...vledgments

...graphs

...Boris Starosta/Stock Connection

...er 1 **xxii** Sylvain Coffie/Tony Stone Images; **29** Culver Pictures; ...lett-Packard Company

...ter 2 **54** Ron Stanford/Tony Stone Images; **74** John ...l/Bruce Coleman Inc.; **89** Jet Propulsion Labs

...pter 3 **94** Michael W. Davidson/Photo Researchers; **132** From ...C Physics, 2/e, 1965; D.C. Heath & Company, with Education ...elopment Center, Inc., Newton, MA. Reprinted with permission; ...8 Jenny Thomas*; **171** Steve Wewerka*

...hapter 4 **176** Peter Poulides/Tony Stone Images; **193** Focus on ...ports; **231** NASA

Chapter 5 **246** Ralph Mercer/Tony Stone Images; **260** German Information Center; **277** Corbis-Bettmann; **285** Archive Photos; **297** Marshall Henrichs*; **301** Jenny Thomas*

Chapter 6 **302** Ralph Oberlander/Stock, Boston; **341** Norman Kent Productions; **357** Jenny Thomas*

Chapter 7 **362** Douglas Pulsipher/Third Coast Stock; **400** Jenny Thomas*; **403** Corbis-Bettmann; **412** Keith Burke*

Chapter 8 **416** NASA; **430** Hank Morgan/Rainbow

Chapter 9 **456** Dan McCoy/Rainbow; **467** From *PSSC Physics*, 2/e, 1965; D.C. Heath & Co. with Education Development Center, Inc., Newton, MA. Reprinted with permission; **499** Frank Siteman/Stock, Boston; **511** Peter Yates*

Chapter 10 **512** Roger Tully/Tony Stone Images; **543** Agence France Presse/Corbis-Bettmann; **550** From *PSSC Physics*, 2/e, 1965; D.C. Heath & Co. with Education Development Center, Inc., Newton, MA. Reprinted with permission; **572** Corbis-Bettmann

*Photographed expressly for Addison Wesley Longman

Illustrations

Technical illustrative and technical artwork for this edition was produced by Tech-Graphics.

Text

Chapter 2 **75** Figure 2.20 from A. J. Lotka, *Elements of Mathematical Biology*, p. 69. (New York: Dover, 1956). Reprinted by permission.

Printed in the United States of America

ISBN 0-201-44138-1

1 2 3 4 5 6 7 8 9 10–VH–02 01 00 99 98

CALCULUS

SECOND EDITION

Ross L. Finney

Franklin D. Demana
The Ohio State University

Bert K. Waits
The Ohio State University

Daniel Kennedy
Baylor School

With

Marianne H. Lepp
GTE Laboratories

 **ADDISON-WESLEY**

An imprint of Addison Wesley Longman, Inc.

Reading, Massachusetts • Menlo Park, California • New York • Harlow, England
Don Mills, Ontario • Sydney • Mexico City • Madrid • Amsterdam

About the Authors

Ross L. Finney

Ross Finney received his undergraduate degree and Ph.D. from the University of Michigan at Ann Arbor. He taught at the University of Illinois at Urbana-Champaign from 1966 to 1980 and at the Massachusetts Institute of Technology (MIT) from 1980 to 1990. Dr. Finney worked as a consultant for the Educational Development Center in Newton, Massachusetts. He directed the Undergraduate Mathematics and Its Applications Project (UMAP) from 1977 to 1984 and was the founding editor of the *UMAP Journal*. In 1984, he traveled with a Mathematical Association of America (MAA) delegation to China on a teacher education project through People to People International. Currently, he is involved in a number of mathematical organizations.

Dr. Finney has coauthored a number of Addison Wesley Longman textbooks, including *Calculus; Calculus and Analytic Geometry; Elementary Differential Equations with Linear Algebra;* and *Calculus for Engineers and Scientists.*

Franklin D. Demana

Frank Demana received his master's degree in mathematics and his Ph.D. from Michigan State University. Currently, he is Professor Emeritus of Mathematics at The Ohio State University. As an active supporter of the use of technology to teach and learn mathematics, he is cofounder of the national Teachers Teaching with Technology (T^3) professional development program. He has been the director and codirector of more than $10 million of National Science Foundation (NSF) and foundational grant activities over the past 15 years. Along with frequent presentations at professional meetings, he has published a variety of articles in the areas of computer and calculator-enhanced mathematics instruction. Dr. Demana is also the cofounder (with Bert Waits) of the annual International Conference on Technology in Collegiate Mathematics (ICTCM). He is corecipient of the 1997 Glenn Gilbert National Leadership Award presented by the National Council of Supervisors of Mathematics.

Dr. Demana coauthored *Essential Algebra: A Calculator Approach; Transition to College Mathematics; Precalculus: A Graphing Approach; College Algebra and Trigonometry: A Graphing Approach; College Algebra: A Graphing Approach; Precalculus: Functions and Graphs;* and *Intermediate Algebra: A Graphing Approach.*

Bert K. Waits

Bert Waits received his Ph.D. from The Ohio State University and is currently Professor Emeritus of Mathematics there. Dr. Waits is cofounder of the national Teachers Teaching with Technology (T^3) professional development program, and has been codirector or principal investigator on several large NSF projects. Dr. Waits has published articles in more than 50 nationally recognized professional journals. He frequently gives invited lectures, workshops, and minicourses at national meetings of the MAA and the National Council of Teachers of Mathematics (NCTM) on how to use computer technology to enhance the teaching and learning of mathematics. He has given invited presentations at the International Congress on Mathematical Education (ICME 6, 7, and 8) in Budapest (1988), Quebec (1992), and Seville (1996). Dr. Waits is corecipient of the 1997 Glenn Gilbert National Leadership Award presented by the National Council of Supervisors of Mathematics, and is the cofounder (with Frank Demana) of the ICTCM.

Dr. Waits coauthored *Precalculus: A Graphing Approach; College Algebra and Trigonometry: A Graphing Approach; College Algebra: A Graphing Approach; Precalculus: Functions and Graphs;* and *Intermediate Algebra: A Graphing Approach.*

Daniel Kennedy

Dan Kennedy received his undergraduate degree from the College of the Holy Cross and his master's and Ph.D. in mathematics from the University of North Carolina at Chapel Hill. Since 1973 he has taught mathematics at the Baylor School in Chattanooga, Tennessee, where he holds the Cartter Lupton Distinguished Professorship. Dr. Kennedy became an Advanced Placement Calculus reader in 1978, which led to an increasing level of involvement with the program as workshop consultant, table leader, and exam leader. He joined the Advanced Placement Calculus Test Development Committee in 1986, then in 1990 became the first high school teacher in 35 years to chair that committee. It was during his tenure as chair that the program moved to require graphing calculators and laid the early groundwork for the recent major reform of the Advanced Placement Calculus curriculum. The author of the 1997 *Teacher's Guide—AP® Calculus,* Dr. Kennedy has conducted more than 50 workshops and institutes for high school calculus teachers. His articles on mathematics teaching have appeared in the *Mathematics Teacher* and the *American Mathematical Monthly,* and he is a frequent speaker on calculus curriculum reform at professional and civic meetings. Dr. Kennedy was named a Tandy Technology Scholar in 1992 and a Presidential Award winner in 1995.

Preface

This is the lean and lively technology-based calculus book college instructors have been waiting for. Without distraction from inessential material, students get right into the calculus to see how it works from the very beginning. This text combines an appropriate use of technology with standard analytic techniques to provide a balanced approach to the study and implementation of calculus. Technology is fully integrated, rather than just added. The text encourages group work, mathematical modeling (problem solving), conceptual understanding, and facility with technology, and fosters an appreciation of calculus both as a coherent body of knowledge and as a human accomplishment.

Audience and Prerequisites

This book is a complete revision of the first edition and explores all concepts necessary for the standard calculus sequence. Its purpose, in addition to making it possible to learn calculus, is to teach students how to use calculus effectively and to show how knowing calculus can pay off in any profession they decide to enter. The applications described within the text are real, and their presentations are self-contained; students will not need any experience in the fields from which the applications are drawn. The prerequisites are exposure to algebra, geometry, and trigonometry. Chapter 1 reviews this material.

Philosophy

We believe in using the full array of calculator functions from the beginning, and Chapter 1 introduces them with appropriate examples and applications. In addition to the usual linear, polynomial, power, and rational functions, there are exponential functions, parametrizations, logarithmic functions, inverses of familiar functions, trigonometric functions, and inverse trigonometric functions. Chapter 1 also introduces the capability that most graphing calculators now have for regression analysis, and shows how to use this capability to summarize and visualize numerical data and to evaluate the predictions of analytic models.

We explore calculus through the interpretation of graphs and tables as well as the application of analytic methods. Derivatives are interpreted both as rates of change and as local linear approximations. Local linearity is a recurring theme. The definite integral is introduced both as a description of the net effect of a rate of change over an interval of time and as a limit of Riemann sums. There are applications to such diverse topics as pollution control, medicine, diesel oil consumption, household electricity, economics, projectile motion, population dynamics, and work done pumping. Please see the Index of Applications on page 771 for a complete listing of all applications by subject.

Chapter 6 focuses on the use of differential equations to model real-world problems. We interpret differential equations using slope fields and solve initial value problems analytically and numerically. In presenting infinite series, the emphasis (as it was historically and still is in applied fields) is on the utility of power series and the calculus connection, and not on sequences and series of constants. Convergence and divergence are interpreted graphically, numerically, and analytically, and Lagrange error bounds are used to confirm the accuracies of Taylor-polynomial approximations.

Technology is integrated throughout the book to provide a balanced approach to teaching and learning calculus. The presentation uses graphical, numerical, and algebraic/analytic techniques and encourages written and oral communication. Students are expected to use a multi-representational approach to investigating problems, to write about their conclusions, and often to work in groups to communicate mathematics orally. This book reflects what we have learned about the appropriate use of technology in the classroom during the last decade.

Rich and varied problems have been a hallmark of Finney/Thomas books, and to these we have added a wide variety of new graphical and data-based problems. Enhanced versions of the visualizations and technological explorations pioneered by Demana and Waits have been incorporated throughout.

The Rule of Four—A Balanced Approach

A principal feature of this edition is the balance attained among graphical, numerical, and algebraic/analytic techniques and communication (the rule of four). We believe that students must value all of these methods of representation, understand how they are connected in a given problem, and learn how to choose the one or ones most appropriate for solving a particular problem. Suggestions from numerous instructors have helped us shape this modern, balanced, technological approach to the discipline of calculus.

In support of the rule of four, we use a variety of techniques to solve problems and to communicate their solutions. For instance, we obtain solutions algebraically or analytically, support our results graphically or numerically with technology, and then interpret the results in the original problem context. We have written exercises that ask students to solve problems by one method and then to support or confirm their solutions using another method. We want students to understand that technology can be used to support, but not prove, results, and that algebraic or analytic techniques are needed to prove results. We want students to understand that it is *mathematics* that provides the foundation that allows us to use technology to solve problems.

Applications

The text includes a rich array of interesting applications in biology, business, chemistry, economics, engineering, finance, physics, the social sciences, and statistics. For a complete listing of all applications organized by subject, please see the Index of Applications on page 771. Many applications are based on real data from cited sources. Students are first exposed to functions as mechanisms for modeling data and then learn how various functions can model real-life problems. They learn to analyze and model data, represent data graphically, interpret from graphs, and fit curves. Additionally, the tabular representations of data presented in the text highlight the concept that a function is a correspondence between numerical variables, helping students build the connection between the numbers and the graphs.

Explorations

We want students to be actively involved in understanding calculus concepts and solving problems. Stepped Explorations in the text provide guided investigations of many key concepts. These Explorations help build problem-solving ability by guiding students to develop a mathematical model of a problem, solve the mathematical model, support or confirm the solution, and interpret the solution. The ability to communicate their understanding is just as important to the students' learning process as reading and studying, not only in mathematics but in every academic pursuit. Students gain an entirely new perspective on their knowledge when they explain what they know in writing.

Graphing Utilities

This book assumes familiarity with a graphing utility that will produce the graph of a function within an arbitrary viewing window, find the zeros of a function, compute the derivative of a function numerically, and compute definite integrals numerically. Students are expected to recognize that a graph is reasonable, identify all of the important characteristics of a graph, interpret those characteristics, and confirm them using analytic methods. Toward that end, most graphs appearing in this book resemble students' actual grapher output or suggest hand-drawn sketches. This is one of the first calculus textbooks to take full advantage of graphing calculators, philosophically restructuring the course to teach new things in new ways to achieve new understanding, while abandoning some old things and old ways that no longer serve a purpose.

Exercise Sets and Key Terms

The exercise sets were revised extensively for this edition, with many new exercises added. There are more than 3800 exercises, including more than 560 Quick Review Exercises. The exercises involve, as appropriate,

- algebraic and analytic manipulation
- data analysis
- exploration
- extending ideas
- graphical and numerical representations of functions
- group activities
- interpretations of graphs
- titled applications
- writing to learn

Each exercise set begins with a set of Quick Review Exercises, which can be used to introduce lessons, support examples, and review prerequisite skills. The exercises that follow are graded from routine to challenging. Additional blocks of exercises, entitled Explorations and Extending the Ideas, may be used in a variety of ways, including group work. Each chapter concludes with a set of Review Exercises, followed by a list of Key Terms with appropriate page references. A closer look at the features of this revision can be found on page xvi.

Acknowledgments

The authors would like to express their gratitude to all those who have used the first edition of the text, with special thanks to the professors who have commented on this edition. In particular, we thank Marie Aratari, Oakland Community College; Bill Ardis, Collin County Community College; Dan Burkett, Indiana University of Pennsylvania; Stephen Merrill, Marquette University; Cheryl Slayden, Pellissippi State Technical Community College; and Randall Westhoff, Bemidji State University. We are particularly indebted to Judith Broadwin, who consulted with us throughout the development of this revision, for her dedication and excellent advice.

In Conclusion

We are excited by the new directions our text has taken, and hope you will contact us with your comments.

Ross L. Finney
Franklin D. Demana
Bert K. Waits
Daniel Kennedy
Math@awl.com

Contents

C H A P T E R

1

Prerequisites for Calculus

$x = 3 \cos t,\ y = 4 \sin t$

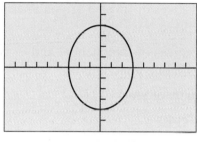

[−9, 9] by [−6, 6]

C H A P T E R

2

Limits and Continuity

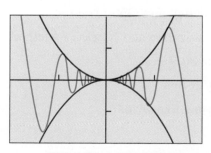

[−0.2, 0.2] by [−0.02, 0.02]

CHAPTER

3
Derivatives

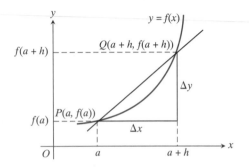

CHAPTER

4
Applications of Derivatives

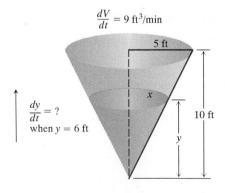

CHAPTER

5

The Definite Integral

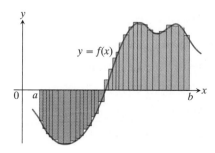

CHAPTER

6

Differential Equations and Mathematical Modeling

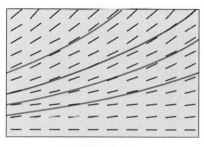

[0, 20] by [0, 300]

CHAPTER
7
Applications of Definite Integrals

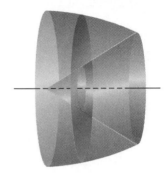

CHAPTER
8
L'Hôpital's Rule, Improper Integrals, and Partial Fractions

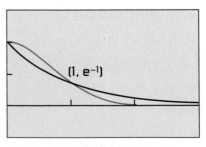

$[0, 3]$ by $[-0.5, 1.5]$

CHAPTER 9
Infinite Series

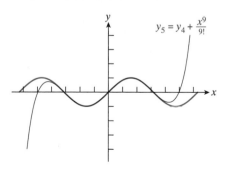

CHAPTER 10
Parametric, Vector, and Polar Functions

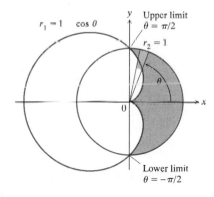

Supplements

The following products are designed to accompany *Calculus,* Second Edition:

For the Student

Technology Resource Manual, Volume I
    For TI Calculators

Technology Resource Manual, Volume II
    For Casio and Hewlett-Packard Calculators

For the Instructor

Teacher's Guide
    Bibliography • Teaching Notes • Exploration Extensions • Assignment
    Guides • Alternate Assessment Ideas • Answer Key

Assessment
    Quizzes • Tests • Semester and Final Tests • Alternate Assessment

Transparencies
    50 4-color transparencies

Solutions Manual
    Complete solutions for all problems in the Student Edition

Technology Resources

Teacher Resource Planner CD-ROM
    All Supplements • Calculator Programs

Calculus Testworks
    Electronic Test Bank

To the Student

Working with graphical, numerical, analytical, and verbal forms of functions

You are about to embark on what we hope will be the beginning of your adventures into higher level mathematics. All of your previous math courses will serve as the foundation for calculus. Your study of algebra has taught you how to work with functions, equations, and graphs. Your study of geometry has taught you how to work with figures, areas, volumes, and coordinatization.

Theorems and definitions continue to be important. Some of the proofs will be provided and some will be developed graphically. You will be writing proofs in some of the exercises.

Theorems, Definitions, and Rules will always be boxed and shaded in yellow.

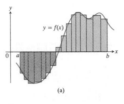

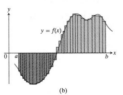

Figure 5.15 The curve of Figure 5.12 with rectangles from finer partitions of $[a, b]$. Finer partitions create more rectangles, with shorter bases.

Definition The Definite Integral as a Limit of Riemann Sums

Let f be a function defined on a closed interval $[a, b]$. For any partition P of $[a, b]$, let the numbers c_k be chosen arbitrarily in the subintervals $[x_{k-1}, x_k]$.

If there exists a number I such that

$$\lim_{\|P\| \to 0} \sum_{k=1}^{n} f(c_k) \Delta x_k = I$$

no matter how P and the c_k's are chosen, then f is **integrable** on $[a, b]$ and I is the **definite integral** of f over $[a, b]$.

Theorem 4 (continued) The Fundamental Theorem of Calculus, Part 2

If f is continuous at every point of $[a, b]$, and if F is any antiderivative of f on $[a, b]$, then

$$\int_a^b f(x)\, dx = F(b) - F(a).$$

This part of the Fundamental Theorem is also called the **Integral Evaluation Theorem**.

Proof Part 1 of the Fundamental Theorem tells us that an antiderivative of f exists, namely

$$G(x) = \int_a^x f(t)\, dt.$$

Thus, if F is *any* antiderivative of f, then $F(x) = G(x) + C$ for some constant C (by Corollary 3 of the Mean Value Theorem for Derivatives, Section 4.2).

Evaluating $F(b) - F(a)$, we have

$$F(b) - F(a) = [G(b) + C] - [G(a) + C]$$
$$= G(b) - G(a)$$
$$= \int_a^b f(t)\, dt - \int_a^a f(t)\, dt$$
$$= \int_a^b f(t)\, dt - 0$$
$$= \int_a^b f(t)\, dt. \qquad \blacksquare$$

Numerous **Examples** show how the theorems and definitions are applied.

Justification for each step in an Example often appears in red type to the right of the step.

Example 1 APPLYING THE FUNDAMENTAL THEOREM

Find

$$\frac{d}{dx} \int_{-\pi}^{x} \cos t\, dt \quad \text{and} \quad \frac{d}{dx} \int_{0}^{x} \frac{1}{1 + t^2}\, dt$$

by using the Fundamental Theorem.

Solution

$$\frac{d}{dx} \int_{-\pi}^{x} \cos t\, dt = \cos x \qquad \text{Eq. 1 with } f(t) = \cos t$$

$$\frac{d}{dx} \int_{0}^{x} \frac{1}{1 + t^2}\, dt = \frac{1}{1 + x^2}. \qquad \text{Eq. 1 with } f(t) = \frac{1}{1 + t^2}$$

Green boxes help summarize important skills and strategies.

Graph Viewing Skills

1. Recognize that the graph is reasonable.
2. See all the important characteristics of the graph.
3. Interpret those characteristics.
4. Recognize grapher failure.

Linking derivatives and integrals with the Fundamental Theorem of Calculus

In calculus, you will study limits, derivatives, and integrals, and applications of all of these ideas. Your study will be based on a balanced approach. You will be asked to solve graphically, support numerically, confirm analytically, and solve algebraically, all while applying calculus to problem situations.

Solving graphically gives insight into the behavior of functions.

Often you will **confirm** the graphical results by using the algebraic techniques based on the theorems, definitions, and rules you have acquired throughout your math courses.

You will **analyze** graphs and use **numerical** tables to **support** your conjectures.

You will participate actively in your learning. Working through the **Explorations** will give you insights into calculus. This will help you gain a depth of understanding that will help you get the most out of this, and future, math courses.

Many **Explorations** are designed for you to work in groups. Working together will give you the opportunity to reinforce each other's ideas and verify your work.

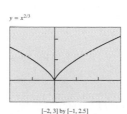

$y = x^{2/3}$

[-2, 3] by [-1, 2.5]

Figure 4.5 (Example 3)

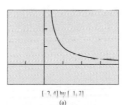

[-2, 4] by [-1, 2]

(a)

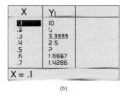

X	Y1
.1	10
.2	5
.3	3.3333
.4	2.5
.5	2
.6	1.6667
.7	1.4286

X = .1

(b)

Figure 3.17 (a) The graph of NDER ln (x) and (b) a table of values. What graph could this be? (Example 4)

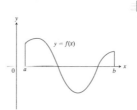

y

$y = f(x)$

0 a b x

Figure 5.27 The graph of the function in Exploration 2.

Example 3 FINDING ABSOLUTE EXTREMA

Find the absolute maximum and minimum values of $f(x) = x^{2/3}$ on the interval $[-2, 3]$.

Solution

Solve Graphically

Figure 4.5 suggests that f has an absolute maximum value of about 2 at $x = 3$ and an absolute minimum value of 0 at $x = 0$.

Confirm Analytically

We evaluate the function at the critical points and endpoints and take the largest and smallest of the resulting values.

The first derivative

$$f'(x) = \frac{2}{3} x^{-1/3} = \frac{2}{3\sqrt[3]{x}}$$

has no zeros but is undefined at $x = 0$. The values of f at this one critical point and at the endpoints are

Critical point value: $f(0) = 0$;

Endpoint values: $f(-2) = (-2)^{2/3} = \sqrt[3]{4}$;

$f(3) = (3)^{2/3} = \sqrt[3]{9}$.

We can see from this list that the function's absolute maximum value is $\sqrt[3]{9} \approx 2.08$, and occurs at the right endpoint $x = 3$. The absolute minimum value is 0, and occurs at the interior point $x = 0$.

Example 4 GRAPHING A DERIVATIVE USING NDER

Let $f(x) = \ln x$. Use NDER to graph $y = f'(x)$. Can you guess what function $f'(x)$ is by analyzing its graph?

Solution The graph is shown in Figure 3.17a. The shape of the graph suggests, and the table of values in Figure 3.17b supports, the conjecture that this is the graph of $y = 1/x$. We will prove in Section 3.9 (using analytic methods) that this is indeed the case.

Differentiability Implies Continuity

We began this section with a look at the typical ways that a function could fail to have a derivative at a point. As one example, we indicated graphically that a discontinuity in the graph of f would cause one or both of the one-sided derivatives to be nonexistent. It is actually not difficult to give an analytic proof that continuity is an essential condition for the derivative to exist, so we include that as a theorem here.

Exploration 2 **Finding the Derivative of an Integral**

Suppose we are given the graph of a continuous function f, as in Figure 5.27. *Work in groups of two or three.*

1. Copy the graph of f onto your own paper. Choose any x greater than a in the interval $[a, b]$ and mark it on the x-axis.

2. Using only *vertical line segments*, shade in the region between the graph of f and the x-axis from a to x. (Some shading might be below the x-axis.)

3. Your shaded region represents a definite integral. Explain why this integral can be written as $\int_a^x f(t)\, dt$. (Why don't we write it as $\int_a^x f(x)\, dx$?)

4. Compare your picture with others produced by your group. Notice how your integral (a real number) depends on which x you chose in the interval $[a, b]$. The integral is therefore a *function of x* on $[a, b]$. Call it F.

5. Recall that $F'(x)$ is the limit of $\Delta F/\Delta x$ as Δx gets smaller and smaller. Represent ΔF in your picture by drawing *one more vertical shading segment* to the right of the last one you drew in step 2. ΔF is the (signed) *area* of your vertical segment.

6. Represent Δx in your picture by moving x to beneath your newly-drawn segment. That small change in Δx is the *thickness* of your vertical segment.

7. What is now the *height* of your vertical segment?

8. Can you see why Newton and Leibniz concluded that $F'(x) = f(x)$?

Creating a mathematical model for physical situations

Each chapter begins with a problem that can be solved using the mathematics in the chapter. These problems are intended to pique your interest. Problem solving is continued throughout the chapter to develop the concepts in application contexts.

Practical applications and situations (many with sources cited) are included. You will learn how to evaluate a problem critically, formulate a strategy, solve the problem, and analyze and interpret the results.

You will model problems using various methods, including logistical regression, linearizations, and solving differential equations.

Differential calculus includes the study of rates of change of functions and approximations.

We model rates of change using differential equations, and use slope fields to interpret differential equations. We also solve differential equations using numerical methods such as Euler's method.

Another important calculus topic is integral calculus.

Integrals are used to model physical, economic, and social situations. We use integrals of rates of change to get accumulated change, and often use Riemann sums to approximate accumulated change.

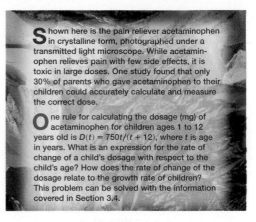

S hown here is the pain reliever acetaminophen in crystalline form, photographed under a transmitted light microscope. While acetaminophen relieves pain with few side effects, it is toxic in large doses. One study found that only 30% of parents who gave acetaminophen to their children could accurately calculate and measure the correct dose.

O ne rule for calculating the dosage (mg) of acetaminophen for children ages 1 to 12 years old is $D(t) = 750t/(t + 12)$, where t is age in years. What is an expression for the rate of change of a child's dosage with respect to the child's age? How does the rate of change of the dosage relate to the growth rate of children? This problem can be solved with the information covered in Section 3.4.

Example 2   MODELING BEAR POPULATION

A national park is known to be capable of supporting no more than 100 grizzly bears. Ten bears are in the park at present. We model the population with a logistic differential equation with $k = 0.1$.

(a) Draw and describe a slope field for the differential equation.

(b) Find a logistic growth model $P(t)$ for the population and draw its graph.

(c) When will the bear population reach 50?

Solution   (a) The carrying capacity is 100, so $M = 100$. The model we seek is a solution to the following differential equation.

$$\frac{dP}{dt} = \frac{0.1}{100}P(100 - P) \quad \text{Eq. 1 with } k = 0.1, M = 100$$

$$= 0.001P(100 - P)$$

Figure 6.12 shows a slope field for this differential equation. There appears to be a horizontal asymptote at $P = 100$. The solution curves fall toward this level from above, and rise toward it from below.

[0, 150] by [0, 150]

Figure 6.12 A slope field for the logistic differential equation

$$\frac{dP}{dt} = 0.001P(100 - P).$$

(Example 2)

Example 5   POTATO CONSUMPTION

From 1970 to 1980, the rate of potato consumption in a particular country was $C(t) = 2.2 + 1.1^t$ millions of bushels per year, with t being years since the beginning of 1970. How many bushels were consumed from the beginning of 1972 to the end of 1973?

Solution We seek the cumulative effect of the consumption rate for $2 \le t \le 4$.

Step 1:
Riemann sum. We partition $[2, 4]$ into subintervals of length Δt and let t_k be a time in the kth subinterval. The amount consumed during this interval is approximately

$$C(t_k) \, \Delta t \text{ million bushels.}$$

The consumption for $2 \le t \le 4$ is approximately

$$\sum C(t_k) \, \Delta t \text{ million bushels.}$$

Step 2:
Definite integral. The amount consumed from $t = 2$ to $t = 4$ is the limit of these sums as the norms of the partitions go to zero.

$$\int_2^4 C(t) \, dt = \int_2^4 (2.2 + 1.1^t) \, dt \text{ million bushels}$$

Step 3:
Evaluate. Evaluating numerically, we obtain

$$\text{NINT } (2.2 + 1.1^t, t, 2, 4) \approx 7.066 \text{ million bushels.}$$

Interpreting data

We hope that you will find the applications in this text relevant and interesting. Examples provide problems from a wide variety of subjects, including medicine, health, nature, economics, and physics. Exercises also include numerous opportunities to apply calculus to many situations.

Applications come from many sources, including newspapers, government data, and experimental data obtained with CBL™ software.

You will be given data from up-to-date sources and asked to interpret this data mathematically. You will be asked to find regression equations that, in some cases, provide a good approximation for the mathematical model of the data. From these equations you will be able to predict the behavior for new values of the function.

Graphing calculator art shows what will appear as you work with your grapher. The window is provided to help you duplicate the graph.

Explorations using the graphing calculator provide simulations of motion, an important calculus application.

Angiography

An opaque dye is injected into a partially blocked artery to make the inside visible under X-rays. This reveals the location and severity of the blockage.

Angioplasty

A balloon-tipped catheter is inflated inside the artery to widen it at the blockage site.

Table 6.3 Experimental Data

Time (sec)	T (°C)	$T - T_s$ (°C)
2	64.8	60.3
5	49.0	44.5
10	31.4	26.9
15	22.0	17.5
20	16.5	12.0
25	14.2	9.7
30	12.0	7.5

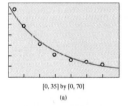

[0, 35] by [0, 70]
(a)

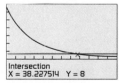

Intersection
X = 38.227514 Y = 8

[0, 60] by [−20, 70]
(b)

Figure 6.9 (Example 5)

Example 12 UNCLOGGING ARTERIES

In the late 1830s, the French physiologist Jean Poiseuille ("pwa-ZOY") discovered the formula we use today to predict how much the radius of a partially clogged artery has to be expanded to restore normal flow. His formula,

$$V = kr^4,$$

says that the volume V of fluid flowing through a small pipe or tube in a unit of time at a fixed pressure is a constant times the fourth power of the tube's radius r. How will a 10% increase in r affect V?

Solution The differentials of r and V are related by the equation

$$dV = \frac{dV}{dr}\,dr = 4kr^3\,dr.$$

The relative change in V is

$$\frac{dV}{V} = \frac{4kr^3 dr}{kr^4} = 4\,\frac{dr}{r}.$$

The relative change in V is 4 times the relative change in r, so a 10% increase in r will produce a 40% increase in the flow.

Example 5 USING NEWTON'S LAW OF COOLING

A temperature probe (thermometer) is removed from a cup of coffee and placed in water that has a temperature of $T_s = 4.5$°C. Temperature readings T, as recorded in Table 6.3, are taken after 2 sec, 5 sec, and every 5 sec thereafter. Estimate

(a) the coffee's temperature at the time the temperature probe was removed.

(b) the time when the temperature probe reading will be 8°C.

Solution

Model

According to Newton's Law of Cooling, $T - T_s = (T_0 - T_s)e^{-kt}$, where $T_s = 4.5$ and T_0 is the temperature of the coffee (probe reading) at $t = 0$.

We use exponential regression to find that

$$T - 4.5 = 61.66(0.9277^t)$$

is a model for the $(t, T - T_s) = (t, T - 4.5)$ data.

Thus,

$$T = 4.5 + 61.66(0.9277^t)$$

is a model for the (t, T) data.

Figure 6.9a shows the graph of the model superimposed on a scatter plot of the (t, T) data.

(a) At time $t = 0$, when the probe was removed, the temperature was

$$T = 4.5 + 61.66(0.9277^0) \approx 66.16°C.$$

(b) Solve Graphically

Figure 6.9b shows that the graphs of

$$y = 8 \quad \text{and} \quad y = T = 4.5 + 61.66(0.9277^t)$$

intersect at about $t = 38$.

Interpret

The temperature of the coffee was about 66.2°C when the temperature probe was removed. The temperature probe will reach 8°C about 38 sec after it is removed from the coffee and placed in the water.

Exploration 3 Seeing Motion on a Graphing Calculator

The graphs in Figure 3.26b give us plenty of information about the flight of the rock in Example 4, but neither graph shows the path of the rock in flight. We can simulate the moving rock by graphing the parametric equations

$$x_1(t) = 3(t < 5) + 3.1(t \geq 5), \quad y_1(t) = 160t - 16t^2$$

in dot mode.

This will show the upward flight of the rock along the vertical line $x = 3$, and the downward flight of the rock along the line $x = 3.1$.

1. To see the flight of the rock from beginning to end, what should we use for tMin and tMax in our graphing window?

2. Set xMin = 0, xMax = 6, and yMin = −10. Use the results from Example 4 to determine an appropriate value for yMax.

Studying a human accomplishment enhanced by technology

Calculus has been developed through the centuries by individuals such as Leibniz and Agnesi, and it has evolved over time. This evolution has been forwarded dramatically by advances in technology. The graphing calculator will play an important part in your study by enabling you to explore functions in ways that were not available to the individuals who first discovered calculus.

This book assumes you are familiar with a graphing utility that will produce the graph of a function within an arbitrary viewing window, find the zeros of a function, compute the derivative of a function numerically, and compute definite integrals numerically.

Graphing technology manuals that give more specific information about each specific calculator are available. The textbook uses special programs such as SLOPEFLD. These programs are given in the *Technology Resource Manuals.*

$y = 100e^{0.056x}$

[0, 20] by [0, 300]

Figure 6.2 The solution of the initial value problem in Example 1.

Exploration 1 Constructing a Slope Field

1. Let $h = (\text{Xmax} - \text{Xmin})/10$ and $k = (\text{Ymax} - \text{Ymin})/10$. Explain why

$$\left(\text{Xmin} + (2i - 1)\frac{h}{2},\ \text{Ymin} + (2j - 1)\frac{k}{2}\right)$$

for $i, j = 1, 2, ..., 10$, represents 100 points in the viewing window [Xmin, Xmax] by [Ymin, Ymax].

2. Show that the horizontal distance between adjacent points with the same y-coordinate in part 1 is h.

3. Show that the vertical distance between adjacent points with the same x-coordinate in part 1 is k.

4. Consider the differential equation $dy/dx = 2/(x + 1)$ and the viewing window [0, 10] by [0, 10]. Find the slope (given by the differential equation) and construct a short line segment at each point in the third column ($i = 3$) of the points given by part 1. What do the line segments have in common?

5. Use the same differential equation and viewing window as in part 4. Find the slope and construct a short line segment at each point in the fourth row ($j = 4$) of the points given by part 1.

6. Explain how you could complete the slope field using the information in parts 4 and 5.

7. Let L and W be positive integers, $h = (\text{Xmax} - \text{Xmin})/W$, and $k = (\text{Ymax} - \text{Ymin})/L$. If $i = 1, 2, ..., W$ and $j = 1, 2, ..., L$, how many points are produced in part 1?

Gottfried Wilhelm Leibniz (1646–1716)

The method of limits used in this book was not discovered until nearly a century after Newton and Leibniz, the discoverers of calculus, had died.

To Leibniz, the key idea was the *differential*, an infinitely small quantity that was almost like zero, but which—unlike zero—could be used in the denominator of a fraction. Thus, Leibniz thought of the derivative dy/dx as the quotient of two differentials, dy and dx.

Maria Agnesi (1718–1799)

The first text to include differential and integral calculus along with analytic geometry, infinite series, and differential equations was written in the 1740s by the Italian mathematician Maria Gaetana Agnesi. Agnesi, a gifted scholar and linguist whose Latin essay defending higher education for women was published when she was only nine years old, was a well-published scientist by age 20, and an honorary faculty member of the University of Bologna by age 30.

Calculus at Work features introduce you to individuals who are using calculus in their jobs. Some of the applications of calculus they encounter are mentioned throughout the text.

Calculus at Work

I have a Bachelor's and Master's degree in Aerospace Engineering from the University of California at Davis. I started my professional career as a Facility Engineer managing productivity and maintenance projects in the Unitary Project Wind Tunnel facility at NASA Ames Research Center. I used calculus and differential equations in fluid mechanic analyses of the tunnels. I then moved to the position of Test Manager, still using some fluid mechanics and other mechanical engineering analysis tools to solve problems. For example, the lift and drag forces acting on an airplane wing can be determined by integrating the known pressure distribution on the wing.

I am currently a NASA On-Site Systems Engineer for the Lunar Prospector spacecraft project, at Lockheed Martin Missiles and Space in Sunnyvale, California. Differential equations and integration are used to design some of the flight hardware for the spacecraft. I work on ensuring that the different systems of the spacecraft are adequately integrated together to meet the specified design requirements. This often means doing some analysis to determine if the systems will function properly and within the constraints of the space environment. Some of these analyses require use of differential equations and integration to determine the most exact results, within some margin of error.

Ross Shaw
NASA Ames Research Center
Sunnyvale, CA

Solutions to problems must be checked for reasonableness and accuracy.

Each set of exercises is carefully graded to run from routine at the beginning to more challenging at the end. Answers for most odd-numbered problems appear in the back of the text. Complete solutions are included for representative problems.

Quick Review exercises help you review skills from previous courses or chapters that you will need in the rest of the exercise set.

Many of the exercises are similar to Examples worked out in the lesson, but you will also be expected to apply the principles to new problems.

Explorations are opportunities to discover exciting mathematics on your own or in *groups of two or three*. These exercises will often ask you to use critical thinking to discover ideas.

Many of the exercises ask you to practice your writing skills by **Writing to Learn.** Communicating your ideas can give you a new perspective on your own knowledge that can benefit your study of any discipline.

Extending the Ideas exercises go beyond what is presented in the textbook. These exercises are challenging extensions of the book's material. You may be asked to work on these problems in groups or to solve them as projects.

Review Exercises at the end of each chapter prepare you for Chapter Tests. These exercises cover the entire chapter and include many additional applications.

Quick Review 4.2

In Exercises 1 and 2, find exact solutions to the inequality.

1. $2x^2 - 6 < 0$ **2.** $3x^2 - 6 > 0$

In Exercises 3–5, let $f(x) = \sqrt{8 - 2x^2}$.

3. Find the domain of f.   **4.** Where is f continuous?

5. Where is f differentiable?

In Exercises 6–8, let $f(x) = \dfrac{x}{x^2 - 1}$.

6. Find the domain of f.   **7.** Where is f continuous?

8. Where is f differentiable?

In Exercises 9 and 10, find C so that the graph of the function f passes through the specified point.

9. $f(x) = -2x + C$, $(-2, 7)$

10. $g(x) = x^2 + 2x + C$, $(1, -1)$

In Exercises 35–38, *work in groups of two or three* and sketch a graph of a differentiable function $y = f(x)$ that has the given properties.

35. (a) local minimum at $(1, 1)$, local maximum at $(3, 3)$

   (b) local minima at $(1, 1)$ and $(3, 3)$

   (c) local maxima at $(1, 1)$ and $(3, 3)$

36. $f(2) = 3$, $f'(2) = 0$, and

   (a) $f'(x) > 0$ for $x < 2$, $f'(x) < 0$ for $x > 2$.

   (b) $f'(x) < 0$ for $x < 2$, $f'(x) > 0$ for $x > 2$.

   (c) $f'(x) < 0$ for $x \neq 2$.

   (d) $f'(x) > 0$ for $x \neq 2$.

37. $f'(-1) = f'(1) = 0$, $f'(x) > 0$ on $(-1, 1)$,

   $f'(x) < 0$ for $x < -1$, $f'(x) > 0$ for $x > 1$.

38. A local minimum value that is greater than one of its local maximum values.

39. *Speeding* A trucker handed in a ticket at a toll booth showing that in 2 h she had covered 159 mi on a toll road with speed limit 65 mph. The trucker was cited for speeding. Why?

40. *Temperature Change* It took 20 sec for the temperature to rise from 0°F to 212°F when a thermometer was taken from a freezer and placed in boiling water. Explain why at some moment in that interval the mercury was rising at exactly 10.1°F/sec.

44. *Diving* (a) With what velocity will you hit the water if you step off from a 10-m diving platform?

   (b) With what velocity will you hit the water if you dive off the platform with an upward velocity of 2 m/sec?

45. Writing to Learn The function

$$f(x) = \begin{cases} x, & 0 \le x < 1 \\ 0, & x = 1 \end{cases}$$

is zero at $x = 0$ and at $x = 1$. Its derivative is equal to 1 at every point between 0 and 1, so f' is never zero between 0 and 1, and the graph of f has no tangent parallel to the chord from $(0, 0)$ to $(1, 0)$. Explain why this does not contradict the Mean Value Theorem.

46. Writing to Learn Explain why there is a zero of $y = \cos x$ between every two zeros of $y = \sin x$.

24. *Oil Flow* Oil flows through a cylindrical pipe of radius 3 inches, but friction from the pipe slows the flow toward the outer edge. The speed at which the oil flows at a distance r inches from the center is $8(10 - r^2)$ inches per second.

   (a) In a plane cross section of the pipe, a thin ring with thickness Δr at a distance r inches from the center approximates a rectangular strip when you straighten it out. What is the area of the strip (and hence the approximate area of the ring)?

   (b) Explain why we know that oil passes through this ring at approximately $8(10 - r^2)(2\pi r)\, \Delta r$ cubic inches per second.

   (c) Set up and evaluate a definite integral that will give the rate (in cubic inches per second) at which oil is flowing through the pipe.

Explorations

51. *Analyzing Derivative Data* Assume that f is continuous on $[-2, 2]$ and differentiable on $(-2, 2)$. The table gives some values of $f'(x)$.

x	$f'(x)$	x	$f'(x)$
-2	7	0.25	-4.81
-1.75	4.19	0.5	-4.25
-1.5	1.75	0.75	3.31
-1.25	-0.31	1	-2
-1	-2	1.25	-0.31
-0.75	-3.31	1.5	1.75
-0.5	-4.25	1.75	4.19
-0.25	-4.81	2	7
0	-5		

   (a) Estimate where f is increasing, decreasing, and has local extrema.

   (b) Find a quadratic regression equation for the data in the table and superimpose its graph on a scatter plot of the data.

   (c) Use the model in (b) for f' and find a formula for f that satisfies $f(0) = 0$.   ∎

Extending the Ideas

53. *Geometric Mean* The **geometric mean** of two positive numbers a and b is $\sqrt{ab}$. Show that for $f(x) = 1/x$ on any interval $[a, b]$ of positive numbers, the value of c in the conclusion of the Mean Value Theorem is $c = \sqrt{ab}$.

54. *Arithmetic Mean* The **arithmetic mean** of two numbers a and b is $(a + b)/2$. Show that for $f(x) = x^2$ on any interval $[a, b]$, the value of c in the conclusion of the Mean Value Theorem is $c = (a + b)/2$.

Prerequisites for Calculus

Exponential functions are used to model situations in which growth or decay change dramatically. Such situations are found in nuclear power plants, which contain rods of plutonium-239, an extremely toxic radioactive isotope.

Operating at full capacity for one year, a 1,000 megawatt power plant discharges about 435 lb of plutonium-239. With a half-life of 24,400 years, how much of the isotope will remain after 1,000 years? This question can be answered with the mathematics covered in Section 1.3.

Chapter 1 Overview

This chapter reviews the most important things you need to know to start learning calculus. It also introduces the use of a graphing utility as a tool to investigate mathematical ideas, to support analytic work, and to solve problems with numerical and graphical methods. The emphasis is on functions and graphs, the main building blocks of calculus.

Functions and parametric equations are the major tools for describing the real world in mathematical terms, from temperature variations to planetary motions, from brain waves to business cycles, and from heartbeat patterns to population growth. Many functions have particular importance because of the behavior they describe. Trigonometric functions describe cyclic, repetitive activity; exponential, logarithmic, and logistic functions describe growth and decay; and polynomial functions can approximate these and most other functions.

1.1 Lines

Increments • Slope of a Line • Parallel and Perpendicular Lines • Equations of Lines • Applications

Increments

One reason calculus has proved to be so useful is that it is the right mathematics for relating the rate of change of a quantity to the graph of the quantity. Explaining that relationship is one goal of this book. It all begins with the slopes of lines.

When a particle in the plane moves from one point to another, the net changes or *increments* in its coordinates are found by subtracting the coordinates of its starting point from the coordinates of its stopping point.

Definition Increments

If a particle moves from the point (x_1, y_1) to the point (x_2, y_2), the **increments** in its coordinates are

$$\Delta x = x_2 - x_1 \quad \text{and} \quad \Delta y = y_2 - y_1.$$

The symbols Δx and Δy are read "delta x" and "delta y." The letter Δ is a Greek capital d for "difference." Neither Δx nor Δy denotes multiplication; Δx is not "delta times x" nor is Δy "delta times y."

Increments can be positive, negative, or zero, as shown in Example 1.

Example 1   FINDING INCREMENTS

The coordinate increments from $(4, -3)$ to $(2, 5)$ are

$$\Delta x = 2 - 4 = -2, \quad \Delta y = 5 - (-3) = 8.$$

From $(5, 6)$ to $(5, 1)$, the increments are

$$\Delta x = 5 - 5 = 0, \quad \Delta y = 1 - 6 = -5.$$

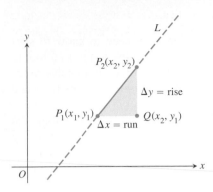

Figure 1.1 The slope of line L is
$$m = \frac{\text{rise}}{\text{run}} = \frac{\Delta y}{\Delta x}.$$

Slope of a Line

Each nonvertical line has a *slope,* which we can calculate from increments in coordinates.

Let L be a nonvertical line in the plane and $P_1(x_1, y_1)$ and $P_2(x_2, y_2)$ two points on L (Figure 1.1). We call $\Delta y = y_2 - y_1$ the **rise** from P_1 to P_2 and $\Delta x = x_2 - x_1$ the **run** from P_1 to P_2. Since L is not vertical, $\Delta x \neq 0$ and we define the slope of L to be the amount of rise per unit of run. It is conventional to denote the slope by the letter m.

Definition Slope

Let $P_1(x_1, y_1)$ and $P_2(x_2, y_2)$ be points on a nonvertical line, L. The **slope** of L is
$$m = \frac{\text{rise}}{\text{run}} = \frac{\Delta y}{\Delta x} = \frac{y_2 - y_1}{x_2 - x_1}.$$

A line that goes uphill as x increases has a positive slope. A line that goes downhill as x increases has a negative slope. A horizontal line has slope zero since all of its points have the same y-coordinate, making $\Delta y = 0$. For vertical lines, $\Delta x = 0$ and the ratio $\Delta y/\Delta x$ is undefined. We express this by saying that vertical lines *have no slope.*

Parallel and Perpendicular Lines

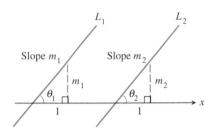

Figure 1.2 If $L_1 \parallel L_2$, then $\theta_1 = \theta_2$ and $m_1 = m_2$. Conversely, if $m_1 = m_2$, then $\theta_1 = \theta_2$ and $L_1 \parallel L_2$.

Parallel lines form equal angles with the x-axis (Figure 1.2). Hence, nonvertical parallel lines have the same slope. Conversely, lines with equal slopes form equal angles with the x-axis and are therefore parallel.

If two nonvertical lines L_1 and L_2 are perpendicular, their slopes m_1 and m_2 satisfy $m_1 m_2 = -1$, so each slope is the *negative reciprocal* of the other:
$$m_1 = -\frac{1}{m_2}, \qquad m_2 = -\frac{1}{m_1}.$$

The argument goes like this: In the notation of Figure 1.3, $m_1 = \tan \phi_1 = a/h$, while $m_2 = \tan \phi_2 = -h/a$. Hence, $m_1 m_2 = (a/h)(-h/a) = -1$.

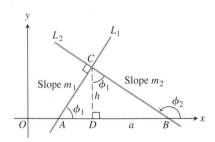

Figure 1.3 $\triangle ADC$ is similar to $\triangle CDB$. Hence ϕ_1 is also the upper angle in $\triangle CDB$, where $\tan \phi_1 = a/h$.

Example 2 DETERMINING PERPENDICULARITY FROM SLOPE

Let L be a line with slope 3/4. Then any line with slope $-4/3$ will be perpendicular to L.

Equations of Lines

The vertical line through the point (a, b) has equation $x = a$ since every x-coordinate on the line has the value a. Similarly, the horizontal line through (a, b) has equation $y = b$.

Example 3 FINDING EQUATIONS OF VERTICAL AND HORIZONTAL LINES

The vertical and horizontal lines through the point $(2, 3)$ have equations $x = 2$ and $y = 3$, respectively (Figure 1.4).

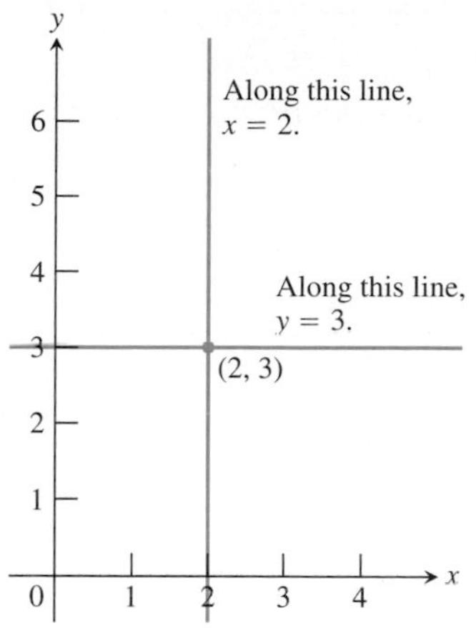

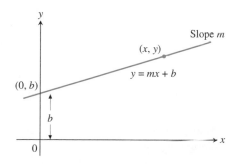

Figure 1.4 The standard equations for the vertical and horizontal lines through the point (2, 3) are $x = 2$ and $y = 3$. (Example 3)

We can write an equation for any nonvertical line L if we know its slope m and the coordinates of one point $P_1(x_1, y_1)$ on it. If $P(x, y)$ is *any* other point on L, then

$$\frac{y - y_1}{x - x_1} = m,$$

so that

$$y - y_1 = m(x - x_1) \quad \text{or} \quad y = m(x - x_1) + y_1.$$

Definition Point-slope Equation

The equation

$$y = m(x - x_1) + y_1$$

is the **point-slope equation** of the line through the point (x_1, y_1) with slope m.

Example 4 USING THE POINT-SLOPE EQUATION

Write an equation for the line through the point $(2, 3)$ with slope $-3/2$.

Solution We substitute $x_1 = 2$, $y_1 = 3$, and $m = -3/2$ into the point-slope equation and obtain

$$y = -\frac{3}{2}(x - 2) + 3 \quad \text{or} \quad y = -\frac{3}{2}x + 6.$$

Example 5 USING THE POINT-SLOPE EQUATION

Write an equation for the line through $(-2, -1)$ and $(3, 4)$.

Solution The line's slope is

$$m = \frac{4 - (-1)}{3 - (-2)} = \frac{5}{5} = 1.$$

We can use this slope with either of the two given points in the point-slope equation. For $(x_1, y_1) = (-2, -1)$, we obtain

$$y = 1 \cdot (x - (-2)) + (-1)$$
$$y = x + 2 + (-1)$$
$$y = x + 1.$$

The y-coordinate of the point where a nonvertical line intersects the y-axis is the **y-intercept** of the line. Similarly, the x-coordinate of the point where a nonhorizontal line intersects the x-axis is the **x-intercept** of the line. A line with slope m and y-intercept b passes through $(0, b)$ (Figure 1.5), so

$$y = m(x - 0) + b, \quad \text{or, more simply,} \quad y = mx + b.$$

Figure 1.5 A line with slope m and y-intercept b.

> **Definition** Slope-intercept Equation
>
> The equation
>
> $$y = mx + b$$
>
> is the **slope-intercept equation** of the line with slope m and y-intercept b.

Example 6 WRITING EQUATIONS FOR LINES

Write an equation for the line through the point $(-1, 2)$ that is **(a)** parallel, and **(b)** perpendicular to the line $L: y = 3x - 4$.

Solution The line L, $y = 3x - 4$, has slope 3.

(a) The line $y = 3(x + 1) + 2$, or $y = 3x + 5$, passes through the point $(-1, 2)$, and is parallel to L because it has slope 3.

(b) The line $y = (-1/3)(x + 1) + 2$, or $y = (-1/3)x + 5/3$, passes through the point $(-1, 2)$, and is perpendicular to L because it has slope $-1/3$.

If A and B are not both zero, the graph of the equation $Ax + By = C$ is a line. Every line has an equation in this form, even lines with undefined slopes.

> **Definition** General Linear Equation
>
> The equation
>
> $$Ax + By = C \quad (A \text{ and } B \text{ not both } 0)$$
>
> is a **general linear equation** in x and y.

Although the general linear form helps in the quick identification of lines, the slope-intercept form is the one to enter into a calculator for graphing.

Example 7 ANALYZING AND GRAPHING A GENERAL LINEAR EQUATION

Find the slope and y-intercept of the line $8x + 5y = 20$. Graph the line.

Solution Solve the equation for y to put the equation in slope-intercept form:

$$8x + 5y = 20$$
$$5y = -8x + 20$$
$$y = -\frac{8}{5}x + 4$$

This form reveals the slope ($m = -8/5$) and y-intercept ($b = 4$), and puts the equation in a form suitable for graphing (Figure 1.6).

$y = -\dfrac{8}{5}x + 4$

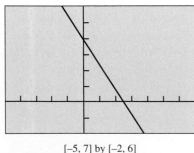

[–5, 7] by [–2, 6]

Figure 1.6 The line $8x + 5y = 20$. (Example 7)

Applications

Many important variables are related by linear equations. For example, the relationship between Fahrenheit temperature and Celsius temperature is linear, a fact we use to advantage in the next example.

Example 8 TEMPERATURE CONVERSION

Find the relationship between Fahrenheit and Celsius temperature. Then find the Celsius equivalent of 90°F and the Fahrenheit equivalent of −5°C.

Solution Because the relationship between the two temperature scales is linear, it has the form $F = mC + b$. The freezing point of water is $F = 32°$ or $C = 0°$, while the boiling point is $F = 212°$ or $C = 100°$. Thus,

$$32 = m \cdot 0 + b \quad \text{and} \quad 212 = m \cdot 100 + b,$$

so $b = 32$ and $m = (212 - 32)/100 = 9/5$. Therefore,

$$F = \frac{9}{5}C + 32, \quad \text{or} \quad C = \frac{5}{9}(F - 32).$$

These relationships let us find equivalent temperatures. The Celsius equivalent of 90°F is

$$C = \frac{5}{9}(90 - 32) \approx 32.2°.$$

The Fahrenheit equivalent of −5°C is

$$F = \frac{9}{5}(-5) + 32 = 23°.$$

Some graphing utilities have a feature that enables them to approximate the relationship between variables with a linear equation. We use this feature in Example 9.

It can be difficult to see patterns or trends in lists of paired numbers. For this reason, we sometimes begin by plotting the pairs (such a plot is called a **scatter plot**) to see whether the corresponding points lie close to a curve of some kind. If they do, and if we can find an equation $y = f(x)$ for the curve, then we have a formula that

1. summarizes the data with a simple expression, and

2. lets us predict values of y for other values of x.

The process of finding a curve to fit data is called **regression analysis** and the curve is called a **regression curve.**

There are many useful types of regression curves—power, exponential, logarithmic, sinusoidal, and so on. In the next example, we use the calculator's linear regression feature to fit the data in Table 1.1 with a line.

Table 1.1 World Population

Year	Population (millions)
1986	4936
1987	5023
1988	5111
1989	5201
1990	5329
1991	5422

Source: Statistical Office of the United Nations, *Monthly Bulletin of Statistics*, 1991.

Example 9 REGRESSION ANALYSIS—PREDICTING WORLD POPULATION

Starting with the data in Table 1.1, build a linear model for the growth of the world population. Use the model to predict the world population in the year 2010.

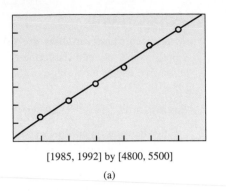

[1985, 1992] by [4800, 5500]

(a)

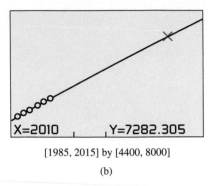

[1985, 2015] by [4400, 8000]

(b)

Figure 1.7 (Example 9)

Why Not Round the Decimals in Equation 1 Even More?

If we do, our final calculation will be way off. Using $y = 98x - 190{,}157$, for instance, gives $y = 6823$ when $x = 2010$, as compared to $y = 7282$, a shortfall of 459 million. The rule is: *Retain all decimal places while working a problem. Round only at the end.* We rounded the coefficients in Equation 1 enough to make it readable, but not enough to hurt the outcome. However, we knew how much we could safely round *only from first having done the entire calculation with numbers unrounded.*

Rounding Rule

Round your answer as appropriate, but do not round the numbers in the calculations that lead to it.

Solution

Model

Upon entering the data into the grapher, we find the regression equation to be approximately

$$y = 98.2286x - 190{,}157.181, \tag{1}$$

where x represents the year and y the population *in millions.*

Figure 1.7a shows the scatter plot for Table 1.1 together with a graph of the regression line just found. You can see how well the line fits the data.

Solve Graphically

Our goal is to predict the population in the year 2010. Reading from the graph in Figure 1.7b, we conclude that when x is 2010, y is approximately 7282.

Confirm Algebraically

Evaluating Equation 1 for $x = 2010$ gives

$$y = 98.2286(2010) - 190{,}157.181$$

$$\approx 7282.$$

Interpret

The linear regression equation suggests that the world population in the year 2010 will be about 7282 million, or approximately 7.3 billion.

Regression Analysis

Regression analysis has four steps:

1. Plot the data (scatter plot).
2. Find the regression equation. For a line, it has the form $y = mx + b$.
3. Superimpose the graph of the regression equation on the scatter plot to see the fit.
4. Use the regression equation to predict y-values for particular values of x.

Quick Review 1.1

1. Find the value of y that corresponds to $x = 3$ in $y = -2 + 4(x - 3)$.

2. Find the value of x that corresponds to $y = 3$ in $y = 3 - 2(x + 1)$.

In Exercises 3 and 4, find the value of m that corresponds to the values of x and y.

3. $x = 5$, $y = 2$, $m = \dfrac{y - 3}{x - 4}$

4. $x = -1$, $y = -3$, $m = \dfrac{2 - y}{3 - x}$

In Exercises 5 and 6, determine whether the ordered pair is a solution to the equation.

5. $3x - 4y = 5$

    (a) $(2, 1/4)$ **(b)** $(3, -1)$

6. $y = -2x + 5$

    (a) $(-1, 7)$ **(b)** $(-2, 1)$

In Exercises 7 and 8, find the distance between the points.

7. $(1, 0)$, $(0, 1)$

8. $(2, 1)$, $(1, -1/3)$

In Exercises 9 and 10, solve for y in terms of x.

9. $4x - 3y = 7$

10. $-2x + 5y = -3$

Section 1.1 Exercises

In Exercises 1–4, find the coordinate increments from A to B.

1. $A(1, 2)$, $B(-1, -1)$ **2.** $A(-3, 2)$, $B(-1, -2)$

3. $A(-3, 1)$, $B(-8, 1)$ **4.** $A(0, 4)$, $B(0, -2)$

In Exercises 5–8, let L be the line determined by points A and B.

  (a) Plot A and B.   **(b)** Find the slope of L.

  (c) Draw the graph of L.

5. $A(1, -2)$, $B(2, 1)$ **6.** $A(-2, -1)$, $B(1, -2)$

7. $A(2, 3)$, $B(-1, 3)$ **8.** $A(1, 2)$, $B(1, -3)$

In Exercise 9–12, write an equation for **(a)** the vertical line and **(b)** the horizontal line through the point P.

9. $P(2, 3)$ **10.** $P(-1, 4/3)$

11. $P(0, -\sqrt{2})$ **12.** $P(-\pi, 0)$

In Exercises 13–16, write the point-slope equation for the line through the point P with slope m.

13. $P(1, 1)$, $m = 1$ **14.** $P(-1, 1)$, $m = -1$

15. $P(0, 3)$, $m = 2$ **16.** $P(-4, 0)$, $m = -2$

In Exercises 17–20, write a general linear equation for the line through the two points.

17. $(0, 0)$, $(2, 3)$ **18.** $(1, 1)$, $(2, 1)$

19. $(-2, 0)$, $(-2, -2)$ **20.** $(-2, 1)$, $(2, -2)$

In Exercises 21–24, write the slope-intercept equation for the line with slope m and y-intercept b.

21. $m = 3$, $b = -2$ **22.** $m = -1$, $b = 2$

23. $m = -1/2$, $b = -3$ **24.** $m = 1/3$, $b = -1$

In Exercises 25 and 26, the line contains the origin and the point in the upper right corner of the grapher screen. Write an equation for the line.

25.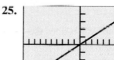

26.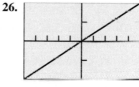

    $[-10, 10]$ by $[-25, 25]$ $[-5, 5]$ by $[-2, 2]$

In Exercises 27–30, find the **(a)** slope and **(b)** y-intercept, and **(c)** graph the line.

27. $3x + 4y = 12$ **28.** $x + y = 2$

29. $\dfrac{x}{3} + \dfrac{y}{4} = 1$ **30.** $y = 2x + 4$

In Exercises 31–34, write an equation for the line through P that is **(a)** parallel to L, **(b)** perpendicular to L.

31. $P(0, 0)$, $L: y = -x + 2$

32. $P(-2, 2)$, $L: 2x + y = 4$

33. $P(-2, 4)$, $L: x = 5$

34. $P(-1, 1/2)$, $L: y = 3$

In Exercises 35 and 36, a table of values is given for the linear function $f(x) = mx + b$. Determine m and b.

35.

x	$f(x)$
1	2
3	9
5	16

36.

x	$f(x)$
2	-1
4	-4
6	-7

In Exercises 37 and 38, find the value of x or y for which the line through A and B has the given slope m.

37. $A(-2, 3)$, $B(4, y)$, $m = -2/3$

38. $A(-8, -2)$, $B(x, 2)$, $m = 2$

In Exercises 39 and 40, use linear regression analysis.

39. Table 1.2 lists the ages and weights of nine girls.

Table 1.2 Girls' Ages and Weights

Age (months)	Weight (pounds)
19	22
21	23
24	25
27	28
29	31
31	28
34	32
38	34
43	39

(a) Find the linear regression equation for the data.

(b) Find the slope of the regression line. What does the slope represent?

(c) Superimpose the graph of the linear regression equation on a scatter plot of the data.

(d) Use the regression equation to predict the approximate weight of a 30-month-old girl.

40. Table 1.3 shows the mean annual compensation of construction workers.

Table 1.3 Construction Workers' Average Annual Compensation

Year	Annual Compensation (dollars)
1980	22,033
1985	27,581
1988	30,466
1989	31,465
1990	32,836

Source: U.S. Bureau of Economic Analysis.

(a) Find the linear regression equation for the data.

(b) Find the slope of the regression line. What does the slope represent?

(c) Superimpose the graph of the linear regression equation on a scatter plot of the data.

(d) Use the regression equation to predict the construction workers' average annual compensation in the year 2000.

41. *Revisiting Example 5* Show that you get the same equation in Example 5 if you use the point $(3, 4)$ to write the equation.

42. Writing to Learn *x- and y-intercepts*

(a) Explain why c and d are the x-intercept and y-intercept, respectively, of the line

$$\frac{x}{c} + \frac{y}{d} = 1.$$

(b) How are the x-intercept and y-intercept related to c and d in the line

$$\frac{x}{c} + \frac{y}{d} = 2?$$

43. *Parallel and Perpendicular Lines* For what value of k are the two lines $2x + ky = 3$ and $x + y = 1$ **(a)** parallel? **(b)** perpendicular?

In Exercises 44–46, *work in groups of two or three to solve the problem.*

44. *Insulation* By measuring slopes in the figure below, find the temperature change in degrees per inch for the following materials.

(a) gypsum wallboard

(b) fiberglass insulation

(c) wood sheathing

(d) Writing to Learn Which of the materials in (a)–(c) is the best insulator? the poorest? Explain.

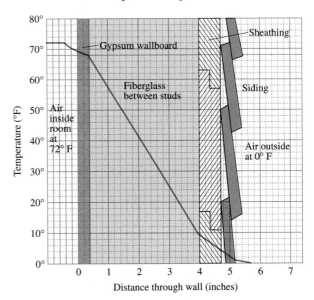

45. *Pressure under Water* The pressure p experienced by a diver under water is related to the diver's depth d by an equation of the form $p = kd + 1$ (k a constant). When $d = 0$ meters, the pressure is 1 atmosphere. The pressure at 100 meters is 10.94 atmospheres. Find the pressure at 50 meters.

46. *Modeling Distance Traveled* A car starts from point P at time $t = 0$ and travels at 45 mph.

(a) Write an expression $d(t)$ for the distance the car travels from P.

(b) Graph $y = d(t)$.

(c) What is the slope of the graph in (b)? What does it have to do with the car?

(d) Writing to Learn Create a scenario in which t could have negative values.

(e) Writing to Learn Create a scenario in which the y-intercept of $y = d(t)$ could be 30.

Extending the Ideas

47. The median price of existing single-family homes has increased consistently during the past two decades. However, the data in Table 1.4 show that there have been differences in various parts of the country.

Year	Table 1.4 Median Price of Single-Family Homes	
	Northeast (dollars)	Midwest (dollars)
1970	25,200	20,100
1975	39,300	30,100
1980	60,800	51,900
1985	88,900	58,900
1990	141,200	74,000

Source: National Association of Realtors®, *Home Sales Yearbook* (Washington, DC, 1990).

(a) Find the linear regression equation for home cost in the Northeast.

(b) What does the slope of the regression line represent?

(c) Find the linear regression equation for home cost in the Midwest.

(d) Where is the median price increasing more rapidly, in the Northeast or the Midwest?

48. *Fahrenheit versus Celsius* We found a relationship between Fahrenheit temperature and Celsius temperature in Example 8.

(a) Is there a temperature at which a Fahrenheit thermometer and a Celsius thermometer give the same reading? If so, what is it?

(b) Writing to Learn Graph $y_1 = (9/5)x + 32$, $y_2 = (5/9)(x - 32)$, and $y_3 = x$ in the same viewing window. Explain how this figure is related to the question in (a).

49. *Parallelogram* Three different parallelograms have vertices at $(-1, 1)$, $(2, 0)$, and $(2, 3)$. Draw the three and give the coordinates of the missing vertices.

50. *Parallelogram* Show that if the midpoints of consecutive sides of any quadrilateral are connected, the result is a parallelogram.

51. *Tangent Line* Consider the circle of radius 5 centered at $(0, 0)$. Find an equation of the line tangent to the circle at the point $(3, 4)$.

52. *Distance From a Point to a Line* This activity investigates how to find the distance from a point $P(a, b)$ to a line $L: Ax + By = C$.

We suggest that students *work in groups of two or three.*

(a) Write an equation for the line M through P perpendicular to L.

(b) Find the coordinates of the point Q in which M and L intersect.

(c) Find the distance from P to Q.

1.2 Functions and Graphs

Functions • Domains and Ranges • Viewing and Interpreting Graphs • Even Functions and Odd Functions—Symmetry • Functions Defined in Pieces • Absolute Value Function • Composite Functions

Functions

The values of one variable often depend on the values for another:

- The temperature at which water boils depends on elevation (the boiling point drops as you go up).

- The amount by which your savings will grow in a year depends on the interest rate offered by the bank.

- The area of a circle depends on the circle's radius.

In each of these examples, the value of one variable quantity depends on the value of another. For example, the boiling temperature of water, *b*, depends on the elevation, *e;* the amount of interest, *I*, depends on the interest rate, *r*. We call *b* and *I* **dependent variables** because they are determined by the values of the variables *e* and *r* on which they depend. The variables *e* and *r* are **independent variables.**

A rule that assigns to each element in one set a unique element in another set is called a *function.* The sets may be sets of any kind and do not have to be the same. A function is like a machine that assigns a unique output to every allowable input. The inputs make up the **domain** of the function; the outputs make up the **range** (Figure 1.8).

Figure 1.8 A "machine" diagram for a function.

> **Definition Function**
>
> A **function** from a set *D* to a set *R* is a rule that assigns a unique element in *R* to each element in *D*.

In this definition, *D* is the domain of the function and *R* is a set *containing* the range (Figure 1.9).

Euler invented a symbolic way to say "*y* is a function of *x*":

$$y = f(x),$$

which we read as "*y* equals *f* of *x*." This notation enables us to give different functions different names by changing the letters we use. To say that the boiling point of water is a function of elevation, we can write $b = f(e)$. To say that the area of a circle is a function of the circle's radius, we can write $A = A(r)$, giving the function the same name as the dependent variable.

The notation $y = f(x)$ gives a way to denote specific values of a function. The value of *f* at *a* can be written as $f(a)$, read "*f* of *a*."

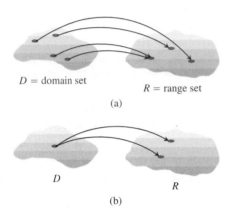

Figure 1.9 (a) A function from a set *D* to a set *R*. (b) *Not* a function. The assignment is not unique.

Example 1 THE CIRCLE-AREA FUNCTION

The domain of the circle-area function $A(r) = \pi r^2$ is the set of all possible radii—the set of all positive real numbers. The range is also the set of all positive real numbers.

The value of the function at $r = 2$ is

$$A(2) = \pi(2)^2 = 4\pi.$$

The area of a circle of radius 2 is 4π.

Domains and Ranges

In Example 1, the domain of the function is restricted by context: the independent variable is a radius and must be positive. When we define a function $y = f(x)$ with a formula and the domain is not stated explicitly or restricted by context, the domain is assumed to be the largest set of *x*-values for which the formula gives real *y*-values—the so-called **natural domain.** If we want to restrict the domain, we must say so. The domain of $y = x^2$ is understood to be the entire set of real numbers. We must write "$y = x^2, x > 0$" if we want to restrict the function to positive values of *x*.

Leonhard Euler (1707–1783)

Leonhard Euler, the dominant mathematical figure of his century and the most prolific mathematician ever, was also an astronomer, physicist, botanist, and chemist, and an expert in oriental languages. His work was the first to give the function concept the prominence that it has in mathematics today. Euler's collected books and papers fill 72 volumes. This does not count his enormous correspondence to approximately 300 addresses. His introductory algebra text, written originally in German (Euler was Swiss), is still available in English translation.

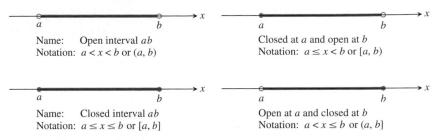

Name: Open interval ab
Notation: $a < x < b$ or (a, b)

Closed at a and open at b
Notation: $a \leq x < b$ or $[a, b)$

Name: Closed interval ab
Notation: $a \leq x \leq b$ or $[a, b]$

Open at a and closed at b
Notation: $a < x \leq b$ or $(a, b]$

Figure 1.10 Open and closed finite intervals.

Figure 1.11 Half-open finite intervals.

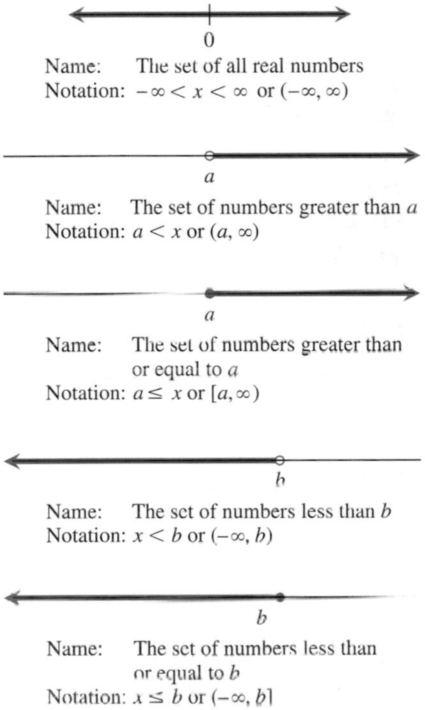

The domains and ranges of many real-valued functions of a real variable are intervals or combinations of intervals. The intervals may be open, closed, or half-open (Figures 1.10 and 1.11) and finite or infinite (Figure 1.12).

The endpoints of an interval make up the interval's **boundary** and are called **boundary points.** The remaining points make up the interval's **interior** and are called **interior points.** Closed intervals contain their boundary points. Open intervals contain no boundary points. Every point of an open interval is an interior point of the interval.

Example 2 IDENTIFYING DOMAIN AND RANGE OF A FUNCTION

Verify the domains of these functions.

Function	Domain (x)	Range (y)
$y = x^2$	$(-\infty, \infty)$	$[0, \infty)$
$y = 1/x$	$(-\infty, 0) \cup (0, \infty)$	$(-\infty, 0) \cup (0, \infty)$
$y = \sqrt{x}$	$[0, \infty)$	$[0, \infty)$
$y = \sqrt{4 - x}$	$(-\infty, 4]$	$[0, \infty)$
$y = \sqrt{1 - x^2}$	$[-1, 1]$	$[0, 1]$

Solution The formula $y = x^2$ gives a real y value for any real number x, so the domain is $(-\infty, \infty)$.

The formula $y = 1/x$ gives a real y-value for every real x-value except $x = 0$. *We cannot divide any number by* 0.

The formula $y = \sqrt{x}$ gives a real y-value only when x is positive or zero.

The formula $y = \sqrt{4 - x}$ gives a real y-value only when $4 - x$ is greater than or equal to zero. So, $0 \leq 4 - x$, or $x \leq 4$.

The formula $y = \sqrt{1 - x^2}$ gives a real y-value for every value of x in the closed interval from -1 to 1. Outside this interval $1 - x^2$ is negative and its square root is not a real number. The domain is $[-1, 1]$.

Viewing and Interpreting Graphs

The points (x, y) in the plane whose coordinates are the input-output pairs of a function $y = f(x)$ make up the function's **graph.** The graph of the function $y = x + 2$, for example, is the set of points with coordinates (x, y) for which y equals $x + 2$.

0

Name: The set of all real numbers
Notation: $-\infty < x < \infty$ or $(-\infty, \infty)$

a

Name: The set of numbers greater than a
Notation: $a < x$ or (a, ∞)

a

Name: The set of numbers greater than or equal to a
Notation: $a \leq x$ or $[a, \infty)$

b

Name: The set of numbers less than b
Notation: $x < b$ or $(-\infty, b)$

b

Name: The set of numbers less than or equal to b
Notation: $x \leq b$ or $(-\infty, b]$

Figure 1.12 Infinite intervals—rays on the number line and the number line itself. The symbol ∞ (infinity) is used merely for convenience; it does not mean there is a number ∞.

$y = \dfrac{1}{\sqrt{4-x^2}}$

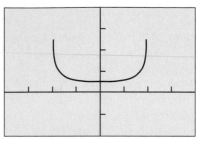

[−4, 4] by [−2, 4]

(a)

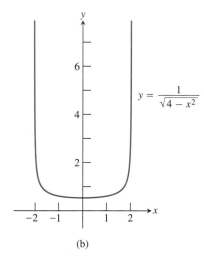

$y = \dfrac{1}{\sqrt{4 - x^2}}$

(b)

Figure 1.13 (a) Grapher failure. (b) A more accurate graph of $y = 1/\sqrt{4 - x^2}$. (Example 3)

Graphing $y = x^{2/3}$ — Possible Grapher Failure

On some graphing calculators you need to enter this function as $y = (x^2)^{1/3}$ or $y = (x^{1/3})^2$ to obtain a correct graph. Try graphing this function on your grapher.

Graphing with pencil and paper requires that you develop graph *drawing* skills. Graphing with a grapher requires that you develop graph *viewing* skills.

> **Graph Viewing Skills**
>
> **1.** Recognize that the graph is reasonable.
> **2.** See all the important characteristics of the graph.
> **3.** Interpret those characteristics.
> **4.** Recognize grapher failure.

Being able to recognize that a graph is reasonable comes with experience. You need to know the basic functions, their graphs, and how changes in their equations affect the graphs.

Grapher failure occurs when the graph produced by a grapher is less than precise—or even incorrect—usually due to the limitations of the screen resolution of the grapher.

Example 3   RECOGNIZING GRAPHER FAILURE

Find the domain and range of $y = f(x) = 1/\sqrt{4 - x^2}$.

Solution   The graph of f in Figure 1.13a seems to suggest that the domain of f is an interval between −2 and 2, and that the range is also a finite interval. The latter observation is the result of grapher failure; in this case, we can recognize the failure using algebra.

Solve Algebraically

The expression $4 - x^2$ must be greater than zero.

$$4 - x^2 > 0$$
$$x^2 < 4$$

Thus, $-2 < x < 2$, and the domain is $(-2, 2)$.

The smallest value of f is $1/2$ and occurs when $x = 0$. The values of f get very large as x approaches 2 from the left or −2 from the right, as suggested by the following table. (The values of f are rounded to three decimal places.)

x	±1.99	±1.999	±1.9999	±1.99999
$f(x)$	5.006	15.813	50.001	158.114

The range of f is $[0.5, \infty)$.

Figure 1.14 shows graphs of *power functions* that arise frequently in calculus. Knowing the general shapes of these graphs will help you recognize grapher failure. We will review other functions as the chapter continues.

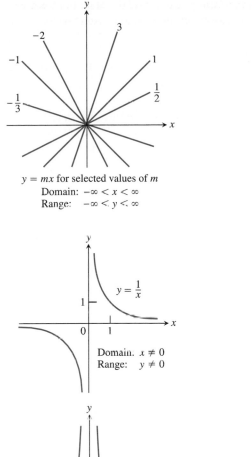

$y = mx$ for selected values of m
Domain: $-\infty < x < \infty$
Range: $-\infty < y < \infty$

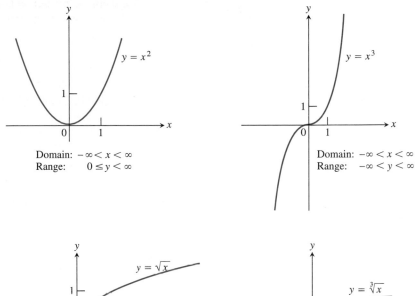

Domain: $-\infty < x < \infty$
Range: $0 \le y < \infty$

Domain: $-\infty < x < \infty$
Range: $-\infty < y < \infty$

$y = \dfrac{1}{x}$

Domain: $x \ne 0$
Range: $y \ne 0$

$y = \sqrt{x}$

Domain: $0 \le x < \infty$
Range: $0 \le y < \infty$

$y = \sqrt[3]{x}$

Domain: $-\infty < x < \infty$
Range: $-\infty < y < \infty$

$y = \dfrac{1}{x^2}$

Domain: $x \ne 0$
Range: $y > 0$

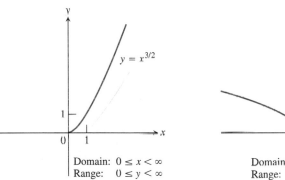

$y = x^{3/2}$

Domain: $0 \le x < \infty$
Range: $0 \le y < \infty$

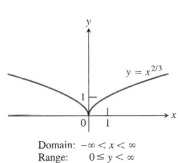

$y = x^{2/3}$

Domain: $-\infty < x < \infty$
Range: $0 \le y < \infty$

Figure 1.14 Useful power functions.

Even Functions and Odd Functions—Symmetry

The graphs of *even* and *odd* functions have important symmetry properties.

Definitions **Even Function, Odd Function**

A function $y = f(x)$ is an

> **even function of x** if $f(-x) = f(x)$,
>
> **odd function of x** if $f(-x) = -f(x)$,

for every x in the function's domain.

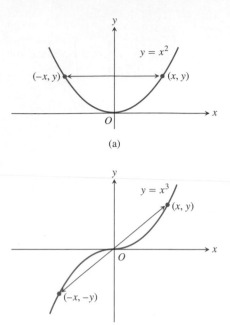

(a)

(b)

Figure 1.15 (a) The graph of $y = x^2$ (an even function) is symmetric about the y-axis. (b) The graph of $y = x^3$ (an odd function) is symmetric about the origin.

The names even and odd come from powers of x. If y is an even power of x, as in $y = x^2$ or $y = x^4$, it is an even function of x (because $(-x)^2 = x^2$ and $(-x)^4 = x^4$). If y is an odd power of x, as in $y = x$ or $y = x^3$, it is an odd function of x (because $(-x)^1 = -x$ and $(-x)^3 = -x^3$).

The graph of an even function is **symmetric about the y-axis.** Since $f(-x) = f(x)$, a point (x, y) lies on the graph if and only if the point $(-x, y)$ lies on the graph (Figure 1.15a).

The graph of an odd function is **symmetric about the origin.** Since $f(-x) = -f(x)$, a point (x, y) lies on the graph if and only if the point $(-x, -y)$ lies on the graph (Figure 1.15b). Equivalently, a graph is symmetric about the origin if a rotation of $180°$ about the origin leaves the graph unchanged.

Example 4 RECOGNIZING EVEN AND ODD FUNCTIONS

$f(x) = x^2$ Even function: $(-x)^2 = x^2$ for all x; symmetry about y-axis.

$f(x) = x^2 + 1$ Even function: $(-x)^2 + 1 = x^2 + 1$ for all x; symmetry about y-axis (Figure 1.16a).

$f(x) = x$ Odd function: $(-x) = -x$ for all x; symmetry about the origin.

$f(x) = x + 1$ Not odd: $f(-x) = -x + 1$, but $-f(x) = -x - 1$. The two are not equal. Not even: $(-x) + 1 \neq x + 1$ for all $x \neq 0$ (Figure 1.16b).

It is useful in graphing to recognize even and odd functions. Once we know the graph of either type of function on one side of the y-axis, we know its graph on both sides.

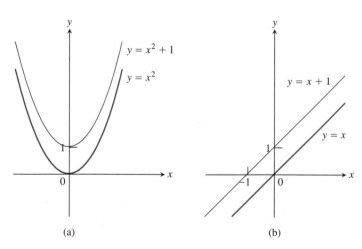

(a)

(b)

Figure 1.16 (a) When we add the constant term 1 to the function $y = x^2$, the resulting function $y = x^2 + 1$ is still even and its graph is still symmetric about the y-axis. (b) When we add the constant term 1 to the function $y = x$, the resulting function $y = x + 1$ is no longer odd. The symmetry about the origin is lost. (Example 4)

$$y = \begin{cases} -x, & x < 0 \\ x^2, & 0 \le x \le 1 \\ 1, & x > 1 \end{cases}$$

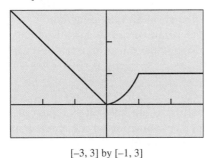

[−3, 3] by [−1, 3]

Figure 1.17 The graph of a piecewise defined function. (Example 5)

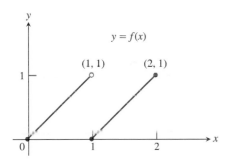

Figure 1.18 The segment on the left contains (0, 0) but not (1, 1). The segment on the right contains both of its endpoints. (Example 6)

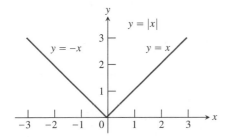

Figure 1.19 The absolute value function has domain $(-\infty, \infty)$ and range $[0, \infty)$.

Functions Defined in Pieces

While some functions are defined by single formulas, others are defined by applying different formulas to different parts of their domains.

Example 5 GRAPHING PIECEWISE DEFINED FUNCTIONS

Graph $y = f(x) = \begin{cases} -x, & x < 0 \\ x^2, & 0 \le x \le 1 \\ 1, & x > 1. \end{cases}$

Solution The values of f are given by three separate formulas: $y = -x$ when $x < 0$, $y = x^2$ when $0 \le x \le 1$, and $y = 1$ when $x > 1$. However, the function is *just one function*, whose domain is the entire set of real numbers (Figure 1.17).

Example 6 WRITING FORMULAS FOR PIECEWISE FUNCTIONS

Write a formula for the function $y = f(x)$ whose graph consists of the two line segments in Figure 1.18.

Solution We find formulas for the segments from (0, 0) to (1, 1) and from (1, 0) to (2, 1) and piece them together in the manner of Example 5.

Segment from (0, 0) to (1, 1) The line through (0, 0) and (1, 1) has slope $m = (1 - 0)/(1 - 0) = 1$ and y-intercept $b = 0$. Its slope-intercept equation is $y = x$. The segment from (0, 0) to (1, 1) that includes the point (0, 0) but not the point (1, 1) is the graph of the function $y = x$ restricted to the half-open interval $0 \le x < 1$, namely,

$$y = x, \quad 0 \le x < 1.$$

Segment from (1, 0) to (2, 1) The line through (1, 0) and (2, 1) has slope $m = (1 - 0)/(2 - 1) = 1$ and passes through the point (1, 0). The corresponding point-slope equation for the line is

$$y = 1(x - 1) + 0, \quad \text{or} \quad y = x - 1.$$

The segment from (1, 0) to (2, 1) that includes both endpoints is the graph of $y = x - 1$ restricted to the closed interval $1 \le x \le 2$, namely,

$$y = x - 1, \quad 1 \le x \le 2.$$

Piecewise formula Combining the formulas for the two pieces of the graph, we obtain

$$f(x) = \begin{cases} x, & 0 \le x < 1 \\ x - 1, & 1 \le x \le 2. \end{cases}$$

Absolute Value Function

The **absolute value function** $y = |x|$ is defined piecewise by the formula

$$|x| = \begin{cases} -x, & x < 0 \\ x, & x \ge 0. \end{cases}$$

The function is even, and its graph (Figure 1.19) is symmetric about the y-axis.

$y = |x - 2| - 1$

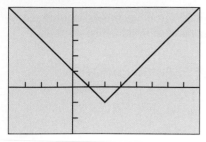

[–4, 8] by [–3, 5]

Figure 1.20 The lowest point of the graph of $f(x) = |x - 2| - 1$ is $(2, -1)$. (Example 7)

Figure 1.21 Two functions can be composed when a portion of the range of the first lies in the domain of the second.

Example 7 USING TRANSFORMATIONS

Find the domain, range, and draw the graph of $f(x) = |x - 2| - 1$.

Solution The graph of f is the graph of the absolute value function shifted 2 units horizontally to the right and 1 unit vertically downward (Figure 1.20). The domain of f is $(-\infty, \infty)$ and the range is $[-1, \infty)$.

Composite Functions

Suppose that some of the outputs of a function g can be used as inputs of a function f. We can then link g and f to form a new function whose inputs x are inputs of g and whose outputs are the numbers $f(g(x))$, as in Figure 1.21. We say that the function $f(g(x))$ (read "f of g of x") is the **composite of g and f**. It is made by *composing* g and f in the order of first g, then f. The usual "stand-alone" notation for this composite is $f \circ g$, which is read as "f of g." Thus, the value of $f \circ g$ at x is $(f \circ g)(x) = f(g(x))$.

Example 8 COMPOSING FUNCTIONS

Find a formula for $f(g(x))$ if $g(x) = x^2$ and $f(x) = x - 7$. Then find $f(g(2))$.

Solution To find $f(g(x))$, we replace x in the formula $f(x) = x - 7$ by the expression given for $g(x)$.

$$f(x) = x - 7$$

$$f(g(x)) = g(x) - 7 = x^2 - 7$$

We then find the value of $f(g(2))$ by substituting 2 for x.

$$f(g(2)) = (2)^2 - 7 = -3$$

Exploration 1 Composing Functions

Some graphers allow a function such as y_1 to be used as the independent variable of another function. With such a grapher, we can compose functions.

1. Enter the functions $y_1 = f(x) = 4 - x^2$, $y_2 = g(x) = \sqrt{x}$, $y_3 = y_2(y_1(x))$, and $y_4 = y_1(y_2(x))$. Which of y_3 and y_4 corresponds to $f \circ g$? to $g \circ f$?

2. Graph y_1, y_2, and y_3 and make conjectures about the domain and range of y_3.

3. Graph y_1, y_2, and y_4 and make conjectures about the domain and range of y_4.

4. Confirm your conjectures algebraically by finding formulas for y_3 and y_4.

Quick Review 1.2

In Exercises 1–6, solve for *x*.

1. $3x - 1 \le 5x + 3$

2. $x(x - 2) > 0$

3. $|x - 3| \le 4$

4. $|x - 2| \ge 5$

5. $x^2 < 16$

6. $9 - x^2 \ge 0$

In Exercises 7 and 8, describe how the graph of *f* can be transformed to the graph of *g*.

7. $f(x) = x^2$, $g(x) = (x + 2)^2 - 3$

8. $f(x) = |x|$, $g(x) = |x - 5| + 2$

In Exercises 9–12, find all real solutions to the equations.

9. $f(x) = x^2 - 5$

   (a) $f(x) = 4$ **(b)** $f(x) = -6$

10. $f(x) = 1/x$

   (a) $f(x) = -5$ **(b)** $f(x) = 0$

11. $f(x) = \sqrt{x + 7}$

   (a) $f(x) = 4$ **(b)** $f(x) = 1$

12. $f(x) = \sqrt[3]{x - 1}$

   (a) $f(x) = -2$ **(b)** $f(x) = 3$

Section 1.2 Exercises

In Exercises 1–4, write a formula that expresses the first variable as a function of the second.

1. the area of a circle as a function of its diameter

2. the height of an equilateral triangle as a function of its side length

3. the surface area of a cube as a function of the length of the cube's edges

4. the volume of a sphere as a function of the sphere's radius

In Exercises 5–18, do the following for the function.

   (a) Find the domain.   **(b)** Find the range.

   (c) Draw its graph.

   (d) Determine any symmetries discussed in this section that are characteristics of the graph.

5. $y = 4 - x^2$

6. $y = x^2 - 9$

7. $y = 2 + \sqrt{x - 1}$

8. $y = -\sqrt{-x}$

9. $y = 2\sqrt{3 - x}$

10. $y = \dfrac{1}{x - 2}$

11. $y = \sqrt[3]{x - 3}$

12. $y = \sqrt[3]{1 - x^2}$

13. $y = \sqrt[4]{-x}$

14. $y = 1 + \dfrac{1}{x}$

15. $y = \sqrt{4 - x^2}$

16. $y = x^{2/3}$

17. $y = 1 + \dfrac{1}{x^2}$

18. $y = x^{3/2}$

In Exercises 19–28, determine whether the function is even, odd, or neither. Try to answer without writing anything (except the answer).

19. $y = x^4$

20. $y = x + x^2$

21. $y = x + 2$

22. $y = x^2 - 3$

23. $y = \sqrt{x^2 + 2}$

24. $y = x + x^3$

25. $y = \dfrac{x^3}{x^2 - 1}$

26. $y = \sqrt[3]{2 - x}$

27. $y = \dfrac{1}{x - 1}$

28. $y = \dfrac{1}{x^2 - 1}$

In Exercises 29–34, **(a)** draw the graph of the function. Then find its **(b)** domain and **(c)** range.

29. $f(x) = -|3 - x| + 2$

30. $f(x) = 2|x + 4| - 3$

31. $f(x) = \begin{cases} 3 - x, & x \le 1 \\ 2x, & 1 < x \end{cases}$

32. $f(x) = \begin{cases} 1, & x < 0 \\ \sqrt{x}, & x \ge 0 \end{cases}$

33. $f(x) = \begin{cases} 4 - x^2, & x < 1 \\ (3/2)x + 3/2, & 1 \le x \le 3 \\ x + 3, & x > 3 \end{cases}$

34. $f(x) = \begin{cases} x^2, & x < 0 \\ x^3, & 0 \le x \le 1 \\ 2x - 1, & x > 1 \end{cases}$

35. Writing to Learn The *vertical line test* to determine whether a curve is the graph of a function states: If every vertical line in the *xy*-plane intersects a given curve in at most one point, then the curve is the graph of a function. Explain why this is true.

36. Writing to Learn For a curve to be *symmetric about the x-axis*, the point (x, y) must lie on the curve if and only if the point $(x, -y)$ lies on the curve. Explain why a curve that is symmetric about the *x*-axis is not the graph of a function, unless the function is $y > 0$.

In Exercises 37–40, use the vertical line test (see Exercise 35) to determine whether the curve is the graph of a function.

37.

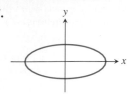

38.

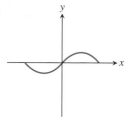

39.

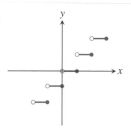

40.

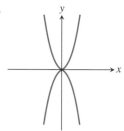

In Exercises 41–48, write a piecewise formula for the function.

41.

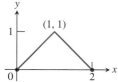

42.

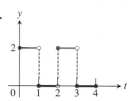

43.

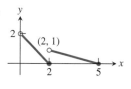

44.

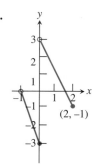

45.

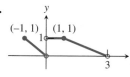

46.

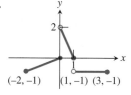

47.

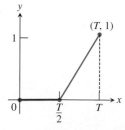

48.

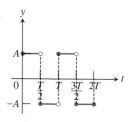

In Exercises 49 and 50, find

(a) $f(g(x))$ (b) $g(f(x))$ (c) $f(g(0))$

(d) $g(f(0))$ (e) $g(g(-2))$ (f) $f(f(x))$

49. $f(x) = x + 5$, $g(x) = x^2 - 3$

50. $f(x) = x + 1$, $g(x) = x - 1$

Explorations

In Exercises 51–54, (a) graph $f \circ g$ and $g \circ f$ and make a conjecture about the domain and range of each function.
(b) Then confirm your conjectures by finding formulas for $f \circ g$ and $g \circ f$.

51. $f(x) = x - 7$, $g(x) = \sqrt{x}$

52. $f(x) = 1 - x^2$, $g(x) = \sqrt{x}$

53. $f(x) = x^2 - 3$, $g(x) = \sqrt{x + 2}$

54. $f(x) = \dfrac{2x - 1}{x + 3}$, $g(x) = \dfrac{3x + 1}{2 - x}$ ■

In Exercises 55–58, use a graph to determine the domain and range of the function. Confirm your answer algebraically.

55. $f(x) = \dfrac{1}{\sqrt{x^2 - 4}}$ **56.** $f(x) = \dfrac{2}{\sqrt[4]{9 - x^2}}$

57. $f(x) = \dfrac{2}{\sqrt[3]{9 - x^2}}$ **58.** $f(x) = \dfrac{1}{\sqrt[3]{x^2 - 1}}$

In Exercises 59–62, a portion of the graph of a function defined on $[-2, 2]$ is shown. *Work in groups of two or three* to complete each graph assuming that the graph is (a) even, (b) odd.

59.

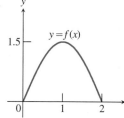

60.

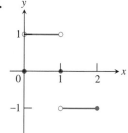

61.

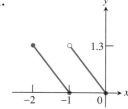

62.

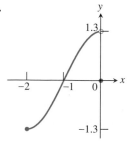

63. Copy and complete the following table.

	$g(x)$	$f(x)$	$(f \circ g)(x)$
(a)	?	$\sqrt{x-5}$	$\sqrt{x^2-5}$
(b)	?	$1 + 1/x$	x
(c)	$1/x$	?	x
(d)	$\sqrt{x}$	?	$|x|$

64. *Mutual Fund* Table 1.5 shows how an initial investment of $10,000 on December 31, 1977, grew over time.

Table 1.5 Merrill Lynch Capital Fund

Year	Amount (dollars)
1990	74,240
1991	92,570
1992	97,225
1993	110,551
1994	111,561
1995	148,226

Source: Data provided by Morningstar, Inc., as reported in *USA Today,* February 1, 1996.

(a) Find the power regression equation for the data in Table 1.5. Let $x = 0$ represent 1970, $x = 1$ represent 1971, and so forth.

(b) Superimpose the graph of the power regression equation on a scatter plot of the data.

(c) Use the graph of the power regression equation to predict the amount to which the initial $10,000 investment will grow in the year 2000. Confirm algebraically.

(d) Now use linear regression to predict the amount to which the initial $10,000 investment will grow in the year 2000.

65. *The Cone Problem* Begin with a circular piece of paper with a 4-in. radius as shown in (a). Cut out a sector with an arc length of x. Join the two edges of the remaining portion to form a cone with radius r and height h, as shown in (b).

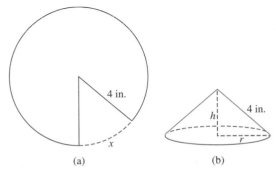

(a) (b)

(a) Explain why the circumference of the base of the cone is $8\pi - x$.

(b) Express the radius r as a function of x.

(c) Express the height h as a function of x.

(d) Express the volume V of the cone as a function of x.

66. *Industrial Costs* Dayton Power and Light, Inc., has a power plant on the Miami River where the river is 800 ft wide. To lay a new cable from the plant to a location in the city 2 mi downstream on the opposite side costs $180 per foot across the river and $100 per foot along the land.

(a) Suppose that the cable goes from the plant to a point Q on the opposite side that is x ft from the point P directly opposite the plant. Write a function $C(x)$ that gives the cost of laying the cable in terms of the distance x.

(b) Generate a table of values to determine if the least expensive location for point Q is less than 2000 ft or greater than 2000 ft from point P.

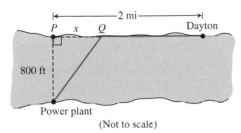

(Not to scale)

Extending the Ideas

67. Enter $y_1 = \sqrt{x}$, $y_2 = \sqrt{1 - x}$ and $y_3 = y_1 + y_2$ on your grapher.

(a) Graph y_3 in $[-3, 3]$ by $[-1, 3]$.

(b) Compare the domain of the graph of y_3 with the domains of the graphs of y_1 and y_2.

(c) Replace y_3 by

$$y_1 - y_2, \quad y_2 - y_1, \quad y_1 \cdot y_2, \quad y_1/y_2, \quad \text{and} \quad y_2/y_1,$$

in turn, and repeat the comparison of part (b).

(d) Based on your observations in (b) and (c), what would you conjecture about the domains of sums, differences, products, and quotients of functions?

68. *Even and Odd Functions*

(a) Must the product of two even functions always be even? Give reasons for your answer.

(b) Can anything be said about the product of two odd functions? Give reasons for your answer.

1.3 Exponential Functions

Exponential Growth • Exponential Decay • Applications • The Number e

Exponential Growth

Table 1.6 shows the growth of $100 invested in 1996 at an interest rate of 5.5%, compounded annually.

Table 1.6 Savings Account Growth

Year	Amount (dollars)	Increase (dollars)
1996	100	
1997	$100(1.055) = 105.50$	5.50
1998	$100(1.055)^2 = 111.30$	5.80
1999	$100(1.055)^3 = 117.42$	6.12
2000	$100(1.055)^4 = 123.88$	6.46

After the first year, the value of the account is always 1.055 times its value in the previous year. After n years, the value is $y = 100 \cdot (1.055)^n$.

Compound interest provides an example of *exponential growth* and is modeled by a function of the form $y = P \cdot a^x$, where P is the initial investment and a is equal to 1 plus the interest rate expressed as a decimal.

The equation $y = P \cdot a^x$, $a > 0$, $a \neq 1$, identifies a family of functions called *exponential functions*.

Exploration 1 **Exponential Functions**

1. Graph the function $y = a^x$ for $a = 2, 3, 5$, in a $[-5, 5]$ by $[-2, 5]$ viewing window.
2. For what values of x is it true that $2^x < 3^x < 5^x$?
3. For what values of x is it true that $2^x > 3^x > 5^x$?
4. For what values of x is it true that $2^x = 3^x = 5^x$?
5. Graph the function $y = (1/a)^x = a^{-x}$ for $a = 2, 3, 5$.
6. Repeat parts 2–4 for the functions in part 5.

Definition **Exponential Function**

Let a be a positive real number other than 1. The function

$$f(x) = a^x$$

is the **exponential function with base a.**

$y = 2^x$

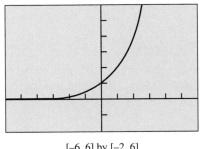

[–6, 6] by [–2, 6]

(a)

$y = 2^{-x}$

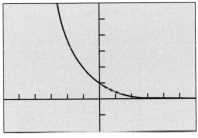

[–6, 6] by [–2, 6]

(b)

Figure 1.22 A graph of (a) $y = 2^x$ and (b) $y = 2^{-x}$.

The domain of $f(x) = a^x$ is $(-\infty, \infty)$ and the range is $(0, \infty)$. If $a > 1$, the graph of f looks like the graph of $y = 2^x$ in Figure 1.22a. If $0 < a < 1$, the graph of f looks like the graph of $y = 2^{-x}$ in Figure 1.22b.

Exponential functions obey the rules for exponents.

Rules for Exponents

If $a > 0$ and $b > 0$, the following hold for all real numbers x and y.

1. $a^x \cdot a^y = a^{x+y}$

2. $\dfrac{a^x}{a^y} = a^{x-y}$

3. $(a^x)^y = (a^y)^x = a^{xy}$

4. $a^x \cdot b^x = (ab)^x$

5. $\left(\dfrac{a}{b}\right)^x = \dfrac{a^x}{b^x}$

Population growth can sometimes be modeled with an exponential function. In Section 1.1 we gave some values for the population of the world in Table 1.1, which we repeat here as Table 1.7. We have also divided the population in one year by the population in the previous year to get an idea of how the population is growing. These ratios are in the third column.

Table 1.7 World Population

Year	Population (millions)	Ratio
1986	4936	$5023/4936 \approx 1.0176$
1987	5023	$5111/5023 \approx 1.0175$
1988	5111	$5201/5111 \approx 1.0176$
1989	5201	$5329/5201 \approx 1.0246$
1990	5329	$5422/5329 \approx 1.0175$
1991	5422	

Source: Statistical Office of the United Nations, *Monthly Bulletin of Statistics*, 1991.

Example 1 PREDICTING WORLD POPULATION

Use the data in Table 1.7 and an exponential model to predict the population of the world in the year 2010.

Solution Based on the third column of Table 1.7, we might be willing to conjecture that the population of the world in any year is about 1.018 times the population in the previous year.

If we start with the population in 1986, then according to the model the population (in millions) in 2010 would be about

$$4936(1.018)^{24} \approx 7573.9,$$

or about 7.6 billion people.

Exponential Decay

Exponential functions can also model phenomena that produce a decrease over time, such as happens with radioactive decay. The **half-life** of a radioactive substance is the amount of time it takes for half of the substance to change from its original radioactive state to a nonradioactive state by emitting energy in the form of radiation.

Example 2 MODELING RADIOACTIVE DECAY

Suppose the half-life of a certain radioactive substance is 20 days and that there are 5 grams present initially. When will there be only 1 gram of the substance remaining?

Solution

Model

The number of grams remaining after 20 days is

$$5\left(\frac{1}{2}\right) = \frac{5}{2}.$$

The number of grams remaining after 40 days is

$$5\left(\frac{1}{2}\right)\left(\frac{1}{2}\right) = 5\left(\frac{1}{2}\right)^2 = \frac{5}{4}.$$

The function $y = 5(1/2)^{t/20}$ models the mass in grams of the radioactive substance after t days.

Solve Graphically

Figure 1.23 shows that the graphs of $y_1 = 5(1/2)^{t/20}$ and $y_2 = 1$ (for 1 gram) intersect when t is approximately 46.44.

Interpret

There will be 1 gram of the radioactive substance left after approximately 46.44 days, or about 46 days 10.5 hours.

Compound interest investments, population growth, and radioactive decay are all examples of *exponential growth and decay.*

$y = 5\left(\frac{1}{2}\right)^{t/20}, y = 1$

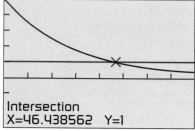

Intersection
X=46.438562   Y=1

[0, 80] by [−3, 5]

Figure 1.23 (Example 2)

> **Definitions   Exponential Growth, Exponential Decay**
>
> The function $y = k \cdot a^x$, $k > 0$ is a model for **exponential growth** if $a > 1$, and a model for **exponential decay** if $0 < a < 1$.

Applications

Most graphers have the exponential growth and decay model $y = k \cdot a^x$ built in as an exponential regression equation. We use this feature in Example 3 to analyze the U.S. population from the data in Table 1.8.

Table 1.8 U.S. Population

Year	Population (millions)
1880	50.2
1890	63.0
1900	76.0
1910	92.0
1920	105.7
1930	122.8
1940	131.7
1950	151.3
1960	179.3
1970	203.3

Source: The Statesman's Yearbook, 129th ed. (London: The Macmillan Press, Ltd., 1992).

Example 3 PREDICTING THE U.S. POPULATION

Use the population data in Table 1.8 to estimate the population for the year 1990. Compare the result with the actual 1990 population of approximately 250 million.

Solution

Model

Let $x = 0$ represent 1880, $x = 1$ represent 1890, and so on. We enter the data into the grapher and find the exponential regression equation to be

$$f(x) = 55.05(1.16063^x).$$

Figure 1.24 shows the graph of f superimposed on the scatter plot of the data.

Solve Graphically

The year 1990 is represented by $x = 11$. Reading from the curve, we find

$$f(11) \approx 283.4.$$

The exponential model estimates the 1990 population to be 283.4 million, an overestimate of approximately 33 million, or about 13%.

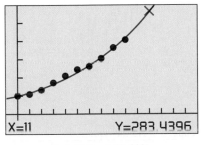

$y = 55.05(1.16063^x)$

X=11   Y=283.4396

[−1, 15] by [−50, 300]

Figure 1.24 (Example 3)

Example 4 INTERPRETING EXPONENTIAL REGRESSION

What *annual* rate of growth can we infer from the exponential regression equation in Example 3?

Solution Let r be the annual rate of growth of the U.S. population, expressed as a decimal. Because the time increments we used were 10-year intervals, we have

$$(1 + r)^{10} \approx 1.16063$$

$$r \approx \sqrt[10]{1.16063} - 1$$

$$r \approx 0.015.$$

The annual rate of growth is about 1.5%. This is surprisingly close to the 1.8% rate in Example 1.

The Number e

Many natural, physical, and economic phenomena are best modeled by an exponential function whose base is the famous number e, which is 2.718281828 to nine decimal places. We can define e to be the number that the function $f(x) = (1 + 1/x)^x$ approaches as x approaches infinity. The graph and table in Figure 1.25 strongly suggest that such a number exists.

The exponential functions $y = e^x$ and $y = e^{-x}$ are frequently used as models of exponential growth or decay. For example, interest **compounded continuously** uses the model $y = P \cdot e^{rt}$, where P is the initial investment, r is the interest rate as a decimal, and t is time in years.

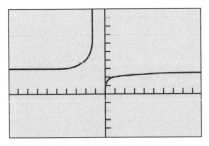

$y = (1 + 1/x)^x$

[−10, 10] by [−5, 10]

X	Y₁	
1000	2.7169	
2000	2.7176	
3000	2.7178	
4000	2.7179	
5000	2.718	
6000	2.7181	
7000	2.7181	

$Y_1 = (1 + 1/X)^X$

Figure 1.25 A graph and table of values for $f(x) = (1 + 1/x)^x$ both suggest that as $x \to \infty$, $f(x) \to e \approx 2.718$.

Quick Review 1.3

In Exercises 1–3, evaluate the expression. Round your answers to 3 decimal places.

1. $5^{2/3}$

2. $3^{\sqrt{2}}$

3. $3^{-1.5}$

In Exercises 4–6, solve the equation. Round your answers to 4 decimal places.

4. $x^3 = 17$

5. $x^5 = 24$

6. $x^{10} = 1.4567$

In Exercises 7 and 8, find the value of investing P dollars for n years with the interest rate r compounded annually.

7. $P = \$500, \quad r = 4.75\%, \quad n = 5$ years

8. $P = \$1000, \quad r = 6.3\%, \quad n = 3$ years

In Exercises 9 and 10, simplify the exponential expression.

9. $\dfrac{(x^{-3}y^2)^2}{(x^4y^3)^3}$

10. $\left(\dfrac{a^3b^{-2}}{c^4}\right)^2\left(\dfrac{a^4c^{-2}}{b^3}\right)^{-1}$

Section 1.3 Exercises

In Exercises 1–6 , match the function with its graph. Try to do it without using your grapher.

1. $y = 2^x$

2. $y = 3^{-x}$

3. $y = -3^{-x}$

4. $y = -0.5^{-x}$

5. $y = 2^{-x} - 2$

6. $y = 1.5^x - 2$

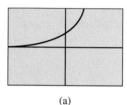

(a)

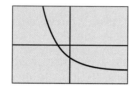

(b)

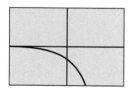

(c)

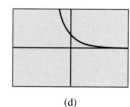

(d)

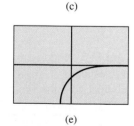

(e)

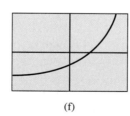

(f)

In Exercises 7–10, graph the function. State its domain, range, and intercepts.

7. $y = -2^x + 3$

8. $y = e^x + 3$

9. $y = 3 \cdot e^{-x} - 2$

10. $y = -2^{-x} - 1$

In Exercises 11–14, rewrite the exponential expression to have the indicated base.

11. 9^{2x}, base 3

12. 16^{3x}, base 2

13. $(1/8)^{2x}$, base 2

14. $(1/27)^x$, base 3

In Exercises 15–18, use graphs to solve the equations.

15. $2^x = 5$

16. $e^x = 4$

17. $3^x - 0.5 = 0$

18. $3 - 2^{-x} = 0$

In Exercises 19–22, copy and *work in groups of two or three* to complete the table for the function.

19. $y = 2x - 3$

x	y	Change (Δy)
1	?	
		?
2	?	
		?
3	?	
		?
4	?	

20. $y = -3x + 4$

x	y	Change (Δy)
1	?	
		?
2	?	
		?
3	?	
		?
4	?	

21. $y = x^2$

x	y	Change (Δy)
1	?	
		?
2	?	
		?
3	?	
		?
4	?	

22. $y = 3e^x$

x	y	Ratio (y_i/y_{i-1})
1	?	
		?
2	?	
		?
3	?	
		?
4	?	

In Exercises 23–34, use an exponential model to solve the problem.

23. *Population Growth* The population of Knoxville is 500,000 and is increasing at the rate of 3.75% each year. Approximately when will the population reach 1 million?

24. *Population Growth* The population of Silver Run in the year 1890 was 6250. Assume the population increased at a rate of 2.75% per year.

  (a) Estimate the population in 1915 and 1940.

  (b) Approximately when did the population reach 50,000?

25. *Radioactive Decay* The half-life of phosphorus-32 is about 14 days. There are 6.6 grams present initially.

  (a) Express the amount of phosphorus-32 remaining as a function of time t.

  (b) When will there be 1 gram remaining?

26. *Finding Time* If John invests $2300 in a savings account with a 6% interest rate compounded annually, how long will it take until John's account has a balance of $4150?

27. *Doubling Your Money* Determine how much time is required for an investment to double in value if interest is earned at the rate of 6.25% compounded annually.

28. *Doubling Your Money* Determine how much time is required for an investment to double in value if interest is earned at the rate of 6.25% compounded monthly.

29. *Doubling Your Money* Determine how much time is required for an investment to double in value if interest is earned at the rate of 6.25% compounded continuously.

30. *Tripling Your Money* Determine how much time is required for an investment to triple in value if interest is earned at the rate of 5.75% compounded annually.

31. *Tripling Your Money* Determine how much time is required for an investment to triple in value if interest is earned at the rate of 5.75% compounded daily.

32. *Tripling Your Money* Determine how much time is required for an investment to triple in value if interest is earned at the rate of 5.75% compounded continuously.

33. *Cholera Bacteria* Suppose that a colony of bacteria starts with 1 bacterium and doubles in number every half hour. How many bacteria will the colony contain at the end of 24 h?

34. *Eliminating a Disease* Suppose that in any given year, the number of cases of a disease is reduced by 20%. If there are 10,000 cases today, how many years will it take

  (a) to reduce the number of cases to 1000?

  (b) to eliminate the disease; that is, to reduce the number of cases to less than 1?

35. Writing to Learn Explain how the change Δy is related to the slopes of the lines in Exercises 19 and 20. If the changes in x are constant for a linear function, what would you conclude about the corresponding changes in y?

36. *Bacteria Growth* The number of bacteria in a petri dish culture after t hours is

$$B = 100e^{0.693t}.$$

  (a) What was the initial number of bacteria present?

  (b) How many bacteria are present after 6 hours?

  (c) Approximately when will the number of bacteria be 200? Estimate the doubling time of the bacteria.

37. Table 1.9 gives some data about the population of Mexico.

Table 1.9 Population of Mexico

Year	Population (millions)
1950	25.8
1960	34.9
1970	48.2
1980	66.8
1990	81.1

Source: The Statesman's Yearbook, 129th ed. (London: The Macmillan Press, Ltd., 1992).

  (a) Let $x = 0$ represent 1900, $x = 1$ represent 1901, and so forth. Find an exponential regression equation for the data and superimpose its graph on a scatter plot of the data.

  (b) Use the exponential regression equation to estimate the population of Mexico in 1900. How close is the estimate to the actual population in 1900 of 13,607,272?

  (c) Use the exponential regression equation to estimate the annual rate of growth of the population of Mexico.

38. Table 1.10 gives population data for South Africa.

Table 1.10 Population of South Africa

Year	Population (millions)
1904	5.2
1911	6.0
1921	6.9
1936	9.6
1946	11.4
1951	12.7
1960	16.0
1970	18.3
1980	20.6

Source: The Statesman's Yearbook, 129th ed. (London: The Macmillan Press, Ltd., 1992).

  (a) Let $x = 0$ represent 1900, $x = 1$ represent 1901, and so forth. Find an exponential regression equation for the data and superimpose its graph on a scatter plot of the data.

  (b) Use the exponential regression equation to estimate the population of South Africa in 1990.

  (c) Use the exponential regression equation to estimate the annual rate of growth of the population of South Africa.

39. World Population *(continuation of Example 1)* Use 1.018 and the population in 1991 to estimate the population of the world in the year 2010.

Exploration

40. Let $y_1 = x^2$ and $y_2 = 2^x$.

(a) Graph y_1 and y_2 in $[-5, 5]$ by $[-2, 10]$. How many times do you think the two graphs cross?

(b) Compare the corresponding changes in y_1 and y_2 as x changes from 1 to 2, 2 to 3, and so on. How large must x be for the changes in y_2 to overtake the changes in y_1?

(c) Solve for x: $x^2 = 2^x$.

(d) Solve for x: $x^2 < 2^x$.

Extending the Ideas

In Exercises 41 and 42, assume that the graph of the exponential function $f(x) = k \cdot a^x$ passes through the two points. Find the values of a and k.

41. $(1, 4.5), (-1, 0.5)$ **42.** $(1, 1.5), (-1, 6)$

■

1.4 Parametric Equations

Relations • Circles • Ellipses • Lines and Other Curves

Relations

A **relation** is a set of ordered pairs (x, y) of real numbers. The **graph of a relation** is the set of points in the plane that correspond to the ordered pairs of the relation. If x and y are *functions* of a third variable t, called a *parameter*, then we can use the *parametric mode* of a grapher to obtain a graph of the relation.

$x = \sqrt{t}, y = t$

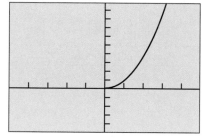

[–5, 5] by [–5, 10]

Figure 1.26 You must choose a *smallest* and *largest* value for t in parametric mode. Here we used 0 and 10, respectively. (Example 1)

Example 1 GRAPHING HALF A PARABOLA

Describe the graph of the relation determined by

$$x = \sqrt{t}, \quad y = t, \quad t \geq 0.$$

Solution Set $x_1 = \sqrt{t}$, $y_1 = t$, and use the parametric mode of the grapher to draw the graph in Figure 1.26. The graph appears to be the right half of the parabola $y = x^2$. Notice that there is no information about t on the graph itself.

Confirm Algebraically

Both x and y will be greater than or equal to zero because $t \geq 0$. Eliminating t we find that for every value of t,

$$y = t = (\sqrt{t})^2 = x^2.$$

Thus, the relation is the function $y = x^2$, $x \geq 0$.

Definitions Parametric Curve, Parametric Equations

If x and y are given as functions

$$x = f(t), \quad y = g(t)$$

over an interval of t-values, then the set of points $(x, y) = (f(t), g(t))$ defined by these equations is a **parametric curve.** The equations are **parametric equations** for the curve.

The variable t is a **parameter** for the curve and its domain I is the **parameter interval**. If I is a closed interval, $a \le t \le b$, the point $(f(a), g(a))$ is the **initial point** of the curve and the point $(f(b), g(b))$ is the **terminal point** of the curve. When we give parametric equations and a parameter interval for a curve, we say that we have **parametrized** the curve. The equations and interval constitute a **parametrization** of the curve.

In Example 1, the parameter interval is $[0, \infty)$, so $(0, 0)$ is the initial point and there is no terminal point.

A grapher can draw a parametrized curve only over a closed interval, so the portion it draws has endpoints even when the curve being graphed does not. Keep this in mind when you graph.

Circles

In applications, t often denotes time, an angle, or the distance a particle has traveled along its path from its starting point. In fact, parametric graphing can be used to simulate the motion of the particle.

Exploration 1 **Parametrizing Circles**

Let $x = a \cos t$ and $y = a \sin t$.

1. Let $a = 1, 2,$ or 3 and graph the parametric equations in a *square viewing window* using the parameter interval $[0, 2\pi]$. How does changing a affect this graph?

2. Let $a = 2$ and graph the parametric equations using the following parameter intervals: $[0, \pi/2]$, $[0, \pi]$, $[0, 3\pi/2]$, $[2\pi, 4\pi]$, and $[0, 4\pi]$. Describe the role of the length of the parameter interval.

3. Let $a = 3$ and graph the parametric equations using the intervals $[\pi/2, 3\pi/2]$, $[\pi, 2\pi]$, $[3\pi/2, 3\pi]$, and $[\pi, 5\pi]$. What are the initial point and terminal point in each case?

4. Graph $x = 2 \cos(-t)$ and $y = 2 \sin(-t)$ using the parameter intervals $[0, 2\pi]$, $[\pi, 3\pi]$, and $[\pi/2, 3\pi/2]$. In each case, describe how the graph is traced.

For $x = a \cos t$ and $y = a \sin t,$ we have

$$x^2 + y^2 = a^2 \cos^2 t + a^2 \sin^2 t = a^2(\cos^2 t + \sin^2 t) = a^2(1) = a^2,$$

using the identity $\cos^2 t + \sin^2 t = 1$. Thus, the curves in Exploration 1 were either circles or portions of circles, each with center at the origin.

Ellipses

Parametrizations of ellipses are similar to parametrizations of circles. Recall that the standard form of an ellipse centered at $(0, 0)$ is

$$\frac{x^2}{a^2} + \frac{y^2}{b^2} = 1.$$

Example 2 GRAPHING AN ELLIPSE

Graph the parametric curve $x = 3 \cos t$, $y = 4 \sin t$, $0 \le t \le 2\pi$.

Find a Cartesian equation for a curve that contains the parametric curve. What portion of the graph of the Cartesian equation is traced by the parametric curve?

Solution Figure 1.27 suggests that the curve is an ellipse. The Cartesian equation is

$$\left(\frac{x}{3}\right)^2 + \left(\frac{y}{4}\right)^2 = \cos^2 t + \sin^2 t = 1,$$

so the parametrized curve lies along an ellipse with major axis endpoints $(0, \pm 4)$ and minor axis endpoints $(\pm 3, 0)$. As t increases from 0 to 2π, the point $(x, y) = (3 \cos t, 4 \sin t)$ starts at $(3, 0)$ and traces the entire ellipse once counterclockwise.

$x = 3 \cos t, y = 4 \sin t$

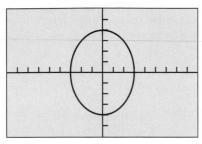

[–9, 9] by [–6, 6]

Figure 1.27 A graph of the parametric equations $x = 3 \cos t$, $y = 4 \sin t$ for $0 \le t \le 2\pi$. (Example 2)

Exploration 2 **Parametrizing Ellipses**

Let $x = a \cos t$ and $y = b \sin t$.

1. Let $a = 2$ and $b = 3$. Then graph using the parameter interval $[0, 2\pi]$. Repeat, changing b to 4, 5, and 6.
2. Let $a = 3$ and $b = 4$. Then graph using the parameter interval $[0, 2\pi]$. Repeat, changing a to 5, 6, and 7.
3. Based on parts 1 and 2, how do you identify the axis that contains the major axis of the ellipse? the minor axis?
4. Let $a = 4$ and $b = 3$. Then graph using the parameter intervals $[0, \pi/2]$, $[0, \pi]$, $[0, 3\pi/2]$, and $[0, 4\pi]$. Describe the role of the length of the parameter interval.
5. Graph $x = 5 \cos (-t)$ and $y = 2 \sin (-t)$ using the parameter intervals $[0, 2\pi]$, $[\pi, 3\pi]$, and $[\pi/2, 3\pi/2]$. Describe how the graph is traced. What are the initial point and terminal point in each case?

For $x = a \cos t$ and $y = b \sin t$, we have $(x/a)^2 + (y/b)^2 = \cos^2 t + \sin^2 t = 1$. Thus, the curves in Exploration 2 were either ellipses or portions of ellipses, each with center at the origin.

In the exercises you will see how to graph hyperbolas parametrically.

Lines and Other Curves

Lines, line segments, and many other curves can be defined parametrically.

$x = 3t, y = 2 - 2t$

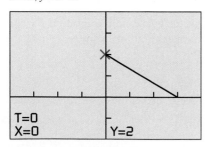

T=0
X=0   Y=2

[–4, 4] by [–2, 4]

Figure 1.28 The graph of the line segment $x = 3t$, $y = 2 - 2t$, $0 \le t \le 1$, with trace on the initial point $(0, 2)$. (Example 3)

Example 3 GRAPHING A LINE SEGMENT

Draw and identify the graph of the parametric curve determined by

$$x = 3t, \quad y = 2 - 2t, \quad 0 \le t \le 1.$$

Solution The graph (Figure 1.28) appears to be a line segment with endpoints $(0, 2)$ and $(3, 0)$.

$x = 2 \cot t,\ y = 2 \sin^2 t$

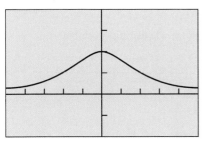

[−5, 5] by [−2, 4]

Figure 1.29 The witch of Agnesi. (Exploration 3)

Maria Agnesi (1718–1799)

The first text to include differential and integral calculus along with analytic geometry, infinite series, and differential equations was written in the 1740s by the Italian mathematician Maria Gaetana Agnesi, a gifted scholar and linguist whose Latin essay defending higher education for women was published when she was only nine years old, was a well-published scientist by age 20, and an honorary faculty member of the University of Bologna by age 30.

Today, Agnesi is remembered chiefly for a bell-shaped curve called *the witch of Agnesi.* This name, found only in English texts, is the result of a mistranslation. Agnesi's own name for the curve was *versiera* or "turning curve." John Colson, a noted Cambridge mathematician, probably confused versiera with *avversiera*, which means "wife of the devil" and translated it into "witch."

Confirm Algebraically

When $t = 0$, the equations give $x = 0$ and $y = 2$. When $t = 1$, they give $x = 3$ and $y = 0$. When we substitute $t = x/3$ into the y equation, we obtain

$$y = 2 - 2\left(\frac{x}{3}\right) = -\frac{2}{3}x + 2.$$

Thus, the parametric curve traces the segment of the line $y = -(2/3)x + 2$ from the point $(0, 2)$ to $(3, 0)$.

If we change the parameter interval $[0, 1]$ in Example 3 to $(-\infty, \infty)$, the parametrization will trace the entire line $y = -(2/3)x + 2$.

The bell-shaped curve in Exploration 3 is the famous witch of Agnesi. You will find more information about this curve in Exercise 41.

Exploration 3 **Graphing the Witch of Agnesi**

The witch of Agnesi is the curve

$$x = 2 \cot t, \quad y = 2 \sin^2 t, \quad 0 < t < \pi.$$

1. Draw the curve using the window in Figure 1.29. What did you choose as a closed parameter interval for your grapher? In what direction is the curve traced? How far to the left and right of the origin do you think the curve extends?

2. Graph the same parametric equations using the parameter intervals $(-\pi/2, \pi/2)$, $(0, \pi/2)$, and $(\pi/2, \pi)$. In each case, describe the curve you see and the direction in which it is traced by your grapher.

3. What happens if you replace $x = 2 \cot t$ by $x = -2 \cot t$ in the original parametrization? What happens if you use $x = 2 \cot (\pi - t)$?

Example 4 PARAMETRIZING A LINE SEGMENT

Find a parametrization for the line segment with endpoints $(-2, 1)$ and $(3, 5)$.

Solution Using $(-2, 1)$ we create the parametric equations

$$x = -2 + at, \quad y = 1 + bt.$$

These represent a line, as we can see by solving each equation for t and equating to obtain

$$\frac{x + 2}{a} = \frac{y - 1}{b}.$$

This line goes through the point $(-2, 1)$ when $t = 0$. We determine a and b so that the line goes through $(3, 5)$ when $t = 1$.

$$3 = -2 + a \quad \Rightarrow \quad a = 5 \quad x = 3 \text{ when } t = 1.$$

$$5 = 1 + b \quad \Rightarrow \quad b = 4 \quad y = 5 \text{ when } t = 1.$$

Therefore,

$$x = -2 + 5t, \quad y = 1 + 4t, \quad 0 \le t \le 1$$

is a parametrization of the line segment with initial point $(-2, 1)$ and terminal point $(3, 5)$.

Quick Review 1.4

In Exercises 1–3, write an equation for the line.

1. the line through the points (1, 8) and (4, 3)

2. the horizontal line through the point (3, −4)

3. the vertical line through the point (2, −3)

In Exercises 4–6, find the *x*- and *y*-intercepts of the graph of the relation.

4. $\dfrac{x^2}{9} + \dfrac{y^2}{16} = 1$

5. $\dfrac{x^2}{16} - \dfrac{y^2}{9} = 1$

6. $2y^2 = x + 1$

In Exercises 7 and 8, determine whether the given points lie on the graph of the relation.

7. $2x^2y + y^2 = 3$

(a) (1, 1)   (b) (−1, −1)   (c) (1/2, −2)

8. $9x^2 - 18x + 4y^2 = 27$

(a) (1, 3)   (b) (1, −3)   (c) (−1, 3)

9. Solve for *t*.

(a) $2x + 3t = -5$ (b) $3y - 2t = -1$

10. For what values of *a* is each equation true?

(a) $\sqrt{a^2} = a$ (b) $\sqrt{a^2} = \pm a$ (c) $\sqrt{4a^2} = 2|a|$

Section 1.4 Exercises

In Exercises 1–4, match the parametric equations with their graph. State the approximate dimensions of the viewing window. Give a parameter interval that traces the curve exactly once.

1. $x = 3 \sin (2t)$, $y = 1.5 \cos t$

2. $x = \sin^3 t$, $y = \cos^3 t$

3. $x = 7 \sin t - \sin (7t)$, $y = 7 \cos t - \cos (7t)$

4. $x = 12 \sin t - 3 \sin (6t)$, $y = 12 \cos t + 3 \cos (6t)$

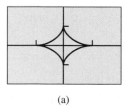

(a)

(b)

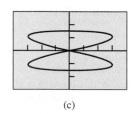

(c)

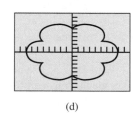

(d)

Explorations

5. *Hyperbolas* Let $x = a \sec t$ and $y = b \tan t$.

(a) **Writing to Learn** Let $a = 1, 2,$ or 3, $b = 1, 2,$ or 3, and graph using the parameter interval $(-\pi/2, \pi/2)$. Explain what you see, and describe the role of *a* and *b* in these parametric equations. (Caution: If you get what appear to be asymptotes, try using the approximation $[-1.57, 1.57]$ for the parameter interval.)

(b) Let $a = 2$, $b = 3$, and graph in the parameter interval $(\pi/2, 3\pi/2)$. Explain what you see.

(c) **Writing to Learn** Let $a = 2$, $b = 3$, and graph using the parameter interval $(-\pi/2, 3\pi/2)$. Explain why you must be careful about graphing in this interval or any interval that contains $\pm\pi/2$.

(d) Use algebra to explain why

$$\left(\frac{x}{a}\right)^2 - \left(\frac{y}{b}\right)^2 = 1.$$

(e) Let $x = a \tan t$ and $y = b \sec t$. Repeat (a), (b), and (d) using an appropriate version of (d).

6. *Transformations* Let $x = (2 \cos t) + h$ and $y = (2 \sin t) + k$.

(a) **Writing to Learn** Let $k = 0$ and $h = -2, -1, 1,$ and 2, in turn. Graph using the parameter interval $[0, 2\pi]$. Describe the role of *h*.

(b) **Writing to Learn** Let $h = 0$ and $k = -2, -1, 1,$ and 2, in turn. Graph using the parameter interval $[0, 2\pi]$. Describe the role of *k*.

(c) Find a parametrization for the circle with radius 5 and center at (2, −3).

(d) Find a parametrization for the ellipse centered at (−3, 4) with semimajor axis of length 5 parallel to the *x*-axis and semiminor axis of length 2 parallel to the *y*-axis. ■

In Exercises 7–26, a parametrization is given for a curve.

(a) Graph the curve. What are the initial and terminal points, if any? Indicate the direction in which the curve is traced.

(b) Find a Cartesian equation for a curve that contains the parametrized curve. What portion of the graph of the Cartesian equation is traced by the parametrized curve?

7. $x = \cos t$, $y = \sin t$, $0 \le t \le \pi$

8. $x = \sin (2\pi t)$, $y = \cos (2\pi t)$, $0 \le t \le 1$

9. $x = \cos(\pi - t), \quad y = \sin(\pi - t), \quad 0 \le t \le \pi$

10. $x = 4\cos t, \quad y = 2\sin t, \quad 0 \le t \le 2\pi$

11. $x = 4\sin t, \quad y = 2\cos t, \quad 0 \le t \le \pi$

12. $x = 4\sin t, \quad y = 5\cos t, \quad 0 \le t \le 2\pi$

13. $x = 3t, \quad y = 9t^2, \quad -\infty < t < \infty$

14. $x = -\sqrt{t}, \quad y = t, \quad t \ge 0$

15. $x = t, \quad y = \sqrt{t}, \quad t \ge 0$

16. $x = (\sec^2 t) - 1, \quad y = \tan t, \quad -\pi/2 < t < \pi/2$

17. $x = -\sec t, \quad y = \tan t, \quad -\pi/2 < t < \pi/2$

18. $x = \tan t, \quad y = -2\sec t, \quad -\pi/2 < t < \pi/2$

19. $x = 2t - 5, \quad y = 4t - 7, \quad -\infty < t < \infty$

20. $x = 1 - t, \quad y = 1 + t, \quad -\infty < t < \infty$

21. $x = t, \quad y = 1 - t, \quad 0 \le t \le 1$

22. $x = 3 - 3t, \quad y = 2t, \quad 0 \le t \le 1$

23. $x = 4 - \sqrt{t}, \quad y = \sqrt{t}, \quad 0 \le t$

24. $x = t^2, \quad y = \sqrt{4 - t^2}, \quad 0 \le t \le 2$

25. $x = \sin t, \quad y = \cos 2t, \quad -\infty < t < \infty$

26. $x = t^2 - 3, \quad y = t, \quad t \le 0$

In Exercises 27–32, find a parametrization for the curve.

27. the line segment with endpoints $(-1, -3)$ and $(4, 1)$

28. the line segment with endpoints $(-1, 3)$ and $(3, -2)$

29. the lower half of the parabola $x - 1 = y^2$

30. the left half of the parabola $y = x^2 + 2x$

31. the ray (half line) with initial point $(2, 3)$ that passes through the point $(-1, -1)$

32. the ray (half line) with initial point $(-1, 2)$ that passes through the point $(0, 0)$

In Exercises 33–36, refer to the graph of
$$x = 3 - |t|, \quad y = t - 1, \quad -5 \le t \le 5,$$
shown in the figure. *Work in groups of two or three to find the values of t that produce the graph in the given quadrant.*

33. Quadrant I

34. Quadrant II

35. Quadrant III

36. Quadrant IV

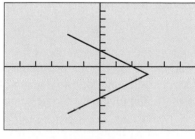

[−6, 6] by [−8, 8]

In Exercises 37 and 38, find a parametrization for the part of the graph that lies in Quadrant I.

37. $y = x^2 + 2x + 2$ **38.** $y = \sqrt{x + 3}$

39. *Circles* Find parametrizations to model the motion of a particle that starts at $(a, 0)$ and traces the circle $x^2 + y^2 = a^2$, $a > 0$, as indicated.

(a) once clockwise   **(b)** once counterclockwise

(c) twice clockwise   **(d)** twice counterclockwise

40. *Ellipses* Find parametrizations to model the motion of a particle that starts at $(-a, 0)$ and traces the ellipse
$$\left(\frac{x}{a}\right)^2 + \left(\frac{y}{b}\right)^2 = 1, \quad a > 0, b > 0,$$
as indicated.

(a) once clockwise   **(b)** once counterclockwise

(c) twice clockwise   **(d)** twice counterclockwise

Extending the Ideas

41. *The Witch of Agnesi* The bell-shaped witch of Agnesi can be constructed as follows. Start with the circle of radius 1, centered at the point $(0, 1)$ as shown in the figure.

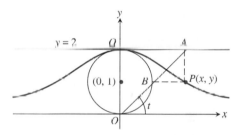

Choose a point A on the line $y = 2$, and connect it to the origin with a line segment. Call the point where the segment crosses the circle B. Let P be the point where the vertical line through A crosses the horizontal line through B. The witch is the curve traced by P as A moves along the line $y = 2$.

Find a parametrization for the witch by expressing the coordinates of P in terms of t, the radian measure of the angle that segment OA makes with the positive x-axis. The following equalities (which you may assume) will help:

(i) $x = AQ$

(ii) $y = 2 - AB \sin t$

(iii) $AB \cdot AO = (AQ)^2$

42. *Parametrizing Lines and Segments*

(a) Show that $x = x_1 + (x_2 - x_1)t, \quad y = y_1 + (y_2 - y_1)t,$ $-\infty < t < \infty$ is a parametrization for the line through the points (x_1, y_1) and (x_2, y_2).

(b) Find a parametrization for the line segment with endpoints (x_1, y_1) and (x_2, y_2).

1.5 Functions and Logarithms

One-to-One Functions • Inverses • Finding Inverses •
Logarithmic Functions • Properties of Logarithms • Applications

One-to-One Functions

As you know, a function is a rule that assigns a single value in its range to each point in its domain. Some functions assign the same output to more than one input. For example, $f(x) = x^2$ assigns the output 4 to both 2 and -2. Other functions never output a given value more than once. For example, the cubes of different numbers are always different.

If each output value of a function is associated with exactly one input value, the function is *one-to-one*.

Definition One-to-One Function

A function $f(x)$ is **one-to-one** on a domain D if $f(a) \neq f(b)$ whenever $a \neq b$.

The graph of a one-to-one function $y = f(x)$ can intersect any horizontal line at most once (the *horizontal line test*). If it intersects such a line more than once it assumes the same y-value more than once, and is therefore not one-to-one (Figure 1.30).

$y = |x|$

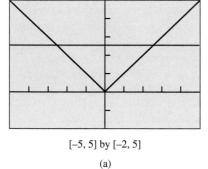

[−5, 5] by [−2, 5]

(a)

$y = \sqrt{x}$

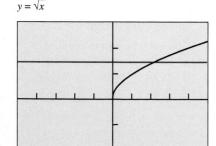

[−5, 5] by [−2, 3]

(b)

Figure 1.31 (a) The graph of $f(x) = |x|$ and a horizontal line. (b) The graph of $g(x) = \sqrt{x}$ and a horizontal line. (Example 1)

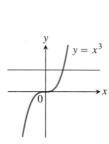

One-to-one: Graph meets each
horizontal line once.

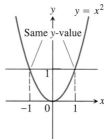

Not one-to-one: Graph meets some
horizontal lines more than once.

Figure 1.30 Using the horizontal line test, we see that $y = x^3$ is one-to-one and $y = x^2$ is not.

Example 1   USING THE HORIZONTAL LINE TEST

Determine whether the functions are one-to-one.

$$\textbf{(a)}\, f(x) = |x| \quad \textbf{(b)}\, g(x) = \sqrt{x}$$

Solution As Figure 1.31a suggests, each horizontal line $y = c,\ c > 0$, intersects the graph of $f(x) = |x|$ twice. So f is not one-to-one. As Figure 1.31b suggests, each horizontal line intersects the graph of $g(x) = \sqrt{x}$ either once or not at all. The function g is one-to-one.

Inverses

Since each output of a one-to-one function comes from just one input, a one-to-one function can be reversed to send outputs back to the inputs from which they came. The function defined by reversing a one-to-one function f is the **inverse of f.** The functions in Tables 1.11 and 1.12 are inverses of one another. The symbol for the inverse of f is f^{-1}, read "f inverse." The -1 in f^{-1} is not an exponent; $f^{-1}(x)$ does not mean $1/f(x)$.

Table 1.11 Rental Charge versus Time	
Time x (hours)	Charge y (dollars)
1	5.00
2	7.50
3	10.00
4	12.50
5	15.00
6	17.50

Table 1.12 Time versus Rental Charge	
Charge x (dollars)	Time y (hours)
5.00	1
7.50	2
10.00	3
12.50	4
15.00	5
17.50	6

As Tables 1.11 and 1.12 suggest, composing a function with its inverse in either order sends each output back to the input from which it came. In other words, the result of composing a function and its inverse in either order is the **identity function,** the function that assigns each number to itself. This gives a way to test whether two functions f and g are inverses of one another. Compute $f \circ g$ and $g \circ f$. If $(f \circ g)(x) = (g \circ f)(x) = x$, then f and g are inverses of one another; otherwise they are not. The functions $f(x) = x^3$ and $g(x) = x^{1/3}$ are inverses of one another because $(x^3)^{1/3} = x$ and $(x^{1/3})^3 = x$ for every number x.

Exploration 1 **Testing For Inverses Graphically**

For each of the function pairs below,

(a) Graph f and g together in a square window.
(b) Graph $f \circ g$.
(c) Graph $g \circ f$.

What can you conclude from the graphs?

1. $f(x) = x^3$, $g(x) = x^{1/3}$
2. $f(x) = x$, $g(x) = 1/x$
3. $f(x) = 3x$, $g(x) = x/3$
4. $f(x) = e^x$, $g(x) = \ln x$

Finding Inverses

How do we find the graph of the inverse of a function? Suppose, for example, that the function is the one pictured in Figure 1.32a. To read the graph, we start at the point x on the x-axis, go up to the graph, and then move over to the *y*-axis to read the value of *y*. If we start with *y* and want to find the *x* from which it came, we reverse the process (Figure 1.32b).

The graph of *f* is already the graph of f^{-1}, although the latter graph is not drawn in the usual way with the domain axis horizontal and the range axis vertical. For f^{-1}, the input-output pairs are reversed. To display the graph of f^{-1} in the usual way, we have to reverse the pairs by reflecting the graph in the 45° line $y = x$ (Figure 1.32c) and interchanging the letters *x* and *y* (Figure 1.32d). This puts the independent variable, now called *x,* on the horizontal axis and the dependent variable, now called *y,* on the vertical axis.

The fact that the graphs of *f* and f^{-1} are reflections of each other across the line $y = x$ is to be expected because the input-output pairs (a, b) of *f* have been reversed to produce the input-output pairs (b, a) of f^{-1}.

The pictures in Figure 1.32 tell us how to express f^{-1} as a function of *x* algebraically.

Writing f^{-1} as a Function of *x*

1. Solve the equation $y = f(x)$ for *x* in terms of *y*.

2. Interchange *x* and *y*. The resulting formula will be $y = f^{-1}(x)$.

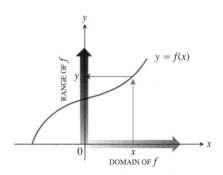

(a) To find the value of *f* at *x,* we start at *x,* go up to the curve, and then and over to the *y*-axis.

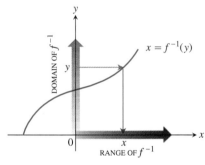

(b) The graph of *f* is also the graph of f^{-1}. To find the *x* that gave *y,* we start at *y* and go over to the curve and down to the *x*-axis. The domain of f^{-1} is the range of *f*. The range of f^{-1} is the domain of *f*.

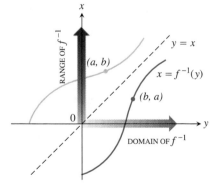

(c) To draw the graph of f^{-1} in the usual way, we reflect the system in the line $y = x$.

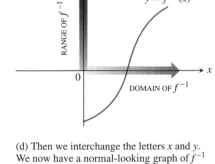

(d) Then we interchange the letters *x* and *y*. We now have a normal-looking graph of f^{-1} as a function of *x*.

Figure 1.32 The graph of $y = f^{-1}(x)$.

Example 2 FINDING THE INVERSE FUNCTION

Show that the function $y = f(x) = -2x + 4$ is one-to-one and find its inverse function.

Solution Every horizontal line intersects the graph of f exactly once, so f is one-to-one and has an inverse.

Step 1:
Solve for x in terms of y:
$$y = -2x + 4$$
$$x = -\frac{1}{2}y + 2$$

Step 2:
Interchange x and y:
$$y = -\frac{1}{2}x + 2$$

The inverse of the function $f(x) = -2x + 4$ is the function $f^{-1}(x) = -(1/2)x + 2$. We can verify that both composites are the identity function.

$$f^{-1}(f(x)) = -\frac{1}{2}(-2x + 4) + 2 = x - 2 + 2 = x$$

$$f(f^{-1}(x)) = -2(-\frac{1}{2}x + 2) + 4 = x - 4 + 4 = x$$

We can use parametric graphing to graph the inverse of a function without finding an explicit rule for the inverse, as illustrated in Example 3.

Graphing $y = f(x)$ and $y = f^{-1}(x)$ Parametrically

We can graph any function $y = f(x)$ as
$$x_1 = t, \quad y_1 = f(t).$$
Interchanging t and $f(t)$ produces parametric equations for the inverse:
$$x_2 = f(t), \quad y_2 = t.$$

Example 3 GRAPHING THE INVERSE PARAMETRICALLY

(a) Graph the one-to-one function $f(x) = x^2$, $x \geq 0$, together with its inverse and the line $y = x$, $x \geq 0$.

(b) Express the inverse of f as a function of x.

Solution

(a) We can graph the three functions parametrically as follows.

Graph of f: $x_1 = t, \quad y_1 = t^2, \quad t \geq 0$

Graph of f^{-1}: $x_2 = t^2, \quad y_2 = t$

Graph of $y = x$: $x_3 = t, \quad y_3 = t$

Figure 1.33 shows the three graphs.

(b) Next we find a formula for $f^{-1}(x)$.

Step 1:
Solve for x in terms of y.
$$y = x^2$$
$$\sqrt{y} = \sqrt{x^2}$$
$$\sqrt{y} = x \qquad \text{Because } x \geq 0$$

Step 2:
Interchange x and y.
$$\sqrt{x} = y$$

Thus, $f^{-1}(x) = \sqrt{x}$.

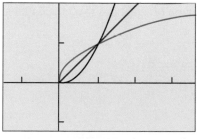

[–1.5, 3] by [–1, 2]

Figure 1.33 The graphs of f and f^{-1} are reflections of each other across the line $y = x$. (Example 3)

Logarithmic Functions

If a is any positive real number other than 1, the base a exponential function $f(x) = a^x$ is one-to-one. It therefore has an inverse. Its inverse is called the *base a logarithm function.*

Definition Base *a* Logarithm Function

The **base *a* logarithm function** $y = \log_a x$ is the inverse of the base a exponential function $y = a^x$ $(a > 0, a \neq 1)$.

The domain of $\log_a x$ is $(0, \infty)$, the range of a^x. The range of $\log_a x$ is $(-\infty, \infty)$, the domain of a^x.

Because we have no technique for solving for x in terms of y in the equation $y = a^x$, we do not have an explicit formula for the logarithm function as a function of x. However, the graph of $y = \log_a x$ can be obtained by reflecting the graph of $y = a^x$ across the line $y = x$, or by using parametric graphing (Figure 1.34).

Logarithms with base e and base 10 are so important in applications that calculators have special keys for them. They also have their own special notation and names:

$$\log_e x = \ln x,$$

$$\log_{10} x = \log x.$$

The function $y = \ln x$ is called the **natural logarithm function** and $y = \log x$ is often called the **common logarithm function.**

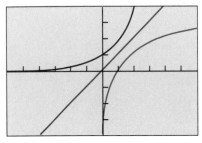

[–6, 6] by [–4, 4]

Figure 1.34 The graphs of $y = 2^x$ ($x_1 = t$, $y_1 = 2^t$), its inverse $y = \log_2 x$ ($x_2 = 2^t$, $y_2 = t$), and $y = x$ ($x_3 = t, y_3 = t$).

Properties of Logarithms

Because a^x and $\log_a x$ are inverses of each other, composing them in either order gives the identity function. This gives two useful properties.

Inverse Properties for a^x and $\log_a x$

1. **Base a:** $a^{\log_a x} = x,$ $\log_a a^x = x,$ $a > 1, x > 0$
2. **Base e:** $e^{\ln x} = x,$ $\ln e^x = x,$ $x > 0$

These properties help us with the solution of equations that contain logarithms and exponential functions.

Example 4   USING THE INVERSE PROPERTIES

Solve for x: **(a)** $\ln x = 3t + 5$ **(b)** $e^{2x} = 10$

Solution

(a) $\ln x = 3t + 5$

$\quad e^{\ln x} = e^{3t+5}$ Exponentiate both sides.

$\quad\quad x = e^{3t+5}$ Inverse Property

(b) $e^{2x} = 10$

$\ln e^{2x} = \ln 10$ Take logarithms of both sides.

$2x = \ln 10$ Inverse Property

$x = \dfrac{1}{2} \ln 10 \approx 1.15$

The logarithm function has the following useful arithmetic properties.

Properties of Logarithms

For any real numbers $x > 0$ and $y > 0$,

1. *Product Rule:* $\log_a xy = \log_a x + \log_a y$

2. *Quotient Rule:* $\log_a \dfrac{x}{y} = \log_a x - \log_a y$

3. *Power Rule:* $\log_a x^y = y \log_a x$

Exploration 2   **Supporting the Product Rule**

Let $y_1 = \ln (ax)$, $y_2 = \ln x$, and $y_3 = y_1 - y_2$.

1. Graph y_1 and y_2 for $a = 2, 3, 4,$ and 5. How do the graphs of y_1 and y_2 appear to be related?

2. Support your finding by graphing y_3.

3. Confirm your finding algebraically.

The following formula allows us to evaluate $\log_a x$ for any base $a > 0$, $a \neq 1$, and to obtain its graph using the natural logarithm function on our grapher.

Change of Base Formula

$$\log_a x = \frac{\ln x}{\ln a}$$

$y = \dfrac{\ln x}{\ln 2}$

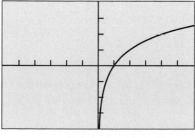

[–6, 6] by [–4, 4]

Figure 1.35 The graph of $f(x) = \log_2 x$ using $f(x) = (\ln x)/(\ln 2)$. (Example 5)

Example 5   GRAPHING A BASE *a* LOGARITHM FUNCTION

Graph $f(x) = \log_2 x$.

Solution We use the change of base formula to rewrite $f(x)$.

$$f(x) = \log_2 x = \frac{\ln x}{\ln 2}.$$

Figure 1.35 gives the graph of f.

Applications

In Section 1.3 we used graphical methods to solve exponential growth and decay problems. Now we can use the properties of logarithms to solve the same problems algebraically.

Example 6 FINDING TIME

Sarah invests $1000 in an account that earns 5.25% interest compounded annually. How long will it take the account to reach $2500?

Solution

Model

The amount in the account at any time t in years is $1000(1.0525)^t$, so we need to solve the equation

$$1000(1.0525)^t = 2500.$$

Solve Algebraically

$(1.0525)^t = 2.5$	Divide by 1000.
$\ln (1.0525)^t = \ln 2.5$	Take logarithms of both sides.
$t \ln 1.0525 = \ln 2.5$	Power Rule

$$t = \frac{\ln 2.5}{\ln 1.0525} \approx 17.9$$

Interpret

The amount in Sarah's account will be $2500 in about 17.9 years, or about 17 years and 11 months.

Example 7 ESTIMATING OIL PRODUCTION

Table 1.13 shows the number of metric tons of oil produced by Indonesia for three different years.

Find the natural logarithm regression equation for the data in Table 1.13 and use it to estimate the number of metric tons of oil produced by Indonesia in 1982 and 2000.

Solution

Model

We let $x = 60$ represent 1960, $x = 70$ represent 1970, and so forth. We compute the natural logarithm regression equation to be

$$f(x) = -474.31 + 121.1346 \ln x.$$

Solve Graphically

Figure 1.36 shows the graph of f superimposed on the scatter plot of the data. The year 1982 is represented by 82 and the year 2000 by 100. Reading from the graph, we find

$$f(82) \approx 59.5 \quad \text{and} \quad f(100) \approx 83.5.$$

Interpret

The natural logarithm regression model estimates the 1982 oil production to have been 59.5 million metric tons, and the 2000 oil production to be 83.5 million metric tons.

Table 1.13 Indonesia's Oil Production

Year	Metric Tons (millions)
1960	20.56
1970	42.10
1990	70.10

Source: The Statesman's Yearbook, 129th ed. (London: The Macmillan Press, Ltd., 1992).

$y = -474.31 + 121.1346 \ln x$

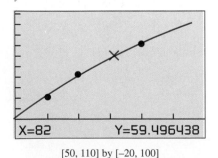

X=82   Y=59.496438

[50, 110] by [−20, 100]

Figure 1.36 The value of f at $x = 82$ is about 59.5. (Example 7)

Quick Review 1.5

In Exercises 1–4, let $f(x) = \sqrt[3]{x - 1}$, $g(x) = x^2 + 1$, and evaluate the expression.

1. $(f \circ g)(1)$

2. $(g \circ f)(-7)$

3. $(f \circ g)(x)$

4. $(g \circ f)(x)$

In Exercises 5 and 6, choose parametric equations and a parameter interval to represent the function on the interval specified.

5. $y = \dfrac{1}{x - 1}$, $x \geq 2$

6. $y = x$, $x < -3$

In Exercises 7–10, find the points of intersection of the two curves. Round your answers to 2 decimal places.

7. $y = 2x - 3$, $y = 5$

8. $y = -3x + 5$, $y = -3$

9. (a) $y = 2^x$, $y = 3$ (b) $y = 2^x$, $y = -1$

10. (a) $y = e^{-x}$, $y = 4$ (b) $y = e^{-x}$, $y = -1$

Section 1.5 Exercises

In Exercises 1–6, determine whether the function is one-to-one.

1.

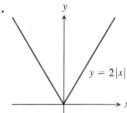

$y = 2|x|$

2.

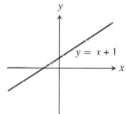

$y = x + 1$

3.

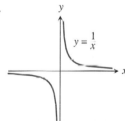

$y = \dfrac{1}{x}$

4.

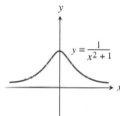

$y = \dfrac{1}{x^2 + 1}$

5.

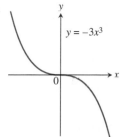

$y = -3x^3$

6.

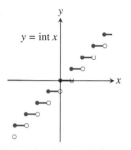

$y = \text{int } x$

In Exercises 7–12, determine whether the function has an inverse function.

7. $y = \dfrac{3}{x - 2} - 1$

8. $y = x^2 + 5x$

9. $y = x^3 - 4x + 6$

10. $y = x^3 + x$

11. $y = \ln x^2$

12. $y = 2^{3-x}$

In Exercises 13–24, find f^{-1} and verify that
$$(f \circ f^{-1})(x) = (f^{-1} \circ f)(x) = x.$$

13. $f(x) = 2x + 3$

14. $f(x) = 5 - 4x$

15. $f(x) = x^3 - 1$

16. $f(x) = x^2 + 1$, $x \geq 0$

17. $f(x) = x^2$, $x \leq 0$

18. $f(x) = x^{2/3}$, $x \geq 0$

19. $f(x) = -(x - 2)^2$, $x \leq 2$

20. $f(x) = x^2 + 2x + 1$, $x \geq -1$

21. $f(x) = \dfrac{1}{x^2}$, $x > 0$

22. $f(x) = \dfrac{1}{x^3}$

23. $f(x) = \dfrac{2x + 1}{x + 3}$

24. $f(x) = \dfrac{x + 3}{x - 2}$

In Exercises 25–32, use parametric graphing to graph f, f^{-1}, and $y = x$.

25. $f(x) = e^x$

26. $f(x) = 3^x$

27. $f(x) = 2^{-x}$

28. $f(x) = 3^{-x}$

29. $f(x) = \ln x$

30. $f(x) = \log x$

31. $f(x) = \sin^{-1} x$

32. $f(x) = \tan^{-1} x$

In Exercises 33–36, draw the graph and determine the domain and range of the function.

33. $y = 2 \ln (3 - x) - 4$

34. $y = -3 \log (x + 2) + 1$

35. $y = \log_2 (x + 1)$

36. $y = \log_3 (x - 4)$

In Exercises 37–40, solve the equation algebraically. Support your solution graphically.

37. $(1.045)^t = 2$

38. $e^{0.05t} = 3$

39. $e^x + e^{-x} = 3$

40. $2^x + 2^{-x} = 5$

In Exercises 41 and 42, solve for y.

41. $\ln y = 2t + 4$

42. $\ln (y - 1) - \ln 2 = x + \ln x$

In Exercises 43 and 44, find a formula for f^{-1} and verify that $(f \circ f^{-1})(x) = (f^{-1} \circ f)(x) = x$.

43. $f(x) = \dfrac{100}{1 + 2^{-x}}$ **44.** $f(x) = \dfrac{50}{1 + 1.1^{-x}}$

45. *Self-inverse* Prove that the function f is its own inverse.

(a) $f(x) = \sqrt{1 - x^2}, \quad x \geq 0$ (b) $f(x) = 1/x$

46. *Radioactive Decay* The half-life of a certain radioactive substance is 12 hours. There are 8 grams present initially.

(a) Express the amount of substance remaining as a function of time t.

(b) When will there be 1 gram remaining?

47. *Doubling Your Money* Determine how much time is required for a \$500 investment to double in value if interest is earned at the rate of 4.75% compounded annually.

48. *Population Growth* The population of Glenbrook is 375,000 and is increasing at the rate of 2.25% per year. Predict when the population will be 1 million.

In Exercises 49 and 50, let $x = 60$ represent 1960, $x = 70$ represent 1970, and so forth.

49. *Oil Production*

(a) Find a natural logarithm regression equation for the data in Table 1.14.

(b) Estimate the number of metric tons of oil produced by Saudi Arabia in 1975.

(c) Predict when Saudi Arabian oil production will reach 400 million metric tons.

Table 1.14 Saudi Arabian Oil Production

Year	Metric Tons (millions)
1960	61.09
1970	176.85
1990	321.93

Source: *The Statesman's Yearbook,* 129th ed. (London: The Macmillan Press, Ltd., 1992).

50. *Oil Production*

(a) Find a natural logarithm regression equation for the data in Table 1.15.

(b) Estimate the number of metric tons of oil produced by Canada in 1985.

(c) Predict when Canadian oil production will reach 120 metric tons.

Table 1.15 Canadian Oil Production

Year	Metric Tons (millions)
1960	27.48
1970	69.95
1990	92.24

Source: *The Statesman's Yearbook,* 129th ed. (London: The Macmillan Press, Ltd., 1992).

51. *Inverse Functions* We suggest that students *work in groups of two or three.*

Let $y = f(x) = mx + b, \quad m \neq 0$.

(a) **Writing to Learn** Give a convincing argument that f is a one-to-one function.

(b) Find a formula for the inverse of f. How are the slopes of f and f^{-1} related?

(c) If the graphs of two functions are parallel lines with a nonzero slope, what can you say about the graphs of the inverses of the functions?

(d) If the graphs of two functions are perpendicular lines with a nonzero slope, what can you say about the graphs of the inverses of the functions?

Exploration

52. *Supporting the Quotient Rule* Let $y_1 = \ln (x/a)$, $y_2 = \ln x$, $y_3 = y_2 - y_1$, and $y_4 = e^{y_3}$.

(a) Graph y_1 and y_2 for $a = 2, 3, 4,$ and 5. How are the graphs of y_1 and y_2 related?

(b) Graph y_3 for $a = 2, 3, 4,$ and 5. Describe the graphs.

(c) Graph y_4 for $a = 2, 3, 4,$ and 5. Compare the graphs to the graph of $y = a$.

(d) Use $e^{y_3} = e^{y_2 - y_1} = a$ to solve for y_1.   ■

Extending the Ideas

53. *One-to-One Functions* If f is a one-to-one function, prove that $g(x) = -f(x)$ is also one-to-one.

54. *One-to-One Functions* If f is a one-to-one function and $f(x)$ is never zero, prove that $g(x) = 1/f(x)$ is also one-to-one.

55. *Domain and Range* Suppose that $a \neq 0, b \neq 1,$ and $b > 0$. Determine the domain and range of the function.

(a) $y = a(b^{c-x}) + d$ (b) $y = a \log_b (x - c) + d$

56. *Inverse Functions* We suggest that students *work in groups of two or three.*

Let $f(x) = \dfrac{ax + b}{cx + d}, \quad c \neq 0, \quad ad - bc \neq 0$.

(a) **Writing to Learn** Give a convincing argument that f is one-to-one.

(b) Find a formula for the inverse of f.

(c) Find the horizontal and vertical asymptotes of f.

(d) Find the horizontal and vertical asymptotes of f^{-1}. How are they related to those of f?

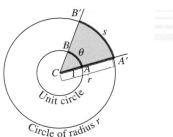

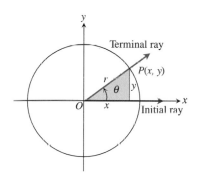

Figure 1.37 The radian measure of angle *ACB* is the length θ of arc *AB* on the unit circle centered at *C*. The value of θ can be found from any other circle, however, as the ratio s/r.

Figure 1.38 An angle θ in standard position.

1.6 Trigonometric Functions

Radian Measure • Graphs of Trigonometric Functions • Periodicity • Even and Odd Trigonometric Functions • Transformations of Trigonometric Graphs • Applications • Inverse Trigonometric Functions

Radian Measure

The **radian measure** of the angle *ACB* at the center of the unit circle (Figure 1.37) equals the length of the arc that *ACB* cuts from the unit circle.

When an angle of measure θ is placed in *standard position* at the center of a circle of radius *r* (Figure 1.38), the six basic trigonometric functions of θ are defined as follows:

$$\textbf{sine:}\ \sin\theta = \frac{y}{r} \qquad \textbf{cosecant:}\ \csc\theta = \frac{r}{y}$$

$$\textbf{cosine:}\ \cos\theta = \frac{x}{r} \qquad \textbf{secant:}\ \sec\theta = \frac{r}{x}$$

$$\textbf{tangent:}\ \tan\theta = \frac{y}{x} \qquad \textbf{cotangent:}\ \cot\theta = \frac{x}{y}$$

Graphs of Trigonometric Functions

When we graph trigonometric functions in the coordinate plane, we usually denote the independent variable (radians) by *x* instead of θ. Figure 1.39 shows sketches of the six trigonometric functions. It is a good exercise for you to compare these with what you see in a grapher viewing window. (Some graphers have a "trig viewing window.")

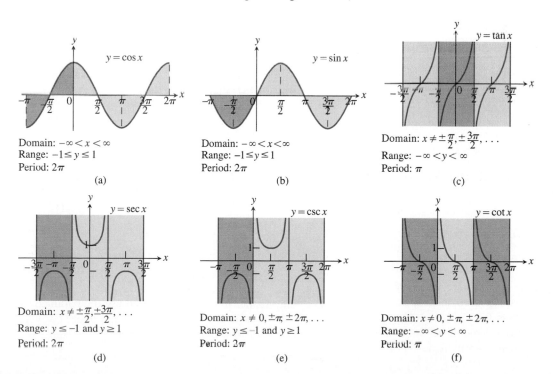

Domain: $-\infty < x < \infty$
Range: $-1 \le y \le 1$
Period: 2π
(a)

Domain: $-\infty < x < \infty$
Range: $-1 \le y \le 1$
Period: 2π
(b)

Domain: $x \ne \pm\frac{\pi}{2}, \pm\frac{3\pi}{2}, \ldots$
Range: $-\infty < y < \infty$
Period: π
(c)

Domain: $x \ne \pm\frac{\pi}{2}, \pm\frac{3\pi}{2}, \ldots$
Range: $y \le -1$ and $y \ge 1$
Period: 2π
(d)

Domain: $x \ne 0, \pm\pi, \pm 2\pi, \ldots$
Range: $y \le -1$ and $y \ge 1$
Period: 2π
(e)

Domain: $x \ne 0, \pm\pi, \pm 2\pi, \ldots$
Range: $-\infty < y < \infty$
Period: π
(f)

Figure 1.39 Graphs of the (a) cosine, (b) sine, (c) tangent, (d) secant, (e) cosecant, and (f) cotangent functions using radian measure.

Exploration 1 **Unwrapping Trigonometric Functions**

Set your grapher in *radian mode,* parametric mode, and *simultaneous mode* (all three). Enter the parametric equations

$$x_1 = \cos t, \quad y_1 = \sin t \quad \text{and} \quad x_2 = t, \quad y_2 = \sin t.$$

1. Graph for $0 \le t \le 2\pi$ in the window $[-1.5, 2\pi]$ by $[-2.5, 2.5]$. Describe the two curves. (You may wish to make the viewing window square.)

2. Use trace to compare the y-values of the two curves.

3. Repeat part 2 in the window $[-1.5, 4\pi]$ by $[-5, 5]$, using the parameter interval $0 \le t \le 4\pi$.

4. Let $y_2 = \cos t$. Use trace to compare the x-values of curve 1 (the unit circle) with the y-values of curve 2 using the parameter intervals $[0, 2\pi]$ and $[0, 4\pi]$.

5. Set $y_2 = \tan t$, $\csc t$, $\sec t$, and $\cot t$. Graph each in the window $[-1.5, 2\pi]$ by $[-2.5, 2.5]$ using the interval $0 \le t \le 2\pi$. How is a y-value of curve 2 related to the corresponding point on curve 1? (Use trace to explore the curves.)

A grapher can find trigonometric values, as the next exploration demonstrates.

Exploration 2 **Finding Sines and Cosines**

Set your grapher in radian mode and enter the parametric equations

$$x_1 = \cos t, \quad y_1 = \sin t.$$

1. Graph in a square viewing window using the parameter interval $0 \le t \le 2\pi$ with t-step = 0.1. Use trace to read x and y for different values of t as suggested by Figure 1.40.

2. Try to match the entries in Table 1.16.

Table 1.16 Some Values of Sine and Cosine

t (radians)	$\cos t$	$\sin t$
0.5	0.87758256	0.47942554
1	0.54030231	0.84147098
2	-0.41614684	0.90929743
3.5	-0.93645669	-0.35078323
6.2	0.99654210	-0.08308940

3. Set your grapher in degree mode and use the parameter interval $0 \le t \le 360$ with t-step = 15. Graph and trace. For what values of t are $\cos t$ and $\sin t$ being computed?

$x = \cos t, y = \sin t$

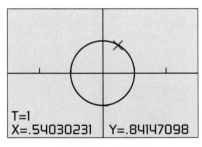

T=1
X=.54030231   Y=.84147098

$[-3, 3]$ by $[-2, 2]$

Figure 1.40 A graph of the unit circle $x^2 + y^2 = 1$ obtained in parametric mode. The values of $x = \cos 1$ and $y = \sin 1$ are shown using trace. (Exploration 2)

Angle Convention: Use Radians

From now on in this book it is assumed that all angles are measured in radians unless degrees or some other unit is stated explicitly. When we talk about the angle $\pi/3$, we mean $\pi/3$ radians (which is 60°), not $\pi/3$ degrees. When you do calculus, keep your calculator in radian mode.

Periodicity

When an angle of measure θ and an angle of measure $\theta + 2\pi$ are in standard position, their terminal rays coincide. The two angles therefore have the same trigonometric function values:

$$\cos(\theta + 2\pi) = \cos\theta \quad \sin(\theta + 2\pi) = \sin\theta \quad \tan(\theta + 2\pi) = \tan\theta$$

$$\sec(\theta + 2\pi) = \sec\theta \quad \csc(\theta + 2\pi) = \csc\theta \quad \cot(\theta + 2\pi) = \cot\theta \qquad (1)$$

Similarly, $\cos(\theta - 2\pi) = \cos\theta$, $\sin(\theta - 2\pi) = \sin\theta$, and so on.

We see the values of the trigonometric functions repeat at regular intervals. We describe this behavior by saying that the six basic trigonometric functions are *periodic*.

Periods of Trigonometric Functions

Period π: $\tan(x + \pi) = \tan x$
          $\cot(x + \pi) = \cot x$

Period 2π: $\sin(x + 2\pi) = \sin x$
          $\cos(x + 2\pi) = \cos x$
          $\sec(x + 2\pi) = \sec x$
          $\csc(x + 2\pi) = \csc x$

> **Definition Periodic Function, Period**
>
> A function $f(x)$ is **periodic** if there is a positive number p such that $f(x + p) = f(x)$ for every value of x. The smallest such value of p is the **period** of f.

As we can see in Figure 1.39, the functions $\cos x$, $\sin x$, $\sec x$, and $\csc x$ are periodic with period 2π. The functions $\tan x$ and $\cot x$ are periodic with period π.

The importance of periodic functions stems from the fact that much of the behavior we study in science is periodic (Figure 1.41). Brain waves and heartbeats are periodic, as are household voltage and electric current. The electromagnetic field that heats food in a microwave oven is periodic, as are cash flows in seasonal businesses and the behavior of rotational machinery. The seasons are periodic; so is the weather. The phases of the moon are periodic, as are the motions of the planets. There is strong evidence that the ice ages are periodic, with a period of 90,000–100,000 years.

Why are trigonometric functions so important in the study of things periodic? The answer lies in a surprising and beautiful theorem from advanced calculus that says that every periodic function we want to use in mathematical modeling can be written as an algebraic combination of sines and cosines. Thus, once we learn the calculus of sines and cosines, we will know everything we need to know to model the mathematical behavior of most periodic phenomena.

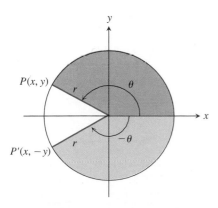

Figure 1.41 This compact patient monitor shows several periodic functions associated with the human body. This device dynamically monitors electrocardiogram (ECG) and respiration, and blood pressure.

Even and Odd Trigonometric Functions

The graphs in Figure 1.39 suggest that $\cos x$ and $\sec x$ are even functions because their graphs are symmetric about the y-axis. The other four basic trigonometric functions are odd.

Example 1 CONFIRMING EVEN AND ODD

Show that cosine is an even function and sine is odd.

Solution From Figure 1.42 it follows that

$$\cos(-\theta) = \frac{x}{r} = \cos\theta, \quad \sin(-\theta) = \frac{-y}{r} = -\sin\theta,$$

so cosine is an even function and sine is odd.

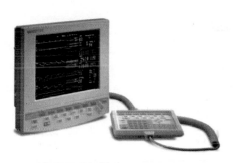

Figure 1.42 Angles of opposite sign. (Example 1)

We can use the results of Example 1 to establish the *parity* of the other four basic trigonometric functions. For example,

$$\sec(-\theta) = \frac{1}{\cos(-\theta)} = \frac{1}{\cos\theta} = \sec\theta,$$

$$\tan(-\theta) = \frac{\sin(-\theta)}{\cos(-\theta)} = \frac{-\sin\theta}{\cos\theta} = -\tan\theta,$$

so secant is an even function and tangent is odd. Similar steps will show that cosecant and cotangent are odd functions.

Transformations of Trigonometric Graphs

The rules for shifting, stretching, shrinking, and reflecting the graph of a function apply to the trigonometric functions. The following diagram will remind you of the controlling parameters.

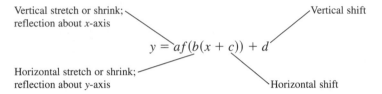

Vertical stretch or shrink; reflection about *x*-axis

Vertical shift

$$y = af(b(x + c)) + d$$

Horizontal stretch or shrink; reflection about *y*-axis

Horizontal shift

Applications

Example 2   FINDING A TRIGONOMETRIC MODEL

The builders of the Trans-Alaska Pipeline used insulated pads to keep the pipeline heat from melting the permanently frozen soil beneath. To design the pads, it was necessary to take into account the variation in air temperature throughout the year. The variation was represented in the calculations by a general sine function or **sinusoid** of the form

$$f(x) = A\sin\left[\frac{2\pi}{B}(x - C)\right] + D,$$

where $|A|$ is the *amplitude*, $|B|$ is the *period*, C is the *horizontal shift*, and D is the *vertical shift* (Figure 1.43).

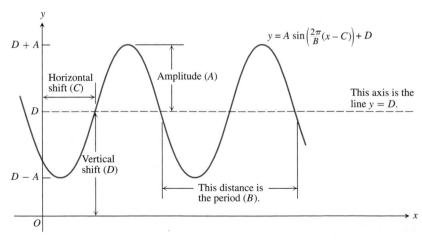

Figure 1.43 The general sine curve $y = A\sin[(2\pi/B)(x - C)] + D$, shown for A, B, C, and D positive. (Example 2)

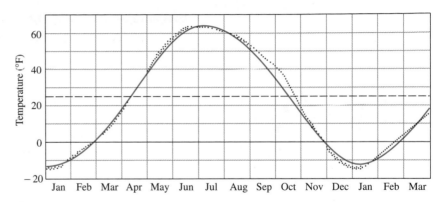

Figure 1.44 Normal mean air temperatures for Fairbanks, Alaska, plotted as data points (red). The approximating sine function is $f(x) = 37 \sin[(2\pi/365)(x - 101)] + 25$. *Source:* "Is the Curve of Temperature Variation a Sine Curve?" by B. M. Lando and C. A. Lando, *The Mathematics Teacher,* 7:6, Fig. 2, p. 535 (September 1977).

Figure 1.44 shows how we can use such a function to represent temperature data. The data points in the figure are plots of the mean daily air temperatures for Fairbanks, Alaska, based on records of the National Weather Service from 1941 to 1970. The sine function used to fit the data is

$$f(x) = 37 \sin\left[\frac{2\pi}{365}(x - 101)\right] + 25,$$

where f is temperature in degrees Fahrenheit and x is the number of the day counting from the beginning of the year. The fit is remarkably good.

Musical notes are pressure waves in the air. The wave behavior can be modeled with great accuracy by general sine curves. New devices called Calculator Based Laboratory™ (CBL) systems can record these waves with the aid of a microphone. The data in Table 1.17 give pressure displacement versus time in seconds of a musical note produced by a tuning fork and recorded with a CBL system.

Table 1.17 Tuning Fork Data

Time	Pressure	Time	Pressure	Time	Pressure
0.00091	−0.080	0.00271	−0.141	0.00453	0.749
0.00108	0.200	0.00289	−0.309	0.00471	0.581
0.00125	0.480	0.00307	−0.348	0.00489	0.346
0.00144	0.693	0.00325	−0.248	0.00507	0.077
0.00162	0.816	0.00344	−0.041	0.00525	−0.164
0.00180	0.844	0.00362	0.217	0.00543	−0.320
0.00198	0.771	0.00379	0.480	0.00562	−0.354
0.00216	0.603	0.00398	0.681	0.00579	−0.248
0.00234	0.368	0.00416	0.810	0.00598	−0.035
0.00253	0.099	0.00435	0.827		

$y = 0.6 \sin (2488.6x - 2.832) + 0.266$

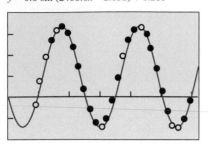

[0, 0.0062] by [−0.5, 1]

Figure 1.45 A sinusoidal regression model for the tuning fork data in Table 1.17. (Example 3)

$x = t,\ y = \sin t,\ -\dfrac{\pi}{2} \le t \le \dfrac{\pi}{2}$

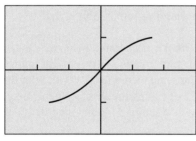

[−3, 3] by [−2, 2]

(a)

$x = \sin t,\ y = t,\ -\dfrac{\pi}{2} \le t \le \dfrac{\pi}{2}$

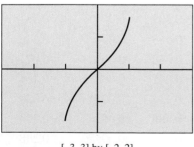

[−3, 3] by [−2, 2]

(b)

Figure 1.46 (a) A restricted sine function and (b) its inverse. (Example 4)

Example 3 FINDING THE FREQUENCY OF A MUSICAL NOTE

Consider the tuning fork data in Table 1.17 on the previous page.

(a) Find a sinusoidal regression equation (general sine curve) for the data and superimpose its graph on a scatter plot of the data.

(b) The *frequency* of a musical note, or wave, is measured in cycles per second, or hertz (1 Hz = 1 cycle per second). The frequency is the reciprocal of the *period* of the wave, which is measured in seconds per cycle. Estimate the frequency of the note produced by the tuning fork.

Solution

(a) The sinusoidal regression equation produced by our calculator is approximately

$$y = 0.6 \sin (2488.6x - 2.832) + 0.266.$$

Figure 1.45 shows its graph together with a scatter plot of the tuning fork data.

(b) The period is $\dfrac{2\pi}{2488.6}$ sec, so the frequency is $\dfrac{2488.6}{2\pi} \approx 396$ Hz.

Interpretation

The tuning fork is vibrating at a frequency of about 396 Hz. On the pure tone scale, this is the note G above middle C. It is a few cycles per second different from the frequency of the G we hear on a piano's tempered scale, 392 Hz.

Inverse Trigonometric Functions

None of the six basic trigonometric functions graphed in Figure 1.39 is one-to-one. These functions do not have inverses. However, in each case the domain can be restricted to produce a new function that does have an inverse, as illustrated in Example 4.

Example 4 RESTRICTING THE DOMAIN OF THE SINE

Show that the function $y = \sin x,\ -\pi/2 \le x \le \pi/2$, is one-to-one, and graph its inverse.

Solution Figure 1.46a shows the graph of this restricted sine function using the parametric equations

$$x_1 = t, \qquad y_1 = \sin t, \qquad -\frac{\pi}{2} \le t \le \frac{\pi}{2}.$$

This restricted sine function is one-to-one because it does not repeat any output values. It therefore has an inverse, which we graph in Figure 1.46b by interchanging the ordered pairs using the parametric equations

$$x_2 = \sin t, \qquad y_2 = t, \qquad -\frac{\pi}{2} \le t \le \frac{\pi}{2}.$$

The inverse of the restricted sine function of Example 4 is called the *inverse sine function*. The inverse sine of x is the angle whose sine is x. It is denoted by $\sin^{-1} x$ or arcsin x. Either notation is read "arcsine of x" or "the inverse sine of x."

The domains of the other basic trigonometric functions can also be restricted to produce a function with an inverse. The domains and ranges of the resulting inverse functions become parts of their definitions.

Definitions **Inverse Trigonometric Functions**

Function	Domain	Range
$y = \cos^{-1} x$	$-1 \le x \le 1$	$0 \le y \le \pi$
$y = \sin^{-1} x$	$-1 \le x \le 1$	$-\dfrac{\pi}{2} \le y \le \dfrac{\pi}{2}$
$y = \tan^{-1} x$	$-\infty < x < \infty$	$-\dfrac{\pi}{2} < y < \dfrac{\pi}{2}$
$y = \sec^{-1} x$	$\lvert x \rvert \ge 1$	$0 \le y \le \pi, y \ne \dfrac{\pi}{2}$
$y = \csc^{-1} x$	$\lvert x \rvert \ge 1$	$-\dfrac{\pi}{2} \le y \le \dfrac{\pi}{2}, y \ne 0$
$y = \cot^{-1} x$	$-\infty < x < \infty$	$0 < y < \pi$

The graphs of the six inverse trigonometric functions are shown in Figure 1.47.

Domain: $-1 \le x \le 1$
Range: $0 \le y \le \pi$

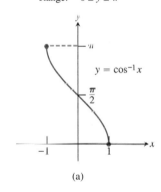

(a)

Domain: $-1 \le x \le 1$
Range: $-\dfrac{\pi}{2} \le y \le \dfrac{\pi}{2}$

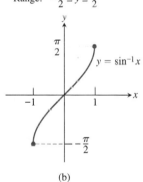

(b)

Domain: $-\infty < x < \infty$
Range: $-\dfrac{\pi}{2} < y < \dfrac{\pi}{2}$

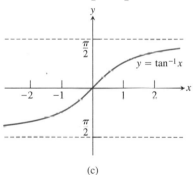

(c)

Domain: $x \le -1$ or $x \ge 1$
Range: $0 \le y \le \pi, y \ne \dfrac{\pi}{2}$

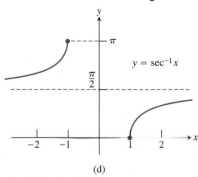

(d)

Domain: $x \le -1$ or $x \ge 1$
Range: $-\dfrac{\pi}{2} \le y \le \dfrac{\pi}{2}, y \ne 0$

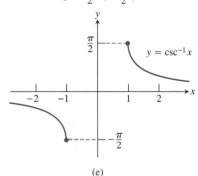

(e)

Domain: $-\infty < x < \infty$
Range: $0 < y < \pi$

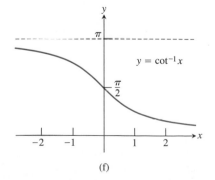

(f)

Figure 1.47 Graphs of (a) $y = \cos^{-1} x$, (b) $y = \sin^{-1} x$, (c) $y = \tan^{-1} x$, (d) $y = \sec^{-1} x$, (e) $y = \csc^{-1} x$, and (f) $y = \cot^{-1} x$.

Example 5   USING THE INVERSE TRIGONOMETRIC FUNCTIONS

Solve for x.

(a) $\sin x = 0.7$ in $0 \le x < 2\pi$

(b) $\tan x = -2$ in $-\infty < x < \infty$

Solution

(a) Notice that $x = \sin^{-1}(0.7) \approx 0.775$ is in the first quadrant, so 0.775 is one solution of this equation. The angle $\pi - x$ is in the second quadrant and has sine equal to 0.7. Thus two solutions in this interval are

$$\sin^{-1}(0.7) \approx 0.775 \quad \text{and} \quad \pi - \sin^{-1}(0.7) \approx 2.366.$$

(b) The angle $x = \tan^{-1}(-2) \approx -1.107$ is in the fourth quadrant and is the only solution to this equation in the interval $-\pi/2 < x < \pi/2$ where $\tan x$ is one-to-one. Since $\tan x$ is periodic with period π, the solutions to this equation are of the form

$$\tan^{-1}(-2) + k\pi \approx -1.107 + k\pi$$

where k is any integer.

Quick Review 1.6

In Exercises 1–4, convert from radians to degrees or degrees to radians.

1. $\pi/3$

2. -2.5

3. $-40°$

4. $45°$

In Exercises 5–7, solve the equation graphically in the given interval.

5. $\sin x = 0.6, \quad 0 \le x < 2\pi$

6. $\cos x = -0.4, \quad 0 \le x < 2\pi$

7. $\tan x = 1, \quad -\dfrac{\pi}{2} \le x < \dfrac{3\pi}{2}$

8. Show that $f(x) = 2x^2 - 3$ is an even function. Explain why its graph is symmetric about the y-axis.

9. Show that $f(x) = x^3 - 3x$ is an odd function. Explain why its graph is symmetric about the origin.

10. Give one way to restrict the domain of the function $f(x) = x^4 - 2$ to make the resulting function one-to-one.

Section 1.6 Exercises

In Exercises 1–4, the angle lies at the center of a circle and subtends an arc of the circle. Find the missing angle measure, circle radius, or arc length.

Angle	Radius	Arc Length
1. $5\pi/8$	2	?
2. $175°$	?	10
3. ?	14	7
4. ?	6	$3\pi/2$

In Exercises 5 and 6, choose an appropriate viewing window to display two complete periods of each trigonometric function in radian mode.

5. (a) $y = \sec x$ **(b)** $y = \csc x$ **(c)** $y = \cot x$

6. (a) $y = \sin x$ **(b)** $y = \cos x$ **(c)** $y = \tan x$

In Exercises 7–10, give the measure of the angle in radians and degrees. Give exact answers whenever possible.

7. $\sin^{-1}(0.5)$

8. $\sin^{-1}\left(-\dfrac{\sqrt{2}}{2}\right)$

9. $\tan^{-1}(-5)$

10. $\cos^{-1}(0.7)$

In Exercises 11–16, specify **(a)** the period, **(b)** the amplitude, and **(c)** identify the viewing window that is shown.

11. $y = 1.5 \sin 2x$

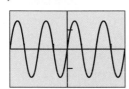

12. $y = 2 \cos 3x$

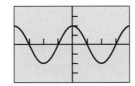

13. $y = -3 \cos 2x$

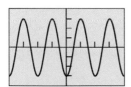

14. $y = 5 \sin \dfrac{x}{2}$

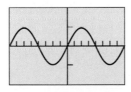

15. $y = -4 \sin \dfrac{\pi}{3} x$

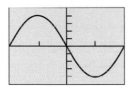

16. $y = \cos \pi x$

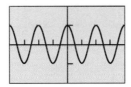

In Exercises 17–20, determine **(a)** the period, **(b)** the domain, and **(c)** the range, and **(d)** draw the graph of the function.

17. $y = 3 \csc (3x + \pi) - 2$

18. $y = 2 \sin (4x + \pi) + 3$

19. $y = -3 \tan (3x + \pi) + 2$

20. $y = 2 \sin \left(2x + \dfrac{\pi}{3} \right)$

In Exercises 21–24, use the given information to find the values of the six trigonometric functions at the angle θ. Give exact answers.

21. $\theta = \sin^{-1} \left(\dfrac{8}{17} \right)$

22. $\theta = \tan^{-1} \left(-\dfrac{5}{12} \right)$

23. The point $P(-3, 4)$ is on the terminal side of θ.

24. The point $P(-2, 2)$ is on the terminal side of θ.

In Exercises 25–30, solve the equation in the specified interval.

25. $\tan x = 2.5, \quad 0 \le x < 2\pi$

26. $\cos x = -0.7, \quad 2\pi \le x < 4\pi$

27. $\csc x = 2, \quad 0 < x < 2\pi$

28. $\sec x = -3, \quad -\pi \le x < \pi$

29. $\sin x = -0.5, \quad -\infty < x < \infty$

30. $\cot x = -1, \quad -\infty < x < \infty$

In Exercises 31 and 32, evaluate the expression.

31. $\sin \left(\cos^{-1} \left(\dfrac{7}{11} \right) \right)$

32. $\tan \left(\sin^{-1} \left(\dfrac{9}{13} \right) \right)$

33. We suggest that students *work in groups of two or three.* A musical note like that produced with a tuning fork or pitch meter is a pressure wave. Table 1.18 gives frequencies (in Hz) of musical notes on the tempered scale. The pressure versus time tuning fork data in Table 1.19 were collected using a CBL™ and a microphone.

Table 1.18 Frequencies of Notes

Note	Frequency (Hz)
C	262
C$^\sharp$ or D$^\flat$	277
D	294
D$^\sharp$ or E$^\flat$	311
E	330
F	349
F$^\sharp$ or G$^\flat$	370
G	392
G$^\sharp$ or A$^\flat$	415
A	440
A$^\sharp$ or B$^\flat$	466
B	494
C (next octave)	524

Source: CBL™ System Experimental Workbook, Texas Instruments, Inc., 1994.

Table 1.19 Tuning Fork Data

Time (s)	Pressure	Time (s)	Pressure
0.0002368	1.29021	0.0049024	−1.06632
0.0005664	1.50851	0.0051520	0.09235
0.0008256	1.51971	0.0054112	1.44694
0.0010752	1.51411	0.0056608	1.51411
0.0013344	1.47493	0.0059200	1.51971
0.0015840	0.45619	0.0061696	1.51411
0.0018432	−0.89280	0.0064288	1.43015
0.0020928	−1.51412	0.0066784	0.19871
0.0023520	−1.15588	0.0069408	−1.06072
0.0026016	−0.04758	0.0071904	−1.51412
0.0028640	1.36858	0.0074496	−0.97116
0.0031136	1.50851	0.0076992	0.23229
0.0033728	1.51971	0.0079584	1.46933
0.0036224	1.51411	0.0082080	1.51411
0.0038816	1.45813	0.0084672	1.51971
0.0041312	0.32185	0.0087168	1.50851
0.0043904	−0.97676	0.0089792	1.36298
0.0046400	−1.51971		

(a) Find a sinusoidal regression equation for the data in Table 1.19 and superimpose its graph on a scatter plot of the data.

(b) Determine the frequency of and identify the musical note produced by the tuning fork.

34. *Temperature Data* Table 1.20 gives the average monthly temperatures for St. Louis for a 12-month period starting with January. Model the monthly temperature with an equation of the form

$$y = a \sin [b(t - h)] + k,$$

y in degrees Fahrenheit, t in months, as follows:

(a) Find the value of b assuming that the period is 12 months.

(b) How is the amplitude a related to the difference $80° - 30°$?

(c) Use the information in (b) to find k.

(d) Find h, and write an equation for y.

(e) Superimpose a graph of y on a scatter plot of the data.

Table 1.20 Temperature Data for St. Louis

Time (months)	Temperature (°F)
1	34
2	30
3	39
4	44
5	58
6	67
7	78
8	80
9	72
10	63
11	51
12	40

35. *Temperatures in Fairbanks, Alaska* Find the **(a)** amplitude, **(b)** period, **(c)** horizontal shift, and **(d)** vertical shift of the model

$$f(x) = 37 \sin \left[\frac{2\pi}{365}(x - 101) \right] + 25$$

used in Example 2.

36. *Temperatures in Fairbanks, Alaska* Use the equation of Exercise 35 to approximate the answers to the following questions about the temperatures in Fairbanks, Alaska, shown in Figure 1.44 on page 45. Assume that the year has 365 days.

(a) What are the highest and lowest mean daily temperatures?

(b) What is the average of the highest and lowest mean daily temperatures? Why is this average the vertical shift of the function?

37. *Even-Odd*

(a) Show that $\cot x$ is an odd function of x.

(b) Show that the quotient of an even function and an odd function is an odd function.

38. *Even-Odd*

(a) Show that $\csc x$ is an odd function of x.

(b) Show that the reciprocal of an odd function is odd.

39. *Even-Odd* Show that the product of an even function and an odd function is an odd function.

40. *Finding the Period* Give a convincing argument that the period of $\tan x$ is π.

41. *Sinusoidal Regression* Table 1.21 gives the values of the function

$$f(x) = a \sin (bx + c) + d$$

accurate to two decimals.

(a) Find a sinusoidal regression equation for the data.

(b) Rewrite the equation with a, b, c, and d rounded to the nearest integer.

Table 1.21 Values of a Function

x	$f(x)$
1	3.42
2	0.73
3	0.12
4	2.16
5	4.97
6	5.97

Exploration

42. *Trigonometric Identities* Let $f(x) = \sin x + \cos x$.

(a) Graph $y = f(x)$. Describe the graph.

(b) Use the graph to identify the amplitude, period, horizontal shift, and vertical shift.

(c) Use the formula

$$\sin \alpha \cos \beta + \cos \alpha \sin \beta = \sin (\alpha + \beta)$$

for the sine of the sum of two angles to confirm your answers. ∎

Extending the Ideas

43. **Exploration** Let $y = \sin (ax) + \cos (ax)$.

Use the symbolic manipulator of a computer algebra system (CAS) to help you with the following:

(a) Express y as a sinusoid for $a = 2, 3, 4$, and 5.

(b) Conjecture another formula for y for a equal to any positive integer n.

(c) Check your conjecture with a CAS.

(d) Use the formula for the sine of the sum of two angles (see Exercise 42c) to confirm your conjecture.

44. Exploration Let $y = a \sin x + b \cos x$.

Use the symbolic manipulator of a computer algebra system (CAS) to help you with the following:

(a) Express y as a sinusoid for the following pairs of values:
$a = 2, b = 1$; $a = 1, b = 2$; $a = 5, b = 2$;
$a = 2, b = 5$; $a = 3, b = 4$.

(b) Conjecture another formula for y for any pair of positive integers. Try other values if necessary.

(c) Check your conjecture with a CAS.

(d) Use the following formulas for the sine or cosine of a sum or difference of two angles to confirm your conjecture.

$$\sin \alpha \cos \beta \pm \cos \alpha \sin \beta = \sin (\alpha \pm \beta)$$
$$\cos \alpha \cos \beta \pm \sin \alpha \sin \beta = \cos (\alpha \pm \beta)$$

In Exercises 45 and 46, show that the function is periodic and find its period.

45. $y = \sin^3 x$　　　　　　**46.** $y = |\tan x|$

In Exercises 47 and 48, graph one period of the function.

47. $f(x) = \sin (60x)$　　　　**48.** $f(x) = \cos (60\pi x)$

Chapter 1　Key Terms

absolute value function (p. 15)
base a logarithm function (p. 36)
boundary of an interval (p. 11)
boundary points (p. 11)
change of base formula (p. 37)
closed interval (p. 11)
common logarithm function (p. 36)
composing (p. 16)
composite function (p. 16)
compounded continuously (p. 23)
cosecant function (p. 41)
cosine function (p. 41)
cotangent function (p. 41)
dependent variable (p. 10)
domain (p. 10)
even function (p. 13)
exponential decay (p. 22)
exponential function base a (p. 20)
exponential growth (p. 22)
function (p. 10)
general linear equation (p. 4)
graph of a function (p. 11)
graph of a relation (p. 26)
grapher failure (p. 12)
half-life (p. 21)
half-open interval (p. 11)
identity function (p. 33)
increments (p. 1)

independent variable (p. 10)
initial point of parametrized curve (p. 27)
interior of an interval (p. 11)
interior points of an interval (p. 11)
inverse cosecant function (p. 47)
inverse cosine function (p. 47)
inverse cotangent function (p. 47)
inverse function (p. 33)
inverse properties for a^x and $\log_a x$ (p. 36)
inverse secant function (p. 47)
inverse sine function (p. 47)
inverse tangent function (p. 47)
linear regression (p. 5)
natural domain (p. 10)
natural logarithm function (p. 36)
odd function (p. 13)
one-to-one function (p. 32)
open interval (p. 11)
parallel lines (p. 2)
parameter (p. 27)
parameter interval (p. 27)
parametric curve (p. 26)
parametric equations (p. 26)
parametrization of a curve (p. 27)
parametrize (p. 27)
period of a function (p. 43)
periodic function (p. 43)
perpendicular lines (p. 2)

piecewise defined function (p. 15)
point-slope equation (p. 3)
power function (p. 12)
power rule for logarithms (p. 37)
product rule for logarithms (p. 37)
quotient rule for logarithms (p. 37)
radian measure (p. 41)
range (p. 10)
regression analysis (p. 5)
regression curve (p. 5)
relation (p. 26)
rise (p. 2)
run (p. 2)
scatter plot (p. 5)
secant function (p. 41)
sine function (p. 40)
sinusoid (p. 44)
sinusoidal regression (p. 46)
slope (p. 2)
slope-intercept equation (p. 4)
symmetry about the origin (p. 14)
symmetry about the y-axis (p. 14)
tangent function (p. 41)
terminal point of parametrized curve (p. 27)
witch of Agnesi (p. 29)
x-intercept (p. 3)
y-intercept (p. 3)

Chapter 1 Review Exercises

In Exercises 1–14, write an equation for the specified line.

1. through $(1, -6)$ with slope 3

2. through $(-1, 2)$ with slope $-1/2$

3. the vertical line through $(0, -3)$

4. through $(-3, 6)$ and $(1, -2)$

5. the horizontal line through $(0, 2)$

6. through $(3, 3)$ and $(-2, 5)$

7. with slope -3 and y-intercept 3

8. through $(3, 1)$ and parallel to $2x - y = -2$

9. through $(4, -12)$ and parallel to $4x + 3y = 12$

10. through $(-2, -3)$ and perpendicular to $3x - 5y = 1$

11. through $(-1, 2)$ and perpendicular to $\frac{1}{2}x + \frac{1}{3}y = 1$

12. with x-intercept 3 and y-intercept -5

13. the line $y = f(x)$, where f has the following values:

x	-2	2	4
$f(x)$	4	2	1

14. through $(4, -2)$ with x-intercept -3

In Exercises 15–18, determine whether the graph of the function is symmetric about the y-axis, the origin, or neither.

15. $y = x^{1/5}$

16. $y = x^{2/5}$

17. $y = x^2 - 2x - 1$

18. $y = e^{-x^2}$

In Exercises 19–26, determine whether the function is even, odd, or neither.

19. $y = x^2 + 1$

20. $y = x^5 - x^3 - x$

21. $y = 1 - \cos x$

22. $y = \sec x \tan x$

23. $y = \dfrac{x^4 + 1}{x^3 - 2x}$

24. $y = 1 - \sin x$

25. $y = x + \cos x$

26. $y = \sqrt{x^4 - 1}$

In Exercises 27–38, find the (a) domain and (b) range, and (c) graph the function.

27. $y = |x| - 2$

28. $y = -2 + \sqrt{1 - x}$

29. $y = \sqrt{16 - x^2}$

30. $y = 3^{2-x} + 1$

31. $y = 2e^{-x} - 3$

32. $y = \tan(2x - \pi)$

33. $y = 2\sin(3x + \pi) - 1$

34. $y = x^{2/5}$

35. $y = \ln(x - 3) + 1$

36. $y = -1 + \sqrt[3]{2 - x}$

37. $y = \begin{cases} \sqrt{-x}, & -4 \le x \le 0 \\ \sqrt{x}, & 0 < x \le 4 \end{cases}$

38. $y = \begin{cases} -x - 2, & -2 \le x \le -1 \\ x, & -1 < x \le 1 \\ -x + 2, & 1 < x \le 2 \end{cases}$

In Exercises 39 and 40, write a piecewise formula for the function.

39.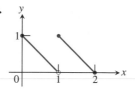

40.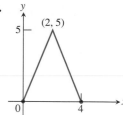

In Exercises 41 and 42, find

(a) $(f \circ g)(-1)$ (b) $(g \circ f)(2)$

(c) $(f \circ f)(x)$ (d) $(g \circ g)(x)$

41. $f(x) = \dfrac{1}{x}$, $g(x) = \dfrac{1}{\sqrt{x + 2}}$

42. $f(x) = 2 - x$, $g(x) = \sqrt[3]{x + 1}$

In Exercises 43 and 44, (a) write a formula for $f \circ g$ and $g \circ f$ and find the (b) domain and (c) range of each.

43. $f(x) = 2 - x^2$, $g(x) = \sqrt{x + 2}$

44. $f(x) = \sqrt{x}$, $g(x) = \sqrt{1 - x}$

In Exercises 45–48, a parametrization is given for a curve.

(a) Graph the curve. Identify the initial and terminal points, if any. Indicate the direction in which the curve is traced.

(b) Find a Cartesian equation for a curve that contains the parametrized curve. What portion of the graph of the Cartesian equation is traced by the parametrized curve?

45. $x = 5\cos t$, $y = 2\sin t$, $0 \le t \le 2\pi$

46. $x = 4\cos t$, $y = 4\sin t$, $\pi/2 \le t < 3\pi/2$

47. $x = 2 - t$, $y = 11 - 2t$, $-2 \le t \le 4$

48. $x = 1 + t$, $y = \sqrt{4 - 2t}$, $t \le 2$

In Exercises 49–52, give a parametrization for the curve.

49. the line segment with endpoints $(-2, 5)$ and $(4, 3)$

50. the line through $(-3, -2)$ and $(4, -1)$

51. the ray with initial point $(2, 5)$ that passes through $(-1, 0)$

52. $y = x(x - 4)$, $x \le 2$

In Exercises 53 and 54, *work in groups of two or three* to (a) find f^{-1} and show that $(f \circ f^{-1})(x) = (f^{-1} \circ f)(x) = x$.
(b) graph f and f^{-1} in the same viewing window.

53. $f(x) = 2 - 3x$

54. $f(x) = (x + 2)^2$, $x \ge -2$

In Exercises 55 and 56, find the measure of the angle in radians and degrees.

55. $\sin^{-1}(0.6)$ **56.** $\tan^{-1}(-2.3)$

57. Find the six trigonometric values of $\theta = \cos^{-1}(3/7)$. Give exact answers.

58. Solve the equation $\sin x = -0.2$ in the following intervals.

(a) $0 \leq x < 2\pi$ (b) $-\infty < x < \infty$

59. Solve for x: $e^{-0.2x} = 4$

60. The graph of f is shown. Draw the graph of each function.

(a) $y = f(-x)$ (b) $y = -f(x)$

(c) $y = -2f(x + 1) + 1$ (d) $y = 3f(x - 2) - 2$

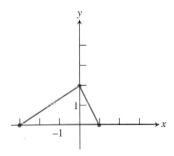

61. A portion of the graph of a function defined on $[-3, 3]$ is shown. Complete the graph assuming that the function is

(a) even.   (b) odd.

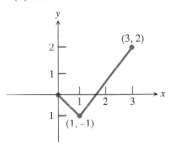

62. *Depreciation* Smith Hauling purchased an 18-wheel truck for $100,000. The truck depreciates at the constant rate of $10,000 per year for 10 years.

(a) Write an expression that gives the value y after x years.

(b) When is the value of the truck $55,000?

63. *Drug Absorption* A drug is administered intravenously for pain. The function

$$f(t) = 90 - 52 \ln(1 + t), \quad 0 \leq t \leq 4$$

gives the number of units of the drug in the body after t hours.

(a) What was the initial number of units of the drug administered?

(b) How much is present after 2 hours?

(c) Draw the graph of f.

64. *Finding Time* If Joenita invests $1500 in a retirement account that earns 8% compounded annually, how long will it take this single payment to grow to $5000?

65. *Guppy Population* The number of guppies in Susan's aquarium doubles every day. There are four guppies initially.

(a) Write the number of guppies as a function of time t.

(b) How many guppies were present after 4 days? after 1 week?

(c) When will there be 2000 guppies?

(d) **Writing to Learn** Give reasons why this might not be a good model for the growth of Susan's guppy population.

66. *Doctoral Degrees* Table 1.22 shows the number of doctoral degrees earned in the given academic year by Hispanic students. Let $x = 0$ represent 1970–71, $x = 1$ represent 1971–72, and so forth.

Table 1.22 Doctorates Earned by Hispanic Americans

Year	Number of Degrees
1976–77	520
1980–81	460
1984–85	680
1988–89	630
1990–91	730
1991–92	810
1992–93	830

Source: U.S. Department of Education, as reported in the *Chronicle of Higher Education*, April 28, 1995.

(a) Find the linear regression equation for the data and superimpose its graph on a scatter plot of the data.

(b) Use the regression equation to predict the number of doctoral degrees that will be earned by Hispanic Americans in the academic year 2000–01.

(c) **Writing to Learn** Find the slope of the regression line. What does the slope represent?

67. *Estimating Population Growth* Use the data in Table 1.23 about the population of New York State. Let $x = 60$ represent 1960, $x = 70$ represent 1970, and so forth.

Table 1.23 Population of New York State

Year	Population (millions)
1960	16.78
1980	17.56
1990	17.99

Source: The Statesman's Yearbook, 129th ed. (London: The Macmillan Press, Ltd., 1992).

(a) Find the exponential regression equation for the data.

(b) Use the regression equation to predict when the population will be 25 million.

(c) What annual rate of growth can we infer from the regression equation?

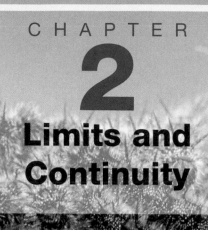

C H A P T E R

2

Limits and Continuity

An Economic Injury Level (EIL) is a measurement of the fewest number of insect pests that will cause economic damage to a crop or forest. It has been estimated that monitoring pest populations and establishing EILs can reduce pesticide use by 30–50%.

Accurate population estimates are crucial for determining EILs. A population density of one insect pest can be approximated by

$$D(t) = \frac{t^2}{90} + \frac{t}{3}$$

pests per plant, where t is the number of days since initial infestation. What is the rate of change of this population density when the population density is equal to the EIL of 20 pests per plant? Section 2.4 can help answer this question.

Chapter 2 Overview

The concept of limit is one of the ideas that distinguish calculus from algebra and trigonometry.

In this chapter, we show how to define and calculate limits of function values. The calculation rules are straightforward and most of the limits we need can be found by substitution, graphical investigation, numerical approximation, algebra, or some combination of these.

One of the uses of limits is to test functions for continuity. Continuous functions arise frequently in scientific work because they model such an enormous range of natural behavior. They also have special mathematical properties, not otherwise guaranteed.

2.1 Rates of Change and Limits

Average and Instantaneous Speed • Definition of Limit • Properties of Limits • One-sided and Two-sided Limits • Sandwich Theorem

Average and Instantaneous Speed

A moving body's **average speed** during an interval of time is found by dividing the distance covered by the elapsed time. The unit of measure is length per unit time—kilometers per hour, feet per second, or whatever is appropriate to the problem at hand.

Free Fall

Near the surface of the earth, all bodies fall with the same constant acceleration. The distance a body falls after it is released from rest is a constant multiple of the square of the time fallen. At least, that is what happens when a body falls in a vacuum, where there is no air to slow it down. The square-of-time rule also holds for dense, heavy objects like rocks, ball bearings, and steel tools during the first few seconds of fall through air, before the velocity builds up to where air resistance begins to matter. When air resistance is absent or insignificant and the only force acting on a falling body is the force of gravity, we call the way the body falls *free fall*.

Example 1 FINDING AN AVERAGE SPEED

A rock breaks loose from the top of a tall cliff. What is its average speed during the first 2 seconds of fall?

Solution Experiments show that a dense solid object dropped from rest to fall freely near the surface of the earth will fall

$$y = 16t^2$$

feet in the first t seconds. The average speed of the rock over any given time interval is the distance traveled, Δy, divided by the length of the interval Δt. For the first 2 seconds of fall, from $t = 0$ to $t = 2$, we have

$$\frac{\Delta y}{\Delta t} = \frac{16(2)^2 - 16(0)^2}{2 - 0} = 32\frac{\text{ft}}{\text{sec}}.$$

Example 2 FINDING AN INSTANTANEOUS SPEED

Find the speed of the rock in Example 1 at the instant $t = 2$.

Solution

Solve Numerically

We can calculate the average speed of the rock over the interval from time $t = 2$ to any slightly later time $t = 2 + h$ as

$$\frac{\Delta y}{\Delta t} = \frac{16(2 + h)^2 - 16(2)^2}{h}. \tag{1}$$

Table 2.1 Average Speeds over Short Time Intervals Starting at $t = 2$

$$\frac{\Delta y}{\Delta t} = \frac{16(2 + h)^2 - 16(2)^2}{h}$$

Length of Time Interval, h (sec)	Average Speed for Interval $\Delta y/\Delta t$ (ft/sec)
1	80
0.1	65.6
0.01	64.16
0.001	64.016
0.0001	64.0016
0.00001	64.00016

We cannot use this formula to calculate the speed at the exact instant $t = 2$ because that would require taking $h = 0$, and $0/0$ is undefined. However, we can get a good idea of what is happening at $t = 2$ by evaluating the formula at values of h *close* to 0. When we do, we see a clear pattern (Table 2.1). As h approaches 0, the average speed approaches the limiting value 64 ft/sec.

Confirm Algebraically

If we expand the numerator of Equation 1 and simplify, we find that

$$\frac{\Delta y}{\Delta t} = \frac{16(2 + h)^2 - 16(2)^2}{h} = \frac{16(4 + 4h + h^2) - 64}{h}$$

$$= \frac{64h + 16h^2}{h} = 64 + 16h.$$

For values of h different from 0, the expressions on the right and left are equivalent and the average speed is $64 + 16h$ ft/sec. We can now see why the average speed has the limiting value $64 + 16(0) = 64$ ft/sec as h approaches 0.

Definition of Limit

As in the preceding example, most limits of interest in the real world can be viewed as numerical limits of values of functions. And this is where a graphing utility and calculus come in. A calculator can suggest the limits, and calculus can give the mathematics for confirming the limits analytically.

Limits give us a language for describing how the outputs of a function behave as the inputs approach some particular value. In Example 2, the average speed was not defined at $h = 0$ but approached the limit 64 as h approached 0. We were able to see this numerically and to confirm it algebraically by eliminating h from the denominator. But we cannot always do that. For instance, we can see both graphically and numerically (Figure 2.1) that the values of $f(x) = (\sin x)/x$ approach 1 as x approaches 0.

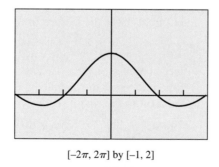

$[-2\pi, 2\pi]$ by $[-1, 2]$

(a)

(b)

Figure 2.1 (a) A graph and (b) table of values for $f(x) = (\sin x)/x$ that suggest the limit of f as x approaches 0 is 1.

We cannot eliminate the x from the denominator of $(\sin x)/x$ to confirm the observation algebraically. We need to use a theorem about limits to make that confirmation, as you will see in Exercise 63.

Definition Limit

Let c and L be real numbers. The function f **has limit L as x approaches c** if, given any positive number ε, there is a positive number δ such that for all x,

$$0 < |x - c| < \delta \Rightarrow |f(x) - L| < \varepsilon.$$

We write

$$\lim_{x \to c} f(x) = L.$$

The sentence $\lim_{x \to c} f(x) = L$ is read, "The limit of f of x as x approaches c equals L." The notation means that the values $f(x)$ of the function f approach or equal L as the values of x approach *(but do not equal)* c. Appendix 3 provides practice applying the definition of limit.

We saw in Example 2 that $\lim_{h \to 0} (64 + 16h) = 64$.

As suggested in Figure 2.1,

$$\lim_{x \to 0} \frac{\sin x}{x} = 1.$$

Figure 2.2 illustrates the fact that the existence of a limit as $x \to c$ never depends on how the function may or may not be defined at c. The function f has limit 2 as $x \to 1$ even though f is not defined at 1. The function g has limit 2 as $x \to 1$ even though $g(1) \neq 2$. The function h is the only one whose limit as $x \to 1$ equals its value at $x = 1$.

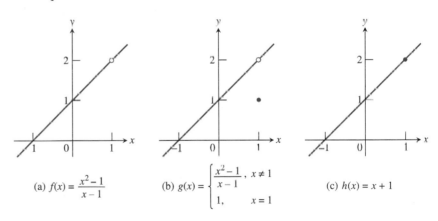

(a) $f(x) = \dfrac{x^2 - 1}{x - 1}$

(b) $g(x) = \begin{cases} \dfrac{x^2 - 1}{x - 1}, & x \neq 1 \\ 1, & x = 1 \end{cases}$

(c) $h(x) = x + 1$

Figure 2.2 $\lim_{x \to 1} f(x) = \lim_{x \to 1} g(x) = \lim_{x \to 1} h(x) = 2$

Properties of Limits

By applying six basic facts about limits, we can calculate many unfamiliar limits from limits we already know. For instance, from knowing that

$$\lim_{x \to c} (k) = k \qquad \text{Limit of the function with constant value } k$$

and

$$\lim_{x \to c} (x) = c, \qquad \text{Limit of the identity function at } x - c$$

we can calculate the limits of all polynomial and rational functions. The facts are listed in Theorem 1.

Theorem 1 Properties of Limits

If L, M, c, and k are real numbers and

$$\lim_{x \to c} f(x) = L \quad \text{and} \quad \lim_{x \to c} g(x) = M, \text{ then}$$

1. *Sum Rule:* $\qquad\qquad\qquad\qquad \lim_{x \to c} (f(x) + g(x)) = L + M$

   The limit of the sum of two functions is the sum of their limits.

2. *Difference Rule:* $\qquad\qquad\quad\; \lim_{x \to c} (f(x) - g(x)) = L - M$

   The limit of the difference of two functions is the difference of their limits.

3. *Product Rule:* $\qquad\qquad\qquad \lim_{x \to c} (f(x) \cdot g(x)) = L \cdot M$

   The limit of a product of two functions is the product of their limits.

4. *Constant Multiple Rule:* $\qquad \lim_{x \to c} (k \cdot f(x)) = k \cdot L$

   The limit of a constant times a function is the constant times the limit of the function.

5. *Quotient Rule:* $\qquad\qquad\qquad \lim_{x \to c} \dfrac{f(x)}{g(x)} = \dfrac{L}{M}, M \neq 0$

   The limit of a quotient of two functions is the quotient of their limits, provided the limit of the denominator is not zero.

6. *Power Rule:* If r and s are integers, $s \neq 0$, then

   $$\lim_{x \to c} (f(x))^{r/s} = L^{r/s}$$

   provided that $L^{r/s}$ is a real number.

   The limit of a rational power of a function is that power of the limit of the function, provided the latter is a real number.

Here are some examples of how Theorem 1 can be used to find limits of polynomial and rational functions.

Example 3 USING PROPERTIES OF LIMITS

Use the observations $\lim_{x \to c} k = k$ and $\lim_{x \to c} x = c,$ and the properties of limits to find the following limits.

(a) $\lim_{x \to c} (x^3 + 4x^2 - 3)$ **(b)** $\lim_{x \to c} \dfrac{x^4 + x^2 - 1}{x^2 + 5}$

Solution

(a) $\lim_{x \to c} (x^3 + 4x^2 - 3) = \lim_{x \to c} x^3 + \lim_{x \to c} 4x^2 - \lim_{x \to c} 3$ ⟶ Sum & Difference Rules

$$= c^3 + 4c^2 - 3 \qquad\qquad \text{Product \& Multiple Rules}$$

(b) $\lim_{x \to c} \dfrac{x^4 + x^2 - 1}{x^2 + 5} = \dfrac{\lim_{x \to c} (x^4 + x^2 - 1)}{\lim_{x \to c} (x^2 + 5)}$ ⟶ Quotient Rule

$$= \dfrac{\lim_{x \to c} x^4 + \lim_{x \to c} x^2 - \lim_{x \to c} 1}{\lim_{x \to c} x^2 + \lim_{x \to c} 5} \qquad \text{Sum \& Difference Rules}$$

$$= \frac{c^4 + c^2 - 1}{c^2 + 5} \qquad \text{Product Rule}$$

Example 3 shows the remarkable strength of Theorem 1. From the two simple observations that $\lim_{x \to c} k = k$ and $\lim_{x \to c} x = c,$ we can immediately work our way to limits of polynomial functions and most rational functions using substitution.

Theorem 2 Polynomial and Rational Functions

1. If $f(x) = a_n x^n + a_{n-1} x^{n-1} + \cdots + a_0$ is any polynomial function and c is any real number, then

$$\lim_{x \to c} f(x) = f(c) = a_n c^n + a_{n-1} c^{n-1} + \cdots + a_0.$$

2. If $f(x)$ and $g(x)$ are polynomials and c is any real number, then

$$\lim_{x \to c} \frac{f(x)}{g(x)} = \frac{f(c)}{g(c)}, \qquad \text{provided that } g(c) \neq 0.$$

Example 4   USING THEOREM 2

(a) $\lim_{x \to 3} \left[x^2(2 \quad x) \right] = (3)^2(2 - 3) = -9$

(b) $\lim_{x \to 2} \frac{x^2 + 2x + 4}{x + 2} = \frac{(2)^2 + 2(2) + 4}{2 + 2} = \frac{12}{4} = 3$

As with polynomials, limits of many familiar functions can be found by substitution at points where they are defined. This includes trigonometric functions, exponential and logarithmic functions, and composites of these functions. Feel free to use these properties.

Example 5   USING THE PRODUCT RULE

Determine $\lim_{x \to 0} \dfrac{\tan x}{x}.$

Solution

Solve Graphically

The graph of $f(x) = (\tan x)/x$ in Figure 2.3 suggests that the limit exists and is about 1.

Confirm Analytically

Using the analytic result of Exercise 63, we have

$$\lim_{x \to 0} \frac{\tan x}{x} = \lim_{x \to 0} \left(\frac{\sin x}{x} \cdot \frac{1}{\cos x} \right) \qquad \tan x = \frac{\sin x}{\cos x}$$

$$= \lim_{x \to 0} \frac{\sin x}{x} \cdot \lim_{x \to 0} \frac{1}{\cos x} \qquad \text{Product Rule}$$

$$= 1 \cdot \frac{1}{\cos 0} = 1 \cdot \frac{1}{1} = 1.$$

Sometimes we can use a graph to discover that limits do not exist, as illustrated by Example 6.

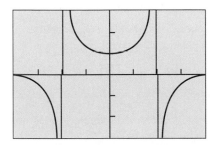

$[-\pi, \pi]$ by $[-3, 3]$

Figure 2.3 The graph of $f(x) = (\tan x)/x$ suggests that $f(x) \to 1$ as $x \to 0$. (Example 5)

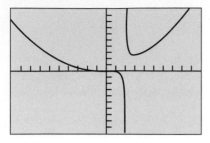

[–10, 10] by [–60, 60]

Figure 2.4 The graph of
$$f(x) = (x^3 - 1)/(x - 2)$$
obtained using parametric graphing to produce a more accurate graph. (Example 6)

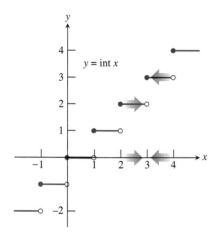

Figure 2.5 At each integer, the greatest integer function $y = \text{int } x$ has different right-hand and left-hand limits. (Example 7)

On the Far Side

If f is not defined to the left of $x = c$, then f does not have a left-hand limit at c. Similarly, if f is not defined to the right of $x = c$, then f does not have a right-hand limit at c.

Example 6 EXPLORING A NONEXISTENT LIMIT

Use a graph to show that
$$\lim_{x \to 2} \frac{x^3 - 1}{x - 2}$$
does not exist.

Solution Notice that the denominator is 0 when x is replaced by 2, so we cannot use substitution to determine the limit. The graph in Figure 2.4 of $f(x) = (x^3 - 1)/(x - 2)$ strongly suggests that as $x \to 2$ from either side, the absolute values of the function values get very large. This, in turn, suggests that the limit does not exist.

One-sided and Two-sided Limits

Sometimes the values of a function f tend to different limits as x approaches a number c from opposite sides. When this happens, we call the limit of f as x approaches c from the right the **right-hand limit** of f at c and the limit as x approaches c from the left the **left-hand limit** of f at c. Here is the notation we use:

right-hand: $\lim_{x \to c^+} f(x)$ *The limit of f as x approaches c from the right.*

left-hand: $\lim_{x \to c^-} f(x)$ *The limit of f as x approaches c from the left.*

Example 7 FUNCTION VALUES APPROACH TWO NUMBERS

The greatest integer function $f(x) = \text{int } x$ has different right-hand and left-hand limits at each integer, as we can see in Figure 2.5. For example,

$$\lim_{x \to 3^+} \text{int } x = 3 \quad \text{and} \quad \lim_{x \to 3^-} \text{int } x = 2.$$

The limit of int x as x approaches an integer n from the right is n, while the limit as x approaches n from the left is $n - 1$.

We sometimes call $\lim_{x \to c} f(x)$ the **two-sided limit** of f at c to distinguish it from the *one-sided* right-hand and left-hand limits of f at c. Theorem 3 shows how these limits are related.

Theorem 3 One-sided and Two-sided Limits

A function $f(x)$ has a limit as x approaches c if and only if the right-hand and left-hand limits at c exist and are equal. In symbols,

$$\lim_{x \to c} f(x) = L \Leftrightarrow \lim_{x \to c^+} f(x) = L \quad \text{and} \quad \lim_{x \to c^-} f(x) = L.$$

Thus, the greatest integer function $f(x) = \text{int } x$ of Example 7 does not have a limit as $x \to 3$ even though each one-sided limit exists.

Example 8 EXPLORING RIGHT- AND LEFT-HAND LIMITS

All the following statements about the function $y = f(x)$ graphed in Figure 2.6 are true.

At $x = 0$: $\lim_{x \to 0^+} f(x) = 1.$

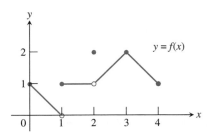

Figure 2.6 The graph of the function

$$f(x) = \begin{cases} -x + 1, & 0 \leq x < 1 \\ 1, & 1 \leq x < 2 \\ 2, & x = 2 \\ x - 1, & 2 < x \leq 3 \\ -x + 5, & 3 < x \leq 4. \end{cases}$$

(Example 8)

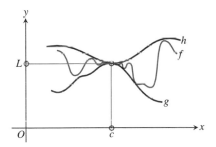

Figure 2.7 Sandwiching f between g and h forces the limiting value of f to be between the limiting values of g and h.

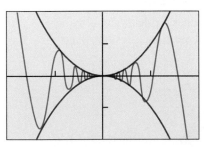

[−0.2, 0.2] by [−0.02, 0.02]

Figure 2.8 The graphs of $y_1 = x^2$, $y_2 = x^2 \sin(1/x)$, and $y_3 = -x^2$. Notice that $y_3 \leq y_2 \leq y_1$. (Example 9)

At $x = 1$: $\quad \lim_{x \to 1^-} f(x) = 0$ even though $f(1) = 1$,

$$\lim_{x \to 1^+} f(x) = 1,$$

f has no limit as $x \to 1$. (The right- and left-hand limits at 1 are not equal, so $\lim_{x \to 1} f(x)$ does not exist.)

At $x = 2$: $\quad \lim_{x \to 2^-} f(x) = 1$,

$$\lim_{x \to 2^+} f(x) = 1,$$

$$\lim_{x \to 2} f(x) = 1 \text{ even though } f(2) = 2.$$

At $x = 3$: $\quad \lim_{x \to 3^-} f(x) = \lim_{x \to 3^+} f(x) = 2 = f(3) = \lim_{x \to 3} f(x)$.

At $x = 4$: $\quad \lim_{x \to 4^-} f(x) = 1$.

At noninteger values of c between 0 and 4, f has a limit as $x \to c$.

Sandwich Theorem

If we cannot find a limit directly, we may be able to find it indirectly with the Sandwich Theorem. The theorem refers to a function f whose values are sandwiched between the values of two other functions, g and h. If g and h have the same limit as $x \to c$, then f has that limit too, as suggested by Figure 2.7.

Theorem 4 The Sandwich Theorem

If $g(x) \leq f(x) \leq h(x)$ for all $x \neq c$ in some interval about c, and

$$\lim_{x \to c} g(x) = \lim_{x \to c} h(x) = L,$$

then

$$\lim_{x \to c} f(x) = L.$$

Example 9 USING THE SANDWICH THEOREM

Show that $\lim_{x \to 0} \left[x^2 \sin(1/x) \right] = 0$.

Solution We know that the values of the sine function lie between -1 and 1. So, it follows that

$$\left| x^2 \sin \frac{1}{x} \right| = |x^2| \cdot \left| \sin \frac{1}{x} \right| \leq |x^2| \cdot 1 = x^2$$

and

$$-x^2 \leq x^2 \sin \frac{1}{x} \leq x^2.$$

Because $\lim_{x \to 0} (-x^2) = \lim_{x \to 0} x^2 = 0$, the Sandwich Theorem gives

$$\lim_{x \to 0} \left(x^2 \sin \frac{1}{x} \right) = 0.$$

The graphs in Figure 2.8 support this result.

Quick Review 2.1

In Exercises 1–4, find $f(2)$.

1. $f(x) = 2x^3 - 5x^2 + 4$

2. $f(x) = \dfrac{4x^2 - 5}{x^3 + 4}$

3. $f(x) = \sin\left(\pi \dfrac{x}{2}\right)$

4. $f(x) = \begin{cases} 3x - 1, & x < 2 \\ \dfrac{1}{x^2 - 1}, & x \geq 2 \end{cases}$

In Exercises 5–8, write the inequality in the form $a < x < b$.

5. $|x| < 4$

6. $|x| < c^2$

7. $|x - 2| < 3$

8. $|x - c| < d^2$

In Exercises 9 and 10, write the fraction in reduced form.

9. $\dfrac{x^2 - 3x - 18}{x + 3}$

10. $\dfrac{2x^2 - x}{2x^2 + x - 1}$

Section 2.1 Exercises

In Exercises 1–6, use the graph to estimate the limits and value of the function, or explain why the limits do not exist.

1.

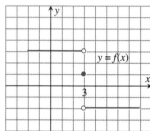

(a) $\lim\limits_{x \to 3^-} f(x)$ **(b)** $\lim\limits_{x \to 3^+} f(x)$ **(c)** $\lim\limits_{x \to 3} f(x)$ **(d)** $f(3)$

2.

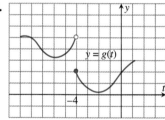

(a) $\lim\limits_{t \to -4^-} g(t)$ **(b)** $\lim\limits_{t \to -4^+} g(t)$ **(c)** $\lim\limits_{t \to -4} g(t)$ **(d)** $g(-4)$

3.

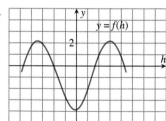

(a) $\lim\limits_{h \to 0^-} f(h)$ **(b)** $\lim\limits_{h \to 0^+} f(h)$ **(c)** $\lim\limits_{h \to 0} f(h)$ **(d)** $f(0)$

4.

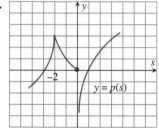

(a) $\lim\limits_{s \to -2^-} p(s)$ **(b)** $\lim\limits_{s \to -2^+} p(s)$ **(c)** $\lim\limits_{s \to -2} p(s)$ **(d)** $p(-2)$

5.

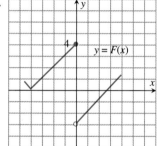

(a) $\lim\limits_{x \to 0^-} F(x)$ **(b)** $\lim\limits_{x \to 0^+} F(x)$ **(c)** $\lim\limits_{x \to 0} F(x)$ **(d)** $F(0)$

6.

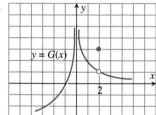

(a) $\lim\limits_{x \to 2^-} G(x)$ **(b)** $\lim\limits_{x \to 2^+} G(x)$ **(c)** $\lim\limits_{x \to 2} G(x)$ **(d)** $G(2)$

In Exercises 7–16, determine the limit by substitution. Support graphically.

7. $\lim\limits_{x \to -1/2} 3x^2(2x - 1)$

8. $\lim\limits_{x \to -4} (x + 3)^{1998}$

9. $\lim\limits_{x \to 1} (x^3 + 3x^2 - 2x - 17)$

10. $\lim\limits_{y \to 2} \dfrac{y^2 + 5y + 6}{y + 2}$

11. $\lim\limits_{y \to -3} \dfrac{y^2 + 4y + 3}{y^2 - 3}$

12. $\lim\limits_{x \to 1/2}$ int x

13. $\lim\limits_{x \to -2} (x - 6)^{2/3}$

14. $\lim\limits_{x \to 2} \sqrt{x + 3}$

15. $\lim\limits_{x \to 0} e^x \cos x$

16. $\lim\limits_{x \to \pi/2} \ln(\sin x)$

In Exercises 17–20, explain why you cannot use substitution to determine the limit. Find the limit if it exists.

17. $\lim\limits_{x \to -2} \sqrt{x - 2}$

18. $\lim\limits_{x \to 0} \dfrac{1}{x^2}$

19. $\lim\limits_{x \to 0} \dfrac{|x|}{x}$

20. $\lim\limits_{x \to 0} \dfrac{(4 + x)^2 - 16}{x}$

In Exercises 21–30, determine the limit graphically. Confirm algebraically.

21. $\lim\limits_{x \to 1} \dfrac{x - 1}{x^2 - 1}$

22. $\lim\limits_{t \to 2} \dfrac{t^2 - 3t + 2}{t^2 - 4}$

23. $\lim\limits_{x \to 0} \dfrac{5x^3 + 8x^2}{3x^4 - 16x^2}$

24. $\lim\limits_{x \to 0} \dfrac{\dfrac{1}{2 + x} - \dfrac{1}{2}}{x}$

25. $\lim\limits_{x \to 0} \dfrac{(2 + x)^3 - 8}{x}$

26. $\lim\limits_{x \to 0} \dfrac{\sin 2x}{x}$

27. $\lim\limits_{x \to 0} \dfrac{\sin x}{2x^2 - x}$

28. $\lim\limits_{x \to 0} \dfrac{x + \sin x}{x}$

29. $\lim\limits_{x \to 0} \dfrac{\sin^2 x}{x}$

30. $\lim\limits_{x \to 0} \dfrac{3 \sin 4x}{\sin 3x}$

In Exercises 31 and 32, which of the statements are true about the function $y = f(x)$ graphed there, and which are false?

31.

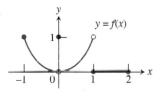

(a) $\lim\limits_{x \to -1^+} f(x) = 1$

(b) $\lim\limits_{x \to 0^-} f(x) = 0$

(c) $\lim\limits_{x \to 0^-} f(x) = 1$

(d) $\lim\limits_{x \to 0^-} f(x) = \lim\limits_{x \to 0^+} f(x)$

(e) $\lim\limits_{x \to 0} f(x)$ exists

(f) $\lim\limits_{x \to 0} f(x) = 0$

(g) $\lim\limits_{x \to 0} f(x) = 1$

(h) $\lim\limits_{x \to 1} f(x) = 1$

(i) $\lim\limits_{x \to 1} f(x) = 0$

(j) $\lim\limits_{x \to 2^-} f(x) = 2$

32.

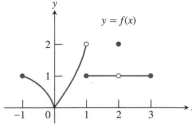

(a) $\lim\limits_{x \to -1^+} f(x) = 1$

(b) $\lim\limits_{x \to 2} f(x)$ does not exist.

(c) $\lim\limits_{x \to 2} f(x) = 2$

(d) $\lim\limits_{x \to 1^-} f(x) = 2$

(e) $\lim\limits_{x \to 1^+} f(x) = 1$

(f) $\lim\limits_{x \to 1} f(x)$ does not exist.

(g) $\lim\limits_{x \to 0^+} f(x) = \lim\limits_{x \to 0^-} f(x)$

(h) $\lim\limits_{x \to c} f(x)$ exists at every c in $(-1, 1)$.

(i) $\lim\limits_{x \to c} f(x)$ exists at every c in $(1, 3)$.

In Exercises 33–36, match the function with the table.

33. $y_1 = \dfrac{x^2 + x - 2}{x - 1}$

34. $y_1 = \dfrac{x^2 - x - 2}{x - 1}$

35. $y_1 = \dfrac{x^2 - 2x + 1}{x - 1}$

36. $y_1 = \dfrac{x^2 + x - 2}{x + 1}$

X	Y₁
.7	-.4765
.8	-.3111
.9	-.1526
1	0
1.1	.14762
1.2	.29091
1.3	.43043

X = .7

(a)

X	Y₁
.7	7.3667
.8	10.8
.9	20.9
1	ERROR
1.1	-18.9
1.2	-8.8
1.3	-5.367

X = .7

(b)

X	Y₁
.7	2.7
.8	2.8
.9	2.9
1	ERROR
1.1	3.1
1.2	3.2
1.3	3.3

X = .7

(c)

X	Y₁
.7	-.3
.8	-.2
.9	-.1
1	ERROR
1.1	.1
1.2	.2
1.3	.3

X = .7

(d)

In Exercises 37–44, determine the limit.

37. $\lim\limits_{x \to 0^+}$ int x

38. $\lim\limits_{x \to 0^-}$ int x

39. $\lim\limits_{x \to 0.01}$ int x

40. $\lim\limits_{x \to 2^-}$ int x

41. $\lim\limits_{x \to 0^+} \dfrac{x}{|x|}$

42. $\lim\limits_{x \to 0^-} \dfrac{x}{|x|}$

43. Assume that $\lim\limits_{x \to 4} f(x) = 0$ and $\lim\limits_{x \to 4} g(x) = 3$.

(a) $\lim\limits_{x \to 4} (g(x) + 3)$

(b) $\lim\limits_{x \to 4} x f(x)$

(c) $\lim\limits_{x \to 4} g^2(x)$

(d) $\lim\limits_{x \to 4} \dfrac{g(x)}{f(x) - 1}$

44. Assume that $\lim\limits_{x \to b} f(x) = 7$ and $\lim\limits_{x \to b} g(x) = -3$.

(a) $\lim\limits_{x \to b} (f(x) + g(x))$

(b) $\lim\limits_{x \to b} (f(x) \cdot g(x))$

(c) $\lim\limits_{x \to b} 4 g(x)$

(d) $\lim\limits_{x \to b} \dfrac{f(x)}{g(x)}$

In Exercises 45–48, complete parts (a), (b), and (c) for the piecewise-defined function.

(a) Draw the graph of f.

(b) Determine $\lim\limits_{x \to c^+} f(x)$ and $\lim\limits_{x \to c^-} f(x)$.

(c) Writing to Learn Does $\lim\limits_{x \to c} f(x)$ exist? If so, what is it? If not, explain.

45. $c = 2, f(x) = \begin{cases} 3 - x, & x < 2 \\ \dfrac{x}{2} + 1, & x > 2 \end{cases}$

46. $c = 2, f(x) = \begin{cases} 3 - x, & x < 2 \\ 2, & x = 2 \\ x/2, & x > 2 \end{cases}$

47. $c = 1, f(x) = \begin{cases} \dfrac{1}{x - 1}, & x < 1 \\ x^3 - 2x + 5, & x \geq 1 \end{cases}$

48. $c = -1, f(x) = \begin{cases} 1 - x^2, & x \neq -1 \\ 2, & x = -1 \end{cases}$

In Exercises 49–52, complete parts (a)–(d) for the piecewise-defined function.

(a) Draw the graph of f.

(b) At what points c in the domain of f does $\lim_{x \to c} f(x)$ exist?

(c) At what points c does only the left-hand limit exist?

(d) At what points c does only the right-hand limit exist?

49. $f(x) = \begin{cases} \sin x, & -2\pi \leq x < 0 \\ \cos x, & 0 \leq x \leq 2\pi \end{cases}$

50. $f(x) = \begin{cases} \cos x, & -\pi \leq x < 0 \\ \sec x, & 0 \leq x \leq \pi \end{cases}$

51. $f(x) = \begin{cases} \sqrt{1 - x^2}, & 0 \leq x < 1 \\ 1, & 1 \leq x < 2 \\ 2, & x = 2 \end{cases}$

52. $f(x) = \begin{cases} x, & -1 \leq x < 0, \text{ or } 0 < x \leq 1 \\ 1, & x = 0 \\ 0, & x < -1, \text{ or } x > 1 \end{cases}$

In Exercises 53–56, find the limit graphically. Use the Sandwich Theorem to confirm your answer.

53. $\lim_{x \to 0} x \sin x$

54. $\lim_{x \to 0} x^2 \sin x$

55. $\lim_{x \to 0} x^2 \sin \dfrac{1}{x^2}$

56. $\lim_{x \to 0} x^2 \cos \dfrac{1}{x^2}$

57. *Free Fall* A water balloon dropped from a window high above the ground falls $y = 4.9t^2$ m in t sec. Find the balloon's

(a) average speed during the first 3 sec of fall.

(b) speed at the instant $t = 3$.

58. *Free Fall on a Small Airless Planet* A rock released from rest to fall on a small airless planet falls $y = gt^2$ m in t sec, g a constant. Suppose that the rock falls to the bottom of a crevasse 20 m below and reaches the bottom in 4 sec.

(a) Find the value of g.

(b) Find the average speed for the fall.

(c) With what speed did the rock hit the bottom?

Explorations

In Exercises **59–62**, complete the following tables and state what you believe $\lim_{x \to 0} f(x)$ to be.

(a)

x	-0.1	-0.01	-0.001	-0.0001	...
$f(x)$	?	?	?	?	

(b)

x	0.1	0.01	0.001	0.0001	...
$f(x)$	?	?	?	?	

59. $f(x) = x \sin \dfrac{1}{x}$

60. $f(x) = \sin \dfrac{1}{x}$

61. $f(x) = \dfrac{10^x - 1}{x}$

62. $f(x) = x \sin (\ln |x|)$ ■

63. *Proof that $\lim_{\theta \to 0} (\sin \theta)/\theta = 1$ when θ is Measured in Radians* Work in groups of two or three. The plan is to show that the right- and left-hand limits are both 1.

(a) To show that the right-hand limit is 1, explain why we can restrict our attention to $0 < \theta < \pi/2$.

(b) Use the figure to show that

$$\text{area of } \triangle OAP = \frac{1}{2} \sin \theta,$$

$$\text{area of sector } OAP = \frac{\theta}{2},$$

$$\text{area of } \triangle OAT = \frac{1}{2} \tan \theta.$$

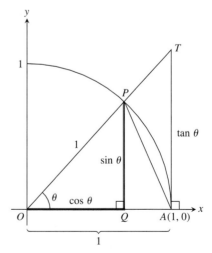

(c) Use (b) and the figure to show that for $0 < \theta < \pi/2$,

$$\frac{1}{2} \sin \theta < \frac{1}{2}\theta < \frac{1}{2} \tan \theta.$$

(d) Show that for $0 < \theta < \pi/2$ the inequality of (c) can be written in the form

$$1 < \frac{\theta}{\sin \theta} < \frac{1}{\cos \theta}.$$

(e) Show that for $0 < \theta < \pi/2$ the inequality of (d) can be written in the form

$$\cos \theta < \frac{\sin \theta}{\theta} < 1.$$

(f) Use the Sandwich Theorem to show that

$$\lim_{\theta \to 0^+} \frac{\sin \theta}{\theta} = 1.$$

(g) Show that $(\sin \theta)/\theta$ is an even function.

(h) Use (g) to show that

$$\lim_{\theta \to 0^-} \frac{\sin \theta}{\theta} = 1.$$

(i) Finally, show that

$$\lim_{\theta \to 0} \frac{\sin \theta}{\theta} = 1.$$

Extending the Ideas

64. *Controlling Outputs* Let $f(x) = \sqrt{3x - 2}$.

(a) Show that $\lim_{x \to 2} f(x) = 2 = f(2)$.

(b) Use a graph to estimate values for a and b so that $1.8 < f(x) < 2.2$ provided $a < x < b$.

(c) Use a graph to estimate values for a and b so that $1.99 < f(x) < 2.01$ provided $a < x < b$.

65. *Controlling Outputs* Let $f(x) = \sin x$.

(a) Find $f(\pi/6)$.

(b) Use a graph to estimate an interval (a, b) about $x = \pi/6$ so that $0.3 < f(x) < 0.7$ provided $a < x < b$.

(c) Use a graph to estimate an interval (a, b) about $x = \pi/6$ so that $0.49 < f(x) < 0.51$ provided $a < x < b$.

66. *Limits and Geometry* Let $P(a, a^2)$ be a point on the parabola $y = x^2, a > 0$. Let O be the origin and $(0, b)$ the y-intercept of the perpendicular bisector of line segment OP. Find $\lim_{P \to O} b$.

2.2

Limits Involving Infinity

Finite Limits as $x \to \pm\infty$ • Sandwich Theorem Revisited • Infinite Limits as $x \to a$ • End Behavior Models • "Seeing" Limits as $x \to \pm\infty$

Finite Limits as $x \to \pm\infty$

The symbol for infinity (∞) does not represent a real number. We use ∞ to describe the behavior of a function when the values in its domain or range outgrow all finite bounds. For example, when we say "the limit of f as x approaches infinity" we mean the limit of f as x moves increasingly far to the right on the number line. When we say "the limit of f as x approaches negative infinity ($-\infty$)" we mean the limit of f as x moves increasingly far to the left. (The limit in each case may or may not exist.)

Looking at $f(x) = 1/x$ (Figure 2.9), we observe

(a) as $x \to \infty$, $(1/x) \to 0$ and we write $\lim_{x \to \infty} (1/x) = 0$,

(b) as $x \to -\infty$, $(1/x) \to 0$ and we write $\lim_{x \to -\infty} (1/x) = 0$.

We say that the line $y = 0$ is a *horizontal asymptote* of the graph of f.

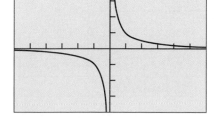

[–6, 6] by [–4, 4]

Figure 2.9 The graph of $f(x) = 1/x$.

> **Definition Horizontal Asymptote**
>
> The line $y = b$ is a **horizontal asymptote** of the graph of a function $y = f(x)$ if either
>
> $$\lim_{x \to \infty} f(x) = b \quad \text{or} \quad \lim_{x \to -\infty} f(x) = b.$$

The graph of $f(x) = 2 + (1/x)$ has the single horizontal asymptote $y = 2$ because

$$\lim_{x \to \infty} \left(2 + \frac{1}{x} \right) = 2 \quad \text{and} \quad \lim_{x \to -\infty} \left(2 + \frac{1}{x} \right) = 2.$$

A function can have more than one horizontal asymptote, as Example 1 demonstrates.

Example 1   LOOKING FOR HORIZONTAL ASYMPTOTES

Investigate the graph of $f(x) = x/\sqrt{x^2 + 1}$.

Solution

Solve Graphically

Figure 2.10a shows the graph for $-10 \le x \le 10$. The graph climbs rapidly toward the line $y = 1$ as x moves away from the origin to the right. On our calculator screen, the graph soon becomes indistinguishable from the line. Similarly, as x moves away from the origin to the left, the graph drops rapidly toward the line $y = -1$ and soon appears to overlap the line.

Confirm Numerically

The table in Figure 2.10b confirms the rapid approach of $f(x)$ toward 1 as $x \to \infty$. Since f is an odd function of x, we can expect its values to approach -1 in a similar way as $x \to -\infty$.

[–10, 10] by [–1.5, 1.5]

(a)

X	Y1
0	0
1	.70711
2	.89443
3	.94868
4	.97014
5	.98058
6	.98639

Y1 ▤ X/√(X² + 1)

(b)

Figure 2.10 (a) The graph of $f(x) = x/\sqrt{x^2 + 1}$ has two horizontal asymptotes, $y = -1$ and $y = 1$. (b) Selected values of f. (Example 1)

Sandwich Theorem Revisited

The Sandwich Theorem also holds for limits as $x \to \pm\infty$.

Example 2   FINDING A LIMIT AS x APPROACHES ∞

Find $\lim_{x \to \infty} f(x)$ for $f(x) = \dfrac{\sin x}{x}$.

Solution

Solve Graphically and Numerically

The graph and table of values in Figure 2.11 suggest that $y = 0$ is the horizontal asymptote of f.

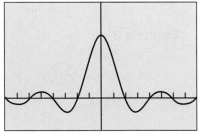

[–4π, 4π] by [–0.5, 1.5]

(a)

X	Y1
100	–.0051
200	–.0044
300	–.0033
400	–.0021
500	–9E–4
600	7.4E–5
700	7.8E–4

Y1 ▤ sin(X)/X

(b)

Figure 2.11 (a) The graph of $f(x) = (\sin x)/x$ oscillates about the x-axis. The amplitude of the oscillations decreases toward zero as $x \to \pm\infty$. (b) A table of values for f that suggests $f(x) \to 0$ as $x \to \infty$. (Example 2)

Confirm Analytically

We know that $-1 \le \sin x \le 1$. So, for $x > 0$ we have

$$-\frac{1}{x} \le \frac{\sin x}{x} \le \frac{1}{x}.$$

Therefore, by the Sandwich Theorem,

$$0 = \lim_{x \to \infty}\left(-\frac{1}{x}\right) = \lim_{x \to \infty}\frac{\sin x}{x} = \lim_{x \to \infty}\frac{1}{x} = 0.$$

Since $(\sin x)/x$ is an even function of x, we can also conclude that

$$\lim_{x \to -\infty}\frac{\sin x}{x} = 0.$$

Limits at infinity have properties similar to those of finite limits.

Theorem 5 Properties of Limits as $x \to \pm\infty$

If L, M, and k are real numbers and

$$\lim_{x \to \pm\infty} f(x) = L \quad \text{and} \quad \lim_{x \to \pm\infty} g(x) = M, \text{ then}$$

1. *Sum Rule:* $\qquad\qquad\qquad\qquad \lim_{x \to \pm\infty}(f(x) + g(x)) = L + M$

2. *Difference Rule:* $\qquad\qquad\quad \lim_{x \to \pm\infty}(f(x) - g(x)) = L - M$

3. *Product Rule:* $\qquad\qquad\qquad \lim_{x \to \pm\infty}(f(x) \cdot g(x)) = L \cdot M$

4. *Constant Multiple Rule:* $\qquad\; \lim_{x \to \pm\infty}(k \cdot f(x)) = k \cdot L$

5. *Quotient Rule:* $\qquad\qquad\qquad \lim_{x \to \pm\infty}\frac{f(x)}{g(x)} = \frac{L}{M}, M \ne 0$

6. *Power Rule:* If r and s are integers, $s \ne 0$, then

$$\lim_{x \to \pm\infty}(f(x))^{r/s} = L^{r/s}$$

provided that $L^{r/s}$ is a real number.

We can use Theorem 5 to find limits at infinity of functions with complicated expressions, as illustrated in Example 3.

Example 3 USING THEOREM 5

Find $\displaystyle\lim_{x \to \infty}\frac{5x + \sin x}{x}$.

Solution Notice that

$$\frac{5x + \sin x}{x} = \frac{5x}{x} + \frac{\sin x}{x} = 5 + \frac{\sin x}{x}.$$

So,

$$\lim_{x \to \infty}\frac{5x + \sin x}{x} = \lim_{x \to \infty} 5 + \lim_{x \to \infty}\frac{\sin x}{x} \qquad \text{Sum Rule}$$

$$= 5 + 0 = 5. \qquad\qquad\qquad \text{Known values}$$

Exploration 1 **Exploring Theorem 5**

We must be careful how we apply Theorem 5.

1. (Example 3 again) Let $f(x) = 5x + \sin x$ and $g(x) = x$. Do the limits as $x \to \infty$ of f and g exist? Can we apply the Quotient Rule to $\lim_{x \to \infty} f(x)/g(x)$? Explain. Does the limit of the quotient exist?

2. Let $f(x) = \sin^2 x$ and $g(x) = \cos^2 x$. Describe the behavior of f and g as $x \to \infty$. Can we apply the Sum Rule to $\lim_{x \to \infty} (f(x) + g(x))$? Explain. Does the limit of the sum exist?

3. Let $f(x) = \ln(2x)$ and $g(x) = \ln(x + 1)$. Find the limits as $x \to \infty$ of f and g. Can we apply the Difference Rule to $\lim_{x \to \infty} (f(x) - g(x))$? Explain. Does the limit of the difference exist?

4. Based on parts 1–3, what advice might you give about applying Theorem 5?

Infinite Limits as $x \to a$

If the values of a function $f(x)$ outgrow all positive bounds as x approaches a finite number a, we say that $\lim_{x \to a} f(x) = \infty$. If the values of f become large and negative, exceeding all negative bounds as $x \to a$, we say that $\lim_{x \to a} f(x) = -\infty$.

Looking at $f(x) = 1/x$ (Figure 2.9, page 65), we observe that

$$\lim_{x \to 0^+} 1/x = \infty \quad \text{and} \quad \lim_{x \to 0^-} 1/x = -\infty.$$

We say that the line $x = 0$ is a *vertical asymptote* of the graph of f.

Definition **Vertical Asymptote**

The line $x = a$ is a **vertical asymptote** of the graph of a function $y = f(x)$ if either

$$\lim_{x \to a^+} f(x) = \pm\infty \quad \text{or} \quad \lim_{x \to a^-} f(x) = \pm\infty.$$

Example 4 **FINDING VERTICAL ASYMPTOTES**

Find the vertical asymptotes of $f(x) = \dfrac{1}{x^2}$.

Solution The values of the function approach ∞ on either side of $x = 0$.

$$\lim_{x \to 0^+} \frac{1}{x^2} = \infty \quad \text{and} \quad \lim_{x \to 0^-} \frac{1}{x^2} = \infty.$$

The line $x = 0$ is the only vertical asymptote.

We can also say that $\lim_{x \to 0} (1/x^2) = \infty$. We can make no such statement about $1/x$.

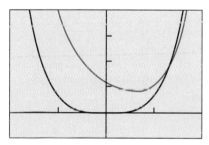

[−2π, 2π] by [−5, 5]

Figure 2.12 The graph of $f(x) = \tan x$ has a vertical asymptote at
$$\ldots, -\frac{3\pi}{2}, -\frac{\pi}{2}, \frac{\pi}{2}, \frac{3\pi}{2}, \ldots. \text{ (Example 5)}$$

$y = 3x^4 - 2x^3 + 3x^2 - 5x + 6$

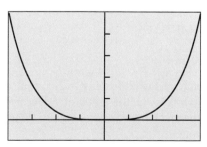

[−2, 2] by [−5, 20]

(a)

[−20, 20] by [−100000, 500000]

(b)

Figure 2.13 The graphs of f and g, (a) distinct for $|x|$ small, are (b) nearly identical for $|x|$ large. (Example 6)

Example 5 FINDING VERTICAL ASYMPTOTES

The graph of $f(x) = \tan x = (\sin x)/(\cos x)$ has infinitely many vertical asymptotes, one at each point where the cosine is zero. If a is an odd multiple of $\pi/2$, then

$$\lim_{x \to a^+} \tan x = -\infty \quad \text{and} \quad \lim_{x \to a^-} \tan x = \infty,$$

as suggested by Figure 2.12.

You might think that the graph of a quotient always has a vertical asymptote where the denominator is zero, but that need not be the case. For example, we observed in Section 2.1 that $\lim_{x \to 0} (\sin x)/x = 1$.

End Behavior Models

For numerically large values of x, we can sometimes model the behavior of a complicated function by a simpler one that acts virtually in the same way.

Example 6 MODELING FUNCTIONS FOR $|x|$ LARGE

Let $f(x) = 3x^4 - 2x^3 + 3x^2 - 5x + 6$ and $g(x) = 3x^4$. Show that while f and g are quite different for numerically small values of x, they are virtually identical for $|x|$ large.

Solution

Solve Graphically

The graphs of f and g (Figure 2.13a), quite different near the origin, are virtually identical on a larger scale (Figure 2.13b).

Confirm Analytically

We can test the claim that g models f for numerically large values of x by examining the ratio of the two functions as $x \to \pm\infty$. We find that

$$\lim_{x \to \pm\infty} \frac{f(x)}{g(x)} = \lim_{x \to \pm\infty} \frac{3x^4 - 2x^3 + 3x^2 - 5x + 6}{3x^4}$$

$$= \lim_{x \to \pm\infty} \left(1 - \frac{2}{3x} + \frac{1}{x^2} - \frac{5}{3x^3} + \frac{2}{x^4}\right)$$

$$= 1,$$

convincing evidence that f and g behave alike for $|x|$ large.

Definition End Behavior Model

The function g is

(a) a **right end behavior model** for f if and only if $\displaystyle \lim_{x \to \infty} \frac{f(x)}{g(x)} = 1$.

(b) a **left end behavior model** for f if and only if $\displaystyle \lim_{x \to -\infty} \frac{f(x)}{g(x)} = 1$.

A function's right and left end behavior models need not be the same function.

Example 7 FINDING END BEHAVIOR MODELS

Let $f(x) = x + e^{-x}$. Show that $g(x) = x$ is a right end behavior model for f while $h(x) = e^{-x}$ is a left end behavior model for f.

Solution On the right,

$$\lim_{x \to \infty} \frac{f(x)}{g(x)} = \lim_{x \to \infty} \frac{x + e^{-x}}{x} = \lim_{x \to \infty} \left(1 + \frac{e^{-x}}{x}\right) = 1 \text{ because } \lim_{x \to \infty} \frac{e^{-x}}{x} = 0.$$

On the left,

$$\lim_{x \to -\infty} \frac{f(x)}{h(x)} = \lim_{x \to -\infty} \frac{x + e^{-x}}{e^{-x}} = \lim_{x \to -\infty} \left(\frac{x}{e^{-x}} + 1\right) = 1 \text{ because } \lim_{x \to -\infty} \frac{x}{e^{-x}} = 0.$$

The graph of f in Figure 2.14 supports these end behavior conclusions.

If one function provides both a left and right end behavior model, it is simply called an **end behavior model.** Thus, $g(x) = 3x^4$ is an end behavior model for $f(x) = 3x^4 - 2x^3 + 3x^2 - 5x + 6$ (Example 6).

In general, $g(x) = a_n x^n$ is an end behavior model for the polynomial function $f(x) = a_n x^n + a_{n-1}x^{n-1} + \cdots + a_0$, $a_n \neq 0$. In the large, all polynomials behave like monomials. This is the key to the end behavior of rational functions, as illustrated in Example 8.

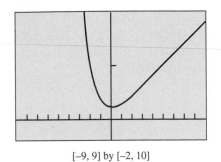

[–9, 9] by [–2, 10]

Figure 2.14 The graph of $f(x) = x + e^{-x}$ looks like the graph of $g(x) = x$ to the right of the y-axis, and like the graph of $h(x) = e^{-x}$ to the left of the y-axis. (Example 7)

Example 8 FINDING END BEHAVIOR MODELS

Find an end behavior model for

(a) $f(x) = \dfrac{2x^5 + x^4 - x^2 + 1}{3x^2 - 5x + 7}$. **(b)** $g(x) = \dfrac{2x^3 - x^2 + x - 1}{5x^3 + x^2 + x - 5}$.

Solution

(a) Notice that $2x^5$ is an end behavior model for the numerator of f, and $3x^2$ is one for the denominator. This makes

$$\frac{2x^5}{3x^2} = \frac{2}{3}x^3$$

an end behavior model for f.

(b) Similarly, $2x^3$ is an end behavior model for the numerator of g, and $5x^3$ is one for the denominator of g. This makes

$$\frac{2x^3}{5x^3} = \frac{2}{5}$$

an end behavior model for g.

Notice in Example 8b that the end behavior model for g, $y = 2/5$, is also a horizontal asymptote of the graph of g, while in 8a, the graph of f does not have a horizontal asymptote. We can use the end behavior model of a rational function to identify any horizontal asymptote.

Example 9 FINDING A HORIZONTAL ASYMPTOTE

Does the graph of

$$f(x) = \frac{4x^2 - 3x + 5}{2x^3 + x - 1}$$

have a horizontal asymptote? If so, what is it?

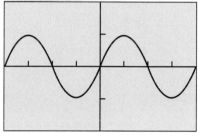

[–10, 10] by [–1, 1]

(a)

[–2π, 2π] by [–2, 2]

(b)

Figure 2.15 The graphs of (a) $f(x) =$ sin $(1/x)$ and (b) $g(x) = f(1/x) = \sin x$. (Example 10)

Solution An end behavior model for f is

$$\frac{4x^2}{2x^3} = \frac{2}{x}.$$

So, $\lim_{x \to \pm\infty} f(x) = \lim_{x \to \pm\infty} 2/x = 0$ and $y = 0$ is the horizontal asymptote of f.

We can see from Examples 8 and 9 that a rational function always has a simple power function as an end behavior model.

"Seeing" Limits as $x \to \pm\infty$

We can investigate the graph of $y = f(x)$ as $x \to \pm\infty$ by investigating the graph of $y = f(1/x)$ as $x \to 0$.

Example 10 USING SUBSTITUTION

Find $\lim\limits_{x \to \infty} \sin(1/x)$.

Solution Figure 2.15a suggests that the limit is 0. Indeed, replacing $\lim_{x \to \infty} \sin(1/x)$ by the equivalent $\lim_{x \to 0^+} \sin x = 0$ (Figure 2.15b), we find

$$\lim_{x \to \infty} \sin 1/x = \lim_{x \to 0^+} \sin x = 0.$$

Quick Review 2.2

In Exercises 1–4, find f^{-1} and graph f, f^{-1}, and $y = x$ in the same square viewing window.

1. $f(x) = 2x - 3$

2. $f(x) = e^x$

3. $f(x) = \tan^{-1} x$

4. $f(x) = \cot^{-1} x$

In Exercises 5 and 6, find the quotient $q(x)$ and remainder $r(x)$ when $f(x)$ is divided by $g(x)$.

5. $f(x) = 2x^3 - 3x^2 + x - 1,\quad g(x) = 3x^3 + 4x - 5$

6. $f(x) = 2x^5 - x^3 + x - 1,\quad g(x) = x^3 - x^2 + 1$

In Exercises 7–10, write a formula for **(a)** $f(-x)$ and **(b)** $f(1/x)$. Simplify where possible.

7. $f(x) = \cos x$

8. $f(x) = e^{-x}$

9. $f(x) = \dfrac{\ln x}{x}$

10. $f(x) = \left(x + \dfrac{1}{x}\right) \sin x$

Section 2.2 Exercises

In Exercises 1–8, use graphs and tables to find **(a)** $\lim_{x \to \infty} f(x)$ and **(b)** $\lim_{x \to -\infty} f(x)$. **(c)** Identify all horizontal asymptotes.

1. $f(x) = \cos\left(\dfrac{1}{x}\right)$

2. $f(x) = \dfrac{\sin 2x}{x}$

3. $f(x) = \dfrac{e^{-x}}{x}$

4. $f(x) = \dfrac{3x^3 - x + 1}{x + 3}$

5. $f(x) = \dfrac{3x + 1}{|x| + 2}$

6. $f(x) = \dfrac{2x - 1}{|x| - 3}$

7. $f(x) = \dfrac{x}{|x|}$

8. $f(x) = \dfrac{|x|}{|x| + 1}$

In Exercises 9–16, use graphs and tables to find the limits.

9. $\lim\limits_{x \to 2^+} \dfrac{1}{x - 2}$

10. $\lim\limits_{x \to 2^-} \dfrac{x}{x - 2}$

11. $\lim\limits_{x \to -3^-} \dfrac{1}{x + 3}$

12. $\lim\limits_{x \to -3^+} \dfrac{x}{x + 3}$

13. $\lim\limits_{x \to 0^+} \dfrac{\text{int } x}{x}$

14. $\lim\limits_{x \to 0^-} \dfrac{\text{int } x}{x}$

15. $\lim\limits_{x \to 0^+} \csc x$

16. $\lim\limits_{x \to (\pi/2)^+} \sec x$

In Exercises 17–22, **(a)** find the vertical asymptotes of the graph of $f(x)$. **(b)** Describe the behavior of $f(x)$ to the left and right of each vertical asymptote.

17. $f(x) = \dfrac{1}{x^2 - 4}$

18. $f(x) = \dfrac{x^2 - 1}{2x + 4}$

19. $f(x) = \dfrac{x^2 - 2x}{x + 1}$

20. $f(x) = \dfrac{1 - x}{2x^2 - 5x - 3}$

21. $f(x) = \cot x$

22. $f(x) = \sec x$

In Exercises 23–28, find $\lim\limits_{x \to \infty} y$ and $\lim\limits_{x \to -\infty} y$.

23. $y = \left(2 - \dfrac{x}{x + 1}\right)\left(\dfrac{x^2}{5 + x^2}\right)$

24. $y = \left(\dfrac{2}{x} + 1\right)\left(\dfrac{5x^2 - 1}{x^2}\right)$

25. $y = \dfrac{\cos (1/x)}{1 + (1/x)}$

26. $y = \dfrac{2x + \sin x}{x}$

27. $y = \dfrac{\sin x}{2x^2 + x}$

28. $y = \dfrac{x \sin x + 2 \sin x}{2x^2}$

In Exercises 29–32, match the function with the graph of its end behavior model.

29. $y = \dfrac{2x^3 - 3x^2 + 1}{x + 3}$

30. $y = \dfrac{x^5 - x^4 + x + 1}{2x^2 + x - 3}$

31. $y = \dfrac{2x^4 - x^3 + x^2 - 1}{2 - x}$

32. $y = \dfrac{x^4 - 3x^3 + x^2 - 1}{1 - x^2}$

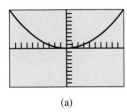

(a)

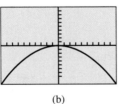

(b)

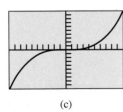

(c)

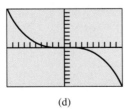

(d)

In Exercises 33–38, **(a)** find a power function end behavior model for f. **(b)** Identify any horizontal asymptotes.

33. $f(x) = 3x^2 - 2x + 1$

34. $f(x) = -4x^3 + x^2 - 2x - 1$

35. $f(x) = \dfrac{x - 2}{2x^2 + 3x - 5}$

36. $f(x) = \dfrac{3x^2 - x + 5}{x^2 - 4}$

37. $f(x) = \dfrac{4x^3 - 2x + 1}{x - 2}$

38. $f(x) = \dfrac{-x^4 + 2x^2 + x - 3}{x^2 - 4}$

In Exercises 39–42, find **(a)** a simple basic function as a right end behavior model and **(b)** a simple basic function as a left end behavior model for the function.

39. $y = e^x - 2x$

40. $y = x^2 + e^{-x}$

41. $y = x + \ln |x|$

42. $y = x^2 + \sin x$

In Exercises 43–46, use the graph of $y = f(1/x)$ to find $\lim\limits_{x \to \infty} f(x)$ and $\lim\limits_{x \to -\infty} f(x)$.

43. $f(x) = xe^x$

44. $f(x) = x^2 e^{-x}$

45. $f(x) = \dfrac{\ln |x|}{x}$

46. $f(x) = x \sin \dfrac{1}{x}$

In Exercises 47 and 48, find the limit of $f(x)$ as **(a)** $x \to -\infty$, **(b)** $x \to \infty$, **(c)** $x \to 0^-$, and **(d)** $x \to 0^+$.

47. $f(x) = \begin{cases} 1/x, & x < 0 \\ -1, & x \geq 0 \end{cases}$

48. $f(x) = \begin{cases} \dfrac{x - 2}{x - 1}, & x \leq 0 \\ 1/x^2, & x > 0 \end{cases}$

In Exercises 49 and 50, *work in groups of two or three* and sketch a graph of a function $y = f(x)$ that satisfies the stated conditions. Include any asymptotes.

49. $\lim\limits_{x \to 1} f(x) = 2$, $\quad \lim\limits_{x \to 5^-} f(x) = \infty$, $\quad \lim\limits_{x \to 5^+} f(x) = \infty$,

$\lim\limits_{x \to \infty} f(x) = -1$, $\quad \lim\limits_{x \to -2^+} f(x) = -\infty$,

$\lim\limits_{x \to -2^-} f(x) = \infty$, $\quad \lim\limits_{x \to -\infty} f(x) = 0$

50. $\lim\limits_{x \to 2} f(x) = -1$, $\quad \lim\limits_{x \to 4^+} f(x) = -\infty$, $\quad \lim\limits_{x \to 4^-} f(x) = \infty$,

$\lim\limits_{x \to \infty} f(x) = \infty$, $\quad \lim\limits_{x \to -\infty} f(x) = 2$

51. *End Behavior Models* Work in groups of two or three. Suppose that $g_1(x)$ is a right end behavior model for $f_1(x)$ and that $g_2(x)$ is a right end behavior model for $f_2(x)$. Explain why this makes $g_1(x)/g_2(x)$ a right end behavior model for $f_1(x)/f_2(x)$.

52. *Writing to Learn* Let L be a real number, $\lim\limits_{x \to c} f(x) = L$, and $\lim\limits_{x \to c} g(x) = \infty$ or $-\infty$. Can $\lim\limits_{x \to c} (f(x) + g(x))$ be determined? Explain.

53. Table 2.2 gives the average amount per grant of emergency financial assistance for veterans in Franklin County, Ohio. Let $x = 0$ represent 1980, $x = 1$ represent 1981, and so forth.

Table 2.2 Emergency Financial Assistance	
Year	Average Amount (dollars)
1986	145.38
1987	155.00
1988	230.43
1989	420.70
1990	494.55
1991	555.00
1992	508.77
1993	460.92
1994	453.77

Source: Franklin County Veterans Service Commission as reported by Mary Stephens in *The Columbus [Ohio] Dispatch*, May 31, 1995.

(a) Find a cubic regression equation and superimpose its graph on a scatter plot of the data.

(b) Find a quartic regression equation and superimpose its graph on a scatter plot of the data.

(c) Use each regression equation to predict the average amount per grant of financial assistance for veterans in the year 2000.

(d) Writing to Learn Find an end behavior model for each regression equation. What does each predict about the future sizes of grants? Does this seem reasonable? Explain.

Exploration

54. *Exploring Properties of Limits* Find the limits of f, g, and fg as $x \to c$.

(a) $f(x) = \dfrac{1}{x}$, $g(x) = x$, $c = 0$

(b) $f(x) = -\dfrac{2}{x^3}$, $g(x) = 4x^3$, $c = 0$

(c) $f(x) = \dfrac{3}{x - 2}$, $g(x) = (x - 2)^3$, $c = 2$

(d) $f(x) = \dfrac{5}{(3 - x)^4}$, $g(x) = (x - 3)^2$, $c = 3$

(e) Writing to Learn Suppose that $\lim_{x \to c} f(x) = 0$ and $\lim_{x \to c} g(x) = \infty$. Based on your observations in (a)–(d), what can you say about $\lim_{x \to c} (f(x) \cdot g(x))$? ∎

Extending the Ideas

55. *The Greatest Integer Function*

(a) Show that

$$\frac{x - 1}{x} < \frac{\text{int } x}{x} \le 1 \ (x > 0) \quad \text{and} \quad \frac{x - 1}{x} > \frac{\text{int } x}{x} \ge 1 \ (x < 0).$$

(b) Determine $\displaystyle \lim_{x \to \infty} \frac{\text{int } x}{x}$.   **(c)** Determine $\displaystyle \lim_{x \to -\infty} \frac{\text{int } x}{x}$.

56. *Sandwich Theorem* Use the Sandwich Theorem to confirm the limit as $x \to \infty$ found in Exercise 3.

57. Writing to Learn Explain why there is no value L for which $\lim_{x \to \infty} \sin x = L$.

In Exercises 58–60, find the limit. Give a convincing argument that the value is correct.

58. $\displaystyle \lim_{x \to \infty} \frac{\ln x^2}{\ln x}$ **59.** $\displaystyle \lim_{x \to \infty} \frac{\ln x}{\log x}$

60. $\displaystyle \lim_{x \to \infty} \frac{\ln (x + 1)}{\ln x}$

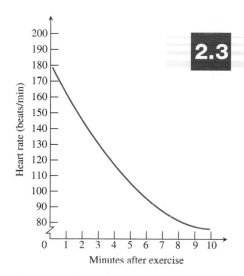

Figure 2.16 How the heartbeat returns to a normal rate after running.

| **2.3** | # Continuity |

Continuity at a Point • Continuous Functions • Algebraic Combinations • Composites • Intermediate Value Theorem for Continuous Functions

Continuity at a Point

When we plot function values generated in the laboratory or collected in the field, we often connect the plotted points with an unbroken curve to show what the function's values are likely to have been at the times we did not measure (Figure 2.16). In doing so, we are assuming that we are working with a *continuous function,* a function whose outputs vary continuously with the inputs and do not jump from one value to another without taking on the values in between. Any function $y = f(x)$ whose graph can be sketched in one continuous motion without lifting the pencil is an example of a continuous function.

Figure 2.17 The laser was developed as a result of an understanding of the nature of the atom.

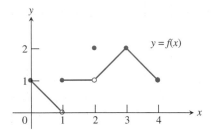

Figure 2.18 The function is continuous on $[0, 4]$ except at $x = 1$ and $x = 2$. (Example 1)

Continuous functions are the functions we use to find a planet's closest point of approach to the sun or the peak concentration of antibodies in blood plasma. They are also the functions we use to describe how a body moves through space or how the speed of a chemical reaction changes with time. In fact, so many physical processes proceed continuously that throughout the eighteenth and nineteenth centuries it rarely occurred to anyone to look for any other kind of behavior. It came as a surprise when the physicists of the 1920s discovered that light comes in particles and that heated atoms emit light at discrete frequencies (Figure 2.17). As a result of these and other discoveries, and because of the heavy use of discontinuous functions in computer science, statistics, and mathematical modeling, the issue of continuity has become one of practical as well as theoretical importance.

To understand continuity, we need to consider a function like the one in Figure 2.18, whose limits we investigated in Example 8, Section 2.1.

Example 1 INVESTIGATING CONTINUITY

Find the points at which the function f in Figure 2.18 is continuous, and the points at which f is discontinuous.

Solution The function f is continuous at every point in its domain $[0, 4]$ except at $x = 1$ and $x = 2$. At these points there are breaks in the graph. Note the relationship between the limit of f and the value of f at each point of the function's domain.

Points at which f is continuous:

At $x = 0$, $\qquad\qquad \lim_{x \to 0^+} f(x) = f(0)$.

At $x = 4$, $\qquad\qquad \lim_{x \to 4^-} f(x) = f(4)$.

At $0 < c < 4$, $c \neq 1, 2$, $\quad \lim_{x \to c} f(x) = f(c)$.

Points at which f is discontinuous:

At $x = 1$, $\qquad \lim_{x \to 1} f(x)$ does not exist.

At $x = 2$, $\qquad \lim_{x \to 2} f(x) = 1$, but $1 \neq f(2)$.

At $c < 0$, $c > 4$, $\quad$ these points are not in the domain of f.

To define continuity at a point in a function's domain, we need to define continuity at an interior point (which involves a two-sided limit) and continuity at an endpoint (which involves a one-sided limit). (Figure 2.19)

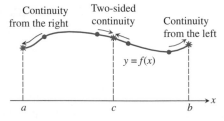

Figure 2.19 Continuity at points a, b, and c for a function $y = f(x)$ that is continuous on the interval $[a, b]$.

Definition Continuity at a Point

Interior Point: A function $y = f(x)$ is **continuous at an interior point** c of its domain if

$$\lim_{x \to c} f(x) = f(c).$$

Endpoint: A function $y = f(x)$ is **continuous at a left endpoint** a or is **continuous at a right endpoint** b of its domain if

$$\lim_{x \to a^+} f(x) = f(a) \quad \text{or} \quad \lim_{x \to b^-} f(x) = f(b), \quad \text{respectively.}$$

If a function f is not continuous at a point c, we say that f is **discontinuous** at c and c is a **point of discontinuity** of f. Note that c need not be in the domain of f.

Example 2   FINDING POINTS OF CONTINUITY & DISCONTINUITY

Find the points of continuity and the points of discontinuity of the greatest integer function (Figure 2.20).

Solution For the function to be continuous at $x = c$, the limit as $x \to c$ must exist and must equal the value of the function at $x = c$. The greatest integer function is discontinuous at every integer. For example,

$$\lim_{x \to 3^-} \text{int } x = 2 \quad \text{and} \quad \lim_{x \to 3^+} \text{int } x = 3$$

so the limit as $x \to 3$ does not exist. Notice that int $3 = 3$. In general, if n is any integer,

$$\lim_{x \to n^-} \text{int } x = n - 1 \quad \text{and} \quad \lim_{x \to n^+} \text{int } x = n,$$

so the limit as $x \to n$ does not exist.

The greatest integer function is continuous at every other real number. For example,

$$\lim_{x \to 1.5} \text{int } x = 1 = \text{int } 1.5.$$

In general, if $n - 1 < c < n$, n an integer, then

$$\lim_{x \to c} \text{int } x = n - 1 = \text{int } c.$$

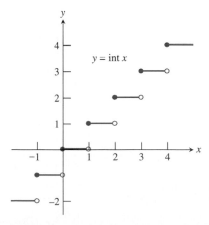

Figure 2.20 The function int x is continuous at every noninteger point. (Example 2)

Figure 2.21 is a catalog of discontinuity types. The function in (a) is continuous at $x = 0$. The function in (b) would be continuous if it had $f(0) = 1$. The function in (c) would be continuous if $f(0)$ were 1 instead of 2. The discontinuities in (b) and (c) are **removable.** Each function has a limit as $x \to 0$, and we can remove the discontinuity by setting $f(0)$ equal to this limit.

The discontinuities in (d)–(f) of Figure 2.21 are more serious: $\lim_{x \to 0} f(x)$ does not exist and there is no way to improve the situation by changing f at 0. The step function in (d) has a **jump discontinuity:** the one-sided limits exist but have different values. The function $f(x) = 1/x^2$ in (e) has an **infinite discontinuity.** The function in (f) has an **oscillating discontinuity:** it oscillates too much to have a limit as $x \to 0$.

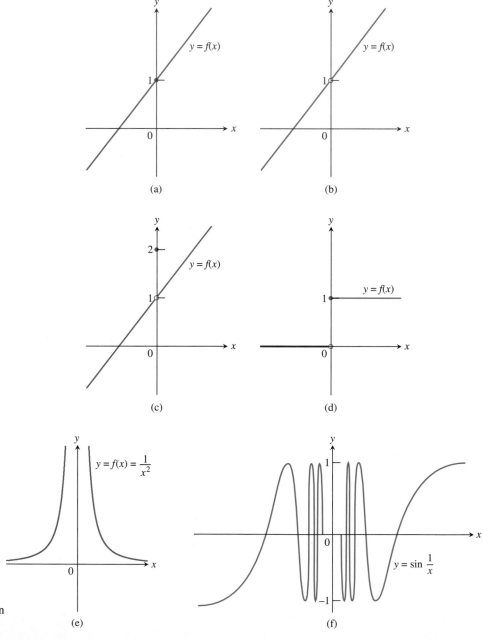

Figure 2.21 The function in (a) is continuous at $x = 0$. The functions in (b)–(f) are not.

Exploration 1 **Removing a Discontinuity**

Let $f(x) = \dfrac{x^3 - 7x - 6}{x^2 - 9}$.

1. Factor the denominator. What is the domain of f?

2. Investigate the graph of f around $x = 3$ to see that f has a removable discontinuity at $x = 3$.

3. How should f be defined at $x = 3$ to remove the discontinuity? Use zoom-in and tables as necessary.

4. Show that $(x - 3)$ is a factor of the numerator of f, and remove all common factors. Now compute the limit as $x \to 3$ of the reduced form for f.

5. Show that the *extended function*

$$g(x) = \begin{cases} \dfrac{x^3 - 7x - 6}{x^2 - 9}, & x \neq 3 \\ 10/3, & x = 3 \end{cases}$$

is continuous at $x = 3$. The function g is the **continuous extension** of the original function f to include $x = 3$.

Continuous Functions

A function is **continuous on an interval** if and only if it is continuous at every point of the interval. A **continuous function** is one that is continuous at every point of its domain. A continuous function need not be continuous on every interval. For example, $y = 1/x$ is not continuous on $[-1, 1]$.

Example 3   IDENTIFYING CONTINUOUS FUNCTIONS

The reciprocal function $y = 1/x$ (Figure 2.22) is a continuous function because it is continuous at every point of its domain. However, it has a point of discontinuity at $x = 0$ because it is not defined there.

Polynomial functions f are continuous at every real number c because $\lim_{x \to c} f(x) = f(c)$. Rational functions are continuous at every point of their domains. They have points of discontinuity at the zeros of their denominators. The absolute value function $y = |x|$ is continuous at every real number. The exponential functions, logarithmic functions, trigonometric functions, and radical functions like $y = \sqrt[n]{x}$ (n a positive integer greater than 1) are continuous at every point of their domains. All of these functions are continuous functions.

Algebraic Combinations

As you may have guessed, algebraic combinations of continuous functions are continuous wherever they are defined.

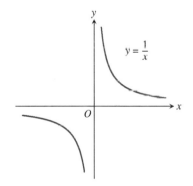

Figure 2.22 The function $y = 1/x$ is continuous at every value of x except $x = 0$. It has a point of discontinuity at $x = 0$. (Example 3)

Theorem 6 Properties of Continuous Functions

If the functions f and g are continuous at $x = c$, then the following combinations are continuous at $x = c$.

1. *Sums:* $f + g$
2. *Differences:* $f - g$
3. *Products:* $f \cdot g$
4. *Constant multiples:* $k \cdot f$, for any number k
5. *Quotients:* f/g, provided $g(c) \neq 0$

Composites

All composites of continuous functions are continuous. This means composites like

$$y = \sin(x^2) \quad \text{and} \quad y = |\cos x|$$

are continuous at every point at which they are defined. The idea is that if $f(x)$ is continuous at $x = c$ and $g(x)$ is continuous at $x = f(c)$, then $g \circ f$ is continuous at $x = c$ (Figure 2.23). In this case, the limit as $x \to c$ is $g(f(c))$.

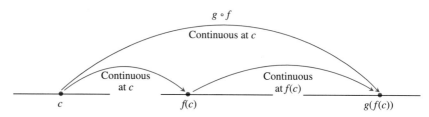

Figure 2.23 Composites of continuous functions are continuous.

Theorem 7 Composite of Continuous Functions

If f is continuous at c and g is continuous at $f(c)$, then the composite $g \circ f$ is continuous at c.

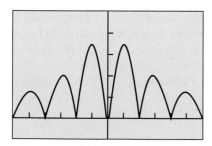

$[-3\pi, 3\pi]$ by $[-0.1, 0.5]$

Figure 2.24 The graph suggests that $y = |(x \sin x)/(x^2 + 2)|$ is continuous. (Example 4)

Example 4 USING THEOREM 7

Show that $y = \left| \dfrac{x \sin x}{x^2 + 2} \right|$ is continuous.

Solution The graph (Figure 2.24) of $y = |(x \sin x)/(x^2 + 2)|$ suggests that the function is continuous at every value of x. By letting

$$g(x) = |x| \quad \text{and} \quad f(x) = \frac{x \sin x}{x^2 + 2},$$

we see that y is the composite $g \circ f$.

We know that the absolute value function g is continuous. The function f is continuous by Theorem 6. Their composite is continuous by Theorem 7.

Intermediate Value Theorem for Continuous Functions

Functions that are continuous on intervals have properties that make them particularly useful in mathematics and its applications. One of these is the *intermediate value property*. A function is said to have the **intermediate value property** if it never takes on two values without taking on all the values in between.

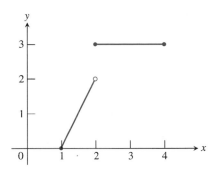

Figure 2.25 The function

$$f(x) = \begin{cases} 2x - 2, & 1 \le x < 2 \\ 3, & 2 \le x \le 4 \end{cases}$$

does not take on all values between $f(1) = 0$ and $f(4) = 3$; it misses all the values between 2 and 3.

Grapher Failure

In connected mode, a grapher may conceal a function's discontinuities by portraying the graph as a connected curve when it is not. To see what we mean, graph $y = $ int (x) in a $[-10, 10]$ by $[-10, 10]$ window in both connected and dot modes. A knowledge of where to expect discontinuities will help you recognize this form of grapher failure.

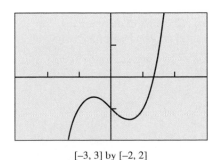

$[-3, 3]$ by $[-2, 2]$

Figure 2.26 The graph of $f(x) = x^3 - x - 1$. (Example 5)

> **Theorem 8 The Intermediate Value Theorem for Continuous Functions**
>
> A function $y = f(x)$ that is continuous on a closed interval $[a, b]$ takes on every value between $f(a)$ and $f(b)$. In other words, if y_0 is between $f(a)$ and $f(b)$, then $y_0 = f(c)$ for some c in $[a, b]$.
>
>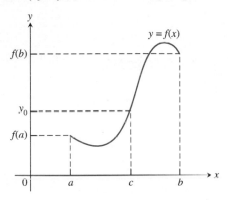

The continuity of f on the interval is essential to Theorem 8. If f is discontinuous at even one point of the interval, the theorem's conclusion may fail, as it does for the function graphed in Figure 2.25.

A Consequence for Graphing: Connectivity Theorem 8 is the reason why the graph of a function continuous on an interval cannot have any breaks. The graph will be **connected,** a single, unbroken curve, like the graph of sin x. It will not have jumps like those in the graph of the greatest integer function int x, or separate branches like we see in the graph of $1/x$.

Most graphers can plot points *(dot mode)*. Some can turn on pixels between plotted points to suggest an unbroken curve *(connected mode)*. For functions, the connected format basically assumes that outputs *vary continuously* with inputs and do not jump from one value to another without taking on all values in between.

Example 5 USING THEOREM 8

Is any real number exactly 1 less than its cube?

Solution We answer this question by applying the Intermediate Value Theorem in the following way. Any such number must satisfy the equation $x = x^3 - 1$ or, equivalently, $x^3 - x - 1 = 0$. Hence, we are looking for a zero value of the continuous function $f(x) = x^3 - x - 1$ (Figure 2.26). The function changes sign between 1 and 2, so there must be a point c between 1 and 2 where $f(c) = 0$.

Quick Review 2.3

1. Find $\lim\limits_{x \to -1} \dfrac{3x^2 - 2x + 1}{x^3 + 4}$.

2. Let $f(x) = \text{int } x$. Find each limit.

(a) $\lim\limits_{x \to -1^-} f(x)$ (b) $\lim\limits_{x \to -1^+} f(x)$ (c) $\lim\limits_{x \to -1} f(x)$ (d) $f(-1)$

3. Let $f(x) = \begin{cases} x^2 - 4x + 5, & x < 2 \\ 4 - x, & x \geq 2. \end{cases}$

Find each limit.

(a) $\lim\limits_{x \to 2^-} f(x)$ (b) $\lim\limits_{x \to 2^+} f(x)$ (c) $\lim\limits_{x \to 2} f(x)$ (d) $f(2)$

In Exercises 4–6, find the remaining functions in the list of functions: f, g, $f \circ g$, $g \circ f$.

4. $f(x) = \dfrac{2x - 1}{x + 5}$, $g(x) = \dfrac{1}{x} + 1$

5. $f(x) = x^2$, $(g \circ f)(x) = \sin x^2$,   domain of $g = [0, \infty)$

6. $g(x) = \sqrt{x - 1}$, $(g \circ f)(x) = 1/x$, $x > 0$

7. Use factoring to solve $2x^2 + 9x - 5 = 0$.

8. Use graphing to solve $x^3 + 2x - 1 = 0$.

In Exercises 9 and 10, let

$$f(x) = \begin{cases} 5 - x, & x \leq 3 \\ -x^2 + 6x - 8, & x > 3. \end{cases}$$

9. Solve the equation $f(x) = 4$.

10. Find a value of c for which the equation $f(x) = c$ has no solution.

Section 2.3 Exercises

In Exercises 1–10, find the points of discontinuity of the function. Identify each type of discontinuity.

1. $y = \dfrac{1}{(x + 2)^2}$

2. $y = \dfrac{x + 1}{x^2 - 4x + 3}$

3. $y = \dfrac{1}{x^2 + 1}$

4. $y = |x - 1|$

5. $y = \sqrt{2x + 3}$

6. $y = \sqrt[3]{2x - 1}$

7. $y = |x|/x$

8. $y = \cot x$

9. $y = e^{1/x}$

10. $y = \ln(x + 1)$

In Exercises 11–18, use the function f defined and graphed below to answer the questions.

$$f(x) = \begin{cases} x^2 - 1, & -1 \leq x < 0 \\ 2x, & 0 < x < 1 \\ 1, & x = 1 \\ -2x + 4, & 1 < x < 2 \\ 0, & 2 < x < 3 \end{cases}$$

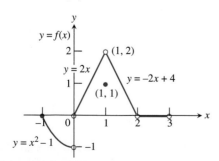

11. (a) Does $f(-1)$ exist?

(b) Does $\lim\limits_{x \to -1^+} f(x)$ exist?

(c) Does $\lim\limits_{x \to -1^+} f(x) = f(-1)$?

(d) Is f continuous at $x = -1$?

12. (a) Does $f(1)$ exist?

(b) Does $\lim\limits_{x \to 1} f(x)$ exist?

(c) Does $\lim\limits_{x \to 1} f(x) = f(1)$?

(d) Is f continuous at $x = 1$?

13. (a) Is f defined at $x = 2$? (Look at the definition of f.)

(b) Is f continuous at $x = 2$?

14. At what values of x is f continuous?

15. What value should be assigned to $f(2)$ to make the extended function continuous at $x = 2$?

16. What new value should be assigned to $f(1)$ to make the new function continuous at $x = 1$?

17. Writing to Learn Is it possible to extend f to be continuous at $x = 0$? If so, what value should the extended function have there? If not, why not?

18. Writing to Learn Is it possible to extend f to be continuous at $x = 3$? If so, what value should the extended function have there? If not, why not?

In Exercises 19–24, **(a)** find each point of discontinuity. **(b)** Which of the discontinuities are removable? not removable? Give reasons for your answers.

19. $f(x) = \begin{cases} 3 - x, & x < 2 \\ \dfrac{x}{2} + 1, & x > 2 \end{cases}$

20. $f(x) = \begin{cases} 3 - x, & x < 2 \\ 2, & x = 2 \\ x/2, & x > 2 \end{cases}$

21. $f(x) = \begin{cases} \dfrac{1}{x - 1}, & x < 1 \\ x^3 - 2x + 5, & x \geq 1 \end{cases}$

22. $f(x) = \begin{cases} 1 - x^2, & x \neq -1 \\ 2, & x = -1 \end{cases}$

23.

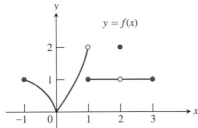

In Exercises 25–30, give a formula for the extended function that is continuous at the indicated point.

25. $f(x) = \dfrac{x^2 - 9}{x + 3}, \quad x = -3$ **26.** $f(x) = \dfrac{x^3 - 1}{x^2 - 1}, \quad x = 1$

27. $f(x) = \dfrac{\sin x}{x}, \quad x = 0$ **28.** $f(x) = \dfrac{\sin 4x}{x}, \quad x = 0$

29. $f(x) = \dfrac{x - 4}{\sqrt{x} - 2}, \quad x = 4$

30. $f(x) = \dfrac{x^3 - 4x^2 - 11x + 30}{x^2 - 4}, \quad x = 2$

In Exercises 31–34, *work in groups of two or three.* Verify that the function is continuous and state its domain. Indicate which theorems you are using, and which functions you are assuming to be continuous.

31. $y = \dfrac{1}{\sqrt{x + 2}}$ **32.** $y = x^2 + \sqrt[3]{4 - x}$

33. $y = |x^2 - 4x|$ **34.** $y = \begin{cases} \dfrac{x^2 - 1}{x - 1}, & x \neq 1 \\ 2, & x = 1 \end{cases}$

In Exercises 35–38, sketch a possible graph for a function f that has the stated properties.

35. $f(3)$ exists but $\lim_{x \to 3} f(x)$ does not.

36. $f(-2)$ exists, $\lim_{x \to -2^+} f(x) = f(-2)$, but $\lim_{x \to -2} f(x)$ does not exist.

37. $f(4)$ exists, $\lim_{x \to 4} f(x)$ exists, but f is not continuous at $x = 4$.

38. $f(x)$ is continuous for all x except $x = 1$, where f has a nonremovable discontinuity.

39. *Solving Equations* Is any real number exactly 1 less than its fourth power? Give any such values accurate to 3 decimal places.

40. *Solving Equations* Is any real number exactly 2 more than its cube? Give any such values accurate to 3 decimal places.

41. *Continuous Function* Find a value for a so that the function

$$f(x) = \begin{cases} x^2 - 1, & x < 3 \\ 2ax, & x \geq 3 \end{cases}$$

is continuous.

42. **Writing to Learn** Explain why the equation $e^{-x} = x$ has at least one solution.

43. *Salary Negotiation* A welder's contract promises a 3.5% salary increase each year for 4 years and Luisa has an initial salary of $36,500.

(a) Show that Luisa's salary is given by

$$y = 36{,}500(1.035)^{\text{int } t},$$

where t is the time, measured in years, since Luisa signed the contract.

(b) Graph Luisa's salary function. At what values of t is it continuous?

44. *Airport Parking* Valuepark charge $1.10 per hour or fraction of an hour for airport parking. The maximum charge per day is $7.25.

(a) Write a formula that gives the charge for x hours with $0 \leq x \leq 24$. (*Hint:* See Exercise 43.)

(b) Graph the function in (a). At what values of x is it continuous?

Exploration

45. Let $f(x) = \left(1 + \dfrac{1}{x}\right)^x$.

(a) Find the domain of f.

(b) Draw the graph of f.

(c) **Writing to Learn** Explain why $x = -1$ and $x = 0$ are points of discontinuity of f.

(d) **Writing to Learn** Are either of the discontinuities in (c) removable? Explain.

(e) Use graphs and tables to estimate $\lim_{x \to \infty} f(x)$.   ■

Extending the Ideas

46. *Continuity at a Point* Show that $f(x)$ is continuous at $x = a$ if and only if

$$\lim_{h \to 0} f(a + h) = f(a).$$

47. *Continuity on Closed Intervals* Let f be continuous and never zero on $[a, b]$. Show that either $f(x) > 0$ for all x in $[a, b]$ or $f(x) < 0$ for all x in $[a, b]$.

48. *Properties of Continuity* Prove that if f is continuous on an interval, then so is $|f|$.

49. *Everywhere Discontinuous* Give a convincing argument that the following function is not continuous at any real number.

$$f(x) = \begin{cases} 1, & \text{if } x \text{ is rational} \\ 0, & \text{if } x \text{ is irrational} \end{cases}$$

2.4 # Rates of Change and Tangent Lines

Average Rates of Change • Tangent to a Curve • Slope of a Curve • Normal to a Curve • Speed Revisited

Average Rates of Change

We encounter average rates of change in such forms as average speed (in miles per hour), growth rates of populations (in percent per year), and average monthly rainfall (in inches per month). The **average rate of change** of a quantity over a period of time is the amount of change divided by the time it takes. In general, the *average rate of change* of a function over an interval is the amount of change divided by the length of the interval.

Example 1 FINDING AVERAGE RATE OF CHANGE

Find the average rate of change of $f(x) = x^3 - x$ over the interval $[1, 3]$.

Solution Since $f(1) = 0$ and $f(3) = 24$, the average rate of change over the interval $[1, 3]$ is

$$\frac{f(3) - f(1)}{3 - 1} = \frac{24 - 0}{2} = 12.$$

Experimental biologists often want to know the rates at which populations grow under controlled laboratory conditions. Figure 2.27 shows how the number of fruit flies *(Drosophila)* grew in a controlled 50-day experiment. The graph was made by counting flies at regular intervals, plotting a point for each count, and drawing a smooth curve through the plotted points.

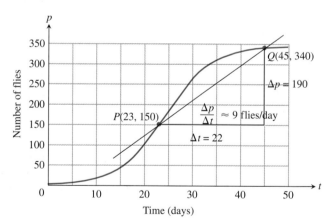

Figure 2.27 Growth of a fruit fly population in a controlled experiment. *Source: Elements of Mathematical Biology.* (Example 2)

Secant to a Curve

A line through two points on a curve is a **secant to the curve.**

Example 2 GROWING *DROSOPHILA* IN A LABORATORY

Use the points $P(23, 150)$ and $Q(45, 340)$ in Figure 2.27 to compute the average rate of change and the slope of the secant line PQ.

Solution There were 150 flies on day 23 and 340 flies on day 45. This gives an increase of $340 - 150 = 190$ flies in $45 - 23 = 22$ days.

The average rate of change in the population p from day 23 to day 45 was

$$\textit{Average rate of change: } \frac{\Delta p}{\Delta t} = \frac{340 - 150}{45 - 23} = \frac{190}{22} \approx 8.6 \text{ flies/day,}$$

or about 9 flies per day.

This average rate of change is also the slope of the secant line through the two points P and Q on the population curve. We can calculate the slope of the secant PQ from the coordinates of P and Q.

$$\textit{Secant slope: } \frac{\Delta p}{\Delta t} = \frac{340 - 150}{45 - 23} = \frac{190}{22} \approx 8.6 \text{ flies/day}$$

As suggested by Example 2, *we can always think of an average rate of change as the slope of a secant line.*

In addition to knowing the average rate at which the population grew from day 23 to day 45, we may also want to know how fast the population was growing on day 23 itself. To find out, we can watch the slope of the secant PQ change as we back Q along the curve toward P. The results for four positions of Q are shown in Figure 2.28.

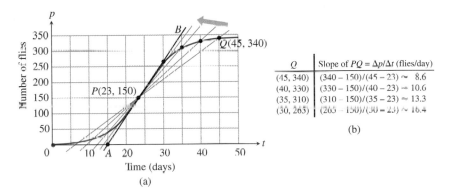

Q	Slope of $PQ = \Delta p/\Delta t$ (flies/day)
(45, 340)	$(340 - 150)/(45 - 23) \approx 8.6$
(40, 330)	$(330 - 150)/(40 - 23) \approx 10.6$
(35, 310)	$(310 - 150)/(35 - 23) \approx 13.3$
(30, 265)	$(265 - 150)/(30 - 23) \approx 16.4$

(b)

Figure 2.28 (a) Four secants to the fruit fly graph of Figure 2.27, through the point $P(23, 150)$. (b) The slopes of the four secants.

In terms of geometry, what we see as Q approaches P along the curve is this: The secant PQ approaches the tangent line AB that we drew by eye at P. This means that within the limitations of our drawing, the slopes of the secants approach the slope of the tangent, which we calculate from the coordinates of A and B to be

$$\frac{350 - 0}{35 - 15} = 17.5 \text{ flies/day.}$$

In terms of population, what we see as Q approaches P is this: The average growth rates for increasingly smaller time intervals approach the slope of the tangent to the curve at P (17.5 flies per day). The slope of the tangent line is therefore the number we take as the rate at which the fly population was growing on day $t = 23$.

Tangent to a Curve

The moral of the fruit fly story would seem to be that we should define the rate at which the value of the function $y = f(x)$ is changing with respect to x at any particular value $x = a$ to be the slope of the tangent to the curve

Why Find Tangents to Curves?

In mechanics, the tangent determines the direction of a body's motion at every point along its path.

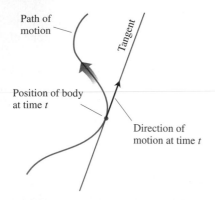

In geometry, the tangents to two curves at a point of intersection determine the angle at which the curves intersect.

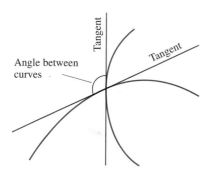

In optics, the tangent determines the angle at which a ray of light enters a curved lens (more about this in Section 3.7). The problem of how to find a tangent to a curve became the dominant mathematical problem of the early seventeenth century and it is hard to overestimate how badly the scientists of the day wanted to know the answer. Descartes went so far as to say that the problem was the most useful and most general problem not only that he knew but that he had any desire to know.

$y = f(x)$ at $x = a$. But how are we to define the tangent line at an arbitrary point P on the curve and find its slope from the formula $y = f(x)$? The problem here is that we know only one point. Our usual definition of slope requires two points.

The solution that mathematician Pierre Fermat found in 1629 proved to be one of that century's major contributions to calculus. We still use his method of defining tangents to produce formulas for slopes of curves and rates of change:

1. We start with what we *can* calculate, namely, the slope of a secant through P and a point Q nearby on the curve.

2. We find the limiting value of the secant slope (if it exists) as Q approaches P along the curve.

3. We define the *slope of the curve* at P to be this number and define the *tangent to the curve* at P to be the line through P with this slope.

Example 3   FINDING SLOPE AND TANGENT LINE

Find the slope of the parabola $y = x^2$ at the point $P(2, 4)$. Write an equation for the tangent to the parabola at this point.

Solution   We begin with a secant line through $P(2, 4)$ and a nearby point $Q(2 + h, (2 + h)^2)$ on the curve (Figure 2.29).

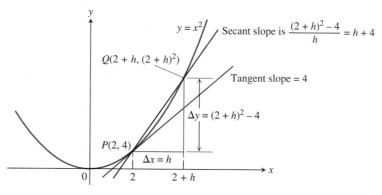

Figure 2.29   The slope of the tangent to the parabola $y = x^2$ at $P(2, 4)$ is 4.

We then write an expression for the slope of the secant line and find the limiting value of this slope as Q approaches P along the curve.

$$\text{Secant slope} = \frac{\Delta y}{\Delta x} = \frac{(2 + h)^2 - 4}{h}$$

$$= \frac{h^2 + 4h + 4 - 4}{h}$$

$$= \frac{h^2 + 4h}{h} = h + 4$$

The limit of the secant slope as Q approaches P along the curve is

$$\lim_{Q \to P} (\text{secant slope}) = \lim_{h \to 0} (h + 4) = 4.$$

Thus, the slope of the parabola at P is 4.

The tangent to the parabola at P is the line through $P(2, 4)$ with slope $m = 4$.

$$y - 4 = 4(x - 2)$$
$$y = 4x - 8 + 4$$
$$y = 4x - 4$$

Slope of a Curve

To find the tangent to a curve $y = f(x)$ at a point $P(a, f(a))$ we use the same dynamic procedure. We calculate the slope of the secant line through P and a point $Q(a + h, f(a + h))$. We then investigate the limit of the slope as $h \to 0$ (Figure 2.30). If the limit exists, it is the slope of the curve at P and we define the tangent at P to be the line through P having this slope.

Definition   Slope of a Curve at a Point

The **slope of the curve** $y = f(x)$ at the point $P(a, f(a))$ is the number

$$m = \lim_{h \to 0} \frac{f(a + h) - f(a)}{h},$$

provided the limit exists.

The **tangent line to the curve** at P is the line through P with this slope.

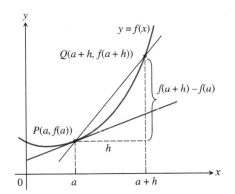

$y = f(x)$

$Q(a + h, f(a + h))$

$f(a + h) - f(a)$

$P(a, f(a))$

h

a $a + h$

Figure 2.30 The tangent slope is

$$\lim_{h \to 0} \frac{f(a + h) - f(a)}{h}.$$

Example 4   EXPLORING SLOPE AND TANGENT

Let $f(x) = 1/x$.

(a) Find the slope of the curve at $x = a$.

(b) Where does the slope equal $-1/4$?

(c) What happens to the tangent to the curve at the point $(a, 1/a)$ for different values of a?

Solution

(a) The slope at $x = a$ is

$$\lim_{h \to 0} \frac{f(a + h) - f(a)}{h} = \lim_{h \to 0} \frac{\dfrac{1}{a + h} - \dfrac{1}{a}}{h}$$

$$= \lim_{h \to 0} \frac{1}{h} \cdot \frac{a - (a + h)}{a(a + h)}$$

$$= \lim_{h \to 0} \frac{-h}{ha(a + h)}$$

$$= \lim_{h \to 0} \frac{-1}{a(a + h)} = -\frac{1}{a^2}.$$

(b) The slope will be $-1/4$ if

$$-\frac{1}{a^2} = -\frac{1}{4}$$

$$a^2 = 4 \qquad \text{Multiply by } -4a^2.$$

$$a = \pm 2.$$

Pierre de Fermat (1601–1665)

The dynamic approach to tangency, invented by Fermat in 1629, proved to be one of the seventeenth century's major contributions to calculus.

Fermat, a skilled linguist and one of his century's greatest mathematicians, tended to confine his writing to professional correspondence and to papers written for personal friends. He rarely wrote completed descriptions of his work, even for his personal use. His name slipped into relative obscurity until the late 1800s, and it was only from a four-volume edition of his works published at the beginning of this century that the true importance of his many achievements became clear.

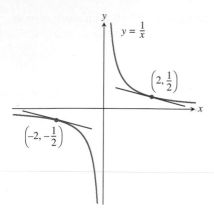

Figure 2.31 The two tangent lines to $y = 1/x$ having slope $-1/4$. (Example 4)

All of these are the same:

1. the slope of $y = f(x)$ at $x = a$

2. the slope of the tangent to $y = f(x)$ at $x = a$

3. the (instantaneous) rate of change of $f(x)$ with respect to x at $x = a$

4. $\displaystyle\lim_{h \to 0} \frac{f(a + h) - f(a)}{h}$

About the Word *Normal*

When analytic geometry was developed in the seventeenth century, European scientists still wrote about their work and ideas in Latin, the one language that all educated Europeans could read and understand. The Latin word *normalis,* which scholars used for *perpendicular,* became *normal* when they discussed geometry in English.

The curve has the slope $-1/4$ at the two points $(2, 1/2)$ and $(-2, -1/2)$ (Figure 2.31).

(c) The slope $-1/a^2$ is always negative. As $a \to 0^+$, the slope approaches $-\infty$ and the tangent becomes increasingly steep. We see this again as $a \to 0^-$. As a moves away from the origin in either direction, the slope approaches 0^- and the tangent becomes increasingly horizontal.

The expression

$$\frac{f(a + h) - f(a)}{h}$$

is the **difference quotient of f at a.** Suppose the difference quotient has a limit as h approaches zero. If we interpret the difference quotient as a secant slope, the limit is the slope of both the curve and the tangent to the curve at the point $x = a$. If we interpret the difference quotient as an average rate of change, the limit is the function's rate of change with respect to x at the point $x = a$. This limit is one of the two most important mathematical objects considered in calculus. We will begin a thorough study of it in Chapter 3.

Normal to a Curve

The **normal line** to a curve at a point is the line perpendicular to the tangent at that point.

Example 5   FINDING A NORMAL LINE

Write an equation for the normal to the curve $f(x) = 4 - x^2$ at $x = 1$.

Solution The slope of the tangent to the curve at $x = 1$ is

$$\lim_{h \to 0} \frac{f(1 + h) - f(1)}{h} = \lim_{h \to 0} \frac{4 - (1 + h)^2 - 3}{h}$$

$$= \lim_{h \to 0} \frac{4 - 1 - 2h - h^2 - 3}{h}$$

$$= \lim_{h \to 0} \frac{-h(2 + h)}{h} = -2.$$

Thus, the slope of the normal is $1/2$, the negative reciprocal of -2. The normal to the curve at $(1, f(1)) = (1, 3)$ is the line through $(1, 3)$ with slope $m = 1/2$.

$$y - 3 = \frac{1}{2}(x - 1)$$

$$y = \frac{1}{2}x - \frac{1}{2} + 3$$

$$y = \frac{1}{2}x + \frac{5}{2}$$

You can support this result by drawing the graphs in a square viewing window.

Speed Revisited

The function $y = 16t^2$ that gave the distance fallen by the rock in Example 1, Section 2.1, was the rock's *position function.* A body's average speed along a coordinate axis (here, the y-axis) for a given period of time is the average rate

Particle Motion

We only have considered objects moving in one direction in this chapter. In Chapter 3, we will deal with more complicated motion.

of change of its *position* $y = f(t)$. Its **instantaneous speed** at any time t is the **instantaneous rate of change** of position with respect to time at time t, or

$$\lim_{h \to 0} \frac{f(t + h) - f(t)}{h}.$$

We saw in Example 1, Section 2.1, that the rock's instantaneous speed at $t = 2$ sec was 64 ft/sec.

Example 6   INVESTIGATING FREE FALL

Find the speed of the falling rock in Example 1, Section 2.1, at $t = 1$ sec.

Solution The position function of the rock is $f(t) = 16t^2$. The average speed of the rock over the interval between $t = 1$ and $t = 1 + h$ sec was

$$\frac{f(1 + h) - f(1)}{h} = \frac{16(1 + h)^2 - 16(1)^2}{h} = \frac{16(h^2 + 2h)}{h} = 16(h + 2).$$

The rock's speed at the instant $t = 1$ was

$$\lim_{h \to 0} 16(h + 2) = 32 \text{ ft/sec.}$$

Quick Review 2.4

In Exercises 1 and 2, find the increments Δx and Δy from point A to point B.

1. $A(-5, 2)$, $B(3, 5)$

2. $A(1, 3)$, $B(a, b)$

In Exercises 3 and 4, find the slope of the line determined by the points.

3. $(-2, 3)$, $(5, -1)$

4. $(-3, -1)$, $(3, 3)$

In Exercises 5–9, write an equation for the specified line.

5. through $(-2, 3)$ with slope $= 3/2$

6. through $(1, 6)$ and $(4, -1)$

7. through $(1, 4)$ and parallel to $y = -\dfrac{3}{4}x + 2$

8. through $(1, 4)$ and perpendicular to $y = -\dfrac{3}{4}x + 2$

9. through $(1, 3)$ and parallel to $2x + 3y = 5$

10. For what value of b will the slope of the line through $(2, 3)$ and $(4, b)$ be 5/3?

Section 2.4 Exercises

In Exercises 1–6, find the average rate of change of the function over each interval.

1. $f(x) = x^3 + 1$

(a) $[2, 3]$ (b) $[-1, 1]$

2. $f(x) = \sqrt{4x + 1}$

(a) $[0, 2]$ (b) $[10, 12]$

3. $f(x) = e^x$

(a) $[-2, 0]$ (b) $[1, 3]$

4. $f(x) = \ln x$

(a) $[1, 4]$ (b) $[100, 103]$

5. $f(x) = \cot t$

(a) $[\pi/4, 3\pi/4]$ (b) $[\pi/6, \pi/2]$

6. $f(x) = 2 + \cos t$

(a) $[0, \pi]$ (b) $[-\pi, \pi]$

In Exercises 7 and 8, a distance-time graph is shown.

(a) Estimate the slopes of the secants PQ_1, PQ_2, PQ_3, and PQ_4, arranging them in order in a table. What is the appropriate unit for these slopes?

(b) Estimate the speed at point P.

7. *Accelerating from a Standstill* The figure shows the distance-time graph for a 1994 Ford® Mustang Cobra™ accelerating from a standstill.

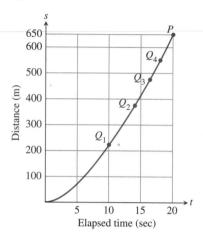

8. *Lunar Data* The figure shows a distance-time graph for a wrench that fell from the top platform of a communication mast on the moon to the station roof 80 m below.

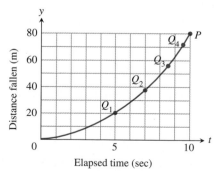

In Exercises 9–12, at the indicated point find

(a) the slope of the curve,

(b) an equation of the tangent, and

(c) an equation of the normal.

(d) Then draw a graph of the curve, tangent line, and normal line in the same square viewing window.

9. $y = x^2$ at $x = -2$

10. $y = x^2 - 4x$ at $x = 1$

11. $y = \dfrac{1}{x - 1}$ at $x = 2$

12. $y = x^2 - 3x - 1$ at $x = 0$

In Exercises 13 and 14, find the slope of the curve at the indicated point.

13. $f(x) = |x|$ at   (a) $x = 2$ (b) $x = -3$

14. $f(x) = |x - 2|$ at $x = 1$

In Exercises 15–18, determine whether the curve has a tangent at the indicated point. If it does, give its slope. If not, explain why not.

15. $f(x) = \begin{cases} 2 - 2x - x^2, & x < 0 \\ 2x + 2, & x \geq 0 \end{cases}$ at $x = 0$

16. $f(x) = \begin{cases} -x, & x < 0 \\ x^2 - x, & x \geq 0 \end{cases}$ at $x = 0$

17. $f(x) = \begin{cases} 1/x, & x \leq 2 \\ \dfrac{4 - x}{4}, & x > 2 \end{cases}$ at $x = 2$

18. $f(x) = \begin{cases} \sin x, & 0 \leq x < 3\pi/4 \\ \cos x, & 3\pi/4 \leq x \leq 2\pi \end{cases}$ at $x = 3\pi/4$

In Exercises 19–22, (a) find the slope of the curve at $x = a$ and (b) **Writing to Learn** describe what happens to the tangent at $x = a$ as a changes.

19. $y = x^2 + 2$ **20.** $y = 2/x$

21. $y = \dfrac{1}{x - 1}$ **22.** $y = 9 - x^2$

23. *Free Fall* An object is dropped from the top of a 100-m tower. Its height above ground after t sec is $100 - 4.9t^2$ m. How fast is it falling 2 sec after it is dropped?

24. *Rocket Launch* At t sec after lift-off, the height of a rocket is $3t^2$ ft. How fast is the rocket climbing after 10 sec?

25. *Area of Circle* What is the rate of change of the area of a circle with respect to the radius when the radius is $r = 3$ in.?

26. *Volume of Sphere* What is the rate of change of the volume of a sphere with respect to the radius when the radius is $r = 2$ in.?

27. *Free Fall on Mars* The equation for free fall at the surface of Mars is $s = 1.86t^2$ m with t in seconds. Assume a rock is dropped from the top of a 200-m cliff. Find the speed of the rock at $t = 1$ sec.

28. *Free Fall on Jupiter* The equation for free fall at the surface of Jupiter is $s = 11.44t^2$ m with t in seconds. Assume a rock is dropped from the top of a 500-m cliff. Find the speed of the rock at $t = 2$ sec.

29. *Horizontal Tangent* At what point is the tangent to $f(x) = x^2 + 4x - 1$ horizontal?

30. *Horizontal Tangent* At what point is the tangent to $f(x) = 3 - 4x - x^2$ horizontal?

31. *Finding Tangents and Normals*

(a) Find an equation for each tangent to the curve $y = 1/(x - 1)$ that has slope -1. (See Exercise 21.)

(b) Find an equation for each normal to the curve $y = 1/(x - 1)$ that has slope 1.

32. *Finding Tangents* Find the equations of all lines tangent to $y = 9 - x^2$ that pass through the point $(1, 12)$.

33. Table 2.3 gives the amount of federal funding for the Immigration and Naturalization Service (INS) for several years.

(a) Find the average rate of change in funding from 1993 to 1995.

(b) Find the average rate of change from 1995 to 1997.

(c) Let $x = 0$ represent 1990, $x = 1$ represent 1991, and so forth. Find the quadratic regression equation for the data and superimpose its graph on a scatter plot of the data.

(d) Compute the average rates of change in (a) and (b) using the regression equation.

(e) Use the regression equation to find how fast the funding was growing in 1997.

Table 2.3 Federal Funding

Year	INS Funding ($ billions)
1993	1.5
1994	1.6
1995	2.1
1996	2.6
1997	3.1

Source: Immigration and Naturalization Service as reported by Bob Laird in *USA Today,* February 18, 1997.

34. Table 2.4 gives the amounts of money earmarked by the U.S. Congress for collegiate academic programs for several years.

(a) Let $x = 0$ represent 1980, $x = 1$ represent 1981, and so forth. Make a scatter plot of the data.

(b) Let P represent the point corresponding to 1997, and Q the point for any one of the previous years. Make a table of the slopes possible for the secant line PQ.

(c) **Writing to Learn** Based on the computations, explain why someone might be hesitant to make a prediction about the rate of change of congressional funding in 1997.

Table 2.4 Congressional Academic Funding

Year	Funding ($ millions)
1988	225
1989	289
1990	270
1991	493
1992	684
1993	763
1994	651
1995	600
1996	296
1997	440

Source: The Chronicle of Higher Education, March 28, 1997.

Explorations

In Exercises **35** and **36,** complete the following for the function.

(a) Compute the difference quotient

$$\frac{f(1 + h) - f(1)}{h}.$$

(b) Use graphs and tables to estimate the limit of the difference quotient in (a) as $h \rightarrow 0$.

(c) Compare your estimate in (b) with the given number.

(d) **Writing to Learn** Based on your computations, do you think the graph of f has a tangent at $x = 1$? If so, estimate its slope. If not, explain why not.

35. $f(x) = e^x,\ e$ **36.** $f(x) = 2^x,\ \ln 4$ ■

In Exercises 37–40, *work in groups of two or three.*

The curve $y = f(x)$ has a **vertical tangent** at $x = a$ if

$$\lim_{h \to 0} \frac{f(a + h) - f(a)}{h} = \infty$$

or if

$$\lim_{h \to 0} \frac{f(a + h) - f(a)}{h} = -\infty.$$

In each case, the right- and left-hand limits are required to be the same: both $+\infty$ or both $-\infty$.

Use graphs to investigate whether the curve has a vertical tangent at $x = 0$.

37. $y = x^{2/5}$

38. $y = x^{3/5}$

39. $y = x^{1/3}$

40. $y = x^{2/3}$

Extending the Ideas

In Exercises 41 and 42, determine whether the graph of the function has a tangent at the origin. Explain your answer.

41. $f(x) = \begin{cases} x^2 \sin \dfrac{1}{x}, & x \neq 0 \\ 0, & x = 0 \end{cases}$

42. $f(x) = \begin{cases} x \sin \dfrac{1}{x}, & x \neq 0 \\ 0, & x = 0 \end{cases}$

43. *Sine Function* Estimate the slope of the curve $y = \sin x$ at $x = 1$. (*Hint:* See Exercises 35 and 36.)

Chapter 2 Key Terms

average rate of change (p. 82)
average speed (p. 55)
connected graph (p. 79)
Constant Multiple Rule for Limits (p. 67)
continuity at a point (p. 75)
continuous at an endpoint (p. 75)
continuous at an interior point (p. 75)
continuous extension (p. 77)
continuous function (p. 77)
continuous on an interval (p. 77)
difference quotient (p. 86)
Difference Rule for Limits (p. 67)
discontinuous (p. 75)
end behavior model (p. 70)
free fall (p. 55)
horizontal asymptote (p. 65)
infinite discontinuity (p. 76)
instantaneous rate of change (p. 87)
instantaneous speed (p. 87)
intermediate value property (p. 79)
Intermediate Value Theorem for
    Continuous Functions (p. 79)

jump discontinuity (p. 76)
left end behavior model (p. 69)
left-hand limit (p. 60)
limit of a function (p. 57)
normal to a curve (p. 86)
oscillating discontinuity (p. 76)
point of discontinuity (p. 75)
Power Rule for Limits (p. 67)
Product Rule for Limits (p. 67)
Quotient Rule for Limits (p. 67)
removable discontinuity (p. 76)
right end behavior model (p. 69)
right-hand limit (p. 60)
Sandwich Theorem (p. 61)
secant to a curve (p. 82)
slope of a curve (p. 85)
Sum Rule for Limits (p. 67)
tangent line to a curve (p. 85)
two-sided limit (p. 60)
vertical asymptote (p. 68)
vertical tangent (p. 90)

Chapter 2   Review Exercises

In Exercises 1–14, find the limits.

1. $\lim\limits_{x \to -2} (x^3 - 2x^2 + 1)$

2. $\lim\limits_{x \to -2} \dfrac{x^2 + 1}{3x^2 - 2x + 5}$

3. $\lim\limits_{x \to 4} \sqrt{1 - 2x}$

4. $\lim\limits_{x \to 5} \sqrt[4]{9 - x^2}$

5. $\lim\limits_{x \to 0} \dfrac{\dfrac{1}{2 + x} - \dfrac{1}{2}}{x}$

6. $\lim\limits_{x \to \pm\infty} \dfrac{2x^2 + 3}{5x^2 + 7}$

7. $\lim\limits_{x \to \pm\infty} \dfrac{x^4 + x^3}{12x^3 + 128}$

8. $\lim\limits_{x \to 0} \dfrac{\sin 2x}{4x}$

9. $\lim\limits_{x \to 0} \dfrac{x \csc x + 1}{x \csc x}$

10. $\lim\limits_{x \to 0} e^x \sin x$

11. $\lim\limits_{x \to 7/2^+} \text{int}\,(2x - 1)$

12. $\lim\limits_{x \to 7/2^-} \text{int}\,(2x - 1)$

13. $\lim\limits_{x \to \infty} e^{-x} \cos x$

14. $\lim\limits_{x \to \infty} \dfrac{x + \sin x}{x + \cos x}$

In Exercises 15–20, determine whether the limit exists on the basis of the graph of $y = f(x)$. The domain of f is the set of real numbers.

15. $\lim\limits_{x \to d} f(x)$

16. $\lim\limits_{x \to c^+} f(x)$

17. $\lim\limits_{x \to c^-} f(x)$

18. $\lim\limits_{x \to c} f(x)$

19. $\lim\limits_{x \to b} f(x)$

20. $\lim\limits_{x \to a} f(x)$

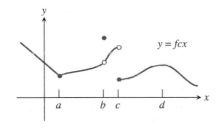

In Exercises 21–24, determine whether the function f used in Exercises 15–20 is continuous at the indicated point.

21. $x = a$

22. $x = b$

23. $x = c$

24. $x = d$

In Exercises 25 and 26, use the graph of the function with domain $-1 \le x \le 3$.

25. Determine

(a) $\lim\limits_{x \to 3^-} g(x)$.

(b) $g(3)$.

(c) whether $g(x)$ is continuous at $x = 3$.

(d) the points of discontinuity of $g(x)$.

(e) **Writing to Learn** whether any points of discontinuity are removable. If so, describe the extended function. If not, explain why not.

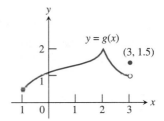

26. Determine

(a) $\lim\limits_{x \to 1^-} k(x)$.

(b) $\lim\limits_{x \to 1^+} k(x)$.

(c) $k(1)$.

(d) whether $k(x)$ is continuous at $x = 1$.

(e) the points of discontinuity of $k(x)$.

(f) **Writing to Learn** whether any points of discontinuity are removable. If so, describe the extended function. If not, explain why not.

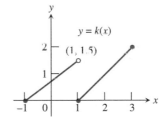

In Exercises 27 and 28, (a) find the vertical asymptotes of the graph of $y = f(x)$, and (b) describe the behavior of $f(x)$ to the left and right of any vertical asymptote.

27. $f(x) = \dfrac{x + 3}{x + 2}$

28. $f(x) = \dfrac{x - 1}{x^2\,(x + 2)}$

In Exercises 29 and 30, answer the questions for the piecewise-defined function.

29. $f(x) = \begin{cases} 1, & x \le -1 \\ -x, & -1 < x < 0 \\ 1, & x = 0 \\ -x, & 0 < x < 1 \\ 1, & x \ge 1 \end{cases}$

(a) Find the right-hand and left-hand limits of f at $x = -1$, 0, and 1.

(b) Does f have a limit as x approaches -1? 0? 1? If so, what is it? If not, why not?

(c) Is f continuous at $x = -1$? 0? 1? Explain.

30. $f(x) = \begin{cases} |x^3 - 4x|, & x < 1 \\ x^2 - 2x - 2, & x \ge 1 \end{cases}$

(a) Find the right-hand and left-hand limits of f at $x = 1$.

(b) Does f have a limit as $x \to 1$? If so, what is it? If not, why not?

(c) At what points is f continuous?

(d) At what points is f discontinuous?

In Exercises 31 and 32, find all points of discontinuity of the function.

31. $f(x) = \dfrac{x + 1}{4 - x^2}$

32. $g(x) = \sqrt[3]{3x + 2}$

In Exercises 33–36, find (a) a power function end behavior model and (b) any horizontal asymptotes.

33. $f(x) = \dfrac{2x + 1}{x^2 - 2x + 1}$

34. $f(x) = \dfrac{2x^2 + 5x - 1}{x^2 + 2x}$

35. $f(x) = \dfrac{x^3 - 4x^2 + 3x + 3}{x - 3}$

36. $f(x) = \dfrac{x^4 - 3x^2 + x - 1}{x^3 - x + 1}$

In Exercises 37 and 38, find (a) a right end behavior model and (b) a left end behavior model for the function.

37. $f(x) = x + e^x$

38. $f(x) = \ln |x| + \sin x$

In Exercises 39 and 40, *work in groups of two or three.* What value should be assigned to k to make f a continuous function?

39. $f(x) = \begin{cases} \dfrac{x^2 + 2x - 15}{x - 3}, & x \ne 3 \\ k, & x = 3 \end{cases}$

40. $f(x) = \begin{cases} \dfrac{\sin x}{2x}, & x \ne 0 \\ k, & x = 0 \end{cases}$

In Exercises 41 and 42, *work in groups of two or three.* Sketch a graph of a function f that satisfies the given conditions.

41. $\lim\limits_{x \to \infty} f(x) = 3$, $\quad \lim\limits_{x \to -\infty} f(x) = \infty$,

$\lim\limits_{x \to 3^+} f(x) = \infty$, $\quad \lim\limits_{x \to 3^-} f(x) = -\infty$

42. $\lim\limits_{x \to 2} f(x)$ does not exist, $\quad \lim\limits_{x \to 2^+} f(x) = f(2) = 3$

43. *Average Rate of Change* Find the average rate of change of $f(x) = 1 + \sin x$ over the interval $[0, \pi/2]$.

44. *Rate of Change* Find the instantaneous rate of change of the volume $V = (1/3)\pi r^2 H$ of a cone with respect to the radius r at $r = a$ if the height H does not change.

45. *Rate of Change* Find the instantaneous rate of change of the surface area $S = 6x^2$ of a cube with respect to the edge length x at $x = a$.

46. *Slope of a Curve* Find the slope of the curve $y = x^2 - x - 2$ at $x = a$.

47. *Tangent and Normal* Let $f(x) = x^2 - 3x$ and $P = (1, f(1))$. Find (a) the slope of the curve $y = f(x)$ at P, (b) an equation of the tangent at P, and (c) an equation of the normal at P.

48. *Horizontal Tangents* At what points, if any, are the tangents to the graph of $f(x) = x^2 - 3x$ horizontal? (See Exercise 47)

49. *Bear Population* The number of bears in a federal wildlife reserve is given by the population equation

$$p(t) = \frac{200}{1 + 7e^{-0.1t}},$$

where t is in years.

(a) **Writing to Learn** Find $p(0)$. Give a possible interpretation of this number.

(b) Find $\lim\limits_{t \to \infty} p(t)$.

(c) **Writing to Learn** Give a possible interpretation of the result in (b).

50. *Taxi Fares* Bluetop Cab charges \$3.20 for the first mile and \$1.35 for each additional mile or part of a mile.

(a) Write a formula that gives the charge for x miles with $0 \leq x \leq 20$.

(b) Graph the function in (a). At what values of x is it discontinuous?

51. *Congressional Academic Funding* Consider Table 2.4 in Exercise 34 of Section 2.4.

(a) Let $x = 0$ represent 1980, $x = 1$ represent 1981, and so forth. Find a cubic and a quartic regression equation for the data.

(b) Find an end behavior model for each regression equation. What does each predict about future funding?

52. *Limit Properties* Assume that

$$\lim_{x \to c} [f(x) + g(x)] = 2,$$

$$\lim_{x \to c} [f(x) - g(x)] = 1,$$

and that $\lim_{x \to c} f(x)$ and $\lim_{x \to c} g(x)$ exist. Find $\lim_{x \to c} f(x)$ and $\lim_{x \to c} g(x)$.

53. Table 2.5 gives the spending for software, equipment, and services to create and run *intranets*—private, Internet-like networks that link a company's operations.

(a) Let $x = 0$ represent 1990, $x = 1$ represent 1991, and so forth. Make a scatter plot of the data.

(b) Let P represent the point corresponding to 2000, and Q the point for any one of the previous years. Make a table of the slopes possible for the secant line PQ.

(c) Predict the rate of change of spending in 2000.

(d) Find a quadratic regression equation for the data, and use it to calculate the rate of change of spending in 2000.

Table 2.5 Intranet Spending

Year	Spending ($ billions)
1995	2.7
1996	4.8
1997	7.8
1998	11.2
1999	15.2
2000	20.1

Source: Killen & Associates as reported by Anne R. Carey and Elys A. McLean in *USA Today*, April 14, 1997.

Derivatives

Shown here is the pain reliever acetaminophen in crystalline form, photographed under a transmitted light microscope. While acetaminophen relieves pain with few side effects, it is toxic in large doses. One study found that only 30% of parents who gave acetaminophen to their children could accurately calculate and measure the correct dose.

One rule for calculating the dosage (mg) of acetaminophen for children ages 1 to 12 years old is $D(t) = 750t/(t + 12)$, where t is age in years. What is an expression for the rate of change of a child's dosage with respect to the child's age? How does the rate of change of the dosage relate to the growth rate of children? This problem can be solved with the information covered in Section 3.4.

Chapter 3 Overview

In Chapter 2, we learned how to find the slope of a tangent to a curve as the limit of the slopes of secant lines. In Example 4 of Section 2.4, we derived a formula for the slope of the tangent at an arbitrary point $(a, 1/a)$ on the graph of the function $f(x) = 1/x$ and showed that it was $-1/a^2$.

This seemingly unimportant result is more powerful than it might appear at first glance, as it gives us a simple way to calculate the instantaneous rate of change of f at any point. The study of rates of change of functions is called *differential calculus,* and the formula $-1/a^2$ was our first look at a *derivative.* The derivative was the 17th-century breakthrough that enabled mathematicians to unlock the secrets of planetary motion and gravitational attraction—of objects changing position over time. We will learn many uses for derivatives in Chapter 4, but first we will concentrate in this chapter on understanding what derivatives are and how they work.

3.1 Derivative of a Function

Definition of Derivative • Notation • Relationships between the Graphs of f and f' • Graphing the Derivative from Data • One-sided Derivatives

Definition of Derivative

In Section 2.4, we defined the slope of a curve $y = f(x)$ at the point where $x = a$ to be

$$m = \lim_{h \to 0} \frac{f(a + h) - f(a)}{h}.$$

When it exists, this limit is called the **derivative of f at a.** In this section, we investigate the derivative as a *function* derived from f by considering the limit at each point of the domain of f.

Definition Derivative

The **derivative** of the function f with respect to the variable x is the function f' whose value at x is

$$f'(x) = \lim_{h \to 0} \frac{f(x + h) - f(x)}{h}, \tag{1}$$

provided the limit exists.

The domain of f', the set of points in the domain of f for which the limit exists, may be smaller than the domain of f. If $f'(x)$ exists, we say that f **has a derivative (is differentiable)** at x. A function that is differentiable at every point of its domain is a **differentiable function.**

Example 1 APPLYING THE DEFINITION

Differentiate (that is, find the derivative of) $f(x) = x^3$.

Solution Applying the definition, we have

$$f'(x) = \lim_{h \to 0} \frac{f(x + h) - f(x)}{h}$$

$$= \lim_{h \to 0} \frac{(x + h)^3 - x^3}{h} \qquad \text{Eq. 1 with } f(x) = x^3, \\ f(x + h) = (x + h)^3$$

$$= \lim_{h \to 0} \frac{(x^3 + 3x^2h + 3xh^2 + h^3) - x^3}{h} \qquad (x + h)^3 \text{ expanded}$$

$$= \lim_{h \to 0} \frac{(3x^2 + 3xh + h^2)h}{h} \qquad x^3\text{s cancelled, } h \text{ factored out}$$

$$= \lim_{h \to 0} (3x^2 + 3xh + h^2)$$

$$= 3x^2.$$

The derivative of $f(x)$ at a point where $x = a$ is found by taking the limit as $h \to 0$ of slopes of secant lines, as shown in Figure 3.1.

By relabeling the picture as in Figure 3.2, we arrive at a useful alternate formula for calculating the derivative. This time, the limit is taken as x approaches a.

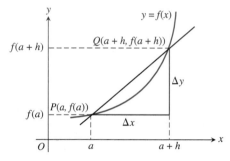

Figure 3.1 The slope of the secant line PQ is

$$\frac{\Delta y}{\Delta x} = \frac{f(a + h) - f(a)}{(a + h) - a} = \frac{f(a + h) - f(a)}{h}.$$

Definition (Alternate) **Derivative at a Point**

The **derivative** of the function f **at the point** $x = a$ is the limit

$$f'(a) = \lim_{x \to a} \frac{f(x) - f(a)}{x - a}, \qquad (2)$$

provided the limit exists.

After we find the derivative of f at a point $x = a$ using the alternate form, we can find the derivative of f as a function by applying the resulting formula to an arbitrary x in the domain of f.

Example 2 APPLYING THE ALTERNATE DEFINITION

Differentiate $f(x) = \sqrt{x}$ using the alternate definition.

Solution At the point $x = a$,

$$f'(a) = \lim_{x \to a} \frac{f(x) - f(a)}{x - a}$$

$$= \lim_{x \to a} \frac{\sqrt{x} - \sqrt{a}}{x - a} \qquad \text{Eq. 2 with } f(x) = \sqrt{x}$$

$$= \lim_{x \to a} \frac{\sqrt{x} - \sqrt{a}}{x - a} \cdot \frac{\sqrt{x} + \sqrt{a}}{\sqrt{x} + \sqrt{a}} \qquad \text{Rationalize...}$$

$$= \lim_{x \to a} \frac{x - a}{(x - a)(\sqrt{x} + \sqrt{a})} \qquad \text{...the numerator.}$$

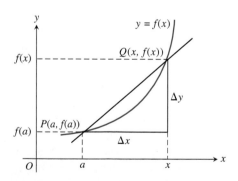

Figure 3.2 The slope of the secant line PQ is

$$\frac{\Delta y}{\Delta x} = \frac{f(x) - f(a)}{x - a}.$$

$$= \lim_{x \to a} \frac{1}{\sqrt{x} + \sqrt{a}}$$ We can now take the limit.

$$= \frac{1}{2\sqrt{a}}.$$

Applying this formula to an arbitrary $x > 0$ in the domain of f identifies the derivative as the function $f'(x) = 1/(2\sqrt{x})$ with domain $(0, \infty)$.

Notation

There are many ways to denote the derivative of a function $y = f(x)$. Besides $f'(x)$, the most common notations are these:

y'	"y prime"	Nice and brief, but does not name the independent variable.
$\dfrac{dy}{dx}$	"$dy\ dx$" or "the derivative of y with respect to x"	Names both variables and uses d for derivative.
$\dfrac{df}{dx}$	"$df\ dx$" or "the derivative of f with respect to x"	Emphasizes the function's name.
$\dfrac{d}{dx}f(x)$	"$d\ dx$ of f at x" or "the derivative of f at x"	Emphasizes the idea that differentiation is an operation performed on f.

Relationships between the Graphs of *f* and *f'*

When we have the explicit formula for $f(x)$, we can derive a formula for $f'(x)$ using methods like those in Examples 1 and 2. We have already seen, however, that functions are encountered in other ways: graphically, for example, or in tables of data.

Because we can think of the derivative at a point in graphical terms as *slope,* we can get a good idea of what the graph of the function f' looks like by *estimating the slopes* at various points along the graph of f.

Example 3   GRAPHING *f'* FROM *f*

Graph the derivative of the function f whose graph is shown in Figure 3.3a. Discuss the behavior of f in terms of the signs and values of f'.

Solution First, we draw a pair of coordinate axes, marking the horizontal axis in x-units and the vertical axis in slope units (Figure 3.3b). Next, we estimate the slope of the graph of f at various points, plotting the corresponding slope values using the new axes. At $A(0, f(0))$, the graph of f has slope 4, so $f'(0) = 4$. At B, the graph of f has slope 1, so $f' = 1$ at B', and so on.

We complete our estimate of the graph of f' by connecting the plotted points with a smooth curve.

Although we do not have a formula for either f or f', the graph of each reveals important information about the behavior of the other. In particular, notice that f is decreasing where f' is negative and increasing where f' is positive. Where f' is zero, the graph of f has a horizontal tangent, changing from increasing to decreasing (point C) or from decreasing to increasing (point F).

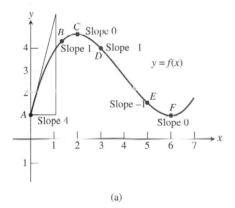

(a)

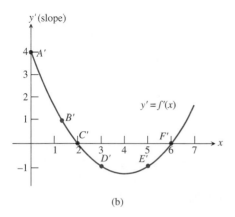

(b)

Figure 3.3 By plotting the slopes at points on the graph of $y = f(x)$, we obtain a graph of $y' = f'(x)$. The slope at point A of the graph of f in part (a) is the y-coordinate of point A' on the graph of f' in part (b), and so on. (Example 3)

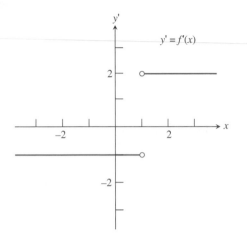

Figure 3.4 The graph of the derivative. (Example 4)

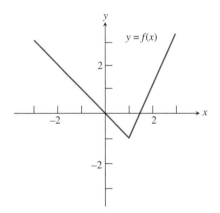

Figure 3.5 The graph of *f*, constructed from the graph of *f′* and two other conditions. (Example 4)

What's happening at *x* = 1?

Notice that *f* is defined at *x* = 1, while *f′* is not. It is the continuity of *f* that enables us to conclude that *f*(1) = −1. Looking at the graph of *f*, can you see why *f′* could not possibly be defined at *x* = 1? We will explore the reason for this in Example 6.

Exploration 1   Reading the Graphs

Suppose that the function *f* in Figure 3.3a on the previous page represents the depth *y* (in inches) of water in a ditch alongside a dirt road as a function of time *x* (in days). How would you answer the following questions?

1. What does the graph in Figure 3.3b represent? What units would you use along the *y′*-axis?

2. Describe as carefully as you can what happened to the water in the ditch over the course of the 7-day period.

3. Can you describe the weather during the 7 days? When was it the wettest? When was it the driest?

4. How does the graph of the derivative help in finding when the weather was wettest or driest?

5. Interpret the significance of point *C* in terms of the water in the ditch. How does the significance of point *C′* reflect that in terms of rate of change?

6. It is tempting to say that it rains right up until the beginning of the second day, but that overlooks a fact about rainwater that is important in flood control. Explain.

Construct your own "real-world" scenario for the function in Example 3, and pose a similar set of questions that could be answered by considering the two graphs in Figure 3.3.

Example 4   GRAPHING *f* FROM *f′*

Sketch the graph of a function *f* that has the following properties:

i. $f(0) = 0$;

ii. the graph of *f′*, the derivative of *f*, is as shown in Figure 3.4;

iii. *f* is continuous for all *x*.

Solution To satisfy property (i), we begin with a point at the origin.

To satisfy property (ii), we consider what the graph of the derivative tells us about slopes. To the left of $x = 1$, the graph of *f* has a constant slope of -1; therefore we draw a line with slope -1 to the left of $x = 1$, making sure that it goes through the origin.

To the right of $x = 1$, the graph of *f* has a constant slope of 2, so it must be a line with slope 2. There are infinitely many such lines but only one— the one that meets the left side of the graph at $(1, -1)$—will satisfy the continuity requirement. The resulting graph is shown in Figure 3.5.

Graphing the Derivative from Data

Discrete points plotted from sets of data do not yield a continuous curve, but we have seen that the shape and pattern of the graphed points (called a scatter plot) can be meaningful nonetheless. It is often possible to fit a curve to the points using regression techniques. If the fit is good, we could use the curve to get a graph of the derivative visually, as in Example 3. However, it is also possible to get a scatter plot of the derivative numerically, directly from the data, by computing the slopes between successive points, as in Example 5.

Example 5 ESTIMATING THE PROBABILITY OF SHARED BIRTHDAYS

Suppose 30 people are in a room. What is the probability that two of them share the same birthday? Ignore the year of birth.

Solution It may surprise you to learn that the probability of a shared birthday among 30 people is at least 0.706, well above two-thirds! In fact, if we assume that no one day is more likely to be a birthday than any other day, the probabilities shown in Table 3.1 are not hard to determine (see Exercise 29).

Table 3.1 Probabilities of Shared Birthdays		Table 3.2 Estimates of Slopes on the Probability Curve	
People in Room (x)	Probability (y)	Midpoint of Interval (x)	Change (slope $\Delta y/\Delta x$)
0	0		
5	0.027	2.5	0.0054
10	0.117	7.5	0.0180
15	0.253	12.5	0.0272
20	0.411	17.5	0.0316
25	0.569	22.5	0.0316
30	0.706	27.5	0.0274
35	0.814	32.5	0.0216
40	0.891	37.5	0.0154
45	0.941	42.5	0.0100
50	0.970	47.5	0.0058
55	0.986	52.5	0.0032
60	0.994	57.5	0.0016
65	0.998	62.5	0.0008
70	0.999	67.5	0.0002

A scatter plot of the data in Table 3.1 is shown in Figure 3.6.

Notice that the probabilities grow slowly at first, then faster, then much more slowly past $x = 45$. At which x are they growing the fastest? To answer the question, we need the graph of the derivative.

Using the data in Table 3.1, we compute the slopes between successive points on the probability plot. For example, from $x = 0$ to $x = 5$ the slope is

$$\frac{0.027 - 0}{5 - 0} = 0.0054.$$

We make a new table showing the slopes, beginning with slope 0.0054 on the interval $[0, 5]$ (Table 3.2). A logical x value to use to represent the interval is its midpoint 2.5.

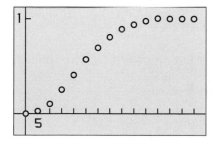

[–5, 75] by [–0.2, 1.1]

Figure 3.6 Scatter plot of the probabilities (y) of shared birthdays among x people, for $x = 0, 5, 10, \ldots, 70$. (Example 5)

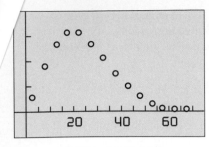

[–5, 75] by [–0.01, 0.04]

Figure 3.7 A scatter plot of the derivative data in Table 3.2. (Example 5)

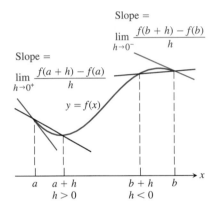

Figure 3.8 Derivatives at endpoints are one-sided limits.

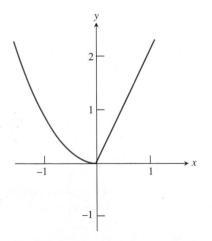

Figure 3.9 A function with different one-sided derivatives at $x = 0$. (Example 6)

A scatter plot of the derivative data in Table 3.2 is shown in Figure 3.7.

From the derivative plot, we can see that the rate of change peaks near $x = 20$. You can impress your friends with your "psychic powers" by predicting a shared birthday in a room of just 25 people (since you will be right about 57% of the time), but the derivative warns you to be cautious: a few less people can make quite a difference. On the other hand, going from 40 people to 100 people will not improve your chances much at all.

Generating shared birthday probabilities: If you know a little about probability, you might try generating the probabilities in Table 3.1. Exploration Exercise 29 at the end of this section shows how to generate them on a calculator.

One-sided Derivatives

A function $y = f(x)$ is **differentiable on a closed interval** $[a, b]$ if it has a derivative at every interior point of the interval, and if the limits

$$\lim_{h \to 0^+} \frac{f(a + h) - f(a)}{h} \quad \text{[the \textbf{right-hand derivative at} } a\text{]}$$

$$\lim_{h \to 0^-} \frac{f(b + h) - f(b)}{h} \quad \text{[the \textbf{left-hand derivative at} } b\text{]}$$

exist at the endpoints. In the right-hand derivative, h is positive and $a + h$ approaches a from the right. In the left-hand derivative, h is negative and $b + h$ approaches b from the left (Figure 3.8).

Right-hand and left-hand derivatives may be defined at any point of a function's domain.

The usual relationship between one-sided and two-sided limits holds for derivatives. Theorem 3, Section 2.1, allows us to conclude that a function has a (two-sided) derivative at a point if and only if the function's right-hand and left-hand derivatives are defined and equal at that point.

Example 6   ONE-SIDED DERIVATIVES CAN DIFFER AT A POINT

Show that the following function has left-hand and right-hand derivatives at $x = 0$, but no derivative there (Figure 3.9).

$$y = \begin{cases} x^2, & x \leq 0 \\ 2x, & x > 0 \end{cases}$$

Solution We verify the existence of the left-hand derivative:

$$\lim_{h \to 0^-} \frac{(0 + h)^2 - 0^2}{h} = \lim_{h \to 0^-} \frac{h^2}{h} = 0.$$

We verify the existence of the right-hand derivative:

$$\lim_{h \to 0^+} \frac{2(0 + h) - 0^2}{h} = \lim_{h \to 0^+} \frac{2h}{h} = 2.$$

Since the left-hand derivative equals zero and the right-hand derivative equals 2, the derivatives are not equal at $x = 0$. The function does not have a derivative at 0.

Quick Review 3.1

In Exercises 1–4, evaluate the indicated limit algebraically.

1. $\lim\limits_{h \to 0} \dfrac{(2 + h)^2 - 4}{h}$

2. $\lim\limits_{x \to 2^+} \dfrac{x + 3}{2}$

3. $\lim\limits_{y \to 0^-} \dfrac{|y|}{y}$

4. $\lim\limits_{x \to 4} \dfrac{2x - 8}{\sqrt{x} - 2}$

5. Find the slope of the line tangent to the parabola $y = x^2 + 1$ at its vertex.

6. By considering the graph of $f(x) = x^3 - 3x^2 + 2$, find the intervals on which f is increasing.

In Exercises 7–10, let

$$f(x) = \begin{cases} x + 2, & x \le 1 \\ (x - 1)^2, & x > 1. \end{cases}$$

7. Find $\lim_{x \to 1^+} f(x)$ and $\lim_{x \to 1^-} f(x)$.

8. Find $\lim_{h \to 0^+} f(1 + h)$.

9. Does $\lim_{x \to 1} f(x)$ exist? Explain.

10. Is f continuous? Explain.

Section 3.1 Exercises

1. If $f(2) = 3$ and $f'(2) = 5$, find an equation of **(a)** the *tangent* line, and **(b)** the *normal* line to the graph of $y = f(x)$ at the point where $x = 2$.

2. Use the definition

$$f'(a) = \lim_{h \to 0} \frac{f(a + h) - f(a)}{h}$$

to find the derivative of $f(x) = (1/x)$ at $x = 3$.

3. Use the definition

$$f'(a) = \lim_{x \to a} \frac{f(x) - f(a)}{x - a}$$

to find the derivative of $f(x) = (1/x)$ at $x = 3$.

4. Find $f'(x)$ if $f(x) = 3x - 12$.

5. Find dy/dx if $y = 7x$.

6. Find $\dfrac{d}{dx}(x^2)$.

In Exercises 7–10, match the graph of the function with the graph of the derivative shown here:

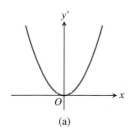

(a)

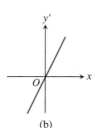

(b)

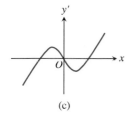

(c)

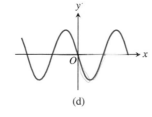

(d)

7.

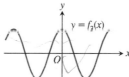

$y = f_1(x)$

8.

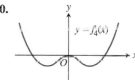

$y = f_2(x)$

9.

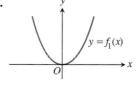

$y = f_3(x)$

10.

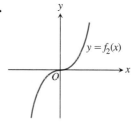

$y = f_4(x)$

11. Find the derivative of the function $y = 2x^2 - 13x + 5$ and use it to find the equation of the line tangent to the curve at $x = 3$.

12. Find the lines that are **(a)** tangent and **(b)** normal to the curve $y = x^3$ at the point $(1, 1)$.

13. Shown below is the graph of $f(x) = x \ln x - x$. From what you know about the graphs of functions (i) through (v), pick out the one that is the *derivative* of f for $x > 0$.

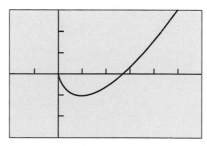

[−2, 6] by [−3, 3]

i. $y = \sin x$ **ii.** $y = \ln x$ **iii.** $y = \sqrt{x}$

iv. $y = x^2$ **v.** $y = 3x - 1$

14. From what you know about the graphs of functions (i) through (v), pick out the one that is *its own derivative*.

i. $y = \sin x$ **ii.** $y = x$ **iii.** $y = \sqrt{x}$

iv. $y = e^x$ **v.** $y = x^2$

15. *Daylight in Fairbanks* The viewing window below shows the number of hours of daylight in Fairbanks, Alaska, on each day for a typical 365-day period from January 1 to December 31. Answer the following questions by estimating slopes on the graph in hours per day. For the purposes of estimation, assume that each month has 30 days.

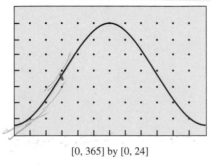

[0, 365] by [0, 24]

(a) On about what date is the amount of daylight increasing at the fastest rate? What is that rate?

(b) Do there appear to be days on which the rate of change in the amount of daylight is zero? If so, which ones?

(c) On what dates is the rate of change in the number of daylight hours positive? negative?

16. *Graphing f' from f* Given the graph of the function f below, sketch a graph of the *derivative* of f.

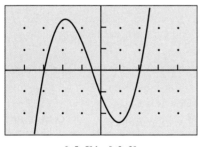

[–5, 5] by [–3, 3]

17. The graphs in Figure 3.10a show the numbers of rabbits and foxes in a small arctic population. They are plotted as functions of time for 200 days. The number of rabbits increases at first, as the rabbits reproduce. But the foxes prey on the rabbits and, as the number of foxes increases, the rabbit population levels off and then drops. Figure 3.10b shows the graph of the derivative of the rabbit population. We made it by plotting slopes, as in Example 3.

(a) What is the value of the derivative of the rabbit population in Figure 3.10 when the number of rabbits is largest? smallest?

(b) What is the size of the rabbit population in Figure 3.10 when its derivative is largest? smallest?

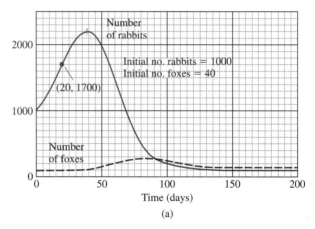

(a)

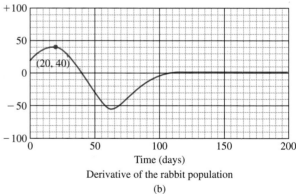

Derivative of the rabbit population

(b)

Figure 3.10 Rabbits and foxes in an arctic predator-prey food chain. *Source: Differentiation* by W. U. Walton et al., Project CALC, Education Development Center, Inc., Newton, MA, 1975, p. 86.

18. The graph of the function $y = f(x)$ shown here is made of line segments joined end to end.

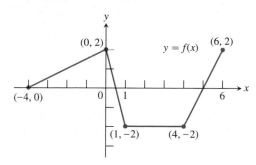

(a) Graph the function's derivative.

(b) At what values of x between $x = -4$ and $x = 6$ is the function not differentiable?

In Exercises 19 and 20, use the data to answer the questions.

19. *A Downhill Skier* Table 3.3 gives the approximate distance traveled by a downhill skier after t seconds for $0 \le t \le 10$. Use the method of Example 5 to sketch a graph of the derivative; then answer the following questions:

(a) What does the derivative represent?

(b) In what units would the derivative be measured?

(c) Can you guess an equation of the derivative by considering its graph?

Table 3.3 Skiing Distances

Time t (seconds)	Distance Traveled (feet)
0	0
1	3.3
2	13.3
3	29.9
4	53.2
5	83.2
6	119.8
7	163.0
8	212.9
9	269.5
10	332.7

20. *A Whitewater River* Bear Creek, a Georgia river known to kayaking enthusiasts, drops more than 770 feet over one stretch of 3.24 miles. By reading a contour map, one can estimate the elevations (y) at various distances (x) downriver from the start of the kayaking route (Table 3.4).

Table 3.4 Elevations along Bear Creek

Distance Downriver (miles)	River Elevation (feet)
0.00	1577
0.56	1512
0.92	1448
1.19	1384
1.30	1319
1.39	1255
1.57	1191
1.74	1126
1.98	1062
2.18	998
2.41	933
2.64	869
3.24	805

(a) Sketch a graph of elevation (y) as a function of distance downriver (x).

(b) Use the technique of Example 5 to get an approximate graph of the derivative, dy/dx.

(c) The average change in elevation over a given distance is called a *gradient*. In this problem, what units of measure would be appropriate for a gradient?

(d) In this problem, what units of measure would be appropriate for the derivative?

(e) How would you identify the most dangerous section of the river (ignoring rocks) by analyzing the graph in (a)? Explain.

(f) How would you identify the most dangerous section of the river by analyzing the graph in (b)? Explain.

21. Writing to Learn Graph $y = \sin x$ and $y = \cos x$ in the same viewing window. Which function could be the derivative of the other? Defend your answer in terms of the behavior of the graphs.

22. Using one-sided derivatives, show that the function

$$f(x) = \begin{cases} x^3, & x \le 1 \\ 3x, & x > 1 \end{cases}$$

does not have a derivative at $x = 1$.

23. In Example 2 of this section we showed that the derivative of $y = \sqrt{x}$ is a function with domain $(0, \infty)$. However, the function $y = \sqrt{x}$ itself has domain $[0, \infty)$, so it could have a right-hand derivative at $x = 0$. Prove that it does not.

24. Writing to Learn Use the concept of the derivative to define what it might mean for two parabolas to be parallel. Construct equations for two such parallel parabolas and graph them. Are the parabolas "everywhere equidistant," and if so, in what sense?

25. *Graphing f from f'* Sketch the graph of a continuous function f with $f(0) = -1$ and

$$f'(x) = \begin{cases} 1, & x < -1 \\ -2, & x > -1. \end{cases}$$

Explorations

26. Let $f(x) = \begin{cases} x^2, & x \le 1 \\ 2x, & x > 1. \end{cases}$

(a) Find $f'(x)$ for $x < 1$.

(b) Find $f'(x)$ for $x > 1$.

(c) Find $\lim_{x \to 1^-} f'(x)$.

(d) Find $\lim_{x \to 1^+} f'(x)$.

(e) Does $\lim_{x \to 1} f'(x)$ exist? Explain.

(f) Use the definition to find the left-hand derivative of f at $x = 1$ if it exists.

(g) Use the definition to find the right-hand derivative of f at $x = 1$ if it exists.

(h) Does $f'(1)$ exist? Explain.

27. *Work in groups of two or three.* Using graphing calculators, have each person in your group do the following:

(a) pick two numbers a and b between 1 and 10;

(b) graph the function $y = (x - a)(x + b)$;

(c) graph the *derivative* of your function (it will be a line with slope 2);

(d) find the y-intercept of your derivative graph.

(e) Compare your answers and determine a simple way to predict the y-intercept, given the values of a and b. Test your result. ■

Extending the Ideas

28. Find the unique value of k that makes the function

$$f(x) = \begin{cases} x^3, & x \le 1 \\ 3x + k, & x > 1 \end{cases}$$

differentiable at $x = 1$.

29. *Generating the Birthday Probabilities* Example 5 of this section concerns the probability that, in a group of n people, at least two people will share a common birthday. You can generate these probabilities on your calculator for values of n from 1 to 365.

Step 1: Set the values of N and P to zero:

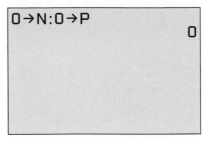

Step 2: Type in this single, multi-step command:

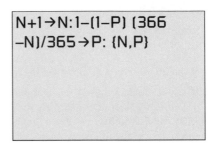

Now each time you press the ENTER key, the command will print a new value of N (the number of people in the room) alongside P (the probability that at least two of them share a common birthday):

```
                        {1  0}
          {2  .002739726}
          {3  .0082041659}
          {4  .0163559125}
          {5  .0271355737}
          {6  .0404624836}
          {7  .0562357031}
```

If you have some experience with probability, try to answer the following questions without looking at the table:

(a) If there are three people in the room, what is the probability that they all have *different* birthdays? (Assume that there are 365 possible birthdays, all of them equally likely.)

(b) If there are three people in the room, what is the probability that at least two of them share a common birthday?

(c) Explain how you can use the answer in (b) to find the probability of a shared birthday when there are *four* people in the room. (This is how the calculator statement in Step 2 generates the probabilities.)

(d) Is it reasonable to assume that all calendar dates are equally likely birthdays? Explain your answer.

3.2 Differentiability

How $f'(a)$ Might Fail to Exist • Differentiability Implies Local Linearity • Derivatives on a Calculator • Differentiability Implies Continuity • Intermediate Value Theorem for Derivatives

How $f'(a)$ Might Fail to Exist

A function will not have a derivative at a point $P(a, f(a))$ where the slopes of the secant lines,

$$\frac{f(x) - f(a)}{x - a},$$

fail to approach a limit as x approaches a. The next figures illustrate four different instances where this occurs. For example, a function whose graph is otherwise smooth will fail to have a derivative at a point where the graph has

1. a *corner,* where the one-sided derivatives differ;

   Example: $f(x) = |x|$

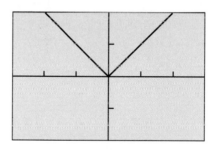

[−3, 3] by [−2, 2]

Figure 3.11 There is a "corner" at $x = 0$.

2. a *cusp,* where the slopes of the secant lines approach ∞ from one side and $-\infty$ from the other (an extreme case of a corner);

   Example: $f(x) = x^{2/3}$

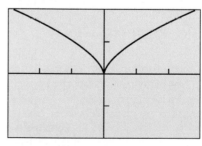

[−3, 3] by [−2, 2]

Figure 3.12 There is a "cusp" at $x = 0$.

3. a *vertical tangent,* where the slopes of the secant lines approach either ∞ or $-\infty$ from both sides (in this example, ∞);

Example: $f(x) = \sqrt[3]{x}$

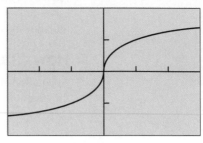

[–3, 3] by [–2, 2]

Figure 3.13 There is a vertical tangent line at $x = 0$.

4. a *discontinuity* (which will cause one or both of the one-sided derivatives to be nonexistent).

Example: The *Unit Step Function*

$$U(x) = \begin{cases} -1, & x < 0 \\ 1, & x \geq 0 \end{cases}$$

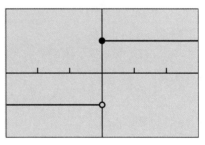

[–3, 3] by [–2, 2]

Figure 3.14 There is a discontinuity at $x = 0$.

In this example, the left-hand derivative fails to exist:

$$\lim_{h \to 0^-} \frac{(-1) - (1)}{h} = \lim_{h \to 0^-} \frac{-2}{h} = \infty.$$

Later in this section we will prove a theorem that states that a function *must* be continuous at a to be differentiable at a. This theorem would provide a quick and easy verification that U is not differentiable at $x = 0$.

Example 1 FINDING WHERE A FUNCTION IS NOT DIFFERENTIABLE

Find all points in the domain of $f(x) = |x - 2| + 3$ where f is not differentiable.

Solution   Think graphically! The graph of this function is the same as that of $y = |x|$, translated 2 units to the right and 3 units up. This puts the corner at the point $(2, 3)$, so this function is not differentiable at $x = 2$.

At every other point, the graph is (locally) a straight line and f has derivative $+1$ or -1 (again, just like $y = |x|$).

How rough can the graph of a continuous function be?

The graph of the absolute value function fails to be differentiable at a single point. If you graph $y = \sin^{-1}(\sin(x))$ on your calculator, you will see a continuous function with an *infinite* number of points of non-differentiability. But can a continuous function fail to be differentiable at *every* point?

The answer, surprisingly enough, is yes, as Karl Weierstrass showed in 1872. One of his formulas (there are many like it) was

$$f(x) = \sum_{n=0}^{\infty} \left(\frac{2}{3}\right)^n \cos(9^n \pi x),$$

a formula that expresses f as an infinite (but converging) sum of cosines with increasingly higher frequencies. By adding wiggles to wiggles infinitely many times, so to speak, the formula produces a function whose graph is too bumpy in the limit to have a tangent anywhere!

Most of the functions we encounter in calculus are differentiable wherever they are defined, which means that they will *not* have corners, cusps, vertical tangent lines, or points of discontinuity within their domains. Their graphs will be unbroken and smooth, with a well-defined slope at each point. Polynomials are differentiable, as are rational functions, trigonometric functions, exponential functions, and logarithmic functions. Composites of differentiable functions are differentiable, and so are sums, products, integer powers, and quotients of differentiable functions, where defined. We will see why all of this is true as the chapter continues.

Differentiability Implies Local Linearity

A good way to think of differentiable functions is that they are **locally linear;** that is, a function that is differentiable at a closely resembles its own tangent line very close to a. In the jargon of graphing calculators, differentiable curves will "straighten out" when we zoom in on them at a point of differentiability. (See Figure 3.15.)

Exploration 1 **Zooming in to "See" Differentiability**

Is either of these functions differentiable at $x = 0$?

(a) $f(x) = |x| + 1$ **(b)** $g(x) = \sqrt{x^2 + 0.0001} + 0.99$

1. We already know that f is not differentiable at $x = 0$; its graph has a corner there. Graph f and zoom in at the point $(0, 1)$ several times. Does the corner show signs of straightening out?

2. Now do the same thing with g. Does the graph of g show signs of straightening out? We will learn a quick way to differentiate g in Section 3.6, but for now suffice it to say that it *is* differentiable at $x = 0$, and in fact has a horizontal tangent there.

3. How many zooms does it take before the graph of g looks exactly like a horizontal line?

4. Now graph f and g *together* in a standard square viewing window. They appear to be identical until you start zooming in. The differentiable function eventually straightens out, while the nondifferentiable function remains impressively unchanged.

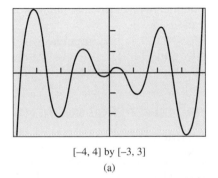

[−4, 4] by [−3, 3]

(a)

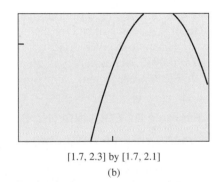

[1.7, 2.3] by [1.7, 2.1]

(b)

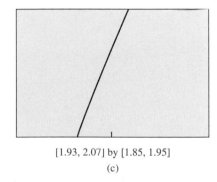

[1.93, 2.07] by [1.85, 1.95]

(c)

Figure 3.15 Three different views of the differentiable function $f(x) = x \cos(3x)$. We have zoomed in here at the point $(2, 1.9)$.

Derivatives on a Calculator

Many graphing utilities can approximate derivatives numerically with good accuracy at most points of their domains.

For small values of h, the difference quotient

$$\frac{f(a + h) - f(a)}{h}$$

is often a good numerical approximation of $f'(a)$. However, as suggested by Figure 3.16, the same value of h will usually yield a *better* approximation if we use the **symmetric difference quotient**

$$\frac{f(a + h) - f(a - h)}{2h},$$

which is what our graphing calculator uses to calculate NDER $f(a)$, the **numerical derivative of f at a point a**. The **numerical derivative of f** as a function is denoted by NDER $f(x)$. Sometimes we will use NDER $(f(x), a)$ for NDER $f(a)$ when we want to emphasize both the function *and* the point.

Although the symmetric difference quotient is not the quotient used in the definition of $f'(a)$, it can be proven that

$$\lim_{h \to 0} \frac{f(a + h) - f(a - h)}{2h}$$

equals $f'(a)$ wherever $f'(a)$ exists.

You might think that an extremely small value of h would be required to give an accurate approximation of $f'(a)$, but in most cases $h = 0.001$ is more than adequate. In fact, your calculator probably assumes such a value for h unless you choose to specify otherwise (consult your *Owner's Manual*). The numerical derivatives we compute in this book will use $h = 0.001$; that is,

$$\text{NDER } f(a) = \frac{f(a + 0.001) - f(a - 0.001)}{0.002}.$$

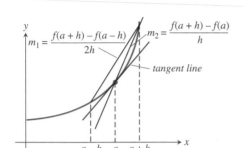

Figure 3.16 The symmetric difference quotient (slope m_1) usually gives a better approximation of the derivative for a given value of h than does the regular difference quotient (slope m_2), which is why the symmetric difference quotient is used in the numerical derivative.

Example 2 COMPUTING A NUMERICAL DERIVATIVE

Compute NDER $(x^3, 2)$, the numerical derivative of x^3 at $x = 2$.

Solution Using $h = 0.001$,

$$\text{NDER } (x^3, 2) = \frac{(2.001)^3 - (1.999)^3}{0.002} = 12.000001.$$

In Example 1 of Section 3.1, we found the derivative of x^3 to be $3x^2$, whose value at $x = 2$ is $3(2)^2 = 12$. The numerical derivative is accurate to 5 decimal places. Not bad for the push of a button.

Example 2 gives dramatic evidence that NDER is very accurate when $h = 0.001$. Such accuracy is usually the case, although it is also possible for NDER to produce some surprisingly inaccurate results, as in Example 3.

Example 3 FOOLING THE SYMMETRIC DIFFERENCE QUOTIENT

Compute NDER $(|x|, 0)$, the numerical derivative of $|x|$ at $x = 0$.

Solution We saw at the start of this section that $|x|$ is not differentiable at $x = 0$ since its right-hand and left-hand derivatives at $x = 0$ are not the same. Nonetheless,

$$\text{NDER} (|x|, 0) = \lim_{h \to 0} \frac{|0 + h| - |0 - h|}{2h}$$

$$= \lim_{h \to 0} \frac{|h| - |h|}{2h}$$

$$= \lim_{h \to 0} \frac{0}{2h}$$

$$= 0.$$

The symmetric difference quotient, which works symmetrically on either side of 0, never detects the corner! Consequently, most graphing utilities will indicate (wrongly) that $y = |x|$ is differentiable at $x = 0$, with derivative 0.

In light of Example 3, it is worth repeating here that NDER $f(a)$ actually does approach $f'(a)$ *when $f'(a)$ exists*, and in fact approximates it quite well (as in Example 2).

An Alternative to NDER

Graphing

$$y = \frac{f(x + 0.001) - f(x - 0.001)}{0.002}$$

is equivalent to graphing $y =$ NDER $f(x)$ (useful if NDER is not readily available on your calculator).

Exploration 2 **Looking at the Symmetric Difference Quotient Analytically**

Let $f(x) = x^2$ and let $h = 0.01$.

1. Find
$$\frac{f(10 + h) - f(10)}{h}.$$
How close is it to $f'(10)$?

2. Find
$$\frac{f(10 + h) - f(10 - h)}{2h}.$$
How close is it to $f'(10)$?

3. Repeat this comparison for $f(x) = x^3$.

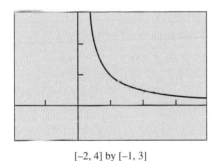

[−2, 4] by [−1, 3]

(a)

X	Y₁	
.1	10	
.2	5	
.3	3.3333	
.4	2.5	
.5	2	
.6	1.6667	
.7	1.4286	
X = .1		

(b)

Figure 3.17 (a) The graph of NDER ln (x) and (b) a table of values. What graph could this be? (Example 4)

Example 4 GRAPHING A DERIVATIVE USING NDER

Let $f(x) = \ln x$. Use NDER to graph $y = f'(x)$. Can you guess what function $f'(x)$ is by analyzing its graph?

Solution The graph is shown in Figure 3.17a. The shape of the graph suggests, and the table of values in Figure 3.17b supports, the conjecture that this is the graph of $y = 1/x$. We will prove in Section 3.9 (using analytic methods) that this is indeed the case.

Differentiability Implies Continuity

We began this section with a look at the typical ways that a function could fail to have a derivative at a point. As one example, we indicated graphically that a discontinuity in the graph of f would cause one or both of the one-sided derivatives to be nonexistent. It is actually not difficult to give an analytic proof that continuity is an essential condition for the derivative to exist, so we include that as a theorem here.

> **Theorem 1 Differentiability Implies Continuity**
>
> If f has a derivative at $x = a$, then f is continuous at $x = a$.

Proof Our task is to show that $\lim_{x \to a} f(x) = f(a)$, or, equivalently, that

$$\lim_{x \to a} [f(x) - f(a)] = 0.$$

Using the Limit Product Rule (and noting that $x - a$ is not zero), we can write

$$\lim_{x \to a} [f(x) - f(a)] = \lim_{x \to a} \left[(x - a) \frac{f(x) - f(a)}{x - a} \right]$$

$$= \lim_{x \to a} (x - a) \cdot \lim_{x \to a} \frac{f(x) - f(a)}{x - a}$$

$$= 0 \cdot f'(a)$$

$$= 0. \qquad \blacksquare$$

The converse of Theorem 1 is false, as we have already seen. A continuous function might have a corner, a cusp, or a vertical tangent line, and hence not be differentiable at a given point.

Intermediate Value Theorem for Derivatives

Not every function can be a derivative. A derivative must have the intermediate value property, as stated in the following theorem (the proof of which can be found in advanced texts).

> **Theorem 2 Intermediate Value Theorem for Derivatives**
>
> If a and b are any two points in an interval on which f is differentiable, then f' takes on every value between $f'(a)$ and $f'(b)$.

Example 5 APPLYING THEOREM 2

Does any function have the Unit Step Function (Figure 3.14) as its derivative?

Solution No. Choose some $a < 0$ and some $b > 0$. Then $U(a) = -1$ and $U(b) = 1$, but U does not take on any value between -1 and 1.

The question of when a function is a derivative of some function is one of the central questions in all of calculus. The answer, found by Newton and Leibniz, would revolutionize the world of mathematics. We will see what that answer is when we reach Chapter 5.

Quick Review 3.2

In Exercises 1–5, tell whether the limit could be used to define $f'(a)$ (assuming that f is differentiable at a).

1. $\lim\limits_{h\to 0} \dfrac{f(a+h) - f(a)}{h}$

2. $\lim\limits_{h\to 0} \dfrac{f(a+h) - f(h)}{h}$

3. $\lim\limits_{x\to a} \dfrac{f(x) - f(a)}{x - a}$

4. $\lim\limits_{x\to a} \dfrac{f(a) - f(x)}{a - x}$

5. $\lim\limits_{h\to 0} \dfrac{f(a+h) + f(a-h)}{h}$

6. Find the domain of the function $y = x^{4/3}$.

7. Find the domain of the function $y = x^{3/4}$.

8. Find the range of the function $y = |x - 2| + 3$.

9. Find the slope of the line $y - 5 = 3.2(x + \pi)$.

10. If $f(x) = 5x$, find
$$\frac{f(3 + 0.001) - f(3 - 0.001)}{0.002}.$$

Section 3.2 Exercises

In Exercises 1–4, compare the right-hand and left-hand derivatives to show that the function is not differentiable at the point P.

1.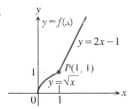

2.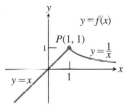

3.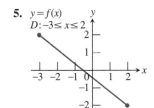

4.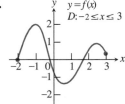

7.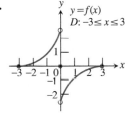

8.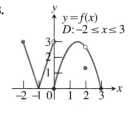

9.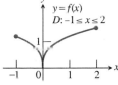

10.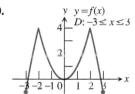

In Exercises 5–10, the graph of a function over a closed interval D is given. At what domain points does the function appear to be

(a) differentiable?

(b) continuous but not differentiable?

(c) neither continuous nor differentiable?

5.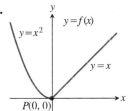

6.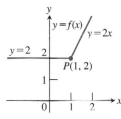

In Exercises 11–16, the function fails to be differentiable at $x = 0$. Tell whether the problem is a corner, a cusp, a vertical tangent, or a discontinuity.

11. $y = \begin{cases} \tan^{-1} x, & x \neq 0 \\ 1, & x = 0 \end{cases}$

12. $y = x^{4/5}$

13. $y = x + \sqrt{x^2} + 2$

14. $y = 3 - \sqrt[3]{x}$

15. $y = 3x - 2|x| - 1$

16. $y = \sqrt[3]{|x|}$

In Exercises 17–22, find all values of x for which the function is differentiable.

17. $f(x) = \dfrac{x^3 - 8}{x^2 - 4x - 5}$

18. $h(x) = \sqrt[3]{3x - 6} + 5$

19. $P(x) = \sin(|x|) - 1$

20. $Q(x) = 3\cos(|x|)$

21. $g(x) = \begin{cases} (x + 1)^2, & x \leq 0 \\ 2x + 1, & 0 < x < 3 \\ (4 - x)^2, & x \geq 3 \end{cases}$

22. $C(x) = x|x|$

23. Writing to Learn Recall that the numerical derivative (NDER) can give meaningless values at points where a function is not differentiable. In this exercise, we consider the numerical derivatives of the functions $1/x$ and $1/x^2$ at $x = 0$.

(a) Explain why neither function is differentiable at $x = 0$.

(b) Find NDER at $x = 0$ for each function.

(c) By analyzing the definition of the symmetric difference quotient, explain why NDER returns wrong responses that are so different from each other for these two functions.

In Exercises 24–28, *work in groups of two or three.* Use NDER to graph the derivative of the function. If possible, identify the derivative function by looking at the graph.

24. $y = 0.25x^4$

25. $y = -\cos x$

26. $y = \dfrac{x|x|}{2}$

27. $y = -\ln |\cos x|$

28. $y = 2x \cos (2x) + 4x \sin x$

29. Let f be the function defined as

$$f(x) = \begin{cases} 3 - x, & x < 1 \\ ax^2 + bx, & x \geq 1 \end{cases}$$

where a and b are constants.

(a) If the function is continuous for all x, what is the relationship between a and b?

(b) Find the unique values for a and b that will make f both continuous and differentiable.

30. Show that the function

$$f(x) = \begin{cases} 0, & -1 \leq x < 0 \\ 1, & 0 \leq x \leq 1 \end{cases}$$

is not the derivative of any function on the interval $-1 \leq x \leq 1$.

Extending the Ideas

31. *Oscillation* There is another way that a function might fail to be differentiable, and that is by *oscillation*. Let

$$f(x) = \begin{cases} x \sin \dfrac{1}{x}, & x \neq 0 \\ 0, & x = 0. \end{cases}$$

(a) Show that f is continuous at $x = 0$.

(b) Show that

$$\frac{f(0 + h) - f(0)}{h} = \sin \frac{1}{h}.$$

(c) Explain why

$$\lim_{h \to 0} \frac{f(0 + h) - f(0)}{h}$$

does not exist.

(d) Does f have either a left-hand or right-hand derivative at $x = 0$?

(e) Now consider the function

$$g(x) = \begin{cases} x^2 \sin \dfrac{1}{x}, & x \neq 0 \\ 0, & x = 0. \end{cases}$$

Use the definition of the derivative to show that g is differentiable at $x = 0$ and that $g'(0) = 0$.

3.3 Rules for Differentiation

Positive Integer Powers, Multiples, Sums, and Differences •
Products and Quotients • Negative Integer Powers of x •
Second and Higher Order Derivatives

Positive Integer Powers, Multiples, Sums, and Differences

The first rule of differentiation is that the derivative of every constant function is the zero function.

Rule 1 Derivative of a Constant Function

If f is the function with the constant value c, then

$$\frac{df}{dx} = \frac{d}{dx}(c) = 0.$$

Proof of Rule 1   If $f(x) = c$ is a function with a constant value c, then

$$\lim_{h \to 0} \frac{f(x + h) - f(x)}{h} = \lim_{h \to 0} \frac{c - c}{h} = \lim_{h \to 0} 0 = 0. \qquad \blacksquare$$

The next rule is a first step toward a rule for differentiating any polynomial.

Rule 2 Power Rule for Positive Integer Powers of x

If n is a positive integer, then

$$\frac{d}{dx}(x^n) = nx^{n-1}.$$

Proof of Rule 2   If $f(x) = x^n$, then $f(x + h) = (x + h)^n$ and the difference quotient for f is

$$\frac{(x + h)^n - x^n}{h}.$$

We can readily find the limit of this quotient as $h \to 0$ if we apply the algebraic identity

$$a^n - b^n = (a - b)(a^{n-1} + a^{n-2}b + \cdots + ab^{n-2} + b^{n-1}) \qquad n \text{ a positive integer}$$

with $a = x + h$ and $b = x$. For then $(a - b) = h$ and the h's in the numerator and denominator of the quotient cancel, giving

$$\frac{f(x + h) - f(x)}{h} = \frac{(x + h)^n - x^n}{h}$$

$$= \frac{h[(x + h)^{n-1} + (x + h)^{n-2}x + \cdots + (x + h)x^{n-2} + x^{n-1}]}{h}$$

$$= \underbrace{(x + h)^{n-1} + (x + h)^{n-2}x + \cdots + (x + h)x^{n-2} + x^{n-1}}.$$

n terms, each with limit x^{n-1} as $h \to 0$

Hence,

$$\frac{d}{dx}(x^n) = \lim_{h \to 0} \frac{f(x + h) - f(x)}{h} = nx^{n-1}. \qquad \blacksquare$$

The Power Rule says: To differentiate x^n, multiply by n and subtract 1 from the exponent. For example, the derivatives of x^2, x^3, and x^4 are $2x^1$, $3x^2$, and $4x^3$, respectively.

Rule 3 The Constant Multiple Rule

If u is a differentiable function of x and c is a constant, then

$$\frac{d}{dx}(cu) = c\frac{du}{dx}.$$

Proof of Rule 3

$$\frac{d}{dx}(cu) = \lim_{h \to 0} \frac{cu(x + h) - cu(x)}{h}$$

$$= c\lim_{h \to 0} \frac{u(x + h) - u(x)}{h} = c\frac{du}{dx} \qquad \blacksquare$$

Denoting Functions by *u* and *v*

The functions we work with when we need a differentiation formula are likely to be denoted by letters like *f* and *g*. When we apply the formula, we do not want to find the formula using these same letters in some other way. To guard against this, we denote the functions in differentiation rules by letters like *u* and *v* that are not likely to be already in use.

Rule 3 says that if a differentiable function is multiplied by a constant, then its derivative is multiplied by the same constant. Combined with Rule 2, it enables us to find the derivative of any monomial quickly; for example, the derivative of $7x^4$ is $7(4x^3) = 28x^3$.

To find the derivatives of polynomials, we need to be able to differentiate sums and differences of monomials. We can accomplish this by applying the Sum and Difference Rule.

Rule 4 The Sum and Difference Rule

If *u* and *v* are differentiable functions of *x*, then their sum and difference are differentiable at every point where *u* and *v* are differentiable. At such points,

$$\frac{d}{dx}(u \pm v) = \frac{du}{dx} \pm \frac{dv}{dx}.$$

Proof of Rule 4

We use the difference quotient for $f(x) = u(x) + v(x)$.

$$\frac{d}{dx}[u(x) + v(x)] = \lim_{h\to 0} \frac{[u(x+h) + v(x+h)] - [u(x) + v(x)]}{h}$$

$$= \lim_{h\to 0} \left[\frac{u(x+h) - u(x)}{h} + \frac{v(x+h) - v(x)}{h}\right]$$

$$= \lim_{h\to 0} \frac{u(x+h) - u(x)}{h} + \lim_{h\to 0} \frac{v(x+h) - v(x)}{h}$$

$$= \frac{du}{dx} + \frac{dv}{dx}$$

The proof of the rule for the difference of two functions is similar. ∎

Example 1 DIFFERENTIATING A POLYNOMIAL

Find $\frac{dp}{dt}$ if $p = t^3 + 6t^2 - \frac{5}{3}t + 16$.

Solution By Rule 4 we can differentiate the polynomial term-by-term, applying Rules 1 through 3 as we go.

$$\frac{dp}{dt} = \frac{d}{dt}(t^3) + \frac{d}{dt}(6t^2) - \frac{d}{dt}\left(\frac{5}{3}t\right) + \frac{d}{dt}(16) \quad \text{Sum and Difference Rule}$$

$$= 3t^2 + 6 \cdot 2t - \frac{5}{3} + 0 \quad \text{Constant and Power Rules}$$

$$= 3t^2 + 12t - \frac{5}{3}$$

Example 2 FINDING HORIZONTAL TANGENTS

Does the curve $y = x^4 - 2x^2 + 2$ have any horizontal tangents? If so, where?

Solution The horizontal tangents, if any, occur where the slope dy/dx is zero. To find these points, we

(a) calculate *dy/dx:*

$$\frac{dy}{dx} = \frac{d}{dx}(x^4 - 2x^2 + 2) = 4x^3 - 4x.$$

(b) solve the equation *dy/dx* = 0 for *x:*

$$4x^3 - 4x = 0$$

$$4x(x^2 - 1) = 0$$

$$x = 0, 1, -1.$$

The curve has horizontal tangents at $x = 0, 1$, and -1. The corresponding points on the curve (found from the equation $y = x^4 - 2x^2 + 2$) are $(0, 2)$, $(1, 1)$, and $(-1, 1)$. You might wish to graph the curve to see where the horizontal tangents go.

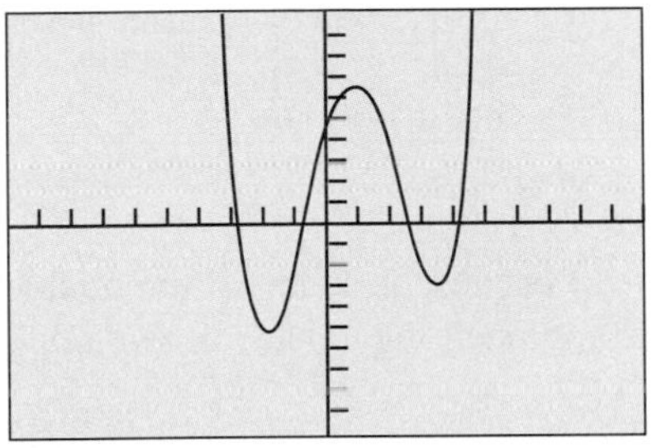

The derivative in Example 2 was easily factored, making an algebraic solution of the equation $dy/dx = 0$ correspondingly simple. When a simple algebraic solution is not possible, the solutions to $dy/dx = 0$ can still be found to a high degree of accuracy by using the SOLVE capability of your calculator.

Example 3 USING CALCULUS AND CALCULATOR

As can be seen in the viewing window $[-10, 10]$ by $[-10, 10]$, the graph of $y = 0.2x^4 - 0.7x^3 - 2x^2 + 5x + 4$ has three horizontal tangents (Figure 3.18). At what points do these horizontal tangents occur?

Solution First we find the derivative

$$\frac{dy}{dx} = 0.8x^3 - 2.1x^2 - 4x + 5.$$

Using the calculator solver, we find that $0.8x^3 - 2.1x^2 - 4x + 5 = 0$ when $x \approx -1.862, 0.9484$, and 3.539. We use the calculator again to evaluate the original function at these x-values and find the corresponding points to be approximately $(-1.862, -5.321)$, $(0.9484, 6.508)$, and $(3.539, -3.008)$.

Products and Quotients

While the derivative of the sum of two functions is the sum of their derivatives and the derivative of the difference of two functions is the difference of their derivatives, the derivative of the product of two functions is *not* the product of their derivatives.

For instance,

$$\frac{d}{dx}(x \cdot x) = \frac{d}{dx}(x^2) = 2x, \quad \text{while} \quad \frac{d}{dx}(x) \cdot \frac{d}{dx}(x) = 1 \cdot 1 = 1.$$

The derivative of a product is actually the sum of *two* products, as we now explain.

[−10, 10] by [−10, 10]

Figure 3.18 The graph of

$$y = 0.2x^4 - 0.7x^3 - 2x^2 + 5x + 4$$

has three horizontal tangents. (Example 3)

On Rounding Calculator Values

Notice in Example 3 that we rounded the *x*-values to four significant digits when we presented the answers. The calculator actually presented many more digits, but there was no practical reason for writing all of them. When we used the calculator to compute the corresponding *y*-values, however, we *used the x-values stored in the calculator,* not the rounded values. We then rounded the *y*-values to four significant digits when we presented the ordered pairs. Significant "round-off errors" can accumulate in a problem if you use rounded intermediate values for doing additional computations, so avoid rounding until the final answer.

You can remember the Product Rule with the phrase "the first times the derivative of the second plus the second times the derivative of the first."

Rule 5 The Product Rule

The product of two differentiable functions u and v is differentiable, and

$$\frac{d}{dx}(uv) = u\frac{dv}{dx} + v\frac{du}{dx}.$$

Gottfried Wilhelm Leibniz (1646–1716)

The method of limits used in this book was not discovered until nearly a century after Newton and Leibniz, the discoverers of calculus, had died.

To Leibniz, the key idea was the *differential,* an infinitely small quantity that was almost like zero, but which—unlike zero—could be used in the denominator of a fraction. Thus, Leibniz thought of the derivative dy/dx as the quotient of two differentials, dy and dx.

The problem was explaining why these differentials sometimes became zero and sometimes did not! See Exercise 40.

Some 17th-century mathematicians were confident that the calculus of Newton and Leibniz would eventually be found to be fatally flawed because of these mysterious quantities. It was only after later generations of mathematicians had found better ways to prove their results that the calculus of Newton and Leibniz was accepted by the entire scientific community.

Proof of Rule 5 We begin, as usual, by applying the definition.

$$\frac{d}{dx}(uv) = \lim_{h \to 0} \frac{u(x+h)v(x+h) - u(x)v(x)}{h}$$

To change the fraction into an equivalent one that contains difference quotients for the derivatives of u and v, we subtract and add $u(x+h)v(x)$ in the numerator. Then,

$$\frac{d}{dx}(uv) = \lim_{h \to 0} \frac{u(x+h)v(x+h) - u(x+h)v(x) + u(x+h)v(x) - u(x)v(x)}{h}$$

$$= \lim_{h \to 0} \left[u(x+h)\frac{v(x+h) - v(x)}{h} + v(x)\frac{u(x+h) - u(x)}{h} \right] \quad \text{Factor and separate.}$$

$$= \lim_{h \to 0} u(x+h) \cdot \lim_{h \to 0} \frac{v(x+h) - v(x)}{h} + v(x) \cdot \lim_{h \to 0} \frac{u(x+h) - u(x)}{h}.$$

As h approaches 0, $u(x+h)$ approaches $u(x)$ because u, being differentiable at x, is continuous at x. The two fractions approach the values of dv/dx and du/dx, respectively, at x. Therefore

$$\frac{d}{dx}(uv) = u\frac{dv}{dx} + v\frac{du}{dx}. \qquad \blacksquare$$

Example 4 DIFFERENTIATING A PRODUCT

Find $f'(x)$ if $f(x) = (x^2 + 1)(x^3 + 3)$.

Solution From the Product Rule with $u = x^2 + 1$ and $v = x^3 + 3$, we find

$$f'(x) = \frac{d}{dx}[(x^2 + 1)(x^3 + 3)] = (x^2 + 1)(3x^2) + (x^3 + 3)(2x)$$

$$= 3x^4 + 3x^2 + 2x^4 + 6x$$

$$= 5x^4 + 3x^2 + 6x.$$

We could also have done Example 4 by multiplying out the original expression and then differentiating the resulting polynomial. That alternate strategy will not work, however, on a product like $x^2 \sin x$, nor in a problem like Example 5.

Example 5 WORKING WITH NUMERICAL VALUES

Let $y = uv$ be the product of the functions u and v. Find $y'(2)$ if

$$u(2) = 3, \quad u'(2) = -4, \quad v(2) = 1, \quad \text{and} \quad v'(2) = 2.$$

Solution From the Product Rule, $y' = (uv)' = uv' + vu'$. In particular,

$$y'(2) = u(2)v'(2) + v(2)u'(2)$$

$$= (3)(2) + (1)(-4)$$

$$= 2.$$

Just as the derivative of the product of two differentiable functions is not the product of their derivatives, the derivative of a quotient of two functions is not the quotient of their derivatives. What happens instead is this:

Using the Quotient Rule

Since order is important in subtraction, be sure to set up the numerator of the Quotient Rule correctly:

> *v* times the derivative of *u*
minus
> *u* times the derivative of *v*.

You can remember the Quotient Rule with the phrase "bottom times the derivative of the top minus the top times the derivative of the bottom, all over the bottom squared."

Rule 6 The Quotient Rule

At a point where $v \neq 0$, the quotient $y = u/v$ of two differentiable functions is differentiable, and

$$\frac{d}{dx}\left(\frac{u}{v}\right) = \frac{v\dfrac{du}{dx} - u\dfrac{dv}{dx}}{v^2}.$$

Proof of Rule 6

$$\frac{d}{dx}\left(\frac{u}{v}\right) = \lim_{h \to 0} \frac{\dfrac{u(x+h)}{v(x+h)} - \dfrac{u(x)}{v(x)}}{h}$$

$$= \lim_{h \to 0} \frac{v(x)u(x+h) - u(x)v(x+h)}{hv(x+h)v(x)}$$

To change the last fraction into an equivalent one that contains the difference quotients for the derivatives of *u* and *v*, we subtract and add $v(x)u(x)$ in the numerator. This allows us to continue with

$$\frac{d}{dx}\left(\frac{u}{v}\right) = \lim_{h \to 0} \frac{v(x)u(x+h) - v(x)u(x) + v(x)u(x) - u(x)v(x+h)}{hv(x+h)v(x)}$$

$$- \lim_{h \to 0} \frac{v(x)\dfrac{u(x+h) - u(x)}{h} - u(x)\dfrac{v(x+h) - v(x)}{h}}{v(x+h)v(x)}.$$

Taking the limits in both the numerator and denominator now gives us the Quotient Rule. ∎

Example 6 SUPPORTING COMPUTATIONS GRAPHICALLY

Differentiate $f(x) = \dfrac{x^2 - 1}{x^2 + 1}$. Support graphically.

Solution We apply the Quotient Rule with $u = x^2 - 1$ and $v = x^2 + 1$:

$$f'(x) = \frac{(x^2+1)\cdot 2x - (x^2-1)\cdot 2x}{(x^2+1)^2} \qquad \frac{v(du/dx) - u(dv/dx)}{v^2}$$

$$= \frac{2x^3 + 2x - 2x^3 + 2x}{(x^2+1)^2}$$

$$= \frac{4x}{(x^2+1)^2}.$$

The graphs of $y_1 = f'(x)$ calculated above and of $y_2 = \text{NDER } f(x)$ are shown in Figure 3.19. The fact that they appear to be identical provides strong graphical support that our calculations are indeed correct.

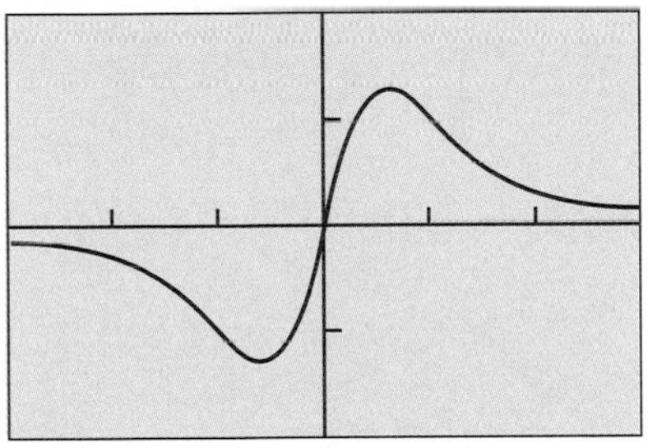

Negative Integer Powers of *x*

The rule for differentiating negative powers of *x* is the same as Rule 2 for differentiating positive powers of *x*, although our proof of Rule 2 does not work for negative values of *n*. We can now extend the Power Rule to negative integer powers by a clever use of the Quotient Rule.

[–3, 3] by [–2, 2]

Figure 3.19 The graph of

$$y = \frac{4x}{(x^2+1)^2}$$

and the graph of

$$y = \text{NDER}\left(\frac{x^2-1}{x^2+1}\right)$$

appear to be the same. (Example 6)

> **Rule 7 Power Rule for Negative Integer Powers of *x***
>
> If *n* is a negative integer and $x \neq 0$, then
>
> $$\frac{d}{dx}(x^n) = nx^{n-1}.$$

Proof of Rule 7 If *n* is a negative integer, then $n = -m$, where *m* is a positive integer. Hence, $x^n = x^{-m} = 1/x^m$, and

$$\frac{d}{dx}(x^n) = \frac{d}{dx}\left(\frac{1}{x^m}\right)$$

$$= \frac{x^m \cdot \dfrac{d}{dx}(1) - 1 \cdot \dfrac{d}{dx}(x^m)}{(x^m)^2}$$

$$= \frac{0 - mx^{m-1}}{x^{2m}}$$

$$= -mx^{-m-1}$$

$$= nx^{n-1}.$$ ■

Example 7 USING THE POWER RULE

Find an equation for the line tangent to the curve

$$y = \frac{x^2 + 3}{2x}$$

at the point (1, 2). Support your answer graphically.

Solution We could find the derivative by the Quotient Rule, but it is easier to first simplify the function as a sum of two powers of *x*.

$$\frac{dy}{dx} = \frac{d}{dx}\left(\frac{x^2}{2x} + \frac{3}{2x}\right)$$

$$= \frac{d}{dx}\left(\frac{1}{2}x + \frac{3}{2}x^{-1}\right)$$

$$= \frac{1}{2} - \frac{3}{2}x^{-2}$$

The slope at $x = 1$ is

$$\frac{dy}{dx}\bigg|_{x=1} = \left[\frac{1}{2} - \frac{3}{2}x^{-2}\right]_{x=1} = \frac{1}{2} - \frac{3}{2} = -1.$$

The line through (1, 2) with slope $m = -1$ is

$$y - 2 = (-1)(x - 1)$$

$$y = -x + 1 + 2$$

$$y = -x + 3.$$

We graph $y = (x^2 + 3)/2x$ and $y = -x + 3$ (Figure 3.20), observing that the line appears to be tangent to the curve at (1, 2). Thus, we have graphical support that our computations are correct.

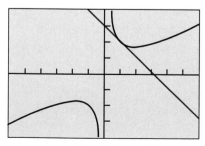

[–6, 6] by [–4, 4]

Figure 3.20 The line $y = -x + 3$ appears to be tangent to the graph of

$$y = \frac{x^2 + 3}{2x}$$

at the point (1, 2). (Example 7)

Second and Higher Order Derivatives

The derivative $y' = dy/dx$ is called the *first derivative* of y with respect to x. The first derivative may itself be a differentiable function of x. If so, its derivative,

$$y'' = \frac{dy'}{dx} = \frac{d}{dx}\left(\frac{dy}{dx}\right) = \frac{d^2y}{dx^2},$$

is called the *second derivative* of y with respect to x. If y'' ("y double-prime") is differentiable, its derivative,

$$y''' = \frac{dy''}{dx} = \frac{d^3y}{dx^3},$$

is called the *third derivative* of y with respect to x. The names continue as you might expect they would, except that the multiple-prime notation begins to lose its usefulness after about three primes. We use

$$y^{(n)} = \frac{d}{dx}y^{(n-1)} \quad \text{"y super n"}$$

to denote the **nth derivative** of y with respect to x. (We also use d^ny/dx^n.) Do not confuse $y^{(n)}$ with the nth power of y, which is y^n.

Technology Tip

HIGHER ORDER DERIVATIVES WITH NDER

Some graphers will allow the *nesting* of the NDER function,

NDER2 f = NDER(NDER f),

but such nesting, in general, is safe only to the second derivative. Beyond that, the error buildup in the algorithm makes the results unreliable.

Example 8 FINDING HIGHER ORDER DERIVATIVES

Find the first four derivatives of $y = x^3 - 5x^2 + 2$.

Solution The first four derivatives are:

First derivative: $y' = 3x^2 - 10x;$

Second derivative: $y'' = 6x - 10;$

Third derivative: $y''' = 6;$

Fourth derivative: $y^{(4)} = 0.$

This function has derivatives of all orders, the fourth and higher order derivatives all being zero.

Quick Review 3.3

In Exercises 1–6, write the expression as a sum of powers of x.

1. $(x^2 - 2)(x^{-1} + 1)$

2. $\left(\dfrac{x}{x^2 + 1}\right)^{-1}$

3. $3x^2 - \dfrac{2}{x} + \dfrac{5}{x^2}$

4. $\dfrac{3x^4 - 2x^3 + 4}{2x^2}$

5. $(x^{-1} + 2)(x^{-2} + 1)$

6. $\dfrac{x^{-1} + x^{-2}}{x^{-3}}$

7. Find the positive roots of the equation

$$2x^3 - 5x^2 - 2x + 6 = 0$$

and evaluate the function $y = 500x^6$ at each root. Round your answers to the nearest integer, but only in the final step.

8. If $f(x) = 7$ for all real numbers x, find

(a) $f(10)$.   (b) $f(0)$.

(c) $f(x + h)$.   (d) $\lim\limits_{x \to a} \dfrac{f(x) - f(a)}{x - a}$.

9. Find the derivatives of these functions with respect to x.

(a) $f(x) = \pi$ (b) $f(x) = \pi^2$ (c) $f(x) = \pi^{15}$

10. Find the derivatives of these functions with respect to x using the definition of the derivative.

(a) $f(x) = \dfrac{x}{\pi}$ (b) $f(x) = \dfrac{\pi}{x}$

Section 3.3 Exercises

In Exercises 1–10, find dy/dx and d^2y/dx^2.

1. $y = -x^2 + 3$

2. $y = \dfrac{x^3}{3} - x$

3. $y = 2x + 1$

4. $y = x^2 + x + 1$

5. $y = \dfrac{x^3}{3} + \dfrac{x^2}{2} + x$

6. $y = 1 - x + x^2 - x^3$

7. $y = x^4 - 7x^3 + 2x^2 + 15$

8. $y = 5x^3 - 3x^5$

9. $y = 4x^{-2} - 8x + 1$

10. $y = \dfrac{x^{-4}}{4} - \dfrac{x^{-3}}{3} + \dfrac{x^{-2}}{2} - x^{-1} + 3$

11. Let $y = (x + 1)(x^2 + 1)$. Find dy/dx **(a)** by applying the Product Rule, and **(b)** by multiplying the factors first and then differentiating.

12. Let $y = (x^2 + 3)/x$. Find dy/dx **(a)** by using the Quotient Rule, and **(b)** by first dividing the terms in the numerator by the denominator and then differentiating.

In Exercises 13–19, find dy/dx.

13. $y = \dfrac{2x + 5}{3x - 2}$

14. $y = \dfrac{x^2 + 5x - 1}{x^2}$

15. $y = \dfrac{(x - 1)(x^2 + x + 1)}{x^3}$

16. $y = (1 - x)(1 + x^2)^{-1}$

17. $y = \dfrac{x^2}{1 - x^3}$

18. $y = \dfrac{\sqrt{x} - 1}{\sqrt{x} + 1}$

19. $y = \dfrac{(x + 1)(x + 2)}{(x - 1)(x - 2)}$

20. Use the definition of derivative (given in Section 3.1, Equation 1) to show that

(a) $\dfrac{d}{dx}(x) = 1.$ **(b)** $\dfrac{d}{dx}(-u) = -\dfrac{du}{dx}.$

21. Use the Product Rule to show that

$$\frac{d}{dx}(c \cdot f(x)) = c \cdot \frac{d}{dx}f(x)$$

for any constant c.

22. Devise a rule for $\dfrac{d}{dx}\left(\dfrac{1}{f(x)}\right)$.

23. Suppose u and v are functions of x that are differentiable at $x = 0$, and that $u(0) = 5$, $u'(0) = -3$, $v(0) = -1$, $v'(0) = 2$. Find the values of the following derivatives at $x = 0$.

(a) $\dfrac{d}{dx}(uv)$ **(b)** $\dfrac{d}{dx}\left(\dfrac{u}{v}\right)$

(c) $\dfrac{d}{dx}\left(\dfrac{v}{u}\right)$ **(d)** $\dfrac{d}{dx}(7v - 2u)$

24. Suppose u and v are functions of x that are differentiable at $x = 2$ and that $u(2) = 3$, $u'(2) = -4$, $v(2) = 1$, and $v'(2) = 2$. Find the values of the following derivatives at $x = 2$.

(a) $\dfrac{d}{dx}(uv)$ **(b)** $\dfrac{d}{dx}\left(\dfrac{u}{v}\right)$

(c) $\dfrac{d}{dx}\left(\dfrac{v}{u}\right)$ **(d)** $\dfrac{d}{dx}(3u - 2v + 2uv)$

25. Which of the following numbers is the slope of the line tangent to the curve $y = x^2 + 5x$ at $x = 3$?

i. 24   **ii.** $-5/2$ **iii.** 11   **iv.** 8

26. Which of the following numbers is the slope of the line $3x - 2y + 12 = 0$?

i. 6   **ii.** 3   **iii.** $3/2$ **iv.** $2/3$

In Exercises 27–32, support your answer graphically.

27. Find an equation of the line perpendicular to the tangent to the curve $y = x^3 - 3x + 1$ at the point $(2, 3)$.

28. Find the tangents to the curve $y = x^3 + x$ at the points where the slope is 4. What is the smallest slope of the curve? At what value of x does the curve have this slope?

29. Find the points on the curve $y = 2x^3 - 3x^2 - 12x + 20$ where the tangent is parallel to the x-axis.

30. Find the x- and y-intercepts of the line that is tangent to the curve $y = x^3$ at the point $(-2, -8)$.

31. Find the tangents to *Newton's serpentine,*

$$y = \frac{4x}{x^2 + 1},$$

at the origin and the point $(1, 2)$.

32. Find the tangent to the *witch of Agnesi,*

$$y = \frac{8}{4 + x^2},$$

at the point $(2, 1)$.

When we work with functions of a single variable in mathematics, we often call the independent variable x and the dependent variable y. Applied fields use many different letters, however. Here are some examples.

33. *Cylinder Pressure* If gas in a cylinder is maintained at a constant temperature T, the pressure P is related to the volume V by a formula of the form

$$P = \frac{nRT}{V - nb} - \frac{an^2}{V^2},$$

in which a, b, n, and R are constants. Find dP/dV.

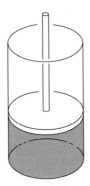

34. *Free Fall* When a rock falls from rest near the surface of the earth, the distance it covers during the first few seconds is given by the equation

$$s = 4.9t^2.$$

In this equation, s is the distance in meters and t is the elapsed time in seconds. Find ds/dt and d^2s/dt^2.

In Exercises 35–39, *work in groups of two or three* to solve the problem.

35. *The Body's Reaction to Medicine* The reaction of the body to a dose of medicine can often be represented by an equation of the form

$$R = M^2 \left(\frac{C}{2} - \frac{M}{3} \right),$$

where C is a positive constant and M is the amount of medicine absorbed in the blood. If the reaction is a change in blood pressure, R is measured in millimeters of mercury. If the reaction is a change in temperature, R is measured in degrees, and so on.

Find dR/dM. This derivative, as a function of M, is called the sensitivity of the body to medicine. In Chapter 4, we shall see how to find the amount of medicine to which the body is most sensitive. *Source: Some Mathematical Models in Biology,* Revised Edition, December 1967, PB-202 364, p. 221; distributed by N.T.I.S., U.S. Department of Commerce.

36. **Writing to Learn** Recall that the area A of a circle with radius r is πr^2 and that the circumference C is $2\pi r$. Notice that $dA/dr = C$. Explain in terms of geometry why the instantaneous rate of change of the area with respect to the radius should equal the circumference.

37. **Writing to Learn** Recall that the volume V of a sphere of radius r is $(4/3)\pi r^3$ and that the surface area A is $4\pi r^2$. Notice that $dV/dr = A$. Explain in terms of geometry why the instantaneous rate of change of the volume with respect to the radius should equal the surface area.

38. *Orchard Farming* An apple farmer currently has 156 trees yielding an average of 12 bushels of apples per tree. He is expanding his farm at a rate of 13 trees per year, while improved husbandry is improving his average annual yield by 1.5 bushels per tree. What is the current (instantaneous) rate of increase of his total annual production of apples? Answer in appropriate units of measure.

39. *Picnic Pavilion Rental* The members of the Blue Boar society always divide the pavilion rental fee for their picnics equally among the members. Currently there are 65 members and the pavilion rents for $250. The pavilion cost is increasing at a rate of $10 per year, while the Blue Boar membership is increasing at a rate of 6 members per year. What is the current (instantaneous) rate of change in each member's share of the pavilion rental fee? Answer in appropriate units of measure.

Extending the Ideas

40. *Leibniz's Proof of the Product Rule* Here's how Leibniz explained the Product Rule in a letter to his colleague John Wallis:

It is useful to consider quantities infinitely small such that when their ratio is sought, they may not be considered zero, but which are rejected as often as they occur with quantities incomparably greater. Thus if we have $x + dx$, dx is rejected. Similarly we cannot have xdx and $dxdx$ standing together, as xdx is incomparably greater than $dxdx$. Hence if we are to differentiate uv, we write

$$d(uv) = (u + du)(v + dv) - uv$$
$$= uv + vdu + udv + dudv - uv$$
$$= vdu + udv.$$

Answer the following questions about Leibniz's proof.

(a) What does Leibniz mean by a quantity being "rejected"?

(b) What happened to $dudv$ in the last step of Leibniz's proof?

(c) Divide both sides of Leibniz's formula

$$[d(uv) = vdu + udv]$$

by the differential dx. What formula results?

(d) Why would the critics of Leibniz's time have objected to dividing both sides of the equation by dx?

(e) Leibniz had a similar simple (but not-so-clean) proof of the Quotient Rule. Can you reconstruct it?

3.4 Velocity and Other Rates of Change

Instantaneous Rates of Change • Motion along a Line • Sensitivity to Change • Derivatives in Economics

Instantaneous Rates of Change

In this section we examine some applications in which derivatives as functions are used to represent the rates at which things change in the world around us. It is natural to think of change as change with respect to time, but other variables can be treated in the same way. For example, a physician may want to know how change in dosage affects the body's response to a drug. An economist may want to study how the cost of producing steel varies with the number of tons produced.

If we interpret the difference quotient

$$\frac{f(x + h) - f(x)}{h}$$

as the average rate of change of the function f over the interval from x to $x + h$, we can interpret its limit as h approaches 0 to be the rate at which f is changing at the point x.

Definition Instantaneous Rate of Change

The **(instantaneous) rate of change** of f with respect to x at a is the derivative

$$f'(a) = \lim_{h \to 0} \frac{f(a + h) - f(a)}{h},$$

provided the limit exists.

It is conventional to use the word *instantaneous* even when x does not represent time. The word, however, is frequently omitted in practice. When we say *rate of change,* we mean *instantaneous rate of change.*

Example 1 ENLARGING CIRCLES

(a) Find the rate of change of the area A of a circle with respect to its radius r.

(b) Evaluate the rate of change of A at $r = 5$ and at $r = 10$.

(c) If r is measured in inches and A is measured in square inches, what units would be appropriate for dA/dr?

Solution The area of a circle is related to its radius by the equation $A = \pi r^2$.

(a) The (instantaneous) rate of change of A with respect to r is

$$\frac{dA}{dr} = \frac{d}{dr}(\pi r^2) = \pi \cdot 2r = 2\pi r.$$

(b) At $r = 5$, the rate is 10π (about 31.4). At $r = 10$, the rate is 20π (about 62.8).

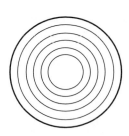

Figure 3.21 The same change in radius brings about a larger change in area as the circles grow radially away from the center. (Example 1, Exploration 1)

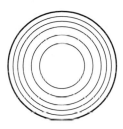

Figure 3.22 Which is the more appropriate model for the growth of rings in a tree—the circles here or those in Figure 3.21? (Exploration 1)

Notice that the rate of change gets bigger as r gets bigger. As can be seen in Figure 3.21, the same change in radius brings about a bigger change in area as the circles grow radially away from the center.

(c) The appropriate units for dA/dr are square inches (of area) per inch (of radius).

Exploration 1 **Growth Rings on a Tree**

The phenomenon observed in Example 1, that the rate of change in area of a circle with respect to its radius gets larger as the radius gets larger, is reflected in nature in many ways. When trees grow, they add layers of wood directly under the inner bark during the growing season, then form a darker, protective layer for protection during the winter. This results in concentric rings that can be seen in a cross-sectional slice of the trunk. The age of the tree can be determined by counting the rings.

1. Look at the concentric rings in Figure 3.21 and Figure 3.22. Which is a better model for the pattern of growth rings in a tree? Is it likely that a tree could find the nutrients and light necessary to increase its amount of growth every year?

2. Considering how trees grow, explain why the change in *area* of the rings remains relatively constant from year to year.

3. If the change in area is constant, and if

$$\frac{dA}{dr} = \frac{\text{change in area}}{\text{change in radius}} = 2\pi r,$$

explain why the change in radius must get smaller as r gets bigger.

Motion along a Line

Suppose that an object is moving along a coordinate line (say an s-axis) so that we know its position s on that line as a function of time t:

$$s = f(t).$$

The **displacement** of the object over the time interval from t to $t + \Delta t$ is

$$\Delta s = f(t + \Delta t) - f(t)$$

Position at time t …   and at time $t + \Delta t$

$s = f(t)$ $s + \Delta s = f(t + \Delta t)$

Figure 3.23 The positions of an object moving along a coordinate line at time t and shortly later at time $t + \Delta t$.

(Figure 3.23) and the **average velocity** of the object over that time interval is

$$v_{av} = \frac{\text{displacement}}{\text{travel time}} = \frac{\Delta s}{\Delta t} = \frac{f(t + \Delta t) - f(t)}{\Delta t}.$$

To find the object's velocity at the exact instant t, we take the limit of the average velocity over the interval from t to $t + \Delta t$ as Δt shrinks to zero. The limit is the derivative of f with respect to t.

> ### Definition Instantaneous Velocity
>
> The **(instantaneous) velocity** is the derivative of the position function $s = f(t)$ with respect to time. At time t the velocity is
>
> $$v(t) = \frac{ds}{dt} = \lim_{\Delta t \to 0} \frac{f(t + \Delta t) - f(t)}{\Delta t}.$$

Example 2 FINDING THE VELOCITY OF AN AUTOMOBILE

Figure 3.24 shows the time-to-distance graph of a 1996 Riley & Scott Mk III-Olds WSC race car. The slope of the secant PQ is the average velocity for the 3-second interval from $t = 2$ to $t = 5$ sec, in this case, about 100 ft/sec or 68 mph. The slope of the tangent at P is the speedometer reading at $t = 2$ sec, about 57 ft/sec or 39 mph. The acceleration for the period shown is a nearly constant 28.5 ft/sec during each second, which is about $0.89g$ where g is the acceleration due to gravity. The race car's top speed is an estimated 190 mph. *Source: Road and Track,* March 1997.

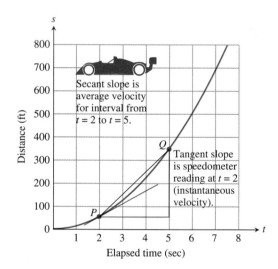

Figure 3.24 The time-to-distance graph for Example 2.

Besides telling how fast an object is moving, velocity tells the direction of motion. When the object is moving forward (when s is increasing), the velocity is positive; when the object is moving backward (when s is decreasing), the velocity is negative.

If we drive to a friend's house and back at 30 mph, the speedometer will show 30 on the way over but will not show -30 on the way back, even though our distance from home is decreasing. The speedometer always shows *speed,* which is the absolute value of velocity. Speed measures the rate of motion regardless of direction.

> ### Definition Speed
>
> **Speed** is the absolute value of velocity.
>
> $$\text{Speed} = |v(t)| = \left| \frac{ds}{dt} \right|$$

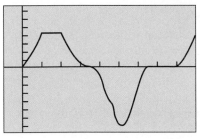

[–4, 36] by [–7.5, 7.5]

Figure 3.25 A student's velocity graph from data recorded by a motion detector. (Example 3)

Example 3 READING A VELOCITY GRAPH

A student walks around in front of a motion detector that records her velocity at 1-second intervals for 36 seconds. She stores the data in her graphing calculator and uses it to generate the time-velocity graph shown in Figure 3.25. Describe her motion as a function of time by reading the velocity graph. When is her *speed* a maximum?

Solution The student moves forward for the first 14 seconds, moves backward for the next 12 seconds, stands still for 6 seconds, and then moves forward again. She achieves her maximum speed at $t \approx 20$, while moving backward.

The rate at which a body's velocity changes is called the body's *acceleration*. The acceleration measures how quickly the body picks up or loses speed.

Definition Acceleration

Acceleration is the derivative of velocity with respect to time. If a body's velocity at time t is $v(t) = ds/dt$, then the body's acceleration at time t is

$$a(t) = \frac{dv}{dt} = \frac{d^2s}{dt^2}.$$

The earliest questions that motivated the discovery of calculus were concerned with velocity and acceleration, particularly the motion of freely falling bodies under the force of gravity. (See Examples 1 and 2 in Section 2.1.) The mathematical description of this type of motion captured the imagination of many great scientists, including Aristotle, Galileo, and Newton. Experimental and theoretical investigations revealed that the distance a body released from rest falls freely is proportional to the square of the amount of time it has fallen. We express this by saying that

$$s = \frac{1}{2}gt^2,$$

where s is distance, g is the acceleration due to Earth's gravity, and t is time. The value of g in the equation depends on the units used to measure s and t. With t in seconds (the usual unit), we have the following values:

Free-fall Constants (Earth)

English units: $g = 32\dfrac{\text{ft}}{\text{sec}^2},$ $s = \dfrac{1}{2}(32)t^2 = 16t^2$ (*s* in feet)

Metric units: $g = 9.8\dfrac{\text{m}}{\text{sec}^2},$ $s = \dfrac{1}{2}(9.8)t^2 = 4.9t^2$ (*s* in meters)

The abbreviation ft/sec^2 is read "feet per second squared" or "feet per second per second," and m/sec^2 is read "meters per second squared."

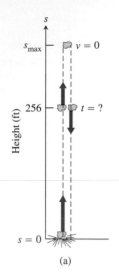

(a)

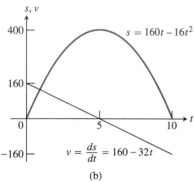

(b)

Figure 3.26 (a) The rock in Example 4. (b) The graphs of s and v as functions of time t, showing that s is largest when $v = ds/dt = 0$. (The graph of s is *not* the path of the rock; it is a plot of height as a function of time.) (Example 4)

Example 4 MODELING VERTICAL MOTION

A dynamite blast propels a heavy rock straight up with a launch velocity of 160 ft/sec (about 109 mph) (Figure 3.26a). It reaches a height of $s = 160t - 16t^2$ ft after t seconds.

(a) How high does the rock go?

(b) What is the velocity and speed of the rock when it is 256 ft above the ground on the way up? on the way down?

(c) What is the acceleration of the rock at any time t during its flight (after the blast)?

(d) When does the rock hit the ground?

Solution In the coordinate system we have chosen, s measures height from the ground up, so velocity is positive on the way up and negative on the way down.

(a) The instant when the rock is at its highest point is the one instant during the flight when the velocity is 0. At any time t, the velocity is

$$v = \frac{ds}{dt} = \frac{d}{dt}(160t - 16t^2) = 160 - 32t \text{ ft/sec.}$$

The velocity is zero when $160 - 32t = 0$, or at $t = 5$ sec.

The maximum height is the height of the rock at $t = 5$ sec. That is,

$$s_{max} = s(5) = 160(5) - 16(5)^2 = 400 \text{ ft.}$$

See Figure 3.26b.

(b) To find the velocity when the height is 256 ft, we determine the two values of t for which $s(t) = 256$ ft.

$$s(t) = 160t - 16t^2 = 256$$
$$16t^2 - 160t + 256 = 0$$
$$16(t^2 - 10t + 16) = 0$$
$$(t - 2)(t - 8) = 0$$
$$t = 2 \text{ sec} \quad \text{or} \quad t = 8 \text{ sec}$$

The velocity of the rock at each of these times is

$$v(2) = 160 - 32(2) = 96 \text{ ft/sec,}$$
$$v(8) = 160 - 32(8) = -96 \text{ ft/sec.}$$

At both instants, the speed of the rock is 96 ft/sec.

(c) At any time during its flight after the explosion, the rock's acceleration is

$$a = \frac{dv}{dt} = \frac{d}{dt}(160 - 32t) = -32 \text{ ft/sec}^2.$$

The acceleration is always downward. When the rock is rising, it is slowing down; when it is falling, it is speeding up.

(d) The rock hits the ground at the positive time for which $s = 0$. The equation $160t - 16t^2 = 0$ has two solutions: $t = 0$ and $t = 10$. The blast initiated the flight of the rock from ground level at $t = 0$. The rock returned to the ground 10 seconds later.

Exploration 2 **Modeling Horizontal Motion**

The position (*x*-coordinate) of a particle moving on the horizontal line $y = 2$ is given by $x(t) = 4t^3 - 16t^2 + 15t$ for $t \geq 0$.

1. Graph the parametric equations $x_1(t) = 4t^3 - 16t^2 + 15t$, $y_1(t) = 2$ in $[-4, 6]$ by $[-3, 5]$. Use TRACE to support that the particle starts at the point $(0, 2)$, moves to the right, then to the left, and finally to the right. At what times does the particle reverse direction?

2. Graph the parametric equations $x_2(t) = x_1(t)$, $y_2(t) = t$ in the same viewing window. Explain how this graph shows the back and forth motion of the particle. Use this graph to find when the particle reverses direction.

3. Graph the parametric equations $x_3(t) = t$, $y_3(t) = x_1(t)$ in the same viewing window. Explain how this graph shows the back and forth motion of the particle. Use this graph to find when the particle reverses direction.

4. Use the methods in parts 1, 2, and 3 to represent and describe the *velocity* of the particle.

Exploration 3 **Seeing Motion on a Graphing Calculator**

The graphs in Figure 3.26b give us plenty of information about the flight of the rock in Example 4, but neither graph shows the path of the rock in flight. We can simulate the moving rock by graphing the parametric equations

$$x_1(t) = 3(t < 5) + 3.1(t \geq 5), \qquad y_1(t) = 160t - 16t^2$$

in dot mode.

This will show the upward flight of the rock along the vertical line $x = 3$, and the downward flight of the rock along the line $x = 3.1$.

1. To see the flight of the rock from beginning to end, what should we use for *t*Min and *t*Max in our graphing window?

2. Set xMin $= 0$, xMax $= 6$, and yMin $= -10$. Use the results from Example 4 to determine an appropriate value for yMax. (You will want the entire flight of the rock to fit within the vertical range of the screen.)

3. Set *t*Step initially at 0.1. (A higher number will make the simulation move faster. A lower number will slow it down.)

4. Can you explain why the grapher actually slows down when the rock would slow down, and speeds up when the rock would speed up?

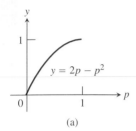

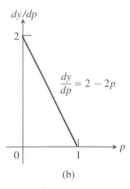

Figure 3.27 (a) The graph of $y = 2p - p^2$ describing the proportion of smooth-skinned peas. (b) The graph of dy/dp. (Example 5)

Sensitivity to Change

When a small change in x produces a large change in the value of a function $f(x)$, we say that the function is relatively **sensitive** to changes in x. The derivative $f'(x)$ is a measure of this sensitivity.

Example 5   GENETIC DATA AND SENSITIVITY TO CHANGE

The Austrian monk Gregor Johann Mendel (1822–1884), working with garden peas and other plants, provided the first scientific explanation of hybridization. His careful records showed that if p (a number between 0 and 1) is the relative frequency of the gene for smooth skin in peas (dominant) and $(1 - p)$ is the relative frequency of the gene for wrinkled skin in peas (recessive), then the proportion of smooth-skinned peas in the next generation will be

$$y = 2p(1 - p) + p^2 = 2p - p^2.$$

The graph of y versus p in Figure 3.27a suggests that the value of y is more sensitive to a change in p when p is small than it is to a change in p when p is large. Indeed, this is borne out by the derivative graph in Figure 3.27b, which shows that dy/dp is close to 2 when p is near 0 and close to 0 when p is near 1.

The implication for genetics is that introducing a few more dominant genes into a highly recessive population will have a more dramatic effect on later generations than will a similar increase in a highly dominant population.

Derivatives in Economics

Engineers use the terms *velocity* and *acceleration* to refer to the derivatives of functions describing motion. Economists, too, have a specialized vocabulary for rates of change and derivatives. They call them *marginals*.

In a manufacturing operation, *the cost of production $c(x)$ is a function of x,* the number of units produced. The *marginal cost of production* is the rate of change of cost with respect to the level of production, so it is dc/dx.

Suppose $c(x)$ represents the dollars needed to produce x tons of steel in one week. It costs more to produce $x + h$ tons per week, and the cost difference divided by h is the average cost of producing each additional ton.

$$\frac{c(x + h) - c(x)}{h} = \begin{cases} \text{the average cost of each of the} \\ \text{additional } h \text{ tons produced} \end{cases}$$

The limit of this ratio as $h \to 0$ is the **marginal cost** of producing more steel per week when the current production is x tons (Figure 3.28).

$$\frac{dc}{dx} = \lim_{h \to 0} \frac{c(x + h) - c(x)}{h} = \text{marginal cost of production}$$

Sometimes the marginal cost of production is loosely defined to be the extra cost of producing one more unit,

$$\frac{\Delta c}{\Delta x} = \frac{c(x + 1) - c(x)}{1},$$

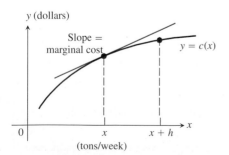

Figure 3.28 Weekly steel production: $c(x)$ is the cost of producing x tons per week. The cost of producing an additional h tons per week is $c(x + h) - c(x)$.

which is approximated by the value of dc/dx at x. This approximation is acceptable if the slope of c does not change quickly near x, for then the difference quotient is close to its limit dc/dx even if $\Delta x = 1$ (Figure 3.29). The approximation works best for large values of x.

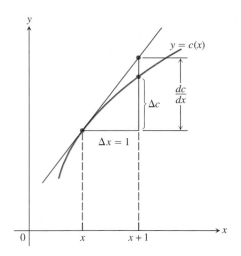

Figure 3.29 Because dc/dx is the slope of the tangent at x, the marginal cost dc/dx approximates the extra cost Δc of producing $\Delta x = 1$ more unit.

Example 6   MARGINAL COST AND MARGINAL REVENUE

Suppose it costs

$$c(x) = x^3 - 6x^2 + 15x$$

dollars to produce x radiators when 8 to 10 radiators are produced, and that

$$r(x) = x^3 - 3x^2 + 12x$$

gives the dollar revenue from selling x radiators. Your shop currently produces 10 radiators a day. Find the marginal cost and **marginal revenue.**

Solution The marginal cost of producing one more radiator a day when 10 are being produced is $c'(10)$.

$$c'(x) = \frac{d}{dx}(x^3 - 6x^2 + 15x) = 3x^2 - 12x + 15$$

$$c'(10) = 3(100) - 12(10) + 15 = 195 \text{ dollars}$$

The marginal revenue is

$$r'(x) = \frac{d}{dx}(x^3 - 3x^2 + 12x) = 3x^2 - 6x + 12,$$

so,

$$r'(10) = 3(100) - 6(10) + 12 = 252 \text{ dollars.}$$

Quick Review 3.4

In Exercises 1–10, answer the questions about the graph of the quadratic function $y = f(x) = -16x^2 + 160x - 256$ by analyzing the equation algebraically. Then support your answers graphically.

1. Does the graph open upward or downward?

2. What is the y-intercept?

3. What are the x-intercepts?

4. What is the range of the function?

5. What point is the vertex of the parabola?

6. At what x-values does $f(x) = 80$?

7. For what x-value does $dy/dx = 100$?

8. On what interval is $dy/dx > 0$?

9. Find $\lim_{h \to 0} \dfrac{f(3 + h) - f(3)}{h}$.

10. Find d^2y/dx^2 at $x = 7$.

Section 3.4 Exercises

1. Find the instantaneous rate of change of the volume V of a cube with respect to a side s.

2. *Particle Motion* A particle moves along a line so that its position at any time $t \geq 0$ is given by the function

$$s(t) = t^2 - 3t + 2,$$

where s is measured in meters and t is measured in seconds.

(a) Find the displacement during the first 5 seconds.

(b) Find the average velocity during the first 5 seconds.

(c) Find the instantaneous velocity when $t = 4$.

(d) Find the acceleration of the particle when $t = 4$.

(e) At what values of t does the particle change direction?

(f) Where is the particle when s is a minimum?

3. *Lunar Projectile Motion* A rock thrown vertically upward from the surface of the moon at a velocity of 24 m/sec (about 86 km/h) reaches a height of $s = 24t - 0.8t^2$ meters in t seconds.

(a) Find the rock's velocity and acceleration as functions of time. (The acceleration in this case is the acceleration of gravity on the moon.)

(b) How long did it take the rock to reach its highest point?

(c) How high did the rock go?

(d) When did the rock reach half its maximum height?

(e) How long was the rock aloft?

4. *Free Fall* The equations for free fall near the surfaces of Mars and Jupiter (s in meters, t in seconds) are: Mars, $s = 1.86t^2$; Jupiter, $s = 11.44t^2$. How long would it take a rock falling from rest to reach a velocity of 16.6 m/sec (about 60 km/h) on each planet?

5. *Projectile Motion* On Earth, in the absence of air, the rock in Exercise 3 would reach a height of $s = 24t - 4.9t^2$ meters in t seconds. How high would the rock go?

6. *Speeding Bullet* A bullet fired straight up from the moon surface would reach a height of $s = 832t - 2.6t^2$ ft after t sec. On Earth, in the absence of air, its height would be $s = 832t - 16t^2$ ft after t sec. How long would it take the bullet to get back down in each case?

7. *Parametric Graphing* Devise a grapher simulation of the problem situation in Exercise 6. Use it to support the answers obtained analytically.

8. *Bacterium Population* When a bactericide was added to a nutrient broth in which bacteria were growing, the bacterium population continued to grow for a while but then stopped growing and began to decline. The size of the population at time t (hours) was $b(t) = 10^6 + 10^4 t - 10^3 t^2$. Find the growth rates at $t = 0$, $t = 5$, and $t = 10$ hours.

9. *Draining a Tank* The number of gallons of water in a tank t minutes after the tank has started to drain is $Q(t) = 200(30 - t)^2$. How fast is the water running out at the end of 10 min? What is the average rate at which the water flows out during the first 10 min?

10. *Marginal Cost* Suppose that the dollar cost of producing x washing machines is $c(x) = 2000 + 100x - 0.1x^2$.

(a) Find the average cost of producing 100 washing machines.

(b) Find the marginal cost when 100 machines are produced.

(c) Show that the marginal cost when 100 washing machines are produced is approximately the cost of producing one more washing machine after the first 100 have been made, by calculating the latter cost directly.

11. *Marginal Revenue* Suppose the weekly revenue in dollars from selling x custom-made office desks is

$$r(x) = 2000\left(1 - \frac{1}{x + 1}\right).$$

(a) Draw the graph of r. What values of x make sense in this problem situation?

(b) Find the marginal revenue when x desks are sold.

(c) Use the function $r'(x)$ to estimate the increase in revenue that will result from increasing sales from 5 desks a week to 6 desks a week.

(d) **Writing to Learn** Find the limit of $r'(x)$ as $x \to \infty$. How would you interpret this number?

12. *Particle Motion* The position of a body at time t sec is $s = t^3 - 6t^2 + 9t$ m. Find the body's acceleration each time the velocity is zero.

13. *Finding Speed* A body's velocity at time t sec is $v = 2t^3 - 9t^2 + 12t - 5$ m/sec. Find the body's speed each time the acceleration is zero.

14. *Finding f from f'* Let $f'(x) = 3x^2$.

(a) Compute the derivatives of $g(x) = x^3$, $h(x) = x^3 - 2$, and $t(x) = x^3 + 3$.

(b) Graph the numerical derivatives of g, h, and t.

(c) Describe a *family* of functions, $f(x)$, that have the property that $f'(x) = 3x^2$.

(d) Is there a function f such that $f'(x) = 3x^2$ and $f(0) = 0$? If so, what is it?

(e) Is there a function f such that $f'(x) = 3x^2$ and $f(0) = 3$? If so, what is it?

15. *Finding Profit* The monthly profit (in thousands of dollars) of a software company is given by

$$P(x) = \frac{10}{1 + 50 \cdot 2^{5-0.1x}},$$

where x is the number of software packages sold.

(a) Graph $P(x)$.

(b) What values of x make sense in the problem situation?

(c) Use NDER to graph $P'(x)$. For what values of x is P relatively sensitive to changes in x?

(d) What is the profit when the marginal profit is greatest?

(e) What is the marginal profit when 50 units are sold? 100 units, 125 units, 150 units, 175 units, and 300 units?

(f) What is $\lim_{x \to \infty} P(x)$? What is the maximum profit possible?

(g) **Writing to Learn** Is there a practical explanation to the maximum profit answer? Explain your reasoning.

Explorations

16. *Launching a Rocket* When a model rocket is launched, the propellant burns for a few seconds, accelerating the rocket upward. After burnout, the rocket coasts upward for a while and then begins to fall. A small explosive charge pops out a parachute shortly after the rocket starts downward. The parachute slows the rocket to keep it from breaking when it lands. This graph shows velocity data from the flight.

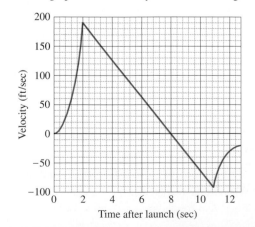

Use the graph to answer the following.

(a) How fast was the rocket climbing when the engine stopped?

(b) For how many seconds did the engine burn?

(c) When did the rocket reach its highest point? What was its velocity then?

(d) When did the parachute pop out? How fast was the rocket falling then?

(e) How long did the rocket fall before the parachute opened?

(f) When was the rocket's acceleration greatest? When was the acceleration constant?

17. *Pisa by Parachute* *(continuation of Exercise 16)* A few years ago, Mike McCarthy parachuted 179 ft from the top of the Tower of Pisa. Make a rough sketch to show the shape of the graph of his downward velocity during the jump.

18. *Fruit Flies* *(Example 2, Section 2.4 continued) Work in groups of two or three.* Populations starting out in closed environments grow slowly at first, when there are relatively few members, then more rapidly as the number of reproducing individuals increases and resources are still abundant, then slowly again as the population reaches the carrying capacity of the environment.

(a) Use the graphical technique of Section 3.1, Example 3, to graph the derivative of the fruit fly population introduced in Section 2.4. The graph of the population is reproduced below. What units should be used on the horizontal and vertical axes for the derivative's graph?

(b) During what days does the population seem to be increasing fastest? slowest?

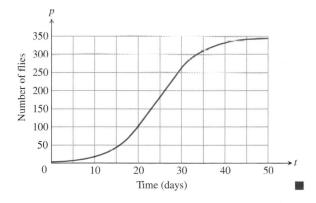

Time (days)

19. In step 1 of Exploration 2, at what time is the particle at the point (5, 2)?

20. *Particle Motion* *Work in groups of two or three.* The position (*x*-coordinate) of a particle moving on the line $y = 2$ is given by $x(t) = 2t^3 - 13t^2 + 22t - 5$ where t is time in seconds.

(a) Describe the motion of the particle for $t \geq 0$.

(b) When does the particle speed up? slow down?

(c) When does the particle change direction?

(d) When is the particle at rest?

(e) Describe the velocity and speed of the particle.

(f) When is the particle at the point (5, 2)?

In Exercises 21 and 22, the coordinates *s* of a moving body for various values of *t* are given. *Work in groups of two or three.*
(a) Plot *s* versus *t* on coordinate paper, and sketch a smooth curve through the given points. **(b)** Assuming that this smooth curve represents the motion of the body, estimate the velocity at $t = 1.0$, $t = 2.5$, and $t = 3.5$.

21.

t (sec)	0	0.5	1.0	1.5	2.0	2.5	3.0	3.5	4.0
s (ft)	12.5	26	36.5	44	48.5	50	48.5	44	36.5

22.

t (sec)	0	0.5	1.0	1.5	2.0	2.5	3.0	3.5	4.0
s (ft)	3.5	−4	−8.5	−10	−8.5	−4	3.5	14	27.5

23. *Particle Motion* The accompanying figure shows the velocity $v = ds/dt = f(t)$ (m/sec) of a body moving along a coordinate line.

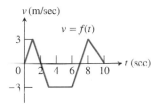

(a) When does the body reverse direction?

(b) When (approximately) is the body moving at a constant speed?

(c) Graph the body's speed for $0 \leq t \leq 10$.

(d) Graph the acceleration, where defined.

24. *Particle Motion* A particle P moves on the number line shown in part (a) of the accompanying figure. Part (b) shows the position of P as a function of time t.

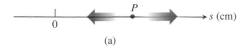

(a)

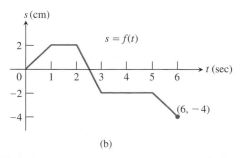

(b)

(a) When is P moving to the left? moving to the right? standing still?

(b) Graph the particle's velocity and speed (where defined).

25. *Particle Motion* The accompanying figure shows the velocity $v = f(t)$ of a particle moving on a coordinate line.

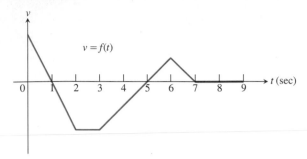

(a) When does the particle move forward? move backward? speed up? slow down?

(b) When is the particle's acceleration positive? negative? zero?

(c) When does the particle move at its greatest speed?

(d) When does the particle stand still for more than an instant?

26. *Moving Truck* The graph here shows the position s of a truck traveling on a highway. The truck starts at $t = 0$ and returns 15 hours later at $t = 15$.

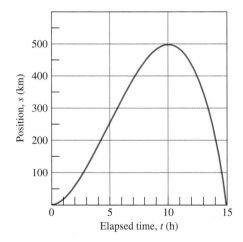

(a) Use the technique described in Section 3.1, Example 3, to graph the truck's velocity $v = ds/dt$ for $0 \le t \le 15$. Then repeat the process, with the velocity curve, to graph the truck's acceleration dv/dt.

(b) Suppose $s = 15t^2 - t^3$. Graph ds/dt and d^2s/dt^2, and compare your graphs with those in (a).

27. *Falling Objects* The multiflash photograph in Figure 3.30 shows two balls falling from rest. The vertical rulers are marked in centimeters. Use the equation $s = 490t^2$ (the free-fall equation for s in centimeters and t in seconds) to answer the following questions. *Work in groups of two or three.*

(a) How long did it take the balls to fall the first 160 cm? What was their average velocity for the period?

(b) How fast were the balls falling when they reached the 160-cm mark? What was their acceleration then?

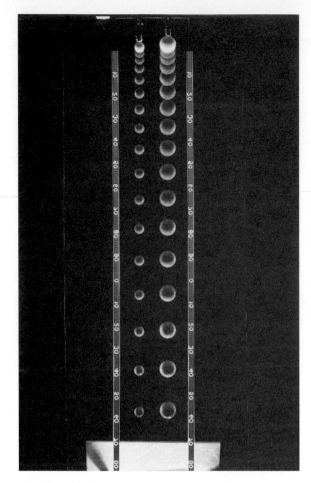

Figure 3.30 Two balls falling from rest. (Exercise 27)

(c) About how fast was the light flashing (flashes per second)?

28. **Writing to Learn** The graphs in Figure 3.31 show as functions of time t the position s, velocity $v = ds/dt$, and acceleration $a = d^2s/dt^2$ of a body moving along a coordinate line. Which graph is which? Give reasons for your answers.

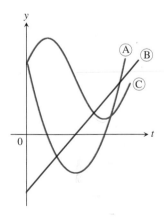

Figure 3.31 The graphs for Exercise 28.

29. Writing to Learn The graphs in Figure 3.32 show as functions of time t the position s, the velocity $v = ds/dt$, and the acceleration $a = d^2s/dt^2$ of a body moving along a coordinate line. Which graph is which? Give reasons for your answers.

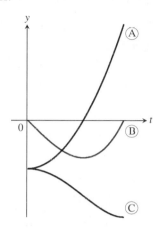

Figure 3.32 The graphs for Exercise 29.

30. Draining a Tank It takes 12 hours to drain a storage tank by opening the valve at the bottom. The depth y of fluid in the tank t hours after the valve is opened is given by the formula

$$y = 6\left(1 - \frac{t}{12}\right)^2 \text{ m.}$$

(a) Find the rate dy/dt (m/h) at which the water level is changing at time t.

(b) When is the fluid level in the tank falling fastest? slowest? What are the values of dy/dt at these times?

(c) Graph y and dy/dt together and discuss the behavior of y in relation to the signs and values of dy/dt.

31. Inflating a Balloon The volume $V = (4/3)\pi r^3$ of a spherical balloon changes with the radius.

(a) At what rate does the volume change with respect to the radius when $r = 2$ ft?

(b) By approximately how much does the volume increase when the radius changes from 2 to 2.2 ft?

32. Airplane Takeoff Suppose that the distance an aircraft travels along a runway before takeoff is given by $D = (10/9)t^2$, where D is measured in meters from the starting point and t is measured in seconds from the time the brakes are released. If the aircraft will become airborne when its speed reaches 200 km/h, how long will it take to become airborne, and what distance will it have traveled by that time?

33. Volcanic Lava Fountains Although the November 1959 Kilauea Iki eruption on the island of Hawaii began with a line of fountains along the wall of the crater, activity was later confined to a single vent in the crater's floor, which at one point shot lava 1900 ft straight into the air (a world record). What was the lava's exit velocity in feet per second? in miles per hour? (*Hint:* If v_0 is the exit velocity of a particle of lava, its height t seconds later will be $s = v_0 t - 16t^2$ feet. Begin by finding the time at which $ds/dt = 0$. Neglect air resistance.)

34. Writing to Learn Suppose you are looking at a graph of velocity as a function of time. How can you estimate the acceleration at a given point in time?

35. Writing to Learn Explain how the Sum and Difference Rule (Rule 4 in Section 3.3) can be used to derive a formula for *marginal profit* in terms of marginal revenue and marginal cost.

36. Thoroughbred Racing A racehorse is running a 10-furlong race. (A furlong is 220 yards, although we will use furlongs and seconds as our units in this exercise.) As the horse passes each furlong marker (F), a steward records the time elapsed (t) since the beginning of the race, as shown in the table below:

F	0	1	2	3	4	5	6	7	8	9	10
t	0	20	33	46	59	73	86	100	112	124	135

(a) How long does it take the horse to finish the race?

(b) What is the average speed of the horse over the first 5 furlongs?

(c) What is the approximate speed of the horse as it passes the 3-furlong marker?

(d) During which portion of the race is the horse running the fastest?

(e) During which portion of the race is the horse accelerating the fastest?

Extending the Ideas

37. Even and Odd Functions

(a) Show that if f is an even function, then f' is an odd function.

(b) Show that if f is an odd function, then f' is an even function.

38. Extended Product Rule Derive a formula for the derivative of the product fgh of three differentiable functions.

3.5 Derivatives of Trigonometric Functions

Derivative of the Sine Function • Derivative of the Cosine Function • Simple Harmonic Motion • Jerk • Derivatives of the Other Basic Trigonometric Functions

Derivative of the Sine Function

Trigonometric functions are important because so many of the phenomena we want information about are periodic (heart rhythms, earthquakes, tides, weather). It is known that continuous periodic functions can always be expressed in terms of sines and cosines, so the derivatives of sines and cosines play a key role in describing periodic change. This section introduces the derivatives of the six basic trigonometric functions.

Exploration 1 **Making a Conjecture with NDER**

In the window $[-2\pi, 2\pi]$ by $[-4, 4]$, graph $y_1 = \sin x$ and $y_2 = \text{NDER}(\sin x)$ (Figure 3.33).

1. When the graph of $y_1 = \sin x$ is increasing, what is true about the graph of $y_2 = \text{NDER}(\sin x)$?

2. When the graph of $y_1 = \sin x$ is decreasing, what is true about the graph of $y_2 = \text{NDER}(\sin x)$?

3. When the graph of $y_1 = \sin x$ stops increasing and starts decreasing, what is true about the graph of $y_2 = \text{NDER}(\sin x)$?

4. At the places where $\text{NDER}(\sin x) = \pm 1$, what appears to be the slope of the graph of $y_1 = \sin x$?

5. Make a conjecture about what function the derivative of sine might be. Test your conjecture by graphing your function and $\text{NDER}(\sin x)$ in the same viewing window.

6. Now let $y_1 = \cos x$ and $y_2 = \text{NDER}(\cos x)$. Answer questions (1) through (5) *without* looking at the graph of $\text{NDER}(\cos x)$ until you are ready to test your conjecture about what function the derivative of cosine might be.

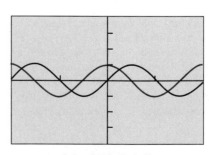

$[-2\pi, 2\pi]$ by $[-4, 4]$

Figure 3.33 Sine and its derivative. What is the derivative? (Exploration 1)

If you conjectured that the derivative of the sine function is the cosine function, then you are right. We will confirm this analytically, but first we appeal to technology one more time to evaluate two limits needed in the proof:

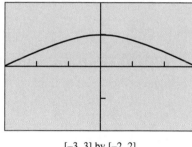

[−3, 3] by [−2, 2]

(a)

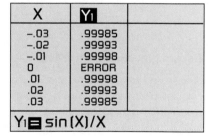

X	Y₁	
−.03	.99985	
−.02	.99993	
−.01	.99998	
0	ERROR	
.01	.99998	
.02	.99993	
.03	.99985	

Y₁ = sin (X)/X

(b)

Figure 3.34 (a) Graphical and (b) tabular support that $\lim\limits_{h \to 0} \dfrac{\sin (h)}{h} = 1$.

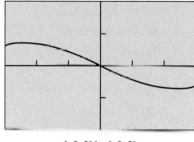

[−3, 3] by [−2, 2]

(a)

X	Y₁	
−.03	.015	
−.02	.01	
−.01	.005	
0	ERROR	
.01	−.005	
.02	−.01	
.03	−.015	

Y₁ = (cos(X)−1)/X

(b)

Figure 3.35 (a) Graphical and (b) tabular support that $\lim\limits_{h \to 0} \dfrac{\cos (h) - 1}{h} = 0$.

Confirm Analytically

(Also, see Section 2.1, Exercise 63.) Now, let $y = \sin x$. Then

$$\frac{dy}{dx} = \lim_{h \to 0} \frac{\sin (x + h) - \sin x}{h}$$

$$= \lim_{h \to 0} \frac{\sin x \cos h + \cos x \sin h - \sin x}{h} \qquad \text{Angle sum identity}$$

$$= \lim_{h \to 0} \frac{(\sin x)(\cos h - 1) + \cos x \sin h}{h}$$

$$= \lim_{h \to 0} \sin x \cdot \lim_{h \to 0} \frac{(\cos h - 1)}{h} + \lim_{h \to 0} \cos x \cdot \lim_{h \to 0} \frac{\sin h}{h}$$

$$= \sin x \cdot 0 + \cos x \cdot 1$$

$$= \cos x.$$

In short, the derivative of the sine is the cosine.

$$\frac{d}{dx} \sin x = \cos x$$

Now that we know that the sine function is differentiable, we know that sine and its derivative obey all the rules for differentiation. We also know that $\sin x$ is continuous. The same holds for the other trigonometric functions in this section. Each one is differentiable at every point in its domain, so each one is continuous at every point in its domain, and the differentiation rules apply for each one.

Radian Measure in Calculus

In case you have been wondering why calculus uses radian measure when the rest of the world seems to measure angles in degrees, you are now ready to understand the answer. The derivative of $\sin x$ is $\cos x$ *only if* x is measured in radians! If you look at the analytic confirmation, you will note that the derivative comes down to

$$\cos x \text{ times } \lim_{h \to 0} \frac{\sin h}{h}.$$

We saw that

$$\lim_{h \to 0} \frac{\sin h}{h} = 1$$

in Figure 3.34, but only because the graph in Figure 3.34 is in *radian mode*. If you look at the limit of the same function in *degree* mode you will get a very different limit (and hence a different derivative for sine). See Exercise 25.

Derivative of the Cosine Function

If you conjectured in Exploration 1 that the derivative of the cosine function is the negative of the sine function, you were correct. You can confirm this analytically in Exercise 14.

$$\frac{d}{dx}\cos x = -\sin x$$

Example 1 REVISITING THE DIFFERENTIATION RULES

Find the derivatives of **(a)** $y = x^2 \sin x$ and **(b)** $u = \cos x/(1 - \sin x)$.

Solution

(a)
$$\frac{dy}{dx} = x^2 \cdot \frac{d}{dx}(\sin x) + \sin x \cdot \frac{d}{dx}(x^2) \quad \text{Product Rule}$$

$$= x^2 \cos x + 2x \sin x$$

(b)
$$\frac{du}{dx} = \frac{(1 - \sin x) \cdot \dfrac{d}{dx}(\cos x) - \cos x \cdot \dfrac{d}{dx}(1 - \sin x)}{(1 - \sin x)^2} \quad \text{Quotient Rule}$$

$$= \frac{(1 - \sin x)(-\sin x) - \cos x\,(0 - \cos x)}{(1 - \sin x)^2}$$

$$= \frac{-\sin x + \sin^2 x + \cos^2 x}{(1 - \sin x)^2}$$

$$= \frac{1 - \sin x}{(1 - \sin x)^2} \qquad \sin^2 x + \cos^2 x = 1$$

$$= \frac{1}{1 - \sin x}$$

Simple Harmonic Motion

The motion of a weight bobbing up and down on the end of a spring is an example of **simple harmonic motion**. Example 2 describes a case in which there are no opposing forces like friction or buoyancy to slow down the motion.

Example 2 THE MOTION OF A WEIGHT ON A SPRING

A weight hanging from a spring (Figure 3.36) is stretched 5 units beyond its rest position $(s = 0)$ and released at time $t = 0$ to bob up and down. Its position at any later time t is

$$s = 5 \cos t.$$

What are its velocity and acceleration at time t? Describe its motion.

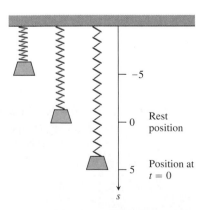

Figure 3.36 The weighted spring in Example 2.

Solution We have:

Position: $\quad s = 5 \cos t;$

Velocity: $\quad v = \dfrac{ds}{dt} = \dfrac{d}{dt}(5 \cos t) = -5 \sin t;$

Acceleration: $\quad a = \dfrac{dv}{dt} = \dfrac{d}{dt}(-5 \sin t) = -5 \cos t.$

Notice how much we can learn from these equations:

1. As time passes, the weight moves down and up between $s = -5$ and $s = 5$ on the *s*-axis. The amplitude of the motion is 5. The period of the motion is 2π.

2. The velocity $v = -5 \sin t$ attains its greatest magnitude, 5, when $\cos t = 0$, as the graphs show in Figure 3.37. Hence the speed of the weight, $|v| = 5 |\sin t|$, is greatest when $\cos t = 0$, that is, when $s = 0$ (the rest position). The speed of the weight is zero when $\sin t = 0$. This occurs when $s = 5 \cos t = \pm 5$, at the endpoints of the interval of motion.

3. The acceleration value is always the exact opposite of the position value. When the weight is above the rest position, gravity is pulling it back down; when the weight is below the rest position, the spring is pulling it back up.

4. The acceleration, $a = -5 \cos t$, is zero only at the rest position where $\cos t = 0$ and the force of gravity and the force from the spring offset each other. When the weight is anywhere else, the two forces are unequal and acceleration is nonzero. The acceleration is greatest in magnitude at the points farthest from the rest position, where $\cos t = \pm 1$.

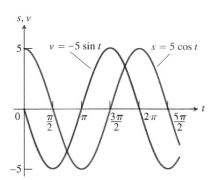

Figure 3.37 Graphs of the position and velocity of the weight in Example 2.

Jerk

A sudden change in acceleration is called a "jerk." When a ride in a car or a bus is jerky, it is not that the accelerations involved are necessarily large but that the changes in acceleration are abrupt. Jerk is what spills your soft drink. The derivative responsible for jerk is the *third* derivative of position.

Definition Jerk

Jerk is the derivative of acceleration. If a body's position at time t is $s(t)$, the body's jerk at time t is

$$j(t) = \frac{da}{dt} = \frac{d^3 s}{dt^3}.$$

Recent tests have shown that motion sickness comes from accelerations whose changes in magnitude or direction take us by surprise. Keeping an eye on the road helps us to see the changes coming. A driver is less likely to become sick than a passenger who is reading in the back seat.

Example 3   A COUPLE OF JERKS

(a) The jerk caused by the constant acceleration of gravity ($g = -32$ ft/sec^2) is zero:

$$j = \frac{d}{dt}(g) = 0.$$

This explains why we don't experience motion sickness while just sitting around.

(b) The jerk of the simple harmonic motion in Example 2 is

$$j = \frac{da}{dt} = \frac{d}{dt}(-5 \cos t)$$

$$= 5 \sin t.$$

It has its greatest magnitude when $\sin t = \pm 1$. This does not occur at the extremes of the displacement, but at the rest position, where the acceleration changes direction and sign.

Derivatives of the Other Basic Trigonometric Functions

Because $\sin x$ and $\cos x$ are differentiable functions of x, the related functions

$$\tan x = \frac{\sin x}{\cos x}, \quad \sec x = \frac{1}{\cos x},$$

$$\cot x = \frac{\cos x}{\sin x}, \quad \csc x = \frac{1}{\sin x}$$

are differentiable at every value of x for which they are defined. Their derivatives (Exercises 15 and 16) are given by the following formulas.

$$\frac{d}{dx}\tan x = \sec^2 x \qquad \frac{d}{dx}\sec x = \sec x \tan x$$

$$\frac{d}{dx}\cot x = -\csc^2 x \qquad \frac{d}{dx}\csc x = -\csc x \cot x$$

Example 4   FINDING TANGENT AND NORMAL LINES

Find equations for the lines that are tangent and normal to the graph of

$$f(x) = \frac{\tan x}{x}$$

at $x = 2$. Support graphically.

Solution

Solve Numerically

Since we will be using a calculator approximation for $f(2)$ anyway, this is a good place to use NDER.

We compute $(\tan 2)/2$ on the calculator and store it as k. The slope of the tangent line at $(2, k)$ is

$$\text{NDER}\left(\frac{\tan x}{x}, 2\right),$$

which we compute and store as m. The equation of the tangent line is $y - k = m(x - 2)$, or

$$y = mx + k - 2m.$$

Only after we have found m and $k - 2m$ do we round the coefficients, giving the tangent line as

$$y = 3.43x - 7.96.$$

The equation of the normal line is

$$y - k = -\frac{1}{m}(x - 2), \text{ or}$$

$$y = -\frac{1}{m}x + k + \frac{2}{m}.$$

Again we wait until the end to round the coefficients, giving the normal line as

$$y = -0.291x - 0.51.$$

Confirm Graphically

Figure 3.38, showing the original function and the two lines, supports our computations.

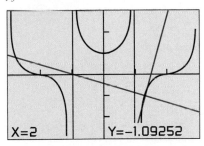

$y_1 = \tan(x)/x$
$y_2 = 3.43x - 7.96$
$y_3 = -0.291x - 0.51$

X=2   Y=-1.09252

$[-3\pi/2, 3\pi/2]$ by $[-3, 3]$

Figure 3.38 Graphical support for Example 4.

Example 5 A TRIGONOMETRIC SECOND DERIVATIVE

Find y'' if $y = \sec x$.

Solution $y = \sec x$

$$y' = \sec x \tan x$$

$$y'' = \frac{d}{dx}(\sec x \tan x)$$

$$= \sec x \frac{d}{dx}(\tan x) + \tan x \frac{d}{dx}(\sec x)$$

$$= \sec x (\sec^2 x) + \tan x (\sec x \tan x)$$

$$= \sec^3 x + \sec x \tan^2 x$$

Quick Review 3.5

1. Convert 135 degrees to radians.

2. Convert 1.7 radians to degrees.

3. Find the exact value of $\sin(\pi/3)$ without a calculator.

4. State the domain and the range of the cosine function.

5. State the domain and the range of the tangent function.

6. If $\sin a = -1$, what is $\cos a$?

7. If $\tan a = -1$, what are two possible values of $\sin a$?

8. Verify the identity:
$$\frac{1 - \cos h}{h} = \frac{\sin^2 h}{h(1 + \cos h)}.$$

9. Find an equation of the line tangent to the curve $y = 2x^3 - 7x^2 + 10$ at the point $(3, 1)$.

10. A particle moves along a line with velocity $v = 2t^3 - 7t^2 + 10$ for time $t \geq 0$. Find the acceleration of the particle at $t = 3$.

Section 3.5 Exercises

In Exercises 1–10, find dy/dx. Use your grapher to support your analysis if you are unsure of your answer.

1. $y = 1 + x - \cos x$

2. $y = 2 \sin x - \tan x$

3. $y = \dfrac{1}{x} + 5 \sin x$

4. $y = x \sec x$

5. $y = 4 - x^2 \sin x$

6. $y = 3x + x \tan x$

7. $y = \dfrac{4}{\cos x}$

8. $y = \dfrac{x}{1 + \cos x}$

9. $y = \dfrac{\cot x}{1 + \cot x}$

10. $y = \dfrac{\cos x}{1 + \sin x}$

11. Find an equation of the line tangent to the graph of $y = \sin x + 3$ at $x = \pi$.

12. Find an equation of the line normal to the graph of $y = (\tan x)/x$ at $x = \pi/4$.

13. Find an equation of the line tangent to the graph of $y = x^2 \sin x$ at $x = 3$.

14. Use the definition of the derivative to prove that $(d/dx) (\cos x) = -\sin x$. (You will need the limits found at the beginning of this section.)

15. Assuming that $(d/dx) (\sin x) = \cos x$ and $(d/dx) (\cos x) = -\sin x$, prove each of the following.

(a) $\dfrac{d}{dx} \tan x = \sec^2 x$

(b) $\dfrac{d}{dx} \sec x = \sec x \tan x$

16. Assuming that $(d/dx) (\sin x) = \cos x$ and $(d/dx) (\cos x) = -\sin x$, prove each of the following.

(a) $\dfrac{d}{dx} \cot x = -\csc^2 x$

(b) $\dfrac{d}{dx} \csc x = -\csc x \cot x$

17. Show that the graphs of $y = \sec x$ and $y = \cos x$ have horizontal tangents at $x = 0$.

18. Show that the graphs of $y = \tan x$ and $y = \cot x$ have no horizontal tangents.

19. Find equations for the lines that are tangent and normal to the curve $y = \sqrt{2} \cos x$ at the point $(\pi/4, 1)$.

20. Find the points on the curve $y = \tan x$, $-\pi/2 < x < \pi/2$, where the tangent is parallel to the line $y = 2x$.

In Exercises 21 and 22, find an equation for **(a)** the tangent to the curve at P and **(b)** the horizontal tangent to the curve at Q.

21.

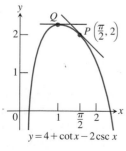

$y = 4 + \cot x - 2 \csc x$

22.

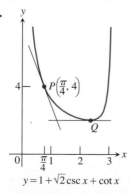

$y = 1 + \sqrt{2} \csc x + \cot x$

In Exercises 23 and 24, a body is moving in simple harmonic motion with position $s = f(t)$ (s in meters, t in seconds). *Work in groups of two or three.*

(a) Find the body's velocity, speed, acceleration, and jerk at time t.

(b) Find the body's velocity, speed, acceleration, and jerk at time $t = \pi/4$ sec.

(c) Describe the motion of the body.

23. $s = 2 - 2 \sin t$

24. $s = \sin t + \cos t$

Exploration

25. *Radians vs. Degrees* What happens to the derivatives of $\sin x$ and $\cos x$ if x is measured in degrees instead of radians? To find out, take the following steps.

(a) With your grapher in degree mode, graph

$$f(h) = \frac{\sin h}{h}$$

and estimate $\lim_{h \to 0} f(h)$. Compare your estimate with $\pi/180$. Is there any reason to believe the limit should be $\pi/180$?

(b) With your grapher in degree mode, estimate

$$\lim_{h \to 0} \frac{\cos h - 1}{h}.$$

(c) Now go back to the derivation of the formula for the derivative of $\sin x$ in the text and carry out the steps of the derivation using degree-mode limits. What formula do you obtain for the derivative?

(d) Derive the formula for the derivative of $\cos x$ using degree-mode limits.

(e) The disadvantages of the degree-mode formulas become apparent as you start taking derivatives of higher order. What are the second and third degree-mode derivatives of $\sin x$ and $\cos x$? ∎

26. Find y'' if $y = \csc x$.

27. Find y'' if $y = \theta \tan \theta$.

28. **Writing to Learn** Is there a value of b that will make

$$g(x) = \begin{cases} x + b, & x < 0 \\ \cos x, & x \geq 0 \end{cases}$$

continuous at $x = 0$? differentiable at $x = 0$? Give reasons for your answers.

29. Find $\dfrac{d^{999}}{dx^{999}} (\cos x)$.

30. Find $\dfrac{d^{725}}{dx^{725}} (\sin x)$.

31. *Local Linearity* This is the graph of the function $y = \sin x$ close to the origin. Since $\sin x$ is differentiable, this graph resembles a line. Find an equation for this line.

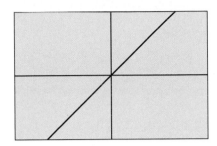

32. *(Continuation of Exercise 31)* For values of x close to 0, the linear equation found in Exercise 31 gives a good approximation of $\sin x$.

(a) Use this fact to estimate $\sin (0.12)$.

(b) Find $\sin (0.12)$ with a calculator. How close is the approximation in (a)?

33. Use the identity $\sin 2x = 2 \sin x \cos x$ to find the derivative of $\sin 2x$. Then use the identity $\cos 2x = \cos^2 x - \sin^2 x$ to express that derivative in terms of $\cos 2x$.

34. Use the identity $\cos 2x = \cos x \cos x - \sin x \sin x$ to find the derivative of $\cos 2x$. Express the derivative in terms of $\sin 2x$.

Extending the Ideas

35. Use analytic methods to show that

$$\lim_{h \to 0} \frac{\cos h - 1}{h} = 0.$$

(*Hint:* Multiply numerator and denominator by $(\cos h + 1)$.)

36. Find A and B in $y = A \sin x + B \cos x$ so that $y'' - y = \sin x$.

3.6 Chain Rule

Derivative of a Composite Function • "Outside-Inside" Rule • Repeated Use of the Chain Rule • Slopes of Parametrized Curves • Power Chain Rule

Derivative of a Composite Function

We now know how to differentiate $\sin x$ and $x^2 - 4$, but how do we differentiate a composite like $\sin (x^2 - 4)$? The answer is with the Chain Rule, which is probably the most widely used differentiation rule in mathematics. This section describes the rule and how to use it.

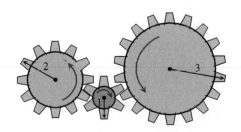

Figure 3.39 When gear A makes x turns, gear B makes u turns and gear C makes y turns. By comparing circumferences or counting teeth, we see that $y = u/2$ and $u = 3x$, so $y = 3x/2$. Thus $dy/du = 1/2$, $du/dx = 3$, and $dy/dx = 3/2 = (dy/du)(du/dx)$.

Example 1   RELATING DERIVATIVES

The function $y = 6x - 10 = 2(3x - 5)$ is the composite of the functions $y = 2u$ and $u = 3x - 5$. How are the derivatives of these three functions related?

Solution We have

$$\frac{dy}{dx} = 6, \quad \frac{dy}{du} = 2, \quad \frac{du}{dx} = 3.$$

Since $6 = 2 \cdot 3$,

$$\frac{dy}{dx} = \frac{dy}{du} \cdot \frac{du}{dx}.$$

Is it an accident that $dy/dx = dy/du \cdot du/dx$?

If we think of the derivative as a rate of change, our intuition allows us to see that this relationship is reasonable. For $y = f(u)$ and $u = g(x)$, if y changes twice as fast as u and u changes three times as fast as x, then we expect y to change six times as fast as x. This is much like the effect of a multiple gear train (Figure 3.39).

Let us try again on another function.

Example 2 RELATING DERIVATIVES

The polynomial $y = 9x^4 + 6x^2 + 1 = (3x^2 + 1)^2$ is the composite of $y = u^2$ and $u = 3x^2 + 1$. Calculating derivatives, we see that

$$\frac{dy}{du} \cdot \frac{du}{dx} = 2u \cdot 6x$$

$$= 2(3x^2 + 1) \cdot 6x$$

$$= 36x^3 + 12x.$$

Also,

$$\frac{dy}{dx} = \frac{d}{dx}(9x^4 + 6x^2 + 1)$$

$$= 36x^3 + 12x.$$

Once again,

$$\frac{dy}{du} \cdot \frac{du}{dx} = \frac{dy}{dx}.$$

The derivative of the composite function $f(g(x))$ at x is the derivative of f at $g(x)$ times the derivative of g at x (Figure 3.40). This is known as the Chain Rule.

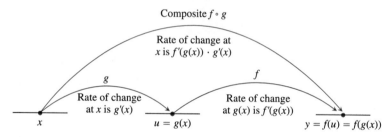

Figure 3.40 Rates of change multiply: the derivative of $f \circ g$ at x is the derivative of f at the point $g(x)$ times the derivative of g at x.

Rule 8 The Chain Rule

If f is differentiable at the point $u = g(x)$, and g is differentiable at x, then the composite function $(f \circ g)(x) = f(g(x))$ is differentiable at x, and

$$(f \circ g)'(x) = f'(g(x)) \cdot g'(x).$$

In Leibniz notation, if $y = f(u)$ and $u = g(x)$, then

$$\frac{dy}{dx} = \frac{dy}{du} \cdot \frac{du}{dx},$$

where dy/du is evaluated at $u = g(x)$.

It would be tempting to try to prove the Chain Rule by writing

$$\frac{\Delta y}{\Delta x} = \frac{\Delta y}{\Delta u} \cdot \frac{\Delta u}{\Delta x}$$

(a true statement about fractions with nonzero denominators) and taking the limit as $\Delta x \to 0$. This is essentially what is happening, and it would work as a proof if we knew that Δu, the change in u, was nonzero; but we do not know this. A small change in x could conceivably produce no change in u. An airtight proof of the Chain Rule can be constructed through a different approach, but we will omit it here.

Example 3 APPLYING THE CHAIN RULE

An object moves along the x-axis so that its position at any time $t \geq 0$ is given by $x(t) = \cos(t^2 + 1)$. Find the velocity of the object as a function of t.

Solution We know that the velocity is dx/dt. In this instance, x is a composite function: $x = \cos(u)$ and $u = t^2 + 1$. We have

$$\frac{dx}{du} = -\sin(u) \qquad x = \cos(u)$$

$$\frac{du}{dt} = 2t. \qquad u = t^2 + 1$$

By the Chain Rule,

$$\frac{dx}{dt} = \frac{dx}{du} \cdot \frac{du}{dt}$$

$$= -\sin(u) \cdot 2t$$

$$= -\sin(t^2 + 1) \cdot 2t$$

$$= -2t \sin(t^2 + 1).$$

"Outside-Inside" Rule

It sometimes helps to think about the Chain Rule this way: If $y = f(g(x))$, then

$$\frac{dy}{dx} = f'(g(x)) \cdot g'(x).$$

In words, differentiate the "outside" function f and evaluate it at the "inside" function $g(x)$ left alone; then multiply by the derivative of the "inside function."

Example 4 DIFFERENTIATING FROM THE OUTSIDE IN

Differentiate $\sin(x^2 + x)$ with respect to x.

Solution

$$\frac{d}{dx}\sin\underbrace{(x^2 + x)}_{\text{inside}} = \cos\underbrace{(x^2 + x)}_{\substack{\text{inside} \\ \text{left alone}}} \cdot \underbrace{(2x + 1)}_{\substack{\text{derivative of} \\ \text{the inside}}}$$

Repeated Use of the Chain Rule

We sometimes have to use the Chain Rule two or more times to find a derivative. Here is an example:

Example 5 A THREE-LINK "CHAIN"

Find the derivative of $g(t) = \tan(5 - \sin 2t)$.

Solution Notice here that tan is a function of $5 - \sin 2t$, while sin is a function of $2t$, which is itself a function of t. Therefore, by the Chain Rule,

$$g'(t) = \frac{d}{dt}(\tan(5 - \sin 2t))$$

$$= \sec^2(5 - \sin 2t) \cdot \frac{d}{dt}(5 - \sin 2t) \qquad \text{Derivative of } \tan u \text{ with } u = 5 - \sin 2t$$

$$= \sec^2(5 - \sin 2t) \cdot \left(0 - \cos 2t \cdot \frac{d}{dt}(2t)\right) \qquad \text{Derivative of } 5 - \sin u \text{ with } u = 2t$$

$$= \sec^2(5 - \sin 2t) \cdot (-\cos 2t) \cdot 2$$

$$= -2(\cos 2t)\sec^2(5 - \sin 2t).$$

Slopes of Parametrized Curves

A parametrized curve $(x(t), y(t))$ is *differentiable at t* if x and y are differentiable at t. At a point on a differentiable parametrized curve where y is also a differentiable function of x, the derivatives dy/dt, dx/dt, and dy/dx are related by the Chain Rule:

$$\frac{dy}{dt} = \frac{dy}{dx} \cdot \frac{dx}{dt}.$$

If $dx/dt \neq 0$, we may divide both sides of this equation by dx/dt to solve for dy/dx.

Finding *dy/dx* Parametrically

If all three derivatives exist and $dx/dt \neq 0$,

$$\frac{dy}{dx} = \frac{dy/dt}{dx/dt}. \tag{3}$$

Example 6   DIFFERENTIATING WITH A PARAMETER

Find the line tangent to the right-hand hyperbola branch defined parametrically by

$$x = \sec t, \qquad y = \tan t, \qquad -\frac{\pi}{2} < t < \frac{\pi}{2}$$

at the point $(\sqrt{2}, 1)$, where $t = \pi/4$ (Figure 3.41).

Solution All three of the derivatives in Equation 3 exist and $dx/dt = \sec t \tan t \neq 0$ at the indicated point. Therefore, Equation 3 applies and

$$\frac{dy}{dx} = \frac{dy/dt}{dx/dt}$$

$$= \frac{\sec^2 t}{\sec t \tan t}$$

$$= \frac{\sec t}{\tan t}$$

$$= \csc t.$$

Setting $t = \pi/4$ gives

$$\left.\frac{dy}{dx}\right|_{t=\pi/4} = \csc(\pi/4) = \sqrt{2}.$$

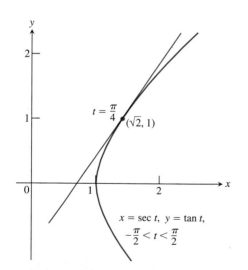

Figure 3.41 The hyperbola branch in Example 6. Eq. 3 applies for every point on the graph except $(1, 0)$. Can you state why Eq. 3 fails at $(1, 0)$?

$x = \sec t, \ y = \tan t,$

$-\frac{\pi}{2} < t < \frac{\pi}{2}$

$t = \frac{\pi}{4}$

$(\sqrt{2}, 1)$

The equation of the tangent line is

$$y - 1 = \sqrt{2}(x - \sqrt{2})$$
$$y = \sqrt{2}x - 2 + 1$$
$$y = \sqrt{2}x - 1.$$

Power Chain Rule

If f is a differentiable function of u, and u is a differentiable function of x, then substituting $y = f(u)$ into the Chain Rule formula

$$\frac{dy}{dx} = \frac{dy}{du} \cdot \frac{du}{dx}$$

leads to the formula

$$\frac{d}{dx}f(u) = f'(u)\frac{du}{dx}.$$

Here's an example of how it works: If n is an integer and $f(u) = u^n$, the Power Rules (Rules 2 and 7) tell us that $f'(u) = nu^{n-1}$. If u is a differentiable function of x, then we can use the Chain Rule to extend this to the **Power Chain Rule:**

$$\frac{d}{dx}u^n = nu^{n-1}\frac{du}{dx}. \qquad \frac{d}{du}(u^n) = nu^{n-1}$$

Example 7   FINDING SLOPE

(a) Find the slope of the line tangent to the curve $y = \sin^5 x$ at the point where $x = \pi/3$.

(b) Show that the slope of every line tangent to the curve $y = 1/(1 - 2x)^3$ is positive.

Solution

(a) $\quad \dfrac{dy}{dx} = 5 \sin^4 x \cdot \dfrac{d}{dx}\sin x \quad$ Power Chain Rule with $u = \sin x$, $n = 5$

$$= 5 \sin^4 x \cos x$$

The tangent line has slope

$$\left.\frac{dy}{dx}\right|_{x=\pi/3} = 5\left(\frac{\sqrt{3}}{2}\right)^4\left(\frac{1}{2}\right) = \frac{45}{32}.$$

(b) $\quad \dfrac{dy}{dx} = \dfrac{d}{dx}(1 - 2x)^{-3}$

$$= -3(1 - 2x)^{-4} \cdot \frac{d}{dx}(1 - 2x) \qquad \text{Power Chain Rule with } u = (1 - 2x),\ n = -3$$

$$= -3(1 - 2x)^{-4} \cdot (-2)$$

$$= \frac{6}{(1 - 2x)^4}$$

At any point (x, y) on the curve, $x \neq 1/2$ and the slope of the tangent line is

$$\frac{dy}{dx} = \frac{6}{(1 - 2x)^4},$$

the quotient of two positive numbers.

Example 8 RADIANS VERSUS DEGREES

It is important to remember that the formulas for the derivatives of both $\sin x$ and $\cos x$ were obtained under the assumption that x is measured in radians, *not* degrees. The Chain Rule gives us new insight into the difference between the two. Since $180° = \pi$ radians, $x° = \pi x/180$ radians. By the Chain Rule,

$$\frac{d}{dx}\sin\,(x°) = \frac{d}{dx}\sin\left(\frac{\pi x}{180}\right) = \frac{\pi}{180}\cos\left(\frac{\pi x}{180}\right) = \frac{\pi}{180}\cos\,(x°).$$

See Figure 3.42. Similarly, the derivative of $\cos\,(x°)$ is $-(\pi/180)\sin\,(x°)$.

The factor $\pi/180$, annoying in the first derivative, would compound with repeated differentiation. We see at a glance the compelling reason for the use of radian measure.

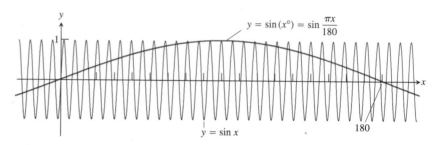

Figure 3.42 $\sin\,(x°)$ oscillates only $\pi/180$ times as often as $\sin x$ oscillates. Its maximum slope is $\pi/180$. (Example 8)

Quick Review 3.6

In Exercises 1–5, let $f(x) = \sin x$, $g(x) = x^2 + 1$, and $h(x) = 7x$. Write a simplified expression for the composite function.

1. $f(g(x))$

2. $f(g(h(x)))$

3. $(g \circ h)(x)$

4. $(h \circ g)(x)$

5. $f\left(\dfrac{g(x)}{h(x)}\right)$

In Exercises 6–10, let $f(x) = \cos x$, $g(x) = \sqrt{x + 2}$, and $h(x) = 3x^2$. Write the given function as a composite of two or more of f, g, and h. For example, $\cos 3x^2$ is $f(h(x))$.

6. $\sqrt{\cos x + 2}$

7. $\sqrt{3\cos^2 x + 2}$

8. $3\cos x + 6$

9. $\cos 27x^4$

10. $\cos \sqrt{2 + 3x^2}$

Section 3.6 Exercises

In Exercises 1–20, find dy/dx. If you are unsure of your answer, use NDER to support your computation.

1. $y = \sin (3x + 1)$

2. $y = \sin (7 - 5x)$

3. $y = \cos (\sqrt{3}x)$

4. $y = \tan (2x - x^3)$

5. $y = 5 \cot \left(\dfrac{2}{x}\right)$

6. $y = \left(\dfrac{\sin x}{1 + \cos x}\right)^2$

7. $y = \cos (\sin x)$

8. $y = \sec (\tan x)$

9. $y = (x + \sqrt{x})^{-2}$

10. $y = (\csc x + \cot x)^{-1}$

11. $y = \sin^{-5} x - \cos^3 x$

12. $y = x^3(2x - 5)^4$

13. $y = \sin^3 x \tan 4x$

14. $y = 4\sqrt{\sec x + \tan x}$

15. $y = \dfrac{3}{\sqrt{2x + 1}}$

16. $y = \dfrac{x}{\sqrt{1 + x^2}}$

17. $y = \sin^2 (3x - 2)$

18. $y = (1 + \cos 2x)^2$

19. $y = (1 + \cos^2 7x)^3$

20. $y = \sqrt{\tan 5x}$

In Exercises 21–24, find ds/dt.

21. $s = \cos\left(\dfrac{\pi}{2} - 3t\right)$

22. $s = t \cos (\pi - 4t)$

23. $s = \dfrac{4}{3\pi} \sin 3t + \dfrac{4}{5\pi} \cos 5t$

24. $s = \sin\left(\dfrac{3\pi}{2}t\right) + \cos\left(\dfrac{7\pi}{4}t\right)$

In Exercises 25–28 find $dr/d\theta$.

25. $r = \tan(2 - \theta)$

26. $r = \sec 2\theta \tan 2\theta$

27. $r = \sqrt{\theta} \sin \theta$

28. $r = 2\theta\sqrt{\sec \theta}$

In Exercises 29–32, find y''.

29. $y = \tan x$

30. $y = \cot x$

31. $y = \cot(3x - 1)$

32. $y = 9\tan(x/3)$

In Exercises 33–38, find the value of $(f \circ g)'$ at the given value of x.

33. $f(u) = u^5 + 1, \quad u = g(x) = \sqrt{x}, \quad x = 1$

34. $f(u) = 1 - \dfrac{1}{u}, \quad u = g(x) - \dfrac{1}{1 - x}, \quad x = -1$

35. $f(u) = \cot\dfrac{\pi u}{10}, \quad u = g(x) = 5\sqrt{x}, \quad x = 1$

36. $f(u) = u + \dfrac{1}{\cos^2 u}, \quad u = g(x) = \pi x, \quad x = \dfrac{1}{4}$

37. $f(u) = \dfrac{2u}{u^2 + 1}, \quad u = g(x) = 10x^2 + x + 1, \quad x = 0$

38. $f(u) = \left(\dfrac{u - 1}{u + 1}\right)^2, \quad u = g(x) = \dfrac{1}{x^2} - 1, \quad x = -1$

What happens if you can write a function as a composite in different ways? Do you get the same derivative each time? The Chain Rule says you should. Try it with the functions in Exercises 39 and 40.

39. Find dy/dx if $y = \cos(6x + 2)$ by writing y as a composite with

   (a) $y = \cos u$ and $u = 6x + 2$.

   (b) $y = \cos 2u$ and $u = 3x + 1$.

40. Find dy/dx if $y = \sin(x^2 + 1)$ by writing y as a composite with

   (a) $y = \sin(u + 1)$ and $u = x^2$.

   (b) $y - \sin u$ and $u = x^2 + 1$.

In Exercises 41–48, find the equation of the line tangent to the curve at the point defined by the given value of t.

41. $x = 2\cos t, \quad y = 2\sin t, \quad t = \pi/4$

42. $x = \sin 2\pi t, \quad y = \cos 2\pi t, \quad t = -1/6$

43. $x = \sec^2 t - 1, \quad y = \tan t, \quad t = -\pi/4$

44. $x = \sec t, \quad y = \tan t, \quad t = \pi/6$

45. $x = t, \quad y = \sqrt{t}, \quad t = 1/4$

46. $x = 2t^2 + 3, \quad y = t^4, \quad t = -1$

47. $x = t - \sin t, \quad y = 1 - \cos t, \quad t = \pi/3$

48. $x = \cos t, \quad y = 1 + \sin t, \quad t = \pi/2$

49. Let $x = t^2 + t$, and let $y = \sin t$.

   (a) Find dy/dx as a function of t.

   (b) Find $\dfrac{d}{dt}\left(\dfrac{dy}{dx}\right)$ as a function of t.

   (c) Find $\dfrac{d}{dx}\left(\dfrac{dy}{dx}\right)$ as a function of t.

   Use the Chain Rule and your answer from (b).

   (d) Which of the expressions in parts (b) and (c) is d^2y/dx^2?

50. A circle of radius 2 and center $(0, 0)$ can be parametrized by the equations $x = 2\cos t$ and $y = 2\sin t$. Show that for any value of t, the line tangent to the circle at $(2\cos t, 2\sin t)$ is perpendicular to the radius.

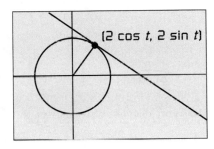

(2 cos *t*, 2 sin *t*)

51. Let $s - \cos \theta$. Evaluate ds/dt when $\theta = 3\pi/2$ and $d\theta/dt - 5$.

52. Let $y = x^2 + 7x - 5$. Evaluate dy/dt when $x = 1$ and $dx/dt = 1/3$.

53. What is the largest value possible for the slope of the curve $y = \sin(x/2)$?

54. Write an equation for the tangent to the curve $y = \sin mx$ at the origin.

55. Find the lines that are tangent and normal to the curve $y = 2\tan(\pi x/4)$ at $x = 1$. Support your answer graphically.

56. *Working with Numerical Values* Suppose that functions f and g and their derivatives have the following values at $x = 2$ and $x = 3$.

x	$f(x)$	$g(x)$	$f'(x)$	$g'(x)$
2	8	2	1/3	−3
3	3	−4	2π	5

Evaluate the derivatives with respect to x of the following combinations at the given value of x.

   (a) $2f(x)$ at $x = 2$

   (b) $f(x) + g(x)$ at $x = 3$

   (c) $f(x) \cdot g(x)$ at $x = 3$

   (d) $f(x)/g(x)$ at $x = 2$ **(e)** $f(g(x))$ at $x = 2$

   (f) $\sqrt{f(x)}$ at $x = 2$ **(g)** $1/g^2(x)$ at $x = 3$

   (h) $\sqrt{f^2(x) + g^2(x)}$ at $x = 2$

57. *Working with Numerical Values* Suppose that the functions f and g and their derivatives with respect to x have the following values at $x = 0$ and $x = 1$.

x	$f(x)$	$g(x)$	$f'(x)$	$g'(x)$
0	1	1	5	1/3
1	3	−4	−1/3	−8/3

Evaluate the derivatives with respect to x of the following combinations at the given value of x.

(a) $5f(x) - g(x), \quad x = 1$ **(b)** $f(x)g^3(x), \quad x = 0$

(c) $\dfrac{f(x)}{g(x) + 1}, \quad x = 1$ **(d)** $f(g(x)), \quad x = 0$

(e) $g(f(x)), \quad x = 0$

(f) $(g(x) + f(x))^{-2}, \quad x = 1$

(g) $f(x + g(x)), \quad x = 0$

58. *Orthogonal Curves* Two curves are said to cross at right angles if their tangents are perpendicular at the crossing point. The technical word for "crossing at right angles" is **orthogonal.** Show that the curves $y = \sin 2x$ and $y = -\sin(x/2)$ are orthogonal at the origin. Draw both graphs and both tangents in a square viewing window.

59. *Writing to Learn* Explain why the Chain Rule formula

$$\frac{dy}{dx} = \frac{dy}{du} \cdot \frac{du}{dx}$$

is not simply the well-known rule for multiplying fractions.

60. *Running Machinery Too Fast* Suppose that a piston is moving straight up and down and that its position at time t seconds is

$$s = A \cos(2\pi bt),$$

with A and b positive. The value of A is the amplitude of the motion, and b is the frequency (number of times the piston moves up and down each second). What effect does doubling the frequency have on the piston's velocity, acceleration, and jerk? (Once you find out, you will know why machinery breaks when you run it too fast.)

Figure 3.43 The internal forces in the engine get so large that they tear the engine apart when the velocity is too great.

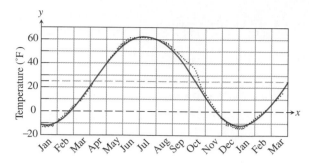

Figure 3.44 Normal mean air temperatures at Fairbanks, Alaska, plotted as data points, and the approximating sine function (Exercise 61).

61. *Temperatures in Fairbanks, Alaska* Work in groups of two or three. The graph in Figure 3.44 shows the average Fahrenheit temperature in Fairbanks, Alaska, during a typical 365-day year. The equation that approximates the temperature on day x is

$$y = 37 \sin\left[\frac{2\pi}{365}(x - 101)\right] + 25.$$

(a) On what day is the temperature increasing the fastest?

(b) About how many degrees per day is the temperature increasing when it is increasing at its fastest?

62. *Constant Acceleration* The position of a particle moving along a coordinate line is $s = \sqrt{1 + 4t}$, with s in meters and t in seconds. Find the particle's velocity and acceleration at $t = 6$ sec.

63. *Particle Motion* Suppose the velocity of a falling body is $v = k\sqrt{s}$ m/sec (k a constant) at the instant the body has fallen s meters from its starting point. Show that the body's acceleration is constant.

64. *Falling Meteorite* The velocity of a heavy meteorite entering the earth's atmosphere is inversely proportional to $\sqrt{s}$ when it is s kilometers from the earth's center. Show that the meteorite's acceleration is inversely proportional to s^2.

65. *Particle Acceleration* A particle moves along the x-axis with velocity $dx/dt = f(x)$. Show that the particle's acceleration is $f(x)f'(x)$.

66. *Temperature and the Period of a Pendulum* For oscillations of small amplitude (short swings), we may safely model the relationship between the period T and the length L of a simple pendulum with the equation

$$T = 2\pi\sqrt{\frac{L}{g}},$$

where g is the constant acceleration of gravity at the pendulum's location. If we measure g in centimeters per second squared, we measure L in centimeters and T in seconds. If the pendulum is made of metal, its length will vary with temperature, either increasing or decreasing at a rate that is roughly proportional to L. In symbols, with u being temperature and k the proportionality constant,

$$\frac{dL}{du} = kL.$$

Assuming this to be the case, show that the rate at which the period changes with respect to temperature is $kT/2$.

67. **Writing to Learn** *Chain Rule* Suppose that $f(x) = x^2$ and $g(x) = |x|$. Then the composites

$$(f \circ g)(x) = |x|^2 = x^2 \quad \text{and} \quad (g \circ f)(x) = |x^2| = x^2$$

are both differentiable at $x = 0$ even though g itself is not differentiable at $x = 0$. Does this contradict the Chain Rule? Explain.

68. *Tangents* Suppose that $u = g(x)$ is differentiable at $x = 1$ and that $y = f(u)$ is differentiable at $u = g(1)$. If the graph of $y = f(g(x))$ has a horizontal tangent at $x = 1$, can we conclude anything about the tangent to the graph of g at $x = 1$ or the tangent to the graph of f at $u = g(1)$? Give reasons for your answer.

Explorations

69. *The Derivative of sin 2x* Graph the function $y = 2 \cos 2x$ for $2 \le x \le 3.5$. Then, on the same screen, graph

$$y = \frac{\sin 2(x + h) - \sin 2x}{h}$$

for $h = 1.0, 0.5$, and 0.2. Experiment with other values of h, including negative values. What do you see happening as $h \to 0$? Explain this behavior.

70. *The Derivative of cos* (x^2) Graph $y = -2x \sin (x^2)$ for $-2 \le x \le 3$. Then, on screen, graph

$$y = \frac{\cos [(x + h)^2] - \cos (x^2)}{h}$$

for $h = 1.0, 0.7$, and 0.3. Experiment with other values of h. What do you see happening as $h \to 0$? Explain this behavior. ∎

Extending the Ideas

71. *Absolute Value Functions* Let u be a differentiable function of x.

   (a) Show that $\dfrac{d}{dx}|u| = u' \dfrac{u}{|u|}$.

   (b) Use part (a) to find the derivatives of $f(x) = |x^2 - 9|$ and $g(x) = |x| \sin x$.

72. *Geometric and Arithmetic Mean* The geometric mean of u and v is $G = \sqrt{uv}$ and the arithmetic mean is $A = (u + v)/2$. Show that if $u = x$, $v = x + c$, c a real number, then

$$\frac{dG}{dx} = \frac{A}{G}.$$

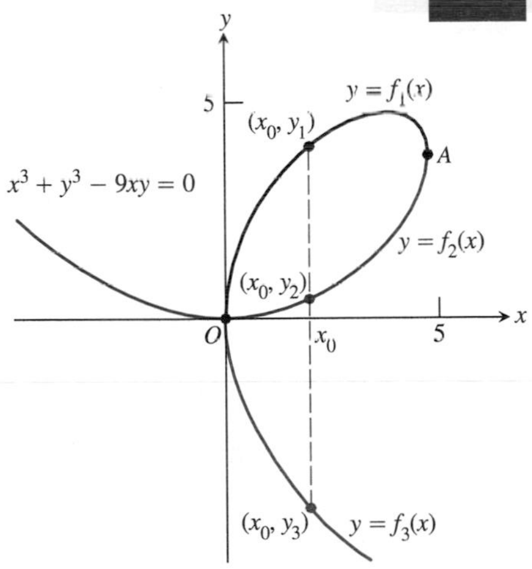

Figure 3.45 The graph of

$$x^3 + y^3 - 9xy = 0$$

(called a *folium*). Although not the graph of a function, it can be seen as the union of the graphs of three separate functions. This particular curve dates to Descartes in 1638.

3.7 Implicit Differentiation

Implicitly Defined Functions • Lenses, Tangents, and Normal Lines • Derivatives of Higher Order • Rational Powers of Differentiable Functions

Implicitly Defined Functions

The graph of the equation $x^3 + y^3 - 9xy = 0$ (Figure 3.45) has a well-defined slope at nearly every point because it is the union of the graphs of the functions $y = f_1(x)$, $y = f_2(x)$, and $y = f_3(x)$, which are differentiable except at O and A. But how do we find the slope when we cannot conveniently solve the equation to find the functions? The answer is to treat y as a differentiable function of x and differentiate both sides of the equation with respect to x, using the differentiation rules for sums, products, and quotients, and the Chain Rule. Then solve for dy/dx in terms of x and y *together* to obtain a formula that calculates the slope at any point (x, y) on the graph from the values of x and y.

The process by which we find dy/dx is called **implicit differentiation.** The phrase derives from the fact that the equation $x^3 + y^3 - 9xy = 0$ defines the functions $f_1, f_2,$ and f_3 implicitly (i.e., hidden inside the equation), without giving us *explicit* formulas to work with.

Example 1 DIFFERENTIATING IMPLICITLY

Find dy/dx if $y^2 = x$.

Solution To find dy/dx, we simply differentiate both sides of the equation $y^2 = x$ with respect to x, treating y as a differentiable function of x and applying the Chain Rule:

$$y^2 = x$$

$$2y\frac{dy}{dx} = 1 \qquad \frac{d}{dx}(y^2) = \frac{d}{dy}(y^2) \cdot \frac{dy}{dx}$$

$$\frac{dy}{dx} = \frac{1}{2y}.$$

In the previous example we differentiated with respect to x, and yet the derivative we obtained appeared as a function of y. Not only is this acceptable, it is actually quite useful. Figure 3.46, for example, shows that the curve has two different tangent lines when $x = 4$: one at the point $(4, 2)$ and the other at the point $(4, -2)$. Since the formula for dy/dx depends on y, our single formula gives the slope in both cases.

Implicit differentiation will frequently yield a derivative that is expressed in terms of both x and y, as in Example 2.

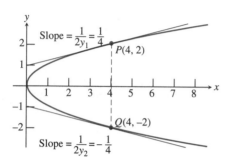

Figure 3.46 The derivative found in Example 1 gives the slope for the tangent lines at both P and Q, because it is a function of y.

Example 2 FINDING SLOPE ON A CIRCLE

Find the slope of the circle $x^2 + y^2 = 25$ at the point $(3, -4)$.

Solution The circle is not the graph of a single function of x, but it is the union of the graphs of two differentiable functions, $y_1 = \sqrt{25 - x^2}$ and $y_2 = -\sqrt{25 - x^2}$ (Figure 3.47). The point $(3, -4)$ lies on the graph of y_2, so it is possible to find the slope by calculating explicitly:

$$\frac{dy_2}{dx}\bigg|_{x=3} = -\frac{-2x}{2\sqrt{25 - x^2}}\bigg|_{x=3} = -\frac{-6}{2\sqrt{25 - 9}} = \frac{3}{4}.$$

But we can also find this slope more easily by differentiating both sides of the equation of the circle implicitly with respect to x:

$$\frac{d}{dx}(x^2 + y^2) = \frac{d}{dx}(25) \qquad \text{Differentiate both sides with respect to } x.$$

$$2x + 2y\frac{dy}{dx} = 0$$

$$\frac{dy}{dx} = -\frac{x}{y}.$$

The slope at $(3, -4)$ is

$$-\frac{x}{y}\bigg|_{(3, -4)} = -\frac{3}{-4} = \frac{3}{4}.$$

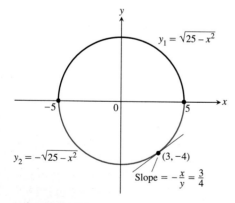

Figure 3.47 The circle combines the graphs of two functions. The graph of y_2 is the lower semicircle and passes through $(3, -4)$. (Example 2)

The implicit solution, besides being computationally easier, yields a formula for dy/dx that applies at any point on the circle (except, of course, $(\pm 5, 0)$, where slope is undefined). The explicit solution derived from the formula for y_2 applies only to the lower half of the circle.

To calculate the derivatives of other implicitly defined functions, we proceed as in Examples 1 and 2. We treat y as a differentiable function of x and apply the usual rules to differentiate both sides of the defining equation.

Example 3 SOLVING FOR dy/dx

Show that the slope dy/dx is defined at every point on the graph of $2y = x^2 + \sin y$.

Solution First we need to know dy/dx, which we find by implicit differentiation:

$$2y = x^2 + \sin y$$

$$\frac{d}{dx}(2y) = \frac{d}{dx}(x^2 + \sin y) \qquad \text{Differentiate both sides with respect to } x \ldots$$

$$= \frac{d}{dx}(x^2) + \frac{d}{dx}(\sin y)$$

$$2\frac{dy}{dx} = 2x + \cos y \frac{dy}{dx} \qquad \ldots \text{treating } y \text{ as a function of } x \text{ and using the Chain Rule.}$$

$$2\frac{dy}{dx} - (\cos y)\frac{dy}{dx} = 2x \qquad \text{Collect terms with } dy/dx \ldots$$

$$(2 - \cos y)\frac{dy}{dx} = 2x \qquad \ldots \text{and factor out } dy/dx.$$

$$\frac{dy}{dx} = \frac{2x}{2 - \cos y}. \qquad \text{Solve for } dy/dx \text{ by dividing.}$$

The formula for dy/dx is defined at every point (x, y), except for those points at which $\cos y = 2$. Since $\cos y$ cannot be greater than 1, this never happens.

Implicit Differentiation Process

1. Differentiate both sides of the equation with respect to x.
2. Collect the terms with dy/dx on one side of the equation.
3. Factor out dy/dx.
4. Solve for dy/dx.

Lenses, Tangents, and Normal Lines

In the law that describes how light changes direction as it enters a lens, the important angles are the angles the light makes with the line perpendicular to the surface of the lens at the point of entry (angles A and B in Figure 3.48). This line is called the *normal to the surface* at the point of entry. In a profile view of a lens like the one in Figure 3.48, the normal is a line perpendicular to the tangent to the profile curve at the point of entry.

The profiles of lenses are often described by quadratic curves like the one on the next page. When they are, we can use implicit differentiation to find the tangents and normals.

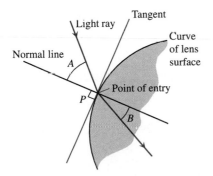

Figure 3.48 The profile of a lens, showing the bending (refraction) of a ray of light as it passes through the lens surface.

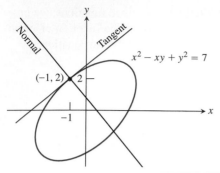

Figure 3.49 Tangent and normal lines to the ellipse $x^2 - xy + y^2 = 7$ at the point $(-1, 2)$. (Example 4)

Example 4 TANGENT AND NORMAL TO AN ELLIPSE

Find the tangent and normal to the ellipse $x^2 - xy + y^2 = 7$ at the point $(-1, 2)$. (See Figure 3.49.)

Solution We first use implicit differentiation to find dy/dx:

$$x^2 - xy + y^2 = 7$$

$$\frac{d}{dx}(x^2) - \frac{d}{dx}(xy) + \frac{d}{dx}(y^2) = \frac{d}{dx}(7) \qquad \text{Differentiate both sides with respect to } x \dots$$

$$2x - \left(x\frac{dy}{dx} + y\frac{dx}{dx}\right) + 2y\frac{dy}{dx} = 0 \qquad \dots \text{ treating } xy \text{ as a product and } y \text{ as a function of } x.$$

$$(2y - x)\frac{dy}{dx} = y - 2x \qquad \text{Collect terms.}$$

$$\frac{dy}{dx} = \frac{y - 2x}{2y - x}. \qquad \text{Solve for } dy/dx.$$

We then evaluate the derivative at $x = -1$, $y = 2$ to obtain

$$\left.\frac{dy}{dx}\right|_{(-1, 2)} = \left.\frac{y - 2x}{2y - x}\right|_{(-1, 2)}$$

$$= \frac{2 - 2(-1)}{2(2) - (-1)}$$

$$= \frac{4}{5}.$$

The tangent to the curve at $(-1, 2)$ is

$$y - 2 = \frac{4}{5}(x - (-1))$$

$$y = \frac{4}{5}x + \frac{14}{5}.$$

The normal to the curve at $(-1, 2)$ is

$$y - 2 = -\frac{5}{4}(x + 1)$$

$$y = -\frac{5}{4}x + \frac{3}{4}.$$

Derivatives of Higher Order

Implicit differentiation can also be used to find derivatives of higher order. Here is an example.

Example 5 FINDING A SECOND DERIVATIVE IMPLICITLY

Find d^2y/dx^2 if $2x^3 - 3y^2 = 8$.

Solution To start, we differentiate both sides of the equation with respect to x in order to find $y' = dy/dx$.

$$\frac{d}{dx}(2x^3 - 3y^2) = \frac{d}{dx}(8)$$

$$6x^2 - 6yy' = 0$$

$$x^2 - yy' = 0$$

$$y' = \frac{x^2}{y}, \text{ when } y \neq 0$$

We now apply the Quotient Rule to find y''.

$$y'' = \frac{d}{dx}\left(\frac{x^2}{y}\right) = \frac{2xy - x^2 y'}{y^2} = \frac{2x}{y} - \frac{x^2}{y^2} \cdot y'$$

Finally, we substitute $y' = x^2/y$ to express y'' in terms of x and y.

$$y'' = \frac{2x}{y} - \frac{x^2}{y^2}\left(\frac{x^2}{y}\right) = \frac{2x}{y} - \frac{x^4}{y^3}, \text{ when } y \neq 0$$

Exploration 1 **An Unexpected Derivative**

Consider the set of all points (x, y) satisfying the equation $x^2 - 2xy + y^2 = 4$. What does the graph of the equation look like? You can find out in two ways in this Exploration.

1. Use implicit differentiation to find dy/dx. Are you surprised by this derivative?

2. Knowing the derivative, what do you conjecture about the graph?

3. What are the possible values of y when $x = 0$? Does this information enable you to refine your conjecture about the graph?

4. The original equation can be written as $(x - y)^2 - 4 = 0$. By factoring the expression on the left, write two equations whose graphs combine to give the graph of the original equation. Then sketch the graph.

5. Explain why your graph is consistent with the derivative found in part 1.

Rational Powers of Differentiable Functions

We know that the Power Rule

$$\frac{d}{dx}x^n = nx^{n-1}$$

holds for any integer n (Rules 2 and 7 of this chapter). We can now prove that it holds when n is any rational number.

Rule 9 Power Rule for Rational Powers of *x*

If n is any rational number, then

$$\frac{d}{dx}x^n = nx^{n-1}.$$

If $n < 1$, then the derivative does not exist at $x = 0$.

Proof   Let p and q be integers with $q > 0$ and suppose that $y = \sqrt[q]{x^p} = x^{p/q}$. Then

$$y^q = x^p.$$

Since p and q are integers (for which we already have the Power Rule), we can differentiate both sides of the equation with respect to x and obtain

$$qy^{q-1}\frac{dy}{dx} = px^{p-1}.$$

If $y \neq 0$, we can divide both sides of the equation by qy^{q-1} to solve for dy/dx, obtaining

$$\frac{dy}{dx} = \frac{px^{p-1}}{qy^{q-1}}$$

$$= \frac{p}{q} \cdot \frac{x^{p-1}}{(x^{p/q})^{q-1}} \qquad y = x^{p/q}$$

$$= \frac{p}{q} \cdot \frac{x^{p-1}}{x^{p-p/q}} \qquad \frac{p}{q}(q-1) = p - \frac{p}{q}$$

$$= \frac{p}{q} \cdot x^{(p-1)-(p-p/q)} \qquad \text{A law of exponents}$$

$$= \frac{p}{q} \cdot x^{(p/q)-1}.$$

This proves the rule.   ∎

By combining this result with the Chain Rule, we get an extension of the Power Chain Rule to rational powers of u:

If n is a rational number and u is a differentiable function of x, then u^n is a differentiable function of x and

$$\frac{d}{dx}u^n = nu^{n-1}\frac{du}{dx},$$

provided that $u \neq 0$ if $n < 1$.

The restriction that $u \neq 0$ when $n < 1$ is necessary because 0 might be in the domain of u^n but not in the domain of u^{n-1}, as we see in the first two parts of Example 6.

Example 6   USING THE RATIONAL POWER RULE

(a) $\dfrac{d}{dx}(\sqrt{x}) = \dfrac{d}{dx}(x^{1/2}) = \dfrac{1}{2}x^{-1/2} = \dfrac{1}{2\sqrt{x}}$

Notice that $\sqrt{x}$ is defined at $x = 0$, but $1/(2\sqrt{x})$ is not.

(b) $\dfrac{d}{dx}(x^{2/3}) = \dfrac{2}{3}(x^{-1/3}) = \dfrac{2}{3x^{1/3}}$

The original function is defined for all real numbers, but the derivative is undefined at $x = 0$. Recall Figure 3.12, which showed that this function's graph has a *cusp* at $x = 0$.

(c) $\dfrac{d}{dx}(\cos x)^{-1/5} = -\dfrac{1}{5}(\cos x)^{-6/5} \cdot \dfrac{d}{dx}(\cos x)$

$$= -\frac{1}{5}(\cos x)^{-6/5}(-\sin x)$$

$$= \frac{1}{5}\sin x(\cos x)^{-6/5}$$

Quick Review 3.7

In Exercises 1–5, sketch the curve defined by the equation and find two functions y_1 and y_2 whose graphs will combine to give the curve.

1. $x - y^2 = 0$

2. $4x^2 + 9y^2 = 36$

3. $x^2 - 4y^2 = 0$

4. $x^2 + y^2 = 9$

5. $x^2 + y^2 = 2x + 3$

In Exercises 6–8, solve for y' in terms of y and x.

6. $x^2 y' - 2xy = 4x - y$

7. $y' \sin x - x \cos x = xy' + y$

8. $x(y^2 - y') = y'(x^2 - y)$

In Exercises 9–10, find an expression for the function using rational powers rather than radicals.

9. $\sqrt{x}(x - \sqrt[3]{x})$

10. $\dfrac{x + \sqrt[3]{x^2}}{\sqrt{x^3}}$

Section 3.7 Exercises

In Exercises 1–20, find dy/dx.

1. $y = x^{9/4}$

2. $y = x^{-3/5}$

3. $y = \sqrt[3]{x}$

4. $y = \sqrt[4]{x}$

5. $y = (2x + 5)^{-1/2}$

6. $y = (1 - 6x)^{2/3}$

7. $y = x\sqrt{x^2 + 1}$

8. $y = \dfrac{x}{\sqrt{x^2 + 1}}$

9. $x^2 y + xy^2 = 6$

10. $x^3 + y^3 = 18xy$

11. $y^2 = \dfrac{x - 1}{x + 1}$

12. $x^2 = \dfrac{x - y}{x + y}$

13. $y = \sqrt{1 - \sqrt{x}}$

14. $y = 3(2x^{-1/2} + 1)^{-1/3}$

15. $y = 3(\csc x)^{3/2}$

16. $y = |\sin (x + 5)]^{5/4}$

17. $x = \tan y$

18. $x = \sin y$

19. $x + \tan (xy) = 0$

20. $x + \sin y = xy$

21. Which of the following could be true if $f''(x) = x^{-1/3}$?

   (a) $f(x) = \dfrac{3}{2}x^{2/3} - 3$

   (b) $f(x) = \dfrac{9}{10}x^{5/3} - 7$

   (c) $f'''(x) = -\dfrac{1}{3}x^{-4/3}$

   (d) $f'(x) = \dfrac{3}{2}x^{2/3} + 6$

22. Which of the following could be true if $g''(t) = 1/t^{3/4}$?

   (a) $g'(t) = 4\sqrt[4]{t} - 4$

   (b) $g'''(t) = -4/\sqrt[4]{t}$

   (c) $g(t) = t - 7 + (16/5)t^{5/4}$

   (d) $g'(t) = (1/4)t^{1/4}$

In Exercises 23–26, use implicit differentiation to find dy/dx and then d^2y/dx^2.

23. $x^2 + y^2 = 1$

24. $x^{2/3} + y^{2/3} = 1$

25. $y^2 = x^2 + 2x$

26. $y^2 + 2y = 2x + 1$

In Exercises 27–36, find the lines that are **(a)** tangent and **(b)** normal to the curve at the given point.

27. $x^2 + xy - y^2 = 1$, $(2, 3)$

28. $x^2 + y^2 = 25$, $(3, -4)$

29. $x^2 y^2 = 9$, $(-1, 3)$

30. $y^2 - 2x - 4y - 1 = 0$, $(-2, 1)$

31. $6x^2 + 3xy + 2y^2 + 17y - 6 = 0$, $(-1, 0)$

32. $x^2 - \sqrt{3}xy + 2y^2 = 5$, $(\sqrt{3}, 2)$

33. $2xy + \pi \sin y = 2\pi$, $(1, \pi/2)$

34. $x \sin 2y = y \cos 2x$, $(\pi/4, \pi/2)$

35. $y = 2 \sin (\pi x - y)$, $(1, 0)$

36. $x^2 \cos^2 y - \sin y = 0$, $(0, \pi)$

37. *The Eight Curve* **(a)** Find the slopes of the figure-eight-shaped curve

$$y^4 = y^2 - x^2$$

at the two points shown on the graph that follows.

(b) Use parametric mode and the two pairs of parametric equations

$$x_1(t) = \sqrt{t^2 - t^4}, \quad y_1(t) = t,$$
$$x_2(t) = -\sqrt{t^2 - t^4}, \quad y_2(t) = t,$$

to graph the curve. Specify a window and a parameter interval.

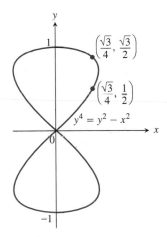

38. *The Cissoid of Diocles (dates from about 200 B.C.)*

   (a) Find equations for the tangent and normal to the cissoid of Diocles,

$$y^2(2 - x) = x^3,$$

at the point $(1, 1)$ as pictured below.

   (b) Explain how to reproduce the graph on a grapher.

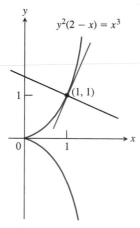

39. (a) Confirm that $(-1, 1)$ is on the curve defined by $x^3 y^2 = \cos(\pi y)$.

   (b) Use part (a) to find the slope of the line tangent to the curve at $(-1, 1)$.

40. *Work in groups of two or three.*

   (a) Show that the relation

$$y^3 - xy = -1$$

cannot be a function of x by showing that there is more than one possible y-value when $x = 2$.

   (b) On a small enough square with center $(2, 1)$, the part of the graph of the relation within the square will define a function $y = f(x)$. For this function, find $f'(2)$ and $f''(2)$.

41. Find the two points where the curve $x^2 + xy + y^2 = 7$ crosses the x-axis, and show that the tangents to the curve at these points are parallel. What is the common slope of these tangents?

42. Find points on the curve $x^2 + xy + y^2 = 7$ **(a)** where the tangent is parallel to the x-axis and **(b)** where the tangent is parallel to the y-axis. (In the latter case, dy/dx is not defined, but dx/dy is. What value does dx/dy have at these points?)

43. *Orthogonal Curves*   Two curves are *orthogonal* at a point of intersection if their tangents at that point cross at right angles. Show that the curves $2x^2 + 3y^2 = 5$ and $y^2 = x^3$ are orthogonal at $(1, 1)$ and $(1, -1)$. Use parametric mode to draw the curves and to show the tangent lines.

44. The position of a body moving along a coordinate line at time t is $s = (4 + 6t)^{3/2}$, with s in meters and t in seconds. Find the body's velocity and acceleration when $t = 2$ sec.

45. The velocity of a falling body is $v = 8\sqrt{s - t} + 1$ feet per second at the instant t(sec) the body has fallen s feet from its starting point. Show that the body's acceleration is 32 ft/sec².

46. *The Devil's Curve (Gabriel Cramer [the Cramer of Cramer's Rule], 1750)*   Find the slopes of the devil's curve $y^4 - 4y^2 = x^4 - 9x^2$ at the four indicated points.

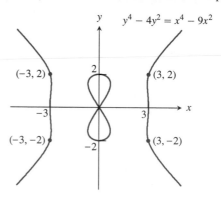

47. *The Folium of Descartes*   (See Figure 3.45 on page 149)

   (a) Find the slope of the folium of Descartes, $x^3 + y^3 - 9xy = 0$ at the points $(4, 2)$ and $(2, 4)$.

   (b) At what point other than the origin does the folium have a horizontal tangent?

   (c) Find the coordinates of the point A in Figure 3.45, where the folium has a vertical tangent.

48. The line that is normal to the curve $x^2 + 2xy - 3y^2 = 0$ at $(1, 1)$ intersects the curve at what other point?

49. Find the normals to the curve $xy + 2x - y = 0$ that are parallel to the line $2x + y = 0$.

50. Show that if it is possible to draw these three normals from the point $(a, 0)$ to the parabola $x = y^2$ shown here, then a must be greater than $1/2$. One of the normals is the x-axis. For what value of a are the other two normals perpendicular?

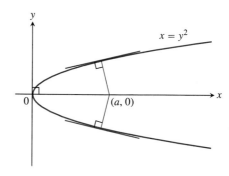

Extending the Ideas

51. *Finding Tangents*

(a) Show that the tangent to the ellipse

$$\frac{x^2}{a^2} + \frac{y^2}{b^2} = 1$$

at the point (x_1, y_1) has equation

$$\frac{x_1 x}{a^2} + \frac{y_1 y}{b^2} = 1.$$

(b) Find an equation for the tangent to the hyperbola

$$\frac{x^2}{a^2} - \frac{y^2}{b^2} = 1$$

at the point (x_1, y_1).

52. *End Behavior Model* Consider the hyperbola

$$\frac{x^2}{a^2} - \frac{y^2}{b^2} = 1.$$

Show that

(a) $y = \pm\dfrac{b}{a}\sqrt{x^2 - a^2}$.

(b) $g(x) = (b/a)|x|$ is an end behavior model for

$$f(x) = (b/a)\sqrt{x^2 - a^2}.$$

(c) $g(x) = -(b/a)|x|$ is an end behavior model for

$$f(x) = -(b/a)\sqrt{x^2 - a^2}.$$

3.8 Derivatives of Inverse Trigonometric Functions

Derivatives of Inverse Functions • Derivative of the Arcsine • Derivative of the Arctangent • Derivative of the Arcsecant • Derivatives of the Other Three

Derivatives of Inverse Functions

In Section 1.5 we learned that the graph of the inverse of a function f can be obtained by reflecting the graph of f across the line $y = x$. If we combine that with our understanding of what makes a function differentiable, we can gain some quick insights into the differentiability of inverse functions.

As Figure 3.50 suggests, the reflection of a continuous curve with no cusps or corners will be another continuous curve with no cusps or corners. Indeed, if there is a tangent line to the graph of f at the point $(a, f(a))$, then that line will reflect across $y = x$ to become a tangent line to the graph of f^{-1} at the point $(f(a), a)$. We can even see geometrically that the *slope* of the reflected tangent line (when it exists and is not zero) will be the *reciprocal* of the slope of the original tangent line, since a change in y becomes a change in x in the reflection, and a change in x becomes a change in y.

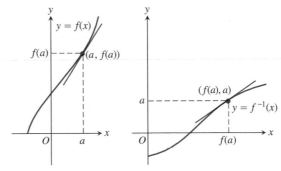

The slopes are reciprocal: $\left.\dfrac{df^{-1}}{dx}\right|_{f(a)} = \dfrac{1}{\left.\dfrac{df}{dx}\right|_a}$

Figure 3.50 The graphs of a function and its inverse. Notice that the tangent lines have reciprocal slopes.

All of this serves as an introduction to the following theorem, which we will assume as we proceed to find derivatives of inverse functions. Although the essentials of the proof are illustrated in the geometry of Figure 3.50, a careful analytic proof is more appropriate for an advanced calculus text and will be omitted here.

Theorem 3 Derivatives of Inverse Functions

If f is differentiable at every point of an interval I and df/dx is never zero on I, then f has an inverse and f^{-1} is differentiable at every point of the interval $f(I)$.

Exploration 1 **Finding a Derivative on an Inverse Graph Geometrically**

Let $f(x) = x^5 + 2x - 1$. Since the point $(1, 2)$ is on the graph of f, it follows that the point $(2, 1)$ is on the graph of f^{-1}. Can you find

$$\frac{df^{-1}}{dx}(2),$$

the value of df^{-1}/dx at 2, without knowing a formula for f^{-1}?

1. Graph $f(x) = x^5 + 2x - 1$. A function must be one-to-one to have an inverse function. Is this function one-to-one?

2. Find $f'(x)$. How could this derivative help you to conclude that f has an inverse?

3. Reflect the graph of f across the line $y = x$ to obtain a graph of f^{-1}.

4. Sketch the tangent line to the graph of f^{-1} at the point $(2, 1)$. Call it L.

5. Reflect the line L across the line $y = x$. At what point is the reflection of L tangent to the graph of f?

6. What is the slope of the reflection of L?

7. What is the slope of L?

8. What is $\dfrac{df^{-1}}{dx}(2)$?

Derivative of the Arcsine

We know that the function $x = \sin y$ is differentiable in the open interval $-\pi/2 < y < \pi/2$ and that its derivative, the cosine, is positive there. Theorem 3 therefore assures us that the inverse function $y = \sin^{-1}(x)$ (the *arcsine* of x) is differentiable throughout the interval $-1 < x < 1$. We cannot expect it to be differentiable at $x = -1$ or $x = 1$, however, because the tangents to the graph are vertical at these points (Figure 3.51).

We find the derivative of $y = \sin^{-1}(x)$ as follows:

$$y = \sin^{-1} x$$

$$\sin y = x \qquad \text{Inverse function relationship}$$

$$\frac{d}{dx}(\sin y) = \frac{d}{dx}x \qquad \text{Differentiate both sides.}$$

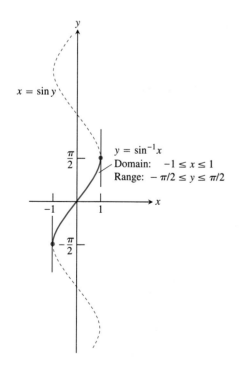

$y = \sin^{-1}x$
Domain: $-1 \leq x \leq 1$
Range: $-\pi/2 \leq y \leq \pi/2$

Figure 3.51 The graph of $y = \sin^{-1}x$ has vertical tangents $x = -1$ and $x = 1$.

$$\cos y \frac{dy}{dx} = 1 \qquad \text{Implicit differentiation}$$

$$\frac{dy}{dx} = \frac{1}{\cos y}$$

The division in the last step is safe because $\cos y \neq 0$ for $-\pi/2 < y < \pi/2$. In fact, $\cos y$ is *positive* for $-\pi/2 < y < \pi/2$, so we can replace $\cos y$ with $\sqrt{1 - (\sin y)^2}$, which is $\sqrt{1 - x^2}$. Thus

$$\frac{d}{dx}(\sin^{-1} x) = \frac{1}{\sqrt{1 - x^2}}.$$

If u is a differentiable function of x with $|u| < 1$, we apply the Chain Rule to get

$$\frac{d}{dx}\sin^{-1} u = \frac{1}{\sqrt{1 - u^2}} \frac{du}{dx}, \qquad |u| < 1.$$

Example 1 APPLYING THE FORMULA

$$\frac{d}{dx}(\sin^{-1} x^2) = \frac{1}{\sqrt{1 - (x^2)^2}} \cdot \frac{d}{dx}(x^2) = \frac{2x}{\sqrt{1 - x^4}}$$

Derivative of the Arctangent

Although the function $y = \sin^{-1}(x)$ has a rather narrow domain of $[-1, 1]$, the function $y = \tan^{-1} x$ is defined for all real numbers. It is also differentiable for all real numbers, as we will now see. The differentiation proceeds exactly as with the arcsine function above.

$$y = \tan^{-1} x$$

$$\tan y = x \qquad \text{Inverse function relationship}$$

$$\frac{d}{dx}(\tan y) = \frac{d}{dx}x$$

$$\sec^2 y \frac{dy}{dx} = 1 \qquad \text{Implicit differentiation}$$

$$\frac{dy}{dx} = \frac{1}{\sec^2 y}$$

$$= \frac{1}{1 + (\tan y)^2} \qquad \text{Trig identity: } \sec^2 y = 1 + \tan^2 y$$

$$= \frac{1}{1 + x^2}$$

The derivative is defined for all real numbers. If u is a differentiable function of x, we get the Chain Rule form:

$$\frac{d}{dx}\tan^{-1} u = \frac{1}{1 + u^2} \frac{du}{dx}.$$

Example 2 A MOVING PARTICLE

A particle moves along the x-axis so that its position at any time $t \geq 0$ is $x(t) = \tan^{-1} \sqrt{t}$. What is the velocity of the particle when $t = 16$?

Solution

$$v(t) = \frac{d}{dt}\tan^{-1}\sqrt{t} = \frac{1}{1 + (\sqrt{t})^2} \cdot \frac{d}{dt}\sqrt{t} = \frac{1}{1 + t} \cdot \frac{1}{2\sqrt{t}}$$

When $t = 16$, the velocity is

$$v(16) = \frac{1}{1 + 16} \cdot \frac{1}{2\sqrt{16}} = \frac{1}{136}.$$

Derivative of the Arcsecant

We find the derivative of $y = \sec^{-1} x$, $|x| > 1$, beginning as we did with the other inverse trigonometric functions.

$$y = \sec^{-1} x$$

$$\sec y = x \qquad \text{Inverse function relationship}$$

$$\frac{d}{dx}(\sec y) = \frac{d}{dx}x$$

$$\sec y \tan y \frac{dy}{dx} = 1$$

$$\frac{dy}{dx} = \frac{1}{\sec y \tan y} \qquad \text{Since } |x| > 1, y \text{ lies in } (0, \pi/2) \cup (\pi/2, \pi)$$
$$\text{and } \sec y \tan y \neq 0.$$

To express the result in terms of x, we use the relationships

$$\sec y = x \quad \text{and} \quad \tan y = \pm\sqrt{\sec^2 y - 1} = \pm\sqrt{x^2 - 1}$$

to get

$$\frac{dy}{dx} = \pm\frac{1}{x\sqrt{x^2 - 1}}.$$

Can we do anything about the $\pm$ sign? A glance at Figure 3.52 shows that the slope of the graph $y = \sec^{-1} x$ is always positive. That must mean that

$$\frac{d}{dx}\sec^{-1} x = \begin{cases} +\dfrac{1}{x\sqrt{x^2 - 1}} & \text{if } x > 1 \\[2mm] -\dfrac{1}{x\sqrt{x^2 - 1}} & \text{if } x < -1. \end{cases}$$

With the absolute value symbol we can write a single expression that eliminates the "$\pm$" ambiguity:

$$\frac{d}{dx}\sec^{-1} x = \frac{1}{|x|\sqrt{x^2 - 1}}.$$

If u is a differentiable function of x with $|u| > 1$, we have the formula

$$\frac{d}{dx}\sec^{-1} u = \frac{1}{|u|\sqrt{u^2 - 1}} \frac{du}{dx}, \qquad |u| > 1.$$

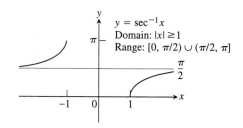

$y = \sec^{-1}x$
Domain: $|x| \geq 1$
Range: $[0, \pi/2) \cup (\pi/2, \pi]$

Figure 3.52 The slope of the curve $y = \sec^{-1} x$ is positive for both $x < -1$ and $x > 1$.

Example 3 USING THE FORMULA

$$\frac{d}{dx}\sec^{-1}(5x^4) = \frac{1}{|5x^4|\sqrt{(5x^4)^2 - 1}}\frac{d}{dx}(5x^4)$$

$$= \frac{1}{5x^4\sqrt{25x^8 - 1}}(20x^3)$$

$$= \frac{4}{x\sqrt{25x^8 - 1}}$$

Derivatives of the Other Three

We could use the same technique to find the derivatives of the other three inverse trigonometric functions: arccosine, arccotangent, and arccosecant, but there is a much easier way, thanks to the following identities.

Inverse Function – Inverse Cofunction Identities

$$\cos^{-1} x = \pi/2 - \sin^{-1} x$$

$$\cot^{-1} x = \pi/2 - \tan^{-1} x$$

$$\csc^{-1} x = \pi/2 - \sec^{-1} x$$

It follows easily that the derivatives of the inverse cofunctions are the negatives of the derivatives of the corresponding inverse functions (see Exercises 24–26).

You have probably noticed by now that most calculators do not have buttons for $\cot^{-1}$, $\sec^{-1}$, or $\csc^{-1}$. They are not needed because of the following identities:

Calculator Conversion Identities

$$\sec^{-1} x = \cos^{-1}(1/x)$$

$$\cot^{-1} x = \pi/2 - \tan^{-1} x$$

$$\csc^{-1} x = \sin^{-1}(1/x)$$

Notice that we do not use $\tan^{-1}(1/x)$ as an identity for $\cot^{-1} x$. A glance at the graphs of $y = \tan^{-1}(1/x)$ and $y = \pi/2 - \tan^{-1} x$ reveals the problem (Figure 3.53).

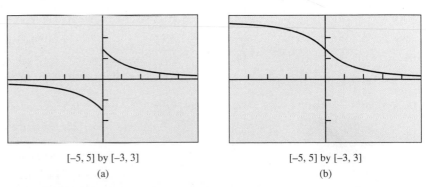

[–5, 5] by [–3, 3]	[–5, 5] by [–3, 3]
(a)	(b)

Figure 3.53 The graphs of (a) $y = \tan^{-1}(1/x)$ and (b) $y = \pi/2 - \tan^{-1} x$. The graph in (b) is the same as the graph of $y = \cot^{-1} x$.

We cannot replace $\cot^{-1} x$ by the function $y = \tan^{-1}(1/x)$ in the identity for the inverse functions and inverse cofunctions, and so it is not the function we want for $\cot^{-1} x$. The ranges of the inverse trigonometric functions have been chosen in part to make the two sets of identities above hold.

Example 4 A TANGENT LINE TO THE ARCCOTANGENT CURVE

Find an equation for the line tangent to the graph of $y = \cot^{-1} x$ at $x = -1$.

Solution First, we note that

$$\cot^{-1}(-1) = \pi/2 - \tan^{-1}(-1) = \pi/2 - (-\pi/4) = 3\pi/4.$$

The slope of the tangent line is

$$\frac{dy}{dx}\bigg|_{x=-1} = -\frac{1}{1+x^2}\bigg|_{x=-1} = -\frac{1}{1+(-1)^2} = -\frac{1}{2}.$$

So the tangent line has equation $y - 3\pi/4 = (-1/2)(x + 1)$.

Quick Review 3.8

In Exercises 1–5, give the *domain* and *range* of the function, and evaluate the function at $x = 1$.

1. $y = \sin^{-1} x$ **2.** $y = \cos^{-1} x$

3. $y = \tan^{-1} x$ **4.** $y = \sec^{-1} x$

5. $y = \tan(\tan^{-1} x)$

In Exercises 6–10, find the inverse of the given function.

6. $y = 3x - 8$ **7.** $y = \sqrt[3]{x + 5}$

8. $y = \dfrac{8}{x}$ **9.** $y = \dfrac{3x - 2}{x}$

10. $y = \arctan(x/3)$

Section 3.8 Exercises

In Exercises 1–18, find the derivative of y with respect to the appropriate variable.

1. $y = \cos^{-1}(x^2)$ **2.** $y = \cos^{-1}(1/x)$

3. $y = \sin^{-1}\sqrt{2}t$ **4.** $y = \sin^{-1}(1 - t)$

5. $y = \sec^{-1}(2s + 1)$ **6.** $y = \sec^{-1} 5s$

7. $y = \csc^{-1}(x^2 + 1), \quad x > 0$ **8.** $y = \csc^{-1} x/2$

9. $y = \sec^{-1}\dfrac{1}{t}, \quad 0 < t < 1$ **10.** $y = \sin^{-1}\dfrac{3}{t^2}$

11. $y = \cot^{-1}\sqrt{t}$ **12.** $y = \cot^{-1}\sqrt{t - 1}$

13. $y = s\sqrt{1 - s^2} + \cos^{-1} s$ **14.** $y = \sqrt{s^2 - 1} - \sec^{-1} s$

15. $y = \tan^{-1}\sqrt{x^2 - 1} + \csc^{-1} x, \quad x > 1$

16. $y = \cot^{-1}\dfrac{1}{x} - \tan^{-1} x$ **17.** $y = x\sin^{-1} x + \sqrt{1 - x^2}$

18. $y = \dfrac{1}{\sin^{-1}(2x)}$

19. (a) Find an equation for the line tangent to the graph of $y = \tan x$ at the point $(\pi/4, 1)$.

    (b) Find an equation for the line tangent to the graph of $y = \tan^{-1} x$ at the point $(1, \pi/4)$.

20. Let $f(x) = x^5 + 2x^3 + x - 1$.

    (a) Find $f(1)$ and $f'(1)$.

    (b) Find $f^{-1}(3)$ and $(f^{-1})'(3)$.

21. Let $f(x) = \cos x + 3x$.

    (a) Show that f has a differentiable inverse.

    (b) Find $f(0)$ and $f'(0)$.

    (c) Find $f^{-1}(1)$ and $(f^{-1})'(1)$.

In Excercises 22 and 23, *work in groups of two or three.*

22. Graph the function $f(x) = \sin^{-1}(\sin x)$ in the viewing window $[-2\pi, 2\pi]$ by $[-4, 4]$. Then answer the following questions:

    (a) What is the domain of f?

    (b) What is the range of f?

(c) At which points is f not differentiable?

(d) Sketch a graph of $y = f'(x)$ without using NDER or computing the derivative.

(e) Find $f'(x)$ algebraically. Can you reconcile your answer with the graph in (d)?

23. A particle moves along the x-axis so that its position at any time $t \geq 0$ is given by $x = \arctan t$.

(a) Prove that the particle is always moving to the right.

(b) Prove that the particle is always decelerating.

(c) What is the limiting position of the particle as t approaches infinity?

In Exercises 24–26, use the inverse function–inverse cofunction identities to derive the formula for the derivative of the function.

24. arccosine

25. arccotangent

26. arccosecant

Explorations

In Exercises 27–32, find **(a)** the right end behavior model, **(b)** the left end behavior model, and **(c)** any horizontal tangents for the function if they exist.

27. $y = \tan^{-1} x$

28. $y = \cot^{-1} x$

29. $y = \sec^{-1} x$

30. $y = \csc^{-1} x$

31. $y = \sin^{-1} x$

32. $y = \cos^{-1} x$ ■

Extending the Ideas

33. *Identities* Confirm the following identities for $x > 0$.

(a) $\cos^{-1} x + \sin^{-1} x = \pi/2$

(b) $\tan^{-1} x + \cot^{-1} x = \pi/2$

(c) $\sec^{-1} x + \csc^{-1} x = \pi/2$

34. *Proof Without Words* The figure gives a proof without words that $\tan^{-1} 1 + \tan^{-1} 2 + \tan^{-1} 3 = \pi$. Explain what is going on.

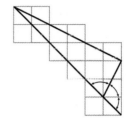

35. *(Continuation of Exercise 34)* Here is a way to construct $\tan^{-1} 1$, $\tan^{-1} 2$, and $\tan^{-1} 3$ by folding a square of paper. Try it and explain what is going on.

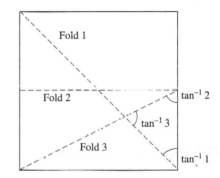

3.9

Derivatives of Exponential and Logarithmic Functions

Derivative of e^x • Derivative of a^x • Derivative of $\ln x$ • Derivative of $\log_a x$ • Power Rule for Arbitrary Real Powers

Derivative of e^x

At the end of the brief review of exponential functions in Section 1.3, we mentioned that the function $y = e^x$ was a particularly important function for modeling exponential growth. The number e was defined in that section to be the limit of $(1 + 1/x)^x$ as $x \to \infty$. This intriguing number shows up in other interesting limits as well, but the one with the most interesting implications for the *calculus* of exponential functions is this one:

$$\lim_{h \to 0} \frac{e^h - 1}{h} = 1.$$

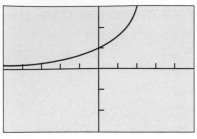

[−4.9, 4.9] by [−2.9, 2.9]

(a)

X	Y₁	
−.03	.98515	
−.02	.99007	
−.01	.99502	
0	ERROR	
.01	1.005	
.02	1.0101	
.03	1.0152	

X=0

(b)

Figure 3.54 (a) The graph and (b) the table support the conclusion that

$$\lim_{h\to 0} \frac{e^h - 1}{h} = 1.$$

Is any other function its own derivative?

The zero function is also its own derivative, but this hardly seems worth mentioning. (Its value is always 0 and its slope is always 0.) In addition to e^x, however, we can also say that any constant *multiple* of e^x is its own derivative:

$$\frac{d}{dx}(c \cdot e^x) = c \cdot e^x.$$

The next obvious question is whether there are still *other* functions that are their own derivatives, and this time the answer is no. The only functions that satisfy the condition $dy/dx = y$ are functions of the form $y = ke^x$ (and notice that the zero function can be included in this category). We will prove this significant fact in Chapter 6.

(The graph and the table in Figure 3.54 provide strong support for this limit being 1. A formal algebraic proof that begins with our limit definition of e would require some rather subtle limit arguments, so we will not include one here.)

The fact that the limit is 1 creates a remarkable relationship between the function e^x and its derivative, as we will now see.

$$\begin{aligned} \frac{d}{dx}(e^x) &= \lim_{h\to 0} \frac{e^{x+h} - e^x}{h} \\ &= \lim_{h\to 0} \frac{e^x \cdot e^h - e^x}{h} \\ &= \lim_{h\to 0} \left(e^x \cdot \frac{e^h - 1}{h}\right) \\ &= e^x \cdot \lim_{h\to 0} \left(\frac{e^h - 1}{h}\right) \\ &= e^x \cdot 1 \\ &= e^x \end{aligned}$$

In other words, the derivative of this particular function is itself!

$$\frac{d}{dx}(e^x) = e^x$$

If u is a differentiable function of x, then we have

$$\frac{d}{dx}e^u = e^u \frac{du}{dx}.$$

We will make extensive use of this formula when we study exponential growth and decay in Chapter 6.

Example 1 HOW FAST DOES A FLU SPREAD?

The spread of a flu in a certain school is modeled by the equation

$$P(t) = \frac{100}{1 + e^{3-t}},$$

where $P(t)$ is the total number of students infected t days after the flu was first noticed. Many of them may already be well again at time t.

(a) Estimate the initial number of students infected with the flu.

(b) How fast is the flu spreading after 3 days?

(c) When will the flu spread at its maximum rate? What is this rate?

Solution The graph of P as a function of t is shown in Figure 3.55.

(a) $P(0) = 100/(1 + e^3) = 5$ students (to the nearest whole number).

(b) To find the rate at which the flu spreads, we find dP/dt. To find dP/dt, we need to invoke the Chain Rule twice:

$$\frac{dP}{dt} = \frac{d}{dt}(100(1 + e^{3-t})^{-1}) = 100 \cdot (-1)(1 + e^{3-t})^{-2} \cdot \frac{d}{dt}(1 + e^{3-t})$$

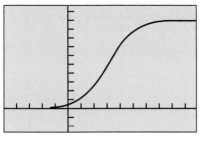

[−5, 10] by [−20, 120]

Figure 3.55 The graph of

$$P(t) = \frac{100}{1 + e^{3-t}},$$

modeling the spread of a flu. (Example 1)

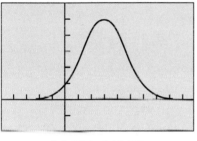

[−5, 10] by [−10, 30]

Figure 3.56 The graph of dP/dt, the rate of spread of the flu in Example 1. The graph of P is shown in Figure 3.55

$$= -100(1 + e^{3-t})^{-2} \cdot (0 + e^{3-t} \cdot \frac{d}{dt}(3 - t))$$

$$= -100(1 + e^{3-t})^{-2}(e^{3-t} \cdot (-1))$$

$$= \frac{100e^{3-t}}{(1 + e^{3-t})^2}$$

At $t = 3$, then, $dP/dt = 100/4 = 25$. The flu is spreading to 25 students per day.

(c) We could estimate when the flu is spreading the fastest by seeing where the graph of $y = P(t)$ has the steepest upward slope, but we can answer both the "when" and the "what" parts of this question most easily by finding the maximum point on the graph of the derivative (Figure 3.56).

We see by tracing on the curve that the maximum rate occurs at about 3 days, when (as we have just calculated) the flu is spreading at a rate of 25 students per day.

Derivative of a^x

What about an exponential function with a base other than e? We will assume that the base is positive and different from 1, since negative numbers to arbitrary real powers are not always real numbers, and $y = 1^x$ is a constant function.

If $a > 0$ and $a \neq 1$, we can use the properties of logarithms to write a^x in terms of e^x. The formula for doing so is

$$a^x = e^{x \ln a}. \qquad e^{x \ln a} = e^{\ln(a^x)} = a^x$$

We can then find the derivative of a^x with the Chain Rule.

$$\frac{d}{dx}a^x - \frac{d}{dx}e^{x \ln a} = e^{x \ln a} \cdot \frac{d}{dx}(x \ln a) = e^{x \ln a} \cdot \ln a = a^x \ln a$$

Thus, if u is a differentiable function of x, we get the following rule.

For $a > 0$ and $a \neq 1$,

$$\frac{d}{dx}(a^u) = a^u \ln a \frac{du}{dx}.$$

Example 2   REVIEWING THE ALGEBRA OF LOGARITHMS

At what point on the graph of the function $y = 2^t - 3$ does the tangent line have slope 21?

Solution The slope is the derivative:

$$\frac{d}{dt}(2^t - 3) = 2^t \cdot \ln 2 - 0 = 2^t \ln 2.$$

We want the value of t for which $2^t \ln 2 = 21$. We could use the solver on the calculator, but we will use logarithms for the sake of review.

$$2^t \ln 2 = 21$$

$$2^t = \frac{21}{\ln 2}$$

$$\ln 2^t = \ln\left(\frac{21}{\ln 2}\right) \qquad \text{Logarithm of both sides}$$

$$t \cdot \ln 2 = \ln 21 - \ln(\ln 2) \qquad \text{Properties of logarithms}$$

$$t = \frac{\ln 21 - \ln(\ln 2)}{\ln 2}$$

$$t \approx 4.921$$

$$y = 2^t - 3 \approx 27.297 \qquad \text{Using the stored value of } t$$

The point is approximately $(4.9, 27.3)$.

Exploration 1 **Leaving Milk on the Counter**

A glass of cold milk from the refrigerator is left on the counter on a warm summer day. Its temperature y (in degrees Fahrenheit) after sitting on the counter t minutes is

$$y = 72 - 30(0.98)^t.$$

Answer the following questions by interpreting y and dy/dt.

1. What is the temperature of the refrigerator? How can you tell?
2. What is the temperature of the room? How can you tell?
3. When is the milk warming up the fastest? How can you tell?
4. Determine algebraically when the temperature of the milk reaches 55°F.
5. At what rate is the milk warming when its temperature is 55°F? Answer with an appropriate unit of measure.

Derivative of ln x

Now that we know the derivative of e^x, it is relatively easy to find the derivative of its inverse function, $\ln x$.

$$y = \ln x$$

$$e^y = x \qquad \text{Inverse function relationship}$$

$$\frac{d}{dx}(e^y) = \frac{d}{dx}(x) \qquad \text{Differentiate implicitly.}$$

$$e^y \frac{dy}{dx} = 1$$

$$\frac{dy}{dx} = \frac{1}{e^y} = \frac{1}{x}$$

If u is a differentiable function of x and $u > 0$,

$$\frac{d}{dx}\ln u = \frac{1}{u}\frac{du}{dx}.$$

This equation answers what was once a perplexing problem: Is there a function with derivative x^{-1}? All of the other power functions follow the Power Rule,

$$\frac{d}{dx}x^n = nx^{n-1}.$$

However, this formula is not much help if one is looking for a function with x^{-1} as its derivative! Now we know why: The function we should be looking for is not a power function at all; it is the natural logarithm function.

Example 3 A TANGENT THROUGH THE ORIGIN

A line with slope m passes through the origin and is tangent to the graph of $y = \ln x$. What is the value of m?

Solution This problem is a little harder than it looks, since we do not know the point of tangency. However, we do know two important facts about that point:

1. it has coordinates $(a, \ln a)$ for some positive a, and

2. the tangent line there has slope $m = 1/a$ (Figure 3.57).

Since the tangent line passes through the origin, its slope is

$$m = \frac{\ln a - 0}{a - 0} = \frac{\ln a}{a}.$$

Setting these two formulas for m equal to each other, we have

$$\frac{\ln a}{a} = \frac{1}{a}$$

$$\ln a = 1$$

$$e^{\ln a} = e^1$$

$$a = e$$

$$m = \frac{1}{e}.$$

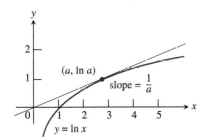

Figure 3.57 The tangent line intersects the curve at some point $(a, \ln a)$, where the slope of the curve is $1/a$. (Example 3)

Derivative of $\log_a x$

To find the derivative of $\log_a x$ for an arbitrary base $(a > 0, a \neq 1)$, we use the change-of-base formula for logarithms to express $\log_a x$ in terms of natural logarithms, as follows:

$$\log_a x = \frac{\ln x}{\ln a}.$$

The rest is easy:

$$\frac{d}{dx}\log_a x = \frac{d}{dx}\left(\frac{\ln x}{\ln a}\right)$$

$$= \frac{1}{\ln a} \cdot \frac{d}{dx}\ln x \quad \text{Since ln } a \text{ is a constant}$$

$$= \frac{1}{\ln a} \cdot \frac{1}{x}$$

$$= \frac{1}{x \ln a}.$$

So, if u is a differentiable function of x and $u > 0$, the formula is as follows.

> For $a > 0$ and $a \neq 1$,
> $$\frac{d}{dx}\log_a u = \frac{1}{u \ln a}\frac{du}{dx}.$$

Example 4 GOING THE LONG WAY WITH THE CHAIN RULE

Find dy/dx if $y = \log_a a^{\sin x}$.

Solution Carefully working from the outside in, we apply the Chain Rule to get:

$$\frac{d}{dx}(\log_a a^{\sin x}) = \frac{1}{a^{\sin x} \ln a} \cdot \frac{d}{dx}(a^{\sin x}) \qquad \log_a u, \ u = a^{\sin x}$$

$$= \frac{1}{a^{\sin x} \ln a} \cdot a^{\sin x} \ln a \cdot \frac{d}{dx}(\sin x) \quad a^u, \ u = \sin x$$

$$= \frac{a^{\sin x} \ln a}{a^{\sin x} \ln a} \cdot \cos x$$

$$= \cos x.$$

We could have saved ourselves a lot of work in Example 4 if we had noticed at the beginning that $\log_a a^{\sin x}$, being the composite of inverse functions, is equal to $\sin x$. It is always a good idea to simplify functions *before* differentiating, wherever possible. On the other hand, it is comforting to know that all these rules do work if applied correctly.

Power Rule for Arbitrary Real Powers

We are now ready to prove the Power Rule in its final form. As long as $x > 0$, we can write any real power of x as a power of e, specifically

$$x^n = e^{n \ln x}.$$

This enables us to differentiate x^n for any real power n, as follows:

$$\frac{d}{dx}(x^n) = \frac{d}{dx}(e^{n \ln x})$$

$$= e^{n \ln x} \cdot \frac{d}{dx}(n \ln x) \quad e^u, \ u = n \ln x$$

$$= e^{n \ln x} \cdot \frac{n}{x}$$

$$= x^n \cdot \frac{n}{x}$$

$$= nx^{n-1}.$$

The Chain Rule extends this result to the Power Rule's final form.

Rule 10 **Power Rule for Arbitrary Real Powers**

If u is a positive differentiable function of x and n is any real number, then u^n is a differentiable function of x, and

$$\frac{d}{dx}u^n = nu^{n-1}\frac{du}{dx}.$$

Example 5 USING THE POWER RULE IN ALL ITS POWER

(a) If $y = x^{\sqrt{2}}$, then
$$\frac{dy}{dx} = \sqrt{2}\,x^{(\sqrt{2}-1)}.$$

(b) If $y = (2 + \sin 3x)^{\pi}$, then
$$\frac{d}{dx}(2 + \sin 3x)^{\pi} = \pi(2 + \sin 3x)^{\pi-1}(\cos 3x) \cdot 3$$
$$= 3\pi(2 + \sin 3x)^{\pi-1}(\cos 3x).$$

Example 6 FINDING DOMAIN

If $f(x) = \ln(x - 3)$, find $f'(x)$. State the domain of f'.

Solution The domain of f is $(3, \infty)$ and
$$f'(x) = \frac{1}{x - 3}.$$

The domain of f' appears to be all $x \neq 3$. However, since f is not defined for $x < 3$, neither is f'. Thus,
$$f'(x) = \frac{1}{x - 3}, \quad x > 3.$$

That is, the domain of f' is $(3, \infty)$.

Sometimes the properties of logarithms can be used to simplify the differentiation process, even if we must introduce the logarithms ourselves as a step in the process. Example 7 shows a clever way to differentiate $y = x^x$ for $x > 0$.

Example 7 LOGARITHMIC DIFFERENTIATION

Find dy/dx for $y = x^x$, $x > 0$.

$$y = x^x$$
$$\ln y = \ln x^x \qquad \text{Logs of both sides}$$
$$\ln y = x \ln x \qquad \text{Property of logs}$$
$$\frac{d}{dx}(\ln y) = \frac{d}{dx}(x \ln x) \qquad \text{Differentiate implicitly.}$$
$$\frac{1}{y}\frac{dy}{dx} = 1 \cdot \ln x + x \cdot \frac{1}{x}$$
$$\frac{dy}{dx} = y(\ln x + 1)$$
$$\frac{dy}{dx} = x^x(\ln x + 1)$$

Quick Review 3.9

1. Write $\log_5 8$ in terms of natural logarithms.

2. Write 7^x as a power of e.

In Exercises 3–7, simplify the expression using properties of exponents and logarithms.

3. $\ln(e^{\tan x})$

4. $\ln(x^2 - 4) - \ln(x + 2)$

5. $\log_2(8^{x-5})$

6. $(\log_4 x^{15})/(\log_4 x^{12})$

7. $3 \ln x - \ln 3x + \ln(12x^2)$

In Exercises 8–10, solve the equation algebraically using logarithms. Give an *exact* answer, such as $(\ln 2)/3$, and also an approximate answer to the nearest hundredth.

8. $3^x = 19$

9. $5^t \ln 5 = 18$

10. $3^{x+1} = 2^x$

Section 3.9 Exercises

In Exercises 1–40, find dy/dx. Remember that you can use NDER to support your computations.

1. $y = 2e^x$

2. $y = e^{2x}$

3. $y = e^{-x}$

4. $y = e^{-5x}$

5. $y = e^{2x/3}$

6. $y = e^{-x/4}$

7. $y = xe^2 - e^x$

8. $y = x^2 e^x - xe^x$

9. $y = e^{\sqrt{x}}$

10. $y = e^{(x^2)}$

11. $y = x^\pi$

12. $y = x^{1+\sqrt{2}}$

13. $y = x^{-\sqrt{2}}$

14. $y = x^{1-e}$

15. $y = 8^x$

16. $y = 9^{-x}$

17. $y = 3^{\csc x}$

18. $y = 3^{\cot x}$

19. $y = x^{\ln x}$

20. $y = x^{(1/\ln x)}$

21. $y = \ln(x^2)$

22. $y = (\ln x)^2$

23. $y = \ln(1/x)$

24. $y = \ln(10/x)$

25. $y = \ln(x + 2)$

26. $y = \ln(2x + 2)$

27. $y = \ln(2 - \cos x)$

28. $y = \ln(x^2 + 1)$

29. $y = \ln(\ln x)$

30. $y = x \ln x - x$

31. $y = \log_4 x^2$

32. $y = \log_5 \sqrt{x}$

33. $y = \log_2(3x + 1)$

34. $y = \log_{10} \sqrt{x + 1}$

35. $y = \log_2(1/x)$

36. $y = 1/\log_2 x$

37. $y = \ln 2 \cdot \log_2 x$

38. $y = \log_3(1 + x \ln 3)$

39. $y = \log_{10} e^x$

40. $y = \ln 10^x$

41. Find an equation for a line that is tangent to the graph of $y = e^x$ and goes through the origin.

42. Find an equation for a line that is normal to the graph of $y = xe^x$ and goes through the origin.

In Exercises 43–46, use the technique of logarithmic differentiation to find dy/dx. *Work in groups of two or three.*

43. $y = (\sin x)^x, \quad 0 < x < \pi/2$

44. $y = x^{\tan x}, \quad x > 0$

45. $y = \sqrt[5]{\dfrac{(x - 3)^4(x^2 + 1)}{(2x + 5)^3}}$

46. $y = \dfrac{x\sqrt{x^2 + 1}}{(x + 1)^{2/3}}$

47. *Radioactive Decay* The amount A (in grams) of radioactive plutonium remaining in a 20-gram sample after t days is given by the formula

$$A = 20 \cdot (1/2)^{t/140}.$$

At what rate is the plutonium decaying when $t = 2$ days? Answer in appropriate units.

48. For any positive constant k, the derivative of $\ln(kx)$ is $1/x$. Prove this fact

(a) by using the Chain Rule.

(b) by using a property of logarithms and differentiating.

49. Let $f(x) = 2^x$.

(a) Find $f'(0)$.

(b) Use the definition of the derivative to write $f'(0)$ as a limit.

(c) Deduce the exact value of

$$\lim_{h \to 0} \frac{2^h - 1}{h}.$$

(d) What is the exact value of

$$\lim_{h \to 0} \frac{7^h - 1}{h}?$$

50. **Writing to Learn** The graph of $y = \ln x$ looks as though it might be approaching a horizontal asymptote. Write an argument based on the graph of $y = e^x$ to explain why it does not.

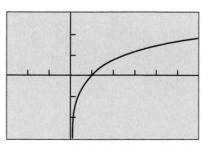

$[-3, 6]$ by $[-3, 3]$

Exploration

51. Let $y_1 = a^x$, $y_2 = \text{NDER } y_1$, $y_3 = y_2/y_1$, and $y_4 = e^{y_3}$.

(a) Describe the graph of y_4 for $a = 2,\ 3,\ 4,\ 5$. Generalize your description to an arbitrary $a > 1$.

(b) Describe the graph of y_3 for $a = 2,\ 3,\ 4,\ 5$. Compare a table of values for y_3 for $a = 2,\ 3,\ 4,\ 5$ with $\ln a$. Generalize your description to an arbitrary $a > 1$.

(c) Explain how parts (a) and (b) support the statement

$$\frac{d}{dx} a^x = a^x \quad \text{if and only if} \quad a = e.$$

(d) Show algebraically that $y_1 = y_2$ if and only if $a = e$. ∎

Extending the Ideas

52. *Orthogonal Families of Curves* Prove that all curves in the family

$$y = -\frac{1}{2}x^2 + k$$

(k any constant) are perpendicular to all curves in the family $y = \ln x + c$ (c any constant) at their points of intersection. (See figure below.)

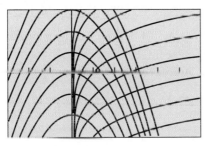

[−3, 6] by [−3, 3]

53. *Which is Bigger, π^e or e^π?* Calculators have taken some of the mystery out of this once-challenging question. (Go ahead and check; you will see that it is a surprisingly close call.) You can answer the question without a calculator, though, by using the result from Example 3 of this section.

Recall from that example that the line through the origin tangent to the graph of $y = \ln x$ has slope $1/e$.

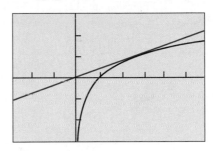

[−3, 6] by [−3, 3]

(a) Find an equation for this tangent line.

(b) Give an argument based on the graphs of $y = \ln x$ and the tangent line to explain why $\ln x < x/e$ for all positive $x \neq e$.

(c) Show that $\ln (x^e) < x$ for all positive $x \neq e$.

(d) Conclude that $x^e < e^x$ for all positive $x \neq e$.

(e) So which is bigger, π^e or e^π?

Calculus at Work

I work at Ramsey County Hospital and other community hospitals in the Minneapolis area, both with patients and in a laboratory. I have wanted to be a physician since I was about 12 years old, and I began attending medical school when I was 30 years old. I am now working in the field of internal medicine.

Cardiac patients are common in my field, especially in the diagnostic stages. One of the machines that is sometimes used in the emergency room to diagnose problems is called a Swan-Ganz catheter, named after its inventors Harold James Swan and William Ganz. The catheter is inserted into the pulmonary artery and then is hooked up to a cardiac monitor. A program measures cardiac output by looking at changes of slope in the curve. This information alerts me to left-sided heart failure.

Lupe Bolding, M.D.
Ramsey County Hospital
Minneapolis, MN

Chapter 3   Key Terms

acceleration (p. 125)
average velocity (p. 123)
Chain Rule (p. 142)
Constant Multiple Rule (p. 113)
Derivative of a Constant Function (p. 112)
derivative of f at a (p. 95)
differentiable function (p. 95)
differentiable on a closed interval (p. 100)
displacement (p. 123)
free-fall constants (p. 125)
implicit differentiation (p. 149)
instantaneous rate of change (p. 122)
instantaneous velocity (p. 124)
Intermediate Value Theorem for
    Derivatives (p. 110)

inverse function–inverse
    cofunction identities (p. 161)
jerk (p. 137)
left-hand derivative (p. 100)
local linearity (p. 107)
logarithmic differentiation (p. 169)
marginal cost (p. 128)
marginal revenue (p. 129)
nth derivative (p. 119)
numerical derivative (NDER) (p. 108)
orthogonal curves (p. 148)
orthogonal families (p. 171)
Power Chain Rule (p. 145)
Power Rule for Arbitrary Real
    Powers of u (p. 168)

Power Rule for Negative Integer
    Powers of x (p. 118)
Power Rule for Positive Integer
    Powers of x (p. 113)
Power Rule for Rational Powers
    of u (p. 153)
Product Rule (p. 115)
Quotient Rule (p. 117)
right-hand derivative (p. 100)
sensitivity to change (p. 128)
simple harmonic motion (p. 136)
speed (p. 124)
Sum and Difference Rule (p. 114)
symmetric difference quotient (p. 108)
velocity (p. 124)

Chapter 3   Review Exercises

In Exercises 1–30, find the derivative of the function.

1. $y = x^5 - \dfrac{1}{8}x^2 + \dfrac{1}{4}x$

2. $y = 3 - 7x^3 + 3x^7$

3. $y = 2 \sin x \cos x$

4. $y = \dfrac{2x + 1}{2x - 1}$

5. $s = \cos (1 - 2t)$

6. $s = \cot \dfrac{2}{t}$

7. $y = \sqrt{x} + 1 + \dfrac{1}{\sqrt{x}}$

8. $y = x\sqrt{2x + 1}$

9. $r = \sec (1 + 3\theta)$

10. $r = \tan^2 (3 - \theta^2)$

11. $y = x^2 \csc 5x$

12. $y = \ln \sqrt{x}$

13. $y = \ln (1 + e^x)$

14. $y = xe^{-x}$

15. $y = e^{(1 + \ln x)}$

16. $y = \ln (\sin x)$

17. $r = \ln (\cos^{-1} x)$

18. $r = \log_2 (\theta^2)$

19. $s = \log_5 (t - 7)$

20. $s = 8^{-t}$

21. $y = x^{\ln x}$

22. $y = \dfrac{(2x)2^x}{\sqrt{x^2 + 1}}$

23. $y = e^{\tan^{-1} x}$

24. $y = \sin^{-1}\sqrt{1 - u^2}$

25. $y = t \sec^{-1} t - \dfrac{1}{2}\ln t$

26. $y = (1 + t^2) \cot^{-1} 2t$

27. $y = z \cos^{-1} z - \sqrt{1 - z^2}$

28. $y = 2\sqrt{x - 1} \csc^{-1}\sqrt{x}$

29. $y = \csc^{-1} (\sec x), 0 \le x \le 2\pi$

30. $r = \left(\dfrac{1 + \sin \theta}{1 - \cos \theta}\right)^2$

In Exercises 31–34, find all values of x for which the function is differentiable.

31. $y = \ln x^2$

32. $y = \sin x - x \cos x$

33. $y = \sqrt{\dfrac{1 - x}{1 + x^2}}$

34. $y = (2x - 7)^{-1}(x + 5)$

In Exercises 35–38, find dy/dx.

35. $xy + 2x + 3y = 1$

36. $5x^{4/5} + 10y^{6/5} = 15$

37. $\sqrt{xy} = 1$

38. $y^2 = \dfrac{x}{x + 1}$

In Exercises 39–42, find d^2y/dx^2 by implicit differentiation.

39. $x^3 + y^3 = 1$

40. $y^2 = 1 - \dfrac{2}{x}$

41. $y^3 + y = 2 \cos x$

42. $x^{1/3} + y^{1/3} = 4$

In Exercises 43 and 44, find all derivatives of the function.

43. $y = \dfrac{x^4}{2} - \dfrac{3}{2}x^2 - x$

44. $y = \dfrac{x^5}{120}$

In Exercises 45–48, find an equation for the **(a)** tangent and **(b)** normal to the curve at the indicated point.

45. $y = \sqrt{x^2 - 2x}, \quad x = 3$

46. $y = 4 + \cot x - 2 \csc x, \quad x = \pi/2$

47. $x^2 + 2y^2 = 9, \quad (1, 2)$ **48.** $x + \sqrt{xy} = 6, \quad (4, 1)$

In Exercises 49–52, find an equation for the line tangent to the curve at the point defined by the given value of t.

49. $x = 2 \sin t, \quad y = 2 \cos t, \quad t = 3\pi/4$

50. $x = 3 \cos t,$ $y = 4 \sin t,$ $t = 3\pi/4$

51. $x = 3 \sec t,$ $y = 5 \tan t,$ $t = \pi/6$

52. $x = \cos t,$ $y = t + \sin t,$ $t = -\pi/4$

53. Writing to Learn

(a) Graph the function

$$f(x) = \begin{cases} x, & 0 \le x \le 1 \\ 2 - x, & 1 < x \le 2. \end{cases}$$

(b) Is f continuous at $x = 1$? Explain.

(c) Is f differentiable at $x = 1$? Explain.

54. Writing to Learn For what values of the constant m is

$$f(x) = \begin{cases} \sin 2x, & x \le 0 \\ mx, & x > 0 \end{cases}$$

(a) continuous at $x = 0$? Explain.

(b) differentiable at $x = 0$? Explain.

In Exercises 55–58, determine where the function is (a) differentiable, (b) continuous but not differentiable, and (c) neither continuous nor differentiable.

55. $f(x) = x^{4/5}$ **56.** $g(x) = \sin (x^2 + 1)$

57. $f(x) = \begin{cases} 2x - 3, & -1 \le x < 0 \\ x - 3, & 0 \le x \le 4 \end{cases}$

58. $g(x) = \begin{cases} \dfrac{x - 1}{x}, & -2 \le x < 0 \\ \dfrac{x + 1}{x}, & 0 \le x \le 2 \end{cases}$

In Exercises 59 and 60, use the graph of f to sketch the graph of f'.

59. *Sketching f' from f*

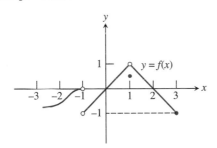

60. *Sketching f' from f*

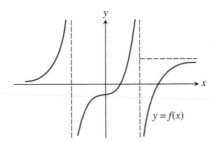

61. *Recognizing Graphs* The following graphs show the distance traveled, velocity, and acceleration for each second of a 2-minute automobile trip. Which graph shows

(a) distance? (b) velocity? (c) acceleration?

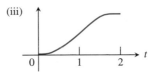

62. *Sketching f from f'* Sketch the graph of a continuous function f with $f(0) = 5$ and

$$f'(x) = \begin{cases} -2, & x < 2 \\ -0.5, & x > 2. \end{cases}$$

63. *Sketching f from f'* Sketch the graph of a continuous function f with $f(-1) = 2$ and

$$f'(x) = \begin{cases} -2, & x < 1 \\ 1, & 1 < x < 4 \\ -1, & 4 < x < 6. \end{cases}$$

64. Which of the following statements could be true if $f''(x) = x^{1/3}$?

i. $f(x) = \dfrac{9}{28}x^{7/3} + 9$ **ii.** $f'(x) = \dfrac{9}{28}x^{7/3} - 2$

iii. $f'(x) = \dfrac{3}{4}x^{4/3} + 6$ **iv.** $f(x) = \dfrac{3}{4}x^{4/3} - 4$

A. i only **B.** iii only

C. ii and iv only **D.** i and iii only

65. *Derivative from Data* The following data give the coordinates of a moving body for various values of t.

t (sec)	0	0.5	1	1.5	2	2.5	3	3.5	4
s (ft)	10	38	58	70	74	70	58	38	10

(a) Make a scatter plot of the (t, s) data and sketch a smooth curve through the points.

(b) Compute the average velocity between consecutive points of the table.

(c) Make a scatter plot of the data in (b) using the midpoints of the t values to represent the data. Then sketch a smooth curve through the points.

(d) **Writing to Learn** Why does the curve in (c) approximate the graph of ds/dt?

66. *Working with Numerical Values* Suppose that a function f and its first derivative have the following values at $x = 0$ and $x = 1$.

x	$f(x)$	$f'(x)$
0	9	-2
1	-3	$1/5$

Find the first derivative of the following combinations at the given value of x.

(a) $\sqrt{x}f(x)$, $x = 1$
(b) $\sqrt{f(x)}$, $x = 0$

(c) $f(\sqrt{x})$, $x = 1$
(d) $f(1 - 5\tan x)$, $x = 0$

(e) $\dfrac{f(x)}{2 + \cos x}$, $x = 0$
(f) $10\sin\left(\dfrac{\pi x}{2}\right)f^2(x)$, $x = 1$

67. *Working with Numerical Values* Suppose that functions f and g and their first derivatives have the following values at $x = -1$ and $x = 0$.

x	$f(x)$	$g(x)$	$f'(x)$	$g'(x)$
-1	0	-1	2	1
0	-1	-3	-2	4

Find the first derivative of the following combinations at the given value of x.

(a) $3f(x) - g(x)$, $x = -1$
(b) $f^2(x)g^3(x)$, $x = 0$

(c) $g(f(x))$, $x = -1$
(d) $f(g(x))$, $x = -1$

(e) $\dfrac{f(x)}{g(x) + 2}$, $x = 0$
(f) $g(x + f(x))$, $x = 0$

68. Find the value of dw/ds at $s = 0$ if $w = \sin(\sqrt{r} - 2)$ and $r = 8\sin(s + \pi/6)$.

69. Find the value of dr/dt at $t = 0$ if $r = (\theta^2 + 7)^{1/3}$ and $\theta^2 t + \theta = 1$.

70. *Particle Motion* The position at time $t \geq 0$ of a particle moving along the s-axis is

$$s(t) = 10\cos(t + \pi/4).$$

(a) Give parametric equations that can be used to simulate the motion of the particle.

(b) What is the particle's initial position $(t = 0)$?

(c) What points reached by the particle are farthest to the left and right of the origin?

(d) When does the particle first reach the origin? What are its velocity, speed, and acceleration then?

71. *Vertical Motion* On Earth, if you shoot a paper clip 64 ft straight up into the air with a rubber band, the paper clip will be $s(t) = 64t - 16t^2$ feet above your hand at t sec after firing.

(a) Find ds/dt and d^2s/dt^2.

(b) How long does it take the paper clip to reach its maximum height?

(c) With what velocity does it leave your hand?

(d) On the moon, the same force will send the paper clip to a height of $s(t) = 64t - 2.6t^2$ ft in t sec. About how long will it take the paper clip to reach its maximum height, and how high will it go?

72. *Free Fall* Suppose two balls are falling from rest at a certain height in centimeters above the ground. Use the equation $s = 490t^2$ to answer the following questions.

(a) How long does it take the balls to fall the first 160 cm? What is their average velocity for the period?

(b) How fast are the balls falling when they reach the 160-cm mark? What is their acceleration then?

73. *Filling a Bowl* If a hemispherical bowl of radius 10 in. is filled with water to a depth of x in., the volume of water is given by $V = \pi[10 - (x/3)]x^2$. Find the rate of increase of the volume per inch increase of depth.

74. *Marginal Revenue* A bus will hold 60 people. The fare charged (p dollars) is related to the number x of people who use the bus by the formula $p = [3 - (x/40)]^2$.

(a) Write a formula for the total revenue per trip received by the bus company.

(b) What number of people per trip will make the marginal revenue equal to zero? What is the corresponding fare?

(c) **Writing to Learn** Do you think the bus company's fare policy is good for its business?

75. *Searchlight* The figure shows a boat 1 km offshore sweeping the shore with a searchlight. The light turns at a constant rate, $d\theta/dt = -0.6$ rad/sec.

(a) How fast is the light moving along the shore when it reaches point A?

(b) How many revolutions per minute is 0.6 rad/sec?

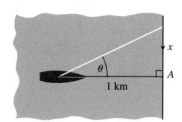

76. *Horizontal Tangents* The graph of $y = \sin(x - \sin x)$ appears to have horizontal tangents at the x-axis. Does it?

77. *Fundamental Frequency of a Vibrating Piano String* We measure the frequencies at which wires vibrate in cycles (trips back and forth) per sec. The unit of measure is a *hertz:* 1 cycle per sec. Middle A on a piano has a frequency 440 hertz. For any given wire, the fundamental frequency y is a function of four variables:

r: the radius of the wire;

l: the length;

d: the density of the wire;

T: the tension (force) holding the wire taut.

With r and l in centimeters, d in grams per cubic centimeter, and T in dynes (it takes about 100,000 dynes to lift an apple), the fundamental frequency of the wire is

$$y = \frac{1}{2rl} \sqrt{\frac{T}{\pi d}}.$$

If we keep all the variables fixed except one, then y can be alternatively thought of as four different functions of one variable, $y(r)$, $y(l)$, $y(d)$, and $y(T)$. How would changing each variable affect the string's fundamental frequency? To find out, calculate $y'(r)$, $y'(l)$, $y'(d)$, and $y'(T)$.

78. *Spread of Measles* The spread of measles in a certain school is given by

$$P(t) = \frac{200}{1 + e^{5-t}},$$

where t is the number of days since the measles first appeared, and $P(t)$ is the total number of students who have caught the measles to date.

(a) Estimate the initial number of students infected with measles.

(b) About how many students in all will get the measles?

(c) When will the rate of spread of measles be greatest? What is this rate?

79. Graph the function $f(x) = \tan^{-1}(\tan 2x)$ in the window $[-\pi, \pi]$ by $[-4, 4]$. Then answer the following questions.

(a) What is the domain of f?

(b) What is the range of f?

(c) At which points is f not differentiable?

(d) Describe the graph of f'.

80. If $x^2 - y^2 = 1$, find d^2y/dx^2 at the point $(2, \sqrt{3})$.

An automobile's gas mileage is a function of many variables, including road surface, tire type, velocity, fuel octane rating, road angle, and the speed and direction of the wind. If we look only at velocity's effect on gas mileage, the mileage of a certain car can be approximated by:

$$m(v) = 0.00015v^3 - 0.032v^2 + 1.8v + 1.7$$

(where v is velocity)

At what speed should you drive this car to obtain the best gas mileage? The ideas in Section 4.1 will help you find the answer.

Chapter 4 Overview

In the past, when virtually all graphing was done by hand—often laboriously—derivatives were the key tool used to sketch the graph of a function. Now we can graph a function quickly, and usually correctly, using a grapher. However, confirmation of much of what we see and conclude true from a grapher view must still come from calculus.

This chapter shows how to draw conclusions from derivatives about the extreme values of a function and about the general shape of a function's graph. We will also see how a tangent line captures the shape of a curve near the point of tangency, how to deduce rates of change we cannot measure from rates of change we already know, and how to find a function when we know only its first derivative and its value at a single point. The key to recovering functions from derivatives is the Mean Value Theorem, a theorem whose corollaries provide the gateway to *integral calculus,* which we begin in Chapter 5.

4.1 Extreme Values of Functions

Absolute (Global) Extreme Values • Local (Relative) Extreme Values • Finding Extreme Values

Absolute (Global) Extreme Values

One of the most useful things we can learn from a function's derivative is whether the function assumes any maximum or minimum values on a given interval and where these values are located if it does. Once we know how to find a function's extreme values, we will be able to answer such questions as "What is the most effective size for a dose of medicine?" and "What is the least expensive way to pipe oil from an offshore well to a refinery down the coast?" We will see how to answer questions like these in Section 4.4.

Definition Absolute Extreme Values

Let f be a function with domain D. Then $f(c)$ is the

(a) absolute maximum value on D if and only if $f(x) \leq f(c)$ for all x in D.

(b) absolute minimum value on D if and only if $f(x) \geq f(c)$ for all x in D.

Absolute (or **global**) maximum and minimum values are also called **absolute extrema** (plural of the Latin *extremum*). We often omit the term "absolute" or "global" and just say maximum and minimum.

Example 1 shows that extreme values can occur at interior points or endpoints of intervals.

Example 1 EXPLORING EXTREME VALUES

On $[-\pi/2, \pi/2]$, $f(x) = \cos x$ takes on a maximum value of 1 (once) and a minimum value of 0 (twice). The function $g(x) = \sin x$ takes on a maximum value of 1 and a minimum value of -1 (Figure 4.1).

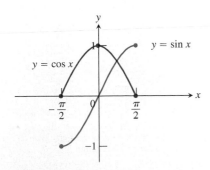

Figure 4.1 (Example 1)

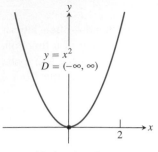

(a) abs min only

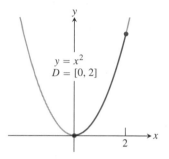

(b) abs max and min

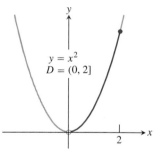

(c) abs max only

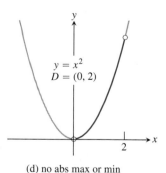

(d) no abs max or min

Figure 4.2 (Example 2)

Functions with the same defining rule can have different extrema, depending on the domain.

Example 2   EXPLORING ABSOLUTE EXTREMA

The absolute extrema of the following functions on their domains can be seen in Figure 4.2.

	Function Rule	Domain D	Absolute Extrema on D
(a)	$y = x^2$	$(-\infty, \infty)$	No absolute maximum. Absolute minimum of 0 at $x = 0$.
(b)	$y = x^2$	$[0, 2]$	Absolute maximum of 4 at $x = 2$. Absolute minimum of 0 at $x = 0$.
(c)	$y = x^2$	$(0, 2]$	Absolute maximum of 4 at $x = 2$. No absolute minimum.
(d)	$y = x^2$	$(0, 2)$	No absolute extrema.

Example 2 shows that a function may fail to have a maximum or minimum value. This cannot happen with a continuous function on a finite closed interval.

Theorem 1   The Extreme Value Theorem

If f is continuous on a closed interval $[a, b]$, then f has both a maximum value and a minimum value on the interval. (Figure 4.3)

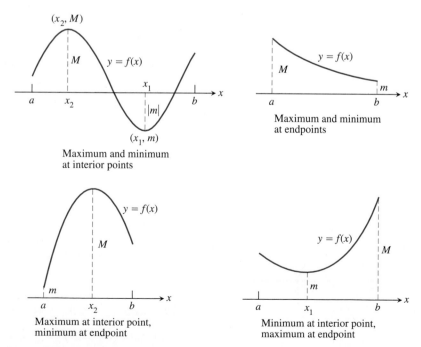

Figure 4.3 Some possibilities for a continuous function's maximum and minimum on a closed interval $[a, b]$.

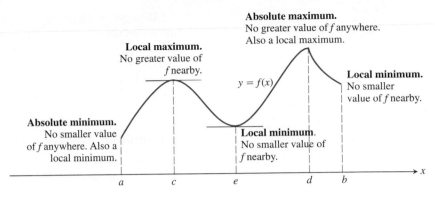

Figure 4.4 Classifying extreme values.

Local (Relative) Extreme Values

Figure 4.4 shows a graph with five points where a function has extreme values on its domain $[a, b]$. The function's absolute minimum occurs at a even though at e the function's value is smaller than at any other point *nearby*. The curve rises to the left and falls to the right around c, making $f(c)$ a maximum locally. The function attains its absolute maximum at d.

Definition Local Extreme Values

Let c be an interior point of the domain of the function f. Then $f(c)$ is a

(a) local maximum value at c if and only if $f(x) \leq f(c)$ for all x in some open interval containing c.

(b) local minimum value at c if and only if $f(x) \geq f(c)$ for all x in some open interval containing c.

Local extrema are also called **relative extrema.** We can extend the definitions of local extrema to endpoints of intervals. A function f has a local maximum or local minimum *at an endpoint* c if the appropriate inequality holds for all x in some half-open domain interval containing c.

An absolute extremum is also a local extremum, because being an extreme value overall makes it an extreme value in its immediate neighborhood. Hence, *a list of local extrema will automatically include absolute extrema if there are any.*

Finding Extreme Values

The interior domain points where the function in Figure 4.4 has local extreme values are points where either f' is zero or f' does not exist. This is generally the case, as we see from the following theorem.

Theorem 2 Local Extreme Values

If a function f has a local maximum value or a local minimum value at an interior point c of its domain, and if f' exists at c, then

$$f'(c) = 0.$$

Because of Theorem 2, we usually need to look at only a few points to find a function's extrema. These consist of the interior domain points where $f' = 0$ or f' does not exist (the domain points covered by the theorem) and the domain endpoints (the domain points not covered by the theorem). At all other domain points, $f' > 0$ or $f' < 0$.

The following definition helps us summarize these findings.

Definition   Critical Point

A point in the domain of a function f at which $f' = 0$ or f' does not exist is a **critical point** of f.

Thus, in summary, extreme values occur only at critical points and endpoints.

Example 3   FINDING ABSOLUTE EXTREMA

Find the absolute maximum and minimum values of $f(x) = x^{2/3}$ on the interval $[-2, 3]$.

Solution

Solve Graphically

Figure 4.5 suggests that f has an absolute maximum value of about 2 at $x = 3$ and an absolute minimum value of 0 at $x = 0$.

Confirm Analytically

We evaluate the function at the critical points and endpoints and take the largest and smallest of the resulting values.

The first derivative

$$f'(x) = \frac{2}{3}x^{-1/3} = \frac{2}{3\sqrt[3]{x}}$$

has no zeros but is undefined at $x = 0$. The values of f at this one critical point and at the endpoints are

Critical point value: $f(0) = 0$;

Endpoint values: $f(-2) = (-2)^{2/3} = \sqrt[3]{4}$;

$$f(3) = (3)^{2/3} = \sqrt[3]{9}.$$

We can see from this list that the function's absolute maximum value is $\sqrt[3]{9} \approx 2.08$, and occurs at the right endpoint $x = 3$. The absolute minimum value is 0, and occurs at the interior point $x = 0$.

In Example 4, we investigate the function whose graph was drawn in Example 3 of Section 1.2.

Example 4   FINDING EXTREME VALUES

Find the extreme values of $f(x) = \dfrac{1}{\sqrt{4 - x^2}}$.

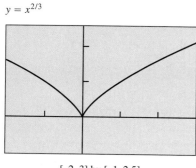

$y = x^{2/3}$

[-2, 3] by [-1, 2.5]

Figure 4.5 (Example 3)

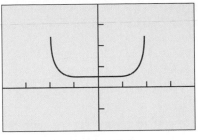

[–4, 4] by [–2, 4]

Figure 4.6 The graph of

$$f(x) = \frac{1}{\sqrt{4 - x^2}}.$$

(Example 4)

Solution
Solve Graphically
Figure 4.6 suggests that f has an absolute minimum of about 0.5 at $x = 0$. There also appear to be local maxima at $x = -2$ and $x = 2$. However, f is not defined at these points and there do not appear to be maxima anywhere else.

Confirm Analytically
The function f is defined only for $4 - x^2 > 0$, so its domain is the open interval $(-2, 2)$. The domain has no endpoints, so all the extreme values must occur at critical points. We rewrite the formula for f to find f':

$$f(x) = \frac{1}{\sqrt{4 - x^2}} = (4 - x^2)^{-1/2}.$$

Thus,

$$f'(x) = -\frac{1}{2}(4 - x^2)^{-3/2}(-2x) = \frac{x}{(4 - x^2)^{3/2}}.$$

The only critical point in the domain $(-2, 2)$ is $x = 0$. The value

$$f(0) = \frac{1}{\sqrt{4 - 0^2}} = \frac{1}{2}$$

is therefore the sole candidate for an extreme value.

To determine whether $1/2$ is an extreme value of f, we examine the formula

$$f(x) = \frac{1}{\sqrt{4 - x^2}}.$$

As x moves away from 0 on either side, the denominator gets smaller, the values of f increase, and the graph rises. We have a minimum value at $x = 0$, and the minimum is absolute.

The function has no maxima, either local or absolute. This does not violate Theorem 1 (The Extreme Value Theorem) because here f is defined on an *open* interval. To invoke Theorem 1's guarantee of extreme points, the interval must be closed.

While a function's extrema can occur only at critical points and endpoints, not every critical point or endpoint signals the presence of an extreme value. Figure 4.7 illustrates this for interior points. Exercise 53 describes a function that fails to assume an extreme value at an endpoint of its domain.

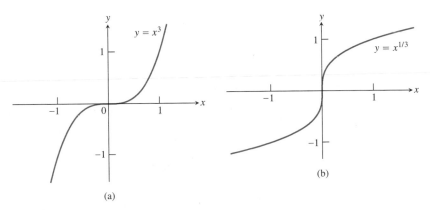

Figure 4.7 Critical points without extreme values. (a) $y' = 3x^2$ is 0 at $x = 0$, but $y = x^3$ has no extremum there. (b) $y' = (1/3)x^{-2/3}$ is undefined at $x = 0$, but $y = x^{1/3}$ has no extremum there.

Example 5 FINDING EXTREME VALUES

Find the extreme values of

$$f(x) = \begin{cases} 5 - 2x^2, & x \le 1 \\ x + 2, & x > 1. \end{cases}$$

Solution

Solve Graphically

The graph in Figure 4.8 suggests that $f'(0) = 0$ and that $f'(1)$ does not exist. There appears to be a local maximum value of 5 at $x = 0$ and a local minimum value of 3 at $x = 1$.

Confirm Analytically

For $x \ne 1$, the derivative is

$$f'(x) = \begin{cases} \dfrac{d}{dx}(5 - 2x^2) = -4x, & x < 1 \\[2mm] \dfrac{d}{dx}(x + 2) = 1, & x > 1. \end{cases}$$

The only point where $f' = 0$ is $x = 0$. What happens at $x = 1$?

At $x = 1$, the right- and left-hand derivatives are respectively

$$\lim_{h \to 0^+} \frac{f(1 + h) - f(1)}{h} = \lim_{h \to 0^+} \frac{(1 + h) + 2 - 3}{h} = \lim_{h \to 0^+} \frac{h}{h} = 1,$$

$$\lim_{h \to 0^-} \frac{f(1 + h) - f(1)}{h} = \lim_{h \to 0^-} \frac{5 - 2(1 + h)^2 - 3}{h}$$

$$= \lim_{h \to 0^-} \frac{-2h(2 + h)}{h} = -4.$$

Since these one-sided derivatives differ, f has no derivative at $x = 1$, and 1 is a second critical point of f.

The domain $(-\infty, \infty)$ has no endpoints, so the only values of f that might be local extrema are those at the critical points:

$$f(0) = 5 \quad \text{and} \quad f(1) = 3.$$

From the formula for f, we see that the values of f immediately to either side of $x = 0$ are less than 5, so 5 is a local maximum. Similarly, the values of f immediately to either side of $x = 1$ are greater than 3, so 3 is a local minimum.

Most graphing calculators have built-in methods to find the coordinates of points where extreme values occur. We must, of course, be sure that we use correct graphs to find these values. The calculus that you learn in this chapter should make you feel more confident about working with graphs.

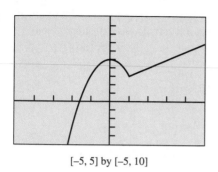

[–5, 5] by [–5, 10]

Figure 4.8 The function in Example 5.

Example 6 USING GRAPHICAL METHODS

Find the extreme values of $f(x) = \ln \left| \dfrac{x}{1 + x^2} \right|$.

Solution

Solve Graphically

The domain of f is the set of all nonzero real numbers. Figure 4.9 suggests that f is an even function with a maximum value at two points. The coordinates found in this window suggest an extreme value of about -0.69 at approximately $x = 1$. Because f is even, there is another extreme of the same value at approximately $x = -1$. The figure also suggests a minimum value at $x = 0$, but f is not defined there.

Confirm Analytically

The derivative

$$f'(x) = \frac{1 - x^2}{x(1 + x^2)}$$

is defined at every point of the function's domain. The critical points where $f'(x) = 0$ are $x = 1$ and $x = -1$. The corresponding values of f are both $\ln (1/2) = -\ln 2 \approx -0.69$.

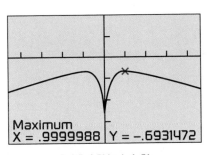

Maximum
X = .9999988 Y = -.6931472

$[-4.5, 4.5]$ by $[-4, 2]$

Figure 4.9 The function in Example 6.

Exploration 1 Finding Extreme Values

Let $f(x) = \left| \dfrac{x}{x^2 + 1} \right|$, $-2 \le x \le 2$.

1. Determine graphically the extreme values of f and where they occur. Find f' at these values of x.
2. Graph f and f' (or NDER $(f(x), x, x)$) in the same viewing window. Comment on the relationship between the graphs.
3. Find a formula for $f'(x)$.

Quick Review 4.1

In Exercises 1–8, find the first derivative of the function.

1. $f(x) = \sqrt{4 - x}$ **2.** $f(x) = x^{3/4}$

3. $f(x) = \dfrac{2}{\sqrt{9 - x^2}}$ **4.** $f(x) = \dfrac{1}{\sqrt[3]{x^2 - 1}}$

5. $g(x) = \ln (x^2 + 1)$ **6.** $g(x) = \cos (\ln x)$

7. $h(x) = e^{2x}$ **8.** $h(x) = e^{\ln x}$

In Exercises 9 and 10, find the limit for

$$f(x) = \frac{2}{\sqrt{9 - x^2}}.$$

9. $\displaystyle\lim_{x \to 3^-} f(x)$ **10.** $\displaystyle\lim_{x \to -3^+} f(x)$

In Exercises 11 and 12, let

$$f(x) = \begin{cases} x^3 - 2x, & x \le 2 \\ x + 2, & x > 2. \end{cases}$$

11. Find **(a)** $f'(1)$, **(b)** $f'(3)$, **(c)** $f'(2)$.

12. (a) Find the domain of f'.

     (b) Write a formula for $f'(x)$.

Section 4.1 Exercises

In Exercises 1–6, identify each x-value at which any absolute extreme value occurs. Explain how your answer is consistent with the Extreme Value Theorem.

1.
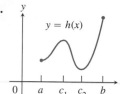

2.
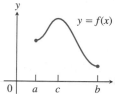

3.
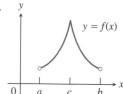

4.
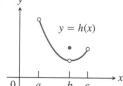

5.
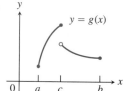

6.
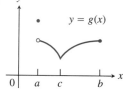

In Exercises 7–10, find the extreme values and where they occur.

7.
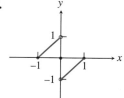

8.
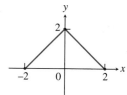

9.
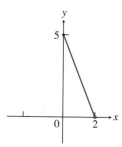

10.
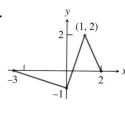

In Exercises 11–18, use analytic methods to find the extreme values of the function on the interval and where they occur.

11. $f(x) = \dfrac{1}{x} + \ln x, \quad 0.5 \le x \le 4$

12. $g(x) = e^{-x}, \quad -1 \le x \le 1$

13. $h(x) = \ln (x + 1), \quad 0 \le x \le 3$

14. $k(x) = e^{-x^2}, \quad -\infty < x < \infty$

15. $f(x) = \sin \left(x + \dfrac{\pi}{4} \right), \quad 0 \le x \le \dfrac{7\pi}{4}$

16. $g(x) = \sec x, \quad -\dfrac{\pi}{2} < x < \dfrac{3\pi}{2}$

17. $f(x) = x^{2/5}, \quad -3 \le x < 1$

18. $f(x) = x^{3/5}, \quad -2 < x \le 3$

In Exercises 19–30, find the extreme values of the function and where they occur.

19. $y = 2x^2 - 8x + 9$

20. $y = x^3 - 2x + 4$

21. $y = x^3 + x^2 - 8x + 5$

22. $y = x^3 - 3x^2 + 3x - 2$

23. $y = \sqrt{x^2 - 1}$

24. $y = \dfrac{1}{x^2 - 1}$

25. $y = \dfrac{1}{\sqrt{1 - x^2}}$

26. $y = \dfrac{1}{\sqrt[3]{1 - x^2}}$

27. $y = \sqrt{3 + 2x - x^2}$

28. $y = \dfrac{3}{2}x^4 + 4x^3 - 9x^2 + 10$

29. $y = \dfrac{x}{x^2 + 1}$

30. $y = \dfrac{x + 1}{x^2 + 2x + 2}$

In Exercises 31–34, *work in groups of two or three* to find the extreme values of the function on the interval and where they occur.

31. $f(x) = |x - 2| + |x + 3|, \quad -5 \le x \le 5$

32. $g(x) = |x - 1| - |x - 5|, \quad -2 \le x \le 7$

33. $h(x) = |x + 2| - |x - 3|, \quad -\infty < x < \infty$

34. $k(x) = |x + 1| + |x - 3|, \quad -\infty < x < \infty$

Explorations

In Exercises **35** and **36**, give reasons for your answers.

35. Writing to Learn Let $f(x) = (x - 2)^{2/3}$.

(a) Does $f'(2)$ exist?

(b) Show that the only local extreme value of f occurs at $x = 2$.

(c) Does the result in (b) contradict the Extreme Value Theorem?

(d) Repeat (a) and (b) for $f(x) = (x - a)^{2/3}$, replacing 2 by a.

36. Writing to Learn Let $f(x) = |x^3 - 9x|$.

(a) Does $f'(0)$ exist?

(b) Does $f'(3)$ exist?

(c) Does $f'(-3)$ exist?

(d) Determine all extrema of f. ∎

In Exercises 37–44, find the derivative at each critical point and determine the local extreme values.

37. $y = x^{2/3}(x + 2)$

38. $y = x^{2/3}(x^2 - 4)$

39. $y = x\sqrt{4 - x^2}$

40. $y = x^2\sqrt{3 - x}$

41. $y = \begin{cases} 4 - 2x, & x \le 1 \\ x + 1, & x > 1 \end{cases}$

42. $y = \begin{cases} 3 - x, & x < 0 \\ 3 + 2x - x^2, & x \ge 0 \end{cases}$

43. $y = \begin{cases} -x^2 - 2x + 4, & x \le 1 \\ -x^2 + 6x - 4, & x > 1 \end{cases}$

44. $y = \begin{cases} -\dfrac{1}{4}x^2 - \dfrac{1}{2}x + \dfrac{15}{4}, & x \le 1 \\ x^3 - 6x^2 + 8x, & x > 1 \end{cases}$

In Exercises 45–48, match the table with a graph.

45.

x	$f'(x)$
a	0
b	0
c	5

46.

x	$f'(x)$
a	0
b	0
c	-5

47.

x	$f'(x)$
a	does not exist
b	0
c	-2

48.

x	$f'(x)$
a	does not exist
b	does not exist
c	-1.7

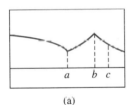

(a)

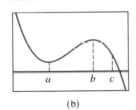

(b)

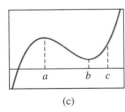

(c)

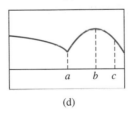

(d)

49. Writing to Learn The function

$$V(x) = x(10 - 2x)(16 - 2x), \quad 0 < x < 5,$$

models the volume of a box.

(a) Find the extreme values of V.

(b) Interpret any values found in (a) in terms of volume of the box.

50. Writing to Learn The function

$$P(x) = 2x + \frac{200}{x}, \quad 0 < x < \infty,$$

models the perimeter of a rectangle of dimensions x by $100/x$.

(a) Find any extreme values of P.

(b) Give an interpretation in terms of perimeter of the rectangle for any values found in (a).

Extending the Ideas

51. *Cubic Functions* Consider the cubic function

$$f(x) = ax^3 + bx^2 + cx + d.$$

(a) Show that f can have 0, 1, or 2 critical points. Give examples and graphs to support your argument.

(b) How many local extreme values can f have?

52. *Proving Theorem 2* Assume that the function f has a local maximum value at the interior point c of its domain.

(a) Show that there is an open interval containing c such that $f(x) - f(c) \le 0$ for all x in the open interval.

(b) Writing to Learn Now explain why we may say

$$\lim_{x \to c^+} \frac{f(x) - f(c)}{x - c} \le 0.$$

(c) Writing to Learn Now explain why we may say

$$\lim_{x \to c^-} \frac{f(x) - f(c)}{x - c} \ge 0.$$

(d) Writing to Learn Explain how (b) and (c) allow us to conclude $f'(c) = 0$.

(e) Writing to Learn Give a similar argument if f has a local minimum value at an interior point.

In Exercise 53, *work in groups of two or three.*

53. *Functions with No Extreme Values at Endpoints*

(a) Graph the function

$$f(x) = \begin{cases} \sin \dfrac{1}{x}, & x > 0 \\ 0, & x = 0. \end{cases}$$

Explain why $f(0) = 0$ is not a local extreme value of f.

(b) Construct a function of your own that fails to have an extreme value at a domain endpoint.

Mean Value Theorem • Physical Interpretation • Increasing and Decreasing Functions • Other Consequences

Mean Value Theorem

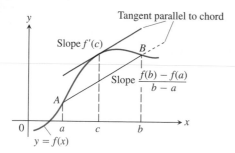

The Mean Value Theorem connects the average rate of change of a function over an interval with the instantaneous rate of change of the function at a point within the interval. Its powerful corollaries lie at the heart of some of the most important applications of the calculus.

The theorem says that somewhere between points A and B on a differentiable curve, there is at least one tangent line parallel to chord AB (Figure 4.10).

Figure 4.10 Figure for the Mean Value Theorem.

> ### Theorem 3 Mean Value Theorem for Derivatives
>
> If $y = f(x)$ is continuous at every point of the closed interval $[a, b]$ and differentiable at every point of its interior (a, b), then there is at least one point c in (a, b) at which
>
> $$f'(c) = \frac{f(b) - f(a)}{b - a}.$$

The hypotheses of Theorem 3 cannot be relaxed. If they fail at even one point, the graph may fail to have a tangent parallel to the chord. For instance, the function $f(x) = |x|$ is continuous on $[-1, 1]$ and differentiable at every point of the interior $(-1, 1)$ except $x = 0$. The graph has no tangent parallel to chord AB (Figure 4.11a). The function $g(x) = \text{int}\,(x)$ is differentiable at every point of $(1, 2)$ and continuous at every point of $[1, 2]$ except $x = 2$. Again, the graph has no tangent parallel to chord AB (Figure 4.11b).

The Mean Value Theorem is an *existence theorem*. It tells us the number c exists without telling how to find it. We can sometimes satisfy our curiosity about the value of c but the real importance of the theorem lies in the surprising conclusions we can draw from it.

Rolle's Theorem

The first version of the Mean Value Theorem was proved by French mathematician Michel Rolle (1652–1719). His version had $f(a) = f(b) = 0$ and was proved only for polynomials, using algebra and geometry.

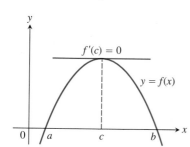

Rolle distrusted calculus and spent most of his life denouncing it. It is ironic that he is known today only for an unintended contribution to a field he tried to suppress.

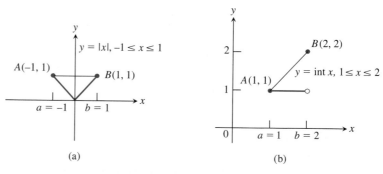

(a)   (b)

Figure 4.11 No tangent parallel to chord AB.

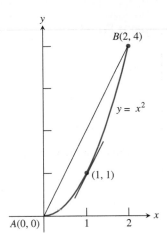

Figure 4.12 (Example 1)

Example 1 EXPLORING THE MEAN VALUE THEOREM

Show that the function $f(x) = x^2$ satisfies the hypotheses of the Mean Value Theorem on the interval $[0, 2]$. Then find a solution c to the equation

$$f'(c) = \frac{f(b) - f(a)}{b - a}$$

on this interval.

Solution The function $f(x) = x^2$ is continuous on $[0, 2]$ and differentiable on $(0, 2)$. Since $f(0) = 0$ and $f(2) = 4$, the Mean Value Theorem guarantees a point c in the interval $(0, 2)$ for which

$$f'(c) = \frac{f(b) - f(a)}{b - a}$$

$$2c = \frac{f(2) - f(0)}{2 - 0} = 2 \quad f'(x) = 2x$$

$$c = 1.$$

Interpret

The tangent line to $f(x) = x^2$ at $x = 1$ has slope 2 and is parallel to the chord joining $A(0, 0)$ and $B(2, 4)$ (Figure 4.12).

Example 2 APPLYING THE MEAN VALUE THEOREM

Let $f(x) = \sqrt{1 - x^2}$, $A = (-1, f(-1))$, and $B = (1, f(1))$. Find a tangent to f in the interval $(-1, 1)$ that is parallel to the secant AB.

Solution The function f (Figure 4.13) is continuous on the interval $[-1, 1]$ and

$$f'(x) = \frac{-x}{\sqrt{1 - x^2}}$$

is defined on the interval $(-1, 1)$. The function is not differentiable at $x = -1$ and $x = 1$, but it does not need to be for the theorem to apply. Since $f(-1) = f(1) = 0$, the tangent we are looking for is horizontal. We find that $f' = 0$ at $x = 0$, where the graph has the horizontal tangent $y = 1$.

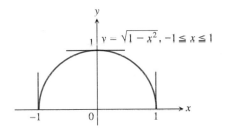

Figure 4.13 (Example 2)

Physical Interpretation

If we think of the difference quotient $(f(b) - f(a))/(b - a)$ as the average change in f over $[a, b]$ and $f'(c)$ as an instantaneous change, then the Mean Value Theorem says that the instantaneous change at some interior point must equal the average change over the entire interval.

Example 3 INTERPRETING THE MEAN VALUE THEOREM

If a car accelerating from zero takes 8 sec to go 352 ft, its average velocity for the 8-sec interval is $352/8 = 44$ ft/sec, or 30 mph. At some point during the acceleration, the theorem says, the speedometer must read exactly 30 mph (Figure 4.14).

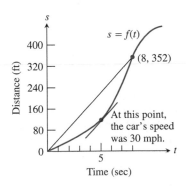

Figure 4.14 (Example 3)

Increasing and Decreasing Functions

Our first use of the Mean Value Theorem will be its application to increasing and decreasing functions.

Monotonic Functions

A function that is always increasing on an interval or always decreasing on an interval is said to be **monotonic** there.

Definitions Increasing Function, Decreasing Function

Let f be a function defined on an interval I and let x_1 and x_2 be any two points in I.

1. f **increases** on I if $\quad x_1 < x_2 \quad \Rightarrow \quad f(x_1) < f(x_2)$.
2. f **decreases** on I if $\quad x_1 < x_2 \quad \Rightarrow \quad f(x_1) > f(x_2)$.

The Mean Value Theorem allows us to identify exactly where graphs rise and fall. Functions with positive derivatives are increasing functions; functions with negative derivatives are decreasing functions.

Corollary 1 Increasing and Decreasing Functions

Let f be continuous on $[a, b]$ and differentiable on (a, b).

1. If $f' > 0$ at each point of (a, b), then f increases on $[a, b]$.
2. If $f' < 0$ at each point of (a, b), then f decreases on $[a, b]$.

Proof Let x_1 and x_2 be any two points in $[a, b]$ with $x_1 < x_2$. The Mean Value Theorem applied to f on $[x_1, x_2]$ gives

$$f(x_2) - f(x_1) = f'(c)(x_2 - x_1)$$

for some c between x_1 and x_2. The sign of the right-hand side of this equation is the same as the sign of $f'(c)$ because $x_2 - x_1$ is positive. Therefore,

(a) $f(x_1) < f(x_2)$ if $f' > 0$ on (a, b) (f is increasing), or

(b) $f(x_1) > f(x_2)$ if $f' < 0$ on (a, b) (f is decreasing). ∎

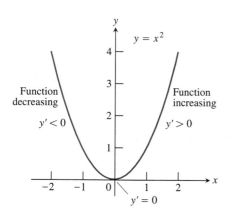

Figure 4.15 (Example 4)

Example 4 DETERMINING WHERE GRAPHS RISE OR FALL

The function $y = x^2$ (Figure 4.15) is

(a) decreasing on $(-\infty, 0]$ because $y' = 2x < 0$ on $(-\infty, 0)$.

(b) increasing on $[0, \infty)$ because $y' = 2x > 0$ on $(0, \infty)$.

Example 5 DETERMINING WHERE GRAPHS RISE OR FALL

Where is the function $f(x) = x^3 - 4x$ increasing and where is it decreasing?

Solution

Solve Graphically

The graph of f in Figure 4.16 suggests that f is increasing from $-\infty$ to the x-coordinate of the local maximum, decreasing between the two local extrema, and increasing again from the x-coordinate of the local minimum to ∞. This information is supported by the superimposed graph of $f'(x) = 3x^2 - 4$.

Confirm Analytically

The function is increasing where $f'(x) > 0$.

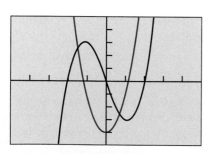

[−5, 5] by [−5, 5]

Figure 4.16 By comparing the graphs of $f(x) = x^3 - 4x$ and $f'(x) = 3x^2 - 4$ we can relate the increasing and decreasing behavior of f to the sign of f'. (Example 5)

$$3x^2 - 4 > 0$$

$$x^2 > \frac{4}{3}$$

$$x < -\sqrt{\frac{4}{3}} \quad \text{or} \quad x > \sqrt{\frac{4}{3}}$$

The function is decreasing where $f'(x) < 0$.

$$3x^2 - 4 < 0$$

$$x^2 < \frac{4}{3}$$

$$-\sqrt{\frac{4}{3}} < x < \sqrt{\frac{4}{3}}$$

Interpret

To describe the intervals we use the approximation $\sqrt{4/3} \approx 1.15$. The function increases on $(-\infty, -1.15]$ and $[1.15, \infty)$. It decreases on $[-1.15, 1.15]$.

In Example 5, the exact interval on which the function f decreases is $[-\sqrt{4/3}, \sqrt{4/3}]$. We reported it as $[-1.15, 1.15]$. As we have been doing all along, we continue to display results to a number of decimal places that seems appropriate to the problem.

Other Consequences

We know that constant functions have the zero function as their derivative. We can now use the Mean Value Theorem to show conversely that the only functions with the zero function as derivative are constant functions.

Corollary 2 Functions with $f' = 0$ are Constant

If $f'(x) = 0$ at each point of an interval I, then there is a constant C for which $f(x) = C$ for all x in I.

Proof Our plan is to show that $f(x_1) = f(x_2)$ for any two points x_1 and x_2 in I. We can assume the points are numbered so that $x_1 < x_2$. Since f is differentiable at every point of $[x_1, x_2]$, it is continuous at every point as well. Thus, f satisfies the hypotheses of the Mean Value Theorem on $[x_1, x_2]$. Therefore, there is a point c between x_1 and x_2 for which

$$f'(c) = \frac{f(x_2) - f(x_1)}{x_2 - x_1}.$$

Because $f'(c) = 0$, it follows that $f(x_1) = f(x_2)$. ∎

We can use Corollary 2 to show that if two functions have the same derivative, they differ by a constant.

Corollary 3 Functions with the Same Derivative Differ by a Constant

If $f'(x) = g'(x)$ at each point of an interval I, then there is a constant C such that $f(x) = g(x) + C$ for all x in I.

Proof Let $h = f - g$. Then for each point x in I,

$$h'(x) = f'(x) - g'(x) = 0.$$

It follows from Corollary 2 that there is a constant C such that $h(x) = C$ for all x in I. Thus, $h(x) = f(x) - g(x) = C$, or $f(x) = g(x) + C$. ∎

We know that the derivative of $f(x) = x^2$ is $2x$ on the interval $(-\infty, \infty)$. So, any other function $g(x)$ with derivative $2x$ on $(-\infty, \infty)$ must have the formula $g(x) = x^2 + C$ for some constant C.

Example 6 APPLYING COROLLARY 3

Find the function $f(x)$ whose derivative is $\sin x$ and whose graph passes through the point $(0, 2)$.

Solution Since f has the same derivative as $g(x) = -\cos x$, we know that $f(x) = -\cos x + C$ for some constant C. To identify C, we use the condition that the graph must pass through $(0, 2)$. This is equivalent to saying that

$$f(0) = 2$$

$$-\cos(0) + C = 2 \qquad f(x) = -\cos x + C$$

$$-1 + C = 2$$

$$C = 3.$$

The formula for f is $f(x) = -\cos x + 3$.

In Example 6 we were given a derivative and asked to find a function with that derivative. This type of function is so important that it has a name.

Definition Antiderivative

A function $F(x)$ is an **antiderivative** of a function $f(x)$ if $F'(x) = f(x)$ for all x in the domain of f. The process of finding an antiderivative is **antidifferentiation.**

We know that if f has one antiderivative F then it has infinitely many antiderivatives, each differing from F by a constant. Corollary 3 says these are all there are. In Example 6, we found the particular antiderivative of $\sin x$ whose graph passed through the point $(0, 2)$.

Example 7 FINDING VELOCITY AND POSITION

Find the velocity and position functions of a freely falling body for each of the following sets of conditions:

(a) The acceleration is 9.8 m/sec^2 and the body falls from rest.

(b) The acceleration is 9.8 m/sec^2 and the body is propelled downward with an initial velocity of 1 m/sec.

Solution

(a) *Falling from rest.* We measure distance fallen in meters and time in seconds, and assume that the body is released from rest at time $t = 0$.

Velocity: We know that the velocity $v(t)$ is an antiderivative of the constant function 9.8. We also know that $g(t) = 9.8t$ is an antiderivative of 9.8. By Corollary 3,

$$v(t) = 9.8t + C$$

for some constant C. Since the body falls from rest, $v(0) = 0$. Thus,

$$9.8(0) + C = 0 \quad \text{and} \quad C = 0.$$

The body's velocity function is $v(t) = 9.8t$.

Position: We know that the position $s(t)$ is an antiderivative of 9.8t. We also know that $h(t) = 4.9t^2$ is an antiderivative of 9.8t. By Corollary 3,

$$s(t) = 4.9t^2 + C$$

for some constant C. Since $s(0) = 0$,

$$4.9(0)^2 + C = 0 \quad \text{and} \quad C = 0.$$

The body's position function is $s(t) = 4.9t^2$.

(b) *Propelled downward.* We measure distance fallen in meters and time in seconds, and assume that the body is propelled downward with velocity of 1 m/sec at time $t = 0$.

Velocity: The velocity function still has the form $9.8t + C$, but instead of being zero, the initial velocity (velocity at $t = 0$) is now 1 m/sec. Thus,

$$9.8(0) + C = 1 \quad \text{and} \quad C = 1.$$

The body's velocity function is $v(t) = 9.8t + 1$.

Position: We know that the position $s(t)$ is an antiderivative of $9.8t + 1$. We also know that $k(t) = 4.9t^2 + t$ is an antiderivative of $9.8t + 1$. By Corollary 3,

$$s(t) = 4.9t^2 + t + C$$

for some constant C. Since $s(0) = 0$,

$$4.9(0)^2 + 0 + C = 0 \quad \text{and} \quad C = 0.$$

The body's position function is $s(t) = 4.9t^2 + t$.

Quick Review 4.2

In Exercises 1 and 2, find exact solutions to the inequality.

1. $2x^2 - 6 < 0$ **2.** $3x^2 - 6 > 0$

In Exercises 3–5, let $f(x) = \sqrt{8 - 2x^2}$.

3. Find the domain of f. **4.** Where is f continuous?

5. Where is f differentiable?

In Exercises 6–8, let $f(x) = \dfrac{x}{x^2 - 1}$.

6. Find the domain of f. **7.** Where is f continuous?

8. Where is f differentiable?

In Exercises 9 and 10, find C so that the graph of the function f passes through the specified point.

9. $f(x) = -2x + C, \quad (-2, 7)$

10. $g(x) = x^2 + 2x + C, \quad (1, -1)$

Section 4.2 Exercises

In Exercises 1–8, use analytic methods to find **(a)** the local extrema, **(b)** the intervals on which the function is increasing, and **(c)** the intervals on which the function is decreasing.

1. $f(x) = 5x - x^2$

2. $g(x) = x^2 - x - 12$

3. $h(x) = \dfrac{2}{x}$

4. $k(x) = \dfrac{1}{x^2}$

5. $f(x) = e^{2x}$

6. $f(x) = e^{-0.5x}$

7. $y = 4 - \sqrt{x+2}$

8. $y = x^4 - 10x^2 + 9$

In Exercises 9–14, find **(a)** the local extrema, **(b)** the intervals on which the function is increasing, and **(c)** the intervals on which the function is decreasing.

9. $f(x) = x\sqrt{4-x}$

10. $g(x) = x^{1/3}(x+8)$

11. $h(x) = \dfrac{-x}{x^2+4}$

12. $k(x) = \dfrac{x}{x^2-4}$

13. $f(x) = x^3 - 2x - 2\cos x$

14. $g(x) = 2x + \cos x$

In Exercises 15–18, **(a)** show that the function f satisfies the hypotheses of the Mean Value Theorem on the given interval $[a, b]$. **(b)** Find each value of c in (a, b) that satisfies the equation

$$f'(c) = \frac{f(b) - f(a)}{b - a}.$$

15. $f(x) = x^2 + 2x - 1, \quad [0, 1]$

16. $f(x) = x^{2/3}, \quad [0, 1]$

17. $f(x) = \sin^{-1} x, \quad [-1, 1]$

18. $f(x) = \ln(x - 1), \quad [2, 4]$

In Exercises 19 and 20, the interval $a \le x \le b$ is given. Let $A = (a, f(a))$ and $B = (b, f(b))$. Write an equation for

(a) the secant line AB.

(b) a tangent line to f in the interval (a, b) that is parallel to AB.

19. $f(x) = x + \dfrac{1}{x}, \quad 0.5 \le x \le 2$

20. $f(x) = \sqrt{x - 1}, \quad 1 \le x \le 3$

In Exercises 21–24, **(a)** show that the function f does not satisfy the hypotheses of the Mean Value Theorem on the given interval $[a, b]$. **(b)** Graph f together with the line through the points $A(a, f(a))$ and $B(b, f(b))$. **(c)** Find any values of c in (a, b) that satisfy the equation

$$f'(c) = \frac{f(b) - f(a)}{b - a}.$$

21. $f(x) = x^{1/3}, \quad [-1, 1]$ **22.** $f(x) = |x - 1|, \quad [0, 3]$

23. $f(x) = 1 - |x|, \quad [-1, 1]$

24. $f(x) = \begin{cases} \cos x, & -\pi \le x < 0, \\ 1 + \sin x, & 0 \le x \le \pi, \end{cases}$ on $[-\pi, \pi]$

In Exercises 25–30, find all possible functions f with the given derivative.

25. $f'(x) = x$

26. $f'(x) = 2$

27. $f'(x) = 3x^2 - 2x + 1$

28. $f'(x) = \sin x$

29. $f'(x) = e^x$

30. $f'(x) = \dfrac{1}{x-1}, \quad x > 1$

In Exercises 31–34, find the function with the given derivative whose graph passes through the point P.

31. $f'(x) = -\dfrac{1}{x^2}, \quad x > 0, \quad P(2, 1)$

32. $f'(x) = \dfrac{1}{4x^{3/4}}, \quad P(1, -2)$

33. $f'(x) = \dfrac{1}{x+2}, \quad x > -2, \quad P(-1, 3)$

34. $f'(x) = 2x + 1 - \cos x, \quad P(0, 3)$

In Exercises 35–38, *work in groups of two or three* and sketch a graph of a differentiable function $y = f(x)$ that has the given properties.

35. **(a)** local minimum at $(1, 1)$, local maximum at $(3, 3)$

(b) local minima at $(1, 1)$ and $(3, 3)$

(c) local maxima at $(1, 1)$ and $(3, 3)$

36. $f(2) = 3, \quad f'(2) = 0, \quad$ and

(a) $f'(x) > 0$ for $x < 2, \quad f'(x) < 0$ for $x > 2$.

(b) $f'(x) < 0$ for $x < 2, \quad f'(x) > 0$ for $x > 2$.

(c) $f'(x) < 0$ for $x \ne 2$.

(d) $f'(x) > 0$ for $x \ne 2$.

37. $f'(-1) = f'(1) = 0, \quad f'(x) > 0$ on $(-1, 1)$, $f'(x) < 0$ for $x < -1, \quad f'(x) > 0$ for $x > 1$.

38. A local minimum value that is greater than one of its local maximum values.

39. *Speeding* A trucker handed in a ticket at a toll booth showing that in 2 h she had covered 159 mi on a toll road with speed limit 65 mph. The trucker was cited for speeding. Why?

40. *Temperature Change* It took 20 sec for the temperature to rise from 0°F to 212°F when a thermometer was taken from a freezer and placed in boiling water. Explain why at some moment in that interval the mercury was rising at exactly 10.1°F/sec.

41. *Triremes* Classical accounts tell us that a 170-oar trireme (ancient Greek or Roman warship) once covered 184 sea miles in 24 h. Explain why at some point during this feat the trireme's speed exceeded 7.5 knots (sea miles per hour).

42. *Running a Marathon* A marathoner ran the 26.2-mi New York City Marathon in 2.2 h. Show that at least twice, the marathoner was running at exactly 11 mph.

43. *Free Fall* On the moon, the acceleration due to gravity is 1.6 m/sec².

(a) If a rock is dropped into a crevasse, how fast will it be going just before it hits bottom 30 sec later?

(b) How far below the point of release is the bottom of the crevasse?

(c) If instead of being released from rest, the rock is thrown into the crevasse from the same point with a downward velocity of 4 m/sec, when will it hit the bottom and how fast will it be going when it does?

44. *Diving* (a) With what velocity will you hit the water if you step off from a 10-m diving platform?

(b) With what velocity will you hit the water if you dive off the platform with an upward velocity of 2 m/sec?

45. **Writing to Learn** The function

$$f(x) = \begin{cases} x, & 0 \le x < 1 \\ 0, & x = 1 \end{cases}$$

is zero at $x = 0$ and at $x = 1$. Its derivative is equal to 1 at every point between 0 and 1, so f' is never zero between 0 and 1, and the graph of f has no tangent parallel to the chord from $(0, 0)$ to $(1, 0)$. Explain why this does not contradict the Mean Value Theorem.

46. **Writing to Learn** Explain why there is a zero of $y = \cos x$ between every two zeros of $y = \sin x$.

47. *Unique Solution* Assume that f is continuous on $[a, b]$ and differentiable on (a, b). Also assume that $f(a)$ and $f(b)$ have opposite signs and $f' \neq 0$ between a and b. Show that $f(x) = 0$ exactly once between a and b.

In Exercises 48 and 49, show that the equation has exactly one solution in the interval. (*Hint:* See Exercise 47.)

48. $x^4 + 3x + 1 = 0$, $-2 \le x \le -1$

49. $x + \ln (x + 1) = 0$, $0 \le x \le 3$

50. *Parallel Tangents* Assume that f and g are differentiable on $[a, b]$ and that $f(a) = g(a)$ and $f(b) = g(b)$. Show that there is at least one point between a and b where the tangents to the graphs of f and g are parallel or the same line. Illustrate with a sketch.

Explorations

51. *Analyzing Derivative Data* Assume that f is continuous on $[-2, 2]$ and differentiable on $(-2, 2)$. The table gives some values of $f'(x)$.

x	$f'(x)$	x	$f'(x)$
-2	7	0.25	-4.81
-1.75	4.19	0.5	-4.25
-1.5	1.75	0.75	-3.31
-1.25	-0.31	1	-2
-1	-2	1.25	-0.31
-0.75	-3.31	1.5	1.75
-0.5	-4.25	1.75	4.19
-0.25	-4.81	2	7
0	-5		

(a) Estimate where f is increasing, decreasing, and has local extrema.

(b) Find a quadratic regression equation for the data in the table and superimpose its graph on a scatter plot of the data.

(c) Use the model in (b) for f' and find a formula for f that satisfies $f(0) = 0$.

52. *Analyzing Motion Data* Priya's distance D in meters from a motion detector is given by the data in Table 4.1.

Table 4.1 Motion Detector Data

t (sec)	D (m)	t (sec)	D (m)
0.0	3.36	4.5	3.59
0.5	2.61	5.0	4.15
1.0	1.86	5.5	3.99
1.5	1.27	6.0	3.37
2.0	0.91	6.5	2.58
2.5	1.14	7.0	1.93
3.0	1.69	7.5	1.25
3.5	2.37	8.0	0.67
4.0	3.01		

(a) Estimate when Priya is moving toward the motion detector; away from the motion detector.

(b) **Writing to Learn** Give an interpretation of any local extreme values in terms of this problem situation.

(c) Find a cubic regression equation $D = f(t)$ for the data in Table 4.1 and superimpose its graph on a scatter plot of the data.

(d) Use the model in (c) for f to find a formula for f'. Use this formula to estimate the answers to (a). ■

Extending the Ideas

53. *Geometric Mean* The **geometric mean** of two positive numbers a and b is $\sqrt{ab}$. Show that for $f(x) = 1/x$ on any interval $[a, b]$ of positive numbers, the value of c in the conclusion of the Mean Value Theorem is $c = \sqrt{ab}$.

54. *Arithmetic Mean* The **arithmetic mean** of two numbers a and b is $(a + b)/2$. Show that for $f(x) = x^2$ on any interval $[a, b]$, the value of c in the conclusion of the Mean Value Theorem is $c = (a + b)/2$.

55. *Upper Bounds* Show that for any numbers a and b, $|\sin b - \sin a| \le |b - a|$.

56. *Sign of f'* Assume that f is differentiable on $a \le x \le b$ and that $f(b) < f(a)$. Show that f' is negative at some point between a and b.

57. *Monotonic Functions* Show that monotonic functions are one-to-one.

4.3 Connecting f' and f'' with the Graph of f

First Derivative Test for Local Extrema • Concavity • Points of Inflection • Second Derivative Test for Local Extrema • Learning about Functions from Derivatives

First Derivative Test for Local Extrema

As we see once again in Figure 4.17, a function f may have local extrema at some critical points while failing to have local extrema at others. The key is the sign of f' in a critical point's immediate vicinity. As x moves from left to right, the values of f increase where $f' > 0$ and decrease where $f' < 0$.

At the points where f has a minimum value, we see that $f' < 0$ on the interval immediately to the left and $f' > 0$ on the interval immediately to the right. (If the point is an endpoint, there is only the interval on the appropriate side to consider.) This means that the curve is falling (values decreasing) on the left of the minimum value and rising (values increasing) on its right. Similarly, at the points where f has a maximum value, $f' > 0$ on the interval immediately to the left and $f' < 0$ on the interval immediately to the right. This means that the curve is rising (values increasing) on the left of the maximum value and falling (values decreasing) on its right.

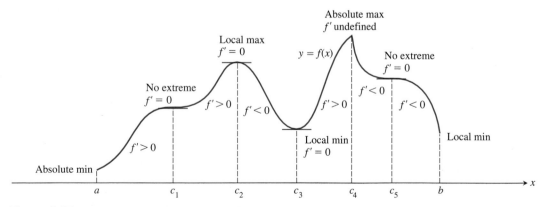

Figure 4.17 A function's first derivative tells how the graph rises and falls.

Theorem 4 First Derivative Test for Local Extrema

The following test applies to a continuous function $f(x)$.

At a critical point c:

1. If f' changes sign from positive to negative at c ($f' > 0$ for $x < c$ and $f' < 0$ for $x > c$), then f has a local maximum value at c.

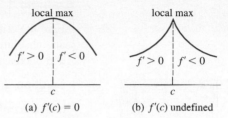

2. If f' changes sign from negative to positive at c ($f' < 0$ for $x < c$ and $f' > 0$ for $x > c$), then f has a local minimum value at c.

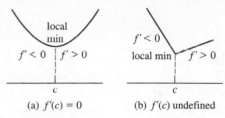

3. If f' does not change sign at c (f' has the same sign on both sides of c), then f has no local extreme value at c.

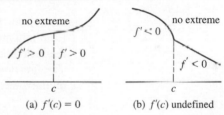

At a left endpoint a:

If $f' < 0$ ($f' > 0$) for $x > a$, then f has a local maximum (minimum) value at a.

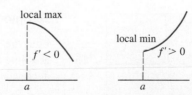

At a right endpoint b:

If $f' < 0$ ($f' > 0$) for $x < b$, then f has a local minimum (maximum) value at b.

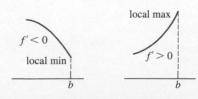

Here is how we apply the first derivative test to find the local extrema of a function. The critical points of a function f partition the x-axis into intervals on which f' is either positive or negative. We determine the sign of f' in each interval by evaluating f' for one value of x in the interval. Then we apply Theorem 4 as shown in Examples 1 and 2.

Example 1   USING THE FIRST DERIVATIVE TEST

Find the critical points of $f(x) = x^3 - 12x - 5$. Find the function's local and absolute extreme values. Identify the intervals on which f is increasing and decreasing.

Solution   Figure 4.18 suggests that f has two critical points.

Solve Analytically

Since f is continuous and differentiable for all real numbers, the critical points occur only at the zeros of f'.

$$f'(x) = 3x^2 - 12 = 0$$
$$x^2 = 4$$
$$x = -2, 2$$

The zeros of f' partition the x-axis into intervals as follows.

Intervals	$-\infty < x < -2$	$-2 < x < 2$	$2 < x < \infty$
Sign of f'	$+$	$-$	$+$
Behavior of f	increasing	decreasing	increasing

We can see from the table that there is a local maximum at $x = -2$ and a local minimum at $x = 2$. The local maximum value is $f(-2) = 11$, and the local minimum value is $f(2) = -21$. There are no absolute extrema. The function increases on the intervals $(-\infty, -2]$ and $[2, \infty)$, and decreases on the interval $[-2, 2]$.

If a function increases, then decreases, and finally increases again like the function in Example 1, you might conjecture that it is a polynomial function. Example 2 shows that this is not necessarily true.

Example 2   USING THE FIRST DERIVATIVE TEST

Find the critical points of $f(x) = (x^2 - 3)e^x$. Find the function's local and absolute extreme values. Identify the intervals on which f is increasing and decreasing.

Solution   This time it is a little harder to see one of the extrema (Figure 4.19).

Solve Analytically

The function f is continuous and differentiable for all real numbers, so the critical points occur only at the zeros of f'.

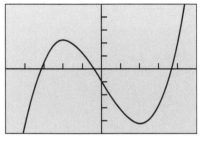

[–5, 5] by [–25, 25]

Figure 4.18 The graph of $f(x) = x^3 - 12x - 5$. (Example 1)

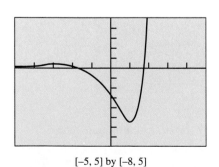

[–5, 5] by [–8, 5]

Figure 4.19 The graph of $f(x) = (x^2 - 3)e^x$. (Example 2)

Using the Product Rule we find

$$f'(x) = (x^2 - 3) \cdot \frac{d}{dx}e^x + \frac{d}{dx}(x^2 - 3) \cdot e^x$$

$$= (x^2 - 3) \cdot e^x + (2x) \cdot e^x$$

$$= (x^2 + 2x - 3)e^x.$$

Since e^x is never zero, the first derivative is zero if and only if

$$x^2 + 2x - 3 = 0$$

$$(x + 3)(x - 1) = 0.$$

The zeros $x = -3$ and $x = 1$ partition the x-axis into intervals as follows.

Intervals	$x < -3$	$-3 < x < 1$	$1 < x$
Sign of f'	+	−	+
Behavior of f	increasing	decreasing	increasing

We can see from the table that there is a local maximum (about 0.299) at $x = -3$, and a local minimum (about -5.437) at $x = 1$. The local minimum value is also an absolute minimum because $f(x) > 0$ for $|x| > \sqrt{3}$. There is no absolute maximum. The function increases on $(-\infty, -3]$ and $[1, \infty)$, and decreases on $[-3, 1]$.

Concavity

As you can see in Figure 4.20, the function $y = x^3$ rises as x increases, but the portions defined on the intervals $(-\infty, 0)$ and $(0, \infty)$ *turn* in different ways. Looking at tangents as we scan from left to right, we see that the slope y' of the curve decreases on the interval $(-\infty, 0)$ and then increases on the interval $(0, \infty)$. The curve $y = x^3$ is *concave down* on $(-\infty, 0)$ and *concave up* on $(0, \infty)$. The curve lies below the tangents where it is concave down, and above the tangents where it is concave up.

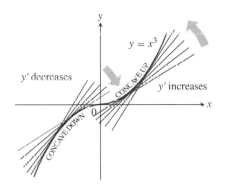

Figure 4.20 The graph of $y = x^3$ is concave down on $(-\infty, 0)$ and concave up on $(0, \infty)$.

Definition   Concavity

The graph of a differentiable function $y = f(x)$ is

(a) concave up on an open interval I if y' is increasing on I.

(b) concave down on an open interval I if y' is decreasing on I.

If a function $y = f(x)$ has a second derivative, then we can conclude that y' increases if $y'' > 0$ and y' decreases if $y'' < 0$.

Concavity Test

The graph of a twice-differentiable function $y = f(x)$ is

(a) concave up on any interval where $y'' > 0$.

(b) concave down on any interval where $y'' < 0$.

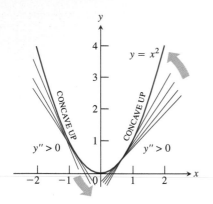

Figure 4.21 The graph of $y = x^2$ is concave up on any interval. (Example 3)

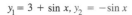

$y_1 = 3 + \sin x$, $y_2 = -\sin x$

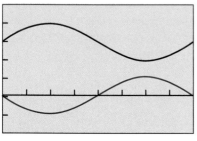

$[0, 2\pi]$ by $[-2, 5]$

Figure 4.22 Using the graph of y'' to determine the concavity of y. (Example 4)

Example 3 APPLYING THE CONCAVITY TEST

The curve $y = x^2$ (Figure 4.21) is concave up on $(-\infty, \infty)$ because its second derivative $y'' = 2$ is always positive.

Example 4 DETERMINING CONCAVITY

Determine the concavity of $y = 3 + \sin x$ on $[0, 2\pi]$.

Solution The graph of $y = 3 + \sin x$ is concave down on $(0, \pi)$, where $y'' = -\sin x$ is negative. It is concave up on $(\pi, 2\pi)$, where $y'' = -\sin x$ is positive (Figure 4.22).

Points of Inflection

The curve $y = 3 + \sin x$ in Example 4 changes concavity at the point $(\pi, 3)$. We call $(\pi, 3)$ a *point of inflection* of the curve.

> **Definition Point of Inflection**
>
> A point where the graph of a function has a tangent line and where the concavity changes is a **point of inflection.**

A point on a curve where y'' is positive on one side and negative on the other is a point of inflection. At such a point, y'' is either zero (because derivatives have the intermediate value property) or undefined. If y is a twice-differentiable function, $y'' = 0$ at a point of inflection and y' has a local maximum or minimum.

To study the motion of a body moving along a line, we often graph the body's position as a function of time. One reason for doing so is to reveal where the body's acceleration, given by the second derivative, changes sign. On the graph, these are the points of inflection.

Example 5 STUDYING MOTION ALONG A LINE

A particle is moving along a horizontal line with position function

$$s(t) = 2t^3 - 14t^2 + 22t - 5, \quad t \geq 0.$$

Find the velocity and acceleration, and describe the motion of the particle.

Solution

Solve Analytically

The velocity is

$$v(t) = s'(t) = 6t^2 - 28t + 22 = 2(t - 1)(3t - 11),$$

and the acceleration is

$$a(t) = v'(t) = s''(t) = 12t - 28 = 4(3t - 7).$$

When the function $s(t)$ is increasing, the particle is moving to the right; when $s(t)$ is decreasing, the particle is moving to the left. Figure 4.23 shows the graphs of the position, velocity, and acceleration of the particle.

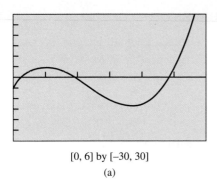
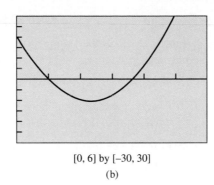
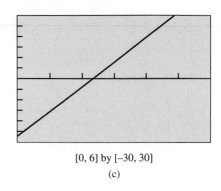

[0, 6] by [−30, 30]

(a)

[0, 6] by [−30, 30]

(b)

[0, 6] by [−30, 30]

(c)

Figure 4.23 The graph of (a) $s(t) = 2t^3 - 14t^2 + 22t - 5$, $t \geq 0$, (b) $s'(t) = 6t^2 - 28t + 22$, and (c) $s''(t) = 12t - 28$. (Example 5)

Notice that the first derivative $(v - s')$ is zero when $t = 1$ and $t = 11/3$.

Intervals	$0 < t < 1$	$1 < t < 11/3$	$11/3 < t$
Sign of $v = s'$	+	−	+
Behavior of s	increasing	decreasing	increasing
Particle motion	right	left	right

The particle is moving to the right in the time intervals $[0, 1)$ and $(11/3, \infty)$, and moving to the left in $(1, 11/3)$.

The acceleration $a(t) = s''(t) = 12t - 28 = 4(3t - 7)$ is zero when $t = 7/3$.

Intervals	$0 < t < 7/3$	$7/3 < t$
Sign of $a = s''$	−	+
Graph of s	concave down	concave up

The accelerating force is directed toward the left during the time interval $[0, 7/3)$, is momentarily zero at $t = 7/3$, and is directed toward the right thereafter.

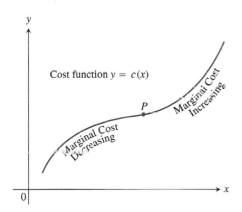

Figure 4.24 On a typical cost curve, a point of inflection separates an interval of decreasing marginal cost from an interval of increasing marginal cost.

Inflection points have applications in some areas of economics. If $y = c(x)$ (Figure 4.24) is the total cost of producing x units of something, the point of inflection at P is then the point at which the marginal cost (the cost of producing one more unit) changes from decreasing to increasing; i.e., it is where the marginal cost reaches a minimum.

The growth of an individual company, of a population, in sales of a new product, or of salaries often follows a *logistic* or *life cycle curve* like the one shown in Figure 4.25. For example, sales of a new product will generally grow slowly at first, then experience a period of rapid growth. Eventually, sales growth slows down again. The function f in Figure 4.25 is increasing. Its rate of increase, f', is at first increasing ($f'' > 0$) up to the point of inflection, and then its rate of increase, f', is decreasing ($f'' < 0$). This is, in a sense, the opposite of what happens in Figure 4.20.

Some graphers have the logistic curve as a built-in regression model. We use this feature in Example 6.

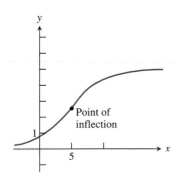

Figure 4.25 A logistic curve

$$y = \frac{c}{1 + ae^{-bx}}.$$

Table 4.2 NFL Average Salaries

Year	Average Salary (dollars)
1983	141,000
1984	206,000
1985	217,000
1986	220,000
1987	220,000
1988	250,000
1989	319,000
1990	365,000
1991	425,000
1992	492,000
1993	683,000
1994	636,000
1995	714,000
1996	795,000

Source: NFLPA as reported by Gordon Forbes in *USA Today,* May 7, 1997.

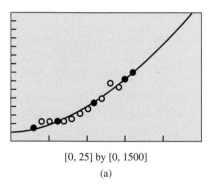

[0, 25] by [0, 1500]

(a)

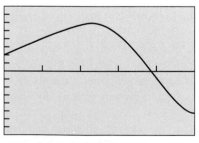

[0, 25] by [−8, 8]

(b)

Figure 4.26 The graph of (a)
$$y = \frac{2171}{1 + 24.72e^{-0.168x}}$$
and (b) y''. (Example 6)

Example 6 USING LOGISTIC REGRESSION

Table 4.2 shows the average salaries for National Football League (NFL) players.

(a) Find the logistic regression equation for the data.

(b) Use the regression equation to predict when the rate of salary increases (f') will start to decrease, and the average salary at that time.

(c) Is there a ceiling to the salaries? If so, what is it?

Solution

(a) We let $x = 0$ represent 1980, $x = 1$ represent 1981, and so forth. We enter the salaries in thousands, using 141 for the salary in 1983, and so on. The logistic regression equation is approximately

$$y = \frac{2171}{1 + 24.72e^{-0.168x}}.$$

Its graph is superimposed on a scatter plot of the data in Figure 4.26a.

(b) We need to find the point of inflection, so we need to solve the equation

$$y'' = 0.$$

Using the graph of y'' in Figure 4.26b we find that $y'' = 0$ when $x \approx 19.097$. The corresponding salary is $y(19.097) \approx 1086$. So, the rate of increase should start to decrease in 1999, and the average player salary at that time should be approximately $1,086,000.

(c) Notice that

$$\lim_{x \to \infty} \frac{2171}{1 + 24.72e^{-0.168x}} = 2171,$$

so the average salaries have a ceiling of about $2,171,000.

Second Derivative Test for Local Extrema

Instead of looking for sign changes in y' at critical points, we can sometimes use the following test to determine the presence of local extrema.

Theorem 5   Second Derivative Test for Local Extrema

1. If $f'(c) = 0$ and $f''(c) < 0$, then f has a local maximum at $x = c$.
2. If $f'(c) = 0$ and $f''(c) > 0$, then f has a local minimum at $x = c$.

This test requires us to know f'' *only at c itself* and not in an interval about c. This makes the test easy to apply. That's the good news. The bad news is that the test fails if $f''(c) = 0$ or if $f''(c)$ fails to exist. When this happens, go back to the first derivative test for local extreme values.

In Example 7, we apply the second derivative test to the function in Example 1.

Example 7 USING THE SECOND DERIVATIVE TEST

Find the extreme values of $f(x) = x^3 - 12x - 5$.

Solution We have

$$f'(x) = 3x^2 - 12 = 3(x^2 - 4)$$
$$f''(x) = 6x.$$

Testing the critical points $x = \pm 2$ (there are no endpoints), we find

$$f''(-2) = -12 < 0 \Rightarrow f \text{ has a local maximum at } x = -2 \text{ and}$$
$$f''(2) = 12 > 0 \Rightarrow f \text{ has a local minimum at } x = 2.$$

Example 8   USING f′ AND f″ TO GRAPH f

Let $f'(x) = 4x^3 - 12x^2$.

(a) Identify where the extrema of f occur.

(b) Find the intervals on which f is increasing and the intervals on which f is decreasing.

(c) Find where the graph of f is concave up and where it is concave down.

(d) Sketch a possible graph for f.

Solution f is continuous since f′ exists. The domain of f′ is $(-\infty, \infty)$, so the domain of f is also $(-\infty, \infty)$. Thus, the critical points of f occur only at the zeros of f′. Since

$$f'(x) = 4x^3 - 12x^2 = 4x^2(x - 3),$$

the first derivative is zero at $x = 0$ and $x = 3$.

Intervals	$x < 0$	$0 < x < 3$	$3 < x$
Sign of f′	−		+
Behavior of f	decreasing	decreasing	increasing

(a) Using the first derivative test and the table above we see that there is no extremum at $x = 0$ and a local minimum at $x = 3$.

(b) Using the table above we see that f is decreasing in $(-\infty, 0]$ and $[0, 3]$, and increasing in $[3, \infty)$.

(c) $f''(x) = 12x^2 - 24x = 12x(x - 2)$ is zero at $x = 0$ and $x = 2$.

Intervals	$x < 0$	$0 < x < 2$	$2 < x$
Sign of f″	+	−	+
Behavior of f	concave up	concave down	concave up

We see that f is concave up on the intervals $(-\infty, 0)$ and $(2, \infty)$, and concave down on $(0, 2)$.

(d) Summarizing the information in the two tables above we obtain.

$x < 0$	$0 < x < 2$	$2 < x < 3$	$x < 3$
decreasing	decreasing	decreasing	increasing
concave up	concave down	concave up	concave up

Figure 4.27 shows one possibility for the graph of f.

Note

The second derivative test does not apply at $x = 0$ because $f''(0) = 0$. We need the first derivative test to see that there is no local extremum at $x = 0$.

Figure 4.27 The graph for f has no extremum but has points of inflection where $x = 0$ and $x = 2$, and a local minimum where $x = 3$. (Example 8)

Exploration 1   **Finding *f* from *f′***

Let $f'(x) = 4x^3 - 12x^2$.

1. Find three different functions with derivative equal to $f'(x)$. How are the graphs of the three functions related?

2. Compare their behavior with the behavior found in Example 8.

Learning about Functions from Derivatives

We have seen in Example 8 and Exploration 1 that we are able to recover almost everything we need to know about a differentiable function $y = f(x)$ by examining y'. We can find where the graph rises and falls and where any local extrema are assumed. We can differentiate y' to learn how the graph bends as it passes over the intervals of rise and fall. We can determine the shape of the function's graph. The only information we cannot get from the derivative is how to place the graph in the xy-plane. As we discovered in Section 4.2, the only additional information we need to position the graph is the value of f at one point.

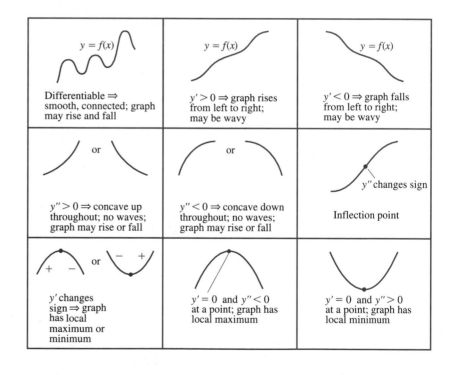

Exploration 2 **Finding f from f' and f"**

A function f is continuous on its domain $[-2, 4]$, $f(-2) = 5$, $f(4) = 1$, and f' and f'' have the following properties.

x	$-2 < x < 0$	$x = 0$	$0 < x < 2$	$x = 2$	$2 < x < 4$
f'	$+$	does not exist	$-$	0	$-$
f''	$+$	does not exist	$+$	0	$-$

1. Find where all absolute extrema of f occur.
2. Find where the points of inflection of f occur.
3. Sketch a possible graph of f.

Quick Review 4.3

In Exercises 1 and 2, factor the expression and use sign charts to solve the inequality.

1. $x^2 - 9 < 0$

2. $x^3 - 4x > 0$

In Exercises 3–6, find the domains of f and f'.

3. $f(x) = xe^x$

4. $f(x) = x^{3/5}$

5. $f(x) = \dfrac{x}{x - 2}$

6. $f(x) = x^{2/5}$

In Exercises 7–10, find the horizontal asymptotes of the function's graph.

7. $y = (4 - x^2)e^x$

8. $y = (x^2 - x)e^{-x}$

9. $y = \dfrac{200}{1 + 10e^{-0.5x}}$

10. $y = \dfrac{750}{2 + 5e^{-0.1x}}$

Section 4.3 Exercises

In Exercises 1 and 2, use the graph of the function f to estimate where (**a**) f' and (**b**) f'' are 0, positive, and negative.

1.

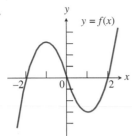

2.

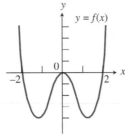

In Exercises 3–6, use the graph of f' to estimate the intervals on which the function f is (**a**) increasing or (**b**) decreasing. (**c**) Estimate where f has local extreme values.

3.

4.

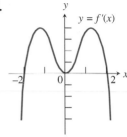

5. The domain of f' is $[0, 4) \cup (4, 6]$.

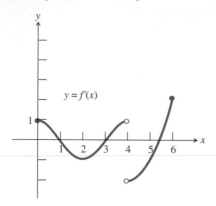

6. The domain of f' is $[0, 1) \cup (1, 2) \cup (2, 3]$.

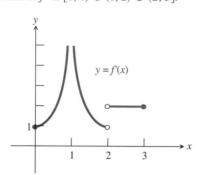

In Exercises 7–12, use analytic methods to find the intervals on which the function is

   (a) increasing,   **(b)** decreasing,

   (c) concave up,   **(d)** concave down.

Then find any

   (e) local extreme values,   **(f)** inflection points.

Support your answers graphically.

  7. $y = x^2 - x - 1$ **8.** $y = -2x^3 + 6x^2 - 3$

  9. $y = 2x^4 - 4x^2 + 1$ **10.** $y = xe^{1/x}$

 11. $y = x\sqrt{8 - x^2}$ **12.** $y = \begin{cases} 3 - x^2, & x < 0 \\ x^2 + 1, & x \geq 0 \end{cases}$

In Exercises 13–28, find the intervals on which the function is

   (a) increasing,   **(b)** decreasing,

   (c) concave up,   **(d)** concave down.

Then find any

   (e) local extreme values,   **(f)** inflection points.

 13. $y = 4x^3 + 21x^2 + 36x - 20$ **14.** $y = -x^4 + 4x^3 - 4x + 1$

 15. $y = 2x^{1/5} + 3$ **16.** $y = 5 - x^{1/3}$

 17. $y = \dfrac{5e^x}{e^x + 3e^{0.8x}}$ **18.** $y = \dfrac{8e^{-x}}{2e^{-x} + 5e^{-1.5x}}$

 19. $y = \begin{cases} 2x, & x < 1 \\ 2 - x^2, & x \geq 1 \end{cases}$ **20.** $y = e^x, \quad 0 \leq x \leq 2\pi$

21. $y = xe^{1/x^2}$ **22.** $y = x^2\sqrt{9 - x^2}$

23. $y = \tan^{-1} x$ **24.** $y = x^{3/4}(5 - x)$

25. $y = x^{1/3}(x - 4)$ **26.** $y = x^{1/4}(x + 3)$

27. $y = \dfrac{x^3 - 2x^2 + x - 1}{x - 2}$ **28.** $y = \dfrac{x}{x^2 + 1}$

In Exercises 29 and 30, use the derivative of the function $y = f(x)$ to find the points at which f has a

   (a) local maximum,   **(b)** local minimum, or

   (c) point of inflection.

29. $y' = (x - 1)^2(x - 2)$

30. $y' = (x - 1)^2(x - 2)(x - 4)$

Exercises 31 and 32 show the graphs of the first and second derivatives of a function $y = f(x)$. *Work in groups of two or three.* Copy the figure and add a sketch of a possible graph of f that passes through the point P.

31.

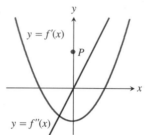

32.

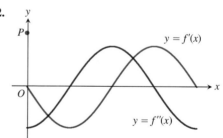

In Exercises 33 and 34, *work in groups of two or three.*

   (a) Find the absolute extrema of f and where they occur.

   (b) Find any points of inflection.

   (c) Sketch a possible graph of f.

33. f is continuous on $[0, 3]$ and satisfies the following.

x	0	1	2	3
f	0	2	0	-2
f'	3	0	does not exist	-3
f''	0	-1	does not exist	0

x	$0 < x < 1$	$1 < x < 2$	$2 < x < 3$
f	$+$	$+$	$-$
f'	$+$	$-$	$-$
f''	$-$	$-$	$-$

34. f is an even function, continuous on $[-3, 3]$, and satisfies the following.

x	0	1	2
f	2	0	-1
f'	does not exist	0	does not exist
f''	does not exist	0	does not exist

x	$0 < x < 1$	$1 < x < 2$	$2 < x < 3$
f	+	$-$	$-$
f'	$-$	$-$	+
f''	+	$-$	$-$

(d) What can you conclude about $f(3)$ and $f(-3)$?

In Exercises 35 and 36, *work in groups of two or three*. Sketch a possible graph of a continuous function f that has the given properties.

35. Domain $[0, 6]$, graph of f' given in Exercise 5, and $f(0) = 2$.

36. Domain $[0, 3]$, graph of f' given in Exercise 6, and $f(0) = -3$.

In Exercises 37–40, a particle is moving along a line with position function $s(t)$. Find the **(a)** velocity and **(b)** acceleration, and **(c)** describe the motion of the particle for $t \geq 0$.

37. $s(t) = t^2 - 4t + 3$ **38.** $s(t) = 6 - 2t - t^2$

39. $s(t) = t^3 - 3t + 3$ **40.** $s(t) = 3t^2 - 2t^3$

In Exercises 41 and 42, the graph of the position function $y = s(t)$ of a particle moving along a line is given. At approximately what times is the particle's **(a)** velocity equal to zero? **(b)** acceleration equal to zero?

41.

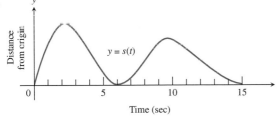

42.

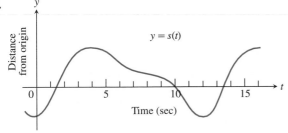

43. Writing to Learn If $f(x)$ is a differentiable function and $f'(c) = 0$ at an interior point c of f's domain, must f have a local maximum or minimum at $x = c$? Explain.

44. Writing to Learn If $f(x)$ is a twice-differentiable function and $f''(c) = 0$ at an interior point c of f's domain, must f have an inflection point at $x = c$? Explain.

45. *Connecting f and f'* Sketch a smooth curve $y = f(x)$ through the origin with the properties that $f'(x) < 0$ for $x < 0$ and $f'(x) > 0$ for $x > 0$.

46. *Connecting f and f''* Sketch a smooth curve $y = f(x)$ through the origin with the properties that $f''(x) < 0$ for $x < 0$ and $f''(x) > 0$ for $x > 0$.

47. *Connecting f, f', and f''* Sketch a continuous curve $y = f(x)$ with the following properties. Label coordinates where possible.

$f(-2) = 8$	$f'(x) > 0$ for $|x| > 2$
$f(0) = 4$	$f'(x) < 0$ for $|x| < 2$
$f(2) = 0$	$f''(x) < 0$ for $x < 0$
$f'(2) = f'(-2) = 0$	$f''(x) > 0$ for $x > 0$

48. *Using Behavior to Sketch* Sketch a continuous curve $y = f(x)$ with the following properties. Label coordinates where possible.

x	y	Curve
$x < 2$		falling, concave up
2	1	horizontal tangent
$2 < x < 4$		rising, concave up
4	4	inflection point
$4 < x < 6$		rising, concave down
6	7	horizontal tangent
$x > 6$		falling, concave down

49. Table 4.3 gives the sales of prepaid phone cards for several years (the values for 1997 and 1998 were estimated). Let $x = 0$ represent 1992, $x = 1$ represent 1993, and so forth.

(a) Find the logistic regression equation and superimpose its graph on a scatter plot of the data.

(b) Use the regression equation to predict when the rate of increase in sales will start to decrease, and to predict the sales at that time.

(c) At what amount does the regression equation predict that sales will stabilize?

Table 4.3 Prepaid Phone Card Sales

Year	Sales (millions of dollars)
1992	12
1993	114
1994	348
1995	792
1996	1110
1997	1520
1998	1900

Source: Atlantic ACM as reported by Grant Jerding in *USA Today*, April 16, 1997.

Exploration

50. *Graphs of Cubics* There is almost no leeway in the locations of the inflection point and the extrema of $f(x) = ax^3 + bx^2 + cx + d,\ a \neq 0,$ because the one inflection point occurs at $x = -b/(3a)$ and the extrema, if any, must be located symmetrically about this value of x. Check this out by examining **(a)** the cubic in Exercise 13 and **(b)** the cubic in Exercise 8. Then **(c)** prove the general case. ■

Extending the Ideas

In Exercises 51 and 52, feel free to use a CAS (computer algebra system), if you have one, to solve the problem.

51. *Logistic Functions* Let $f(x) = c/(1 + ae^{-bx})$ with $a > 0$, $abc \neq 0$.

    (a) Show that f is increasing on the interval $(-\infty, \infty)$ if $abc > 0$, and decreasing if $abc < 0$.

    (b) Show that the point of inflection of f occurs at $x = (\ln|a|)/b$.

52. *Quartic Polynomial Functions* Let $f(x) = ax^4 + bx^3 + cx^2 + dx + e$ with $a \neq 0$.

    (a) Show that the graph of f has 0 or 2 points of inflection.

    (b) Write a condition that must be satisfied by the coefficients if the graph of f has 0 or 2 points of inflection.

4.4 Modeling and Optimization

Examples from Business and Industry • Examples from Mathematics • Examples from Economics • Modeling Discrete Phenomena with Differentiable Functions

Examples from Business and Industry

To *optimize* something means to maximize or minimize some aspect of it. What is the size of the most profitable production run? What is the least expensive shape for an oil can? What is the stiffest rectangular beam we can cut from a 12-inch log? We usually answer such questions by finding the greatest or smallest value of some function that we have used to model the situation.

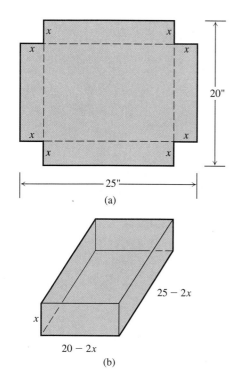

Figure 4.28 An open box made by cutting the corners from a piece of tin. (Example 1)

Example 1 FABRICATING A BOX

An open-top box is to be made by cutting congruent squares of side length x from the corners of a 20- by 25-inch sheet of tin and bending up the sides (Figure 4.28). How large should the squares be to make the box hold as much as possible? What is the resulting maximum volume?

Solution

Model

The height of the box is x, and the other two dimensions are $(20 - 2x)$ and $(25 - 2x)$. Thus, the volume of the box is

$$V(x) = x(20 - 2x)(25 - 2x).$$

Solve Graphically

Because $2x$ cannot exceed 20, we have $0 \leq x \leq 10$. Figure 4.29 suggests that the maximum value of V is about 820.53 and occurs at $x \approx 3.68$.

Confirm Analytically

Expanding, we obtain $V(x) = 4x^3 - 90x^2 + 500x$. The first derivative of V is

$$V'(x) = 12x^2 - 180x + 500.$$

$y = x(20 - 2x)(25 - 2x)$

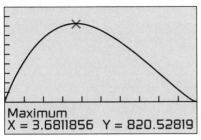

```
Maximum
X = 3.6811856  Y = 820.52819
```

[0, 10] by [−300, 1000]

Figure 4.29 We chose the −300 in −300 < y < 1000 so that the coordinates of the local maximum at the bottom of the screen would not interfere with the graph. (Example 1)

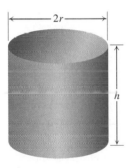

Figure 4.30 This one-liter can uses the least material when $h = 2r$. (Example 2)

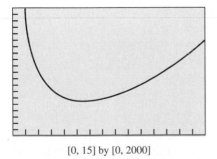

[0, 15] by [0, 2000]

Figure 4.31 The graph of $A = 2\pi r^2 + 2000/r, r > 0$. (Example 2)

The two solutions of the quadratic equation $V'(x) = 0$ are

$$c_1 = \frac{180 - \sqrt{180^2 - 48(500)}}{24} \approx 3.68 \text{ and}$$

$$c_2 = \frac{180 + \sqrt{180^2 - 48(500)}}{24} \approx 11.32.$$

Only c_1 is in the domain [0, 10] of V. The values of V at this one critical point and the two endpoints are

Critical point value: $\quad V(c_1) \approx 820.53$

Endpoint values: $\quad V(0) = 0, \quad V(10) = 0.$

Interpret

Cutout squares that are about 3.68 in. on a side give the maximum volume, about 820.53 in³.

Example 2 DESIGNING A CAN

You have been asked to design a one-liter oil can shaped like a right circular cylinder (Figure 4.30). What dimensions will use the least material?

Solution

Volume of can: If r and h are measured in centimeters, then the volume of the can in cubic centimeters is

$$\pi r^2 h = 1000. \qquad \text{1 liter = 1000 cm}^3$$

Surface area of can: $\quad A = 2\underbrace{\pi r^2}_{\substack{\text{circular} \\ \text{ends}}} + \underbrace{2\pi rh}_{\substack{\text{cylinder} \\ \text{wall}}}$

How can we interpret the phrase "least material"? One possibility is to ignore the thickness of the material and the waste in manufacturing. Then we ask for dimensions r and h that make the total surface area as small as possible while satisfying the constraint $\pi r^2 h = 1000$. (Exercise 17 describes one way to take waste into account.)

Model

To express the surface area as a function of one variable, we solve for one of the variables in $\pi r^2 h = 1000$ and substitute that expression into the surface area formula. Solving for h is easier,

$$h = \frac{1000}{\pi r^2}.$$

Thus,

$$A = 2\pi r^2 + 2\pi rh$$

$$= 2\pi r^2 + 2\pi r \left(\frac{1000}{\pi r^2} \right)$$

$$= 2\pi r^2 + \frac{2000}{r}.$$

Solve Analytically

Our goal is to find a value of $r > 0$ that minimizes the value of A. Figure 4.31 suggests that such a value exists.

Notice from the graph that for small r (a tall thin container, like a piece of pipe), the term $2000/r$ dominates and A is large. For large r (a short wide container, like a pizza pan), the term $2\pi r^2$ dominates and A again is large.

Since A is differentiable on $r > 0$, an interval with no endpoints, it can have a minimum value only where its first derivative is zero.

$$\frac{dA}{dr} = 4\pi r - \frac{2000}{r^2}$$

$$0 = 4\pi r - \frac{2000}{r^2} \qquad \text{Set } dA/dr = 0.$$

$$4\pi r^3 = 2000 \qquad \text{Multiply by } r^2.$$

$$r = \sqrt[3]{\frac{500}{\pi}} \approx 5.42 \qquad \text{Solve for } r.$$

Something happens at $r = \sqrt[3]{500/\pi}$, but what?

If the domain of A were a closed interval, we could find out by evaluating A at this critical point and the endpoints and comparing the results. But the domain is an open interval, so we must learn what is happening at $r = \sqrt[3]{500/\pi}$ by referring to the shape of A's graph. The second derivative

$$\frac{d^2A}{dr^2} = 4\pi + \frac{4000}{r^3}$$

is positive throughout the domain of A. The graph is therefore concave up and the value of A at $r = \sqrt[3]{500/\pi}$ an absolute minimum.

The corresponding value of h (after a little algebra) is

$$h = \frac{1000}{\pi r^2} = 2\sqrt[3]{\frac{500}{\pi}} = 2r.$$

Interpret

The one-liter can that uses the least material has height equal to the diameter, with $r \approx 5.42$ cm and $h \approx 10.84$ cm.

Strategy for Solving Max-Min Problems

1. **Understand the Problem** Read the problem carefully. Identify the information you need to solve the problem.

2. **Develop a Mathematical Model of the Problem** Draw pictures and label the parts that are important to the problem. Introduce a variable to represent the quantity to be maximized or minimized. Using that variable, write a function whose extreme value gives the information sought.

3. **Graph the Function** Find the domain of the function. Determine what values of the variable make sense in the problem.

4. **Identify the Critical Points and Endpoints** Find where the derivative is zero or fails to exist.

5. **Solve the Mathematical Model** If unsure of the result, support or confirm your solution with another method.

6. **Interpret the Solution** Translate your mathematical result into the problem setting and decide whether the result makes sense.

Examples from Mathematics

Example 3 USING THE STRATEGY

Find two numbers whose sum is 20 and whose product is as large as possible.

Solution

Model

If one number is x, the other is $(20 - x)$, and their product is
$f(x) = x(20 - x)$.

Solve Graphically

We can see from the graph of f in Figure 4.32 that there is a maximum. From what we know about parabolas, the maximum occurs at $x = 10$.

Interpret

The two numbers we seek are $x = 10$ and $20 - x = 10$.

Sometimes we find it helpful to use both analytic and graphical methods together, as in Example 4.

[–5, 25] by [–100, 150]

Figure 4.32 The graph of $f(x) = x(20 - x)$ with domain $(-\infty, \infty)$ has an absolute maximum of 100 at $x = 10$. (Example 3)

Example 4 INSCRIBING RECTANGLES

A rectangle is to be inscribed under one arch of the sine curve (Figure 4.33). What is the largest area the rectangle can have, and what dimensions give that area?

Solution

Model

Let $(x, \sin x)$ be the coordinates of point P in Figure 4.33. From what we know about the sine function the x-coordinate of point Q is $(\pi - x)$. Thus,

$$\pi - 2x = \text{length of rectangle}$$

and

$$\sin x = \text{height of rectangle.}$$

The area of the rectangle is

$$A(x) = (\pi - 2x) \sin x.$$

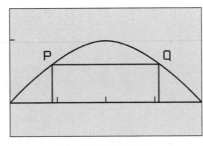

[0, π] by [–0.5, 1.5]

Figure 4.33 A rectangle inscribed under one arch of $y = \sin x$. (Example 4)

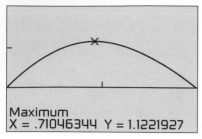

Maximum
X = .71046344 Y = 1.1221927

[0, π/2] by [−1, 2]

(a)

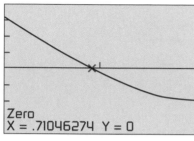

Zero
X = .71046274 Y = 0

[0, π/2] by [−4, 4]

(b)

Figure 4.34 The graph of (a) $A(x) = (\pi - 2x)\sin x$ and (b) A' in the interval $0 \le x \le \pi/2$. (Example 4)

Solve Analytically and Graphically

We can assume that $0 \le x \le \pi/2$. Notice that $A = 0$ at the endpoints $x = 0$ and $x = \pi/2$. Since A is differentiable, the only critical points occur at the zeros of the first derivative,

$$A'(x) = -2\sin x + (\pi - 2x)\cos x.$$

It is not possible to solve the equation $A'(x) = 0$ using algebraic methods. We can use the graph of A (Figure 4.34a) to find the maximum value and where it occurs. Or, we can use the graph of A' (Figure 4.34b) to find where the derivative is zero, and then evaluate A at this value of x to find the maximum value. The two x-values appear to be the same, as they should.

Interpret

The rectangle has a maximum area of about 1.12 square units when $x \approx 0.71$. At this point, the rectangle is $\pi - 2x \approx 1.72$ units long by $\sin x \approx 0.65$ unit high.

Exploration 1 Constructing Cones

A cone of height h and radius r is constructed from a flat, circular disk of radius 4 in. by removing a sector AOC of arc length x in. and then connecting the edges OA and OC. What arc length x will produce the cone of maximum volume, and what is that volume?

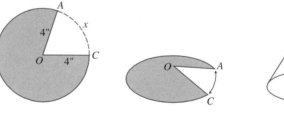
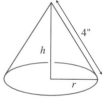

NOT TO SCALE

1. Show that

$$r = \frac{8\pi - x}{2\pi}, \qquad h = \sqrt{16 - r^2}, \qquad \text{and}$$

$$V(x) = \frac{\pi}{3}\left(\frac{8\pi - x}{2\pi}\right)^2 \sqrt{16 - \left(\frac{8\pi - x}{2\pi}\right)^2}.$$

2. Show that the natural domain of V is $0 \le x \le 16\pi$. Graph V over this domain.

3. Explain why the restriction $0 \le x \le 8\pi$ makes sense in the problem situation. Graph V over this domain.

4. Use graphical methods to find where the cone has its maximum volume, and what that volume is.

5. Confirm your findings in part 4 analytically. (*Hint:* Use $V(x) = (1/3)\pi r^2 h$, $h^2 + r^2 = 16$, and the Chain Rule.)

Examples from Economics

Here we want to point out two more places where calculus makes a contribution to economic theory. The first has to do with maximizing profit. The second has to do with minimizing average cost.

Suppose that

$$r(x) = \text{the revenue from selling } x \text{ items,}$$

$$c(x) = \text{the cost of producing the } x \text{ items,}$$

$$p(x) = r(x) - c(x) = \text{the profit from selling } x \text{ items.}$$

The marginal revenue, marginal cost, and marginal profit at this production level (x items) are

$$\frac{dr}{dx} = \text{marginal revenue,}$$

$$\frac{dc}{dx} = \text{marginal cost,}$$

$$\frac{dp}{dx} = \text{marginal profit.}$$

The first observation is about the relationship of p to these derivatives.

Marginal Analysis

Because differentiable functions are locally linear, we can use the marginals to approximate the extra revenue, cost, or profit resulting from selling or producing one more item. Using these approximations is referred to as *marginal analysis*.

Theorem 6 Maximum Profit

Maximum profit (if any) occurs at a production level at which marginal revenue equals marginal cost.

Proof We assume that $r(x)$ and $c(x)$ are differentiable for all $x > 0$, so if $p(x) = r(x) - c(x)$ has a maximum value, it occurs at a production level at which $p'(x) = 0$. Since $p'(x) = r'(x) - c'(x)$, $p'(x) = 0$ implies that

$$r'(x) - c'(x) = 0 \quad \text{or} \quad r'(x) = c'(x). \qquad \blacksquare$$

Figure 4.35 gives more information about this situation.

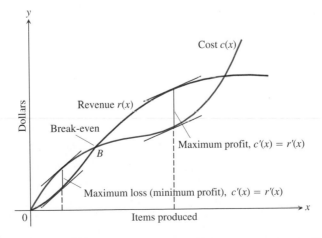

Figure 4.35 The graph of a typical cost function starts concave down and later turns concave up. It crosses the revenue curve at the break-even point B. To the left of B, the company operates at a loss. To the right, the company operates at a profit, the maximum profit occurring where $r'(x) = c'(x)$. Farther to the right, cost exceeds revenue (perhaps because of a combination of market saturation and rising labor and material costs) and production levels become unprofitable again.

What guidance do we get from this observation? We know that a production level at which $p'(x) = 0$ need not be a level of maximum profit. It might be a level of minimum profit, for example. But if we are making financial projections for our company, we should look for production levels at which marginal cost seems to equal marginal revenue. If there is a most profitable production level, it will be one of these.

Example 5   MAXIMIZING PROFIT

Suppose that $r(x) = 9x$ and $c(x) = x^3 - 6x^2 + 15x$, where x represents thousands of units. Is there a production level that maximizes profit? If so, what is it?

Solution   Notice that $r'(x) = 9$ and $c'(x) = 3x^2 - 12x + 15$.

$$3x^2 - 12x + 15 = 9 \quad \text{Set } c'(x) = r'(x).$$

$$3x^2 - 12x + 6 = 0$$

The two solutions of the quadratic equation are

$$x_1 = \frac{12 - \sqrt{72}}{6} = 2 - \sqrt{2} \approx 0.586 \quad \text{and}$$

$$x_2 = \frac{12 + \sqrt{72}}{6} = 2 + \sqrt{2} \approx 3.414.$$

The possible production levels for maximum profit are $x \approx 0.586$ thousand units or $x \approx 3.414$ thousand units. The graphs in Figure 4.36 show that maximum profit occurs at about $x = 3.414$ and maximum loss occurs at about $x = 0.586$.

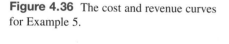

Figure 4.36 The cost and revenue curves for Example 5.

Another way to look for optimal production levels is to look for levels that minimize the average cost of the units produced. Theorem 7 helps us find them.

Theorem 7   Minimizing Average Cost

The production level (if any) at which average cost is smallest is a level at which the average cost equals the marginal cost.

Proof   We assume that $c(x)$ is differentiable.

$$c(x) = \text{cost of producing } x \text{ items}, \quad x > 0$$

$$\frac{c(x)}{x} = \text{average cost of producing } x \text{ items}$$

If the average cost can be minimized, it will be a production level at which

$$\frac{d}{dx}\left(\frac{c(x)}{x}\right) = 0$$

$$\frac{xc'(x) - c(x)}{x^2} = 0 \qquad \text{Quotient Rule}$$

$$xc'(x) - c(x) = 0 \qquad \text{Multiply by } x^2.$$

$$\underbrace{c'(x)}_{\substack{\text{marginal} \\ \text{cost}}} = \underbrace{\frac{c(x)}{x}}_{\substack{\text{average} \\ \text{cost}}}.$$

Again we have to be careful about what Theorem 7 does and does not say. It does not say that there is a production level of minimum average cost—it says where to look to see if there is one. Look for production levels at which average cost and marginal cost are equal. Then check to see if any of them gives a minimum average cost.

Example 6 MINIMIZING AVERAGE COST

Suppose $c(x) = x^3 - 6x^2 + 15x$, where x represents thousands of units. Is there a production level that minimizes average cost? If so, what is it?

Solution We look for levels at which average cost equals marginal cost.

Marginal cost: $c'(x) = 3x^2 - 12x + 15$

Average cost: $\dfrac{c(x)}{x} = x^2 - 6x + 15$

$$3x^2 - 12x + 15 = x^2 - 6x + 15 \qquad \text{Marginal cost = Average cost}$$

$$2x^2 - 6x = 0$$

$$2x(x - 3) = 0$$

$$x = 0 \quad \text{or} \quad x = 3$$

Since $x > 0$, the only production level that might minimize average cost is $x = 3$ thousand units.

We use the second derivative test.

$$\frac{d}{dx}\left(\frac{c(x)}{x}\right) = 2x - 6$$

$$\frac{d^2}{dx^2}\left(\frac{c(x)}{x}\right) = 2 > 0$$

The second derivative is positive for all $x > 0$, so $x = 3$ gives an absolute minimum.

Modeling Discrete Phenomena with Differentiable Functions

In case you are wondering how we can use differentiable functions $c(x)$ and $r(x)$ to describe the cost and revenue that comes from producing a number of items x that can only be an integer, here is the rationale.

When x is large, we can reasonably fit the cost and revenue data with smooth curves $c(x)$ and $r(x)$ that are defined not only at integer values of x but at the values in between just as we do when we use regression equations. Once we have these differentiable functions, which are supposed to behave like the real cost and revenue when x is an integer, we can apply calculus to draw conclusions about their values. We then translate these mathematical conclusions into inferences about the real world that we hope will have predictive value. When they do, as is the case with the economic theory here, we say that the functions give a good model of reality.

What do we do when our calculus tells us that the best production level is a value of x that isn't an integer, as it did in Example 5? We use the nearest convenient integer. For $x \approx 3.414$ thousand units in Example 5, we might use 3414, or perhaps 3410 or 3420 if we ship in boxes of 10.

Quick Review 4.4

1. Use the first derivative test to identify the local extrema of $y = x^3 - 6x^2 + 12x - 8$.

2. Use the second derivative test to identify the local extrema of $y = 2x^3 + 3x^2 - 12x - 3$.

3. Find the volume of a cone with radius 5 cm and height 8 cm.

4. Find the dimensions of a right circular cylinder with volume 1000 cm^3 and surface area 600 cm^2.

In Exercises 5–8, rewrite the expression as a trigonometric function of the angle α.

5. $\sin(-\alpha)$

6. $\cos(-\alpha)$

7. $\sin(\pi - \alpha)$

8. $\cos(\pi - \alpha)$

In Exercises 9 and 10, use substitution to find the exact solutions of the system of equations.

9. $\begin{cases} x^2 + y^2 = 4 \\ y = \sqrt{3}x \end{cases}$

10. $\begin{cases} \dfrac{x^2}{4} + \dfrac{y^2}{9} = 1 \\ y = x + 3 \end{cases}$

Section 4.4 Exercises

In Exercises 1–10, solve the problem analytically. Support your answer graphically.

1. *Finding Numbers* The sum of two nonnegative numbers is 20. Find the numbers if

   (a) the sum of their squares is as large as possible; as small as possible.

   (b) one number plus the square root of the other is as large as possible; as small as possible.

2. *Maximizing Area* What is the largest possible area for a right triangle whose hypotenuse is 5 cm long, and what are its dimensions?

3. *Maximizing Perimeter* What is the smallest perimeter possible for a rectangle whose area is 16 in^2, and what are its dimensions?

4. *Finding Area* Show that among all rectangles with an 8-m perimeter, the one with largest area is a square.

5. *Inscribing Rectangles* The figure shows a rectangle inscribed in an isosceles right triangle whose hypotenuse is 2 units long.

   (a) Express the y-coordinate of P in terms of x. (*Hint:* Write an equation for the line AB.)

   (b) Express the area of the rectangle in terms of x.

   (c) What is the largest area the rectangle can have, and what are its dimensions?

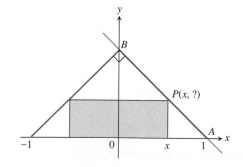

6. *Largest Rectangle* A rectangle has its base on the x-axis and its upper two vertices on the parabola $y = 12 - x^2$. What is the largest area the rectangle can have, and what are its dimensions?

7. *Optimal Dimensions* You are planning to make an open rectangular box from an 8- by 15-in. piece of cardboard by cutting congruent squares from the corners and folding up the sides. What are the dimensions of the box of largest volume you can make this way, and what is its volume?

8. *Closing Off the First Quadrant* You are planning to close off a corner of the first quadrant with a line segment 20 units long running from $(a, 0)$ to $(0, b)$. Show that the area of the triangle enclosed by the segment is largest when $a = b$.

9. *The Best Fencing Plan* A rectangular plot of farmland will be bounded on one side by a river and on the other three sides by a single-strand electric fence. With 800 m of wire at your disposal, what is the largest area you can enclose, and what are its dimensions?

10. *The Shortest Fence* A 216-m^2 rectangular pea patch is to be enclosed by a fence and divided into two equal parts by another fence parallel to one of the sides. What dimensions for the outer rectangle will require the smallest total length of fence? How much fence will be needed?

11. *Designing a Tank* Your iron works has contracted to design and build a 500-ft^3, square-based, open-top, rectangular steel holding tank for a paper company. The tank is to be made by welding thin stainless steel plates together along their edges. As the production engineer, your job is to find dimensions for the base and height that will make the tank weigh as little as possible.

    (a) What dimensions do you tell the shop to use?

    (b) **Writing to Learn** Briefly describe how you took weight into account.

12. *Catching Rainwater* A 1125-ft^3 open-top rectangular tank with a square base x ft on a side and y ft deep is to be built with its top flush with the ground to catch runoff water. The costs associated with the tank involve not only the material from which the tank is made but also an excavation charge proportional to the product xy.

(a) If the total cost is

$$c = 5(x^2 + 4xy) + 10xy,$$

what values of x and y will minimize it?

(b) Writing to Learn Give a possible scenario for the cost function in (a).

13. *Designing a Poster* You are designing a rectangular poster to contain 50 in^2 of printing with a 4-in. margin at the top and bottom and a 2-in. margin at each side. What overall dimensions will minimize the amount of paper used?

14. *Vertical Motion* The height of an object moving vertically is given by

$$s = -16t^2 + 96t + 112,$$

with s in ft and t in sec. Find **(a)** the object's velocity when $t = 0$, **(b)** its maximum height and when it occurs, and **(c)** its velocity when $s = 0$.

15. *Finding an Angle* Two sides of a triangle have lengths a and b, and the angle between them is θ. What value of θ will maximize the triangle's area? (*Hint:* $A = (1/2) ab \sin \theta$.)

16. *Designing a Can* What are the dimensions of the lightest open-top right circular cylindrical can that will hold a volume of 1000 cm^3? Compare the result here with the result in Example 2.

17. *Designing a Can* You are designing a 1000-cm^3 right circular cylindrical can whose manufacture will take waste into account. There is no waste in cutting the aluminum for the side, but the top and bottom of radius r will be cut from squares that measure $2r$ units on a side. The total amount of aluminum used up by the can will therefore be

$$A = 8r^2 + 2\pi rh$$

rather than the $A = 2\pi r^2 + 2\pi rh$ in Example 2. In Example 2 the ratio of h to r for the most economical can was 2 to 1. What is the ratio now?

In Exercises 18 and 19, *work in groups of two or three.*

18. *Designing a Box with Lid* A piece of cardboard measures 10- by 15-in. Two equal squares are removed from the corners of a 10-in. side as shown in the figure. Two equal rectangles are removed from the other corners so that the tabs can be folded to form a rectangular box with lid.

(a) Write a formula $V(x)$ for the volume of the box.

(b) Find the domain of V for the problem situation and graph V over this domain.

(c) Use a graphical method to find the maximum volume and the value of x that gives it.

(d) Confirm your result in (c) analytically.

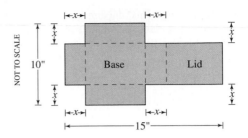

19. *Designing a Suitcase* A 24- by 36-in. sheet of cardboard is folded in half to form a 24- by 18-in. rectangle as shown in the figure. Then four congruent squares of side length x are cut from the corners of the folded rectangle. The sheet is unfolded, and the six tabs are folded up to form a box with sides and a lid.

(a) Write a formula $V(x)$ for the volume of the box.

(b) Find the domain of V for the problem situation and graph V over this domain.

(c) Use a graphical method to find the maximum volume and the value of x that gives it.

(d) Confirm your result in (c) analytically.

(e) Find a value of x that yields a volume of 1120 in^3.

(f) Writing to Learn Write a paragraph describing the issues that arise in part (b).

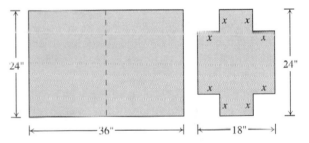

The sheet is then unfolded.

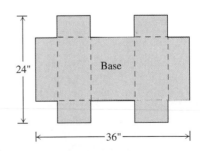

20. *Quickest Route* Jane is 2 mi offshore in a boat and wishes to reach a coastal village 6 mi down a straight shoreline from the point nearest the boat. She can row 2 mph and can walk 5 mph. Where should she land her boat to reach the village in the least amount of time?

21. *Inscribing Rectangles* A rectangle is to be inscribed under the arch of the curve $y = 4 \cos (0.5x)$ from $x = -\pi$ to $x = \pi$. What are the dimensions of the rectangle with largest area, and what is the largest area?

22. *Maximizing Volume* Find the dimensions of a right circular cylinder of maximum volume that can be inscribed in a sphere of radius 10 cm. What is the maximum volume?

Exploration

23. The figure shows the graph of $f(x) = xe^{-x}, x \geq 0$.

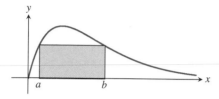

(a) Find where the absolute maximum of f occurs.

(b) Let $a > 0$ and $b > 0$ be given as shown in the figure. Complete the following table where A is the area of the rectangle in the figure.

a	b	A
0.1		
0.2		
0.3		
$\vdots$		
1		

(c) Draw a scatter plot of the data (a, A).

(d) Find the quadratic, cubic, and quartic regression equations for the data in (b), and superimpose their graphs on a scatter plot of the data.

(e) Use each of the regression equations in (d) to estimate the maximum possible value of the area of the rectangle. ■

24. *Cubic Polynomial Functions*
Let $f(x) = ax^3 + bx^2 + cx + d, \ a \neq 0$.

(a) Show that f has either 0 or 2 local extrema.

(b) Give an example of each possibility in (a).

25. *Shipping Packages* The U.S. Postal Service will accept a box for domestic shipment only if the sum of its length and girth (distance around), as shown in the figure, does not exceed 108 in. What dimensions will give a box with a square end the largest possible volume?

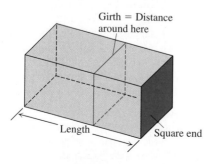

26. *Constructing Cylinders* Compare the answers to the following two construction problems.

(a) A rectangular sheet of perimeter 36 cm and dimensions x cm by y cm is to be rolled into a cylinder as shown in part (a) of the figure. What values of x and y give the largest volume?

(b) The same sheet is to be revolved about one of the sides of length y to sweep out the cylinder as shown in part (b) of the figure. What values of x and y give the largest volume?

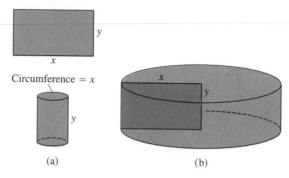
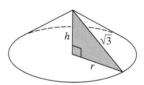

(a)　　　　　　　　　(b)

27. *Constructing Cones* A right triangle whose hypotenuse is $\sqrt{3}$ m long is revolved about one of its legs to generate a right circular cone. Find the radius, height, and volume of the cone of greatest volume that can be made this way.

28. *Finding Parameter Values* What value of a makes $f(x) = x^2 + (a/x)$ have **(a)** a local minimum at $x = 2$? **(b)** a point of inflection at $x = 1$?

29. *Finding Parameter Values* Show that $f(x) = x^2 + (a/x)$ cannot have a local maximum for any value of a.

30. *Finding Parameter Values* What values of a and b make $f(x) = x^3 + ax^2 + bx$ have **(a)** a local maximum at $x = -1$ and a local minimum at $x = 3$? **(b)** a local minimum at $x = 4$ and a point of inflection at $x = 1$?

31. *Inscribing a Cone* Find the volume of the largest right circular cone that can be inscribed in a sphere of radius 3.

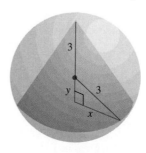

32. *Strength of a Beam* The strength S of a rectangular wooden beam is proportional to its width times the square of its depth.

(a) Find the dimensions of the strongest beam that can be cut from a 12-in. diameter cylindrical log.

(b) **Writing to Learn** Graph S as a function of the beam's width w, assuming the proportionality constant to be $k = 1$. Reconcile what you see with your answer in (a).

(c) **Writing to Learn** On the same screen, graph S as a function of the beam's depth d, again taking $k = 1$. Compare the graphs with one another and with your answer in (a). What would be the effect of changing to some other value of k? Try it.

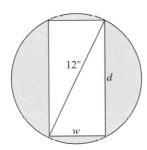

33. *Stiffness of a Beam* The stiffness S of a rectangular beam is proportional to its width times the cube of its depth.

(a) Find the dimensions of the stiffest beam that can be cut from a 12-in. diameter cylindrical log.

(b) **Writing to Learn** Graph S as a function of the beam's width w, assuming the proportionality constant to be $k = 1$. Reconcile what you see with your answer in (a).

(c) **Writing to Learn** On the same screen, graph S as a function of the beam's depth d, again taking $k = 1$. Compare the graphs with one another and with your answer in (a). What would be the effect of changing to some other value of k? Try it.

34. *Frictionless Cart* A small frictionless cart, attached to the wall by a spring, is pulled 10 cm from its rest position and released at time $t = 0$ to roll back and forth for 4 sec. Its position at time t is $s = 10 \cos \pi t$.

(a) What is the cart's maximum speed? When is the cart moving that fast? Where is it then? What is the magnitude of the acceleration then?

(b) Where is the cart when the magnitude of the acceleration is greatest? What is the cart's speed then?

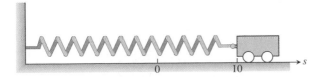

35. *Electrical Current* Suppose that at any time t (sec) the current i (amp) in an alternating current circuit is $i = 2 \cos t + 2 \sin t$. What is the peak (largest magnitude) current for this circuit?

36. *Calculus and Geometry* How close does the curve $y = \sqrt{x}$ come to the point $(3/2, 0)$? (*Hint:* If you minimize the *square* of the distance, you can avoid square roots.)

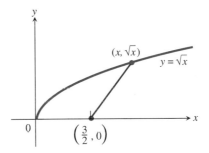

37. *Calculus and Geometry* How close does the semicircle $y = \sqrt{16 - x^2}$ come to the point $(1, \sqrt{3})$?

38. **Writing to Learn** Is the function $f(x) = x^2 - x + 1$ ever negative? Explain.

39. **Writing to Learn** You have been asked to determine whether the function $f(x) = 3 + 4 \cos x + \cos 2x$ is ever negative.

(a) Explain why you need to consider values of x only in the interval $[0, 2\pi]$.

(b) Is f ever negative? Explain.

40. *Vertical Motion* Two masses hanging side by side from springs have positions $s_1 = 2 \sin t$ and $s_2 = \sin 2t$, respectively, with s_1 and s_2 in meters and t in seconds.

(a) At what times in the interval $t > 0$ do the masses pass each other? (*Hint:* $\sin 2t = 2 \sin t \cos t$.)

(b) When in the interval $0 \leq t \leq 2\pi$ is the vertical distance between the masses the greatest? What is this distance? (*Hint:* $\cos 2t = 2 \cos^2 t - 1$.)

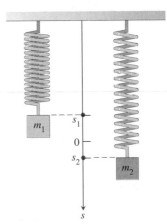

41. *Motion on a Line* The positions of two particles on the s-axis are $s_1 = \sin t$ and $s_2 = \sin (t + \pi/3)$, with s_1 and s_2 in meters and t in seconds.

(a) At what time(s) in the interval $0 \le t \le 2\pi$ do the particles meet?

(b) What is the farthest apart that the particles ever get?

(c) When in the interval $0 \le t \le 2\pi$ is the distance between the particles changing the fastest?

42. *Finding an Angle* The trough in the figure is to be made to the dimensions shown. Only the angle θ can be varied. What value of θ will maximize the trough's volume?

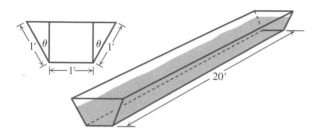

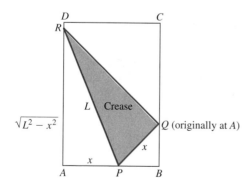

43. *Paper Folding* *Work in groups of two or three.* A rectangular sheet of 8 1/2- by 11-in. paper is placed on a flat surface. One of the corners is placed on the opposite longer edge, as shown in the figure, and held there as the paper is smoothed flat. The problem is to make the length of the crease as small as possible. Call the length L. Try it with paper.

(a) Show that $L^2 = 2x^3/(2x - 8.5)$.

(b) What value of x minimizes L^2?

(c) What is the minimum value of L?

44. *Sensitivity to Medicine* *(continuation of Exercise 35, Section 3.3)* Find the amount of medicine to which the body is most sensitive by finding the value of M that maximizes the derivative dR/dM.

45. *Selling Backpacks* It costs you c dollars each to manufacture and distribute backpacks. If the backpacks sell at x dollars each, the number sold is given by

$$n = \frac{a}{x - c} + b(100 - x),$$

where a and b are certain positive constants. What selling price will bring a maximum profit?

46. *Fermat's Principle in Optics* Fermat's principle in optics states that light always travels from one point to another along a path that minimizes the travel time. Light from a source A is reflected by a plane mirror to a receiver at point B, as shown in the figure. Show that for the light to obey Fermat's principle, the angle of incidence must equal the angle of reflection, both measured from the line normal to the reflecting surface. (This result can also be derived without calculus. There is a purely geometric argument, which you may prefer.)

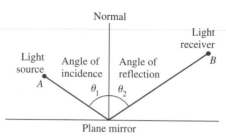

47. *Tin Pest* When metallic tin is kept below 13.2°C, it slowly becomes brittle and crumbles to a gray powder. Tin objects eventually crumble to this gray powder spontaneously if kept in a cold climate for years. The Europeans who saw tin organ pipes in their churches crumble away years ago called the change *tin pest* because it seemed to be contagious. And indeed it was, for the gray powder is a catalyst for its own formation.

A *catalyst* for a chemical reaction is a substance that controls the rate of reaction without undergoing any permanent change in itself. An *autocatalytic reaction* is one whose product is a catalyst for its own formation. Such a reaction may proceed slowly at first if the amount of catalyst present is small and slowly again at the end, when most of the original substance is used up. But in between, when both the substance and its catalyst product are abundant, the reaction proceeds at a faster pace.

In some cases it is reasonable to assume that the rate $v = dx/dt$ of the reaction is proportional both to the amount of the original substance present and to the amount of product. That is, v may be considered to be a function of x alone, and

$$v = kx(a - x) = kax - kx^2,$$

where

$x =$ the amount of product,

$a =$ the amount of substance at the beginning,

$k =$ a positive constant.

At what value of x does the rate v have a maximum? What is the maximum value of v?

48. How We Cough When we cough, the trachea (windpipe) contracts to increase the velocity of the air going out. This raises the question of how much it should contract to maximize the velocity and whether it really contracts that much when we cough.

Under reasonable assumptions about the elasticity of the tracheal wall and about how the air near the wall is slowed by friction, the average flow velocity v (in cm/sec) can be modeled by the equation

$$v = c(r_0 - r)r^2, \quad \frac{r_0}{2} \le r \le r_0,$$

where r_0 is the rest radius of the trachea in cm and c is a positive constant whose value depends in part on the length of the trachea.

(a) Show that v is greatest when $r = (2/3)r_0$, that is, when the trachea is about 33% contracted. The remarkable fact is that X-ray photographs confirm that the trachea contracts about this much during a cough.

(b) Take r_0 to be 0.5 and c to be 1, and graph v over the interval $0 \le r \le 0.5$. Compare what you see to the claim that v is a maximum when $r = (2/3)r_0$.

49. Tour Service You operate a tour service that offers the following rates:

• $200 per person if 50 people (the minimum number to book the tour) go on the tour.

• For each additional person, up to a maximum of 80 people total, the rate per person is reduced by $2.

It costs $6000 (a fixed cost) plus $32 per person to conduct the tour. How many people does it take to maximize your profit?

50. Wilson Lot Size Formula One of the formulas for inventory management says that the average weekly cost of ordering, paying for, and holding merchandise is

$$A(q) = \frac{km}{q} + cm + \frac{hq}{2},$$

where q is the quantity you order when things run low (shoes, radios, brooms, or whatever the item might be), k is the cost of placing an order (the same, no matter how often you order), c is the cost of one item (a constant), m is the number of items sold each week (a constant), and h is the weekly holding cost per item (a constant that takes into account things such as space, utilities, insurance, and security).

(a) Your job, as the inventory manager for your store, is to find the quantity that will minimize $A(q)$. What is it? (The formula you get for the answer is called the *Wilson lot size formula*.)

(b) Shipping costs sometimes depend on order size. When they do, it is more realistic to replace k by $k + bq$, the sum of k and a constant multiple of q. What is the most economical quantity to order now?

51. Production Level Show that if $r(x) = 6x$ and $c(x) = x^3 - 6x^2 + 15x$ are your revenue and cost functions, then the best you can do is break even (have revenue equal cost).

52. Production Level Suppose $c(x) = x^3 - 20x^2 + 20,000x$ is the cost of manufacturing x items. Find a production level that will minimize the average cost of making x items.

Extending the Ideas

53. Airplane Landing Path An airplane is flying at altitude H when it begins its descent to an airport runway that is at horizontal ground distance L from the airplane, as shown in the figure. Assume that the landing path of the airplane is the graph of a cubic polynomial function $y = ax^3 + bx^2 + cx + d$ where $y(-L) = H$ and $y(0) = 0$.

(a) What is dy/dx at $x = 0$?

(b) What is dy/dx at $x = -L$?

(c) Use the values for dy/dx at $x = 0$ and $x = -L$ together with $y(0) = 0$ and $y(-L) = H$ to show that

$$y(x) = H\left[2\left(\frac{x}{L}\right)^3 + 3\left(\frac{x}{L}\right)^2\right].$$

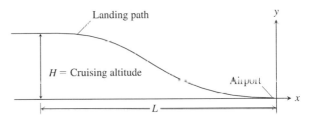

In Exercises 54 and 55, you might find it helpful to use a CAS and to *work in groups of two or three*.

54. Generalized Cone Problem A cone of height h and radius r is constructed from a flat, circular disk of radius a in. as described in Exploration 1.

(a) Find a formula for the volume V of the cone in terms of x and a.

(b) Find r and h in the cone of maximum volume for $a = 4, 5, 6, 8$.

(c) Writing to Learn Find a simple relationship between r and h that is independent of a for the cone of maximum volume. Explain how you arrived at your relationship.

55. *Circumscribing an Ellipse* Let $P(x, a)$ and $Q(-x, a)$ be two points on the upper half of the ellipse

$$\frac{x^2}{100} + \frac{(y - 5)^2}{25} = 1$$

centered at $(0, 5)$. A triangle RST is formed by using the tangent lines to the ellipse at Q and P as shown in the figure.

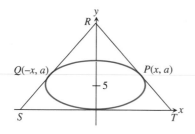

(a) Show that the area of the triangle is

$$A(x) = -f'(x)\left[x - \frac{f(x)}{f'(x)}\right]^2,$$

where $y = f(x)$ is the function representing the upper half of the ellipse.

(b) What is the domain of A? Draw the graph of A. How are the asymptotes of the graph related to the problem situation?

(c) Determine the height of the triangle with minimum area. How is it related to the y-coordinate of the center of the ellipse?

(d) Repeat parts (a)–(c) for the ellipse

$$\frac{x^2}{C^2} + \frac{(y - B)^2}{B^2} = 1$$

centered at $(0, B)$. Show that the triangle has minimum area when its height is $3B$.

<table>
<tr><td>**4.5**</td><td>## Linearization and Newton's Method</td></tr>
</table>

Linear Approximation • Newton's Method • Differentials •
Estimating Change with Differentials • Absolute, Relative, and
Percentage Change • Sensitivity to Change

Linear Approximation

The line $y = mx + b$ tangent to a curve $y = f(x)$ lies close to the curve near the point of tangency. You can observe this phenomenon by zooming in on the two graphs at the point of tangency, or by looking at tables of values for the difference $y = f(x) - (mx + b)$ near the x-coordinate of the point of tangency. Exploration 1 gives more details.

Exploration 1 **Approximating with Tangent Lines**

Let $f(x) = x^2$.

1. Show that the line tangent to the graph of f at the point $(1, 1)$ is $y = 2x - 1$.

2. Set $y_1 = x^2$ and $y_2 = 2x - 1$. Zoom in on the two graphs at $(1, 1)$.

3. For a more dramatic view, set $y_3 = y_1 - y_2$. Then turn off y_1 and y_2, zoom in on the point $(1, 0)$, and watch the difference collapse to zero. Explain why this gives a graphical measure of how well the tangent line approximates the function near $x = 1$.

4. Look at tables of values for y_3 with Δ Table $= 0.1, 0.01, 0.001$, and 0.0001. Explain why this gives a numerical measure of how well the tangent line approximates the function near $x = 1$.

Figure 4.37 The tangent to the curve $y = f(x)$ at $x = a$ is the line $y = f(a) + f'(a)(x - a)$.

In general, the tangent to $y = f(x)$ at $x = a$ (Figure 4.37) passes through the point $(a, f(a))$, so its point-slope equation is

$$y = f(a) + f'(a)(x - a).$$

Thus, the tangent is the graph of the linear function

$$L(x) = f(a) + f'(a)(x - a).$$

For as long as the line remains close to the graph of f, $L(x)$ gives a good approximation to $f(x)$.

Definition   Linearization

If f is differentiable at $x = a$, then the approximating function

$$L(x) = f(a) + f'(a)(x - a)$$

is the **linearization** of f at a.

The approximation $f(x) \approx L(x)$ is the **standard linear approximation** of f at a. The point $x = a$ is the **center** of the approximation.

Example 1   FINDING A LINEARIZATION

Find the linearization of $f(x) = \sqrt{1 + x}$ at $x = 0$ (Figure 4.38).

Solution Since

$$f'(x) = \frac{1}{2}(1 + x)^{-1/2},$$

we have $f(0) = 1,\ f'(0) = 1/2,$ and

$$L(x) = f(a) + f'(a)(x - a) = 1 + \frac{1}{2}(x - 0) = 1 + \frac{x}{2}.$$

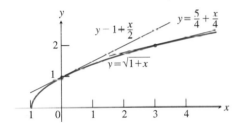

Figure 4.38 The graph of $f(x) = \sqrt{1 + x}$ and its linearization at $x = 0$ and $x = 3$. (Example 1)

Look at how accurate the approximation $\sqrt{1 + x} \approx 1 + (x/2)$ is for values of x near 0.

Approximation	$|$True Value $-$ Approximation$|$
$\sqrt{1.2} \approx 1 + \dfrac{0.2}{2} = 1.10$	$< 10^{-2}$
$\sqrt{1.05} \approx 1 + \dfrac{0.05}{2} = 1.025$	$< 10^{-3}$
$\sqrt{1.005} \approx 1 + \dfrac{0.005}{2} = 1.00250$	$< 10^{-5}$

As we move away from zero we lose accuracy. For example, for $x = 2$ the linearization gives 2 as the approximation for $\sqrt{3}$, which is not even accurate to one decimal place.

Do not be misled by the preceding calculations into thinking that whatever we do with a linearization is better done with a calculator. In practice, we would never use a linearization to find a particular square root. The utility of a linearization is its ability to replace a complicated formula by a simpler one over an entire interval of values. If we have to work with $\sqrt{1 + x}$ for x close to 0 and can tolerate the small amount of error involved, we can work with $1 + (x/2)$ instead. Of course, we then need to know how much error there is. We will have a full story on error in Chapter 9.

In Exercise 7 you will provide the details for the linearization in Example 2.

Example 2 FINDING A LINEARIZATION

The most important linear approximation for roots and powers is

$$(1 + x)^k \approx 1 + kx, \quad (x \approx 0, \text{ any number } k).$$

Example 3 APPLYING EXAMPLE 2

The following approximations are consequences of Example 2.

$$\sqrt{1 + x} \approx 1 + \frac{1}{2}x \qquad k = 1/2$$

$$\frac{1}{1 - x} = (1 - x)^{-1} \approx 1 + (-1)(-x) = 1 + x \qquad k = -1; \text{ replace } x \text{ by } -x.$$

$$\sqrt[3]{1 + 5x^4} = (1 + 5x^4)^{1/3} \approx 1 + \frac{1}{3}(5x^4) = 1 + \frac{5}{3}x^4 \qquad k = 1/3; \text{ replace } x \text{ by } 5x^4.$$

$$\frac{1}{\sqrt{1 - x^2}} = (1 - x^2)^{-1/2} \approx 1 + \left(-\frac{1}{2}\right)(-x^2) = 1 + \frac{1}{2}x^2 \qquad k = -1/2; \text{ replace } x \text{ by } -x^2.$$

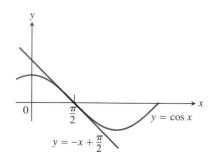

Example 4 FINDING A LINEARIZATION

Find the linearization of $f(x) = \cos x$ at $x = \pi/2$ (Figure 4.39).

Solution Since $f(\pi/2) = \cos(\pi/2) = 0$, $f'(x) = -\sin x$, and $f'(\pi/2) = -\sin(\pi/2) = -1$, we have

$$L(x) = f(a) + f'(a)(x - a)$$

$$= 0 + (-1)\left(x - \frac{\pi}{2}\right)$$

$$= -x + \frac{\pi}{2}.$$

Figure 4.39 The graph of $f(x) = \cos x$ and its linearization at $x = \pi/2$. Near $x = \pi/2$, $\cos x \approx -x + (\pi/2)$. (Example 4)

Newton's Method

Newton's method is a numerical technique for approximating a zero of a function with zeros of its linearizations. Under favorable circumstances, the zeros of the linearizations *converge* rapidly to an accurate approximation. Many calculators use the method because it applies to a wide range of functions and usually gets results in only a few steps. Here is how it works.

To find a solution of an equation $f(x) = 0$, we begin with an initial estimate x_1, found either by looking at a graph or simply guessing. Then we use the tangent to the curve $y = f(x)$ at $(x_1, f(x_1))$ to approximate the curve (Figure 4.40). The point where the tangent crosses the x-axis is the next approximation x_2. The number x_2 is usually a better approximation to the solution than is x_1. The point where the tangent to the curve at $(x_2, f(x_2))$ crosses the x-axis is the next approximation x_3. We continue on, using each approximation to generate the next, until we are close enough to the zero to stop.

There is a formula for finding the $(n + 1)$st approximation x_{n+1} from the nth approximation x_n. The point-slope equation for the tangent to the curve at $(x_n, f(x_n))$ is

$$y - f(x_n) = f'(x_n)(x - x_n).$$

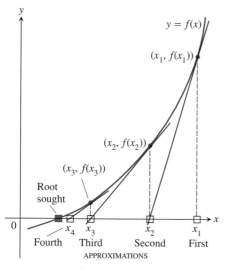

Figure 4.40 Usually the approximations rapidly approach an actual zero of $y = f(x)$.

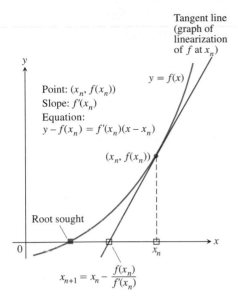

Point: $(x_n, f(x_n))$
Slope: $f'(x_n)$
Equation:
$y - f(x_n) = f'(x_n)(x - x_n)$

$(x_n, f(x_n))$

Root sought

$x_{n+1} = x_n - \dfrac{f(x_n)}{f'(x_n)}$

Figure 4.41 From x_n we go up to the curve and follow the tangent line down to find x_{n+1}.

We can find where it crosses the x-axis by setting $y = 0$ (Figure 4.41).

$$0 - f(x_n) = f'(x_n)(x - x_n)$$

$$-f(x_n) = f'(x_n) \cdot x - f'(x_n) \cdot x_n$$

$$f'(x_n) \cdot x = f'(x_n) \cdot x_n - f(x_n)$$

$$x = x_n - \frac{f(x_n)}{f'(x_n)} \qquad \text{If } f'(x_n) \neq 0$$

This value of x is the next approximation x_{n+1}. Here is a summary of Newton's method.

Procedure for Newton's Method

1. Guess a first approximation to a solution of the equation $f(x) = 0$. A graph of $y = f(x)$ may help.

2. Use the first approximation to get a second, the second to get a third, and so on, using the formula

$$x_{n+1} = x_n - \frac{f(x_n)}{f'(x_n)}.$$

Example 5   USING NEWTON'S METHOD

Use Newton's method to solve $x^3 + 3x + 1 = 0$.

Solution Let $f(x) = x^3 + 3x + 1$, then $f'(x) = 3x^2 + 3$ and

$$x_{n+1} = x_n - \frac{f(x_n)}{f'(x_n)} = x_n - \frac{x_n^3 + 3x_n + 1}{3x_n^2 + 3}.$$

The graph of f in Figure 4.42 suggests that $x_1 = -0.3$ is a good first approximation to the zero of f in the interval $-1 \leq x \leq 0$. Then,

$$x_1 = -0.3,$$

$$x_2 = -0.322324159,$$

$$x_3 = -0.3221853603,$$

$$x_4 = -0.3221853546.$$

The x_n for $n \geq 5$ all appear to equal x_4 on the calculator we used to do the computations. We conclude that the solution to the equation $x^3 + 3x + 1 = 0$ is about -0.3221853546.

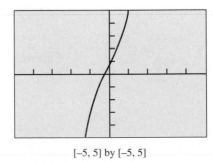

[-5, 5] by [-5, 5]

Figure 4.42 Our graphing calculator's root finder reports -0.3221853546 as the zero of $f(x) = x^3 + 3x + 1$. (Example 5)

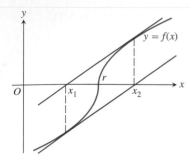

Figure 4.43 The graph of the function

$$f(x) = \begin{cases} -\sqrt{r - x}, & x < r \\ \sqrt{x - r}, & x \geq r. \end{cases}$$

If $x_1 = r - h$, then $x_2 = r + h$. Successive approximations go back and forth between these two values, and Newton's method fails to converge.

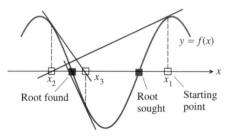

Figure 4.44 Newton's method may miss the zero you want if you start too far away.

Newton's method does not work if $f'(x_1) = 0$. In that case, choose a new starting point.

Newton's method does not always converge. For instance (see Figure 4.43), successive approximations $r - h$ and $r + h$ can go back and forth between these two values, and no amount of iteration will bring us any closer to the zero r.

If Newton's method does converge, it converges to a zero of f. However, the method may converge to a zero that is different from the expected one if the starting value is not close enough to the zero sought. Figure 4.44 shows how this might happen.

Differentials

We sometimes use the notation dy/dx to represent the derivative y' of y with respect to x. Contrary to its appearance, it is not a ratio. We now introduce two new variables dx and dy with the property that if their ratio exists, it will be equal to the derivative.

Definition **Differentials**

Let $y = f(x)$ be a differentiable function. The **differential dx** is an independent variable. The **differential dy** is

$$dy = f'(x)\, dx.$$

Unlike the independent variable dx, the variable dy is always a dependent variable. It depends on both x and dx.

Example 6 FINDING THE DIFFERENTIAL dy

Find dy if

(a) $y = x^5 + 37x$.   **(b)** $y = \sin 3x$.

Solution

(a) $dy = (5x^4 + 37)\, dx$ **(b)** $dy = (3 \cos 3x)\, dx$

If $dx \neq 0$, then the quotient of the differential dy by the differential dx is equal to the derivative $f'(x)$ because

$$\frac{dy}{dx} = \frac{f'(x)\,dx}{dx} = f'(x).$$

We sometimes write

$$df = f'(x)\,dx$$

in place of $dy = f'(x)\,dx$, calling df the **differential of f**. For instance, if $f(x) = 3x^2 - 6$, then

$$df = d(3x^2 - 6) = 6x\,dx.$$

Every differentiation formula like

$$\frac{d(u+v)}{dx} = \frac{du}{dx} + \frac{dv}{dx} \quad \text{or} \quad \frac{d(\sin u)}{dx} = \cos u \frac{du}{dx}$$

has a corresponding differential form like

$$d(u+v) = du + dv \quad \text{or} \quad d(\sin u) = \cos u\,du.$$

Example 7 FINDING DIFFERENTIALS OF FUNCTIONS

(a) $d(\tan 2x) = \sec^2(2x)\,d(2x) = 2\sec^2 2x\,dx$

(b) $d\left(\dfrac{x}{x+1}\right) = \dfrac{(x+1)\,dx - x\,d(x+1)}{(x+1)^2} = \dfrac{x\,dx + dx - x\,dx}{(x+1)^2} = \dfrac{dx}{(x+1)^2}$

Estimating Change with Differentials

Suppose we know the value of a differentiable function $f(x)$ at a point a and we want to predict how much this value will change if we move to a nearby point $a + dx$. If dx is small, f and its linearization L at a will change by nearly the same amount (Figure 4.45). Since the values of L are simple to calculate, calculating the change in L offers a practical way to estimate the change in f.

In the notation of Figure 4.45, the change in f is

$$\Delta f = f(a + dx) - f(a).$$

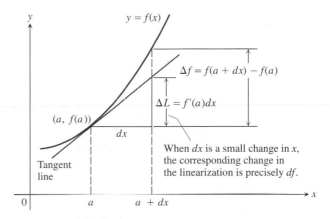

Figure 4.45 Approximating the change in the function f by the change in the linearization of f.

The corresponding change in L is

$$\Delta L = L(a + dx) - L(a)$$

$$= \underbrace{f(a) + f'(a)[(a + dx) - a]}_{L(a + dx)} - \underbrace{f(a)}_{L(a)}$$

$$= f'(a)\, dx.$$

Thus, the differential $df = f'(x)\, dx$ has a geometric interpretation: The value of df at $x = a$ is ΔL, the change in the linearization of f corresponding to the change dx.

Differential Estimate of Change

Let $f(x)$ be differentiable at $x = a$. The approximate change in the value of f when x changes from a to $a + dx$ is

$$df = f'(a)\, dx.$$

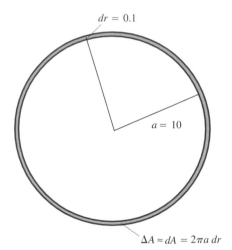

Figure 4.46 When dr is small compared with a, as it is when $dr = 0.1$ and $a = 10$, the differential $dA = 2\pi a\, dr$ gives a good estimate of ΔA. (Example 8)

Example 8 ESTIMATING CHANGE WITH DIFFERENTIALS

The radius r of a circle increases from $a = 10$ m to 10.1 m (Figure 4.46). Use dA to estimate the increase in the circle's area A. Compare this estimate with the true change ΔA.

Solution Since $A = \pi r^2$, the estimated increase is

$$dA = A'(a)\, dr = 2\pi a\, dr = 2\pi(10)(0.1) = 2\pi \text{ m}^2.$$

The true change is

$$\Delta A = \pi(10.1)^2 - \pi(10)^2 = (102.01 - 100)\pi = \underbrace{(2\pi}_{dA} + \underbrace{0.01\pi)}_{\text{error}} \text{ m}^2.$$

Absolute, Relative, and Percentage Change

As we move from a to a nearby point $a + dx$, we can describe the change in f in three ways:

	True	**Estimated**
Absolute change	$\Delta f = f(a + dx) - f(a)$	$df = f'(a)\, dx$
Relative change	$\dfrac{\Delta f}{f(a)}$	$\dfrac{df}{f(a)}$
Percentage change	$\dfrac{\Delta f}{f(a)} \times 100$	$\dfrac{df}{f(a)} \times 100$

Example 9 COMPUTING PERCENTAGE CHANGE

The estimated percentage change in the area of the circle in Example 8 is

$$\frac{dA}{A(a)} \times 100 = \frac{2\pi}{100\pi} \times 100 = 2\%.$$

The true percentage change is

$$\frac{\Delta A}{A(a)} \times 100 = \frac{2.01\pi}{100\pi} \times 100 = 2.01\%.$$

Usually it is not possible or easy to compute the exact (true) change as we did in Example 9. This is sometimes due to uncertainty in measurement, as shown in Example 10.

Example 10 ESTIMATING THE EARTH'S SURFACE AREA

Suppose the earth were a perfect sphere and we determined its radius to be 3959 ± 0.1 miles. What effect would the tolerance of ± 0.1 mi have on our estimate of the earth's surface area?

Solution The surface area of a sphere of radius r is $S = 4\pi r^2$. The uncertainty in the calculation of S that arises from measuring r with a tolerance of dr miles is

$$dS = \left(\frac{dS}{dr}\right) dr = 8\pi r \, dr.$$

With $r = 3959$ and $dr = 0.1$, our estimate of S could be off by as much as

$$dS = 8\pi(3959)(0.1) \approx 9950 \text{ mi}^2,$$

to the nearest square mile, which is about the area of the state of Maryland.

Note

If we underestimated the radius of the earth by 528 ft during a calculation of the earth's surface area, we would leave out an area the size of the state of Maryland.

Example 11 DETERMINING TOLERANCE

About how accurately should we measure the radius r of a sphere to calculate the surface area $S = 4\pi r^2$ within 1% of its true value?

Solution We want any inaccuracy in our measurement to be small enough to make the corresponding increment ΔS in the surface area satisfy the inequality

$$|\Delta S| \leq \frac{1}{100} S = \frac{4\pi r^2}{100}.$$

We replace ΔS in this inequality by its approximation

$$dS = \left(\frac{dS}{dr}\right) dr = 8\pi r \, dr.$$

This gives

$$|8\pi r \, dr| \leq \frac{4\pi r^2}{100}, \quad \text{or} \quad |dr| \leq \frac{1}{8\pi r} \cdot \frac{4\pi r^2}{100} = \frac{1}{2} \cdot \frac{r}{100}.$$

We should measure r with an error dr that is no more than 0.5% of the true value.

Angiography

An opaque dye is injected into a partially blocked artery to make the inside visible under X-rays. This reveals the location and severity of the blockage.

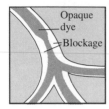

Opaque dye
Blockage

Angioplasty

A balloon-tipped catheter is inflated inside the artery to widen it at the blockage site.

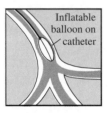

Inflatable balloon on catheter

Example 12 UNCLOGGING ARTERIES

In the late 1830s, the French physiologist Jean Poiseuille ("pwa-ZOY") discovered the formula we use today to predict how much the radius of a partially clogged artery has to be expanded to restore normal flow. His formula,

$$V = kr^4,$$

says that the volume V of fluid flowing through a small pipe or tube in a unit of time at a fixed pressure is a constant times the fourth power of the tube's radius r. How will a 10% increase in r affect V?

Solution The differentials of r and V are related by the equation

$$dV = \frac{dV}{dr}\,dr = 4kr^3\,dr.$$

The relative change in V is

$$\frac{dV}{V} = \frac{4kr^3 dr}{kr^4} = 4\frac{dr}{r}.$$

The relative change in V is 4 times the relative change in r, so a 10% increase in r will produce a 40% increase in the flow.

Sensitivity to Change

The equation $df = f'(x)\,dx$ tells how *sensitive* the output of f is to a change in input at different values of x. The larger the value of f' at x, the greater the effect of a given change dx.

Example 13 FINDING DEPTH OF A WELL

You want to calculate the depth of a well from the equation $s = 16t^2$ by timing how long it takes a heavy stone you drop to splash into the water below. How sensitive will your calculations be to a 0.1 sec error in measuring the time?

Solution The size of ds in the equation

$$ds = 32t\,dt$$

depends on how big t is. If $t = 2$ sec, the error caused by $dt = 0.1$ is only

$$ds = 32(2)(0.1) = 6.4 \text{ ft.}$$

Three seconds later at $t = 5$ sec, the error caused by the same dt is

$$ds = 32(5)(0.1) = 16 \text{ ft.}$$

Quick Review 4.5

In Exercises 1 and 2, find dy/dx.

1. $y = \sin(x^2 + 1)$

2. $y = \dfrac{x + \cos x}{x + 1}$

In Exercises 3 and 4, solve the equation graphically.

3. $xe^{-x} + 1 = 0$

4. $x^3 + 3x + 1 = 0$

In Exercises 5 and 6, let $f(x) = xe^{-x} + 1$. Write an equation for the line tangent to f at $x = c$.

5. $c = 0$

6. $c = -1$

7. Find where the tangent line in **(a)** Exercise 5 and **(b)** Exercise 6 crosses the x-axis.

8. Let $g(x)$ be the function whose graph is the tangent line to the graph of $f(x) = x^3 - 4x + 1$ at $x = 1$. Complete the table.

x	$f(x)$	$g(x)$
0.7		
0.8		
0.9		
1		
1.1		
1.2		
1.3		

In Exercises 9 and 10, graph $y = f(x)$ and its tangent line at $x = c$.

9. $c = 1.5$, $f(x) = \sin x$

10. $c = 4$, $f(x) = \begin{cases} -\sqrt{3 - x}, & x < 3 \\ \sqrt{x - 3}, & x \geq 3 \end{cases}$

Section 4.5 Exercises

In Exercises 1–6, **(a)** find the linearization $L(x)$ of $f(x)$ at $x = a$. **(b)** How accurate is the approximation $L(a + 0.1) \approx f(a + 0.1)$? See the comparisons following Example 1.

1. $f(x) = x^3 - 2x + 3$, $a = 2$

2. $f(x) = \sqrt{x^2 + 9}$, $a = -4$

3. $f(x) = x + \dfrac{1}{x}$, $a = 1$

4. $f(x) = \ln(x + 1)$, $a = 0$

5. $f(x) = \tan x$, $a = \pi$

6. $f(x) = \cos^{-1} x$, $a = 0$

7. Show that the linearization of $f(x) = (1 + x)^k$ at $x = 0$ is $L(x) = 1 + kx$.

8. Use the linear approximation $(1 + x)^k \approx 1 + kx$ to find an approximation for the function $f(x)$ for values of x near zero.

(a) $f(x) = (1 - x)^6$

(b) $f(x) = \dfrac{2}{1 - x}$

(c) $f(x) = \dfrac{1}{\sqrt{1 + x}}$

(d) $f(x) = \sqrt{2 + x^2}$

(e) $f(x) = (4 + 3x)^{1/3}$

(f) $f(x) = \sqrt[3]{\left(1 - \dfrac{1}{2 + x}\right)^2}$

9. Writing to Learn Find the linearization of $f(x) = \sqrt{x + 1} + \sin x$ at $x = 0$. How is it related to the individual linearizations for $\sqrt{x + 1}$ and $\sin x$?

10. Use the linearization $(1 + x)^k \approx 1 + kx$ to approximate the following. State how accurate your approximation is.

(a) $(1.002)^{100}$

(b) $\sqrt[3]{1.009}$

In Exercises 11–14, choose a linearization with center not at $x = a$ but at a nearby value at which the function and its derivative are easy to evaluate. State the linearization and the center.

11. $f(x) = 2x^2 + 4x - 3$, $a = -0.9$

12. $f(x) = \sqrt[3]{x}$, $a = 8.5$

13. $f(x) = \dfrac{x}{x + 1}$, $a = 1.3$

14. $f(x) = \cos x$, $a = 1.7$

In Exercises 15–18, use Newton's method to estimate all real solutions of the equation. Make your answers accurate to 6 decimal places.

15. $x^3 + x - 1 = 0$

16. $x^4 + x - 3 = 0$

17. $x^2 - 2x + 1 = \sin x$

18. $x^4 - 2 = 0$

In Exercises 19–26, **(a)** find dy, and **(b)** evaluate dy for the given value of x and dx.

19. $y = x^3 - 3x$, $x = 2$, $dx = 0.05$

20. $y = \dfrac{2x}{1 + x^2}$, $x = -2$, $dx = 0.1$

21. $y = x^2 \ln x$, $x = 1$, $dx = 0.01$

22. $y = x\sqrt{1 - x^2}$, $x = 0$, $dx = -0.2$

23. $y = e^{\sin x}$, $x = \pi$, $dx = -0.1$

24. $y = 3 \csc \left(1 - \dfrac{x}{3}\right)$, $x = 1$, $dx = 0.1$

25. $y + xy - x = 0$, $x = 0$, $dx = 0.01$

26. $y = \sec(x^2 - 1)$, $x = 1.5$, $dx = 0.05$

In Exercises 27–30, the function f changes value when x changes from a to $a + dx$. Find

    (a) the absolute change $\Delta f = f(a + dx) - f(a)$.

    (b) the estimated change $df = f'(a)\, dx$.

    (c) the approximation error $|\Delta f - df|$.

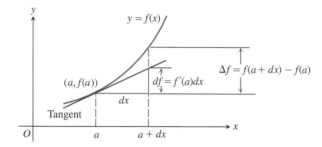

27. $f(x) = x^2 + 2x$, $a = 0$, $dx = 0.1$

28. $f(x) = x^3 - x$, $a = 1$, $dx = 0.1$

29. $f(x) = x^{-1}$, $a = 0.5$, $dx = 0.05$

30. $f(x) = x^4$, $a = 1$, $dx = 0.01$

In Exercises 31–36, write a differential formula that estimates the given change in volume or surface area.

31. *Volume* The change in the volume $V = (4/3)\pi r^3$ of a sphere when the radius changes from a to $a + dr$

32. *Surface Area* The change in the surface area $S = 4\pi r^2$ of a sphere when the radius changes from a to $a + dr$

$V = \dfrac{4}{3}\pi r^3$, $S = 4\pi r^2$

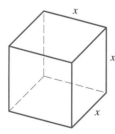

$V = x^3$, $S = 6x^2$

33. *Volume* The change in the volume $V = x^3$ of a cube when the edge lengths change from a to $a + dx$

34. *Surface Area* The change in the surface area $S = 6x^2$ of a cube when the edge lengths change from a to $a + dx$

35. *Volume* The change in the volume $V = \pi r^2 h$ of a right circular cylinder when the radius changes from a to $a + dr$ and the height does not change

36. *Surface Area* The change in the lateral surface area $S = 2\pi rh$ of a right circular cylinder when the height changes from a to $a + dh$ and the radius does not change

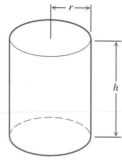

$V = \pi r^2 h$, $S = 2\pi rh$

37. *Linear Approximation* Let f be a function with $f(0) = 1$ and $f'(x) = \cos(x^2)$.

    (a) Find the linearization of f at $x = 0$.

    (b) Estimate the value of f at $x = 0.1$.

    (c) **Writing to Learn** Do you think the actual value of f at $x = 0.1$ is greater than or less than the estimate in (b)? Explain.

38. *Expanding Circle* The radius of a circle is increased from 2.00 to 2.02 m.

    (a) Estimate the resulting change in area.

    (b) Estimate as a percentage of the circle's original area.

39. *Growing Tree* The diameter of a tree was 10 in. During the following year, the circumference increased 2 in. About how much did the tree's diameter increase? the tree's cross section area?

40. *Percentage Error* The edge of a cube is measured as 10 cm with an error of 1%. The cube's volume is to be calculated from this measurement. Estimate the percentage error in the volume calculation.

41. *Percentage Error* About how accurately should you measure the side of a square to be sure of calculating the area to within 2% of its true value?

42. *Estimating Volume* Estimate the volume of material in a cylindrical shell with height 30 in., radius 6 in., and shell thickness 0.5 in.

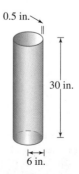

43. *Estimating Volume* A surveyor is standing 30 ft from the base of a building. She measures the angle of elevation to the top of the building to be 75°. How accurately must the angle be measured for the percentage error in estimating the height of the building to be less than 4%?

44. *Tolerance* The height and radius of a right circular cylinder are equal, so the cylinder's volume is $V = \pi h^3$. The volume is to be calculated with an error of no more than 1% of the true value. Find approximately the greatest error that can be tolerated in the measurement of h, expressed as a percentage of h.

45. *Tolerance* **(a)** About how accurately must the interior diameter of a 10-m high cylindrical storage tank be measured to calculate the tank's volume to within 1% of its true value?

(b) About how accurately must the tank's exterior diameter be measured to calculate the amount of paint it will take to paint the side of the tank to within 5% of the true amount?

46. *Minting Coins* A manufacturer contracts to mint coins for the federal government. How much variation dr in the radius of the coins can be tolerated if the coins are to weigh within 1/1000 of their ideal weight? Assume the thickness does not vary.

47. *The Effect of Flight Maneuvers on the Heart* The amount of work done by the heart's main pumping chamber, the left ventricle, is given by the equation

$$W = PV + \frac{V\delta v^2}{2g},$$

where W is the work per unit time, P is the average blood pressure, V is the volume of blood pumped out during the unit of time, δ ("delta") is the density of the blood, v is the average velocity of the exiting blood, and g is the acceleration of gravity.

When P, V, δ, and v remain constant, W becomes a function of g, and the equation takes the simplified form

$$W = a + \frac{b}{g} \ (a, \ b \text{ constant}).$$

As a member of NASA's medical team, you want to know how sensitive W is to apparent changes in g caused by flight maneuvers, and this depends on the initial value of g. As part of your investigation, you decide to compare the effect on W of a given change dg on the moon, where $g = 5.2$ ft/sec², with the effect the same change dg would have on Earth, where $g = 32$ ft/sec². Use the simplified equation above to find the ratio of dW_{moon} to dW_{Earth}.

48. *Measuring Acceleration of Gravity* When the length L of a clock pendulum is held constant by controlling its temperature, the pendulum's period T depends on the acceleration of gravity g. The period will therefore vary slightly as the clock is moved from place to place on the earth's surface, depending on the change in g. By keeping track of ΔT, we can estimate the variation in g from the equation $T = 2\pi(L/g)^{1/2}$ that relates T, g, and L.

(a) With L held constant and g as the independent variable, calculate dT and use it to answer (b) and (c).

(b) **Writing to Learn** If g increases, will T increase or decrease? Will a pendulum clock speed up or slow down? Explain.

(c) A clock with a 100-cm pendulum is moved from a location where $g = 980$ cm/sec² to a new location. This increases the period by $dT = 0.001$ sec. Find dg and estimate the value of g at the new location.

49. *Newton's Method* Suppose your first guess in using Newton's method is lucky in the sense that x_1 is a root of $f(x) = 0$. What happens to x_2 and later approximations?

50. *Oscillation* Show that if $h > 0$, applying Newton's method to

$$f(x) = \begin{cases} \sqrt{x}, & x \geq 0 \\ \sqrt{-x}, & x < 0 \end{cases}$$

leads to $x_2 = -h$ if $x_1 = h$, and to $x_2 = h$ if $x_1 = -h$. Draw a picture that shows what is going on.

51. *Approximations that Get Worse and Worse* Apply Newton's method to $f(x) = x^{1/3}$ with $x_1 = 1$, and calculate x_2, x_3, x_4, and x_5. Find a formula for $|x_n|$. What happens to $|x_n|$ as $n \to \infty$? Draw a picture that shows what is going on.

Exploration

52. *Quadratic Approximations*

(a) Let $Q(x) = b_0 + b_1(x - a) + b_2(x - a)^2$ be a quadratic approximation to $f(x)$ at $x = a$ with the properties:

i. $Q(a) = f(a)$,

ii. $Q'(a) = f'(a)$,

iii. $Q''(a) = f''(a)$.

Determine the coefficients b_0, b_1, and b_2.

(b) Find the quadratic approximation to $f(x) = 1/(1 - x)$ at $x = 0$.

(c) Graph $f(x) = 1/(1 - x)$ and its quadratic approximation at $x = 0$. Then zoom in on the two graphs at the point $(0, 1)$. Comment on what you see.

(d) Find the quadratic approximation to $g(x) = 1/x$ at $x = 1$. Graph g and its quadratic approximation together. Comment on what you see.

(e) Find the quadratic approximation to $h(x) = \sqrt{1 + x}$ at $x = 0$. Graph h and its quadratic approximation together. Comment on what you see.

(f) What are the linearizations of f, g, and h at the respective points in (b), (d), and (e)?

■

Extending the Ideas

53. *Formulas for Differentials* Verify the following formulas.

  (a) $d(c) = 0$ (c a constant)

  (b) $d(cu) = c\,du$ (c a constant)

  (c) $d(u + v) = du + dv$

  (d) $d(u \cdot v) = u\,dv + v\,du$

  (e) $d\left(\dfrac{u}{v}\right) = \dfrac{v\,du - u\,dv}{v^2}$

  (f) $d(u^n) = nu^{n-1}\,du$

54. *Linearization* Show that the approximation of $\tan x$ by its linearization at the origin must improve as $x \to 0$ by showing that

$$\lim_{x \to 0} \frac{\tan x}{x} = 1.$$

55. *The Linearization is the Best Linear Approximation*
Suppose that $y = f(x)$ is differentiable at $x = a$ and that $g(x) = m(x - a) + c$ (m and c constants). If the error $E(x) = f(x) - g(x)$ were small enough near $x = a$, we might think of using g as a linear approximation of f instead of the linearization $L(x) = f(a) + f'(a)(x - a)$. Show that if we impose on g the conditions

  i. $E(a) = 0$, The error is zero at $x = a$.

  ii. $\displaystyle\lim_{x \to a} \frac{E(x)}{x - a} = 0$, The error is negligible when compared with $(x - a)$.

then $g(x) = f(a) + f'(a)(x - a)$. Thus, the linearization gives the only linear approximation whose error is both zero at $x = a$ and negligible in comparison with $(x - a)$.

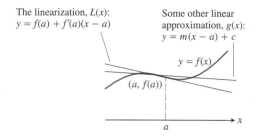

The linearization, $L(x)$:
$y = f(a) + f'(a)(x - a)$

Some other linear approximation, $g(x)$:
$y = m(x - a) + c$

$y = f(x)$

$(a, f(a))$

x

a

4.6 Related Rates

Related Rate Equations • Solution Strategy • Simulating Related Motion

Related Rate Equations

Suppose that a particle $P(x, y)$ is moving along a curve C in the plane so that its coordinates x and y are differentiable functions of time t. If D is the distance from the origin to P, then using the Chain Rule we can find an equation that relates dD/dt, dx/dt, and dy/dt.

$$D = \sqrt{x^2 + y^2}$$

$$\frac{dD}{dt} = \frac{1}{2}(x^2 + y^2)^{-1/2}\left(2x\frac{dx}{dt} + 2y\frac{dy}{dt}\right)$$

Any equation involving two or more variables that are differentiable functions of time t can be used to find an equation that relates their corresponding rates.

Example 1 FINDING RELATED RATE EQUATIONS

Assume that the radius r and height h of a cone are differentiable functions of t and let V be the volume of the cone. Find an equation that relates dV/dt, dr/dt, and dh/dt.

Solution $V = \dfrac{\pi}{3}r^2 h$ Cone volume formula

$$\frac{dV}{dt} = \frac{\pi}{3}\left(r^2 \cdot \frac{dh}{dt} + 2r\frac{dr}{dt} \cdot h\right) = \frac{\pi}{3}\left(r^2\frac{dh}{dt} + 2rh\frac{dr}{dt}\right)$$

Solution Strategy

How fast is a balloon rising at a given instant? How fast does the water level drop when a tank is drained at a certain rate? Questions like these ask us to calculate a rate that may be difficult to measure from a rate that we know, or is easy to measure. We begin by writing an equation that relates the variables involved and then differentiate it to get an equation that relates the rate we seek to the rates we know.

Example 2 A RISING BALLOON

A hot-air balloon rising straight up from a level field is tracked by a range finder 500 ft from the lift-off point. At the moment the range finder's elevation angle is $\pi/4$, the angle is increasing at the rate of 0.14 rad/min. How fast is the balloon rising at that moment?

Solution We answer the question in six steps.

Step 1:
Draw a picture and name the variables and constants (Figure 4.47). The variables in the picture are

θ = the angle in radians the range finder makes with the ground,

y = the height in feet of the balloon.

We let t represent time in minutes and assume θ and y are differentiable functions of t.

The one constant in the picture is the distance from the range finder to the lift-off point (500 ft). There is no need to give it a special symbol.

Step 2:
Write down the additional numerical information.
$$\frac{d\theta}{dt} = 0.14 \text{ rad/min} \quad \text{when} \quad \theta = \frac{\pi}{4}$$

Step 3:
Write down what we are to find. We want dy/dt when $\theta = \pi/4$.

Step 4:
Write an equation that relates the variables y and θ.
$$\frac{y}{500} = \tan \theta \quad \text{or} \quad y = 500 \tan \theta$$

Step 5:
Differentiate with respect to t using the Chain Rule. The result tells how dy/dt (which we want) is related to $d\theta/dt$ (which we know).
$$\frac{dy}{dt} = 500(\sec^2 \theta)\frac{d\theta}{dt}$$

Step 6:
Evaluate with $\theta = \pi/4$ and $d\theta/dt = 0.14$ to find dy/dt.
$$\frac{dy}{dt} = 500(\sqrt{2})^2(0.14) = 140 \quad \sec\frac{\pi}{4} = \sqrt{2}$$

Interpret
At the moment in question, the balloon is rising at the rate of 140 ft/min.

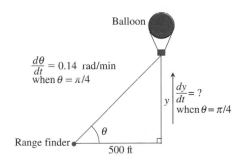

Balloon

$\frac{d\theta}{dt} = 0.14$ rad/min
when $\theta = \pi/4$

$\frac{dy}{dt} = ?$
when $\theta = \pi/4$

y

θ

Range finder

500 ft

Figure 4.47 (Example 2)

The units in the formula for *dy/dt* (Step 6)

A radian is the ratio of two lengths, meaning its units are length/length. In Example 2, the units are ft/ft. Radians per minute are (ft/ft)/min = ft/(ft × min). The secant of an angle is also a ratio of lengths. The units of the secant in Example 2 are ft/ft. For the equation

$$\frac{dy}{dt} = 500 \times (\sec \theta)^2 \times \frac{d\theta}{dt} = 140,$$

we therefore have

$$\frac{dy}{dt} = \text{ft} \times \left(\frac{\text{ft}}{\text{ft}}\right)^2 \times \frac{\text{ft}}{\text{ft} \times \text{min}} = \frac{\text{ft}}{\text{min}}.$$

Because the units of radians (length/length) cancel, we usually regard radians as *dimensionless* and don't write anything in. The trigonometric ratios (length/length) are dimensionless, too. With this understanding, the units of radians/minute, instead of being (length/length)/minute, are 1/minute or "per minute," the secant's units do not appear at all, and the units of

$$500 \times (\sqrt{2})^2 \times (0.14) = 140$$

are calculated as

$$\text{ft} \times \frac{1}{\text{min}} = \frac{\text{ft}}{\text{min}}.$$

Related Rate Problem Strategy

1. *Draw a picture and name the variables and constants.* Use *t* for time. Assume all variables are differentiable functions of *t*.

2. *Write down the numerical information* (in terms of the symbols you have chosen).

3. *Write down what we are asked to find* (usually a rate, expressed as a derivative).

4. *Write an equation that relates the variables.* You may have to combine two or more equations to get a single equation that relates the variable whose rate you want to the variables whose rates you know.

5. *Differentiate with respect to t.* Then express the rate you want in terms of the rate and variables whose values you know.

6. *Evaluate.* Use known values to find the unknown rate.

Example 3   A HIGHWAY CHASE

A police cruiser, approaching a right-angled intersection from the north, is chasing a speeding car that has turned the corner and is now moving straight east. When the cruiser is 0.6 mi north of the intersection and the car is 0.8 mi to the east, the police determine with radar that the distance between them and the car is increasing at 20 mph. If the cruiser is moving at 60 mph at the instant of measurement, what is the speed of the car?

Solution We carry out the steps of the strategy.

Step 1:

Picture and variables. We picture the car and cruiser in the coordinate plane, using the positive *x*-axis as the eastbound highway and the positive *y*-axis as the southbound highway (Figure 4.48). We let *t* represent time and set

$$x = \text{position of car at time } t,$$

$$y = \text{position of cruiser at time } t, \text{ and}$$

$$s = \text{distance between car and cruiser at time } t.$$

We assume *x*, *y*, and *s* are differentiable functions of *t*.

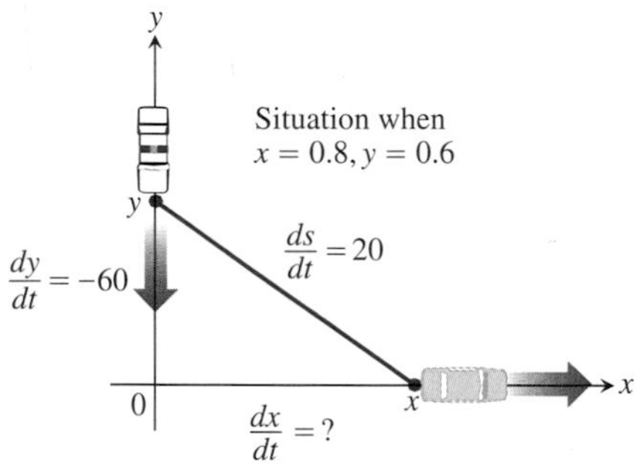

Step 2:

Numerical information. At the instant in question,

$$x = 0.8 \text{ mi}, \quad y = 0.6 \text{ mi}, \quad \frac{dy}{dt} = -60 \text{ mph}, \quad \frac{ds}{dt} = 20 \text{ mph}.$$

Note *dy/dt* is negative because *y* is decreasing.

Step 3:

To find: dx/dt

Step 4:

How the variables are related: $s^2 = x^2 + y^2$
(We could also use $s = \sqrt{x^2 + y^2}$.)

Situation when $x = 0.8, y = 0.6$

$$\frac{dy}{dt} = -60$$

$$\frac{ds}{dt} = 20$$

$$\frac{dx}{dt} = ?$$

Figure 4.48 (Example 3)

Step 5:

Differentiate with respect to t.

$$2s\frac{ds}{dt} = 2x\frac{dx}{dt} + 2y\frac{dy}{dt}$$

$$\frac{ds}{dt} = \frac{1}{s}\left(x\frac{dx}{dt} + y\frac{dy}{dt}\right)$$

$$= \frac{1}{\sqrt{x^2 + y^2}}\left(x\frac{dx}{dt} + y\frac{dy}{dt}\right)$$

Step 6:

Evaluate. Use $x = 0.8$, $y = 0.6$, $dy/dt = -60$, $ds/dt = 20$, and solve for dx/dt.

$$20 = \frac{1}{\sqrt{(0.8)^2 + (0.6)^2}}\left(0.8\frac{dx}{dt} + (0.6)(-60)\right)$$

$$\frac{dx}{dt} = \frac{20\sqrt{(0.8)^2 + (0.6)^2} + (0.6)(60)}{0.8} = 70$$

Interpret

At the moment in question, the car's speed is 70 mph.

Example 4   FILLING A CONICAL TANK

Water runs into a conical tank at the rate of 9 ft³/min. The tank stands point down and has a height of 10 ft and a base radius of 5 ft. How fast is the water level rising when the water is 6 ft deep?

Solution We carry out the steps of the strategy.

Step 1:

Picture and variables. Figure 4.49 on the next page shows a partially filled conical tank. The variables in the problem are

V = volume (ft³) of the water in the tank at time t (min),

x = radius (ft) of the surface of the water at time t, and

y = depth (ft) of water in tank at time t.

We assume V, x, and y are differentiable functions of t. The constants are the dimensions of the tank.

Step 2:

Numerical information. At the time in question,

$$y = 6 \text{ ft} \quad \text{and} \quad \frac{dV}{dt} = 9 \text{ ft}^3/\text{min}.$$

Step 3:

To find: dy/dt

Step 4:

How the variables are related: The water forms a cone with volume

$$V = \frac{1}{3}\pi x^2 y.$$

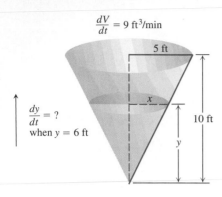

$$\frac{dV}{dt} = 9 \text{ ft}^3/\text{min}$$

5 ft

$$\frac{dy}{dt} = ?$$
when $y = 6$ ft

x

10 ft

y

Figure 4.49 (Example 4)

This equation involves x as well as V and y. Because no information is given about x and dx/dt at the time in question, we need to eliminate x. The similar triangles in Figure 4.49 give us a way to express x in terms of y:

$$\frac{x}{y} = \frac{5}{10} \quad \text{or} \quad x = \frac{y}{2}.$$

Therefore,

$$V = \frac{1}{3}\pi\left(\frac{y}{2}\right)^2 y = \frac{\pi}{12}y^3.$$

Step 5:

Differentiate with respect to t.

$$\frac{dV}{dt} = \frac{\pi}{12} \cdot 3y^2 \frac{dy}{dt} = \frac{\pi}{4}y^2 \frac{dy}{dt}$$

Step 6:

Evaluate. Use $y = 6$ and $dV/dt = 9$ to solve for dy/dt.

$$9 = \frac{\pi}{4}(6)^2 \frac{dy}{dt}$$

$$\frac{dy}{dt} = \frac{1}{\pi} \approx 0.32 \text{ ft/min}$$

Interpret

At the moment in question, the water level is rising at about 0.32 ft/min.

Simulating Related Motion

Parametric mode on a grapher can be used to simulate the motion of moving objects when the motion of each can be expressed as a function of time.

Exploration 1 **Sliding Ladder**

A 13-ft ladder is leaning against a wall. Suppose that the base of the ladder slides away from the wall at the constant rate of 3 ft/sec.

1. Explain why the motion of the two ends of the ladder can be represented by the parametric equations:

$$x_1(t) = 3t, \quad y_1(t) = 0$$
$$x_2(t) = 0, \quad y_2(t) = \sqrt{13^2 - (3t)^2}.$$

2. What values of t make sense in this problem situation?

3. Use simultaneous mode. Give a viewing window that shows the action. (*Hint:* It may be helpful to "hide" the coordinate axes if your grapher has this feature. If not, adjust the parametric equations to move the action away from the axes.)

4. Use analytic methods to find the rates at which the top of the ladder is moving down the wall at $t = 0.5$, 1, 1.5, and 2 sec. In theory, how fast is the top of the ladder moving as it hits the ground?

Quick Review 4.6

In Exercises 1 and 2, find the distance between the points *A* and *B*.

1. $A(0, 5)$, $B(7, 0)$

2. $A(0, a)$, $B(b, 0)$

In Exercises 3–6, find dy/dx.

3. $2xy + y^2 = x + y$

4. $x \sin y = 1 - xy$

5. $x^2 = \tan y$

6. $\ln (x + y) = 2x$

In Exercises 7 and 8, find a parametrization for the line segment with endpoints *A* and *B*.

7. $A(-2, 1)$, $B(4, -3)$

8. $A(0, -4)$, $B(5, 0)$

In Exercises 9 and 10, let $x = 2 \cos t$, $y = 2 \sin t$. Find a parameter interval that produces the indicated portion of the graph.

9. The portion in the second and third quadrants, including the points on the axes.

10. The portion in the fourth quadrant, including the points on the axes.

Section 4.6 Exercises

In Exercises 1–41, assume all variables are differentiable functions of *t*.

1. *Area* The radius *r* and area *A* of a circle are related by the equation $A = \pi r^2$. Write an equation that relates dA/dt to dr/dt.

2. *Surface Area* The radius *r* and surface area *S* of a sphere are related by the equation $S = 4\pi r^2$. Write an equation that relates dS/dt to dr/dt.

3. *Volume* The radius *r*, height *h*, and volume *V* of a right circular cylinder are related by the equation $V = \pi r^2 h$.

(a) How is dV/dt related to dh/dt if *r* is constant?

(b) How is dV/dt related to dr/dt if *h* is constant?

(c) How is dV/dt related to dr/dt and dh/dt if neither *r* nor *h* is constant?

4. *Electrical Power* The power *P* (watts) of an electric circuit is related to the circuit's resistance *R* (ohms) and current *I* (amperes) by the equation $P = RI^2$.

(a) How is dP/dt related to dR/dt and dI/dt?

(b) How is dR/dt related to dI/dt if *P* is constant?

5. *Diagonals* If *x*, *y*, and *z* are lengths of the edges of a rectangular box, the common length of the box's diagonals is $s = \sqrt{x^2 + y^2 + z^2}$. How is ds/dt related to dx/dt, dy/dt, and dz/dt?

6. *Area* If *a* and *b* are the lengths of two sides of a triangle, and θ the measure of the included angle, the area *A* of the triangle is $A = (1/2) ab \sin \theta$. How is dA/dt related to da/dt, db/dt, and $d\theta/dt$?

7. *Changing Voltage* The voltage *V* (volts), current *I* (amperes), and resistance *R* (ohms) of an electric circuit like the one shown here are related by the equation $V = IR$. Suppose that *V* is increasing at the rate of 1 volt/sec while *I* is decreasing at the rate of 1/3 amp/sec. Let *t* denote time in sec.

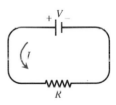

(a) What is the value of dV/dt?

(b) What is the value of dI/dt?

(c) Write an equation that relates dR/dt to dV/dt and dI/dt.

(d) **Writing to Learn** Find the rate at which *R* is changing when $V = 12$ volts and $I = 2$ amp. Is *R* increasing, or decreasing? Explain.

8. *Heating a Plate* When a circular plate of metal is heated in an oven, its radius increases at the rate of 0.01 cm/sec. At what rate is the plate's area increasing when the radius is 50 cm?

9. *Changing Dimensions in a Rectangle* The length ℓ of a rectangle is decreasing at the rate of 2 cm/sec while the width *w* is increasing at the rate of 2 cm/sec. When $\ell = 12$ cm and $w = 5$ cm, find the rates of change of

(a) the area,   (b) the perimeter, and

(c) the length of a diagonal of the rectangle.

(d) **Writing to Learn** Which of these quantities are decreasing, and which are increasing? Explain.

10. *Changing Dimensions in a Rectangular Box* Suppose that the edge lengths x, y, and z of a closed rectangular box are changing at the following rates:

$$\frac{dx}{dt} = 1 \text{ m/sec}, \quad \frac{dy}{dt} = -2 \text{ m/sec}, \quad \frac{dz}{dt} = 1 \text{ m/sec}.$$

Find the rates at which the box's **(a)** volume, **(b)** surface area, and **(c)** diagonal length $s = \sqrt{x^2 + y^2 + z^2}$ are changing at the instant when $x = 4$, $y = 3$, and $z = 2$.

11. *Air Traffic Control* An airplane is flying at an altitude of 7 mi and passes directly over a radar antenna as shown in the figure. When the plane is 10 mi from the antenna ($s = 10$), the radar detects that the distance s is changing at the rate of 300 mph. What is the speed of the airplane at that moment?

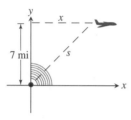

12. *Filling a Trough* A trough is 15 ft long and 4 ft across the top as shown in the figure. Its ends are isosceles triangles with height 3 ft. Water runs into the trough at the rate of 2.5 ft³/min. How fast is the water level rising when it is 2 ft deep?

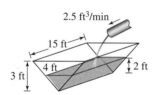

13. *Sliding Ladder* A 13-ft ladder is leaning against a house (see figure) when its base starts to slide away. By the time the base is 12 ft from the house, the base is moving at the rate of 5 ft/sec.

(a) How fast is the top of the ladder sliding down the wall at that moment?

(b) At what rate is the area of the triangle formed by the ladder, wall, and ground changing at that moment?

(c) At what rate is the angle θ between the ladder and the ground changing at that moment?

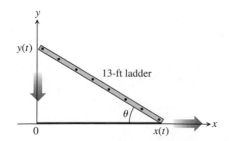

14. *Flying a Kite* Inge flies a kite at a height of 300 ft, the wind carrying the kite horizontally away at a rate of 25 ft/sec. How fast must she let out the string when the kite is 500 ft away from her?

15. *Boring a Cylinder* The mechanics at Lincoln Automotive are reboring a 6-in. deep cylinder to fit a new piston. The machine they are using increases the cylinder's radius one-thousandth of an inch every 3 min. How rapidly is the cylinder volume increasing when the bore (diameter) is 3.800 in.?

16. *Growing Sand Pile* Sand falls from a conveyor belt at the rate of 10 m³/min onto the top of a conical pile. The height of the pile is always three-eighths of the base diameter. How fast are the **(a)** height and **(b)** radius changing when the pile is 4 m high? Give your answer in cm/min.

17. *Draining Conical Reservoir* Water is flowing at the rate of 50 m³/min from a concrete conical reservoir (vertex down) of base radius 45 m and height 6 m. **(a)** How fast is the water level falling when the water is 5 m deep? **(b)** How fast is the radius of the water's surface changing at that moment? Give your answer in cm/min.

18. *Draining Hemispherical Reservoir* Water is flowing at the rate of 6 m³/min from a reservoir shaped like a hemispherical bowl of radius 13 m, shown here in profile. Answer the following questions given that the volume of water in a hemispherical bowl of radius R is $V = (\pi/3)y^2(3R - y)$ when the water is y units deep.

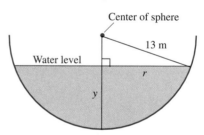

(a) At what rate is the water level changing when the water is 8 m deep?

(b) What is the radius r of the water's surface when the water is y m deep?

(c) At what rate is the radius r changing when the water is 8 m deep?

19. *Growing Raindrop* Suppose that a droplet of mist is a perfect sphere and that, through condensation, the droplet picks up moisture at a rate proportional to its surface area. Show that under these circumstances the droplet's radius increases at a constant rate.

20. *Inflating Balloon* A spherical balloon is inflated with helium at the rate of 100π ft³/min.

(a) How fast is the balloon's radius increasing at the instant the radius is 5 ft?

(b) How fast is the surface area increasing at that instant?

21. *Hauling in a Dinghy* A dinghy is pulled toward a dock by a rope from the bow through a ring on the dock 6 ft above the bow as shown in the figure. The rope is hauled in at the rate of 2 ft/sec.

(a) How fast is the boat approaching the dock when 10 ft of rope are out?

(b) At what rate is angle θ changing at that moment?

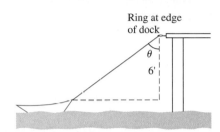

Ring at edge
of dock

θ

6'

22. *Rising Balloon* A balloon is rising vertically above a level, straight road at a constant rate of 1 ft/sec. Just when the balloon is 65 ft above the ground, a bicycle moving at a constant rate of 17 ft/sec passes under it. How fast is the distance between the bicycle and balloon increasing 3 sec later (see figure)?

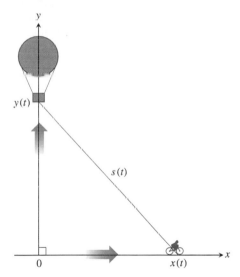

y

$y(t)$

$s(t)$

0

$x(t)$

x

23. *Cost, Revenue, and Profit* A company can manufacture x items at a cost of $c(x)$ dollars, a sales revenue of $r(x)$ dollars, and a profit of $p(x) = r(x) - c(x)$ dollars (all amounts in thousands). Find dc/dt, dr/dt, and dp/dt for the following values of x and dx/dt.

(a) $r(x) = 9x$, $c(x) = x^3 - 6x^2 + 15x$,
and $dx/dt = 0.1$ when $x = 2$.

(b) $r(x) = 70x$, $c(x) = x^3 - 6x^2 + 45/x$,
and $dx/dt = 0.05$ when $x = 1.5$.

24. *Making Coffee* Coffee is draining from a conical filter into a cylindrical coffeepot at the rate of 10 in³/min.

(a) How fast is the level in the pot rising when the coffee in the cone is 5 in. deep?

(b) How fast is the level in the cone falling at that moment?

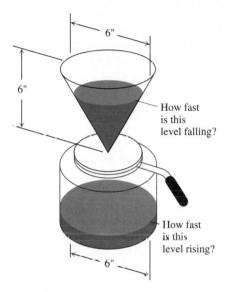

6"

6"

How fast
is this
level falling?

How fast
is this
level rising?

6"

25. *Cardiac Output* *Work in groups of two or three.* In the late 1860s, Adolf Fick, a professor of physiology in the Faculty of Medicine in Würtzberg, Germany, developed one of the methods we use today for measuring how much blood your heart pumps in a minute. Your cardiac output as you read this sentence is probably about 7 liters a minute. At rest it is likely to be a bit under 6 L/min. If you are a trained marathon runner running a marathon, your cardiac output can be as high as 30 L/min.

Your cardiac output can be calculated with the formula

$$y = \frac{Q}{D},$$

where Q is the number of milliliters of CO_2 you exhale in a minute and D is the difference between the CO_2 concentration (mL/L) in the blood pumped to the lungs and the CO_2 concentration in the blood returning from the lungs. With $Q = 233$ mL/min and $D = 97 - 56 = 41$ mL/L,

$$y = \frac{233 \text{ mL/min}}{41 \text{ mL/L}} \approx 5.68 \text{ L/min},$$

fairly close to the 6 L/min that most people have at basal (resting) conditions. (Data courtesy of J. Kenneth Herd, M.D., Quillan College of Medicine, East Tennessee State University.)

Suppose that when $Q = 233$ and $D = 41$, we also know that D is decreasing at the rate of 2 units a minute but that Q remains unchanged. What is happening to the cardiac output?

26. *Particle Motion* A particle moves along the parabola $y = x^2$ in the first quadrant in such a way that its x-coordinate (in meters) increases at a constant rate of 10 m/sec. How fast is the angle of inclination θ of the line joining the particle to the origin changing when $x = 3$?

27. *Particle Motion* A particle moves from right to left along the parabolic curve $y = \sqrt{-x}$ in such a way that its x-coordinate (in meters) decreases at the rate of 8 m/sec. How fast is the angle of inclination θ of the line joining the particle to the origin changing when $x = -4$?

28. *Particle Motion* A particle $P(x, y)$ is moving in the coordinate plane in such a way that $dx/dt = -1$ m/sec and $dy/dt = -5$ m/sec. How fast is the particle's distance from the origin changing as it passes through the point $(5, 12)$?

29. *Moving Shadow* A man 6 ft tall walks at the rate of 5 ft/sec toward a streetlight that is 16 ft above the ground. At what rate is the length of his shadow changing when he is 10 ft from the base of the light?

30. *Moving Shadow* A light shines from the top of a pole 50 ft high. A ball is dropped from the same height from a point 30 ft away from the light as shown in the figure. How fast is the shadow of the ball moving along the ground $1/2$ sec later? (Assume the ball falls a distance $s = 16t^2$ in t sec.)

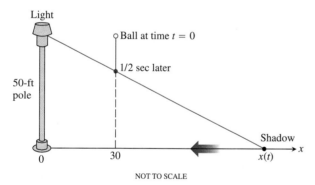

NOT TO SCALE

31. *Moving Race Car* You are videotaping a race from a stand 132 ft from the track, following a car that is moving at 180 mph (264 ft/sec) as shown in the figure. About how fast will your camera angle θ be changing when the car is right in front of you? a half second later?

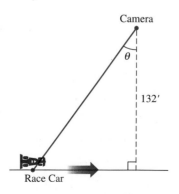

32. *Melting Ice* A spherical iron ball is coated with a layer of ice of uniform thickness. If the ice melts at the rate of 8 mL/min, how fast is the outer surface area of ice decreasing when the outer diameter (ball plus ice) is 20 cm?

33. *Speed Trap* A highway patrol airplane flies 3 mi above a level, straight road at a constant rate of 120 mph. The pilot sees an oncoming car and with radar determines that at the instant the line-of-sight distance from plane to car is 5 mi the line-of-sight distance is decreasing at the rate of 160 mph. Find the car's speed along the highway.

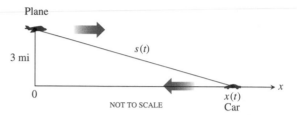

NOT TO SCALE

34. *Building's Shadow* On a morning of a day when the sun will pass directly overhead, the shadow of an 80-ft building on level ground is 60 ft long as shown in the figure. At the moment in question, the angle θ the sun makes with the ground is increasing at the rate of 0.27°/min. At what rate is the shadow length decreasing? Express your answer in in./min, to the nearest tenth. (Remember to use radians.)

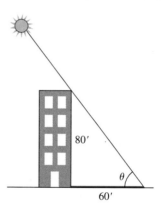

35. *Walkers* A and B are walking on straight streets that meet at right angles. A approaches the intersection at 2 m/sec and B moves away from the intersection at 1 m/sec as shown in the figure. At what rate is the angle θ changing when A is 10 m from the intersection and B is 20 m from the intersection? Express your answer in degrees per second to the nearest degree.

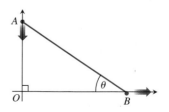

36. *Moving Ships* Two ships are steaming away from a point O along routes that make a 120° angle. Ship A moves at 14 knots (nautical miles per hour; a nautical mile is 2000 yards). Ship B moves at 21 knots. How fast are the ships moving apart when $OA = 5$ and $OB = 3$ nautical miles?

In Exercises 37 and 38, a particle is moving along the curve $y = f(x)$.

37. Let $y = f(x) = \dfrac{10}{1 + x^2}$.

If $dx/dt = 3$ cm/sec, find dy/dt at the point where

(a) $x = -2$.   **(b)** $x = 0$.   **(c)** $x = 20$.

38. Let $y = f(x) = x^3 - 4x$.

If $dx/dt = -2$ cm/sec, find dy/dt at the point where

(a) $x = -3$.   **(b)** $x = 1$.   **(c)** $x = 4$.

Extending the Ideas

39. *Motion along a Circle* A wheel of radius 2 ft makes 8 revolutions about its center every second.

(a) Explain how the parametric equations

$$x = 2 \cos \theta, \quad y = 2 \sin \theta$$

can be used to represent the motion of the wheel.

(b) Express θ as a function of time t.

(c) Find the rate of horizontal movement and the rate of vertical movement of a point on the edge of the wheel when it is at the position given by $\theta = \pi/4$, $\pi/2$, and π.

40. *Ferris Wheel* A Ferris wheel with radius 30 ft makes one revolution every 10 sec.

(a) Assume that the center of the Ferris wheel is located at the point $(0, 40)$, and write parametric equations to model its motion. (*Hint:* See Exercise 39.)

(b) At $t = 0$ the point P on the Ferris wheel is located at $(30, 40)$. Find the rate of horizontal movement, and the rate of vertical movement of the point P when $t = 5$ sec and $t = 8$ sec.

41. *Industrial Production* **(a)** Economists often use the expression "rate of growth" in relative rather than absolute terms. For example, let $u = f(t)$ be the number of people in the labor force at time t in a given industry. (We treat this function as though it were differentiable even though it is an integer-valued step function.)

Let $v = g(t)$ be the average production per person in the labor force at time t. The total production is then $y = uv$. If the labor force is growing at the rate of 4% per year ($du/dt = 0.04u$) and the production per worker is growing at the rate of 5% per year ($dv/dt = 0.05v$), find the rate of growth of the total production, y.

(b) Suppose that the labor force in (a) is decreasing at the rate of 2% per year while the production per person is increasing at the rate of 3% per year. Is the total production increasing, or is it decreasing, and at what rate?

<table>
<tr><td>Chapter 4</td><td>Key Terms</td></tr>
</table>

absolute change (p. 226)	extrema (p. 177)	marginal analysis (p. 211)
absolute maximum value (p. 177)	Extreme Value Theorem (p. 178)	marginal cost and revenue (p. 211)
absolute minimum value (p. 177)	first derivative test (p. 194)	Mean Value Theorem (p. 186)
antiderivative (p. 190)	first derivative test for	monotonic function (p. 188)
antidifferentiation (p. 190)	local extrema (p. 194)	Newton's method (p. 222)
arithmetic mean (p. 194)	geometric mean (p. 194)	optimization (p. 206)
average cost (p. 212)	global maximum value (p. 177)	percentage change (p. 226)
concave down (p. 197)	global minimum value (p. 177)	point of inflection (p. 198)
concave up (p. 197)	increasing function (p. 188)	quadratic approximation (p. 231)
concavity test (p. 197)	linear approximation (p. 220)	related rates (p. 232)
critical point (p. 180)	linearization (p. 221)	relative change (p. 226)
decreasing function (p. 188)	local maximum value (p. 179)	relative extrema (p. 179)
differential (p. 224)	local minimum value (p. 179)	second derivative test for
differential estimate of change (p. 226)	logistic curve (p. 199)	local extrema (p. 200)
differential of a function (p. 225)	logistic regression (p. 200)	standard linear approximation (p. 221)

Chapter 4 Review Exercises

In Exercises 1 and 2, use analytic methods to find the global extreme values of the function on the interval and state where they occur.

1. $y = x\sqrt{2 - x}, \quad -2 \le x \le 2$

2. $y = x^3 - 9x^2 - 21x - 11, \quad -\infty < x < \infty$

In Exercises 3 and 4, use analytic methods. Find the intervals on which the function is

    (a) increasing,   **(b)** decreasing,

    (c) concave up,   **(d)** concave down.

Then find any

    (e) local extreme values,   **(f)** inflection points.

3. $y = x^2 e^{1/x^2}$ **4.** $y = x\sqrt{4 - x^2}$

In Exercises 5–16, find the intervals on which the function is

    (a) increasing,   **(b)** decreasing,

    (c) concave up,   **(d)** concave down.

Then find any

    (e) local extreme values,   **(f)** inflection points.

5. $y = 1 + x - x^2 - x^4$ **6.** $y = e^{x-1} - x$

7. $y = \dfrac{1}{\sqrt[4]{1 - x^2}}$ **8.** $y = \dfrac{x}{x^3 - 1}$

9. $y = \cos^{-1} x$ **10.** $y = \dfrac{x}{x^2 + 2x + 3}$

11. $y = \ln |x|, \quad -2 \le x \le 2, \quad x \ne 0$

12. $y = \sin 3x + \cos 4x, \quad 0 \le x \le 2\pi$

13. $y = \begin{cases} e^{-x}, & x \le 0 \\ 4x - x^3, & x > 0 \end{cases}$

14. $y = -x^5 + \dfrac{7}{3}x^3 + 5x^2 + 4x + 2$

15. $y = x^{4/5}(2 - x)$

16. $y = \dfrac{5 - 4x + 4x^2 - x^3}{x - 2}$

In Exercises 17 and 18, use the derivative of the function $y = f(x)$ to find the points at which f has a

    (a) local maximum,   **(b)** local minimum, or

    (c) point of inflection.

17. $y' = 6(x + 1)(x - 2)^2$ **18.** $y' = 6(x + 1)(x - 2)$

In Exercises 19–22, find all possible functions with the given derivative.

19. $f'(x) = x^{-5} + e^{-x}$ **20.** $f'(x) = \sec x \tan x$

21. $f'(x) = \dfrac{2}{x} + x^2 + 1, \ x > 0$ **22.** $f'(x) = \sqrt{x} + \dfrac{1}{\sqrt{x}}$

In Exercises 23 and 24, find the function with the given derivative whose graph passes through the point P.

23. $f'(x) = \sin x + \cos x, \quad P(\pi, 3)$

24. $f'(x) = x^{1/3} + x^2 + x + 1, \quad P(1, 0)$

In Exercises 25 and 26, the velocity v or acceleration a of a particle is given. Find the particle's position s at time t.

25. $v = 9.8t + 5, \quad s = 10 \quad$ when $\quad t = 0$

26. $a = 32, \quad v = 20 \quad$ and $\quad s = 5 \quad$ when $\quad t = 0$

In Exercises 27–30, find the linearization $L(x)$ of $f(x)$ at $x = a$.

27. $f(x) = \tan x, \quad a = -\pi/4$

28. $f(x) = \sec x, \quad a = \pi/4$

29. $f(x) = \dfrac{1}{1 + \tan x}, \quad a = 0$

30. $f(x) = e^x + \sin x, \quad a = 0$

In Exercises 31–34, use the graph to answer the questions.

31. Identify any global extreme values of f and the values of x at which they occur.

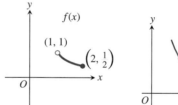   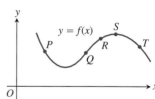

Figure for Exercise 31   Figure for Exercise 32

32. At which of the five points on the graph of $y = f(x)$ shown here

    (a) are y' and y'' both negative?

    (b) is y' negative and y'' positive?

33. Estimate the intervals on which the function $y = f(x)$ is **(a)** increasing; **(b)** decreasing. **(c)** Estimate any local extreme values of the function and where they occur.

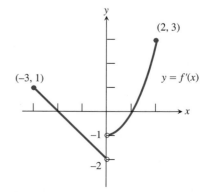

34. Here is the graph of the fruit fly population from Section 2.4, Example 1. On approximately what day did the population's growth rate change from increasing to decreasing?

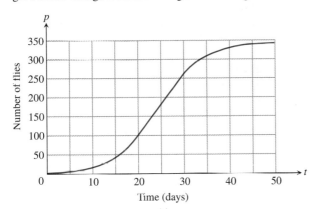

35. *Connecting f and f′* The graph of f' is shown in Exercise 33. Sketch a possible graph of f given that it is continuous with domain $[-3, 2]$ and $f(-3) = 0$.

36. *Connecting f, f′, and f″* The function f is continuous on $[0, 3]$ and satisfies the following.

x	0	1	2	3
f	0	−2	0	3
f'	−3	0	does not exist	4
f''	0	1	does not exist	0

x	$0 < x < 1$	$1 < x < 2$	$2 < x < 3$
f	−	−	+
f'	−	+	+
f''	+	+	+

(a) Find the absolute extrema of f and where they occur.

(b) Find any points of inflection.

(c) Sketch a possible graph of f.

37. *Mean Value Theorem* Let $f(x) = x \ln x$.

(a) **Writing to Learn** Show that f satisfies the hypotheses of the Mean Value Theorem on the interval $[a, b] = [0.5, 3]$.

(b) Find the value(s) of c in (a, b) for which

$$f'(c) = \frac{f(b) - f(a)}{b - a}.$$

(c) Write an equation for the secant line AB where $A = (a, f(a))$ and $B = (b, f(b))$.

(d) Write an equation for the tangent line that is parallel to the secant line AB.

38. *Motion along a Line* A particle is moving along a line with position function $s(t) = 3 + 4t - 3t^2 - t^3$. Find the (a) velocity and (b) acceleration, and (c) describe the motion of the particle for $t \geq 0$.

39. *Approximating Functions* Let f be a function with $f'(x) = \sin x^2$ and $f(0) = -1$.

(a) Find the linearization of f at $x = 0$.

(b) Approximate the value of f at $x = 0.1$.

(c) **Writing to Learn** Is the actual value of f at $x = 0.1$ greater than or less than the approximation in (b)?

40. *Differentials* Let $y = x^2 e^{-x}$. Find (a) dy and (b) evaluate dy for $x = 1$ and $dx = 0.01$.

41. *Logistic Regression* Table 4.4 gives the numbers of on-line debit card transactions for several years. Let $x = 0$ represent 1990, $x = 1$ represent 1991, and so on.

(a) Find the logistic regression equation and superimpose its graph on a scatter plot of the data.

(b) Use the regression equation to predict when the rate of increase in debit card transactions will start to decrease, and the number of transactions at that time.

(c) What is the ceiling of the number of debit card transactions predicted by the regression equation?

Table 4.4 Debit Card Transactions

Year	Number (millions)
1991	211
1992	280
1993	386
1994	551
1995	684
1996	905

Source: "Debit Card News" as reported by Julie Stacey in *USA Today,* May 30, 1997.

42. *Newton's Method* Use Newton's method to estimate all real solutions to $2 \cos x - \sqrt{1 + x} = 0$. State your answers accurate to 6 decimal places.

43. *Rocket Launch* A rocket lifts off the surface of Earth with a constant acceleration of 20 m/sec². How fast will the rocket be going 1 min later?

44. *Launching on Mars* The acceleration of gravity near the surface of Mars is 3.72 m/sec². If a rock is blasted straight up from the surface with an initial velocity of 93 m/sec (about 208 mph), how high does it go?

45. *Area of Sector* If the perimeter of the circular sector shown here is fixed at 100 ft, what values of r and s will give the sector the greatest area?

46. *Area of Triangle* An isosceles triangle has its vertex at the origin and its base parallel to the *x*-axis with the vertices above the axis on the curve $y = 27 - x^2$. Find the largest area the triangle can have.

47. *Storage Bin* Find the dimensions of the largest open-top storage bin with a square base and vertical sides that can be made from 108 ft^2 of sheet steel. (Neglect the thickness of the steel and assume that there is no waste.)

48. *Designing a Vat* You are to design an open-top rectangular stainless-steel vat. It is to have a square base and a volume of 32 ft^3; to be welded from quarter-inch plate, and weigh no more than necessary. What dimensions do you recommend?

49. *Inscribing a Cylinder* Find the height and radius of the largest right circular cylinder that can be put into a sphere of radius $\sqrt{3}$ as described in the figure.

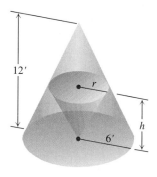

50. *Cone in a Cone* The figure shows two right circular cones, one upside down inside the other. The two bases are parallel, and the vertex of the smaller cone lies at the center of the larger cone's base. What values of *r* and *h* will give the smaller cone the largest possible volume?

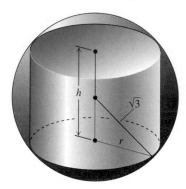

51. *Box with Lid* Repeat Exercise 18 of Section 4.4 but this time remove the two equal squares from the corners of a 15-in. side.

52. *Inscribing a Rectangle* A rectangle is inscribed under one arch of $y = 8 \cos (0.3x)$ with its base on the *x*-axis and its upper two vertices on the curve symmetric about the *y*-axis. What is the largest area the rectangle can have?

53. *Oil Refinery* A drilling rig 12 mi offshore is to be connected by a pipe to a refinery onshore, 20 mi down the coast from the rig as shown in the figure. If underwater pipe costs \$40,000 per mile and land-based pipe costs \$30,000 per mile, what values of *x* and *y* give the least expensive connection?

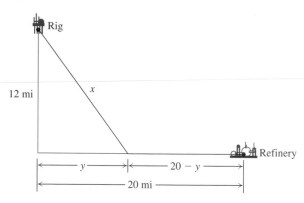

54. *Designing an Athletic Field* An athletic field is to be built in the shape of a rectangle *x* units long capped by semicircular regions of radius *r* at the two ends. The field is to be bounded by a 400-m running track. What values of *x* and *r* will give the rectangle the largest possible area?

55. *Manufacturing Tires* Your company can manufacture *x* hundred grade A tires and *y* hundred grade B tires a day, where $0 \le x \le 4$ and

$$y = \frac{40 - 10x}{5 - x}.$$

Your profit on a grade A tire is twice your profit on a grade B tire. What is the most profitable number of each kind to make?

56. *Particle Motion* The positions of two particles on the *s*-axis are $s_1 = \cos t$ and $s_2 = \cos (t + \pi/4)$.

(a) What is the farthest apart the particles ever get?

(b) When do the particles collide?

57. *Open-top Box* An open-top rectangular box is constructed from a 10- by 16-in. piece of cardboard by cutting squares of equal side length from the corners and folding up the sides. Find analytically the dimensions of the box of largest volume and the maximum volume. Support your answers graphically.

58. *Changing Area* The radius of a circle is changing at the rate of $-2/\pi$ m/sec. At what rate is the circle's area changing when $r = 10$ m?

59. *Particle Motion* The coordinates of a particle moving in the plane are differentiable functions of time *t* with $dx/dt = -1$ m/sec and $dy/dt = -5$ m/sec. How fast is the particle approaching the origin as it passes through the point (5, 12)?

60. *Changing Cube* The volume of a cube is increasing at the rate of 1200 cm³/min at the instant its edges are 20 cm long. At what rate are the edges changing at that instant?

61. *Particle Motion* A point moves smoothly along the curve $y = x^{3/2}$ in the first quadrant in such a way that its distance from the origin increases at the constant rate of 11 units per second. Find dx/dt when $x = 3$.

62. *Draining Water* Water drains from the conical tank shown in the figure at the rate of 5 ft³/min.

(a) What is the relation between the variables h and r?

(b) How fast is the water level dropping when $h = 6$ ft?

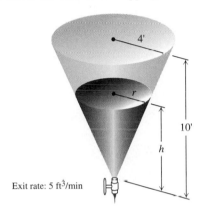

Exit rate: 5 ft³/min

63. *Stringing Telephone Cable* As telephone cable is pulled from a large spool to be strung from the telephone poles along a street, it unwinds from the spool in layers of constant radius as suggested in the figure. If the truck pulling the cable moves at a constant rate of 6 ft/sec, use the equation $s = r\theta$ to find how fast (in rad/sec) the spool is turning when the layer of radius 1.2 ft is being unwound.

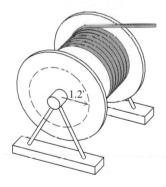

64. *Throwing Dirt* You sling a shovelful of dirt up from the bottom of a 17-ft hole with an initial velocity of 32 ft/sec. Is that enough speed to get the dirt out of the hole, or had you better duck?

65. *Estimating Change* Write a formula that estimates the change that occurs in the volume of a right circular cone (see figure) when the radius changes from a to $a + dr$ and the height does not change.

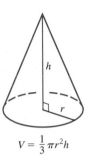

$$V = \frac{1}{3}\pi r^2 h$$

66. *Controlling Error*

(a) How accurately should you measure the edge of a cube to be reasonably sure of calculating the cube's surface area with an error of no more than 2%?

(b) Suppose the edge is measured with the accuracy required in part (a). About how accurately can the cube's volume be calculated from the edge measurement? To find out, estimate the percentage error in the volume calculation that might result from using the edge measurement.

67. *Compounding Error* The circumference of the equator of a sphere is measured as 10 cm with a possible error of 0.4 cm. This measurement is then used to calculate the radius. The radius is then used to calculate the surface area and volume of the sphere. Estimate the percentage errors in the calculated values of (a) the radius, (b) the surface area, and (c) the volume.

68. *Finding Height* To find the height of a lamppost (see figure), you stand a 6-ft pole 20 ft from the lamp and measure the length a of its shadow, finding it to be 15 ft, give or take an inch. Calculate the height of the lamppost using the value $a = 15$, and estimate the possible error in the result.

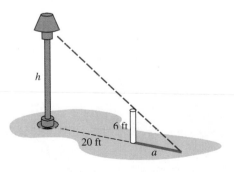

69. *Decreasing Function* Show that the function $y = \sin^2 x - 3x$ decreases on every interval in its domain.

The Definite Integral

The 1995 Reader's Digest Sweepstakes grand prize winner is being paid a total of $5,010,000 over 30 years. If invested, the winnings plus the interest earned generate an amount defined by:

$$A = e^{rT}\int_0^T mPe^{-rt}\,dt$$

(r = interest rate, P = size of payment, T = term in years, m = number of payments per year.)

Would the prize have a different value if it were paid in 15 annual installments of $334,000 instead of 30 annual installments of $167,000? Section 5.4 can help you compare the total amounts.

Chapter 5 Overview

We have seen how the need to calculate instantaneous rates of change led the discoverers of calculus to an investigation of the slopes of tangent lines and, ultimately, to the derivative—to what we call *differential* calculus. But they knew that derivatives revealed only half the story. In addition to a calculation method (a "calculus") to describe how functions were changing at a given instant, they also needed a method to describe how those instantaneous changes could accumulate over an interval to produce the function. That is why they were also investigating *areas under curves,* an investigation that ultimately led to the second main branch of calculus, called *integral* calculus.

Once they had the calculus for finding slopes of tangent lines and the calculus for finding areas under curves—two geometric operations that would seem to have nothing at all to do with each other—the challenge for Newton and Leibniz was to prove the connection that they knew intuitively had to be there. The discovery of this connection (called the Fundamental Theorem of Calculus) brought differential and integral calculus together to become the single most powerful insight mathematicians had ever acquired for understanding how the universe worked.

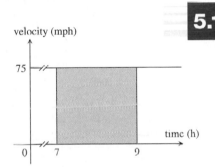

Figure 5.1 The distance traveled by a 75 mph train in 2 hours is 150 miles, which corresponds to the area of the shaded rectangle.

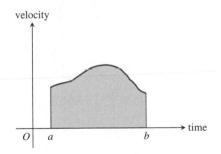

Figure 5.2 If the velocity varies over the time interval [a, b], does the shaded region give the distance traveled?

5.1 Estimating with Finite Sums

Distance Traveled • Rectangular Approximation Method (RAM) • Volume of a Sphere • Cardiac Output

Distance Traveled

We know why a mathematician pondering motion problems might have been led to consider slopes of curves, but what do those same motion problems have to do with areas under curves? Consider the following problem from a typical elementary school textbook:

> A train moves along a track at a steady rate of 75 miles per hour from 7:00 A.M. to 9:00 A.M. What is the total distance traveled by the train?

Applying the well-known formula *distance = rate × time,* we find that the answer is 150 miles. Simple. Now suppose that you are Isaac Newton trying to make a connection between this formula and the graph of the velocity function.

You might notice that the distance traveled by the train (150 miles) is exactly the *area* of the rectangle whose base is the time interval [7, 9] and whose height at each point is the value of the constant velocity function $v = 75$ (Figure 5.1). This is no accident, either, since *the distance traveled* and *the area* in this case are both found by multiplying the rate (75) by the change in time (2).

This same connection between distance traveled and rectangle area could be made no matter how fast the train was going or how long or short the time interval was. But what if the train had a velocity v that *varied* as a function of time? The graph (Figure 5.2) would no longer be a horizontal line, so the region under the graph would no longer be rectangular.

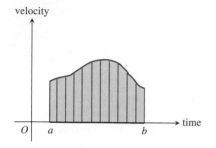

velocity

O   a   b   time

Figure 5.3 The region is partitioned into vertical strips. If the strips are narrow enough, they are almost indistinguishable from rectangles. The sum of the areas of these "rectangles" will give the total area and can be interpreted as distance traveled.

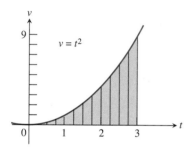

Figure 5.4 The region under the parabola $v = t^2$ from $t = 0$ to $t = 3$ is partitioned into 12 thin strips, each with base $\Delta t = 1/4$. The strips have curved tops. (Example 1)

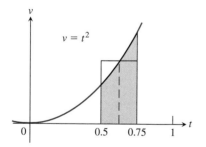

Figure 5.5 The area of the shaded region is approximated by the area of the rectangle whose height is the function value at the midpoint of the interval. (Example 1)

Would the area of this irregular region still give the total distance traveled over the time interval? Newton and Leibniz (and, actually, many others who had considered this question) thought that it obviously would, and that is why they were interested in a calculus for finding areas under curves. They imagined the time interval being partitioned into many tiny subintervals, each one so small that the velocity over it would essentially be constant. Geometrically, this was equivalent to slicing the irregular region into narrow strips, each of which would be nearly indistinguishable from a narrow rectangle (Figure 5.3).

They argued that, just as the total area could be found by summing the areas of the (essentially rectangular) strips, the total distance traveled could be found by summing the small distances traveled over the tiny time intervals.

Example 1 FINDING DISTANCE TRAVELED WHEN VELOCITY VARIES

A particle starts at $x = 0$ and moves along the x-axis with velocity $v(t) = t^2$ for time $t \geq 0$. Where is the particle at $t = 3$?

Solution We graph v and partition the time interval $[0, 3]$ into subintervals of length Δt. (Figure 5.4 shows twelve subintervals of length 3/12 each.)

Notice that the region under the curve is partitioned into thin strips with bases of length 1/4 and *curved* tops that slope upward from left to right. You might not know how to find the area of such a strip, but you can get a good approximation of it by finding the area of a suitable rectangle. In Figure 5.5, we use the rectangle whose height is the y-coordinate of the function at the midpoint of its base.

The area of this narrow rectangle approximates the distance traveled over the time subinterval. Adding all the areas (distances) gives an approximation of the total area under the curve (total distance traveled) from $t = 0$ to $t = 3$ (Figure 5.6).

Computing this sum of areas is straightforward. Each rectangle has a base of length $\Delta t = 1/4$, while the height of each rectangle can be found by evaluating the function at the midpoint of the subinterval. Table 5.1 shows the computations for the first four rectangles.

Table 5.1

Subinterval	$\left[0, \dfrac{1}{4}\right]$	$\left[\dfrac{1}{4}, \dfrac{1}{2}\right]$	$\left[\dfrac{1}{2}, \dfrac{3}{4}\right]$	$\left[\dfrac{3}{4}, 1\right]$
Midpoint m_i	$\dfrac{1}{8}$	$\dfrac{3}{8}$	$\dfrac{5}{8}$	$\dfrac{7}{8}$
Height $= (m_i)^2$	$\dfrac{1}{64}$	$\dfrac{9}{64}$	$\dfrac{25}{64}$	$\dfrac{49}{64}$
Area $= (1/4)(m_i)^2$	$\dfrac{1}{256}$	$\dfrac{9}{256}$	$\dfrac{25}{256}$	$\dfrac{49}{256}$

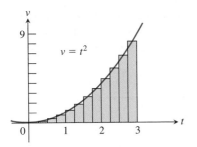

Figure 5.6 These rectangles have approximately the same areas as the strips in Figure 5.4. Each rectangle has height m_i^2, where m_i is the midpoint of its base. (Example 1)

Approximation by Rectangles

Approximating irregularly-shaped regions by regularly-shaped regions for the purpose of computing areas is not new. Archimedes used the idea more than 2200 years ago to find the area of a circle, demonstrating in the process that π was located between 3.140845 and 3.142857. He also used approximations to find the area under a parabolic arch, anticipating the answer to an important seventeenth-century question nearly 2000 years before anyone thought to ask it. The fact that we still measure the area of anything—even a circle—in "square units" is obvious testimony to the historical effectiveness of using rectangles for approximating areas.

Continuing in this manner, we derive the area $(1/4)(m_i)^2$ for each of the twelve subintervals and add them:

$$\frac{1}{256} + \frac{9}{256} + \frac{25}{256} + \frac{49}{256} + \frac{81}{256} + \frac{121}{256} + \frac{169}{256} + \frac{225}{256} +$$

$$\frac{289}{256} + \frac{361}{256} + \frac{441}{256} + \frac{529}{256} = \frac{2300}{256} \approx 8.98.$$

Since this number approximates the area and hence the total distance traveled by the particle, we conclude that the particle has moved approximately 9 units in 3 seconds. If it starts at $x = 0$, then it is very close to $x = 9$ when $t = 3$.

To make it easier to talk about approximations with rectangles, we now introduce some new terminology.

Rectangular Approximation Method (RAM)

In Example 1 we used the *Midpoint Rectangular Approximation Method (MRAM)* to approximate the area under the curve. The name suggests the choice we made when determining the heights of the approximating rectangles: We evaluated the function at the midpoint of each subinterval. If instead we had evaluated the function at the left-hand endpoint we would have obtained the *LRAM* approximation, and if we had used the right-hand endpoints we would have obtained the *RRAM* approximation. Figure 5.7 shows what the three approximations look like graphically when we approximate the area under the curve $y = x^2$ from $x = 0$ to $x = 3$ with six subintervals.

No matter which RAM approximation we compute, we are adding products of the form $f(x_i) \cdot \Delta x$, or, in this case, $(x_i)^2 \cdot (3/6)$.

LRAM:

$$\left(0\right)^2\left(\frac{1}{2}\right) + \left(\frac{1}{2}\right)^2\left(\frac{1}{2}\right) + \left(1\right)^2\left(\frac{1}{2}\right) + \left(\frac{3}{2}\right)^2\left(\frac{1}{2}\right) + \left(2\right)^2\left(\frac{1}{2}\right) + \left(\frac{5}{2}\right)^2\left(\frac{1}{2}\right) = 6.875$$

MRAM:

$$\left(\frac{1}{4}\right)^2\left(\frac{1}{2}\right) + \left(\frac{3}{4}\right)^2\left(\frac{1}{2}\right) + \left(\frac{5}{4}\right)^2\left(\frac{1}{2}\right) + \left(\frac{7}{4}\right)^2\left(\frac{1}{2}\right) + \left(\frac{9}{4}\right)^2\left(\frac{1}{2}\right) + \left(\frac{11}{4}\right)^2\left(\frac{1}{2}\right) = 8.9375$$

RRAM:

$$\left(\frac{1}{2}\right)^2\left(\frac{1}{2}\right) + \left(1\right)^2\left(\frac{1}{2}\right) + \left(\frac{3}{2}\right)^2\left(\frac{1}{2}\right) + \left(2\right)^2\left(\frac{1}{2}\right) + \left(\frac{5}{2}\right)^2\left(\frac{1}{2}\right) + \left(3\right)^2\left(\frac{1}{2}\right) = 11.375$$

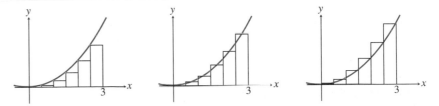

Figure 5.7 LRAM, MRAM, and RRAM approximations to the area under the graph of $y = x^2$ from $x = 0$ to $x = 3$.

As we can see from Figure 5.7, LRAM is smaller than the true area and RRAM is larger. MRAM appears to be the closest of the three approximations. However, observe what happens as the number n of subintervals increases:

n	$LRAM_n$	$MRAM_n$	$RRAM_n$
6	6.875	8.9375	11.375
12	7.90625	8.984375	10.15625
24	8.4453125	8.99609375	9.5703125
48	8.720703125	8.999023438	9.283203125
100	8.86545	8.999775	9.13545
1000	8.9865045	8.99999775	9.0135045

We computed the numbers in this table using a graphing calculator and a summing program called RAM. A version of this program for most graphing calculators can be found in the *Technology Resource Manual* that accompanies this textbook. *All three sums* approach the same number (in this case, 9).

Example 2 ESTIMATING AREA UNDER THE GRAPH OF A NONNEGATIVE FUNCTION

Figure 5.8 shows the graph of $f(x) = x^2 \sin x$ on the interval $[0, 3]$. Estimate the area under the curve from $x = 0$ to $x = 3$.

Solution We apply our RAM program to get the numbers in this table.

n	$LRAM_n$	$MRAM_n$	$RRAM_n$
5	5.15480	5.89668	5.91685
10	5.52574	5.80685	5.90677
25	5.69079	5.78150	5.84320
50	5.73615	5.77788	5.81235
100	5.75701	5.77697	5.79511
1000	5.77476	5.77667	5.77857

It is not necessary to compute all three sums each time just to approximate the area, but we wanted to show again how all three sums approach the same number. With 1000 subintervals, all three agree in the first three digits. (The *exact* area is $-7 \cos 3 + 6 \sin 3 - 2$, which is 5.77666752456 to twelve digits.)

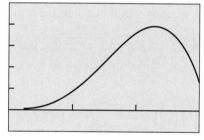

[0, 3] by [–1, 5]

Figure 5.8 The graph of $y = x^2 \sin x$ over the interval $[0, 3]$. (Example 2)

You might think from the previous two RAM tables that LRAM is always a little low and RRAM a little high, with MRAM somewhere in between. That, however, depends on n and on the shape of the curve.

1. Graph $y = 5 - 4 \sin (x/2)$ in the window $[0, 3]$ by $[0, 5]$. Copy the graph on paper and sketch the rectangles for the LRAM, MRAM, and RRAM sums with $n = 3$. Order the three approximations from greatest to smallest.

2. Graph $y = 2 \sin (5x) + 3$ in the same window. Copy the graph on paper and sketch the rectangles for the LRAM, MRAM, and RRAM sums with $n = 3$. Order the three approximations from greatest to smallest.

3. If a positive, continuous function is increasing on an interval, what can we say about the relative sizes of LRAM, MRAM, and RRAM? Explain.

4. If a positive, continuous function is decreasing on an interval, what can we say about the relative sizes of LRAM, MRAM, and RRAM? Explain.

Volume of a Sphere

Although the visual representation of RAM approximation focuses on area, remember that our original motivation for looking at sums of this type was to find distance traveled by an object moving with a nonconstant velocity. The connection between Examples 1 and 2 is that in each case, we have a function f defined on a closed interval and estimate what we want to know with a sum of function values multiplied by interval lengths. Many other physical quantities can be estimated this way.

Example 3 ESTIMATING THE VOLUME OF A SPHERE

Estimate the volume of a solid sphere of radius 4.

Solution We picture the sphere as if its surface were generated by revolving the graph of the function $f(x) = \sqrt{16 - x^2}$ about the x-axis (Figure 5.9a). We partition the interval $-4 \leq x \leq 4$ into n subintervals of equal length $\Delta x = 8/n$. We then slice the sphere with planes perpendicular to the x-axis at the partition points, cutting it like a round loaf of bread into n parallel slices of width Δx. When n is large, each slice can be approximated by a cylinder, a familiar geometric shape of known volume, $\pi r^2 h$. In our case, the cylinders lie on their sides and h is Δx while r varies according to where we are on the x-axis (Figure 5.9b). A logical radius to choose for each cylinder is $f(m_i) = \sqrt{16 - m_i^2}$, where m_i is the midpoint of the interval where the ith slice intersects the x-axis (Figure 5.9c).

We can now approximate the volume of the sphere by using MRAM to sum the cylinder volumes,

$$\pi r^2 h = \pi(\sqrt{16 - m_i^2})^2 \Delta x.$$

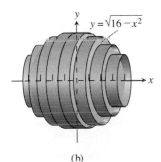

(a)

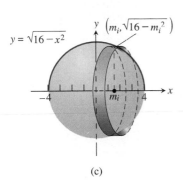

(b)

(c)

Figure 5.9 (a) The semicircle $y = \sqrt{16 - x^2}$ revolved about the x-axis to generate a sphere. (b) Slices of the solid sphere approximated with cylinders (drawn for $n = 8$). (c) The typical approximating cylinder has radius $f(m_i) = \sqrt{16 - m_i^2}$. (Example 3)

Keeping Track of Units

Notice in Example 3 that we are summing products of the form $\pi(16 - x^2) \times$ (a cross section area, measured in square units) times Δx (a length, measured in units). The products are therefore measured in cubic units, which are the correct units for volume.

The function we use in the RAM program is $\pi(\sqrt{16 - x^2})^2 = \pi(16 - x^2)$. The interval is $[-4, 4]$.

Number of Slices (n)	MRAM$_n$
10	269.42299
25	268.29704
50	268.13619
100	268.09598
1000	268.08271

The value for $n = 1000$ compares *very* favorably with the true volume,

$$V = \frac{4}{3}\pi r^3 = \frac{4}{3}\pi(4)^3 = \frac{256\pi}{3} \approx 268.0825731.$$

Even for $n = 10$ the difference between the MRAM approximation and the true volume is a small percentage of V:

$$\frac{|\text{MRAM}_{10} - V|}{V} = \frac{\text{MRAM}_{10} - 256\pi/3}{256\pi/3} \le 0.005.$$

That is, the error percentage is about one half of one percent!

Cardiac Output

So far we have seen applications of the RAM process to finding distance traveled and volume. These applications hint at the usefulness of this technique. To suggest its versatility we will present an application from human physiology.

The number of liters of blood your heart pumps in a fixed time interval is called your *cardiac output*. For a person at rest, the rate might be 5 or 6 liters per minute. During strenuous exercise the rate might be as high as 30 liters per minute. It might also be altered significantly by disease. How can a physician measure a patient's cardiac output without interrupting the flow of blood?

One technique is to inject a dye into a main vein near the heart. The dye is drawn into the right side of the heart and pumped through the lungs and out the left side of the heart into the aorta, where its concentration can be measured every few seconds as the blood flows past. The data in Table 5.2 and the plot in Figure 5.10 (obtained from the data) show the response of a healthy, resting patient to an injection of 5.6 mg of dye.

Table 5.2 Dye Concentration Data

Seconds after Injection t	Dye Concentration (adjusted for recirculation) c
5	0
7	3.8
9	8.0
11	6.1
13	3.6
15	2.3
17	1.45
19	0.91
21	0.57
23	0.36
25	0.23
27	0.14
29	0.09
31	0

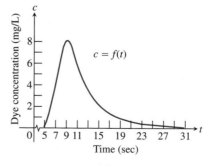

Figure 5.10 The dye concentration data from Table 5.2, plotted and fitted with a smooth curve. Time is measured with $t = 0$ at the time of injection. The dye concentration is zero at the beginning while the dye passes through the lungs. It then rises to a maximum at about $t = 9$ sec and tapers to zero by $t = 31$ sec.

The graph shows dye concentration (measured in milligrams of dye per liter of blood) as a function of time (in seconds). How can we use this graph to obtain the cardiac output (measured in liters of blood per second)? The trick is to divide the *number of mg of dye* by the *area under the dye concentration curve.* You can see why this works if you consider what happens to the units:

$$\frac{\text{mg of dye}}{\text{units of area under curve}} = \frac{\text{mg of dye}}{\dfrac{\text{mg of dye}}{\text{L of blood}} \cdot \text{sec}}$$

$$= \frac{\text{mg of dye}}{\text{sec}} \cdot \frac{\text{L of blood}}{\text{mg of dye}}$$

$$= \frac{\text{L of blood}}{\text{sec}}.$$

So you are now ready to compute like a cardiologist.

Example 4 COMPUTING CARDIAC OUTPUT FROM DYE CONCENTRATION

Estimate the cardiac output of the patient whose data appear in Table 5.2 and Figure 5.10. Give the estimate in liters per minute.

Solution We have seen that we can obtain the cardiac output by dividing the amount of dye (5.6 mg for our patient) by the area under the curve in Figure 5.10. Now we need to find the area. Our geometry formulas do not apply to this irregularly shaped region, and the RAM program is useless without a formula for the function. Nonetheless, we can draw the MRAM rectangles ourselves and estimate their heights from the graph. In Figure 5.11 each rectangle has a base 2 units long and a height $f(m_i)$ equal to the height of the curve above the midpoint of the base.

The area of each rectangle, then, is $f(m_i)$ times 2, and the sum of the rectangular areas is the MRAM estimate for the area under the curve:

$$\text{Area} \approx f(6) \cdot 2 + f(8) \cdot 2 + f(10) \cdot 2 + \cdots + f(28) \cdot 2$$

$$\approx 2 \cdot (1.4 + 6.3 + 7.5 + 4.8 + 2.8 + 1.9 + 1.1 + 0.7 + 0.5 + 0.3 + 0.2 + 0.1)$$

$$= 2 \cdot (27.6) = 55.2 \; (\text{mg/L}) \cdot \text{sec.}$$

Dividing 5.6 mg by this figure gives an estimate for cardiac output in liters per second. Multiplying by 60 converts the estimate to liters per minute:

$$\frac{5.6 \text{ mg}}{55.2 \text{ mg} \cdot \text{sec/L}} \cdot \frac{60 \text{ sec}}{1 \text{ min}} \approx 6.09 \text{ L/min.}$$

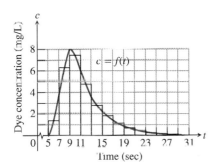

Figure 5.11 The region under the concentration curve of Figure 5.10 is approximated with rectangles. We ignore the portion from $t = 29$ to $t = 31$; its concentration is negligible. (Example 4)

Quick Review 5.1

As you answer the questions in Exercises 1–10, try to associate the answers with area, as in Figure 5.1.

1. A train travels at 80 mph for 5 hours. How far does it travel?

2. A truck travels at an average speed of 48 mph for 3 hours. How far does it travel?

3. Beginning at a standstill, a car maintains a constant acceleration of 10 ft/sec^2 for 10 seconds. What is its velocity after 10 seconds? Give your answer in ft/sec and then convert it to mi/h.

4. In a vacuum, light travels at a speed of 300,000 kilometers per second. How many kilometers does it travel in a year? (This distance equals one *light-year*.)

5. A long distance runner ran a race in 5 hours, averaging 6 mph for the first 3 hours and 5 mph for the last 2 hours. How far did she run?

6. A pump working at 20 gallons/minute pumps for an hour. How many gallons are pumped?

7. At 8:00 P.M. the temperature began dropping at a rate of 1 degree Celsius per hour. Twelve hours later it began rising at a rate of 1.5 degrees per hour for six hours. What was the net change in temperature over the 18-hour period?

8. Water flows over a spillway at a steady rate of 300 cubic feet per second. How many cubic feet of water pass over the spillway in one day?

9. A city has a population density of 350 people per square mile in an area of 50 square miles. What is the population of the city?

10. A hummingbird in flight beats its wings at a rate of 70 times per second. How many times does it beat its wings in an hour if it is in flight 70% of the time?

Section 5.1 Exercises

Exercises 1–4 refer to the region R enclosed between the graph of the function $y = 2x - x^2$ and the x-axis for $0 \le x \le 2$.

1. **(a)** Sketch the region R.

   (b) Partition $[0, 2]$ into 4 subintervals and show the four rectangles that LRAM uses to approximate the area of R. Compute the LRAM sum without a calculator.

2. Repeat Exercise 1(b) for RRAM and MRAM.

3. Using a calculator program, find the RAM sums that complete the following table.

n	LRAM$_n$	MRAM$_n$	RRAM$_n$
10			
50			
100			
500			

4. Make a conjecture about the area of the region R.

In Exercises 5–8, use RAM to estimate the area of the region enclosed between the graph of f and the x-axis for $a \le x \le b$.

5. $f(x) = x^2 - x + 3$, $a = 0$, $b = 3$

6. $f(x) = \dfrac{1}{x}$, $a = 1$, $b = 3$

7. $f(x) = e^{-x^2}$, $a = 0$, $b = 2$

8. $f(x) = \sin x$, $a = 0$, $b = \pi$

9. *Cardiac Output* The following table gives dye concentrations for a dye-concentration cardiac-output determination like the one in Example 4. The amount of dye injected in this patient was 5 mg instead of 5.6 mg. Use rectangles to estimate the area under the dye concentration curve and then go on to estimate the patient's cardiac output.

Seconds after Injection t	Dye Concentration (adjusted for recirculation) c
2	0
4	0.6
6	1.4
8	2.7
10	3.7
12	4.1
14	3.8
16	2.9
18	1.7
20	1.0
22	0.5
24	0

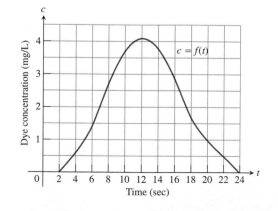

10. *Distance Traveled*   The table below shows the velocity of a model train engine moving along a track for 10 sec. Estimate the distance traveled by the engine, using 10 subintervals of length 1 with **(a)** left-endpoint values (LRAM) and **(b)** right-endpoint values (RRAM).

Time (sec)	Velocity (in./sec)	Time (sec)	Velocity (in./sec)
0	0	6	11
1	12	7	6
2	22	8	2
3	10	9	6
4	5	10	0
5	13		

11. *Distance Traveled Upstream*   You are walking along the bank of a tidal river watching the incoming tide carry a bottle upstream. You record the velocity of the flow every 5 minutes for an hour, with the results shown in the table below. About how far upstream does the bottle travel during that hour? Find the **(a)** LRAM and **(b)** RRAM estimates using 12 subintervals of length 5.

Time (min)	Velocity (m/sec)	Time (min)	Velocity (m/sec)
0	1	35	1.2
5	1.2	40	1.0
10	1.7	45	1.8
15	2.0	50	1.5
20	1.8	55	1.2
25	1.6	60	0
30	1.4		

12. *Length of a Road*   You and a companion are driving along a twisty stretch of dirt road in a car whose speedometer works but whose odometer (mileage counter) is broken. To find out how long this particular stretch of road is, you record the car's velocity at 10-sec intervals, with the results shown in the table below. (The velocity was converted from mi/h to ft/sec using 30 mi/h = 44 ft/sec.) Estimate the length of the road by averaging the LRAM and RRAM sums.

Time (sec)	Velocity (ft/sec)	Time (sec)	Velocity (ft/sec)
0	0	70	15
10	44	80	22
20	15	90	35
30	35	100	44
40	30	110	30
50	44	120	35
60	35		

13. *Distance from Velocity Data*   The table below gives data for the velocity of a vintage sports car accelerating from 0 to 142 mi/h in 36 sec (10 thousandths of an hour.)

Time (h)	Velocity (mi/h)	Time (h)	Velocity (mi/h)
0.0	0	0.006	116
0.001	40	0.007	125
0.002	62	0.008	132
0.003	82	0.009	137
0.004	96	0.010	142
0.005	108		

(a) Use rectangles to estimate how far the car traveled during the 36 sec it took to reach 142 mi/h.

(b) Roughly how many seconds did it take the car to reach the halfway point? About how fast was the car going then?

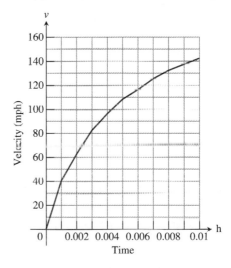

14. *(Continuation of Example 3)*   Use the slicing technique of Example 3 to find the MRAM sums that approximate the volume of a sphere of radius 5. Use $n = 10, 20, 40, 80,$ and 160.

15. *(Continuation of Exercise 14)*   Use a geometry formula to find the volume V of the sphere in Exercise 14 and find **(a)** the error and **(b)** the percentage error in the MRAM approximation for each value of n given.

16. *Volume of a Solid Hemisphere* To estimate the volume of a solid hemisphere of radius 4, imagine its axis of symmetry to be the interval $[0, 4]$ on the x-axis. Partition $[0, 4]$ into eight subintervals of equal length and approximate the solid with cylinders based on the circular cross sections of the hemisphere perpendicular to the x-axis at the subintervals' left endpoints. (See the accompanying profile view.)

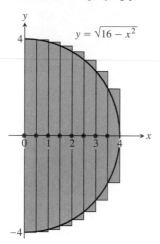

$$y = \sqrt{16 - x^2}$$

(a) Writing to Learn Find the sum S_8 of the volumes of the cylinders. Do you expect S_8 to overestimate V? Give reasons for your answer.

(b) Express $|V - S_8|$ as a percentage of V to the nearest percent.

17. Repeat Exercise 16 using cylinders based on cross sections at the right endpoints of the subintervals.

18. *Volume of Water in a Reservoir* A reservoir shaped like a hemispherical bowl of radius 8 m is filled with water to a depth of 4 m.

(a) Find an estimate S of the water's volume by approximating the water with eight circumscribed solid cylinders.

(b) It can be shown that the water's volume is $V = (320\pi)/3$ m^3. Find the error $|V - S|$ as a percentage of V to the nearest percent.

19. *Volume of Water in a Swimming Pool* A rectangular swimming pool is 30 ft wide and 50 ft long. The table below shows the depth $h(x)$ of the water at 5-ft intervals from one end of the pool to the other. Estimate the volume of water in the pool using **(a)** left-endpoint values and **(b)** right-endpoint values.

Position (ft) x	Depth (ft) $h(x)$	Position (ft) x	Depth (ft) $h(x)$
0	6.0	30	11.5
5	8.2	35	11.9
10	9.1	40	12.3
15	9.9	45	12.7
20	10.5	50	13.0
25	11.0		

20. *Volume of a Nose Cone* The nose "cone" of a rocket is a *paraboloid* obtained by revolving the curve $y = \sqrt{x}$, $0 \le x \le 5$ about the x-axis, where x is measured in feet. Estimate the volume V of the nose cone by partitioning $[0, 5]$ into five subintervals of equal length, slicing the cone with planes perpendicular to the x-axis at the subintervals' left endpoints, constructing cylinders of height 1 based on cross sections at these points, and finding the volumes of these cylinders. (See the accompanying figure.)

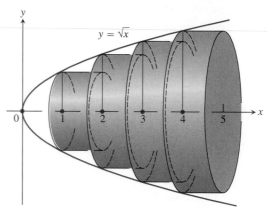

$$y = \sqrt{x}$$

21. *Volume of a Nose Cone* Repeat Exercise 20 using cylinders based on cross sections at the *midpoints* of the subintervals.

22. *Free Fall with Air Resistance* An object is dropped straight down from a helicopter. The object falls faster and faster but its acceleration (rate of change of its velocity) decreases over time because of air resistance. The acceleration is measured in ft/sec^2 and recorded every second after the drop for 5 sec, as shown in the table below.

t	0	1	2	3	4	5
a	32.00	19.41	11.77	7.14	4.33	2.63

(a) Use LRAM$_5$ to find an upper estimate for the speed when $t = 5$.

(b) Use RRAM$_5$ to find a lower estimate for the speed when $t = 5$.

(c) Use upper estimates for the speed during the first second, second second, and third second to find an upper estimate for the distance fallen when $t = 3$.

23. *Distance Traveled by a Projectile* An object is shot straight upward from sea level with an initial velocity of 400 ft/sec.

(a) Assuming gravity is the only force acting on the object, give an upper estimate for its velocity after 5 sec have elapsed. Use $g = 32$ ft/sec^2 for the gravitational constant.

(b) Find a lower estimate for the height attained after 5 sec.

24. *Water Pollution* Oil is leaking out of a tanker damaged at sea. The damage to the tanker is worsening as evidenced by the increased leakage each hour, recorded in the table below.

Time (h)	0	1	2	3	4
Leakage (gal/h)	50	70	97	136	190

Time (h)	5	6	7	8
Leakage (gal/h)	265	369	516	720

(a) Give an upper and lower estimate of the total quantity of oil that has escaped after 5 hours.

(b) Repeat (a) for the quantity of oil that has escaped after 8 hours.

(c) The tanker continues to leak 720 gal/h after the first 8 hours. If the tanker originally contained 25,000 gal of oil, approximately how many more hours will elapse in the worst case before all of the oil has leaked? in the best case?

25. *Air Pollution* A power plant generates electricity by burning oil. Pollutants produced by the burning process are removed by scrubbers in the smokestacks. Over time the scrubbers become less efficient and eventually must be replaced when the amount of pollutants released exceeds government standards. Measurements taken at the end of each month determine the rate at which pollutants are released into the atmosphere as recorded in the table below.

Month	Jan	Feb	Mar	Apr	May	Jun
Pollutant Release Rate (tons/day)	0.20	0.25	0.27	0.34	0.45	0.52

Month	Jul	Aug	Sep	Oct	Nov	Dec
Pollutant Release Rate (tons/day)	0.63	0.70	0.81	0.85	0.89	0.95

(a) Assuming a 30-day month and that new scrubbers allow only 0.05 ton/day released, give an upper estimate of the total tonnage of pollutants released by the end of June. What is a lower estimate?

(b) In the best case, approximately when will a total of 125 tons of pollutants have been released into the atmosphere?

26. **Writing to Learn** The graph shows the sales record for a company over a 10-year period. If sales are measured in millions of units per year, explain what information can be obtained from the area of the region, and why.

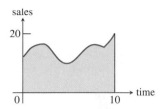

Exploration

27. *Area of a Circle* *Work in groups of two or three.* Inscribe a regular n-sided polygon inside a circle of radius 1 and compute the area of the polygon for the following values of n.

(a) 4 (square) **(b)** 8 (octagon) **(c)** 16

(d) Compare the areas in parts (a), (b), and (c) with the area of the circle. ■

Extending the Ideas

28. *Rectangular Approximation Methods* Prove or disprove the following statement: MRAM_n is always the average of LRAM_n and RRAM_n.

29. *Rectangular Approximation Methods* Show that if f is a nonnegative function on the interval $[a, b]$ and the line $x = (a + b)/2$ is a line of symmetry of the graph of $y = f(x)$, then $\text{LRAM}_n f = \text{RRAM}_n f$ for every positive integer n.

30. *(Continuation of Exercise 27)*

(a) Inscribe a regular n-sided polygon inside a circle of radius 1 and compute the area of one of the n congruent triangles formed by drawing radii to the vertices of the polygon.

(b) Compute the limit of the area of the inscribed polygon as $n \rightarrow \infty$.

(c) Repeat the computations in (a) and (b) for a circle of radius r.

| 5.2 | **Definite Integrals** |

Riemann Sums • Terminology and Notation of Integration •
Definite Integral and Area • Constant Functions • Integrals on a
Calculator • Discontinuous Integrable Functions

Riemann Sums

In the preceding section, we estimated distances, areas, and volumes with finite
sums. The terms in the sums were obtained by multiplying selected function
values by the lengths of intervals. In this section we move beyond finite sums
to see what happens in the limit, as the terms become infinitely small and their
number infinitely large.

Sigma notation enables us to express a large sum in compact form:

$$\sum_{k=1}^{n} a_k = a_1 + a_2 + a_3 + \cdots + a_{n-1} + a_n.$$

The Greek capital letter Σ (sigma) stands for "sum." The index k tells us where
to begin the sum (at the number below the Σ) and where to end (at the number
above). If the symbol ∞ appears above the Σ, it indicates that the terms go on
indefinitely.

The sums in which we will be interested are called *Riemann* ("*ree*-mahn")
sums, after Georg Friedrich Bernhard Riemann (1826–1866). LRAM, MRAM,
and RRAM in the previous section are all examples of Riemann sums—not
because they estimated area, but because they were constructed in a particular
way. We now describe that construction formally, in a more general context that
does not confine us to nonnegative functions.

We begin with an arbitrary continuous function $f(x)$ defined on a closed
interval $[a, b]$. Like the function graphed in Figure 5.12, it may have negative
values as well as positive values.

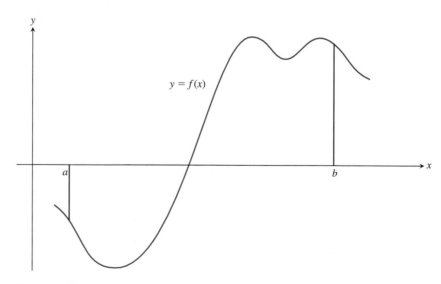

Figure 5.12 The graph of a typical function $y = f(x)$ over a closed interval $[a, b]$.

We then partition the interval $[a, b]$ into n subintervals by choosing $n - 1$ points, say $x_1, x_2, \cdots, x_{n-1}$, between a and b subject only to the condition that

$$a < x_1 < x_2 < \cdots < x_{n-1} < b.$$

To make the notation consistent, we denote a by x_0 and b by x_n. The set

$$P = \{x_0, x_1, x_2, \cdots, x_n\}$$

is called a *partition* of $[a, b]$.

The partition P determines n closed *subintervals,* as shown in Figure 5.13. The kth subinterval is $[x_{k-1}, x_k]$, which has length $\Delta x_k = x_k - x_{k-1}$.

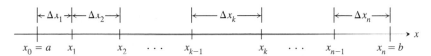

Figure 5.13 The partition $P = \{a = x_0, x_1, x_2, \cdots, x_n = b\}$ divides $[a, b]$ into n subintervals of lengths $\Delta x_1, \Delta x_2, \cdots, \Delta x_n$. The kth subinterval has length Δx_k.

In each subinterval we select some number. Denote the number chosen from the kth subinterval by c_k.

Then, on each subinterval we stand a vertical rectangle that reaches from the x-axis to touch the curve at $(c_k, f(c_k))$. These rectangles could lie either above or below the x-axis (Figure 5.14).

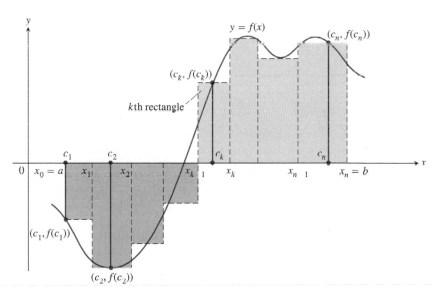

Figure 5.14 Rectangles extending from the x-axis to intersect the curve at the points $(c_k, f(c_k))$. The rectangles approximate the region between the x-axis and the graph of the function.

On each subinterval, we form the product $f(c_k) \cdot \Delta x_k$. This product can be positive, negative, or zero, depending on $f(c_k)$.

Finally, we take the sum of these products:

$$S_n = \sum_{k=1}^{n} f(c_k) \cdot \Delta x_k.$$

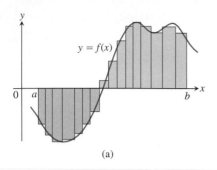

(a)

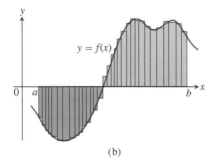

(b)

Figure 5.15 The curve of Figure 5.12 with rectangles from finer partitions of $[a, b]$. Finer partitions create more rectangles, with shorter bases.

Georg Riemann (1826–1866)

The mathematicians of the 17th and 18th centuries blithely assumed the existence of limits of Riemann sums (as we admittedly did in our RAM explorations of the last section), but the existence was not established mathematically until Georg Riemann proved Theorem 1 in 1854. You can find a current version of Riemann's proof in most advanced calculus books.

This sum, which depends on the partition P and the choice of the numbers c_k, is a **Riemann sum for f on the interval $[a, b]$.**

As the partitions of $[a, b]$ become finer and finer, we would expect the rectangles defined by the partitions to approximate the region between the x-axis and the graph of f with increasing accuracy (Figure 5.15).

Just as LRAM, MRAM, and RRAM in our earlier examples converged to a common value in the limit, *all* Riemann sums for a given function on $[a, b]$ converge to a common value, as long as the lengths of the subintervals all tend to zero. This latter condition is assured by requiring the longest subinterval length (called the **norm** of the partition and denoted by $\|P\|$) to tend to zero.

Definition   The Definite Integral as a Limit of Riemann Sums

Let f be a function defined on a closed interval $[a, b]$. For any partition P of $[a, b]$, let the numbers c_k be chosen arbitrarily in the subintervals $[x_{k-1}, x_k]$.

If there exists a number I such that

$$\lim_{\|P\|\to 0} \sum_{k=1}^{n} f(c_k)\Delta x_k = I$$

no matter how P and the c_k's are chosen, then f is **integrable** on $[a, b]$ and I is the **definite integral** of f over $[a, b]$.

Despite the potential for variety in the sums $\sum f(c_k)\Delta x_k$ as the partitions change and as the c_k's are chosen arbitrarily in the intervals of each partition, the sums always have the same limit as $\|P\|\to 0$ as long as f is *continuous* on $[a, b]$.

Theorem 1   The Existence of Definite Integrals

All continuous functions are integrable. That is, if a function f is continuous on an interval $[a, b]$, then its definite integral over $[a, b]$ exists.

Because of Theorem 1, we can get by with a simpler construction for definite integrals of continuous functions. Since we know for these functions that the Riemann sums tend to the same limit for *all* partitions in which $\|P\|\to 0$, we need only to consider the limit of the so-called **regular partitions,** in which all the subintervals have the same length.

The Definite Integral of a Continuous Function on $[a, b]$

Let f be continuous on $[a, b]$, and let $[a, b]$ be partitioned into n subintervals of equal length $\Delta x = (b - a)/n$. Then the definite integral of f over $[a, b]$ is given by

$$\lim_{n\to\infty} \sum_{k=1}^{n} f(c_k)\Delta x,$$

where each c_k is chosen arbitrarily in the kth subinterval.

Terminology and Notation of Integration

Leibniz's clever choice of notation for the derivative, dy/dx, had the advantage of retaining an identity as a "fraction" even though both numerator and denominator had tended to zero. Although not really fractions, derivatives can *behave* like fractions, so the notation makes profound results like the Chain Rule

$$\frac{dy}{dx} = \frac{dy}{du} \cdot \frac{du}{dx}$$

seem almost simple.

The notation that Leibniz introduced for the definite integral was equally inspired. In his derivative notation, the Greek letters ("Δ" for "difference") switch to Roman letters ("d" for "differential") in the limit,

$$\lim_{\Delta x \to 0} \frac{\Delta y}{\Delta x} = \frac{dy}{dx}.$$

In his definite integral notation, the Greek letters again become Roman letters in the limit,

$$\lim_{n \to \infty} \sum_{k=1}^{n} f(c_k) \Delta x = \int_a^b f(x)\, dx.$$

Notice that the difference Δx has again tended to zero, becoming a differential dx. The Greek "Σ" has become an elongated Roman "S," so that the integral can retain its identity as a "sum." The c_k's have become so crowded together in the limit that we no longer think of a choppy selection of x values between a and b, but rather of a continuous, unbroken sampling of x values from a to b. It is as if we were summing *all* products of the form $f(x)\, dx$ as x goes from a to b, so we can abandon the k and the n used in the finite sum expression.

The symbol

$$\int_a^b f(x)\, dx$$

is read as "the integral from a to b of f of x dee x," or sometimes as "the integral from a to b of f of x with respect to x." The component parts also have names:

Upper limit of integration

The function is the **integrand.**

Integral sign

x is the **variable of integration.**

$$\int_a^b f(x)\, dx$$

Lower limit of integration

Integral of f from a to b

When you find the value of the integral, you have **evaluated** the integral.

The value of the definite integral of a function over any particular interval depends on the function and not on the letter we choose to represent its independent variable. If we decide to use t or u instead of x, we simply write the integral as

$$\int_a^b f(t)\, dt \quad \text{or} \quad \int_a^b f(u)\, du \quad \text{instead of} \quad \int_a^b f(x)\, dx.$$

No matter how we represent the integral, it is the same *number,* defined as a limit of Riemann sums. Since it does not matter what letter we use to run from *a* to *b,* the variable of integration is called a **dummy variable.**

Example 1 USING THE NOTATION

The interval $[-1, 3]$ is partitioned into n subintervals of equal length $\Delta x = 4/n$. Let m_k denote the midpoint of the kth subinterval. Express the limit

$$\lim_{n \to \infty} \sum_{k=1}^{n} (3(m_k)^2 - 2m_k + 5)\,\Delta x$$

as an integral.

Solution Since the midpoints m_k have been chosen from the subintervals of the partition, this expression is indeed a limit of Riemann sums. (The points chosen did not have to be midpoints; they could have been chosen from the subintervals in any arbitrary fashion.) The function being integrated is $f(x) = 3x^2 - 2x + 5$ over the interval $[-1, 3]$. Therefore,

$$\lim_{n \to \infty} \sum_{k=1}^{n} (3(m_k)^2 - 2m_k + 5)\,\Delta x = \int_{-1}^{3} (3x^2 - 2x + 5)\,dx.$$

Definite Integral and Area

If an integrable function $y = f(x)$ is nonnegative throughout an interval $[a, b]$, each nonzero term $f(c_k)\Delta x_k$ is the area of a rectangle reaching from the x-axis up to the curve $y = f(x)$. (See Figure 5.16.)

The Riemann sum

$$\sum f(c_k)\,\Delta x_k,$$

which is the sum of the areas of these rectangles, gives an estimate of the area of the region between the curve and the x-axis from a to b. Since the rectangles give an increasingly good approximation of the region as we use partitions with smaller and smaller norms, we call the limiting value the area under the curve.

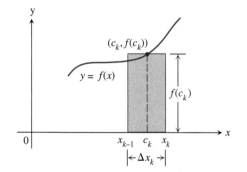

Figure 5.16 A term of a Riemann sum $\sum f(c_k)\Delta x_k$ for a nonnegative function f is either zero or the area of a rectangle such as the one shown.

Definition Area Under a Curve (as a Definite Integral)

If $y = f(x)$ is nonnegative and integrable over a closed interval $[a, b]$, then the **area under the curve $y = f(x)$ from a to b** is the integral of f from a to b,

$$A = \int_{a}^{b} f(x)\,dx.$$

This definition works both ways: We can use integrals to calculate areas *and* we can use areas to calculate integrals.

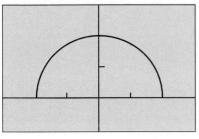

[–3, 3] by [–1, 3]

Figure 5.17 A square viewing window on $y = \sqrt{4 - x^2}$. The graph is a semicircle because $y = \sqrt{4 - x^2}$ is the same as $y^2 = 4 - x^2$, or $x^2 + y^2 = 4$, with $y \geq 0$. (Example 2)

Example 2 REVISITING AREA UNDER A CURVE

Evaluate the integral $\int_{-2}^{2} \sqrt{4 - x^2} \, dx$.

Solution We recognize $f(x) = \sqrt{4 - x^2}$ as a function whose graph is a semicircle of radius 2 centered at the origin (Figure 5.17).

The area between the semicircle and the x-axis from -2 to 2 can be computed using the geometry formula

$$\text{Area} = \frac{1}{2} \cdot \pi r^2 = \frac{1}{2} \cdot \pi (2)^2 = 2\pi.$$

Because the area is also the value of the integral of f from -2 to 2,

$$\int_{-2}^{2} \sqrt{4 - x^2} \, dx = 2\pi.$$

If an integrable function $y = f(x)$ is nonpositive, the nonzero terms $f(c_k)\Delta x_k$ in the Riemann sums for f over an interval $[a, b]$ are negatives of rectangle areas. The limit of the sums, the integral of f from a to b, is therefore the *negative* of the area of the region between the graph of f and the x-axis (Figure 5.18).

$$\int_{a}^{b} f(x) \, dx = -(\text{the area}) \quad \text{if} \quad f(x) \leq 0.$$

Or, turning this around,

$$\text{Area} = -\int_{a}^{b} f(x) \, dx \quad \text{when} \quad f(x) \leq 0.$$

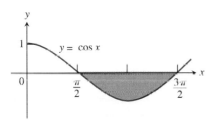

Figure 5.18 Because $f(x) = \cos x$ is nonpositive on $[\pi/2, 3\pi/2]$, the integral of f is a negative number. The area of the shaded region is the opposite of this integral,

$$\text{Area} = -\int_{\pi/2}^{3\pi/2} \cos x \, dx.$$

If an integrable function $y = f(x)$ has both positive and negative values on an interval $[a, b]$, then the Riemann sums for f on $[a, b]$ add areas of rectangles that lie above the x-axis to the negatives of areas of rectangles that lie below the x-axis, as in Figure 5.19. The resulting cancellations mean that the limiting value is a number whose magnitude is less than the total area between the curve and the x-axis. The value of the integral is the area above the x-axis minus the area below.

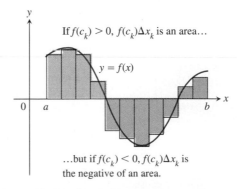

Figure 5.19 An integrable function f with negative as well as positive values.

Net Area

Sometimes $\int_a^b f(x)\,dx$ is called the *net area* of the region determined by the curve $y = f(x)$ and the x-axis between $x = a$ and $x = b$.

For any integrable function,

$$\int_a^b f(x)\,dx = (\text{area above the } x\text{-axis}) - (\text{area below the } x\text{-axis}).$$

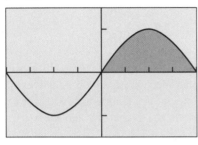

$y = \sin x$

$[-\pi, \pi]$ by $[-1.5, 1.5]$

Figure 5.20

$\int_0^\pi \sin x\,dx = 2.$ (Exploration 1)

Exploration 1 Finding Integrals by Signed Areas

It is a fact (which we will revisit) that $\int_0^\pi \sin x\,dx = 2$ (Figure 5.20). With that information, what you know about integrals and areas, what you know about graphing curves, and sometimes a bit of intuition, determine the values of the following integrals. Give as convincing an argument as you can for each value, based on the graph of the function.

1. $\displaystyle\int_\pi^{2\pi} \sin x\,dx$ 2. $\displaystyle\int_0^{2\pi} \sin x\,dx$ 3. $\displaystyle\int_0^{\pi/2} \sin x\,dx$

4. $\displaystyle\int_0^\pi (2 + \sin x)\,dx$ 5. $\displaystyle\int_0^\pi 2\sin x\,dx$ 6. $\displaystyle\int_2^{\pi+2} \sin(x-2)\,dx$

7. $\displaystyle\int_{-\pi}^\pi \sin u\,du$ 8. $\displaystyle\int_0^{2\pi} \sin(x/2)\,dx$ 9. $\displaystyle\int_0^\pi \cos x\,dx$

10. Suppose k is *any* positive number. Make a conjecture about $\int_{-k}^k \sin x\,dx$ and support your conjecture.

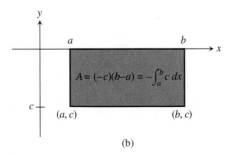

(a)

(b)

Figure 5.21 (a) If c is a positive constant, then $\int_a^b c\,dx$ is the area of the rectangle shown. (b) If c is negative, then $\int_a^b c\,dx$ is the opposite of the area of the rectangle.

Constant Functions

Integrals of constant functions are easy to evaluate. Over a closed interval, they are simply the constant times the length of the interval (Figure 5.21).

Theorem 2 The Integral of a Constant

If $f(x) = c,$ where c is a constant, on the interval $[a, b]$, then

$$\int_a^b f(x)\,dx = \int_a^b c\,dx = c(b - a).$$

Proof A constant function is continuous, so the integral exists, and we can evaluate it as a limit of Riemann sums with subintervals of equal length $(b - a)/n$. Any such sum looks like

$$\sum_{k=1}^n f(c_k) \cdot \Delta x, \quad \text{which is} \quad \sum_{k=1}^n c \cdot \frac{b - a}{n}.$$

Then

$$\sum_{k=1}^{n} c \cdot \frac{b-a}{n} = c \cdot (b-a) \sum_{k=1}^{n} \frac{1}{n}$$

$$= c(b-a) \cdot n\left(\frac{1}{n}\right)$$

$$= c(b-a).$$

Since the sum is *always* $c(b-a)$ for any value of n, it follows that the limit of the sums, the integral to which they converge, is also $c(b-a)$. ∎

Example 3 REVISITING THE TRAIN PROBLEM

A train moves along a track at a steady 75 miles per hour from 7:00 A.M. to 9:00 A.M. Express its total distance traveled as an integral. Evaluate the integral using Theorem 2.

Solution (See Figure 5.22.)

$$\text{Distance traveled} = \int_{7}^{9} 75 \, dt = 75 \cdot (9-7) = 150$$

Since the 75 is measured in miles/hour and the $(9-7)$ is measured in hours, the 150 is measured in miles. The train traveled 150 miles.

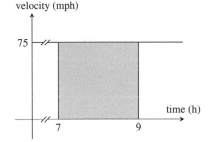

Figure 5.22 The area of the rectangle is a special case of Theorem 2. (Example 3)

Integrals on a Calculator

You do not have to know much about your calculator to realize that finding the limit of a Riemann sum is exactly the kind of thing that it does best. We have seen how effectively it can approximate areas using MRAM, but most modern calculators have sophisticated built-in programs that converge to integrals with much greater speed and precision than that. We will assume that your calculator has such a numerical integration capability, which we will denote as **NINT.** In particular, we will use NINT $(f(x), x, a, b)$ to denote a calculator (or computer) approximation of $\int_{a}^{b} f(x) \, dx$. When we write

$$\int_{a}^{b} f(x) \, dx = \text{NINT} \, (f(x), x, a, b),$$

we do so with the understanding that the right-hand side of the equation is an approximation of the left-hand side.

Example 4 USING NINT

Evaluate the following integrals numerically.

(a) $\int_{-1}^{2} x \sin x \, dx$ **(b)** $\int_{0}^{1} \frac{4}{1+x^2} \, dx$ **(c)** $\int_{0}^{5} e^{-x^2} \, dx$

Solution

(a) NINT $(x \sin x, x, -1, 2) \approx 2.04$

(b) NINT $(4/(1 + x^2), x, 0, 1) \approx 3.14$

(c) NINT $(e^{-x^2}, x, 0, 5) \approx 0.89$

Bounded Functions

We say a function is *bounded* on a given domain if its range is confined between some minimum value m and some maximum value M. That is, given any x in the domain, $m \le f(x) \le M$. Equivalently, the graph of $y = f(x)$ lies between the horizontal lines $y = m$ and $y = M$.

$y = |x|/x$

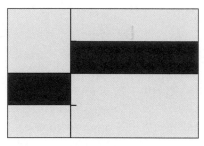

[–1, 2] by [–2, 2]

Figure 5.23 A discontinuous integrable function:

$$\int_{-1}^{2} \frac{|x|}{x}\, dx =$$

$-$(area below x-axis) + (area above x-axis). (Example 5)

A Nonintegrable Function

How "bad" does a function have to be before it is *not* integrable? One way to defeat integrability is to be unbounded (like $y = 1/x$ near $x = 0$), which can prevent the Riemann sums from tending to a finite limit. Another, more subtle, way is to be bounded but badly discontinuous, like the *characteristic function of the rationals*:

$$f(x) = \begin{cases} 1 & \text{if } x \text{ is rational} \\ 0 & \text{if } x \text{ is irrational.} \end{cases}$$

No matter what partition we take of the closed interval [0, 1], every subinterval contains both rational and irrational numbers. That means that we can always form a Riemann sum with all rational c_k's (a Riemann sum of 1) or all irrational c_k's (a Riemann sum of 0). The sums can therefore never tend toward a unique limit.

We will eventually be able to confirm that the exact value for Example 4a is $-2\cos 2 + \sin 2 - \cos 1 + \sin 1$. You might want to conjecture for yourself what the exact answer to Example 4b might be. As for Example 4c, no explicit *exact* value has ever been found for this integral! The best we can do in this case (and in many like it) is to approximate the integral numerically. Here, technology is not only useful, it is essential.

Discontinuous Integrable Functions

Theorem 1 guarantees that all continuous functions are integrable. But some functions with discontinuities are also integrable. For example, a bounded function (see margin note) that has a finite number of points of discontinuity on an interval $[a, b]$ will still be integrable on the interval if it is continuous everywhere else.

Example 5   INTEGRATING A DISCONTINUOUS FUNCTION

Find $\displaystyle\int_{-1}^{2} \frac{|x|}{x}\, dx.$

Solution This function has a discontinuity at $x = 0$, where the graph jumps from $y = -1$ to $y = 1$. The graph, however, determines two rectangles, one below the x-axis and one above (Figure 5.23).

Using the idea of net area, we have

$$\int_{-1}^{2} \frac{|x|}{x}\, dx = -1 + 2 = 1.$$

Exploration 2 **More Discontinuous Integrands**

1. Explain why the function

$$f(x) = \frac{x^2 - 4}{x - 2}$$

is not continuous on [0, 3]. What kind of discontinuity occurs?

2. Use areas to show that

$$\int_{0}^{3} \frac{x^2 - 4}{x - 2}\, dx = 10.5.$$

3. Use areas to show that

$$\int_{0}^{5} \text{int}\,(x)\, dx = 10.$$

Quick Review 5.2

In Exercises 1–3, evaluate the sum.

1. $\displaystyle\sum_{n-1}^{5} n^2$

2. $\displaystyle\sum_{k=0}^{4} (3k - 2)$

3. $\displaystyle\sum_{j=0}^{4} 100\,(j + 1)^2$

In Exercises 4–6, write the sum in sigma notation.

4. $1 + 2 + 3 + \cdots + 98 + 99$

5. $0 + 2 + 4 + \cdots + 48 + 50$

6. $3(1)^2 + 3(2)^2 + \cdots + 3(500)^2$

In Exercises 7 and 8, write the expression as a single sum in sigma notation.

7. $\displaystyle 2\sum_{x=1}^{50} x^2 + 3\sum_{x=1}^{50} x$

8. $\displaystyle\sum_{k=0}^{8} x^k + \sum_{k=9}^{20} x^k$

9. Find $\displaystyle\sum_{k=0}^{n} (-1)^k$ if n is odd.

10. Find $\displaystyle\sum_{k=0}^{n} (-1)^k$ if n is even.

Section 5.2 Exercises

In Exercises 1–6, express the limit as a definite integral.

1. $\displaystyle\lim_{\|P\|\to 0} \sum_{k=1}^{n} c_k^2 \Delta x_k$, where P is any partition of $[0, 2]$

2. $\displaystyle\lim_{\|P\|\to 0} \sum_{k=1}^{n} (c_k^2 - 3c_k)\,\Delta x_k$, where P is any partition of $[-7, 5]$

3. $\displaystyle\lim_{\|P\|\to 0} \sum_{k=1}^{n} \frac{1}{c_k}\,\Delta x_k$, where P is any partition of $[1, 4]$

4. $\displaystyle\lim_{\|P\|\to 0} \sum_{k=1}^{n} \frac{1}{1 - c_k}\,\Delta x_k$, where P is any partition of $[2, 3]$

5. $\displaystyle\lim_{\|P\|\to 0} \sum_{k=1}^{n} \sqrt{4 - c_k^2}\,\Delta x_k$, where P is any partition of $[0, 1]$

6. $\displaystyle\lim_{\|P\|\to 0} \sum_{k=1}^{n} (\sin^3 c_k)\,\Delta x_k$, where P is any partition of $[-\pi, \pi]$

In Exercises 7–12, evaluate the integral.

7. $\displaystyle\int_{-2}^{1} 5\,dx$

8. $\displaystyle\int_{3}^{7} (-20)\,dx$

9. $\displaystyle\int_{0}^{3} (-160)\,dt$

10. $\displaystyle\int_{-4}^{-1} \frac{\pi}{2}\,d\theta$

11. $\displaystyle\int_{-2.1}^{3.4} 0.5\,ds$

12. $\displaystyle\int_{\sqrt{2}}^{\sqrt{18}} \sqrt{2}\,dr$

In Exercises 13–22, use the graph of the integrand and areas to evaluate the integral.

13. $\displaystyle\int_{-2}^{4} \left(\frac{x}{2} + 3\right) dx$

14. $\displaystyle\int_{1/2}^{3/2} (-2x + 4)\,dx$

15. $\displaystyle\int_{-3}^{3} \sqrt{9 - x^2}\,dx$

16. $\displaystyle\int_{-4}^{0} \sqrt{16 - x^2}\,dx$

17. $\displaystyle\int_{-2}^{1} |x|\,dx$

18. $\displaystyle\int_{-1}^{1} (1 - |x|)\,dx$

19. $\displaystyle\int_{-1}^{1} (2 - |x|)\,dx$

20. $\displaystyle\int_{-1}^{1} (1 + \sqrt{1 - x^2})\,dx$

21. $\displaystyle\int_{\pi}^{2\pi} \theta\,d\theta$

22. $\displaystyle\int_{\sqrt{2}}^{5\sqrt{2}} r\,dr$

In Exercises 23–28, use areas to evaluate the integral.

23. $\displaystyle\int_{0}^{b} x\,dx, \quad b > 0$

24. $\displaystyle\int_{0}^{b} 4x\,dx, \quad b > 0$

25. $\displaystyle\int_{a}^{b} 2s\,ds, \quad 0 < a < b$

26. $\displaystyle\int_{a}^{b} 3t\,dt, \quad 0 < a < b$

27. $\displaystyle\int_{a}^{2a} x\,dx, \quad a > 0$

28. $\displaystyle\int_{a}^{\sqrt{3}a} x\,dx, \quad a > 0$

In Exercises 29–38, *work in groups of two or three.* Use graphs, your knowledge of area, and the fact that

$$\int_{0}^{1} x^3\,dx - \frac{1}{4}$$

to evaluate the integral.

29. $\displaystyle\int_{-1}^{1} x^3\,dx$

30. $\displaystyle\int_{0}^{1} (x^3 + 3)\,dx$

31. $\displaystyle\int_{2}^{3} (x - 2)^3\,dx$

32. $\displaystyle\int_{-1}^{1} |x|^3\,dx$

33. $\displaystyle\int_{0}^{1} (1 - x^3)\,dx$

34. $\displaystyle\int_{-1}^{2} (|x| - 1)^3\,dx$

35. $\displaystyle\int_{0}^{2} \left(\frac{x}{2}\right)^3 dx$

36. $\displaystyle\int_{-8}^{8} x^3\,dx$

37. $\displaystyle\int_0^1 (x^3 - 1)\, dx$

38. $\displaystyle\int_0^1 \sqrt[3]{x}\, dx$

In Exercises 39–42, use NINT to evaluate the expression.

39. $\displaystyle\int_0^5 \frac{x}{x^2 + 4}\, dx$

40. $3 + 2\displaystyle\int_0^{\pi/3} \tan x\, dx$

41. Find the area enclosed between the x-axis and the graph of $y = 4 - x^2$ from $x = -2$ to $x = 2$.

42. Find the area enclosed between the x-axis and the graph of $y = x^2 e^{-x}$ from $x = -1$ to $x = 3$.

In Exercises 43–46, **(a)** find the points of discontinuity of the integrand on the interval of integration, and **(b)** use area to evaluate the integral.

43. $\displaystyle\int_{-2}^3 \frac{x}{|x|}\, dx$

44. $\displaystyle\int_{-6}^5 2\text{ int }(x - 3)\, dx$

45. $\displaystyle\int_{-3}^4 \frac{x^2 - 1}{x + 1}\, dx$

46. $\displaystyle\int_{-5}^6 \frac{9 - x^2}{x - 3}\, dx$

Extending the Ideas

47. Writing to Learn The function

$$f(x) = \begin{cases} \dfrac{1}{x^2}, & 0 < x \le 1 \\ 0, & x = 0 \end{cases}$$

is defined on $[0, 1]$ and has a single point of discontinuity at $x = 0$.

(a) What happens to the graph of f as x approaches 0 from the right?

(b) The function f is not integrable on $[0, 1]$. Give a convincing argument based on Riemann sums to explain why it is not.

48. It can be shown by mathematical induction (see Appendix 2) that

$$\sum_{k=1}^n k^2 = \frac{n(n + 1)(2n + 1)}{6}.$$

Use this fact to give a formal proof that

$$\int_0^1 x^2\, dx = \frac{1}{3}$$

by following the steps given below.

(a) Partition $[0, 1]$ into n subintervals of length $1/n$. Show that the RRAM Riemann sum for the integral is

$$\sum_{k=1}^n \left(\left(\frac{k}{n}\right)^2 \cdot \frac{1}{n}\right).$$

(b) Show that this sum can be written as

$$\frac{1}{n^3} \cdot \sum_{k=1}^n k^2.$$

(c) Show that the sum can therefore be written as

$$\frac{n(n + 1)(2n + 1)}{6n^3}.$$

(d) Show that

$$\lim_{n \to \infty} \sum_{k=1}^n \left(\left(\frac{k}{n}\right)^2 \cdot \frac{1}{n}\right) = \frac{1}{3}.$$

(e) Explain why the equation in (d) proves that

$$\int_0^1 x^2\, dx = \frac{1}{3}.$$

5.3 # Definite Integrals and Antiderivatives

Properties of Definite Integrals • Average Value of a Function • Mean Value Theorem for Definite Integrals • Connecting Differential and Integral Calculus

Properties of Definite Integrals

In defining $\int_a^b f(x)$ as a limit of sums $\sum c_k \Delta x_k$, we moved from left to right across the interval $[a, b]$. What would happen if we integrated in the *opposite direction*? The integral would become $\int_b^a f(x)\, dx$—again a limit of sums of the form $\sum f(c_k)\Delta x_k$—but this time each of the Δx_k's would be negative as the x-values *decreased* from b to a. This would change the signs of all the terms in each Riemann sum, and ultimately the sign of the definite integral. This suggests the rule

$$\int_b^a f(x)\, dx = -\int_a^b f(x)\, dx.$$

Since the original definition did not apply to integrating backwards over an interval, we can treat this rule as a logical extension of the definition.

Although $[a, a]$ is technically not an interval, another logical extension of the definition is that $\int_a^a f(x)\, dx = 0$.

These are the first two rules in Table 5.3. The others are inherited from rules that hold for Riemann sums. However, the limit step required to *prove* that these rules hold in the limit (as the norms of the partitions tend to zero) places their mathematical verification beyond the scope of this course. You can see intuitively why each rule makes sense, however.

Table 5.3 Rules for Definite Integrals

1. *Order of Integration:* $\quad \displaystyle\int_b^a f(x)\, dx = -\int_a^b f(x)\, dx \qquad$ A definition

2. *Zero:* $\quad \displaystyle\int_a^a f(x)\, dx = 0 \qquad$ Also a definition

3. *Constant Multiple:* $\quad \displaystyle\int_a^b kf(x)\, dx = k\int_a^b f(x)\, dx \qquad$ Any number k

$\qquad\qquad\qquad\quad\ \ \displaystyle\int_a^b -f(x)\, dx = -\int_a^b f(x)\, dx \qquad k = -1$

4. *Sum and Difference:* $\quad \displaystyle\int_a^b (f(x) \pm g(x))\, dx = \int_a^b f(x)\, dx \pm \int_a^b g(x)\, dx$

5. *Additivity:* $\quad \displaystyle\int_a^b f(x)\, dx + \int_b^c f(x)\, dx = \int_a^c f(x)\, dx$

6. *Max-Min Inequality:* If max f and min f are the maximum and minimum values of f on $[a, b]$, then

$$\min f \cdot (b - a) \le \int_a^b f(x)\, dx \le \max f \cdot (b - a).$$

7. *Domination:* $f(x) \ge g(x)$ on $[a, b] \Rightarrow \displaystyle\int_a^b f(x)\, dx \ge \int_a^b g(x)\, dx$

$\qquad\qquad\quad\ f(x) \ge 0$ on $[a, b] \Rightarrow \displaystyle\int_a^b f(x)\, dx \ge 0 \quad g = 0$

Example 1 USING THE RULES FOR DEFINITE INTEGRALS

Suppose

$$\int_{-1}^{1} f(x)\,dx = 5, \qquad \int_{1}^{4} f(x)\,dx = -2, \qquad \text{and} \qquad \int_{-1}^{1} h(x)\,dx = 7.$$

Find each of the following integrals, if possible.

(a) $\displaystyle\int_{4}^{1} f(x)\,dx$ **(b)** $\displaystyle\int_{-1}^{4} f(x)\,dx$ **(c)** $\displaystyle\int_{-1}^{1} [2f(x) + 3h(x)]\,dx$

(d) $\displaystyle\int_{0}^{1} f(x)\,dx$ **(e)** $\displaystyle\int_{-2}^{2} h(x)\,dx$ **(f)** $\displaystyle\int_{-1}^{4} [f(x) + h(x)]\,dx$

Solution

(a) $\displaystyle\int_{4}^{1} f(x)\,dx = -\int_{1}^{4} f(x)\,dx = -(-2) = 2$

(b) $\displaystyle\int_{-1}^{4} f(x)\,dx = \int_{-1}^{1} f(x)\,dx + \int_{1}^{4} f(x)\,dx = 5 + (-2) = 3$

(c) $\displaystyle\int_{-1}^{1} [2f(x) + 3h(x)]\,dx = 2\int_{-1}^{1} f(x)\,dx + 3\int_{-1}^{1} h(x)\,dx = 2(5) + 3(7) = 31$

(d) Not enough information given. (We cannot assume, for example, that integrating over half the interval would give half the integral!)

(e) Not enough information given. (We have no information about the function h outside the interval $[-1, 1]$.)

(f) Not enough information given (same reason as in (e)).

Example 2 FINDING BOUNDS FOR AN INTEGRAL

Show that the value of $\int_{0}^{1} \sqrt{1 + \cos x}\,dx$ is less than $3/2$.

Solution The Max-Min Inequality for definite integrals (Rule 6) says that $\min f \cdot (b - a)$ is a *lower bound* for the value of $\int_{a}^{b} f(x)\,dx$ and that $\max f \cdot (b - a)$ is an *upper bound*. The maximum value of $\sqrt{1 + \cos x}$ on $[0, 1]$ is $\sqrt{2}$, so

$$\int_{0}^{1} \sqrt{1 + \cos x}\,dx \le \sqrt{2} \cdot (1 - 0) = \sqrt{2}.$$

Since $\int_{0}^{1} \sqrt{1 + \cos x}\,dx$ is bounded above by $\sqrt{2}$ (which is 1.414...), it is less than $3/2$.

Average Value of a Function

The *average* of n numbers is the sum of the numbers divided by n. How would we define the average value of an arbitrary function f over a closed interval $[a, b]$? As there are infinitely many values to consider, adding them and then dividing by infinity is not an option.

Consider, then, what happens if we take a large *sample* of n numbers from regular subintervals of the interval $[a, b]$. One way would be to take some number c_k from each of the n subintervals of length

$$\Delta x = \frac{b - a}{n}.$$

The average of the n sampled values is

$$\frac{f(c_1) + f(c_2) + \cdots + f(c_n)}{n} = \frac{1}{n} \cdot \sum_{k=1}^{n} f(c_k)$$

$$= \frac{\Delta x}{b - a} \sum_{k=1}^{n} f(c_k) \qquad \frac{1}{n} = \frac{\Delta x}{b - a}$$

$$= \frac{1}{b - a} \cdot \sum_{k=1}^{n} f(c_k) \Delta x.$$

Does this last sum look familiar? It is $1/(b - a)$ times a Riemann sum for f on $[a, b]$. That means that when we consider this averaging process as $n \to \infty$, we find it *has a limit*, namely $1/(b - a)$ times the integral of f over $[a, b]$. We are led by this remarkable fact to the following definition.

Definition **Average (Mean) Value**

If f is integrable on $[a, b]$, its **average (mean) value** on $[a, b]$ is

$$av(f) = \frac{1}{b - a} \int_a^b f(x)\, dx.$$

Example 3 APPLYING THE DEFINITION

Find the average value of $f(x) = 4 - x^2$ on $[0, 3]$. Does f actually take on this value at some point in the given interval?

Solution

$$av(f) = \frac{1}{b - a} \int_a^b f(x)\, dx$$

$$= \frac{1}{3 - 0} \int_0^3 (4 - x^2)\, dx$$

$$= \frac{1}{3 - 0} \cdot 3 \qquad\qquad \text{Using NINT}$$

$$= 1$$

The average value of $f(x) = 4 - x^2$ over the interval $[0, 3]$ is 1. The function assumes this value when $4 - x^2 = 1$ or $x = \pm\sqrt{3}$. Since $x = \sqrt{3}$ lies in the interval $[0, 3]$, the function does assume its average value in the given interval (Figure 5.24).

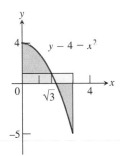

Figure 5.24 The rectangle with base $[0, 3]$ and with height equal to 1 (the average value of the function $f(x) = 4 - x^2$) has area equal to the net area between f and the x-axis from 0 to 3. (Example 3)

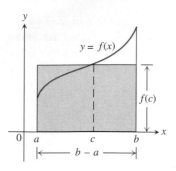

Figure 5.25 The value $f(c)$ in the Mean Value Theorem is, in a sense, the average (or *mean*) height of f on $[a, b]$. When $f \geq 0$, the area of the shaded rectangle

$$f(c)(b - a) = \int_a^b f(x) \, dx,$$

is the area under the graph of f from a to b.

Mean Value Theorem for Definite Integrals

It was no mere coincidence that the function in Example 3 took on its average value at some point in the interval. Look at the graph in Figure 5.25 and imagine rectangles with base $(b - a)$ and heights ranging from the minimum of f (a rectangle too small to give the integral) to the maximum of f (a rectangle too large). Somewhere in between there is a "just right" rectangle, and its topside will intersect the graph of f if f is continuous. The statement that a continuous function on a closed interval *always* assumes its average value at least once in the interval is known as the Mean Value Theorem for Definite Integrals.

Theorem 3 The Mean Value Theorem for Definite Integrals

If f is continuous on $[a, b]$, then at some point c in $[a, b]$,

$$f(c) = \frac{1}{b - a} \int_a^b f(x) \, dx.$$

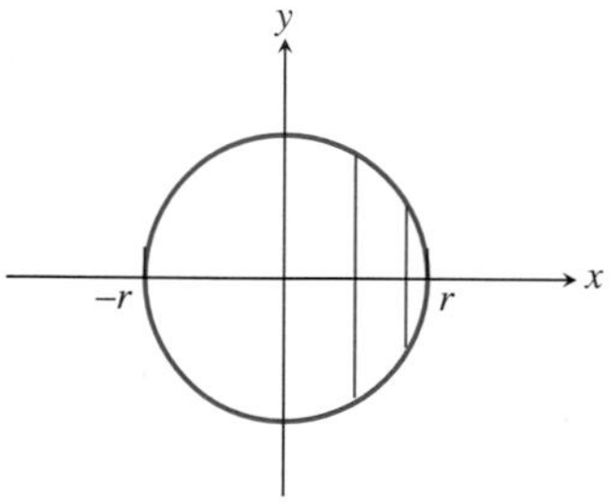

Exploration 1   How Long Is the Average Chord of a Circle?

Suppose we have a circle of radius r centered at the origin. We want to know the average length of the chords perpendicular to the diameter $[-r, r]$ on the x-axis.

1. Show that the length of the chord at x is $2\sqrt{r^2 - x^2}$ (Figure 5.26).

2. Set up an integral expression for the average value of $2\sqrt{r^2 - x^2}$ over the interval $[-r, r]$.

3. Evaluate the integral by identifying its value as an area.

4. So, what is the average length of a chord of a circle of radius r?

5. Explain how we can use the Mean Value Theorem for Definite Integrals (Theorem 3) to show that the function assumes the value in step 4.

Figure 5.26 Chords perpendicular to the diameter $[-r, r]$ in a circle of radius r centered at the origin. (Exploration 1)

Connecting Differential and Integral Calculus

Before we move on to the next section, let us pause for a moment of historical perspective that can help you to appreciate the power of the theorem that you are about to encounter. In Example 3 we used NINT to find the integral, and in Section 5.2, Example 2 we were fortunate that we could use our knowledge of the area of a circle. The area of a circle has been around for a long time, but NINT has not; so how did people evaluate definite integrals when they could not apply some known area formula? For example, in Exploration 1 of the previous section we used the fact that

$$\int_0^\pi \sin x \, dx = 2.$$

Would Newton and Leibniz have known this fact? How?

They did know that *quotients of infinitely small quantities,* as they put it, could be used to get velocity functions from position functions, and that *sums of infinitely thin "rectangle areas"* could be used to get position functions from velocity functions. In some way, then, there had to be a connection between these two seemingly different processes. Newton and Leibniz were able to picture that connection, and it led them to the Fundamental Theorem of Calculus. Can you picture it? Try Exploration 2.

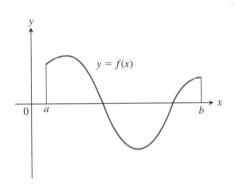

Figure 5.27 The graph of the function in Exploration 2.

Exploration 2 **Finding the Derivative of an Integral**

Suppose we are given the graph of a continuous function f, as in Figure 5.27. *Work in groups of two or three.*

1. Copy the graph of f onto your own paper. Choose any x greater than a in the interval $[a, b]$ and mark it on the x-axis.

2. Using only *vertical line segments,* shade in the region between the graph of f and the x-axis from a to x. (Some shading might be below the x-axis.)

3. Your shaded region represents a definite integral. Explain why this integral can be written as $\int_a^x f(t)\, dt$. (Why don't we write it as $\int_a^x f(x)\, dx$?)

4. Compare your picture with others produced by your group. Notice how your integral (a real number) depends on which x you chose in the interval $[a, b]$. The integral is therefore a *function of x* on $[a, b]$. Call it F.

5. Recall that $F'(x)$ is the limit of $\Delta F/\Delta x$ as Δx gets smaller and smaller. Represent ΔF in your picture by drawing *one more vertical shading segment* to the right of the last one you drew in step 2. ΔF is the (signed) *area* of your vertical segment.

6. Represent Δx in your picture by moving x to beneath your newly-drawn segment. That small change in Δx is the *thickness* of your vertical segment.

7. What is now the *height* of your vertical segment?

8. Can you see why Newton and Leibniz concluded that $F'(x) = f(x)$?

If all went well in Exploration 2, you concluded that the derivative with respect to x of the integral of f from a to x is simply f. Specifically,

$$\frac{d}{dx}\int_a^x f(t)\, dt = f(x).$$

This means that the integral is an *antiderivative* of f, a fact we can exploit in the following way.

If F is any antiderivative of f, then

$$\int_a^x f(t)\, dt = F(x) + C$$

for some constant C. Setting x in this equation equal to a gives

$$\int_a^a f(t)\,dt = F(a) + C$$

$$0 = F(a) + C$$

$$C = -F(a).$$

Putting it all together,

$$\int_a^x f(t)\,dt = F(x) - F(a).$$

The implications of this last equation were enormous for the discoverers of calculus. It meant that they could evaluate the definite integral of f from a to any number x simply by computing $F(x) - F(a)$, where F is *any antiderivative of f.*

Example 4 FINDING AN INTEGRAL USING ANTIDERIVATIVES

Find $\int_0^\pi \sin x\,dx$ using the formula $\int_a^x f(t)\,dt = F(x) - F(a)$.

Solution Since $F(x) = -\cos x$ is an antiderivative of $\sin x$, we have

$$\int_0^\pi \sin x\,dx = -\cos(\pi) - (-\cos(0))$$

$$= -(-1) - (-1)$$

$$= 2.$$

This explains how we obtained the value for Exploration 1 of the previous section.

Quick Review 5.3

In Exercises 1–10, find dy/dx.

1. $y = -\cos x$

2. $y = \sin x$

3. $y = \ln(\sec x)$

4. $y = \ln(\sin x)$

5. $y = \ln(\sec x + \tan x)$

6. $y = x \ln x - x$

7. $y = \dfrac{x^{n+1}}{n+1}$ $\quad(n \neq -1)$

8. $y = \dfrac{1}{2^x + 1}$

9. $y = xe^x$

10. $y = \tan^{-1} x$

Section 5.3 Exercises

The exercises in this section are designed to reinforce your understanding of the definite integral from the algebraic and geometric points of view. For this reason, you should not use the numerical integration capability of your calculator (NINT) except perhaps to support an answer.

1. Suppose that f and g are continuous functions and that

$$\int_1^2 f(x)\,dx = -4, \quad \int_1^5 f(x)\,dx = 6, \quad \int_1^5 g(x)\,dx = 8.$$

Use the rules in Table 5.3 to find each integral.

(a) $\displaystyle\int_2^2 g(x)\,dx$

(b) $\displaystyle\int_5^1 g(x)\,dx$

(c) $\displaystyle\int_1^2 3f(x)\,dx$

(d) $\displaystyle\int_2^5 f(x)\,dx$

(e) $\displaystyle\int_1^5 [f(x) - g(x)]\,dx$

(f) $\displaystyle\int_1^5 [4f(x) - g(x)]\,dx$

2. Suppose that f and h are continuous functions and that

$$\int_1^9 f(x)\,dx = -1, \quad \int_7^9 f(x)\,dx = 5, \quad \int_7^9 h(x)\,dx = 4.$$

Use the rules in Table 5.3 to find each integral.

(a) $\displaystyle\int_1^9 -2f(x)\,dx$ (b) $\displaystyle\int_7^9 [f(x) + h(x)]\,dx$

(c) $\displaystyle\int_7^9 [2f(x) - 3h(x)]\,dx$ (d) $\displaystyle\int_9^1 f(x)\,dx$

(e) $\displaystyle\int_1^7 f(x)\,dx$ (f) $\displaystyle\int_9^7 [h(x) - f(x)]\,dx$

3. Suppose that $\int_1^2 f(x)\,dx = 5$. Find each integral.

(a) $\displaystyle\int_1^2 f(u)\,du$ (b) $\displaystyle\int_1^2 \sqrt{3}\,f(z)\,dz$

(c) $\displaystyle\int_2^1 f(t)\,dt$ (d) $\displaystyle\int_1^2 [-f(x)]\,dx$

4. Suppose that $\int_{-3}^0 g(t)\,dt = \sqrt{2}$. Find each integral.

(a) $\displaystyle\int_0^{-3} g(t)\,dt$ (b) $\displaystyle\int_{-3}^0 g(u)\,du$

(c) $\displaystyle\int_{-3}^0 [-g(x)]\,dx$ (d) $\displaystyle\int_{-3}^0 \frac{g(r)}{\sqrt{2}}\,dr$

5. Suppose that f is continuous and that

$$\int_0^3 f(z)\,dz - 3 \quad \text{and} \quad \int_0^4 f(z)\,dz = 7.$$

Find each integral.

(a) $\displaystyle\int_3^4 f(z)\,dz$ (b) $\displaystyle\int_4^3 f(t)\,dt$

6. Suppose that h is continuous and that

$$\int_{-1}^1 h(r)\,dr = 0 \quad \text{and} \quad \int_{-1}^3 h(r)\,dr = 6.$$

Find each integral.

(a) $\displaystyle\int_1^3 h(r)\,dr$ (b) $\displaystyle-\int_3^1 h(u)\,du$

In Exercises 7–16, evaluate the integral.

7. $\displaystyle\int_3^1 7\,dx$ **8.** $\displaystyle\int_0^2 5x\,dx$

9. $\displaystyle\int_3^5 \frac{x}{8}\,dx$ **10.** $\displaystyle\int_0^2 (2t - 3)\,dt$

11. $\displaystyle\int_0^{\sqrt{2}} (t - \sqrt{2})\,dt$ **12.** $\displaystyle\int_2^1 \left(1 + \frac{z}{2}\right)dz$

13. $\displaystyle\int_{-1}^1 \frac{dx}{1 + x^2}$ **14.** $\displaystyle\int_{-1/2}^{1/2} \frac{dx}{\sqrt{1 - x^2}}$

15. $\displaystyle\int_0^2 e^x\,dx$ **16.** $\displaystyle\int_0^3 \frac{3\,dx}{x + 1}$

In Exercises 17–20, find the total shaded area.

17.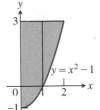

18.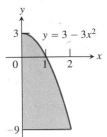

19.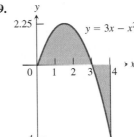

20.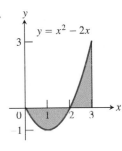

In Exercises 21–24, graph the function over the interval. Then (a) integrate the function over the interval and (b) find the area of the region between the graph and the x-axis.

21. $y = x^2 - 6x + 8$, $[0, 3]$

22. $y = -x^2 + 5x - 4$, $[0, 2]$

23. $y = 2x - x^2$, $[0, 3]$ **24.** $y = x^2 - 4x$, $[0, 5]$

In Exercises 25–28, find the average value of the function on the interval. At what point(s) in the interval does the function assume its average value?

25. $y = x^2 - 1$, $[0, \sqrt{3}]$ **26.** $y = -\dfrac{x^2}{2}$, $[0, 3]$

27. $y = -3x^2 - 1$, $[0, 1]$ **28.** $y = (x - 1)^2$, $[0, 3]$

In Exercises 29–32, find the average value of the function on the interval without integrating, by appealing to the geometry of the region between the graph and the x-axis.

29. $f(x) = \begin{cases} x + 4, & -4 \le x \le -1, \\ -x + 2, & -1 < x \le 2, \end{cases}$ on $[-4, 2]$

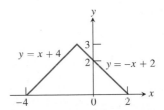

30. $f(t) = 1 - \sqrt{1 - t^2}, \quad [-1, 1]$

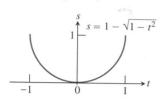

31. $f(t) = \sin t, \quad [0, 2\pi]$

32. $f(\theta) = \tan \theta, \quad \left[-\dfrac{\pi}{4}, \dfrac{\pi}{4} \right]$

In Exercises 33 and 34, *work in groups of two or three.*

33. Use the Max-Min Inequality to find upper and lower bounds for the value of

$$\int_0^1 \frac{1}{1 + x^4} \, dx.$$

34. *(Continuation of Exercise 33)* Use the Max-Min Inequality to find upper and lower bounds for the values of

$$\int_0^{0.5} \frac{1}{1 + x^4} \, dx \quad \text{and} \quad \int_{0.5}^1 \frac{1}{1 + x^4} \, dx.$$

Add these to arrive at an improved estimate for

$$\int_0^1 \frac{1}{1 + x^4} \, dx.$$

35. Show that the value of $\int_0^1 \sin(x^2) \, dx$ cannot possibly be 2.

36. Show that the value of $\int_0^1 \sqrt{x + 8} \, dx$ lies between $2\sqrt{2} \approx 2.8$ and 3.

37. *Integrals of Nonnegative Functions* Use the Max-Min Inequality to show that if f is integrable then

$$f(x) \geq 0 \text{ on } [a, b] \quad \Rightarrow \quad \int_a^b f(x) \, dx \geq 0.$$

38. *Integrals of Nonpositive Functions* Show that if f is integrable then

$$f(x) \leq 0 \text{ on } [a, b] \quad \Rightarrow \quad \int_a^b f(x) \, dx \leq 0.$$

39. Writing to Learn If $av(f)$ really is a typical value of the integrable function $f(x)$ on $[a, b]$, then the number $av(f)$ should have the same integral over $[a, b]$ that f does. Does it? That is, does

$$\int_a^b av(f) \, dx = \int_a^b f(x) \, dx?$$

Give reasons for your answer.

40. Writing to Learn A driver averaged 30 mph on a 150-mile trip and then returned over the same 150 miles at the rate of 50 mph. He figured that his average speed was 40 mph for the entire trip.

(a) What was his total distance traveled?

(b) What was his total time spent for the trip?

(c) What was his average speed for the trip?

(d) Explain the error in the driver's reasoning.

41. Writing to Learn A dam released 1000 m³ of water at 10 m³/min and then released another 1000 m³ at 20 m³/min. What was the average rate at which the water was released? Give reasons for your answer.

42. Use the inequality $\sin x \leq x$, which holds for $x \geq 0$, to find an upper bound for the value of $\int_0^1 \sin x \, dx$.

43. The inequality $\sec x \geq 1 + (x^2/2)$ holds on $(-\pi/2, \pi/2)$. Use it to find a lower bound for the value of $\int_0^1 \sec x \, dx$.

Exploration

44. *Comparing Area Formulas* Consider the region in the first quadrant under the curve $y = (h/b)\, x$ from $x = 0$ to $x = b$ (see figure).

(a) Use a geometry formula to calculate the area of the region.

(b) Find all antiderivatives of y.

(c) Use an antiderivative of y to evaluate $\int_0^b y(x) \, dx$.

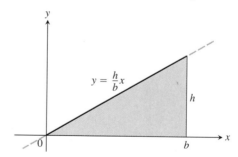

Extending the Ideas

45. *Graphing Calculator Challenge* If $k > 1$, and if the average value of x^k on $[0, k]$ is k, what is k? Check your result with a CAS if you have one available.

46. Show that if $F'(x) = G'(x)$ on $[a, b]$, then

$$F(b) - F(a) = G(b) - G(a).$$

5.4 Fundamental Theorem of Calculus

Fundamental Theorem, Part 1 • Graphing the Function $\int_a^x f(t)\,dt$ • Fundamental Theorem, Part 2 • Area Connection • More Applications

Fundamental Theorem, Part 1

This section presents the discovery by Newton and Leibniz of the astonishing connection between integration and differentiation. This connection started the mathematical development that fueled the scientific revolution for the next 200 years, and is still regarded as the most important computational discovery in the history of mathematics: The Fundamental Theorem of Calculus.

The Fundamental Theorem comes in two parts, both of which were previewed in Exploration 2 of the previous section. The first part says that the definite integral of a continuous function is a differentiable function of its upper limit of integration. Moreover, it tells us what that derivative is. The second part says that the definite integral of a continuous function from a to b can be found from any one of the function's antiderivatives F as the number $F(b) - F(a)$.

Sir Isaac Newton (1642–1727)

Sir Isaac Newton is considered to be one of the most influential mathematicians of all time. Moreover, by the age of 25, he had also made revolutionary advances in optics, physics, and astronomy.

> **Theorem 4 The Fundamental Theorem of Calculus, Part 1**
>
> If f is continuous on $[a, b]$, then the function
>
> $$F(x) = \int_a^x f(t)\,dt$$
>
> has a derivative at every point x in $[a, b]$, and
>
> $$\frac{dF}{dx} = \frac{d}{dx}\int_a^x f(t)\,dt = f(x).$$

Proof The geometric exploration at the end of the previous section contained the idea of the proof, but it glossed over the necessary limit arguments. Here we will be more precise.

Apply the definition of the derivative directly to the function F. That is,

$$\frac{dF}{dx} = \lim_{h\to 0}\frac{F(x+h) - F(x)}{h}$$

$$= \lim_{h\to 0}\frac{\displaystyle\int_a^{x+h} f(t)\,dt - \int_a^x f(t)\,dt}{h}$$

$$= \lim_{h\to 0}\frac{\displaystyle\int_x^{x+h} f(t)\,dt}{h} \qquad \text{Rules for integrals, Section 5.3}$$

$$= \lim_{h\to 0}\left[\frac{1}{h}\int_x^{x+h} f(t)\,dt\right].$$

The expression in brackets in the last line is the average value of f from x to $x + h$. We know from the Mean Value Theorem for Definite Integrals (Theorem 3, Section 5.3) that f, being continuous, takes on its average value at least once in the interval; that is,

$$\frac{1}{h} \int_x^{x+h} f(t)\, dt = f(c) \qquad \text{for some } c \text{ between } x \text{ and } x + h.$$

We can therefore continue our proof, letting $(1/h) \int_x^{x+h} f(t)\, dt = f(c)$,

$$\frac{dF}{dx} = \lim_{h \to 0} \frac{1}{h} \int_x^{x+h} f(t)\, dt$$

$$= \lim_{h \to 0} f(c), \qquad \text{where } c \text{ lies between } x \text{ and } x + h.$$

What happens to c as h goes to zero? As $x + h$ gets closer to x, it carries c along with it like a bead on a wire, forcing c to approach x. Since f is continuous, this means that $f(c)$ approaches $f(x)$:

$$\lim_{h \to 0} f(c) = f(x).$$

Putting it all together,

$$\frac{dF}{dx} = \lim_{h \to 0} \frac{F(x + h) - F(x)}{h} \qquad \text{Definition of derivatives}$$

$$= \lim_{h \to 0} \frac{\int_x^{x+h} f(t)\, dt}{h} \qquad \text{Rules for integrals}$$

$$= \lim_{h \to 0} f(c) \qquad \text{for some } c \text{ between } x \text{ and } x + h.$$

$$= f(x). \qquad \text{Because } f \text{ is continuous}$$

This concludes the proof. ∎

It is difficult to overestimate the power of the equation

$$\frac{d}{dx} \int_a^x f(t)\, dt = f(x). \tag{1}$$

It says that every continuous function f is the derivative of some other function, namely $\int_a^x f(t)\, dt$. It says that every continuous function has an antiderivative. And it says that the processes of integration and differentiation are inverses of one another. If any equation deserves to be called the Fundamental Theorem of Calculus, this equation is surely the one.

Example 1 APPLYING THE FUNDAMENTAL THEOREM

Find

$$\frac{d}{dx} \int_{-\pi}^x \cos t\, dt \qquad \text{and} \qquad \frac{d}{dx} \int_0^x \frac{1}{1 + t^2}\, dt$$

by using the Fundamental Theorem.

Solution

$$\frac{d}{dx}\int_{-\pi}^{x}\cos t\,dt = \cos x \qquad \text{Eq. 1 with } f(t) = \cos t$$

$$\frac{d}{dx}\int_{0}^{x}\frac{1}{1+t^2}\,dt = \frac{1}{1+x^2}. \qquad \text{Eq. 1 with } f(t) = \frac{1}{1+t^2}$$

Example 2 THE FUNDAMENTAL THEOREM WITH THE CHAIN RULE

Find dy/dx if $y = \int_{1}^{x^2}\cos t\,dt$.

Solution The upper limit of integration is not x but x^2. This makes y a composite of

$$y = \int_{1}^{u}\cos t\,dt \quad \text{and} \quad u = x^2.$$

We must therefore apply the Chain Rule when finding dy/dx.

$$\frac{dy}{dx} = \frac{dy}{du}\cdot\frac{du}{dx}$$

$$= \left(\frac{d}{du}\int_{1}^{u}\cos t\,dt\right)\cdot\frac{du}{dx}$$

$$= \cos u \cdot \frac{du}{dx}$$

$$= \cos(x^2)\cdot 2x$$

$$= 2x\cos x^2$$

Example 3 VARIABLE LOWER LIMITS OF INTEGRATION

Find dy/dx.

(a) $y = \int_{x}^{5} 3t\sin t\,dt$ **(b)** $y = \int_{2x}^{x^2}\frac{1}{2+e^t}\,dt$

Solution The rules for integrals set these up for the Fundamental Theorem.

(a) $\dfrac{d}{dx}\displaystyle\int_{x}^{5} 3t\sin t\,dt = \dfrac{d}{dx}\left(-\displaystyle\int_{5}^{x} 3t\sin t\,dt\right)$

$$= -\frac{d}{dx}\int_{5}^{x} 3t\sin t\,dt$$

$$= -3x\sin x$$

(b) $\dfrac{d}{dx}\displaystyle\int_{2x}^{x^2}\dfrac{1}{2+e^t}\,dt = \dfrac{d}{dx}\left(\displaystyle\int_{0}^{x^2}\dfrac{1}{2+e^t}\,dt - \displaystyle\int_{0}^{2x}\dfrac{1}{2+e^t}\,dt\right)$

$$= \frac{1}{2+e^{x^2}}\frac{d}{dx}(x^2) - \frac{1}{2+e^{2x}}\frac{d}{dx}(2x) \qquad \text{Chain Rule}$$

$$= \frac{1}{2+e^{x^2}}\cdot 2x - \frac{1}{2+e^{2x}}\cdot 2$$

$$= \frac{2x}{2+e^{x^2}} - \frac{2}{2+e^{2x}}$$

Example 4 CONSTRUCTING A FUNCTION WITH A GIVEN DERIVATIVE AND VALUE

Find a function $y = f(x)$ with derivative

$$\frac{dy}{dx} = \tan x$$

that satisfies the condition $f(3) = 5$.

Solution The Fundamental Theorem makes it easy to construct a function with derivative $\tan x$:

$$y = \int_3^x \tan t \, dt.$$

Since $y(3) = 0$, we have only to add 5 to this function to construct one with derivative $\tan x$ whose value at $x = 3$ is 5:

$$f(x) = \int_3^x \tan t \, dt + 5.$$

Although the solution to the problem in Example 4 satisfies the two required conditions, you might question whether it is in a useful form. Not many years ago, this form might have posed a computation problem. Indeed, for such problems much effort has been expended over the centuries trying to find solutions that do not involve integrals. We will see some in Chapter 6, where we will learn (for example) how to write the solution in Example 4 as

$$y = \ln \left| \frac{\cos 3}{\cos x} \right| + 5.$$

However, now that computers and calculators are capable of evaluating integrals, the form given in Example 4 is not only useful, but in some ways preferable. It is certainly easier to find and is always available.

Graphing the Function $\int_a^x f(t) \, dt$

Consider for a moment the two forms of the function we have just been discussing,

$$F(x) = \int_3^x \tan t \, dt + 5 \quad \text{and} \quad F(x) = \ln \left| \frac{\cos 3}{\cos x} \right| + 5.$$

With which expression is it easier to evaluate, say, $F(4)$? From the time of Newton almost to the present, there has been no contest: the expression on the right. At least it provides something to compute, and there have always been tables or slide rules or calculators to facilitate that computation. The expression on the left involved at best a tedious summing process and almost certainly an increased opportunity for error.

Today we can find $F(4)$ from either expression on the same machine. The choice is between NINT $(\tan x, x, 3, 4) + 5$ and $\ln (\text{abs}(\cos(3)/\cos(4))) + 5$. Both calculations give 5.415135083 in approximately the same amount of time.

We can even use NINT to graph the function. This modest technology feat would have absolutely dazzled the mathematicians of the 18th and 19th centuries, who knew how the solutions of differential equations, such as $dy/dx = \tan x$, could be written as integrals, but for whom integrals were of no practical use computationally unless they could be written in exact form. Since so few integrals could, in fact, be written in exact form, NINT would have spared generations of scientists much frustration.

Nevertheless, one must not proceed blindly into the world of calculator computation. Exploration 1 will demonstrate the need for caution.

Exploration 1 **Graphing NINT *f***

Let us use NINT to attempt to graph the function we just discussed,

$$F(x) = \int_3^x \tan t \, dt + 5.$$

1. Graph the function $y = F(x)$ in the window $[-10, 10]$ by $[-10, 10]$. You will probably wait a long time and see no graph. Break out of the graphing program if necessary.

2. Recall that the graph of the function $y = \tan x$ has vertical asymptotes. Where do they occur on the interval $[-10, 10]$?

3. When attempting to graph the function $F(x) = \int_3^x \tan t \, dt + 5$ on the interval $[-10, 10]$, your grapher begins by trying to find $F(-10)$. Explain why this might cause a problem for your calculator.

4. Set your viewing window so that your calculator graphs only over the domain of the continuous branch of the tangent function that contains the point $(3, \tan 3)$.

5. What is the domain in step 4? Is it an open interval or a closed interval?

6. What is the domain of $F(x)$? Is it an open interval or a closed interval?

7. Your calculator graphs over the closed interval $[x_{\min}, x_{\max}]$. Find a viewing window that will give you a good look at the graph of F and produce the graph on your calculator.

8. Describe the graph of F.

Graphing NINT *f*

Some graphers can graph the numerical integral $y = \text{NINT} (f(x), x, a, x)$ directly as a function of x. Others will require a tool-box program such as the one called NINT-GRAF provided in the *Technology Resource Manual*.

You have probably noticed that your grapher moves slowly when graphing NINT. This is because it must compute each value as a limit of sums—comparatively slow work even for a microprocessor. Here are some ways to speed up the process:

1. Change the *tolerance* on your grapher. The smaller the tolerance, the more accurate the calculator will try to be when finding the limiting value of each sum (and the longer it will take to do so). The default value is usually quite small (like 0.00001), but a value as large as 1 can be used for graphing in a typical viewing window.

2. Change the *x-resolution*. The default resolution is 1, which means that the grapher will compute a function value for every vertical column of pixels. At resolution 2 it computes only every second value, and so on. With higher resolutions, some graph smoothness is sacrificed for speed.

3. Switch to parametric mode. To graph $y = \text{NINT} (f(x), x, a, x)$ in parametric mode, let $x(t) = t$ and let $y(t) = \text{NINT} (f(t), t, a, t)$. You can then control the speed of the grapher by changing the *t*-step. (Choosing a bigger *t*-step has the same effect as choosing a larger *x*-resolution.)

Exploration 2 **The Effect of Changing *a* in $\int_a^x f(t)\, dt$**

The first part of the Fundamental Theorem of Calculus asserts that the derivative of $\int_a^x f(t)\, dt$ is $f(x)$, regardless of the value of a.

1. Graph NDER (NINT $(x^2, x, 0, x)$).
2. Graph NDER (NINT $(x^2, x, 5, x)$).
3. Without graphing, tell what the x-intercept of NINT $(x^2, x, 0, x)$ is. Explain.
4. Without graphing, tell what the x-intercept of NINT $(x^2, x, 5, x)$ is. Explain.
5. How does changing a affect the graph of $y = (d/dx)\int_a^x f(t)\, dt$?
6. How does changing a affect the graph of $y = \int_a^x f(t)\, dt$?

Fundamental Theorem, Part 2

The second part of the Fundamental Theorem of Calculus shows how to evaluate definite integrals directly from antiderivatives.

Theorem 4 (continued) **The Fundamental Theorem of Calculus, Part 2**

If f is continuous at every point of $[a, b]$, and if F is any antiderivative of f on $[a, b]$, then

$$\int_a^b f(x)\, dx = F(b) - F(a).$$

This part of the Fundamental Theorem is also called the **Integral Evaluation Theorem.**

Proof Part 1 of the Fundamental Theorem tells us that an antiderivative of f exists, namely

$$G(x) = \int_a^x f(t)\, dt.$$

Thus, if F is *any* antiderivative of f, then $F(x) = G(x) + C$ for some constant C (by Corollary 3 of the Mean Value Theorem for Derivatives, Section 4.2).

Evaluating $F(b) - F(a)$, we have

$$F(b) - F(a) = [G(b) + C] - [G(a) + C]$$
$$= G(b) - G(a)$$
$$= \int_a^b f(t)\, dt - \int_a^a f(t)\, dt$$
$$= \int_a^b f(t)\, dt - 0$$
$$= \int_a^b f(t)\, dt. \qquad \blacksquare$$

At the risk of repeating ourselves: It is difficult to overestimate the power of the simple equation

$$\int_a^b f(x)\, dx = F(b) - F(a).$$

It says that any definite integral of any continuous function f can be calculated without taking limits, without calculating Riemann sums, and often without effort—so long as an antiderivative of f can be found. If you can imagine what it was like before this theorem (and before computing machines), when approximations by tedious sums were the only alternative for solving many real-world problems, then you can imagine what a miracle calculus was thought to be. If any equation deserves to be called the Fundamental Theorem of Calculus, this equation is surely the (second) one.

Example 5 EVALUATING AN INTEGRAL

Evaluate $\int_{-1}^3 (x^3 + 1)\, dx$ using an antiderivative.

Solution

Solve Analytically

A simple antiderivative of $x^3 + 1$ is $(x^4/4) + x$. Therefore,

$$\int_{-1}^3 \left(x^3 + 1\right) dx = \left[\frac{x^4}{4} + x\right]_{-1}^3$$

$$= \left(\frac{81}{4} + 3\right) - \left(\frac{1}{4} - 1\right)$$

$$= 24.$$

Support Numerically

NINT $(x^3 + 1, x, -1, 3) = 24.$

Area Connection

In Section 5.2 we saw that the definite integral could be interpreted as the net area between the graph of a function and the x-axis. We can therefore compute areas using antiderivatives, but we must again be careful to distinguish net area (in which area below the x-axis is counted as negative) from total area. The unmodified word "area" will be taken to mean *total area*.

Example 6 FINDING AREA USING ANTIDERIVATIVES

Find the area of the region between the curve $y = 4 - x^2$, $0 \le x \le 3$, and the x-axis.

Solution The curve crosses the x-axis at $x = 2$, partitioning the interval $[0, 3]$ into two subintervals, on each of which $f(x) = 4 - x^2$ will not change sign.

Integral Evaluation Notation

The usual notation for $F(b) \quad F(a)$ is

$$F(x)\Big]_a^b \quad \text{or} \quad \left[F(x)\right]_a^b,$$

depending on whether F has one or more terms. This notation provides a compact "recipe" for the evaluation, allowing us to show the antiderivative in an intermediate step.

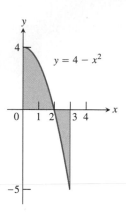

Figure 5.28 The function $f(x) = 4 - x^2$ changes sign only at $x = 2$ on the interval $[0, 3]$. (Example 6)

We can see from the graph (Figure 5.28) that $f(x) > 0$ on $[0, 2)$ and $f(x) < 0$ on $(2, 3]$.

Over $[0, 2]$: $\displaystyle\int_0^2 (4 - x^2)\, dx = \left[4x - \frac{x^3}{3}\right]_0^2 = \frac{16}{3}.$

Over $[2, 3]$: $\displaystyle\int_2^3 (4 - x^2)\, dx = \left[4x - \frac{x^3}{3}\right]_2^3 = -\frac{7}{3}.$

The area of the region is $\left|\dfrac{16}{3}\right| + \left|-\dfrac{7}{3}\right| = \dfrac{23}{3}.$

How to Find Total Area Analytically

To find the area between the graph of $y = f(x)$ and the x-axis over the interval $[a, b]$ analytically,

1. partition $[a, b]$ with the zeros of f,
2. integrate f over each subinterval,
3. add the absolute values of the integrals.

We can find area numerically by using NINT to integrate the *absolute value* of the function over the given interval. There is no need to partition. By taking absolute values, we automatically reflect the negative portions of the graph across the x-axis to count all area as positive (Figure 5.29).

Example 7 FINDING AREA USING NINT

Find the area of the region between the curve $y = x \cos 2x$ and the x-axis over the interval $-3 \le x \le 3$ (Figure 5.29).

Solution Rounded to two decimal places, we have

$$\text{NINT}\,(|x \cos 2x|, x, -3, 3) = 5.43.$$

How to Find Total Area Numerically

To find the area between the graph of $y = f(x)$ and the x-axis over the interval $[a, b]$ numerically, evaluate

$$\text{NINT}\,(|f(x)|, x, a, b).$$

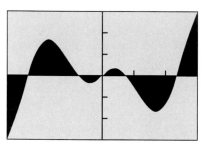

[–3, 3] by [–3, 3]

(a)

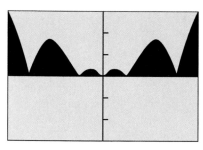

[–3, 3] by [–3, 3]

(b)

Figure 5.29 The graphs of (a) $y = x \cos 2x$ and (b) $y = |x \cos 2x|$ over $[-3, 3]$. The shaded regions have the same area.

More Applications

We close the section with two applications from the world of economics.

Example 8 DETERMINING COST FROM MARGINAL COST

The fixed cost of starting a manufacturing run and producing the first 10 units is $200. After that, the marginal cost at x units output is

$$\frac{dc}{dx} = \frac{1000}{x^2}.$$

Find the total cost of producing the first 100 units.

Solution If $c(x)$ is the cost of x units, then

$$\underbrace{c(100)}_{\substack{\text{Cost of} \\ \text{100 units}}} = \underbrace{200}_{\substack{\text{Startup and} \\ \text{first 10 units}}} + \underbrace{c(100) - c(10)}_{\substack{\text{Cost of units} \\ \text{11 through 100}}}$$

$$= 200 + \int_{10}^{100} \frac{dc}{dx}\, dx \qquad \text{Since } \int_{10}^{100}\frac{dc}{dx}dx = c(100) - c(10)$$

$$= 200 + \int_{10}^{100} \frac{1000}{x^2}\, dx$$

$$- 200 + 1000\left[-\frac{1}{x}\right]_{10}^{100}$$

$$= 290.$$

The total cost of producing the first 100 units is $290.

The notion of a function's average value is used in economics to study things like average daily inventory. An **inventory function** $I(x)$ gives the number of radios, shoes, tires, or whatever product a firm has on hand on day x. The average value of I over a time period $a \le x \le b$ is the firm's average daily inventory for the period.

Definition Average Daily Inventory

If $I(x)$ is the number of items on hand on day x, the **average daily inventory** of the items for the period $a \le x \le b$ is

$$av(I) - \frac{1}{b - a}\int_a^b I(x)\, dx.$$

If h is the dollar cost of holding one item per day, the **average daily holding cost** for the period $a \le x \le b$ is $av(I) \cdot h$.

Example 9 FINDING AVERAGE DAILY INVENTORY

Suppose a wholesaler receives a shipment of 1200 cases of boxes of chocolates every 30 days. The chocolate is sold to retailers at a steady rate, and x days after the shipment arrives, the inventory of cases still on hand is $I(x) = 1200 - 40x$. Find the average daily inventory. Also, find the average daily holding cost if the cost of holding one case is 3 cents a day.

Solution The average daily inventory is

$$av(I) = \frac{1}{30 - 0}\int_0^{30}(1200 - 40x)\, dx$$

$$= \frac{1}{30}\left[1200x - 20x^2\right]_0^{30}$$

$$= 600.$$

The average daily holding cost, then, is $(600)(0.03) = 18$ dollars a day.

Figure 5.30 Maintaining a steady inventory to ship to the wholesaler. (Example 9)

Quick Review 5.4

In Exercises 1–10, find dy/dx.

1. $y = \sin(x^2)$

2. $y = (\sin x)^2$

3. $y = \sec^2 x - \tan^2 x$

4. $y = \ln(3x) - \ln(7x)$

5. $y = 2^x$

6. $y = \sqrt{x}$

7. $y = \dfrac{\cos x}{x}$

8. $y = \sin t$ and $x = \cos t$

9. $xy + x = y^2$

10. $dx/dy = 3x$

Section 5.4 Exercises

In Exercises 1–14, evaluate each integral using Part 2 of the Fundamental Theorem. Support your answer with NINT if you are unsure.

1. $\displaystyle\int_{1/2}^{3} \left(2 - \frac{1}{x}\right) dx$

2. $\displaystyle\int_{2}^{-1} 3^x \, dx$

3. $\displaystyle\int_{0}^{1} (x^2 + \sqrt{x}) \, dx$

4. $\displaystyle\int_{0}^{5} x^{3/2} \, dx$

5. $\displaystyle\int_{1}^{32} x^{-6/5} \, dx$

6. $\displaystyle\int_{-2}^{-1} \frac{2}{x^2} \, dx$

7. $\displaystyle\int_{0}^{\pi} \sin x \, dx$

8. $\displaystyle\int_{0}^{\pi} (1 + \cos x) \, dx$

9. $\displaystyle\int_{0}^{\pi/3} 2 \sec^2 \theta \, d\theta$

10. $\displaystyle\int_{\pi/6}^{5\pi/6} \csc^2 \theta \, d\theta$

11. $\displaystyle\int_{\pi/4}^{3\pi/4} \csc x \cot x \, dx$

12. $\displaystyle\int_{0}^{\pi/3} 4 \sec x \tan x \, dx$

13. $\displaystyle\int_{-1}^{1} (r + 1)^2 \, dr$

14. $\displaystyle\int_{0}^{4} \frac{1 - \sqrt{u}}{\sqrt{u}} \, du$

In Exercises 15–18, find the total area of the region between the curve and the x-axis.

15. $y = 2 - x, \quad 0 \le x \le 3$

16. $y = 3x^2 - 3, \quad -2 \le x \le 2$

17. $y = x^3 - 3x^2 + 2x, \quad 0 \le x \le 2$

18. $y = x^3 - 4x, \quad -2 \le x \le 2$

In Exercises 19–24, answer the following questions about the integral.

    (a) Writing to Learn Can the Fundamental Theorem of Calculus, Part 2 (Theorem 4) be used to evaluate the integral? Explain.

    (b) Does the integral have a value? If so, what is it? Explain.

19. $\displaystyle\int_{-2}^{3} \frac{x^2 - 1}{x + 1} \, dx$

20. $\displaystyle\int_{0}^{5} \frac{9 - x^2}{3x - 9} \, dx$

21. $\displaystyle\int_{0}^{2\pi} \tan x \, dx$

22. $\displaystyle\int_{0}^{2} \frac{x + 1}{x^2 - 1} \, dx$

23. $\displaystyle\int_{-1}^{2} \frac{\sin x}{x} \, dx$

24. $\displaystyle\int_{-2}^{3} \frac{1 - \cos x}{x^2} \, dx$

In Exercises 25–28, find the area of the shaded region.

25.

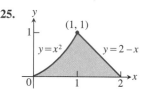

26.

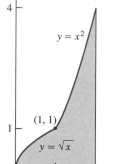

27.

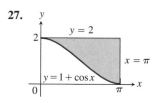

28.

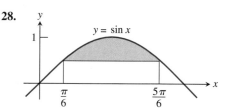

In Exercises 29–34, use NINT to solve the problem.

29. Evaluate $\displaystyle\int_{0}^{10} \frac{1}{3 + 2 \sin x} \, dx$.

30. Evaluate $\displaystyle\int_{-0.8}^{0.8} \frac{2x^4 - 1}{x^4 - 1} \, dx$.

31. Find the average value of $\sqrt{\cos x}$ on the interval $[-1, 1]$.

32. Find the area of the semielliptical region between the x-axis and the graph of $y = \sqrt{8 - 2x^2}$.

33. For what value of x does $\displaystyle\int_{0}^{x} e^{-t^2} \, dt = 0.6$?

34. Find the area of the region in the first quadrant enclosed by the coordinate axes and the graph of $x^3 + y^3 = 1$.

In Exercises 35 and 36, find K so that

$$\int_a^x f(t)\,dt + K = \int_b^x f(t)\,dt.$$

35. $f(x) = x^2 - 3x + 1;\quad a = -1;\quad b = 2$

36. $f(x) = \sin^2 x;\quad a = 0;\quad b = 2$

In Exercises 37–42, find dy/dx.

37. $y = \displaystyle\int_0^x \sqrt{1 + t^2}\,dt$

38. $y = \displaystyle\int_x^1 \frac{1}{t}\,dt, \quad x > 0$

39. $y = \displaystyle\int_0^{\sqrt{x}} \sin(t^2)\,dt$

40. $y = \displaystyle\int_0^{2x} \cos t\,dt$

41. $y = \displaystyle\int_{x^2}^{x^3} \cos(2t)\,dt$

42. $y = \displaystyle\int_{\sin x}^{\cos x} t^2\,dt$

In Exercises 43–46, choose which of the following functions has the given derivative and numerical value. Confirm your answer.

(a) $y = \displaystyle\int_1^x e^{-t^2}\,dt - 3$

(b) $y = \displaystyle\int_0^x \sec t\,dt + 4$

(c) $y = \displaystyle\int_{-1}^x \sec t\,dt + 4$

(d) $y = \displaystyle\int_\pi^x e^{-t^2}\,dt - 3$

43. $\dfrac{dy}{dx} = e^{-x^2}, \quad y(\pi) = -3$

44. $\dfrac{dy}{dx} = \sec x, \quad y(-1) = 4$

45. $\dfrac{dy}{dx} = \sec x, \quad y(0) = 4$

46. $\dfrac{dy}{dx} = e^{-x^2}, \quad y(1) = -3$

47. *Identifying a Zero* For what value of x is $\int_a^x f(t)\,dt$ sure to be zero?

48. Suppose $\int_1^x f(t)\,dt = x^2 - 2x + 1$. Find $f(x)$.

49. *Linearization* Find the linearization of

$$f(x) = 2 + \int_0^x \frac{10}{1+t}\,dt \quad \text{at} \quad x = 0.$$

50. Find $f(4)$ if $\int_0^x f(t)\,dt = x\cos \pi x$.

51. *Finding Area* Show that if k is a positive constant, then the area between the x-axis and one arch of the curve $y = \sin kx$ is always $2/k$.

52. *Archimedes' Area Formula for Parabolas* Archimedes (287–212 B.C.), inventor, military engineer, physicist, and the greatest mathematician of classical times, discovered that the area under a parabolic arch like the one shown here is always two-thirds the base times the height.

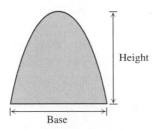

(a) Find the area under the parabolic arch

$$y = 6 - x - x^2, \quad -3 \le x \le 2.$$

(b) Find the height of the arch.

(c) Show that the area is two-thirds the base times the height.

In Exercises 53–55, *work in groups of two or three.*

53. Let

$$H(x) = \int_0^x f(t)\,dt,$$

where f is the continuous function with domain $[0, 12]$ graphed here.

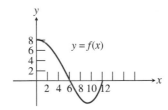

(a) Find $H(0)$.

(b) On what interval is H increasing? Explain.

(c) On what interval is the graph of H concave up? Explain.

(d) Is $H(12)$ positive or negative? Explain.

(e) Where does H achieve its maximum value? Explain.

(f) Where does H achieve its minimum value? Explain.

In Exercises 54 and 55, f is the differentiable function whose graph is shown in the figure. The position at time t (sec) of a particle moving along a coordinate axis is

$$s = \int_0^t f(x)\,dx$$

meters. Use the graph to answer the questions. Give reasons for your answers.

54.

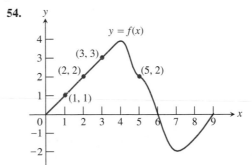

(a) What is the particle's velocity at time $t = 5$?

(b) Is the acceleration of the particle at time $t = 5$ positive or negative?

(c) What is the particle's position at time $t = 3$?

(d) At what time during the first 9 sec does s have its largest value?

(e) Approximately when is the acceleration zero?

(f) When is the particle moving toward the origin? away from the origin?

(g) On which side of the origin does the particle lie at time $t = 9$?

55.

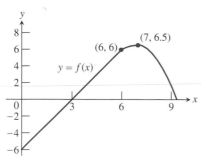

(a) What is the particle's velocity at time $t = 3$?

(b) Is the acceleration of the particle at time $t = 3$ positive or negative?

(c) What is the particle's position at time $t = 3$?

(d) When does the particle pass through the origin?

(e) Approximately when is the acceleration zero?

(f) When is the particle moving toward the origin? away from the origin?

(g) On which side of the origin does the particle lie at time $t = 9$?

Exploration

56. *The Sine Integral Function*   The sine integral function

$$\text{Si}\,(x) = \int_0^x \frac{\sin t}{t}\,dt$$

is one of the many useful functions in engineering that are defined as integrals. Although the notation does not show it, the function being integrated is

$$f(t) = \begin{cases} \dfrac{\sin t}{t}, & t \neq 0 \\ 1, & t = 0, \end{cases}$$

the continuous extension of $(\sin t)/t$ to the origin.

(a) Show that $\text{Si}\,(x)$ is an odd function of x.

(b) What is the value of $\text{Si}\,(0)$?

(c) Find the values of x at which $\text{Si}\,(x)$ has a local extreme value.

(d) Use NINT to graph $\text{Si}\,(x)$.   ∎

57. *Cost from Marginal Cost*   The marginal cost of printing a poster when x posters have been printed is

$$\frac{dc}{dx} = \frac{1}{2\sqrt{x}}$$

dollars. Find

(a) $c(100) - c(1)$, the cost of printing posters 2 to 100.

(b) $c(400) - c(100)$, the cost of printing posters 101 to 400.

58. *Revenue from Marginal Revenue*   Suppose that a company's marginal revenue from the manufacture and sale of eggbeaters is

$$\frac{dr}{dx} = 2 - \frac{2}{(x+1)^2},$$

where r is measured in thousands of dollars and x in thousands of units. How much money should the company expect from a production run of $x = 3$ thousand eggbeaters? To find out, integrate the marginal revenue from $x = 0$ to $x = 3$.

59. *Average Daily Holding Cost*   Solon Container receives 450 drums of plastic pellets every 30 days. The inventory function (drums on hand as a function of days) is $I(x) = 450 - x^2/2$.

(a) Find the average daily inventory.

(b) If the holding cost for one drum is \$0.02 per day, find the average daily holding cost.

60. Suppose that f has a negative derivative for all values of x and that $f(1) = 0$. Which of the following statements must be true of the function

$$h(x) = \int_0^x f(t)\,dt?$$

Give reasons for your answers.

(a) h is a twice-differentiable function of x.

(b) h and dh/dx are both continuous.

(c) The graph of h has a horizontal tangent at $x = 1$.

(d) h has a local maximum at $x = 1$.

(e) h has a local minimum at $x = 1$.

(f) The graph of h has an inflection point at $x = 1$.

(g) The graph of dh/dx crosses the x-axis at $x = 1$.

Extending the Ideas

61. Writing to Learn   If f is an odd continuous function, give a graphical argument to explain why $\int_0^x f(t)\,dt$ is even.

62. Writing to Learn   If f is an even continuous function, give a graphical argument to explain why $\int_0^x f(t)\,dt$ is odd.

63. Writing to Learn   Explain why we can conclude from Exercises 61 and 62 that every even continuous function is the derivative of an odd continuous function and vice versa.

64. Give a convincing argument that the equation

$$\int_0^x \frac{\sin t}{t}\,dt = 1$$

has exactly one solution. Give its approximate value.

5.5 Trapezoidal Rule

Trapezoidal Approximations • Other Algorithms • Error Analysis

Trapezoidal Approximations

You probably noticed in Section 5.1 that MRAM was generally more efficient in approximating integrals than either LRAM or RRAM, even though all three RAM approximations approached the same limit. All three RAM approximations, however, depend on the areas of rectangles. Are there other geometric shapes with known areas that can do the job more efficiently? The answer is yes, and the most obvious one is the trapezoid.

As shown in Figure 5.31, if $[a, b]$ is partitioned into n subintervals of equal length $h = (b - a)/n$, the graph of f on $[a, b]$ can be approximated by a straight line segment over each subinterval.

The region between the curve and the x-axis is then approximated by the trapezoids, the area of each trapezoid being the length of its horizontal "altitude" times the average of its two vertical "bases." That is,

$$\int_a^b f(x)\,dx \approx h \cdot \frac{y_0 + y_1}{2} + h \cdot \frac{y_1 + y_2}{2} + \cdots + h \cdot \frac{y_{n-1} + y_n}{2}$$

$$= h\left(\frac{y_0}{2} + y_1 + y_2 + \cdots + y_{n-1} + \frac{y_n}{2}\right)$$

$$= \frac{h}{2}\left(y_0 + 2y_1 + 2y_2 + \cdots + 2y_{n-1} + y_n\right),$$

where

$$y_0 = f(a), \quad y_1 = f(x_1), \quad \cdots, \quad y_{n-1} = f(x_{n-1}), \quad y_n = f(b).$$

This is algebraically equivalent to finding the numerical average of LRAM and RRAM; indeed, that is how some texts define the Trapezoidal Rule.

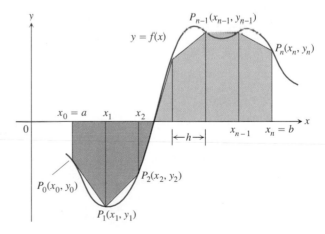

Figure 5.31 The trapezoidal rule approximates short stretches of the curve $y = f(x)$ with line segments. To approximate the integral of f from a to b, we add the "signed" areas of the trapezoids made by joining the ends of the segments to the x axis.

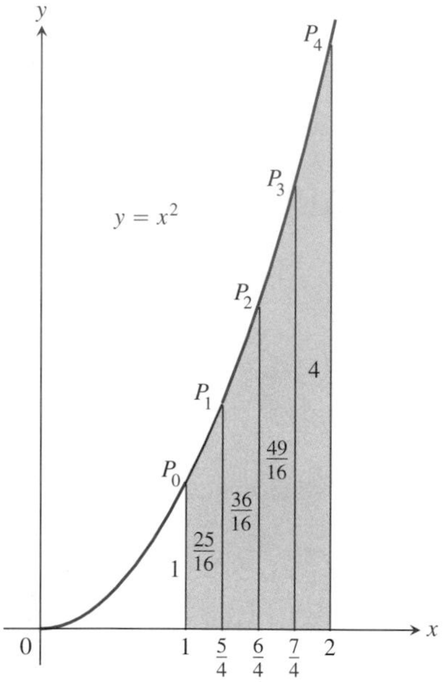

Figure 5.32 The trapezoidal approximation of the area under the graph of $y = x^2$ from $x = 1$ to $x = 2$ is a slight overestimate. (Example 1)

Table 5.4

x	$y = x^2$
1	1
$\dfrac{5}{4}$	$\dfrac{25}{16}$
$\dfrac{6}{4}$	$\dfrac{36}{16}$
$\dfrac{7}{4}$	$\dfrac{49}{16}$
2	4

The Trapezoidal Rule

To approximate $\int_a^b f(x)\, dx$, use

$$T = \frac{h}{2}\left(y_0 + 2y_1 + 2y_2 + \cdots + 2y_{n-1} + y_n\right),$$

where $[a, b]$ is partitioned into n subintervals of equal length $h = (b - a)/n$.

Equivalently,

$$T = \frac{\text{LRAM}_n + \text{RRAM}_n}{2},$$

where LRAM_n and RRAM_n are the Riemann sums using the left and right endpoints, respectively, for f for the partition.

Example 1   APPLYING THE TRAPEZOIDAL RULE

Use the Trapezoidal Rule with $n = 4$ to estimate $\int_1^2 x^2\, dx$. Compare the estimate with the value of $\text{NINT}\,(x^2, x, 1, 2)$ and with the exact value.

Solution Partition $[1, 2]$ into four subintervals of equal length (Figure 5.32). Then evaluate $y = x^2$ at each partition point (Table 5.4).

Using these y values, $n = 4$, and $h = (2 - 1)/4 = 1/4$ in the Trapezoidal Rule, we have

$$T = \frac{h}{2}\left(y_0 + 2y_1 + 2y_2 + 2y_3 + y_4\right)$$

$$= \frac{1}{8}\left(1 + 2\left(\frac{25}{16}\right) + 2\left(\frac{36}{16}\right) + 2\left(\frac{49}{16}\right) + 4\right)$$

$$= \frac{75}{32} = 2.34375.$$

The value of $\text{NINT}\,(x^2, x, 1, 2)$ is 2.333333333.

The exact value of the integral is

$$\int_1^2 x^2\, dx = \frac{x^3}{3}\Bigg]_1^2 = \frac{8}{3} - \frac{1}{3} = \frac{7}{3}.$$

The T approximation overestimates the integral by about half a percent of its true value of 7/3. The percentage error is $(2.34375 - 7/3)/(7/3) \approx 0.00446$.

   We could have predicted that the Trapezoidal Rule would overestimate the integral in Example 1 by considering the geometry of the graph in Figure 5.32. Since the parabola is concave *up,* the approximating segments lie above the curve, giving each trapezoid slightly more area than the corresponding strip under the curve. In Figure 5.31 we see that the straight segments lie *under* the curve on those intervals where the curve is concave *down,* causing the Trapezoidal Rule to *underestimate* the integral on those intervals. The interpretation of "area" changes where the curve lies below the x-axis but it is still the case that the higher y-values give the greater signed area. So we can always say that T overestimates the integral where the graph is concave up and underestimates the integral where the graph is concave down.

Example 2 AVERAGING TEMPERATURES

An observer measures the outside temperature every hour from noon until midnight, recording the temperatures in the following table.

Time	N	1	2	3	4	5	6	7	8	9	10	11	M
Temp	63	65	66	68	70	69	68	68	65	64	62	58	55

What was the average temperature for the 12-hour period?

Solution We are looking for the average value of a continuous function (temperature) for which we know values at discrete times that are one unit apart. We need to find

$$av(f) = \frac{1}{b-a}\int_a^b f(x)\,dx,$$

without having a formula for $f(x)$. The integral, however, can be approximated by the Trapezoidal Rule, using the temperatures in the table as function values at the points of a 12-subinterval partition of the 12-hour interval (making $h = 1$).

$$T = \frac{h}{2}\Big(y_0 + 2y_1 + 2y_2 + \cdots + 2y_{11} + y_{12}\Big)$$

$$= \frac{1}{2}\Big(63 + 2\cdot 65 + 2\cdot 66 + \cdots + 2\cdot 58 + 55\Big)$$

$$= 782$$

Using T to approximate $\int_a^b f(x)\,dx$, we have

$$av(f) \approx \frac{1}{b-a}\cdot T = \frac{1}{12}\cdot 782 \approx 65.17.$$

Rounding to be consistent with the data given, we estimate the average temperature as 65 degrees.

Other Algorithms

LRAM, MRAM, RRAM, and the Trapezoidal Rule all give reasonable approximations to the integral of a continuous function over a closed interval. The Trapezoidal Rule is more efficient, giving a better approximation for small values of n, which makes it a faster algorithm for numerical integration.

Indeed, the only shortcoming of the Trapezoidal Rule seems to be that it depends on approximating curved arcs with straight segments. You might think that an algorithm that approximates the curve with *curved* pieces would be even more efficient (and hence faster for machines), and you would be right. All we need to do is find a geometric figure with a straight base, straight sides, and a curved top that has a known area. You might not know one, but the ancient Greeks did; it is one of the things they knew about parabolas.

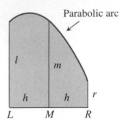

Parabolic arc

Figure 5.33 The area under the parabolic arc can be computed from the length of the base LR and the lengths of the altitudes constructed at L, R and midpoint M. (Exploration 1)

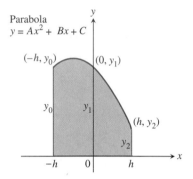

Figure 5.34 A convenient coordinatization of Figure 5.33. The parabola has equation $y = Ax^2 + Bx + C$, and the midpoint of the base is at the origin. (Exploration 1)

What's in a Name?

The formula that underlies Simpson's Rule (see Exploration 1) was discovered long before Thomas Simpson (1720–1761) was born. Just as Pythagoras did not discover the Pythagorean Theorem, Simpson did not discover Simpson's Rule. It is another of history's beautiful quirks that one of the ablest mathematicians of eighteenth-century England is remembered not for his successful textbooks and his contributions to mathematical analysis, but for a rule that was never his, that he never laid claim to, and that bears his name only because he happened to mention it in one of his books.

Exploration 1 **Area Under a Parabolic Arc**

The area A_P of a figure having a horizontal base, vertical sides, and a parabolic top (Figure 5.33) can be computed by the formula

$$A_P = \frac{h}{3}\left(l + 4m + r\right),$$

where h is half the length of the base, l and r are the lengths of the left and right sides, and m is the altitude at the midpoint of the base. This formula, once a profound discovery of ancient geometers, is readily verified today with calculus.

1. Coordinatize Figure 5.33 by centering the base at the origin, as shown in Figure 5.34. Let $y = Ax^2 + Bx + C$ be the equation of the parabola. Using this equation, show that $y_0 = Ah^2 - Bh + C$, $y_1 = C$, and $y_2 = Ah^2 + Bh + C$.

2. Show that $y_0 + 4y_1 + y_2 = 2Ah^2 + 6C$.

3. Integrate to show that the area A_P is

$$\frac{h}{3}(2Ah^2 + 6C).$$

4. Combine these results to derive the formula

$$A_P = \frac{h}{3}\left(y_0 + 4y_1 + y_2\right).$$

This last formula leads to an efficient rule for approximating integrals numerically. Partition the interval of integration into an even number of subintervals, apply the formula for A_P to successive interval pairs, and add the results. This algorithm is known as Simpson's Rule.

Simpson's Rule

To approximate $\int_a^b f(x)\, dx$, use

$$S = \frac{h}{3}\left(y_0 + 4y_1 + 2y_2 + 4y_3 + \cdots + 2y_{n-2} + 4y_{n-1} + y_n\right),$$

where $[a, b]$ is partitioned into an *even* number n of subintervals of equal length $h = (b - a)/n$.

Example 3 APPLYING SIMPSON'S RULE

Use Simpson's Rule with $n = 4$ to approximate $\int_0^2 5x^4\, dx$.

Solution Partition $[0, 2]$ into four subintervals and evaluate $y = 5x^4$ at the partition points (Table 5.5).

Table 5.5	
x	$y = 5x^4$
0	0
$\dfrac{1}{2}$	$\dfrac{5}{16}$
1	5
$\dfrac{3}{2}$	$\dfrac{405}{16}$
2	80

Then apply Simpson's Rule with $n = 4$ and $h = 1/2$:

$$S = \frac{h}{3}\left(y_0 + 4y_1 + 2y_2 + 4y_3 + y_4\right)$$

$$= \frac{1}{6}\left(0 + 4\left(\frac{5}{16}\right) + 2\left(5\right) + 4\left(\frac{405}{16}\right) + 80\right)$$

$$= \frac{385}{12}.$$

This estimate differs from the exact value (32) by only 1/12, a percentage error of less than three-tenths of one percent—and this was with just 4 subintervals.

There are still other algorithms for approximating definite integrals, most of them involving fancy numerical analysis designed to make the calculations more efficient for high-speed computers. Some are kept secret by the companies that design the machines. In any case, we will not deal with them here.

Error Analysis

After finding that the trapezoidal approximation in Example 1 overestimated the integral, we pointed out that this could have been predicted from the concavity of the curve we were approximating.

Knowing something about the error in an approximation is more than just an interesting sidelight. Despite what your years of classroom experience might have suggested, exact answers are not always easy to find in mathematics. It is fortunate that for all *practical* purposes exact answers are also rarely necessary. (For example, a carpenter who computes the need for a board of length $\sqrt{34}$ feet will happily settle for an approximation when cutting the board.)

Suppose that an exact answer really can *not* be found, but that we know that an approximation within 0.001 unit is good enough. How can we tell that our approximation is within 0.001 if we do not know the exact answer? This is where knowing something about the error is critical.

Since the Trapezoidal Rule approximates curves with straight lines, it seems reasonable that the error depends on how "curvy" the graph is. This suggests that the error depends on the second derivative. It is also apparent that the error depends on the length h of the subintervals. It can be shown that if f'' is continuous the error in the trapezoidal approximation, denoted E_T, satisfies the inequality

$$|E_T| \le \frac{b - a}{12}\,h^2 M_{f''},$$

where $[a, b]$ is the interval of integration, h is the length of each subinterval, and $M_{f''}$ is the maximum value of $|f''|$ on $[a, b]$.

It can also be shown that the error E_S in Simpson's Rule depends on h and the *fourth* derivative. It satisfies the inequality

$$|E_S| \le \frac{b - a}{180}\,h^4 M_{f^{(4)}},$$

where $[a, b]$ is the interval of integration, h is the length of each subinterval, and $M_{f^{(4)}}$ is the maximum value of $|f^{(4)}|$ on $[a, b]$, provided that $f^{(4)}$ is continuous.

For comparison's sake, if all the assumptions hold, we have the following *error bounds.*

Error Bounds

If T and S represent the approximations to $\int_a^b f(x)\,dx$ given by the Trapezoidal Rule and Simpson's Rule, respectively, then the errors E_T and E_S satisfy

$$|E_T| \le \frac{b-a}{12}\, h^2 M_{f''} \quad \text{and} \quad |E_S| \le \frac{b-a}{180}\, h^4 M_{f^{(4)}}.$$

If we disregard possible differences in magnitude between $M_{f''}$ and $M_{f^{(4)}}$, we notice immediately that $(b-a)/180$ is one-fifteenth the size of $(b-a)/12$, giving S an obvious advantage over T as an approximation. That, however, is almost insignificant when compared to the fact that the trapezoid error varies as the *square* of h, while Simpson's error varies as the *fourth power* of h. (Remember that h is already a small number in most partitions.)

Table 5.6 shows T and S values for approximations of $\int_1^2 1/x\,dx$ using various values of n. Notice how Simpson's Rule dramatically improves over the Trapezoidal Rule. In particular, notice that when we double the value of n (thereby halving the value of h), the T error is divided by 2 *squared,* while the S error is divided by 2 *to the fourth.*

Table 5.7 Approximations of $\int_1^5 (\sin x)/x\,dx$

Method	Subintervals	Value
LRAM	50	0.6453898
RRAM	50	0.5627293
MRAM	50	0.6037425
TRAP	50	0.6040595
SIMP	50	0.6038481
NINT	Tol = 0.00001	0.6038482

Table 5.6 Trapezoidal Rule Approximations (T_n) and Simpson's Rule Approximations (S_n) of ln 2 $= \int_1^2 (1/x)\,dx$

n	T_n	|Error| less than …	S_n	|Error| less than …
10	.6937714032	.0006242227	.6931502307	.0000030502
20	.6933033818	.0001562013	.6931473747	.0000001942
30	.6932166154	.0000694349	.6931472190	.0000000385
40	.6931862400	.0000390595	.6931471927	.0000000122
50	.6931721793	.0000249988	.6931471856	.0000000050
100	.6931534305	.0000062500	.6931471809	.0000000004

This has a dramatic effect as h gets very small. The Simpson approximation for $n = 50$ rounds accurately to seven places, and for $n = 100$ agrees to nine decimal places (billionths)!

We close by showing you the values (Table 5.7) we found for $\int_1^5 (\sin x)/x\,dx$ by six different calculator methods. The exact value of this integral to six decimal places is 0.603848, so both Simpson's method with 50 subintervals and NINT give results accurate to at least six places (millionths).

Quick Review 5.5

In Exercises 1–10, tell whether the curve is concave up or concave down on the given interval.

1. $y = \cos x$ on $[-1, 1]$

2. $y = x^4 - 12x - 5$ on $[8, 17]$

3. $y = 4x^3 - 3x^2 + 6$ on $[-8, 0]$

4. $y = \sin(x/2)$ on $[48\pi, 50\pi]$

5. $y = e^{2x}$ on $[-5, 5]$

6. $y = \ln x$ on $[100, 200]$

7. $y = \dfrac{1}{x}$ on $[3, 6]$

8. $y = \csc x$ on $[0, \pi]$

9. $y = 10^{10} - 10x^{10}$ on $[10, 10^{10}]$

10. $y = \sin x - \cos x$ on $[1, 2]$

Section 5.5 Exercises

In Exercises 1–6, **(a)** use the Trapezoidal Rule with $n = 4$ to approximate the value of the integral. **(b)** Use the concavity of the function to predict whether the approximation is an overestimate or an underestimate. Finally, **(c)** find the integral's exact value to check your answer.

1. $\displaystyle\int_0^2 x \, dx$

2. $\displaystyle\int_0^2 x^2 \, dx$

3. $\displaystyle\int_0^2 x^3 \, dx$

4. $\displaystyle\int_1^2 \dfrac{1}{x} \, dx$

5. $\displaystyle\int_0^4 \sqrt{x} \, dx$

6. $\displaystyle\int_0^\pi \sin x \, dx$

7. *Volume of Water in a Swimming Pool* A rectangular swimming pool is 30 ft wide and 50 ft long. The table below shows the depth $h(x)$ of the water at 5-ft intervals from one end of the pool to the other. Estimate the volume of water in the pool using the Trapezoidal Rule with $n = 10$, applied to the integral

$$V = \int_0^{50} 30 \cdot h(x) \, dx.$$

Position (ft) x	Depth (ft) $h(x)$	Position (ft) x	Depth (ft) $h(x)$
0	6.0	30	11.5
5	8.2	35	11.9
10	9.1	40	12.3
15	9.9	45	12.7
20	10.5	50	13.0
25	11.0		

8. *Stocking a Fish Pond* As the fish and game warden of your township, you are responsible for stocking the town pond with fish before the fishing season. The average depth of the pond is 20 feet. Using a scaled map, you measure distances across the pond at 200-foot intervals, as shown in the diagram.

(a) Use the Trapezoidal Rule to estimate the volume of the pond.

(b) You plan to start the season with one fish per 1000 cubic feet. You intend to have at least 25% of the opening day's fish population left at the end of the season. What is the maximum number of licenses the town can sell if the average seasonal catch is 20 fish per license?

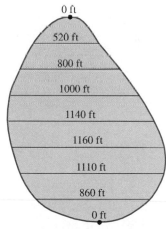

0 ft

520 ft

800 ft

1000 ft

1140 ft

1160 ft

1110 ft

860 ft

0 ft

Vertical spacing = 200 ft

9. *Ford® Mustang Cobra™* The accompanying table shows time-to-speed data for a 1994 Ford Mustang Cobra accelerating from rest to 130 mph. How far had the Mustang traveled by the time it reached this speed? (Use trapezoids to estimate the area under the velocity curve, but be careful: the time intervals vary in length.)

Speed Change	Time (sec)
Zero to 30 mph	2.2
40 mph	3.2
50 mph	4.5
60 mph	5.9
70 mph	7.8
80 mph	10.2
90 mph	12.7
100 mph	16.0
110 mph	20.6
120 mph	26.2
130 mph	37.1

Source: Car and Driver, April 1994.

10. Consider the integral $\int_{-1}^{3} (x^3 - 2x)\, dx$.

(a) Use Simpson's Rule with $n = 4$ to approximate its value.

(b) Find the exact value of the integral. What is the error, $|E_S|$?

(c) Explain how you could have predicted what you found in (b) from knowing the error-bound formula.

(d) Writing to Learn Is it possible to make a general statement about using Simpson's Rule to approximate integrals of cubic polynomials? Explain.

11. Writing to Learn In Example 2 (before rounding) we found the average temperature to be 65.17 degrees when we used the integral approximation, yet the average of the 13 discrete temperatures is only 64.69 degrees. Considering the shape of the temperature curve, explain why you would expect the average of the 13 discrete temperatures to be less than the average value of the temperature function on the entire interval.

12. *(Continuation of Exercise 11)*

(a) In the Trapezoidal Rule, every function value in the sum is doubled except for the two endpoint values. Show that if you double the endpoint values, you get 70.08 for the average temperature.

(b) Explain why it makes more sense to not double the endpoint values if we are interested in the average temperature over the entire 12-hour period.

In Exercises 13–16, use a calculator program to find the Simpson's Rule approximations with $n = 50$ and $n = 100$.

13. $\int_{-1}^{1} 2\sqrt{1 - x^2}\, dx$ The exact value is π.

14. $\int_{0}^{1} \sqrt{1 + x^4}\, dx$ An integral that came up in Newton's research

15. $\int_{0}^{\pi/2} \frac{\sin x}{x}\, dx$

16. $\int_{0}^{\pi/2} \sin(x^2)\, dx$ An integral associated with the diffraction of light

17. Consider the integral $\int_{0}^{\pi} \sin x\, dx$.

(a) Use a calculator program to find the Trapezoidal Rule approximations for $n = 10,\ 100,$ and 1000.

(b) Record the errors with as many decimal places of accuracy as you can.

(c) What pattern do you see?

(d) Writing to Learn Explain how the error bound for E_T accounts for the pattern.

18. *(Continuation of Exercise 17)* Repeat Exercise 17 with Simpson's Rule and E_S.

Explorations

In Exercises 19 and 20, *work in groups of two or three.*

19. Consider the integral $\int_{-1}^{1} \sin(x^2)\, dx$.

(a) Find f'' for $f(x) = \sin(x^2)$.

(b) Graph $y = f''(x)$ in the viewing window $[-1, 1]$ by $[-3, 3]$.

(c) Explain why the graph in (b) suggests that $|f''(x)| \le 3$ for $-1 \le x \le 1$.

(d) Show that the error estimate for the Trapezoidal Rule in this case becomes

$$|E_T| \le \frac{h^2}{2}.$$

(e) Show that the Trapezoidal Rule error will be less than or equal to 0.01 if $h \le 0.1$.

(f) How large must n be for $h \le 0.1$?

20. Consider the integral $\int_{-1}^{1} \sin(x^2)\, dx$.

(a) Find $f^{(4)}$ for $f(x) = \sin(x^2)$. (You may want to check your work with a CAS if you have one available.)

(b) Graph $y = f^{(4)}(x)$ in the viewing window $[-1, 1]$ by $[-30, 10]$.

(c) Explain why the graph in (b) suggests that $|f^{(4)}(x)| \le 30$ for $-1 \le x \le 1$.

(d) Show that the error estimate for Simpson's Rule in this case becomes

$$|E_S| \le \frac{h^4}{3}.$$

(e) Show that the Simpson's Rule error will be less than or equal to 0.01 if $h \le 0.4$.

(f) How large must n be for $h \le 0.4$?

21. *Aerodynamic Drag* A vehicle's aerodynamic drag is determined in part by its cross section area, so, all other things being equal, engineers try to make this area as small as possible. Use Simpson's Rule to estimate the cross section area of the body of James Worden's solar-powered Solectria® automobile at M.I.T. from the diagram below.

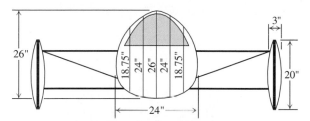

22. *Wing Design* The design of a new airplane requires a gasoline tank of constant cross section area in each wing. A scale drawing of a cross section is shown here. The tank must hold 5000 lb of gasoline, which has a density of 42 lb/ft^3. Estimate the length of the tank.

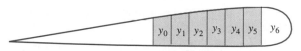

$y_0 = 1.5$ ft, $y_1 = 1.6$ ft, $y_2 = 1.8$ ft, $y_3 = 1.9$ ft,
$y_4 = 2.0$ ft, $y_5 = y_6 = 2.1$ ft   Horizontal spacing = 1 ft

Extending the Ideas

23. Using the definitions, prove that, in general,
$$T_n = \frac{\text{LRAM}_n + \text{RRAM}_n}{2}.$$

24. Using the definitions, prove that, in general,
$$S_{2n} = \frac{\text{MRAM}_n + 2T_{2n}}{3}.$$

Chapter 5 Key Terms

Chapter 5   Review Exercises

Exercises 1–6 refer to the region R in the first quadrant enclosed by the x-axis and the graph of the function $y = 4x - x^3$.

1. Sketch R and partition it into four subregions, each with a base of length $\Delta x = 1/2$.

2. Sketch the rectangles and compute (by hand) the area for the $LRAM_4$ approximation.

3. Sketch the rectangles and compute (by hand) the area for the $MRAM_4$ approximation.

4. Sketch the rectangles and compute (by hand) the area for the $RRAM_4$ approximation.

5. Sketch the trapezoids and compute (by hand) the area for the T_4 approximation.

6. Find the exact area of R by using the Fundamental Theorem of Calculus.

7. Use a calculator program to compute the RAM approximations in the following table for the area under the graph of $y = 1/x$ from $x = 1$ to $x = 5$.

n	$LRAM_n$	$MRAM_n$	$RRAM_n$
10			
20			
30			
50			
100			
1000			

8. *(Continuation of Exercise 7)* Use the Fundamental Theorem of Calculus to determine the value to which the sums in the table are converging.

9. Suppose

$$\int_{-2}^{2} f(x)\, dx = 4, \quad \int_{2}^{5} f(x)\, dx = 3, \quad \int_{-2}^{5} g(x)\, dx = 2.$$

Which of the following statements are true, and which, if any, are false?

(a) $\int_{5}^{2} f(x)\, dx = -3$ (b) $\int_{-2}^{5} [f(x) + g(x)]\, dx = 9$

(c) $f(x) \le g(x)$ on the interval $-2 \le x \le 5$

10. The region under one arch of the curve $y = \sin x$ is revolved around the x-axis to form a solid. (a) Use the method of Example 3, Section 5.1, to set up a Riemann sum that approximates the volume of the solid. (b) Find the volume using NINT.

11. The accompanying graph shows the velocity (m/sec) of a body moving along the s-axis during the time interval from $t = 0$ to $t = 10$ sec. (a) About how far did the body travel during those 10 seconds?

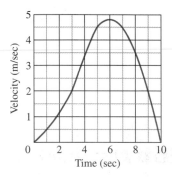

(b) Sketch a graph of position (s) as a function of time (t) for $0 \le t \le 10$, assuming $s(0) = 0$.

12. The interval $[0, 10]$ is partitioned into n subintervals of length $\Delta x = 10/n$. We form the following Riemann sums, choosing each c_k in the kth subinterval. Write the limit as $n \to \infty$ of each Riemann sum as a definite integral.

(a) $\displaystyle\sum_{k=1}^{n} (c_k)^3 \Delta x$ (b) $\displaystyle\sum_{k=1}^{n} c_k(\sin c_k)\Delta x$

(c) $\displaystyle\sum_{k=1}^{n} c_k(3c_k - 2)^2 \Delta x$ (d) $\displaystyle\sum_{k=1}^{n} (1 + c_k^2)^{-1}\Delta x$

(e) $\displaystyle\sum_{k=1}^{n} \pi(9 - \sin^2(\pi c_k/10))\Delta x$

In Exercises 13 and 14, find the total area between the curve and the x-axis.

13. $y = 4 - x, \quad 0 \le x \le 6$ 14. $y = \cos x, \quad 0 \le x \le \pi$

In Exercises 15–24, evaluate the integral analytically by using the Integral Evaluation Theorem (Part 2 of the Fundamental Theorem, Theorem 4).

15. $\displaystyle\int_{-2}^{2} 5\, dx$ 16. $\displaystyle\int_{2}^{5} 4x\, dx$

17. $\displaystyle\int_{0}^{\pi/4} \cos x\, dx$ 18. $\displaystyle\int_{-1}^{1} (3x^2 - 4x + 7)\, dx$

19. $\displaystyle\int_{0}^{1} (8s^3 - 12s^2 + 5)\, ds$ 20. $\displaystyle\int_{1}^{2} \frac{4}{x^2}\, dx$

21. $\displaystyle\int_{1}^{27} y^{-4/3}\, dy$ 22. $\displaystyle\int_{1}^{4} \frac{dt}{t\sqrt{t}}$

23. $\displaystyle\int_{0}^{\pi/3} \sec^2 \theta\, d\theta$ 24. $\displaystyle\int_{1}^{e} (1/x)\, dx$

In Exercises 25–29, evaluate the integral.

25. $\int_0^1 \dfrac{36}{(2x+1)^3}\,dx$ **26.** $\int_1^2 \left(x + \dfrac{1}{x^2}\right) dx$

27. $\int_{-\pi/3}^0 \sec x \tan x \, dx$ **28.** $\int_{-1}^1 2x \sin(1 - x^2)\, dx$

29. $\int_0^2 \dfrac{2}{y+1}\, dy$

In Exercises 30–32, evaluate the integral by interpreting it as area and using formulas from geometry.

30. $\int_0^2 \sqrt{4 - x^2}\, dx$ **31.** $\int_{-4}^8 |x|\, dx$

32. $\int_{-8}^8 2\sqrt{64 - x^2}\, dx$

33. *Oil Consumption on Pathfinder Island* A diesel generator runs continuously, consuming oil at a gradually increasing rate until it must be temporarily shut down to have the filters replaced.

Day	Oil Consumption Rate (liters/hour)
Sun	0.019
Mon	0.020
Tue	0.021
Wed	0.023
Thu	0.025
Fri	0.028
Sat	0.031
Sun	0.035

(a) Give an upper estimate and a lower estimate for the amount of oil consumed by the generator during that week.

(b) Use the Trapezoidal Rule to estimate the amount of oil consumed by the generator during that week.

34. *Rubber-Band–Powered Sled* A sled powered by a wound rubber band moves along a track until friction and the unwinding of the rubber band gradually slow it to a stop. A speedometer in the sled monitors its speed, which is recorded at 3-second intervals during the 27-second run.

Time (sec)	Speed (ft/sec)
0	5.30
3	5.25
6	5.04
9	4.71
12	4.25
15	3.66
18	2.94
21	2.09
24	1.11
27	0

(a) Give an upper estimate and a lower estimate for the distance traveled by the sled.

(b) Use the Trapezoidal Rule to estimate the distance traveled by the sled.

35. Writing to Learn Your friend knows how to compute integrals but never could understand what difference the "*dx*" makes, claiming that it is irrelevant. How would you explain to your friend why it is necessary?

36. The function
$$f(x) = \begin{cases} x^2, & x \ge 0 \\ x - 2, & x < 0 \end{cases}$$
is discontinuous at 0, but integrable on $[-4, 4]$. Find $\int_{-4}^4 f(x)\, dx$.

37. Show that $0 \le \int_0^1 \sqrt{1 + \sin^2 x}\, dx \le \sqrt{2}$.

38. Find the average value of
(a) $y = \sqrt{x}$ over the interval $[0, 4]$.
(b) $y = a\sqrt{x}$ over the interval $[0, a]$.

In Exercises 39–42, find dy/dx.

39. $y = \int_2^x \sqrt{2 + \cos^3 t}\, dt$ **40.** $y = \int_2^{7x^2} \sqrt{2 + \cos^3 t}\, dt$

41. $y = \int_x^1 \dfrac{6}{3 + t^4}\, dt$ **42.** $y = \int_x^{2x} \dfrac{1}{t^2 + 1}\, dt$

43. *Printing Costs* Including start-up costs, it costs a printer $50 to print 25 copies of a newsletter, after which the marginal cost at x copies is
$$\frac{dc}{dx} = \frac{2}{\sqrt{x}} \text{ dollars per copy.}$$
Find the total cost of printing 2500 newsletters.

44. *Average Daily Inventory* Rich Wholesale Foods, a manufacturer of cookies, stores its cases of cookies in an air-conditioned warehouse for shipment every 14 days. Rich tries to keep 600 cases on reserve to meet occasional peaks in demand, so a typical 14-day inventory function is $I(t) = 600 + 600t$, $0 \le t \le 14$. The holding cost for each case is 4¢ per day. Find Rich's average daily inventory and average daily holding cost.

45. Solve for x: $\int_0^x (t^3 - 2t + 3)\, dt = 4$.

46. Suppose $f(x)$ has a positive derivative for all values of x and that $f(1) = 0$. Which of the following statements must be true of
$$g(x) = \int_0^x f(t)\, dt?$$

(a) g is a differentiable function of x.
(b) g is a continuous function of x.
(c) The graph of g has a horizontal tangent line at $x = 1$.
(d) g has a local maximum at $x = 1$.
(e) g has a local minimum at $x = 1$.
(f) The graph of g has an inflection point at $x = 1$.
(g) The graph of dg/dx crosses the x-axis at $x = 1$.

47. Suppose $F(x)$ is an antiderivative of $f(x) = \sqrt{1 + x^4}$. Express $\int_0^1 \sqrt{1 + x^4}\, dx$ in terms of F.

48. Express the function $y(x)$ with

$$\frac{dy}{dx} = \frac{\sin x}{x} \quad \text{and} \quad y(5) = 3$$

as a definite integral.

49. Show that $y = x^2 + \int_1^x 1/t\, dt + 1$ satisfies both of the following conditions:

    i. $y'' = 2 - \dfrac{1}{x^2}$

    ii. $y = 2$ and $y' = 3$ when $x = 1$.

50. Writing to Learn Which of the following is the graph of the function whose derivative is $dy/dx = 2x$ and whose value at $x = 1$ is 4? Explain your answer.

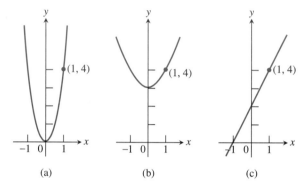

    (a)   (b)   (c)

51. Fuel Efficiency An automobile computer gives a digital readout of fuel consumption in gallons per hour. During a trip, a passenger recorded the fuel consumption every 5 minutes for a full hour of travel.

time	gal/h	time	gal/h
0	2.5	35	2.5
5	2.4	40	2.4
10	2.3	45	2.3
15	2.4	50	2.4
20	2.4	55	2.4
25	2.5	60	2.3
30	2.6		

(a) Use the Trapezoidal Rule to approximate the total fuel consumption during the hour.

(b) If the automobile covered 60 miles in the hour, what was its fuel efficiency (in miles per gallon) for that portion of the trip?

52. *Skydiving* Skydivers A and B are in a helicopter hovering at 6400 feet. Skydiver A jumps and descends for 4 sec before opening her parachute. The helicopter then climbs to 7000 feet and hovers there. Forty-five seconds after A leaves the aircraft, B jumps and descends for 13 sec before opening her parachute. Both skydivers descend at 16 ft/sec with parachutes open. Assume that the skydivers fall freely (with acceleration -32 ft/sec^2) before their parachutes open.

(a) At what altitude does A's parachute open?

(b) At what altitude does B's parachute open?

(c) Which skydiver lands first?

53. *Relating Simpson's Rule, MRAM, and T* The figure below shows an interval of length $2h$ with a trapezoid, a midpoint rectangle, and a parabolic region on it.

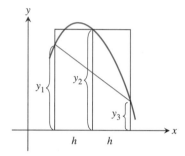

(a) Show that the area of the trapezoid plus twice the area of the rectangle equals

$$h(y_1 + 4y_2 + y_3).$$

(b) Use the result in (a) to prove that

$$S_{2n} = \frac{2 \cdot \text{MRAM}_n + T_n}{3}.$$

54. The graph of a function f consists of a semicircle and two line segments as shown below.

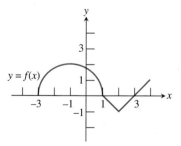

Let $g(x) = \int_1^x f(t)\, dt$.

(a) Find $g(1)$.   **(b)** Find $g(3)$.   **(c)** Find $g(-1)$.

(d) Find all values of x on the open interval $(-3, 4)$ at which g has a relative maximum.

(e) Write an equation for the line tangent to the graph of g at $x = -1$.

(f) Find the x-coordinate of each point of inflection of the graph of g on the open interval $(-3, 4)$.

(g) Find the range of g.

55. What is the total area under the curve $y = e^{-x^2/2}$?

The graph approaches the x-axis as an asymptote both to the left and the right, but quickly enough so that the total area is a finite number. In fact,

$$\text{NINT} (e^{-x^2/2}, x, -10, 10)$$

computes all but a negligible amount of the area.

(a) Find this number on your calculator. Verify that NINT $(e^{-x^2/2}, x, -20, 20)$ does not increase the number enough for the calculator to distinguish the difference.

(b) This area has an interesting relationship to π. Perform various (simple) algebraic operations on the number to discover what it is.

56. *Filling a Swamp* A town wants to drain and fill the small polluted swamp shown below. The swamp averages 5 ft deep. About how many cubic yards of dirt will it take to fill the area after the swamp is drained?

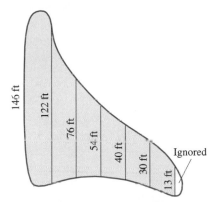

Horizontal spacing − 20 ft

57. *Household Electricity* We model the voltage V in our homes with the sine function

$$V = V_{max} \sin (120 \pi t),$$

which expresses V in volts as a function of time t in seconds. The function runs through 60 cycles each second. The number V_{max} is the *peak voltage*.

To measure the voltage effectively, we use an instrument that measures the square root of the average value of the square of the voltage over a 1-second interval:

$$V_{rms} = \sqrt{(V^2)_{av}} \,.$$

The subscript "rms" stands for "root mean square." It turns out that

$$V_{rms} = \frac{V_{max}}{\sqrt{2}}. \tag{1}$$

The familiar phrase "115 volts ac" means that the rms voltage is 115. The peak voltage, obtained from Equation 1 as $V_{max} = 115\sqrt{2}$, is about 163 volts.

(a) Find the average value of V^2 over a 1-sec interval. Then find V_{rms}, and verify Equation 1.

(b) The circuit that runs your electric stove is rated 240 volts rms. What is the peak value of the allowable voltage?

Calculus at Work

I have a degree in Mechanical Engineering with a minor in Psychology. I am a Research and Development Engineer at Komag, which designs and manufactures hard disks in Santa Clara, California. My job is to test the durability and reliability of the disks, measuring the rest friction between the read/write heads and the disk surface, which is called "Contact-Start-Stop" testing.

I use calculus to evaluate the moment of inertia of different disk stacks, which consist of disks on a spindle, separated by spacer rings. Because the rings vary in size as well as material, the mass of each

ring must be determined. For such problems, I refer to my college calculus textbook and its tables of summations and integrals. For instance, I use:

Moment of Inertia =

$$\left[\sum_{i=1}^{n} M_i L_i^2 \right] + \frac{1}{3} M_{rod} L_{rod}^2$$

where i = components 1 to n;
M_i = mass of component i such as the disk and/or ring stack;
L_i = distance of component i from a reference point;
M_{rod} = mass of the spindle that rotates; L_{rod} = length of the spindle.

Andrea Woo
Komag
Santa Clara, CA

Differential Equations and Mathematical Modeling

One way to measure how light in the ocean diminishes as water depth increases involves using a Secchi disk. This white disk is 30 centimeters in diameter, and is lowered into the ocean until it disappears from view. The depth of this point (in meters), divided into 1.7, yields the coefficient k used in the equation $I_x = I_0 e^{-kx}$. This equation estimates the intensity I_x of light at depth x using I_0, the intensity of light at the surface.

In an ocean experiment, if the Secchi disk disappears at 55 meters, at what depth will only 1% of surface radiation remain? Section 6.4 will help you answer this question.

Chapter 6 Overview

One of the early accomplishments of calculus was predicting the future position of a planet from its present position and velocity. Today this is just one of a number of occasions on which we deduce everything we need to know about a function from one of its known values and its rate of change. From this kind of information, we can tell how long a sample of radioactive polonium will last; whether, given current trends, a population will grow or become extinct; and how large major league baseball salaries are likely to be in the year 2010. In this chapter, we examine the analytic, graphical, and numerical techniques on which such predictions are based.

6.1 Antiderivatives and Slope Fields

Solving Initial Value Problems • Antiderivatives and Indefinite Integrals • Properties of Indefinite Integrals • Applications

Solving Initial Value Problems

An equation like

$$\frac{dy}{dx} = y \ln x$$

Order

The **order** of a differential equation is the order of the highest derivative involved in the equation.

containing a derivative is a **differential equation.** The problem of finding a function y of x when we are given its derivative and its value at a particular point is called an **initial value problem.** The value of f for one value of x is the **initial condition** of the problem. When we find all the functions y that satisfy the differential equation we have **solved the differential equation.** When we then find the **particular solution** that fulfills the initial condition, we have **solved the initial value problem.**

Table 6.1 Initial Value Problems from Earlier Chapters

Original Problem...	... as an Initial Value Problem
(a) Find the function $y = f(x)$ whose derivative is $\sin x$ and whose graph passes through the point $(0, 2)$. (Section 4.2, Example 6)	Solve the initial value problem: Differential equation: $\dfrac{dy}{dx} = \sin x,$ Initial condition: $y(0) = 2.$
(b) Find the function $y = f(x)$ with derivative $dy/dx = \tan x$ satisfying the condition $f(3) = 5$. (Section 5.4, Example 4)	Solve the initial value problem: Differential equation: $\dfrac{dy}{dx} = \tan x,$ Initial condition: $y(3) = 5.$

Suppose that y_0 dollars are invested in an account at an interest rate of 5.6% per year, compounded *continuously.* This means that at any given instant the amount y in the account is increasing at a rate that is 5.6% of the current balance. If t is the time in years since the account was opened, we can model the situation with the following initial value problem:

Differential equation: $\dfrac{dy}{dt} = 0.056y,$ At any instant the rate of change of y is 5.6% of y.

Initial condition: $y(0) = y_0.$ There are y_0 dollars in the account when $t = 0$.

Example 1   COMPOUNDING INTEREST CONTINUOUSLY

Suppose \$100 is invested in an account that pays 5.6% interest compounded continuously. Find a formula for the amount in the account at any time t.

Solution   We let $t = 0$ when the initial \$100 is deposited in the account.

Model

We can model this problem with the initial value problem

$$\frac{dy}{dt} = 0.056y \quad \text{and} \quad y(0) = 100.$$

Solve Analytically

We are looking for a function whose derivative is a constant multiple of itself. We know that exponential functions have this property. In fact, the function $y(t) = Ce^{0.056t}$ is a solution of the differential equation for any real number C because

$$\frac{dy}{dt} = C(0.056 \cdot e^{0.056t}) = 0.056(Ce^{0.056t}) = 0.056y(t).$$

In Section 6.4 we will see that there are no other solutions.

Applying the initial condition we find

$$y(t) = Ce^{0.056t}$$
$$y(0) = Ce^{(0.056)(0)}$$
$$100 = C.$$

Interpret

The amount in the account at any time t is $y(t) = 100e^{0.056t}$. At the end of the first year, for instance, the balance will be $y(1) = 100e^{0.056(1)} = \$105.76$.

Each member of the family of functions $y(t) = Ce^{0.056t}$ is a solution of the equation

$$\frac{dy}{dt} = 0.056y. \tag{1}$$

We can use our grapher to obtain a reasonably accurate visualization of these solutions by using a slope field program (see *Technology Resource Manual*) that plots short line segments of slope $0.056y$ at selected points (x, y) in a viewing window (Figure 6.1a). In the window of Figure 6.1a, line segments are constructed at 100 points in the plane. The line segments are arranged in 10 columns and 10 rows. (See Exploration 1 for more details.) This is equivalent to plotting short pieces of the linearizations of the solution curves that pass through these points. The plot is called a *slope field* or *direction field* for the differential equation. Figure 6.1b shows the graphs of four solutions of Equation 1.

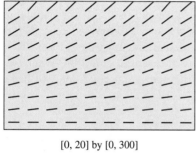

[0, 20] by [0, 300]

(a)

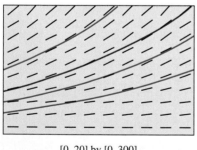

[0, 20] by [0, 300]

(b)

Figure 6.1 Notice how the slope field (a) gives a visualization of a solution curve (b) by indicating the direction of the curve as we proceed from left to right across the viewing window.

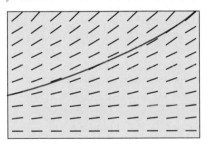

$y = 100e^{0.056x}$

[0, 20] by [0, 300]

Figure 6.2 The solution of the initial value problem in Example 1.

The solution $y(t) = 100e^{0.056t}$ in Example 1 is the one passing through the point (0, 100) (Figure 6.2).

Definition Slope Field or Direction Field

A **slope field** or **direction field** for the first order differential equation

$$\frac{dy}{dx} = f(x, y)$$

is a plot of short line segments with slopes $f(x, y)$ for a lattice of points (x, y) in the plane.

Here $f(x, y)$ is a function of the two variables x and y, for example, $f(x, y) = -2xy/(1 + x^2)$. Slope fields enable us to graph solution curves without solving the differential equation analytically (handy if it is an equation we cannot solve). This is especially useful when the formula given for dy/dx involves both x and y, as in the formulas we get from implicit differentiation.

Visual Support from a Slope Field

A slope field can be used to check that we have solved an initial value problem correctly.

Exploration 1 **Constructing a Slope Field**

1. Let $h = (\text{Xmax} - \text{Xmin})/10$ and $k = (\text{Ymax} - \text{Ymin})/10$. Explain why

$$\left(\text{Xmin} + (2i - 1)\frac{h}{2}, \text{Ymin} + (2j - 1)\frac{k}{2}\right)$$

   for $i, j = 1, 2, \ldots, 10$, represents 100 points in the viewing window [Xmin, Xmax] by [Ymin, Ymax].

2. Show that the horizontal distance between adjacent points with the same y-coordinate in part 1 is h.

3. Show that the vertical distance between adjacent points with the same x-coordinate in part 1 is k.

4. Consider the differential equation $dy/dx = 2/(x + 1)$ and the viewing window [0, 10] by [0, 10]. Find the slope (given by the differential equation) and construct a short line segment at each point in the third column ($i = 3$) of the points given by part 1. What do the line segments have in common?

5. Use the same differential equation and viewing window as in part 4. Find the slope and construct a short line segment at each point in the fourth row ($j = 4$) of the points given by part 1.

6. Explain how you could complete the slope field using the information in parts 4 and 5.

7. Let L and W be positive integers, $h = (\text{Xmax} - \text{Xmin})/W$, and $k = (\text{Ymax} - \text{Ymin})/L$. If $i = 1, 2, \ldots, W$ and $j = 1, 2, \ldots, L$, how many points are produced in part 1?

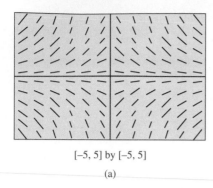

[−5, 5] by [−5, 5]

(a)

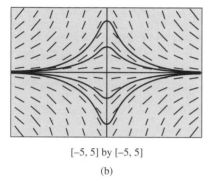

[−5, 5] by [−5, 5]

(b)

Figure 6.3 Notice that the slope $y' = -2xy/(1 + x^2)$ is zero when $x = 0$. (Example 2)

Example 2 PORTRAYING SOLUTIONS WITH A SLOPE FIELD

Plot the solution curves of the differential equation

$$y' = -\frac{2xy}{1 + x^2}.$$

Solution With our grapher in *differential equation graphing mode,* we generate the slope field (Figure 6.3a). We can see from the way the field "flows" how the solution curves will go. Then, to be specific, we add the solutions through the points $(0, 0)$, $(0, \pm 2)$ and $(0, \pm 4)$ (Figure 6.3b). These are the solutions satisfying the initial conditions $y(0) = 0$, $y(0) = \pm 2$, and $y(0) = \pm 4$.

We can tell from the concavity of the curves where the second derivatives of the solutions are positive, where they are negative, and approximately where they are zero, observations that can be significant in an application. We can also see that the solutions approach 0 as $x \rightarrow \pm \infty$. And we learn all this without touching the equation except to enter it into the machine.

Antiderivatives and Indefinite Integrals

Recall that a function $F(x)$ is an antiderivative of a function $f(x)$ if $F'(x) = f(x)$.

Definition Indefinite Integral

The set of all antiderivatives of a function $f(x)$ is the **indefinite integral** of f with respect to x and is denoted by

$$\int f(x)\, dx.$$

As in Chapter 5, the symbol $\int$ is an **integral sign,** the function f is the **integrand** of the integral, and x is the **variable of integration.**

According to Corollary 3 of the Mean Value Theorem for Derivatives in Section 4.2, once we have found one antiderivative F of a function f, all the other antiderivatives differ from F by constants. We indicate this in integral notation in the following way.

$$\int f(x)\, dx = F(x) + C$$

The equation is read, "The indefinite integral of f with respect to x is $F(x) + C$." The constant C is the **constant of integration** and is an **arbitrary constant.** When we find $F(x) + C$ we have **integrated** f or **evaluated** the integral.

Example 3 FINDING INDEFINITE INTEGRALS ANALYTICALLY

Evaluate $\int 2x\, dx$.

Solution

$$\int 2x\, dx = x^2 + C$$

an antiderivative of $2x$

the arbitrary constant

The formula $x^2 + C$ generates all the antiderivatives of the function $2x$. The functions $x^2 + 1$, $x^2 - \pi$, and $x^2 + \sqrt{2}$ are all antiderivatives of $2x$, as you can check by differentiation.

Many indefinite integrals needed in scientific work are found by reversing derivative formulas. You will see what we mean if you look at Table 6.2, which lists standard integral forms side by side with their derivative-formula sources.

Table 6.2 Integral Formulas

Indefinite Integral	Reversed Derivative Formula
1. (a) $\displaystyle\int x^n\, dx = \frac{x^{n+1}}{n+1} + C, \ n \neq -1$	$\dfrac{d}{dx}\left(\dfrac{x^{n+1}}{n+1}\right) = x^n, \ n \neq -1$
(b) $\displaystyle\int \frac{dx}{x} = \ln|x| + C$	$\dfrac{d}{dx}\ln|x| = \dfrac{1}{x}$
2. $\displaystyle\int e^{kx}\, dx = \frac{e^{kx}}{k} + C$	$\dfrac{d}{dx}\left(\dfrac{e^{kx}}{k}\right) = e^{kx}$
3. $\displaystyle\int \sin kx\, dx = -\frac{\cos kx}{k} + C$	$\dfrac{d}{dx}\left(-\dfrac{\cos kx}{k}\right) = \sin kx$
4. $\displaystyle\int \cos kx\, dx = \frac{\sin kx}{k} + C$	$\dfrac{d}{dx}\left(\dfrac{\sin kx}{k}\right) = \cos kx$
5. $\displaystyle\int \sec^2 x\, dx = \tan x + C$	$\dfrac{d}{dx}\tan x = \sec^2 x$
6. $\displaystyle\int \csc^2 x\, dx = -\cot x + C$	$\dfrac{d}{dx}(-\cot x) = \csc^2 x$
7. $\displaystyle\int \sec x \tan x\, dx = \sec x + C$	$\dfrac{d}{dx}\sec x = \sec x \tan x$
8. $\displaystyle\int \csc x \cot x\, dx = -\csc x + C$	$\dfrac{d}{dx}(-\csc x) = \csc x \cot x$

Example 4   USING INTEGRAL FORMULAS

(a) $\displaystyle\int x^5\,dx = \frac{x^6}{6} + C$ Formula 1a with $n = 5$

(b) $\displaystyle\int \frac{1}{\sqrt{x}}\,dx = \int x^{-1/2}\,dx = 2x^{1/2} + C = 2\sqrt{x} + C$ Formula 1a with $n = -1/2$

(c) $\displaystyle\int e^{-3x}\,dx = \frac{e^{-3x}}{-3} + C = -\frac{1}{3}e^{-3x} + C$ Formula 2 with $k = -3$

(d) $\displaystyle\int \cos\frac{x}{2}\,dx = \frac{\sin(1/2)x}{1/2} + C = 2\sin\frac{x}{2} + C$ Formula 4 with $k = 1/2$

When we cannot find a ready-made antiderivative like the ones in Table 6.2, we can construct one of our own using Part 1 of the Fundamental Theorem of Calculus from Section 5.4. This is the part that says that if $f(x)$ is continuous for $a \le x \le b$ then

$$\frac{d}{dx}\int_a^x f(t)\,dt = f(x)$$

at every point x in $[a, b]$. The integral is the antiderivative we seek.

$y = \text{NINT}\,(x\sin x, x, 0, x)$

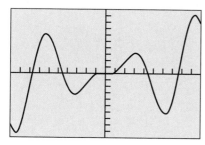

[–10, 10] by [–10, 10]

Figure 6.4 An antiderivative of $x \sin x$. (Example 5)

Example 5   USING THE FUNDAMENTAL THEOREM

Evaluate $\displaystyle\int x \sin x\,dx$.

Solution The function $f(x) = x \sin x$ is continuous for all x, so

$$F(x) = \int_0^x t \sin t\,dt$$

is an antiderivative of f. Therefore

$$\int x \sin x\,dx = F(x) + C = \int_0^x t \sin t\,dt + C.$$

Figure 6.4 shows $F(x) = \int_0^x t \sin t\,dt$ graphed as $\text{NINT}\,(x\sin x, x, 0, x)$. All other antiderivatives have graphs that are vertical shifts of this graph.

Properties of Indefinite Integrals

Here are the standard arithmetic rules for indefinite integration.

Properties of Indefinite Integrals

Let k be a real number.

1. *Constant Multiple Rule:* $\displaystyle\int kf(x)\,dx = k\int f(x)\,dx$

If $k = -1$, then: $\displaystyle\int -f(x)\,dx = -\int f(x)\,dx$

2. *Sum and Difference Rule:* $\displaystyle\int (f(x) \pm g(x))\,dx = \int f(x)\,dx \pm \int g(x)\,dx$

Just as the Sum and Difference Rule for differentiation enables us to differentiate expressions term by term, the Sum and Difference Rule for integration enables us to integrate expressions term by term. When we do so, we combine the individual constants of integration into a single arbitrary constant at the end.

Example 6 INTEGRATING TERM BY TERM

Evaluate $\int (x^2 - 2x + 5)\, dx$.

Solution If we recognize that $(x^3/3) - x^2 + 5x$ is an antiderivative of $x^2 - 2x + 5$, we can evaluate the integral as

$$\int (x^2 - 2x + 5)\, dx = \overbrace{\frac{x^3}{3} - x^2 + 5x}^{\text{antiderivative}} + \overset{\text{arbitrary constant}}{C.}$$

If we do not recognize the antiderivative right away, we can generate it term by term with the Sum and Difference Rule.

$$\int (x^2 - 2x + 5)\, dx = \int x^2\, dx - \int 2x\, dx + \int 5\, dx$$

$$= \frac{x^3}{3} + C_1 - x^2 + C_2 + 5x + C_3$$

This formula is more complicated than it needs to be. If we combine C_1, C_2, and C_3 into a single constant $C = C_1 + C_2 + C_3$, the formula simplifies to

$$\frac{x^3}{3} - x^2 + 5x + C$$

and still gives *all* the antiderivatives. For this reason we recommend that you go right to the final form even if you elect to integrate term by term. Write

$$\int (x^2 - 2x + 5)\, dx = \int x^2\, dx - \int 2x\, dx + \int 5\, dx$$

$$= \frac{x^3}{3} - x^2 + 5x + C.$$

Find the simplest antiderivative you can for each part and add the constant at the end.

Applications

Torricelli's Law says that if you drain a tank like the one in Figure 6.5, the rate at which the water runs out is a constant times the square root of the water's depth x. The constant depends on the size of the exit valve. In Example 7 we assume the constant is $1/2$.

Example 7 DRAINING A TANK

A right circular cylindrical tank with radius 5 ft and height 16 ft that was initially full of water is being drained at the rate of $0.5\sqrt{x}$ ft³/min. Find a formula for the depth and the amount of water in the tank at any time t. How long will it take the tank to empty?

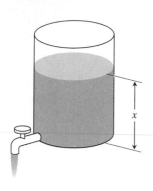

Figure 6.5 The rate at which water runs out is $0.5\sqrt{x}$ ft^3/min. (Example 7)

Solution The volume of a right circular cylinder with radius r and height h is $V = \pi r^2 h$.

Model

The volume of water in the tank (Figure 6.5) is

$$V = \pi r^2 h = \pi (5)^2 x = 25\pi x.$$

Differential equation: $\qquad \dfrac{dV}{dt} = -25\pi \dfrac{dx}{dt} \qquad$ Negative because V is decreasing

$$0.5\sqrt{x} = -25\pi \frac{dx}{dt} \qquad \text{Torricelli's Law}$$

$$\frac{dx}{dt} = -\frac{\sqrt{x}}{50\pi}$$

Initial condition: $\qquad x(0) = 16 \qquad$ The water is 16 ft deep when $t = 0$.

Solve Analytically

We first solve the differential equation.

$$x^{-1/2} \frac{dx}{dt} = -\frac{1}{50\pi} \qquad \text{Terms rearranged}$$

$$\int x^{-1/2} \frac{dx}{dt}\, dt = -\int \frac{1}{50\pi}\, dt \qquad \begin{array}{l}\text{Integrate both sides with} \\ \text{respect to } t.\end{array}$$

$$\int x^{-1/2}\, dx = -\int \frac{1}{50\pi}\, dt \qquad \begin{array}{l} dx = (dx/dt)\, dt, \text{ property} \\ \text{of differentials}\end{array}$$

$$2x^{1/2} = -\frac{1}{50\pi} t + C \qquad \begin{array}{l}\text{Example 6 (constants of} \\ \text{integration combined as one)}\end{array}$$

The initial condition $x(0) = 16$ determines the value of C.

$$2(16)^{1/2} = -\frac{1}{50\pi}(0) + C$$

$$C = 8$$

With $C = 8$, we have

$$2x^{1/2} = -\frac{1}{50\pi} t + 8 \qquad \text{or} \qquad x^{1/2} = 4 - \frac{t}{100\pi}.$$

The formulas we seek are

$$x = \left(4 - \frac{t}{100\pi}\right)^2 \qquad \text{and} \qquad V = 25\pi x = 25\pi\left(4 - \frac{t}{100\pi}\right)^2.$$

Interpret

At any time t, the water in the tank is $(4 - t/(100\pi))^2$ ft deep and the amount of water is $25\pi(4 - t/(100\pi))^2$ ft^3. At $t = 0$, we have $x = 16$ ft and $V = 400\pi$ ft^3, as required. The tank will empty ($V = 0$) in $t = 400\pi$ minutes, which is about 21 hours.

In Example 8 we have to integrate twice, starting with a second derivative, to find the function we seek.

Example 8 FINDING A PROJECTILE'S HEIGHT

A heavy projectile is fired straight up from a platform 3 m above the ground with an initial velocity of 160 m/sec. Assume that the only force affecting the projectile during its flight is from gravity, which produces a downward acceleration of 9.8 m/sec^2. If $t = 0$ when the projectile is fired, find a formula for the projectile's

(a) velocity as a function of time t.

(b) height above the ground as a function of time t.

Solution

Model

To model the problem we draw Figure 6.6 and let s represent the projectile's height above the ground at time t. We assume s to be a twice-differentiable function of t and represent the projectile's velocity and acceleration with the derivatives

$$v = \frac{ds}{dt} \quad \text{and} \quad a = \frac{dv}{dt} = \frac{d^2s}{dt^2}.$$

Since gravity acts in the direction of decreasing s in Figure 6.6, the initial value problem to solve is

Differential equation: $\dfrac{d^2s}{dt^2} = -9.8,$

Initial conditions: $\dfrac{ds}{dt}(0) = 160 \quad \text{and} \quad s(0) = 3.$

Solve Analytically

(a) Integrate both sides of the differential equation with respect to t to find ds/dt.

$$\int \frac{d^2s}{dt^2}\, dt = \int (-9.8)\, dt$$

$$\frac{ds}{dt} = -9.8t + C_1$$

We apply the first initial condition to find C_1.

$$160 = -9.8(0) + C_1 \qquad \frac{ds}{dt}(0) = 160$$

$$C_1 = 160$$

Thus,

$$v = \frac{ds}{dt} = -9.8t + 160.$$

(b) We integrate with respect to t again to find s.

$$\int \frac{ds}{dt}\, dt = \int (-9.8t + 160)\, dt$$

$$s = -4.9t^2 + 160t + C_2$$

We apply the second initial condition to find C_2.

$$3 = -4.9(0)^2 + 160(0) + C_2 \qquad s(0) = 3$$

$$C_2 = 3$$

Therefore,

$$s = -4.9t^2 + 160t + 3.$$

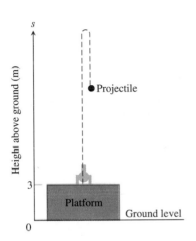

Figure 6.6 (Example 8)

When we solve an initial value problem involving a first derivative, we have one arbitrary constant, as in Examples 1 and 7. When we find a function from its second derivative, we have to deal with two constants, one from each antidifferentiation, as in Example 8. To find a function from its third derivative would require finding the values of three constants, and so on. In each case, the values of the constants are determined by the problem's initial conditions. Each time we find an antiderivative, we need an initial condition to tell us the value of C.

Quick Review 6.1

In Exercises 1–4, determine the amount in the account one year later if $100 is invested at 6% interest compounded k times per year.

1. $k = 1$ (annually)

2. $k = 4$ (quarterly)

3. $k = 12$ (monthly)

4. $k = 365$ (daily)

In Exercises 5–8, find dy/dx.

5. $y = \sin 3x$

6. $y = \tan \dfrac{5}{2} x$

7. $y = Ce^{2x}$, $\quad C$ a constant

8. $y = \ln (x + 2)$

In Exercises 9 and 10, use NINT to graph the function.

9. Graph $\int_1^x dt/t$ for $x > 0$. How does the graph compare with the graph of $y = \ln x$ for $x > 0$?

10. Graph $\int_{-1}^x dt/t$ for $x < 0$. How does the graph compare with the graph of $y = \ln (-x)$ for $x < 0$?

Section 6.1 Exercises

In Exercises 1–6, find an antiderivative for the function. Confirm your answer by differentiation.

1. $x^2 - 2x + 1$

2. $-3x^{-4}$

3. $x^2 - 4\sqrt{x}$

4. $8 + \csc x \cot x$

5. e^{4x}

6. $\dfrac{1}{x + 3}$

In Exercises 7–24, evaluate the integral.

7. $\displaystyle\int (x^5 - 6x + 3)\, dx$

8. $\displaystyle\int (-x^{-3} + x - 1)\, dx$

9. $\displaystyle\int \left(e^{t/2} - \dfrac{5}{t^2}\right) dt$

10. $\displaystyle\int \dfrac{4}{3}\sqrt[3]{t}\, dt$

11. $\displaystyle\int \left(x^3 - \dfrac{1}{x^3}\right) dx$

12. $\displaystyle\int \left(\sqrt[3]{x} + \dfrac{1}{\sqrt[3]{x}}\right) dx$

13. $\displaystyle\int \dfrac{1}{3}x^{-2/3}\, dx$

14. $\displaystyle\int (3 \sin x - \sin 3x)\, dx$

15. $\displaystyle\int \dfrac{\pi}{2} \cos \dfrac{\pi x}{2}\, dx$

16. $\displaystyle\int 2 \sec t \tan t\, dt$

17. $\displaystyle\int \left(\dfrac{2}{x - 1} + \dfrac{1}{x}\right) dx$

18. $\displaystyle\int \left(\dfrac{1}{x - 2} + \sin 5x - e^{-2x}\right) dx$

19. $\displaystyle\int 5 \sec^2 5r\, dr$

20. $\displaystyle\int \csc^2 7t\, dt$

21. $\displaystyle\int \cos^2 x\, dx$ (*Hint:* $\cos^2 x = \dfrac{1 + \cos 2x}{2}$)

22. $\displaystyle\int \sin^2 x\, dx$ (*Hint:* $\sin^2 x = \dfrac{1 - \cos 2x}{2}$)

23. $\displaystyle\int \tan^2 \theta\, d\theta$

24. $\displaystyle\int \cot^2 t\, dt$

In Exercises 25 and 26, **(a)** determine which graph shows the solution of the initial value problem without actually solving the problem. **(b) Writing to Learn** Explain why you eliminated two of the possibilities.

25. $\dfrac{dy}{dx} = \dfrac{1}{1 + x^2}$, $\quad y(0) = \dfrac{\pi}{2}$

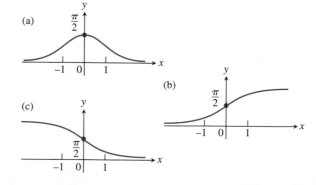

(a)

(b)

(c)

26. $\dfrac{dy}{dx} = -x$, $y(-1) = 1$

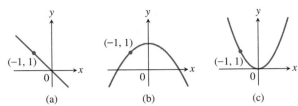

(a)   (b)   (c)

In Exercises 27–30, solve the initial value problem. Support your answer by overlaying your solution on a slope field for the differential equation.

27. $\dfrac{dy}{dx} - 2x - 1$, $y(2) = 0$ **28.** $\dfrac{dy}{dx} = \dfrac{1}{x^2} + x$, $y(2) = 1$

29. $\dfrac{dy}{dx} = \sec^2 x$, $y(\pi/4) = -1$ **30.** $\dfrac{dy}{dx} = x^{-2/3}$, $y(-1) = -5$

In Exercises 31–38, solve the initial value problem.

31. $\dfrac{dy}{dx} = 9x^2 - 4x + 5$, $y(-1) = 0$

32. $\dfrac{dy}{dx} = \cos x + \sin x$, $y(\pi) = 1$

33. $\dfrac{dy}{dt} = 2e^{-t}$, $y(\ln 2) = 0$

34. $\dfrac{dy}{dx} = \dfrac{1}{x}$, $y(e^3) = 0$

35. $\dfrac{d^2y}{d\theta^2} - \sin \theta$, $y(0) = -3$, $y'(0) = 0$

36. $\dfrac{d^2y}{dx^2} = 2 - 6x$, $y(0) = 1$, $y'(0) = 4$

37. $\dfrac{d^3y}{dt^3} = \dfrac{1}{t^3}$, $y(1) = 1$, $y'(1) = 3$, $y''(1) = 2$

38. $\dfrac{d^4y}{d\theta^4} = \sin \theta + \cos \theta$, $y(0) = -3$, $y'(0) = -1$,

$y''(0) = -1$, $y^3(0) = -3$

In Exercises 39–42, the velocity $v = ds/dt$ or acceleration $a = dv/dt$ of a body moving along a coordinate line is given. Find the body's position s at time t.

39. $v = 9.8t + 5$, $s(0) = 10$ **40.** $v = \sin \pi t$, $s(1) = 0$

41. $a = 32$, $s(0) = 0$, $v(0) = 20$

42. $a = \cos t$, $s(0) = 1$, $v(0) = -1$

In Exercises 43 and 44, *work in groups of two or three*. Draw a possible graph for the function f with the given slope field that satisfies the stated condition.

43. $f(0) = 0$

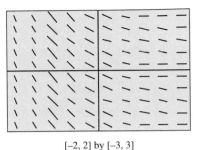

[-2, 2] by [-3, 3]

44. $f(-1) = -2$

[-2, 3] by [-3, 3]

In Exercises 45–48, confirm the integration formula by differentiation.

45. $\displaystyle\int \dfrac{dx}{1 + x^2} = \tan^{-1} x + C$ **46.** $\displaystyle\int \dfrac{dx}{\sqrt{1 - x^2}} = \sin^{-1} x + C$

47. $\displaystyle\int \dfrac{dx}{|x|\sqrt{x^2 - 1}} = \sec^{-1} x + C$

48. $\displaystyle\int \dfrac{dx}{\sqrt{1 - x^2}} = -\cos^{-1} x + C$

Exploration

49. Let $\dfrac{dy}{dx} = x - \dfrac{1}{x^2}$.

(a) Find a solution to the differential equation in the interval $(0, \infty)$ that satisfies $y(1) = 2$.

(b) Find a solution to the differential equation in the interval $(-\infty, 0)$ that satisfies $y(-1) = 1$.

(c) Show that the following piecewise function is a solution to the differential equation for any values of C_1 and C_2.

$$y = \begin{cases} \dfrac{1}{x} + \dfrac{x^2}{2} + C_1, & x < 0 \\[2mm] \dfrac{1}{x} + \dfrac{x^2}{2} + C_2, & x > 0 \end{cases}$$

(d) Choose values for C_1 and C_2 so that the solution in part (c) agrees with the solutions in parts (a) and (b).

(e) Choose values for C_1 and C_2 so that the solution in part (c) satisfies $y(2) = -1$ and $y(-2) = 2$.

50. *Marginal Revenue* Suppose that the marginal revenue when x thousand units are sold is

$$\frac{dr}{dx} = 3x^2 - 6x + 12$$

dollars per unit. Find the revenue function $r(x)$ if $r(0) = 0$.

51. *Marginal Cost* Suppose that the marginal cost of manufacturing an item when x thousand units are produced is

$$\frac{dc}{dx} = 3x^2 - 12x + 15$$

dollars per item. Find the cost function $c(x)$ if $c(0) = 400$.

52. *Indefinite Integration* Suppose that

$$f(x) = \frac{d}{dx}(x^2 e^x) \quad \text{and} \quad g(x) = \frac{d}{dx}(x \sin x).$$

Evaluate each integral.

(a) $\int f(x)\, dx$

(b) $\int g(x)\, dx$

(c) $\int [-f(x)]\, dx$

(d) $\int [-g(x)]\, dx$

(e) $\int [f(x) + g(x)]\, dx$

(f) $\int [f(x) - g(x)]\, dx$

(g) $\int [x + f(x)]\, dx$

(h) $\int [g(x) - 4]\, dx$

53. *Stopping a Car in Time* You are driving along a highway at a steady 60 mph (88 ft/sec) when you see an accident ahead and slam on the brakes. What constant deceleration is required to stop your car in 242 ft? To find out, do the following.

(a) Solve the initial value problem

Differential equation: $\dfrac{d^2 s}{dt^2} = -k \quad (k \text{ constant})$,

Initial conditions: $\dfrac{ds}{dt} = 88 \text{ and } s = 0 \text{ when } t = 0$,

measuring time and distance from when the brakes are applied.

(b) Find the value of t that makes $ds/dt = 0$. (The answer will involve k.)

(c) Find the value of k that makes $s = 242$ for the value of t you found in (b).

54. *Stopping a Motorcycle* The State of Illinois Cycle Rider Safety Program requires riders to be able to brake from 30 mph (44 ft/sec) to 0 in 45 ft. What constant deceleration does it take to do that? (*Hint:* See Exercise 53.)

55. *The Hammer and the Feather* When Apollo 15 astronaut David Scott dropped a hammer and a feather on the moon to demonstrate that in a vacuum all bodies fall with the same (constant) acceleration, he dropped them from about 4 ft above the ground. The television footage of the event shows the hammer and feather falling more slowly than on Earth, where, in a vacuum, they would have taken

only half a second to fall the 4 ft. How long did it take the hammer and feather to fall 4 ft on the moon? To find out, solve the following initial value problem for s as a function of t. Then find the value of t that makes s equal to 0.

Differential equation: $\dfrac{d^2 s}{dt^2} = -5.2 \text{ ft/sec}^2$

Initial conditions: $\dfrac{ds}{dt} = 0 \text{ and } s = 4 \text{ when } t = 0$

56. *Motion with Constant Acceleration* The standard equation for the position s of a body moving with a constant acceleration a along a coordinate line is

$$s = \frac{a}{2} t^2 + v_0 t + s_0,$$

where v_0 and s_0 are the body's velocity and position at time $t = 0$. Derive this equation by setting up and solving an initial value problem.

57. *Draining a Tank* A conical tank with radius 4 ft and height 10 ft that was initially full of water is being drained at the rate of $(1/6)\sqrt{h}$ (see figure). Find a formula for the depth and the amount of water in the tank at any time t.

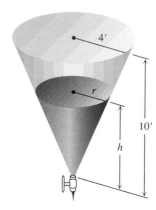

58. *Continuous Compound Interest* Suppose $500 is invested in an account that pays 4.75% interest compounded continuously.

(a) Find a formula for the amount in the account at any time t.

(b) When will the amount in the account be double the initial value?

59. *Continuous Compound Interest* Suppose $1200 is invested in an account that pays 6.25% interest compounded continuously.

(a) Find a formula for the amount in the account at any time t.

(b) When will the amount in the account be triple the initial value?

In Exercises 60 and 61, **(a)** use Part 1 of the Fundamental Theorem of Calculus to evaluate the indefinite integral. **(b)** Find the antiderivative whose graph passes through the point $(0, 1)$.

60. $\int x^2 \cos x\, dx$

61. $\int x e^x\, dx$

Extending the Ideas

62. *Solving Differential Equations* Let

$$\frac{d^2y}{dx^2} = 6x.$$

(a) Find a solution to the differential equation that is continuous for $-\infty < x < \infty$ and whose graph passes through the point $(0, 1)$ and has a horizontal tangent there.

(b) Writing to Learn How many functions like this are there? How do you know?

In Exercises 63–66, *work in groups of two or three.* Generate a slope field for the differential equation in a viewing window large enough to display the listed points. Then add to your display the solution curves through these points. What can you learn from the concavity of the solution curves?

63. $y' = xy;$ $(1, 1),\ (-1, 2),\ (0, -2),\ (-2, -1)$

64. $y' = y;$ $(0, 1),\ (0, 2),\ (0, -1)$

65. $y' = 2(y - 4);$ $(0, 1),\ (0, 4),\ (0, 5)$

66. $y' = y^2;$ $(0, 1),\ (0, 2),\ (0, -1),\ (0, 0)$

67. *Solving Differential Equations* Let

$$\frac{dy}{dx} = \frac{1}{x}.$$

(a) Show that $y = \ln x + C$ is a solution to the differential equation in the interval $(0, \infty)$.

(b) Show that $y = \ln(-x) + C$ is a solution to the differential equation in the interval $(-\infty, 0)$.

(c) Writing to Learn Explain why $y = \ln|x| + C$ is a solution to the differential equation in the domain $(-\infty, 0) \cup (0, \infty)$.

(d) Show that the function

$$y = \begin{cases} \ln(-x) + C_1, & x < 0 \\ \ln x + C_2, & x > 0 \end{cases}$$

is a solution to the differential equation for any values of C_1 and C_2.

6.2 Integration by Substitution

Power Rule in Integral Form • Trigonometric Integrands • Substitution in Indefinite Integrals • Substitution in Definite Integrals • Separable Differential Equations

Power Rule in Integral Form

When u is a differentiable function of x and n is a real number different from -1, the Chain Rule tells us that

$$\frac{d}{dx}\left(\frac{u^{n+1}}{n + 1}\right) = u^n \frac{du}{dx}.$$

By reversing this derivative formula, we obtain the integral formula

$$\int \left(u^n \frac{du}{dx}\right) dx = \frac{u^{n+1}}{n + 1} + C.$$

When the integrand of this integral is rewritten in the simpler *differential* form $u^n\, du$, we obtain the *Power Rule for Integration.*

Power Rule for Integration

If u is any differentiable function of x, then

$$\int u^n\, du = \frac{u^{n+1}}{n + 1} + C, \qquad n \neq -1.$$

A change of variables can often turn an unfamiliar integral into one that we can evaluate. The method for doing this is called the **substitution method of integration.** In Example 1 we use substitution to put the integral in the form

$$\int u^n\, du.$$

Example 1 USING SUBSTITUTION

Evaluate $\int (x + 2)^5 \, dx$.

Solution

$$\int (x + 2)^5 \, dx = \int u^5 du \qquad u = x + 2, \, du = d(x + 2) = dx$$

$$= \frac{u^6}{6} + C \qquad \text{Power Rule}$$

$$= \frac{(x + 2)^6}{6} + C \qquad u = x + 2$$

Example 2 USING SUBSTITUTION

Evaluate $\int \sqrt{4x - 1} \, dx$.

Solution

$$\int \sqrt{4x - 1} \, dx = \int u^{1/2} \cdot \frac{1}{4} \, du \qquad \begin{array}{l} u = 4x - 1, \, du = d(4x - 1) = 4dx, \\ dx = du/4 \end{array}$$

$$= \frac{1}{4} \int u^{1/2} \, du \qquad \text{Constant Multiple Rule}$$

$$= \frac{1}{4} \cdot \frac{u^{3/2}}{3/2} + C \qquad \text{Power Rule}$$

$$= \frac{1}{6} u^{3/2} + C \qquad \text{Simplify.}$$

$$= \frac{1}{6} (4x - 1)^{3/2} + C \qquad u = 4x - 1$$

Exploration 1 **Supporting Indefinite Integrals Graphically**

Consider the integral $\int \sqrt{1 + x^2} \cdot 2x \, dx$.

1. Use the substitution $u = 1 + x^2$ to show that

$$\int \sqrt{1 + x^2} \cdot 2x \, dx = \frac{2}{3} \left(1 + x^2 \right)^{3/2} + C.$$

2. Explain why

$$y_1 = \frac{2}{3} \left(1 + x^2 \right)^{3/2} \qquad \text{and} \qquad y_2 = \int_0^x \sqrt{1 + t^2} \cdot 2t \, dt$$

are both antiderivatives of $2x\sqrt{1 + x^2}$.

3. Graph y_1 and y_2 in the same viewing window. Find a value of C for which $y_1 = y_2 + C$. Explain why this shift is possible.

4. If the factor $2x$ is removed from the integrand, the resulting integral has no easily obtained explicit form. You can understand why by trying to find a function whose derivative is $\sqrt{1 + x^2}$. Graph $\int_a^x \sqrt{1 + t^2} \, dt$ for $a = -4, -2, 0, 2, 4$ in $[-10, 10]$ by $[-30, 30]$ to see what five antiderivatives of $\sqrt{1 + x^2}$ look like.

5. Explain why the five curves in part 4 are solution curves of the differential equation $dy/dx = \sqrt{1 + x^2}$.

Trigonometric Integrands

When u is a differentiable function of x, the six trigonometric functions of u are also differentiable functions of x. Their derivative formulas, obtained using the Chain Rule, can be reversed to obtain integral formulas. For example,

$$\frac{d}{dx} \sin u = \cos u \frac{du}{dx} \quad \text{gives} \quad \int \cos u \, du = \sin u + C.$$

In this way we obtain the six formulas.

Trigonometric Integral Formulas

If u is any differentiable function of x, then

1. $\displaystyle\int \cos u \, du = \sin u + C.$ **2.** $\displaystyle\int \sin u \, du = -\cos u + C.$

3. $\displaystyle\int \sec^2 u \, du = \tan u + C.$ **4.** $\displaystyle\int \csc^2 u \, du = -\cot u + C.$

5. $\displaystyle\int \sec u \tan u \, du = \sec u + C.$ **6.** $\displaystyle\int \csc u \cot u \, du = -\csc u + C.$

Example 3   USING SUBSTITUTION

Evaluate $\int \cos (7x + 5) \, dx.$

Solution

$$\int \cos (7x + 5) \, dx = \int \cos u \cdot \frac{1}{7} \, du \qquad \begin{aligned} &u = 7x + 5, \, du = 7dx, \\ &dx = du/7 \end{aligned}$$

$$= \frac{1}{7} \int \cos u \, du \qquad \text{Constant Multiple Rule}$$

$$= \frac{1}{7} \sin u + C \qquad \text{Formula 1}$$

$$= \frac{1}{7} \sin (7x + 5) + C \quad u = 7x + 5$$

Example 4   USING IDENTITIES AND SUBSTITUTION

Evaluate $\displaystyle\int \frac{1}{\cos^2 2x} \, dx.$

Solution

$$\int \frac{1}{\cos^2 2x} \, dx = \int \sec^2 2x \, dx \qquad \frac{1}{\cos 2x} = \sec 2x$$

$$= \int \sec^2 u \cdot \frac{1}{2} \, du \quad u = 2x, \, du = 2dx, \, dx = (1/2) \, du$$

$$= \frac{1}{2} \int \sec^2 u \, du$$

$$= \frac{1}{2} \tan u + C \qquad \text{Formula 3}$$

$$= \frac{1}{2} \tan 2x + C \qquad u = 2x$$

Confirm Analytically

$$\frac{d}{dx}\left(\frac{1}{2} \tan 2x + C\right) = \frac{1}{2}(\sec^2 2x)(2) = \sec^2 2x = \frac{1}{\cos^2 2x}$$

The integral of the tangent function involves the logarithm function as shown in Example 5. The same is true for the cotangent, secant, and cosecant functions, as you will see if you do Exercises 23, 29, and 30.

Example 5 USING SUBSTITUTION

Evaluate $\int \tan x \, dx$.

Solution

$$\int \tan x \, dx = \int \frac{\sin x}{\cos x} \, dx \qquad \tan x = \frac{\sin x}{\cos x}$$

$$= \int \frac{-du}{u} \qquad u = \cos x, \, du = -\sin x \, dx$$

$$= -\int \frac{du}{u}$$

$$= -\ln|u| + C \qquad \text{Integrate.}$$

$$= \ln \frac{1}{|u|} + C \qquad \ln \frac{1}{|u|} = -\ln|u|$$

$$= \ln \frac{1}{|\cos x|} + C \qquad u = \cos x$$

$$= \ln|\sec x| + C \qquad \frac{1}{\cos x} = \sec x$$

The Substitution Method

Take these steps to evaluate the integral

$$\int f(g(x)) \cdot g'(x) \, dx,$$

when f and g' are continuous functions.

1. Substitute $u = g(x)$ and $du = g'(x) \, dx$ to obtain the integral

$$\int f(u) \, du.$$

2. Integrate with respect to u.
3. Replace u by $g(x)$ in the result.

Substitution in Indefinite Integrals

The substitutions in the preceding Examples and Exploration are all instances of the following general method.

$$\int f(g(x)) \cdot g'(x) \, dx = \int f(u) \, du \qquad \text{1. Substitute } u = g(x), \, du = g'(x) \, dx.$$

$$= F(u) + C \qquad \text{2. Evaluate by finding an antiderivative } F(u) \text{ of } f(u).$$

$$= F(g(x)) + C \qquad \text{3. Replace } u \text{ by } g(x).$$

These three steps are the steps of the substitution method of integration. The method works because $F(g(x))$ is an antiderivative of $f(g(x)) \cdot g'(x)$ whenever F is an antiderivative of f.

Example 6 USING THE SUBSTITUTION METHOD

Evaluate $\int (x^2 + 2x - 3)^2(x + 1)\,dx$.

Solution Let $u = x^2 + 2x - 3$, then $du = (2x + 2)\,dx = 2(x + 1)\,dx$ and $(x + 1)\,dx = (1/2)\,du$.

$$\int (x^2 + 2x - 3)^2(x + 1)\,dx = \int u^2 \cdot \frac{1}{2}du = \frac{1}{2}\int u^2\,du \qquad \text{Step 1}$$

$$= \frac{1}{2} \cdot \frac{u^3}{3} + C = \frac{1}{6}u^3 + C \qquad \text{Step 2}$$

$$= \frac{1}{6}(x^2 + 2x - 3)^3 + C \qquad \text{Step 3}$$

Example 7 USING THE SUBSTITUTION METHOD

$$\int \sin^4 x \cos x\,dx = \int u^4\,du \qquad u = \sin x,\ du = \cos x\,dx$$

$$= \frac{u^5}{5} + C \qquad \text{Integrate.}$$

$$= \frac{\sin^5 x}{5} + C \qquad u = \sin x$$

Substitution in Definite Integrals

To evaluate a *definite* integral by substitution, we can avoid Step 3 of the substitution method (replacing u by $g(x)$). Here is the alternative procedure.

Substitution in Definite Integrals

Substitute $u = g(x)$, $du = g'(x)\,dx$, and integrate with respect to u from $u = g(a)$ to $u = g(b)$.

$$\int_a^b f(g(x)) \cdot g'(x)\,dx = \int_{g(a)}^{g(b)} f(u)\,du$$

Example 8 EVALUATING DEFINITE INTEGRALS

Evaluate $\int_0^{\pi/4} \tan x \sec^2 x\,dx$.

Solution Let $u = \tan x$, then $du = \sec^2 x\,dx$. Notice that

$$u(0) = \tan 0 = 0 \quad \text{and} \quad u\left(\frac{\pi}{4}\right) = \tan \frac{\pi}{4} = 1.$$

Thus,

$$\int_0^{\pi/4} \tan x \sec^2 x\,dx = \int_0^1 u\,du$$

$$= \frac{u^2}{2}\Bigg]_0^1$$

$$= \frac{1}{2} - 0 = \frac{1}{2}.$$

Exploration 2 compares two methods for evaluating a definite integral.

Exploration 2 **Two Routes to the Integral**

Evaluate $\int_{-1}^{1} 3x^2\sqrt{x^3 + 1}\, dx$ by the two methods below.

1. Substitute $u = x^3 + 1$, $du = 3x^2 dx$ and integrate from $u(-1)$ to $u(1)$.

2. Substitute $u = x^3 + 1$, $du = 3x^2 dx$. Then find an antiderivative with respect to u. Replace u with $x^3 + 1$. Evaluate using the original x-limits of integration, $x = -1$ and 1.

Separable Differential Equations

A differential equation $y' = f(x, y)$ is **separable** if f can be expressed as a product of a function of x and a function of y. The differential equation then has the form

$$\frac{dy}{dx} = g(x)h(y). \tag{1}$$

If $h(y) \neq 0$, we can **separate the variables** by dividing both sides of the equation by $h(y)$, obtaining, in succession,

$$\frac{1}{h(y)}\frac{dy}{dx} = g(x)$$

$$\int \frac{1}{h(y)}\frac{dy}{dx}dx = \int g(x)\, dx \qquad \text{Integrate both sides with respect to } x.$$

$$\int \frac{1}{h(y)}dy = \int g(x)\, dx. \qquad dy = (dy/dx)\, dx \tag{2}$$

With x and y now separated, the integrated equation provides the solution we seek by expressing y either explicitly or implicitly as a function of x, up to an arbitrary constant.

In practice, we go directly from Equation 1 to Equation 2 without writing the intervening steps.

Example 9 SOLVING BY SEPARATION OF VARIABLES

Solve the differential equation

$$\frac{dy}{dx} = 2x(1 + y^2)e^{x^2}.$$

Solution Since $h(y) = 1 + y^2$ is never zero, we can solve the equation by separating the variables. Dividing both sides by $(1 + y^2)$ and integrating as above, we have

$$\frac{dy}{dx} = 2x(1 + y^2)e^{x^2}$$

$$\int \frac{1}{1 + y^2}dy = \int 2xe^{x^2}\, dx. \qquad \text{Separate the variables and integrate.}$$

On the left, $\displaystyle\int \frac{1}{1 + y^2}\, dy = \tan^{-1} y + C.$

On the right, $\displaystyle\int 2xe^{x^2}\, dx = \int e^u\, du$ $u = x^2,\ du = 2x\, dx$

$$= e^u + C$$
$$= e^{x^2} + C.$$

Combining the constants of integration, we have

$$\tan^{-1} y = e^{x^2} + C.$$ *C* now represents the com- bined constants of integration.

The equation $\tan^{-1} y = e^{x^2} + C$ gives *y* as an implicit function of *x*. In this case, we can solve for *y* as an explicit function of *x* by taking the tangent of both sides:

$$\tan (\tan^{-1} y) = \tan (e^{x^2} + C)$$
$$y = \tan (e^{x^2} + C).$$

Quick Review 6.2

In Exercises 1 and 2, evaluate the definite integral.

1. $\displaystyle\int_0^2 x^4\, dx$

2. $\displaystyle\int_1^5 \sqrt{x - 1}\, dx$

In Exercises 3–10, find dy/dx.

3. $y = \displaystyle\int_2^x 3^t\, dt$

4. $y = \displaystyle\int_0^x 3^t\, dt$

5. $y = (x^3 - 2x^2 + 3)^4$

6. $y = \sin^2 (4x - 5)$

7. $y = \ln \cos x$

8. $y = \ln \sin x$

9. $y = \ln (\sec x + \tan x)$

10. $y = \ln (\csc x + \cot x)$

Section 6.2 Exercises

In Exercises 1–8, use the indicated substitution to evaluate the integral. Confirm your answer by differentiation.

1. $\displaystyle\int \sin 3x\, dx, \quad u = 3x$

2. $\displaystyle\int x \cos (2x^2)\, dx, \quad u = 2x^2$

3. $\displaystyle\int \sec 2x \tan 2x\, dx, \quad u = 2x$

4. $\displaystyle\int 28(7x - 2)^3\, dx, \quad u = 7x - 2$

5. $\displaystyle\int \frac{dx}{x^2 + 9}, \quad u = \frac{x}{3}$

6. $\displaystyle\int \frac{9r^2\, dr}{\sqrt{1 - r^3}}, \quad u = 1 - r^3$

7. $\displaystyle\int \left(1 - \cos \frac{t}{2}\right)^2 \sin \frac{t}{2}\, dt, \quad u = 1 - \cos \frac{t}{2}$

8. $\displaystyle\int 8(y^4 + 4y^2 + 1)^2 (y^3 + 2y)\, dy, \quad u = y^4 + 4y^2 + 1$

In Exercises 9–30, use substitution to evaluate the integral.

9. $\displaystyle\int \frac{dx}{(1 - x)^2}$

10. $\displaystyle\int \sec^2 (x + 2)\, dx$

11. $\displaystyle\int \sqrt{\tan x} \sec^2 x\, dx$

12. $\int \sec\left(\theta + \dfrac{\pi}{2}\right) \tan\left(\theta + \dfrac{\pi}{2}\right) d\theta$

13. $\int_e^6 \dfrac{dx}{x \ln x}$

14. $\int_{-\pi/4}^{\pi/4} \tan^2 x \sec^2 x \, dx$

15. $\int \cos(3z + 4) \, dz$

16. $\int \sqrt{\cot x} \, \csc^2 x \, dx$

17. $\int \dfrac{\ln^6 x}{x} \, dx$

18. $\int \tan^7 \dfrac{x}{2} \sec^2 \dfrac{x}{2} \, dx$

19. $\int s^{1/3} \cos(s^{4/3} - 8) \, ds$

20. $\int \dfrac{dx}{\sin^2 3x}$

21. $\int \dfrac{\sin(2t + 1)}{\cos^2(2t + 1)} \, dt$

22. $\int \dfrac{6 \cos t}{(2 + \sin t)^2} \, dt$

23. $\int_{\pi/4}^{3\pi/4} \cot x \, dx$

24. $\int_0^7 \dfrac{dx}{x + 2}$

25. $\int_{-1}^3 \dfrac{x \, dx}{x^2 + 1}$

26. $\int_0^5 \dfrac{40 \, dx}{x^2 + 25}$

27. $\int \dfrac{dx}{\cot 3x}$

28. $\int \dfrac{dx}{\sqrt{5x + 8}}$

29. $\int \sec x \, dx$ (*Hint:* Multiply the integrand by

$$\frac{\sec x + \tan x}{\sec x + \tan x}$$

and then use a substitution to integrate the result.)

30. $\int \csc x \, dx$ (*Hint:* Multiply the integrand by

$$\frac{\csc x + \cot x}{\csc x + \cot x}$$

and then use a substitution to integrate the result.)

In Exercises 31–38, make a u-substitution and integrate from $u(a)$ to $u(b)$.

31. $\int_0^3 \sqrt{y + 1} \, dy$

32. $\int_0^1 r\sqrt{1 - r^2} \, dr$

33. $\int_{-\pi/4}^0 \tan x \sec^2 x \, dx$

34. $\int_{-1}^1 \dfrac{5r}{(4 + r^2)^2} \, dr$

35. $\int_0^1 \dfrac{10\sqrt{\theta}}{(1 + \theta^{3/2})^2} \, d\theta$

36. $\int_{-\pi}^{\pi} \dfrac{\cos x}{\sqrt{4 + 3 \sin x}} \, dx$

37. $\int_0^1 \sqrt{t^5 + 2t} \, (5t^4 + 2) \, dt$

38. $\int_0^{\pi/6} \cos^{-3} 2\theta \sin 2\theta \, d\theta$

In Exercises 39–42, solve the differential equation by separation of variables.

39. $\dfrac{dy}{dx} = (y + 5)(x + 2)$

40. $\dfrac{dy}{dx} = x\sqrt{y} \cos^2 \sqrt{y}$

41. $\dfrac{dy}{dx} = (\cos x)e^{y + \sin x}$

42. $\dfrac{dy}{dx} = e^{x - y}$

In Exercises 43 and 44, solve the initial value problem by separation of variables.

43. $\dfrac{dy}{dx} = -2xy^2$, $y(1) = 0.25$

44. $\dfrac{dy}{dx} = \dfrac{4\sqrt{y} \ln x}{x}$, $y(e) = 1$

Explorations

45. *Constant of Integration* Consider the integral

$$\int \sqrt{x + 1} \, dx.$$

(a) Show that $\displaystyle\int \sqrt{x + 1} \, dx = \dfrac{2}{3}(x + 1)^{3/2} + C$.

(b) Writing to Learn Explain why

$$y_1 = \int_0^x \sqrt{t + 1} \, dt \quad \text{and} \quad y_2 = \int_3^x \sqrt{t + 1} \, dt$$

are antiderivatives of $\sqrt{x + 1}$.

(c) Use a table of values for $y_1 - y_2$ to find the value of C for which $y_1 = y_2 + C$.

(d) Writing to Learn Give a convincing argument that

$$C = \int_0^3 \sqrt{x + 1} \, dx.$$

46. *Making Connections* Work in groups of two or three. Suppose that

$$\int f(x) \, dx = F(x) + C.$$

(a) Explain how you can use the derivative of $F(x) + C$ to confirm the integration is correct.

(b) Explain how you can use a slope field of f and the graph of $y = F(x)$ to support your evaluation of the integral.

(c) Explain how you can use the graphs of $y_1 = F(x)$ and $y_2 = \int_0^x f(t) \, dt$ to support your evaluation of the integral.

(d) Explain how you can use a table of values for $y_1 - y_2$, y_1 and y_2 defined as in (c), to support your evaluation of the integral.

(e) Explain how you can use graphs of f and NDER of $F(x)$ to support your evaluation of the integral.

(f) Illustrate (a)–(e) for $f(x) = \dfrac{x}{\sqrt{x^2 + 1}}$. ∎

Two Routes to the Integral In Exercises 47 and 48, make a substitution $u = \cdots$ (an expression in x), $du = \cdots$. Then

    (a) integrate with respect to u from $u(a)$ to $u(b)$.

    (b) find an antiderivative with respect to u, replace u by the expression in x, then evaluate from a to b.

47. $\displaystyle\int_0^1 \frac{x^3}{\sqrt{x^4 + 9}}\, dx$

48. $\displaystyle\int_{\pi/6}^{\pi/3} (1 - \cos 3x) \sin 3x\, dx$

49. Show that

$$y = \ln\left|\frac{\cos 3}{\cos x}\right| + 5$$

is the solution to the initial value problem

$$\frac{dy}{dx} = \tan x, \quad f(3) = 5.$$

(See the discussion following Example 4, Section 5.4.)

Extending the Ideas

50. *Connecting Antiderivatives* Consider the integral

$$\int \csc^2 2\theta \cot 2\theta\, d\theta.$$

    (a) Use the substitution $u = \cot 2\theta$ to write the integral in the form $F_1(\theta) + C$. Identify F_1.

    (b) Use the substitution $u = \csc 2\theta$ to write the integral in the form $F_2(\theta) + C$. Identify F_2.

    (c) Confirm that each integration is correct.

    (d) Determine a constant b so that $F_1 = F_2 + b$.

51. *Connecting Antiderivatives* Consider the integral

$$\int 2 \sin x \cos x\, dx.$$

    (a) Use the substitution $u = \sin x$ to evaluate the integral.

    (b) Use the substitution $u = \cos x$ to evaluate the integral.

    (c) Rewrite the integrand as $\sin 2x$ and evaluate the integral.

    (d) Confirm that each integration is correct.

6.3 Integration by Parts

Product Rule in Integral Form • Repeated Use • Solving for the Unknown Integral • Tabular Integration

Product Rule in Integral Form

When u and v are differentiable functions of x, the Product Rule for differentiation tells us that

$$\frac{d}{dx}(uv) = u\frac{dv}{dx} + v\frac{du}{dx}.$$

Integrating both sides with respect to x and rearranging leads to the integral equation

$$\int \left(u\frac{dv}{dx}\right) dx = \int \left(\frac{d}{dx}(uv)\right) dx - \int \left(v\frac{du}{dx}\right) dx$$

$$= uv - \int \left(v\frac{du}{dx}\right) dx.$$

When this equation is written in the simpler differential notation we obtain the following formula.

Integration by Parts Formula

$$\int u\, dv = uv - \int v\, du$$

This formula expresses one integral, $\int u\, dv$, in terms of a second integral, $\int v\, du$. With a proper choice of u and v, the second integral may be easier to evaluate than the first. This is the reason for the importance of the formula. When faced with an integral that we cannot handle analytically, we can replace it by one with which we might have more success.

Example 1 USING INTEGRATION BY PARTS

Evaluate $\int x \cos x \, dx$.

Solution We use the formula $\int u \, dv = uv - \int v \, du$ with

$$u = x, \quad dv = \cos x \, dx.$$

To complete the formula, we take the differential of u and find the simplest antiderivative of $\cos x$.

$$du = dx \quad v = \sin x$$

Then,

$$\int x \cos x \, dx = x \sin x - \int \sin x \, dx = x \sin x + \cos x + C.$$

Let's examine the choices available for u and v in Example 1.

Example 2 INVESTIGATING INTEGRATION BY PARTS

What are the choices for u and dv when we apply integration by parts to

$$\int x \cos x \, dx = \int u \, dv?$$

Which choices lead to a successful evaluation of the original integral?

Solution There are four possible choices.

1. $u = 1$ and $dv = x \cos x \, dx$ **2.** $u = x$ and $dv = \cos x \, dx$

3. $u = x \cos x$ and $dv = dx$ **4.** $u = \cos x$ and $dv = x \, dx$

Choice 1 won't do because we still don't know how to integrate $dv = x \cos x \, dx$ to get v.

Choice 2 works well as we saw in Example 1.

Choice 3 leads to $u = x \cos x, \quad\quad dv = dx,$

$$du = (\cos x - x \sin x) \, dx, \quad v = x,$$

and the new integral

$$\int v \, du = \int (x \cos x - x^2 \sin x) \, dx.$$

This is worse than the integral we started with.

Choice 4 leads to $u = \cos x, \quad\quad dv = x \, dx,$

$$du = -\sin x \, dx, \quad v = x^2/2,$$

and the new integral

$$\int v \, du = -\int \frac{x^2}{2} \sin x \, dx.$$

This, too, is worse.

LIPET

If you are wondering what to choose for *u,* here is what we usually do. Our first choice is a natural logarithm (L), if there is one. If there isn't, we look for an inverse trigonometric function (I). If there isn't one of these either, look for a polynomial (P). Barring that, look for an exponential (E) or a trigonometric function (T). That's the preference order: **L I P E T.**

In general, we want *u* to be something that simplifies when differentiated, and *dv* to be something that remains manageable when integrated.

The goal of integration by parts is to go from an integral $\int u\, dv$ that we don't see how to evaluate to an integral $\int v\, du$ that we can evaluate. Keep in mind that integration by parts does not always work.

Example 3 FINDING AREA

Find the area of the region bounded by the curve $y = xe^{-x}$ and the x-axis from $x = 0$ to $x = 3$.

Solution The region is shaded in Figure 6.7. Its area is

$$\int_0^3 xe^{-x}\, dx.$$

Solve Analytically

We use the formula $\int u\, dv = uv - \int v\, du$ with

$$u = x, \qquad dv = e^{-x}\, dx,$$
$$du = dx, \qquad v = -e^{-x}.$$

Then

$$\int xe^{-x}\, dx = -xe^{-x} - \int (-e^{-x})\, dx$$

$$= -xe^{-x} + \int e^{-x}\, dx$$

$$= -xe^{-x} - e^{-x} + C.$$

Thus,

$$\int_0^3 xe^{-x}\, dx = \left[-xe^{-x} - e^{-x} \right]_0^3$$

$$= (-3e^{-3} - e^{-3}) - (-e^0) = 1 - 4e^{-3} \approx 0.80.$$

Support Numerically

NINT $(xe^{-x}, x, 0, 3)$ gives the same value.

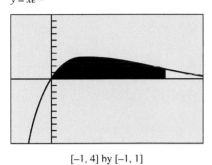

$y = xe^{-x}$

[−1, 4] by [−1, 1]

Figure 6.7 The region in Example 3.

| Exploration 1 | Evaluating and Checking Integrals |

Consider $\int \ln x\, dx$.

1. Use integration by parts with $u = \ln x$ and $dv = dx$ to show that the integral is $x \ln x - x + C$.

2. Differentiate $x \ln x - x$ to confirm the result in part 1.

3. Use a slope field of the differential equation $dy/dx = \ln x$ and the graph of $y = x \ln x - x$ to support the result in part 1. Explain the connection.

4. Use the graphs of $y = x \ln x - x$ and $y = \int_1^x \ln t\, dt$ to support the result in part 1. Explain the connection.

Repeated Use

Sometimes we have to use integration by parts more than once to evaluate an integral.

Example 4 REPEATED USE OF INTEGRATION BY PARTS

Evaluate $\int x^2 e^x \, dx$.

Solution With $u = x^2$, $dv = e^x \, dx$, $du = 2x \, dx$, and $v = e^x$, we have

$$\int x^2 e^x \, dx = x^2 e^x - 2 \int x e^x \, dx.$$

The new integral is less complicated than the original because the exponent on x is reduced by one. To evaluate the integral on the right, we integrate by parts again with $u = x$, $dv = e^x \, dx$. Then $du = dx$, $v = e^x$, and

$$\int x e^x \, dx = x e^x - \int e^x \, dx = x e^x - e^x + C.$$

Hence,

$$\int x^2 e^x \, dx = x^2 e^x - 2 \int x e^x \, dx$$

$$= x^2 e^x - 2x e^x + 2e^x + C.$$

The technique of Example 4 works for any integral $\int x^n e^x \, dx$ in which n is a positive integer, because differentiating x^n will eventually lead to zero and integrating e^x is easy. We will say more on this later in this section when we discuss *tabular integration*.

Solving for the Unknown Integral

Integrals like the one in the next example occur in electrical engineering. Their evaluation requires two integrations by parts, followed by solving for the unknown integral.

Example 5 SOLVING FOR THE UNKNOWN INTEGRAL

Evaluate $\int e^x \cos x \, dx$.

Solution Let $u = e^x$, $dv = \cos x \, dx$. Then $du = e^x \, dx$, $v = \sin x$, and

$$\int e^x \cos x \, dx = e^x \sin x - \int e^x \sin x \, dx.$$

The second integral is like the first, except it has $\sin x$ in place of $\cos x$. To evaluate it, we use integration by parts with

$$u = e^x, \quad dv = \sin x \, dx, \quad v = -\cos x, \quad du = e^x \, dx.$$

Then

$$\int e^x \cos x \, dx = e^x \sin x - \left(-e^x \cos x - \int (-\cos x)(e^x \, dx) \right)$$

$$= e^x \sin x + e^x \cos x - \int e^x \cos x \, dx.$$

The unknown integral now appears on both sides of the equation. Adding the integral to both sides gives

$$2 \int e^x \cos x \, dx = e^x \sin x + e^x \cos x + C.$$

Dividing by 2 and renaming the constant of integration gives

$$\int e^x \cos x \, dx = \frac{e^x \sin x + e^x \cos x}{2} + C.$$

When making repeated use of integration by parts in circumstances like Example 5, once a choice for u and dv is made, it is usually not a good idea to switch choices in the second stage of the problem. Doing so may result in undoing the work. For example, if we had switched to the substitution $u = \sin x$, $dv = e^x dx$ in the second integration, we would have obtained

$$\int e^x \cos x \, dx = e^x \sin x - \left(e^x \sin x - \int e^x \cos x \, dx \right)$$

$$= \int e^x \cos x \, dx,$$

undoing the first integration by parts.

Tabular Integration

We have seen that integrals of the form $\int f(x) g(x) dx$, in which f can be differentiated repeatedly to become zero and g can be integrated repeatedly without difficulty, are natural candidates for integration by parts. However, if many repetitions are required, the calculations can be cumbersome. In situations like this, there is a way to organize the calculations that saves a great deal of work. It is **tabular integration,** as shown in Examples 6 and 7.

Example 6 USING TABULAR INTEGRATION

Evaluate $\int x^2 e^x \, dx$.

Solution With $f(x) = x^2$ and $g(x) = e^x$, we list:

$f(x)$ and its derivatives		$g(x)$ and its integrals
x^2	$(+)$	e^x
$2x$	$(-)$	e^x
2	$(+)$	e^x
0		e^x

We combine the products of the functions connected by the arrows according to the operation signs above the arrows to obtain

$$\int x^2 e^x \, dx = x^2 e^x - 2x e^x + 2e^x + C.$$

Compare this with the result in Example 4.

Example 7 USING TABULAR INTEGRATION

Evaluate $\int x^3 \sin x \, dx$.

Solution With $f(x) = x^3$ and $g(x) = \sin x$, we list:

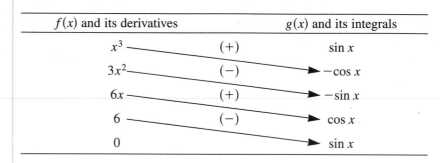

$f(x)$ and its derivatives		$g(x)$ and its integrals
x^3	$(+)$	$\sin x$
$3x^2$	$(-)$	$-\cos x$
$6x$	$(+)$	$-\sin x$
6	$(-)$	$\cos x$
0		$\sin x$

Again we combine the products of the functions connected by the arrows according to the operation signs above the arrows to obtain

$$\int x^3 \sin x \, dx = -x^3 \cos x + 3x^2 \sin x + 6x \cos x - 6 \sin x + C.$$

Quick Review 6.3

In Exercises 1–4, find dy/dx.

1. $y = x^3 \sin 2x$

2. $y = e^{2x} \ln(3x + 1)$

3. $y = \tan^{-1} 2x$

4. $y = \sin^{-1}(x + 3)$

In Exercises 5 and 6, solve for x in terms of y.

5. $y = \tan^{-1} 3x$

6. $y = \cos^{-1}(x + 1)$

7. Find the area under the arch of the curve $y = \sin \pi x$ from $x = 0$ to $x = 1$.

8. Solve the differential equation $\dfrac{dy}{dx} = e^{2x}$.

9. Solve the initial value problem $\dfrac{dy}{dx} = x + \sin x, \quad y(0) = 2$.

10. Use differentiation to confirm the integration formula

$$\int e^x \sin x \, dx = \frac{1}{2} e^x (\sin x - \cos x).$$

Section 6.3 Exercises

In Exercises 1–4, evaluate the integral. Confirm your answer by differentiation.

1. $\displaystyle\int x \sin x \, dx$

2. $\displaystyle\int x^2 \cos x \, dx$

3. $\displaystyle\int y \ln y \, dy$

4. $\displaystyle\int \tan^{-1} y \, dy$

In Exercises 5–8, evaluate the integral. Support your answer by superimposing the graph of one of the antiderivatives on a slope field of the integrand.

5. $\displaystyle\int x \sec^2 x \, dx$

6. $\displaystyle\int \sin^{-1} \theta \, d\theta$

7. $\displaystyle\int t^2 \sin t \, dt$

8. $\displaystyle\int t \csc^2 t \, dt$

In Exercises 9–14, evaluate the integral.

9. $\displaystyle\int x^3 \ln x \, dx$

10. $\displaystyle\int x^4 e^{-x} \, dx$

11. $\displaystyle\int (x^2 - 5x)e^x \, dx$

12. $\displaystyle\int x^3 e^{-2x} \, dx$

13. $\displaystyle\int e^y \sin y \, dy$

14. $\displaystyle\int e^{-y} \cos y \, dy$

In Exercises 15–18, evaluate the integral analytically. Support your answer using NINT.

15. $\displaystyle\int_0^{\pi/2} x^2 \sin 2x \, dx$

16. $\displaystyle\int_0^{\pi/2} x^3 \cos 2x \, dx$

17. $\displaystyle\int_{-2}^{3} e^{2x} \cos 3x \, dx$

18. $\displaystyle\int_{-3}^{2} e^{-2x} \sin 2x \, dx$

In Exercises 19–22, solve the differential equation.

19. $\dfrac{dy}{dx} = x^2 \, e^{4x}$

20. $\dfrac{dy}{dx} = x^2 \ln x$

21. $\dfrac{dy}{d\theta} = \theta \sec^{-1}\theta, \quad \theta > 1$

22. $\dfrac{dy}{d\theta} = \theta \sec\theta \tan\theta$

23. *Finding Area* Find the area of the region enclosed by the x-axis and the curve $y = x \sin x$ for

(a) $0 \le x \le \pi$,   (b) $\pi \le x \le 2\pi$,   (c) $0 \le x \le 2\pi$.

24. *Finding Area* Find the area of the region enclosed by the y-axis and the curves $y = x^2$ and $y = (x^2 + x + 1)e^{-x}$.

25. *Average Value* A retarding force, symbolized by the dashpot in the figure, slows the motion of the weighted spring so that the mass's position at time t is

$$y = 2e^{-t} \cos t, \quad t \ge 0.$$

Find the average value of y over the interval $0 \le t \le 2\pi$.

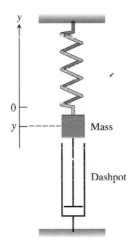

Exploration

26. Consider the integral $\displaystyle\int x^n e^x \, dx$. Use integration by parts to evaluate the integral if

(a) $n = 1$.   (b) $n = 2$.   (c) $n = 3$.

(d) Conjecture the value of the integral for any positive integer n.

(e) **Writing to Learn** Give a convincing argument that your conjecture in (d) is true.

In Exercises 27–30, evaluate the integral by using a substitution prior to integration by parts.

27. $\displaystyle\int \sin \sqrt{x} \, dx$

28. $\displaystyle\int e^{\sqrt{3x + 9}} \, dx$

29. $\displaystyle\int x^7 e^{x^2} \, dx$

30. $\displaystyle\int \sin (\ln r) \, dr$

In Exercises 31–34, use integration by parts to establish the *reduction formula*.

31. $\displaystyle\int x^n \cos x \, dx = x^n \sin x - n \int x^{n-1} \sin x \, dx$

32. $\displaystyle\int x^n \sin x \, dx = -x^n \cos x + n \int x^{n-1} \cos x \, dx$

33. $\displaystyle\int x^n e^{ax} \, dx = \dfrac{x^n \, e^{ax}}{a} - \dfrac{n}{a} \int x^{n-1} e^{ax} \, dx, \quad a \ne 0$

34. $\displaystyle\int (\ln x)^n \, dx = x(\ln x)^n - n \int (\ln x)^{n-1} \, dx$

Extending the Ideas

35. *Integrating Inverse Functions* Assume that the function f has an inverse.

(a) Show that $\displaystyle\int f^{-1}(x)dx = \int yf'(y) \, dy$. (*Hint:* Use the substitution $y = f^{-1}(x)$.)

(b) Use integration by parts on the second integral in (a) to show that

$$\int f^{-1}(x) \, dx = \int yf'(y) \, dy = xf^{-1}(x) - \int f(y) \, dy.$$

36. *Integrating Inverse Functions* Assume that the function f has an inverse. Use integration by parts directly to show that

$$\int f^{-1}(x) \, dx = xf^{-1}(x) - \int x\left(\dfrac{d}{dx} f^{-1}(x)\right)dx.$$

In Exercises 37–40, evaluate the integral using

(a) the technique of Exercise 35.

(b) the technique of Exercise 36.

(c) Show that the expressions (with $C = 0$) obtained in parts (a) and (b) are the same.

37. $\displaystyle\int \sin^{-1} x \, dx$

38. $\displaystyle\int \tan^{-1} x \, dx$

39. $\displaystyle\int \cos^{-1} x \, dx$

40. $\displaystyle\int \log_2 x \, dx$

6.4 Exponential Growth and Decay

Law of Exponential Change • Continuously Compounded Interest • Radioactivity • Newton's Law of Cooling • Resistance Proportional to Velocity

Law of Exponential Change

Suppose we are interested in a quantity y (population, radioactive element, money, whatever) that increases or decreases at a rate proportional to the amount present. If we also know the amount present at time $t = 0$, say y_0, we can find y as a function of t by solving the following initial value problem.

Differential equation: $\dfrac{dy}{dt} = ky$

Initial condition: $y = y_0$ when $t = 0$

If y is positive and increasing, then k is positive and the rate of growth is proportional to what has already been accumulated. If y is positive and decreasing, then k is negative and the rate of decay is proportional to the amount still left.

The constant function $y = 0$ is a solution of the differential equation but we usually aren't interested in that solution. To find nonzero solutions we separate the variables and integrate.

$$\frac{dy}{y} = k \, dt$$

$$\ln |y| = kt + C \qquad \text{Integrate.}$$

$$e^{\ln |y|} = e^{kt+C} \qquad \text{Exponentiate.}$$

$$|y| = e^C \cdot e^{kt} \qquad e^{\ln u} = u, \ e^{a+b} = e^a \cdot e^b$$

$$y = \pm e^C e^{kt} \qquad |y| = r \Rightarrow y = \pm r$$

$$y = Ae^{kt} \qquad \text{Let } A = \pm e^C.$$

By allowing A to take on the value 0 in addition to all the possible values of $\pm e^C$, we can include the solution $y = 0$. This solves the differential equation.

To solve the initial value problem we set $t = 0$ and $y = y_0$ and solve for A.

$$y_0 = Ae^{(k)(0)} = A$$

The solution of the initial value problem is $y = y_0 e^{kt}$.

Law of Exponential Change

If y changes at a rate proportional to the amount present $(dy/dt = ky)$ and $y = y_0$ when $t = 0$, then

$$y = y_0 e^{kt},$$

where $k > 0$ represents growth and $k < 0$ represents decay.

The number k is the **rate constant** of the equation.

Continuously Compounded Interest

Suppose that A_0 dollars are invested at a fixed annual interest rate r (expressed as a decimal). If interest is added to the account k times a year, the amount of money present after t years is

$$A(t) = A_0 \left(1 + \frac{r}{k}\right)^{kt}.$$

The interest might be added ("compounded," bankers say) monthly ($k = 12$), weekly ($k = 52$), daily ($k = 365$), or even more frequently, by the hour or by the minute.

If, instead of being added at discrete intervals, the interest is added continuously at a rate proportional to the amount in the account, we can model the growth of the account with the initial value problem.

Differential equation: $\dfrac{dA}{dt} = rA$

Initial condition: $A(0) = A_0$

The amount of money in the account after t years is then

$$A(t) = A_0 e^{rt}.$$

Interest paid according to this formula is said to be **compounded continuously.** The number r is the **continuous interest rate.**

It can be shown that

$$\lim_{k \to \infty} A_0 \left(1 + \frac{r}{k}\right)^{kt} = A_0 e^{rt}.$$

We will see how this limit is evaluated in Section 8.1, Exercise 49.

Example 1   COMPOUNDING INTEREST CONTINUOUSLY

Suppose you deposit \$800 in an account that pays 6.3% annual interest. How much will you have 8 years later if the interest is **(a)** compounded continuously? **(b)** compounded quarterly?

Solution Here $A_0 = 800$ and $r = 0.063$. The amount in the account to the nearest cent after 8 years is

(a) $A(8) = 800 e^{(0.063)(8)} = 1324.26.$

(b) $A(8) = 800 \left(1 + \dfrac{0.063}{4}\right)^{(4)(8)} = 1319.07.$

You might have expected to generate more than an additional \$5.19 with interest compounded continuously.

For radium-226, which used to be painted on watch dials to make them glow at night (a dangerous practice for the painters, who licked their brush-tips), t is measured in years and $k = 4.3 \times 10^{-4}$. For radon-222 gas, t is measured in days and $k = 0.18$. The decay of radium in the earth's crust is the source of the radon we sometimes find in our basements.

Radioactivity

When an atom emits some of its mass as radiation, the remainder of the atom reforms to make an atom of some new element. This process of radiation and change is **radioactive decay,** and an element whose atoms go spontaneously through this process is **radioactive.** Radioactive carbon-14 decays into nitrogen. Radium, through a number of intervening radioactive steps, decays into lead.

Experiments have shown that at any given time the rate at which a radioactive element decays (as measured by the number of nuclei that change per unit of time) is approximately proportional to the number of radioactive nuclei present. Thus, the decay of a radioactive element is described by the equation $dy/dt = -ky, \ k > 0$. If y_0 is the number of radioactive nuclei present at time zero, the number still present at any later time t will be

$$y = y_0 e^{-kt}, \quad k > 0.$$

Convention

It is conventional to use $-k$ $(k > 0)$ here instead of k $(k < 0)$ to emphasize that y is decreasing.

The **half-life** of a radioactive element is the time required for half of the radioactive nuclei present in a sample to decay. Example 2 shows the surprising fact that the half-life is a constant that depends only on the radioactive substance and not on the number of radioactive nuclei present in the sample.

Example 2 FINDING HALF-LIFE

Find the half-life of a radioactive substance with decay equation $y = y_0 e^{-kt}$ and show that the half-life depends only on k.

Solution

Model

The half-life is the solution to the equation

$$y_0 e^{-kt} = \frac{1}{2} y_0.$$

Solve Algebraically

$$e^{-kt} = \frac{1}{2} \qquad\qquad \text{Divide by } y_0.$$

$$-kt = \ln \frac{1}{2} \qquad\qquad \text{Take ln of both sides.}$$

$$t = -\frac{1}{k} \ln \frac{1}{2} = \frac{\ln 2}{k} \quad \ln\frac{1}{a} = -\ln a$$

Interpret

This value of t is the half-life of the element. It depends only on the value of k. Note that the number y_0 does not appear.

Carbon-14 Dating

The decay of radioactive elements can sometimes be used to date events from earth's past. The ages of rocks more than 2 billion years old have been measured by the extent of the radioactive decay of uranium (half-life 4.5 billion years!). In a living organism, the ratio of radioactive carbon, carbon-14, to ordinary carbon stays fairly constant during the lifetime of the organism, being approximately equal to the ratio in the organism's surroundings at the time. After the organism's death, however, no new carbon is ingested, and the proportion of carbon-14 decreases as the carbon-14 decays. It is possible to estimate the ages of fairly old organic remains by comparing the proportion of carbon-14 they contain with the proportion assumed to have been in the organism's environment at the time it lived. Archeologists have dated shells (which contain $CaCO_3$), seeds, and wooden artifacts this way. The estimate of 15,500 years for the age of the cave paintings at Lascaux, France, is based on carbon-14 dating. After generations of controversy, the Shroud of Turin, long believed by many to be the burial cloth of Christ, was shown by carbon-14 dating in 1988 to have been made after A.D. 1200.

> **Half-life**
>
> The **half-life** of a radioactive substance with rate constant k $(k > 0)$ is
>
> $$\text{half-life} = \frac{\ln 2}{k}.$$

Example 3 USING CARBON-14 DATING

Scientists who do carbon-14 dating use 5700 years for its half-life. Find the age of a sample in which 10% of the radioactive nuclei originally present have decayed.

Solution Because half-life $= 5700 = (\ln 2)/k$, we have $k = (\ln 2)/5700$.

Model

We need to find the value of t for which

$$y_0 e^{-kt} = 0.9 y_0 \quad\text{or}\quad e^{-kt} = 0.9,$$

where $k = (\ln 2)/5700$.

Solve Algebraically

$$e^{-kt} = 0.9$$

$$-kt = \ln 0.9 \quad \text{ln of both sides}$$

$$t = -\frac{1}{k}\ln 0.9 = -\frac{5700}{\ln 2}\ln 0.9 \approx 866$$

Interpret

The sample is about 866 years old.

Newton's Law of Cooling

Soup left in a cup cools to the temperature of the surrounding air. A hot silver ingot immersed in water cools to the temperature of the surrounding water. In situations like these, the rate at which an object's temperature is changing at any given time is roughly proportional to the difference between its temperature and the temperature of the surrounding medium. This observation is *Newton's Law of Cooling,* although it applies to warming as well, and there is an equation for it.

If T is the temperature of the object at time t, and T_s is the surrounding temperature, then

$$\frac{dT}{dt} = -k(T - T_s). \tag{1}$$

Since $dT = d(T - T_s)$, Equation 1 can be written as

$$\frac{d}{dt}(T - T_s) = -k(T - T_s).$$

Its solution, by the law of exponential change, is

$$T - T_s = (T_0 - T_s)e^{-kt},$$

where T_0 is the temperature at time $t = 0$. This equation also bears the name **Newton's Law of Cooling.**

Example 4 USING NEWTON'S LAW OF COOLING

A hard-boiled egg at 98°C is put in a pan under running 18°C water to cool. After 5 minutes, the egg's temperature is found to be 38°C. How much longer will it take the egg to reach 20°?

Solution

Model

Using Newton's Law of Cooling with $T_s = 18$ and $T_0 = 98$, we have

$$T - 18 = (98 - 18)e^{-kt} \quad \text{or} \quad T = 18 + 80e^{-kt}.$$

To find k we use the information that $T = 38$ when $t = 5$.

$$38 = 18 + 80e^{-5k}$$

$$e^{-5k} = \frac{1}{4}$$

$$-5k = \ln\frac{1}{4} = -\ln 4$$

$$k = \frac{1}{5}\ln 4$$

The egg's temperature at time t is $T = 18 + 80e^{-(0.2\ln 4)t}$.

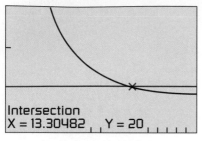

Intersection
X = 13.30482   Y = 20

[0, 20] by [10, 40]

Figure 6.8 The egg will reach 20°C about 13.3 min after being placed in the pan to cool. (Example 4)

Table 6.3 Experimental Data

Time (sec)	T (°C)	$T - T_s$ (°C)
2	64.8	60.3
5	49.0	44.5
10	31.4	26.9
15	22.0	17.5
20	16.5	12.0
25	14.2	9.7
30	12.0	7.5

Solve Graphically

We can now use a grapher to find the time when the egg's temperature is 20°C. Figure 6.8 shows that the graphs of

$$y = 20 \quad \text{and} \quad y = T = 18 + 80e^{-(0.2 \ln 4)t}$$

intersect at about $t = 13.3$.

Interpret

The egg's temperature will reach 20°C in about 13.3 min after it is put in the pan under running water to cool. Because it took 5 min to reach 38°C, it will take slightly more than 8 additional minutes to reach 20°C.

The next example shows how to use exponential regression to fit a function to real data. A CBL™ temperature probe was used to collect the data.

Example 5   USING NEWTON'S LAW OF COOLING

A temperature probe (thermometer) is removed from a cup of coffee and placed in water that has a temperature of $T_s = 4.5$°C. Temperature readings T, as recorded in Table 6.3, are taken after 2 sec, 5 sec, and every 5 sec thereafter. Estimate

(a) the coffee's temperature at the time the temperature probe was removed.

(b) the time when the temperature probe reading will be 8°C.

Solution

Model

According to Newton's Law of Cooling, $T - T_s = (T_0 - T_s)e^{-kt}$, where $T_s = 4.5$ and T_0 is the temperature of the coffee (probe reading) at $t = 0$.

We use exponential regression to find that

$$T - 4.5 = 61.66(0.9277^t)$$

is a model for the $(t, T - T_s) = (t, T - 4.5)$ data.

Thus,

$$T = 4.5 + 61.66(0.9277^t)$$

is a model for the (t, T) data.

Figure 6.9a shows the graph of the model superimposed on a scatter plot of the (t, T) data.

(a) At time $t = 0$, when the probe was removed, the temperature was

$$T = 4.5 + 61.66(0.9277^0) \approx 66.16°C.$$

(b) Solve Graphically

Figure 6.9b shows that the graphs of

$$y = 8 \quad \text{and} \quad y = T = 4.5 + 61.66(0.9277^t)$$

intersect at about $t = 38$.

Interpret

The temperature of the coffee was about 66.2°C when the temperature probe was removed. The temperature probe will reach 8°C about 38 sec after it is removed from the coffee and placed in the water.

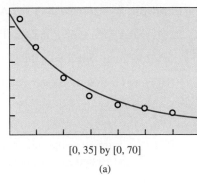

[0, 35] by [0, 70]

(a)

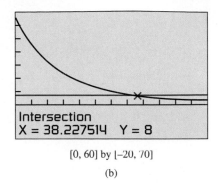

Intersection
X = 38.227514   Y = 8

[0, 60] by [−20, 70]

(b)

Figure 6.9 (Example 5)

Resistance Proportional to Velocity

In some cases it is reasonable to assume that, other forces being absent, the resistance encountered by a moving object, such as a car coasting to a stop, is proportional to the object's velocity. The slower the object moves, the less its forward progress is resisted by the air through which it passes. To describe this in mathematical terms, we picture the object as a mass m moving along a coordinate line with position function s and velocity v at time t. The resisting force opposing the motion is

$$\text{Force} = \text{mass} \times \text{acceleration} = m\frac{dv}{dt}.$$

We can express the assumption that the resisting force is proportional to velocity by writing

$$m\frac{dv}{dt} = -kv \quad \text{or} \quad \frac{dv}{dt} = -\frac{k}{m}v \quad (k > 0).$$

This is a differential equation of exponential change. The solution to the differential equation with initial condition $v = v_0$ at $t = 0$ is

$$v = v_0 e^{-(k/m)t}. \tag{2}$$

In Exploration 1 we fix the values of v_0 and k and investigate what happens as the mass m increases.

Exploration 1 **Slowing Down More Slowly**

An object is traveling along a coordinate line with velocity $v = 100e^{-(0.5/m)t}$.

1. Graph v for $m = 1, 2, 4, 6$ in the window [0, 20] by [0, 120]. Compare and describe the motion in each case.

2. Explain why $s(t) = \int_0^t v(u)\,du$ gives the distance traveled by the object at time t. What is $s(0)$?

3. Graph $s(t)$ for $m = 1, 2, 4, 6$ in the window [0, 70] by [0, 1500]. Compare the total distances traveled.

Example 6   COASTING TO A STOP

For a 50-kg ice skater, the k in Equation 2 is about 2.5 kg/sec.

(a) How long will it take the skater to coast from 7 m/sec to 1 m/sec?

(b) How far will the skater coast before coming to a complete stop?

Solution

Model

If $t = 0$ represents the time when the skater's velocity is 7 m/sec, then $v_0 = 7$ and Equation 2 becomes

$$v = 7e^{-(2.5/50)t} = 7e^{-0.05t}.$$

(a) Solve Algebraically

We want the value of t when $v = 1$.

$$1 = 7e^{-0.05t}$$

$$e^{-0.05t} = \frac{1}{7}$$

$$-0.05t = \ln\frac{1}{7} = -\ln 7$$

$$t = \frac{\ln 7}{0.05} \approx 38.9 \text{ sec}$$

(b) Solve Analytically

The integral of velocity is distance, so

$$s = \int 7e^{-0.05t}\, dt = \frac{7e^{-0.05t}}{-0.05} + C = -140e^{-0.05t} + C.$$

We assume $s = 0$ when $t = 0$, so

$$0 = -140e^0 + C$$

$$C = 140.$$

Thus,

$$s = 140 - 140e^{-0.05t} = 140(1 - e^{-0.05t}).$$

Notice that the limit of $s(t)$ as $t \to \infty$ is 140.

Interpret

Mathematically, s never quite reaches 140 m. However, for all practical purposes, the skater will come to a complete stop after traveling 140 m.

In Exercise 31, you will show that the total distance coasted by a moving object encountering resistance proportional to its velocity is $v_0 m/k$. Further, you will show that if $s(0) = 0$, then

$$s(t) = \frac{v_0 m}{k}\left(1 - e^{-(k/m)t}\right). \tag{3}$$

Example 7 COASTING ON IN-LINE SKATES

The data in Table 6.4 were collected with a motion detector and a CBL™ by Valerie Sharritts, a mathematics teacher at St. Francis DeSales High School in Columbus, Ohio. The table shows the distance s (meters) coasted on in-line skates in t seconds by her daughter Ashley when she was 10 years old.

Table 6.4 Ashley Sharritts Skating Data

t (sec)	s (m)	t (sec)	s (m)	t (sec)	s (m)
0	0	2.24	3.05	4.48	4.77
0.16	0.31	2.40	3.22	4.64	4.82
0.32	0.57	2.56	3.38	4.80	4.84
0.48	0.80	2.72	3.52	4.96	4.86
0.64	1.05	2.88	3.67	5.12	4.88
0.80	1.28	3.04	3.82	5.28	4.89
0.96	1.50	3.20	3.96	5.44	4.90
1.12	1.72	3.36	4.08	5.60	4.90
1.28	1.93	3.52	4.18	5.76	4.91
1.44	2.09	3.68	4.31	5.92	4.90
1.60	2.30	3.84	4.41	6.08	4.91
1.76	2.53	4.00	4.52	6.24	4.90
1.92	2.73	4.16	4.63	6.40	4.91
2.08	2.89	4.32	4.69	6.56	4.91

Find a model for Ashley's position given by the data in Table 6.4 in the form of Equation 3. Her initial velocity was $v_0 = 2.75$ m/sec, her mass $m = 39.92$ kg (she weighed 88 lb), and her total coasting distance was 4.91 m.

Solution We use the total coasting distance to determine k.

$$\frac{v_0 m}{k} = \text{coasting distance}$$

$$\frac{(2.75)(39.92)}{k} = 4.91$$

$$k \approx 22.36$$

Thus, the distance function is

$$s(t) = 4.91(1 - e^{-(22.36/39.92)t}).$$

Figure 6.10 shows the graph of $s(t)$ superimposed on a scatter plot of the position data in Table 6.4.

In Exercise 33 of this section and in Review Exercise 61, we will investigate the skating data of two friends of the Sharritts: Kelly Schmitzer and Johnathon Krueger.

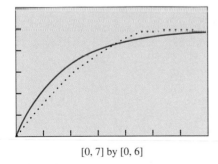

[0, 7] by [0, 6]

Figure 6.10 Notice how well the model obtained using mathematics fits the data obtained experimentally. (Example 7)

Quick Review 6.4

In Exercises 1 and 2, rewrite the equation in exponential form or logarithmic form.

1. $\ln a = b$

2. $e^c = d$

In Exercises 3–8, solve the equation.

3. $\ln(x + 3) = 2$

4. $100e^{2x} = 600$

5. $0.85^x = 2.5$

6. $2^{k+1} = 3^k$

7. $1.1^t = 10$

8. $e^{-2t} = \dfrac{1}{4}$

In Exercises 9 and 10, solve for y.

9. $\ln(y + 1) = 2x - 3$

10. $\ln|y + 2| = 3t - 1$

Section 6.4 Exercises

In Exercises 1–4, find the solution of the differential equation $dy/dt = ky$, k a constant, that satisfies the given conditions.

1. $k = 1.5$, $y(0) = 100$

2. $k = -0.5$, $y(0) = 200$

3. $y(0) = 50$, $y(5) = 100$

4. $y(0) = 60$, $y(10) = 30$

In Exercises 5–8, complete the table for an investment if interest is compounded continuously.

	Initial Deposit ($)	Annual Rate (%)	Doubling Time (yr)	Amount in 30 yr ($)
5.	1000	8.6		
6.	2000		15	
7.		5.25		2898.44
8.	1200			10,405.37

In Exercises 9 and 10, find the amount of time required for a $2000 investment to double if the annual interest rate r is compounded **(a)** annually, **(b)** monthly, **(c)** quarterly, and **(d)** continuously.

9. $r = 4.75\%$

10. $r = 8.25\%$

11. *Growth of Cholera Bacteria* Suppose that the cholera bacteria in a colony grows unchecked according to the Law of Exponential Change. The colony starts with 1 bacterium and doubles in number every half hour.

(a) How many bacteria will the colony contain at the end of 24 h?

(b) Writing to Learn Use (a) to explain why a person who feels well in the morning may be dangerously ill by evening even though, in an infected person, many bacteria are destroyed.

12. *Bacteria Growth* A colony of bacteria is grown under ideal conditions in a laboratory so that the population increases exponentially with time. At the end of 3 h there are 10,000 bacteria. At the end of 5 h there are 40,000 bacteria. How many bacteria were present initially?

13. *Radon-222* The decay equation for radon-222 gas is known to be $y = y_0 e^{-0.18t}$, with t in days. About how long will it take the amount of radon in a sealed sample of air to decay to 90% of its original value?

14. *Polonium-210* The number of radioactive atoms remaining after t days in a sample of polonium-210 that starts with y_0 radioactive atoms is $y = y_0 e^{-0.005t}$.

(a) Find the element's half-life.

(b) Your sample will not be useful to you after 95% of the radioactive nuclei present on the day the sample arrives have disintegrated. For about how many days after the sample arrives will you be able to use the polonium?

In Exercises 15 and 16, find the exponential function $y = y_0 e^{kt}$ whose graph passes through the two points.

15.

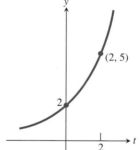

16.

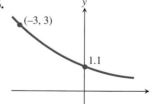

17. *Mean Life of Radioactive Nuclei* Physicists using the radioactive decay equation $y = y_0 e^{-kt}$ call the number $1/k$ the *mean life* of a radioactive nucleus. The mean life of a radon-222 nucleus is about $1/0.18 \approx 5.6$ days. The mean life of a carbon-14 nucleus is more than 8000 years. Show that 95% of the radioactive nuclei originally present in any sample will disintegrate within three mean lifetimes, that is, by time $t = 3/k$. Thus, the mean life of a nucleus gives a quick way to estimate how long the radioactivity of a sample will last.

18. *Finding the Original Temperature of a Beam* An aluminum beam was brought from the outside cold into a machine shop where the temperature was held at 65°F. After 10 min, the beam warmed to 35°F and after another 10 min its temperature was 50°F. Use Newton's Law of Cooling to estimate the beam's initial temperature.

19. *Cooling Soup* Suppose that a cup of soup cooled from 90°C to 60°C in 10 min in a room whose temperature was 20°C. Use Newton's Law of Cooling to answer the following questions.

(a) How much longer would it take the soup to cool to 35°C?

(b) Instead of being left to stand in the room, the cup of 90°C soup is put into a freezer whose temperature is −15°C. How long will it take the soup to cool from 90°C to 35°C?

20. *Cooling Silver* The temperature of an ingot of silver is 60°C above room temperature right now. Twenty minutes ago, it was 70°C above room temperature. How far above room temperature will the silver be

(a) 15 minutes from now?

(b) 2 hours from now?

(c) When will the silver be 10°C above room temperature?

21. *Dating Crater Lake* The charcoal from a tree killed in the volcanic eruption that formed Crater Lake in Oregon contained 44.5% of the carbon-14 found in living matter. About how old is Crater Lake?

22. *Carbon-14 Dating Measurement Sensitivity* To see the effect of a relatively small error in the estimate of the amount of carbon-14 in a sample being dated, answer the following questions about this hypothetical situation.

(a) A fossilized bone found in central Illinois in the year A.D. 2000 contains 17% of its original carbon-14 content. Estimate the year the animal died.

(b) Repeat part (a) assuming 18% instead of 17%.

(c) Repeat part (a) assuming 16% instead of 17%.

23. *Coasting Bicycle* A 66-kg cyclist on a 7-kg bicycle starts coasting on level ground at 9 m/sec. The k in Equation 2 is about 3.9 kg/sec.

(a) About how far will the cyclist coast before reaching a complete stop?

(b) How long will it take the cyclist's speed to drop to 1 m/sec?

24. *Coasting Battleship* Suppose an Iowa class battleship has mass around 51,000 metric tons (51,000,000 kg) and a k value in Equation 2 of about 59,000 kg/sec. Assume the ship loses power when it is moving at a speed of 9 m/sec.

(a) About how far will the ship coast before it is dead in the water?

(b) About how long will it take the ship's speed to drop to 1 m/sec?

25. *The Inversion of Sugar* The processing of raw sugar has an "inversion" step that changes the sugar's molecular structure. Once the process has begun, the rate of change of the amount of raw sugar is proportional to the amount of raw sugar remaining. If 1000 kg of raw sugar reduces to 800 kg of raw sugar during the first 10 h, how much raw sugar will remain after another 14 h?

26. *Oil Depletion* Suppose the amount of oil pumped from one of the canyon wells in Whittier, California, decreases at the continuous rate of 10% per year. When will the well's output fall to one-fifth of its present level?

27. *Atmospheric Pressure* Earth's atmospheric pressure p is often modeled by assuming that the rate dp/dh at which p changes with the altitude h above sea level is proportional to p. Suppose that the pressure at sea level is 1013 millibars (about 14.7 lb/in^2) and that the pressure at an altitude of 20 km is 90 millibars.

(a) Solve the initial value problem

$$\text{Differential equation:} \quad \frac{dp}{dh} = kp,$$

$$\text{Initial condition:} \quad p = p_0 \text{ when } h = 0,$$

to express p in terms of h. Determine the values of p_0 and k from the given altitude-pressure data.

(b) What is the atmospheric pressure at $h = 50$ km?

(c) At what altitude does the pressure equal 900 millibars?

28. *First Order Chemical Reactions* In some chemical reactions the rate at which the amount of a substance changes with time is proportional to the amount present. For the change of δ-glucono lactone into gluconic acid, for example,

$$\frac{dy}{dt} = -0.6y$$

when y is measured in grams and t is measured in hours. If there are 100 grams of a δ-glucono lactone present when $t = 0$, how many grams will be left after the first hour?

29. *Discharging Capacitor Voltage* Suppose that electricity is draining from a capacitor at a rate proportional to the voltage V across its terminals and that, if t is measured in seconds,

$$\frac{dV}{dt} = -\frac{1}{40}V.$$

(a) Solve this differential equation for V, using V_0 to denote the value of V when $t = 0$.

(b) How long will it take the voltage to drop to 10% of its original value?

30. *John Napier's Answer* John Napier (1550–1617), the Scottish laird who invented logarithms, was the first person to answer the question, "What happens if you invest an amount of money at 100% yearly interest, compounded continuously?"

(a) **Writing to Learn** What does happen? Explain.

(b) How long does it take to triple your money?

(c) **Writing to Learn** How much can you earn in a year?

31. *Coasting Object* Assume that the resistance encountered by a moving object is proportional to the object's velocity so that its velocity is $v = v_0 e^{-(k/m)t}$, as given by Equation 2.

(a) Integrate the velocity function with respect to t to obtain the distance function s. Assume that $s(0) = 0$ and show that

$$s(t) = \frac{v_0 m}{k}\left(1 - e^{-(k/m)t}\right).$$

(b) Show that the total coasting distance traveled by the object as it coasts to a complete stop is $v_0 m/k$.

32. *Benjamin Franklin's Will* The Franklin Technical Institute of Boston owes its existence to a provision in a codicil to Benjamin Franklin's will. In part the codicil reads:

> I wish to be useful even after my Death, if possible, in forming and advancing other young men that may be serviceable to their Country in both Boston and Philadelphia. To this end I devote Two thousand Pounds Sterling, which I give, one thousand thereof to the Inhabitants of the Town of Boston in Massachusetts, and the other thousand to the inhabitants of the City of Philadelphia, in Trust and for the Uses, Interests and Purposes hereinafter mentioned and declared.

Franklin's plan was to lend money to young apprentices at 5% interest with the provision that each borrower should pay each year along

> ... with the yearly Interest, one tenth part of the Principal, which sums of Principal and Interest shall be again let to fresh Borrowers. ... If this plan is executed and succeeds as projected without interruption for one hundred Years, the Sum will then be one hundred and thirty-one thousand Pounds of which I would have the Managers of the Donation to the Inhabitants of the Town of Boston, then lay out at their discretion one hundred thousand Pounds in Public Works. ... The remaining thirty-one thousand Pounds, I would have continued to be let out on Interest in the manner above directed for another hundred Years. ... At the end of this second term if no unfortunate accident has prevented the operation the sum will be Four Millions and Sixty-one Thousand Pounds.

It was not always possible to find as many borrowers as Franklin had planned, but the managers of the trust did the best they could. At the end of 100 years from the receipt of the Franklin gift, in January 1894, the fund had grown from 1000 pounds to almost 90,000 pounds. In 100 years the original capital had multiplied about 90 times instead of the 131 times Franklin had imagined.

(a) What annual rate of interest, compounded continuously for 100 years, would have multiplied Benjamin Franklin's original capital by 90?

(b) In Benjamin Franklin's estimate that the original 1000 pounds would grow to 131,000 in 100 years, he was using an annual rate of 5% and compounding once each year. What rate of interest per year when compounded continuously for 100 years would multiply the original amount by 131?

33. *Coasting to a Stop* Table 6.5 shows the distance s (meters) coasted on in-line skates in terms of time t (seconds) by Kelly Schmitzer. Find a model for her position in the form of Equation 3. Her initial velocity was $v_0 = 0.80$ m/sec, her mass $m = 49.90$ kg (110 lb), and her total coasting distance was 1.32 m.

Table 6.5 Kelly Schmitzer Skating Data

t (sec)	s (m)	t (sec)	s (m)	t (sec)	s (m)
0	0	1.5	0.89	3.1	1.30
0.1	0.07	1.7	0.97	3.3	1.31
0.3	0.22	1.9	1.05	3.5	1.32
0.5	0.36	2.1	1.11	3.7	1.32
0.7	0.49	2.3	1.17	3.9	1.32
0.9	0.60	2.5	1.22	4.1	1.32
1.1	0.71	2.7	1.25	4.3	1.32
1.3	0.81	2.9	1.28	4.5	1.32

34. *Temperature Experiment* A temperature probe is removed from a cup of coffee and placed in water whose temperature (T_s) is 10°C. The data in Table 6.6 were collected over the next 30 sec with a CBL™ temperature probe.

Table 6.6 Experimental Data

Time (sec)	T (°C)	$T - T_s$ (°C)
2	80.47	70.47
5	69.39	59.39
10	49.66	39.66
15	35.26	25.26
20	28.15	18.15
25	23.56	13.56
30	20.62	10.62

(a) Find an exponential regression equation for the $(t, T - T_s)$ data.

(b) Use the regression equation in (a) to find a model for the (t, T) data. Superimpose the graph of the model on a scatter plot of the (t, T) data.

(c) Estimate when the temperature probe will read 12°C.

(d) Estimate the coffee's temperature when the temperature probe was removed.

Explorations

35. *Newton's Law of Cooling* Let T be the temperature of an object at time t, T_s the surrounding temperature, and T_0 the value of T at time zero.

(a) Solve the equation

$$\frac{dT}{dt} = -k(T - T_s)$$

by separation of variables to derive Newton's Law of Cooling.

(b) Find $\lim_{t\to\infty} T$ and identify any horizontal asymptotes.

36. *Rules of 70 and 72* The rules state that it takes about $70/i$ or $72/i$ years for money to double at i percent, compounded continuously, using whichever of 70 or 72 is easier to divide by i.

(a) Show that it takes $t = (\ln 2)/r$ years for money to double if it is invested at annual interest rate r (in decimal form) compounded continuously.

(b) Graph the functions

$$y_1 = \frac{\ln 2}{r}, \quad y_2 = \frac{70}{100r}, \quad \text{and} \quad y_3 = \frac{72}{100r}$$

in the $[0, 0.1]$ by $[0, 100]$ viewing window.

(c) Writing to Learn Explain why these two rules of thumb for mental computation are reasonable.

(d) Use the rules to estimate how long it takes to double money at 5% compounded continuously.

(e) Invent a rule for estimating the number of years needed to triple your money. ■

Extending the Ideas

37. *Continuously Compounded Interest*

(a) Use tables to give a numerical argument that

$$\lim_{x\to\infty} \left(1 + \frac{1}{x}\right)^x = e.$$

Support your argument graphically.

(b) For several different values of r, give numerical and graphical evidence that

$$\lim_{x\to\infty} \left(1 + \frac{r}{x}\right)^x = e^r.$$

(c) Writing to Learn Explain why compounding interest over smaller and smaller periods of time leads to the concept of interest compounded continuously.

38. *Skydiving* If a body of mass m falling from rest under the action of gravity encounters an air resistance proportional to the square of the velocity, then the body's velocity $v(t)$ is modeled by the initial value problem

Differential equation: $m\dfrac{dv}{dt} = mg - kv^2$,

Initial condition: $v(0) = 0$,

where t represents time in seconds, g is the acceleration due to gravity, and k is a constant that depends on the body's aerodynamic properties and the density of the air. (We assume that the fall is short enough so that variation in the air's density will not affect the outcome.)

(a) Show that the function

$$v(t) = \sqrt{\frac{mg}{k}} \, \frac{e^{at} - e^{-at}}{e^{at} + e^{-at}},$$

where $a = \sqrt{gk/m}$, is a solution of the initial value problem.

(b) Find the body's limiting velocity, $\lim_{t\to\infty} v(t)$.

(c) For a 160-lb skydiver ($mg = 160$), and with time in seconds and distance in feet, a typical value for k is 0.005. What is the diver's limiting velocity in feet per second? in miles per hour?

Skydivers can vary their limiting velocities by changing the amount of body area opposing the fall. Their velocities can vary from 94 to 321 miles per hour.

6.5 Population Growth

Exponential Model • Logistic Growth Model • Logistic Regression

Exponential Model

Strictly speaking, the number of individuals in a population is a discontinuous function of time because it takes on only whole number values. However, one common way to model a population is with a differentiable function P growing at a rate proportional to the size of the population. Thus, for some constant k,

$$\frac{dP}{dt} = kP.$$

Notice that

$$\frac{dP/dt}{P} = k$$

is constant. This rate is called the **relative growth rate.** As we saw in Section 6.4, we can represent the population by the model $P = P_0 e^{kt}$, where P_0 is the size of the population at time $t = 0$.

In Table 1.7 of Section 1.3 we gave the world population for the years 1986 to 1991, which we repeat here as Table 6.7.

Table 6.7 World Population

Year	Population (millions)	Ratio
1986	4936	
1987	5023	$5023/4936 \approx 1.0176$
1988	5111	$5111/5023 \approx 1.0175$
1989	5201	$5201/5111 \approx 1.0176$
1990	5329	$5329/5201 \approx 1.0246$
1991	5422	$5422/5329 \approx 1.0175$

Source: Statistical Office of the United Nations, *Monthly Bulletin of Statistics,* 1991.

Example 1 PREDICTING WORLD POPULATION

Find an initial value problem model for world population and use it to predict the population in the year 2010. Graph the model and the data.

Solution We let $t = 0$ represent 1986, $t = 1$ represent 1987, and so forth. The year 2010 will be represented by $t = 24$.

If we approximate the ratios in Table 6.7 by 1.018, then the relative rate k of growth satisfies $e^k = 1.018$, so k is about 0.0178. We obtain the initial value problem

Differential equation: $\quad \dfrac{dP}{dt} = 0.0178P,$

Initial condition: $\quad P(0) = 4936.$

The solution to this initial value problem gives us the model $P = 4936e^{0.0178t}$.

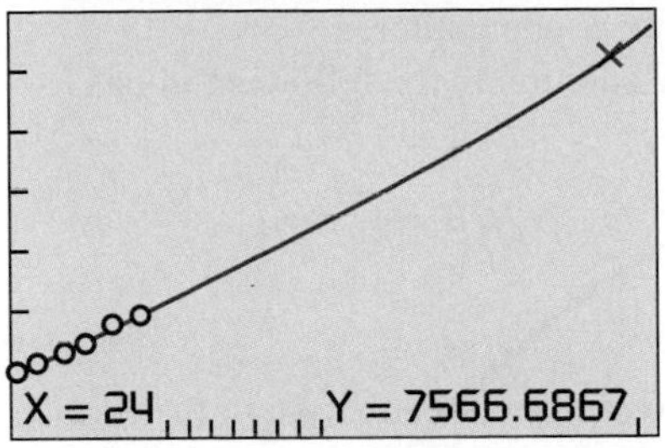

[−1, 25] by [4500, 8000]

Figure 6.11 Notice that the value of the model $P = 4936e^{0.0178t}$ is 7566.6867 when $t = 24$. (Example 1)

The year 2010 is represented by $t = 24$, so

$$P(24) \approx 7567.$$

Figure 6.11 shows the graph of the model superimposed on a scatter plot of the data.

Interpret

This model predicts the world population in the year 2010 to be about 7567 million, or 7.6 billion, slightly less than the prediction in Example 1 of Section 1.3.

Logistic Growth Model

The exponential model for population growth assumes unlimited growth. This is realistic only for a short period of time when the initial population is small. A more realistic assumption is that the relative growth rate is positive but decreases as the population increases due to environmental or economic factors. In other words there is a maximum population M, the **carrying capacity,** that the environment is capable of sustaining in the long run. If we assume the relative growth rate is proportional to $1 - (P/M)$ with positive proportionality constant k, then

$$\frac{dP/dt}{P} = k\left(1 - \frac{P}{M}\right) \quad \text{or} \quad \frac{dP}{dt} = \frac{k}{M}P(M - P). \tag{1}$$

The solution to this **logistic differential equation** is called the **logistic growth model.** Notice that the rate of growth is proportional to both P and $(M - P)$, where M is the assumed maximum population. If P were to exceed M, the growth rate would be negative (as $k > 0$, $M > 0$) and the population would be decreasing.

Example 2 MODELING BEAR POPULATION

A national park is known to be capable of supporting no more than 100 grizzly bears. Ten bears are in the park at present. We model the population with a logistic differential equation with $k = 0.1$.

(a) Draw and describe a slope field for the differential equation.

(b) Find a logistic growth model $P(t)$ for the population and draw its graph.

(c) When will the bear population reach 50?

Solution (a) The carrying capacity is 100, so $M = 100$. The model we seek is a solution to the following differential equation.

$$\frac{dP}{dt} = \frac{0.1}{100}P(100 - P) \quad \text{Eq. 1 with } k = 0.1, M = 100$$

$$= 0.001P(100 - P)$$

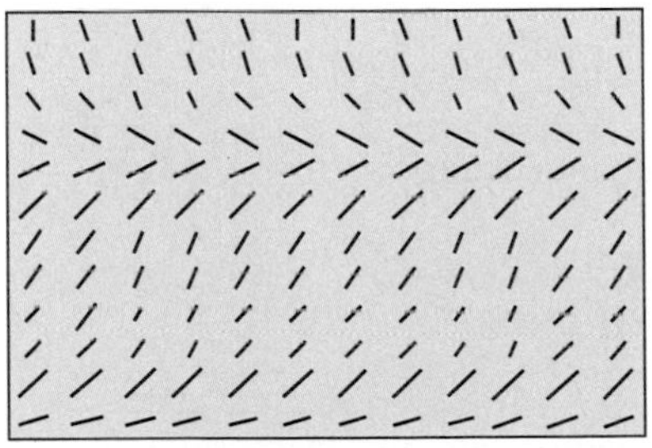

Figure 6.12 shows a slope field for this differential equation. There appears to be a horizontal asymptote at $P = 100$. The solution curves fall toward this level from above, and rise toward it from below.

[0, 150] by [0, 150]

Figure 6.12 A slope field for the logistic differential equation

$$\frac{dP}{dt} = 0.001P(100 - P).$$

(Example 2)

(b) We can assume that $t = 0$ when the bear population is 10, so $P(0) = 10$. The logistic growth model we seek is the solution to the following initial value problem.

Differential equation: $\quad\dfrac{dP}{dt} = 0.001P(100 - P)$

Initial condition: $\quad P(0) = 10$

To prepare for integration we rewrite the differential equation in the form

$$\frac{1}{P(100 - P)} \frac{dP}{dt} = 0.001.$$

Using partial fractions, we rewrite $1/(P(100 - P))$ as

$$\frac{1}{P(100 - P)} = \frac{1}{100}\left(\frac{1}{P} + \frac{1}{100 - P}\right).$$

Substituting this expression into the differential equation and multiplying both sides by 100 we obtain

$$\left(\frac{1}{P} + \frac{1}{100 - P}\right)\frac{dP}{dt} = 0.1$$

$$\ln|P| - \ln|100 - P| = 0.1t + C \qquad \text{Integrate with respect to } t.$$

$$\ln\left|\frac{P}{100 - P}\right| = 0.1t + C$$

$$\ln\left|\frac{100 - P}{P}\right| = -0.1t - C \qquad \ln\frac{a}{b} = -\ln\frac{b}{a}$$

$$\left|\frac{100 - P}{P}\right| = e^{-0.1t - C} \qquad \text{Exponentiate.}$$

$$\frac{100 - P}{P} = (\pm e^{-C})e^{-0.1t}$$

$$\frac{100}{P} - 1 = Ae^{-0.1t} \qquad \text{Let } A = \pm e^{-C}.$$

$$P = \frac{100}{1 + Ae^{-0.1t}}. \qquad \text{Solve for } P.$$

This is the general solution to the differential equation. When $t = 0$, $P = 10$, and we obtain

$$10 = \frac{100}{1 + Ae^0}$$

$$1 + A = 10$$

$$A = 9.$$

Thus, the logistic growth model is

$$P = \frac{100}{1 + 9e^{-0.1t}}.$$

Its graph (Figure 6.13) is superimposed on the slope field from Figure 6.12.

Partial fractions are reviewed in more detail in Section 8.4.

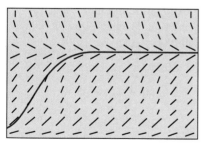

[0, 150] by [0, 150]

Figure 6.13 The graph of

$$P = \frac{100}{1 + 9e^{-0.1t}}$$

superimposed on a slope field for

$$\frac{dP}{dt} = 0.001P(100 - P).$$

(Example 2)

(c) For this model, the bear population will reach 50 when

$$50 = \frac{100}{1 + 9e^{-0.1t}}$$

$$1 + 9e^{-0.1t} = 2$$

$$e^{-0.1t} = \frac{1}{9}$$

$$e^{0.1t} = 9$$

$$t = \frac{\ln 9}{0.1} \approx 22 \text{ years.}$$

The solution of the general logistic differential equation

$$\frac{dP}{dt} = \frac{k}{M}P(M - P)$$

(Equation 1) can be obtained as in Example 2. In Exercise 25, we ask you to show that the solution is

$$P = \frac{M}{1 + Ae^{-kt}}.$$

The value of A is determined by an appropriate initial condition.

Logistic Regression

The logistic growth model is built in on many graphers as the logistic regression equation. It is not a good idea to use logistic regression with the data in Table 6.7, though, because the time interval is too short. In the next example we consider U.S. population data from 1880 to 1990.

Example 3 USING LOGISTIC REGRESSION

(a) Find the logistic regression equation for the U.S. population data in Table 6.8 and superimpose its graph on a scatter plot of the data.

(b) Use the model in (a) to find the carrying capacity predicted by the regression equation.

(c) Determine when the rate of growth predicted by the regression equation changes from increasing to decreasing. Estimate the population at this time.

Solution **(a)** We let $x = 0$ represent 1880, $x = 10$ represent 1890, and so forth. The logistic regression equation obtained from the grapher we use is approximately

$$y = \frac{540.7}{1 + 8.972e^{-0.01856x}}.$$

See Figure 6.14.

(b) The form of the logistic model allows us to read that the carrying capacity is 540.7 million. This number is also the limit of y as $x \to \infty$.

$$\lim_{x \to \infty} y = \lim_{x \to \infty} \frac{540.7}{1 + 8.972e^{-0.01856x}} = 540.7$$

Table 6.8 U.S. Population, 1880–1990

Year	Population (millions)
1880	50.2
1890	63.0
1900	76.0
1910	92.0
1920	105.7
1930	122.8
1940	131.7
1950	151.3
1960	179.3
1970	203.3
1980	226.5
1990	248.7

Source: The Statesman's Yearbook, 133rd ed. (London: The Macmillan Press, Ltd., 1996).

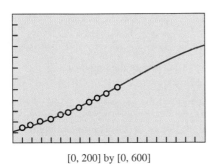

[0, 200] by [0, 600]

Figure 6.14 The graph of $y = 540.7/(1 + 8.972e^{-0.01856x})$ superimposed on a scatter plot of the data in Table 6.8. (Example 3)

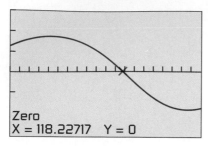

Zero
X = 118.22717 Y = 0

[0, 200] by [−0.03, 0.03]

Figure 6.15 The graph of the second derivative of $y = 540.7/(1 + 8.972e^{-0.01856x})$ obtained using NDER twice. (Example 3)

(c) In the long run, the population will stabilize at 540.7 million. If the first derivative of a twice-differentiable function changes from increasing to decreasing, it must do so at a point of inflection. We look for a zero of y''. Figure 6.15 shows that $y'' = 0$ when $x \approx 118$, or in 1998. The population at this time, $y^{(118)}$, is predicted to be about 270 million.

Suppose we have reason to believe that the maximum population found in Example 3 is too high. Example 4 shows how we can use a projected value of the population to find a new regression model with smaller carrying capacity.

Example 4 USING LOGISTIC REGRESSION

Use the data in Table 6.8 and the estimate that the U.S. population will be 350 million in 2080 to find a new logistic growth model for the population. Graph the data and the model. What is the new carrying capacity?

Solution We add the year 2080 with population 350 million to the data in Table 6.8 and compute the new logistic regression model to be about

$$y = \frac{378}{1 + 6.48e^{-0.02232x}}.$$

The new carrying capacity is 378 million (Figure 6.16).

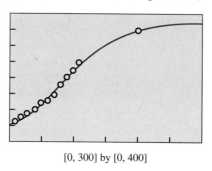

[0, 300] by [0, 400]

Figure 6.16 (Example 4)

Quick Review 6.5

In Exercises 1–8, let

$$f(x) = \frac{50}{1 + 5e^{-0.1x}}.$$

1. Find where f is continuous.

2. Find $\lim_{x \to \infty} f(x)$; $\lim_{x \to -\infty} f(x)$.

3. Find the horizontal asymptotes of the graph of f.

4. Find the domain of f' and the domain of f''.

5. Draw the graph of f and find its zeros.

6. Draw the graph of f' and find the intervals on which f is
(a) increasing, (b) decreasing.

7. Draw the graph of f'' and find where the graph of f is
(a) concave up, (b) concave down.

8. Find the points of inflection of the graph of f.

In Exercises 9 and 10, find the values of A and B to complete the partial fraction decomposition.

9. $\dfrac{x - 12}{x^2 - 4x} = \dfrac{A}{x} + \dfrac{B}{x - 4}$

10. $\dfrac{2x + 16}{x^2 + x - 6} = \dfrac{A}{x + 3} + \dfrac{B}{x - 2}$

Section 6.5 Exercises

In Exercises 1–4, a population is described.

(a) Write a differential equation for the population.

(b) Find a formula for the population P in terms of t.

(c) Superimpose the graph of the population function on a slope field for the differential equation.

1. The relative growth rate of Clairsville is 0.025 and its current population is $P(0) = 75{,}000$.

2. The relative growth rate of Blairsville is 1.9% and its current population is $P(0) = 110{,}000$.

3. logistic growth with $k = 0.05$, $M = 200$, and $P(0) = 10$

4. logistic growth with $k = 0.02$, $M = 150$, and $P(0) = 15$

In Exercises 5 and 6, find the relative growth rate of the population represented by the differential equation.

5. $\dfrac{dP}{dt} = -0.3P$

6. $\dfrac{dP}{dt} = 0.075P$

In Exercises 7 and 8, find k and the carrying capacity for the population represented by the logistic differential equation.

7. $\dfrac{dP}{dt} = 0.04P - 0.0004P^2$

8. $\dfrac{50}{P}\dfrac{dP}{dt} = 2 - \dfrac{P}{250}$

In Exercises 9–12, match the differential equation with its slope field.

9. $\dfrac{dy}{dx} = 0.065y$

10. $\dfrac{dy}{dx} = 0.06y\left(1 - \dfrac{y}{100}\right)$

11. $\dfrac{dy}{dx} = \dfrac{y}{x}$ y

12. $\dfrac{dy}{dx} = 0.06y\left(1 - \dfrac{y}{150}\right)$

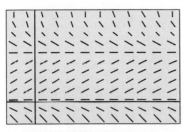

[−20, 100] by [−50, 200]

(a)

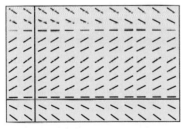

[−20, 100] by [−50, 200]

(b)

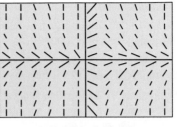

[−10, 10] by [−10, 10]

(c)

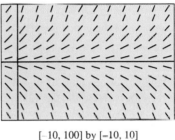

[−10, 100] by [−10, 10]

(d)

In Exercises 13 and 14, a population function is given.

(a) Show that the function is a solution of a logistic differential equation. Identify k and the carrying capacity.

(b) **Writing to Learn** Estimate $P(0)$. Explain its meaning in the context of the problem.

13. *Rabbit Population* A population of rabbits is given by the formula
$$P(t) = \frac{1000}{1 + e^{4.8 - 0.7t}},$$
where t is the number of months after a few rabbits are released.

14. *Spread of Measles* The number of students infected by measles in a certain school is given by the formula
$$P(t) = \frac{200}{1 + e^{5.3 - t}},$$
where t is the number of days after students are first exposed to an infected student.

15. *U.S. Population* The Museum of Science in Boston displays a running total of the U.S. population. On May 11, 1993, the total was increasing at the rate of 1 person every 14 sec. The displayed population figure at 3:45 P.M. that day was 257,313,431. Assume that the relative growth rate is constant.

(a) What is the relative growth rate per year (365 days)?

(b) At this rate, what will the U.S. population be at 3:45 P.M. Boston time on May 11, 2001?

16. *Spread of a Disease*   Assume that the relative rate of spread of a certain disease is constant. Suppose also that in the course of any given year the number of cases of the disease is reduced by 20%.

(a) If there are 10,000 cases today, how many years will it take to reduce the number to 1000?

(b) How long will it take to eradicate the disease, that is, to reduce the number of cases to less than 1?

17. *Guppy Population*   A 2000-gallon tank can support no more than 150 guppies. Six guppies are introduced into the tank. Assume that the rate of growth of the population is

$$\frac{dP}{dt} = 0.0015P(150 - P),$$

where time t is in weeks.

(a) Find a formula for the guppy population in terms of t.

(b) How long will it take for the guppy population to be 100? 125?

18. *Gorilla Population*   A certain wild animal preserve can support no more than 250 lowland gorillas. Twenty-eight gorillas were known to be in the preserve in 1970. Assume that the rate of growth of the population is

$$\frac{dP}{dt} = 0.0004P(250 - P),$$

where time t is in years.

(a) Find a formula for the gorilla population in terms of t.

(b) How long will it take for the gorilla population to reach the carrying capacity of the preserve?

19. *Tooth Size*   The analysis of tooth shrinkage by C. Loring Brace and colleagues at the University of Michigan's Museum of Anthropology indicates that human tooth size is continuing to decrease and that the evolutionary process did not come to a halt some 30,000 years ago as many scientists contend. In northern Europe, for example, tooth size reduction now has a constant relative rate of 1% per 1000 years.

(a) Find a model for northern European tooth size y in terms of time t in years.

(b) In about how many years will northern European tooth size be 90% of its present size?

(c) What will be northern European tooth size 20,000 years from now as a percentage of our present tooth size?
Source: LSA Magazine, Spring 1989, Vol. 12, p. 19, Ann Arbor, MI.

20. *Honeybee Population*   A population of honeybees grows at an annual rate equal to 1/4 of the number present when there are no more than 10,000 bees. If there are more than 10,000 bees but fewer than 50,000 bees, the growth rate is equal to 1/12 of the number present. If there are 5000 bees now, when will there be 25,000 bees?

21. *Continuous Compounding*   You have $1000 with which to open an account to which you plan also to add $1000 per year. All funds in the account will earn 10% annual interest compounded continuously. If the added deposits are also credited to your account continuously, the number of dollars x in your account at time t (years) will satisfy the initial value problem.

Differential equation: $\dfrac{dx}{dt} = 1000 + 0.10x$

Initial condition: $x(0) = 1000$

(a) Solve the initial value problem for x as a function of t.

(b) About how many years will it take for the amount in your account to reach $100,000?

22. *Continuous Price Discounting*   To encourage buyers to place 100-unit orders, your firm's sales department applies a continuous discount that makes the unit price a function $p(x)$ of the number of units ordered. The discount decreases the price at the rate of $0.01 per unit ordered. The price per unit for a 100-unit order is $p(100) = \$20.09$.

(a) Find $p(x)$ by solving the following initial value problem.

Differential equation: $\dfrac{dp}{dx} = -\dfrac{1}{100}p$

Initial condition: $p(100) = 20.09$

(b) Find the unit price $p(10)$ for a 10-unit order and the unit price $p(90)$ for a 90-unit order.

(c) The sales department has asked you to find out whether it is discounting so much that the firm's revenue, $r(x) = x \cdot p(x)$, will actually be less for a 100-unit order than, say, for a 90-unit order. Reassure them by showing that r has its maximum value at $x = 100$.

In Exercises 23 and 24, use the census data provided.

(a) Find the logistic regression equation for the data and superimpose its graph on a scatter plot of the data.

(b) Find the carrying capacity predicted by the regression equation.

(c) Find when the rate of growth predicted by the regression equation changes from increasing to decreasing. Estimate the population at this time.

23. *State of New York*   Let $x = 0$ represent 1910, $x = 20$ represent 1930, and so forth.

Table 6.9 New York Population

Year	Population (millions)
1910	9.1
1930	12.6
1960	16.8
1980	17.6
1990	18.0

Source: The Statesman's Yearbook, 133rd ed. (London: The Macmillan Press, Ltd., 1996).

24. *State of Florida* Let $x = 0$ represent 1950, $x = 10$ represent 1960, and so forth.

Table 6.10 Florida Population

Year	Population (millions)
1950	2.8
1960	5.0
1970	6.8
1980	9.7
1990	12.9

Source: The Statesman's Yearbook, 133rd ed. (London: The Macmillan Press, Ltd., 1996).

25. *Logistic Differential Equation* Show that the solution of the differential equation (Equation 1)

$$\frac{dP}{dt} = \frac{k}{M}P(M - P) \quad \text{is} \quad P = \frac{M}{1 + Ae^{-kt}},$$

where A is an arbitrary constant.

26. *Tinkering with Carrying Capacity* Suppose you decide to estimate the population of Florida to be 16 million in the year 2020. Include this data point with the data in Table 6.10.

(a) Find a logistic regression equation for the new data.

(b) What is the new carrying capacity?

In Exercises 27–30, solve the initial value problem.

27. $\dfrac{dy}{dx} = (\cos x)\, e^{\sin x}, \quad y(0) = 0$

28. $\dfrac{dy}{dx} = -2(y - 3), \quad y(0) = 5$

29. $\dfrac{dy}{dx} = \dfrac{x}{y}, \quad y(0) = 2$ **30.** $\dfrac{dy}{dx} = y\sqrt{x}, \quad y(0) = 1$

Exploration

31. *Extinct Populations* One theory states that if the size of a population falls below a minimum m, the population will become extinct. This condition leads to the *extended* logistic differential equation

$$\frac{dP}{dt} = kP\left(1 - \frac{P}{M}\right)\left(1 - \frac{m}{P}\right)$$

$$= \frac{k}{M}(M - P)(P - m),$$

with $k > 0$ the proportionality constant and M the population maximum.

(a) Show that dP/dt is positive for $m < P < M$ and negative if $P < m$ or $P > M$.

(b) Let $m = 100$, $M = 1200$, and assume that $m < P < M$. Show that the differential equation can be rewritten in the form

$$\left[\frac{1}{1200 - P} + \frac{1}{P - 100}\right]\frac{dP}{dt} = \frac{11}{12}k.$$

Use a procedure similar to that used in Example 2 to solve this differential equation.

(c) Find the solution to (b) that satisfies $P(0) = 300$.

(d) Superimpose the graph of the solution in (c) with $k = 0.1$ on a slope field of the differential equation.

(e) Solve the general extended differential equation with the restriction $m < P < M$. ∎

Extending the Ideas

32. *Variable Inflation Rate* When the inflation rate k varies with time instead of being constant, the formula $p = p_0 e^{kt}$ no longer gives the solution of the differential equation $dp/dt = kp$.

(a) Show that the solution of the differential equation satisfying $p(0) = p_0$ is

$$p(t) = p_0 e^{\int_0^t k(u)\,du}.$$

(b) **Writing to Learn** Suppose that $p_0 = 100$ and $k(t) = 0.04/(1 + t)$, a rate of inflation that starts at 4% at time $t = 0$ but decreases steadily as the years pass. Find $p(9)$ and give a meaning to its value.

(c) Find the price of a \$100 item at $t = 9$ years if the annual rate of inflation is a constant 4%.

(d) Suppose that $p_0 = 100$ and $k(t) = 0.04 + 0.004t$, a rate of inflation that starts at 4% when $t = 0$ but increases steadily as the years pass. Find $p(9)$.

33. *Catastrophic Solution* Let k and P_0 be positive constants.

(a) Solve the initial value problem.

$$\frac{dP}{dt} = kP^2, \quad P(0) = P_0$$

(b) Show that the graph of the solution in (a) has a vertical asymptote at a positive value of t.

Numerical Methods

Euler's Method • Numerical Solutions • Graphical Solutions • Improved Euler's Method

Euler's Method

If we are given a differential equation $dy/dx = f(x, y)$ and an initial condition $y(x_0) = y_0$, we can approximate the solution $y = y(x)$ by its linearization

$$L(x) = y(x_0) + y'(x_0)(x - x_0) \quad \text{or} \quad L(x) = y_0 + f(x_0, y_0)(x - x_0).$$

The function $L(x)$ will give a good approximation to the solution $y(x)$ in a short interval about x_0 (Figure 6.17). The basis of Euler's method is to patch together a string of linearizations to approximate the curve over a longer stretch. Here is how the method works.

We know the point (x_0, y_0) lies on the solution curve. Suppose we specify a new value for the independent variable to be $x_1 = x_0 + dx$. If the increment dx is small, then

$$y_1 = L(x_1) = y_0 + f(x_0, y_0) \, dx$$

is a good approximation to the exact solution value $y = y(x_1)$. So from the point (x_0, y_0), which lies *exactly* on the solution curve, we have obtained the point (x_1, y_1), which lies very close to the point $(x_1, y(x_1))$ on the solution curve (Figure 6.18).

Using the point (x_1, y_1) and the slope $f(x_1, y_1)$ of the solution curve through (x_1, y_1), we take a second step. Setting $x_2 = x_1 + dx$, we use the linearization of the solution curve through (x_1, y_1) to calculate

$$y_2 = y_1 + f(x_1, y_1) \, dx.$$

This gives the next approximation (x_2, y_2) to values along the solution curve $y = y(x)$ (Figure 6.19). Continuing in this fashion, we take a third step from the point (x_2, y_2) with slope $f(x_2, y_2)$ to obtain the third approximation

$$y_3 = y_2 + f(x_2, y_2) \, dx,$$

and so on. We are literally building an approximation to one of the solutions by following the direction of the slope field of the differential equation.

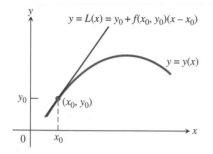

Figure 6.17 The linearization of $y = y(x)$ at $x = x_0$.

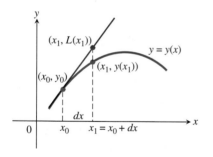

Figure 6.18 The first Euler step approximates $y(x_1)$ with $L(x_1)$.

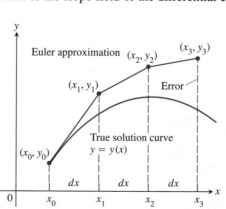

Figure 6.19 Three steps in the Euler approximation to the solution of the initial value problem $y' = f(x, y)$, $y(x_0) = y_0$. As we take more steps, the errors involved usually accumulate, but not in the exaggerated way shown here.

The steps in Figure 6.19 are drawn large to illustrate the construction process, so the approximation looks crude. In practice, dx would be small enough to make the red curve hug the blue one and give a good approximation throughout.

Example 1   USING EULER'S METHOD

Find the first three approximations y_1, y_2, y_3 using Euler's method for the initial value problem

$$y' = 1 + y, \quad y(0) = 1,$$

starting at $x_0 = 0$ with $dx = 0.1$.

Solution   We have $x_0 = 0$, $y_0 = 1$, $x_1 = x_0 + dx = 0.1$, $x_2 = x_0 + 2dx = 0.2$, and $x_3 = x_0 + 3dx = 0.3$.

First:
$$\begin{aligned} y_1 &= y_0 + f(x_0, y_0)\, dx \\ &= y_0 + (1 + y_0)\, dx \\ &= 1 + (1 + 1)(0.1) = 1.2 \end{aligned}$$

Second:
$$\begin{aligned} y_2 &= y_1 + f(x_1, y_1)\, dx \\ &= y_1 + (1 + y_1)\, dx \\ &= 1.2 + (1 + 1.2)(0.1) = 1.42 \end{aligned}$$

Third:
$$\begin{aligned} y_3 &= y_2 + f(x_2, y_2)\, dx \\ &= y_2 + (1 + y_2)\, dx \\ &= 1.42 + (1 + 1.42)(0.1) = 1.662 \end{aligned}$$

Numerical Solutions

If we do not require or cannot immediately find an *exact* solution for an initial value problem $y' = f(x, y)$, $y(x_0) = y_0$, we can use a grapher to generate a table of approximate numerical values of y for values of x in an appropriate interval. Such a table is a **numerical solution** of the initial value problem and the method by which we generate the table is a **numerical method.**

The step-by-step process used in Example 1 can be continued easily. If we use equally spaced values for the independent variable in the table and generate n of them, we first set

$$x_1 = x_0 + dx,$$
$$x_2 = x_1 + dx,$$
$$\vdots$$
$$x_n = x_{n-1} + dx.$$

Then we calculate the approximations to the solution,

$$y_1 = y_0 + f(x_0, y_0)\, dx,$$
$$y_2 = y_1 + f(x_1, y_1)\, dx,$$
$$\vdots$$
$$y_n = y_{n-1} + f(x_{n-1}, y_{n-1})\, dx.$$

The number of steps n can be as large as we like, but errors can accumulate if n is too large.

This process can be easily programmed. We use the program EULERT (EULER Table) to generate numerical solutions to initial value problems. It allows us to choose n and the step size dx.

In Exercise 5, you will show that the exact solution to the initial value problem of Example 1 is $y = 2e^x - 1$. We will use this information in Example 2.

Example 2 INVESTIGATING ACCURACY OF EULER'S METHOD

Use Euler's Method to solve

$$y' = 1 + y, \quad y(0) = 1,$$

on the interval $0 \le x \le 1$, starting at $x_0 = 0$ and taking **(a)** $dx = 0.1$ and **(b)** $dx = 0.05$. Compare the approximations to the values of the exact solution $y = 2e^x - 1$.

Solution **(a)** We use EULERT to generate the approximate values in Table 6.11. The "error" column is obtained by subtracting the unrounded Euler values from the unrounded values found using the exact solution. All entries are then rounded to 4 decimal places.

Table 6.11 Euler Solution of $y' = 1 + y$, $y(0) = 1$,
Step Size $dx = 0.1$

x	y (Euler)	y (exact)	Error
0	1	1	0
0.1	1.2	1.2103	0.0103
0.2	1.42	1.4428	0.0228
0.3	1.662	1.6997	0.0377
0.4	1.9282	1.9836	0.0554
0.5	2.2210	2.2974	0.0764
0.6	2.5431	2.6442	0.1011
0.7	2.8974	3.0275	0.1301
0.8	3.2872	3.4511	0.1639
0.9	3.7159	3.9192	0.2033
1.0	4.1875	4.4366	0.2491

By the time we reach $x = 1$ (after 10 steps), the error is about 5.6% of the exact solution.

(b) One way to try to reduce the error is to decrease the step size. Table 6.12 shows the results and their comparisons with the exact solutions when we decrease the step size to 0.05, doubling the number of steps to 20. As in Table 6.11, all computations are performed before rounding. This time when we reach $x = 1$ the relative error is only about 2.9%.

Table 6.12 Euler Solution of $y' = 1 + y$, $y(0) = 1$, Step Size $dx = 0.05$

x	y (Euler)	y (exact)	Error
0	1	1	0
0.05	1.1	1.1025	0.0025
0.10	1.205	1.2103	0.0053
0.15	1.3153	1.3237	0.0084
0.20	1.4310	1.4428	0.0118
0.25	1.5526	1.5681	0.0155
0.30	1.6802	1.6997	0.0195
0.35	1.8142	1.8381	0.0239
0.40	1.9549	1.9836	0.0287
0.45	2.1027	2.1366	0.0340
0.50	2.2578	2.2974	0.0397
0.55	2.4207	2.4665	0.0458
0.60	2.5917	2.6442	0.0525
0.65	2.7713	2.8311	0.0598
0.70	2.9599	3.0275	0.0676
0.75	3.1579	3.2340	0.0761
0.80	3.3657	3.4511	0.0853
0.85	3.5840	3.6793	0.0953
0.90	3.8132	3.9192	0.1060
0.95	4.0539	4.1714	0.1175
1.00	4.3066	4.4366	0.1300

It might be tempting to reduce the step size even further to obtain greater accuracy. However, each additional calculation not only requires additional grapher time but more importantly adds to the buildup of round-off errors due to the approximate representations of numbers inside the grapher.

The analysis of error and the investigation of methods to reduce it when making numerical calculations are important, but appropriate for a more advanced course. There are numerical methods more accurate than Euler's method, as you will see in your further study of differential equations.

Graphical Solutions

Investigating a graph is usually more instructive than analyzing a large data table. If we plot the data pairs in a numerical solution of an initial value problem we obtain a **graphical solution,** as shown in Example 3.

Example 3   VISUALIZING EULER'S APPROXIMATIONS

Figure 6.20 gives a visualization of the numerical solution shown as Table 6.11 by superimposing the graph of the exact solution on a scatter plot of the data points in the table.

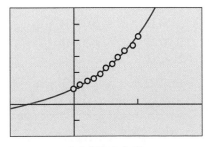

[−1, 2] by [−2, 6]

Figure 6.20 The graph of $y = 2e^x - 1$ superimposed on a scatter plot of the Euler approximations shown in Table 6.11. (Example 3)

Improved Euler's Method

We can improve on Euler's method by taking an average of two slopes. We first estimate y_n as in the original Euler method, but denote it by z_n. We then take the average of $f(x_{n-1}, y_{n-1})$ and $f(x_n, z_n)$ in place of $f(x_{n-1}, y_{n-1})$ in the next step. Thus,

$$z_n = y_{n-1} + f(x_{n-1}, y_{n-1})\, dx,$$

$$y_n = y_{n-1} + \left[\frac{f(x_{n-1}, y_{n-1}) + f(x_n, z_n)}{2} \right] dx.$$

We use the program IMPEULT (IMProved EULer Table) to perform the computations.

Example 4 INVESTIGATING THE ACCURACY OF IMPROVED EULER'S METHOD

Use the improved Euler's method to solve

$$y' = 1 + y, \quad y(0) = 1$$

on the interval $0 \le x \le 1$, starting at $x_0 = 0$ and taking $dx = 0.1$. Compare the approximations to the values of the exact solution $y = 2e^x - 1$.

Solution We use IMPEULT to generate the approximate values in Table 6.13. The "error" column is obtained by subtracting the unrounded improved Euler values from the unrounded values found using the exact solution. All entries are then rounded to 4 decimal places.

Table 6.13 Improved Euler Solution of $y' = 1 + y$, $y(0) = 1$, Step Size $dx = 0.1$

x	y (IMPEULT)	y (exact)	Error
0	1	1	0
0.1	1.21	1.2103	0.0003
0.2	1.4421	1.4428	0.0008
0.3	1.6985	1.6997	0.0013
0.4	1.9818	1.9836	0.0018
0.5	2.2949	2.2974	0.0025
0.6	2.6409	2.6442	0.0034
0.7	3.0231	3.0275	0.0044
0.8	3.4456	3.4511	0.0055
0.9	3.9124	3.9192	0.0068
1.0	4.4282	4.4366	0.0084

By the time we reach $x = 1$ (after 10 steps), the relative error is about 0.19%.

By comparing Tables 6.11 and 6.13, we see that the improved Euler's method is considerably more accurate than the regular Euler's method, at least for the initial value problem $y' = 1 + y$, $y(0) = 1$.

Instead of making a scatter plot of the Euler approximations, we can use the program EULERG (EULER Graph), a modification of EULERT, to plot the points from the Euler approximation directly to obtain a graphical solution.

Example 5 FINDING GRAPHICAL SOLUTIONS

Find a graphical solution for the initial value problem

$$y' = x - y, \quad y(0) = 1.$$

Compare with the graph of the exact solution, $y = x - 1 + 2e^{-x}$.

Solution Figure 6.21a shows the EULERG solution of the initial value problem starting with $x = 0$ with step size 0.1. Figure 6.21b shows the graph of the exact solution superimposed on the Euler graph.

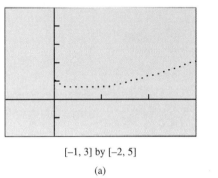
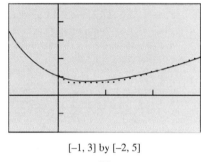

[−1, 3] by [−2, 5]   [−1, 3] by [−2, 5]

(a)   (b)

Figure 6.21 (b) The graph of the exact solution $y = x - 1 + 2e^{-x}$ of the initial value problem $y' = x - y, y(0) = 1$ agrees with (a) the graph of the Euler solution. (Example 5)

Quick Review 6.6

In Exercises 1 and 2, let $f(x) = x^3 - 3x$.

1. Find $f'(2)$.

2. Find the linearization $L(x)$ of f at $x = 2$.

In Exercises 3 and 4, let $f(x) = \tan x$.

3. Find $f'(\pi/4)$.

4. Find the linearization $L(x)$ of f at $x = \pi/4$.

In Exercises 5 and 6, let $f(x) = 0.1x^2 + (5/x)$.

5. Find $f'(4)$.

6. Find the linearization $L(x)$ of f at $x = 4$.

In Exercises 7–10, use the approximation $L(x) \approx f(x)$ found in Exercise 6. Find the **(a)** error and **(b)** percentage error in the approximation at $x = a$.

7. $a = 4.1$ **8.** $a = 4.2$

9. $a = 4.5$ **10.** $a = 3.5$

Section 6.6 Exercises

In Exercises 1–4, use differentiation and substitution to show that the function $y = f(x)$ is the exact solution of the initial value problem.

1. $y' = x - y, \quad y(0) = 1, \quad f(x) = x - 1 + 2e^{-x}$

2. $y' = x - y, \quad y(0) = -2, \quad f(x) = x - 1 - e^{-x}$

3. $y' = 2y + \sin x, \quad y(0) = 0, \quad f(x) = \dfrac{e^{2x} - 2 \sin x - \cos x}{5}$

4. $y' = y - e^{2x} + 1, \quad y(0) = -1, \quad f(x) = e^x - e^{2x} - 1$

In Exercises 5–8, use analytic methods to find the exact solution of the initial value problem.

5. $y' = 1 + y$, $y(0) = 1$ **6.** $y' = x(1 - y)$, $y(-2) = 0$

7. $y' = 2y(x + 1)$, $y(-2) = 2$

8. $y' = y^2(1 + 2x)$, $y(-1) = -1$

In Exercises 9 and 10, use Euler's method to solve the initial value problem on the interval $0 \le x \le 1$ starting at $x_0 = 0$ with $dx = 0.1$. Compare the approximations to the values of the exact solution.

9. $y' = 2y + \sin x$, $y(0) = 0$
    (See Exercise 3 for the exact solution.)

10. $y' = x - y$, $y(0) = -2$
    (See Exercise 2 for the exact solution.)

In Exercises 11 and 12, use the improved Euler's method to solve the initial value problem on the interval $-2 \le x \le 0$ starting at $x_0 = -2$ with $dx = 0.1$. Compare the approximations to the values of the exact solution.

11. $y' = 2y(x + 1)$, $y(-2) = 2$
    (See Exercise 7 for the exact solution.)

12. $y' = x(1 - y)$, $y(-2) = 0$
    (See Exercise 6 for the exact solution.)

In Exercises 13 and 14, *work in groups of two or three.* Use Euler's method and the improved Euler's method to solve the initial value problem on the interval $0 \le x \le 2$ starting at $x_0 = 0$ with $dx = 0.1$. Compare the accuracy of the two methods.

13. $y' = x - y$, $y(0) = 1$
    (See Exercise 1 for the exact solution.)

14. $y' = y - e^{2x} + 1$, $y(0) = -1$
    (See Exercise 4 for the exact solution.)

In Exercises 15 and 16, **(a)** find the exact solution of the initial value problem. Then compare the accuracy of the approximation to $y(x^*)$ using Euler's method starting at x_0 with step size **(b)** 0.2, **(c)** 0.1, and **(d)** 0.05.

15. $y' = 2y^2(x - 1)$, $y(2) = -1/2$, $x_0 = 2$, $x^* = 3$

16. $y' = y - 1$, $y(0) = 3$, $x_0 = 0$, $x^* = 1$

In Exercises 17 and 18, compare the accuracy of the approximation to $y(x^*)$ using the improved Euler's method starting at x_0 with step size **(a)** 0.2, **(b)** 0.1, and **(c)** 0.05. **(d) Writing to Learn** Explain what happens to the error.

17. $y' = 2y^2(x - 1)$, $y(2) = -1/2$, $x_0 = 2$, $x^* = 3$
    (See Exercise 15 for the exact solution.)

18. $y' = y - 1$, $y(0) = 3$, $x_0 = 0$, $x^* = 1$
    (See Exercise 16 for the exact solution.)

In Exercises 19–22, superimpose the graph of the exact solution on a scatter plot of the Euler or improved Euler approximations from the given exercise.

19. Exercise 9   **20.** Exercise 10

21. Exercise 11   **22.** Exercise 12

In Exercises 23 and 24, use the indicated method to solve the initial value problem $y' = x + y$, $y(0) = 1$, on the interval $-1 \le x \le 0$ starting at $x_0 = 0$ with $dx = -0.1$. Compare the approximations to the values of the exact solution $y = 2e^x - x - 1$.

23. Euler's method   **24.** improved Euler's method

In Exercises 25–28, use the indicated method to solve the initial value problem graphically starting at $x_0 = 0$ with **(a)** $dx = 0.1$ and **(b)** $dx = 0.05$.

25. Euler's method, $y' = y + e^x - 2$, $y(0) = 2$

26. Euler's method, $y' = \cos(2x - y)$, $y(0) = 2$

27. improved Euler's method,

$$y' = y\left(\frac{1}{2} - \ln|y|\right), \quad y(0) = \frac{1}{3}$$

28. improved Euler's method, $y' = \sin(2x - y)$ $y(0) = 1$

29. Use Euler's method with $dx = 0.05$ to estimate $y(1)$ if $y' = y$ and $y(0) = 1$. What is the exact value of $y(1)$?

30. Use the improved Euler's method with $dx = 0.05$ to estimate $y(1)$ if $y' = 3y$ and $y(0) = 1$. What is the exact value of $y(1)$?

Extending the Ideas

The calculator programs RUNKUTG (G for "graph") and RUNKUTT (the second T for "table") solve initial value problems using a numerical method called the **Runge-Kutta method.** As implemented by the calculator, the method uses a weighted average of four intermediate calculations to go from each solution step to the next. With

$$k_1 = f(x_{n-1}, y_{n-1})\,dx,$$

$$k_2 = f\left(x_{n-1} + \frac{dx}{2}, y_{n-1} + \frac{k_1}{2}\right)dx,$$

$$k_3 = f\left(x_{n-1} + \frac{dx}{2}, y_{n-1} + \frac{k_2}{2}\right)dx,$$

$$k_4 = f(x_{n-1} + dx, y_{n-1} + k_3)\,dx,$$

the programs calculate y_n from y_{n-1} as

$$y_n = y_{n-1} + \frac{1}{6}(k_1 + 2k_2 + 2k_3 + k_4).$$

31. Use the Runge-Kutta program RUNKUTT to solve the initial value problem $y' = 1 + y$, $y(0) = 1$, on the interval $0 \le x \le 1$ starting at $x_0 = 0$ with $dx = 0.1$. Compare the approximations to the values of the exact solution $y = 2e^x - 1$.

32. Use the Runge-Kutta program RUNKUTG to solve the following initial value problems graphically starting at $x_0 = 0$ with $dx = 0.1$.

(a) $y' = x - y$, $y(0) = 1$ **(b)** $y' = x - y$, $y(0) = -2$

Calculus at Work

I have a Bachelor's and Master's degree in Aerospace Engineering from the University of California at Davis. I started my professional career as a Facility Engineer managing productivity and maintenance projects in the Unitary Project Wind Tunnel facility at NASA Ames Research Center. I used calculus and differential equations in fluid mechanic analyses of the tunnels. I then moved to the position of Test Manager, still using some fluid mechanics and other mechanical engineering analysis tools to solve problems. For example, the lift and drag forces acting on an airplane wing can be determined by integrating the known pressure distribution on the wing.

I am currently a NASA On-Site Systems Engineer for the Lunar Prospector spacecraft project, at Lockheed Martin Missiles and Space in Sunnyvale, California. Differential equations and integration are used to design some of the flight hardware for the spacecraft. I work on ensuring that the different systems of the spacecraft are adequately integrated together to meet the specified design requirements. This often means doing some analysis to determine if the systems will function properly and within the constraints of the space environment. Some of these analyses require use of differential equations and integration to determine the most exact results, within some margin of error.

Ross Shaw

NASA Ames Research Center
Sunnyvale, CA

Chapter 6 Key Terms

Chapter 6 Review Exercises

In Exercises 1–8, evaluate the integral analytically. Then use NINT to support your result.

1. $\displaystyle\int_0^{\pi/3} \sec^2 \theta \, d\theta$

2. $\displaystyle\int_1^2 \left(x + \frac{1}{x^2} \right) dx$

3. $\displaystyle\int_0^1 \frac{36 \, dx}{(2x + 1)^3}$

4. $\displaystyle\int_{-1}^1 2x \sin (1 - x^2) \, dx$

5. $\displaystyle\int_0^{\pi/2} 5 \sin^{3/2} x \cos x \, dx$

6. $\displaystyle\int_{1/2}^4 \frac{x^2 + 3x}{x} \, dx$

7. $\displaystyle\int_0^{\pi/4} e^{\tan x} \sec^2 x \, dx$

8. $\displaystyle\int_1^e \frac{\sqrt{\ln r}}{r} \, dr$

In Exercises 9–20, evaluate the integral.

9. $\displaystyle\int \frac{\cos x}{2 - \sin x} \, dx$

10. $\displaystyle\int \frac{dx}{\sqrt[3]{3x + 4}}$

11. $\displaystyle\int \frac{t \, dt}{t^2 + 5}$

12. $\displaystyle\int \frac{1}{\theta^2} \sec \frac{1}{\theta} \tan \frac{1}{\theta} \, d\theta$

13. $\displaystyle\int \frac{\tan (\ln y)}{y} \, dy$

14. $\displaystyle\int e^x \sec (e^x) \, dx$

15. $\displaystyle\int \frac{dx}{x \ln x}$

16. $\displaystyle\int \frac{dt}{t \sqrt{t}}$

17. $\displaystyle\int x^3 \cos x \, dx$

18. $\displaystyle\int x^4 \ln x \, dx$

19. $\displaystyle\int e^{3x} \sin x \, dx$

20. $\displaystyle\int x^2 e^{-3x} \, dx$

In Exercises 21–28, solve the initial value problem analytically. Support your solution by overlaying its graph on a slope field of the differential equation.

21. $\dfrac{dy}{dx} = 1 + x + \dfrac{x^2}{2}, \quad y(0) = 1$

22. $\dfrac{dy}{dx} = \left(x + \dfrac{1}{x} \right)^2, \quad y(1) = 1$

23. $\dfrac{dy}{dt} = \dfrac{1}{t + 4}, \quad y(-3) = 2$

24. $\dfrac{dy}{d\theta} = \csc 2\theta \cot 2\theta, \quad y(\pi/4) = 1$

25. $\dfrac{d^2 y}{dx^2} = 2x - \dfrac{1}{x^2}, \quad x > 0, \quad y'(1) = 1, \quad y(1) = 0$

26. $\dfrac{d^3 r}{dt^3} = -\cos t, \quad r''(0) = r'(0) = r(0) = -1$

27. $\dfrac{dy}{dx} = y + 2, \quad y(0) = 2$

28. $\dfrac{dy}{dx} = (2x + 1)(y + 1), \quad y(-1) = 1$

In Exercises 29–32, assume that $1 - \sqrt{x}$ is an antiderivative of f and $x + 2$ is an antiderivative of g. Find the indefinite integral of the function.

29. $-f(x)$

30. $x + f(x)$

31. $2f(x) - g(x)$

32. $g(x) - 4$

In Exercises 33 and 34, match the indefinite integral with the graph of one of the antiderivatives of the integrand.

33. $\displaystyle\int \frac{\sin x}{x} \, dx$

34. $\displaystyle\int e^{-x^2} \, dx$

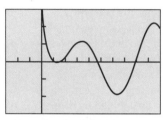

[−3, 10] by [−3, 3]

(a)

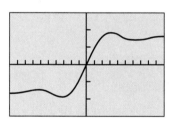

[−10, 10] by [−3, 3]

(b)

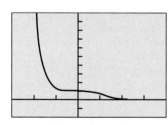

[−3, 4] by [−2, 10]

(c)

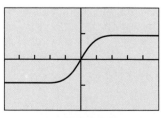

[−5, 5] by [−2, 2]

(d)

35. Writing to Learn The figure shows the graph of the function $y = f(x)$ that is the solution of one of the following initial value problems. Which one? How do you know?

i. $dy/dx = 2x, \quad y(1) = 0$

ii. $dy/dx = x^2, \quad y(1) = 1$

iii. $dy/dx = 2x + 2, \quad y(1) = 1$

iv. $dy/dx = 2x, \quad y(1) = 1$

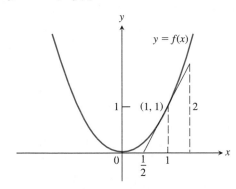

36. Writing to Learn Does the following initial value problem have a solution? Explain.

$$\frac{d^2 y}{dx^2} = 0, \quad y'(0) = 1, \quad y(0) = 0$$

37. Moving Particle The acceleration of a particle moving along a coordinate line is

$$\frac{d^2 s}{dt^2} = 2 + 6t \text{ m/sec}^2.$$

At $t = 0$ the velocity is 4 m/sec.

(a) Find the velocity as a function of time t.

(b) How far does the particle move during the first second of its trip, from $t = 0$ to $t = 1$?

38. Sketching Solutions Draw a possible graph for the function $y = f(x)$ with slope field given in the figure that satisfies the initial condition $y(0) = 0$.

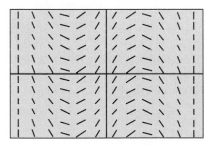

[−10, 10] by [−10, 10]

In Exercises 39 and 40, use the stated method to solve the initial value problem on the given interval starting at x_0 with $dx = 0.1$.

39. Euler; $y' - y + \cos x, \quad y(0) - 0; \quad 0 \le x \le 2; \quad x_0 - 0$

40. improved Euler; $y' = (2 - y)(2x + 3), \quad y(-3) = 1;$
$-3 \le x \le -1; \quad x_0 = -3$

In Exercises 41 and 42, use the stated method with $dx = 0.05$ to estimate $y(c)$ where y is the solution to the given initial value problem.

41. improved Euler; $c = 3; \quad \dfrac{dy}{dx} = \dfrac{x - 2y}{x + 1}, \quad y(0) = 1$

42. Euler; $c = 4; \quad \dfrac{dy}{dx} = \dfrac{x^2 - 2y + 1}{x}, \quad y(1) = 1$

In Exercises 43 and 44, use the stated method to solve the initial value problem graphically, starting at $x_0 = 0$ with **(a)** $dx = 0.1$ and **(b)** $dx = -0.1$.

43. Euler; $\dfrac{dy}{dx} = \dfrac{1}{e^{x+y+2}}, \quad y(0) = -2$

44. improved Euler; $\dfrac{dy}{dx} = -\dfrac{x^2 + y}{e^y + x}, \quad y(0) = 0$

45. *Californium-252* What costs $27 million per gram and can be used to treat brain cancer, analyze coal for its sulfur content, and detect explosives in luggage? The answer is californium-252, a radioactive isotope so rare that only about 8 g of it have been made in the western world since its discovery by Glenn Seaborg in 1950. The half-life of the isotope is 2.645 years—long enough for a useful service life and short enough to have a high radioactivity per unit mass. One microgram of the isotope releases 170 million neutrons per second.

(a) What is the value of k in the decay equation for this isotope?

(b) What is the isotope's mean life? (See Exercise 17, Section 6.4.)

46. *Cooling a Pie* A deep-dish apple pie, whose internal temperature was 220°F when removed from the oven, was set out on a 40°F breezy porch to cool. Fifteen minutes later, the pie's internal temperature was 180°F. How long did it take the pie to cool from there to 70°F?

47. *Finding Temperature* A pan of warm water (46°C) was put into a refrigerator. Ten minutes later, the water's temperature was 39°C; 10 minutes after that, it was 33°C. Use Newton's Law of Cooling to estimate how cold the refrigerator was.

48. *Art Forgery* A painting attributed to Vermeer (1632–1675), which should contain no more than 96.2% of its original carbon-14, contains 99.5% instead. About how old is the forgery?

49. *Carbon-14* What is the age of a sample of charcoal in which 90% of the carbon-14 that was originally present has decayed?

50. *Appreciation* A violin made in 1785 by John Betts, one of England's finest violin makers, cost $250 in 1924 and sold for $7500 in 1988. Assuming a constant relative rate of appreciation, what was that rate?

51. *Working Underwater* The intensity $L(x)$ of light x feet beneath the surface of the ocean satisfies the differential equation

$$\frac{dL}{dx} = -kL,$$

where k is a constant. As a diver you know from experience that diving to 18 ft in the Caribbean Sea cuts the intensity in half. You cannot work without artificial light when the intensity falls below a tenth of the surface value. About how deep can you expect to work without artificial light?

52. *Transport through a Cell Membrane* Under certain conditions, the result of the movement of a dissolved substance across a cell's membrane is described by the equation

$$\frac{dy}{dt} = k\frac{A}{V}(c - y).$$

In this equation, y is the concentration of the substance inside the cell, and dy/dt is the rate with which y changes over time. The letters k, A, V, and c stand for constants, k being the *permeability coefficient* (a property of the membrane), A the surface area of the membrane, V the cell's volume, and c the concentration of the substance outside the cell. The equation says that the rate at which the concentration changes within the cell is proportional to the difference between it and the outside concentration.

(a) Solve the equation for $y(t)$, using $y_0 = y(0)$.

(b) Find the steady-state concentration, $\lim_{t\to\infty} y(t)$.

53. *Logistic Equation* The spread of flu in a certain school is given by the formula

$$P(t) = \frac{150}{1 + e^{4.3-t}},$$

where t is the number of days after students are first exposed to infected students.

(a) Show that the function is a solution of a logistic differential equation. Identify k and the carrying capacity.

(b) Writing to Learn Estimate $P(0)$. Explain its meaning in the context of the problem.

(c) Estimate the number of days it will take for a total of 125 students to become infected.

54. *Confirming a Solution* Show that

$$y = \int_0^x \sin(t^2)\, dt + x^3 + x + 2$$

is the solution of the initial value problem.

    Differential equation: $y'' = 2x\cos(x^2) + 6x$

    Initial conditions: $y'(0) = 1,\quad y(0) = 2$

55. *Finding an Exact Solution* Use analytic methods to find the exact solution to

$$\frac{dP}{dt} = 0.002P\left(1 - \frac{P}{800}\right),\quad P(0) = 50.$$

56. *Supporting a Solution* Give two ways to provide graphical support for the integral formula

$$\int x^2 \ln x\, dx = \frac{x^3}{3}\ln x - \frac{x^3}{9} + C.$$

57. *Doubling Time* Find the amount of time required for $10,000 to double if the 6.3% annual interest is compounded **(a)** annually, **(b)** continuously.

58. *Constant of Integration* Let

$$f(x) = \int_0^x u(t)\, dt \quad\text{and}\quad g(x) = \int_3^x u(t)\, dt.$$

(a) Show that f and g are antiderivatives of $u(x)$.

(b) Find a constant C so that $f(x) = g(x) + C$.

59. *England and Wales Population* Table 6.14 gives the combined population of England and Wales for several years. Let $x = 0$ represent 1800, $x = 1$ represent 1801, and so forth.

Table 6.14 England and Wales Population

Year	Population (millions)	Year	Population (millions)
1801	8.9	1901	32.5
1811	10.2	1911	36.1
1821	12.0	1921	37.9
1831	13.9	1931	40.0
1841	15.9	1941	
1851	17.9	1951	43.8
1861	20.1	1961	46.1
1871	22.7	1971	48.7
1881	26.0	1981	49.0
1891	29.0	1991	49.2

Source: The Statesman's Yearbook, 133rd ed. (London: The Macmillan Press, Ltd., 1996).

(a) Find the logistic regression equation for the data and superimpose its graph on a scatter plot of the data.

(b) Find the carrying capacity predicted by the regression equation.

(c) Find when the rate of growth predicted by the regression equation changes from increasing to decreasing. Estimate the population at that time.

60. *Temperature Experiment* A temperature probe is removed from a cup of hot chocolate and placed in water whose temperature (T_s) is 0°C. The data in Table 6.15 were collected over the next 30 sec with a CBL™ temperature probe.

Table 6.15 Experimental Data

Time (sec)	T (°C)
2	74.68
5	61.99
10	34.89
15	21.95
20	15.36
25	11.89
30	10.02

(a) Find an exponential regression equation for the (t, T) data. Superimpose its graph on a scatter plot of the data.

(b) Estimate when the temperature probe will read 40°C.

(c) Estimate the hot chocolate's temperature when the temperature probe was removed.

61. *Coasting to a Stop* Table 6.16 shows the distance s (meters) coasted on in-line skates in t seconds by Johnathon Krueger. Find a model for his position in the form of Equation 3 of Section 6.4. His initial velocity was $v_0 = 0.86$ m/sec, his mass $m = 30.84$ kg (he weighed 68 lb), and his total coasting distance 0.97 m.

Table 6.16 Johnathon Krueger Skating Data

t (sec)	s (m)	t (sec)	s (m)	t (sec)	s (m)
0	0	0.93	0.61	1.86	0.93
0.13	0.08	1.06	0.68	2.00	0.94
0.27	0.19	1.20	0.74	2.13	0.95
0.40	0.28	1.33	0.79	2.26	0.96
0.53	0.36	1.46	0.83	2.39	0.96
0.67	0.45	1.60	0.87	2.53	0.97
0.80	0.53	1.73	0.90	2.66	0.97

CHAPTER 7

Applications of Definite Integrals

The art of pottery developed independently in many ancient civilizations and still exists in modern times. The desired shape of the side of a pottery vase can be described by:

$$y = 5.0 + 2 \sin (x/4)\,(0 \le x \le 8\pi)$$

where x is the height and y is the radius at height x (in inches).

A base for the vase is preformed and placed on a potter's wheel. How much clay should be added to the base to form this vase if the inside radius is always 1 inch less than the outside radius? Section 7.3 contains the needed mathematics.

Chapter 7 Overview

By this point it should be apparent that finding the limits of Riemann sums is not just an intellectual exercise; it is a natural way to calculate mathematical or physical quantities that appear to be irregular when viewed as a whole, but which can be fragmented into regular pieces. We calculate values for the regular pieces using known formulas, then sum them to find a value for the irregular whole. This approach to problem solving was around for thousands of years before calculus came along, but it was tedious work and the more accurate you wanted to be the more tedious it became.

With calculus it became possible to get *exact* answers for these problems with almost no effort, because in the limit these sums became definite integrals and definite integrals could be evaluated with antiderivatives. With calculus, the challenge became one of fitting an integrable function to the situation at hand (the "modeling" step) and then finding an antiderivative for it.

Today we can finesse the antidifferentiation step (occasionally an insurmountable hurdle for our predecessors) with programs like NINT, but the modeling step is no less crucial. Ironically, it is the modeling step that is thousands of years old. Before either calculus or technology can be of assistance, we must still break down the irregular whole into regular parts and set up a function to be integrated. We have already seen how the process works with area, volume, and average value, for example. Now we will focus more closely on the underlying modeling step: how to set up the function to be integrated.

7.1 Integral as Net Change

Linear Motion Revisited • General Strategy • Consumption over Time • Net Change from Data • Work

Linear Motion Revisited

In many applications, the integral is viewed as net change over time. The classic example of this kind is distance traveled, a problem we discussed in Chapter 5.

Example 1   INTERPRETING A VELOCITY FUNCTION

Figure 7.1 shows the velocity

$$\frac{ds}{dt} = v(t) = t^2 - \frac{8}{(t+1)^2} \quad \frac{cm}{sec}$$

of a particle moving along a horizontal s-axis for $0 \le t \le 5$. Describe the motion.

Solution

Solve Graphically

The graph of v (Figure 7.1) starts with $v(0) = -8$, which we interpret as saying that the particle has an initial velocity of 8 cm/sec to the left. It slows to a halt at about $t = 1.25$ sec, after which it moves to the right ($v > 0$) with increasing speed, reaching a velocity of $v(5) \approx 24.8$ cm/sec at the end.

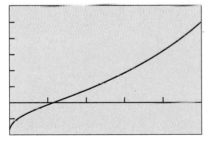

[0, 5] by [−10, 30]

Figure 7.1 The velocity function in Example 1.

Reminder from Section 3.4

A change in position is a **displacement.** If $s(t)$ is a body's position at time t, the displacement over the time interval from t to $t + \Delta t$ is $s(t + \Delta t) - s(t)$. The displacement may be positive, negative, or zero, depending on the motion.

Example 2 FINDING POSITION FROM DISPLACEMENT

Suppose the initial position of the particle in Example 1 is $s(0) = 9$. What is the particle's position at **(a)** $t = 1$ sec? **(b)** $t = 5$ sec?

Solution

Solve Analytically

(a) The position at $t = 1$ is the initial position $s(0)$ plus the displacement (the amount, Δs, that the position changed from $t = 0$ to $t = 1$). When velocity is constant during a motion, we can find the displacement (change in position) with the formula

$$\text{Displacement} = \text{rate of change} \times \text{time.}$$

But in our case the velocity varies, so we resort instead to partitioning the time interval $[0, 1]$ into subintervals of length Δt so short that the velocity is effectively constant on each subinterval. If t_k is any time in the kth subinterval, the particle's velocity throughout that interval will be close to $v(t_k)$. The change in the particle's position during the brief time this constant velocity applies is

$$v(t_k)\, \Delta t. \quad \text{rate of change} \times \text{time}$$

If $v(t_k)$ is negative, the displacement is negative and the particle will move left. If $v(t_k)$ is positive, the particle will move right. The sum

$$\sum v(t_k)\, \Delta t$$

of all these small position changes approximates the displacement for the time interval $[0, 1]$.

The sum $\sum v(t_k)\, \Delta t$ is a Riemann sum for the continuous function $v(t)$ over $[0, 1]$. As the norms of the partitions go to zero, the approximations improve and the sums converge to the integral of v over $[0, 1]$, giving

$$\text{Displacement} = \int_0^1 v(t)\, dt$$

$$= \int_0^1 \left(t^2 - \frac{8}{(t + 1)^2} \right) dt$$

$$= \left[\frac{t^3}{3} + \frac{8}{t + 1} \right]_0^1$$

$$= \frac{1}{3} + \frac{8}{2} - 8 = -\frac{11}{3}.$$

During the first second of motion, the particle moves $11/3$ cm to the left. It starts at $s(0) = 9$, so its position at $t = 1$ is

$$\text{New position} = \text{initial position} + \text{displacement} = 9 - \frac{11}{3} = \frac{16}{3}.$$

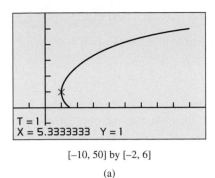

T = 1
X = 5.3333333   Y = 1

[–10, 50] by [–2, 6]

(a)

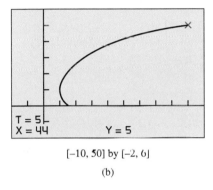

T = 5
X = 44   Y = 5

[–10, 50] by [–2, 6]

(b)

Figure 7.2 Using TRACE and the parametrization in Example 2 you can "see" the left and right motion of the particle.

(b) If we model the displacement from $t = 0$ to $t = 5$ in the same way, we arrive at

$$\text{Displacement} = \int_0^5 v(t)\, dt = \left[\frac{t^3}{3} + \frac{8}{t + 1} \right]_0^5 = 35.$$

The motion has the net effect of displacing the particle 35 cm to the right of its starting point. The particle's final position is

$$\text{Final position} = \text{initial position} + \text{displacement}$$

$$= s(0) + 35 = 9 + 35 = 44.$$

Support Graphically

The position of the particle at any time t is given by

$$s(t) = \int_0^t \left[u^2 - \frac{8}{(u + 1)^2} \right] du + 9,$$

because $s'(t) = v(t)$ and $s(0) = 9$. Figure 7.2 shows the graph of $s(t)$ given by the parametrization

$$x(t) = \text{NINT}\,(v(u), u, 0, t), \quad y(t) = t, \quad 0 < t \le 5.$$

(a) Figure 7.2a supports that the position of the particle at $t = 1$ is 16/3.

(b) Figure 7.2b shows the position of the particle is 44 at $t = 5$. Therefore, the displacement is $44 - 9 = 35$.

The reason for our method in Example 2 was to illustrate the *modeling step* that will be used throughout this chapter. We can also solve Example 2 using the techniques of Chapter 6 as shown in Exploration 1.

Exploration 1   Revisiting Example 2

The velocity of a particle moving along a horizontal s-axis for $0 \le t \le 5$ is

$$\frac{ds}{dt} = t^2 - \frac{8}{(t + 1)^2}.$$

1. Use the indefinite integral of ds/dt to find the solution of the initial value problem

$$\frac{ds}{dt} = t^2 - \frac{8}{(t + 1)^2}, \quad s(0) = 9.$$

2. Determine the position of the particle at $t = 1$. Compare your answer with the answer to Example 2a.

3. Determine the position of the particle at $t = 5$. Compare your answer with the answer to Example 2b.

We know now that the particle in Example 1 was at $s(0) = 9$ at the beginning of the motion and at $s(5) = 44$ at the end. But it did not travel from 9 to 44 directly—it began its trip by moving to the left (Figure 7.2). How much distance did the particle actually travel? We find out in Example 3.

Example 3 CALCULATING TOTAL DISTANCE TRAVELED

Find the *total distance traveled* by the particle in Example 1.

Solution

Solve Analytically

We partition the time interval as in Example 2 but record every position shift as *positive* by taking absolute values. The Riemann sum approximating total distance traveled is

$$\sum |v(t_k)|\, \Delta t,$$

and we are led to the integral

$$\text{Total distance traveled} = \int_0^5 |v(t)|\, dt = \int_0^5 \left| t^2 - \frac{8}{(t+1)^2} \right|\, dt.$$

Evaluate Numerically

We have

$$\text{NINT}\left(\left| t^2 - \frac{8}{(t+1)^2} \right|, t, 0, 5 \right) \approx 42.59.$$

What we learn from Examples 2 and 3 is this: Integrating velocity gives displacement (net area between the velocity curve and the time axis). Integrating the *absolute value* of velocity gives total distance traveled (total area between the velocity curve and the time axis).

General Strategy

The idea of fragmenting net effects into finite sums of easily estimated small changes is not new. We used it in Section 5.1 to estimate cardiac output, volume, and air pollution. What *is* new is that we can now identify many of these sums as Riemann sums and express their limits as integrals. The advantages of doing so are twofold. First, we can evaluate one of these integrals to get an accurate result in less time than it takes to crank out even the crudest estimate from a finite sum. Second, the integral itself becomes a formula that enables us to solve similar problems without having to repeat the modeling step.

The strategy that we began in Section 5.1 and have continued here is the following:

Strategy for Modeling with Integrals

1. *Approximate what you want to find as a Riemann sum* of values of a continuous function multiplied by interval lengths. If $f(x)$ is the function and $[a, b]$ the interval, and you partition the interval into subintervals of length Δx, the approximating sums will have the form $\sum f(c_k)\, \Delta x$ with c_k a point in the kth subinterval.

2. *Write a definite integral*, here $\int_a^b f(x)\, dx$, to express the limit of these sums as the norms of the partitions go to zero.

3. *Evaluate the integral* numerically or with an antiderivative.

Example 4 MODELING THE EFFECTS OF ACCELERATION

A car moving with initial velocity of 5 mph accelerates at the rate of

$$a(t) = 2.4t$$

mph per second for 8 seconds.

(a) How fast is the car going when the 8 seconds are up?

(b) How far did the car travel during those 8 seconds?

Solution

(a) We first model the effect of the acceleration on the car's velocity.

Step 1:

Approximate the net change in velocity as a Riemann sum. When acceleration is constant,

velocity change = acceleration × time applied.   rate of change × time

To apply this formula, we partition $[0, 8]$ into short subintervals of length Δt. On each subinterval the acceleration is nearly constant, so if t_k is any point in the kth subinterval, the change in velocity imparted by the acceleration in the subinterval is approximately

$$a(t_k)\,\Delta t \text{ mph.} \quad \frac{\text{mph}}{\text{sec}} \times \text{sec}$$

The net change in velocity for $0 \le t \le 8$ is approximately

$$\sum a(t_k)\,\Delta t \text{ mph.}$$

Step 2:

Write a definite integral. The limit of these sums as the norms of the partitions go to zero is

$$\int_0^8 a(t)\,dt.$$

Step 3:

Evaluate the integral. Using an antiderivative, we have

$$\text{Net velocity change} = \int_0^8 2.4t\,dt - 1.2t^2 \Big]_0^8 = 76.8 \text{ mph.}$$

So, how fast is the car going when the 8 seconds are up? Its initial velocity is 5 mph and the acceleration adds another 76.8 mph for a total of 81.8 mph.

(b) There is nothing special about the upper limit 8 in the preceding calculation. Applying the acceleration for any length of time t adds

$$\int_0^t 2.4u\,du \text{ mph} \quad u \text{ is just a dummy variable here.}$$

to the car's velocity, giving

$$v(t) = 5 + \int_0^t 2.4u\,du = 5 + 1.2t^2 \text{ mph.}$$

The distance traveled from $t = 0$ to $t = 8$ sec is

$$\int_0^8 |v(t)|\, dt = \int_0^8 (5 + 1.2t^2)\, dt \qquad \text{Extension of Example 3}$$

$$= \left[5t + 0.4t^3 \right]_0^8$$

$$= 244.8 \text{ mph} \times \text{seconds}.$$

Miles-per-hour second is not a distance unit that we normally work with! To convert to miles we multiply by hours/second $= 1/3600$, obtaining

$$244.8 \times \frac{1}{3600} = 0.068 \text{ mile.} \qquad \frac{\text{mi}}{\text{h}} \times \text{sec} \times \frac{\text{h}}{\text{sec}} = \text{mi}$$

The car traveled 0.068 mi during the 8 seconds of acceleration.

Consumption over Time

The integral is a natural tool to calculate net change and total accumulation of more quantities than just distance and velocity. Integrals can be used to calculate growth, decay, and, as in the next example, consumption. Whenever you want to find the cumulative effect of a varying rate of change, integrate it.

Example 5   POTATO CONSUMPTION

From 1970 to 1980, the rate of potato consumption in a particular country was $C(t) = 2.2 + 1.1^t$ millions of bushels per year, with t being years since the beginning of 1970. How many bushels were consumed from the beginning of 1972 to the end of 1973?

Solution We seek the cumulative effect of the consumption rate for $2 \le t \le 4$.

Step 1:

Riemann sum. We partition $[2, 4]$ into subintervals of length Δt and let t_k be a time in the kth subinterval. The amount consumed during this interval is approximately

$$C(t_k)\, \Delta t \ \text{ million bushels.}$$

The consumption for $2 \le t \le 4$ is approximately

$$\sum C(t_k)\, \Delta t \ \text{ million bushels.}$$

Step 2:

Definite integral. The amount consumed from $t = 2$ to $t = 4$ is the limit of these sums as the norms of the partitions go to zero.

$$\int_2^4 C(t)\, dt = \int_2^4 (2.2 + 1.1^t)\, dt \ \text{ million bushels}$$

Step 3:

Evaluate. Evaluating numerically, we obtain

$$\text{NINT } (2.2 + 1.1^t, t, 2, 4) \approx 7.066 \ \text{ million bushels.}$$

Net Change from Data

Many real applications begin with data, not a fully modeled function. In the next example, we are given data on the rate at which a pump operates in consecutive 5-minute intervals and asked to find the total amount pumped.

Example 6   FINDING GALLONS PUMPED FROM RATE DATA

A pump connected to a generator operates at a varying rate, depending on how much power is being drawn from the generator to operate other machinery. The rate (gallons per minute) at which the pump operates is recorded at 5-minute intervals for one hour as shown in Table 7.1. How many gallons were pumped during that hour?

Solution   Let $R(t)$, $0 \le t \le 60$, be the pumping rate as a continuous function of time for the hour. We can partition the hour into short subintervals of length Δt on which the rate is nearly constant and form the sum $\sum R(t_k) \Delta t$ as an approximation to the amount pumped during the hour. This reveals the integral formula for the number of gallons pumped to be

$$\text{Gallons pumped} = \int_0^{60} R(t)\, dt.$$

We have no formula for R in this instance, but the 13 equally spaced values in Table 7.1 enable us to estimate the integral with the Trapezoidal Rule:

$$\int_0^{60} R(t)\, dt \approx \frac{60}{2 \cdot 12}\left[58 + 2(60) + 2(65) + \cdots + 2(63) + 63 \right]$$

$$= 3582.5.$$

The total amount pumped during the hour is about 3580 gal.

Table 7.1 Pumping Rates

Time (min)	Rate (gal/min)
0	58
5	60
10	65
15	64
20	58
25	57
30	55
35	55
40	59
45	60
50	60
55	63
60	63

Work

In everyday life, *work* means an activity that requires muscular or mental effort. In science, the term refers specifically to a force acting on a body and the body's subsequent displacement. When a body moves a distance d along a straight line as a result of the action of a force of constant magnitude F in the direction of motion, the **work** done by the force is

$$W = Fd.$$

The equation $W = Fd$ is the **constant-force formula** for work.

The units of work are force $\times$ distance. In the metric system, the unit is the newton-meter, which, for historical reasons, is called a joule (see margin note). In the U.S. customary system, the most common unit of work is the **foot-pound.**

Hooke's Law for springs says that the force it takes to stretch or compress a spring x units from its natural (unstressed) length is a constant times x. In symbols,

$$F = kx,$$

where k, measured in force units per unit length, is a characteristic of the spring called the **force constant.**

Joules

The joule, abbreviated J and pronounced "jewel," is named after the English physicist James Prescott Joule (1818–1889). The defining equation is

1 joule = (1 newton)(1 meter).

In symbols, 1 J = 1 N · m.

It takes a force of about 1 N to lift an apple from a table. If you lift it 1 m you have done about 1 J of work on the apple. If you eat the apple, you will have consumed about 80 food calories, the heat equivalent of nearly 335,000 joules. If this energy were directly useful for mechanical work (it's not), it would enable you to lift 335,000 more apples up 1 m.

Example 7 A BIT OF WORK

It takes a force of 10 N to stretch a spring 2 m beyond its natural length. How much work is done in stretching the spring 4 m from its natural length?

Solution We let $F(x)$ represent the force in newtons required to stretch the spring x meters from its natural length. By Hooke's Law, $F(x) = kx$ for some constant k. We are told that

$$F(2) = 10 = k \cdot 2, \qquad \begin{array}{l}\text{The force required to stretch}\\\text{the spring 2 m is 10 newtons.}\end{array}$$

so $k = 5$ N/m and $F(x) = 5x$ for this particular spring.

We construct an integral for the work done in applying F over the interval from $x = 0$ to $x = 4$.

Step 1:

Riemann sum. We partition the interval into subintervals on each of which F is so nearly constant that we can apply the constant-force formula for work. If x_k is any point in the kth subinterval, the value of F throughout the interval is approximately $F(x_k) = 5x_k$. The work done by F across the interval is approximately $5x_k \, \Delta x$, where Δx is the length of the interval. The sum

$$\sum F(x_k) \, \Delta x = \sum 5x_k \, \Delta x$$

approximates the work done by F from $x = 0$ to $x = 4$.

Steps 2 and 3:

Integrate. The limit of these sums as the norms of the partitions go to zero is

Numerically, work is the area under the force graph.

$$\int_0^4 F(x) \, dx = \int_0^4 5x \, dx = 5\frac{x^2}{2}\Big]_0^4 = 40 \text{ N} \cdot \text{m}.$$

We will revisit work in Section 7.5.

Quick Review 7.1

In Exercises 1–10, find all values of x (if any) at which the function changes sign on the given interval. Sketch a number line graph of the interval, and indicate the sign of the function on each subinterval.

Example: $f(x) = x^2 - 1$ on $[-2, 3]$ changes sign at $x = \pm 1$.

1. $\sin 2x$ on $[-3, 2]$

2. $x^2 - 3x + 2$ on $[-2, 4]$

3. $x^2 - 2x + 3$ on $[-4, 2]$

4. $2x^3 - 3x^2 + 1$ on $[-2, 2]$

5. $x \cos 2x$ on $[0, 4]$

6. xe^{-x} on $[0, \infty)$

7. $\dfrac{x}{x^2 + 1}$ on $[-5, 30]$

8. $\dfrac{x^2 - 2}{x^2 - 4}$ on $[-3, 3]$

9. $\sec (1 + \sqrt{1 - \sin^2 x})$ on $(-\infty, \infty)$

10. $\sin (1/x)$ on $[0.1, 0.2]$

Section 7.1 Exercises

In Exercises 1–8, the function $v(t)$ is the velocity in m/sec of a particle moving along the *x*-axis. Use analytic methods to do each of the following:

(a) Determine when the particle is moving to the right, to the left, and stopped.

(b) Find the particle's displacement for the given time interval.

(c) Find the total distance traveled by the particle.

1. $v(t) = 5 \cos t, \quad 0 \le t \le 2\pi$

2. $v(t) = 6 \sin 3t, \quad 0 \le t \le \pi/2$

3. $v(t) = 49 - 9.8t, \quad 0 \le t \le 10$

4. $v(t) = 6t^2 - 18t + 12, \quad 0 \le t \le 2$

5. $v(t) = 5 \sin^2 t \cos t, \quad 0 \le t \le 2\pi$

6. $v(t) = \sqrt{4 - t}, \quad 0 \le t \le 4$

7. $v(t) = e^{\sin t} \cos t, \quad 0 \le t \le 2\pi$

8. $v(t) = \dfrac{t}{1 + t^2}, \quad 0 \le t \le 3$

9. An automobile accelerates from rest at $1 + 3\sqrt{t}$ mph/sec for 9 seconds.

(a) What is its velocity after 9 seconds?

(b) How far does it travel in those 9 seconds?

10. A particle travels with velocity

$$v(t) = (t - 2) \sin t \text{ m/sec}$$

for $0 \le t \le 4$ sec.

(a) What is the particle's displacement?

(b) What is the total distance traveled?

11. *Projectile* Recall that the acceleration due to Earth's gravity is 32 ft/sec². From ground level, a projectile is fired straight upward with velocity 90 feet per second.

(a) What is its velocity after 3 seconds?

(b) When does it hit the ground?

(c) When it hits the ground, what is the net distance it has traveled?

(d) When it hits the ground, what is the total distance it has traveled?

In Exercises 12–16, a particle moves along the *x*-axis (units in cm). Its initial position at $t = 0$ sec is $x(0) = 15$. The figure shows the graph of the particle's velocity $v(t)$. The numbers are the *areas* of the enclosed regions.

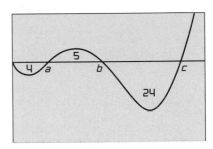

12. What is the particle's displacement between $t = 0$ and $t = c$?

13. What is the total distance traveled by the particle in the same time period?

14. Give the positions of the particle at times a, b, and c.

15. Approximately where does the particle achieve its greatest positive acceleration on the interval $[0, b]$?

16. Approximately where does the particle achieve its greatest positive acceleration on the interval $[0, c]$?

In Exercises 17–20, the graph of the velocity of a particle moving on the *x*-axis is given. The particle starts at $x = 2$ when $t = 0$.

(a) Find where the particle is at the end of the trip.

(b) Find the total distance traveled by the particle.

17.

18.

19.

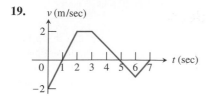

20.

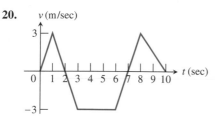

21. *U.S. Oil Consumption* The rate of consumption of oil in the United States during the 1980s (in billions of barrels per year) is modeled by the function $C = 27.08 \cdot e^{t/25}$, where t is the number of years after January 1, 1980. Find the total consumption of oil in the United States from January 1, 1980 to January 1, 1990.

22. *Home Electricity Use* The rate at which your home consumes electricity is measured in kilowatts. If your home consumes electricity at the rate of 1 kilowatt for 1 hour, you will be charged for 1 "kilowatt-hour" of electricity. Suppose that the average consumption rate for a certain home is modeled by the function $C(t) = 3.9 - 2.4 \sin(\pi t/12)$, where $C(t)$ is measured in kilowatts and t is the number of hours past midnight. Find the average daily consumption for this home, measured in kilowatt-hours.

23. *Population Density* Population density measures the number of people per square mile inhabiting a given living area. Washerton's population density, which decreases as you move away from the city center, can be approximated by the function $10,000(2 - r)$ at a distance r miles from the city center.

(a) If the population density approaches zero at the edge of the city, what is the city's radius?

(b) A thin ring around the center of the city has thickness Δr and radius r. If you straighten it out, it suggests a rectangular strip. Approximately what is its area?

(c) **Writing to Learn** Explain why the population of the ring in (b) is approximately

$$10,000(2 - r)(2\pi r) \, \Delta r.$$

(d) Estimate the total population of Washerton by setting up and evaluating a definite integral.

24. *Oil Flow* Oil flows through a cylindrical pipe of radius 3 inches, but friction from the pipe slows the flow toward the outer edge. The speed at which the oil flows at a distance r inches from the center is $8(10 - r^2)$ inches per second.

(a) In a plane cross section of the pipe, a thin ring with thickness Δr at a distance r inches from the center approximates a rectangular strip when you straighten it

out. What is the area of the strip (and hence the approximate area of the ring)?

(b) Explain why we know that oil passes through this ring at approximately $8(10 - r^2)(2\pi r) \, \Delta r$ cubic inches per second.

(c) Set up and evaluate a definite integral that will give the rate (in cubic inches per second) at which oil is flowing through the pipe.

25. Writing to Learn As a school project, Anna accompanies her mother on a trip to the grocery store and keeps a log of the car's speed at 10-second intervals. Explain how she can use the data to estimate the distance from her home to the store. What is the connection between this process and the definite integral?

In Exercises 26 and 27, *work in groups of two or three.*

26. *Bagel Sales* From 1985 to 1995, the Konigsberg Bakery noticed a consistent increase in annual sales of its bagels. The annual sales (in thousands of bagels) are shown below.

Year	Sales (thousands)
1985	5
1986	8.9
1987	16
1988	26.3
1989	39.8
1990	56.5
1991	76.4
1992	99.5
1993	125.8
1994	155.3
1995	188

(a) What was the total number of bagels sold over the 11-year period? (This is not a calculus question!)

(b) Use quadratic regression to model the annual bagel sales (in thousands) as a function $B(x)$, where x is the number of years after 1985.

(c) Integrate $B(x)$ over the interval $[0, 11]$ to find total bagel sales for the 11-year period.

(d) Explain graphically why the answer in (a) is smaller than the answer in (c).

27. *(Continuation of Exercise 26)*

(a) Integrate $B(x)$ over the interval $[-0.5, 10.5]$ to find total bagel sales for the 11-year period.

(b) Explain graphically why the answer in (a) is better than the answer in 26(c).

28. *Filling Milk Cartons* A machine fills milk cartons with milk at an approximately constant rate, but backups along the assembly line cause some variation. The rates (in cases per hour) are recorded at hourly intervals during a 10-hour period, from 8:00 A.M. to 6:00 P.M.

Time	Rate (cases/h)
8	120
9	110
10	115
11	115
12	119
1	120
2	120
3	115
4	112
5	110
6	121

Use the Trapezoidal Rule with $n = 10$ to determine approximately how many cases of milk were filled by the machine over the 10-hour period.

29. *Hooke's Law* A certain spring requires a force of 6 N to stretch it 3 cm beyond its natural length.

(a) What force would be required to stretch the string 9 cm beyond its natural length?

(b) What would be the work done in stretching the string 9 cm beyond its natural length?

30. *Hooke's Law* Hooke's Law also applies to *compressing* springs; that is, it requires a force of kx to compress a spring a distance x from its natural length. Suppose a 10,000-lb force compressed a spring from its natural length of 12 inches to a length of 11 inches. How much work was done in compressing the spring

(a) the first half-inch? **(b)** the second half-inch?

Extending the Ideas

31. *Inflation* Although the economy is continuously changing, we analyze it with discrete measurements. The following table records the *annual* inflation rate as measured each month for 13 consecutive months. Use the Trapezoidal Rule with $n = 12$ to find the overall inflation rate for the year.

Month	Annual Rate
January	0.04
February	0.04
March	0.05
April	0.06
May	0.05
June	0.04
July	0.04
August	0.05
September	0.04
October	0.06
November	0.06
December	0.05
January	0.05

32. *Inflation Rate* The table below shows the *monthly* inflation rate (in *thousandths*) for energy prices for thirteen consecutive months. Use the Trapezoidal Rule with $n = 12$ to approximate the *annual* inflation rate for the 12-month period running from the middle of the first month to the middle of the last month.

Month	Monthly Rate (in thousandths)
January	3.6
February	4.0
March	3.1
April	2.8
May	2.8
June	3.2
July	3.3
August	3.1
September	3.2
October	3.4
November	3.4
December	3.9
January	4.0

33. *Center of Mass* Suppose we have a finite collection of masses in the coordinate plane, the mass m_k located at the point (x_k, y_k) as shown in the figure.

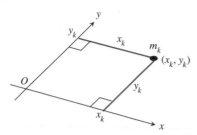

Each mass m_k has **moment $m_k y_k$ with respect to the x-axis** and **moment $m_k x_k$ about the y-axis.** The moments of the entire system with respect to the two axes are

$$\text{Moment about } x\text{-axis: } M_x = \sum m_k y_k,$$

$$\text{Moment about } y\text{-axis: } M_y = \sum m_k x_k.$$

The **center of mass** is $(\overline{x}, \overline{y})$ where

$$\overline{x} = \frac{M_y}{M} = \frac{\sum m_k x_k}{\sum m_k} \quad \text{and} \quad \overline{y} = \frac{M_x}{M} = \frac{\sum m_k y_k}{\sum m_k}.$$

Suppose we have a thin, flat plate occupying a region in the plane.

(a) Imagine the region cut into thin strips parallel to the y-axis. Show that

$$\overline{x} = \frac{\int x \, dm}{\int dm},$$

where $dm = \delta \, dA$, $\delta =$ density (mass per unit area), and $A =$ area of the region.

(b) Imagine the region cut into thin strips parallel to the x-axis. Show that

$$\overline{y} = \frac{\int y \, dm}{\int dm},$$

where $dm = \delta \, dA$, $\delta =$ density, and $A =$ area of the region.

In Exercises 34 and 35, use Exercise 33 to find the center of mass of the region with given density.

34. the region bounded by the parabola $y = x^2$ and the line $y = 4$ with constant density δ

35. the region bounded by the lines $y = x$, $y = -x$, $x = 2$ with constant density δ

7.2 Areas in the Plane

Area Between Curves • Area Enclosed by Intersecting Curves • Boundaries with Changing Functions • Integrating with Respect to y • Saving Time with Geometry Formulas

Area Between Curves

We know how to find the area of a region between a curve and the x-axis but many times we want to know the area of a region that is bounded above by one curve, $y = f(x)$, and below by another, $y = g(x)$ (Figure 7.3).

We find the area as an integral by applying the first two steps of the modeling strategy developed in Section 7.1.

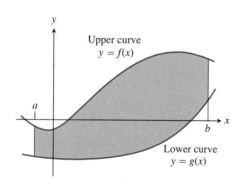

Figure 7.3 The region between $y = f(x)$ and $y = g(x)$ and the lines $x = a$ and $x = b$.

1. We partition the region into vertical strips of equal width Δx and approximate each strip with a rectangle with base parallel to $[a, b]$ (Figure 7.4). Each rectangle has area

$$[f(c_k) - g(c_k)] \, \Delta x$$

for some c_k in its respective subinterval (Figure 7.5). This expression will be nonnegative even if the region lies below the x-axis. We approximate the area of the region with the Riemann sum

$$\sum [f(c_k) - g(c_k)] \, \Delta x.$$

2. The limit of these sums as $\Delta x \to 0$ is

$$\int_a^b [f(x) - g(x)] \, dx.$$

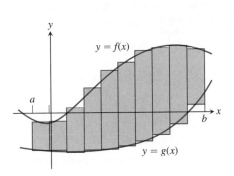

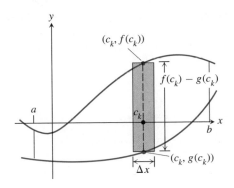

Figure 7.4 We approximate the region with rectangles perpendicular to the x-axis.

Figure 7.5 The area of a typical rectangle is $[f(c_k) - g(c_k)]\,\Delta x$.

This approach to finding area captures the properties of area, so it can serve as a definition.

Definition Area Between Curves

If f and g are continuous with $f(x) \geq g(x)$ throughout $[a, b]$, then the **area between the curves** $y = f(x)$ **and** $y = g(x)$ **from a to b** is the integral of $[f - g]$ from a to b,

$$A = \int_a^b [f(x) - g(x)]\,dx.$$

Example 1 APPLYING THE DEFINITION

Find the area of the region between $y = \sec^2 x$ and $y = \sin x$ from $x = 0$ to $x = \pi/4$.

Solution We graph the curves (Figure 7.6) to find their relative positions in the plane, and see that $y = \sec^2 x$ lies *above* $y = \sin x$ on $[0, \pi/4]$. The area is therefore

$$A = \int_0^{\pi/4} [\sec^2 x - \sin x]\,dx$$

$$= \Big[\tan x + \cos x \Big]_0^{\pi/4}$$

$$= \frac{\sqrt{2}}{2} \text{ units squared.}$$

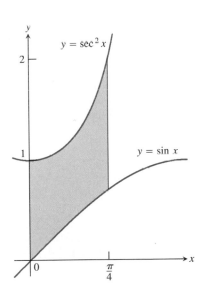

Figure 7.6 The region in Example 1.

$y_1 = 2k - k \sin kx$
$y_2 = k \sin kx$

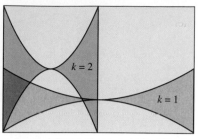

[0, π] by [0, 4]

Figure 7.7 Two members of the family of butterfly-shaped regions described in Exploration 1.

Exploration 1 **A Family of Butterflies**

For each positive integer k, let A_k denote the area of the butterfly-shaped region enclosed between the graphs of $y = k \sin kx$ and $y = 2k - k \sin kx$ on the interval $[0, \pi/k]$. The regions for $k = 1$ and $k = 2$ are shown in Figure 7.7.

1. Find the areas of the two regions in Figure 7.7.
2. Make a conjecture about the areas A_k for $k \geq 3$.
3. Set up a definite integral that gives the area A_k. Can you make a simple u-substitution that will transform this integral into the definite integral that gives the area A_1?
4. What is $\lim_{k \to \infty} A_k$?
5. If P_k denotes the perimeter of the kth butterfly-shaped region, what is $\lim_{k \to \infty} P_k$? (You can answer this question without an explicit formula for P_k.)

Area Enclosed by Intersecting Curves

When a region is enclosed by intersecting curves, the intersection points give the limits of integration.

Example 2 AREA OF AN ENCLOSED REGION

Find the area of the region enclosed by the parabola $y = 2 - x^2$ and the line $y = -x$.

Solution We graph the curves to view the region (Figure 7.8).

The limits of integration are found by solving the equation

$$2 - x^2 = -x$$

either algebraically or by calculator. The solutions are $x = -1$ and $x = 2$.

Since the parabola lies above the line on $[-1, 2]$, the area integrand is $2 - x^2 - (-x)$.

$$A = \int_{-1}^{2} [2 - x^2 - (-x)] \, dx$$

$$= \left[2x - \frac{x^3}{3} + \frac{x^2}{2} \right]_{-1}^{2}$$

$$= \frac{9}{2} \text{ units squared}$$

$y_1 = 2 - x^2$
$y_2 = -x$

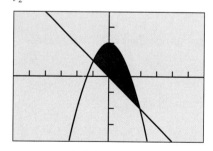

[-6, 6] by [-4, 4]

Figure 7.8 The region in Example 2.

Example 3 USING A CALCULATOR

Find the area of the region enclosed by the graphs of $y = 2 \cos x$ and $y = x^2 - 1$.

Solution The region is shown in Figure 7.9.

$y_1 = 2\cos x$
$y_2 = x^2 - 1$

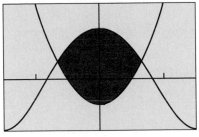

[−3, 3] by [−2, 3]

Figure 7.9 The region in Example 3.

Finding Intersections by Calculator

The coordinates of the points of intersection of two curves are sometimes needed for other calculations. To take advantage of the accuracy provided by calculators, use them to solve for the values and *store* the ones you want.

Using a calculator, we solve the equation

$$2\cos x = x^2 - 1$$

to find the *x*-coordinates of the points where the curves intersect. These are the limits of integration. The solutions are $x = \pm 1.265423706$. We store the negative value as *A* and the positive value as *B*. The area is

$$\text{NINT} \left(2\cos x - (x^2 - 1), x, A, B\right) \approx 4.994907788.$$

This is the final calculation, so we are now free to round. The area is about 4.99.

Boundaries with Changing Functions

If a boundary of a region is defined by more than one function, we can partition the region into subregions that correspond to the function changes and proceed as usual.

Example 4 FINDING AREA USING SUBREGIONS

Find the area of the region *R* in the first quadrant that is bounded above by $y = \sqrt{x}$ and below by the *x*-axis and the line $y = x - 2$.

Solution The region is shown in Figure 7.10.

While it appears that no single integral can give the area of *R* (the bottom boundary is defined by two different curves), we can split the region at $x = 2$ into two regions *A* and *B*. The area of *R* can be found as the sum of the areas of *A* and *B*.

$$\text{Area of } R = \underbrace{\int_0^2 \sqrt{x}\, dx}_{\text{area of } A} + \underbrace{\int_2^4 [\sqrt{x} - (x-2)]\, dx}_{\text{area of } B}$$

$$= \frac{2}{3} x^{3/2} \bigg]_0^2 + \left[\frac{2}{3} x^{3/2} - \frac{x^2}{2} + 2x\right]_2^4$$

$$= \frac{10}{3} \text{ units squared}$$

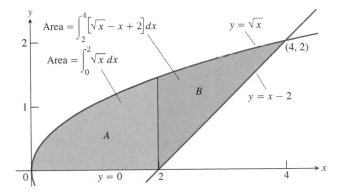

Figure 7.10 Region *R* split into subregions *A* and *B*. (Example 4)

Integrating with Respect to *y*

Sometimes the boundaries of a region are more easily described by functions of *y* than by functions of *x*. We can use approximating rectangles that are horizontal rather than vertical and the resulting basic formula has *y* in place of *x*.

For regions like these

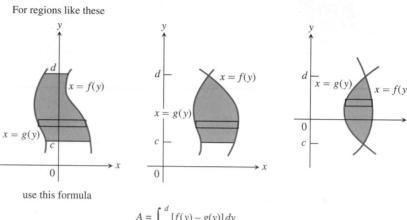

use this formula

$$A = \int_c^d [f(y) - g(y)]\,dy.$$

Example 5 INTEGRATING WITH RESPECT TO *y*

Find the area of the region in Example 4 by integrating with respect to *y*.

Solution We remarked in solving Example 4 that "it appears that no single integral can give the area of *R*," but notice how appearances change when we think of our rectangles being summed over *y*. The interval of integration is [0, 2], and the rectangles run between the same two curves on the entire interval. There is no need to split the region (Figure 7.11).

We need to solve for *x* in terms of *y* in both equations:

$$y = x - 2 \quad \text{becomes} \quad x = y + 2,$$

$$y = \sqrt{x} \quad \text{becomes} \quad x = y^2, \quad y \geq 0.$$

We must still be careful to subtract the lower number from the higher number when forming the integrand. In this case, the higher numbers are the higher *x*-values, which are on the line $x = y + 2$ because the line lies to the *right* of the parabola. So,

$$\text{Area of } R = \int_0^2 (y + 2 - y^2)\,dy$$

$$= \left[\frac{y^2}{2} + 2y - \frac{y^3}{3}\right]_0^2$$

$$= \frac{10}{3} \text{ units squared.}$$

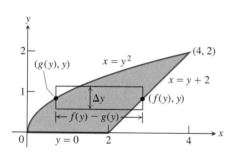

Figure 7.11 It takes two integrations to find the area of this region if we integrate with respect to *x*. It takes only one if we integrate with respect to *y*. (Example 5)

$y_1 = x^3, \; y_2 = \sqrt{x+2}, \; y_3 = -\sqrt{x+2}$

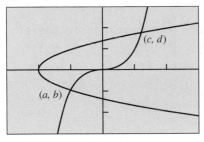

[–3, 3] by [–3, 3]

Figure 7.12 The region in Example 6.

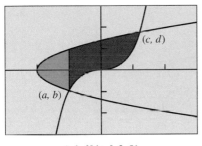

[–3, 3] by [–3, 3]

Figure 7.13 If we integrate with respect to x in Example 6, we must split the region at $x = a$.

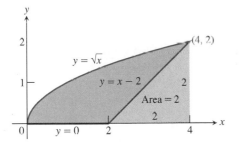

Figure 7.14 The area of the blue region is the area under the parabola $y = \sqrt{x}$ minus the area of the triangle. (Example 7)

Example 6 MAKING THE CHOICE

Find the area of the region enclosed by the graphs of $y = x^3$ and $x = y^2 - 2$.

Solution We can produce a graph of the region on a calculator by graphing the three curves $y = x^3$, $y = \sqrt{x+2}$, and $y = -\sqrt{x+2}$ (Figure 7.12).

This conveniently gives us all of our bounding curves as functions of x. If we integrate in terms of x, however, we need to split the region at $x = a$ (Figure 7.13).

On the other hand, we can integrate from $y = b$ to $y = d$ and handle the entire region at once. We solve the cubic for x in terms of y:

$$y = x^3 \quad \text{becomes} \quad x = y^{1/3}.$$

To find the limits of integration, we solve $y^{1/3} = y^2 - 2$. It is easy to see that the lower limit is $b = -1$, but a calculator is needed to find that the upper limit $d = 1.793003715$. We store this value as D.

The cubic lies to the right of the parabola, so

$$\text{Area} = \text{NINT}\,(y^{1/3} - (y^2 - 2), \, y, \, -1, \, \text{D}) = 4.214939673.$$

The area is about 4.21.

Saving Time with Geometry Formulas

Here is yet another way to handle Example 4.

Example 7 USING GEOMETRY

Find the area of the region in Example 4 by subtracting the area of the triangular region from the area under the square root curve.

Solution Figure 7.14 illustrates the strategy, which enables us to integrate with respect to x without splitting the region.

$$\begin{aligned}
\text{Area} &= \int_0^4 \sqrt{x}\,dx - \frac{1}{2}\,(2)(2) \\[4pt]
&= \frac{2}{3}x^{3/2}\bigg]_0^4 - 2 \\[4pt]
&= \frac{10}{3} \text{ units squared}
\end{aligned}$$

The moral behind Examples 4, 5, and 7 is that you often have options for finding the area of a region, some of which may be easier than others. You can integrate with respect to x or with respect to y, you can partition the region into subregions, and sometimes you can even use traditional geometry formulas. Sketch the region first and take a moment to determine the best way to proceed.

Quick Review 7.2

In Exercises 1–5, find the area between the *x*-axis and the graph of the given function over the given interval.

1. $y = \sin x$ over $[0, \pi]$

2. $y = e^{2x}$ over $[0, 1]$

3. $y = \sec^2 x$ over $[-\pi/4, \pi/4]$

4. $y = 4x - x^3$ over $[0, 2]$

5. $y = \sqrt{9 - x^2}$ over $[-3, 3]$

In Exercises 6–10, find the *x*- and *y*-coordinates of all points where the graphs of the given functions intersect. If the curves never intersect, write "NI."

6. $y = x^2 - 4x$ and $y = x + 6$

7. $y = e^x$ and $y = x + 1$

8. $y = x^2 - \pi x$ and $y = \sin x$

9. $y = \dfrac{2x}{x^2 + 1}$ and $y = x^3$

10. $y = \sin x$ and $y = x^3$

Section 7.2 Exercises

In Exercises 1–8, find the area of the shaded region analytically.

1.

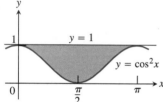

2.

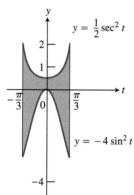

3.

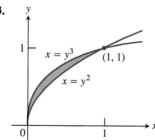

4.

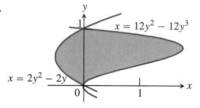

5.

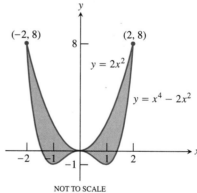

NOT TO SCALE

6.

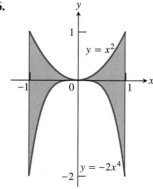

7.

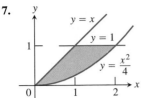

8.

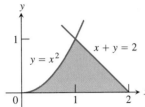

In Exercises 9 and 10, find the total shaded area.

9.

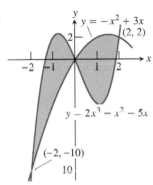

10.

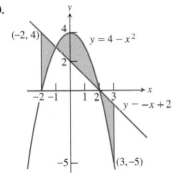

In Exercises 11–30, find the area of the regions enclosed by the lines and curves.

11. $y = x^2 - 2$ and $y = 2$

12. $y = 2x - x^2$ and $y = -3$

13. $y = 7 - 2x^2$ and $y = x^2 + 4$

14. $y = x^4 - 4x^2 + 4$ and $y = x^2$

15. $y = x\sqrt{a^2 - x^2}$, $a > 0$, and $y = 0$

16. $y = \sqrt{|x|}$ and $5y = x + 6$
(How many intersection points are there?)

17. $y = |x^2 - 4|$ and $y = (x^2/2) + 4$

18. $x = y^2$ and $x = y + 2$

19. $y^2 - 4x = 4$ and $4x - y = 16$

20. $x - y^2 = 0$ and $x + 2y^2 = 3$

21. $x + y^2 = 0$ and $x + 3y^2 = 2$

22. $4x^2 + y = 4$ and $x^4 - y = 1$

23. $x + y^2 = 3$ and $4x + y^2 = 0$

24. $y = 2 \sin x$ and $y = \sin 2x$, $0 \le x \le \pi$

25. $y = 8 \cos x$ and $y = \sec^2 x$, $-\pi/3 \le x \le \pi/3$

26. $y = \cos (\pi x/2)$ and $y = 1 - x^2$

27. $y = \sin (\pi x/2)$ and $y = x$

28. $y = \sec^2 x$, $y = \tan^2 x$, $x = -\pi/4$, $x = \pi/4$

29. $x = \tan^2 y$ and $x = -\tan^2 y$, $-\pi/4 \le y \le \pi/4$

30. $x = 3 \sin y \sqrt{\cos y}$ and $x = 0$, $0 \le y \le \pi/2$

31. Find the area of the propeller-shaped region enclosed by the curve $x - y^3 = 0$ and the line $x - y = 0$.

32. Find the area of the region in the first quadrant bounded by the line $y = x$, the line $x = 2$, the curve $y = 1/x^2$, and the x-axis.

33. Find the area of the "triangular" region in the first quadrant bounded on the left by the y-axis and on the right by the curves $y = \sin x$ and $y = \cos x$.

34. Find the area of the region between the curve $y = 3 - x^2$ and the line $y = -1$ by integrating with respect to (a) x, (b) y.

35. The region bounded below by the parabola $y = x^2$ and above by the line $y = 4$ is to be partitioned into two subsections of equal area by cutting across it with the horizontal line $y = c$.

(a) Sketch the region and draw a line $y = c$ across it that looks about right. In terms of c, what are the coordinates of the points where the line and parabola intersect? Add them to your figure.

(b) Find c by integrating with respect to y. (This puts c in the limits of integration.)

(c) Find c by integrating with respect to x. (This puts c into the integrand as well.)

36. Find the area of the region in the first quadrant bounded on the left by the y-axis, below by the line $y = x/4$, above left by the curve $y = 1 + \sqrt{x}$, and above right by the curve $y = 2/\sqrt{x}$.

In Exercises 37–40, *work in groups of two or three.*

37. The figure here shows triangle *AOC* inscribed in the region cut from the parabola $y = x^2$ by the line $y = a^2$. Find the limit of the ratio of the area of the triangle to the area of the parabolic region as *a* approaches zero.

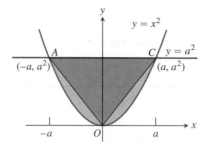

38. Suppose the area of the region between the graph of a positive continuous function *f* and the *x*-axis from $x = a$ to $x = b$ is 4 square units. Find the area between the curves $y = f(x)$ and $y = 2f(x)$ from $x = a$ to $x = b$.

39. **Writing to Learn** Which of the following integrals, if either, calculates the area of the shaded region shown here? Give reasons for your answer.

$$\textbf{i.} \int_{-1}^{1} (x - (-x)) \, dx = \int_{-1}^{1} 2x \, dx$$

$$\textbf{ii.} \int_{-1}^{1} (-x - (x)) \, dx = \int_{-1}^{1} -2x \, dx$$

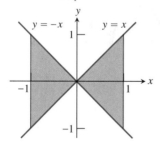

40. **Writing to Learn** Is the following statement true, sometimes true, or never true? The area of the region between the graphs of the continuous functions $y = f(x)$ and $y = g(x)$ and the vertical lines $x = a$ and $x = b \ (a < b)$ is

$$\int_a^b [f(x) - g(x)] \, dx.$$

Give reasons for your answer.

41. Find the area of the propeller-shaped region enclosed between the graphs of

$$y = \frac{2x}{x^2 + 1} \quad \text{and} \quad y = x^3.$$

42. Find the area of the propeller-shaped region enclosed between the graphs of $y = \sin x$ and $y = x^3$.

43. Find the positive value of *k* such that the area of the region enclosed between the graph of $y = k \cos x$ and the graph of $y = kx^2$ is 2.

Exploration

44. *Area of Ellipse* *Work in groups of two or three.*

An ellipse with major axis of length 2*a* and minor axis of length 2*b* can be coordinatized with its center at the origin and its major axis horizontal, in which case it is defined by the equation

$$\frac{x^2}{a^2} + \frac{y^2}{b^2} = 1.$$

(a) Find the equations that define the upper and lower semiellipses as functions of *x*.

(b) Write an integral expression that gives the area of the ellipse.

(c) With your group, use NINT to find the areas of ellipses for various lengths of *a* and *b*.

(d) There is a simple formula for the area of an ellipse with major axis of length 2*a* and minor axis of length 2*b*. Can you tell what it is from the areas you and your group have found?

(e) Work with your group to write a *proof* of this area formula by showing that it is the exact value of the integral expression in (b). ∎

Extending the Ideas

45. *Cavalieri's Theorem* Bonaventura Cavalieri (1598–1647) discovered that if two plane regions can be arranged to lie over the same interval of the *x*-axis in such a way that they have identical vertical cross sections at every point (see figure), then the regions have the same area. Show that this theorem is true.

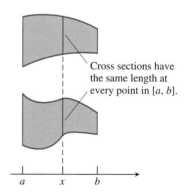

Cross sections have the same length at every point in [*a*, *b*].

46. Find the area of the region enclosed by the curves

$$y = \frac{x}{x^2 + 1} \quad \text{and} \quad y = mx, \quad 0 < m < 1.$$

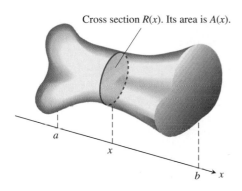

Cross section $R(x)$. Its area is $A(x)$.

Figure 7.15 The cross section of an arbitrary solid at point x.

Volumes

Volume as an Integral • Square Cross Sections • Circular Cross Sections • Cylindrical Shells • Other Cross Sections

Volume as an Integral

In Section 5.1, Example 3, we estimated the volume of a sphere by partitioning it into thin slices that were nearly cylindrical and summing the cylinders' volumes using MRAM. MRAM sums are Riemann sums, and had we known how at the time, we could have continued on to express the volume of the sphere as a definite integral.

Starting the same way, we can now find the volumes of a great many solids by integration. Suppose we want to find the volume of a solid like the one in Figure 7.15. The cross section of the solid at each point x in the interval $[a, b]$ is a region $R(x)$ of area $A(x)$. If A is a continuous function of x, we can use it to define and calculate the volume of the solid as an integral in the following way.

We partition $[a, b]$ into subintervals of length Δx and slice the solid, as we would a loaf of bread, by planes perpendicular to the x-axis at the partition points. The kth slice, the one between the planes at x_{k-1} and x_k, has approximately the same volume as the cylinder between the two planes based on the region $R(x_k)$ (Figure 7.16).

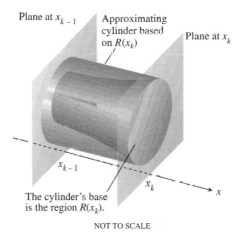

Plane at x_{k-1} Approximating cylinder based on $R(x_k)$ Plane at x_k

x_{k-1}

x_k

x

The cylinder's base is the region $R(x_k)$.

NOT TO SCALE

Figure 7.16 Enlarged view of the slice of the solid between the planes at $x_k - 1$ and x_k

The volume of the cylinder is

$$V_k = \text{base area} \times \text{height} = A(x_k) \times \Delta x.$$

The sum

$$\sum V_k = \sum A(x_k) \times \Delta x$$

approximates the volume of the solid.

This is a Riemann sum for $A(x)$ on $[a, b]$. We expect the approximations to improve as the norms of the partitions go to zero, so we define their limiting integral to be the *volume of the solid*.

Definition Volume of a Solid

The **volume** of a solid of known integrable cross section area $A(x)$ from $x = a$ to $x = b$ is the integral of A from a to b,

$$V = \int_a^b A(x)\, dx.$$

To apply this formula, we proceed as follows.

How to Find Volume by the Method of Slicing

1. Sketch the solid and a typical cross section.
2. Find a formula for $A(x)$.
3. Find the limits of integration.
4. Integrate $A(x)$ to find the volume.

Bonaventura Cavalieri (1598–1647)

Cavalieri, a student of Galileo, discovered that if two plane regions can be arranged to lie over the same interval of the x-axis in such a way that they have identical vertical cross sections at every point, then the regions have the same area. This theorem and a letter of recommendation from Galileo were enough to win Cavalieri a chair at the University of Bologna in 1629. The solid geometry version in Example 1, which Cavalieri never proved, was named after him by later geometers.

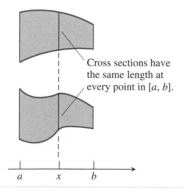

Cross sections have the same length at every point in $[a, b]$.

Example 1 CAVALIERI'S VOLUME THEOREM

Cavalieri's volume theorem says that solids with equal altitudes and identical cross section areas at each height have the same volume (Figure 7.17). This follows immediately from the definition of volume, because the cross section area function $A(x)$ and the interval $[a, b]$ are the same for both solids.

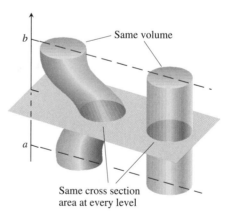

Same volume

Same cross section area at every level

Figure 7.17 *Cavalieri's volume theorem:* These solids have the same volume. You can illustrate this yourself with stacks of coins. (Example 1)

Square Cross Sections

Let us apply the volume formula to a solid with square cross sections.

Example 2   A SQUARE-BASED PYRAMID

A pyramid 3 m high has congruent triangular sides and a square base that is 3 m on each side. Each cross section of the pyramid parallel to the base is a square. Find the volume of the pyramid.

Solution We follow the steps for the method of slicing.

1. *Sketch.* We draw the pyramid with its vertex at the origin and its altitude along the interval $0 \le x \le 3$. We sketch a typical cross section at a point x between 0 and 3 (Figure 7.18).

2. *Find a formula for $A(x)$.* The cross section at x is a square x meters on a side, so
$$A(x) = x^2.$$

3. *Find the limits of integration.* The squares go from $x = 0$ to $x = 3$.

4. *Integrate to find the volume.*
$$V = \int_0^3 A(x)\, dx = \int_0^3 x^2 = \frac{x^3}{3} \Big]_0^3 = 9 \text{ m}^3$$

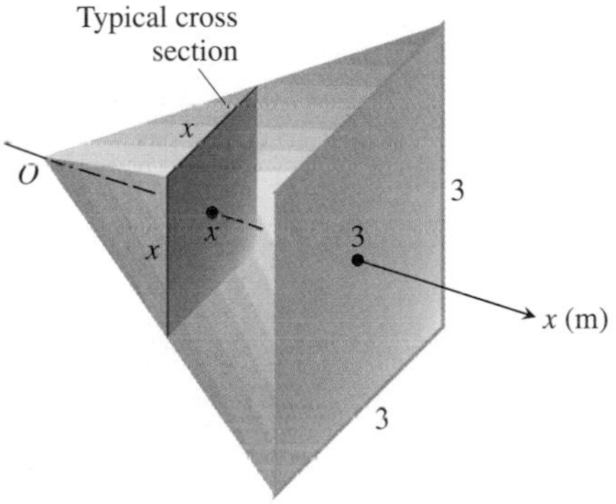

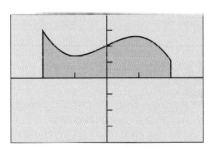

Figure 7.18 A cross section of the pyramid in Example 2.

Circular Cross Sections

The only thing that changes when the cross sections of a solid are circular is the formula for $A(x)$. Many such solids are **solids of revolution,** as in the next example.

Example 3   A SOLID OF REVOLUTION

The region between the graph of $f(x) = 2 + x \cos x$ and the x-axis over the interval $[-2, 2]$ is revolved about the x-axis to generate a solid. Find the volume of the solid.

Solution Revolving the region (Figure 7.19) about the x-axis generates the vase-shaped solid in Figure 7.20. The cross section at a typical point x is circular, with radius equal to $f(x)$. Its area is
$$A(x) = \pi(f(x))^2.$$

The volume of the solid is

$$V = \int_{-2}^2 A(x)\, dx$$
$$\approx \text{NINT}\,(\pi(2 + x \cos x)^2, x, -2, 2)$$
$$\approx 52.43 \text{ units cubed.}$$

[−3, 3] by [−4, 4]

Figure 7.19 The region in Example 3.

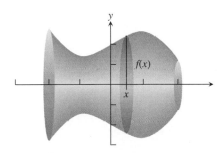

Figure 7.20 The region in Figure 7.19 is revolved about the x-axis to generate a solid. A typical cross section is circular, with radius $f(x) = 2 + x \cos x$. (Example 3)

Example 4 WASHER CROSS SECTIONS

The region in the first quadrant enclosed by the y-axis and the graphs of $y = \cos x$ and $y = \sin x$ is revolved about the x-axis to form a solid. Find its volume.

Solution The region is shown in Figure 7.21.

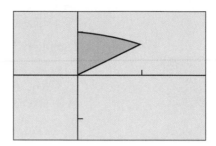

[$-\pi/4$, $\pi/2$] by [-1.5, 1.5]

Figure 7.21 The region in Example 4.

We revolve it about the x-axis to generate a solid with a cone-shaped cavity in its center (Figure 7.22).

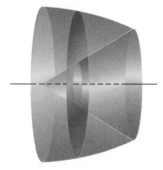

Figure 7.22 The solid generated by revolving the region in Figure 7.21 about the x-axis. A typical cross section is a washer: a circular region with a circular region cut out of its center. (Example 4)

This time each cross section perpendicular to the *axis of revolution* is a *washer,* a circular region with a circular region cut from its center. The area of a washer can be found by subtracting the inner area from the outer area (Figure 7.23).

CAUTION!

The area of a washer is $\pi R^2 - \pi r^2$, which you can simplify to $\pi(R^2 - r^2)$, but *not* to $\pi(R - r)^2$. No matter how tempting it is to make the latter simplification, it's wrong. Don't do it.

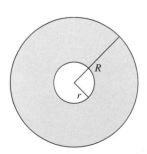

Figure 7.23 The area of a washer is $\pi R^2 - \pi r^2$. (Example 4)

In our region the cosine curve defines the outer radius, and the curves intersect at $x = \pi/4$. The volume is

$$V = \int_0^{\pi/4} \pi(\cos^2 x - \sin^2 x)\, dx$$

$$= \pi \int_0^{\pi/4} \cos 2x\, dx \qquad \text{identity: } \cos^2 x - \sin^2 x = \cos 2x$$

$$= \pi \left[\frac{\sin 2x}{2} \right]_0^{\pi/4} = \frac{\pi}{2} \text{ units cubed.}$$

We could have done the integration in Example 4 with NINT, but we wanted to demonstrate how a trigonometric identity can be useful under unexpected circumstances in calculus. The double-angle identity turned a difficult integrand into an easy one and enabled us to get an exact answer by antidifferentiation.

Cylindrical Shells

There is another way to find volumes of solids of rotation that can be useful when the axis of revolution is perpendicular to the axis containing the natural interval of integration. Instead of summing volumes of thin slices, we sum volumes of thin cylindrical shells that grow outward from the axis of revolution like tree rings.

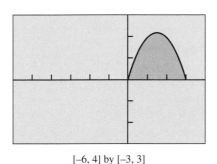

[–6, 4] by [–3, 3]

Figure 7.24 The graph of the region in Exploration 1, before revolution.

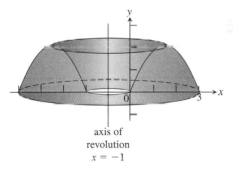

axis of
revolution
$x = -1$

Figure 7.25 The region in Figure 7.24 is revolved about the line $x = -1$ to form a solid cake. The natural interval of integration is along the x-axis, perpendicular to the axis of revolution. (Exploration 1)

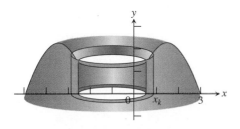

Figure 7.26 Cutting the cake into thin cylindrical slices, working from the inside out. Each slice occurs at some x_k between 0 and 3 and has thickness Δx. (Exploration 1)

Exploration 1 Volume by Cylindrical Shells

The region enclosed by the x-axis and the parabola $y = f(x) = 3x - x^2$ is revolved about the line $x = -1$ to generate the shape of a cake (Figures 7.24, 7.25). (Such a cake is often called a bundt cake.) What is the volume of the cake?

Integrating with respect to y would be awkward here, as it is not easy to get the original parabola in terms of y. (Try finding the volume by washers and you will soon see what we mean.) To integrate with respect to x, you can do the problem by *cylindrical shells*, which requires that you cut the cake in a rather unusual way.

1. Instead of cutting the usual wedge shape, cut a *cylindrical* slice by cutting straight down all the way around close to the inside hole. Then cut another cylindrical slice around the enlarged hole, then another, and so on. The radii of the cylinders gradually increase, and the heights of the cylinders follow the contour of the parabola: smaller to larger, then back to smaller (Figure 7.26). Each slice is sitting over a subinterval of the x-axis of length Δx. Its radius is approximately $(1 + x_k)$. What is its height?

2. If you unroll the cylinder at x_k and flatten it out, it becomes (essentially) a rectangular slab with thickness Δx. Show that the volume of the slab is approximately $2\pi(x_k + 1)(3x_k - x_k^2)\Delta x$.

3. $\sum 2\pi(x_k + 1)(3x_k - x_k^2)\Delta x$ is a Riemann sum. What is the limit of these Riemann sums as $\Delta x \to 0$?

4. Evaluate the integral you found in step 3 to find the volume of the cake!

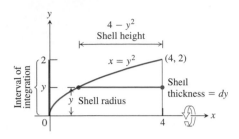

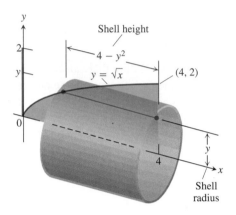

Figure 7.27 The region, shell dimensions, and interval of integration in Example 5.

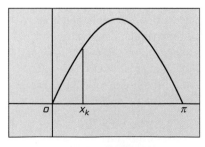

Figure 7.28 The shell swept out by the line segment in Figure 7.27.

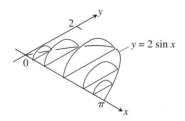

[−1, 3.5] by [−0.8, 2.2]

Figure 7.29 The base of the paperweight in Example 6. The segment perpendicular to the x-axis at x_k is the diameter of a semicircle that is perpendicular to the base.

Example 5 FINDING VOLUMES USING CYLINDRICAL SHELLS

The region bounded by the curve $y = \sqrt{x}$, the x-axis, and the line $x = 4$ is revolved about the x-axis to generate a solid. Find the volume of the solid.

Solution

1. Sketch the region and draw a line segment across it parallel to the axis of revolution (Figure 7.27). Label the segment's length (shell height) and distance from the axis of revolution (shell radius). The width of the segment is the shell thickness dy. (We drew the shell in Figure 7.28, but you need not do that.)

2. Identify the limits of integration: y runs from 0 to 2.

3. Integrate to find the volume.

$$V = \int_0^2 2\pi \left(\begin{matrix} \text{shell} \\ \text{radius} \end{matrix} \right) \left(\begin{matrix} \text{shell} \\ \text{height} \end{matrix} \right) dy$$

$$= \int_0^2 2\pi(y)(4 - y^2)\, dy = 8\pi$$

Other Cross Sections

The method of cross-section slicing can be used to find volumes of a wide variety of unusually shaped solids, so long as the cross sections have areas that we can describe with some formula. Admittedly, it does take a special artistic talent to *draw* some of these solids, but a crude picture is usually enough to suggest how to set up the integral.

Example 6 A MATHEMATICIAN'S PAPERWEIGHT

A mathematician has a paperweight made so that its base is the shape of the region between the x-axis and one arch of the curve $y = 2 \sin x$ (linear units in inches). Each cross section cut perpendicular to the x-axis (and hence to the xy-plane) is a semicircle whose diameter runs from the x-axis to the curve. (Think of the cross section as a semicircular fin sticking up out of the plane.) Find the volume of the paperweight.

Solution The paperweight is not easily drawn, but we know what it looks like. Its base is the region in Figure 7.29, and the cross sections perpendicular to the base are semicircular fins like those in Figure 7.30.

The semicircle at each point x has

$$\text{radius} = \frac{2 \sin x}{2} = \sin x \quad \text{and area} \quad A(x) = \frac{1}{2}\pi(\sin x)^2.$$

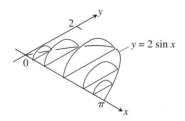

Figure 7.30 Cross sections perpendicular to the region in Figure 7.29 are semicircular. (Example 6)

The volume of the paperweight is

$$V = \int_0^\pi A(x)\, dx$$

$$= \frac{\pi}{2} \int_0^\pi (\sin x)^2\, dx$$

$$\approx \frac{\pi}{2}\, \text{NINT}\, ((\sin x)^2,\, x,\, 0,\, \pi)$$

$$\approx \frac{\pi}{2}(1.570796327).$$

The number in parentheses looks like half of π, an observation that can be confirmed analytically, and which we support numerically by dividing by π to get 0.5. The volume of the paperweight is

$$\frac{\pi}{2} \cdot \frac{\pi}{2} = \frac{\pi^2}{4} \approx 2.47\ \text{in}^3.$$

Exploration 2   **Surface Area**

We know how to find the volume of a solid of revolution, but how would we find the *surface area*? As before, we partition the solid into thin slices, but now we wish to form a Riemann sum of approximations to *surface areas of slices* (rather than of volumes of slices).

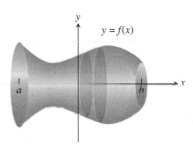

A typical slice has a surface area that can be approximated by $2\pi \cdot f(x) \cdot \Delta s$, where Δs is the tiny *slant height* of the slice. We will see in Section 7.4, when we study arc length, that $\Delta s = \sqrt{\Delta x^2 + \Delta y^2}$, and that this can be written as $\Delta s = \sqrt{1 + (f'(x_k))^2}\, \Delta x$.

Thus, the surface area is approximated by the Riemann sum

$$\sum_{k=1}^{n} 2\pi f(x_k) \sqrt{1 + (f'(x_k))^2}\, \Delta x.$$

1. Write the limit of the Riemann sums as a definite integral from a to b. When will the limit exist?

2. Use the formula from part 1 to find the surface area of the solid generated by revolving a single arch of the curve $y = \sin x$ about the x-axis.

3. The region enclosed by the graphs of $y^2 = x$ and $x = 4$ is revolved about the x-axis to form a solid. Find the surface area of the solid.

Quick Review 7.3

In Exercises 1–10, give a formula for the area of the plane region in terms of the single variable x.

1. a square with sides of length x

2. a square with diagonals of length x

3. a semicircle of radius x

4. a semicircle of diameter x

5. an equilateral triangle with sides of length x

6. an isosceles right triangle with legs of length x

7. an isosceles right triangle with hypotenuse x

8. an isosceles triangle with two sides of length $2x$ and one side of length x

9. a triangle with sides $3x$, $4x$, and $5x$

10. a regular hexagon with sides of length x

Section 7.3 Exercises

In Exercises 1 and 2, find a formula for the area $A(x)$ of the cross sections of the solid that are perpendicular to the x-axis.

1. The solid lies between planes perpendicular to the x-axis at $x = -1$ and $x = 1$. The cross sections perpendicular to the x-axis between these planes run from the semicircle $y = -\sqrt{1 - x^2}$ to the semicircle $y = \sqrt{1 - x^2}$.

    (a) The cross sections are circular disks with diameters in the xy-plane.

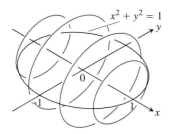

    (b) The cross sections are squares with bases in the xy-plane.

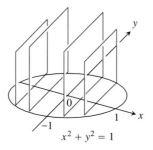

    (c) The cross sections are squares with diagonals in the xy-plane. (The length of a square's diagonal is $\sqrt{2}$ times the length of its sides.)

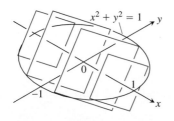

    (d) The cross sections are equilateral triangles with bases in the xy-plane.

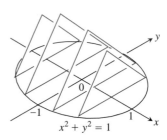

2. The solid lies between planes perpendicular to the x-axis at $x = 0$ and $x = 4$. The cross sections perpendicular to the x-axis between these planes run from $y = -\sqrt{x}$ to $y = \sqrt{x}$.

    (a) The cross sections are circular disks with diameters in the xy-plane.

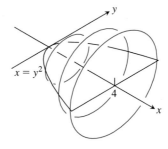

    (b) The cross sections are squares with bases in the xy-plane.

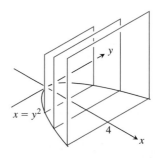

(c) The cross sections are squares with diagonals in the xy-plane.

(d) The cross sections are equilateral triangles with bases in the xy-plane.

In Exercises 3–10, find the volume of the solid analytically.

3. The solid lies between planes perpendicular to the x-axis at $x = 0$ and $x = 4$. The cross sections perpendicular to the axis on the interval $0 \le x \le 4$ are squares whose diagonals run from $y = -\sqrt{x}$ to $y = \sqrt{x}$.

4. The solid lies between planes perpendicular to the x-axis at $x = -1$ and $x = 1$. The cross sections perpendicular to the x-axis are circular disks whose diameters run from the parabola $y = x^2$ to the parabola $y = 2 - x^2$.

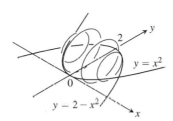

5. The solid lies between planes perpendicular to the x-axis at $x = -1$ and $x = 1$. The cross sections perpendicular to the x-axis between these planes are squares whose bases run from the semicircle $y = -\sqrt{1 - x^2}$ to the semicircle $y = \sqrt{1 - x^2}$.

6. The solid lies between planes perpendicular to the x-axis at $x = -1$ and $x = 1$. The cross sections perpendicular to the x-axis between these planes are squares whose diagonals run from the semicircle $y = -\sqrt{1 - x^2}$ to the semicircle $y = \sqrt{1 - x^2}$.

7. The base of a solid is the region between the curve $y = 2\sqrt{\sin x}$ and the interval $[0, \pi]$ on the x-axis. The cross sections perpendicular to the x-axis are

(a) equilateral triangles with bases running from the x-axis to the curve as shown in the figure.

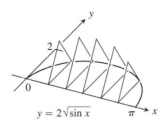

$y = 2\sqrt{\sin x}$

(b) squares with bases running from the x-axis to the curve.

8. The solid lies between planes perpendicular to the x-axis at $x = -\pi/3$ and $x = \pi/3$. The cross sections perpendicular to the x-axis are

(a) circular disks with diameters running from the curve $y = \tan x$ to the curve $y = \sec x$.

(b) squares whose bases run from the curve $y = \tan x$ to the curve $y = \sec x$.

9. The solid lies between planes perpendicular to the y-axis at $y = 0$ and $y = 2$. The cross sections perpendicular to the y-axis are circular disks with diameters running from the y-axis to the parabola $x = \sqrt{5}y^2$.

10. The base of the solid is the disk $x^2 + y^2 \le 1$. The cross sections by planes perpendicular to the y-axis between $y = -1$ and $y = 1$ are isosceles right triangles with one leg in the disk.

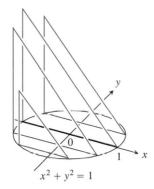

$x^2 + y^2 = 1$

11. *A Twisted Solid*   A square of side length s lies in a plane perpendicular to a line L. One vertex of the square lies on L. As this square moves a distance h along L, the square turns one revolution about L to generate a corkscrew-like column with square cross sections.

(a) Find the volume of the column.

(b) *Writing to Learn*   What will the volume be if the square turns twice instead of once? Give reasons for your answer.

12. **Writing to Learn**   A solid lies between planes perpendicular to the x-axis at $x = 0$ and $x = 12$. The cross sections by planes perpendicular to the x-axis are circular disks whose diameters run from the line $y = x/2$ to the line $y = x$ as shown in the figure. Explain why the solid has the same volume as a right circular cone with base radius 3 and height 12.

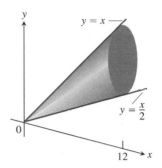

$y = x$

$y = \dfrac{x}{2}$

In Exercises 13–16, find the volume of the solid generated by revolving the shaded region about the given axis.

13. about the *x*-axis   **14.** about the *y*-axis

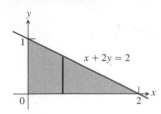   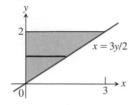

15. about the *y*-axis   **16.** about the *x*-axis

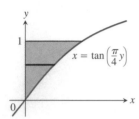   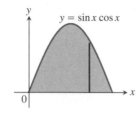

In Exercises 17–26, find the volume of the solid generated by revolving the region bounded by the lines and curves about the *x*-axis.

17. $y = x^2, \quad y = 0, \quad x = 2$

18. $y = x^3, \quad y = 0, \quad x = 2$

19. $y = \sqrt{9 - x^2}, \quad y = 0$

20. $y = x - x^2, \quad y = 0$

21. $y = x, \quad y = 1, \quad x = 0$

22. $y = 2x, \quad y = x, \quad x = 1$

23. $y = x^2 + 1, \quad y = x + 3$

24. $y = 4 - x^2, \quad y = 2 - x$

25. $y = \sec x, \quad y = \sqrt{2}, \quad -\pi/4 \le x \le \pi/4$

26. $y = -\sqrt{x}, \quad y = -2, \quad x = 0$

In Exercises 27 and 28, find the volume of the solid generated by revolving the region about the given line.

27. the region in the first quadrant bounded above by the line $y = \sqrt{2}$, below by the curve $y = \sec x \tan x$, and on the left by the *y*-axis, about the line $y = \sqrt{2}$

28. the region in the first quadrant bounded above by the line $y = 2$, below by the curve $y = 2 \sin x$, $0 \le x \le \pi/2$, and on the left by the *y*-axis, about the line $y = 2$

In Exercises 29–34, find the volume of the solid generated by revolving the region about the *y*-axis.

29. the region enclosed by $x = \sqrt{5}y^2$, $x = 0$, $y = -1$, $y = 1$

30. the region enclosed by $x = y^{3/2}$, $x = 0$, $y = 2$

31. the region enclosed by the triangle with vertices $(1, 0)$, $(2, 1)$, and $(1, 1)$

32. the region enclosed by the triangle with vertices $(0, 1)$, $(1, 0)$, and $(1, 1)$

33. the region in the first quadrant bounded above by the parabola $y = x^2$, below by the *x*-axis, and on the right by the line $x = 2$

34. the region bounded above by the curve $y = \sqrt{x}$ and below by the line $y = x$

In Exercises 35–38, *work in groups of two or three.*

35. Find the volume of the solid generated by revolving the region bounded by $y = \sqrt{x}$ and the lines $y = 2$ and $x = 0$ about

(a) the *x*-axis.   (b) the *y*-axis.

(c) the line $y = 2$.   (d) the line $x = 4$.

36. Find the volume of the solid generated by revolving the triangular region bounded by the lines $y = 2x$, $y = 0$, and $x = 1$ about

(a) the line $x = 1$.   (b) the line $x = 2$.

37. Find the volume of the solid generated by revolving the region bounded by the parabola $y = x^2$ and the line $y = 1$ about

(a) the line $y = 1$.   (b) the line $y = 2$.

(c) the line $y = -1$.

38. By integration, find the volume of the solid generated by revolving the triangular region with vertices $(0, 0)$, $(b, 0)$, $(0, h)$ about

(a) the *x*-axis.   (b) the *y*-axis.

In Exercises 39–42, use the cylindrical shell method to find the volume of the solid generated by revolving the region bounded by the curves about the *y*-axis.

39. $y = x, \quad y = -x/2, \quad x = 2$

40. $y = x^2, \quad y = 2 - x, \quad x = 0, \quad$ for $x \ge 0$

41. $y = \sqrt{x}, \quad y = 0, \quad x = 4$

42. $y = 2x - 1, \quad y = \sqrt{x}, \quad x = 0$

In Exercises 43 and 44, use the cylindrical shell method to find the volume of the solid generated by revolving the shaded region about the indicated axis.

43. (a) the *x*-axis   (b) the line $y = 1$

(c) the line $y = 8/5$ (d) the line $y = -2/5$

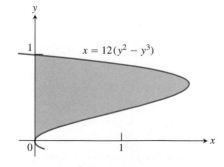

44. (a) the *x*-axis **(b)** the line $y = 2$

 (c) the line $y = 5$ **(d)** the line $y = -5/8$

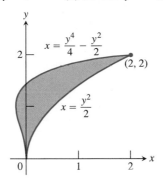

$$x = \frac{y^4}{4} - \frac{y^2}{2}$$

$(2, 2)$

$$x = \frac{y^2}{2}$$

45. Find the volume of the solid generated by revolving the region in the first quadrant bounded by $y = x^3$ and $y = 4x$ about **(a)** the *x*-axis, **(b)** the line $y = 8$.

46. Find the volume of the solid generated by revolving the region bounded by $y = 2x - x^2$ and $y = x$ about **(a)** the *y*-axis, **(b)** the line $x = 1$.

47. The region in the first quadrant that is bounded above by the curve $y = 1/\sqrt{x}$, on the left by the line $x = 1/4$, and below by the line $y = 1$ is revolved about the *y*-axis to generate a solid. Find the volume of the solid by **(a)** the washer method and **(b)** the cylindrical shell method.

48. Let $f(x) = \begin{cases} (\sin x)/x, & 0 < x \le \pi \\ 1, & x = 0. \end{cases}$

 (a) Show that $x f(x) = \sin x, \quad 0 \le x \le \pi$.

 (b) Find the volume of the solid generated by revolving the shaded region about the *y*-axis.

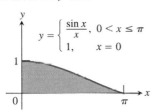

$$y = \begin{cases} \dfrac{\sin x}{x}, & 0 < x \le \pi \\ 1, & x = 0 \end{cases}$$

49. *Designing a Plumb Bob* Having been asked to design a brass plumb bob that will weigh in the neighborhood of 190 g, you decide to shape it like the solid of revolution shown here.

 (a) Find the plumb bob's volume.

 (b) If you specify a brass that weighs 8.5 g/cm³, how much will the plumb bob weigh to the nearest gram?

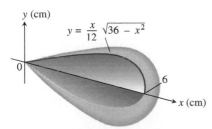

y (cm)

$$y = \frac{x}{12}\sqrt{36 - x^2}$$

6

x (cm)

Explorations

50. *Max-Min* The arch $y = \sin x, \ 0 \le x \le \pi$, is revolved about the line $y = c, \ 0 \le c \le 1$, to generate the solid in the figure.

 (a) Find the value of *c* that minimizes the volume of the solid. What is the minimum volume?

 (b) What value of *c* in $[0, 1]$ maximizes the volume of the solid?

 (c) Writing to Learn Graph the solid's volume as a function of *c*, first for $0 \le c \le 1$ and then on a larger domain. What happens to the volume of the solid as *c* moves away from $[0, 1]$? Does this make sense physically? Give reasons for your answers.

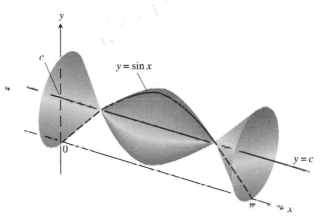

$y = \sin x$

$y = c$

51. *A Vase* We wish to estimate the volume of a flower vase using only a calculator, a string, and a ruler. We measure the height of the vase to be 6 inches. We then use the string and the ruler to find circumferences of the vase (in inches) at half-inch intervals. (We list them from the top down to correspond with the picture of the vase.)

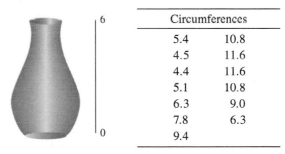

	Circumferences	
	5.4	10.8
	4.5	11.6
	4.4	11.6
	5.1	10.8
	6.3	9.0
	7.8	6.3
	9.4	

 (a) Find the areas of the cross sections that correspond to the given circumferences.

 (b) Express the volume of the vase as an integral with respect to *y* over the interval $[0, 6]$.

 (c) Approximate the integral using the Trapezoidal Rule with $n = 12$. ■

52. *Volume of a Bowl* A bowl has a shape that can be generated by revolving the graph of $y = x^2/2$ between $y = 0$ and $y = 5$ about the y-axis.

(a) Find the volume of the bowl.

(b) If we fill the bowl with water at a constant rate of 3 cubic units per second, how fast will the water level in the bowl be rising when the water is 4 units deep?

53. *The Classical Bead Problem* A round hole is drilled through the center of a spherical solid of radius r. The resulting cylindrical hole has height 4 cm.

(a) What is the volume of the solid that remains?

(b) What is unusual about the answer?

54. Writing to Learn Explain how you could estimate the volume of a solid of revolution by measuring the shadow cast on a table parallel to its axis of revolution by a light shining directly above it.

55. *Same Volume about Each Axis* The region in the first quadrant enclosed between the graph of $y = ax - x^2$ and the x-axis generates the same volume whether it is revolved about the x-axis or the y-axis. Find the value of a.

56. *(Continuation of Exploration 2)* Let $x = g(y) > 0$ have a continuous first derivative on $[c, d]$. Show that the area of the surface generated by revolving the curve $x = g(y)$ about the y-axis is

$$S = \int_c^d 2\pi\, g(y) \sqrt{1 + (g'(y))^2}\; dy.$$

In Exercises 57–64, find the area of the surface generated by revolving the curve about the indicated axis.

57. $x = \sqrt{y}$, $0 \le y \le 2$; y-axis

58. $x = y^3/3$, $0 \le y \le 1$; y-axis

59. $x = y^{1/2} - (1/3)^{3/2}$, $1 \le y \le 3$; y-axis

60. $x = \sqrt{2y - 1}$, $(5/8) \le y \le 1$; y-axis

61. $y = x^2$, $0 \le x \le 2$; x-axis

62. $y = 3x - x^2$, $0 \le x \le 3$; x-axis

63. $y = \sqrt{2x - x^2}$, $0.5 \le x \le 1.5$; x-axis

64. $y = \sqrt{x + 1}$, $1 \le x \le 5$; x-axis

Extending the Ideas

65. *Volume of a Hemisphere* Derive the formula $V = (2/3)\pi R^3$ for the volume of a hemisphere of radius R by comparing its cross sections with the cross sections of a solid right circular cylinder of radius R and height R from which a solid right circular cone of base radius R and height R has been removed as suggested by the figure.

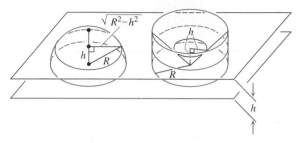

66. *Volume of a Torus* The disk $x^2 + y^2 \le a^2$ is revolved about the line $x = b$ $(b > a)$ to generate a solid shaped like a doughnut, called a *torus*. Find its volume. (*Hint:* $\int_{-a}^a \sqrt{a^2 - y^2}\; dy = \pi a^2/2$, since it is the area of a semicircle of radius a.)

67. *Filling a Bowl*

(a) *Volume* A hemispherical bowl of radius a contains water to a depth h. Find the volume of water in the bowl.

(b) *Related Rates* Water runs into a sunken concrete hemispherical bowl of radius 5 m at a rate of 0.2 m³/sec. How fast is the water level in the bowl rising when the water is 4 m deep?

68. *Consistency of Volume Definitions* The volume formulas in calculus are consistent with the standard formulas from geometry in the sense that they agree on objects to which both apply.

(a) As a case in point, show that if you revolve the region enclosed by the semicircle $y = \sqrt{a^2 - x^2}$ and the x-axis about the x-axis to generate a solid sphere, the calculus formula for volume at the beginning of the section will give $(4/3)\pi a^3$ for the volume just as it should.

(b) Use calculus to find the volume of a right circular cone of height h and base radius r.

7.4 Lengths of Curves

A Sine Wave • Length of a Smooth Curve • Vertical Tangents, Corners, and Cusps

Group Exploration

Later in this section we will use an integral to find the length of the sine wave with great precision. But there are ways to get good approximations without integrating. Take five minutes to come up with a written estimate of the curve's length. No fair looking ahead.

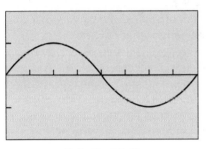

[0, 2π] by [2, 2]

Figure 7.31 One wave of a sine curve has to be longer than 2π.

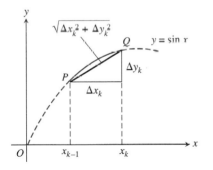

Figure 7.32 The line segment approximating the arc PQ of the sine curve above the subinterval $[x_{k-1}, x_k]$. (Example 1)

A Sine Wave

How long is a sine wave (Figure 7.31)?

The usual meaning of *wavelength* refers to the fundamental period, which for $y = \sin x$ is 2π. But how long is the curve itself? If you straightened it out like a piece of string along the positive x-axis with one end at 0, where would the other end be?

Example 1 THE LENGTH OF A SINE WAVE

What is the length of the curve $y = \sin x$ from $x = 0$ to $x = 2\pi$?

Solution We answer this question with integration, following our usual plan of breaking the whole into measurable parts. We partition $[0, 2\pi]$ into intervals so short that the pieces of curve (call them "arcs") lying directly above the intervals are nearly straight. That way, each arc is nearly the same as the line segment joining its two ends and we can take the length of the segment as an approximation to the length of the arc.

Figure 7.32 shows the segment approximating the arc above the subinterval $[x_{k-1}, x_k]$. The length of the segment is $\sqrt{\Delta x_k^2 + \Delta y_k^2}$. The sum

$$\sum \sqrt{\Delta x_k^2 + \Delta y_k^2}$$

over the entire partition approximates the length of the curve. All we need now is to find the limit of this sum as the norms of the partitions go to zero. That's the usual plan, but this time there is a problem. Do you see it?

The problem is that the sums as written are not Riemann sums. They do not have the form $\sum f(c_k)\,\Delta x$. We can rewrite them as Riemann sums if we multiply and divide each square root by Δx_k.

$$\sum \sqrt{\Delta x_k^2 + \Delta y_k^2} = \sum \frac{\sqrt{(\Delta x_k)^2 + (\Delta y_k)^2}}{\Delta x_k} \Delta x_k$$

$$= \sum \sqrt{1 + \left(\frac{\Delta y_k}{\Delta x_k}\right)^2} \, \Delta x_k$$

This is better, but we still need to write the last square root as a function evaluated at some c_k in the kth subinterval. For this, we call on the Mean Value Theorem for differentiable functions (Section 4.2), which says that since $\sin x$ is continuous on $[x_{k-1}, x_k]$ and is differentiable on (x_{k-1}, x_k) there is a point c_k in $[x_{k-1}, x_k]$ at which $\Delta y_k/\Delta x_k = \sin' c_k$ (Figure 7.33). That gives us

$$\sum \sqrt{1 + (\sin' c_k)^2} \, \Delta x_k,$$

which *is* a Riemann sum.

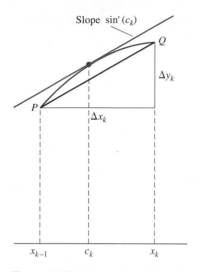

Figure 7.33 The portion of the sine curve above $[x_{k-1}, x_k]$. At some c_k in the interval, $\sin'(c_k) = \Delta y_k/\Delta x_k$, the slope of segment PQ. (Example 1)

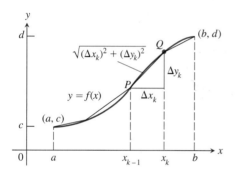

Figure 7.34 The graph of f, approximated by line segments.

Now we take the limit as the norms of the subdivisions go to zero and find that the length of one wave of the sine function is

$$\int_0^{2\pi} \sqrt{1 + (\sin' x)^2}\, dx = \int_0^{2\pi} \sqrt{1 + \cos^2 x}\, dx \approx 7.64. \quad \text{Using NINT}$$

How close was your estimate?

Length of a Smooth Curve

We are almost ready to define the length of a curve as a definite integral, using the procedure of Example 1. We first call attention to two properties of the sine function that came into play along the way.

We obviously used *differentiability* when we invoked the Mean Value Theorem to replace $\Delta y_k/\Delta x_k$ by $\sin'(c_k)$ for some c_k in the interval $[x_{k-1}, x_k]$. Less obviously, we used the continuity of the derivative of sine in passing from $\sum \sqrt{1 + (\sin'(c_k))^2}\, \Delta x_k$ to the Riemann integral. The requirement for finding the length of a curve by this method, then, is that the function have a continuous first derivative. We call this property **smoothness**. A function with a continuous first derivative is **smooth** and its graph is a **smooth curve.**

Let us review the process, this time with a general smooth function $f(x)$. Suppose the graph of f begins at the point (a, c) and ends at (b, d), as shown in Figure 7.34. We partition the interval $a \le x \le b$ into subintervals so short that the arcs of the curve above them are nearly straight. The length of the segment approximating the arc above the subinterval $[x_{k-1}, x_k]$ is $\sqrt{\Delta x_k{}^2 + \Delta y_k{}^2}$. The sum $\sum \sqrt{\Delta x_k{}^2 + \Delta y_k{}^2}$ approximates the length of the curve. We apply the Mean Value Theorem to f on each subinterval to rewrite the sum as a Riemann sum,

$$\sum \sqrt{\Delta x_k{}^2 + \Delta y_k{}^2} = \sum \sqrt{1 + \left(\frac{\Delta y_k}{\Delta x_k}\right)^2}\, \Delta x_k$$

$$= \sum \sqrt{1 + (f'(c_k))^2}\, \Delta x_k. \quad \begin{array}{l}\text{For some point} \\ c_k \text{ in } (x_{k-1}, x_k)\end{array}$$

Passing to the limit as the norms of the subdivisions go to zero gives the length of the curve as

$$L = \int_a^b \sqrt{1 + (f'(x))^2}\, dx = \int_a^b \sqrt{1 + \left(\frac{dy}{dx}\right)^2}\, dx.$$

We could as easily have transformed $\sum \sqrt{\Delta x_k{}^2 + \Delta y_k{}^2}$ into a Riemann sum by dividing and multiplying by Δy_k, giving a formula that involves x as a function of y (say, $x = g(y)$) on the interval $[c, d]$:

$$L \approx \sum \frac{\sqrt{(\Delta x_k)^2 + (\Delta y_k)^2}}{\Delta y_k}\, \Delta y_k = \sum \sqrt{1 + \left(\frac{\Delta x_k}{\Delta y_k}\right)^2}\, \Delta y_k$$

$$= \sum \sqrt{1 + (g'(c_k))^2}\, \Delta y_k. \quad \begin{array}{l}\text{For some } c_k \\ \text{in } (y_{k-1}, y_k)\end{array}$$

The limit of these sums, as the norms of the subdivisions go to zero, gives another reasonable way to calculate the curve's length,

$$L = \int_c^d \sqrt{1 + (g'(y))^2}\, dy = \int_c^d \sqrt{1 + \left(\frac{dx}{dy}\right)^2}\, dy.$$

Putting these two formulas together, we have the following definition for the length of a smooth curve.

Definition   Arc Length: Length of a Smooth Curve

If a smooth curve begins at (a, c) and ends at (b, d), $a < b$, $c < d$, then the **length (arc length) of the curve** is

$$L = \int_a^b \sqrt{1 + \left(\frac{dy}{dx}\right)^2}\, dx \qquad \text{if } y \text{ is a smooth function of } x \text{ on } [a, b];$$

$$L = \int_c^d \sqrt{1 + \left(\frac{dx}{dy}\right)^2}\, dy \qquad \text{if } x \text{ is a smooth function of } y \text{ on } [c, d].$$

Example 2   APPLYING THE DEFINITION

Find the *exact* length of the curve

$$y = \frac{4\sqrt{2}}{3}x^{3/2} - 1 \qquad \text{for} \qquad 0 \leq x \leq 1.$$

Solution

$$\frac{dy}{dx} = \frac{4\sqrt{2}}{3} \cdot \frac{3}{2}x^{1/2} = 2\sqrt{2}\, x^{1/2},$$

which is continuous on $[0, 1]$. Therefore,

$$\begin{aligned}
L &= \int_0^1 \sqrt{1 + \left(\frac{dy}{dx}\right)^2}\, dx \\
&= \int_0^1 \sqrt{1 + \left(2\sqrt{2}x^{1/2}\right)^2}\, dx \\
&= \int_0^1 \sqrt{1 + 8x}\, dx \\
&= \frac{2}{3} \cdot \frac{1}{8}(1 + 8x)^{3/2}\Bigg]_0^1 \\
&= \frac{13}{6}.
\end{aligned}$$

We asked for an exact length in Example 2 to take advantage of the rare opportunity it afforded of taking the antiderivative of an arc length integrand. When you add 1 to the square of the derivative of an arbitrary smooth function and then take the square root of that sum, the result is rarely antidifferentiable by reasonable methods. We know a few more functions that give "nice" integrands, but we are saving those for the exercises.

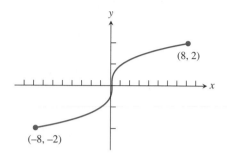

Figure 7.35 The graph of $y = x^{1/3}$ has a vertical tangent line at the origin where dy/dx does not exist. (Example 3)

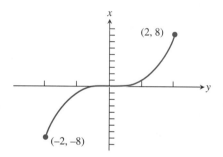

Figure 7.36 The curve in Figure 7.35 plotted with x as a function of y. The tangent at the origin is now horizontal. (Example 3)

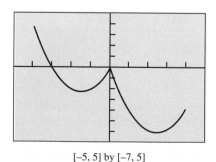

[-5, 5] by [-7, 5]

Figure 7.37 The graph of

$$y = x^2 - 4|x| - x, \quad -4 \leq x \leq 4,$$

has a corner at x 5 0 where neither dy/dx nor dx/dy exists. We find the lengths of the two smooth pieces and add them together. (Example 4)

Vertical Tangents, Corners, and Cusps

Sometimes a curve has a vertical tangent, corner, or cusp where the derivative we need to work with is undefined. We can sometimes get around such a difficulty in ways illustrated by the following examples.

Example 3 A VERTICAL TANGENT

Find the length of the curve $y = x^{1/3}$ between $(-8, -2)$ and $(8, 2)$.

Solution The derivative

$$\frac{dy}{dx} = \frac{1}{3}x^{-2/3} = \frac{1}{3x^{2/3}}$$

is not defined at $x = 0$. Graphically, there is a vertical tangent at $x = 0$ where the derivative becomes infinite (Figure 7.35). If we change to x as a function of y, the tangent at the origin will be horizontal (Figure 7.36) and the derivative will be zero instead of undefined. Solving $y = x^{1/3}$ for x gives $x = y^3$, and we have

$$L = \int_{-2}^{2} \sqrt{1 + \left(\frac{dx}{dy}\right)^2} \, dy = \int_{-2}^{2} \sqrt{1 + (3y^2)^2} \, dy \approx 17.26. \quad \text{Using NINT}$$

What happens if you fail to notice that dy/dx is undefined at $x = 0$ and ask your calculator to compute

$$\text{NINT}\left(\sqrt{1 + \left((1/3) \, x^{-2/3}\right)^2}, x, -8, 8\right)?$$

This actually depends on your calculator. If, in the process of its calculations, it tries to evaluate the function at $x = 0$, then some sort of domain error will result. If it tries to find convergent Riemann sums near $x = 0$, it might get into a long, futile loop of computations that you will have to interrupt. Or it might actually produce an answer—in which case you hope it would be sufficiently bizarre for you to realize that it should not be trusted.

Example 4 GETTING AROUND A CORNER

Find the length of the curve $y = x^2 - 4|x| - x$ from $x = -4$ to $x = 4$.

Solution We should always be alert for abrupt slope changes when absolute value is involved. We graph the function to check (Figure 7.37).

There is clearly a corner at $x = 0$ where neither dy/dx nor dx/dy can exist. To find the length, we split the curve at $x = 0$ to write the function *without* absolute values:

$$x^2 - 4|x| - x = \begin{cases} x^2 + 3x & \text{if} \quad x < 0, \\ x^2 - 5x & \text{if} \quad x \geq 0. \end{cases}$$

Then,

$$L = \int_{-4}^{0} \sqrt{1 + (2x + 3)^2} \, dx + \int_{0}^{4} \sqrt{1 + (2x - 5)^2} \, dx$$

$$\approx 19.56. \quad \text{By NINT}$$

Finally, cusps are handled the same way corners are: split the curve into smooth pieces and add the lengths of those pieces.

Quick Review 7.4

In Exercises 1–5, simplify the function.

1. $\sqrt{1 + 2x + x^2}$ on $[1, 5]$

2. $\sqrt{1 - x + \dfrac{x^2}{4}}$ on $[-3, -1]$

3. $\sqrt{1 + (\tan x)^2}$ on $[0, \pi/3]$

4. $\sqrt{1 + (x/4 - 1/x)^2}$ on $[4, 12]$

5. $\sqrt{1 + \cos 2x}$ on $[0, \pi/2]$

In Exercises 6–10, identify all values of x for which the function fails to be differentiable.

6. $f(x) = |x - 4|$

7. $f(x) = 5x^{2/3}$

8. $f(x) = \sqrt[5]{x + 3}$

9. $f(x) = \sqrt{x^2 - 4x + 4}$

10. $f(x) = 1 + \sqrt[3]{\sin x}$

Section 7.4 Exercises

In Exercises 1–10,

    (a) set up an integral for the length of the curve;

    (b) graph the curve to see what it looks like;

    (c) use NINT to find the length of the curve.

1. $y = x^2, \quad -1 \le x \le 2$

2. $y = \tan x, \quad -\pi/3 \le x \le 0$

3. $x = \sin y, \quad 0 \le y \le \pi$

4. $x = \sqrt{1 - y^2}, \quad -1/2 \le y \le 1/2$

5. $y^2 + 2y = 2x + 1, \quad$ from $(-1, -1)$ to $(7, 3)$

6. $y = \sin x - x \cos x, \quad 0 \le x \le \pi$

7. $y = \int_0^x \tan t \, dt, \quad 0 \le x \le \pi/6$

8. $x = \int_0^y \sqrt{\sec^2 t - 1} \, dt, \quad -\pi/3 \le y \le \pi/4$

9. $y = \sec x, \quad -\pi/3 \le x \le \pi/3$

10. $y = (e^x + e^{-x})/2, \quad 3 \le x \le 3$

In Exercises 11–18, find the exact length of the curve analytically by antidifferentiation. You will need to simplify the integrand algebraically before finding an antiderivative.

11. $y = (1/3)(x^2 + 2)^{3/2} \quad$ from $x = 0$ to $x = 3$

12. $y = x^{3/2} \quad$ from $x = 0$ to $x = 4$

13. $x = (y^3/3) + 1/(4y) \quad$ from $y = 1$ to $y = 3$
    (*Hint:* $1 + (dx/dy)^2$ is a perfect square.)

14. $x = (y^4/4) + 1/(8y^2) \quad$ from $y = 1$ to $y = 2$
    (*Hint:* $1 + (dx/dy)^2$ is a perfect square.)

15. $x = (y^3/6) + 1/(2y) \quad$ from $y = 1$ to $y = 2$
    (*Hint:* $1 + (dx/dy)^2$ is a perfect square.)

16. $y = (x^3/3) + x^2 + x + 1/(4x + 4), \quad 0 \le x \le 2$

17. $x = \int_0^y \sqrt{\sec^4 t - 1} \, dt, \quad -\pi/4 \le y \le \pi/4$

18. $y = \int_{-2}^x \sqrt{3t^4 - 1} \, dt, \quad -2 \le x \le -1$

In Exercises 19 and 20, *work in groups of two or three.*

19. **(a)** Find a curve through the point $(1, 1)$ whose length integral is

$$L = \int_1^4 \sqrt{1 + \frac{1}{4x}} \, dx.$$

    (b) Writing to Learn How many such curves are there? Give reasons for your answer.

20. **(a)** Find a curve through the point $(0, 1)$ whose length integral is

$$L = \int_1^2 \sqrt{1 + \frac{1}{y^4}} \, dy.$$

    (b) Writing to Learn How many such curves are there? Give reasons for your answer.

21. Find the length of the curve

$$y = \int_0^x \sqrt{\cos 2t} \, dt$$

from $x = 0$ to $x = \pi/4$.

22. *The Length of an Astroid* The graph of the equation $x^{2/3} + y^{2/3} = 1$ is one of the family of curves called *astroids* (not "asteroids") because of their starlike appearance (see figure). Find the length of this particular astroid by finding the length of half the first quadrant portion, $y = (1 - x^{2/3})^{3/2}, \quad \sqrt{2}/4 \le x \le 1$, and multiplying by 8.

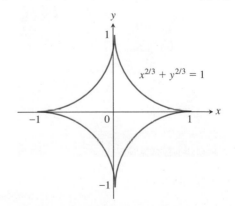

$x^{2/3} + y^{2/3} = 1$

23. *Fabricating Metal Sheets* Your metal fabrication company is bidding for a contract to make sheets of corrugated steel roofing like the one shown here. The cross sections of the corrugated sheets are to conform to the curve

$$y = \sin\left(\frac{3\pi}{20}x\right), \quad 0 \le x \le 20 \text{ in.}$$

If the roofing is to be stamped from flat sheets by a process that does not stretch the material, how wide should the original material be? Give your answer to two decimal places.

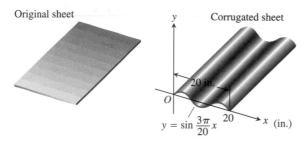

Original sheet Corrugated sheet

20 in.

$y = \sin\frac{3\pi}{20}x$ x (in.)

24. *Tunnel Construction* Your engineering firm is bidding for the contract to construct the tunnel shown here. The tunnel is 300 ft long and 50 ft wide at the base. The cross section is shaped like one arch of the curve $y = 25\cos(\pi x/50)$. Upon completion, the tunnel's inside surface (excluding the roadway) will be treated with a waterproof sealer that costs \$1.75 per square foot to apply. How much will it cost to apply the sealer?

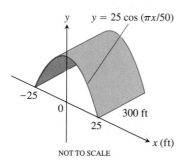

$y = 25\cos(\pi x/50)$

-25

300 ft

25

x (ft)

NOT TO SCALE

Exploration

25. *Modeling Running Tracks* Two lanes of a running track are modeled by the semiellipses as shown. The equation for lane 1 is $y = \sqrt{100 - 0.2x^2}$, and the equation for lane 2 is $y = \sqrt{150 - 0.2x^2}$. The starting point for lane 1 is at the negative x-intercept $(-\sqrt{500}, 0)$. The finish points for both lanes are the positive x-intercepts. Where should the starting point be placed on lane 2 so that the two lane lengths will be equal (running clockwise)?

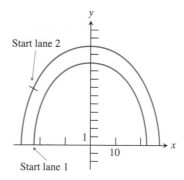

Start lane 2

1

10

x

Start lane 1

In Exercises 26 and 27, find the length of the curve.

26. $f(x) = x^{1/3} + x^{2/3}, \quad 0 \le x \le 2$

27. $f(x) = \dfrac{x-1}{4x^2 + 1}, \quad -\dfrac{1}{2} \le x \le 1$

In Exercises 28–30, find the length of the nonsmooth curve.

28. $y = x^3 + 5|x|$ from $x = -2$ to $x = 1$

29. $\sqrt{x} + \sqrt{y} = 1$

30. $y = \sqrt[4]{x}$ from $x = 0$ to $x = 16$

31. Writing to Learn Explain geometrically why it does not work to use short *horizontal* line segments to approximate the lengths of small arcs when we search for a Riemann sum that leads to the formula for arc length.

32. Writing to Learn A curve is totally contained inside the square with vertices $(0, 0)$, $(1, 0)$, $(1, 1)$, and $(0, 1)$. Is there any limit to the possible length of the curve? Explain.

Extending the Ideas

33. *Using Tangent Fins to Find Arc Length* Assume f is smooth on $[a, b]$ and partition the interval $[a, b]$ in the usual way. In each subinterval $[x_{k-1}, x_k]$ construct the *tangent fin* at the point $(x_{k-1}, f(x_{k-1}))$ as shown in the figure.

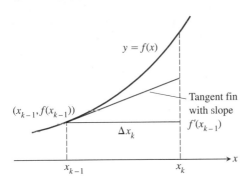

(a) Show that the length of the kth tangent fin over the interval $[x_{k-1}, x_k]$ equals

$$\sqrt{(\Delta x_k)^2 + (f'(x_{k-1}) \Delta x_k)^2}.$$

(b) Show that

$$\lim_{n \to \infty} \sum_{k=1}^{n} (\text{length of } k\text{th tangent fin}) = \int_a^b \sqrt{1 + (f'(x))^2}\, dx,$$

which is the length L of the curve $y = f(x)$ from $x = a$ to $x = b$.

34. Is there a smooth curve $y = f(x)$ whose length over the interval $0 \le x \le a$ is always $a\sqrt{2}$? Give reasons for your answer.

7.5	# Applications from Science and Statistics

Work Revisited • Fluid Force and Fluid Pressure • Normal Probabilities

Our goal in this section is to hint at the diversity of ways in which the definite integral can be used. The contexts may be new to you, but we will explain what you need to know as we go along.

Work Revisited

Recall from Section 7.1 that *work* is defined as force (in the direction of motion) times displacement. A familiar example is to move against the force of gravity to lift an object. The object has to move, incidentally, before "work" is done, no matter how tired you get *trying*.

4.4 newtons ≈ 1 lb

(1 newton)(1 meter) = 1 N • m = 1 Joule

Example 1   WORK DONE LIFTING

A leaky bucket weighs 22 newtons (N) empty. It is lifted from the ground at a constant rate to a point 20 m above the ground by a rope weighing 0.4 N • m. The bucket starts with 70 N (approximately 7.1 liters) of water, but it leaks at a constant rate and just finishes draining as the bucket reaches the top. Find the amount of work done

(a) lifting the bucket alone;

(b) lifting the water alone;

(c) lifting the rope alone;

(d) lifting the bucket, water, and rope together.

Figure 7.38 The work done by a constant 22-N force lifting a bucket 20 m is 440 N • m. (Example 1)

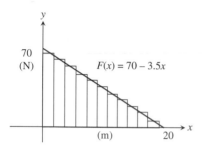

Figure 7.39 The force required to lift the water varies with distance but the work still corresponds to the area under the force graph. (Example 1)

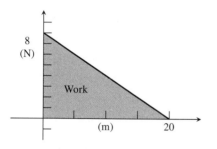

Figure 7.40 The work done lifting the rope to the top corresponds to the area of another triangle. (Example 1)

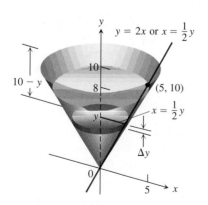

Figure 7.41 The conical tank in Example 2.

Solution

(a) *The bucket alone.* This is easy because the bucket's weight is constant. To lift it, you must exert a force of 22 N through the entire 20-meter interval.

$$\text{Work} = (22\ \text{N}) \times (20\ \text{m}) = 440\ \text{N} \cdot \text{m} = 440\ \text{J}$$

Figure 7.38 shows the graph of force vs. distance applied. The work corresponds to the area under the force graph.

(b) *The water alone.* The force needed to lift the water is equal to the water's weight, which decreases steadily from 70 N to 0 N over the 20-m lift. When the bucket is x m off the ground, the water weighs

$$F(x) = \underbrace{70}_{\substack{\text{original weight} \\ \text{of water}}} \underbrace{\left(\frac{20 - x}{20}\right)}_{\substack{\text{proportion left} \\ \text{at elevation } x}} = 70\left(1 - \frac{x}{20}\right) = 70 - 3.5x\ \text{N}.$$

The work done is (Figure 7.39)

$$W = \int_a^b F(x)\, dx$$

$$= \int_0^{20} (70 - 3.5x)\, dx = \left[70x - 1.75x^2\right]_0^{20} = 1400 - 700 = 700\ \text{J}.$$

(c) *The rope alone.* The force needed to lift the rope is also variable, starting at $(0.4)(20) = 8$ N when the bucket is on the ground and ending at 0 N when the bucket and rope are all at the top. As with the leaky bucket, the rate of decrease is constant. At elevation x meters, the $(20 - x)$ meters of rope still there to lift weigh $F(x) = (0.4)(20 - x)$ N. Figure 7.40 shows the graph of F. The work done lifting the rope is

$$\int_0^{20} F(x)\, dx = \int_0^{20} (0.4)(20 - x)\, dx$$

$$= \left[8x - 0.2x^2\right]_0^{20} = 160 - 80 = 80\ \text{N} \cdot \text{m} = 80\ \text{J}.$$

(d) *The bucket, water, and rope together.* The total work is

$$440 + 700 + 80 = 1220\ \text{J}.$$

Example 2   WORK DONE PUMPING

The conical tank in Figure 7.41 is filled to within 2 ft of the top with olive oil weighing 57 lb/ft^3. How much work does it take to pump the oil to the rim of the tank?

Solution We imagine the oil partitioned into thin slabs by planes perpendicular to the y-axis at the points of a partition of the interval $[0, 8]$. (The 8 represents the top of the oil, not the top of the tank.)

The typical slab between the planes at y and $y + \Delta y$ has a volume of about

$$\Delta V = \pi(\text{radius})^2(\text{thickness}) = \pi\left(\frac{1}{2}y\right)^2 \Delta y = \frac{\pi}{4}y^2\, \Delta y\ \text{ft}^3.$$

The force $F(y)$ required to lift this slab is equal to its weight,

$$F(y) = 57\, \Delta V = \frac{57\pi}{4} y^2\, \Delta y \quad \text{lb.} \quad \text{Weight} = \left(\begin{array}{c}\text{weight per}\\ \text{unit volume}\end{array}\right) \times \text{volume}$$

The distance through which $F(y)$ must act to lift this slab to the level of the rim of the cone is about $(10 - y)$ ft, so the work done lifting the slab is about

$$\Delta W = \frac{57\pi}{4}(10 - y)y^2\, \Delta y \quad \text{ft} \cdot \text{lb.}$$

The work done lifting all the slabs from $y = 0$ to $y = 8$ to the rim is approximately

$$W \approx \sum \frac{57\pi}{4}(10 - y)\, y^2\, \Delta y \quad \text{ft} \cdot \text{lb.}$$

This is a Riemann sum for the function $(57\pi/4)(10 - y)y^2$ on the interval from $y = 0$ to $y = 8$. The work of pumping the oil to the rim is the limit of these sums as the norms of the partitions go to zero.

$$W = \int_0^8 \frac{57\pi}{4}(10 - y)y^2\, dy = \frac{57\pi}{4}\int_0^8 (10y^2 - y^3)\, dy$$

$$= \frac{57\pi}{4}\left[\frac{10y^3}{3} - \frac{y^4}{4}\right]_0^8 \approx 30{,}561 \text{ ft} \cdot \text{lb}$$

Fluid Force and Fluid Pressure

We make dams thicker at the bottom than at the top (Figure 7.42) because the pressure against them increases with depth. It is a remarkable fact that the pressure at any point on a dam depends only on how far below the surface the point lies and not on how much water the dam is holding back. In any liquid, the **fluid pressure** p (force per unit area) at depth h is

$$p = wh, \quad \text{Dimensions check: } \frac{\text{lb}}{\text{ft}^2} = \frac{\text{lb}}{\text{ft}^3} \times \text{ft, for example}$$

where w is the *weight density* (weight per unit volume) of the liquid.

Figure 7.42 To withstand the increasing pressure, dams are built thicker toward the bottom.

Typical Weight-densities (lb/ft³)

Gasoline	42
Mercury	849
Milk	64.5
Molasses	100
Seawater	64
Water	62.4

Example 3 THE GREAT MOLASSES FLOOD OF 1919

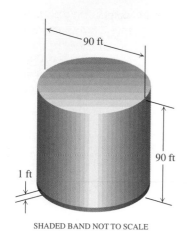

SHADED BAND NOT TO SCALE

Figure 7.43 The molasses tank of Example 3.

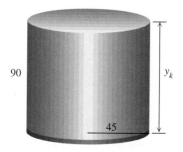

Figure 7.44 The 1-ft band at the bottom of the tank wall can be partitioned into thin strips on which the pressure is approximately constant. (Example 3)

At 1:00 P.M. on January 15, 1919 (an unseasonably warm day), a 90-ft-high, 90-foot-diameter cylindrical metal tank in which the Puritan Distilling Company stored molasses at the corner of Foster and Commercial streets in Boston's North End exploded. Molasses flooded the streets 30 feet deep, trapping pedestrians and horses, knocking down buildings, and oozing into homes. It was eventually tracked all over town and even made its way into the suburbs via trolley cars and people's shoes. It took weeks to clean up.

(a) Given that the tank was full of molasses weighing 100 lb/ft³, what was the total force exerted by the molasses on the bottom of the tank at the time it ruptured?

(b) What was the total force against the bottom foot-wide band of the tank wall (Figure 7.43)?

Solution

(a) At the bottom of the tank, the molasses exerted a constant pressure of

$$p = wh = \left(100 \frac{\text{lb}}{\text{ft}^3}\right)\left(90 \text{ ft}\right) = 9000 \frac{\text{lb}}{\text{ft}^2}.$$

Since the area of the base was $\pi(45)^2$, the total force on the base was

$$\left(9000 \frac{\text{lb}}{\text{ft}^2}\right)\left(2025 \pi \text{ ft}^2\right) \approx 18{,}225{,}000 \text{ lb}.$$

(b) We partition the band from depth 89 ft to depth 90 ft into narrower bands of width Δy and choose a depth y_k in each one. The pressure at this depth y_k is $p = wh = 100 y_k$ lb/ft² (Figure 7.44). The force against each narrow band is approximately

$$\text{pressure} \times \text{area} = (100 y_k)(90\pi \, \Delta y) = 9000\pi y_k \, \Delta y \text{ lb}.$$

Adding the forces against all the bands in the partition and passing to the limit as the norms go to zero, we arrive at

$$F = \int_{89}^{90} 9000\pi y \, dy = 9000\pi \int_{89}^{90} y \, dy \approx 2{,}530{,}553 \text{ lb}$$

for the force against the bottom foot of tank wall.

Normal Probabilities

Suppose you find an old clock in the attic. What is the probability that it has stopped somewhere between 2:00 and 5:00?

If you imagine time being measured continuously over a 12-hour interval, it is easy to conclude that the answer is 1/4 (since the interval from 2:00 to 5:00 contains one-fourth of the time), and that is correct. Mathematically, however, the situation is not quite that clear because both the 12-hour interval and the 3-hour interval contain an *infinite* number of times. In what sense does the ratio of one infinity to another infinity equal 1/4?

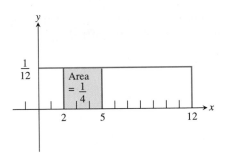

Figure 7.45 The probability that the clock has stopped between 2:00 and 5:00 can be represented as an area of 1/4. The rectangle over the entire interval has area 1.

Improper Integrals

More information about improper integrals like $\int_{-\infty}^{\infty} f(x)\, dx$ can be found in Section 8.3. (You will not need that information here.)

The easiest way to resolve that question is to look at area. We represent the total probability of the 12-hour interval as a rectangle of area 1 sitting above the interval (Figure 7.45).

Not only does it make perfect sense to say that the rectangle over the time interval [2, 5] has an area that is one-fourth the area of the total rectangle, the area actually *equals* 1/4, since the total rectangle has area 1. That is why mathematicians represent probabilities as areas, and that is where definite integrals enter the picture.

Definition Probability Density Function (pdf)

A **probability density function** is a function $f(x)$ with domain all reals such that

$$f(x) \geq 0 \ \text{ for all } x \quad \text{ and } \quad \int_{-\infty}^{\infty} f(x)\, dx = 1.$$

Then the probability associated with an interval $[a, b]$ is

$$\int_{a}^{b} f(x)\, dx.$$

Probabilities of events, such as the clock stopping between 2:00 and 5:00, are integrals of an appropriate pdf.

Example 4 PROBABILITY OF THE CLOCK STOPPING

Find the probability that the clock stopped between 2:00 and 5:00.

Solution The pdf of the clock is

$$f(t) = \begin{cases} 1/12, & 0 \leq t \leq 12 \\ 0, & \text{otherwise.} \end{cases}$$

The probability that the clock stopped at some time t with $2 < t < 5$ is

$$\int_{2}^{5} f(t)\, dt = \frac{1}{4}.$$

By far the most useful kind of pdf is the *normal* kind. ("Normal" here is a technical term, referring to a curve with the shape in Figure 7.46.) The **normal curve,** often called the "bell curve," is one of the most significant curves in applied mathematics because it enables us to describe entire populations based on the statistical measurements taken from a reasonably-sized sample. The measurements needed are the *mean* (μ) and the *standard deviation* (σ), which your calculators will approximate for you from the data. The symbols on the calculator will probably be $\overline{x}$ and s (see your *Owner's Manual*), but go ahead and use them as μ and σ, respectively. Once you have the numbers, you can find the curve by using the following remarkable formula discovered by Karl Friedrich Gauss.

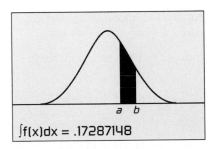

∫f(x)dx = .17287148

Figure 7.46 A normal probability density function. The probability associated with the interval $[a, b]$ is the area under the curve, as shown.

> **Definition** **Normal Probability Density Function (pdf)**
>
> The **normal probability density function (Gaussian curve)** for a population with mean μ and standard deviation σ is
>
> $$f(x) = \frac{1}{\sigma\sqrt{2\pi}} e^{-(x-\mu)^2/(2\sigma^2)}.$$

The mean μ represents the average value of the variable x. The standard deviation σ measures the "scatter" around the mean. For a normal curve, the mean and standard deviation tell you where most of the probability lies. The rule of thumb, illustrated in Figure 7.47, is this:

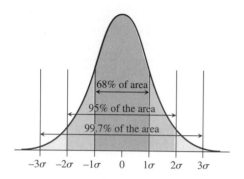

Figure 7.47 The 68-95-99.7 rule for normal distributions.

> **The 68-95-99.7 Rule for Normal Distributions**
>
> Given a normal curve,
>
> • 68% of the area will lie within σ of the mean μ,
> • 95% of the area will lie within 2σ of the mean μ,
> • 99.7% of the area will lie within 3σ of the mean μ.

Even with the 68-95-99.7 rule, the area under the curve can spread quite a bit, depending on the size of σ. Figure 7.48 shows three normal pdfs with mean $\mu = 2$ and standard deviations equal to 0.5, 1, and 2.

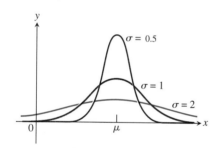

Figure 7.48 Normal pdf curves with mean $\mu = 2$ and $\sigma = 0.5$, 1, and 2.

Example 5   A TELEPHONE HELP LINE

Suppose a telephone help line takes a mean of 2 minutes to answer calls. If the standard deviation is $\sigma = 0.5$, then 68% of the calls are answered in the range of 1.5 to 2.5 minutes and 99.7% of the calls are answered in the range of 0.5 to 3.5 minutes. If $\sigma = 2$, then 68% of the calls are answered in the range of 0 (immediately) to 4 minutes. That is, you are just as likely to have a 4-minute wait as an immediate response.

Example 6   WEIGHTS OF SPINACH BOXES

Suppose that frozen spinach boxes marked as "10 ounces" of spinach have a mean weight of 10.3 ounces and a standard deviation of 0.2 ounce.

(a) What percentage of *all* such spinach boxes can be expected to weigh between 10 and 11 ounces?

(b) What percentage would we expect to weigh less than 10 ounces?

(c) What is the probability that a box weighs *exactly* 10 ounces?

Solution Assuming that some person or machine is *trying* to pack 10 ounces of spinach into these boxes, we expect that most of the weights will be around 10, with probabilities tailing off for boxes being heavier or lighter. We expect, in other words, that a normal pdf will model these probabilities. First, we define $f(x)$ using the formula:

$$f(x) = \frac{1}{0.2\sqrt{2\pi}} e^{-(x-10.3)^2/(0.08)}.$$

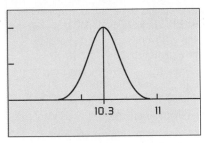

[9, 11.5] by [–1, 2.5]

Figure 7.49 The normal pdf for the spinach weights in Example 6. The mean is at the center.

The graph (Figure 7.49) has the look we are expecting.

(a) For an arbitrary box of this spinach, the probability that it weighs between 10 and 11 ounces is the area under the curve from 10 to 11, which is

$$\text{NINT}\,(f(x),\ x,\ 10,\ 11) \approx 0.933.$$

So without doing any more measuring, we can predict that about 93.3% of all such spinach boxes will weigh between 10 and 11 ounces.

(b) For the probability that a box weighs less than 10 ounces, we use the entire area under the curve to the left of $x = 10$. The curve actually approaches the x-axis as an asymptote, but you can see from the graph (Figure 7.49) that $f(x)$ approaches zero quite quickly. Indeed, $f(9)$ is only slightly larger than a billionth. So getting the area from 9 to 10 should do it:

$$\text{NINT}\,(f(x),\ x,\ 9,\ 10) \approx 0.067.$$

We would expect only about 6.7% of the boxes to weigh less than 10 ounces.

(c) This would be the integral from 10 to 10, which is zero. This zero probability might seem strange at first, but remember that we are assuming a continuous, unbroken interval of possible spinach weights, and 10 is but one of an infinite number of them.

Quick Review 7.5

In Exercises 1–5, find the definite integral by **(a)** antiderivatives and **(b)** using NINT.

1. $\displaystyle\int_{0}^{1} e^{-x}\,dx$

2. $\displaystyle\int_{0}^{1} e^{x}\,dx$

3. $\displaystyle\int_{\pi/4}^{\pi/2} \sin x\,dx$

4. $\displaystyle\int_{0}^{3} (x^2 + 2)\,dx$

5. $\displaystyle\int_{1}^{2} \frac{x^2}{x^3 + 1}\,dx$

In Exercises 6–10 find, but do not evaluate, the definite integral that is the limit as the norms of the partitions go to zero of the Riemann sums on the closed interval $[0, 7]$.

6. $\displaystyle\sum 2\pi(x_k + 2)(\sin x_k)\,\Delta x$

7. $\displaystyle\sum (1 - x_k^2)(2\pi x_k)\,\Delta x$

8. $\displaystyle\sum \pi(\cos x_k)^2\,\Delta x$

9. $\displaystyle\sum \pi\left(\frac{y_k}{2}\right)^2 (10 - y_k)\,\Delta y$

10. $\displaystyle\sum \frac{\sqrt{3}}{4}(\sin^2 x_k)\,\Delta x$

Section 7.5 Exercises

In Exercises 1–4, find the work done by the force of $F(x)$ newtons along the x-axis from $x = a$ meters to $x = b$ meters.

1. $F(x) = xe^{-x/3}, \quad a = 0, \quad b = 5$

2. $F(x) = x \sin(\pi x/4), \quad a = 0, \quad b = 3$

3. $F(x) = x\sqrt{9 - x^2}, \quad a = 0, \quad b = 3$

4. $F(x) = e^{\sin x} \cos x + 2, \quad a = 0, \quad b = 10$

5. *Leaky Bucket* The workers in Example 1 changed to a larger bucket that held 50 L (490 N) of water, but the new bucket had an even larger leak so that it too was empty by the time it reached the top. Assuming the water leaked out at a steady rate, how much work was done lifting the water to a point 20 meters above the ground? (Do not include the rope and bucket.)

6. *Leaky Bucket* The bucket in Exercise 5 is hauled up more quickly so that there is still 10 L (98 N) of water left when the bucket reaches the top. How much work is done lifting the water this time? (Do not include the rope and bucket.)

7. *Leaky Sand Bag* A bag of sand originally weighing 144 lb was lifted at a constant rate. As it rose, sand leaked out at a constant rate. The sand was half gone by the time the bag had been lifted 18 ft. How much work was done lifting the sand this far? (Neglect the weights of the bag and lifting equipment.)

8. *Stretching a Spring* A spring has a natural length of 10 in. An 800-lb force stretches the spring to 14 in.

  (a) Find the force constant.

  (b) How much work is done in stretching the spring from 10 in. to 12 in.?

  (c) How far beyond its natural length will a 1600-lb force stretch the spring?

9. *Subway Car Springs* It takes a force of 21,714 lb to compress a coil spring assembly on a New York City Transit Authority subway car from its free height of 8 in. to its fully compressed height of 5 in.

  (a) What is the assembly's force constant?

  (b) How much work does it take to compress the assembly the first half inch? the second half inch? Answer to the nearest inch-pound.
(*Source:* Data courtesy of Bombardier, Inc., Mass Transit Division, for spring assemblies in subway cars delivered to the New York City Transit Authority from 1985 to 1987.)

10. *Bathroom Scale* A bathroom scale is compressed 1/16 in. when a 150-lb person stands on it. Assuming the scale behaves like a spring that obeys Hooke's Law,

  (a) how much does someone who compresses the scale 1/8 in. weigh?

  (b) how much work is done in compressing the scale 1/8 in.?

11. *Hauling a Rope* A mountain climber is about to haul up a 50-m length of hanging rope. How much work will it take if the rope weighs 0.624 N/m?

Exploration

12. *Compressing Gas* Suppose that gas in a circular cylinder of cross section area A is being compressed by a piston (see figure).

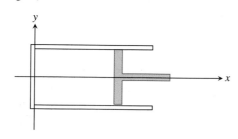

(a) If p is the pressure of the gas in pounds per square inch and V is the volume in cubic inches, show that the work done in compressing the gas from state (p_1, V_1) to state (p_2, V_2) is given by the equation

$$\text{Work} = \int_{(p_1, V_1)}^{(p_2, V_2)} p\, dV \quad \text{in.} \cdot \text{lb},$$

where the force against the piston is pA.

(b) Find the work done in compressing the gas from $V_1 = 243$ in^3 to $V_2 = 32$ in^3 if $p_1 = 50$ lb/in^3 and p and V obey the gas law $pV^{1.4} = $ constant (for adiabatic processes). ∎

In Exercises 13–16, the vertical end of a tank containing water (blue shading) weighing 62.4 lb/ft^3 has the given shape. *Work in groups of two or three.*

  (a) Writing to Learn Explain how to approximate the force against the end of the tank by a Riemann sum.

  (b) Find the force as an integral and evaluate it.

13. semicircle

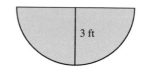

14. semiellipse

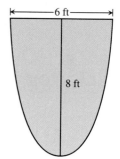

15. triangle

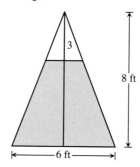

16. parabola

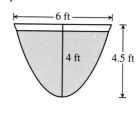

17. *Pumping Water* The rectangular tank shown here, with its top at ground level, is used to catch runoff water. Assume that the water weighs 62.4 lb/ft^3.

(a) How much work does it take to empty the tank by pumping the water back to ground level once the tank is full?

(b) If the water is pumped to ground level with a (5/11)-horsepower motor (work output 250 ft · lb/sec), how long will it take to empty the full tank (to the nearest minute)?

(c) Show that the pump in (b) will lower the water level 10 ft (halfway) during the first 25 min of pumping.

(d) *The Weight of Water* Because of differences in the strength of Earth's gravitational field, the weight of a cubic foot of water at sea level can vary from as little as 62.26 lb at the equator to as much as 62.59 lb near the poles, a variation of about 0.5%. A cubic foot of water that weighs 62.4 lb in Melbourne or New York City will weigh 62.5 lb in Juneau or Stockholm. What are the answers to (a) and (b) in a location where water weighs 62.26 lb/ft^3? 62.5 lb/ft^3?

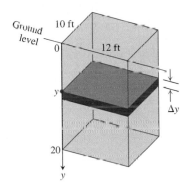

18. *Emptying a Tank* A vertical right cylindrical tank measures 30 ft high and 20 ft in diameter. It is full of kerosene weighing 51.2 lb/ft^3. How much work does it take to pump the kerosene to the level of the top of the tank?

19. *Writing to Learn* The cylindrical tank shown here is to be filled by pumping water from a lake 15 ft below the bottom of the tank. There are two ways to go about this. One is to pump the water through a hose attached to a valve in the bottom of the tank. The other is to attach the hose to the rim of the tank and let the water pour in. Which way will be faster? Give reasons for your answer.

20. *Drinking a Milkshake* The truncated conical container shown here is full of strawberry milkshake that weighs (4/9) oz/in^3. As you can see, the container is 7 in. deep, 2.5 in. across at the base, and 3.5 in. across at the top (a standard size at Brigham's in Boston). The straw sticks up an inch above the top. About how much work does it take to drink the milkshake through the straw (neglecting friction)? Answer in inch-ounces.

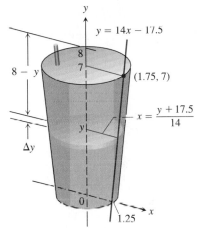

Dimensions in inches

21. *Revisiting Example 2* How much work will it take to pump the oil in Example 2 to a level 3 ft above the cone's rim?

22. *Pumping Milk* Suppose the conical tank in Example 2 contains milk weighing 64.5 lb/ft^3 instead of olive oil. How much work will it take to pump the contents to the rim?

23. **Writing to Learn** You are in charge of the evacuation and repair of the storage tank shown here. The tank is a hemisphere of radius 10 ft and is full of benzene weighing 56 lb/ft^3. A firm you contacted says it can empty the tank for 1/2 cent per foot-pound of work. Find the work required to empty the tank by pumping the benzene to an outlet 2 ft above the tank. If you have budgeted $5000 for the job, can you afford to hire the firm?

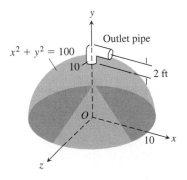

24. *Water Tower* Your town has decided to drill a well to increase its water supply. As the town engineer, you have determined that a water tower will be necessary to provide the pressure needed for distribution, and you have designed the system shown here. The water is to be pumped from a 300-ft well through a vertical 4-in. pipe into the base of a cylindrical tank 20 ft in diameter and 25 ft high. The base of the tank will be 60 ft above ground. The pump is a 3-hp pump, rated at 1650 ft · lb/sec. To the nearest hour, how long will it take to fill the tank the first time? (Include the time it takes to fill the pipe.) Assume water weighs 62.4 lb/ft^3.

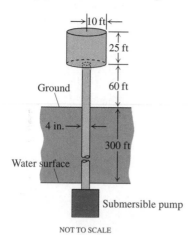

NOT TO SCALE

25. *Fish Tank* A rectangular freshwater fish tank with base 2 × 4 ft and height 2 ft (interior dimensions) is filled to within 2 in. of the top.

(a) Find the fluid force against each end of the tank.

(b) Suppose the tank is sealed and stood on end (without spilling) so that one of the square ends is the base. What does that do to the fluid forces on the rectangular sides?

26. *Milk Carton* A rectangular milk carton measures 3.75 in. by 3.75 in. at the base and is 7.75 in. tall. Find the force of the milk (weighing 64.5 lb/ft^3) on one side when the carton is full.

27. *Heights of Females* The mean height of an adult female in New York City is estimated to be 63.4 inches with a standard deviation of 3.2 inches. What proportion of the adult females in New York City are

(a) less than 63.4 inches tall?

(b) between 63 and 65 inches tall?

(c) taller than 6 feet?

(d) exactly 5 feet tall?

28. *Test Scores* The mean score on a national aptitude test is 498 with a standard deviation of 100 points.

(a) What percentage of the population has scores between 400 and 500?

(b) If we sample 300 test-takers at random, about how many should have scores above 700?

29. *Writing to Learn* Exercises 27 and 28 are subtly different, in that the heights in Exercise 27 are measured *continuously* and the scores in Exercise 28 are measured *discretely*. The discrete probabilities determine rectangles above the individual test scores, so that there actually is a nonzero probability

of scoring, say, 560. The rectangles would look like the figure below, and would have total area 1.

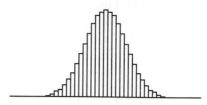

Explain why integration gives a good estimate for the probability, even in the discrete case.

30. **Writing to Learn** Suppose that $f(t)$ is the probability density function for the lifetime of a certain type of lightbulb where t is in hours. What is the meaning of the integral

$$\int_{100}^{800} f(t)\, dt?$$

Extending the Ideas

31. *Putting a Satellite into Orbit* The strength of Earth's gravitational field varies with the distance r from Earth's center, and the magnitude of the gravitational force experienced by a satellite of mass m during and after launch is

$$F(r) = \frac{mMG}{r^2}.$$

Here, $M = 5.975 \times 10^{24}$ kg is Earth's mass, $G = 6.6726 \times 10^{-11}$ N · m^{2}kg^{-2} is the *universal gravitational constant,* and r is measured in meters. The work it takes to lift a 1000-kg satellite from Earth's surface to a circular orbit 35,780 km above Earth's center is therefore given by the integral

$$\text{Work} = \int_{6,370,000}^{35,780,000} \frac{1000MG}{r^2}\, dr \text{ joules.}$$

The lower limit of integration is Earth's radius in meters at the launch site. Evaluate the integral. (This calculation does not take into account energy spent lifting the launch vehicle or energy spent bringing the satellite to orbit velocity.)

32. *Forcing Electrons Together* Two electrons r meters apart repel each other with a force of

$$F = \frac{23 \times 10^{-29}}{r^2} \text{ newton.}$$

(a) Suppose one electron is held fixed at the point $(1, 0)$ on the x-axis (units in meters). How much work does it take to move a second electron along the x-axis from the point $(-1, 0)$ to the origin?

(b) Suppose an electron is held fixed at each of the points $(-1, 0)$ and $(1, 0)$. How much work does it take to move a third electron along the x-axis from $(5, 0)$ to $(3, 0)$?

33. *Kinetic Energy* If a variable force of magnitude $F(x)$ moves a body of mass m along the x-axis from x_1 to x_2, the body's velocity v can be written as dx/dt (where t represents time). Use Newton's second law of motion, $F = m(dv/dt)$, and the Chain Rule

$$\frac{dv}{dt} = \frac{dv}{dx}\frac{dx}{dt} = v\frac{dv}{dx}$$

to show that the net work done by the force in moving the body from x_1 to x_2 is

$$W = \int_{x_1}^{x_2} F(x)\,dx = \frac{1}{2}mv_2^2 - \frac{1}{2}mv_1^2, \qquad (1)$$

where v_1 and v_2 are the body's velocities at x_1 and x_2. In physics the expression $(1/2)mv^2$ is the *kinetic energy* of the body moving with velocity v. Therefore, *the work done by the force equals the change in the body's kinetic energy,* and we can find the work by calculating this change.

Weight vs. Mass

Weight is the force that results from gravity pulling on a mass. The two are related by the equation in Newton's second law,

$$\text{weight} = \text{mass} \times \text{acceleration}.$$

Thus,

$$\text{newtons} = \text{kilograms} \times \text{m}/\text{sec}^2,$$

$$\text{pounds} = \text{slugs} \times \text{ft}/\text{sec}^2.$$

To convert mass to weight, multiply by the acceleration of gravity. To convert weight to mass, divide by the acceleration of gravity.

In Exercises 34–40, use Equation 1 from Exercise 33.

34. *Tennis* A 2-oz tennis ball was served at 160 ft/sec (about 109 mph). How much work was done on the ball to make it go this fast?

35. *Baseball* How many foot-pounds of work does it take to throw a baseball 90 mph? A baseball weighs 5 oz = 0.3125 lb.

36. *Golf* A 1.6-oz golf ball is driven off the tee at a speed of 280 ft/sec (about 191 mph). How many foot-pounds of work are done getting the ball into the air?

37. *Tennis* During the match in which Pete Sampras won the 1990 U.S. Open men's tennis championship, Sampras hit a serve that was clocked at a phenomenal 124 mph. How much work did Sampras have to do on the 2-oz ball to get it to that speed?

38. *Football* A quarterback threw a 14.5-oz football 88 ft/sec (60 mph). How many foot-pounds of work were done on the ball to get it to that speed?

39. *Softball* How much work has to be performed on a 6.5-oz softball to pitch it at 132 ft/sec (90 mph)?

40. *A Ball Bearing* A 2-oz steel ball bearing is placed on a vertical spring whose force constant is $k = 18$ lb/ft. The spring is compressed 3 in. and released. About how high does the ball bearing go?

Calculus at Work

I am working toward becoming an archeaoastronomer and ethnoastronomer of Africa. I have a Bachelor's degree in Physics, a Master's degree in Astronomy, and a Ph.D. in Astronomy and Astrophysics. From 1988 to 1990 I was a member of the Peace Corps, and I taught mathematics to high school students in the Fiji Islands. Calculus is a required course in high schools there.

For my Ph.D. dissertation, I investigated the possibility of the birthrate of stars being related to the composition of star formation clouds. I collected data on the absorption of electromagnetic emissions emanating from these regions. The intensity of emissions graphed versus wavelength produces a flat curve with downward spikes at the characteristic wavelengths of the elements present. An estimate of the area between a spike and the flat curve results in a concentration in molecules/cm^2 of an element. This area is the difference in the integrals of the flat and spike curves. In particular, I was looking for a large concentration of water-ice, which increases the probability of planets forming in a region.

Currently, I am applying for two research grants. One will allow me to use the NASA infrared telescope on Mauna Kea to search for C_3S_2 in comets. The other will help me study the history of astronomy in Tunisia.

Jarita Holbrook
Los Angeles, CA

Chapter 7 Key Terms

arc length (p. 397)
area between curves (p. 375)
Cavalieri's theorems (p. 384)
center of mass (p. 374)
constant-force formula (p. 369)
cylindrical shells (p. 387)
fluid force (p. 403)
fluid pressure (p. 403)
foot-pound (p. 369)
force constant (p. 369)
Gaussian curve (p. 406)
Hooke's Law (p. 369)
inflation rate (p. 373)

joule (p. 369)
length of a curve (p. 397)
mean (p. 405)
moment (p. 374)
net change (p. 363)
newton (p. 369)
normal curve (p. 405)
normal pdf (p. 405)
probability density function (pdf)
(p. 405)
68-95-99.7 rule (p. 406)
smooth curve (p. 396)

smooth function (p. 396)
solid of revolution (p. 385)
standard deviation (p. 405)
surface area (p. 389)
total distance traveled (p. 366)
universal gravitational constant (p. 410)
volume by cylindrical shells (p. 387)
volume by slicing (p. 384)
volume of a solid (p. 383)
weight-density (p. 403)
work (p. 369)

Chapter 7   Review Exercises

In Exercises 1–5, the application involves the accumulation of small changes over an interval to give the net change over that entire interval. Set up an integral to model the accumulation and evaluate it to answer the question.

1. A toy car slides down a ramp and coasts to a stop after 5 sec. Its velocity from $t = 0$ to $t = 5$ is modeled by $v(t) = t^2 - 0.2t^3$ ft/sec. How far does it travel?

2. The fuel consumption of a diesel motor between weekly maintenance periods is modeled by the function $c(t) = 4 + 0.001t^4$ gal/day, $0 \le t \le 7$. How many gallons does it consume in a week?

3. The number of billboards per mile along a 100-mile stretch of an interstate highway approaching a certain city is modeled by the function $B(x) = 21 - e^{0.03x}$, where x is the distance from the city in miles. About how many billboards are along that stretch of highway?

4. A 2-meter rod has a variable density modeled by the function $\rho(x) = 11 - 4x$ g/m, where x is the distance in meters from the base of the rod. What is the total mass of the rod?

5. The electrical power consumption (measured in kilowatts) at a factory t hours after midnight during a typical day is modeled by $E(t) = 300(2 - \cos(\pi t/12))$. How many kilowatt-hours of electrical energy does the company consume in a typical day?

In Exercises 6–19, find the area of the region enclosed by the lines and curves.

6. $y = x$, $y = 1/x^2$, $x = 2$

7. $y = x + 1$, $y = 3 - x^2$

8. $\sqrt{x} + \sqrt{y} = 1$, $x = 0$, $y = 0$

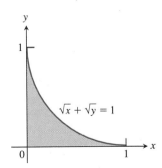

9. $x = 2y^2$, $x = 0$, $y = 3$

10. $4x = y^2 - 4$, $4x = y + 16$

11. $y = \sin x$, $y = x$, $x = \pi/4$

12. $y = 2 \sin x$, $y = \sin 2x$, $0 \le x \le \pi$

13. $y = \cos x$, $y = 4 - x^2$

14. $y = \sec^2 x$, $y = 3 - |x|$

15. *The Necklace* one of the smaller bead-shaped regions enclosed by the graphs of $y = 1 + \cos x$ and $y = 2 - \cos x$

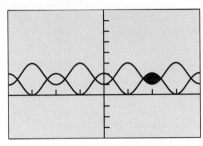

$[-4\pi, 4\pi]$ by $[-4, 8]$

16. one of the larger bead-shaped regions enclosed by the curves in Exercise 15

17. *The Bow Tie* the region enclosed by the graphs of

$$y = x^3 - x \quad \text{and} \quad y = \frac{x}{x^2 + 1}$$

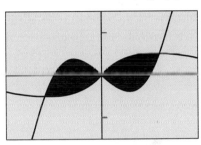

$[-2, 2]$ by $[-1.5, 1.5]$

18. *The Bell* the region enclosed by the graphs of

$$y = 3^{1-x^2} \quad \text{and} \quad y = \frac{x^2 - 3}{10}$$

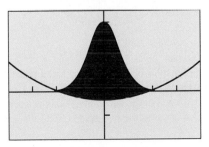

$[-4, 4]$ by $[-2, 3.5]$

19. *The Kissing Fish* the region enclosed between the graphs of $y = x \sin x$ and $y = -x \sin x$ over the interval $[-\pi, \pi]$

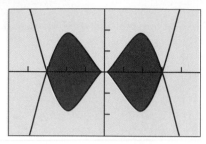

[−5, 5] by [−3, 3]

20. Find the volume of the solid generated by revolving the region bounded by the *x*-axis, the curve $y = 3x^4$, and the lines $x = -1$ and $x = 1$ about the *x*-axis.

21. Find the volume of the solid generated by revolving the region enclosed by the parabola $y^2 = 4x$ and the line $y = x$ about

(a) the *x*-axis.   **(b)** the *y*-axis.

(c) the line $x = 4$.   **(d)** the line $y = 4$.

22. The section of the parabola $y = x^2/2$ from $y = 0$ to $y = 2$ is revolved about the *y*-axis to form a bowl.

(a) Find the volume of the bowl.

(b) Find how much the bowl is holding when it is filled to a depth of *k* units $(0 < k < 2)$.

(c) If the bowl is filled at a rate of 2 cubic units per second, how fast is the depth *k* increasing when $k = 1$?

23. The profile of a football resembles the ellipse shown here (all dimensions in inches). Find the volume of the football to the nearest cubic inch.

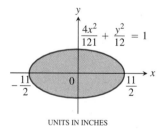

UNITS IN INCHES

24. The base of a solid is the region enclosed between the graphs of $y = \sin x$ and $y = -\sin x$ from $x = 0$ to $x = \pi$. Each cross section perpendicular to the *x*-axis is a semicircle with diameter connecting the two graphs. Find the volume of the solid.

25. The region enclosed by the graphs of $y = e^{x/2}$, $y = 1$, and $x = \ln 3$ is revolved about the *x*-axis. Find the volume of the solid generated.

26. A round hole of radius $\sqrt{3}$ feet is bored through the center of a sphere of radius 2 feet. Find the volume of the piece cut out.

27. Find the length of the arch of the parabola $y = 9 - x^2$ that lies above the *x*-axis.

28. Find the *perimeter* of the bow-tie-shaped region enclosed between the graphs of $y = x^3 - x$ and $y = x - x^3$.

29. A particle travels at 2 units per second along the curve $y = x^3 - 3x^2 + 2$. How long does it take to travel from the local maximum to the local minimum?

30. *Work in groups of two or three.* One of the following statements is true for all $k > 0$ and one is false. Which is which? Explain.

(a) The graphs of $y = k \sin x$ and $y = \sin kx$ have the same length on the interval $[0, 2\pi]$.

(b) The graph of $y = k \sin x$ is *k* times as long as the graph of $y = \sin x$ on the interval $[0, 2\pi]$.

31. Let $F(x) = \int_1^x \sqrt{t^4 - 1} \, dt$. Find the *exact* length of the graph of *F* from $x = 2$ to $x = 5$ without using a calculator.

32. *Rock Climbing* A rock climber is about to haul up 100 N (about 22.5 lb) of equipment that has been hanging beneath her on 40 m of rope weighing 0.8 N/m. How much work will it take to lift

(a) the equipment?   **(b)** the rope?

(c) the rope and equipment together?

33. *Hauling Water* You drove an 800-gallon tank truck from the base of Mt. Washington to the summit and discovered on arrival that the tank was only half full. You had started out with a full tank of water, had climbed at a steady rate, and had taken 50 minutes to accomplish the 4750-ft elevation change. Assuming that the water leaked out at a steady rate, how much work was spent in carrying the water to the summit? Water weighs 8 lb/gal. (Do not count the work done getting you and the truck to the top.)

34. *Stretching a Spring* If a force of 80 N is required to hold a spring 0.3 m beyond its unstressed length, how much work does it take to stretch the spring this far? How much work does it take to stretch the spring an additional meter?

35. **Writing to Learn** It takes a lot more effort to roll a stone up a hill than to roll the stone down the hill, but the weight of the stone and the distance it covers are the same. Does this mean that the same amount of work is done? Explain.

36. *Emptying a Bowl* A hemispherical bowl with radius 8 inches is filled with punch (weighing 0.04 pound per cubic inch) to within 2 inches of the top. How much work is done emptying the bowl if the contents are pumped just high enough to get over the rim?

37. *Fluid Force* The vertical triangular plate shown below is the end plate of a feeding trough full of hog slop, weighing 80 pounds per cubic foot. What is the force against the plate?

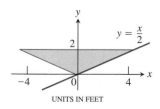

UNITS IN FEET

38. *Fluid Force* A standard olive oil can measures 5.75 in. by 3.5 in. by 10 in. Find the fluid force against the base and each side of the can when it is full. (Olive oil has a weight-density of 57 pounds per cubic foot.)

39. *Volume* A solid lies between planes perpendicular to the *x*-axis at $x = 0$ and at $x = 6$. The cross sections between the planes are squares whose bases run from the *x*-axis up to the curve $\sqrt{x} + \sqrt{y} = \sqrt{6}$. Find the volume of the solid.

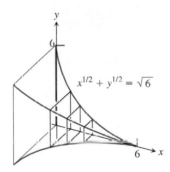

40. *Yellow Perch* A researcher measures the lengths of 3-year-old yellow perch in a fish hatchery and finds that they have a mean length of 17.2 cm with a standard deviation of 3.4 cm. What proportion of 3-year-old yellow perch raised under similar conditions can be expected to reach a length of 20 cm or more?

41. *Work in groups of two or three.* Using as large a sample of classmates as possible, measure the span of each person's fully stretched hand, from the tip of the pinky finger to the tip of the thumb. Based on the mean and standard deviation of your sample, what percentage of students your age would have a finger span of more than 10 inches?

42. *The 68-95-99.7 Rule* **(a)** Verify that for every normal pdf, the proportion of the population lying within one standard deviation of the mean is close to 68%. (*Hint:* Since it is the same for every pdf, you can simplify the function by assuming that $\mu = 0$ and $\sigma = 1$. Then integrate from -1 to 1.)

(b) Verify the two remaining parts of the rule.

43. Writing to Learn Explain why the area under the graph of a probability density function has to equal 1.

In Exercises 44–48, use the cylindrical shell method to find the volume of the solid generated by revolving the region bounded by the curves about the *y*-axis.

44. $y = 2x, \quad y = x/2, \quad x = 1$

45. $y = 1/x, \quad y = 0, \quad x = 1/2, \quad x = 2$

46. $y = \sin x, \quad y = 0, \quad 0 \le x \le \pi$

47. $y = x - 3, \quad y = x^2 - 3x$

48. the bell-shaped region in Exercise 18

49. *Bundt Cake* A bundt cake (see Exploration 1, Section 7.3) has a hole of radius 2 inches and an outer radius of 6 inches at the base. It is 5 inches high, and each cross-sectional slice is parabolic.

(a) Model a typical slice by finding the equation of the parabola with *y*-intercept 5 and *x*-intercepts ± 2.

(b) Revolve the parabolic region about an appropriate line to generate the bundt cake and find its volume.

50. *Finding a Function* Find a function f that has a continuous derivative on $(0, \infty)$ and that has both of the following properties.

   i. The graph of f goes through the point $(1, 1)$.

   ii. The length L of the curve from $(1, 1)$ to any point $(x, f(x))$ is given by the formula $L = \ln x + f(x) - 1$.

In Exercises 51 and 52, find the area of the surface generated by revolving the curve about the indicated axis.

51. $y = \tan x, \quad 0 \le x \le \pi/4; \quad x$-axis

52. $xy = 1, \quad 1 \le y \le 2; \quad y$-axis

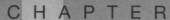

8

L'Hôpital's Rule, Improper Integrals, and Partial Fractions

NASA's Mars Pathfinder collected and transmitted scientific data and photographs back to Earth.

How much work must be done against gravity for the 2000-pound Pathfinder to escape Earth's gravity, that is, to be lifted an infinite distance above the surface of the Earth? Assume that the force due to gravity on an object of weight *w, r* miles from the center of the Earth is:

$$F = 16{,}000{,}000\,w/(r^2)\,(r \geq 4000) \text{ (in pounds)}.$$

The radius of the Earth is approximately 4000 miles. The concepts in Section 8.3 will help you solve this problem.

Chapter 8 Overview

In the late seventeenth century, John Bernoulli discovered a rule for calculating limits of fractions whose numerators and denominators both approach zero. The rule is known today as l'Hôpital's Rule, after Guillaume François Antoine de l'Hôpital (1661–1704), Marquis de St. Mesme, a French nobleman who wrote the first differential calculus text, where the rule first appeared in print. We will also use l'Hôpital's Rule to compare the rates at which functions of x grow as $|x|$ becomes large. In the process, we introduce the little-oh and big-oh notation sometimes used to describe the results of these comparisons.

In Chapter 5 we saw how to evaluate definite integrals of continuous functions and bounded functions with a finite number of discontinuities on finite closed intervals. These ideas are extended to integrals where one or both limits of integration are infinite, and to integrals whose integrands become unbounded on the interval of integration. We also extend the use of partial fractions to integrals whose integrands are rational functions.

8.1 L'Hôpital's Rule

Indeterminate Form 0/0 • Indeterminate Forms ∞/∞, $\infty \cdot 0$, $\infty - \infty$ • Indeterminate Forms 1^{∞}, 0^0, ∞^0

Indeterminate Form 0/0

If functions $f(x)$ and $g(x)$ are both zero at $x = a$, then

$$\lim_{x \to a} \frac{f(x)}{g(x)}$$

cannot be found by substituting $x = a$. The substitution produces 0/0, a meaningless expression known as an **indeterminate form.** Our experience so far has been that limits that lead to indeterminate forms may or may not be hard to find algebraically. It took a lot of analysis in Exercise 63 of Section 2.1 to find $\lim_{x \to 0}$ $(\sin x)/x$. But we have had remarkable success with the limit

$$f'(a) = \lim_{x \to a} \frac{f(x) - f(a)}{x - a},$$

from which we calculate derivatives and which always produces the equivalent of 0/0. L'Hôpital's Rule enables us to draw on our success with derivatives to evaluate limits that otherwise lead to indeterminate forms.

Theorem 1 L'Hôpital's Rule (First Form)

Suppose that $f(a) = g(a) = 0$, that $f'(a)$ and $g'(a)$ exist, and that $g'(a) \neq 0$. Then

$$\lim_{x \to a} \frac{f(x)}{g(x)} = \frac{f'(a)}{g'(a)}.$$

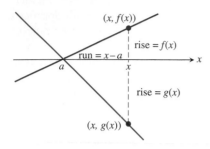

Figure 8.1 A zoom-in view of the graphs of the differentiable functions f and g at $x = a$. (Theorem 1)

Proof
Graphical Argument

If we zoom in on the graphs of f and g at $(a, f(a)) = (a, g(a)) = (a, 0)$, the graphs (Figure 8.1) appear to be straight lines because differentiable functions are locally linear. Let m_1 and m_2 be the slopes of the lines for f and g, respectively. Then for x near a,

$$\frac{f(x)}{g(x)} = \frac{\dfrac{f(x)}{x-a}}{\dfrac{g(x)}{x-a}} = \frac{m_1}{m_2}.$$

As $x \to a$, m_1 and m_2 approach $f'(a)$ and $g'(a)$, respectively. Therefore,

$$\lim_{x\to a}\frac{f(x)}{g(x)} = \lim_{x\to a}\frac{m_1}{m_2} = \frac{f'(a)}{g'(a)}.$$

Confirm Analytically

Working backwards from $f'(a)$ and $g'(a)$, which are themselves limits, we have

$$\frac{f'(a)}{g'(a)} = \frac{\displaystyle\lim_{x\to a}\frac{f(x)-f(a)}{x-a}}{\displaystyle\lim_{x\to a}\frac{g(x)-g(a)}{x-a}} = \lim_{x\to a}\frac{\dfrac{f(x)-f(a)}{x-a}}{\dfrac{g(x)-g(a)}{x-a}}$$

$$= \lim_{x\to a}\frac{f(x)-f(a)}{g(x)-g(a)} = \lim_{x\to a}\frac{f(x)-0}{g(x)-0} = \lim_{x\to a}\frac{f(x)}{g(x)}. \qquad \blacksquare$$

Example 1   USING L'HÔPITAL'S RULE

$$\lim_{x\to 0}\frac{\sqrt{1+x}-1}{x} = \left.\frac{\dfrac{d}{dx}\left(\sqrt{1+x}-1\right)}{\dfrac{d}{dx}(x)}\right|_{x=0} = \left.\frac{1/(2\sqrt{1+x})}{1}\right|_{x=0} = \frac{1}{2}$$

Example 2   WORKING WITH INDETERMINATE FORM 0/0

$$\lim_{x\to 0}\frac{1-\cos x}{x+x^2} \qquad\qquad \frac{0}{0}$$

$$= \left.\frac{\sin x}{(1+2x)}\right|_{x=0} = \frac{0}{1} = 0$$

Sometimes after differentiation the new numerator and denominator both equal zero at $x = a$, as we will see in Example 3. In these cases we apply a stronger form of l'Hôpital's Rule.

Augustin-Louis Cauchy (1789–1857)

An engineer with a genius for mathematics and mathematical modeling, Cauchy created an early modeling of surface wave propagation that is now a classic in hydrodynamics. Cauchy (pronounced "CO-she") invented our notion of continuity and proved the Intermediate Value Theorem for continuous functions. He invented modern limit notation and was the first to prove the convergence of $(1 + 1/n)^n$. His mean value theorem, the subject of Exercise 55, is the key to proving the stronger form of l'Hôpital's Rule. His work advanced not only calculus and mathematical analysis, but also the fields of complex function theory, error theory, differential equations, and celestial mechanics.

Theorem 2   L'Hôpital's Rule (Stronger Form)

Suppose that $f(a) = g(a) = 0$, that f and g are differentiable on an open interval I containing a, and that $g'(x) \neq 0$ on I if $x \neq a$. Then

$$\lim_{x\to a}\frac{f(x)}{g(x)} = \lim_{x\to a}\frac{f'(x)}{g'(x)}.$$

When you apply l'Hôpital's Rule, look for a change from 0/0 into something else. This is where the limit is revealed.

Example 3 APPLYING STRONGER FORM OF L'HÔPITAL'S RULE

$$\lim_{x \to 0} \frac{\sqrt{1+x} - 1 - x/2}{x^2} \qquad \frac{0}{0}$$

$$= \lim_{x \to 0} \frac{(1/2)(1+x)^{-1/2} - 1/2}{2x} \qquad \text{Still } \frac{0}{0}; \text{ differentiate again.}$$

$$= \lim_{x \to 0} \frac{-(1/4)(1+x)^{-3/2}}{2} = -\frac{1}{8} \qquad \text{Not } \frac{0}{0}; \text{ limit is found.}$$

L'Hôpital's Rule applies to one-sided limits as well.

Exploration 1 **Exploring L'Hôpital's Rule Graphically**

Consider the function $f(x) = \dfrac{\sin x}{x}$.

1. Use l'Hôpital's Rule to find $\lim_{x \to 0} f(x)$.

2. Let $y_1 = \sin x$, $y_2 = x$, $y_3 = y_1/y_2$, $y_4 = y_1'/y_2'$. Explain how graphing y_3 and y_4 in the same viewing window provides support for l'Hôpital's Rule in part 1.

3. Let $y_5 = y_3'$. Graph y_3, y_4, and y_5 in the same viewing window. Based on what you see in the viewing window, make a statement about what l'Hôpital's Rule does *not* say.

Example 4 USING L'HÔPITAL'S RULE WITH ONE-SIDED LIMITS

(a) $\lim\limits_{x \to 0^+} \dfrac{\sin x}{x^2}$ $\qquad \dfrac{0}{0}$

$\qquad = \lim\limits_{x \to 0^+} \dfrac{\cos x}{2x} = \infty \qquad \dfrac{1}{0}$

(b) $\lim\limits_{x \to 0^-} \dfrac{\sin x}{x^2}$ $\qquad \dfrac{0}{0}$

$\qquad = \lim\limits_{x \to 0^-} \dfrac{\cos x}{2x} = -\infty \qquad -\dfrac{1}{0}$

These results are supported by the graph in Figure 8.2.

When we reach a point where one of the derivatives approaches 0, as in Example 4, and the other does not, then the limit in question is 0 (if the numerator approaches 0) or infinity (if the denominator approaches 0).

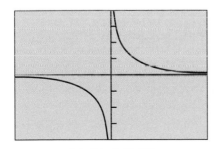

[−1, 1] by [−20, 20]

Figure 8.2 The graph of $f(x) = (\sin x)/x^2$. (Example 4)

Indeterminate Forms ∞/∞, $\infty \cdot 0$, $\infty - \infty$

A version of l'Hôpital's Rule also applies to quotients that lead to the indeterminate form ∞/∞. If $f(x)$ and $g(x)$ both approach infinity as $x \to a$, then

$$\lim_{x \to a} \frac{f(x)}{g(x)} = \lim_{x \to a} \frac{f'(x)}{g'(x)},$$

provided the latter limit exists. The a here (and in the indeterminate form 0/0) may itself be finite or infinite, and may be an endpoint of the interval I of Theorem 2.

Example 5 WORKING WITH INDETERMINATE FORM ∞/∞

Find   **(a)** $\lim\limits_{x \to \pi/2} \dfrac{\sec x}{1 + \tan x}$ **(b)** $\lim\limits_{x \to \infty} \dfrac{\ln x}{2\sqrt{x}}$.

Solution

(a) Solve Analytically

The numerator and denominator are discontinuous at $x = \pi/2$, so we investigate the one-sided limits there. To apply l'Hôpital's Rule we can choose I to be any open interval with $x = \pi/2$ as an endpoint.

$$\lim_{x \to (\pi/2)^-} \frac{\sec x}{1 + \tan x} \qquad \frac{\infty}{\infty} \text{ from the left}$$

$$= \lim_{x \to (\pi/2)^-} \frac{\sec x \tan x}{\sec^2 x} = \lim_{x \to (\pi/2)^-} \sin x = 1$$

The right-hand limit is 1 also, with $(-\infty)/(-\infty)$ as the indeterminate form. Therefore, the two-sided limit is equal to 1.

Support Graphically

The graph of $(\sec x)/(1 + \tan x)$ in Figure 8.3 appears to pass right through the point $(\pi/2, 1)$.

(b) $\lim\limits_{x \to \infty} \dfrac{\ln x}{2\sqrt{x}} = \lim\limits_{x \to \infty} \dfrac{1/x}{1/\sqrt{x}} = \lim\limits_{x \to \infty} \dfrac{1}{\sqrt{x}} = 0$

We can sometimes handle the indeterminate forms $\infty \cdot 0$ and $\infty - \infty$ by using algebra to get 0/0 or ∞/∞ instead. Here again we do not mean to suggest that there is a number $\infty \cdot 0$ or $\infty - \infty$ any more than we mean to suggest that there is a number 0/0 or ∞/∞. These forms are not numbers but descriptions of function behavior.

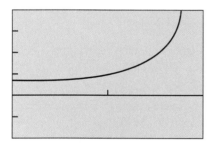

$[\pi/4, 3\pi/4]$ by $[-2, 4]$

Figure 8.3 The graph of $y = (\sec x)/(1 + \tan x)$. (Example 5)

Example 6 WORKING WITH INDETERMINATE FORM $\infty \cdot 0$

Find   **(a)** $\lim\limits_{x \to \infty} \left(x \sin \dfrac{1}{x}\right)$ **(b)** $\lim\limits_{x \to -\infty} \left(x \sin \dfrac{1}{x}\right)$.

Solution Figure 8.4 suggests that the limits exist.

(a) $\lim\limits_{x \to \infty} \left(x \sin \dfrac{1}{x}\right)$ $\infty \cdot 0$

$= \lim\limits_{h \to 0^+} \left(\dfrac{1}{h} \sin h\right)$ Let $h = 1/x$.

$= 1$

$[-5, 5]$ by $[-1, 2]$

Figure 8.4 The graph of $y = x \sin(1/x)$. (Example 6)

(b) Similarly,

$$\lim_{x \to -\infty} \left(x \sin \frac{1}{x} \right) = 1.$$

Example 7   WORKING WITH INDETERMINATE FORM $\infty - \infty$

Find $\lim_{x \to 1} \left(\dfrac{1}{\ln x} - \dfrac{1}{x - 1} \right)$.

Solution   Combining the two fractions converts the indeterminate form $\infty - \infty$ to $0/0$, to which we can apply l'Hôpital's Rule.

$$\lim_{x \to 1} \left(\frac{1}{\ln x} - \frac{1}{x - 1} \right) \qquad \infty - \infty$$

$$= \lim_{x \to 1} \frac{x - 1 - \ln x}{(x - 1) \ln x} \qquad \text{Now } \frac{0}{0}$$

$$= \lim_{x \to 1} \frac{1 - 1/x}{\dfrac{x - 1}{x} + \ln x}$$

$$= \lim_{x \to 1} \frac{x - 1}{x \ln x + x - 1} \qquad \text{Still } \frac{0}{0}$$

$$= \lim_{x \to 1} \frac{1}{2 + \ln x}$$

$$= \frac{1}{2}$$

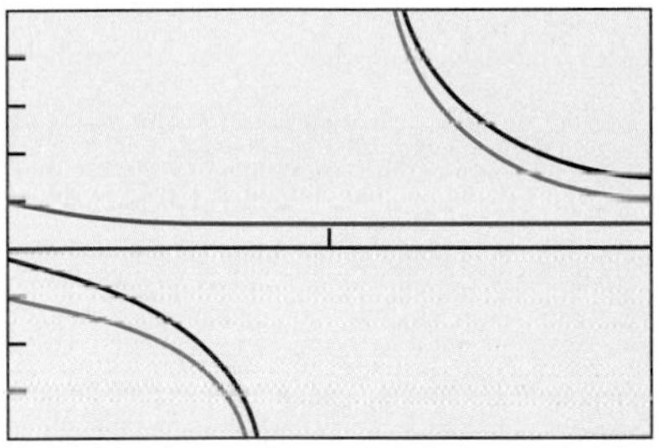

Interpret

The vertical distance between the graphs of $y = 1/(\ln x)$ and $y = 1/(x - 1)$ approaches half a unit as $x \to 1$ from either side (Figure 8.5).

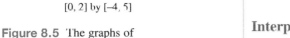

[0, 2] by [−4, 5]

Figure 8.5 The graphs of $y = 1/(\ln x)$, $y = 1/(x - 1)$, and $y = 1/(\ln x) - 1/(x - 1)$. (Example 7)

Indeterminate Forms 1^∞, 0^0, ∞^0

Limits that lead to the indeterminate forms 1^∞, 0^0, and ∞^0 can sometimes be handled by taking logarithms first. We use l'Hôpital's Rule to find the limit of the logarithm and then exponentiate to reveal the original function's behavior.

Since $b = e^{\ln b}$ for every positive number b, we can write $f(x)$ as

$$f(x) = e^{\ln f(x)}$$

for any positive function $f(x)$.

$$\lim_{x \to a} \ln f(x) = L \quad \Rightarrow \quad \lim_{x \to a} f(x) = \lim_{x \to a} e^{\ln f(x)} = e^L$$

Here a can be finite or infinite.

In Section 1.3 we used graphs and tables to investigate the values of $f(x) = (1 + 1/x)^x$ as $x \to \infty$. Now we find this limit with l'Hôpital's Rule.

Example 8   WORKING WITH INDETERMINATE FORM 1^∞

Find $\lim_{x \to \infty} \left(1 + \dfrac{1}{x} \right)^x$.

Solution Let $f(x) = (1 + 1/x)^x$. Then taking logarithms of both sides converts the indeterminate form 1^∞ to $0/0$, to which we can apply l'Hôpital's Rule.

$$\ln f(x) = \ln \left(1 + \frac{1}{x}\right)^x = x \ln \left(1 + \frac{1}{x}\right) = \frac{\ln \left(1 + \frac{1}{x}\right)}{\frac{1}{x}}$$

We apply l'Hôpital's Rule to the last expression above.

$$\lim_{x \to \infty} \ln f(x) = \lim_{x \to \infty} \frac{\ln \left(1 + \frac{1}{x}\right)}{\frac{1}{x}} \qquad \frac{0}{0}$$

$$= \lim_{x \to \infty} \frac{\frac{1}{1 + \frac{1}{x}}\left(-\frac{1}{x^2}\right)}{-\frac{1}{x^2}} \qquad \begin{array}{l}\text{Differentiate numerator} \\ \text{and denominator.}\end{array}$$

$$= \lim_{x \to \infty} \frac{1}{1 + \frac{1}{x}} = 1$$

Therefore,

$$\lim_{x \to \infty} \left(1 + \frac{1}{x}\right)^x = \lim_{x \to \infty} f(x) = \lim_{x \to \infty} e^{\ln f(x)} = e^1 = e.$$

Example 9 WORKING WITH INDETERMINATE FORM 0^0

Determine whether $\lim_{x \to 0^+} x^x$ exists and find its value if it does.

Solution

Investigate Graphically

Figure 8.6 suggests that the limit exists and has a value near 1.

Solve Analytically

The limit leads to the indeterminate form 0^0. To convert the problem to one involving $0/0$, we let $f(x) = x^x$ and take the logarithm of both sides.

$$\ln f(x) = x \ln x = \frac{\ln x}{1/x}$$

Applying l'Hôpital's Rule to $(\ln x)/(1/x)$ we obtain

$$\lim_{x \to 0^+} \ln f(x) = \lim_{x \to 0^+} \frac{\ln x}{1/x} \qquad \frac{-\infty}{\infty}$$

$$= \lim_{x \to 0^+} \frac{1/x}{-1/x^2} \qquad \text{Differentiate.}$$

$$= \lim_{x \to 0^+} (-x) = 0.$$

Therefore,

$$\lim_{x \to 0^+} x^x = \lim_{x \to 0^+} f(x) = \lim_{x \to 0^+} e^{\ln f(x)} = e^0 = 1.$$

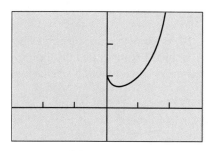

[−3, 3] by [−1, 3]

Figure 8.6 The graph of $y = x^x$. (Example 9)

Example 10 WORKING WITH INDETERMINATE FORM ∞^0

Find $\lim\limits_{x\to\infty} x^{1/x}$.

Solution Let $f(x) = x^{1/x}$. Then

$$\ln f(x) = \frac{\ln x}{x}.$$

Applying l'Hôpital's Rule to $\ln f(x)$ we obtain

$$\lim_{x\to\infty} \ln f(x) = \lim_{x\to\infty} \frac{\ln x}{x} \qquad \frac{\infty}{\infty}$$

$$= \lim_{x\to\infty} \frac{1/x}{1} \qquad \text{Differentiate.}$$

$$= \lim_{x\to\infty} \frac{1}{x} = 0.$$

Therefore,

$$\lim_{x\to\infty} x^{1/x} = \lim_{x\to\infty} f(x) = \lim_{x\to\infty} e^{\ln f(x)} = e^0 = 1.$$

Quick Review 8.1

In Exercises 1 and 2, use tables to estimate the value of the limit.

1. $\lim\limits_{x\to\infty} \left(1 + \dfrac{0.1}{x}\right)^x$

2. $\lim\limits_{x\to 0^+} x^{1/(\ln x)}$

In Exercises 3–8, use graphs or tables to estimate the value of the limit.

3. $\lim\limits_{x\to 0^-} \left(1 - \dfrac{1}{x}\right)^x$

4. $\lim\limits_{r\to -1^-} \left(1 + \dfrac{1}{r}\right)^r$

5. $\lim\limits_{t\to 1} \dfrac{t-1}{\sqrt{t}-1}$

6. $\lim\limits_{x\to\infty} \dfrac{\sqrt{4x^2+1}}{x+1}$

7. $\lim\limits_{x\to 0} \dfrac{\sin 3x}{x}$

8. $\lim\limits_{\theta\to\pi/2} \dfrac{\tan\theta}{2+\tan\theta}$

In Exercises 9 and 10, substitute $x = 1/h$ to express y as a function of h.

9. $y = x\sin\dfrac{1}{x}$

10. $y = \left(1 + \dfrac{1}{x}\right)^x$

Section 8.1 Exercises

In Exercises 1–4, estimate the limit graphically. Use l'Hôpital's Rule to confirm your estimate.

1. $\lim\limits_{x\to 2} \dfrac{x-2}{x^2-4}$

2. $\lim\limits_{x\to 0} \dfrac{\sin 5x}{x}$

3. $\lim\limits_{x\to 0^+} \left(1 + \dfrac{1}{x}\right)^x$

4. $\lim\limits_{x\to\infty} \dfrac{5x^2-3x}{7x^2+1}$

In Exercises 5–8, use l'Hôpital's Rule to evaluate the limit. Support your answer graphically.

5. $\lim\limits_{x\to 1} \dfrac{x^3-1}{4x^3-x-3}$

6. $\lim\limits_{x\to 0} \dfrac{1-\cos x}{x^2}$

7. $\lim\limits_{x\to 0^+} (e^x + x)^{1/x}$

8. $\lim\limits_{x\to\infty} \dfrac{2x^2+3x}{x^3+x+1}$

In Exercises 9 and 10, **(a)** complete the table and estimate the limit. **(b)** Use l'Hôpital's Rule to confirm your estimate.

9. $\lim\limits_{x\to\infty} f(x)$, $f(x) = \dfrac{\ln x^5}{x}$

x	10	10^2	10^3	10^4	10^5
$f(x)$					

10. $\lim\limits_{x\to 0^+} f(x)$, $f(x) = \dfrac{x-\sin x}{x^3}$

x	10^0	10^{-1}	10^{-2}	10^{-3}	10^{-4}
$f(x)$					

In Exercises 11–14, use tables to estimate the limit. Confirm your estimate using l'Hôpital's Rule.

11. $\lim\limits_{\theta \to 0} \dfrac{\sin 3\theta}{\sin 4\theta}$

12. $\lim\limits_{t \to 0} \left(\dfrac{1}{\sin t} - \dfrac{1}{t} \right)$

13. $\lim\limits_{x \to \infty} (1 + x)^{1/x}$

14. $\lim\limits_{x \to \infty} \dfrac{x - 2x^2}{3x^2 + 5x}$

In Exercises 15–42, use l'Hôpital's Rule to evaluate the limit.

15. $\lim\limits_{\theta \to 0} \dfrac{\sin \theta^2}{\theta}$

16. $\lim\limits_{\theta \to \pi/2} \dfrac{1 - \sin \theta}{1 + \cos 2\theta}$

17. $\lim\limits_{t \to 0} \dfrac{\cos t - 1}{e^t - t - 1}$

18. $\lim\limits_{t \to 1} \dfrac{t - 1}{\ln t - \sin \pi t}$

19. $\lim\limits_{x \to \infty} \dfrac{\ln (x + 1)}{\log_2 x}$

20. $\lim\limits_{x \to \infty} \dfrac{\log_2 x}{\log_3 (x + 3)}$

21. $\lim\limits_{y \to 0^+} \dfrac{\ln (y^2 + 2y)}{\ln y}$

22. $\lim\limits_{y \to \pi/2} \left(\dfrac{\pi}{2} - y \right) \tan y$

23. $\lim\limits_{x \to 0^+} x \ln x$

24. $\lim\limits_{x \to \infty} x \tan \dfrac{1}{x}$

25. $\lim\limits_{x \to 0^+} (\csc x - \cot x + \cos x)$

26. $\lim\limits_{x \to \infty} (\ln 2x - \ln (x + 1))$

27. $\lim\limits_{x \to 0^+} (\ln x - \ln \sin x)$

28. $\lim\limits_{x \to 0^+} \left(\dfrac{1}{x} - \dfrac{1}{\sqrt{x}} \right)$

29. $\lim\limits_{x \to 0} (e^x + x)^{1/x}$

30. $\lim\limits_{x \to 0} \left(\dfrac{1}{x^2} \right)^x$

31. $\lim\limits_{x \to \pm\infty} \dfrac{3x - 5}{2x^2 - x + 2}$

32. $\lim\limits_{x \to 0} \dfrac{\sin 7x}{\tan 11x}$

33. $\lim\limits_{x \to \infty} (\ln x)^{1/x}$

34. $\lim\limits_{x \to \infty} (1 + 2x)^{1/(2 \ln x)}$

35. $\lim\limits_{x \to 1} (x^2 - 2x + 1)^{x-1}$

36. $\lim\limits_{x \to (\pi/2)^-} (\cos x)^{\cos x}$

37. $\lim\limits_{x \to 0^+} (1 + x)^{1/x}$

38. $\lim\limits_{x \to 1} x^{1/(x-1)}$

39. $\lim\limits_{x \to 0^+} (\sin x)^x$

40. $\lim\limits_{x \to 0^+} (\sin x)^{\tan x}$

41. $\lim\limits_{x \to 1^+} x^{1/(1-x)}$

42. $\lim\limits_{x \to \infty} \displaystyle\int_x^{2x} \dfrac{dt}{t}$

In Exercises 43 and 44, *work in groups of two or three.*

(a) Writing to Learn Explain why l'Hôpital's Rule does not help you to find the limit.

(b) Use a graph to estimate the limit.

(c) Evaluate the limit analytically using the techniques of Chapter 2.

43. $\lim\limits_{x \to \infty} \dfrac{\sqrt{9x + 1}}{\sqrt{x + 1}}$

44. $\lim\limits_{x \to \pi/2} \dfrac{\sec x}{\tan x}$

Explorations

45. Give an example of two differentiable functions f and g with $\lim_{x \to 3} f(x) = \lim_{x \to 3} g(x) = 0$ that satisfy the following.

(a) $\lim\limits_{x \to 3} \dfrac{f(x)}{g(x)} = 7$

(b) $\lim\limits_{x \to 3} \dfrac{f(x)}{g(x)} = 0$

(c) $\lim\limits_{x \to 3} \dfrac{f(x)}{g(x)} = \infty$

46. Give an example of two differentiable functions f and g with $\lim_{x \to \infty} f(x) = \lim_{x \to \infty} g(x) = \infty$ that satisfy the following.

(a) $\lim\limits_{x \to \infty} \dfrac{f(x)}{g(x)} = 3$

(b) $\lim\limits_{x \to \infty} \dfrac{f(x)}{g(x)} = 0$

(c) $\lim\limits_{x \to \infty} \dfrac{f(x)}{g(x)} = \infty$ ∎

47. *Continuous Extension* Find a value of c that makes the function

$$f(x) = \begin{cases} \dfrac{9x - 3 \sin 3x}{5x^3}, & x \neq 0 \\ c, & x = 0 \end{cases}$$

continuous at $x = 0$. Explain why your value of c works.

48. *Continuous Extension* Let $f(x) = |x|^x$, $x \neq 0$. Show that f has a removable discontinuity at $x = 0$ and extend the definition of f to $x = 0$ so that the extended function is continuous there.

49. *Interest Compounded Continuously*

(a) Show that $\lim\limits_{k \to \infty} A_0 \left(1 + \dfrac{r}{k} \right)^{kt} = A_0 e^{rt}$.

(b) Writing to Learn Explain how the limit in (a) connects interest compounded k times per year with interest compounded continuously.

50. *L'Hôpital's Rule* Let

$$f(x) = \begin{cases} x + 2, & x \neq 0 \\ 0, & x = 0 \end{cases} \quad \text{and} \quad g(x) = \begin{cases} x + 1, & x \neq 0 \\ 0, & x = 0. \end{cases}$$

(a) Show that

$$\lim\limits_{x \to 0} \dfrac{f'(x)}{g'(x)} = 1 \quad \text{but} \quad \lim\limits_{x \to 0} \dfrac{f(x)}{g(x)} = 2.$$

(b) Writing to Learn Explain why this does not contradict l'Hôpital's Rule.

51. *Solid of Revolution* Let $A(t)$ be the area of the region in the first quadrant enclosed by the coordinate axes, the curve $y = e^{-x}$, and the line $x = t > 0$ as shown in the figure. Let $V(t)$ be the volume of the solid generated by revolving the region about the x-axis. Find the following limits.

(a) $\lim\limits_{t \to \infty} A(t)$

(b) $\lim\limits_{t \to \infty} \dfrac{V(t)}{A(t)}$

(c) $\lim\limits_{t \to 0^+} \dfrac{V(t)}{A(t)}$

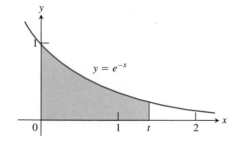

52. *L'Hôpital's Trap* Let $f(x) = \dfrac{1 - \cos x}{x + x^2}$.

(a) Use graphs or tables to estimate $\lim_{x \to 0} f(x)$.

(b) Find the error in the following incorrect application of L'Hôpital's Rule.

$$\lim_{x \to 0} \frac{1 - \cos x}{x + x^2} = \lim_{x \to 0} \frac{\sin x}{1 + 2x}$$

$$= \lim_{x \to 0} \frac{\cos x}{2}$$

$$= \frac{1}{2}$$

53. *Exponential Functions* **(a)** Use the equation

$$a^x = e^{x \ln a}$$

to find the domain of

$$f(x) = \left(1 + \frac{1}{x}\right)^x.$$

(b) Find $\lim_{x \to -1^-} f(x)$.

(c) Find $\lim_{x \to -\infty} f(x)$.

Extending the Ideas

54. *Grapher Precision* Let $f(x) = \dfrac{1 - \cos x^6}{x^{12}}$.

(a) Explain why some graphs of f may give false information about $\lim_{x \to 0} f(x)$. (*Hint:* Try the window $[-1, 1]$ by $[-0.5, 1]$.)

(b) Explain why tables may give false information about $\lim_{x \to 0} f(x)$. (*Hint:* Try tables with increments of 0.01.)

(c) Use l'Hôpital's Rule to find $\lim_{x \to 0} f(x)$.

(d) Writing to Learn This is an example of a function for which graphers do not have enough precision to give reliable information. Explain this statement in your own words.

55. *Cauchy's Mean Value Theorem* Suppose that functions f and g are continuous on $[a, b]$ and differentiable throughout (a, b) and suppose also that $g' \neq 0$ throughout (a, b). Then there exists a number c in (a, b) at which

$$\frac{f'(c)}{g'(c)} = \frac{f(b) - f(a)}{g(b) - g(a)}.$$

Find all values of c in (a, b) that satisfy this property for the following given functions and intervals.

(a) $f(x) = x^3 + 1$, $g(x) = x^2 - x$, $[a, b] = [-1, 1]$

(b) $f(x) = \cos x$, $g(x) = \sin x$, $[a, b] = [0, \pi/2]$

56. *Why 0^∞ and $0^{-\infty}$ Are Not Indeterminate Forms* Assume that $f(x)$ is nonnegative in an open interval containing c and $\lim_{x \to c} f(x) = 0$.

(a) If $\lim_{x \to c} g(x) = \infty$, show that $\lim_{x \to c} f(x)^{g(x)} = 0$.

(b) If $\lim_{x \to c} g(x) = -\infty$, show that $\lim_{x \to c} f(x)^{g(x)} = \infty$.

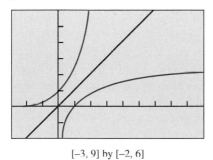

$[-3, 9]$ by $[-2, 6]$

Figure 8.7 The graphs of $y = e^x$, $y = \ln x$, and $y = x$.

8.2 Relative Rates of Growth

Comparing Rates of Growth • Order and Oh-Notation •
Sequential versus Binary Search

Comparing Rates of Growth

We restrict our attention to functions whose values eventually become and remain positive as $x \to \infty$.

The exponential function e^x grows so rapidly and the logarithm function $\ln x$ grows so slowly that they set standards by which we can judge the growth of other functions. The graphs (Figure 8.7) of e^x, $\ln x$, and x suggest how rapidly and slowly e^x and $\ln x$, respectively, grow in comparison to x.

In fact, all the functions a^x, $a > 1$, grow faster (eventually) than any power of x, even $x^{1,000,000}$ (Exercise 35), and hence faster (eventually) than any polynomial function.

To get a feeling for how rapidly the values of e^x grow with increasing x, think of graphing the function on a large blackboard, with the axes scaled in centimeters. At $x = 1$ cm, the graph is $e^1 \approx 3$ cm above the x-axis. At $x = 6$ cm, the graph is $e^6 \approx 403$ cm ≈ 4 m high. (It is about to go through the ceiling if it hasn't done so already.) At $x = 10$ cm, the graph is $e^{10} \approx 22,026$ cm ≈ 220 m high, higher than most buildings. At $x = 24$ cm, the graph is more than halfway to the moon, and at $x = 43$ cm from the origin, the graph is high enough to reach well past the sun's closest stellar neighbor, the red dwarf star Proxima Centauri:

$$e^{43} \approx 4.7 \times 10^{18} \text{ cm}$$

$$= 4.7 \times 10^{13} \text{ km}$$

$$\approx 1.57 \times 10^8 \text{ light-seconds} \qquad \text{Light travels about } 300,000 \text{ km/sec in a vacuum.}$$

$$\approx 5.0 \text{ light-years.}$$

The distance to Proxima Centauri is 4.2 light-years. Yet with $x = 43$ cm from the origin, the graph is still less than 2 feet to the right of the y-axis.

In contrast, the logarithm function $\ln x$ grows more slowly as $x \to \infty$ than any positive power of x, even $x^{1/1,000,000}$ (Exercise 37). Because $\ln x$ and e^x are inverse functions, the calculations above show that with axes scaled in centimeters, you have to go nearly 5 light-years out on the x-axis to find where the graph of $\ln x$ is even 43 cm high.

In fact, all the functions $\log_a x$, $a > 1$, grow slower (eventually) than any positive power of x.

These comparisons of exponential, polynomial, and logarithmic functions can be made precise by defining what it means for a function $f(x)$ to *grow faster* than another function $g(x)$ as $x \to \infty$.

Definitions Faster, Slower, Same-rate Growth as $x \to \infty$

Let $f(x)$ and $g(x)$ be positive for x sufficiently large.

1. f **grows faster** than g (and g **grows slower** than f) as $x \to \infty$ if

$$\lim_{x \to \infty} \frac{f(x)}{g(x)} = \infty, \qquad \text{or, equivalently, if} \qquad \lim_{x \to \infty} \frac{g(x)}{f(x)} = 0.$$

2. f and g **grow at the same rate** as $x \to \infty$ if

$$\lim_{x \to \infty} \frac{f(x)}{g(x)} = L \neq 0. \qquad \text{L finite and not zero}$$

According to these definitions, $y = 2x$ does not grow faster than $y = x$ as $x \to \infty$. The two functions grow at the same rate because

$$\lim_{x \to \infty} \frac{2x}{x} = \lim_{x \to \infty} 2 = 2,$$

which is a finite nonzero limit. The reason for this apparent disregard of common sense is that we want "f grows faster than g" to mean that for large x-values, g is negligible in comparison to f.

If $L = 1$ in part 2 of the definition, then f and g are right end behavior models for each other (Section 2.2). If f grows faster than g, then

$$\lim_{x \to \infty} \frac{f(x) + g(x)}{f(x)} = \lim_{x \to \infty} \left(1 + \frac{g(x)}{f(x)} \right) = 1 + 0 = 1,$$

so f is a right end behavior model for $f + g$. Thus, for large x-values, g can be ignored in the sum $f + g$. This explains why, for large x-values, we can ignore the terms

$$g(x) = a_{n-1}x^{n-1} + \cdots + a_0$$

in

$$f(x) = a_n x^n + a_{n-1}x^{n-1} + \cdots + a_0;$$

that is, why $a_n x^n$ is an end behavior model for

$$a_n x^n + a_{n-1}x^{n-1} + \cdots + a_0.$$

L'Hôpital's Rule can help us to compare rates of growth, as shown in Example 1.

Example 1 COMPARING e^x AND x^2 AS $x \to \infty$

The function e^x grows faster than x^2 as $x \to \infty$ because

$$\underbrace{\lim_{x \to \infty} \frac{e^x}{x^2}}_{\infty/\infty} = \underbrace{\lim_{x \to \infty} \frac{e^x}{2x}}_{\infty/\infty} = \lim_{x \to \infty} \frac{e^x}{2} = \infty. \quad \text{Using l'Hôpital's Rule twice}$$

Exploration 1 **Comparing Rates of Growth as $x \to \infty$**

1. Show that a^x, $a > 1$, grows faster than x^2 as $x \to \infty$.

2. Show that 3^x grows faster than 2^x as $x \to \infty$.

3. If $a > b > 1$, show that a^x grows faster than b^x as $x \to \infty$.

Example 2 COMPARING $\ln x$ WITH x AND x^2 AS $x \to \infty$

Show that $\ln x$ grows slower than **(a)** x and **(b)** x^2 as $x \to \infty$.

Solution

(a) Solve Analytically

$$\lim_{x \to \infty} \frac{\ln x}{x} = \lim_{x \to \infty} \frac{1/x}{1} \quad \text{l'Hôpital's Rule}$$

$$= \lim_{x \to \infty} \frac{1}{x} = 0$$

Support Graphically

Figure 8.8 suggests that the graph of the function $f(x) = (\ln x)/x$ drops dramatically toward the x-axis as x outstrips $\ln x$.

(b) $\displaystyle \lim_{x \to \infty} \frac{\ln x}{x^2} = \lim_{x \to \infty} \left(\frac{\ln x}{x} \cdot \frac{1}{x} \right) = 0 \cdot 0 = 0$

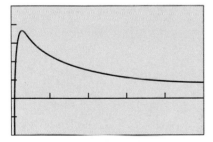

[0, 50] by [–0.2, 0.5]

Figure 8.8 The x-axis is a horizontal asymptote of the function $f(x) = (\ln x)/x$. (Example 2)

Example 3 COMPARING LOGARITHMIC FUNCTIONS AS $x \to \infty$

Let a and b be numbers greater than 1. Show that $\log_a x$ and $\log_b x$ grow at the same rate as $x \to \infty$.

Solution

$$\lim_{x \to \infty} \frac{\log_a x}{\log_b x} = \lim_{x \to \infty} \frac{\ln x / \ln a}{\ln x / \ln b} = \frac{\ln b}{\ln a}$$

The limiting value is finite and nonzero.

Growing at the same rate is a *transitive relation*.

Transitivity of Growing Rates

If f grows at the same rate as g as $x \to \infty$ and g grows at the same rate as h as $x \to \infty$, then f grows at the same rate as h as $x \to \infty$.

The reason is that

$$\lim_{x \to \infty} \frac{f}{g} = L \quad \text{and} \quad \lim_{x \to \infty} \frac{g}{h} = M$$

together imply that

$$\lim_{x \to \infty} \frac{f}{h} = \lim_{x \to \infty} \left(\frac{f}{g} \cdot \frac{g}{h} \right) = LM.$$

If L and M are finite and nonzero, then so is LM.

Example 4 GROWING AT THE SAME RATE AS $x \to \infty$

Show that $f(x) = \sqrt{x^2 + 5}$ and $g(x) = (2\sqrt{x} - 1)^2$ grow at the same rate as $x \to \infty$.

Solution

Solve Analytically

We show that f and g grow at the same rate by showing that they both grow at the same rate as $h(x) = x$.

$$\lim_{x \to \infty} \frac{\sqrt{x^2 + 5}}{x} = \lim_{x \to \infty} \sqrt{1 + \frac{5}{x^2}} = 1$$

$$\lim_{x \to \infty} \frac{(2\sqrt{x} - 1)^2}{x} = \lim_{x \to \infty} \left(\frac{2\sqrt{x} - 1}{\sqrt{x}} \right)^2 = \lim_{x \to \infty} \left(2 - \frac{1}{\sqrt{x}} \right)^2 = 4$$

Thus,

$$\lim_{x \to \infty} \frac{f}{g} = \lim_{x \to \infty} \left(\frac{f}{h} \cdot \frac{h}{g} \right) = 1 \cdot \frac{1}{4} = \frac{1}{4},$$

and f and g grow at the same rate as $x \to \infty$.

Support Graphically

The graph of $y = g/f$ in Figure 8.9 suggests that the quotient g/f is an increasing function with horizontal asymptote $y = 4$. This supports that f and g grow at the same rate.

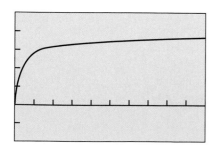

[0, 100] by [−2, 5]

Figure 8.9 The graph of g/f appears to have the line $y = 4$ as a horizontal asymptote. (Example 4)

Order and Oh-Notation

Here we introduce the "little-oh" and "big-oh" notation invented by number theorists a hundred years ago and now commonplace in mathematical analysis and computer science.

Definition *f* of Smaller Order than *g*

Let *f* and *g* be positive for *x* sufficiently large. Then *f* is of **smaller order** than *g* as $x \to \infty$ if

$$\lim_{x \to \infty} \frac{f(x)}{g(x)} = 0.$$

We write $f = o(g)$ and say "*f* is little-oh of *g*."

Saying $f = o(g)$ as $x \to \infty$ is another way to say that *f* grows slower than *g* as $x \to \infty$.

Example 5 SHOWING LITTLE-OH ORDER

$\ln x = o(x)$ as $x \to \infty$ because $\displaystyle \lim_{x \to \infty} \frac{\ln x}{x} = 0.$

$x^2 = o(x^3 + 1)$ as $x \to \infty$ because $\displaystyle \lim_{x \to \infty} \frac{x^2}{x^3 + 1} = 0.$

Definition *f* of at Most the Order of *g*

Let $f(x)$ and $g(x)$ be positive for *x* sufficiently large. Then *f* is of **at most the order** of *g* as $x \to \infty$ if there is a positive integer *M* for which

$$\frac{f(x)}{g(x)} \le M$$

for *x* sufficiently large. We write $f = O(g)$ and say "*f* is big-oh of *g*."

Example 6 SHOWING BIG-OH ORDER

Show that $x + \sin x = O(x)$ as $x \to \infty$.

Solution

Solve Graphically

Figure 8.10 suggests that

$$\frac{x + \sin x}{x} \le 2 \quad \text{for} \quad x > 0.$$

Confirm Algebraically

For $x \ge 1$,

$$\frac{x + \sin x}{x} = 1 + \frac{\sin x}{x} \le 1 + 1 = 2, \quad \text{because} \quad \left| \frac{\sin x}{x} \right| \le \frac{1}{x} \le 1.$$

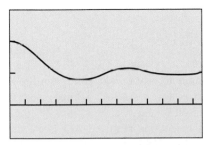

[0, 4π] by [−1, 3]

Figure 8.10 $y = (x + \sin x)/x$ appears to have an absolute maximum of 2 in the interval [0, ∞). (Example 6)

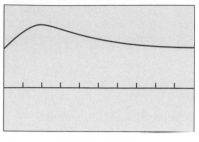

[0, 10] by [−1, 2]

Figure 8.11 The graph of $y = (e^x + x^2)/e^x$ suggests that this function has an absolute maximum less than 2 on the interval $[0, \infty)$. (Example 7)

Note

You would not use a sequential search method to find a word, but you might program a computer to search for a word using this technique.

Figure 8.12 Computer scientists look for the most efficient algorithms when they program searches.

Example 7 SHOWING BIG-OH ORDER

Show that $e^x + x^2 = O(e^x)$ as $x \to \infty$.

Solution Because

$$\lim_{x \to \infty} \frac{e^x + x^2}{e^x} = 1,$$

we must have

$$\frac{e^x + x^2}{e^x} \leq 2$$

for x sufficiently large. This is supported by the graph in Figure 8.11.

Sequential versus Binary Search

Computer scientists sometimes measure the efficiency of an algorithm by counting the number of steps a computer must take to make the algorithm do something (Figure 8.12). (Your graphing calculator works according to algorithms programmed into it.) There can be significant differences in how efficiently algorithms perform, even if they are designed to accomplish the same task. These differences are often described in big-oh notation. Here is an example.

Webster's *Third New International Dictionary* lists about 26,000 words that begin with the letter a. One way to look up a word, or to learn if it is not there, is to read through the list one word at a time until you either find the word or determine that it is not there. This **sequential search** method makes no particular use of the words' alphabetical arrangement. You are sure to get an answer, but it might take about 26,000 steps.

Another way to find the word or to learn that it is not there is to go straight to the middle of the list (give or take a few words). If you do not find the word, then go to the middle of the half that would contain it and forget about the half that would not. (You know which half would contain it because you know the list is ordered alphabetically.) This **binary search** method eliminates roughly 13,000 words in this first step. If you do not find the word on the second try, then jump to the middle of the half that would contain it. Continue this way until you have found the word or divided the list in half so many times that there are no words left. How many times do you have to divide the list to find the word or learn that it is not there? At most 15, because

$$\frac{26,000}{2^{15}} < 1.$$

This certainly beats a possible 26,000 steps.

For a list of length n, a sequential search algorithm takes on the order of n steps to find a word or determine that it is not in the list. A binary search takes on the order of $\log_2 n$ steps. The reason is that if $2^{m-1} < n \leq 2^m$, then $m - 1 < \log_2 n \leq m$, and the number of bisections required to narrow the list to one word will be at most m, the smallest integer greater than or equal to $\log_2 n$.

Big-oh notation provides a compact way to say all this.

> To find an item in a list of length n,
>
> a sequential search takes $O(n)$ steps,
>
> a binary search takes $O(\log_2 n)$ steps.

In our example, there is a big difference between the two (26,000 versus 15), and the difference can only increase with n because n grows faster than $\log_2 n$ as $n \to \infty$.

Quick Review 8.2

In Exercises 1–4, evaluate the limit.

1. $\lim\limits_{x \to \infty} \dfrac{\ln x}{e^x}$

2. $\lim\limits_{x \to \infty} \dfrac{e^x}{x^3}$

3. $\lim\limits_{x \to -\infty} \dfrac{x^2}{e^{2x}}$

4. $\lim\limits_{x \to \infty} \dfrac{x^2}{e^{2x}}$

In Exercises 5 and 6, find an end behavior model (Section 2.2) for the function.

5. $f(x) = -3x^4 + 5x^3 - x + 1$

6. $f(x) = \dfrac{2x^3 - 3x + 1}{x + 2}$

In Exercises 7 and 8, show that g is a right end behavior model for f.

7. $g(x) = x, \quad f(x) = x + \ln x$

8. $g(x) = 2x, \quad f(x) = \sqrt{4x^2 + 5x}$

9. Let $f(x) = \dfrac{e^x + x^2}{e^x}$. Find the

    (a) local extreme values of f and where they occur.

    (b) intervals on which f is increasing.

    (c) intervals on which f is decreasing.

10. Let $f(x) = \dfrac{x + \sin x}{x}$.

    Find the absolute maximum value of f and where it occurs.

Section 8.2 Exercises

In Exercises 1–10, determine whether the function grows faster than e^x, at the same rate as e^x, or slower than e^x as $x \to \infty$.

1. $x^3 - 3x + 1$

2. $\sqrt{1 + x^4}$

3. 4^x

4. $(5/2)^x$

5. e^{x+1}

6. $x \ln x - x$

7. $e^{\cos x}$

8. xe^x

9. x^{1000}

10. $(e^x + e^{-x})/2$

In Exercises 11–16, determine whether the function grows faster than x^2, at the same rate as x^2, or slower than x^2 as $x \to \infty$.

11. $x^2 + 4x$

12. $x^3 + 3$

13. $15x + 3$

14. $\sqrt{x^4 + 5x}$

15. $\ln x$

16. 2^x

In Exercises 17–22, determine whether the function grows faster than $\ln x$, at the same rate as $\ln x$, or slower than $\ln x$ as $x \to \infty$.

17. $\log_2 x^2$

18. $\log \sqrt{x}$

19. $1/\sqrt{x}$

20. e^{-x}

21. $x - \ln x$

22. $5 \ln x$

In Exercises 23 and 24, order the functions from slowest-growing to fastest-growing as $x \to \infty$.

23. $e^x, \quad x^x, \quad (\ln x)^x, \quad e^{x/2}$

24. $2^x, \quad x^2, \quad (\ln 2)^x, \quad e^x$

In Exercises 25–28, show that the three functions grow at the same rate as $x \to \infty$.

25. $f_1(x) = \sqrt{x}, \quad f_2(x) = \sqrt{10x + 1}, \quad f_3(x) = \sqrt{x + 1}$

26. $f_1(x) = x^2, \quad f_2(x) = \sqrt{x^4 + x}, \quad f_3(x) = \sqrt{x^4 - x^3}$

27. $f_1(x) = 3^x, \quad f_2(x) = \sqrt{9^x + 2^x}, \quad f_3(x) = \sqrt{9^x - 4^x}$

28. $f_1(x) = x^3, \quad f_2(x) = \dfrac{x^4 + 2x^2 - 1}{x + 1}, \quad f_3(x) = \dfrac{2x^5 - 1}{x^2 + 1}$

In Exercises 29 and 30, determine whether the statements are true or false as $x \to \infty$. Explain.

29. **(a)** $x = o(x)$
    (b) $x = o(x + 5)$
    (c) $x = O(x + 5)$
    (d) $x = O(2x)$
    (e) $e^x = o(e^{2x})$
    (f) $x + \ln x = O(x)$
    (g) $\ln x = o(\ln 2x)$
    (h) $\sqrt{x^2 + 5} = O(x)$

30. **(a)** $\dfrac{1}{x + 3} = O\left(\dfrac{1}{x}\right)$
    (b) $\dfrac{1}{x} + \dfrac{1}{x^2} = O\left(\dfrac{1}{x}\right)$
    (c) $\dfrac{1}{x} - \dfrac{1}{x^2} = o\left(\dfrac{1}{x}\right)$
    (d) $2 + \cos x = O(2)$
    (e) $e^x + x = O(e^x)$
    (f) $x \ln x = o(x^2)$
    (g) $\ln(\ln x) = O(\ln x)$
    (h) $\ln x = o(\ln(x^2 + 1))$

In Exercises 31–34, only one of the following is true.

    i. $f = o(g)$

    ii. $g = o(f)$

    iii. f and g grow at the same rate.

Use the given graph of f/g to determine which one is true.

31.
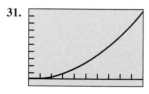
[0, 100] by [–1000, 10000]

32.
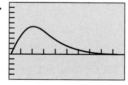
[0, 10] by [–0.5, 1]

33.
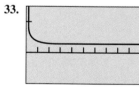
[0, 100] by [–1, 1.5]

34.
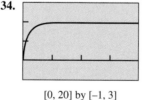
[0, 20] by [–1, 3]

In Exercises 35–37, *work in groups of two or three.*

35. *Comparing Exponential and Power Functions*

    (a) Writing to Learn Explain why e^x grows faster than x^n as $x \to \infty$ for any positive integer n, even $n = 1{,}000{,}000$. (*Hint:* What is the nth derivative of x^n?)

    (b) Writing to Learn Explain why a^x, $a > 1$, grows faster than x^n as $x \to \infty$ for any positive integer n.

36. *Comparing Exponential and Polynomial Functions*

    (a) Writing to Learn Show that e^x grows faster than any polynomial

    $$a_n x^n + a_{n-1}x^{n-1} + \cdots + a_1 x + a_0, \quad a_n > 0,$$

    as $x \to \infty$. Explain.

    (b) Writing to Learn Show that a^x, $a > 1$, grows faster than any polynomial

    $$a_n x^n + a_{n-1}x^{n-1} + \cdots + a_1 x + a_0, \quad a_n > 0,$$

    as $x \to \infty$. Explain.

37. *Comparing Logarithm and Power Functions*

    (a) Writing to Learn Show that $\ln x$ grows slower than $x^{1/n}$ as $x \to \infty$ for any positive integer n, even $n = 1{,}000{,}000$. Explain.

    (b) Writing to Learn Show that for any number $a > 0$, $\ln x$ grows slower than x^a as $x \to \infty$. Explain.

38. *Comparing Logarithm and Polynomial Functions* Show that $\ln x$ grows slower than any nonconstant polynomial

    $$a_n x^n + a_{n-1}x^{n-1} + \cdots + a_1 x + a_0, \quad a_n > 0,$$

    as $x \to \infty$.

39. *Search Algorithms* Suppose you have three different algorithms for solving the same problem and each algorithm provides for a number of steps that is of order of one of the functions listed here.

    $$n \log_2 n, \quad n^{3/2}, \quad n(\log_2 n)^2$$

    Which of the algorithms is likely the most efficient in the long run? Give reasons for your answer.

40. *Sequential and Binary Search* Suppose you are looking for an item in an ordered list one million items long. How many steps might it take to find the item with **(a)** a sequential search? **(b)** a binary search?

41. *Growing at the Same Rate* Show that if functions f and g grow at the same rate as $x \to \infty$, then $f = O(g)$ and $g = O(f)$.

42. *Growing at the Same Rate* Suppose that polynomials $p(x)$ and $q(x)$ grow at the same rate as $x \to \infty$. What can you conclude about

    (a) $\displaystyle\lim_{x \to \infty} \frac{p(x)}{q(x)}$?
    (b) $\displaystyle\lim_{x \to -\infty} \frac{p(x)}{q(x)}$?

Explorations

43. Let

    $$f(x) = a_n x^n + a_{n-1}x^{n-1} + \cdots + a_1 x + a_0$$

    and

    $$g(x) = b_m x^m + b_{m-1}x^{m-1} + \cdots + b_1 x + b_0$$

    be any two polynomial functions with $a_n > 0$, $b_m > 0$.

    (a) Compare the rates of growth of x^5 and x^2 as $x \to \infty$.

    (b) Compare the rates of growth of $5x^3$ and $2x^3$ as $x \to \infty$.

    (c) If x^m grows faster than x^n as $x \to \infty$, what can you conclude about m and n?

    (d) If x^m grows at the same rate as x^n as $x \to \infty$, what can you conclude about m and n?

    (e) If $g(x)$ grows faster than $f(x)$ as $x \to \infty$, what can you conclude about their degrees?

    (f) If $g(x)$ grows at the same rate as $f(x)$ as $x \to \infty$, what can you conclude about their degrees?

44. *Simpson's Rule and Trapezoidal Rule (Section 5.5)*
The definitions in this section can be made more general by lifting the restriction that $x \to \infty$ and considering limits as $x \to a$ for any real number a.

(a) Write definitions for big-oh and little-oh as $x \to a$.

(b) Show that the error E_S in the Simpson's Rule approximation of a definite integral is $O(h^4)$ as $h \to 0$.

(c) Show that the error E_T in the Trapezoidal Rule approximation of a definite integral is $O(h^2)$ as $h \to 0$.

This gives another way to explain the relative accuracies of the two methods. ∎

Extending the Ideas

45. Suppose that the values of the functions $f(x)$ and $g(x)$ eventually become and remain negative as $x \to \infty$. We say that

i. *f* **decreases faster** than *g* as $x \to \infty$ if
$$\lim_{x \to \infty} \frac{f(x)}{g(x)} = \infty.$$

ii. *f* and *g* **decrease at the same rate** as $x \to \infty$ if
$$\lim_{x \to \infty} \frac{f(x)}{g(x)} = L \neq 0.$$

(a) Show that if *f* decreases faster than *g* as $x \to \infty$, then $|f|$ grows faster than $|g|$ as $x \to \infty$.

(b) Show that if *f* and *g* decrease at the same rate as $x \to \infty$, then $|f|$ and $|g|$ grow at the same rate as $x \to \infty$.

46. Suppose that the values of the functions $f(x)$ and $g(x)$ eventually become and remain positive as $x \to -\infty$. We say that

i. *f* **grows faster** than *g* as $x \to -\infty$ if
$$\lim_{x \to -\infty} \frac{f(x)}{g(x)} = \infty.$$

ii. *f* and *g* **grow at the same rate** as $x \to -\infty$ if
$$\lim_{x \to -\infty} \frac{f(x)}{g(x)} = L \neq 0.$$

(a) Show that if *f* grows faster than *g* as $x \to -\infty$, then $f(-x)$ grows faster than $g(-x)$ as $x \to \infty$.

(b) Show that if *f* and *g* grow at the same rate as $x \to -\infty$, then $f(-x)$ and $g(-x)$ grow at the same rate as $x \to \infty$.

8.3 Improper Integrals

Infinite Limits of Integration • The Integral $\int_1^\infty dx/x^p$ • Integrands with Infinite Discontinuities • Tests for Convergence and Divergence • Applications

Infinite Limits of Integration

Consider the infinite region that lies under the curve $y = e^{-x/2}$ in the first quadrant (Figure 8.13a). You might think this region has infinite area, but we will see that it is finite. Here is how we assign a value to the area. First we find the area $A(b)$ of the portion of the region that is bounded on the right by $x = b$ (Figure 8.13b).

$$A(b) = \int_0^b e^{-x/2}\, dx = -2e^{-x/2}\Big]_0^b = -2e^{-b/2} + 2$$

Then we find the limit of $A(b)$ as $b \to \infty$.

$$\lim_{b \to \infty} A(b) = \lim_{b \to \infty} \left(-2e^{-b/2} + 2\right) = 2$$

The area under the curve from 0 to ∞ is

$$\int_0^\infty e^{-x/2}\, dx = \lim_{b \to \infty} \int_0^b e^{-x/2}\, dx = 2.$$

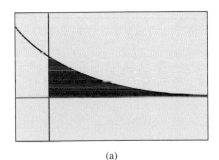

(a)

(b)

Figure 8.13 (a) The area in the first quadrant under the curve $y = e^{-x/2}$ is (b)
$$\lim_{b \to \infty} \int_0^b e^{-x/2}\, dx.$$

Definition **Improper Integrals with Infinite Integration Limits**

Integrals with infinite limits of integration are **improper integrals**.

1. If $f(x)$ is continuous on $[a, \infty)$, then

$$\int_a^\infty f(x)\, dx = \lim_{b \to \infty} \int_a^b f(x)\, dx.$$

2. If $f(x)$ is continuous on $(-\infty, b]$, then

$$\int_{-\infty}^b f(x)\, dx = \lim_{a \to -\infty} \int_a^b f(x)\, dx.$$

3. If $f(x)$ is continuous on $(-\infty, \infty)$, then

$$\int_{-\infty}^\infty f(x)\, dx = \int_{-\infty}^c f(x)\, dx + \int_c^\infty f(x)\, dx,$$

where c is any real number.

In parts 1 and 2, if the limit is finite the improper integral **converges** and the limit is the **value** of the improper integral. If the limit fails to exist, the improper integral **diverges**. In part 3, the integral on the left-hand side of the equation **converges** if both improper integrals on the right-hand side converge, otherwise it **diverges** and has no value. It can be shown that the choice of a in part 3 is unimportant. We can evaluate or determine the convergence or divergence of $\int_{-\infty}^\infty f(x)\, dx$ with any convenient choice.

Example 1 EVALUATING AN IMPROPER INTEGRAL ON $[1, \infty)$

Does the improper integral $\displaystyle\int_1^\infty \frac{dx}{x}$ converge or diverge?

Solution

$$\int_1^\infty \frac{dx}{x} = \lim_{b \to \infty} \int_1^b \frac{dx}{x} \qquad \text{Definition}$$

$$= \lim_{b \to \infty} \ln x \Big]_1^b$$

$$= \lim_{b \to \infty} (\ln b - \ln 1) = \infty$$

Thus, the integral diverges.

Example 2 EVALUATING AN INTEGRAL ON $(-\infty, \infty)$

Evaluate $\displaystyle\int_{-\infty}^\infty \frac{dx}{1 + x^2}$.

Solution According to the definition (part 3) we can write

$$\int_{-\infty}^\infty \frac{dx}{1 + x^2} = \int_{-\infty}^0 \frac{dx}{1 + x^2} + \int_0^\infty \frac{dx}{1 + x^2}.$$

Next, we evaluate each improper integral on the right-hand side of the equation above.

$$\int_{-\infty}^{0} \frac{dx}{1+x^2} = \lim_{a \to -\infty} \int_{a}^{0} \frac{dx}{1+x^2}$$

$$= \lim_{a \to -\infty} \tan^{-1} x \Big]_{a}^{0}$$

$$= \lim_{a \to -\infty} (\tan^{-1} 0 - \tan^{-1} a) = 0 - \left(-\frac{\pi}{2}\right) = \frac{\pi}{2}$$

$$\int_{0}^{\infty} \frac{dx}{1+x^2} = \lim_{b \to \infty} \int_{0}^{b} \frac{dx}{1+x^2}$$

$$= \lim_{b \to \infty} \tan^{-1} x \Big]_{0}^{b}$$

$$= \lim_{b \to \infty} (\tan^{-1} b - \tan^{-1} 0) = \frac{\pi}{2} - 0 = \frac{\pi}{2}$$

Thus,

$$\int_{-\infty}^{\infty} \frac{dx}{1+x^2} = \frac{\pi}{2} + \frac{\pi}{2} = \pi.$$

The Integral $\int_{1}^{\infty} \dfrac{dx}{x^p}$

The function $y = 1/x$ is the boundary between the convergent and divergent improper integrals with integrands of the form $y = 1/x^p$. Example 3 explains.

Example 3   DETERMINING CONVERGENCE

For what values of p does the integral $\int_{1}^{\infty} dx/x^p$ converge? When the integral does converge, what is its value?

Solution Example 1 shows that the integral diverges if $p = 1$. If $p \neq 1$,

$$\int_{1}^{b} \frac{dx}{x^p} = \frac{x^{-p+1}}{-p+1} \bigg]_{1}^{b} = \frac{1}{1-p}(b^{-p+1} - 1) = \frac{1}{1-p}\left(\frac{1}{b^{p-1}} - 1\right).$$

Thus,

$$\int_{1}^{\infty} \frac{dx}{x^p} = \lim_{b \to \infty} \int_{1}^{b} \frac{dx}{x^p}$$

$$= \lim_{b \to \infty} \left[\frac{1}{1-p}\left(\frac{1}{b^{p-1}} - 1\right)\right] = \begin{cases} \dfrac{1}{p-1}, & p > 1 \\ \infty, & p < 1 \end{cases}$$

because

$$\lim_{b \to \infty} \frac{1}{b^{p-1}} = \begin{cases} 0, & p > 1 \\ \infty, & p < 1. \end{cases}$$

Therefore, the integral converges to the value $1/(p-1)$ if $p > 1$. It diverges if $p \leq 1$ (see Example 1).

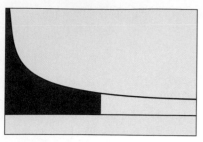

[0, 2] by [−1, 5]

(a)

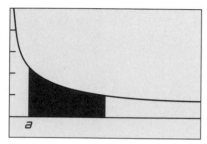

[0, 2] by [−1, 5]

(b)

Figure 8.14 (a) The area under the curve $y = 1/\sqrt{x}$ from $x = 0$ to $x = 1$ is (b)

$$\lim_{a \to 0^+} \int_a^1 (1/\sqrt{x}) \, dx.$$

Integrands with Infinite Discontinuities

Another type of improper integral arises when the integrand has a vertical asymptote — an infinite discontinuity — at a limit of integration or at some point between the limits of integration.

Consider the infinite region in the first quadrant that lies under the curve $y = 1/\sqrt{x}$ from $x = 0$ to $x = 1$ (Figure 8.14a). First we find the area of the portion from a to 1 (Figure 8.14b).

$$\int_a^1 \frac{dx}{\sqrt{x}} = 2\sqrt{x} \, \bigg]_a^1 - 2 - 2\sqrt{a}$$

Then, we find the limit of this area as $a \to 0^+$.

$$\lim_{a \to 0^+} \int_a^1 \frac{dx}{\sqrt{x}} = \lim_{a \to 0^+} (2 - 2\sqrt{a}) = 2$$

The area under the curve from 0 to 1 is

$$\int_0^1 \frac{dx}{\sqrt{x}} = \lim_{a \to 0^+} \int_a^1 \frac{dx}{\sqrt{x}} = 2.$$

Definition Improper Integrals with Infinite Discontinuities

Integrals of functions that become infinite at a point within the interval of integration are **improper integrals.**

1. If $f(x)$ is continuous on $(a, b]$, then

$$\int_a^b f(x) \, dx = \lim_{c \to a^+} \int_c^b f(x) \, dx.$$

2. If $f(x)$ is continuous on $[a, b)$, then

$$\int_a^b f(x) \, dx = \lim_{c \to b^-} \int_a^c f(x) \, dx.$$

3. If $f(x)$ is continuous on $[a, c) \cup (c, b]$, then

$$\int_a^b f(x) \, dx = \int_a^c f(x) \, dx + \int_c^b f(x) \, dx.$$

In parts 1 and 2, if the limit is finite the improper integral **converges** and the limit is the **value** of the improper integral. If the limit fails to exist the improper integral **diverges**. In part 3, the integral on the left-hand side of the equation **converges** if both integrals on the right-hand side have values, otherwise it **diverges**.

Exploration 1 **Investigating** $\displaystyle\int_0^1 \frac{dx}{x^p}$

1. Explain why these integrals are improper if $p > 0$.

2. Show that the integral diverges if $p = 1$.

3. Show that the integral diverges if $p > 1$.

4. Show that the integral converges if $0 < p < 1$.

Example 4 INFINITE DISCONTINUITY AT AN INTERIOR POINT

Evaluate $\displaystyle\int_0^3 \frac{dx}{(x-1)^{2/3}}$.

Solution The integrand has a vertical asymptote at $x = 1$ and is continuous on $[0, 1)$ and $(1, 3]$. Thus, by part 3 of the definition above

$$\int_0^3 \frac{dx}{(x-1)^{2/3}} = \int_0^1 \frac{dx}{(x-1)^{2/3}} + \int_1^3 \frac{dx}{(x-1)^{2/3}}.$$

Next, we evaluate each improper integral on the right-hand side of this equation.

$$\int_0^1 \frac{dx}{(x-1)^{2/3}} = \lim_{c \to 1^-} \int_0^c \frac{dx}{(x-1)^{2/3}}$$

$$= \lim_{c \to 1^-} 3(x-1)^{1/3} \Big]_0^c$$

$$= \lim_{c \to 1^-} [3(c-1)^{1/3} + 3] = 3$$

$$\int_1^3 \frac{dx}{(x-1)^{2/3}} = \lim_{c \to 1^+} \int_c^3 \frac{dx}{(x-1)^{2/3}}$$

$$= \lim_{c \to 1^+} 3(x-1)^{1/3} \Big]_c^3$$

$$= \lim_{c \to 1^+} [3(3-1)^{1/3} - 3(c-1)^{1/3}] = 3\sqrt[3]{2}$$

We conclude that

$$\int_0^3 \frac{dx}{(x-1)^{2/3}} = 3 + 3\sqrt[3]{2}.$$

Example 5 INFINITE DISCONTINUITY AT AN INTERIOR POINT

Evaluate $\displaystyle\int_1^4 \frac{dx}{x-2}$.

Solution The integrand has an infinite discontinuity at $x = 2$ and is continuous on $[1, 2)$ and $(2, 4]$. Thus,

$$\int_1^4 \frac{dx}{x-2} = \int_1^2 \frac{dx}{x-2} + \int_2^4 \frac{dx}{x-2}.$$

Next, we evaluate each improper integral on the right-hand side.

$$\int_1^2 \frac{dx}{x-2} = \lim_{c \to 2^-} \int_1^c \frac{dx}{x-2}$$

$$= \lim_{c \to 2^-} \ln|x-2| \Big]_1^c$$

$$= \lim_{c \to 2^-} (\ln|c-2| - \ln|-1|) = -\infty$$

There is no need to evaluate the second integral on the right-hand side as we can now conclude that the original integral diverges and has no value.

Tests for Convergence and Divergence

When we cannot evaluate an improper integral directly (often the case in practice) we first try to determine whether it converges or diverges. If the integral diverges, that's the end of the story. If it converges, we can then use numerical methods to approximate its value. The principal tests for convergence or divergence are the direct comparison test and the limit comparison test.

Example 6 INVESTIGATING CONVERGENCE

Does the integral $\int_1^\infty e^{-x^2}\,dx$ converge?

Solution

Solve Analytically

By definition,

$$\int_1^\infty e^{-x^2}\,dx = \lim_{b\to\infty} \int_1^b e^{-x^2}\,dx.$$

We cannot evaluate the latter integral directly because there is no simple formula for the antiderivative of e^{-x^2}. We must therefore determine its convergence or divergence some other way. Because $e^{-x^2} > 0$ for all x, $\int_1^b e^{-x^2}\,dx$ is an increasing function of b. Therefore, as $b\to\infty$, the integral either becomes infinite as $b\to\infty$ or it is bounded from above and is forced to converge (have a finite limit).

The two curves $y = e^{-x^2}$ and $y = e^{-x}$ intersect at $(1, e^{-1})$, and $0 < e^{-x^2} \le e^{-x}$ for $x \ge 1$ (Figure 8.15). Thus, for any $b > 1$,

$$0 < \int_1^b e^{-x^2}\,dx \le \int_1^b e^{-x}\,dx = -e^{-b} + e^{-1} < e^{-1} \approx 0.368. \qquad \text{Rounded up to be safe}$$

As an increasing function of b bounded above by 0.368, the integral $\int_1^\infty e^{-x^2}\,dx$ must converge. This does not tell us much about the value of the improper integral, however, except that it is positive and less than 0.368.

Support Graphically

The graph of NINT $(e^{-x^2}, x, 1, x)$ is shown in Figure 8.16. The value of the integral rises rapidly as x first moves away from 1 but changes little past $x = 3$. Values sampled along the curve suggest a limit of about 0.13940 as $x\to\infty$. (Exercise 51 shows how to confirm the accuracy of this estimate.)

The comparison of e^{-x^2} and e^{-x} in Example 6 is a special case of the following test.

Bounded Monotonic Functions

It can be shown that a monotonic function $f(x)$ that is bounded on an infinite interval (a, ∞) must have a finite limit as $x\to\infty$. In Example 6, this is applied to the function

$$f(b) = \int_1^b e^{-x^2}\,dx$$

as $b\to\infty$.

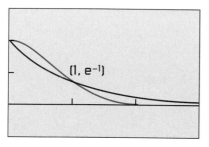

[0, 3] by [–0.5, 1.5]

Figure 8.15 The graph of $y = e^{-x^2}$ lies below the graph of $y = e^{-x}$ for $x > 1$. (Example 6)

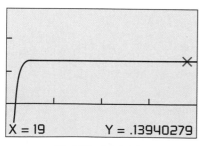

X = 19   Y = .13940279

[0, 20] by [–0.1, 0.3]

Figure 8.16 The graph of NINT $(e^{-x^2}, x, 1, x)$. (Example 6)

Theorem 3 Direct Comparison Test

Let f and g be continuous on $[a, \infty)$ with $0 \le f(x) \le g(x)$ for all $x \ge a$. Then

1. $\displaystyle\int_a^\infty f(x)\,dx$ converges if $\displaystyle\int_a^\infty g(x)\,dx$ converges.

2. $\displaystyle\int_a^\infty g(x)\,dx$ diverges if $\displaystyle\int_a^\infty f(x)\,dx$ diverges.

Example 7   USING THE DIRECT COMPARISON TEST

(a) $\displaystyle\int_1^\infty \frac{\sin^2 x}{x^2}\, dx$ converges because

$$0 \le \frac{\sin^2 x}{x^2} \le \frac{1}{x^2} \quad \text{on} \quad [1, \infty) \quad \text{and} \quad \int_1^\infty \frac{1}{x^2}\, dx \text{ converges.} \qquad \text{Example 3}$$

(b) $\displaystyle\int_1^\infty \frac{1}{\sqrt{x^2 - 0.1}}\, dx$ diverges because

$$\frac{1}{\sqrt{x^2 - 0.1}} \ge \frac{1}{x} \quad \text{on} \quad [1, \infty) \quad \text{and} \quad \int_1^\infty \frac{1}{x}\, dx \text{ diverges.} \qquad \text{Example 1}$$

Theorem 4   Limit Comparison Test

If the positive functions f and g are continuous on $[a, \infty)$ and if

$$\lim_{x \to \infty} \frac{f(x)}{g(x)} = L, \quad 0 < L < \infty,$$

then

$$\int_a^\infty f(x)\, dx \quad \text{and} \quad \int_a^\infty g(x)\, dx \quad \text{both converge or both diverge.}$$

In the language of Section 8.2, Theorem 4 says that if two functions grow at the same rate as $x \to \infty$, then their integrals from a to ∞ behave alike; they both converge or both diverge. This does not mean that their integrals necessarily have the same value, however, as Example 8 shows.

Example 8   USING THE LIMIT COMPARISON TEST

Show that

$$\int_1^\infty \frac{dx}{1 + x^2}$$

converges by comparison with $\int_1^\infty (1/x^2)\, dx$. Find and compare the two integral values.

Solution The functions $f(x) = 1/x^2$ and $g(x) = 1/(1 + x^2)$ are positive and continuous on $[1, \infty)$. Also,

$$\lim_{x \to \infty} \frac{f(x)}{g(x)} = \lim_{x \to \infty} \frac{1/x^2}{1/(1 + x^2)} = \lim_{x \to \infty} \frac{1 + x^2}{x^2}$$

$$= \lim_{x \to \infty} \frac{2x}{2x} = 1. \quad \text{L'Hopital's Rule}$$

Thus, $\int_1^\infty (1/(1 + x^2))\, dx$ converges because $\int_1^\infty (1/x^2)\, dx$ converges (Example 3).

To evaluate the two integrals, we begin with the following from Example 3,

$$\int_1^\infty \frac{dx}{x^2} = \frac{1}{2 - 1} = 1.$$

By definition,

$$\int_1^\infty \frac{dx}{1 + x^2} = \lim_{b \to \infty} \int_1^b \frac{dx}{1 + x^2}$$

$$= \lim_{b \to \infty} \tan^{-1} x \Big]_1^b$$

$$= \lim_{b \to \infty} (\tan^{-1} b - \tan^{-1} 1) = \frac{\pi}{2} - \frac{\pi}{4} = \frac{\pi}{4}.$$

The integrals converge to the different values 1 and $\pi/4$.

Example 9   USING THE LIMIT COMPARISON TEST

Show that $\displaystyle\int_1^\infty \frac{3}{e^x - 5}\, dx$ converges.

Solution From Example 6 it is easy to see that $\int_1^\infty e^{-x}\, dx = \int_1^\infty (1/e^x)\, dx$ converges. Because

$$\lim_{x \to \infty} \frac{1/e^x}{3/(e^x - 5)} = \lim_{x \to \infty} \frac{e^x - 5}{e^x} = \lim_{x \to \infty} \left(1 - \frac{5}{e^x}\right) = 1,$$

the integral $\displaystyle\int_1^\infty \frac{3}{e^x - 5}\, dx$ *also* converges.

Applications

Example 10   FINDING THE VOLUME OF AN INFINITE SOLID

Find the volume of the solid obtained by revolving the curve $y = xe^{-x}$, $0 \le x < \infty$ about the x-axis.

Solution Figure 8.17 shows a portion of the region to be revolved about the x-axis. The area of a typical cross section of the solid is

$$\pi(\text{radius})^2 = \pi y^2 = \pi x^2 e^{-2x}.$$

The volume of the solid is

$$V = \pi \int_0^\infty x^2 e^{-2x}\, dx = \pi \lim_{b \to \infty} \int_0^b x^2 e^{-2x}\, dx.$$

Integrating by parts twice we obtain the following.

$$\int x^2 e^{-2x}\, dx = -\frac{x^2}{2} e^{-2x} + \int x e^{-2x}\, dx \qquad \begin{array}{l} u = x^2,\ dv = e^{-2x}\, dx \\ du = 2x\, dx,\ v = -\dfrac{1}{2} e^{-2x} \end{array}$$

$$= -\frac{x^2}{2} e^{-2x} - \frac{x}{2} e^{-2x} + \frac{1}{2} \int e^{-2x}\, dx \qquad \begin{array}{l} u = x,\ dv = e^{-2x}\, dx \\ du = dx,\ v = -\dfrac{1}{2} e^{-2x} \end{array}$$

$$= -\frac{x^2}{2} e^{-2x} - \frac{x}{2} e^{-2x} - \frac{1}{4} e^{-2x} + C$$

$$= -\frac{2x^2 + 2x + 1}{4e^{2x}} + C$$

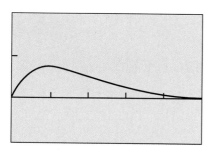

[0, 5] by [–0.5, 1]

Figure 8.17 The graph of $y = xe^{-x}$. (Example 10)

Thus,

$$V = \pi \lim_{b \to \infty} \left[-\frac{2x^2 + 2x + 1}{4e^{2x}} \right]_0^b$$

$$= \pi \lim_{b \to \infty} \left[-\frac{2b^2 + 2b + 1}{4e^{2b}} + \frac{1}{4} \right] = \frac{\pi}{4},$$

and the volume of the solid is $\pi/4$.

Example 11 FINDING CIRCUMFERENCE

Use the arc length formula (Section 7.4) to show that the circumference of the circle $x^2 + y^2 = 4$ is 4π.

Solution One fourth of this circle is given by $y = \sqrt{4 - x^2}, 0 \le x < 2$. Its arc length is

$$L = \int_0^2 \sqrt{1 + (y')^2} \, dx, \quad \text{where} \quad y' = -\frac{x}{\sqrt{4 - x^2}}.$$

The integral is improper because y' is not defined at $x = 2$. We evaluate it as a limit.

$$L = \int_0^2 \sqrt{1 + (y')^2} \, dx = \int_0^2 \sqrt{1 + \frac{x^2}{4 - x^2}} \, dx$$

$$= \int_0^2 \sqrt{\frac{4}{4 - x^2}} \, dx$$

$$= \lim_{b \to 2} \int_0^b \sqrt{\frac{4}{4 - x^2}} \, dx$$

$$= \lim_{b \to 2^-} \int_0^b \sqrt{\frac{1}{1 - (x/2)^2}} \, dx$$

$$= \lim_{b \to 2^-} 2 \sin^{-1} \frac{x}{2} \Big]_0^b$$

$$= \lim_{b \to 2^-} 2 \left[\sin^{-1} \frac{b}{2} - 0 \right] = \pi$$

The circumference of the quarter circle is π; the circumference of the circle is 4π.

Quick Review 8.3

In Exercises 1–4, evaluate the integral.

1. $\int_0^3 \dfrac{dx}{x+3}$

2. $\int_{-1}^1 \dfrac{x\,dx}{x^2+1}$

3. $\int \dfrac{dx}{x^2+4}$

4. $\int \dfrac{dx}{x^4}$

In Exercises 5 and 6, find the domain of the function.

5. $g(x) = \dfrac{1}{\sqrt{9-x^2}}$

6. $h(x) = \dfrac{1}{\sqrt{x-1}}$

In Exercises 7 and 8, confirm the inequality.

7. $\left|\dfrac{\cos x}{x^2}\right| \le \dfrac{1}{x^2}, \quad -\infty < x < \infty$

8. $\dfrac{1}{\sqrt{x^2-1}} \ge \dfrac{1}{x}, \quad x > 1$

In Exercises 9 and 10, show that the functions f and g grow at the same rate as $x \to \infty$.

9. $f(x) = 4e^x - 5, \quad g(x) = 3e^x + 7$

10. $f(x) = \sqrt{2x-1}, \quad g(x) = \sqrt{x+3}$

Section 8.3 Exercises

In Exercises 1–6, **(a)** state why the integral is improper or involves improper integrals. Then, **(b)** determine whether the integral converges or diverges, and **(c)** evaluate the integral if it converges.

1. $\int_0^\infty \dfrac{dx}{x^2+1}$

2. $\int_0^1 \dfrac{dx}{\sqrt{x}}$

3. $\int_{-8}^1 \dfrac{dx}{x^{1/3}}$

4. $\int_{-\infty}^\infty \dfrac{2x\,dx}{(x^2+1)^2}$

5. $\int_0^{\ln 2} x^{-2} e^{1/x}\,dx$

6. $\int_0^{\pi/2} \cot\theta\,d\theta$

In Exercises 7–26, evaluate the integral or state that it diverges.

7. $\int_1^\infty \dfrac{dx}{x^{1.001}}$

8. $\int_{-1}^1 \dfrac{dx}{x^{2/3}}$

9. $\int_0^4 \dfrac{dr}{\sqrt{4-r}}$

10. $\int_0^1 \dfrac{dr}{r^{0.999}}$

11. $\int_0^1 \dfrac{dx}{\sqrt{1-x^2}}$

12. $\int_{-\infty}^2 \dfrac{2\,dx}{x^2+4}$

13. $\int_{-\infty}^{-2} \dfrac{2\,dx}{x^2-1}$

14. $\int_2^\infty \dfrac{3\,dt}{t^2-t}$

15. $\int_0^1 \dfrac{\theta+1}{\sqrt{\theta^2+2\theta}}\,d\theta$

16. $\int_0^2 \dfrac{s+1}{\sqrt{4-s^2}}\,ds$

17. $\int_0^\infty \dfrac{dx}{(1+x)\sqrt{x}}$

18. $\int_1^\infty \dfrac{dx}{x\sqrt{x^2-1}}$

19. $\int_1^2 \dfrac{ds}{s\sqrt{s^2-1}}$

20. $\int_{-1}^\infty \dfrac{d\theta}{\theta^2+5\theta+6}$

21. $\int_0^\infty \dfrac{16\tan^{-1}x}{1+x^2}\,dx$

22. $\int_{-1}^4 \dfrac{dx}{\sqrt{|x|}}$

23. $\int_{-\infty}^0 \theta e^\theta\,d\theta$

24. $\int_0^\infty 2e^{-\theta}\sin\theta\,d\theta$

25. $\int_{-\infty}^\infty e^{-|x|}\,dx$

26. $\int_0^1 x\ln x\,dx$

In Exercises 27–46, use integration, the direct comparison test, or the limit comparison test to determine whether the integral converges or diverges.

27. $\int_0^{\pi/2} \tan\theta\,d\theta$

28. $\int_0^\pi \dfrac{\sin\theta\,d\theta}{\sqrt{\pi-\theta}}$

29. $\int_{-\infty}^\infty 2x\,e^{-x^2}\,dx$

30. $\int_0^4 \dfrac{e^{-\sqrt{x}}}{\sqrt{x}}\,dx$

31. $\int_0^\pi \dfrac{dt}{\sqrt{t+\sin t}}$

32. $\int_4^\infty \dfrac{dx}{\sqrt{x}-1}$

33. $\int_1^\infty \dfrac{dx}{x^3+1}$

34. $\int_0^2 \dfrac{dx}{1-x^2}$

35. $\int_0^2 \dfrac{dx}{1-x}$

36. $\int_{-1}^1 \ln|x|\,dx$

37. $\int_0^\infty \dfrac{d\theta}{1+e^\theta}$

38. $\int_2^\infty \dfrac{dx}{\sqrt{x^2-1}}$

39. $\int_1^\infty \dfrac{\sqrt{x+1}}{x^2}\,dx$

40. $\int_0^\infty \dfrac{dx}{\sqrt{x}}$

41. $\int_\pi^\infty \dfrac{2+\cos x}{x}\,dx$

42. $\int_\pi^\infty \dfrac{1+\sin x}{x^2}\,dx$

43. $\int_{-\infty}^\infty \dfrac{dx}{e^x+e^{-x}}$

44. $\int_{-\infty}^\infty \dfrac{dx}{\sqrt{x^4+1}}$

45. $\int_0^\infty \dfrac{dy}{(1+y^2)(1+\tan^{-1}y)}$

46. $\int_{-\infty}^\infty \dfrac{e^{-y}\,dy}{y^2+1}$

In Exercises 47 and 48, find the area of the region in the first quadrant that lies under the given curve.

47. $y = \dfrac{\ln x}{x^2}$

48. $y = \dfrac{\ln x}{x}$

Explorations

49. *The Infinite Paint Can, or Gabriel's Horn* Consider the region R in the first quadrant bounded above by $y = 1/x$ and on the left by $x = 1$. The region is revolved about the x-axis to form an infinite solid as shown in the figure.

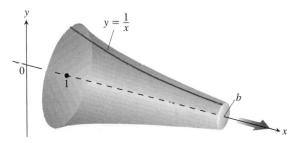

(a) Writing to Learn Explain how Example 1 shows that the region R has infinite area.

(b) Find the surface area of the solid.

(c) Find the volume of the solid.

(d) Why is Gabriel's horn sometimes described as a can that does not hold enough paint to cover its own outside surface?

50. *Normal Probability Distribution Function* In Section 7.5, we encountered the bell-shaped normal distribution curve that is the graph of

$$f(x) = \frac{1}{\sigma\sqrt{2\pi}}\, e^{-\frac{1}{2}\left(\frac{x-\mu}{\sigma}\right)^2},$$

the normal probability density function with mean μ and standard deviation σ. The number μ tells where the distribution is centered, and σ measures the "scatter" around the mean.

From the theory of probability, it is known that

$$\int_{-\infty}^{\infty} f(x)\, dx = 1.$$

In what follows, let $\mu = 0$ and $\sigma = 1$.

(a) Draw the graph of f. Find the intervals on which f is increasing, the intervals on which f is decreasing, and any local extreme values and where they occur.

(b) Evaluate $\displaystyle\int_{-n}^{n} f(x)\, dx$ for $n = 1, 2, 3$.

(c) Give a convincing argument that $\displaystyle\int_{-\infty}^{\infty} f(x)\, dx = 1$.

(*Hint:* Show that $0 < f(x) < e^{-x/2}$ for $x > 1$, and for $b > 1$,

$$\int_{b}^{\infty} e^{-x/2}\, dx \to 0 \quad \text{as} \quad b \to \infty.)$$

51. *Approximating the Value of $\int_{1}^{\infty} e^{-x^2}\, dx$*

(a) Show that $\displaystyle\int_{6}^{\infty} e^{-x^2}\, dx \le \int_{6}^{\infty} e^{-6x}\, dx < 4 \times 10^{-17}.$

(b) Writing to Learn Explain why

$$\int_{1}^{\infty} e^{-x^2}\, dx \approx \int_{1}^{6} e^{-x^2}\, dx$$

with error of at most 4×10^{-17}.

(c) Use the approximation in (b) to estimate the value of $\int_{1}^{\infty} e^{-x^2}\, dx$. Compare this estimate with the value displayed in Figure 8.16.

(d) Writing to Learn Explain why

$$\int_{0}^{\infty} e^{-x^2}\, dx \approx \int_{0}^{3} e^{-x^2}\, dx$$

with error of at most 0.000042. ∎

52. *Work in groups of two or three.*
(a) Show that if f is an even function and the necessary integrals exist, then

$$\int_{-\infty}^{\infty} f(x)\, dx = 2\int_{0}^{\infty} f(x)\, dx.$$

(b) Show that if f is odd and the necessary integrals exist, then

$$\int_{-\infty}^{\infty} f(x)\, dx = 0.$$

53. Writing to Learn

(a) Show that the integral $\displaystyle\int_{0}^{\infty} \frac{2x\, dx}{x^2 + 1}$ diverges.

(b) Explain why we can conclude from (a) that

$$\int_{-\infty}^{\infty} \frac{2x\, dx}{x^2 + 1} \quad \text{diverges.}$$

(c) Show that $\displaystyle\lim_{b \to \infty} \int_{-b}^{b} \frac{2x\, dx}{x^2 + 1} = 0.$

(d) Explain why the result in (c) does not contradict (b).

54. *Finding Perimeter* Find the perimeter of the 4-sided figure $x^{2/3} + y^{2/3} = 1$.

Extending the Ideas

55. Use properties of integrals to give a convincing argument that Theorem 3 is true.

56. Consider the integral

$$f(n + 1) = \int_0^\infty x^n e^{-x} \, dx$$

where $n \geq 0$.

(a) Show that $\int_0^\infty x^n e^{-x} \, dx$ converges for $n = 0, 1, 2$.

(b) Use integration by parts to show that $f(n + 1) = nf(n)$.

(c) Give a convincing argument that $\int_0^\infty x^n e^{-x} \, dx$ converges for all integers $n \geq 0$.

57. Let $f(x) = \displaystyle\int_0^x \frac{\sin t}{t} \, dt$.

(a) Use graphs and tables to investigate the values of $f(x)$ as $x \to \infty$.

(b) Does the integral $\int_0^\infty (\sin x)/x \, dx$ converge? Give a convincing argument.

58. **(a)** Show that we get the same value for the improper integral in Example 2 if we express

$$\int_{-\infty}^\infty \frac{dx}{1 + x^2} = \int_{-\infty}^1 \frac{dx}{1 + x^2} + \int_1^\infty \frac{dx}{1 + x^2},$$

and then evaluate these two integrals.

(b) Show that it doesn't matter what we choose for c in (Improper Integrals with Infinite Integration Limits, part 3)

$$\int_{-\infty}^\infty f(x) \, dx = \int_{-\infty}^c f(x) \, dx + \int_c^\infty f(x) \, dx.$$

8.4 Partial Fractions and Integral Tables

Partial Fractions • General Description of the Method • Integral Tables • Trigonometric Substitutions

Partial Fractions

In Example 2 of Section 6.5, we solved the logistic differential equation

$$\frac{dP}{dt} = 0.001P(100 - P)$$

by rewriting it as

$$\frac{100}{P(100 - P)} \, dP = 0.1 \, dt, \quad \text{Variables separated}$$

expanding the fraction on the left into two basic fractions,

$$\frac{100}{P(100 - P)} = \frac{1}{P} + \frac{1}{100 - P},$$

and integrating both sides to find the solution

$$\ln |P| - \ln |100 - P| = 0.1t + C.$$

This expansion technique is the **method of partial fractions.** Any rational function can be written as a sum of basic fractions, called **partial fractions,** using the method of partial fractions. We can then integrate the rational function by integrating the sum of partial fractions instead.

Example 1 USING PARTIAL FRACTIONS

Use the method of partial fractions to evaluate $\displaystyle\int \frac{5x - 3}{x^2 - 2x - 3} \, dx$.

Solution First we factor the denominator: $x^2 - 2x - 3 = (x + 1)(x - 3)$. Then we determine the values of A and B so that

$$\frac{5x - 3}{x^2 - 2x - 3} = \frac{A}{x + 1} + \frac{B}{x - 3}.$$

Undetermined Coefficients

The A and B in the partial fraction decomposition are referred to as **undetermined coefficients.**

We proceed as follows:

$$5x - 3 = A(x - 3) + B(x + 1) \quad \text{Multiply by } (x + 1)(x - 3).$$

$$= (A + B)x - 3A + B. \quad \text{Combine terms.}$$

Equating coefficients, we obtain the following system of linear equations.

$$A + B = 5$$

$$-3A + B = -3$$

Solving these equations simultaneously yields $A = 2$ and $B = 3$. Therefore,

$$\int \frac{5x - 3}{x^2 - 2x - 3}\, dx = \int \frac{2}{x + 1}\, dx + \int \frac{3}{x - 3}\, dx$$

$$= 2 \ln|x + 1| + 3 \ln|x - 3| + C.$$

General Description of the Method

Success in writing a rational function $f(x)/g(x)$ as a sum of partial fractions depends on two things:

• *The degree of $f(x)$ must be less than the degree of $g(x)$.* That is, the fraction must be *proper*. If it isn't, divide $f(x)$ by $g(x)$ and work with the remainder term. See Example 4 of this section.

• *We must know the factors of $g(x)$.* In theory, any polynomial with real coefficients can be written as a product of real linear factors and real quadratic factors. In practice, the factors may be hard to find.

Here is how we find the partial fractions of a proper fraction $f(x)/g(x)$ when the factors of g are known.

Method of Partial Fractions ($f(x)/g(x)$ Proper)

1. Let $x - r$ be a linear factor of $g(x)$. Suppose $(x - r)^m$ is the highest power of $x - r$ that divides $g(x)$. Then, to this factor, assign the sum of the m partial fractions:

$$\frac{A_1}{x - r} + \frac{A_2}{(x - r)^2} + \cdots + \frac{A_m}{(x - r)^m}.$$

Do this for each distinct linear factor of $g(x)$.

2. Let $x^2 + px + q$ be a quadratic factor of $g(x)$. Suppose $(x^2 + px + q)^n$ is the highest power of this factor that divides $g(x)$. Then, to this factor, assign the sum of the n partial fractions:

$$\frac{B_1x + C_1}{x^2 + px + q} + \frac{B_2x + C_2}{(x^2 + px + q)^2} + \cdots + \frac{B_nx + C_n}{(x^2 + px + q)^n}.$$

Do this for each distinct quadratic factor of $g(x)$ that cannot be factored into linear factors with real coefficients.

3. Set the original fraction $f(x)/g(x)$ equal to the sum of all these partial fractions. Clear the resulting equation of fractions and arrange the terms in decreasing powers of x.

4. Equate the coefficients of corresponding powers of x and solve the resulting equations for the undetermined coefficients.

Example 2 USING A REPEATED LINEAR FACTOR

Express as a sum of partial fractions: $\dfrac{6x + 7}{(x + 2)^2}$.

Solution According to the description above, we must express the fraction as a sum of partial fractions with undetermined coefficients.

$$\frac{6x + 7}{(x + 2)^2} = \frac{A}{x + 2} + \frac{B}{(x + 2)^2}$$

$$6x + 7 = A(x + 2) + B \qquad \text{Multiply both sides by } (x + 2)^2.$$

$$= Ax + (2A + B) \qquad \text{Combine terms.}$$

Equating coefficients of like terms gives

$$A = 6 \quad \text{and} \quad 2A + B = 12 + B = 7, \quad \text{or}$$

$$A = 6 \quad \text{and} \quad B = -5.$$

Therefore,

$$\frac{6x + 7}{(x + 2)^2} = \frac{6}{x + 2} - \frac{5}{(x + 2)^2}.$$

Example 3 USING PARTIAL FRACTIONS

Evaluate $\displaystyle\int \frac{6x + 7}{(x + 2)^2}\, dx$.

Solution

$$\int \frac{6x + 7}{(x + 2)^2}\, dx = \int \left(\frac{6}{x + 2} - \frac{5}{(x + 2)^2} \right) dx \qquad \text{Example 2}$$

$$= 6 \int \frac{dx}{x + 2} - 5 \int (x + 2)^{-2}\, dx$$

$$= 6 \ln |x + 2| + 5(x + 2)^{-1} + C$$

Example 4 INTEGRATING AN IMPROPER FRACTION

Evaluate $\displaystyle\int \frac{2x^3 - 4x^2 - x - 3}{x^2 - 2x - 3}\, dx$.

Solution First we divide the denominator into the numerator to get a polynomial plus a proper fraction.

$$
\begin{array}{r}
2x \\
x^2 - 2x - 3 \overline{) 2x^3 - 4x^2 - x - 3} \\
\underline{2x^3 - 4x^2 - 6x} \\
5x - 3
\end{array}
$$

Then we write the improper fraction as a polynomial plus a proper fraction.

$$\frac{2x^3 - 4x^2 - x - 3}{x^2 - 2x - 3} = 2x + \frac{5x - 3}{x^2 - 2x - 3}$$

Finally, using $\int 2x \, dx = x^2$ and Example 1 we obtain

$$\int \frac{2x^3 - 4x^2 - x - 3}{x^2 - 2x - 3} \, dx = \int 2x \, dx + \int \frac{5x - 3}{x^2 - 2x - 3} \, dx$$

$$= x^2 + 2 \ln |x + 1| + 3 \ln |x - 3| + C.$$

Example 5   SOLVING AN INITIAL VALUE PROBLEM

Find the solution to $dy/dx = 2xy(y^2 + 1)$ that satisfies $y(0) = 1$.

Solution Separating the variables, we rewrite the differential equation as

$$\frac{1}{y(y^2 + 1)} \, dy = 2x \, dx.$$

Integrating both sides gives

$$\int \frac{1}{y(y^2 + 1)} \, dy = \int 2x \, dx = x^2 + C_1.$$

We use partial fractions to rewrite the integrand on the left.

$$\frac{1}{y(y^2 + 1)} = \frac{A}{y} + \frac{By + C}{y^2 + 1}$$

$$1 = A(y^2 + 1) + (By + C)y \quad \text{Multiply by } y(y^2 + 1).$$

$$= (A + B)y^2 + Cy + A$$

Equating coefficients of like terms gives $A + B = 0$, $C = 0$, and $A = 1$. Solving these equations simultaneously, we find $A = 1$, $B = -1$, and $C = 0$.

Accordingly,

$$\int \frac{1}{y(y^2 + 1)} \, dy = \int \frac{1}{y} \, dy - \int \frac{y}{y^2 + 1} \, dy$$

$$= \ln |y| - \frac{1}{2} \ln (y^2 + 1) + C_2.$$

The solution to the differential equation is

$$\ln |y| - \frac{1}{2} \ln (y^2 + 1) = x^2 + C. \quad C = C_1 + C_2$$

Substituting $x = 0$, $y = 1$, we find

$$0 - \frac{1}{2} \ln 2 = C, \quad \text{or} \quad C = -\ln \sqrt{2}.$$

The solution to the initial value problem is

$$\ln |y| - \frac{1}{2} \ln (y^2 + 1) = x^2 - \ln \sqrt{2}.$$

Integral Tables

Every partial fraction that occurs in the decomposition of a rational function can be integrated using analytic methods. This means that we can find the indefinite integral of any rational function analytically. However, rarely do you need to use these techniques. You can usually use a CAS to evaluate the integral, or you can look up the integral in a table like the Brief Table of Integrals in Appendix 7. Example 6 uses this table to evaluate an integral.

Example 6   USING A TABLE OF INTEGRALS

Evaluate $\displaystyle\int \left(\frac{1}{x^2 + 1} + \frac{1}{x^2 - 2x + 5} \right) dx.$

Solution

$$\int \left(\frac{1}{x^2 + 1} + \frac{1}{x^2 - 2x + 5} \right) dx = \int \frac{1}{x^2 + 1}\, dx + \int \frac{1}{x^2 - 2x + 5}\, dx \quad (1)$$

Both integrals on the right-hand side of Equation 1 can be evaluated using Formula 16 from the Brief Table of Integrals.

$$\textbf{16.} \int \frac{dx}{a^2 + x^2} = \frac{1}{a} \tan^{-1} \frac{x}{a} + C$$

You probably remember

$$\int \frac{dx}{1 + x^2} = \tan^{-1} x + C,$$

but if you don't, you can obtain the result by setting $a = 1$ in Formula 16.

To find the value of the second integral, we need to complete the square:

$$x^2 - 2x + 5 = x^2 - 2x + 1 + 4 = (x - 1)^2 + 4.$$

Then,

$$\int \frac{dx}{x^2 - 2x + 5} = \int \frac{dx}{(x - 1)^2 + 4}$$

$$= \int \frac{du}{u^2 + 4} \qquad u = x - 1,\, du = dx$$

$$= \frac{1}{2} \tan^{-1} \frac{u}{2} + C \qquad \text{Formula 16 with } a = 2$$

$$= \frac{1}{2} \tan^{-1} \frac{x - 1}{2} + C.$$

Combining the two integrations gives

$$\int \left(\frac{1}{x^2 + 1} + \frac{1}{x^2 - 2x + 5} \right) dx = \tan^{-1} x + \frac{1}{2} \tan^{-1} \frac{x - 1}{2} + C.$$

The manipulation and substitution in Example 6 are typical of what is needed to use integral tables to evaluate integrals.

Trigonometric Substitutions

Trigonometric substitutions enable us to replace binomials of the form $a^2 + x^2$, $a^2 - x^2$, and $x^2 - a^2$ by single squared terms, and thereby transform a number of integrals into ones we can evaluate directly or find in a table of integrals.

Three of the most common substitutions are $x = a \tan \theta$, $x = a \sin \theta$, and $x = a \sec \theta$. They come from the reference right triangles in Figure 8.18.

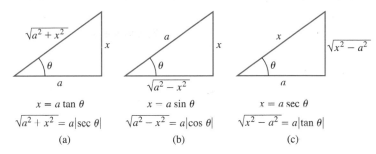

Figure 8.18 Reference triangles for trigonometric substitutions that change binomials into single squared terms.

With $x = a \tan \theta$,

$$a^2 + x^2 = a^2 + a^2 \tan^2 \theta = a^2(1 + \tan^2 \theta) = a^2 \sec^2 \theta.$$

With $x = a \sin \theta$,

$$a^2 - x^2 = a^2 - a^2 \sin^2 \theta = a^2(1 - \sin^2 \theta) = a^2 \cos^2 \theta.$$

With $x = a \sec \theta$,

$$x^2 - a^2 = a^2 \sec^2 \theta - a^2 = a^2(\sec^2 \theta - 1) = a^2 \tan^2 \theta.$$

Trigonometric Substitutions

1. $x = a \tan \theta$ replaces $a^2 + x^2$ with $a^2 \sec^2 \theta$
2. $x = a \sin \theta$ replaces $a^2 - x^2$ with $a^2 \cos^2 \theta$
3. $x = a \sec \theta$ replaces $x^2 - a^2$ with $a^2 \tan^2 \theta$

We want any substitution we use in an integration to be reversible so we can change back to the original variable afterward. For example, if $x = a \tan \theta$, we want to be able to set $\theta = \tan^{-1}(x/a)$ after the integration takes place. If $x = a \sin \theta$, we want to be able to set $\theta = \sin^{-1}(x/a)$ when we're done, and similarly for $x = a \sec \theta$.

For reversibility,

$$x = a \tan \theta \quad \text{requires} \quad \theta = \tan^{-1}\left(\frac{x}{a}\right) \quad \text{with} \quad -\frac{\pi}{2} < \theta < \frac{\pi}{2},$$

$$x = a \sin \theta \quad \text{requires} \quad \theta = \sin^{-1}\left(\frac{x}{a}\right) \quad \text{with} \quad -\frac{\pi}{2} \leq \theta \leq \frac{\pi}{2},$$

$$x = a \sec \theta \quad \text{requires} \quad \theta = \sec^{-1}\left(\frac{x}{a}\right) \quad \text{with} \quad \begin{cases} 0 \leq \theta < \dfrac{\pi}{2}, & \dfrac{x}{a} \geq 1 \\[2mm] \dfrac{\pi}{2} < \theta \leq \pi, & \dfrac{x}{a} \leq -1. \end{cases}$$

Example 7 USING THE SUBSTITUTION $x = a \tan \theta$

Evaluate $\int \sqrt{4 + x^2}\, dx$.

Solution We set

$$x = 2 \tan \theta, \quad dx = 2 \sec^2 \theta\, d\theta, \quad -\frac{\pi}{2} < \theta < \frac{\pi}{2},$$

$$4 + x^2 = 4 + 4 \tan^2 \theta = 4(1 + \tan^2 \theta) = 4 \sec^2 \theta.$$

Then

$$\int \sqrt{4 + x^2}\, dx = \int \sqrt{4 \sec^2 \theta}\, (2 \sec^2 \theta)\, d\theta$$

$$= 4 \int |\sec \theta|\, \sec^2 \theta\, d\theta \quad \sqrt{\sec^2 \theta} = |\sec \theta|$$

$$= 4 \int \sec^3 \theta\, d\theta. \qquad \sec \theta > 0 \text{ for } -\frac{\pi}{2} < \theta < \frac{\pi}{2}$$

Now we are in a position to use Formula 92 from the table of integrals, a *reduction formula*.

92. $\displaystyle \int \sec^n ax\, dx = \frac{\sec^{n-2} ax \tan ax}{a(n-1)} + \frac{n-2}{n-1} \int \sec^{n-2} ax\, dx, \quad n \neq 1$

Applying Formula 92 with $n = 3$, we obtain

$$\int \sec^3 \theta\, d\theta = \frac{\sec \theta \tan \theta}{2} + \frac{1}{2} \int \sec \theta\, d\theta \qquad n = 3, a = 1, x = \theta$$

$$= \frac{\sec \theta \tan \theta}{2} + \frac{1}{2} \ln |\sec \theta + \tan \theta| + C \quad \text{Formula 88, } a = 1$$

$$= \frac{\sec \theta \tan \theta}{2} + \frac{1}{2} \ln (\sec \theta + \tan \theta) + C.$$

The last step is valid because

$$\sec \theta + \tan \theta = \frac{1 + \sin \theta}{\cos \theta} > 0 \quad \text{for} \quad -\frac{\pi}{2} < \theta < \frac{\pi}{2}.$$

Thus,

$$\int \sqrt{4 + x^2}\, dx = 4 \int \sec^3 \theta\, d\theta$$

$$= 2 \sec \theta \tan \theta + 2 \ln (\sec \theta + \tan \theta) + C$$

$$= 2 \cdot \frac{\sqrt{4 + x^2}}{2} \cdot \frac{x}{2} + 2 \ln \left(\frac{\sqrt{4 + x^2}}{2} + \frac{x}{2} \right) + C \quad \begin{array}{l}\text{Figure 8.18a,} \\ a = 2\end{array}$$

$$= \frac{x\sqrt{4 + x^2}}{2} + 2 \ln (\sqrt{4 + x^2} + x) + C.$$

Reduction Formulas

Formula 92 and the formulas in Section 6.3, Exercises 31–34, are called **reduction formulas** because they replace an integral containing some power of a function with an integral of the same form with the power reduced. By applying such a formula repeatedly, we can eventually express the original integral in terms of a power low enough to be evaluated directly.

Example 8 USING THE SUBSTITUTION $x = a \sin \theta$

Evaluate $\displaystyle\int \frac{x^3\, dx}{\sqrt{9 - x^2}}$.

Solution We set

$$x = 3 \sin \theta, \quad dx = 3 \cos \theta\, d\theta, \quad -\frac{\pi}{2} \le \theta \le \frac{\pi}{2}$$

$$9 - x^2 = 9 - 9 \sin^2 \theta = 9(1 - \sin^2 \theta) = 9 \cos^2 \theta.$$

Then

$$\int \frac{x^3\, dx}{\sqrt{9 - x^2}} = \int \frac{27 \sin^3 \theta \cdot 3 \cos \theta\, d\theta}{|3 \cos \theta|}$$

$$= 27 \int \sin^3 \theta\, d\theta \qquad\qquad \cos \theta > 0 \text{ for } -\frac{\pi}{2} < \theta < \frac{\pi}{2}$$

$$= 27 \int (1 - \cos^2 \theta) \sin \theta\, d\theta \qquad\qquad \sin^2 \theta = 1 - \cos^2 \theta$$

$$= -27 \cos \theta + 9 \cos^3 \theta + C$$

$$= -27 \cdot \frac{\sqrt{9 - x^2}}{3} + 9\left(\frac{\sqrt{9 - x^2}}{3}\right)^3 + C \qquad \begin{array}{l}\text{Figure 8.18b,}\\ a = 3\end{array}$$

$$= -9\sqrt{9 - x^2} + \frac{(9 - x^2)^{3/2}}{3} + C.$$

Quick Review 8.4

In Exercises 1 and 2, find the simultaneous solution to the equations.

1. $\begin{aligned} 3A + B &= -5 \\ -2A + 3B &= 7 \end{aligned}$

2. $\begin{aligned} A + 2B - C &= 0 \\ 3A - B + 2C &= 1 \\ A + B + C &= 4 \end{aligned}$

In Exercises 3 and 4, use long division to express the improper fraction (degree of numerator ≥ degree of denominator) as a polynomial plus a proper fraction (degree of numerator < degree of denominator).

3. $\dfrac{2x^3 - 5x^2 - 10x - 7}{x^2 - 3x - 4}$

4. $\dfrac{2x^2 + 11x + 6}{x^2 + 4x + 5}$

In Exercises 5 and 6, factor the polynomials as completely as possible.

5. $x^3 - 3x^2 + x - 3$

6. $y^4 - 5y^2 + 4$

In Exercises 7–10, use the least common denominator to combine the fractions.

7. $\dfrac{2}{x + 3} - \dfrac{3}{x - 2}$

8. $\dfrac{x - 1}{x^2 - 4x + 5} - \dfrac{2}{x + 5}$

9. $\dfrac{t - 1}{t^2 + 2} - \dfrac{3t + 4}{t^2 + 1}$

10. $\dfrac{2}{x - 1} - \dfrac{3}{(x - 1)^2} + \dfrac{1}{(x - 1)^3}$

Section 8.4 Exercises

In Exercises 1–6, express the rational function as a sum of partial fractions.

1. $\dfrac{5x - 7}{x^2 - 3x + 2}$

2. $\dfrac{2x + 2}{x^2 - 2x + 1}$

3. $\dfrac{t + 1}{t^2(t - 1)}$

4. $\dfrac{4}{s^3 - s^2 - 6s}$

5. $\dfrac{x^2 + 8}{x^2 - 5x + 6}$

6. $\dfrac{y^3 + 1}{y^2 + 4}$

In Exercises 7–26, use partial fractions to evaluate the integral.

7. $\displaystyle\int \frac{dx}{1 - x^2}$

8. $\displaystyle\int \frac{dx}{x^2 + 2x}$

9. $\displaystyle\int \frac{y\,dy}{y^2 - 2y - 3}$

10. $\displaystyle\int \frac{y + 4}{y^2 + y}\,dy$

11. $\displaystyle\int \frac{dt}{t^3 + t^2 - 2t}$

12. $\displaystyle\int \frac{t + 3}{2t^3 - 8t}\,dt$

13. $\displaystyle\int \frac{s^3\,ds}{s^2 + 4}$

14. $\displaystyle\int \frac{s^4 + 2s}{s^2 + 1}\,ds$

15. $\displaystyle\int \frac{5x^2\,dx}{x^2 + x + 1}$

16. $\displaystyle\int \frac{x^2\,dx}{(x - 1)(x^2 + 2x + 1)}$

17. $\displaystyle\int \frac{dx}{(x^2 - 1)^2}$

18. $\displaystyle\int \frac{x + 4}{x^2 + 5x - 6}\,dx$

19. $\displaystyle\int \frac{2\,dr}{r^2 - 2r + 2}$

20. $\displaystyle\int \frac{3\,dr}{r^2 - 4r + 5}$

21. $\displaystyle\int \frac{x^2 - 2x - 2}{x^3 - 1}\,dx$

22. $\displaystyle\int \frac{x^2 - 4x + 4}{x^3 + 1}\,dx$

23. $\displaystyle\int \frac{3x^2 - 2x + 12}{(x^2 + 4)^2}\,dx$

24. $\displaystyle\int \frac{x^3 + 2x^2 + 2}{(x^2 + 1)^2}\,dx$

25. $\displaystyle\int_0^1 \frac{\theta}{\theta + 1}\,d\theta$

26. $\displaystyle\int_0^2 \frac{\theta^2}{\theta^2 + 1}\,d\theta$

In Exercises 27–30, solve the initial value problem.

27. $\dfrac{dy}{dx} = e^x(y^2 - y), \quad y(0) = 2$

28. $\dfrac{dy}{d\theta} = (y + 1)^2 \sin\theta, \quad y(\pi/2) = 0$

29. $\dfrac{dy}{dx} = \dfrac{1}{x^2 - 3x + 2}, \quad y(3) = 0$

30. $\dfrac{ds}{dt} = \dfrac{2s + 2}{t^2 + 2t}, \quad s(1) = 1$

In Exercises 31 and 32, *work in groups of two or three.*

31. (a) Writing to Learn Use Formula 18,

$$\int \frac{dx}{a^2 - x^2} = \frac{1}{2a} \ln\left|\frac{x + a}{x - a}\right| + C,$$

from the Brief Table of Integrals to evaluate

$$\int \frac{dx}{5 + 4x - x^2},$$

explaining each step.

(b) Use differentiation to confirm Formula 18.

32. (a) Writing to Learn Use Formula 17,

$$\int \frac{dx}{(a^2 + x^2)^2} = \frac{x}{2a^2(a^2 + x^2)} + \frac{1}{2a^3}\tan^{-1}\frac{x}{a} + C,$$

from the Brief Table of Integrals to evaluate

$$\int \frac{dx}{(x^2 - 2x + 2)^2},$$

explaining each step.

(b) Use differentiation to confirm Formula 17.

In Exercises 33 and 34, find the volume of the solid generated by revolving the shaded region about the indicated axis.

33. the *x*-axis

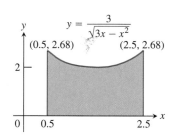

34. the *y*-axis

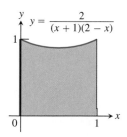

In Exercises 35–42, use trigonometric substitution and the table of integrals (as necessary) to evaluate the integral.

35. $\displaystyle\int \frac{dy}{\sqrt{9 + y^2}}$

36. $\displaystyle\int \sqrt{25 - t^2}\,dt$

37. $\displaystyle\int \frac{dx}{\sqrt{4x^2 - 49}}, \quad x > \frac{7}{2}$ **38.** $\displaystyle\int \frac{dx}{x^2\sqrt{x^2 + 1}}$

39. $\displaystyle\int \frac{x^3\,dx}{\sqrt{1 - x^2}}$ **40.** $\displaystyle\int \frac{2\,dx}{x^3\sqrt{x^2 - 1}}, \quad x > 1$

41. $\displaystyle\int \frac{\sqrt{16 - z^2}}{z}\,dz, \quad 0 < z < 4$ **42.** $\displaystyle\int \frac{8\,dw}{w^2\sqrt{4 - w^2}}$

In Exercises 43 and 44, solve the initial value problem.

43. $\sqrt{x^2 - 9}\,\dfrac{dy}{dx} = 1, \quad x > 3, \quad y(5) = \ln 3$

44. $(x^2 + 1)^2\,\dfrac{dy}{dx} = \sqrt{x^2 + 1}, \quad y(0) = 1$

45. Find the area of the region in the first quadrant that is enclosed by the coordinate axes and the curve $y = \sqrt{9 - x^2}/3$.

46. Find the volume of the solid generated by revolving about the x-axis the region in the first quadrant enclosed by the coordinate axes, the curve $y = 2/(1 + x^2)$, and the line $x = 1$.

Exploration

47. *Social Diffusion* Sociologists sometimes use the phrase "social diffusion" to describe the way information spreads through a population. The information might be a rumor, a cultural fad, or news about a technical innovation. In a sufficiently large population, the number of people x who have the information is treated as a differentiable function of time t, and the rate of diffusion, dx/dt, is assumed to be proportional to the number of people who have the information times the number of people who do not. This leads to the differential equation

$$\frac{dx}{dt} = kx(N - x),$$

where N is the size of the population.

Suppose t is time in days, $k = 1/250$, and two people start a rumor at time $t = 0$ in a population of $N = 1000$ people.

(a) Find x as a function of t.

(b) When will half the population have heard the rumor?

(c) Show that the rumor is spreading the fastest at the time found in (b). ∎

48. *Arc Length* Find the length of the curve $y = \ln(1 - x^2)$ from $x = 0$ to $x = 1/2$.

Extending the Ideas

49. The substitution

$$z = \tan\frac{x}{2}$$

(see figure) reduces the problem of integrating any rational function of $\sin x$ and $\cos x$ to a problem involving a rational function of z. This in turn can be solved by partial fractions.

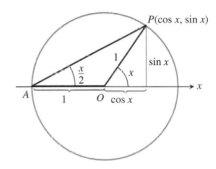

Show that each equation is true.

(a) $\tan\dfrac{x}{2} = \dfrac{\sin x}{1 + \cos x}$ **(b)** $\cos x = \dfrac{1 - z^2}{1 + z^2}$

(c) $\sin x = \dfrac{2z}{1 + z^2}$ **(d)** $dx = \dfrac{2\,dz}{1 + z^2}$

In Exercises 50–53, use the substitution $z = \tan(x/2)$ and the results in Exercise 49 to evaluate the integral.

50. $\displaystyle\int \frac{dx}{1 + \sin x}$ **51.** $\displaystyle\int \frac{dx}{1 - \cos x}$

52. $\displaystyle\int \frac{d\theta}{1 - \sin\theta}$ **53.** $\displaystyle\int \frac{dt}{1 + \sin t + \cos t}$

Chapter 8   Key Terms

at most the order (p. 429)
big-oh (p. 429)
binary search (p. 430)
convergence of improper integral (p. 434)
direct comparison test (p. 438)
divergence of improper integral (p. 434)
grows at the same rate (p. 426)
grows faster (p. 426)

grows slower (p. 426)
improper integral (p. 436)
indeterminate form (p. 417)
l'Hôpital's Rule, first form (p. 417)
l'Hôpital's Rule, stronger form (p. 418)
limit comparison test (p. 439)
little oh (p. 429)
method of partial fractions (p. 444)

monotonic function (p. 438)
partial fractions (p. 444)
reduction formula (p. 450)
sequential search (p. 430)
smaller order (p. 429)
trigonometric substitution (p. 449)
undetermined coefficients (p. 444)
value of improper integral (p. 434)

Chapter 8   Review Exercises

In Exercises 1–14, find the limit.

1. $\lim\limits_{t \to 0} \dfrac{t - \ln(1 + 2t)}{t^2}$

2. $\lim\limits_{t \to 0} \dfrac{\tan 3t}{\tan 5t}$

3. $\lim\limits_{x \to 0} \dfrac{x \sin x}{1 - \cos x}$

4. $\lim\limits_{x \to 1} x^{1/(1-x)}$

5. $\lim\limits_{x \to \infty} x^{1/x}$

6. $\lim\limits_{x \to \infty} \left(1 + \dfrac{3}{x}\right)^x$

7. $\lim\limits_{r \to \infty} \dfrac{\cos r}{\ln r}$

8. $\lim\limits_{\theta \to \pi/2} \left(\theta - \dfrac{\pi}{2}\right) \sec \theta$

9. $\lim\limits_{x \to 1} \left(\dfrac{1}{x - 1} - \dfrac{1}{\ln x}\right)$

10. $\lim\limits_{x \to 0^+} \left(1 + \dfrac{1}{x}\right)^x$

11. $\lim\limits_{\theta \to 0^+} (\tan \theta)^\theta$

12. $\lim\limits_{\theta \to \infty} \theta^2 \sin\left(\dfrac{1}{\theta}\right)$

13. $\lim\limits_{x \to \infty} \dfrac{x^3 - 3x^2 + 1}{2x^2 + x - 3}$

14. $\lim\limits_{x \to \infty} \dfrac{3x^2 - x + 1}{x^4 - x^3 + 2}$

In Exercises 15–26, determine whether f grows faster than, slower than, or at the same rate as g as $x \to \infty$. Give reasons for your answer.

15. $f(x) = x, \quad g(x) = 5x$

16. $f(x) = \log_2 x, \quad g(x) = \log_3 x$

17. $f(x) = x, \quad g(x) = x + \dfrac{1}{x}$

18. $f(x) = \dfrac{x}{100}, \quad g(x) = xe^{-x}$

19. $f(x) = x, \quad g(x) = \tan^{-1} x$

20. $f(x) = \csc^{-1} x, \quad g(x) = \dfrac{1}{x}$

21. $f(x) = x^{\ln x}, \quad g(x) = x^{\log_2 x}$

22. $f(x) = 3^{-x}, \quad g(x) = 2^{-x}$

23. $f(x) = \ln 2x, \quad g(x) = \ln x^2$

24. $f(x) = 10x^3 + 2x^2, \quad g(x) = e^x$

25. $f(x) = \tan^{-1} \dfrac{1}{x}, \quad g(x) = \dfrac{1}{x}$

26. $f(x) = \sin^{-1} \dfrac{1}{x}, \quad g(x) = \dfrac{1}{x^2}$

In Exercises 27 and 28,

  (a) show that f has a removable discontinuity at $x = 0$.

  (b) define f at $x = 0$ so that it is continuous there.

27. $f(x) = \dfrac{2^{\sin x} - 1}{e^x - 1}$

28. $f(x) = x \ln x$

In Exercises 29–38, determine whether the statement is true or false. Give reasons for your answer.

29. $\dfrac{1}{x^2} + \dfrac{1}{x^4} = O\left(\dfrac{1}{x^2}\right)$

30. $\dfrac{1}{x^2} + \dfrac{1}{x^4} = O\left(\dfrac{1}{x^4}\right)$

31. $x = o(x + \ln x)$

32. $\ln(\ln x) = o(\ln x)$

33. $\tan^{-1} x = O(1)$

34. $\dfrac{1}{x^4} = O\left(\dfrac{1}{x^2} + \dfrac{1}{x^4}\right)$

35. $\dfrac{1}{x^4} = o\left(\dfrac{1}{x^2} + \dfrac{1}{x^4}\right)$

36. $\ln x = o(x + 1)$

37. $\ln 2x = O(\ln x)$

38. $\sec^{-1} x = O(1)$

In Exercises 39–48, evaluate the improper integral or state that it diverges.

39. $\displaystyle\int_0^3 \dfrac{dx}{\sqrt{9 - x^2}}$

40. $\displaystyle\int_0^1 \ln x \, dx$

41. $\displaystyle\int_{-1}^1 \dfrac{dy}{y^{2/3}}$

42. $\displaystyle\int_{-2}^0 \dfrac{d\theta}{(\theta + 1)^{3/5}}$

43. $\displaystyle\int_3^\infty \dfrac{2 \, dx}{x^2 - 2x}$

44. $\displaystyle\int_1^\infty \dfrac{3x - 1}{4x^3 - x^2} \, dx$

45. $\displaystyle\int_0^\infty x^2 e^{-x} \, dx$

46. $\displaystyle\int_{-\infty}^0 x \, e^{3x} \, dx$

47. $\displaystyle\int_{-\infty}^\infty \dfrac{4t^3 + t - 1}{t^2(t - 1)(t^2 + 1)} \, dt$

48. $\displaystyle\int_{-\infty}^\infty \dfrac{4 \, dx}{x^2 + 16}$

In Exercises 49–54, determine whether the improper integral converges or diverges. Give reasons for your answer.

49. $\displaystyle\int_0^\infty \dfrac{d\theta}{\sqrt{\theta^2 + 1}}$

50. $\displaystyle\int_0^\infty e^{-u} \cos u \, du$

51. $\displaystyle\int_1^\infty \dfrac{\ln z}{z} \, dz$

52. $\displaystyle\int_1^\infty \dfrac{e^{-t}}{\sqrt{t}} \, dt$

53. $\displaystyle\int_{-\infty}^\infty \dfrac{dx}{e^x + e^{-x}}$

54. $\displaystyle\int_{-\infty}^\infty \dfrac{dx}{x^2(1 + e^x)}$

In Exercises 55–60, evaluate the integral.

55. $\displaystyle\int \dfrac{2x + 1}{x^2 - 7x + 12} \, dx$

56. $\displaystyle\int \dfrac{8 \, dx}{x^3(x + 2)}$

57. $\displaystyle\int \dfrac{3t^2 + 4t + 4}{t^3 + t} \, dt$

58. $\displaystyle\int \dfrac{dt}{t^4 + 4t^2 + 3}$

59. $\displaystyle\int \dfrac{x^3 + 1}{x^3 - x} \, dx$

60. $\displaystyle\int \dfrac{x^3 + 4x^2}{x^2 + 4x + 3} \, dx$

In Exercises 61 and 62, solve the initial value problem.

61. $\dfrac{dy}{dx} = 0.002y(500 - y)$, $y(0) = 20$

62. $\dfrac{dy}{dx} = \dfrac{y^2 + 1}{x + 1}$, $y(0) = \pi/4$

In Exercises 63–66, use trigonometric substitution and the table of integrals (if necessary) to evaluate the integral.

63. $\displaystyle\int \dfrac{3\,dy}{\sqrt{1 + 9y^2}}$

64. $\displaystyle\int \sqrt{1 - 9t^2}\,dt$

65. $\displaystyle\int \dfrac{5\,dx}{\sqrt{25x^2 - 9}}$, $x > \dfrac{3}{5}$

66. $\displaystyle\int \dfrac{4x^2\,dx}{(1 - x^2)^{3/2}}$

67. *Infinite Solid* The infinite region bounded by the coordinate axes and the curve $y = -\ln x$ in the first quadrant (see figure) is revolved about the x axis to generate a solid. Find the volume of the solid.

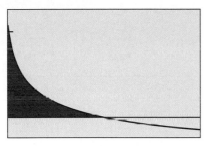

[0, 3] by [-1, 5]

68. *Infinite Region* Find the area of the region in the first quadrant under the curve $y = xe^{-x}$ (see figure).

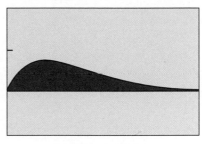

[0, 5] by [-0.5, 1]

69. *Second Order Chemical Reactions* Many chemical reactions are the result of the interaction of two molecules that undergo a change to produce a new product. The rate of the reaction typically depends on the concentrations of the two kinds of molecules. If a is the amount of substance A and b is the amount of substance B at time $t = 0$, and if x is the amount of product at time t, then the rate of formation of the product may be given by the separable differential equation

$$\dfrac{dx}{dt} = k(a - x)(b - x),$$

where k is a constant for the reaction.

Solve this equation to obtain a relation between x and t **(a)** if $a = b$, and **(b)** if $a \neq b$. Assume in each case that $x = 0$ when $t = 0$.

One mathematical constant crucial to the analysis of the world is π. The *p*-series

$$\frac{\pi^2}{6} = 1 + \frac{1}{2^2} + \frac{1}{3^2} + \frac{1}{4^2} + \frac{1}{5^2} + \cdots$$

approximates the value of π. The error, or remainder, of such an approximation is the difference between the actual sum and the *n*th partial sum. For this *p*-series, the remainder is estimated by $R_n \leq 1/n$.

Shown here is a close-up of a high speed microprocessor chip. If a computer adds 1,000,000 terms of the *p*-series in one second, how many places of accuracy will it achieve in 24 hours? Section 9.5 provides a discussion of *p*-series.

Chapter 9 Overview

One consequence of the early and dramatic successes that scientists enjoyed when using calculus to explain natural phenomena was that there suddenly seemed to be no limits, so to speak, on how infinite processes might be exploited. There was still considerable mystery about "infinite sums" and "division by infinitely small quantities" in the years after Newton and Leibniz, but even mathematicians normally insistent on rigorous proof were inclined to throw caution to the wind while things were working. The result was a century of unprecedented progress in understanding the physical universe. (Moreover, we can note happily in retrospect, the proofs eventually followed.)

One infinite process that had puzzled mathematicians for centuries was the summing of infinite series. Sometimes an infinite series of terms added to a number, as in

$$\frac{1}{2} + \frac{1}{4} + \frac{1}{8} + \frac{1}{16} + \cdots = 1.$$

(You can see this by adding the areas in the "infinitely halved" unit square at the right.) But sometimes the infinite sum was infinite, as in

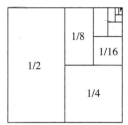

$$\frac{1}{1} + \frac{1}{2} + \frac{1}{3} + \frac{1}{4} + \frac{1}{5} + \cdots = \infty$$

(although this is far from obvious), and sometimes the infinite sum was impossible to pin down, as in

$$1 - 1 + 1 - 1 + 1 - 1 + \cdots$$

(Is it 0? Is it 1? Is it neither?).

Nonetheless, mathematicians like Gauss and Euler successfully used infinite series to derive previously inaccessible results. Laplace used infinite series to prove the stability of the solar system (although that does not stop some people from worrying about it today when they feel that "too many" planets have swung to the same side of the sun). It was years later that careful analysts like Cauchy developed the theoretical foundation for series computations, sending many mathematicians (including Laplace) back to their desks to verify their results.

Our approach in this chapter will be to discover the calculus of infinite series as the pioneers of calculus did: proceeding intuitively, accepting what works and rejecting what does not. Toward the end of the chapter we will return to the crucial question of convergence and take a careful look at it.

9.1 Power Series

Geometric Series • Representing Functions by Series • Differentiation and Integration • Identifying a Series

Geometric Series

The first thing to get straight about an infinite series is that it is not simply an example of addition. Addition of real numbers is a *binary* operation, meaning that we really add numbers two at a time. The only reason that $1 + 2 + 3$

makes sense as "addition" is that we can *group* the numbers and then add them two at a time. The associative property of addition guarantees that we get the same sum no matter how we group them:

$$1 + (2 + 3) = 1 + 5 = 6 \quad \text{and} \quad (1 + 2) + 3 = 3 + 3 = 6.$$

In short, a *finite sum* of real numbers always produces a real number (the result of a finite number of binary additions), but an *infinite sum* of real numbers is something else entirely. That is why we need the following definition.

Definition Infinite Series

An **infinite series** is an expression of the form

$$a_1 + a_2 + a_3 + \cdots + a_n + \cdots, \quad \text{or} \quad \sum_{k=1}^{\infty} a_k.$$

The numbers a_1, a_2, ... are the **terms** of the series; a_n is the **nth term.**

The **partial sums** of the series form a sequence

$$s_1 = a_1$$
$$s_2 = a_1 + a_2$$
$$s_3 = a_1 + a_2 + a_3$$
$$\vdots$$
$$s_n = \sum_{k=1}^{n} a_k$$
$$\vdots$$

of real numbers, each defined as a finite sum. If the sequence of partial sums has a limit S as $n \rightarrow \infty$, we say the series **converges** to the sum S, and we write

$$a_1 + a_2 + a_3 + \cdots + a_n + \cdots = \sum_{k=1}^{\infty} a_k = S.$$

Otherwise, we say the series **diverges.**

Example 1   IDENTIFYING A DIVERGENT SERIES

Does the series $1 - 1 + 1 - 1 + 1 - 1 + \cdots$ converge?

Solution You might be tempted to pair the terms as

$$(1 - 1) + (1 - 1) + (1 - 1) + \cdots.$$

That strategy, however, requires an *infinite* number of pairings, so it cannot be justified by the associative property of addition. This is an infinite series, not a finite sum, so if it has a sum it *has to be* the limit of its sequence of partial sums,

$$1, 0, 1, 0, 1, 0, 1, \ldots.$$

Since this sequence has no limit, the series has no sum. It diverges.

Example 2   IDENTIFYING A CONVERGENT SERIES

Does the series

$$\frac{3}{10} + \frac{3}{100} + \frac{3}{1000} + \cdots + \frac{3}{10^n} + \cdots$$

converge?

Solution Here is the sequence of partial sums, written in decimal form.

$$0.3, \ 0.33, \ 0.333, \ 0.3333, \ \ldots$$

This sequence has a limit $0.\bar{3}$, which we recognize as the fraction $1/3$. The series converges to the sum $1/3$.

The series in Example 2 is a **geometric series** because each term is obtained from its preceding term by multiplying by the same number r—in this case, $r = 1/10$. (The series of areas for the infinitely-halved square at the beginning of this chapter is also geometric.) The convergence of geometric series is one of the few infinite processes with which mathematicians were reasonably comfortable prior to calculus. You may have already seen the following result in a previous course.

The **geometric series**

$$a + ar + ar^2 + ar^3 + \cdots + ar^{n-1} + \cdots = \sum_{n=1}^{\infty} ar^{n-1}$$

converges to the sum $a/(1-r)$ if $|r| < 1$, and diverges if $|r| \geq 1$.

This completely settles the issue for geometric series. We know which ones converge and which ones diverge, and for the convergent ones we know what the sums must be. The interval $-1 < r < 1$ is the **interval of convergence.**

Example 3   ANALYZING GEOMETRIC SERIES

Tell whether each series converges or diverges. If it converges, give its sum.

(a) $\displaystyle\sum_{n=1}^{\infty} 3\left(\frac{1}{2}\right)^{n-1}$

(b) $1 - \dfrac{1}{2} + \dfrac{1}{4} - \dfrac{1}{8} + \cdots + \left(-\dfrac{1}{2}\right)^{n-1} + \cdots$

(c) $\displaystyle\sum_{k=0}^{\infty} \left(\frac{3}{5}\right)^{k}$

(d) $\dfrac{\pi}{2} + \dfrac{\pi^2}{4} + \dfrac{\pi^3}{8} + \cdots$

Solution

(a) First term is $a = 3$ and $r = 1/2$. The series converges to

$$\frac{3}{1 - (1/2)} = 6.$$

(b) First term is $a = 1$ and $r = -1/2$. The series converges to

$$\frac{1}{1 - (-1/2)} = \frac{2}{3}.$$

(c) First term is $a = (3/5)^0 = 1$ and $r = 3/5$. The series converges to

$$\frac{1}{1 - (3/5)} = \frac{5}{2}.$$

(d) In this series, $r = \pi/2 > 1$. The series diverges.

We have hardly begun our study of infinite series, but knowing everything there is to know about the convergence and divergence of an *entire class* of series (geometric) is an impressive start. Like the Renaissance mathematicians, we are ready to explore where this might lead. We are ready to bring in x.

Representing Functions by Series

If $|x| < 1$, then the geometric series formula assures us that

$$1 + x + x^2 + x^3 + \cdots + x^n + \cdots = \frac{1}{1 - x}.$$

Consider this statement for a moment. The expression on the right defines a function whose domain is the set of all numbers $x \neq 1$. The expression on the left defines a function whose domain is the interval of convergence, $|x| < 1$. The equality is understood to hold only on this latter domain, where both sides of the equation are defined. On this domain, the series *represents* the function $1/(1 - x)$.

The partial sums of the infinite series on the left are all polynomials, so we can graph them (Figure 9.1). As expected, we see that the convergence is strong in the interval $(-1, 1)$ but breaks down when $|x| \geq 1$.

Partial Sums

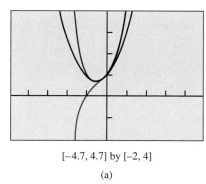

$y = 1/(1-x)$

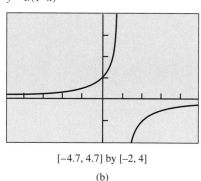

[−4.7, 4.7] by [−2, 4]   [−4.7, 4.7] by [−2, 4]

(a)   (b)

Figure 9.1 (a) Partial sums converging to $1/(1 - x)$ on the interval $(0, 1)$. The partial sums graphed here are $1 + x + x^2$, $1 + x + x^2 + x^3$, and $1 + x + x^2 + x^3 + x^4$. Notice how they resemble the graph of (b) $1/(1 - x)$ on the interval $(-1, 1)$ but are not even close when $|x| \geq 1$.

The expression $\sum_{n=0}^{\infty} x^n$ is like a polynomial in that it is a sum of coefficients times powers of x, but polynomials have *finite* degrees and do not suffer from divergence for the wrong values of x. Just as an infinite series of numbers is not a mere sum, this series of powers of x is not a mere polynomial.

When we set $x = 0$ in the expression

$$\sum_{n=0}^{\infty} c_n x^n = c_0 + c_1 x + c_2 x^2 + \cdots + c_n x^n + \cdots,$$

we get c_0 on the right but $c_0 \cdot 0^0$ on the left. Since 0^0 is not a number, this is a slight flaw in the notation, which we agree to overlook. The same situation arises when we set

$$x = a \quad \text{in} \quad \sum_{n=0}^{\infty} c_n (x - a)^n.$$

In either case, we agree that the expression will equal c_0. (It really *should* equal c_0, so we are not compromising the mathematics; we are clarifying the notation we use to convey the mathematics.)

Definition Power Series

An expression of the form

$$\sum_{n=0}^{\infty} c_n x^n = c_0 + c_1 x + c_2 x^2 + \cdots + c_n x^n + \cdots$$

is a **power series centered at $x = 0$.** An expression of the form

$$\sum_{n=0}^{\infty} c_n (x - a)^n = c_0 + c_1 (x - a) + c_2 (x - a)^2 + \cdots + c_n (x - a)^n + \cdots$$

is a **power series centered at $x = a$.** The term $c_n (x - a)^n$ is the **nth term;** the number a is the **center.**

The geometric series

$$\sum_{n=0}^{\infty} x^n = 1 + x + x^2 + \cdots + x^n + \cdots$$

is a power series centered at $x = 0$. It converges on the interval $-1 < x < 1$, also centered at $x = 0$. This is typical behavior, as we will see in Section 9.4. A power series either converges for all x, converges on a finite interval with the same center as the series, or converges only at the center itself.

We have seen that the power series $\sum_{n=0}^{\infty} x^n$ represents the function $1/(1-x)$ on the domain $(-1, 1)$. Can we find power series to represent other functions?

Exploration 1 Finding Power Series for Other Functions

Given that $1/(1 - x)$ is represented by the power series

$$1 + x + x^2 + \cdots + x^n + \cdots$$

on the interval $(-1, 1)$,

1. find a power series that represents $1/(1 + x)$ on $(-1, 1)$.
2. find a power series that represents $x/(1 + x)$ on $(-1, 1)$.
3. find a power series that represents $1/(1 - 2x)$ on $(-1/2, 1/2)$.
4. find a power series that represents

$$\frac{1}{x} = \frac{1}{1 + (x - 1)}$$

on $(0, 2)$.

Could you have found the intervals of convergence yourself?

5. Find a power series that represents

$$\frac{1}{3x} = \frac{1}{3} \cdot \left(\frac{1}{1 + (x - 1)} \right)$$

and give its interval of convergence.

Differentiation and Integration

So far we have only represented functions by power series that happen to be geometric. The partial sums that converge to those power series, however, are *polynomials,* and we can apply calculus to polynomials. It would seem logical that the calculus of polynomials (the first rules we encountered in Chapter 3) would also apply to power series.

Example 4   FINDING A POWER SERIES BY DIFFERENTIATION

Given that $1/(1 - x)$ is represented by the power series

$$1 + x + x^2 + \cdots + x^n + \cdots$$

on the interval $(-1, 1)$, find a power series to represent $1/(1 - x)^2$.

Solution Notice that $1/(1 - x)^2$ is the derivative of $1/(1 - x)$. To find the power series, we differentiate both sides of the equation

$$\frac{1}{1 - x} = 1 + x + x^2 + x^3 + \cdots + x^n + \cdots .$$

$$\frac{d}{dx}\left(\frac{1}{1 - x}\right) = \frac{d}{dx}(1 + x + x^2 + x^3 + \cdots + x^n + \cdots)$$

$$\frac{1}{(1 - x)^2} = 1 + 2x + 3x^2 + 4x^3 + \cdots + nx^{n-1} + \cdots$$

What about the interval of convergence? Since the original series converges for $-1 < x < 1$, it would seem that the differentiated series ought to converge on the same open interval. Graphs (Figure 9.2) of the partial sums $1 + 2x + 3x^2$, $1 + 2x + 3x^2 + 4x^3$, and $1 + 2x + 3x^2 + 4x^3 + 5x^4$ suggest that this is the case (although such empirical evidence does not constitute a proof).

The basic theorem about differentiating power series is the following.

Partial Sums

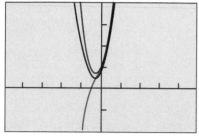

[-4.7, 4.7] by [-2, 4]

(a)

$y = 1/(1-x)^2$

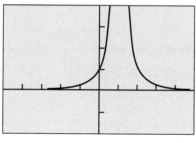

[-4.7, 4.7] by [-2, 4]

(b)

Figure 9.2 (a) The polynomial partial sums of the power series we derived for (b) $1/(1 - x)^2$ seem to converge on the open interval $(-1, 1)$. (Example 4)

Theorem 1 Term-by-Term Differentiation

If $f(x) = \displaystyle\sum_{n=0}^{\infty} c_n(x - a)^n = c_0 + c_1(x - a) + c_2(x - a)^2 + \cdots + c_n(x - a)^n + \cdots$

converges for $|x - a| < R$, then the series

$$\sum_{n=1}^{\infty} nc_n(x - a)^{n-1} = c_1 + 2c_2(x - a) + 3c_3(x - a)^2 + \cdots + nc_n(x - a)^{n-1} + \cdots ,$$

obtained by differentiating the series for f term by term, converges for $|x - a| < R$ and represents $f'(x)$ on that interval. If the series for f converges for all x, then so does the series for f'.

Theorem 1 says that if a power series is differentiated term by term, the new series will converge on the same interval to the derivative of the function represented by the original series. This gives a way to generate new connections between functions and series.

Another way to reveal new connections between functions and series is by integration.

Example 5 FINDING A POWER SERIES BY INTEGRATION

Given that

$$\frac{1}{1+x} = 1 - x + x^2 - x^3 + \cdots + (-x)^n + \cdots, \quad -1 < x < 1$$

(Exploration 1, part 1), find a power series to represent $\ln(1+x)$.

Solution Recall that $1/(1+x)$ is the derivative of $\ln(1+x)$. We can therefore integrate the series for $1/(1+x)$ to obtain a series for $\ln(1+x)$ (no absolute value bars are necessary because $(1+x)$ is positive for $-1 < x < 1$).

$$\frac{1}{1+x} = 1 - x + x^2 - x^3 + \cdots + (-x)^n + \cdots$$

$$= 1 - x + x^2 - x^3 + \cdots + (-1)^n x^n + \cdots$$

$$\int_0^x \frac{1}{1+t}\, dt = \int_0^x (1 - t + t^2 - t^3 + \cdots + (-1)^n t^n + \cdots)\, dt \qquad \text{\textit{t} is a dummy variable.}$$

$$\ln(1+t)\bigg]_0^x = \left[t - \frac{t^2}{2} + \frac{t^3}{3} - \frac{t^4}{4} + \cdots + (-1)^n \frac{t^{n+1}}{n+1} + \cdots \right]_0^x$$

$$\ln(1+x) = x - \frac{x^2}{2} + \frac{x^3}{3} - \frac{x^4}{4} + \cdots + (-1)^n \frac{x^{n+1}}{n+1} + \cdots$$

It would seem logical for the new series to converge where the original series converges, on the open interval $(-1, 1)$. The graphs of the partial sums in Figure 9.3 support this idea.

Partial Sums

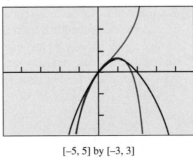

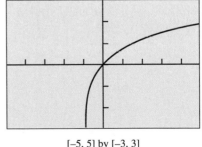

[−5, 5] by [−3, 3]

(a)

[−5, 5] by [−3, 3]

(b)

Figure 9.3 (a) The graphs of the partial sums

$$x - \frac{x^2}{2}, \quad x - \frac{x^2}{2} + \frac{x^3}{3}, \quad \text{and} \quad x - \frac{x^2}{2} + \frac{x^3}{3} - \frac{x^4}{4}$$

closing in on (b) the graph of $\ln(1+x)$ over the interval $(-1, 1)$. (Example 5)

The idea that the integrated series in Example 5 converges to $\ln(1 + x)$ for all x between -1 and 1 is confirmed by the following theorem.

Some calculators have a *sequence mode* that enables you to generate a sequence of partial sums, but you can also do it with simple commands on the home screen. Try entering the two multiple-step commands shown on the first screen below.

```
0 → N: 1 → T
                          1
N+1 → N: T+(−1)^N/(
N+1) → T
                         .5
■
```

```
        .6997694067
         .686611512
        .699598525
        .6867780122
         .69943624
         .68693624
        .699281919
```

If you are successful, then every time you hit ENTER, the calculator will display the next partial sum of the series

$$1 - \frac{1}{2} + \frac{1}{3} - \frac{1}{4} + \cdots + \frac{(-1)^n}{n+1} + \cdots.$$

The second screen shows the result of about 80 ENTERs. The sequence certainly seems to be converging to $\ln 2 = 0.6931471806\ldots$.

Theorem 2 Term-by-Term Integration

If $f(x) = \displaystyle\sum_{n=0}^{\infty} c_n(x - a)^n = c_0 + c_1(x - a) + c_2(x - a)^2 + \cdots + c_n(x - a)^n + \cdots$

converges for $|x - a| < R$, then the series

$$\sum_{n=0}^{\infty} c_n \frac{(x - a)^{n+1}}{n+1} = c_0(x - a) + c_1 \frac{(x - a)^2}{2} + c_2 \frac{(x - a)^3}{3} + \cdots + c_n \frac{(x - a)^{n+1}}{n+1} + \cdots,$$

obtained by integrating the series for f term by term, converges for $|x - a| < R$ and represents $\int_a^x f(t)\, dt$ on that interval. If the series for f converges for all x, then so does the series for the integral.

Theorem 2 says that if a power series is integrated term by term, the new series will converge on the same interval to the integral of the function represented by the original series.

There is still more to be learned from Example 5. The original equation

$$\frac{1}{1 + x} = 1 - x + x^2 - x^3 + \cdots + (-x)^n + \cdots$$

clearly diverges at $x = 1$ (see Example 1). The behavior is not so apparent, however, for the new equation

$$\ln(1 + x) = x - \frac{x^2}{2} + \frac{x^3}{3} - \frac{x^4}{4} + \cdots + (-1)^n \frac{x^{n+1}}{n+1} + \cdots.$$

If we let $x = 1$ on both sides of this equation, we get

$$\ln 2 = 1 - \frac{1}{2} + \frac{1}{3} - \frac{1}{4} + \cdots + \frac{(-1)^n}{n+1} + \cdots,$$

which looks like a reasonable statement. It looks even more reasonable if you look at the partial sums of the series and watch them converge toward $\ln 2$ (see margin note). It would appear that our new series converges at 1 despite the fact that we obtained it from a series that did not! This is all the more reason to take a careful look at convergence later. Meanwhile, we can enjoy the observation that we have created a series that apparently works better than we might have expected and better than Theorem 2 could guarantee.

Exploration 2 **Finding a Power Series for tan⁻¹ x**

1. Find a power series that represents $1/(1 + x^2)$ on $(-1, 1)$.

2. Use the technique of Example 5 to find a power series that represents $\tan^{-1} x$ on $(-1, 1)$.

3. Graph the first four partial sums. Do the graphs suggest convergence on the open interval $(-1, 1)$?

4. Do you think that the series for $\tan^{-1} x$ converges at $x = 1$? Can you support your answer with evidence?

Identifying a Series

So far we have been finding power series to represent functions. Let us now try to find the function that a given power series represents.

Exploration 3 **A Series with a Curious Property**

Define a function f by a power series as follows:

$$f(x) = 1 + x + \frac{x^2}{2!} + \frac{x^3}{3!} + \frac{x^4}{4!} + \cdots + \frac{x^n}{n!} + \cdots .$$

1. Find $f'(x)$.

2. Find $f(0)$.

3. What well-known function do you suppose f is?

4. Use your responses to parts 1 and 2 to set up an initial value problem that the function f must solve. You will need a differential equation and an initial condition.

5. Solve the initial value problem to prove your conjecture in part 3.

6. Graph the first three partial sums. What appears to be the interval of convergence?

7. Graph the next three partial sums. Did you underestimate the interval of convergence?

The correct answer to part 7 in Exploration 3 above is "yes," unless you had the keen insight (or reckless bravado) to answer "all real numbers" in part 6. We will prove the remarkable fact that this series converges *for all x* when we revisit the question of convergence of this series in Section 9.3, Example 4.

Quick Review 9.1

In Exercises 1 and 2, find the first four terms and the 30th term of the sequence
$$\{u_n\}_{n=1}^{\infty} = \{u_1, u_2, \dots, u_n, \dots\}.$$

1. $u_n = \dfrac{4}{n+2}$ **2.** $u_n = \dfrac{(-1)^n}{n}$

In Exercises 3 and 4, the sequences are geometric $(a_{n+1}/a_n = r,$ a constant). Find

     (a) the common ratio r.   **(b)** the tenth term.

     (c) a rule for the nth term.

3. $\{2, 6, 18, 54, \dots\}$

4. $\{8, -4, 2, -1, \dots\}$

In Exercises 5–10,

     (a) graph the sequence $\{a_n\}$.

     (b) determine $\lim\limits_{n\to\infty} a_n$.

5. $a_n = \dfrac{1-n}{n^2}$ **6.** $a_n = \left(1 + \dfrac{1}{n}\right)^n$

7. $a_n = (-1)^n$ **8.** $a_n = \dfrac{1-2n}{1+2n}$

9. $a_n = 2 - \dfrac{1}{n}$ **10.** $a_n = \dfrac{\ln(n+1)}{n}$

Section 9.1 Exercises

1. Replace the $*$ with an expression that will generate the series
$$1 - \frac{1}{4} + \frac{1}{9} - \frac{1}{16} + \cdots.$$

   (a) $\displaystyle\sum_{n=1}^{\infty} (-1)^{n-1}\left(\frac{1}{*}\right)$ **(b)** $\displaystyle\sum_{n=0}^{\infty} (-1)^n\left(\frac{1}{*}\right)$

   (c) $\displaystyle\sum_{n=*}^{\infty} (-1)^n\left(\frac{-1}{(n-2)^2}\right)$

2. Write an expression for the nth term, a_n.

   (a) $\displaystyle\sum_{n=0}^{\infty} a_n = 1 + \frac{1}{3} + \frac{1}{9} + \frac{1}{27} + \frac{1}{81} + \cdots$

   (b) $\displaystyle\sum_{n=1}^{\infty} a_n = 1 - \frac{1}{2} + \frac{1}{3} - \frac{1}{4} + \frac{1}{5} - \cdots$

   (c) $\displaystyle\sum_{n=0}^{\infty} a_n = 5 + 0.5 + 0.05 + 0.005 + 0.0005 + \cdots$

In Exercises 3–6, tell whether the series is the same as
$$\sum_{n=1}^{\infty} \left(-\frac{1}{2}\right)^{n-1}.$$

3. $\displaystyle\sum_{n=1}^{\infty} -\left(\frac{1}{2}\right)^{n-1}$ **4.** $\displaystyle\sum_{n=0}^{\infty} \left(-\frac{1}{2}\right)^n$

5. $\displaystyle\sum_{n=0}^{\infty} (-1)^n\left(\frac{1}{2}\right)^n$ **6.** $\displaystyle\sum_{n=1}^{\infty} \frac{(-1)^n}{2^{n-1}}$

In Exercises 7–16, tell whether the series converges or diverges. If it converges, give its sum.

7. $1 + \dfrac{2}{3} + \left(\dfrac{2}{3}\right)^2 + \left(\dfrac{2}{3}\right)^3 + \cdots + \left(\dfrac{2}{3}\right)^n + \cdots$

8. $1 - 2 + 3 - 4 + 5 - \cdots + (-1)^n(n+1) + \cdots$

9. $\displaystyle\sum_{n=0}^{\infty} \left(\frac{5}{4}\right)\left(\frac{2}{3}\right)^n$ **10.** $\displaystyle\sum_{n=0}^{\infty} \left(\frac{2}{3}\right)\left(\frac{5}{4}\right)^n$

11. $\displaystyle\sum_{n=0}^{\infty} \cos(n\pi)$

12. $3 - 0.3 + 0.03 - 0.003 + 0.0003 - \cdots + 3(-0.1)^n + \cdots$

13. $\displaystyle\sum_{n=0}^{\infty} \sin^n\left(\frac{\pi}{4} + n\pi\right)$

14. $\dfrac{1}{2} + \dfrac{2}{3} + \dfrac{3}{4} + \dfrac{4}{5} + \cdots + \dfrac{n}{n+1} + \cdots$

15. $\displaystyle\sum_{n=1}^{\infty} \left(\frac{e}{\pi}\right)^n$ **16.** $\displaystyle\sum_{n=0}^{\infty} \frac{5^n}{6^{n+1}}$

In Exercises 17–20, find the interval of convergence and the function of x represented by the geometric series.

17. $\displaystyle\sum_{n=0}^{\infty} 2^n x^n$ **18.** $\displaystyle\sum_{n=0}^{\infty} (-1)^n(x+1)^n$

19. $\displaystyle\sum_{n=0}^{\infty} \left(-\frac{1}{2}\right)^n (x-3)^n$ **20.** $\displaystyle\sum_{n=0}^{\infty} 3\left(\frac{x-1}{2}\right)^n$

In Exercises 21 and 22, find the values of x for which the geometric series converges and find the function of x it represents.

21. $\displaystyle\sum_{n=0}^{\infty} \sin^n x$ **22.** $\displaystyle\sum_{n=0}^{\infty} \tan^n x$

23. Writing to Learn Each of the following series diverges in a slightly different way. Explain what is happening to the sequence of partial sums in each case.

(a) $\displaystyle\sum_{n=1}^{\infty} 2n$ **(b)** $\displaystyle\sum_{n=0}^{\infty} (-1)^n$ **(c)** $\displaystyle\sum_{n=1}^{\infty} (-1)^n (2n)$

24. Prove that $\displaystyle\sum_{n=0}^{\infty} \frac{e^{n\pi}}{\pi^{ne}}$ diverges.

25. Solve for *x:* $\displaystyle\sum_{n=0}^{\infty} x^n = 20.$

26. Writing to Learn Explain how it is possible, given any real number at all, to construct an infinite series of non-zero terms that converges to it.

27. Make up a geometric series $\sum ar^{n-1}$ that converges to the number 5 if

   (a) $a = 2$ **(b)** $a = 13/2$

In Exercises 28 and 29, express the repeating decimal as a geometric series and find its sum.

28. $0.\overline{21}$ **29.** $0.\overline{234}$

In Exercises 30–35, express the number as the ratio of two integers.

30. $0.\overline{7} = 0.7777\ldots$

31. $0.\overline{d} = 0.dddd\ldots,$ where *d* is a digit

32. $0.\overline{06} = 0.06666\ldots$

33. $1.\overline{414} = 1.414\ 414\ 414\ldots$

34. $1.24\overline{123} = 1.24\ 123\ 123\ 123\ldots$

35. $3.\overline{142857} = 3.142857\ 142857\ldots$

36. Bouncing Ball A ball is dropped from a height of 4 m. Each time it strikes the pavement after falling from a height of *h* m, it rebounds to a height of $0.6h$ m. Find the total distance the ball travels up and down.

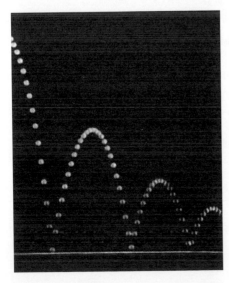

37. *(Continuation of Exercise 36)* Find the total number of seconds that the ball in Exercise 36 travels. (*Hint:* A freely falling ball travels $4.9t^2$ meters in *t* seconds, so it will fall *h* meters in $\sqrt{h/4.9}$ seconds. Bouncing from ground to apex takes the same time as falling from apex to ground.)

38. Summing Areas The figure below shows the first five of an infinite sequence of squares. The outermost square has an area of 4 m². Each of the other squares is obtained by joining the midpoints of the sides of the preceding square. Find the sum of the areas of all the squares.

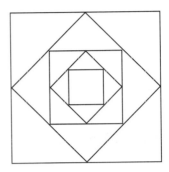

39. *Summing Areas* The accompanying figure shows the first three rows and part of the fourth row of a sequence of rows of semicircles. There are 2^n semicircles in the *n*th row, each of radius $1/(2^n)$. Find the sum of the areas of all the semicircles.

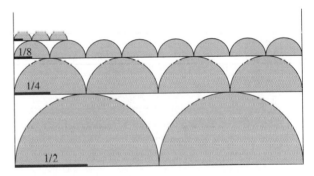

40. *Sum of a Finite Geometric Progression* Let *a* and *r* be real numbers with $r \neq 1$, and let

$$S = a + ar + ar^2 + ar^3 + \cdots + ar^{n-1}.$$

(a) Find $S - rS$.

(b) Use the result in (a) to show that $\displaystyle S = \frac{a - ar^n}{1 - r}.$

41. *Sum of a Convergent Geometric Series* Exercise 40 gives a formula for the *n*th partial sum of an infinite geometric series. Use this formula to show that $\sum_{n=1}^{\infty} ar^{n-1}$ diverges when $|r| \geq 1$ and converges to $a/(1 - r)$ when $|r| < 1$.

In Exercises 42–47, find a power series to represent the given function and identify its interval of convergence. When writing the power series, include a formula for the nth term.

42. $\dfrac{1}{1 + 3x}$

43. $\dfrac{x}{1 - 2x}$

44. $\dfrac{3}{1 - x^3}$

45. $\dfrac{1}{1 + (x - 4)}$

46. $\dfrac{1}{4x} = \dfrac{1}{4}\left(\dfrac{1}{1 + (x - 1)}\right)$

47. $\dfrac{1}{2 - x}$ (*Hint:* Rewrite $2 - x$.)

48. Find the value of b for which $1 + e^b + e^{2b} + e^{3b} + \cdots = 9$.

49. Let S be the series

$$\sum_{n=0}^{\infty} \left(\dfrac{t}{1 + t}\right)^n, \quad t \neq 0.$$

(a) Find the value to which S converges when $t = 1$.

(b) Determine all values of t for which S converges.

(c) Find all values of t that make the sum of the series S greater than 10.

Exploration

50. Let $f(t) = \dfrac{4}{1 + t^2}$ and $G(x) = \displaystyle\int_0^x f(t)\, dt$.

(a) Find the first four nonzero terms and the general term for a power series for $f(t)$ centered at $t = 0$.

(b) Find the first four nonzero terms and the general term for a power series for $G(x)$ centered at $x = 0$.

(c) Find the interval of convergence of the power series in (a).

(d) The interval of convergence of the power series in (b) is almost the same as the interval in (c), but includes two more numbers. What are the numbers? ∎

51. *A Series for* ln *x*   Starting with the power series found for $1/x$ in Exploration 1, Part 4, find a power series for $\ln x$ centered at $x = 1$.

52. *Differentiation*   Use differentiation to find a series for $f(x) = 2/(1 - x)^3$. What is the interval of convergence of your series?

53. *Intervals of Convergence*   *Work in groups of two or three.* How much can the interval of convergence of a power series be changed by integration or differentiation? To be specific, suppose that the power series

$$f(x) = c_0 + c_1 x + c_2 x^2 + \cdots + c_n x^n + \cdots$$

converges for $-1 < x < 1$ and diverges for all other values of x.

(a) Writing to Learn   Could the series obtained by integrating the series for f term by term possibly converge for $-2 < x < 2$? Explain. (*Hint:* Apply Theorem 1, not Theorem 2.)

(b) Writing to Learn   Could the series obtained by differentiating the series for f term by term possibly converge for $-2 < x < 2$? Explain.

Extending the Ideas

The sequence $\{a_n\}$ **converges** to the number L if to every positive number ε there corresponds an integer N such that for all n,

$$n > N \quad \Rightarrow \quad |a_n - L| < \varepsilon.$$

L is the **limit** of the sequence and we write $\lim_{n \to \infty} a_n = L$. If no such number L exists, we say that $\{a_n\}$ **diverges.**

54. *Tail of a Sequence*   Prove that if $\{a_n\}$ is a convergent sequence, then to every positive number ε there corresponds an integer N such that for all m and n,

$$m > N \quad \text{and} \quad n > N \quad \Rightarrow \quad |a_m - a_n| < \varepsilon.$$

(*Hint:* Let $\lim_{n \to \infty} a_n = L$. As the terms approach L, how far apart can they be?)

55. *Uniqueness of Limits*   Prove that limits of sequences are unique. That is, show that if L_1 and L_2 are numbers such that $\lim_{n \to \infty} a_n = L_1$ and $\lim_{n \to \infty} a_n = L_2$, then $L_1 = L_2$.

56. *Limits and Subsequences*   Prove that if two subsequences of a sequence $\{a_n\}$ have different limits $L_1 \neq L_2$, then $\{a_n\}$ diverges.

57. *Limits and Asymptotes*

(a) Show that the sequence with nth term $a_n = (3n + 1)/(n + 1)$ converges.

(b) If $\lim_{n \to \infty} a_n = L$, explain why $y = L$ is a horizontal asymptote of the graph of the function

$$f(x) = \dfrac{3x + 1}{x + 1}$$

obtained by replacing n by x in the nth term.

9.2 Taylor Series

Constructing a Series • Series for sin x and cos x • Beauty Bare • Maclaurin and Taylor Series • Combining Taylor Series • Table of Maclaurin Series

Constructing a Series

A comprehensive understanding of geometric series served us well in Section 9.1, enabling us to find power series to represent certain functions, and functions that are equivalent to certain power series (all of these equivalencies being subject to the condition of convergence). In this section we learn a more general technique for constructing power series, one that makes good use of the tools of calculus.

Let us start by constructing a polynomial.

Exploration 1 **Designing a Polynomial to Specifications**

Construct a polynomial $P(x) = a_0 + a_1x + a_2x^2 + a_3x^3 + a_4x^4$ with the following behavior at $x = 0$:

$$P(0) = 1,$$

$$P'(0) = 2,$$

$$P''(0) = 3,$$

$$P'''(0) = 4, \text{ and}$$

$$P^{(4)}(0) = 5.$$

This task might look difficult at first, but when you try it you will find that the predictability of differentiation when applied to polynomials makes it straightforward. (Be sure to check this out before you move on.)

There is nothing special about the number of derivatives in Exploration 1. We could have prescribed the value of the polynomial and its first n derivatives at $x = 0$ for any n and found a polynomial of degree at most n to match. Our plan now is to use the technique of Exploration 1 to construct polynomials that approximate functions by emulating their behavior at 0.

Example 1 APPROXIMATING ln $(1 + x)$ BY A POLYNOMIAL

Construct a polynomial $P(x) = a_0 + a_1x + a_2x^2 + a_3x^3 + a_4x^4$ that matches the behavior of ln $(1 + x)$ at $x = 0$ through its first four derivatives. That is,

$$P(0) = \ln(1 + x) \qquad \text{at } x = 0,$$

$$P'(0) = (\ln(1 + x))' \qquad \text{at } x = 0,$$

$$P''(0) = (\ln(1 + x))'' \qquad \text{at } x = 0,$$

$$P'''(0) = (\ln(1 + x))''' \qquad \text{at } x = 0, \text{ and}$$

$$P^{(4)}(0) = (\ln(1 + x))^{(4)} \qquad \text{at } x = 0.$$

Solution This is just like Exploration 1, except that first we need to find out what the numbers are.

$$P(0) = \ln(1+x)\Big|_{x=0} = 0$$

$$P'(0) = \frac{1}{1+x}\Big|_{x=0} = 1$$

$$P''(0) = -\frac{1}{(1+x)^2}\Big|_{x=0} = -1$$

$$P'''(0) = \frac{2}{(1+x)^3}\Big|_{x=0} = 2$$

$$P^{(4)}(0) = -\frac{6}{(1+x)^4}\Big|_{x=0} = -6$$

In working through Exploration 1, you probably noticed that the coefficient of the term x^n in the polynomial we seek is $P^{(n)}(0)$ divided by $n!$. The polynomial is

$$P(x) = 0 + x - \frac{x^2}{2} + \frac{x^3}{3} - \frac{x^4}{4}.$$

We have just constructed the fourth order **Taylor polynomial** for the function $\ln(1+x)$ at $x=0$. You might recognize it as the beginning of the power series we discovered for $\ln(1+x)$ in Example 5 of Section 9.1, when we came upon it by integrating a geometric series. If we keep going, of course, we will gradually reconstruct that entire series one term at a time, improving the approximation near $x=0$ with every term we add. The series is called the **Taylor series** generated by the function $\ln(1+x)$ at $x=0$.

You might also recall Figure 9.3, which shows how the polynomial approximations converge nicely to $\ln(1+x)$ near $x=0$, but then gradually peel away from the curve as x gets farther away from 0 in either direction. Given that the coefficients are totally determined by specifying behavior at $x=0$, that is exactly what we ought to expect.

Series for sin x and cos x

We can use the technique of Example 1 to construct Taylor series about $x=0$ for any function, as long as we can keep taking derivatives there. Two functions that are particularly well-suited for this treatment are the sine and cosine.

Example 2   CONSTRUCTING A POWER SERIES FOR sin x

Construct the seventh order Taylor polynomial and the Taylor series for $\sin x$ at $x=0$.

Solution We need to evaluate $\sin x$ and its first seven derivatives at $x=0$. Fortunately, this is not hard to do.

$$\sin(0) = 0$$

$$\sin'(0) = \cos(0) = 1$$

$$\sin''(0) = -\sin(0) = 0$$

$$\sin'''(0) = -\cos(0) = -1$$

$$\sin^{(4)}(0) = \sin(0) = 0$$

$$\sin^{(5)}(0) = \cos(0) = 1$$

$$\vdots$$

The pattern $0, 1, 0, -1$ will keep repeating forever.

The unique seventh order Taylor polynomial that matches all these derivatives at $x = 0$ is

$$P_7(x) = 0 + 1x - 0x^2 - \frac{1}{3!}x^3 + 0x^4 + \frac{1}{5!}x^5 - 0x^6 - \frac{1}{7!}x^7$$

$$= x - \frac{x^3}{3!} + \frac{x^5}{5!} - \frac{x^7}{7!}.$$

P_7 is the seventh order Taylor polynomial for $\sin x$ at $x = 0$. (It also happens to be of seventh *degree,* but that does not always happen. For example, you can see that P_8 for $\sin x$ will be the same polynomial as P_7.)

To form the Taylor series, we just keep on going:

$$x - \frac{x^3}{3!} + \frac{x^5}{5!} - \frac{x^7}{7!} + \frac{x^9}{9!} - \cdots = \sum_{n=0}^{\infty} (-1)^n \frac{x^{2n+1}}{(2n+1)!}.$$

Exploration 2 **A Power Series for the Cosine**

Work in groups of two or three.

1. Construct the sixth order Taylor polynomial and the Taylor series at $x = 0$ for $\cos x$.

2. Compare your method for attacking part 1 with the methods of other groups. Did anyone find a shortcut?

Beauty Bare

Edna St. Vincent Millay, an early twentieth-century American poet, referring to the experience of simultaneously seeing and understanding the geometric under-pinnings of nature, wrote "Euclid alone has looked on Beauty bare." In case you have never experienced that sort of reverie when gazing upon something geometric, we intend to give you that opportunity now.

In Example 2 we constructed a power series for $\sin x$ by matching the behavior of $\sin x$ at $x = 0$. Let us graph the first nine partial sums together with $y = \sin x$ to see how well we did (Figure 9.4).

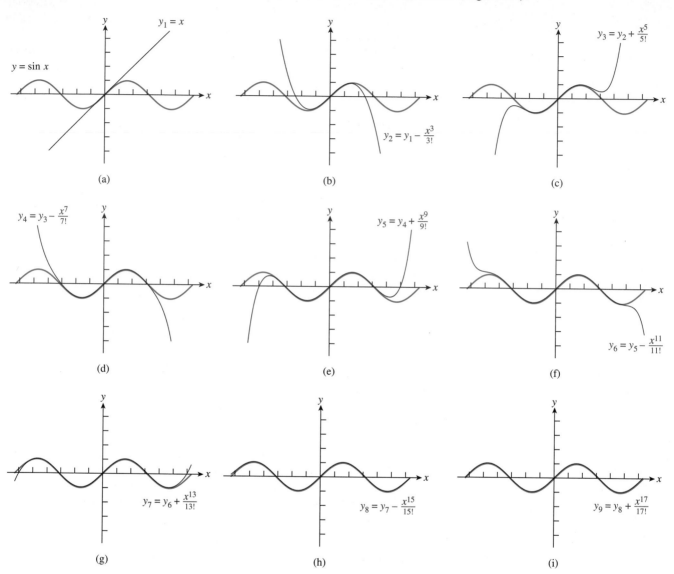

(a)

(b)

(c)

(d)

(e)

(f)

(g)

(h)

(i)

Figure 9.4 $y = \sin x$ and its nine Taylor polynomials $P_1, P_3, \ldots, P_{17}$ for $-2\pi \le x \le 2\pi$. Try graphing these functions in the window $[-2\pi, 2\pi]$ by $[-5, 5]$.

Behold what is occurring here! These polynomials were constructed to mimic the behavior of $\sin x$ near $x = 0$. The *only* information we used to construct the coefficients of these polynomials was information about the sine function and its derivatives at 0. Yet, somehow, the information at $x = 0$ is producing a series whose graph not only looks like sine near the origin, but appears to be a clone of the *entire* sine curve. This is no deception, either; we will show in Section 9.3, Example 3 that the Taylor series for $\sin x$ does, in fact, converge to $\sin x$ over the entire real line. We have managed to construct an entire function by knowing its behavior at a single point! (The same is true about the series for $\cos x$ found in Exploration 2.)

We still must remember that convergence is an infinite process. Even the one-billionth order Taylor polynomial begins to peel away from $\sin x$ as we move away from 0, although imperceptibly at first, and eventually becomes unbounded, as any polynomial must. Nonetheless, we can approximate the sine of *any* number to whatever accuracy we want if we just have the patience to work out enough terms of this series!

This kind of dramatic convergence does not occur for all Taylor series. The Taylor polynomials for $\ln(1 + x)$ do not converge outside the interval from -1 to 1, no matter how many terms we add.

Maclaurin and Taylor Series

If we generalize the steps we followed in constructing the coefficients of the power series in this section so far, we arrive at the following definition.

Definition   Taylor Series Generated by *f* at *x* = 0 (Maclaurin Series)

Let f be a function with derivatives of all orders throughout some open interval containing 0. Then the **Taylor series generated by f at $x = 0$** is

$$f(0) + f'(0)x + \frac{f''(0)}{2!}x^2 + \cdots + \frac{f^{(n)}(0)}{n!}x^n + \cdots = \sum_{k=0}^{\infty} \frac{f^{(k)}(0)}{k!}x^k.$$

This series is also called the **Maclaurin series generated by f.** The partial sum

$$P_n(x) = \sum_{k=0}^{n} \frac{f^{(k)}(0)}{k!}x^k$$

is the **Taylor polynomial of order n for f at $x = 0$.**

We use $f^{(0)}$ to mean f. *Every* power series constructed in this way converges to the function f at $x = 0$, but we have seen that the convergence might well extend to an interval containing 0, or even to the entire real line. When this happens, the Taylor polynomials that form the partial sums of a Taylor series provide good approximations for f near 0.

Example 3   APPROXIMATING A FUNCTION NEAR 0

Find the fourth order Taylor polynomial that approximates $y = \cos 2x$ near $x = 0$.

Solution   The polynomial we want is $P_4(x)$, the Taylor polynomial for $\cos 2x$ at $x = 0$. Before we go cranking out derivatives though, remember that we can use a known power series to generate another, as we did in Section 9.1. We know from Exploration 2 that

$$\cos x = 1 - \frac{x^2}{2!} + \frac{x^4}{4!} - \cdots + (-1)^n \frac{x^{2n}}{(2n)!} + \cdots.$$

Who invented Taylor series?

Brook Taylor (1685–1731) did not invent Taylor series, and Maclaurin series were not developed by Colin Maclaurin (1698–1746). James Gregory was already working with Taylor series when Taylor was only a few years old, and he published the Maclaurin series for $\tan x$, $\sec x$, $\arctan x$, and $\operatorname{arcsec} x$ ten years before Maclaurin was born. Nicolaus Mercator discovered the Maclaurin series for $\ln(1 + x)$ at about the same time.

Taylor was unaware of Gregory's work when he published his book *Methodus Incrementorum Directa et Inversa* in 1715, containing what we now call Taylor series. Maclaurin quoted Taylor's work in a calculus book he wrote in 1742. The book popularized series representations of functions and although Maclaurin never claimed to have discovered them, Taylor series centered at $x = 0$ became known as Maclaurin series. History evened things up in the end. Maclaurin, a brilliant mathematician, was the original discoverer of the rule for solving systems of equations that we call Cramer's rule.

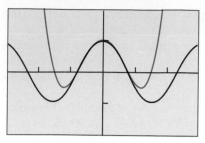

[–3, 3] by [–2, 2]

Figure 9.5 The graphs of
$y = 1 - 2x^2 + (2/3)x^4$ and
$y = \cos 2x$ near $x = 0$. (Example 3)

Therefore,

$$\cos 2x = 1 - \frac{(2x)^2}{2!} + \frac{(2x)^4}{4!} - \cdots + (-1)^n \frac{(2x)^{2n}}{(2n)!} + \cdots.$$

So,

$$P_4(x) = 1 - \frac{(2x)^2}{2!} + \frac{(2x)^4}{4!}$$
$$= 1 - 2x^2 + (2/3)x^4.$$

The graph in Figure 9.5 shows how well the polynomial approximates the cosine near $x = 0$.

These polynomial approximations can be useful in a variety of ways. For one thing, it is easy to do calculus with polynomials. For another thing, polynomials are built using only the two basic operations of addition and multiplication, so computers can handle them easily.

Exploration 3 **Approximating sin 13**

How many terms of the series

$$\sum_{n=0}^{\infty} (-1)^n \frac{x^{2n+1}}{(2n + 1)!}$$

are required to approximate sin 13 accurate to the third decimal place?

1. Find sin 13 on your calculator (radians, of course).

2. Enter these two multiple-step commands on your home screen. They will give you the first order and second order Taylor polynomial approximations for sin 13. Notice that the second order approximation, in particular, is not very good.

```
0 → N: 13 → T
                              13
N+1 → N: T+(-1)^N*1
3^(2N+1)/(2N+1)!
→ T
            -353.1666667
```

3. Continue to hit ENTER. Each time you will add one more term to the Taylor polynomial approximation. Be patient; things will get worse before they get better.

4. How many terms are required before the polynomial approximations stabilize in the thousandths place for $x = 13$?

This strategy for approximation would be of limited practical value if we were restricted to power series at $x = 0$—but we are not. We can match a power series with f in the same way at *any* value $x = a$, provided we can take the derivatives. In fact, we can get a formula for doing that by simply "shifting horizontally" the formula we already have.

Definition **Taylor Series Generated by *f* at *x* = *a***

Let f be a function with derivatives of all orders throughout some open interval containing a. Then the **Taylor series generated by f at $x = a$** is

$$f(a) + f'(a)(x - a) + \frac{f''(a)}{2!}(x - a)^2 + \cdots + \frac{f^{(n)}(a)}{n!}(x - a)^n + \cdots$$

$$= \sum_{k=0}^{\infty} \frac{f^{(k)}(a)}{k!}(x - a)^k.$$

The partial sum

$$P_n(x) = \sum_{k=0}^{n} \frac{f^{(k)}(a)}{k!}(x - a)^k$$

is the **Taylor polynomial of order *n* for *f* at $x = a$.**

Example 4 A TAYLOR SERIES AT $x = 2$

Find the Taylor series generated by $f(x) = e^x$ at $x = 2$.

Solution We first observe that $f(2) = f'(2) = f''(2) = \cdots = f^{(n)}(2) = e^2$. The series, therefore, is

$$e^x = e^2 + e^2(x - 2) + \frac{e^2}{2!}(x - 2)^2 + \cdots + \frac{e^2}{n!}(x - 2)^n + \cdots$$

$$= \sum_{k=0}^{\infty} \left(\frac{e^2}{k!}\right)(x - 2)^k.$$

We illustrate the convergence near $x = 2$ by sketching the graphs of $y = e^x$ and $y = P_3(x)$ in Figure 9.6.

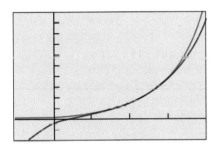

[−1, 4] by [−10, 50]

Figure 9.6 The graphs of $y = e^x$ and $y = P_3(x)$ (the third order Taylor polynomial for e^x at $x = 2$). (Example 4)

Example 5 A TAYLOR POLYNOMIAL FOR A POLYNOMIAL

Find the third order Taylor polynomial for $f(x) = 2x^3 - 3x^2 + 4x - 5$

(a) at $x = 0$.   **(b)** at $x = 1$.

Solution

(a) This is easy. This polynomial is already written in powers of x and is of degree three, so it is its own third order (and fourth order, etc.) Taylor polynomial at $x = 0$.

(b) This would also be easy if we could quickly rewrite the formula for f as a polynomial in powers of $x - 1$, but that would require some messy tinkering. Instead, we apply the Taylor series formula.

$$f(1) = 2x^3 - 3x^2 + 4x - 5 \Big|_{x=1} = -2$$

$$f'(1) = 6x^2 - 6x + 4 \Big|_{x=1} = 4$$

$$f''(1) = 12x - 6 \Big|_{x=1} = 6$$

$$f'''(1) = 12$$

So,

$$P_3(x) = -2 + 4(x - 1) + \frac{6}{2!}(x - 1)^2 + \frac{12}{3!}(x - 1)^3$$

$$= 2(x - 1)^3 + 3(x - 1)^2 + 4(x - 1) - 2.$$

This polynomial function agrees with f at every value of x (as you can verify by multiplying it out) but it is written in powers of $(x - 1)$ instead of x.

Combining Taylor Series

On the intersection of their intervals of convergence, Taylor series can be added, subtracted, and multiplied by constants and powers of x, and the results are once again Taylor series. The Taylor series for $f(x) + g(x)$ is the sum of the Taylor series for $f(x)$ and the Taylor series for $g(x)$ because the nth derivative of $f + g$ is $f^{(n)} + g^{(n)}$, and so on. We can obtain the Maclaurin series for $(1 + \cos 2x)/2$ by substituting $2x$ in the Maclaurin series for $\cos x$, adding 1, and dividing the result by 2. The Maclaurin series for $\sin x + \cos x$ is the term-by-term sum of the series for $\sin x$ and $\cos x$. We obtain the Maclaurin series for $x \sin x$ by multiplying all the terms of the Maclaurin series for $\sin x$ by x.

Table of Maclaurin Series

We conclude the section by listing some of the most useful Maclaurin series, all of which have been derived in one way or another in the first two sections of this chapter. The exercises will ask you to use these series as basic building blocks for constructing other series (e.g., $\tan^{-1} x^2$ or $7xe^x$). We also list the intervals of convergence, although rigorous proofs of convergence are deferred until we develop convergence tests in Sections 9.4 and 9.5.

Maclaurin Series

$$\frac{1}{1-x} = 1 + x + x^2 + \cdots + x^n + \cdots = \sum_{n=0}^{\infty} x^n \quad (|x| < 1)$$

$$\frac{1}{1+x} = 1 - x + x^2 - \cdots + (-x)^n + \cdots = \sum_{n=0}^{\infty}(-1)^n x^n \quad (|x| < 1)$$

$$e^x = 1 + x + \frac{x^2}{2!} + \cdots + \frac{x^n}{n!} + \cdots = \sum_{n=0}^{\infty}\frac{x^n}{n!} \quad (\text{all real } x)$$

$$\sin x = x - \frac{x^3}{3!} + \frac{x^5}{5!} - \cdots + (-1)^n \frac{x^{2n+1}}{(2n+1)!} + \cdots$$
$$= \sum_{n=0}^{\infty}(-1)^n \frac{x^{2n+1}}{(2n+1)!} \quad (\text{all real } x)$$

$$\cos x = 1 - \frac{x^2}{2!} + \frac{x^4}{4!} - \cdots + (-1)^n \frac{x^{2n}}{(2n)!} + \cdots$$
$$= \sum_{n=0}^{\infty}(-1)^n \frac{x^{2n}}{(2n)!} \quad (\text{all real } x)$$

$$\ln(1+x) = x - \frac{x^2}{2} + \frac{x^3}{3} - \cdots + (-1)^{n-1}\frac{x^n}{n} + \cdots$$
$$= \sum_{n=1}^{\infty}(-1)^{n-1}\frac{x^n}{n} \quad (-1 < x \leq 1)$$

$$\tan^{-1} x = x - \frac{x^3}{3} + \frac{x^5}{5} - \cdots + (-1)^n \frac{x^{2n+1}}{2n+1} + \cdots$$
$$= \sum_{n=0}^{\infty}(-1)^n \frac{x^{2n+1}}{2n+1} \quad (|x| \leq 1)$$

Quick Review 9.2

In Exercises 1–5, find a formula for the nth derivative of the function.

1. e^{2x}

2. $\dfrac{1}{x-1}$

3. 3^x

4. $\ln x$

5. x^n

In Exercises 6–10, find dy/dx. (Assume that letters other than x represent constants.)

6. $y = \dfrac{x^n}{n!}$

7. $y = \dfrac{2^n(x-a)^n}{n!}$

8. $y = (-1)^n \dfrac{x^{2n+1}}{(2n+1)!}$

9. $y = \dfrac{(x+a)^{2n}}{(2n)!}$

10. $y = \dfrac{(1-x)^n}{n!}$

Section 9.2 Exercises

In Exercises 1–8, use the table of Maclaurin series on the preceding page. Construct the first three nonzero terms and the general term of the Maclaurin series generated by the function and give the interval of convergence.

1. $\sin 2x$

2. $\ln (1 - x)$

3. $\tan^{-1} x^2$

4. $7x\, e^x$

5. $\cos (x + 2)$
(*Hint:* $\cos (x + 2) = (\cos 2)(\cos x) - (\sin 2)(\sin x)$)

6. $x^2 \cos x$

7. $\dfrac{x}{1 - x^3}$

8. e^{-2x}

In Exercises 9–12, find the Taylor polynomials of orders 0, 1, 2, and 3 generated by f at $x = a$.

9. $f(x) = \dfrac{1}{x}, \quad a = 2$

10. $f(x) = \sin x, \quad a = \pi/4$

11. $f(x) = \cos x, \quad a = \pi/4$

12. $f(x) = \sqrt{x}, \quad a = 4$

In Exercises 13–15, find the Taylor polynomial of order 3 generated by f

  (a) at $x = 0$;   (b) at $x = 1$.

13. $f(x) = x^3 - 2x + 4$

14. $f(x) = 2x^3 + x^2 + 3x - 8$ **15.** $f(x) = x^4$

16. Let f be a function that has derivatives of all orders for all real numbers. Assume $f(0) = 4$, $f'(0) = 5$, $f''(0) = -8$, and $f'''(0) = 6$.

  (a) Write the third order Taylor polynomial for f at $x = 0$ and use it to approximate $f(0.2)$.

  (b) Write the second order Taylor polynomial for f', the derivative of f, at $x = 0$ and use it to approximate $f'(0.2)$.

17. Let f be a function that has derivatives of all orders for all real numbers. Assume $f(1) = 4$, $f'(1) = -1$, $f''(1) = 3$, and $f'''(1) = 2$.

  (a) Write the third order Taylor polynomial for f at $x = 1$ and use it to approximate $f(1.2)$.

  (b) Write the second order Taylor polynomial for f', the derivative of f, at $x = 1$ and use it to approximate $f'(1.2)$.

18. The Maclaurin series for $f(x)$ is

$$f(x) = 1 + \frac{x}{2!} + \frac{x^2}{3!} + \frac{x^3}{4!} + \cdots + \frac{x^n}{(n + 1)!} + \cdots .$$

  (a) Find $f'(0)$ and $f^{(10)}(0)$.

  (b) Let $g(x) = xf(x)$. Write the Maclaurin series for $g(x)$, showing the first three nonzero terms and the general term.

  (c) Write $g(x)$ in terms of a familiar function without using series.

19. (a) Write the first three nonzero terms and the general term of the Taylor series generated by $e^{x/2}$ at $x = 0$.

  (b) Write the first three nonzero terms and the general term of the Taylor series at $x = 0$ for

$$g(x) = \frac{e^x - 1}{x}.$$

  (c) For the function g in part (b), find $g'(1)$ and use it to show that

$$\sum_{n=1}^{\infty} \frac{n}{(n + 1)!} = 1.$$

20. Let

$$f(t) = \frac{2}{1 - t^2} \quad \text{and} \quad G(x) = \int_0^x f(t)\, dt.$$

  (a) Find the first four terms and the general term for the Maclaurin series generated by f.

  (b) Find the first four nonzero terms and the Maclaurin series for G.

21. (a) Find the first four nonzero terms in the Taylor series generated by $f(x) = \sqrt{1 + x}$ at $x = 0$.

  (b) Use the results found in part (a) to find the first four nonzero terms in the Taylor series for $g(x) = \sqrt{1 + x^2}$ at $x = 0$.

  (c) Find the first four nonzero terms in the Taylor series at $x = 0$ for the function h such that $h'(x) = \sqrt{1 + x^2}$ and $h(0) = 5$.

22. Consider the power series

$$\sum_{n=0}^{\infty} a_n x^n, \quad \text{where } a_0 = 1 \text{ and } a_n = \left(\frac{3}{n}\right)a_{n-1} \text{ for } n \geq 1.$$

(This defines the coefficients *recursively*.)

  (a) Find the first four terms and the general term of the series.

  (b) What function f is represented by this power series?

  (c) Find the exact value of $f'(1)$.

23. Use the technique of Exploration 3 to determine the number of terms of the Maclaurin series for $\cos x$ that are needed to approximate the value of $\cos 18$ accurate to within 0.001 of the true value.

24. Writing to Learn Based on what you know about polynomial functions, explain why no Taylor polynomial of any order could actually equal $\sin x$.

25. **Writing to Learn** Your friend has memorized the Maclaurin series for both $\sin x$ and $\cos x$ but is having a hard time remembering which is which. Assuming that your friend knows the trigonometric functions well, what are some tips you could give that would help match $\sin x$ and $\cos x$ with their correct series?

26. What is the coefficient of x^5 in the Maclaurin series generated by $\sin 3x$?

27. What is the coefficient of $(x - 2)^3$ in the Taylor series generated by $\ln x$ at $x = 2$?

28. **Writing to Learn** Review the definition of the *linearization* of a differentiable function f at a in Chapter 4. What is the connection between the linearization of f and Taylor polynomials?

29. *Linearizations at Inflection Points*

(a) As the figure below suggests, linearizations fit particularly well at inflection points. As another example, graph *Newton's serpentine* $f(x) = 4x/(x^2 + 1)$ together with its linearizations at $x = 0$ and $x = \sqrt{3}$.

(b) Show that if the graph of a twice-differentiable function $f(x)$ has an inflection point at $x = a$, then the linearization of f at $x = a$ is also the second order Taylor polynomial of f at $x = a$. This explains why tangent lines fit so well at inflection points.

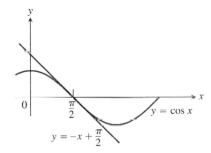

The graph of $f(x) = \cos x$ and its linearization at $\pi/2$. (Exercise 29)

30. According to the table of Maclaurin series, the power series

$$x - \frac{x^3}{3} + \frac{x^5}{5} - \cdots + (-1)^n \frac{x^{2n+1}}{2n+1} + \cdots$$

converges at $x = \pm 1$. To what number does it converge when $x = 1$? To what number does it converge when $x = -1$?

Explorations

31. (a) Using the table of Maclaurin series, find a power series to represent $f(x) = (\sin x)/x$.

(b) The power series you found in part (a) is not quite a Maclaurin series for f, because f is technically not eligible to have a Maclaurin series. Why not?

(c) If we redefine f as follows, then the power series in (a) *will* be a Maclaurin series for f. What is the value of k?

$$f(x) = \begin{cases} \dfrac{\sin x}{x}, & x \neq 0, \\ k, & x = 0 \end{cases}$$

32. *Work in groups of two or three.* Find a function f whose Maclaurin series is

$$1x^1 + 2x^2 + 3x^3 + \cdots + nx^n + \cdots. \qquad \blacksquare$$

Extending the Ideas

33. *The Binomial Series* Let $f(x) = (1 + x)^m$ for some nonzero constant m.

(a) Show that $f'''(x) = m(m - 1)(m - 2)(1 + x)^{m-3}$.

(b) Extend the result of part (a) to show that

$$f^{(k)}(0) = m(m - 1)(m - 2) \cdots (m - k + 1).$$

(c) Find the coefficient of x^k in the Maclaurin series generated by f.

(d) We define the symbol $\dbinom{m}{k}$ as follows:

$$\binom{m}{k} = \frac{m(m - 1)(m - 2) \cdots (m - k + 1)}{k!},$$

with the understanding that

$$\binom{m}{0} = 1 \quad \text{and} \quad \binom{m}{1} = m.$$

With this notation, show that the Maclaurin series generated by $f(x) = (1 + x)^m$ is

$$\sum_{k=0}^{\infty} \binom{m}{k} x^k.$$

This is called the **binomial series.**

34. *(Continuation of Exercise 33)* If m is a positive integer, explain why the Maclaurin series generated by f is a polynomial of degree m. (This means that

$$(1 + x)^m = \sum_{k=0}^{m} \binom{m}{k} x^k.$$

You may recognize this result as the **Binomial Theorem** from algebra.)

Taylor's Theorem

About Taylor Polynomials • The Remainder • Remainder Estimation Theorem • Euler's Formula

About Taylor Polynomials

While there is a certain unspoiled beauty in the exactness of a convergent Taylor series, it is the inexact Taylor polynomials that essentially do all the work. It is satisfying to know, for example, that $\sin x$ can be found *exactly* by summing an infinite Taylor series, but if we want to use that information to find $\sin 3$, we will have to evaluate Taylor polynomials until we arrive at an *approximation* with which we are satisfied. Even a computer must deal with finite sums.

Example 1 APPROXIMATING A FUNCTION TO SPECIFICATIONS

Find a Taylor polynomial that will serve as an adequate substitute for $\sin x$ on the interval $[-\pi, \pi]$.

Solution You do not have to be a professional mathematician to appreciate the imprecision of this problem as written. We are simply unable to proceed until someone decides what an "adequate" substitute is! We will revisit this issue shortly, but for now let us accept the following clarification of "adequate."

By "adequate," we mean that the polynomial should differ from $\sin x$ by less than 0.0001 anywhere on the interval.

Now we have a clear mission: Choose $P_n(x)$ so that $|P_n(x) - \sin x| < 0.0001$ for every x in the interval $[-\pi, \pi]$. How do we do this?

Recall the nine graphs of the partial sums of the Maclaurin series for $\sin x$ in Section 9.2. They show that the approximations get worse as x moves away from 0, suggesting that if we can make $|P_n(\pi) - \sin \pi| < 0.0001$, then P_n will be adequate throughout the interval. Since $\sin \pi = 0$, this means that we need to make $|P_n(\pi)| < 0.0001$.

We evaluate the partial sums at $x = \pi$, adding a term at a time, eventually arriving at the following:

$$\pi - \pi^3/3! + \pi^5/5! - \pi^7/7! + \pi^9/9! - \pi^{11}/11! + \pi^{13}/13!$$
$$2.114256749\text{E}^-5$$

As graphical support that the polynomial $P_{13}(x)$ is adequate throughout the interval, we graph the *absolute error* of the approximation, namely $|P_{13}(x) - \sin x|$, in the window $[-\pi, \pi]$ by $[-0.00004, 0.00004]$ (Figure 9.7).

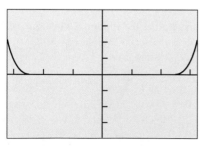

$[-\pi, \pi]$ by $[-0.00004, 0.00004]$

Figure 9.7 The graph shows that $|P_{13}(x) - \sin x| < 0.00010$ throughout the interval $[-\pi, \pi]$. (Example 1)

In practical terms, then, we would like to be able to use Taylor polynomials to approximate functions over the intervals of convergence of the Taylor series, and we would like to keep the error of the approximation within specified bounds. Since the error results from *truncating* the series down to a polynomial (that is, cutting it off after some number of terms), we call it the **truncation error.**

Example 2 TRUNCATION ERROR FOR A GEOMETRIC SERIES

Find a formula for the truncation error if we use $1 + x^2 + x^4 + x^6$ to approximate $1/(1 - x^2)$ over the interval $(-1, 1)$.

Solution We recognize this polynomial as the fourth partial sum of the geometric series for $1/(1 - x^2)$. Since this series converges to $1/(1 - x^2)$ on $(-1, 1)$, the truncation error is the absolute value of the part that we threw away, namely

$$\left| x^8 + x^{10} + \cdots + x^{2n} + \cdots \right|.$$

This is the absolute value of a geometric series with first term x^8 and $r = x^2$. Therefore,

$$\left| x^8 + x^{10} + \cdots + x^{2n} + \cdots \right| = \left| \frac{x^8}{1 - x^2} \right| = \frac{x^8}{1 - x^2}.$$

Figure 9.8 shows that the error is small near 0, but increases as x gets closer to 1 or -1.

You can probably infer from our solution in Example 2 that the truncation error after 5 terms would be $x^{10}/(1 - x^2)$, and after n terms would be $x^{2n}/(1 - x^2)$. Figure 9.9 shows how these errors get closer to 0 on the interval $(-1, 1)$ as n gets larger, and that they still get worse as we approach -1 and 1.

$y = x^8/(1 - x^2)$

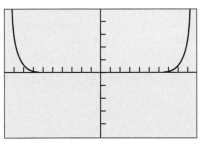

[−1, 1] by [−5, 5]

Figure 9.8 A graph of the truncation error on $(-1, 1)$ if $P_6(x)$ is used to approximate $1/(1 - x^2)$. (Example 2)

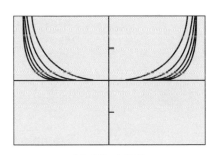

[−1, 1] by [−2, 2]

Figure 9.9 The truncation errors for $n = 2, 4, 6, 8, 10$, when we approximate $1/(1 - x^2)$ by its Taylor polynomials of higher and higher order. (The errors for the higher order polynomials are on the bottom.)

It was fortunate for our error analysis that this series was geometric, since the error was consequently a geometric series itself. This enabled us to write it as a (non-series) function and study it exactly. But how could we handle the error if we were to truncate a *nongeometric* series? That practical question sets the stage for Taylor's Theorem.

The Remainder

Every truncation splits a Taylor series into two equally significant pieces: the Taylor polynomial $P_n(x)$ that gives us the approximation, and the *remainder* $R_n(x)$ that tells us whether the approximation is any good. Taylor's Theorem is about both pieces.

Theorem 3 Taylor's Theorem with Remainder

If f has derivatives of all orders in an open interval I containing a, then for each positive integer n and for each x in I,

$$f(x) = f(a) + f'(a)(x - a) + \frac{f''(a)}{2!}(x - a)^2 + \cdots +$$

$$\frac{f^{(n)}(a)}{n!}(x - a)^n + R_n(x),$$

where

$$R_n(x) = \frac{f^{(n+1)}(c)}{(n + 1)!}(x - a)^{n+1}$$

for some c between a and x.

Pause for a moment to consider how remarkable this theorem is. If we wish to approximate f by a polynomial of degree n over an interval I, the theorem gives us both a formula for the *polynomial* and a formula for the *error* involved in using that approximation over the interval I.

The first equation in Taylor's Theorem is **Taylor's formula.** The function $R_n(x)$ is the **remainder of order n** or the **error term** for the approximation of f by $P_n(x)$ over I. It is also called the **Lagrange form** of the remainder, and bounds on $R_n(x)$ found using this form are **Lagrange error bounds.**

The introduction of $R_n(x)$ finally gives us a mathematically precise way to define what we mean when we say that a Taylor series converges to a function on an interval. If $R_n(x) \to 0$ as $n \to \infty$ for all x in I, we say that the Taylor series generated by f at $x = a$ **converges to f** on I, and we write

$$f(x) = \sum_{k=0}^{\infty} \frac{f^{(k)}(a)}{k!}(x - a)^k.$$

Example 3 PROVING CONVERGENCE OF A MACLAURIN SERIES

Prove that the series

$$\sum_{k=0}^{\infty} (-1)^k \frac{x^{2k+1}}{(2k + 1)!}$$

converges to $\sin x$ for all real x.

Solution We need to consider what happens to $R_n(x)$ as $n \to \infty$.

By Taylor's Theorem,

$$R_n(x) = \frac{f^{(n+1)}(c)}{(n + 1)!}(x - 0)^{n+1},$$

where $f^{(n+1)}(c)$ is the $(n + 1)$st derivative of $\sin x$ evaluated at some c between x and 0. This does not seem at first glance to give us much information, but *for this particular function* we can say something very significant about $f^{(n+1)}(c)$: it lies between -1 and 1 inclusive. Therefore, no matter what x is, we have

$$|R_n(x)| = \left| \frac{f^{(n+1)}(c)}{(n+1)!}(x-0)^{n+1} \right|$$

$$= \frac{|f^{(n+1)}(c)|}{(n+1)!}|x^{n+1}|$$

$$\leq \frac{1}{(n+1)!}|x^{n+1}| = \frac{|x|^{n+1}}{(n+1)!}.$$

What happens to $|x|^{n+1}/(n+1)!$ as $n \to \infty$? The numerator is a product of $n+1$ factors, all of them $|x|$. The denominator is a product of $n+1$ factors, the largest of which eventually exceed $|x|$ and keep on growing as $n \to \infty$. The factorial growth in the denominator, therefore, eventually outstrips the power growth in the numerator, and we have $|x|^{n+1}/(n+1)! \to 0$ for all x. This means that $R_n(x) \to 0$ for all x, which completes the proof.

Exploration 1   Your Turn

Modify the steps of the proof in Example 3 to prove that

$$\sum_{k=0}^{\infty} (-1)^k \frac{x^{2k}}{(2k)!}$$

converges to $\cos x$ for all real x.

Remainder Estimation Theorem

Notice that we were able to use the remainder formula in Taylor's Theorem to verify the convergence of two Taylor series to their generating functions ($\sin x$ and $\cos x$), and yet in neither case did we have to find an actual value for $f^{(n+1)}(c)$. Instead, we were able to put an *upper bound* on $|f^{(n+1)}(c)|$, which was enough to ensure that $R_n(x) \to 0$ for all x. This strategy is so convenient that we state it as a theorem for future reference.

Theorem 4   Remainder Estimation Theorem

If there are positive constants M and r such that $|f^{(n+1)}(t)| \leq Mr^{n+1}$ for all t between a and x, then the remainder $R_n(x)$ in Taylor's Theorem satisfies the inequality

$$|R_n(x)| \leq M \frac{r^{n+1}|x-a|^{n+1}}{(n+1)!}.$$

If these conditions hold for every n and all the other conditions of Taylor's Theorem are satisfied by f, then the series converges to $f(x)$.

It does not matter if M and r are huge; the important thing is that they do not get any *more huge* as $n \to \infty$. This allows the factorial growth to outstrip the power growth and thereby sweep $R_n(x)$ to zero.

Example 4   PROVING CONVERGENCE

Use the Remainder Estimation Theorem to prove that

$$e^x = \sum_{k=0}^{\infty} \frac{x^k}{k!}$$

for all real x.

Solution   We have already seen that this is the Taylor series generated by e^x at $x = 0$, so all that remains is to verify that $R_n(x) \to 0$ for all x. By the Remainder Estimation Theorem, it suffices to find M and r such that $|f^{(n+1)}(t)| = e^t$ is bounded by Mr^{n+1} for t between 0 and an arbitrary x.

We know that e^t is an increasing function on any interval, so it reaches its maximum value at the right-hand endpoint. We can pick M to be that maximum value and simply let $r = 1$. If the interval is $[0, x]$, we let $M = e^x$; if the interval is $[x, 0]$, we let $M = e^0 = 1$. In either case, we have $e^t \le M$ throughout the interval, and the Remainder Estimation Theorem guarantees convergence.

Example 5   ESTIMATING A REMAINDER

The approximation $\ln(1 + x) \approx x - (x^2/2)$ is used when x is small. Use the Remainder Estimation Theorem to get a bound for the maximum error when $|x| \le 0.1$. Support the answer graphically.

Solution   In the notation of the Remainder Estimation Theorem, $f(x) = \ln(1 + x)$, the polynomial is $P_2(x)$, and we need a bound for $|R_2(x)|$. On the interval $[-0.1, 0.1]$, the function $|f^{(3)}(t)| = 2/(1 + t)^3$ is strictly decreasing, achieving its maximum value at the left-hand endpoint, -0.1. We can therefore bound $|f^{(3)}(t)|$ by

$$M = \left| \frac{2}{(1 + (-0.1))^3} \right| = \frac{2000}{729}.$$

We can let $r = 1$.

By the Remainder Estimation Theorem, we may conclude that

$$|R_2(x)| \le \frac{2000}{729} \cdot \frac{|x|^3}{3!} \le \frac{2000}{729} \cdot \frac{|\pm 0.1|^3}{3!} < 4.6 \times 10^{-4}. \quad \text{Rounded up, to be safe}$$

Since $R_2(x) = \ln(1 + x) - (x - x^2/2)$, it is an easy matter to produce a graph to *observe* the behavior of the error on the interval $[-0.1, 0.1]$ (Figure 9.10).

The graph almost appears to have odd-function symmetry, but evaluation shows that $R_2(-0.1) \approx -3.605 \times 10^{-4}$ and $R_2(0.1) \approx 3.102 \times 10^{-4}$. The maximum *absolute* error on the interval is 3.605×10^{-4}, which is indeed less than the bound, 4.6×10^{-4}.

$y = \ln(1 + x) - (x - x^2/2)$

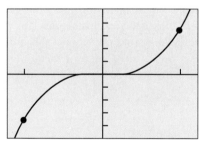

[–0.12, 0.12] by [–0.0005, 0.0005]

Figure 9.10   The graph of the error term $R_2(x)$ in Example 5. Maximum error for $|x| \le 0.1$ occurs at the left-hand endpoint of the interval.

Euler's Formula

We have seen that $\sin x$, $\cos x$, and e^x equal their respective Maclaurin series for all real numbers x. It can also be shown that this is true for all *complex* numbers, although we would need to extend our concept of limit to know what convergence would mean in that context. Accept for the moment that we can substitute complex numbers into these power series, and let us see where that might lead.

We mentioned at the beginning of the chapter that Leonhard Euler had derived some powerful results using infinite series. One of the most impressive was the surprisingly simple relationship he discovered that connects the exponential function e^x to the trigonometric functions $\sin x$ and $\cos x$. You do not need a deep understanding of complex numbers to understand what Euler did, but you do need to recall the powers of $i = \sqrt{-1}$.

$$i^1 = i$$

$$i^2 = -1$$

$$i^3 = -i$$

$$i^4 = 1$$

$$i^5 = i$$

etc.

Now try this exploration!

Exploration 2 **Euler's Formula**

Assume that e^x, $\cos x$, and $\sin x$ equal their Maclaurin series (as in the table in Section 9.2) for complex numbers as well as for real numbers.

1. Find the Maclaurin series for e^{ix}.

2. Use the result of part 1 and the Maclaurin series for $\cos x$ and $\sin x$ to prove that $e^{ix} = \cos x + i \sin x$. This equation is known as **Euler's formula.**

3. Use Euler's formula to prove that $e^{i\pi} + 1 = 0$. This beautiful equation, which brings together the five most celebrated numbers in mathematics in such a stunningly unexpected way, is also widely known as Euler's formula. (There are still others. The prolific Euler had more than his share.)

Quick Review 9.3

In Exercises 1–5, find the smallest number M that bounds $|f|$ from above on the interval I (that is, find the smallest M such that $|f(x)| \le M$ for all x in I).

1. $f(x) = 2 \cos (3x)$, $I = [-2\pi, 2\pi]$

2. $f(x) = x^2 + 3$, $I = [1, 2]$

3. $f(x) = 2^x$, $I = [-3, 0]$

4. $f(x) = \dfrac{x}{x^2 + 1}$, $I = [-2, 2]$

5. $f(x) = \begin{cases} 2 - x^2, & x \le 1, \\ 2x - 1, & x > 1, \end{cases}$ $I = [-3, 3]$

In Exercises 6–10, tell whether the function has derivatives of all orders at the given value of a.

6. $\dfrac{x}{x + 1}$, $a = 0$

7. $|x^2 - 4|$, $a = 2$

8. $\sin x + \cos x$, $a = \pi$

9. e^{-x}, $a = 0$

10. $x^{3/2}$, $a = 0$

Section 9.3 Exercises

In Exercises 1–5, find the Taylor polynomial of order four for the function at $x = 0$, and use it to approximate the value of the function at $x = 0.2$.

1. e^{-2x} **2.** $\cos(\pi x/2)$

3. $5 \sin(-x)$ **4.** $\ln(1 + x^2)$

5. $(1 - x)^{-2}$

In Exercises 6–10, find the Maclaurin series for the function.

6. xe^x **7.** $\sin x - x + \dfrac{x^3}{3!}$

8. $\cos^2 x \quad \left(= \dfrac{1 + \cos 2x}{2} \right)$ **9.** $\sin^2 x$

10. $\dfrac{x^2}{1 - 2x}$

11. For approximately what values of x can you replace $\sin x$ by $x - (x^3/6)$ with an error magnitude no greater than 5×10^{-4}? Give reasons for your answer.

12. If $\cos x$ is replaced by $1 - (x^2/2)$ and $|x| < 0.5$, what estimate can be made of the error? Does $1 - (x^2/2)$ tend to be too large or too small? Support your answer graphically.

13. How close is the approximation $\sin x \approx x$ when $|x| < 10^{-3}$? For which of these values of x is $x < \sin x$? Support your answer graphically.

14. The approximation $\sqrt{1 + x} \approx 1 + (x/2)$ is used when x is small. Estimate the maximum error when $|x| < 0.01$.

15. The approximation $e^x \approx 1 + x + (x^2/2)$ is used when x is small. Use the Remainder Estimation Theorem to estimate the error when $|x| < 0.1$.

16. Hyperbolic sine and cosine The hyperbolic sine and hyperbolic cosine functions, denoted sinh and cosh respectively, are defined as

$$\sinh x = \frac{e^x - e^{-x}}{2} \quad \text{and} \quad \cosh x = \frac{e^x + e^{-x}}{2}.$$

(Appendix A6 gives more information about hyperbolic functions.)

Find the Maclaurin series generated by $\sinh x$ and $\cosh x$.

17. *(Continuation of Exercise 16)* Use the Remainder Estimation Theorem to prove that $\cosh x$ equals its Maclaurin series for all real numbers x.

18. Writing to Learn Review the statement of the Mean Value Theorem (Section 4.2) and explain its relationship to Taylor's Theorem.

Quadratic Approximations Just as we call the Taylor polynomial of order 1 generated by f at $x = a$ the *linearization* of f at a, we call the Taylor polynomial of order 2 generated by f at $x = a$ the *quadratic approximation* of f at a.

In Exercises 19–23, find **(a)** the linearization and **(b)** the quadratic approximation of f at $x = 0$. Then **(c)** graph the function and its linear and quadratic approximations together around $x = 0$ and comment on how the graphs are related.

19. $f(x) = \ln(\cos x)$ **20.** $f(x) = e^{\sin x}$

21. $f(x) = 1/\sqrt{1 - x^2}$ **22.** $f(x) = \sec x$

23. $f(x) = \tan x$

24. Use the Taylor polynomial of order 2 to find the quadratic approximation of $f(x) = (1 + x)^k$ at $x = 0$ (k a constant). If $k = 3$, for approximately what values of x in the interval $[0, 1]$ will the magnitude of the error in the quadratic approximation be less than $1/100$?

25. *A Cubic Approximation of e^x* The approximation

$$e^x \approx 1 + x + \frac{x^2}{2} + \frac{x^3}{6}$$

is used on small intervals about the origin. Estimate the magnitude of the approximation error for $|x| \le 0.1$.

26. *A Cubic Approximation* Use the Taylor polynomial of order 3 to find the *cubic approximation* of $f(x) = 1/(1 - x)$ at $x = 0$. Give an upper bound for the magnitude of the approximation error for $|x| \le 0.1$.

27. Consider the initial value problem,

$$\frac{dy}{dx} = e^{-x^2} \quad \text{and} \quad y = 2 \quad \text{when} \quad x = 0.$$

(a) Can you find a formula for the function y that does not involve any integrals?

(b) Can you represent y by a power series?

(c) For what values of x does this power series actually equal the function y? Give a reason for your answer.

28. (a) Construct the Maclaurin series for $\ln(1 - x)$.

(b) Use this series and the series for $\ln(1 + x)$ to construct a Maclaurin series for

$$\ln \frac{1 + x}{1 - x}.$$

29. *Identifying Graphs* Which well-known functions are approximated on the interval $(-\pi/2, \pi/2)$ by the following Taylor polynomials?

(a) $x + \dfrac{x^3}{3} + \dfrac{2x^5}{15} + \dfrac{17x^7}{315} + \dfrac{62x^9}{2835}$

(b) $1 + \dfrac{x^2}{2} + \dfrac{5x^4}{24} + \dfrac{61x^6}{720} + \dfrac{277x^8}{8064}$

Explorations

30. *Work in groups of two or three.* Try to reinforce each other's ideas and verify your computations at each step.

(a) Use the identity

$$\sin^2 x = \frac{1}{2}(1 - \cos 2x)$$

to obtain the Maclaurin series for $\sin^2 x$.

(b) Differentiate this series to obtain the Maclaurin series for $2 \sin x \cos x$.

(c) Verify that this is the series for $\sin 2x$.

31. *Improving Approximations to π*

(a) Let P be an approximation of π accurate to n decimal places. Check with a calculator to see that $P + \sin P$ gives an approximation correct to $3n$ decimal places!

(b) Use the Remainder Estimation Theorem and the Maclaurin series for $\sin x$ to explain what is happening in part (a). (*Hint:* Let $P = \pi + x$, where x is the error of the estimate. Why should $(P + \sin P) - \pi$ be less than x^3?)

32. *Euler's Identities* Use Euler's formula to show that

(a) $\cos \theta = \dfrac{e^{i\theta} + e^{-i\theta}}{2}$, and

(b) $\sin \theta = \dfrac{e^{i\theta} - e^{-i\theta}}{2i}$.

Extending the Ideas

33. When a and b are real numbers, we define $e^{(a+ib)x}$ with the equation

$$e^{(a+ib)x} = e^{ax} \cdot e^{ibx} = e^{ax}(\cos bx + i \sin bx).$$

Differentiate the right-hand side of this equation to show that

$$\frac{d}{dx}e^{(a+ib)x} = (a + ib)\,e^{(a+ib)x}.$$

Thus, the familiar rule

$$\frac{d}{dx}e^{kx} = ke^{kx}$$

holds for complex values of k as well as for real values.

34. *(Continuation of Exercise 33)*

(a) Confirm the antiderivative formula

$$\int e^{(a+ib)x}\,dx = \frac{a - ib}{a^2 + b^2}\,e^{(a+ib)x} + C$$

by differentiating both sides. (In this case, $C = C_1 + iC_2$ is a complex constant of integration.)

(b) Two complex numbers $a + ib$ and $c + id$ are equal if and only if $a = c$ and $b = d$. Use this fact and the formula in (a) to evaluate $\int e^{ax} \cos bx\,dx$ and $\int e^{ax} \sin bx\,dx$.

■

9.4 # Radius of Convergence

Convergence • *n*th-Term Test • Comparing Nonnegative Series • Ratio Test • Endpoint Convergence

Convergence

Throughout our explorations of infinite series we stressed the importance of convergence. In terms of numbers, the difference between a convergent series and a divergent series could hardly be more stark: a convergent series is a number and may be treated as such; a divergent series is not a number and must not be treated as one.

Recall that the symbol "=" means many different things in mathematics.

1. $1 + 1 = 2$ signifies *equality of real numbers*. It is a true sentence.

2. $2(x - 3) = 2x - 6$ signifies *equivalent expressions*. It is a true sentence.

3. $x^2 + 3 = 7$ is an *equation*. It is an *open sentence*, because it can be true or false, depending on whether x is a solution to the equation.

4. $(x^2 - 1)/(x + 1) = x - 1$ is an *identity*. It is a true sentence (very much like (2)), but with the important qualification that *x must be in the domain of both expressions*. If either side of the equality is undefined, the sentence is meaningless. Substituting -1 into both sides of the equation in (3) gives a sentence that is mathematically false (i.e., $4 = 7$); substituting -1 into both sides of this identity gives a sentence that is meaningless.

Seki Kowa (1642–1708)

Child prodigy, brilliant mathematician, and inspirational teacher, Seki Kowa was born into a samurai warrior family in Fujioka, Kozuke, Japan, and adopted by the family of an accountant. Among his contributions were an improved method of solving higher-degree equations, the use of determinants in solving simultaneous equations, and a form of calculus known in Japan as *yenri*. It is difficult to know the full extent of his work because the samurai code demanded great modesty. Seki Kowa is credited with awakening in Japan a scientific spirit that continues to this day.

$y = \dfrac{1}{1 + x^2}$

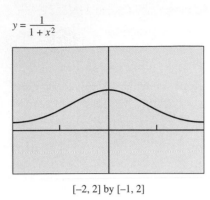

[−2, 2] by [−1, 2]

(a)

Partial Sums

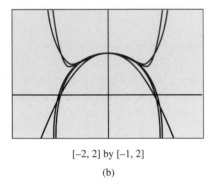

[−2, 2] by [−1, 2]

(b)

Figure 9.11 (a) The graph of $y = 1/(1 + x^2)$ and (b) the graphs of the Taylor polynomials $P_2(x)$, $P_4(x)$, $P_6(x)$, $P_8(x)$, and $P_{10}(x)$. The approximations become better and better, but only over the interval of convergence $(-1, 1)$. (Example 1)

Example 1   THE IMPORTANCE OF CONVERGENCE

Consider the sentence

$$\frac{1}{1 + x^2} = 1 - x^2 + x^4 - x^6 + \cdots + (-1)^n x^{2n} + \cdots .$$

For what values of x is this an identity?

Solution   The function on the left has domain all real numbers. The function on the right can be viewed as a limit of Taylor polynomials. Each Taylor polynomial has domain all real numbers, but the polynomial values *converge* only when $|x| < 1$, so the *series* has domain $(-1, 1)$. If we graph the Taylor polynomials (Figure 9.11) we can see the dramatic convergence to $1/(1 + x^2)$ over the interval $(-1, 1)$. The divergence is just as dramatic for $|x| \geq 1$.

For values of x outside the interval, the statement in this example is meaningless. The Taylor series on the right diverges so it is not a number. The sentence is an identity for x in $(-1, 1)$.

As convincing as these graphs are, they do not *prove* convergence or divergence as $n \to \infty$. The series in Example 1 happens to be geometric, so we do have an analytic proof that it converges for $|x| < 1$ and diverges for $|x| \geq 1$, but for nongeometric series we do not have such undeniable assurance about convergence (yet).

In this section we develop a strategy for finding the interval of convergence of an arbitrary power series and backing it up with proof. We begin by noting that any power series of the form $\sum_{n=0}^{\infty} c_n(x - a)^n$ always converges at $x = a$, thus assuring us of at least one coordinate on the real number line where the series must converge. We have encountered power series that converge for all real numbers (the Maclaurin series for $\sin x$, $\cos x$, and e^x), and we have encountered power series like the series in Example 1 that converge only on a finite interval centered at a. A useful fact about power series is that those are the only possibilities, as the following theorem attests.

Theorem 5   The Convergence Theorem for Power Series

There are three possibilities for $\sum_{n=0}^{\infty} c_n(x - a)^n$ with respect to convergence:

1. There is a positive number R such that the series diverges for $|x - a| > R$ but converges for $|x - a| < R$. The series may or may not converge at either of the endpoints $x = a - R$ and $x = a + R$.

2. The series converges for every x $(R = \infty)$.

3. The series converges at $x = a$ and diverges elsewhere $(R = 0)$.

The number R is the **radius of convergence,** and the set of all values of x for which the series converges is the **interval of convergence.** The radius of convergence completely determines the interval of convergence if R is either zero or infinite. For $0 < R < \infty$, however, there remains the question of what happens at the endpoints of the interval. The table of Maclaurin series at the end of Section 9.2 includes intervals of convergence that are open, half-open, and closed.

We will learn how to find the radius of convergence first, and then we will settle the endpoint question in Section 9.5.

*n*th-Term Test

The most obvious requirement for convergence of a series is that the *n*th term must go to zero as $n \to \infty$. If the partial sums are approaching a limit S, then they also must be getting close to one another, so that for a convergent series $\sum a_n$,

$$\lim_{n \to \infty} a_n = \lim_{n \to \infty} (S_n - S_{n-1}) = S - S = 0.$$

This gives a handy test for divergence:

Theorem 6   The *n*th-Term Test for Divergence

$\sum_{n=1}^{\infty} a_n$ diverges if $\lim_{n \to \infty} a_n$ fails to exist or is different from zero.

Example 2   A SERIES WITH RADIUS OF CONVERGENCE 0

Find the radius of convergence of the series $\sum_{n=0}^{\infty} n! x^n$.

Solution If $|x| \geq 1$, the terms $n! x^n$ grow without bound as $n \to \infty$, $\lim_{n \to \infty} n! x^n$ fails to exist, and the series diverges by the *n*th-Term Test.

If $0 < |x| < 1$, then

$$\lim_{n \to \infty} n! x^n = \lim_{n \to \infty} \frac{n!}{(1/x)^n},$$

a quotient in which both numerator and denominator grow without bound. Recall, however, that factorial growth outstrips power growth, so again the limit fails to exist.

If $x = 0$, the series does converge. The radius of convergence is $R = 0$, and the interval of convergence is $\{0\}$.

Comparing Nonnegative Series

An effective way to show that a series $\sum a_n$ of nonnegative numbers *converges* is to compare it term by term with a known convergent series $\sum c_n$.

Theorem 7   The Direct Comparison Test

Let $\sum a_n$ be a series with no negative terms.

(a) $\sum a_n$ converges if there is a convergent series $\sum c_n$ with $a_n \leq c_n$ for all $n > N$, for some integer N.

(b) $\sum a_n$ diverges if there is a divergent series $\sum d_n$ of nonnegative terms with $a_n \geq d_n$ for all $n > N$, for some integer N.

If we can show that $\sum a_n$, $a_n \geq 0$ is eventually dominated by a convergent series, that will establish the convergence of $\sum a_n$. If we can show that $\sum a_n$ eventually dominates a divergent series of nonnegative terms, that will establish the divergence of $\sum a_n$.

We leave the proof to Exercises 42 and 43.

Example 3 PROVING CONVERGENCE BY COMPARISON

Prove that $\displaystyle\sum_{n=0}^{\infty} \frac{x^{2n}}{(n!)^2}$ converges for all real x.

Solution Let x be any real number. The series

$$\sum_{n=0}^{\infty} \frac{x^{2n}}{(n!)^2}$$

has no negative terms.

For any n, we have

$$\frac{x^{2n}}{(n!)^2} \leq \frac{x^{2n}}{n!} = \frac{(x^2)^n}{n!}.$$

We recognize

$$\sum_{n=0}^{\infty} \frac{(x^2)^n}{n!}$$

as the Taylor series for e^{x^2}, which we know converges to e^{x^2} for all real numbers. Since the e^{x^2} series dominates

$$\sum_{n=0}^{\infty} \frac{x^{2n}}{(n!)^2}$$

term by term, the latter series must also converge for all real numbers by the Direct Comparison Test.

For the Direct Comparison Test to apply, the terms of the unknown series must be *nonnegative*. The fact that $\sum a_n$ is dominated by a convergent positive series means nothing if $\sum a_n$ diverges to $-\infty$. You might think that the requirement of nonnegativity would limit the usefulness of the Direct Comparison Test, but in practice this does not turn out to be the case. We can apply our test to $\sum |a_n|$ (which certainly has no negative terms); if $\sum |a_n|$ converges, then $\sum a_n$ converges.

Definition Absolute Convergence

If the series $\sum |a_n|$ of absolute values converges, then $\sum a_n$ **converges absolutely.**

Theorem 8 Absolute Convergence Implies Convergence

If $\sum |a_n|$ converges, then $\sum a_n$ converges.

Proof For each n,

$$-|a_n| \leq a_n \leq |a_n|, \quad \text{so} \quad 0 \leq a_n + |a_n| \leq 2|a_n|.$$

If $\sum |a_n|$ converges, then $\sum 2|a_n|$ converges, and by the Direct Comparison Test, the nonnegative series $\sum (a_n + |a_n|)$ converges. The equality $a_n = (a_n + |a_n|) - |a_n|$ now allows us to express $\sum a_n$ as the difference of two convergent series:

$$\sum a_n = \sum (a_n + |a_n| - |a_n|) = \sum (a_n + |a_n|) - \sum |a_n|.$$

Therefore, $\sum a_n$ converges. ∎

Example 4 USING ABSOLUTE CONVERGENCE

Show that

$$\sum_{n=0}^{\infty} \frac{(\sin x)^n}{n!}$$

converges for all x.

Solution Let x be any real number. The series

$$\sum_{n=0}^{\infty} \frac{|\sin x|^n}{n!}$$

has no negative terms, and it is term-by-term less than or equal to the series $\sum_{n=0}^{\infty} (1/n!)$, which we know converges to e. Therefore,

$$\sum_{n=0}^{\infty} \frac{|\sin x|^n}{n!}$$

converges by direct comparison. Since

$$\sum_{n=0}^{\infty} \frac{(\sin x)^n}{n!}$$

converges absolutely, it converges.

Ratio Test

Our strategy for finding the radius of convergence for an arbitrary power series will be to check for absolute convergence using a powerful test called the *Ratio Test*.

L'Hôpital's Rule is occasionally helpful in determining the limits that arise here.

Theorem 9 The Ratio Test

Let $\sum a_n$ be a series with positive terms, and with

$$\lim_{n \to \infty} \frac{a_{n+1}}{a_n} = L.$$

Then,

(a) the series *converges* if $L < 1$,

(b) the series *diverges* if $L > 1$,

(c) the test is *inconclusive* if $L = 1$.

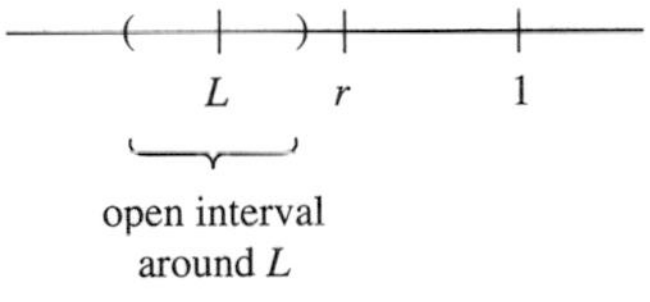

---

open interval
around L

Figure 9.12 Since

$$\lim_{n \to \infty} \frac{a_{n+1}}{a_n} = L,$$

there is some N large enough so that a_{n+1}/a_n lies inside this open interval around L for all $n \geq N$. This guarantees that $a_{n+1}/a_n < r < 1$ for all $n \geq N$.

Proof

(a) $L < 1$:

Choose some number r such that $L < r < 1$. Since

$$\lim_{n \to \infty} \frac{a_{n+1}}{a_n} = L,$$

we know that there is some N large enough so that a_{n+1}/a_n is arbitrarily close to L for all $n \geq N$. In particular, we can guarantee that for some N large enough, $(a_{n+1}/a_n) < r$ for all $n \geq N$. (See Figure 9.12.)

Thus,

$$\frac{a_{N+1}}{a_N} < r \quad \text{so} \quad a_{N+1} < ra_N$$

$$\frac{a_{N+2}}{a_{N+1}} < r \quad \text{so} \quad a_{N+2} < ra_{N+1} < r^2a_N$$

$$\frac{a_{N+3}}{a_{N+2}} < r \quad \text{so} \quad a_{N+3} < ra_{N+2} < r^3a_N$$

$$\vdots$$

This shows that for $n \geq N$ we can dominate $\Sigma\, a_n$ by $a_N(1 + r + r^2 + \cdots)$. Since $0 < r < 1$, this latter series is a convergent geometric series, and so $\Sigma\, a_n$ converges by the Direct Comparison Test.

(b) $L > 1$:

From some index M,

$$\frac{a_{n+1}}{a_n} > 1$$

for all $n \geq M$. In particular,

$$a_M < a_{M+1} < a_{M+2} < \cdots.$$

The terms of the series do not approach 0, so $\Sigma\, a_n$ diverges by the nth-Term Test.

(c) $L = 1$:

In Exploration 1 you will finish the proof by showing that the Ratio Test is inconclusive when $L = 1$. ∎

A Note on Absolute Convergence: The proof of the Ratio Test shows that the convergence of a power series inside its radius of convergence is *absolute* convergence, a stronger result than we first stated in Theorem 5. We will learn more about the distinction between convergence and absolute convergence in Section 9.5.

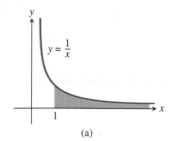

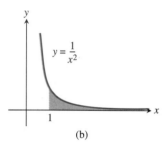

Figure 9.13 Find these areas. (Exploration 1)

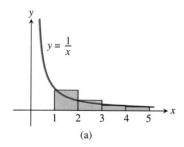

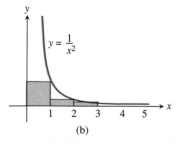

Figure 9.14 The areas of the rectangles form a series in each case. (Exploration 1)

Exploration 1   **Finishing the Proof of the Ratio Test**

Consider

$$\sum_{n=1}^{\infty} \frac{1}{n} \quad \text{and} \quad \sum_{n=1}^{\infty} \frac{1}{n^2}.$$

(We will refer to them hereafter in this exploration as $\Sigma\, 1/n$ and $\Sigma\, 1/n^2$.)

1. Show that the Ratio Test yields $L = 1$ for both series.

2. Use improper integrals to find the areas shaded in Figures 9.13a and 9.13b for $1 \leq x < \infty$.

3. Explain how Figure 9.14a shows that $\Sigma\, 1/n$ diverges, while Figure 9.14b shows that $\Sigma\, 1/n^2$ converges.

4. Explain how this proves the last part of the Ratio Test.

Example 5 FINDING THE RADIUS OF CONVERGENCE

Find the radius of convergence of

$$\sum_{n=0}^{\infty} \frac{nx^n}{10^n}.$$

Solution We check for absolute convergence using the Ratio Test.

$$\lim_{n\to\infty} \frac{|a_{n+1}|}{|a_n|} = \lim_{n\to\infty} \frac{(n+1)\,|x^{n+1}|}{10^{n+1}} \cdot \frac{10^n}{n|x^n|}$$

$$= \lim_{n\to\infty} \left(\frac{n+1}{n}\right) \frac{|x|}{10} = \frac{|x|}{10}$$

Setting $|x|/10 < 1$, we see that the series converges absolutely (and hence converges) for $-10 < x < 10$. The series diverges for $|x| > 10$, which means (by Theorem 5, the Convergence Theorem for Power Series) that it diverges for $x > 10$ and for $x < -10$. The radius of convergence is 10.

Endpoint Convergence

The Ratio Test, which is really a test for absolute convergence, establishes the radius of convergence for $\sum |c_n(x-a)^n|$. Theorem 5 guarantees that this is the same as the radius of convergence of $\sum c_n(x-a)^n$. Therefore, all that remains to be resolved about the convergence of an arbitrary power series is the question of convergence at the endpoints of the convergence interval when the radius of convergence is a finite, nonzero number.

Exploration 2 **Revisiting a Maclaurin Series**

For what values of x does the series

$$x - \frac{x^2}{2} + \frac{x^3}{3} - \cdots + (-1)^{n-1}\frac{x^n}{n} + \cdots$$

converge?

1. Apply the Ratio Test to determine the radius of convergence.

2. Substitute the left-hand endpoint of the interval into the power series. Use Figure 9.14a of Exploration 1 to help you decide whether the resulting series converges or diverges.

3. Substitute the right-hand endpoint of the interval into the power series. You should get

$$1 - \frac{1}{2} + \frac{1}{3} - \frac{1}{4} + \cdots + \frac{(-1)^{n-1}}{n} + \cdots.$$

Chart the progress of the partial sums of this series geometrically on a number line as follows: Start at 0. Go forward 1. Go back 1/2. Go forward 1/3. Go back 1/4. Go forward 1/5, etc.

4. Does the series converge at the right-hand endpoint? Give a convincing argument based on your geometric journey in part 3.

5. Does the series converge *absolutely* at the right-hand endpoint?

The question of convergence of a power series at an endpoint is really a question about the convergence of a series of numbers. If the series is geometric with first term a and common ratio r, then the series converges to $a/(1 - r)$ if $|r| < 1$ and diverges if $|r| \geq 1$. Another type of series whose sums are easily found are **telescoping series**, as illustrated in Example 6.

Telescoping Series

We call the series in Example 6 a *telescoping series* because its partial sums collapse like an old handheld telescope.

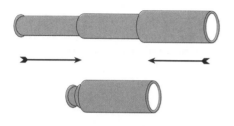

Example 6   SUMMING A TELESCOPING SERIES

Find the sum of $\displaystyle\sum_{n=1}^{\infty} \frac{1}{n(n + 1)}$.

Solution Use partial fractions to rewrite the nth term.

$$\frac{1}{n(n + 1)} = \frac{1}{n} - \frac{1}{n + 1}$$

We compute a few partial sums to find a general formula.

$$s_1 = 1 - \frac{1}{2}$$

$$s_2 = \left(1 - \frac{1}{2}\right) + \left(\frac{1}{2} - \frac{1}{3}\right) = 1 - \frac{1}{3}$$

$$s_3 = \left(1 - \frac{1}{2}\right) + \left(\frac{1}{2} - \frac{1}{3}\right) + \left(\frac{1}{3} - \frac{1}{4}\right) = 1 - \frac{1}{4}$$

We can see that, in general,

$$s_n = 1 - \frac{1}{n},$$

because all the terms between the first and last cancel when the parentheses are removed. Therefore, the sum of the series is

$$S = \lim_{n \to \infty} s_n = 1.$$

The final section of this chapter will formalize some of the strategies used in Exploration 2 and Example 6 and will develop additional tests that can be used to determine series behavior at endpoints.

Quick Review 9.4

In Exercises 1–5, find the limit of the expression as $n \to \infty$. Assume x remains fixed as n changes.

1. $\dfrac{n|x|}{n + 1}$

2. $\dfrac{n^2 |x - 3|}{n(n - 1)}$

3. $\dfrac{|x|^n}{n!}$

4. $\dfrac{(n + 1)^4 x^2}{(2n)^4}$

5. $\dfrac{|2x + 1|^{n+1} 2^n}{2^{n+1} |2x + 1|^n}$

In Exercises 6–10, let a_n be the nth term of the first and b_n the nth term of the second series. Find the smallest positive integer N for which $a_n > b_n$ for all $n \geq N$. Identify a_n and b_n.

6. $\sum 5n$, $\sum n^2$

7. $\sum n^5$, $\sum 5^n$

8. $\sum \ln n$, $\sum \sqrt{n}$

9. $\sum \dfrac{1}{10^n}$, $\sum \dfrac{1}{n!}$

10. $\sum \dfrac{1}{n^2}$, $\sum n^{-3}$

Section 9.4 Exercises

In Exercises 1–16, determine the convergence or divergence of the series. Identify the test (or tests) you use. There may be more than one correct way to determine convergence or divergence of a given series.

1. $\displaystyle\sum_{n=1}^{\infty} \frac{n}{n+1}$

2. $\displaystyle\sum_{n=1}^{\infty} \frac{2^n}{n+1}$

3. $\displaystyle\sum_{n=1}^{\infty} \frac{n^2-1}{2^n}$

4. $\displaystyle\sum_{n=1}^{\infty} -\frac{1}{8^n}$

5. $\displaystyle\sum_{n=1}^{\infty} \frac{2^n}{3^n+1}$

6. $\displaystyle\sum_{n=1}^{\infty} n \sin\left(\frac{1}{n}\right)$

7. $\displaystyle\sum_{n=0}^{\infty} n^2 e^{-n}$

8. $\displaystyle\sum_{n=0}^{\infty} \frac{n^{10}}{10^n}$

9. $\displaystyle\sum_{n=1}^{\infty} \frac{(n+3)!}{3!\,n!\,3^n}$

10. $\displaystyle\sum_{n=1}^{\infty} \left(1+\frac{1}{n}\right)^n$

11. $\displaystyle\sum_{n=0}^{\infty} \frac{(-2)^n}{3^n}$

12. $\displaystyle\sum_{n=1}^{\infty} n!\,e^{-n}$

13. $\displaystyle\sum_{n=1}^{\infty} \frac{3^n}{n^3 2^n}$

14. $\displaystyle\sum_{n=1}^{\infty} \frac{n \ln n}{2^n}$

15. $\displaystyle\sum_{n=1}^{\infty} \frac{n!}{(2n+1)!}$

16. $\displaystyle\sum_{n=1}^{\infty} \frac{n!}{n^n}$ (*Hint:* If you do not recognize L, try recognizing the reciprocal of L.)

17. Give an example to show that the converse of the nth-Term Test is false. That is, $\sum a_n$ might diverge even though $\lim_{n\to\infty} a_n = 0$.

18. Find two convergent series $\sum a_n$ and $\sum b_n$ such that $\sum (a_n/b_n)$ diverges.

In Exercises 19–34, find the *radius* of convergence of the power series.

19. $\displaystyle\sum_{n=0}^{\infty} x^n$

20. $\displaystyle\sum_{n=0}^{\infty} (x+5)^n$

21. $\displaystyle\sum_{n=0}^{\infty} (-1)^n (4x+1)^n$

22. $\displaystyle\sum_{n=1}^{\infty} \frac{(3x-2)^n}{n}$

23. $\displaystyle\sum_{n=0}^{\infty} \frac{(x-2)^n}{10^n}$

24. $\displaystyle\sum_{n=0}^{\infty} \frac{nx^n}{n+2}$

25. $\displaystyle\sum_{n=1}^{\infty} \frac{x^n}{n\sqrt{n}\,3^n}$

26. $\displaystyle\sum_{n=0}^{\infty} \frac{x^{2n+1}}{n!}$

27. $\displaystyle\sum_{n=0}^{\infty} \frac{n(x+3)^n}{5^n}$

28. $\displaystyle\sum_{n=0}^{\infty} \frac{nx^n}{4^n(n^2+1)}$

29. $\displaystyle\sum_{n=0}^{\infty} \frac{\sqrt{n}\,x^n}{3^n}$

30. $\displaystyle\sum_{n=0}^{\infty} n!(x-4)^n$

31. $\displaystyle\sum_{n=0}^{\infty} (-2)^n (n+1)(x-1)^n$

32. $\displaystyle\sum_{n=1}^{\infty} \frac{(4x-5)^{2n+1}}{n^{3/2}}$

33. $\displaystyle\sum_{n=1}^{\infty} \frac{(x+\pi)^n}{\sqrt{n}}$

34. $\displaystyle\sum_{n=0}^{\infty} \frac{(x-\sqrt{2})^{2n+1}}{2^n}$

In Exercises 35–40, find the *interval* of convergence of the series and, within this interval, the sum of the series as a function of x.

35. $\displaystyle\sum_{n=0}^{\infty} \frac{(x-1)^{2n}}{4^n}$

36. $\displaystyle\sum_{n=0}^{\infty} \frac{(x+1)^{2n}}{9^n}$

37. $\displaystyle\sum_{n=0}^{\infty} \left(\frac{\sqrt{x}}{2}-1\right)^n$

38. $\displaystyle\sum_{n=0}^{\infty} (\ln x)^n$

39. $\displaystyle\sum_{n=0}^{\infty} \left(\frac{x^2-1}{3}\right)^n$

40. $\displaystyle\sum_{n=0}^{\infty} \left(\frac{\sin x}{2}\right)^n$

41. Writing to Learn We reviewed in Section 9.1 how to find the interval of convergence for the geometric series $\sum_{n=0}^{\infty} x^n$. Can we find the interval of convergence of a geometric series by using the Ratio Test? Explain.

Explorations

Nondecreasing Sequences As you already know, a nondecreasing (or increasing) function $f(x)$ that is bounded from above on an interval $[a, \infty)$ has a limit as $x\to\infty$ that is less than or equal to the bound. The same is true of sequences of numbers. If $s_1 \le s_2 \le s_3 \le \ldots \le s_n \ldots$ and there is a number M such that $|s_n| \le M$ for all n, then the sequence converges to a limit $S \le M$. You will need this fact as you work through Exercises **42** and **43**. *Work in groups of two or three.*

42. *Proof of the Direct Comparison Test, Part a* Let $\sum a_n$ be a series with no negative terms, and let $\sum c_n$ be a convergent series such that $a_n \le c_n$ for all $n \ge N$, for some integer N.

(a) Show that the partial sums of $\sum a_n$ are bounded above by

$$a_1 + \cdots + a_N + \sum_{n=N+1}^{\infty} c_n.$$

(b) Explain why this shows that $\sum a_n$ must converge.

43. *Proof of the Direct Comparison Test, Part b* Let $\sum a_n$ be a series with no negative terms, and let $\sum d_n$ be a divergent series of nonnegative terms such that $a_n \geq d_n$ for all $n \geq N$, for some integer N.

(a) Show that the partial sums of $\sum d_n$ are bounded above by

$$d_1 + \cdots + d_N + \sum_{n=N+1}^{\infty} a_n.$$

(b) Explain why this leads to a contradiction if we assume that $\sum a_n$ converges.

44. *Work in groups of n students each.* Within your group, have each student make up a power series with radius of convergence equal to one of the numbers $1, 2, \ldots, n$. Then exchange series with another group and match the other group's series with the correct radii of convergence. ■

In Exercises 45–51, find the sum of the telescoping series.

45. $\displaystyle\sum_{n=1}^{\infty} \frac{4}{(4n-3)(4n+1)}$

46. $\displaystyle\sum_{n=1}^{\infty} \frac{6}{(2n-1)(2n+1)}$

47. $\displaystyle\sum_{n=1}^{\infty} \frac{40n}{(2n-1)^2(2n+1)^2}$

48. $\displaystyle\sum_{n=1}^{\infty} \frac{2n+1}{n^2(n+1)^2}$

49. $\displaystyle\sum_{n=1}^{\infty} \left(\frac{1}{\sqrt{n}} - \frac{1}{\sqrt{n+1}} \right)$

50. $\displaystyle\sum_{n=1}^{\infty} \left(\frac{1}{\ln(n+2)} - \frac{1}{\ln(n+1)} \right)$

51. $\displaystyle\sum_{n=1}^{\infty} (\tan^{-1}(n) - \tan^{-1}(n+1))$

Extending the Ideas

52. We can show that the series

$$\sum_{n=0}^{\infty} \frac{n^2}{2^n}$$

converges by the Ratio Test, but what is its sum?

To find out, express $1/(1-x)$ as a geometric series. Differentiate both sides of the resulting equation with respect to x, multiply both sides of the result by x, differentiate again, multiply by x again, and set x equal to $1/2$. What do you get? (*Source:* David E. Dobbs's letter to the editor, *Illinois Mathematics Teacher,* Vol. 33, Issue 4, 1982, p. 27.)

9.5 Testing Convergence at Endpoints

Integral Test • Harmonic Series and *p*-series • Comparison Tests • Alternating Series • Absolute and Conditional Convergence • Intervals of Convergence • A Word of Caution

Integral Test

In Exploration 1 of Section 9.4, you showed that $\sum 1/n$ *diverges* by modeling it as a sum of rectangle areas that contain the area under the curve $y = 1/x$ from 1 to ∞. You also showed that $\sum 1/n^2$ *converges* by modeling it as a sum of rectangle areas contained by the area under the curve $y = 1/x^2$ from 1 to ∞. This area-based convergence test in its general form is known as the *Integral Test*.

Theorem 10 The Integral Test

Let $\{a_n\}$ be a sequence of positive terms. Suppose that $a_n = f(n)$, where f is a continuous, positive, decreasing function of x for all $x \geq N$ (N a positive integer). Then the series $\sum_{n=N}^{\infty} a_n$ and the integral $\int_N^{\infty} f(x)\, dx$ either both converge or both diverge.

Proof We will illustrate the proof for $N = 1$ to keep the notation simple, but the illustration can be shifted horizontally to any value of N without affecting the logic of the proof.

The proof is entirely contained in these two pictures (Figure 9.15):

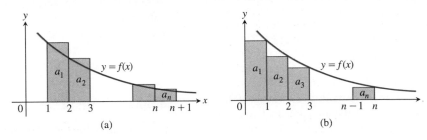

Figure 9.15 (a) The sum $a_1 + a_2 + \cdots + a_n$ provides an upper bound for $\int_1^{n+1} f(x)\,dx$. (b) The sum $a_2 + a_3 + \cdots + a_n$ provides a lower bound for $\int_1^n f(x)\,dx$. (Theorem 10)

We leave it to you (in Exercise 44) to supply the words. ∎

Example 1 APPLYING THE INTEGRAL TEST

Does $\displaystyle\sum_{n=1}^{\infty} \frac{1}{n\sqrt{n}}$ converge?

Solution The Integral Test applies because

$$f(x) = \frac{1}{x\sqrt{x}}$$

is a continuous, positive, decreasing function of x for $x > 1$.

We have

$$\int_1^{\infty} \frac{1}{x\sqrt{x}}\,dx = \lim_{k\to\infty} \int_1^k x^{-3/2}\,dx$$

$$= \lim_{k\to\infty} \left[-2x^{-1/2} \right]_1^k$$

$$= \lim_{k\to\infty} \left(-\frac{2}{\sqrt{k}} + 2 \right)$$

$$= 2.$$

Since the integral converges, so must the series.

Harmonic Series and *p*-series

The Integral Test can be used to settle the question of convergence for any series of the form $\sum_{n=1}^{\infty} (1/n^p)$, p a real constant. (The series in Example 1 had this form, with $p = 3/2$.) Such a series is called a ***p*-series.**

Caution

The series and the integral in the Integral Test need not have the same value in the convergent case. Although the integral converges to 2 in Example 1, the series might have a quite different sum. If you use your calculator to compute or graph partial sums for the series, you can see that the 11th partial sum is already greater than 2. The *Technology Resource Manual* contains two programs, PARTSUMT, which displays partial sums in table form, and PARTSUMG, which displays partial sums graphically.

Exploration 1 The *p*-Series Test

1. Use the Integral Test to prove that $\sum_{n=1}^{\infty}(1/n^p)$ converges if $p > 1$.

2. Use the Integral Test to prove that $\sum_{n=1}^{\infty}(1/n^p)$ diverges if $p < 1$.

3. Use the Integral Test to prove that $\sum_{n=1}^{\infty}(1/n^p)$ diverges if $p = 1$.

What is harmonic about the harmonic series?

The terms in the harmonic series correspond to the nodes on a vibrating string that produce multiples of the fundamental frequency. For example, 1/2 produces the harmonic that is twice the fundamental frequency, 1/3 produces a frequency that is three times the fundamental frequency, and so on. The fundamental frequency is the lowest note or pitch we hear when a string is plucked. (Figure 9.17)

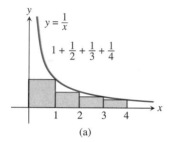

(a)

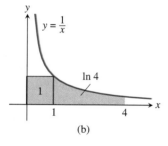

(b)

Figure 9.16 Finding an upper bound for one of the partial sums of the harmonic series. (Example 2)

The p-series with $p = 1$ is the **harmonic series,** and it is probably the most famous divergent series in mathematics. The p-Series Test shows that the harmonic series is just *barely* divergent; if we increase p to 1.000000001, for instance, the series converges!

The slowness with which the harmonic series approaches infinity is most impressive. Consider the following example.

Example 2   THE SLOW DIVERGENCE OF THE HARMONIC SERIES

Approximately how many terms of the harmonic series are required to form a partial sum larger than 20?

Solution Before you set your graphing calculator to the task of finding this number, you might want to estimate how long the calculation might take. The graphs tell the story (Figure 9.16).

Let H_n denote the nth partial sum of the harmonic series. Comparing the two graphs, we see that $H_4 < (1 + \ln 4)$ and (in general) that $H_n \le (1 + \ln n)$. If we wish H_n to be greater than 20, then

$$1 + \ln n > H_n > 20$$
$$1 + \ln n > 20$$
$$\ln n > 19$$
$$n > e^{19}.$$

The exact value of e^{19} rounds up to 178,482,301. It will take *at least* that many terms of the harmonic series to move the partial sums beyond 20. It would take your calculator several weeks to compute a partial sum of this many terms. Nonetheless, the harmonic series really does diverge!

Comparison Tests

The p-Series Test tells everything there is to know about the convergence or divergence of series of the form $\sum (1/n^p)$. This is admittedly a rather narrow class of series, but we can test many other kinds (including those in which the nth term is any rational function of n) by *comparing* them to p-series.

The Direct Comparison Test (Theorem 7, Section 9.4) is one method of comparison, but the *Limit Comparison Test* is another.

Theorem 11 The Limit Comparison Test (LCT)

Suppose that $a_n > 0$ and $b_n > 0$ for all $n \ge N$ (N a positive integer).

1. If $\displaystyle\lim_{n\to\infty} \frac{a_n}{b_n} = c,\ 0 < c < \infty$, then $\sum a_n$ and $\sum b_n$ both converge or both diverge.

2. If $\displaystyle\lim_{n\to\infty} \frac{a_n}{b_n} = 0$ and $\sum b_n$ converges, then $\sum a_n$ converges.

3. If $\displaystyle\lim_{n\to\infty} \frac{a_n}{b_n} = \infty$ and $\sum b_n$ diverges, then $\sum a_n$ diverges.

We omit the proof.

Example 3 USING THE LIMIT COMPARISON TEST

Determine whether the series converge or diverge.

(a) $\dfrac{3}{4} + \dfrac{5}{9} + \dfrac{7}{16} + \dfrac{9}{25} + \cdots = \displaystyle\sum_{n=1}^{\infty} \dfrac{2n + 1}{(n + 1)^2}$

(b) $\dfrac{1}{1} + \dfrac{1}{3} + \dfrac{1}{7} + \dfrac{1}{15} + \cdots = \displaystyle\sum_{n=1}^{\infty} \dfrac{1}{2^n - 1}$

(c) $\dfrac{8}{4} + \dfrac{11}{21} + \dfrac{14}{56} + \dfrac{17}{115} + \cdots = \displaystyle\sum_{n=2}^{\infty} \dfrac{3n + 2}{n^3 - 2n}$

(d) $\sin 1 + \sin \dfrac{1}{2} + \sin \dfrac{1}{3} + \cdots = \displaystyle\sum_{n=1}^{\infty} \sin\left(\dfrac{1}{n}\right)$

Solution

(a) For n large,

$$\frac{2n + 1}{(n + 1)^2}$$

behaves like

$$\frac{2n}{n^2} = \frac{2}{n},$$

so we compare terms of the given series to terms of $\sum (1/n)$ and try the LCT.

$$\lim_{n\to\infty} \frac{a_n}{b_n} = \lim_{n\to\infty} \frac{(2n + 1)/(n + 1)^2}{1/n}$$

$$= \lim_{n\to\infty} \frac{2n + 1}{(n + 1)^2} \cdot \frac{n}{1}$$

$$= \lim_{n\to\infty} \frac{2n^2 + n}{n^2 + 2n + 1} = 2$$

Since the limit is positive and $\sum (1/n)$ diverges,

$$\sum_{n=1}^{\infty} \frac{2n + 1}{(n + 1)^2}$$

also diverges.

(b) For n large, $1/(2^n - 1)$ behaves like $1/2^n$, so we compare the given series to $\sum (1/2^n)$.

$$\lim_{n\to\infty} \frac{a_n}{b_n} = \lim_{n\to\infty} \frac{1}{2^n - 1} \cdot \frac{2^n}{1}$$

$$= \lim_{n\to\infty} \frac{2^n}{2^n - 1}$$

$$= \lim_{n\to\infty} \frac{1}{1 - (1/2^n)} = 1$$

Since $\sum (1/2^n)$ converges (geometric, $r = 1/2$), the LCT guarantees that

$$\sum_{n=1}^{\infty} \frac{1}{2^n - 1}$$

also converges.

Figure 9.17 On a guitar, the second harmonic note is produced when the finger is positioned halfway between the bridge and nut of the string while the string is plucked with the other hand.

(c) For n large,

$$\frac{3n + 2}{n^3 - 2n}$$

behaves like $3/n^2$, so we compare the given series to $\sum (1/n^2)$.

$$\lim_{n \to \infty} \frac{a_n}{b_n} = \lim_{n \to \infty} \frac{3n + 2}{n^3 - 2n} \cdot \frac{n^2}{1}$$

$$= \lim_{n \to \infty} \frac{3n^3 + 2n^2}{n^3 - 2n} = 3$$

Since $\sum (1/n^2)$ converges by the p-Series Test,

$$\sum_{n=2}^{\infty} \frac{3n + 2}{n^3 - 2n}$$

also converges (by the LCT).

(d) Recall that

$$\lim_{x \to 0} \frac{\sin x}{x} = 1,$$

so we try the LCT by comparing the given series to $\sum (1/n)$.

$$\lim_{n \to \infty} \frac{a_n}{b_n} = \lim_{n \to \infty} \frac{\sin (1/n)}{(1/n)} = 1$$

Since $\sum (1/n)$ diverges, $\sum_{n=1}^{\infty} \sin (1/n)$ also diverges.

As Example 3 suggests, applying the Limit Comparison Test has strong connections to analyzing end behavior in functions. In part (c) of Example 3, we could have reached the same conclusion if a_n had been *any* linear polynomial in n divided by *any* cubic polynomial in n, since any such rational function "in the end" will grow like $1/n^2$.

Alternating Series

A series in which the terms are alternately positive and negative is an **alternating series.**

Here are three examples.

$$1 - \frac{1}{2} + \frac{1}{3} - \frac{1}{4} + \frac{1}{5} - \cdots + \frac{(-1)^{n+1}}{n} + \cdots \tag{1}$$

$$-2 + 1 - \frac{1}{2} + \frac{1}{4} - \frac{1}{8} + \cdots + \frac{(-1)^n 4}{2^n} + \cdots \tag{2}$$

$$1 - 2 + 3 - 4 + 5 - 6 + \cdots + (-1)^{n+1} n + \cdots \tag{3}$$

Series 1, called the **alternating harmonic series,** converges, as we will see shortly. (You may have come to this conclusion already in Exploration 2 of Section 9.4.) Series 2, a geometric series with $a = -2$, $r = -1/2$, converges to $-2/[1 + (1/2)] = -4/3$. Series 3 diverges by the nth-Term Test.

We prove the convergence of the alternating harmonic series by applying the following test.

> **Theorem 12 The Alternating Series Test (Leibniz's Theorem)**
>
> The series
>
> $$\sum_{n=1}^{\infty} (-1)^{n+1} u_n = u_1 - u_2 + u_3 - u_4 + \cdots$$
>
> converges if all three of the following conditions are satisfied:
>
> **1.** each u_n is positive;
>
> **2.** $u_n \geq u_{n+1}$ for all $n \geq N$, for some integer N;
>
> **3.** $\lim_{n\to\infty} u_n \to 0$.

Figure 9.18 illustrates the convergence of the partial sums to their limit L.

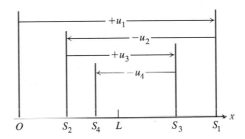

Figure 9.18 Closing in on the sum of a convergent alternating series. (Theorem 12)

The figure that proves the Alternating Series Test actually proves more than the *fact* of convergence; it also shows the *way* that an alternating series converges when it satisfies the conditions of the test. The partial sums keep "overshooting" the limit as they go back and forth on the number line, gradually closing in as the terms tend to zero. If we stop at the nth partial sum, we know that the next term $(\pm u_{n+1})$ will again cause us to overshoot the limit in the positive direction or negative direction, depending on the sign carried by u_{n+1}. This gives us a convenient bound for the truncation error, which we state as another theorem.

> **Theorem 13 The Alternating Series Estimation Theorem**
>
> If the alternating series $\sum_{n=1}^{\infty} (-1)^{n+1} u_n$ satisfies the conditions of Theorem 12, then the truncation error for the nth partial sum is less than u_{n+1} and has the same sign as the unused term.

Example 4 THE ALTERNATING HARMONIC SERIES

Prove that the alternating harmonic series is convergent, but not absolutely convergent. Find a bound for the truncation error after 99 terms.

Solution The terms are strictly alternating in sign and decrease in absolute value from the start:

$$1 > \frac{1}{2} > \frac{1}{3} > \cdots . \quad \text{Also,} \quad \frac{1}{n} \to 0.$$

A Note on the Error Bound

Theorem 13 does not give a *formula* for the truncation error, but a *bound* for the truncation error. The bound might be fairly conservative. For example, the first 99 terms of the alternating harmonic series add to about 0.6981721793, while the series itself has a sum of ln 2 ≈ 0.6931471806. That makes the actual truncation error very close to 0.005, about half the size of the bound of 0.01 given by Theorem 13.

By the Alternating Series Test,

$$\sum_{n=1}^{\infty} \frac{(-1)^{n+1}}{n}$$

converges.

On the other hand, the series $\sum_{n=1}^{\infty} (1/n)$ of absolute values is the harmonic series, which diverges, so the alternating harmonic series is not absolutely convergent.

The Alternating Series Estimation Theorem guarantees that the truncation error after 99 terms is less than $u_{99+1} = 1/(99 + 1) = 1/100$.

Absolute and Conditional Convergence

Because the alternating harmonic series is convergent but not absolutely convergent, we say it is **conditionally convergent** (or **converges conditionally**).

We take it for granted that we can rearrange the terms of a *finite* sum without affecting the sum. We can also rearrange a *finite number* of terms of an infinite series without affecting the sum. But if we rearrange an infinite number of terms of an infinite series, we can be sure of leaving the sum unaltered *only if it converges absolutely.*

Rearrangements of Absolutely Convergent Series

If $\sum a_n$ converges absolutely, and if $b_1, b_2, b_3, \dots, b_n, \dots$ is any rearrangement of the sequence $\{a_n\}$, then $\sum b_n$ converges absolutely and $\sum_{n=1}^{\infty} b_n = \sum_{n=1}^{\infty} a_n$.

On the other hand, consider this:

Rearrangements of Conditionally Convergent Series

If $\sum a_n$ converges conditionally, then the terms can be rearranged to form a divergent series. The terms can also be rearranged to form a series that converges to *any* preassigned sum.

This seems incredible, but it is a logical consequence of the definition of the sum as the *limit of the sequence of partial sums.* A conditionally convergent series consists of positive terms that sum to ∞ and negative terms that sum to −∞, so we can manipulate the partial sums to do virtually anything we wish. We illustrate the technique with the alternating harmonic series.

Example 5 REARRANGING THE ALTERNATING HARMONIC SERIES

Show how to rearrange the terms of

$$\sum_{n=1}^{\infty} \frac{(-1)^{n+1}}{n}$$

to form

(a) a divergent series;   **(b)** a series that converges to π.

Solution The series of positive terms,

$$1 + \frac{1}{3} + \frac{1}{5} + \cdots + \frac{1}{2n+1} + \cdots,$$

diverges to ∞, while the series of negative terms,

$$-\frac{1}{2} - \frac{1}{4} - \frac{1}{6} - \cdots - \frac{1}{2n} - \cdots,$$

diverges to $-\infty$. No matter what finite number of terms we use, the remaining positive terms or negative terms still diverge. So, we build our series as follows:

(a) Start by adding positive terms until the partial sum is greater than 1. Then add negative terms until the partial sum is less than -2. Then add positive terms until the sum is greater than 3. Then add negative terms until the sum is less than -4. Continue in this manner indefinitely, so that the sequence of partial sums swings arbitrarily far in both directions and hence diverges.

(b) Start by adding positive terms until the partial sum is greater than π. Then add negative terms until the partial sum is less than π. Then add positive terms until the sum is greater than π. Continue in this manner indefinitely, always closing in on π. Since the positive and negative terms of the original series both approach zero, the amount by which the partial sums exceed or fall short of π approaches zero.

Intervals of Convergence

Our purpose in this section has been to develop tests for convergence that can be used at the endpoints of the intervals of absolute convergence of power series. There are three possibilities at each endpoint: The series could diverge, it could converge absolutely, or it could converge conditionally.

How to Test a Power Series $\displaystyle\sum_{n=0}^{\infty} c_n(x-a)^n$ for Convergence

1. Use the Ratio Test to find the values of x for which the series converges absolutely. Ordinarily, this is an open interval

$$a - R < x < a + R.$$

In some instances, the series converges for all values of x. In rare cases, the series converges only at $x = a$.

2. If the interval of absolute convergence is finite, test for convergence or divergence at each endpoint. The Ratio Test fails at these points. Use a comparison test, the Integral Test, or the Alternating Series Test.

3. If the interval of absolute convergence is $a - R < x < a + R$, conclude that the series diverges (it does not even converge conditionally) for $|x - a| > R$, because for those values of x the nth term does not approach zero.

Example 6 FINDING INTERVALS OF CONVERGENCE

For what values of x do the following series converge?

(a) $\displaystyle\sum_{n=1}^{\infty} (-1)^{n+1}\frac{x^{2n}}{2n} = \frac{x^2}{2} - \frac{x^4}{4} + \frac{x^6}{6} - \cdots$

(b) $\displaystyle\sum_{n=0}^{\infty} \frac{(10x)^n}{n!} = 1 + 10x + \frac{100x^2}{2!} + \frac{1000x^3}{3!} + \cdots$

(c) $\displaystyle\sum_{n=0}^{\infty} n!(x+1)^n = 1 + (x+1) + 2!(x+1)^2 + 3!(x+1)^3 + \cdots$

(d) $\displaystyle\sum_{n=1}^{\infty} \frac{(x-3)^n}{2n} = \frac{(x-3)}{2} + \frac{(x-3)^2}{4} + \frac{(x-3)^3}{6} + \cdots$

Solution We apply the Ratio Test to find the interval of absolute convergence, then check the endpoints if they exist.

(a) $\displaystyle\lim_{n\to\infty} \left| \frac{u_{n+1}}{u_n} \right| = \lim_{n\to\infty} \frac{x^{2n+2}}{2n+2} \cdot \frac{2n}{x^{2n}}$

$$= \lim_{n\to\infty} \left(\frac{2n}{2n+2} \right) x^2 = x^2$$

The series converges absolutely for $x^2 < 1$, i.e., on the interval $(-1, 1)$. At $x = 1$, the series is

$$\sum \frac{(-1)^{n+1}}{2n},$$

which converges by the Alternating Series Test. (It is half the sum of the alternating harmonic series.) At $x = -1$, the series is the same as at $x = 1$, so it converges. The interval of convergence is $[-1, 1]$.

(b) $\displaystyle\lim_{n\to\infty} \left| \frac{u_{n+1}}{u_n} \right| = \lim_{n\to\infty} \frac{|10x|^{n+1}}{(n+1)!} \cdot \frac{n!}{|10x|^n}$

$$= \lim_{n\to\infty} \frac{|10x|}{n+1} = 0$$

The series converges absolutely for all x.

(c) $\displaystyle\lim_{n\to\infty} \left| \frac{u_{n+1}}{u_n} \right| = \lim_{n\to\infty} \frac{(n+1)!|x+1|^{n+1}}{n!|x+1|^n}$

$$= \lim_{n\to\infty} (n+1)|x+1| = \begin{cases} \infty, & x \neq -1 \\ 0, & x = -1 \end{cases}$$

The series converges only at $x = -1$.

(d) $\displaystyle\lim_{n\to\infty} \left| \frac{u_{n+1}}{u_n} \right| = \lim_{n\to\infty} \frac{|x-3|^{n+1}}{2n+2} \cdot \frac{2n}{|x-3|^n}$

$$= \lim_{n\to\infty} \left(\frac{2n}{2n+2} \right) |x-3| = |x-3|$$

The series converges absolutely for $|x - 3| < 1$, i.e., on the interval $(2, 4)$. At $x = 2$, the series is $\sum (-1)^n/2n$, which converges by the Alternating Series Test. At $x = 4$, the series is $\sum 1/2n$, which diverges by limit comparison with the harmonic series. The interval of convergence is $[2, 4)$.

To facilitate testing convergence at endpoints we can use the following flowchart.

Procedure for Determining Convergence

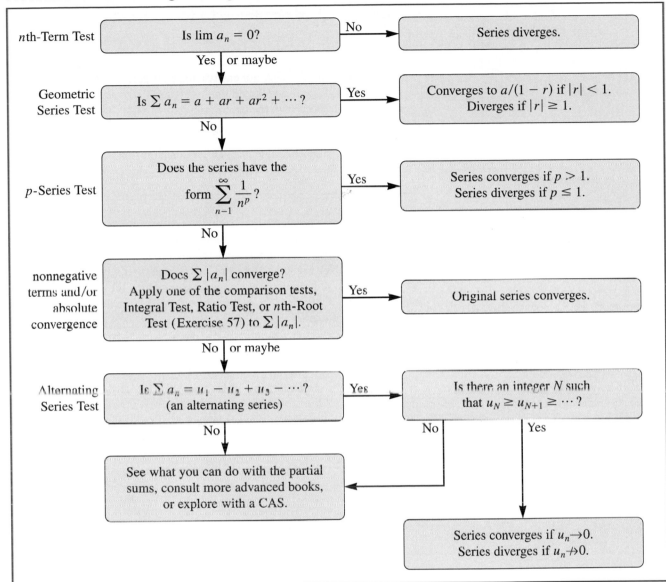

A Word of Caution

Although we can use the tests we have developed to find where a given power series converges, they do not tell us what function that power series is converging *to*. Even if the series is known to be a Maclaurin series generated by a function *f*, we cannot automatically conclude that the series converges *to the function f* on its interval of convergence. That is why it is so important to estimate the error.

For example, we can use the Ratio Test to show that the Maclaurin series for sin *x*, cos *x*, and e^x all converge absolutely for all real numbers. However, the reason we know that they converge to sin *x*, cos *x*, and e^x is that we used the Remainder Estimation Theorem to show that the respective truncation errors went to zero.

The following exploration shows what can happen with a strange function.

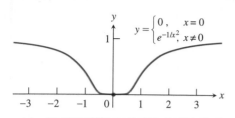

Figure 9.19 The graph of the continuous extension of $y = e^{-1/x^2}$ is so flat at the origin that all of its derivatives there are zero. (Exploration 2)

Exploration 2 **The Maclaurin Series of a Strange Function**

Let $f(x) = \begin{cases} 0, & x = 0 \\ e^{-1/x^2}, & x \neq 0. \end{cases}$

It can be shown (although not easily) that f (Figure 9.19) has derivatives of all orders at $x = 0$ and that $f^{(n)}(0) = 0$ for all n. Use this fact as you proceed with the exploration.

1. Construct the Maclaurin series for f.
2. For what values of x does this series converge?
3. Find all values of x for which the series actually converges to $f(x)$.

If you are surprised by the behavior of the series in Exploration 2, remember that we identified it up front as a strange function. It was fortunate for the early history of calculus that the functions that modeled physical behavior in the Newtonian world were much more predictable, enabling the early theories to enjoy encouraging successes before they could be lost in detail. When the subtleties of convergence emerged later, the theory was prepared to confront them.

Quick Review 9.5

In Exercises 1–5, determine whether the improper integral converges or diverges. Give reasons for your answer. (You do not need to evaluate the integral.)

1. $\int_1^\infty \frac{1}{x^{4/3}} \, dx$

2. $\int_1^\infty \frac{x^2}{x^3 + 1} \, dx$

3. $\int_1^\infty \frac{\ln x}{x} \, dx$

4. $\int_1^\infty \frac{1 + \cos x}{x^2} \, dx$

5. $\int_1^\infty \frac{\sqrt{x}}{x + 1} \, dx$

In Exercises 6–10, determine whether the function is both positive and decreasing on some interval (N, ∞). (You do not need to identify N.)

6. $f(x) = \frac{3}{x}$

7. $f(x) = \frac{7x}{x^2 - 8}$

8. $f(x) = \frac{3 + x^2}{3 - x^2}$

9. $f(x) = \frac{\sin x}{x^5}$

10. $f(x) = \ln(1/x)$

Section 9.5 Exercises

In Exercises 1–16, determine whether the series converges or diverges. There may be more than one correct way to determine convergence or divergence of a given series.

1. $\sum_{n=1}^\infty \frac{5}{n + 1}$

2. $\sum_{n=1}^\infty \frac{3}{\sqrt{n}}$

3. $\sum_{n=2}^\infty \frac{\ln n}{n}$

4. $\sum_{n=1}^\infty \frac{1}{2n - 1}$

5. $\sum_{n=1}^\infty \frac{1}{(\ln 2)^n}$

6. $\sum_{n=1}^\infty \frac{1}{(\ln 3)^n}$

7. $\sum_{n=1}^\infty n \sin\left(\frac{1}{n}\right)$

8. $\sum_{n=0}^\infty \frac{e^n}{1 + e^{2n}}$

9. $\sum_{n=1}^\infty \frac{\sqrt{n}}{n^2 + 1}$

10. $\sum_{n=1}^\infty \frac{5n^3 - 3n}{n^2(n + 2)(n^2 + 5)}$

11. $\displaystyle\sum_{n=1}^{\infty} \frac{3^{n-1}+1}{3^n}$

12. $\displaystyle\sum_{n=2}^{\infty} (-1)^{n+1} \frac{1}{\ln n}$

13. $\displaystyle\sum_{n=1}^{\infty} (-1)^{n+1} \frac{10^n}{n^{10}}$

14. $\displaystyle\sum_{n=1}^{\infty} (-1)^{n+1} \frac{\sqrt{n}+1}{n+1}$

15. $\displaystyle\sum_{n=2}^{\infty} (-1)^{n+1} \frac{\ln n}{\ln n^2}$

16. $\displaystyle\sum_{n=1}^{\infty} \left(\frac{1}{n} - \frac{1}{n^2}\right)$

In Exercises 17–26, determine whether the series converges absolutely, converges conditionally, or diverges. Give reasons for your answer.

17. $\displaystyle\sum_{n=1}^{\infty} (-1)^{n+1}(0.1)^n$

18. $\displaystyle\sum_{n-1}^{\infty} (-1)^{n+1} \frac{1+n}{n^2}$

19. $\displaystyle\sum_{n=1}^{\infty} (-1)^n n^2 \left(\frac{2}{3}\right)^n$

20. $\displaystyle\sum_{n-2}^{\infty} (-1)^{n+1} \frac{1}{n \ln n}$

21. $\displaystyle\sum_{n=1}^{\infty} (-1)^{n+1} \frac{n!}{2^n}$

22. $\displaystyle\sum_{n=1}^{\infty} (-1)^{n+1} \frac{\sin n}{n^2}$

23. $\displaystyle\sum_{n=1}^{\infty} \frac{(-1)^n}{1+\sqrt{n}}$

24. $\displaystyle\sum_{n=1}^{\infty} \frac{\cos n\pi}{n\sqrt{n}}$

25. $\displaystyle\sum_{n=1}^{\infty} \frac{\cos n\pi}{n}$

26. $\displaystyle\sum_{n=1}^{\infty} \frac{(-1)^n}{\sqrt{n}+\sqrt{n+1}}$

In Exercises 27–42, find (a) the *interval* of convergence of the series. For what values of x does the series converge (b) absolutely, (c) conditionally?

27. $\displaystyle\sum_{n=0}^{\infty} x^n$

28. $\displaystyle\sum_{n=0}^{\infty} (x+5)^n$

29. $\displaystyle\sum_{n=0}^{\infty} (-1)^n(4x+1)^n$

30. $\displaystyle\sum_{n=1}^{\infty} \frac{(3x-2)^n}{n}$

31. $\displaystyle\sum_{n=0}^{\infty} \frac{(x-2)^n}{10^n}$

32. $\displaystyle\sum_{n=0}^{\infty} \frac{nx^n}{n+2}$

33. $\displaystyle\sum_{n=1}^{\infty} \frac{x^n}{n\sqrt{n}\,3^n}$

34. $\displaystyle\sum_{n=0}^{\infty} \frac{x^{2n+1}}{n!}$

35. $\displaystyle\sum_{n=0}^{\infty} \frac{n(x+3)^n}{5^n}$

36. $\displaystyle\sum_{n=0}^{\infty} \frac{nx^n}{4^n(n^2+1)}$

37. $\displaystyle\sum_{n=0}^{\infty} \frac{\sqrt{n}\,x^n}{3^n}$

38. $\displaystyle\sum_{n=0}^{\infty} n!(x-4)^n$

39. $\displaystyle\sum_{n=0}^{\infty} (-2)^n(n+1)(x-1)^n$

40. $\displaystyle\sum_{n=1}^{\infty} \frac{(4x-5)^{2n+1}}{n^{3/2}}$

41. $\displaystyle\sum_{n=1}^{\infty} \frac{(x+\pi)^n}{\sqrt{n}}$

42. $\displaystyle\sum_{n=0}^{\infty} (\ln x)^n$

43. Not only do the figures in Example 2 show that the nth partial sum of the harmonic series is less than $1 + \ln n$; they also show that it is *greater* than $\ln(n+1)$. Suppose you had started summing the harmonic series with $S_1 = 1$ at the time

the universe was formed, 13 billion years ago. If you had been able to add a term every *second* since then, about how large would your partial sum be today? (Assume a 365-day year.)

44. Writing to Learn Write out a proof of the Integral Test (Theorem 10) for $N = 1$, explaining what you see in Figure 9.15.

45. *(Continuation of Exercise 44)* Relabel the pictures for an arbitrary N and explain why the same conclusions about convergence can be drawn.

46. In each of the following cases, decide whether the infinite series converges. Justify your answer.

(a) $\displaystyle\sum_{k=1}^{\infty} \frac{1}{\sqrt{2k+7}}$

(b) $\displaystyle\sum_{k=1}^{\infty} \left(1 + \frac{1}{k}\right)^k$

(c) $\displaystyle\sum_{k=1}^{\infty} \frac{\cos k}{k^2 + \sqrt{k}}$

(d) $\displaystyle\sum_{k=3}^{\infty} \frac{18}{k(\ln k)}$

47. Construct a series that diverges more slowly than the harmonic series. Justify your answer.

48. Let $a_k = (-1)^{k+1} \int_0^{1/k} 6(kx)^2 \, dx$.

(a) Evaluate a_k.

(b) Show that $\sum_{k=1}^{\infty} a_k$ converges.

(c) Show that

$$1 \le \sum_{k=1}^{\infty} a_k \le \frac{3}{2}.$$

49. (a) Determine whether the series

$$A = \sum_{n-1}^{\infty} \frac{n}{3n^2 + 1}$$

converges or diverges. Justify your answer.

(b) If S is the series formed by multiplying the nth term in A by the nth term in $\sum_{n=1}^{\infty} (3/n)$, write an expression using summation notation for S and determine whether S converges or diverges.

50. (a) Find the Taylor series generated by $f(x) = \ln(1 + x)$ at $x = 0$. Include an expression for the general term.

(b) For what values of x does the series in part (a) converge?

(c) Use Theorem 13 to find a bound for the error in evaluating $\ln(3/2)$ by using only the first five nonzero terms of the series in part (a).

(d) Use the result found in part (a) to determine the logarithmic function whose Taylor series is

$$\sum_{n-1}^{\infty} \frac{(-1)^{n+1}x^{2n}}{2n}.$$

51. Determine all values of x for which the series

$$\sum_{k=0}^{\infty} \frac{2^k x^k}{\ln(k+2)}$$

converges. Justify your answer.

52. Consider the series $\sum_{n=2}^{\infty} \dfrac{1}{n^p \ln n}$, where $p \geq 0$.

    (a) Show that the series converges for $p > 1$.

    (b) Writing to Learn Determine whether the series converges or diverges for $p = 1$. Show your analysis.

    (c) Show that the series diverges for $0 \leq p < 1$.

53. The Maclaurin series for $1/(1 + x)$ converges for $-1 < x < 1$, but when we integrate it term by term, the resulting series for $\ln|1 + x|$ converges for $-1 < x \leq 1$. Verify the convergence at $x = 1$.

54. The Maclaurin series for $1/(1 + x^2)$ converges for $-1 < x < 1$, but when we integrate it term by term, the resulting series for $\arctan x$ converges for $-1 \leq x \leq 1$. Verify the convergence at $x = 1$ and $x = -1$.

55. **(a)** The series

    $$\frac{1}{3} - \frac{1}{2} + \frac{1}{9} - \frac{1}{4} + \frac{1}{27} - \frac{1}{8} + \cdots + \frac{1}{3^n} - \frac{1}{2^n} + \cdots$$

    fails to satisfy one of the conditions of the Alternating Series Test. Which one?

    (b) Find the sum of the series in (a).

Exploration

56. *Work in groups of n people.* Within your group, have each student construct a series that converges to one of the numbers $1, \ldots, n$. Then exchange your series with another group and try to figure out which number is matched with which series. ∎

Extending the Ideas

Here is a test called the *nth-Root Test*.

> **nth-Root Test**   Let $\sum a_n$ be a series with $a_n \geq 0$ for $n \geq N$, and suppose that $\lim_{n \to \infty} \sqrt[n]{a_n} = L$. Then,
>
> **(a)** the series *converges* if $L < 1$,
>
> **(b)** the series *diverges* if $L > 1$ or L is infinite,
>
> **(c)** the test is *inconclusive* if $L = 1$.

57. Use the *nth*-Root Test and the fact that $\lim_{n \to \infty} \sqrt[n]{n} = 1$ to test the following series for convergence or divergence.

    (a) $\sum_{n=1}^{\infty} \dfrac{n^2}{2^n}$ **(b)** $\sum_{n=1}^{\infty} \left(\dfrac{n}{2n - 1} \right)^n$

    (c) $\sum_{n=1}^{\infty} a_n$, where $a_n = \begin{cases} n/2^n, & n \text{ is odd} \\ 1/2^n, & n \text{ is even} \end{cases}$

58. Use the *nth*-Root Test and whatever else you need to find the intervals of convergence of the following series.

    (a) $\sum_{n=0}^{\infty} \dfrac{(x - 1)^n}{4^n}$ **(b)** $\sum_{n=1}^{\infty} \dfrac{(x - 2)^n}{n \cdot 3^n}$

    (c) $\sum_{n=1}^{\infty} 2^n x^n$ **(d)** $\sum_{n=0}^{\infty} (\ln x)^n$

Chapter 9   Key Terms

Chapter 9 Review Exercises

In Exercises 1–16, find **(a)** the radius of convergence for the series and **(b)** its interval of convergence. Then identify the values of x for which the series converges **(c)** absolutely and **(d)** conditionally.

1. $\displaystyle\sum_{n=0}^{\infty} \frac{(-x)^n}{n!}$

2. $\displaystyle\sum_{n=1}^{\infty} \frac{(x+4)^n}{n3^n}$

3. $\displaystyle\sum_{n=0}^{\infty} \left(\frac{2}{3}\right)^n (x-1)^n$

4. $\displaystyle\sum_{n=1}^{\infty} \frac{(x-1)^{2n-2}}{(2n-1)!}$

5. $\displaystyle\sum_{n=1}^{\infty} \frac{(-1)^{n-1}(3x-1)^n}{n^2}$

6. $\displaystyle\sum_{n=0}^{\infty} (n+1)x^{3n}$

7. $\displaystyle\sum_{n=0}^{\infty} \frac{(n+1)(2x+1)^n}{(2n+1)2^n}$

8. $\displaystyle\sum_{n=1}^{\infty} \frac{x^n}{n^n}$

9. $\displaystyle\sum_{n=1}^{\infty} \frac{x^n}{\sqrt{n}}$

10. $\displaystyle\sum_{n=1}^{\infty} \frac{e^n}{n^e} x^n$

11. $\displaystyle\sum_{n=0}^{\infty} \frac{(n+1)x^{2n-1}}{3^n}$

12. $\displaystyle\sum_{n=0}^{\infty} \frac{(-1)^n(x-1)^{2n+1}}{2n+1}$

13. $\displaystyle\sum_{n=1}^{\infty} \frac{n!}{2^n} x^{2n}$

14. $\displaystyle\sum_{n=2}^{\infty} \frac{(10x)^n}{\ln n}$

15. $\displaystyle\sum_{n=0}^{\infty} (n+1)!x^n$

16. $\displaystyle\sum_{n=0}^{\infty} \left(\frac{x^2-1}{2}\right)^n$

In Exercises 17–22, the series is the value of the Maclaurin series of a function $f(x)$ at a particular point. What function and what point? What is the sum of the series?

17. $1 - \dfrac{1}{4} + \dfrac{1}{16} - \cdots + (-1)^n \dfrac{1}{4^n} + \cdots$

18. $\dfrac{2}{3} - \dfrac{4}{18} + \dfrac{8}{81} - \cdots + (-1)^{n-1} \dfrac{2^n}{n3^n} + \cdots$

19. $\pi - \dfrac{\pi^3}{3!} + \dfrac{\pi^5}{5!} - \cdots + (-1)^n \dfrac{\pi^{2n+1}}{(2n+1)!} + \cdots$

20. $1 - \dfrac{\pi^2}{9 \cdot 2!} + \dfrac{\pi^4}{81 \cdot 4!} - \cdots + (-1)^n \dfrac{\pi^{2n}}{3^{2n}(2n)!} + \cdots$

21. $1 + \ln 2 + \dfrac{(\ln 2)^2}{2!} + \cdots + \dfrac{(\ln 2)^n}{n!} + \cdots$

22. $\dfrac{1}{\sqrt{3}} - \dfrac{1}{9\sqrt{3}} + \dfrac{1}{45\sqrt{3}} - \cdots +$
$(-1)^{n-1}\dfrac{1}{(2n-1)(\sqrt{3})^{2n-1}} + \cdots$

In Exercises 23–36, find a Maclaurin series for the function.

23. $\dfrac{1}{1-6x}$

24. $\dfrac{1}{1+x^3}$

25. $x^9 - 2x^2 + 1$

26. $\dfrac{4x}{1-x}$

27. $\sin \pi x$

28. $-\sin \dfrac{2x}{3}$

29. $-x + \sin x$

30. $\dfrac{e^x + e^{-x}}{2}$

31. $\cos \sqrt{5x}$

32. $e^{(\pi x/2)}$

33. xe^{-x^2}

34. $\tan^{-1} 3x$

35. $\ln(1-2x)$

36. $x \ln(1-x)$

In Exercises 37–40, find the first four nonzero terms and the general term of the Taylor series generated by f at $x=a$.

37. $f(x) = \dfrac{1}{3-x}, \quad a=2$

38. $f(x) = x^3 - 2x^2 + 5, \quad a=-1$

39. $f(x) = \dfrac{1}{x}, \quad a=3$

40. $f(x) = \sin x, \quad a=\pi$

In Exercises 41–52, determine if the series converges absolutely, converges conditionally, or diverges. Give reasons for your answer.

41. $\displaystyle\sum_{n=1}^{\infty} \frac{-5}{n}$

42. $\displaystyle\sum_{n=1}^{\infty} \frac{(-1)^n}{\sqrt{n}}$

43. $\displaystyle\sum_{n=1}^{\infty} \frac{\ln n}{n^3}$

44. $\displaystyle\sum_{n=1}^{\infty} \frac{n+1}{n!}$

45. $\displaystyle\sum_{n=1}^{\infty} \frac{(-1)^n}{\ln(n+1)}$

46. $\displaystyle\sum_{n=2}^{\infty} \frac{1}{n(\ln n)^2}$

47. $\displaystyle\sum_{n=1}^{\infty} \frac{(-3)^n}{n!}$

48. $\displaystyle\sum_{n=1}^{\infty} \frac{2^n 3^n}{n^n}$

49. $\displaystyle\sum_{n=1}^{\infty} \frac{(-1)^n(n^2+1)}{2n^2+n-1}$

50. $\displaystyle\sum_{n=1}^{\infty} \frac{1}{\sqrt{n(n+1)(n+2)}}$

51. $\displaystyle\sum_{n=2}^{\infty} \frac{1}{n\sqrt{n^2-1}}$

52. $\displaystyle\sum_{n=1}^{\infty} \left(\frac{n}{n+1}\right)^n$

In Exercises 53 and 54, find the sum of the series.

53. $\displaystyle\sum_{n=3}^{\infty} \frac{1}{(2n-3)(2n-1)}$

54. $\displaystyle\sum_{n=2}^{\infty} \frac{-2}{n(n+1)}$

55. Let f be a function that has derivatives of all orders for all real numbers. Assume that $f(3) = 1$, $f'(3) = 4$, $f''(3) = 6$, and $f'''(3) = 12$.

(a) Write the third order Taylor polynomial for f at $x=3$ and use it to approximate $f(3.2)$.

(b) Write the second order Taylor polynomial for f' at $x=3$ and use it to approximate $f'(2.7)$.

(c) Does the linearization of f underestimate or overestimate the values of $f(x)$ near $x=3$? Justify your answer.

56. Let

$$P_4(x) = 7 - 3(x - 4) + 5(x - 4)^2 - 2(x - 4)^3 + 6(x - 4)^4$$

be the Taylor polynomial of order 4 for the function f at $x = 4$. Assume f has derivatives of all orders for all real numbers.

(a) Find $f(4)$ and $f'''(4)$.

(b) Write the second order Taylor polynomial for f' at $x = 4$ and use it to approximate $f'(4.3)$.

(c) Write the fourth order Taylor polynomial for $g(x) = \int_4^x f(t)\, dt$ at $x = 4$.

(d) Can the exact value of $f(3)$ be determined from the information given? Justify your answer.

57. (a) Write the first three nonzero terms and the general term of the Taylor series generated by $f(x) = 5 \sin (x/2)$ at $x = 0$.

(b) What is the interval of convergence for the series found in (a)? Show your method.

(c) Writing to Learn What is the minimum number of terms of the series in (a) needed to approximate $f(x)$ on the interval $(-2, 2)$ with an error not exceeding 0.1 in magnitude? Show your method.

58. Let $f(x) = 1/(1 - 2x)$.

(a) Write the first four terms and the general term of the Taylor series generated by $f(x)$ at $x = 0$.

(b) What is the interval of convergence for the series found in part (a)? Show your method.

(c) Find $f(-1/4)$. How many terms of the series are adequate for approximating $f(-1/4)$ with an error not exceeding one percent in magnitude? Justify your answer.

59. Let

$$f(x) = \sum_{n=1}^{\infty} \frac{x^n n^n}{n!}$$

for all x for which the series converges.

(a) Find the radius of convergence of this series.

(b) Use the first three terms of this series to approximate $f(-1/3)$.

(c) Estimate the error involved in the approximation in part (b). Justify your answer.

60. Let $f(x) = 1/(x - 2)$.

(a) Write the first four terms and the general term of the Taylor series generated by $f(x)$ at $x = 3$.

(b) Use the result from part (a) to find the first four terms and the general term of the series generated by $\ln |x - 2|$ at $x = 3$.

(c) Use the series in part (b) to compute a number that differs from $\ln (3/2)$ by less than 0.05. Justify your answer.

61. Let $f(x) = e^{-2x^2}$.

(a) Find the first four nonzero terms and the general term for the power series generated by $f(x)$ at $x = 0$.

(b) Find the interval of convergence of the series generated by $f(x)$ at $x = 0$. Show the analysis that leads to your conclusion.

(c) Writing to Learn Let g be the function defined by the sum of the first four nonzero terms of the series generated by $f(x)$. Show that $|f(x) - g(x)| < 0.02$ for $-0.6 \leq x \leq 0.6$.

62. (a) Find the Maclaurin series generated by $f(x) = x^2/(1 + x)$.

(b) Does the series converge at $x = 1$? Explain.

63. *Evaluating Nonelementary Integrals* Maclaurin series can be used to express nonelementary integrals in terms of series.

(a) Express $\int_0^x \sin t^2\, dt$ as a power series.

(b) According to the Alternating Series Estimation Theorem, how many terms of the series in (a) should you use to estimate $\int_0^1 \sin x^2\, dx$ with an error of less than 0.001?

(c) Use NINT to approximate $\int_0^1 \sin x^2\, dx$.

(d) How close to the answer in (c) do you get if you use four terms of the series in (a)?

64. *Estimating an Integral* Suppose you want a quick noncalculator estimate for the value of $\int_0^1 x^2 e^x\, dx$. There are several ways to get one.

(a) Use the Trapezoidal Rule with $n = 2$ to estimate $\int_0^1 x^2 e^x\, dx$.

(b) Write the first three nonzero terms of the Maclaurin series for $x^2 e^x$ to obtain the fourth order Maclaurin polynomial $P_4(x)$ for $x^2 e^x$. Use $\int_0^1 P_4(x)\, dx$ to obtain another estimate of $\int_0^1 x^2 e^x\, dx$.

(c) Writing to Learn The second derivative of $f(x) = x^2 e^x$ is positive for all $x > 0$. Explain why this enables you to conclude that the Trapezoidal Rule estimate obtained in (a) is too large.

(d) Writing to Learn All the derivatives of $f(x) = x^2 e^x$ are positive for $x > 0$. Explain why this enables you to conclude that all Maclaurin series approximations to $f(x)$ for x in $[0, 1]$ will be too small. (*Hint:* $f(x) = P_n(x) + R_n(x)$.)

(e) Use integration by parts to evaluate $\int_0^1 x^2 e^x\, dx$.

65. *Perpetuities* Suppose you want to give a favorite school or charity $1000 a year forever. This kind of gift is called a *perpetuity*. Assume you can earn 8% annually on your money, i.e., that a payment of a_n today will be worth $a_n (1.08)^n$ in n years.

(a) Show that the amount you must invest today to cover the nth $1000 payment in n years is $1000(1.08)^{-n}$.

(b) Construct an infinite series that gives the amount you must invest today to cover *all* the payments in the perpetuity.

(c) Show that the series in (b) converges and find its sum. This sum is called the *present value* of the perpetuity. What does it represent?

66. (*Continuation of Exercise 65*) Find the present value of a $1000-per-year perpetuity at 6% annual interest.

67. *Expected Payoff* How much would you expect to win playing the following game?

Toss a *fair* coin (heads and tails equally likely). Every time it comes up heads you win a dollar, but the game is over as soon as it comes up tails.

(a) The *expected payoff* of the game is computed by summing all possible payoffs times their respective probabilities. If the probability of tossing the first tail on the nth toss is $(1/2)^n$, express the expected payoff of this game as an infinite series.

(b) Differentiate both sides of

$$\frac{1}{1-x} = 1 + x + x^2 + \cdots + x^n + \cdots$$

to get a series for $1/(1-x)^2$.

(c) Use the series in (b) to get a series for $x^2/(1-x)^2$.

(d) Use the series in (c) to evaluate the expected payoff of the game.

68. *Punching out Triangles* This exercise refers to the "right side up" equilateral triangle with sides of length $2b$ in the accompanying figure. "Upside down" equilateral triangles are removed from the original triangle as the sequence of pictures suggests. The sum of the areas removed from the original triangle forms an infinite series.

(a) Find this infinite series.

(b) Find the sum of this infinite series and hence find the total area removed from the original triangle.

(c) Is every point on the original triangle removed? Explain why or why not.

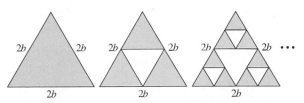

69. *Nicole Oresme's (pronounced "O-rem's") Theorem* Prove Nicole Oresme's theorem that

$$1 + \frac{1}{2} \cdot 2 + \frac{1}{4} \cdot 3 + \cdots + \frac{n}{2^{n-1}} + \cdots = 4.$$

(*Hint:* Differentiate both sides of the equation $1/(1-x) = 1 + \sum_{n=1}^{\infty} x^n$.)

70. (a) Show that

$$\sum_{n=1}^{\infty} \frac{n(n+1)}{x^n} = \frac{2x^2}{(x-1)^3}$$

for $|x| > 1$ by differentiating the identity

$$\sum_{n=1}^{\infty} x^{n+1} - \frac{x^2}{1-x}$$

twice, multiplying the result by x, and then replacing x by $1/x$.

(b) Use part (a) to find the real solution greater than 1 of the equation

$$x = \sum_{n=1}^{\infty} \frac{n(n+1)}{x^n}.$$

Calculus at Work

Roberta M. Johnson
University of Michigan
Ann Arbor, MI

10

Parametric, Vector, and Polar Functions

In 1935, air traffic control was conducted with a system of teletype machines, wall-sized blackboards, large table maps, and movable markers representing airplanes. Today's radar data processing includes an automatic display of aircraft identification, speed, altitude, and velocity vectors.

A DC-10 flying due west at 600 mph enters a region with a steady air current coming from the southwest at 100 mph. How should the pilot adjust the airplane's course and speed to maintain its original velocity vector? This type of problem is covered in Section 10.2.

Chapter 10 Overview

When a body travels in the xy-plane, the parametric equations $x = f(t)$ and $y = g(t)$ can be used to model the body's motion and path. In this chapter, we introduce the *vector* form of parametric equations, which allows us to track the positions of moving bodies with vectors, calculate the directions and magnitudes of their velocities and accelerations, and predict the effects of the forces we see working on them.

One of the principal applications of vector functions is the analysis of motion in space. Planetary motion is best described with polar coordinates (another of Newton's inventions, although James Bernoulli usually gets the credit because he published first), so we prepare for the future by investigating curves, derivatives, and integrals in this new coordinate system.

10.1 Parametric Functions

Derivatives • Parametric Formula for d^2y/dx^2 • Length of a Smooth Curve • Cycloids • Surface Area

Derivatives

If f and g are functions of t, then the curve given by the parametric equations $x = f(t)$, $y = g(t)$ can be treated as the graph of a function of the parameter t, because each value of t produces a unique point on the curve.

You may want to review the material on parametrization in Section 1.4.

> **Definition Derivative at a Point**
>
> A parametrized curve $x = f(t)$, $y = g(t)$, $a \le t \le b$, has a **derivative** at $t = t_0$ if f and g have derivatives at $t = t_0$.

The curve is **differentiable** if it is differentiable at every parameter value. The curve is **smooth** if f' and g' are continuous and not simultaneously zero.

Example 1 USING THE DEFINITION

The parametrization

$$x = \cos t, \qquad y = \sin t, \qquad 0 \le t \le 2\pi,$$

of the unit circle is differentiable at every t in the interval $[0, 2\pi]$ because $\cos t$ and $\sin t$ are differentiable in $[0, 2\pi]$. Thus, the unit circle is a differentiable curve. It is also smooth because $x' = -\sin t$ and $y' = \cos t$ are never zero at the same time.

Parametric Formula for d^2y/dx^2

A parametrized curve can yield one or more ways to define y as a function of x. In Section 3.6, we showed that at a point where both the curve and $y(x)$ are differentiable and $dx/dt \ne 0$, the derivatives dx/dt, dy/dt, and dy/dx are related by the formula

$$\frac{dy}{dx} = \frac{dy/dt}{dx/dt}. \tag{1}$$

This is the formula we use to express dy/dx as a function of t.

If the parametric equations define y as a twice-differentiable function of x, we can apply Equation 1 to the function $dy/dx = y'$ to calculate d^2y/dx^2 as a function of t:

$$\frac{d^2y}{dx^2} = \frac{d}{dx}(y') = \frac{dy'/dt}{dx/dt}.\quad \text{Eq. 1 with } y' \text{ in place of } y$$

Parametric Formula for d^2y/dx^2

If the equations $x = f(t),\; y = g(t)$ define y as a twice-differentiable function of x, then at any point where $dx/dt \neq 0$,

$$\frac{d^2y}{dx^2} = \frac{dy'/dt}{dx/dt}.$$

Finding d^2y/dx^2 in terms of t:

1. Express $y' = dy/dx$ in terms of t.
2. Find dy'/dt.
3. Divide dy'/dt by dx/dt.

Example 2 FINDING d^2y/dx^2 FOR A PARAMETRIZED CURVE

Find d^2y/dx^2 as a function of t if $x = t - t^2,\; y = t - t^3$.

Solution

Step 1: Express $y' = dy/dx$ in terms of t.

$$y' = \frac{dy}{dx} = \frac{dy/dt}{dx/dt} = \frac{1 - 3t^2}{1 - 2t}$$

Step 2: Differentiate y' with respect to t.

$$\frac{dy'}{dt} = \frac{d}{dt}\left(\frac{1 - 3t^2}{1 - 2t}\right) = \frac{2 - 6t + 6t^2}{(1 - 2t)^2}\quad \text{Quotient Rule}$$

Step 3: Divide dy'/dt by dx/dt.

$$\frac{d^2y}{dx^2} = \frac{dy'/dt}{dx/dt} = \frac{(2 - 6t + 6t^2)/(1 - 2t)^2}{1 - 2t} = \frac{2 - 6t + 6t^2}{(1 - 2t)^3}$$

Length of a Smooth Curve

Line segment approximations similar to those in Section 7.4 lead to the following formula for the length of a smooth parametrized curve.

Length (Arc Length) of a Smooth Parametrized Curve

If a smooth curve $x = f(t),\; y = g(t),\; a \le t \le b,$ is traversed exactly once as t increases from a to b, the curve's length is

$$L = \int_a^b \sqrt{\left(\frac{dx}{dt}\right)^2 + \left(\frac{dy}{dt}\right)^2}\, dt.$$

In Exercises 25 and 26, you will see how the length formulas in Section 7.4 are special cases of this parametric formula.

$x = \cos^3 t, \ y = \sin^3 t, \ 0 \le t \le 2\pi$

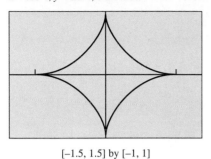

[−1.5, 1.5] by [−1, 1]

Figure 10.1 The astroid in Example 3.

Example 3 FINDING LENGTH

Find the length of the astroid (Figure 10.1)

$$x = \cos^3 t, \qquad y = \sin^3 t, \qquad 0 \le t \le 2\pi.$$

Solution

Solve Analytically

The curve is traced once as t goes from 0 to 2π. Because of the curve's symmetry with respect to the coordinate axes, its length is four times the length of the first quadrant portion. We have

$$\left(\frac{dx}{dt}\right)^2 = \left((3\cos^2 t)(-\sin t)\right)^2 = 9\cos^4 t \sin^2 t$$

$$\left(\frac{dy}{dt}\right)^2 = \left((3\sin^2 t)(\cos t)\right)^2 = 9\sin^4 t \cos^2 t$$

$$\sqrt{\left(\frac{dx}{dt}\right)^2 + \left(\frac{dy}{dt}\right)^2} = \sqrt{9\cos^2 t \sin^2 t \underbrace{(\cos^2 t + \sin^2 t)}_{1}}$$

$$= \sqrt{9\cos^2 t \sin^2 t}$$

$$= 3|\cos t \sin t|.$$

Thus, the length of the first quadrant portion of the curve is

$$\int_0^{\pi/2} 3|\cos t \sin t| \, dt = 3\int_0^{\pi/2} \cos t \sin t \, dt \quad \cos t \sin t \ge 0, \ 0 \le t \le \pi/2$$

$$= \frac{3}{2}\sin^2 t \Big]_0^{\pi/2} \qquad u = \sin t, \ du = \cos t \, dt$$

$$= \frac{3}{2}.$$

The length of the astroid is $4(3/2) = 6$.

Support Numerically

NINT $(3|\cos t \sin t|, t, 0, 2\pi) = 6$.

Cycloids

Suppose that a wheel of radius a rolls along a horizontal line without slipping (see Figure 10.2). The path traced by a point P on the wheel's edge is a **cycloid,** where P is originally at the origin.

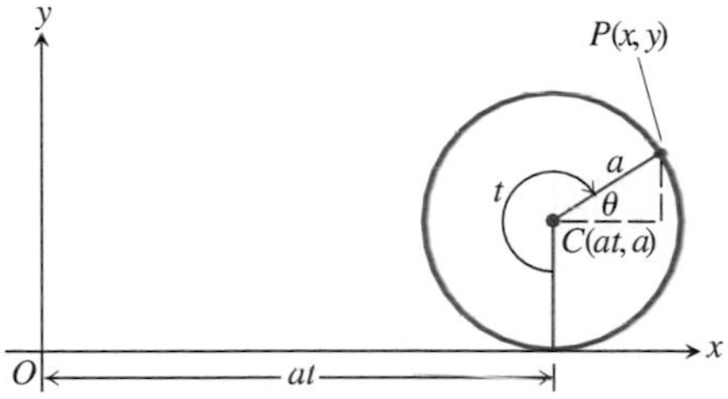

Example 4 FINDING PARAMETRIC EQUATIONS FOR A CYCLOID

Find parametric equations for the path of the point P in Figure 10.2.

Solution We suppose that the wheel rolls to the right, P being at the origin when the turn angle t equals 0. Figure 10.2 shows the wheel after it has turned t radians. The base of the wheel is at distance at from the origin. The wheel's center is at (at, a), and the coordinates of P are

$$x = at + a\cos\theta, \qquad y = a + a\sin\theta.$$

Figure 10.2 The position of $P(x, y)$ on the edge of the wheel when the wheel has turned t radians. (Example 4)

Huygens's Clock

The problem with a pendulum clock whose bob swings in a circular arc is that the frequency of the swing depends on the amplitude of the swing. The wider the swing, the longer it takes the bob to return to center.

This does not happen if the bob can be made to swing in a cycloid. In 1673, Christiaan Huygens (1629–1695), the Dutch mathematician, physicist, and astronomer who discovered the rings of Saturn, designed a pendulum clock whose bob would swing in a cycloid. Driven by a need to make accurate determinations of longitude at sea, he hung the bob from a fine wire constrained by guards that caused it to draw up as it swung away from center. How were the guards shaped? They were cycloids, too.

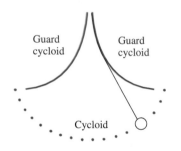

To express θ in terms of t, we observe that $t + \theta = 3\pi/2 + 2k\pi$ for some integer k, so

$$\theta = \frac{3\pi}{2} - t + 2k\pi.$$

Thus,

$$\cos \theta = \cos\left(\frac{3\pi}{2} - t + 2k\pi\right) = -\sin t,$$

$$\sin \theta = \sin\left(\frac{3\pi}{2} - t + 2k\pi\right) = -\cos t.$$

Therefore,

$$x = at - a \sin t = a(t - \sin t),$$
$$y = a - a \cos t = a(1 - \cos t).$$

Exploration 1   Investigating Cycloids

Consider the cycloids with parametric equations

$$x = a(t - \sin t), \quad y = a(1 - \cos t), \quad a > 0.$$

1. Graph the equations for $a = 1, 2,$ and 3.
2. Find the x-intercepts.
3. Show that $y \geq 0$ for all t.
4. Explain why the arches of a cycloid are congruent.
5. What is the maximum value of y? Where is it attained?
6. Describe the graph of a cycloid.

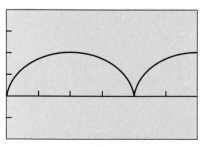

[0, 9π] by [-2, 4]

Figure 10.3 The graph of the cycloid $x = t - \sin t, y = 1 - \cos t, t \geq 0$. (Example 5)

Example 5   FINDING LENGTH

Find the length of one arch of the cycloid

$$x = a(t - \sin t), \quad y = a(1 - \cos t), \quad a > 0.$$

Solution Figure 10.3 shows the first arch of the cycloid and part of the next for $a = 1$. In Exploration 1 you found that the x-intercepts occur at t equal to multiples of 2π and that the arches are congruent.

The length of the first arch is

$$\int_0^{2\pi} \sqrt{\left(\frac{dx}{dt}\right)^2 + \left(\frac{dy}{dt}\right)^2}\, dt.$$

We have

$$\left(\frac{dx}{dt}\right)^2 = [a(1 - \cos t)]^2 = a^2(1 - 2\cos t + \cos^2 t)$$

$$\left(\frac{dy}{dt}\right)^2 = [a \sin t]^2 = a^2 \sin^2 t$$

$$\sqrt{\left(\frac{dx}{dt}\right)^2 + \left(\frac{dy}{dt}\right)^2} = a\sqrt{2 - 2\cos t}. \quad a > 0, \sin^2 t + \cos^2 t = 1$$

Therefore,

$$\int_0^{2\pi} \sqrt{\left(\frac{dx}{dt}\right)^2 + \left(\frac{dy}{dt}\right)^2} \, dt = a \int_0^{2\pi} \sqrt{2 - 2\cos t} \, dt = 8a. \quad \text{Using NINT}$$

The length of one arch of the cycloid is $8a$.

Surface Area

The following formulas give the area of the surface obtained by revolving the smooth parametrized curve $x = f(t)$, $y = g(t)$, $a \le t \le b$, about a coordinate axis.

Surface Area (from a Smooth Parametrized Curve)

If a smooth curve $x = f(t)$, $y = g(t)$, $a \le t \le b$, is traversed exactly once as t increases from a to b, then the areas of the surfaces generated by revolving the curve about the coordinate axes are as follows.

1. Revolution about the x-axis ($y \ge 0$):

$$S = \int_a^b 2\pi y \sqrt{\left(\frac{dx}{dt}\right)^2 + \left(\frac{dy}{dt}\right)^2} \, dt$$

2. Revolution about the y-axis ($x \ge 0$):

$$S = \int_a^b 2\pi x \sqrt{\left(\frac{dx}{dt}\right)^2 + \left(\frac{dy}{dt}\right)^2} \, dt$$

Example 6   FINDING SURFACE AREA

The standard parametrization of the circle of radius 1 centered at the point $(0, 1)$ in the xy-plane is

$$x = \cos t, \quad y = 1 + \sin t, \quad 0 \le t \le 2\pi.$$

Use this parametrization to find the area of the surface swept out by revolving the circle about the x-axis (Figure 10.4).

Solution We evaluate the formula

$$S = \int_a^b 2\pi y \sqrt{\left(\frac{dx}{dt}\right)^2 + \left(\frac{dy}{dt}\right)^2} \, dt.$$

$$S = \int_0^{2\pi} 2\pi(1 + \sin t)\sqrt{(-\sin t)^2 + (\cos t)^2} \, dt$$

$$= 2\pi \int_0^{2\pi} (1 + \sin t) \, dt \qquad \sin^2 t + \cos^2 t = 1$$

$$= 2\pi \left[t - \cos t \right]_0^{2\pi} = 4\pi^2$$

The surface area of the solid in Figure 10.4 is $4\pi^2$.

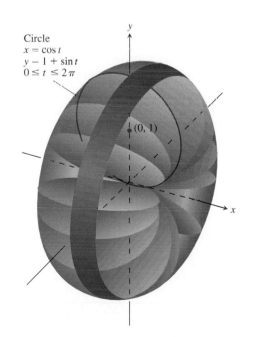

Circle
$x = \cos t$
$y = 1 + \sin t$
$0 \le t \le 2\pi$

$(0, 1)$

y

x

Figure 10.4 The surface in Example 6.

Quick Review 10.1

Exercises 1–4 refer to the parametrized curve $x = \cos t$, $y = \sin t$, $0 \le t \le 3\pi/2$.

1. Identify the initial point of the curve.

2. Identify the terminal point of the curve.

3. Find a Cartesian equation for a curve that contains the graph of the parametrized curve.

4. Which portion of the Cartesian graph in Exercise 3 is traced by the parametrized curve?

In Exercises 5 and 6, find a parametrization for the curve.

5. The portion of the curve $y = x^2 + 1$ between the points $(-1, 2)$ and $(3, 10)$.

6. $(x - 2)^2 + (y - 3)^2 = 4$

Exercises 7–9 refer to the parametrized curve $x = 2 \cos t$, $y = 3 \sin t$, $0 \le t \le 2\pi$.

7. Find dy/dx at the point where $t = 3\pi/4$.

8. Find an equation for the tangent at the point where $t = 3\pi/4$.

9. Find an equation for the normal at the point where $t = 3\pi/4$.

10. Find the length of the curve $y = x^{3/2}$ from $x = 0$ to $x = 3$.

Section 10.1 Exercises

In Exercises 1–6, find **(a)** dy/dx and **(b)** d^2y/dx^2 in terms of t.

1. $x = 4 \sin t$, $y = 2 \cos t$ **2.** $x = \cos t$, $y = \sqrt{3} \cos t$

3. $x = -\sqrt{t + 1}$, $y = \sqrt{3}t$ **4.** $x = 1/t$, $y = -2 + \ln t$

5. $x = t^2 - 3t$, $y = t^3$ **6.** $x = t^2 + t$, $y = t^2 - t$

In Exercises 7–10, find the points at which the tangent to the curve is **(a)** horizontal, **(b)** vertical.

7. $x = 2 + \cos t$, $y = -1 + \sin t$

8. $x = \sec t$, $y = \tan t$

9. $x = 2 - t$, $y = t^3 - 4t$

10. $x = -2 + 3 \cos t$, $y = 1 + 3 \sin t$

In Exercises 11–16, find the length of the curve.

11. $x = \cos t$, $y = t + \sin t$, $0 \le t \le \pi$

12. $x = \dfrac{(2t + 3)^{3/2}}{3}$, $y = t + \dfrac{t^2}{2}$, $0 \le t \le 3$

13. $x = \dfrac{1}{3}t^3$, $y = \dfrac{1}{2}t^2$, $0 \le t \le 1$

14. $x = 8 \cos t + 8t \sin t$, $y = 8 \sin t - 8t \cos t$, $0 \le t \le \pi/2$

15. $x = \ln (\sec t + \tan t) - \sin t$, $y = \cos t$, $0 \le t \le \pi/3$

16. $x = e^t - t^2$, $y = t + e^{-t}$, $-1 \le t \le 2$

In Exercises 17–20, find the area of the surface generated by revolving the curve about the indicated axis.

17. $x = \cos t$, $y = 2 + \sin t$, $0 \le t \le 2\pi$;   *x*-axis

18. $x = (2/3)t^{3/2}$, $y = 2\sqrt{t}$, $0 \le t \le 2$;   *y*-axis

19. $x = t + 1$, $y = t^2 + 2$, $0 \le t \le 3$;   *y*-axis

20. $x = \ln (\sec t + \tan t) - \sin t$, $y = \cos t$, $0 \le t \le \pi/3$;   *x*-axis

Explorations

21. *A Frustum of a Cone*   The line segment joining the points $(0, 1)$ and $(2, 2)$ is revolved about the *x*-axis to generate a frustum of a cone.

(a) Find a parametrization for the line segment joining the two points.

(b) Find the surface area of the frustum using the parametric equations in (a).

(c) Find the (lateral) surface area using the geometry formula

$$\text{Area} = \pi(r_1 + r_2)(\text{slant height}).$$

22. *A Cone*   The line segment joining the origin to the point (h, r) is revolved about the *x*-axis to generate a cone of height h and base radius r.

(a) **Writing to Learn**   Explain why

$$x = ht, \quad y = rt, \quad 0 \le t \le 1$$

gives a parametrization of the line segment joining the two points.

(b) Find the cone's (lateral) surface area using the parametrization in (a).

(c) Use the geometry formula

$$\text{Area} = \pi(r)(\text{slant height})$$

to compute the surface area of the cone.   ∎

23. *Length is Independent of Parametrization* To illustrate the fact that the numbers we get for length do not usually depend on the way we parametrize our curves, calculate the length of the semicircle $y = \sqrt{1 - x^2}$ with these two different parametrizations.

(a) $x = \cos 2t$, $y = \sin 2t$, $0 \le t \le \pi/2$

(b) $x = \sin \pi t$, $y = \cos \pi t$, $-1/2 \le t \le 1/2$

24. *Perimeter of an Ellipse* Find the length of the ellipse

$$x = 3 \cos t, \quad y = 4 \sin t, \quad 0 \le t \le 2\pi.$$

25. *Cartesian Length Formula* The graph of a function $y = f(x)$ over an interval $[a, b]$ automatically has the parametrization

$$x = x, \quad y = f(x), \quad a \le x \le b.$$

The parameter in this case is x itself. Show that for this parametrization, the length formula

$$L = \int_a^b \sqrt{\left(\frac{dx}{dt}\right)^2 + \left(\frac{dy}{dt}\right)^2}\, dt$$

reduces to the Cartesian formula

$$L = \int_a^b \sqrt{1 + \left(\frac{dy}{dx}\right)^2}\, dx$$

derived in Section 7.4.

26. (*Continuation of Exercise 25*) Show that the Cartesian formula

$$L = \int_c^d \sqrt{1 + \left(\frac{dx}{dy}\right)^2}\, dy$$

for the length of the curve $x = g(y)$, $c \le y \le d$, from Section 7.4 is a special case of the parametric length formula

$$L = \int_a^b \sqrt{\left(\frac{dx}{dt}\right)^2 + \left(\frac{dy}{dt}\right)^2}\, dt.$$

27. *Finding a Midpoint* A particle travels from $t = 0$ to $t = 4$ along the curve

$$x = \frac{t^2}{2}, \quad y = \frac{1}{3}(2t + 1)^{3/2}, \quad 0 \le t \le 4.$$

At what point (x, y) has the particle covered half of the length of the curve?

28. *Finding Surface Area* Find the area of the surface generated by revolving the curve

$$x = 3 \sin t, \quad y = 5 + 3 \sin 2t, \quad 0 \le t \le 2\pi$$

about the y-axis.

29. *Finding Surface Area* Find the area of the surface generated by revolving the curve

$$x = 5 + 3 \sin t, \quad y = 3 \sin 2t, \quad 0 \le t \le 2\pi$$

about the x-axis.

30. *Cycloid* Find the area of the surface generated by revolving one arch of the cycloid

$$x = a(t - \sin t), \quad y = a(1 - \cos t)$$

about the x-axis.

Exercises 31 and 32 refer to the region bounded by the x-axis and one arch of the cycloid

$$x = a(t - \sin t), \quad y = a(1 - \cos t)$$

that is shaded in the figure.

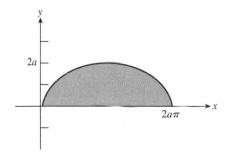

31. Find the area of the shaded region. (*Hint:* $dx = (dx/dt)\, dt$)

32. Find the volume swept out by revolving the region about the x-axis. (*Hint:* $dV = \pi y^2\, dx = \pi y^2 (dx/dt)\, dt$)

33. *Involute of a Circle* Work in groups of two or three. If a string wound around a fixed circle is unwound while being held taut in the plane of the circle, its end P traces an *involute* of the circle as suggested by the diagram below. In the diagram, the circle is the unit circle in the xy-plane, and the initial position of the tracing point is the point $(1, 0)$ on the x-axis. The unwound portion of the string is tangent to the circle at Q, and t is the radian measure of the angle from the positive x-axis to the segment OQ.

(a) Derive parametric equations for the involute by expressing the coordinates x and y of P in terms of t for $t \ge 0$.

(b) Find the length of the involute for $0 \le t \le 2\pi$.

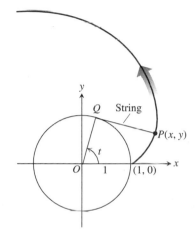

34. *(Continuation of Exercise 33)* Repeat Exercise 33 using the circle of radius a centered at the origin, $x^2 + y^2 = a^2$.

In Exercises 35–38, a projectile is launched over horizontal ground at an angle θ with the horizontal and with initial velocity v_0 ft/sec. Its path is given by the parametric equations

$$x = (v_0 \cos \theta)t, \quad y = (v_0 \sin \theta)t - 16t^2.$$

(a) Find the length of the path traveled by the projectile.

(b) Estimate the maximum height of the projectile.

35. $\theta = 20°, \quad v_0 = 150$ **36.** $\theta = 30°, \quad v_0 = 150$

37. $\theta = 60°, \quad v_0 = 150$ **38.** $\theta = 90°, \quad v_0 = 150$

Extending the Ideas

39. The function $y = f(x)$ is nonnegative on the interval $[a, b]$. Use the parametrization

$$x = x, \quad y = f(x), \quad a \leq x \leq b,$$

to show that the area of the surface generated by revolving the curve $y = f(x)$ from $x = a$ to $x = b$ about the x-axis is

$$S = \int_a^b 2\pi f(x) \sqrt{1 + \left(\frac{dy}{dx}\right)^2} \, dx.$$

In Exercises 40–42, use the formula of Exercise 39 to find the area of the surface generated by revolving the curve $y = f(x)$ from $x = a$ to $x = b$ about the x-axis.

40. $f(x) = e^x, \quad a = 0, \quad b = 3$

41. $f(x) = 1/x, \quad a = 1, \quad b = 4$

42. $f(x) = 2^x + 2^{-x}, \quad a = -2, \quad b = 2$

10.2 Vectors in the Plane

Component Form • Zero Vector • Vector Operations • Angle Between Vectors • Applications

Component Form

Some things we measure are determined by their *magnitudes.* To record mass, length, or time, for example, we need only write down a real number and name an appropriate unit of measure. These are *scalar quantities,* and the associated real numbers are **scalars.**

A quantity such as force, displacement, or velocity has direction as well as magnitude and is represented by a **directed line segment** (Figure 10.5). The arrow points in the direction of the action and its length gives the magnitude of the action in terms of a suitably chosen unit. The directed line segment $\overrightarrow{AB}$ has **initial point** A and **terminal point** B; its **length** is denoted by $|\overrightarrow{AB}|$. Directed line segments that have the same length and direction are **equivalent.**

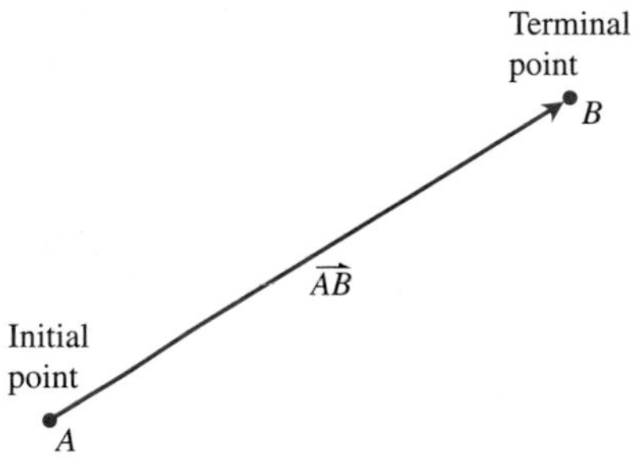

> **Definitions** Vector, Equal Vectors
>
> A **vector** in the plane is represented by a directed line segment. Two vectors are **equal** (or **the same**) if they have the same length and direction.

Thus, the arrows we use when we draw vectors are understood to represent the same vector if they have the same length, are parallel, and point in the same direction (Figure 10.6).

In textbooks, vectors are usually written in lowercase, boldface letters, for example **u**, **v**, and **w**. In handwritten form, it is customary to draw small arrows above the letters, for example $\vec{u}$, $\vec{v}$, and $\vec{w}$.

Figure 10.5 The directed line segment $\overrightarrow{AB}$.

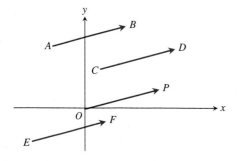

Figure 10.6 The four arrows (directed line segments) shown here have the same length and direction. They therefore represent the same vector, and we write $\overrightarrow{AB} = \overrightarrow{CD} = \overrightarrow{OP} = \overrightarrow{EF}$.

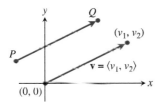

Figure 10.7 Two equal vectors. (Example 1)

Example 1 SHOWING VECTORS ARE EQUAL

Let $A = (0, 0)$, $B = (3, 4)$, $C = (-4, 2)$, and $D = (-1, 6)$. Show that the vectors $\mathbf{u} = \overrightarrow{AB}$ and $\mathbf{v} = \overrightarrow{CD}$ are equal.

Solution We need to show that $\mathbf{u}$ and $\mathbf{v}$ have the same length and direction (Figure 10.7). We use the distance formula to find their lengths.

$$|\mathbf{u}| = |\overrightarrow{AB}| = \sqrt{(3 - 0)^2 + (4 - 0)^2} = 5$$

$$|\mathbf{v}| = |\overrightarrow{CD}| = \sqrt{(-1 - (-4))^2 + (6 - 2)^2} = 5$$

Next we calculate the slopes of the two line segments.

$$\text{slope of } \overrightarrow{AB} = \frac{4 - 0}{3 - 0} = \frac{4}{3}, \quad \text{slope of } \overrightarrow{CD} = \frac{6 - 2}{-1 - (-4)} = \frac{4}{3}$$

The line segments have the same direction because they are parallel and directed toward the upper right. Therefore, $\mathbf{u} = \mathbf{v}$ because they have the same length and direction.

Let $\mathbf{v} = \overrightarrow{PQ}$. There is one directed line segment equivalent to $\overrightarrow{PQ}$ whose initial point is the origin (Figure 10.8). It is the representative of $\mathbf{v}$ in **standard position** and is the vector we normally use to represent $\mathbf{v}$.

Definition Component Form of a Vector

If $\mathbf{v}$ is a vector in the plane equal to the vector with initial point $(0, 0)$ and terminal point (v_1, v_2), then the **component form** of $\mathbf{v}$ is

$$\mathbf{v} = \langle v_1, v_2 \rangle.$$

The coordinates v_1 and v_2 are the **components** of $\mathbf{v}$. The **magnitude (length)** of $\mathbf{v} = \langle v_1, v_2 \rangle$ is

$$|\mathbf{v}| = \sqrt{v_1{}^2 + v_2{}^2}.$$

Two vectors $\langle a, b \rangle$ and $\langle c, d \rangle$ are equal if and only if $a = c$ and $b = d$. The vector $\langle a, b \rangle$ is called the **position vector** of the point (a, b).

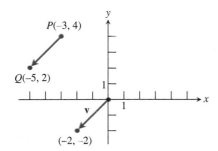

Figure 10.8 The standard position of a vector.

Figure 10.9 The vectors $\overrightarrow{PQ}$ and $\mathbf{v}$ are equal. (Example 2)

Example 2 FINDING COMPONENT FORM

Find the **(a)** component form and **(b)** length of the vector with initial point $P = (-3, 4)$ and terminal point $Q = (-5, 2)$.

Solution

(a) We can move the initial point $(-3, 4)$ of the vector $\overrightarrow{PQ}$ to the origin by translating the vector 3 units right and 4 units down (Figure 10.9). Thus, Q translates to $(-2, -2)$ and the component form of $\overrightarrow{PQ}$ is

$$\mathbf{v} = \langle -2, -2 \rangle.$$

(b) The length of $\mathbf{v} = \overrightarrow{PQ}$ is

$$|\mathbf{v}| = \sqrt{(-2)^2 + (-2)^2} = 2\sqrt{2}.$$

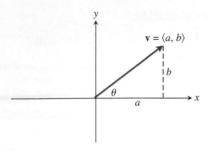

Figure 10.10 The vector $\mathbf{v} = \langle a, b \rangle$ has slope b/a.

Example 2 can be generalized in the following way. The component form of the vector with initial point $P = (p_1, p_2)$ and terminal point $Q = (q_1, q_2)$ is

$$\mathbf{v} = \langle v_1, v_2 \rangle = \langle q_1 - p_1, q_2 - p_2 \rangle,$$

and its length is

$$|\mathbf{v}| = \sqrt{v_1^2 + v_2^2} = \sqrt{(q_1 - p_1)^2 + (q_2 - p_2)^2}.$$

If $|\mathbf{v}| = 1$, then $\mathbf{v}$ is a **unit vector.**

The **slope** of a nonvertical vector is the slope shared by the lines parallel to the vector. Thus, if $a \neq 0$, the vector $\mathbf{v} = \langle a, b \rangle$ has slope b/a (Figure 10.10), and $a = |\mathbf{v}| \cos \theta$, $b = |\mathbf{v}| \sin \theta$.

Example 3 FINDING COMPONENT FORM

Find the component form of the vector $\mathbf{v}$ of length 3 that makes an angle of $40°$ with the positive x-axis.

Solution From Figure 10.10 and the discussion preceding this example, we have

$$\mathbf{v} = \langle 3 \cos 40°, 3 \sin 40° \rangle \approx \langle 2.298, 1.928 \rangle.$$

Zero Vector

It is convenient to define $\langle 0, 0 \rangle$ as the **zero vector,**

$$\mathbf{0} = \langle 0, 0 \rangle.$$

The zero vector has length $|\mathbf{0}| = \sqrt{0^2 + 0^2} = 0$ and is the only vector with no direction.

Vector Operations

Two principal operations involving vectors are *vector addition* and *scalar multiplication*.

Definitions Vector Operations

Let $\mathbf{u} = \langle u_1, u_2 \rangle$, $\mathbf{v} = \langle v_1, v_2 \rangle$ be vectors with k a scalar (real number).

Addition: $\mathbf{u} + \mathbf{v} = \langle u_1, u_2 \rangle + \langle v_1, v_2 \rangle = \langle u_1 + v_1, u_2 + v_2 \rangle$

Subtraction: $\mathbf{u} - \mathbf{v} = \langle u_1, u_2 \rangle - \langle v_1, v_2 \rangle = \langle u_1 - v_1, u_2 - v_2 \rangle$

Scalar multiplication: $k\mathbf{u} = \langle ku_1, ku_2 \rangle$

Negative (opposite): $-\mathbf{u} = (-1)\mathbf{u} = \langle -u_1, -u_2 \rangle$

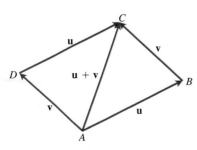

Figure 10.11 The parallelogram law of addition.

Two vectors $\mathbf{u}$ and $\mathbf{v}$ may be added geometrically by placing the initial point of one at the terminal point of the other as shown in Figure 10.11. The sum, called the **resultant vector,** is the diagonal of the resulting parallelogram. This geometric description of addition is sometimes called the **parallelogram law** of addition. In Exercise 33, you will show that the parallelogram law agrees with the component form of addition.

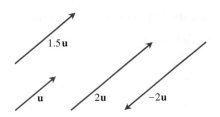

Figure 10.12 Scalar multiples of **u**.

The product $k\mathbf{u}$ of the scalar k and the vector $\mathbf{u}$ can be obtained geometrically by a stretch or shrink of $\mathbf{u}$ by the factor k when $k > 0$ (Figure 10.12). If $k < 0$, then there is a stretch or shrink by the factor $|k|$, and the direction of $k\mathbf{u}$ is opposite that of $\mathbf{u}$.

Example 4 PERFORMING OPERATIONS ON VECTORS

Let $\mathbf{u} = \langle -1, 3 \rangle$ and $\mathbf{v} = \langle 4, 7 \rangle$. Find

(a) $2\mathbf{u} + 3\mathbf{v}$, **(b)** $\mathbf{u} - \mathbf{v}$, **(c)** $\left| \dfrac{1}{2}\mathbf{u} \right|$.

Solution

(a) $2\mathbf{u} + 3\mathbf{v} = 2\langle -1, 3 \rangle + 3\langle 4, 7 \rangle$

$$= \langle 2(-1) + 3(4), 2(3) + 3(7) \rangle = \langle 10, 27 \rangle$$

(b) $\mathbf{u} - \mathbf{v} = \langle -1, 3 \rangle - \langle 4, 7 \rangle$

$$= \langle -1 - 4, 3 - 7 \rangle = \langle -5, -4 \rangle$$

(c) $\left| \dfrac{1}{2}\mathbf{u} \right| = \left| \left\langle -\dfrac{1}{2}, \dfrac{3}{2} \right\rangle \right| = \sqrt{\left(-\dfrac{1}{2}\right)^2 + \left(\dfrac{3}{2}\right)^2} = \dfrac{1}{2}\sqrt{10}$

Vector operations have many of the properties of ordinary arithmetic.

Properties of Vector Operations

Let $\mathbf{u}$, $\mathbf{v}$, $\mathbf{w}$ be vectors and a, b be scalars.

1. $\mathbf{u} + \mathbf{v} = \mathbf{v} + \mathbf{u}$	**2.** $(\mathbf{u} + \mathbf{v}) + \mathbf{w} = \mathbf{u} + (\mathbf{v} + \mathbf{w})$
3. $\mathbf{u} + \mathbf{0} = \mathbf{u}$	**4.** $\mathbf{u} + (-\mathbf{u}) = \mathbf{0}$
5. $0\mathbf{u} = \mathbf{0}$	**6.** $1\mathbf{u} = \mathbf{u}$
7. $a(b\mathbf{u}) = (ab)\mathbf{u}$	**8.** $a(\mathbf{u} + \mathbf{v}) = a\mathbf{u} + a\mathbf{v}$
9. $(a + b)\mathbf{u} = a\mathbf{u} + b\mathbf{u}$	

Angle Between Vectors

When two nonzero vectors $\mathbf{u}$ and $\mathbf{v}$ are placed so their initial points coincide, they form an angle θ of measure $0 \le \theta \le \pi$ (Figure 10.13). This angle is the **angle between u** and **v**.

Theorem 1 gives a formula we can use to determine the angle between two vectors.

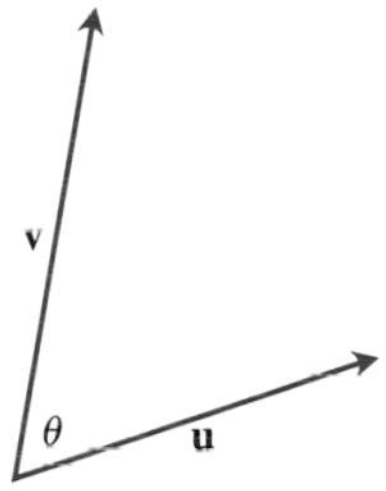

Figure 10.13 The angle between **u** and **v**.

Theorem 1 Angle Between Two Vectors

The angle θ between two nonzero vectors $\mathbf{u} = \langle u_1, u_2 \rangle$ and $\mathbf{v} = \langle v_1, v_2 \rangle$ is given by

$$\theta = \cos^{-1} \frac{u_1 v_1 + u_2 v_2}{|\mathbf{u}||\mathbf{v}|}.$$

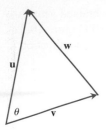

Figure 10.14 The parallelogram law of addition of vectors gives $\mathbf{w} = \mathbf{u} - \mathbf{v}$.

Proof Applying the law of cosines to the triangle in Figure 10.14 we find

$$|\mathbf{w}|^2 = |\mathbf{u}|^2 + |\mathbf{v}|^2 - 2|\mathbf{u}||\mathbf{v}| \cos \theta$$

$$2|\mathbf{u}||\mathbf{v}| \cos \theta = |\mathbf{u}|^2 + |\mathbf{v}|^2 - |\mathbf{w}|^2.$$

Because $\mathbf{w} = \mathbf{u} - \mathbf{v}$, the component form of $\mathbf{w}$ is $\langle u_1 - v_1, u_2 - v_2 \rangle$. Thus,

$$|\mathbf{u}|^2 = \left(\sqrt{u_1^2 + u_2^2} \right)^2 = u_1^2 + u_2^2,$$

$$|\mathbf{v}|^2 = \left(\sqrt{v_1^2 + v_2^2} \right)^2 = v_1^2 + v_2^2,$$

$$|\mathbf{w}|^2 = \left(\sqrt{(u_1 - v_1)^2 + (u_2 - v_2)^2} \right)^2 = (u_1 - v_1)^2 + (u_2 - v_2)^2$$

$$= (u_1^2 - 2u_1v_1 + v_1^2) + (u_2^2 - 2u_2v_2 + v_2^2),$$

and

$$|\mathbf{u}|^2 + |\mathbf{v}|^2 - |\mathbf{w}|^2 = 2(u_1v_1 + u_2v_2).$$

Therefore,

$$2|\mathbf{u}||\mathbf{v}| \cos \theta = |\mathbf{u}|^2 + |\mathbf{v}|^2 - |\mathbf{w}|^2 = 2(u_1v_1 + u_2v_2)$$

$$|\mathbf{u}||\mathbf{v}| \cos \theta = u_1v_1 + u_2v_2$$

$$\cos \theta = \frac{u_1v_1 + u_2v_2}{|\mathbf{u}||\mathbf{v}|}.$$

So

$$\theta = \cos^{-1} \left(\frac{u_1v_1 + u_2v_2}{|\mathbf{u}||\mathbf{v}|} \right). \qquad \blacksquare$$

Definition Dot Product (Inner Product)

The **dot product** (or **inner product**) $\mathbf{u} \cdot \mathbf{v}$ ("$\mathbf{u}$ dot $\mathbf{v}$") of vectors $\mathbf{u} = \langle u_1, u_2 \rangle$ and $\mathbf{v} = \langle v_1, v_2 \rangle$ is the number

$$\mathbf{u} \cdot \mathbf{v} = |\mathbf{u}||\mathbf{v}| \cos \theta = u_1v_1 + u_2v_2.$$

We can use the dot product to rewrite the formula in Theorem 1 for finding the angle between two vectors.

Corollary Angle Between Two Vectors

The angle between nonzero vectors $\mathbf{u}$ and $\mathbf{v}$ is

$$\theta = \cos^{-1} \left(\frac{\mathbf{u} \cdot \mathbf{v}}{|\mathbf{u}||\mathbf{v}|} \right). \qquad (1)$$

Example 5 FINDING AN ANGLE OF A TRIANGLE

Find the angle θ in the triangle ABC determined by the vertices $A = (0, 0)$, $B = (3, 5)$, and $C = (5, 2)$ (Figure 10.15).

Solution The angle θ is the angle between the vectors $\overrightarrow{CA}$ and $\overrightarrow{CB}$. The component forms of these two vectors are

$$\overrightarrow{CA} = \langle -5, -2 \rangle \quad \text{and} \quad \overrightarrow{CB} = \langle -2, 3 \rangle.$$

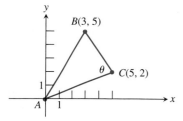

Figure 10.15 The triangle in Example 5.

We use Equation 1.

$$\vec{CA} \cdot \vec{CB} = (-5)(-2) + (-2)(3) = 4$$

$$|\vec{CA}| = \sqrt{(-5)^2 + (-2)^2} = \sqrt{29}$$

$$|\vec{CB}| = \sqrt{(-2)^2 + (3)^2} = \sqrt{13}$$

Thus,

$$\theta = \cos^{-1}\left(\frac{\vec{CA} \cdot \vec{CB}}{|\vec{CA}||\vec{CB}|} \right)$$

$$= \cos^{-1}\left(\frac{4}{(\sqrt{29})(\sqrt{13})} \right)$$

$$\approx 78.1° \ \text{ or } \ 1.36 \text{ radians.}$$

Applications

Suppose the motion of a particle in the plane is represented by parametric equations. The tangent line, suitably directed, models the *direction* of the motion at the point of tangency (Figure 10.16).

A vector is **tangent** or **normal** to a curve at a point P if it is parallel or normal, respectively, to the line that is tangent to the curve at P. Example 6 shows how to find such vectors.

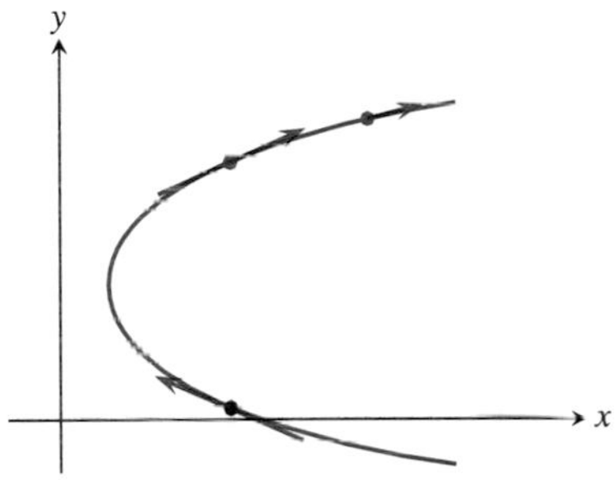

Figure 10.16 The tangent line models the direction of a moving particle at a given point.

Example 6 FINDING VECTORS TANGENT AND NORMAL TO A CURVE

Find unit vectors tangent and normal to the parametrized curve

$$x = \frac{t}{2} + 1, \quad y = \sqrt{t} + 1, \quad t \geq 0,$$

at the point where $t = 4$.

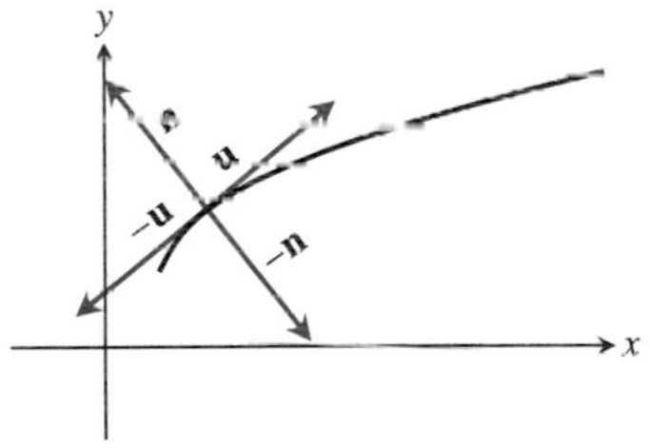

Solution Since $x = y - 3$ when $t = 4$, we seek the unit vectors that are parallel and normal to the curve's tangent line at $(3, 3)$ (Figure 10.17). The slope of the tangent is

$$\frac{dy}{dx} = \frac{dy/dt}{dx/dt} = \frac{(1/2)t^{-1/2}}{1/2}\bigg|_{t=4} = \frac{1}{2}.$$

The vector $\mathbf{v} = \langle 2, 1 \rangle$ has slope $1/2$, as does every nonzero multiple of $\mathbf{v}$. To find a multiple of $\mathbf{v}$ that is a unit vector, we divide $\mathbf{v}$ by its length

$$|\mathbf{v}| = \sqrt{2^2 + 1^2} = \sqrt{5},$$

obtaining

$$\mathbf{u} = \frac{\mathbf{v}}{|\mathbf{v}|} = \left\langle \frac{2}{\sqrt{5}}, \frac{1}{\sqrt{5}} \right\rangle.$$

The vector $\mathbf{u}$ has length 1 and is tangent to the curve at $(3, 3)$ because it has the same direction as $\mathbf{v}$. Of course,

$$-\mathbf{u} = \left\langle -\frac{2}{\sqrt{5}}, -\frac{1}{\sqrt{5}} \right\rangle,$$

which points in the opposite direction, is also tangent to the curve at $(3, 3)$. Unless there are additional requirements, either $\mathbf{u}$ or $-\mathbf{u}$ is acceptable.

Figure 10.17 Unit tangent and normal vectors at the point $(3, 3)$ on the curve $x = (1/2)t + 1$, $y = \sqrt{t} + 1$, $t \geq 0$. (Example 6)

To find unit vectors normal to the curve at (3, 3), we look for unit vectors whose slopes are the negative reciprocal of the slope of **u**. This is quickly done by interchanging the components of **u** and changing the sign of one of them. We obtain

$$\mathbf{n} = \left\langle -\frac{1}{\sqrt{5}}, \frac{2}{\sqrt{5}} \right\rangle \quad \text{and} \quad -\mathbf{n} = \left\langle \frac{1}{\sqrt{5}}, -\frac{2}{\sqrt{5}} \right\rangle.$$

Again, either one will do. The vectors have opposite directions but both are normal to the curve at (3, 3) and have length 1.

An important application of vectors occurs in navigation.

Example 7 FINDING GROUND SPEED AND DIRECTION

A Boeing® 727® airplane, flying due east at 500 mph in still air, encounters a 70-mph tail wind acting in the direction 60° north of east. The airplane holds its compass heading due east but, because of the wind, acquires a new ground speed and direction. What are they?

Solution If **u** = the velocity of the airplane alone and **v** = the velocity of the tail wind, then $|\mathbf{u}| = 500$ and $|\mathbf{v}| = 70$ (Figure 10.18).

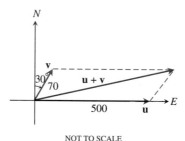

NOT TO SCALE

Figure 10.18 Vectors representing the velocities of the airplane and tail wind in Example 7.

We need to find the magnitude and direction of the *resultant vector* **u** + **v**. If we let the positive *x*-axis represent east and the positive *y*-axis represent north, then the component forms of **u** and **v** are

$$\mathbf{u} = \langle 500, 0 \rangle \quad \text{and} \quad \mathbf{v} = \langle 70 \cos 60°, 70 \sin 60° \rangle = \langle 35, 35\sqrt{3} \rangle.$$

Therefore,

$$\mathbf{u} + \mathbf{v} = \langle 535, 35\sqrt{3} \rangle,$$

$$|\mathbf{u} + \mathbf{v}| = \sqrt{535^2 + (35\sqrt{3})^2} \approx 538.4,$$

and

$$\theta = \tan^{-1}\frac{35\sqrt{3}}{535} \approx 6.5°.$$

Interpret

The new ground speed of the airplane is about 538.4 mph, and its new direction is about 6.5° north of east.

Quick Review 10.2

In Exercises 1–3, let $P = (1, 2)$ and $Q = (5, 3)$.

1. Find the distance between the points P and Q.

2. Find the slope of line segment PQ.

3. If $R = (3, b)$, determine b so that segments PQ and RQ are perpendicular.

In Exercises 4 and 5, determine the missing coordinate so that the four points form a parallelogram $ABCD$.

4. $A = (0, 0)$, $B = (1, 3)$, $C = (5, 3)$, $D = (a, 0)$

5. $A = (1, 1)$, $B = (3, 5)$, $C = (8, b)$, $D = (6, 2)$

In Exercises 6–8, find the measure of the angle θ in **(a)** degrees and **(b)** radians.

6. $\theta = \cos^{-1}(-1/2)$ **7.** $\theta = \sin^{-1}(-1/2)$

8. $\theta = \tan^{-1}(-1)$

9. Use the law of cosines to find the value of c in the triangle.

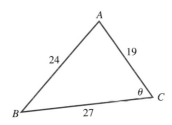

10. Use the law of cosines to find the value of θ in the triangle.

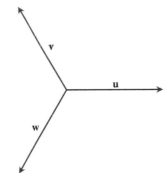

Section 10.2 Exercises

In Exercises 1–8, let $\mathbf{u} = \langle 3, \ 2 \rangle$ and $\mathbf{v} = \langle -2, 5 \rangle$. Find the **(a)** component form and **(b)** magnitude of the vector.

1. $3\mathbf{u}$ **2.** $-2\mathbf{v}$

3. $\mathbf{u} + \mathbf{v}$ **4.** $\mathbf{u} - \mathbf{v}$

5. $2\mathbf{u} - 3\mathbf{v}$ **6.** $-2\mathbf{u} + 5\mathbf{v}$

7. $\dfrac{3}{5}\mathbf{u} + \dfrac{4}{5}\mathbf{v}$ **8.** $-\dfrac{5}{13}\mathbf{u} + \dfrac{12}{13}\mathbf{v}$

In Exercises 9–16, find the component form of the vector.

9. the vector $\overrightarrow{PQ}$, where $P = (1, 3)$ and $Q = (2, -1)$

10. the vector $\overrightarrow{OP}$ where O is the origin and P is the midpoint of segment RS, where $R = (2, -1)$ and $S = (-4, 3)$

11. the vector from the point $A = (2, 3)$ to the origin

12. the sum of $\overrightarrow{AB}$ and $\overrightarrow{CD}$, where $A = (1, -1)$, $B = (2, 0)$, $C = (-1, 3)$, and $D = (-2, 2)$

13. the unit vector that makes an angle $\theta = 2\pi/3$ with the positive x-axis

14. the unit vector that makes an angle $\theta = -3\pi/4$ with the positive x-axis

15. the unit vector obtained by rotating the vector $\langle 0, 1 \rangle$ $120°$ counterclockwise about the origin

16. the unit vector obtained by rotating the vector $\langle 1, 0 \rangle$ $135°$ counterclockwise about the origin

In Exercises 17 and 18, copy vectors $\mathbf{u}$, $\mathbf{v}$, and $\mathbf{w}$ head to tail as needed to sketch the indicated vector.

17.

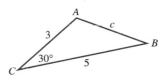

(a) $\mathbf{u} + \mathbf{v}$ **(b)** $\mathbf{u} + \mathbf{v} + \mathbf{w}$

(c) $\mathbf{u} - \mathbf{v}$ **(d)** $\mathbf{u} - \mathbf{w}$

18.

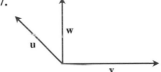

(a) $\mathbf{u} - \mathbf{v}$ **(b)** $\mathbf{u} - \mathbf{v} + \mathbf{w}$

(c) $2\mathbf{u} - \mathbf{v}$ **(d)** $\mathbf{u} + \mathbf{v} + \mathbf{w}$

In Exercises 19–22, find a unit vector in the direction of the given vector.

19. $\langle 3, 4 \rangle$ **20.** $\langle 4, -3 \rangle$

21. $\langle -15, 8 \rangle$ **22.** $\langle -5, -2 \rangle$

In Exercises 23–26, find the unit vectors (four vectors in all) that are tangent and normal to the curve at the given point.

23. $x = \sqrt{t} + 1, \quad y = t + 1 + 2\sqrt{t}, \quad t = 1$

24. $x = \ln(t - 1), \quad y = t - 1, \quad t = 3$

25. $x = 4 \cos t, \quad y = 5 \sin t, \quad t = \pi/3$

26. $x = 3 \cos t, \quad y = 3 \sin t, \quad t = -\pi/4$

27. *Triangle* Find the measures of the angles of the triangle whose vertices are $A = (-1, 0)$, $B = (2, 1)$, and $C = (1, -2)$.

28. *Rectangle* Find the measures of the angles between the diagonals of the rectangle whose vertices are $A = (1, 0)$, $B = (0, 3)$, $C = (3, 4)$, and $D = (4, 1)$.

In Exercises 29–34, *work in groups of two or three.*

29. *Distributive Property* Use components to show that for any vectors **u**, **v**, and **w**,

   (a) $\mathbf{u} \cdot (\mathbf{v} + \mathbf{w}) = \mathbf{u} \cdot \mathbf{v} + \mathbf{u} \cdot \mathbf{w}$

   (b) $(\mathbf{u} + \mathbf{v}) \cdot \mathbf{w} = \mathbf{u} \cdot \mathbf{w} + \mathbf{v} \cdot \mathbf{w}$

30. *Inner Product and Length* Show that $\mathbf{u} \cdot \mathbf{u} = |\mathbf{u}|^2$ for any vector **u**.

31. *Inner Product and Length* Show that $(\mathbf{u} + \mathbf{v}) \cdot (\mathbf{u} - \mathbf{v}) = |\mathbf{u}|^2 - |\mathbf{v}|^2$ for any two vectors **u** and **v**.

32. *Test for Orthogonality* If **u** and **v** are nonzero vectors, show that $\mathbf{u} \cdot \mathbf{v} = 0$ if and only if **u** and **v** are **orthogonal** (perpendicular).

33. *Addition* Show that geometric addition of vectors agrees with component-form addition. (*Hint:* Place Figure 10.11 in a coordinate plane and write component forms for the vectors.)

34. *Subtraction*

   (a) Give a geometric version of subtraction of vectors.

   (b) Show that your geometric version of subtraction agrees with component-form subtraction. (*Hint:* See Exercise 33.)

Explorations

35. *Location on a Line Segment* Let M be a point on the line segment PQ.

   (a) If M is the midpoint of the segment PQ, show that

$$\overrightarrow{OM} = \frac{1}{2}\overrightarrow{OP} + \frac{1}{2}\overrightarrow{OQ}.$$

   (b) If M is one third of the way from P to Q, find a formula for the position vector $\overrightarrow{OM}$ in terms of the position vectors $\overrightarrow{OP}$ and $\overrightarrow{OQ}$.

   (c) If M is two thirds of the way from P to Q, find a formula for the position vector $\overrightarrow{OM}$ in terms of the position vectors $\overrightarrow{OP}$ and $\overrightarrow{OQ}$.

   (d) Writing to Learn If M is between P and Q on the segment PQ, state a formula for the position vector $\overrightarrow{OM}$ in terms of the position vectors $\overrightarrow{OP}$ and $\overrightarrow{OQ}$. Then prove your statement.

36. *Circle* Suppose that AB is the diameter of a circle with center O and that C is a point on one of the two arcs joining A and B, as shown in the figure. Show that $\overrightarrow{CA}$ and $\overrightarrow{CB}$ are orthogonal (perpendicular).

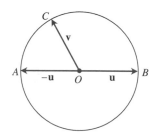

37. *Rhombus* Show that the diagonals of a rhombus (parallelogram with all sides of equal length) are perpendicular.

38. *Square* Show that squares are the only rectangles with perpendicular diagonals.

39. *Parallelogram* Prove the fact, often exploited by carpenters, that a parallelogram is a rectangle if and only if the diagonals are equal in length. ■

40. *Parallelogram* Show that the indicated diagonal of the parallelogram determined by vectors **u** and **v** bisects the angle between **u** and **v** if $|\mathbf{u}| = |\mathbf{v}|$.

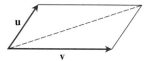

41. Writing to Learn What can you say about the slopes of the (nonzero) vectors **v** and $-\mathbf{v}$?

42. Writing to Learn If $|\mathbf{v}| = 0$, what can you say about the vector **v**?

43. *Velocity* An airplane is flying in the direction 25° west of north at 800 km/h. Find the component form of the velocity of the airplane, assuming that the positive x-axis represents due east and the positive y-axis represents due north.

44. *Velocity* An airplane is flying in the direction 10° east of south at 600 km/h. Find the component form of the velocity of the airplane, assuming that the positive x-axis represents due east and the positive y-axis represents due north.

45. *Navigation* An airplane, flying in the direction 20° east of north at 325 mph in still air, encounters a 40-mph tail wind acting in the direction 40° west of north. The airplane maintains its compass heading but, because of the wind, acquires a new ground speed and direction. What are they?

46. *Inclined Plane* Suppose a box is being towed up an inclined plane as shown in the figure. Find the force **w** needed to make the component of the force parallel to the inclined plane equal to 2.5 lb.

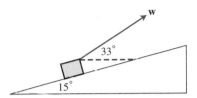

47. *Combining Forces* Juana and Diego Gonzales, ages six and four respectively, own a strong and stubborn puppy named Corporal. It is so hard to take Corporal for a walk that they devise a scheme to use two leashes. If Juana and Diego pull with forces of 23 lb and 18 lb at the angles shown in the figure, how hard is Corporal pulling if the puppy holds the children at a standstill?

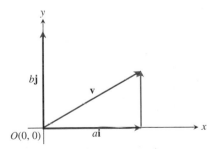

48. *Locating Points* A bird flies from its nest 7 km in the direction northeast, where it stops to rest on a tree. It then flies 8 km in the direction 30° south of west to land atop a telephone pole. Place an *xy*-coordinate system so that the origin is the bird's nest, the *x*-axis points east, and the *y*-axis points north.

(a) At what point is the tree located?

(b) At what point is the telephone pole located?

In Exercises 49 and 50, show that the vectors $\overrightarrow{AB}$ and $\overrightarrow{CD}$ are equal.

49. $A = (0, 0)$, $B = (-3, 4)$, $C = (4, 1)$, $D = (1, 5)$

50. $A = (-4, 3)$, $B = (-2, -2)$, $C = (1, 1)$, $D = (3, -4)$

Extending the Ideas

51. *Vector Operation Properties* Use the component form for vectors to prove the nine properties of vector operations listed in the text.

52. *Linear Combinations of Vectors* Write the vector $\langle 3, 4 \rangle$ as the sum of a vector parallel to $\mathbf{u} = \langle 1, 1 \rangle$ and a vector perpendicular to $\mathbf{u}$.

53. *Equations for Lines* Write an equation for the line in the plane that passes through the point $(-2, 1)$ and is

(a) parallel to the vector $\langle 1, -1 \rangle$.

(b) perpendicular to the vector $\langle 1, -1 \rangle$.

54. *Angle Between Lines* Find the acute angle between the lines $3x - 4y = 3$, $x - y = 7$.

10.3	**Vector-valued Functions**

Standard Unit Vectors • Planar Curves • Limits and Continuity • Derivatives and Motion • Differentiation Rules • Integrals

Standard Unit Vectors

Any vector $\mathbf{v} = \langle a, b \rangle$ in the plane can be written as a *linear combination* of the two **standard unit vectors**

$$\mathbf{i} = \langle 1, 0 \rangle \quad \text{and} \quad \mathbf{j} = \langle 0, 1 \rangle$$

as follows:

$$\mathbf{v} = \langle a, b \rangle = \langle a, 0 \rangle + \langle 0, b \rangle = a\langle 1, 0 \rangle + b\langle 0, 1 \rangle$$
$$= a\mathbf{i} + b\mathbf{j}.$$

The vector **v** is a **linear combination** of the vectors **i** and **j**; the scalar *a* is the **horizontal component** of **v** and the scalar *b* is the **vertical component** of **v** (Figure 10.19).

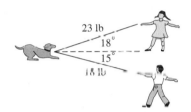

Figure 10.19 **v** is a linear combination of **i** and **j**.

Example 1 EXPRESSING VECTORS AS LINEAR COMBINATIONS OF **i** AND **j**

Let $P = (-1, 5)$ and $Q = (3, 2)$. Write the vector $\mathbf{v} = \overrightarrow{PQ}$ as a linear combination of **i** and **j**.

Solution The component form of **v** is $\langle 3 - (-1), 2 - 5 \rangle = \langle 4, -3 \rangle$. Thus,

$$\mathbf{v} = \langle 4, -3 \rangle = 4\mathbf{i} + (-3)\mathbf{j} = 4\mathbf{i} - 3\mathbf{j}.$$

Planar Curves

When a particle moves through the plane during a time interval *I*, we think of the particle's coordinates as functions defined on *I*:

$$x = f(t), \quad y = g(t), \quad t \in I. \tag{1}$$

The points $(x, y) = (f(t), g(t))$, $t \in I$, make up the curve in the plane that is the particle's **path.** The equations and interval in Equation 1 parametrize the curve. The vector

$$\mathbf{r}(t) = \overrightarrow{OP} = \langle f(t), g(t) \rangle = f(t)\mathbf{i} + g(t)\mathbf{j} \tag{2}$$

from the origin to the particle's **position** $P(f(t), g(t))$ at time *t* is the particle's **position vector.** The functions *f* and *g* are the **component functions** (**components**) of the position vector. We think of the particle's path as the **curve traced by r** during the time interval *I*.

Equation 2 defines **r** as a *vector function* of the real variable *t* on the interval *I*. More generally, a **vector function** or **vector-valued function** on a domain *D* is a rule that assigns a vector in the plane to each element in *D*. The curve traced by a vector function is its **graph.**

We refer to real-valued functions as **scalar functions** to distinguish them from vector functions. The components of **r** are scalar functions of *t*. When we define a vector-valued function by giving its component functions, we assume the vector function's domain to be the common domain of the components.

Example 2 GRAPHING AN ARCHIMEDES SPIRAL

Graph the vector function

$$\mathbf{r}(t) = (t \cos t)\mathbf{i} + (t \sin t)\mathbf{j}, \quad t \geq 0.$$

Solution We can graph the vector function parametrically using

$$x = t \cos t, \quad y = t \sin t, \quad t \geq 0.$$

The curve (Figure 10.20) is a spiral that winds around the origin, getting farther away from the origin as *t* increases. (Enter the equations into your calculator with $t\text{Max} = 50$ and watch the curve develop.)

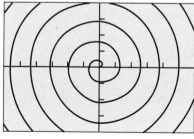

[–30, 30] by [–20, 20]

Figure 10.20 The graph of $\mathbf{r}(t) = (t \cos t)\mathbf{i} + (t \sin t)\mathbf{j}$, $t \geq 0$, is the curve $x = t \cos t$, $y = t \sin t$, $t \geq 0$. (Example 2)

Limits and Continuity

We define limits of vector functions in terms of their scalar components.

Definition Limit

Let $\mathbf{r}(t) = f(t)\mathbf{i} + g(t)\mathbf{j}$. If

$$\lim_{t \to c} f(t) = L_1 \quad \text{and} \quad \lim_{t \to c} g(t) = L_2,$$

then the **limit** of $\mathbf{r}(t)$ as t approaches c is

$$\lim_{t \to c} \mathbf{r}(t) = \mathbf{L} = L_1\mathbf{i} + L_2\mathbf{j}.$$

Example 3 FINDING A LIMIT OF A VECTOR FUNCTION

If $\mathbf{r}(t) = (\cos t)\mathbf{i} + (\sin t)\mathbf{j}$, then

$$\lim_{t \to \pi/4} \mathbf{r}(t) = \left(\lim_{t \to \pi/4} \cos t \right)\mathbf{i} + \left(\lim_{t \to \pi/4} \sin t \right)\mathbf{j} = \frac{\sqrt{2}}{2}\mathbf{i} + \frac{\sqrt{2}}{2}\mathbf{j}.$$

We define continuity for vector functions in the same way we define continuity for scalar functions.

Definition Continuity at a Point

A vector function $\mathbf{r}(t)$ is **continuous at a point** $t = c$ in its domain if

$$\lim_{t \to c} \mathbf{r}(t) = \mathbf{r}(c).$$

A vector function $\mathbf{r}(t)$ is **continuous** if it is continuous at every point in its domain. Since limits of vector functions are defined in terms of components we have the following test for continuity.

Component Test for Continuity at a Point

The vector function $\mathbf{r}(t) = f(t)\mathbf{i} + g(t)\mathbf{j}$ is continuous at $t = c$ if and only if f and g are continuous at $t = c$.

Example 4 FINDING POINTS OF CONTINUITY & DISCONTINUITY

(a) The function

$$\mathbf{r}(t) = (t \cos t)\mathbf{i} + (t \sin t)\mathbf{j}$$

is continuous everywhere because the component functions $t \cos t$ and $t \sin t$ are continuous everywhere.

(b) The function

$$\mathbf{r}(t) = (1/t)\mathbf{i} + (\sin t)\mathbf{j}$$

is not continuous at $t = 0$ because the first component is not continuous at $t = 0$. It is a continuous function, however, because it is continuous on its domain, the set of all nonzero real numbers.

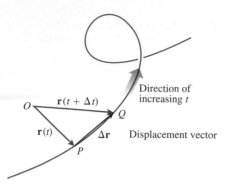

Figure 10.21 Between time t and time $t + \Delta t$, the particle moving along the path shown here undergoes the displacement $\overrightarrow{PQ} = \Delta\mathbf{r}$. The vector sum $\mathbf{r}(t) + \Delta\mathbf{r}$ gives the new position, $\mathbf{r}(t + \Delta t)$.

Derivatives and Motion

Suppose that $\mathbf{r}(t) = f(t)\mathbf{i} + g(t)\mathbf{j}$ is the position vector of a particle moving along a curve in the plane and that f and g are differentiable functions of t. Then (see Figure 10.21) the difference between the particle's positions at time $t + \Delta t$ and time t is $\Delta\mathbf{r} = \mathbf{r}(t + \Delta t) - \mathbf{r}(t)$. In terms of components,

$$\Delta\mathbf{r} = \mathbf{r}(t + \Delta t) - \mathbf{r}(t)$$
$$= [f(t + \Delta t)\mathbf{i} + g(t + \Delta t)\mathbf{j}] - [f(t)\mathbf{i} + g(t)\mathbf{j}]$$
$$= [f(t + \Delta t) - f(t)]\mathbf{i} + [g(t + \Delta t) - g(t)]\mathbf{j}.$$

As Δt approaches zero, three things seem to happen simultaneously. First, Q approaches P along the curve. Second, the secant line PQ seems to approach a limiting position tangent to the curve at P. Third, the quotient $\Delta\mathbf{r}/\Delta t$ approaches the limit

$$\lim_{\Delta t \to 0} \frac{\Delta\mathbf{r}}{\Delta t} = \left[\lim_{\Delta t \to 0} \frac{f(t + \Delta t) - f(t)}{\Delta t}\right]\mathbf{i} + \left[\lim_{\Delta t \to 0} \frac{g(t + \Delta t) - g(t)}{\Delta t}\right]\mathbf{j}$$
$$= \left[\frac{df}{dt}\right]\mathbf{i} + \left[\frac{dg}{dt}\right]\mathbf{j}.$$

We are therefore led by past experience to the following definition.

Definition Derivative at a Point

The vector function $\mathbf{r}(t) = f(t)\mathbf{i} + g(t)\mathbf{j}$ has a **derivative (is differentiable) at t** if f and g have derivatives at t. The derivative is the vector

$$\frac{d\mathbf{r}}{dt} = \lim_{\Delta t \to 0} \frac{\mathbf{r}(t + \Delta t) - \mathbf{r}(t)}{\Delta t} = \frac{df}{dt}\mathbf{i} + \frac{dg}{dt}\mathbf{j}.$$

A vector function $\mathbf{r}$ is **differentiable** if it is differentiable at every point of its domain. The curve traced by $\mathbf{r}$ is **smooth** if $d\mathbf{r}/dt$ is continuous and never $\mathbf{0}$, that is, if f and g have continuous first derivatives that are not simultaneously 0. On a smooth curve there are no sharp corners or cusps.

The vector $d\mathbf{r}/dt$, when different from $\mathbf{0}$, is also a vector *tangent* to the curve. The **tangent line** to the curve at a point $P = (f(a), g(a))$ is defined to be the line through P parallel to $d\mathbf{r}/dt$ at $t = a$.

A curve that is made up of a finite number of smooth curves pieced together in a continuous fashion is **piecewise smooth** (Figure 10.22).

Look once again at Figure 10.21. We drew the figure for Δt positive, so $\Delta\mathbf{r}$ points forward, in the direction of the motion. The vector $\Delta\mathbf{r}/\Delta t$ (not shown), having the same direction as $\Delta\mathbf{r}$, points forward also. Had Δt been negative, $\Delta\mathbf{r}$ would have pointed backwards, against the direction of motion. The quotient $\Delta\mathbf{r}/\Delta t$, however, being a negative scalar multiple of $\Delta\mathbf{r}$, would have once again pointed forward. No matter how $\Delta\mathbf{r}$ points, $\Delta\mathbf{r}/\Delta t$ points forward and we expect the vector $d\mathbf{r}/dt = \lim_{\Delta t \to 0} \Delta\mathbf{r}/\Delta t$, when different from $\mathbf{0}$, to do the same. This means that the derivative $d\mathbf{r}/dt$ is just what we want for modeling a particle's velocity. It points in the direction of motion and gives the rate of change of position with respect to time. For a smooth curve, the velocity is never zero; the particle does not stop or reverse direction.

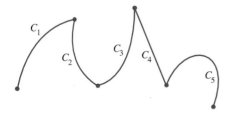

Figure 10.22 A piecewise smooth curve made by connecting five smooth curves end to end in continuous fashion.

> **Definitions** Velocity, Speed, Acceleration, Direction of Motion
>
> If **r** is the position vector of a particle moving along a smooth curve in the plane, then at any time t,
>
> **1.** $\mathbf{v}(t) = \dfrac{d\mathbf{r}}{dt}$ is the particle's **velocity vector** and is tangent to the curve.
>
> **2.** $|\mathbf{v}(t)|$, the magnitude of **v**, is the particle's **speed.**
>
> **3.** $\mathbf{a}(t) = \dfrac{d\mathbf{v}}{dt} = \dfrac{d^2\mathbf{r}}{dt^2}$, the derivative of velocity and the second derivative of position, is the particle's **acceleration vector.**
>
> **4.** $\dfrac{\mathbf{v}}{|\mathbf{v}|}$, a unit vector, is the **direction of motion.**

We can express the velocity of a moving particle as the product of its speed and direction.

$$\text{velocity} = |\mathbf{v}|\left(\frac{\mathbf{v}}{|\mathbf{v}|}\right) = (\text{speed})(\text{direction})$$

Example 5 STUDYING MOTION

The vector $\mathbf{r}(t) = (3\cos t)\mathbf{i} + (3\sin t)\mathbf{j}$ gives the position of a moving particle at time t. Find

(a) the velocity and acceleration vectors.

(b) the velocity, acceleration, speed, and direction of motion at $t = \pi/4$.

(c) $\mathbf{v} \cdot \mathbf{a}$. Interpret this result graphically.

Solution

(a) $\mathbf{v} = \dfrac{d\mathbf{r}}{dt} = (-3\sin t)\mathbf{i} + (3\cos t)\mathbf{j}$

$\mathbf{a} = \dfrac{d\mathbf{v}}{dt} = (-3\cos t)\mathbf{i} - (3\sin t)\mathbf{j}$

(b) At $t = \pi/4$, the particle's velocity and acceleration are

velocity: $\quad \mathbf{v}\left(\dfrac{\pi}{4}\right) = \left(-3\sin\dfrac{\pi}{4}\right)\mathbf{i} + \left(3\cos\dfrac{\pi}{4}\right)\mathbf{j} = -\dfrac{3}{\sqrt{2}}\mathbf{i} + \dfrac{3}{\sqrt{2}}\mathbf{j};$

acceleration: $\quad \mathbf{a}\left(\dfrac{\pi}{4}\right) = \left(-3\cos\dfrac{\pi}{4}\right)\mathbf{i} - \left(3\sin\dfrac{\pi}{4}\right)\mathbf{j} = -\dfrac{3}{\sqrt{2}}\mathbf{i} - \dfrac{3}{\sqrt{2}}\mathbf{j}.$

Its speed and direction are

speed: $\quad \left|\mathbf{v}\left(\dfrac{\pi}{4}\right)\right| = \sqrt{\left(\dfrac{-3}{\sqrt{2}}\right)^2 + \left(\dfrac{3}{\sqrt{2}}\right)^2} = 3;$

direction: $\quad \dfrac{\mathbf{v}(\pi/4)}{|\mathbf{v}(\pi/4)|} = \dfrac{-3/\sqrt{2}}{3}\mathbf{i} + \dfrac{3/\sqrt{2}}{3}\mathbf{j} = -\dfrac{1}{\sqrt{2}}\mathbf{i} + \dfrac{1}{\sqrt{2}}\mathbf{j}.$

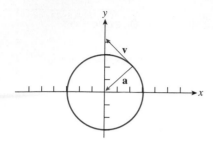

Figure 10.23 At $t = \pi/4$, the velocity vector $-(3/\sqrt{2})\mathbf{i} + (3/\sqrt{2})\mathbf{j}$ is tangent to the circle, and the acceleration vector $-(3/\sqrt{2})\mathbf{i} - (3/\sqrt{2})\mathbf{j}$ is perpendicular to the tangent, pointing toward the center of the circle. (Example 5)

(c) $\mathbf{v} \cdot \mathbf{a} = 9 \sin t \cos t - 9 \sin t \cos t = 0$

Thus, $\mathbf{v}$ and $\mathbf{a}$ are perpendicular for all values of t (Exercise 32, Section 10.2). Figure 10.23 shows the path, a circle with radius 3 centered at the origin, and the velocity and acceleration vectors at $t = \pi/4$.

Example 6 STUDYING MOTION

The vector $\mathbf{r}(t) = (2t^3 - 3t^2)\mathbf{i} + (t^3 - 12t)\mathbf{j}$ gives the position of a moving particle at time t.

(a) Write an equation for the line tangent to the path of the particle at the point where $t = -1$.

(b) Find the coordinates of each point on the path where the horizontal component of the velocity is 0.

Solution

(a) $\mathbf{v}(t) = \dfrac{d\mathbf{r}}{dt} = (6t^2 - 6t)\mathbf{i} + (3t^2 - 12)\mathbf{j}$

At $t = -1$, $\mathbf{r}(-1) = -5\mathbf{i} + 11\mathbf{j}$ and $\mathbf{v}(-1) = 12\mathbf{i} - 9\mathbf{j}$. Thus, we want the equation of the line through $(-5, 11)$ with slope $-9/12 = -3/4$.

$$y - 11 = -\frac{3}{4}(x + 5) \quad \text{or} \quad y = -\frac{3}{4}x + \frac{29}{4}$$

(b) The horizontal component of the velocity is $6t^2 - 6t$. It equals 0 when $t = 0$ and $t = 1$. The point corresponding to $t = 0$ is the origin $(0, 0)$; the point corresponding to $t = 1$ is $(-1, -11)$.

Differentiation Rules

Because the derivatives of vector functions are computed component by component, the rules for differentiating vector functions have the same form as the rules for differentiating scalar functions.

Differentiation Rules for Vector Functions

Let $\mathbf{u}$ and $\mathbf{v}$ be differentiable vector functions of t, and $\mathbf{C}$ a constant vector.

1. *Constant Function Rule:* $\dfrac{d}{dt}\mathbf{C} = \mathbf{0}$

2. *Scalar Multiple Rules:* $\dfrac{d}{dt}(c\mathbf{u}) = c\dfrac{d\mathbf{u}}{dt}$ c any scalar

$\dfrac{d}{dt}(f\mathbf{u}) = \dfrac{df}{dt}\mathbf{u} + f\dfrac{d\mathbf{u}}{dt}$ f any differentiable scalar function

3. *Sum Rule:* $\dfrac{d}{dt}(\mathbf{u} + \mathbf{v}) = \dfrac{d\mathbf{u}}{dt} + \dfrac{d\mathbf{v}}{dt}$

4. *Difference Rule:* $\dfrac{d}{dt}(\mathbf{u} - \mathbf{v}) = \dfrac{d\mathbf{u}}{dt} - \dfrac{d\mathbf{v}}{dt}$

5. *Dot Product Rule:* $\dfrac{d}{dt}(\mathbf{u} \cdot \mathbf{v}) = \dfrac{d\mathbf{u}}{dt} \cdot \mathbf{v} + \mathbf{u} \cdot \dfrac{d\mathbf{v}}{dt}$

6. *Chain Rule:* $\dfrac{d\mathbf{r}}{ds} = \dfrac{d\mathbf{r}}{dt}\dfrac{dt}{ds}$ $\mathbf{r}$ a differentiable function of t, t a differentiable function of s

We will prove the Dot Product Rule but leave the rest for exercises.

Proof of Rule 5 Suppose

$$\mathbf{u} = u_1(t)\mathbf{i} + u_2(t)\mathbf{j} \quad \text{and} \quad \mathbf{v} = v_1(t)\mathbf{i} + v_2(t)\mathbf{j}.$$

Then

$$\frac{d}{dt}(\mathbf{u} \cdot \mathbf{v}) = \frac{d}{dt}(u_1 v_1 + u_2 v_2)$$

$$= u_1' v_1 + u_2' v_2 + u_1 v_1' + u_2 v_2'$$

$$= \quad \mathbf{u}' \cdot \mathbf{v} \quad + \quad \mathbf{u} \cdot \mathbf{v}'. \qquad \blacksquare$$

Integrals

A differentiable vector function $\mathbf{R}(t)$ is an **antiderivative** of a vector function $\mathbf{r}(t)$ on an interval I if $d\mathbf{R}/dt = \mathbf{r}$ at each point t of I. If $\mathbf{R}$ is an antiderivative of $\mathbf{r}$ on I, it can be shown, working one component at a time, that every antiderivative of $\mathbf{r}$ on I has the form $\mathbf{R} + \mathbf{C}$ for some constant vector $\mathbf{C}$ (Exercise 37).

Definition Indefinite Integral

The **indefinite integral** of $\mathbf{r}$ with respect to t is the set of all antiderivatives of $\mathbf{r}$, denoted by $\int \mathbf{r}(t)\, dt$. If $\mathbf{R}$ is any antiderivative of $\mathbf{r}$, then

$$\int \mathbf{r}(t)dt = \mathbf{R}(t) + \mathbf{C}.$$

The usual arithmetic rules for indefinite integrals apply.

Example 7 FINDING ANTIDERIVATIVES

$$\int ((\cos t)\mathbf{i} - 2t\mathbf{j})\, dt = \left(\int \cos t\, dt\right)\mathbf{i} - \left(\int 2t\, dt\right)\mathbf{j} \qquad (3)$$

$$= (\sin t + C_1)\mathbf{i} - (t^2 + C_2)\mathbf{j} \qquad (4)$$

$$= (\sin t)\mathbf{i} - t^2\mathbf{j} + \mathbf{C} \qquad \mathbf{C} = C_1\mathbf{i} + C_2\mathbf{j}$$

As with integration of scalar functions, we recommend that you skip the steps in Equations 3 and 4 and go directly to the final form. Find an antiderivative for each component and add a constant vector at the end.

As with derivatives and indefinite integrals, definite integrals of vector functions are calculated component by component.

Definition Definite Integral

If the components of $\mathbf{r}(t) = f(t)\mathbf{i} + g(t)\mathbf{j}$ are integrable on $[a, b]$, then so is $\mathbf{r}$, and the **definite integral** of $\mathbf{r}$ from a to b is

$$\int_a^b \mathbf{r}(t)\, dt = \left(\int_a^b f(t)\, dt\right)\mathbf{i} + \left(\int_a^b g(t)\, dt\right)\mathbf{j}.$$

Example 8   EVALUATING DEFINITE INTEGRALS

$$\int_0^{\pi} ((\cos t)\mathbf{i} - 2t\mathbf{j})\, dt = \left(\int_0^{\pi} \cos t\, dt\right)\mathbf{i} - \left(\int_0^{\pi} 2t\, dt\right)\mathbf{j}$$

$$= \left(\sin t\Big]_0^{\pi}\right)\mathbf{i} - \left(t^2\Big]_0^{\pi}\right)\mathbf{j} = 0\mathbf{i} - \pi^2\mathbf{j} = -\pi^2\mathbf{j}$$

Example 9   FINDING A PATH

The velocity vector of a particle moving in the plane (scaled in meters) is

$$\frac{d\mathbf{r}}{dt} = \frac{1}{t+1}\mathbf{i} + 2t\mathbf{j}, \quad t \geq 0.$$

(a) Find the particle's position as a vector function of t if $\mathbf{r} = (\ln 2)\mathbf{i}$ when $t = 1$.

(b) Find the distance the particle travels from $t = 0$ to $t = 2$.

Solution

(a) $\mathbf{r} = \left(\int \dfrac{dt}{t+1}\right)\mathbf{i} + \left(\int 2t\, dt\right)\mathbf{j} = (\ln (t+1))\mathbf{i} + t^2\mathbf{j} + \mathbf{C}$

$\mathbf{r}(1) = (\ln 2)\mathbf{i} + \mathbf{j} + \mathbf{C} = (\ln 2)\mathbf{i}$

Thus, $\mathbf{C} = -\mathbf{j}$ and

$$\mathbf{r} = (\ln (t+1))\mathbf{i} + (t^2 - 1)\mathbf{j}.$$

(b) The parametrization

$$x = \ln (t+1), \quad y = t^2 - 1, \quad 0 \leq t \leq 2$$

is smooth, and because x and y are increasing functions of t the path is traversed exactly once as t increases from 0 to 2 (Figure 10.24). The length is

$$L = \int_0^2 \sqrt{\left(\frac{dx}{dt}\right)^2 + \left(\frac{dy}{dt}\right)^2}\, dt = \int_0^2 \sqrt{\left(\frac{1}{t+1}\right)^2 + (2t)^2}\, dt \approx 4.34 \text{ m}.$$

$x = \ln (t + 1),\ y = t^2 - 1$

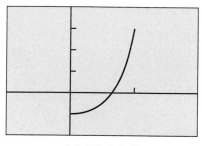

[−1, 2] by [−2, 4]

Figure 10.24. The path of the particle in Example 9 for $0 \leq t \leq 2$.

Quick Review 10.3

In Exercises 1 and 2, let $f(x) = \sqrt{4 - x^2}$.

1. Write an equation for the line tangent to f at $x = 1$.

2. Write an equation for the line normal to f at $x = 1$.

Exercises 3–6 refer to the curve defined parametrically by $x = 4 \cos t,\ y = 5 \sin t$.

3. Find the slope of the tangent to the curve at $t = \pi/2$.

4. Find the slope of the tangent to the curve at $t = \pi$.

5. Write an equation for the tangent to the curve at $t = \pi/6$.

6. Write an equation for the line normal to the curve at $t = \pi/6$.

7. Evaluate $\displaystyle\lim_{x \to 2} \frac{x - 2}{x^2 - 4}$.

8. Find the length of the curve $y = 3x - x^2$ from $x = 0$ to $x = 2$.

9. Find the length of the curve

$$x = t \sin t, \quad y = t \cos t, \quad 0 \leq t \leq 2.$$

10. Find y if $dy/dx = xe^x$ and $y(0) = 2$.

Section 10.3 Exercises

In Exercises 1–4, express each vector as a linear combination of **i** and **j**.

1. the vector $\overrightarrow{PQ}$, where $P = (-1, 4)$ and $Q = (5, 1)$

2. the vector from the point $P = (3, -4)$ to the origin

3. the vectors **(a)** $\overrightarrow{AB} + \overrightarrow{CD}$ and **(b)** $\overrightarrow{AB} - \overrightarrow{CD}$, where $A = (-3, 0)$, $B = (0, 2)$, $C = (4, 0)$, and $D = (0, -3)$

4. the vectors **(a)** $\mathbf{u} + \mathbf{v}$, **(b)** $\mathbf{u} - \mathbf{v}$, **(c)** $3\mathbf{u}$, **(d)** $2\mathbf{u} - 3\mathbf{v}$, where $\mathbf{u} = 5\mathbf{i} - 2\mathbf{j}$ and $\mathbf{v} = 3\mathbf{i} + 4\mathbf{j}$

In Exercises 5–8, $\mathbf{r}(t)$ is the position vector of a particle in the plane at time t.

    (a) Draw the graph of the path of the particle.

    (b) Find the velocity and acceleration vectors.

    (c) Find the particle's speed and direction of motion at the given value of t.

    (d) Write the particle's velocity at that time as the product of its speed and direction.

5. $\mathbf{r}(t) = (2 \cos t)\mathbf{i} + (3 \sin t)\mathbf{j}, \quad t = \pi/2$

6. $\mathbf{r}(t) = (\cos 2t)\mathbf{i} + (2 \sin t)\mathbf{j}, \quad t = 0$

7. $\mathbf{r}(t) = (\sec t)\mathbf{i} + (\tan t)\mathbf{j}, \quad t = \pi/6$

8. $\mathbf{r}(t) = (2 \ln (t + 1))\mathbf{i} + (t^2)\mathbf{j}, \quad t = 1$

In Exercises 9 and 10, find an equation for the line that is **(a)** tangent and **(b)** normal to the curve $\mathbf{r}(t)$ at the point determined by the given value of t.

9. $\mathbf{r}(t) = (\sin t)\mathbf{i} + (t^2 - \cos t)\mathbf{j}, \quad t = 0$

10. $\mathbf{r}(t) = (2 \cos t - 3)\mathbf{i} + (3 \sin t + 1)\mathbf{j}, \quad t = \pi/4$

In Exercises 11–14, evaluate the integral.

11. $\displaystyle\int_{1}^{2} [(6 - 6t)\mathbf{i} + 3\sqrt{t}\,\mathbf{j}]\, dt$

12. $\displaystyle\int_{-\pi/4}^{\pi/4} [(\sin t)\mathbf{i} + (1 + \cos t)\mathbf{j}]\, dt$

13. $\displaystyle\int [(\sec t \tan t)\mathbf{i} + (\tan t)\mathbf{j}]\, dt$

14. $\displaystyle\int \left[\frac{1}{t}\mathbf{i} + \frac{1}{5 - t}\mathbf{j}\right] dt$

In Exercises 15–18, solve the initial value problem for **r** as a vector function of t.

15. $\dfrac{d\mathbf{r}}{dt} = \dfrac{3}{2}(t + 1)^{1/2}\mathbf{i} + e^{-t}\mathbf{j}, \quad \mathbf{r}(0) = \mathbf{0}$

16. $\dfrac{d\mathbf{r}}{dt} = (t^3 + 4t)\mathbf{i} + t\mathbf{j}, \quad \mathbf{r}(0) = \mathbf{i} + \mathbf{j}$

17. $\dfrac{d^2\mathbf{r}}{dt^2} = -32\mathbf{j}, \quad \mathbf{r}(0) = 100\mathbf{i}, \quad \dfrac{d\mathbf{r}}{dt}\Big|_{t=0} = 8\mathbf{i} + 8\mathbf{j}$

18. $\dfrac{d^2\mathbf{r}}{dt^2} = -\mathbf{i} - \mathbf{j}, \quad \mathbf{r}(0) = 10\mathbf{i} + 10\mathbf{j}, \quad \dfrac{d\mathbf{r}}{dt}\Big|_{t=0} = \mathbf{0}$

In Exercises 19–22, $\mathbf{r}(t)$ is the position vector of a particle in the plane at time t. Find the time, or times, in the given time interval when the velocity and acceleration vectors are perpendicular.

19. $\mathbf{r}(t) = (t - \sin t)\mathbf{i} + (1 - \cos t)\mathbf{j}, \quad 0 \leq t \leq 2\pi$

20. $\mathbf{r}(t) = (\sin t)\mathbf{i} + t\mathbf{j}, \quad t \geq 0$

21. $\mathbf{r}(t) = (3 \cos t)\mathbf{i} + (4 \sin t)\mathbf{j}, \quad t \geq 0$

22. $\mathbf{r}(t) = (5 \cos t)\mathbf{i} + (5 \sin t)\mathbf{j}, \quad t \geq 0$

In Exercises 23 and 24, $\mathbf{r}(t)$ is the position vector of a particle in the plane at time t. Find the angle between the velocity and acceleration vectors at the given value of t.

23. $\mathbf{r}(t) = (2 \cos t)\mathbf{i} + (\sin t)\mathbf{j}, \quad t = \pi/4$

24. $\mathbf{r}(t) = (3t + 1)\mathbf{i} + (t^2)\mathbf{j}, \quad t = 0$

In Exercises 25 and 26, **(a)** evaluate the limit, and **(b)** find the values of t for which the vector function is continuous and **(c)** discontinuous.

25. $\displaystyle\lim_{t \to 3} \left[t\mathbf{i} + \frac{t^2 - 9}{t^2 + 3t}\mathbf{j}\right]$

26. $\displaystyle\lim_{t \to 0} \left[\frac{\sin 2t}{t}\mathbf{i} + (\ln (t + 1))\mathbf{j}\right]$

27. *Finding Distance Traveled* The position of a particle in the plane at time t is given by
$$\mathbf{r}(t) = (1 - \cos t)\mathbf{i} + (t - \sin t)\mathbf{j}.$$
Find the distance the particle travels along the path from $t = 0$ to $t = 2\pi/3$.

28. Let C be the path traced by
$$\mathbf{r}(t) = \left(\frac{1}{4}e^{4t} - t\right)\mathbf{i} + (e^{2t})\mathbf{j}, \quad 0 \leq t \leq 2.$$

    (a) Find the initial and terminal point of C.

    (b) Find the length of C.

    (c) Find the area of the surface generated by revolving C about the y-axis.

In Exercises 29–32, *work in groups of two or three.*

29. The position of a particle is given by
$$\mathbf{r}(t) = (\sin t)\mathbf{i} + (\cos 2t)\mathbf{j}.$$

    (a) Find the velocity vector for the particle.

    (b) For what values of t in the interval $0 \leq t \leq 2\pi$ is $d\mathbf{r}/dt$ equal to 0?

    (c) Find a Cartesian equation for a curve that contains the particle's path. What portion of the graph of the Cartesian equation is traced by the particle? Describe the motion as t increases from 0 to 2π.

30. *Revisiting Example 6* The position of a particle is given by $\mathbf{r}(t) = (2t^3 - 3t^2)\mathbf{i} + (t^3 - 12t)\mathbf{j}$.

(a) Find dy/dx in terms of t.

(b) **Writing to Learn** Find the x- and y-coordinates for each critical point of the path (point where dy/dx is zero or does not exist). Does the path have a vertical or horizontal tangent at the critical point? Explain.

31. *Finding a Position Vector* At time $t = 0$, a particle is located at the point $(1, 2)$. It travels in a straight line to the point $(4, 1)$, has speed 2 at $(1, 2)$, and constant acceleration $3\mathbf{i} - \mathbf{j}$. Find an equation for the position vector $\mathbf{r}(t)$ of the particle at time t.

32. The path of a particle for $t > 0$ is given by

$$\mathbf{r}(t) = \left(t + \frac{2}{t}\right)\mathbf{i} + (3t^2)\mathbf{j}.$$

(a) Find the coordinates of each point on the path where the horizontal component of the velocity of the particle is zero.

(b) Find dy/dx when $t = 1$.

(c) Find d^2y/dx^2 when $y = 12$.

Explorations

33. *Flying a Kite* The position of a kite is given by

$$\mathbf{r}(t) = \frac{t}{8}\mathbf{i} - \frac{3}{64}t(t - 160)\mathbf{j},$$

where $t \geq 0$ is measured in seconds and distance is measured in meters.

(a) How long is the kite above ground?

(b) How high is the kite at $t = 40$ sec?

(c) At what rate is the kite's altitude increasing at $t = 40$ sec?

(d) At what time does the kite start to lose altitude?

34. *Colliding Particles* The paths of two particles for $t \geq 0$ are given by

$$\mathbf{r}_1(t) = (t - 3)\mathbf{i} + (t - 3)^2\mathbf{j},$$

$$\mathbf{r}_2(t) = \left(\frac{3t}{2} - 4\right)\mathbf{i} + \left(\frac{3t}{2} - 2\right)\mathbf{j}.$$

(a) Determine the exact time(s) at which the particles collide.

(b) Find the direction of motion of each particle at the time(s) of collision.

35. *A Satellite in Circular Orbit* A satellite of mass m is moving at a constant speed v around a planet of mass M in a circular orbit of radius r_0, as measured from the planet's center of mass. Determine the satellite's orbital period T (the time to complete one full orbit), as follows:

(a) Coordinatize the orbital plane by placing the origin at the planet's center of mass, with the satellite on the x-axis at $t = 0$ and moving counterclockwise, as in the accompanying figure.

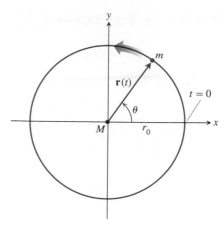

Let $\mathbf{r}(t)$ be the satellite's position vector at time t. Show that $\theta = vt/r_0$ and hence that

$$\mathbf{r}(t) = \left(r_0 \cos \frac{vt}{r_0}\right)\mathbf{i} + \left(r_0 \sin \frac{vt}{r_0}\right)\mathbf{j}.$$

(b) Find the acceleration of the satellite.

(c) According to Newton's law of gravitation, the gravitational force exerted on the satellite by the planet is directed toward the origin and is given by

$$\mathbf{F} = \left(-\frac{GmM}{r_0^2}\right)\frac{\mathbf{r}}{r_0},$$

where G is the universal constant of gravitation. Using Newton's second law, $\mathbf{F} = m\mathbf{a}$, show that $v^2 = GM/r_0$.

(d) Show that the orbital period T satisfies $vT = 2\pi r_0$.

(e) From parts (c) and (d), deduce that

$$T^2 = \frac{4\pi^2}{GM}r_0^3;$$

that is, the square of the period of a satellite in circular orbit is proportional to the cube of the radius from the orbital center.   ∎

36. *Revisiting Example 9* Find a Cartesian equation for a curve that contains the parametrized curve

$$x = \ln(t + 1), \quad y = t^2 - 1, \quad 0 \leq t < \infty.$$

What portion of the graph of the Cartesian equation is traced by the parametrized curve?

37. *Antiderivatives of Vector Functions*

(a) Use Corollary 3 in Section 4.2 (a consequence of the Mean Value Theorem for scalar functions) to show that two vector functions $\mathbf{R}_1(t)$ and $\mathbf{R}_2(t)$ that have identical derivatives on an interval I differ by a constant vector value throughout I.

(b) Use the result in (a) to show that if $\mathbf{R}(t)$ is any antiderivative of $\mathbf{r}(t)$ on I, then every other antiderivative of $\mathbf{r}(t)$ on I equals $\mathbf{R}(t) + \mathbf{C}$ for some constant vector $\mathbf{C}$.

38. *Constant Length* Let **v** be a differentiable vector function of t. Show that if $\mathbf{v} \cdot (d\mathbf{v}/dt) = 0$ for all t, then $|\mathbf{v}|$ is constant.

Extending the Ideas

39. *Constant Function Rule* Prove that if **u** is the vector function with the constant value **C**, then $d\mathbf{u}/dt = \mathbf{0}$.

40. *Scalar Multiple Rules*

(a) Prove that if **u** is a differentiable function of t and c is any real number, then

$$\frac{d(c\mathbf{u})}{dt} = c\frac{d\mathbf{u}}{dt}.$$

(b) Prove that if **u** is a differentiable function of t and f is a differentiable scalar function of t, then

$$\frac{d(f\mathbf{u})}{dt} = \frac{df}{dt}\mathbf{u} + f\frac{d\mathbf{u}}{dt}.$$

41. *Sum and Difference Rules* Prove that if **u** and **v** are differentiable functions of t, then

(a) $\dfrac{d}{dt}(\mathbf{u} + \mathbf{v}) = \dfrac{d\mathbf{u}}{dt} + \dfrac{d\mathbf{v}}{dt}.$

(b) $\dfrac{d}{dt}(\mathbf{u} - \mathbf{v}) = \dfrac{d\mathbf{u}}{dt} - \dfrac{d\mathbf{v}}{dt}.$

42. *Chain Rule* Prove that if $\mathbf{r}(t) = f(t)\mathbf{i} + g(t)\mathbf{j}$ is a differentiable vector function of t and t is a differentiable scalar function of s, then

$$\frac{d\mathbf{r}}{ds} = \frac{d\mathbf{r}}{dt} \cdot \frac{dt}{ds}.$$

43. *Differentiable Vector Functions Are Continuous* Show that if $\mathbf{r}(t) = f(t)\mathbf{i} + g(t)\mathbf{j}$ is differentiable at $t = c$, then **r** is continuous at c as well.

44. *Integration Properties* Establish the following properties of integrable vector functions.

(a) *Constant Scalar Multiple Rule:*

$$\int_a^b k\mathbf{r}(t)\, dt = k\int_a^b \mathbf{r}(t)\, dt$$

for any scalar constant k.

(b) *Sum and Difference Rules:*

$$\int_a^b (\mathbf{r}_1(t) \pm \mathbf{r}_2(t))\, dt = \int_a^b \mathbf{r}_1(t)\, dt \pm \int_a^b \mathbf{r}_2(t)\, dt$$

(c) *Constant Vector Multiple Rule:*

$$\int_a^b \mathbf{C} \cdot \mathbf{r}(t)\, dt = \mathbf{C} \cdot \int_a^b \mathbf{r}(t)\, dt$$

for any constant vector **C**.

45. *Fundamental Theorem of Calculus* The Fundamental Theorem of Calculus for scalar functions of a real variable holds for vector functions of a real variable as well.

(a) Prove this by using the theorem for scalar functions to show that if a vector function $\mathbf{r}(t)$ is continuous for $a \le t \le b$, then

$$\frac{d}{dt}\int_a^t \mathbf{r}(q)\, dq = \mathbf{r}(t)$$

at every point t of $[a, b]$.

(b) Use the conclusion in part (b) of Exercise 37 to show that if **R** is any antiderivative of **r** on $[a, b]$ then

$$\int_a^b \mathbf{r}(t)\, dt = \mathbf{R}(b) - \mathbf{R}(a).$$

10.4 Modeling Projectile Motion

Ideal Projectile Motion • Height, Flight Time, and Range • Ideal Trajectories Are Parabolic • Firing from (x_0, y_0) • Projectile Motion with Wind Gusts • Projectile Motion with Air Resistance

Ideal Projectile Motion

To derive equations for *ideal* projectile motion, we assume that the projectile behaves like a particle moving in a vertical coordinate plane and that the only force acting on the projectile during its flight is the constant force of gravity, which always points straight down. In practice, none of these assumptions really holds. The ground moves beneath the projectile as the earth turns, the air creates a frictional force that varies with the projectile's speed and altitude, and the force (vector) of gravity changes magnitude and direction as the projectile moves along. All this must be taken into account by applying corrections to the predictions of the ideal equations we are about to derive.

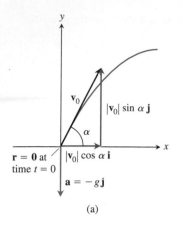

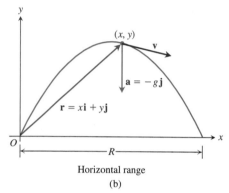

Figure 10.25 (a) Position, velocity, acceleration due to gravity, and launch angle at $t = 0$. (b) Position, velocity, and acceleration at a later time t.

We assume the projectile is launched from the origin at time $t = 0$ into the first quadrant with an initial velocity $\mathbf{v}_0$ (Figure 10.25). If $\mathbf{v}_0$ makes an angle α with the horizontal, then

$$\mathbf{v}_0 = (|\mathbf{v}_0| \cos \alpha)\mathbf{i} + (|\mathbf{v}_0| \sin \alpha)\mathbf{j}.$$

If we use the simpler notation v_0 for the initial speed $|\mathbf{v}_0|$, then

$$\mathbf{v}_0 = (v_0 \cos \alpha)\mathbf{i} + (v_0 \sin \alpha)\mathbf{j}. \tag{1}$$

The projectile's initial position is

$$\mathbf{r}_0 = 0\mathbf{i} + 0\mathbf{j} = \mathbf{0}. \tag{2}$$

Newton's second law of motion says that the force acting on the projectile is equal to the projectile's mass m times its acceleration, or $m(d^2\mathbf{r}/dt^2)$ if $\mathbf{r}$ is the projectile's position vector and t is time. If the force is solely the gravitational force $-mg\mathbf{j}$, then

$$m\frac{d^2\mathbf{r}}{dt^2} = -mg\mathbf{j} \quad \text{and} \quad \frac{d^2\mathbf{r}}{dt^2} = -g\mathbf{j}.$$

We find $\mathbf{r}$ as a function of t by solving the following initial value problem.

Differential equation: $\quad \dfrac{d^2\mathbf{r}}{dt^2} = -g\mathbf{j}$

Initial conditions: $\quad \mathbf{r} = \mathbf{r}_0 \quad$ and $\quad \dfrac{d\mathbf{r}}{dt} = \mathbf{v}_0 \quad$ when $\; t = 0$

The first integration gives

$$\frac{d\mathbf{r}}{dt} = -(gt)\mathbf{j} + \mathbf{v}_0.$$

A second integration gives

$$\mathbf{r} = -\frac{1}{2}gt^2\mathbf{j} + \mathbf{v}_0 t + \mathbf{r}_0. \tag{3}$$

Substituting the values of $\mathbf{v}_0$ and $\mathbf{r}_0$ from Equations 1 and 2 gives

$$\mathbf{r} = -\frac{1}{2}gt^2\mathbf{j} + \underbrace{(v_0 \cos \alpha)t\mathbf{i} + (v_0 \sin \alpha)t\mathbf{j}}_{\mathbf{v}_0 t} + \mathbf{0}$$

or

$$\mathbf{r} = (v_0 \cos \alpha)t\mathbf{i} + \left((v_0 \sin \alpha)t - \frac{1}{2}gt^2\right)\mathbf{j}. \tag{4}$$

Equation 4 is the **vector equation** for ideal projectile motion. The angle α is the projectile's **launch angle (firing angle, angle of elevation)**, and v_0, as we said before, is the projectile's **initial speed.**

Equation 4 gives us a pair of scalar equations,

$$x = (v_0 \cos \alpha)t, \qquad y = (v_0 \sin \alpha)t - \frac{1}{2}gt^2, \tag{5}$$

the **parametric equations** for ideal projectile motion. If time is measured in seconds and distance in meters, g is 9.8 m/sec^2 and Equations 5 give x and y in meters. With feet in place of meters, g is 32 ft/sec^2 and Equations 5 give x and y in feet.

Example 1 FIRING AN IDEAL PROJECTILE

A projectile is fired from the origin over horizontal ground at an initial speed of 500 m/sec and a launch angle of 60°. Where will the projectile be 10 sec later?

Solution We use Equations 5 with $v_0 = 500$, $\alpha = 60°$, $g = 9.8$, and $t = 10$ to find the projectile's coordinates 10 sec after firing.

$$x = (v_0 \cos \alpha)t = (500)\left(\frac{1}{2}\right)10 = 2500$$

$$y = (v_0 \sin \alpha)t - \frac{1}{2}gt^2$$

$$= (500)\left(\frac{\sqrt{3}}{2}\right)10 - \left(\frac{1}{2}\right)(9.8)10^2 \approx 3840.13$$

Interpret
Ten seconds after firing, the projectile is about 3840 m in the air and 2500 m downrange.

Height, Flight Time, and Range

Equations 5 enable us to answer most questions about the ideal motion for a projectile fired from the origin.

The projectile reaches its highest point when its vertical velocity component is zero, that is, when

$$\frac{dy}{dt} = v_0 \sin \alpha - gt = 0, \quad \text{or} \quad t = \frac{v_0 \sin \alpha}{g}.$$

For this value of t, the value of y is

$$y_{\text{max}} = (v_0 \sin \alpha)\left(\frac{v_0 \sin \alpha}{g}\right) - \frac{1}{2}g\left(\frac{v_0 \sin \alpha}{g}\right)^2 = \frac{(v_0 \sin \alpha)^2}{2g}.$$

To find when the projectile lands when fired over horizontal ground, we set y equal to zero in Equations 5 and solve for t

$$(v_0 \sin \alpha)t - \frac{1}{2}gt^2 = 0$$

$$t\left(v_0 \sin \alpha - \frac{1}{2}gt\right) = 0$$

$$t = 0, \quad t = \frac{2v_0 \sin \alpha}{g}$$

Since 0 is the time the projectile is fired, $(2v_0 \sin \alpha)/g$ must be the time when the projectile strikes the ground.

To find the projectile's **range** R, the distance from the origin to the point of impact on horizontal ground, we find the value of x when $t = (2v_0 \sin \alpha)/g$.

$$x = (v_0 \cos \alpha)t$$

$$R = (v_0 \cos \alpha)\left(\frac{2v_0 \sin \alpha}{g}\right) = \frac{v_0^2}{g}(2 \sin \alpha \cos \alpha) = \frac{v_0^2}{g}\sin 2\alpha$$

The range is largest when $\sin 2\alpha = 1$ or $\alpha = 45°$.

Height, Flight Time, and Range for Ideal Projectile Motion

For ideal projectile motion when an object is launched from the origin over a horizontal surface with initial speed v_0 and launch angle α:

$$\text{Maximum height:} \quad y_{\text{max}} = \frac{(v_0 \sin \alpha)^2}{2g} \tag{6}$$

$$\text{Flight time:} \quad t = \frac{2v_0 \sin \alpha}{g} \tag{7}$$

$$\text{Range:} \quad R = \frac{v_0^2}{g} \sin 2\alpha \tag{8}$$

Example 2   INVESTIGATING IDEAL PROJECTILE MOTION

Find the maximum height, flight time, and range of a projectile fired from the origin over horizontal ground at an initial speed of 500 m/sec and a launch angle of 60° (same projectile as Example 1). Then graph the path of the projectile.

Solution

Maximum height (Equation 6): $y_{\text{max}} = \dfrac{(v_0 \sin \alpha)^2}{2g}$

$$= \frac{(500 \sin 60°)^2}{2(9.8)} \approx 9566.33 \text{ m}$$

Flight time (Equation 7): $t = \dfrac{2v_0 \sin \alpha}{g}$

$$= \frac{2(500) \sin 60°}{9.8} \approx 88.37 \text{ sec}$$

Range (Equation 8): $R = \dfrac{v_0^2}{g} \sin 2\alpha$

$$= \frac{(500)^2 \sin 120°}{9.8} \approx 22{,}092.48 \text{ m}$$

Equations 5 give parametric equations for the path of the projectile:

$$x = (v_0 \cos \alpha)t = (500 \cos 60°)t = 250t,$$

$$y = (v_0 \sin \alpha)t - \frac{1}{2}gt^2$$

$$= (500 \sin 60°)t - \frac{1}{2}(9.8)t^2 = (250\sqrt{3})t - 4.9t^2.$$

A graph of these equations (the path of the projectile) is shown in Figure 10.26.

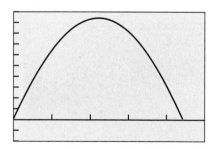

[0, 25000] by [−2000, 10000]

Figure 10.26 The graph of the parametric equations $x = 250t$, $y = (250\sqrt{3})t - 4.9t^2$ for $0 \le t \le 88.4$. (Example 2)

Ideal Trajectories Are Parabolic

It is often claimed that water from a hose traces a parabola in the air, but anyone who looks closely enough will see this is not so. The air slows the water down, and its forward progress is too slow at the end to keep pace with the rate at which it falls.

What is really being claimed is that ideal projectiles move along parabolas, and this we can see from Equations 5. If we substitute $t = x/(v_0 \cos \alpha)$ from the first equation into the second, we obtain the Cartesian-coordinate equation

$$y = -\left(\frac{g}{2v_0^2 \cos^2 \alpha} \right)x^2 + (\tan \alpha)x.$$

This equation has the form $y = ax^2 + bx$, so its graph is a parabola.

Firing from (x_0, y_0)

If we fire our ideal projectile from the point (x_0, y_0) instead of the origin (Figure 10.27), the equations that replace Equations 5 are

$$x = x_0 + (v_0 \cos \alpha)t, \qquad y = y_0 + (v_0 \sin \alpha)t - \frac{1}{2}gt^2, \tag{9}$$

as you will be asked to show in Exercise 19.

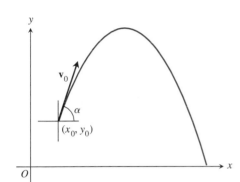

Figure 10.27 The path of a projectile fired from (x_0, y_0) with an initial velocity $\mathbf{v}_0$ at an angle of α degrees with the horizontal.

Example 3   FIRING A FLAMING ARROW

To open the 1992 Summer Olympics in Barcelona, bronze medalist archer Antonio Rebollo lit the Olympic torch with a flaming arrow (Figure 10.28). Suppose that Rebollo shot the arrow at a height of 6 ft above ground level 30 yd from the 70-ft-high cauldron, and he wanted the arrow to reach maximum height exactly 4 ft above the center of the cauldron (Figure 10.29).

Figure 10.28 Spanish archer Antonio Rebollo lights the Olympic torch in Barcelona with a flaming arrow.

(a) Express $y_{\max}$ in terms of the initial speed v_0 and firing angle α.

(b) Use $y_{\max} = 74$ ft (Figure 10.29 on the next page) and the result from part (a) to find the value of $v_0 \sin \alpha$.

(c) Find the value of $v_0 \cos \alpha$.

(d) Find the initial firing angle of the arrow.

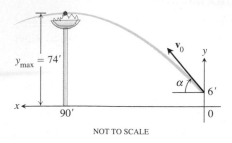

Figure 10.29 Ideal path of the arrow that lit the Olympic torch. (Example 3)

NOT TO SCALE

$y_{max} = 74'$

$90'$

Solution

(a) We use a coordinate system in which the positive x-axis lies along the ground toward the left (to match the second photograph in Figure 10.28) and the coordinates of the flaming arrow at $t = 0$ are $x_0 = 0$ and $y_0 = 6$ (Figure 10.29). We have

$$y = y_0 + (v_0 \sin \alpha)t - \frac{1}{2}gt^2 \quad \text{Eqs. 9}$$

$$= 6 + (v_0 \sin \alpha)t - \frac{1}{2}gt^2. \quad y_0 = 6$$

We find the time when the arrow reaches its highest point by setting $dy/dt = 0$ and solving for t, obtaining

$$t = \frac{v_0 \sin \alpha}{g}.$$

For this value of t, the value of y is

$$y_{max} = 6 + (v_0 \sin \alpha)\left(\frac{v_0 \sin \alpha}{g}\right) - \frac{1}{2}g\left(\frac{v_0 \sin \alpha}{g}\right)^2$$

$$= 6 + \frac{(v_0 \sin \alpha)^2}{2g}.$$

(b) Using $y_{max} = 74$ and $g = 32$, we see from the preceding equation in part (a) that

$$74 = 6 + \frac{(v_0 \sin \alpha)^2}{2(32)}$$

or

$$v_0 \sin \alpha = \sqrt{(68)(64)}.$$

(c) When the arrow reaches y_{max}, the horizontal distance traveled to the center of the cauldron is $x = 90$ ft. We substitute the time to reach y_{max} from part (a) and the horizontal distance $x = 90$ ft into Equations 9 to obtain

$$x = x_0 + (v_0 \cos \alpha)t \qquad \text{Eqs. 9}$$

$$90 = 0 + (v_0 \cos \alpha)t \qquad x = 90, \ x_0 = 0$$

$$= (v_0 \cos \alpha)\left(\frac{v_0 \sin \alpha}{g}\right). \qquad t = (v_0 \sin \alpha)/g$$

Solving this equation for $v_0 \cos \alpha$ and using $g = 32$ and the result from part (b), we have

$$v_0 \cos \alpha = \frac{90g}{v_0 \sin \alpha} = \frac{(90)(32)}{\sqrt{(68)(64)}}.$$

(d) Parts (b) and (c) together tell us that

$$\tan \alpha = \frac{v_0 \sin \alpha}{v_0 \cos \alpha} = \frac{(\sqrt{(68)(64)})^2}{(90)(32)} = \frac{68}{45}$$

or

$$\alpha = \tan^{-1}\left(\frac{68}{45}\right) \approx 56.5°.$$

This is Rebollo's firing angle.

Projectile Motion with Wind Gusts

We begin with an example that shows how to account for a second force acting on a projectile. We also assume that the path of the baseball in Example 4 lies in a vertical plane.

Example 4   HITTING A BASEBALL

A baseball is hit when it is 3 ft above the ground. It leaves the bat with initial speed of 152 ft/sec, making an angle of 20° with the horizontal. At the instant the ball is hit, an instantaneous gust of wind blows in the horizontal direction directly opposite the direction the ball is taking toward the outfield, adding a component of $-8.8\mathbf{i}$ (ft/sec) to the ball's initial velocity (8.8 ft/sec = 6 mph).

(a) Find the vector equation (position vector) and the parametric equations for the path of the baseball.

(b) How high does the baseball go, and when does it reach maximum height?

(c) Assuming the ball is not caught, find its range and flight time.

Solution

Solve Analytically

(a) Using Equation 1 and accounting for the gust of wind, the initial velocity of the baseball is

$$\mathbf{v}_0 = (v_0 \cos \alpha)\mathbf{i} + (v_0 \sin \alpha)\mathbf{j} - 8.8\mathbf{i}$$
$$= (152 \cos 20°)\mathbf{i} + (152 \sin 20°)\mathbf{j} - (8.8)\mathbf{i}$$
$$= (152 \cos 20° - 8.8)\mathbf{i} + (152 \sin 20°)\mathbf{j}.$$

The initial position is $\mathbf{r}_0 = 0\mathbf{i} + 3\mathbf{j}$.

Substituting the values of $\mathbf{v}_0$ and $\mathbf{r}_0$ into Equation 3 gives the position vector of the baseball.

$$\mathbf{r} = -\frac{1}{2}gt^2\mathbf{j} + \mathbf{v}_0 t + \mathbf{r}_0$$
$$= -16t^2\mathbf{j} + (152 \cos 20° - 8.8)t\mathbf{i} + (152 \sin 20°)t\mathbf{j} + 3\mathbf{j}$$
$$= (152 \cos 20° - 8.8)t\mathbf{i} + (3 + (152 \sin 20°)t - 16t^2)\mathbf{j}$$

The corresponding parametric equations are

$$x = (152 \cos 20° - 8.8)t, \quad y = 3 + (152 \sin 20°)t - 16t^2.$$

Solve Graphically

(b) Figure 10.30a shows the path of the baseball as the graph of the parametric equations in part (a). Tracing along the path, we find that the maximum height is about 45.2 ft, and it is reached in about 1.6 sec.

(c) The x-axis represents the ground, so the baseball hits the ground at the x-intercept (Figure 10.30b). Continuing to trace, we see that the horizontal range is about 442.3 ft, and the flight time about 3.3 sec.

$x = (152 \cos 20 - 8.8)t$
$y = 3 + (152 \sin 20)t - 16t^2$

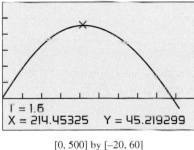

$$T = 1.6$$
$$X = 214.45325 \quad Y = 45.219299$$

[0, 500] by [−20, 60]

(a)

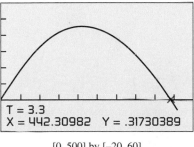

$$T = 3.3$$
$$X = 442.30982 \quad Y = .31730389$$

[0, 500] by [−20, 60]

(b)

Figure 10.30 The (a) maximum height and (b) horizontal range of the baseball of Example 4.

Exploration 1 **Hitting a Home Run**

Suppose the baseball in Example 4 is hit toward the right-centerfield fence, 400 ft from the batter.

1. If the fence is 15 ft high, is the hit a home run (does it clear the fence in flight)? Simulate the fence and the flight of the baseball.

2. Find the range and the flight time if the angle at which the ball comes off the bat is **(a)** 25°, **(b)** 30°, **(c)** 45°.

3. Suppose there is no gust of wind. Simulate this flight and the fence. Is this hit a home run?

4. Repeat part 2 for the flight in part 3.

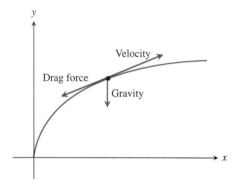

Figure 10.31 Projectile motion with air resistance.

Projectile Motion with Air Resistance

The main force affecting the motion of a projectile, other than gravity, is air resistance. This slowing down force is **drag force** and it acts in a direction *opposite* to the velocity of the projectile (Figure 10.31).

For some high speed projectiles, the drag force is proportional to different powers of the speed over different velocity ranges. For projectiles moving through the air at relatively low speeds, however, the drag force is (very nearly) proportional to the speed (to the first power) and so is called **linear.** Here is what happens for linear drag force.

Projectile Motion with Linear Drag

Equations for the motion of a projectile with linear drag force launched from the origin over a horizontal surface at $t = 0$:

Vector form:

$$\mathbf{r} = \frac{v_0}{k}(1 - e^{-kt})(\cos \alpha)\mathbf{i} +$$
$$\left[\frac{v_0}{k}(1 - e^{-kt})(\sin \alpha) + \frac{g}{k^2}(1 - kt - e^{-kt})\right]\mathbf{j} \qquad (10)$$

Parametric form:

$$x = \frac{v_0}{k}(1 - e^{-kt}) \cos \alpha,$$

$$y = \frac{v_0}{k}(1 - e^{-kt})(\sin \alpha) + \frac{g}{k^2}(1 - kt - e^{-kt}) \qquad (11)$$

where the **drag coefficient** k is a positive constant representing resistance due to air density, v_0 and α are the projectile's initial speed and launch angle, and g is the acceleration of gravity.

Proof Because drag is linear, directly proportional to the particle's speed, and acts in the direction opposite to the velocity vector, we have the following initial value problem to solve for $\mathbf{r} = x\mathbf{i} + y\mathbf{j}$.

Differential equation: $\dfrac{d^2\mathbf{r}}{dt^2} = -g\mathbf{j} - k\mathbf{v} = -g\mathbf{j} - k\dfrac{d\mathbf{r}}{dt}$ (12)

Initial conditions: $\mathbf{r}(0) = \mathbf{0}$

$$\left.\dfrac{d\mathbf{r}}{dt}\right|_{t=0} = \mathbf{v}_0 = (v_0 \cos \alpha)\mathbf{i} + (v_0 \sin \alpha)\mathbf{j}$$

Substituting $\mathbf{r} = x\mathbf{i} + y\mathbf{j}$ into Equation 12 gives

$$\dfrac{d^2\mathbf{r}}{dt^2} = -g\mathbf{j} - k\left(\dfrac{dx}{dt}\mathbf{i} + \dfrac{dy}{dt}\mathbf{j}\right)$$

$$\dfrac{d^2x}{dt^2}\mathbf{i} + \dfrac{d^2y}{dt^2}\mathbf{j} = -k\dfrac{dx}{dt}\mathbf{i} - \left(g + k\dfrac{dy}{dt}\right)\mathbf{j}. \qquad \dfrac{d^2\mathbf{r}}{dt^2} = \dfrac{d^2x}{dt^2}\mathbf{i} + \dfrac{d^2y}{dt^2}\mathbf{j}$$

Comparing components of the two sides, we have

$$\dfrac{d^2x}{dt^2} = -k\dfrac{dx}{dt} \qquad (13)$$

and

$$\dfrac{d^2y}{dt^2} = -g - k\dfrac{dy}{dt}. \qquad (14)$$

We integrate both sides of Equation 13 with respect to t and use $x = 0$ and $dx/dt = v_0 \cos \alpha$ when $t = 0$ to obtain

$$\dfrac{dx}{dt} = -kx + v_0 \cos \alpha. \qquad (15)$$

To solve Equation 15 for x, we first separate the variables (Section 6.2).

$$\dfrac{dx}{v_0 \cos \alpha - kx} = dt$$

Integrating both sides, we obtain

$$-\dfrac{1}{k} \ln |v_0 \cos \alpha - kx| = t + C$$

$$\ln |v_0 \cos \alpha - kx| = -kt + C$$

$$|v_0 \cos \alpha - kx| = Ce^{-kt}$$

$$v_0 \cos \alpha - kx = Ce^{-kt}$$

$$v_0 \cos \alpha - kx = (v_0 \cos \alpha)e^{-kt} \qquad x(0) = 0 \Rightarrow C = v_0 \cos \alpha$$

$$x = \dfrac{v_0}{k}(1 - e^{-kt})(\cos \alpha),$$

the x-component of the parametric form.

To solve Equation 14, we set $w = dy/dt$ and separate the variables.

$$\dfrac{dw}{dt} = -g - kw$$

$$\dfrac{dw}{-g - kw} = dt$$

$$-\dfrac{1}{k} \ln |-g - kw| = t + C$$

$$\ln |-g - kw| = -kt + C$$

$$-g - kw = Ce^{-kt} \qquad (16)$$

Now when $t = 0$,

$$w = \frac{dy}{dt} = v_0 \sin \alpha,$$

so

$$C = -g - kv_0 \sin \alpha.$$

Equation 16 becomes

$$-g - kw = (-g - kv_0 \sin \alpha)e^{-kt}$$

$$-g - k\frac{dy}{dt} = (-g - kv_0 \sin \alpha)e^{-kt}. \quad w = dy/dt \qquad (17)$$

Integrating both sides of Equation 17 with respect to t, solving for the constant of integration, and then solving for y, we obtain

$$y = \frac{v_0}{k}(1 - e^{-kt})(\sin \alpha) + \frac{g}{k^2}(1 - kt - e^{-kt}),$$

the y-component of the parametric form. ∎

Exploration 2 **Hitting a Baseball—Linear Drag**

A baseball is hit as described in Example 4, but this time with linear drag to slow it down instead of wind (a very calm day). For the temperature and humidity on game day, it has been established that the drag coefficient k is 0.05.

1. Find and graph the parametric equations for the path of the baseball.

2. How high does the baseball go, and when does it reach maximum height?

3. Assuming that the ball is not caught, find the range and flight time of the baseball.

Quick Review 10.4

In Exercises 1 and 2, find the component form of the vector **v**.

1.

2.

In Exercises 3–6, let $f(x) = 2x^2 + 11x - 40$ and $g(x) = 20x - x^2$.

3. Find the x- and y-intercepts of the graph of f.

4. Find the coordinates of the vertex of the graph of f.

5. Find the x- and y-intercepts of the graph of g.

6. Find the coordinates of the vertex of the graph of g.

In Exercises 7 and 8, solve the initial value problem.

7. $\dfrac{dy}{dx} = \sin x, \quad y(\pi/2) = 2$

8. $\dfrac{d^2y}{dt^2} = 2t, \quad y(-1) = 5, \quad y'(-1) = 4$

In Exercises 9 and 10, use separation of variables to solve the initial value problem.

9. $\dfrac{dy}{dt} = 16 - y, \quad y(0) = 20$

10. $\dfrac{dy}{dx} = (4 - 2y)x, \quad y(0) = 1$

Section 10.4 Exercises

Projectile flights in the following exercises are to be treated as ideal unless stated otherwise. All launch angles are assumed to be measured from the horizontal. All projectiles are assumed to be launched from the origin over a horizontal surface unless stated otherwise.

1. *Travel Time* A projectile is fired at a speed of 840 m/sec at an angle of 60°. How long will it take to get 21 km downrange?

2. *Finding Muzzle Speed* Find the muzzle speed of a gun whose maximum range is 24.5 km.

3. *Flight Time and Height* A projectile is fired with an initial speed of 500 m/sec at an angle of elevation of 45°.

   (a) When and how far away will the projectile strike?

   (b) How high overhead will the projectile be when it is 5 km downrange?

   (c) What is the greatest height reached by the projectile?

4. *Throwing a Baseball* A baseball is thrown from the stands 32 ft above the field at an angle of 30° up from the horizontal. When and how far away will the ball strike the ground if its initial speed is 32 ft/sec?

5. *Shot Put* An athlete puts a 16-lb shot at an angle of 45° to the horizontal from 6.5 ft above the ground at an initial speed of 44 ft/sec as suggested in the figure. How long after launch and how far from the inner edge of the stopboard does the shot land?

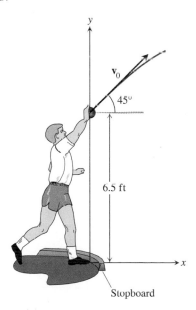

Stopboard

6. *(Continuation of Exercise 5)* Because of its initial elevation, the shot in Exercise 5 would have gone slightly farther if it had been launched at a 40° angle. How much farther? Answer in inches.

7. *Firing Golf Balls* A spring gun at ground level fires a golf ball at an angle of 45°. The ball lands 10 m away.

   (a) What was the ball's initial speed?

   (b) For the same initial speed, find the two firing angles that make the range 6 m.

8. *Beaming Electrons* An electron in a TV tube is beamed horizontally at a speed of 5×10^6 m/sec toward the face of the tube 40 cm away. About how far will the electron drop before it hits?

9. *Finding Golf Ball Speed* Laboratory tests designed to find how far golf balls of different hardness go when hit with a driver showed that a 100-compression ball hit with a club-head speed of 100 mph at a launch angle of 9° carried 248.8 yd. What was the launch speed of the ball? (It was more than 100 mph. At the same time the club head was moving forward, the compressed ball was kicking away from the club face, adding to the ball's forward speed.)

10. **Writing to Learn** A *human cannonball* is to be fired with an initial speed of $v_0 = 80\sqrt{10}/3$ ft/sec. The circus performer (of the right caliber, naturally) hopes to land on a special cushion located 200 ft downrange at the same height as the muzzle of the cannon. The circus is being held in a large room with a flat ceiling 75 ft higher than the muzzle. Can the performer be fired to the cushion without striking the ceiling? If so, what should the cannon's angle of elevation be?

11. **Writing to Learn** A golf ball leaves the ground at a 30° angle at a speed of 90 ft/sec. Will it clear the top of a 30-ft tree that is in the way, 135 ft down the fairway? Explain.

12. *Elevated Green* A golf ball is hit with an initial speed of 116 ft/sec at an angle of elevation of 45° from the tee to a green that is elevated 45 ft above the tee as shown in the diagram. Assuming that the pin, 369 ft downrange, does not get in the way, where will the ball land in relation to the pin?

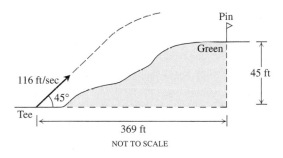

NOT TO SCALE

13. *The Green Monster* A baseball hit by a Boston Red Sox player at a 20° angle from 3 ft above the ground just cleared the left end of the "Green Monster," the left-field wall in Fenway Park (see figure). This wall is 37 ft high and 315 ft from home plate.

    (a) What was the initial speed of the ball?

    (b) How long did it take the ball to reach the wall?

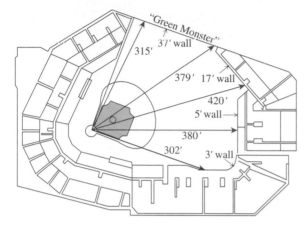

14. *Equal Range Firing Angles* Show that a projectile fired at an angle of α degrees, $0 < \alpha < 90$, has the same range as a projectile fired at the same speed at an angle of $(90 - \alpha)$ degrees. (In models that take air resistance into account, this symmetry is lost.)

15. *Equal Range Firing Angles* What two angles of elevation will enable a projectile to reach a target 16 km downrange on the same level as the gun if the projectile's initial speed is 400 m/sec?

16. *Range and Height vs Speed*

    (a) Show that doubling a projectile's initial speed at a given launch angle multiplies its range by 4.

    (b) By about what percentage should you increase the initial speed to double the height and range?

17. *Shot Put* In Moscow in 1987, Natalya Lisouskaya set a women's world record by putting an 8-lb 13-oz shot 73 ft 10 in. Assuming that she launched the shot at a 40° angle to the horizontal from 6.5 ft above the ground, what was the shot's initial speed?

18. *Height vs Time* Show that a projectile attains three-quarters of its maximum height in half the time it takes to reach the maximum height.

19. *Firing from* (x_0, y_0) Derive the equations

$$x = x_0 + (v_0 \cos \alpha)t,$$

$$y = y_0 + (v_0 \sin \alpha)t - \frac{1}{2}gt^2,$$

(Equations 9 in the text) by solving the following initial value problem for a vector **r** in the plane.

Differential equation: $\dfrac{d^2\mathbf{r}}{dt^2} = -g\mathbf{j}$

Initial conditions: $\mathbf{r}(0) = x_0\mathbf{i} + y_0\mathbf{j}$

$\dfrac{d\mathbf{r}}{dt}(0) = (v_0 \cos \alpha)\mathbf{i} + (v_0 \sin \alpha)\mathbf{j}$

20. *Flaming Arrow* Using the firing angle found in Example 3, find the speed at which the flaming arrow left Rebollo's bow. See Figure 10.29.

21. *Flaming Arrow* The cauldron in Example 3 is 12 ft in diameter. Using Equations 9 and Example 3c, find how long it takes the flaming arrow to cover the horizontal distance to the rim. How high is the arrow at this time?

22. **Writing to Learn** Describe the path of a projectile given by Equations 5 when $\alpha = 90°$.

In Exercises 23–25, *work in groups of two or three.*

23. *Model Train* The multiflash photograph below shows a model train engine moving at a constant speed on a straight horizontal track. As the engine moved along, a marble was fired into the air by a spring in the engine's smokestack. The marble, which continued to move with the same forward speed as the engine, rejoined the engine 1 sec after it was fired. Measure the angle the marble's path made with the horizontal and use the information to find how high the marble went and how fast the engine was moving.

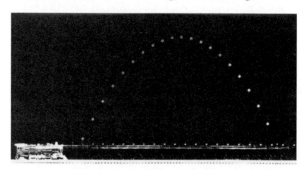

24. **Writing to Learn** The figure shows an experiment with two marbles. Marble A was launched toward marble B with launch angle α and initial speed v_0. At the same instant, marble B was released to fall from rest at $R \tan \alpha$ units directly above a spot R units downrange from A. The marbles were found to collide regardless of the value of v_0. Was this mere coincidence, or must this happen? Give reasons for your answer.

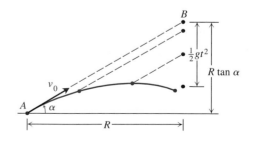

25. *Launching Downhill* An ideal projectile is launched straight down an inclined plane as shown in the figure.

(a) Show that the greatest downhill range is achieved when the initial velocity vector bisects angle *AOR*.

(b) If the projectile were fired uphill instead of down, what launch angle would maximize its range? Give reasons for your answer.

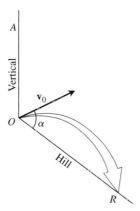

26. *Hitting a Baseball* A baseball is hit when it is 2.5 ft above the ground. It leaves the bat with an initial velocity of 145 ft/sec at a launch angle of 23°. At the instant the ball is hit, an instantaneous gust of wind blows against the ball, adding a component of $-14\mathbf{i}$ (ft/sec) to the ball's initial velocity. A 15-ft-high fence is 300 ft from home plate in the direction of the flight.

(a) Find vector and parametric forms for the path of the baseball.

(b) How high does the baseball go, and when does it reach maximum height?

(c) Find the range and flight time of the baseball, assuming the ball is not caught.

(d) When is the baseball 20 ft high? How far (ground distance) is the baseball from home plate at that height?

(e) Writing to Learn Has the batter hit a home run? Explain.

Explorations

27. *Volleyball* A volleyball is hit when it is 4 ft above the ground and 12 ft from a 6-ft-high net. It leaves the point of impact with an initial velocity of 35 ft/sec at an angle of 27° and slips by the opposing team untouched.

(a) Find vector and parametric forms for the path of the volleyball.

(b) How high does the volleyball go, and when does it reach maximum height?

(c) Find its range and flight time.

(d) When is the volleyball 7 ft above the ground? How far (ground distance) is the volleyball from where it will land?

(e) Writing to Learn Suppose the net is raised to 8 ft. Does this change things? Explain.

28. *Hitting a Baseball* Consider the baseball problem of Exploration 2 again. Assume a drag coefficient of 0.12.

(a) Find vector and parametric forms for the path of the baseball.

(b) How high does the baseball go, and when does it reach maximum height?

(c) Find the range and flight time of the baseball.

(d) When is the baseball 30 ft high? How far (ground distance) is the baseball from home plate at that height?

(e) Writing to Learn A 10-ft-high outfield fence is 340 ft from home plate in the direction of the flight of the baseball. The outfielder can jump and catch any ball up to 11 ft off the ground to stop it from going over the fence. Has the batter hit a home run?

29. *Hitting a Baseball* Consider the baseball problem of Exploration 2 again. This time assume a drag coefficient of 0.08 *and* an instantaneous gust of wind that adds a component of $-17.6\mathbf{i}$ (ft/sec) to the initial velocity at the instant the baseball is hit.

(a) Find vector and parametric forms for the path of the baseball.

(b) How high does the baseball go, and when does it reach maximum height?

(c) Find the range and flight time of the baseball.

(d) When is the baseball 35 ft high? How far (ground distance) is the baseball from home plate at that height?

(e) Writing to Learn A 20-ft-high outfield fence is 380 ft from home plate in the direction of the flight of the baseball. Has the batter hit a home run? If "yes," what change in the horizontal component of the ball's initial velocity would have kept the ball in the park? If "no," what change would have allowed it to be a home run?

30. *Hitting a Baseball* Consider the baseball problem of Exploration 2 again. Assume drag coefficients of 0.01, 0.02, 0.10, 0.15, 0.20, and 0.25.

(a) Graph the paths of the baseballs in the same viewing window.

(b) In each case, how high does the baseball go, and when does it reach maximum height?

(c) In each case, find the range and flight time of the baseball.

(d) Writing to Learn As $k \to 0$, show that the right-hand members of Equations 11 approach the right-hand members of Equations 5. Explain why this makes sense. ∎

Extending the Ideas

31. *Where Trajectories Crest* For a projectile fired from the ground at launch angle α with initial speed v_0, consider α as a variable and v_0 as a fixed constant. For each α, $0 < \alpha < \pi/2$, we obtain a parabolic trajectory as shown in the figure. Show that the points in the plane that give the maximum heights of these parabolic trajectories all lie on the ellipse

$$x^2 + 4\left(y - \frac{v_0{}^2}{4g}\right)^2 = \frac{v_0{}^4}{4g^2},$$

where $x \geq 0$.

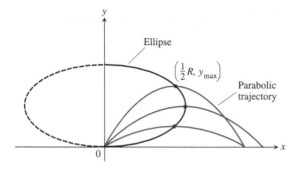

32. *The Linear Drag Equation* Show directly that Equation 10 is the solution to this initial value problem.

Differential equation: $\dfrac{d^2\mathbf{r}}{dt^2} = -g\mathbf{j} - k\mathbf{v} = -g\mathbf{j} - k\dfrac{d\mathbf{r}}{dt}$

Initial conditions: $\mathbf{r}(0) = \mathbf{0}$

$$\left.\frac{d\mathbf{r}}{dt}\right|_{t=0} = \mathbf{v}_0 = (v_0\cos\alpha)\mathbf{i} + (v_0\sin\alpha)\mathbf{j}$$

10.5 Polar Coordinates and Polar Graphs

Polar Coordinates • Polar Graphing • Relating Polar and Cartesian Coordinates

Polar Coordinates

To define polar coordinates, we first fix an **origin** O (called the **pole**) and an **initial ray** from O (Figure 10.32). Then each point P can be located by assigning to it a **polar coordinate pair** (r, θ) in which r gives the directed distance from O to P and θ gives the directed angle from the initial ray to ray OP.

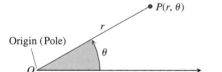

Figure 10.32 To define polar coordinates for the plane, we start with an origin, called the pole, and an initial ray.

Polar Coordinates

$$P(r, \theta)$$

Directed distance from O to P Directed angle from initial ray to ray OP

As in trigonometry, θ is positive when measured counterclockwise and negative when measured clockwise. The angle associated with a given point is not unique. For instance, the point 2 units from the origin along the ray $\theta = \pi/6$ has polar coordinates $r = 2$, $\theta = \pi/6$. It also has coordinates $r = 2$, $\theta = -11\pi/6$ (Figure 10.33).

There are also occasions when we wish to allow r to be negative. This is why we use *directed* distance in defining $P(r, \theta)$. The point $P(2, 7\pi/6)$ can be reached by turning $7\pi/6$ radians counterclockwise from the initial ray and going forward 2 units (Figure 10.35 on the next page). It can also be reached by turning $\pi/6$ radians counterclockwise from the initial ray and going *backwards* 2 units. So the point also has polar coordinates $r = -2$, $\theta = \pi/6$.

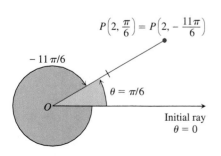

Figure 10.33 Polar coordinates are not unique.

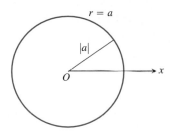

Figure 10.34 The polar equation for this circle is $r = a$.

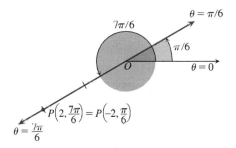

Figure 10.35 Polar coordinates can have negative r values.

Polar Graphing

If we hold r fixed at a constant value $a \neq 0$, the point $P(r, \theta)$ will lie $|a|$ units from the origin O. As θ varies over any interval of length 2π, P traces a circle of radius $|a|$ centered at O (Figure 10.34).

If we hold θ fixed at a constant value $\theta = \alpha$ and let r vary between $-\infty$ and ∞, the point $P(r, \theta)$ traces the line through O that makes an angle of measure α with the initial ray.

Equation	Polar Graph
$r = a$	Circle of radius $|a|$ centered at O
$\theta = \alpha$	Line through O making an angle α with the initial ray

Example 1 FINDING POLAR EQUATIONS FOR GRAPHS

(a) $r = 1$ and $r = -1$ are equations for the circle of radius 1 centered at O.

(b) $\theta = \pi/6$, $\theta = 7\pi/6$, and $\theta = -5\pi/6$ are equations for the line in Figure 10.35.

Equations of the form $r = a$ and $\theta = \alpha$ can be combined to define regions, segments, and rays.

Example 2 IDENTIFYING GRAPHS

Graph the set of points whose polar coordinates satisfy the given conditions.

(a) $1 \leq r \leq 2$ and $0 \leq \theta \leq \dfrac{\pi}{2}$

(b) $-3 \leq r \leq 2$ and $\theta = \dfrac{\pi}{4}$

(c) $r \leq 0$ and $\theta = \dfrac{\pi}{4}$

(d) $\dfrac{2\pi}{3} \leq \theta \leq \dfrac{5\pi}{6}$ (no restriction on r)

Solution The graphs are shown in Figure 10.36.

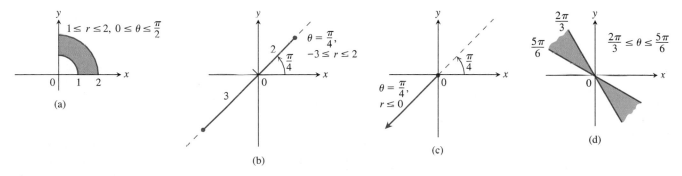

Figure 10.36 The graphs of typical inequalities in r and θ. (Example 2)

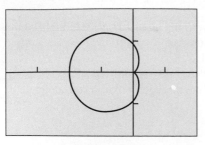

[−4, 2] by [−2, 2]

Figure 10.37 The graph of the cardioid $r = 1 - \cos \theta$ (calculator in polar mode). (Example 3)

The curve in Figure 10.37 is called a **cardioid** because of its heart-like shape.

Example 3 USING A POLAR GRAPHER

Draw the graph of $r = 1 - \cos \theta$. What is the shortest length a θ-interval can have and still produce the graph?

Solution Figure 10.37 shows the graph of

$$r = 1 - \cos \theta, \quad 0 \le \theta \le 2\pi.$$

Any interval of length 2π will produce the graph. Any interval shorter than 2π will leave part of the cardioid untraced.

The graph in Figure 10.37 appears to be symmetric about the x-axis. Figure 10.38 illustrates the standard polar coordinate tests for symmetry.

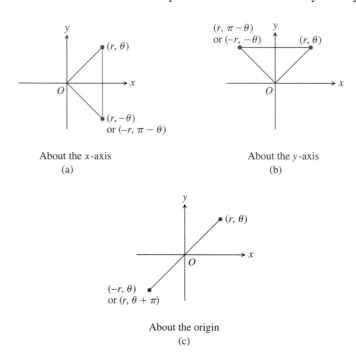

Figure 10.38 Three tests for symmetry.

Symmetry Tests for Polar Graphs

1. **Symmetry about the x-axis:** If the point (r, θ) lies on the graph, the point $(r, -\theta)$ or $(-r, \pi - \theta)$ lies on the graph (Figure 10.38a).

2. **Symmetry about the y-axis:** If the point (r, θ) lies on the graph, the point $(r, \pi - \theta)$ or $(-r, -\theta)$ lies on the graph (Figure 10.38b).

3. **Symmetry about the origin:** If the point (r, θ) lies on the graph, the point $(-r, \theta)$ or $(r, \theta + \pi)$ lies on the graph (Figure 10.38c).

If we replace θ by $-\theta$ in the equation $r = 1 - \cos \theta$ of the cardioid of Figure 10.37 we obtain the same equation

$$r = 1 - \cos (-\theta) = 1 - \cos \theta.$$

Thus, if point (r, θ) lies on the graph of $r = 1 - \cos \theta$, so does the point $(r, -\theta)$. This confirms our conjecture that the graph in Figure 10.37 is symmetric about the x-axis.

Exploration 1 **Investigating Polar Graphs**

Let $r^2 = 4 \cos \theta$.

1. Graph $r^2 = 4 \cos \theta$ in a square viewing window using $r_1 = 2\sqrt{\cos \theta}$ and $r_2 = -2\sqrt{\cos \theta}$.
2. Explain why the origin is part of the graph.
3. What is the shortest length a θ-interval can have and still produce the graph?
4. Show algebraically that the graph is symmetric about the x-axis, the y-axis, and the origin.

Relating Polar and Cartesian Coordinates

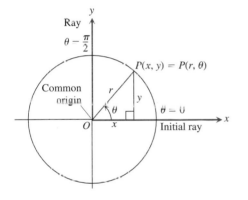

Figure 10.39 The usual way to relate polar and Cartesian coordinates.

When we use both polar and Cartesian coordinates in a plane, we place the two origins together and take the polar initial ray as the positive x-axis. The ray $\theta = \pi/2$, $r > 0$, becomes the positive y-axis (Figure 10.39). The two coordinate systems are then related by the following equations.

Equations Relating Polar and Cartesian Coordinates

$$x = r \cos \theta, \quad y = r \sin \theta, \quad x^2 + y^2 = r^2, \quad \frac{y}{x} = \tan \theta$$

We use these equations and algebra (sometimes a lot of it!) to rewrite polar equations in Cartesian form and vice versa.

Example 4 EQUIVALENT EQUATIONS

Polar equation	Cartesian equivalent
$r \cos \theta = 2$	$x = 2$
$r^2 \cos \theta \sin \theta = 4$	$xy = 4$
$r^2 \cos^2 \theta - r^2 \sin^2 \theta = 1$	$x^2 - y^2 = 1$
$r = 1 + 2r \cos \theta$	$y^2 - 3x^2 - 4x - 1 = 0$
$r = 1 - \cos \theta$	$x^4 + y^4 + 2x^2y^2 + 2x^3 + 2xy^2 - y^2 = 0$

With some curves, we are better off with polar coordinates; with others, we aren't.

Example 5 CONVERTING CARTESIAN TO POLAR

Find a polar equation for the circle $x^2 + (y - 3)^2 = 9$ (Figure 10.40a). Support graphically.

Solution

$$x^2 + y^2 - 6y + 9 = 9 \quad \text{Expand } (y - 3)^2.$$

$$x^2 + y^2 - 6y = 0$$

$$r^2 - 6r \sin \theta = 0 \quad x^2 + y^2 = r^2, \quad y = r \sin \theta$$

$$r = 0 \quad \text{or} \quad r - 6 \sin \theta = 0$$

The equation $r = 6 \sin \theta$ includes the possibility that $r = 0$, as supported by its graph in Figure 10.40b.

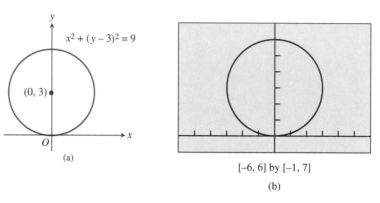

(a)

$[-6, 6]$ by $[-1, 7]$

(b)

Figure 10.40 The graphs of (a) the circle $x^2 + (y - 3)^2 = 9$, and (b) the polar equation $r = 6 \sin \theta$. (Example 5)

Example 6 CONVERTING POLAR TO CARTESIAN

Find a Cartesian equivalent for the polar equation. Identify the graph.

(a) $r^2 = 4r \cos \theta$ **(b)** $r = \dfrac{4}{2 \cos \theta - \sin \theta}$

Solution

(a)

$$r^2 = 4r \cos \theta$$

$$x^2 + y^2 = 4x \qquad r^2 = x^2 + y^2, \quad r \cos \theta = x$$

$$x^2 - 4x + y^2 = 0$$

$$x^2 - 4x + 4 + y^2 = 4 \qquad \text{Completing the square}$$

$$(x - 2)^2 + y^2 = 4$$

The graph of the equivalent Cartesian equation $(x - 2)^2 + y^2 = 4$ is a circle with radius 2 and center $(2, 0)$.

(b)

$$r = \frac{4}{2\cos\theta - \sin\theta}$$

$$r(2\cos\theta - \sin\theta) = 4$$

$$2r\cos\theta - r\sin\theta = 4$$

$$2x - y = 4 \qquad r\cos\theta = x, \quad r\sin\theta = y$$

$$y = 2x - 4$$

The graph of the equivalent Cartesian equation $y = 2x - 4$ is a line with slope 2 and y-intercept -4.

Exploration 2 **Graphing Rose Curves**

The polar curves $r = a\cos n\theta$ and $r = a\sin n\theta,$ where n is an integer and $|n| > 1,$ are *rose curves*.

1. Graph $r = 2\cos n\theta$ for $n = \pm2, \pm4, \pm6.$ Describe the curves.

2. What is the shortest length a θ-interval can have and still produce the graphs in (1)?

3. Based on your observations in (1), describe the graph of $r = 2\cos n\theta$ when n is a nonzero even integer.

4. Graph $r = 2\cos n\theta$ for $n = \pm3, \pm5, \pm7.$ Describe the curves.

5. What is the shortest length a θ-interval can have and still produce the graphs in (4)?

6. Based on your observations in (4), describe the graph of $r = 2\cos n\theta$ when n is an odd integer different from $\pm1.$

Quick Review 10.5

In Exercises 1–3, write a Cartesian equation for the curve.

1. the line through the points $(-2, 4)$ and $(3, -1)$

2. the circle with center $(0, 0)$ and radius 3

3. the circle with center $(-2, 4)$ and radius 2

In Exercises 4–7, determine whether the curve is symmetric about **(a)** the x-axis, **(b)** the y-axis, or **(c)** the origin.

4. $y = x^3 - x$

5. $y = x^2 - x$

6. $y = \cos x$

7. $4x^2 + 9y^2 = 36$

In Exercises 8 and 9, explain how to use a function grapher to graph the curve.

8. $y^2 = x - 2$

9. $x^2 + 3y^2 = 4$

10. Use completing the square to find the center and radius of the circle $x^2 + y^2 - 4x + 6y + 9 = 0.$

Section 10.5 Exercises

In Exercises 1 and 2, determine which polar coordinate pairs name the same point.

1. (a) $(3, 0)$ **(b)** $(-3, 0)$ **(c)** $(2, 2\pi/3)$

    (d) $(2, 7\pi/3)$ **(e)** $(-3, \pi)$ **(f)** $(2, \pi/3)$

    (g) $(-3, 2\pi)$ **(h)** $(-2, -\pi/3)$

2. (a) $(-2, \pi/3)$ **(b)** $(2, -\pi/3)$ **(c)** (r, θ)

    (d) $(r, \theta + \pi)$ **(e)** $(-r, \theta)$ **(f)** $(2, -2\pi/3)$

    (g) $(-r, \theta + \pi)$ **(h)** $(-2, 2\pi/3)$

In Exercises 3 and 4, plot the points with the given polar coordinates and find their Cartesian coordinates.

3. (a) $(\sqrt{2}, \pi/4)$ **(b)** $(1, 0)$

    (c) $(0, \pi/2)$ **(d)** $(-\sqrt{2}, \pi/4)$

4. (a) $(-3, 5\pi/6)$ **(b)** $(5, \tan^{-1}(4/3))$

    (c) $(-1, 7\pi)$. **(d)** $(2\sqrt{3}, 2\pi/3)$

In Exercises 5 and 6, plot the points with the given Cartesian coordinates and find two sets of polar coordinates for each.

5. (a) $(-1, 1)$ **(b)** $(1, -\sqrt{3})$

    (c) $(0, 3)$ **(d)** $(-1, 0)$

6. (a) $(-\sqrt{3}, -1)$ **(b)** $(3, 4)$

    (c) $(0, -2)$ **(d)** $(2, 0)$

In Exercises 7–18, graph the set of points whose polar coordinates satisfy the given equations and inequalities.

7. $r = 2$ **8.** $0 \leq r \leq 2$

9. $r \geq 1$ **10.** $0 \leq \theta \leq \pi/6, \quad r \geq 0$

11. $\theta = 2\pi/3, \quad r \leq -2$ **12.** $\theta = \pi/3, \quad -1 \leq r \leq 3$

13. $0 \leq \theta \leq \pi, \quad r = 1$ **14.** $0 \leq \theta \leq \pi, \quad r = -1$

15. $\theta = \pi/2, \quad r \leq 0$

16. $\pi/4 \leq \theta \leq 3\pi/4, \quad 0 \leq r \leq 1$

17. $-\pi/4 \leq \theta \leq \pi/4, \quad -1 \leq r \leq 1$

18. $0 \leq \theta \leq \pi/2, \quad 1 \leq |r| \leq 2$

In Exercises 19–36, replace the polar equation by an equivalent Cartesian equation. Then identify or describe the graph.

19. $r \sin \theta = 0$ **20.** $r \cos \theta = 0$

21. $r = 4 \csc \theta$ **22.** $r = -3 \sec \theta$

23. $r \cos \theta + r \sin \theta = 1$ **24.** $r^2 = 1$

25. $r^2 = 4r \sin \theta$ **26.** $r = \dfrac{5}{\sin \theta - 2 \cos \theta}$

27. $r^2 \sin 2\theta = 2$ **28.** $r = \cot \theta \csc \theta$

29. $r = \csc \theta \, e^{r \cos \theta}$ **30.** $\cos^2 \theta = \sin^2 \theta$

31. $r \sin \theta = \ln r + \ln \cos \theta$ **32.** $r^2 + 2r^2 \cos \theta \sin \theta = 1$

33. $r^2 = -4r \cos \theta$ **34.** $r = 8 \sin \theta$

35. $r = 2 \cos \theta + 2 \sin \theta$ **36.** $r \sin\left(\theta + \dfrac{\pi}{6}\right) = 2$

In Exercises 37–48, replace the Cartesian equation by an equivalent polar equation. Support graphically.

37. $x = 7$ **38.** $y = 1$

39. $x = y$ **40.** $x - y = 3$

41. $x^2 + y^2 = 4$ **42.** $x^2 - y^2 = 1$

43. $\dfrac{x^2}{9} + \dfrac{y^2}{4} = 1$ **44.** $xy = 2$

45. $y^2 = 4x$ **46.** $x^2 + xy + y^2 = 1$

47. $x^2 + (y - 2)^2 = 4$

48. $(x - 3)^2 + (y + 1)^2 = 4$

In Exercises 49–58, **(a)** graph the polar curve. **(b)** What is the shortest length a θ-interval can have and still produce the graph?

49. $r = 1 + \cos \theta$ **50.** $r = 2 - 2 \cos \theta$

51. $r^2 = -\sin 2\theta$ **52.** $r = 1 - \sin \theta$

53. $r = 1 - 2 \sin 3\theta$ **54.** $r = \sin (\theta/2)$

55. $r = \theta$ **56.** $r = 1 + \sin \theta$

57. $r = 2 \cos 3\theta$ **58.** $r = 1 + 2 \sin \theta$

In Exercises 59–62, *work in groups of two or three.* Determine the symmetries of the curve.

59. $r^2 = 4 \cos 2\theta$ **60.** $r^2 = 4 \sin 2\theta$

61. $r = 2 + \sin \theta$ **62.** $r^2 = -\cos 2\theta$

63. Writing to Learn

    (a) Explain why every vertical line in the plane has a polar equation of the form $r = a \sec \theta$.

    (b) Find an analogous polar equation for horizontal lines. Give reasons for your answer.

Explorations

64. *Rose Curves* Let $r = 2 \sin n\theta$.

    (a) Graph $r = 2 \sin n\theta$ for $n = \pm 2, \pm 4, \pm 6$. Describe the curves.

    (b) What is the smallest length a θ-interval can have and still produce the graphs in (a)?

    (c) Based on your observations in (a), describe the graph of $r = 2 \sin n\theta$ when n is a nonzero even integer.

    (d) Graph $r = 2 \sin n\theta$ for $n = \pm 3, \pm 5, \pm 7$. Describe the curves.

(e) What is the smallest length a θ-interval can have and still produce the graphs in (d)?

(f) Based on your observations in (d), describe the graph of $r = 2 \sin n\theta$ when n is an odd integer different from ± 1.

65. *Relating Polar Equations to Parametric Equations*
Let $r = f(\theta)$ be a polar curve.

(a) Writing to Learn Explain why

$$x = f(t) \cos t, \quad y = f(t) \sin t$$

are parametric equations for the curve.

(b) Use (a) to write parametric equations for the circle $r = 3$. Support your answer by graphing the parametric equations.

(c) Repeat (b) with $r = 1 - \cos \theta$.

(d) Repeat (b) with $r = 3 \sin 2\theta$.

66. *Rotating Curves* Let $r_1(\theta) = 3(1 - \cos \theta)$ and $r_2(\theta) = r_1(\theta - \alpha)$.

(a) Graph r_2 for $\alpha = \pi/6, \pi/4, \pi/3$, and $\pi/2$ and compare with the graph of r_1.

(b) Graph r_2 for $\alpha = -\pi/6, -\pi/4, -\pi/3$, and $-\pi/2$ and compare with the graph of r_1.

(c) Based on your observations in (a) and (b), describe the relationship between the graphs of $r_1 - f(\theta)$ and $r_2 = f(\theta - \alpha)$.

Extending the Ideas

67. *Distance Formula* Show that the distance between two points (r_1, θ_1) and (r_2, θ_2) in polar coordinates is

$$d = \sqrt{r_1^2 + r_2^2 - 2r_1 r_2 \cos(\theta_1 - \theta_2)}.$$

68. Let $r = \dfrac{2}{1 + k \cos \theta}$.

(a) Graph r in a square viewing window for $k = 0.1, 0.3, 0.5, 0.7$, and 0.9. Describe the graphs.

(b) Based on your observations in (a), conjecture what happens to the graphs for $0 < k < 1$ and $k \rightarrow 0^+$.

69. Let $r = \dfrac{2}{1 + k \cos \theta}$.

(a) Graph r in a square viewing window for $k = 1.1, 1.3, 1.5, 1.7$, and 1.9. Describe the graphs.

(b) Based on your observations in (a), conjecture what happens to the graphs for $k > 1$ and $k \rightarrow 1^+$.

70. Let $r = \dfrac{k}{1 + \cos \theta}$.

(a) Graph r in a square viewing window for $k = 1, 3, 5, 7$, and 9. Describe the graphs.

(b) Based on your observations in (a), conjecture what happens to the graphs for $k > 0$ and $k \rightarrow 0^+$.

10.6 Calculus of Polar Curves

Slope • Area in the Plane • Length of a Curve • Area of a Surface of Revolution

Slope

The slope of the tangent line to a polar curve $r = f(\theta)$ is given by dy/dx, not by $r' = df/d\theta$. To see why, think of the graph of f as the graph of the parametric equations (see Section 10.5, Exercise 65)

$$x = r \cos \theta = f(\theta) \cos \theta, \quad y = r \sin \theta = f(\theta) \sin \theta.$$

If f is a differentiable function of θ, then so are x and y and, when $dx/d\theta \neq 0$, we can calculate dy/dx from the parametric formula

$$\frac{dy}{dx} = \frac{dy/d\theta}{dx/d\theta} \qquad \text{Section 10.1, Eq. 1 with } t = \theta$$

$$= \frac{\dfrac{d}{d\theta}(f(\theta) \sin \theta)}{\dfrac{d}{d\theta}(f(\theta) \cos \theta)}$$

$$= \frac{\dfrac{df}{d\theta} \sin \theta + f(\theta) \cos \theta}{\dfrac{df}{d\theta} \cos \theta - f(\theta) \sin \theta}. \qquad \text{Product Rule for derivatives}$$

Slope of the Curve $r = f(\theta)$

$$\frac{dy}{dx}\bigg|_{(r,\,\theta)} = \frac{dy/d\theta}{dx/d\theta} = \frac{f'(\theta)\sin\theta + f(\theta)\cos\theta}{f'(\theta)\cos\theta - f(\theta)\sin\theta}, \tag{1}$$

provided $dx/d\theta \neq 0$ at (r, θ).

We can see from Equation 1 and its derivation that the curve $r = f(\theta)$ has a

1. horizontal tangent at a point where $dy/d\theta = 0$ and $dx/d\theta \neq 0$,

2. vertical tangent at a point where $dx/d\theta = 0$ and $dy/d\theta \neq 0$.

If both derivatives are zero, no conclusion can be drawn without further investigation, as illustrated in Example 1.

Example 1 FINDING HORIZONTAL AND VERTICAL TANGENTS

Find the horizontal and vertical tangents to the cardioid $r = 1 - \cos\theta$, $0 \leq \theta \leq 2\pi$.

Solution The graph in Figure 10.41 suggests that there are at least two horizontal and three vertical tangents.

The parametric form of the equation is

$$x = r\cos\theta = (1 - \cos\theta)\cos\theta = \cos\theta - \cos^2\theta,$$

$$y = r\sin\theta = (1 - \cos\theta)\sin\theta = \sin\theta - \cos\theta\sin\theta.$$

We need to find the zeros of $dy/d\theta$ and $dx/d\theta$.

(a) Zeros of $dy/d\theta$ in $[0, 2\pi]$:

$$\frac{dy}{d\theta} = \cos\theta + \sin^2\theta - \cos^2\theta = \cos\theta + (1 - \cos^2\theta) - \cos^2\theta$$

$$= 1 + \cos\theta - 2\cos^2\theta = (1 + 2\cos\theta)(1 - \cos\theta)$$

Now,

$$1 - \cos\theta = 0 \;\Rightarrow\; \theta = 0, 2\pi,$$

$$1 + 2\cos\theta = 0 \;\Rightarrow\; \theta = 2\pi/3, 4\pi/3.$$

Thus, $dy/d\theta = 0$ in $0 \leq \theta \leq 2\pi$ if $\theta = 0, 2\pi/3, 4\pi/3$, or 2π.

(b) Zeros of $dx/d\theta$ in $[0, 2\pi]$:

$$\frac{dx}{d\theta} = -\sin\theta + 2\cos\theta\sin\theta = (2\cos\theta - 1)\sin\theta$$

Now,

$$2\cos\theta - 1 = 0 \;\Rightarrow\; \theta = \pi/3, 5\pi/3,$$

$$\sin\theta = 0 \;\Rightarrow\; \theta = 0, \pi, 2\pi.$$

Thus, $dx/d\theta = 0$ in $0 \leq \theta \leq 2\pi$ if $\theta = 0, \pi/3, \pi, 5\pi/3$, or 2π.

We can now see that there are horizontal tangents ($dy/d\theta = 0$, $dx/d\theta \neq 0$) at the points where $\theta = 2\pi/3$ and $4\pi/3$, and vertical tangents ($dx/d\theta = 0$, $dy/d\theta \neq 0$) at the points where $\theta = \pi/3, \pi$, and $5\pi/3$.

$r = 1 - \cos\theta, \; 0 \leq \theta \leq 2\pi$

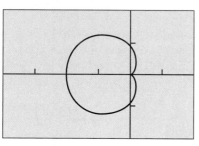

[−4, 2] by [−2, 2]

Figure 10.41 Where are the horizontal and vertical tangents to this cardioid? (Example 1)

At the points where $\theta = 0$ or 2π, the right side of Equation 1 takes the form $0/0$. We can use l'Hôpital's rule (Section 8.1) to see that

$$\lim_{\theta \to 0, 2\pi} \frac{dy/d\theta}{dx/d\theta} = \lim_{\theta \to 0, 2\pi} \frac{1 + \cos \theta - 2 \cos^2 \theta}{2 \cos \theta \sin \theta - \sin \theta}$$

$$= \lim_{\theta \to 0, 2\pi} \frac{-\sin \theta + 4 \cos \theta \sin \theta}{2 \cos^2 \theta - 2 \sin^2 \theta - \cos \theta} = \frac{0}{1} = 0.$$

The curve has a horizontal tangent at the point where $\theta = 0$ or 2π. Summarizing, we have

horizontal tangents at $(0, 0) = (0, 2\pi), \; (1.5, 2\pi/3), \; (1.5, 4\pi/3);$

vertical tangents at $(0.5, \pi/3), \; (2, \pi), \; (0.5, 5\pi/3).$

If the curve $r = f(\theta)$ passes through the origin at $\theta = \theta_0$, then $f(\theta_0) = 0$, and Equation 1 gives

$$\left. \frac{dy}{dx} \right|_{(0, \, \theta_0)} = \frac{f'(\theta_0) \sin \theta_0}{f'(\theta_0) \cos \theta_0} = \tan \theta_0,$$

provided $f'(\theta_0) \neq 0$ (not the case in Example 1). The reason we say "slope at $(0, \theta_0)$" and not just "slope at the origin" is that a polar curve may pass through the origin more than once, with different slopes at different θ values.

Example 2 FINDING TANGENT LINES AT THE POLE (ORIGIN)

Find the lines tangent to the rose curve

$$r = f(\theta) = 2 \sin 3\theta, \qquad 0 \leq \theta \leq \pi,$$

at the pole.

Solution $f(\theta)$ is zero when $\theta = 0, \pi/3, 2\pi/3,$ and π. The derivative $f'(\theta) = 6 \cos 3\theta$ is not zero at these four values of θ. Thus, this curve has tangent lines at the pole (Figure 10.42) with slopes $\tan 0 = \tan \pi = 0$, $\tan (\pi/3) = \sqrt{3}$, and $\tan (2\pi/3) = -\sqrt{3}$. The three corresponding tangent lines are $y = 0$, $y = \sqrt{3}x$, and $y = -\sqrt{3}x$.

Area in the Plane

The region OTS in Figure 10.43 is bounded by the rays $\theta = \alpha$ and $\theta = \beta$ and the curve $r = f(\theta)$. We approximate the region with n nonoverlapping circular sectors based on a partition P of angle TOS. The typical sector has radius $r_k = f(\theta_k)$ and central angle of radian measure $\Delta \theta_k$. Its area is

$$A_k = \frac{1}{2} r_k^2 \, \Delta \theta_k = \frac{1}{2} (f(\theta_k))^2 \, \Delta \theta_k.$$

The area of the region OTS is approximately

$$\sum_{k=1}^{n} A_k = \sum_{k=1}^{n} \frac{1}{2} (f(\theta_k))^2 \, \Delta \theta.$$

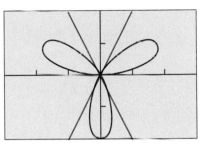

[3, 3] by [−2, 2]

Figure 10.42 The three tangent lines to $r = f(\theta) = 2 \sin 3\theta$, $0 \leq \theta \leq \pi$, are $y = 0$, $y = \sqrt{3}x$, and $y = -\sqrt{3}x$. (Example 2)

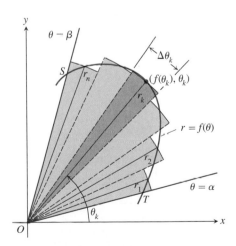

Figure 10.43 To derive a formula for the area of the region OTS, we approximate the region with thin circular sectors.

If f is continuous, we expect the approximations to improve as $\|P\| \to 0$, and we are led to the following formula for the region's area:

$$A = \lim_{\|P\| \to 0} \sum_{k=1}^{n} \frac{1}{2}(f(\theta_k))^2 \, \Delta\theta$$

$$= \int_{\alpha}^{\beta} \frac{1}{2}(f(\theta))^2 \, d\theta.$$

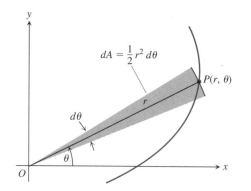

Figure 10.44 The area differential dA.

Area in Polar Coordinates

The **area** of the region **between the origin and the curve** $r = f(\theta)$, $\alpha \le \theta \le \beta$, is

$$A = \int_{\alpha}^{\beta} \frac{1}{2} r^2 \, d\theta.$$

This is the integral of the **area differential** (Figure 10.44),

$$dA = \frac{1}{2} r^2 \, d\theta.$$

Example 3 FINDING AREA

Find the area of the region in the plane enclosed by the cardioid $r = 2(1 + \cos\theta)$.

Solution We graph the cardioid (Figure 10.45) and determine that the *radius OP* sweeps out the region exactly once as θ runs from 0 to 2π.

Solve Analytically
The area is therefore

$$\int_{\theta=0}^{\theta=2\pi} \frac{1}{2} r^2 \, d\theta = \int_0^{2\pi} \frac{1}{2} \cdot 4(1 + \cos\theta)^2 \, d\theta$$

$$= \int_0^{2\pi} 2(1 + 2\cos\theta + \cos^2\theta) \, d\theta$$

$$= \int_0^{2\pi} \left(2 + 4\cos\theta + 2\frac{1 + \cos 2\theta}{2}\right) d\theta$$

$$= \int_0^{2\pi} (3 + 4\cos\theta + \cos 2\theta) \, d\theta$$

$$= \left[3\theta + 4\sin\theta + \frac{\sin 2\theta}{2}\right]_0^{2\pi} = 6\pi - 0 = 6\pi.$$

Support Numerically
NINT $(2(1 + \cos\theta)^2, \theta, 0, 2\pi) = 18.84955592$, which agrees with 6π to eight decimal places.

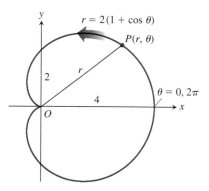

Figure 10.45 The cardioid in Example 3.

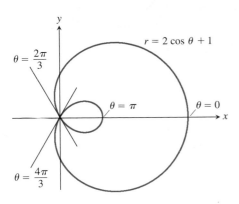

Figure 10.46 The limaçon in Example 4. ("Limaçon (pronounced LEE-ma-sahn) is an old French word for *snail*.)

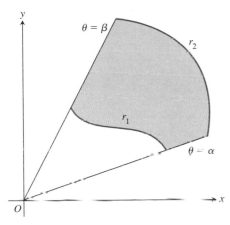

Figure 10.47 The area of the shaded region is calculated by subtracting the area of the region between r_1 and the origin from the area of the region between r_2 and the origin.

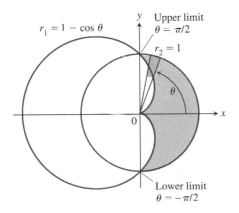

Figure 10.48 The region and limits of integration in Example 5.

Example 4 FINDING AREA

Find the area inside the smaller loop of the limaçon $r = 2 \cos \theta + 1$.

Solution After watching the grapher generate the curve (Figure 10.46), we see that the smaller loop is traced by the point (r, θ) as θ increases from $\theta = 2\pi/3$ to $\theta = 4\pi/3$. The area we seek is

$$A = \int_{2\pi/3}^{4\pi/3} \frac{1}{2} r^2 \, d\theta = \frac{1}{2} \int_{2\pi/3}^{4\pi/3} (2 \cos \theta + 1)^2 \, d\theta.$$

Solve Numerically

$$\frac{1}{2} \, \text{NINT} \, ((2 \cos \theta + 1)^2, \, \theta, \, 2\pi/3, \, 4\pi/3) \approx 0.544.$$

To find the area of a region like the one in Figure 10.47, which lies between two polar curves $r_1 = r_1(\theta)$ and $r_2 = r_2(\theta)$ from $\theta = \alpha$ to $\theta = \beta$, we subtract the integral of $(1/2)r_1^2$ from the integral of $(1/2)r_2^2$. This leads to the following formula.

Area Between Polar Curves

The area of the region $0 \le r_1(\theta) \le r_2(\theta), \; \alpha \le \theta \le \beta$, is

$$A = \int_\alpha^\beta \frac{1}{2} r_2^2 \, d\theta - \int_\alpha^\beta \frac{1}{2} r_1^2 \, d\theta = \int_\alpha^\beta \frac{1}{2} (r_2^2 - r_1^2) \, d\theta. \qquad (2)$$

Example 5 FINDING AREA BETWEEN CURVES

Find the area of the region that lies inside the circle $r = 1$ and outside the cardioid $r = 1 - \cos \theta$.

Solution The region is shown in Figure 10.48. The outer curve is $r_2 = 1$, the inner curve is $r_1 = 1 - \cos \theta$, and θ runs from $-\pi/2$ to $\pi/2$. The area, from Equation 2, is

$$A = \int_{-\pi/2}^{\pi/2} \frac{1}{2} (r_2^2 - r_1^2) \, d\theta$$

$$= 2 \int_0^{\pi/2} \frac{1}{2} (r_2^2 - r_1^2) \, d\theta \qquad \text{Symmetry}$$

$$= \int_0^{\pi/2} (1 - (1 - 2 \cos \theta + \cos^2 \theta)) \, d\theta$$

$$= \int_0^{\pi/2} (2 \cos \theta - \cos^2 \theta) \, d\theta \approx 1.215. \qquad \text{Using NINT}$$

In case you are interested, the exact value is $2 - \pi/4$.

Length of a Curve

We can obtain a polar coordinate formula for the length of a curve $r = f(\theta)$, $\alpha \le \theta \le \beta$, by parametrizing the curve as

$$x = r \cos \theta = f(\theta) \cos \theta, \quad y = r \sin \theta = f(\theta) \sin \theta, \quad \alpha \le \theta \le \beta. \quad (3)$$

The parametric length formula from Section 10.1 then gives the length as

$$L = \int_{\alpha}^{\beta} \sqrt{\left(\frac{dx}{d\theta}\right)^2 + \left(\frac{dy}{d\theta}\right)^2} \, d\theta.$$

This equation becomes

$$L = \int_{\alpha}^{\beta} \sqrt{r^2 + \left(\frac{dr}{d\theta}\right)^2} \, d\theta$$

when Equations 3 are substituted for x and y (Exercise 43).

Length of a Polar Curve

If $r = f(\theta)$ has a continuous first derivative for $\alpha \le \theta \le \beta$ and if the point $P(r, \theta)$ traces the curve $r = f(\theta)$ exactly once as θ runs from α to β, then the length of the curve is

$$L = \int_{\alpha}^{\beta} \sqrt{r^2 + \left(\frac{dr}{d\theta}\right)^2} \, d\theta. \quad (4)$$

Example 6   FINDING LENGTH OF A CARDIOID

Find the length of the cardioid $r = 1 - \cos \theta$.

Solution The graph is shown in Figure 10.49. The point $P(r, \theta)$ traces the curve once counterclockwise as θ runs from 0 to 2π, so these are the values we take for α and β.

Since $r = 1 - \cos \theta$, $dr/d\theta = \sin \theta$, and we have

$$r^2 + \left(\frac{dr}{d\theta}\right)^2 = (1 - \cos \theta)^2 + (\sin \theta)^2$$

$$= 1 - 2 \cos \theta + \underbrace{\cos^2 \theta + \sin^2 \theta}_{1} = 2 - 2 \cos \theta.$$

Therefore,

$$L = \int_{\alpha}^{\beta} \sqrt{r^2 + \left(\frac{dr}{d\theta}\right)^2} \, d\theta$$

$$= \int_{0}^{2\pi} \sqrt{2 - 2 \cos \theta} \, d\theta \approx 8. \quad \text{Using NINT}$$

This is, in fact, the exact value of the integral.

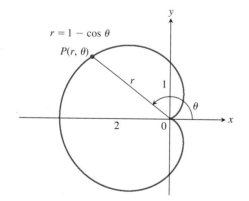

Figure 10.49 Example 6 calculates the length of this cardioid.

$r = 1 - \cos \theta$

Area of a Surface of Revolution

To derive polar coordinate formulas for the area of a surface of revolution, we parametrize the curve $r = f(\theta)$, $\alpha \le \theta \le \beta$, with Equations 3 and apply the parametric surface area equations in Section 10.1.

Area of a Surface of Revolution

If $r = f(\theta)$ has a continuous first derivative for $\alpha \le \theta \le \beta$ and if the point $P(r, \theta)$ traces the curve $r = f(\theta)$ exactly once as θ runs from α to β, then the areas of the surfaces generated by revolving the curve about the x- and y-axes are given by the following formulas.

Revolution about the x-axis ($y \ge 0$):

$$S = \int_{\alpha}^{\beta} 2\pi r \sin\theta \sqrt{r^2 + \left(\frac{dr}{d\theta}\right)^2}\, d\theta \qquad (5)$$

Revolution about the y-axis ($x \ge 0$):

$$S = \int_{\alpha}^{\beta} 2\pi r \cos\theta \sqrt{r^2 + \left(\frac{dr}{d\theta}\right)^2}\, d\theta \qquad (6)$$

Example 7 FINDING SURFACE AREA

Find the area of the surface generated by revolving the right-hand loop of the lemniscate $r^2 = \cos 2\theta$ about the y-axis.

Solution The lemniscate is shown in Figure 10.50a. You can use a grapher to see that the equation $r_1 = \sqrt{\cos 2\theta}$ produces the entire graph. Using $r_1 = \sqrt{\cos 2\theta}$, the point $P(r, \theta)$ traces the right-hand loop of the curve once counterclockwise as θ runs from $-\pi/4$ to $\pi/4$, so these are the values we take for α and β.

We evaluate the area integrand in Equation 6 in stages. Since $r_1 = \sqrt{\cos 2\theta}$,

$$\frac{dr_1}{d\theta} = \frac{1}{2}(\cos 2\theta)^{-1/2}(-2\sin 2\theta) = -\frac{\sin 2\theta}{\sqrt{\cos 2\theta}},$$

and

$$r_1^2 + \left(\frac{dr_1}{d\theta}\right)^2 = \cos 2\theta + \frac{\sin^2 2\theta}{\cos 2\theta}$$

$$= \frac{\cos^2 2\theta + \sin^2 2\theta}{\cos 2\theta} = \frac{1}{\cos 2\theta}.$$

Equation 6 becomes

$$S = \int_{\alpha}^{\beta} 2\pi\, r_1 \cos\theta \sqrt{r_1^2 + \left(\frac{dr_1}{d\theta}\right)^2}\, d\theta$$

$$= \int_{-\pi/4}^{\pi/4} 2\pi \sqrt{\cos 2\theta}\, \cos\theta \sqrt{\frac{1}{\cos 2\theta}}\, d\theta$$

$$= 2\pi \int_{-\pi/4}^{\pi/4} \cos\theta\, d\theta = 2\pi \left[\sin\theta\right]_{-\pi/4}^{\pi/4}$$

$$= 2\pi\sqrt{2} \approx 8.886.$$

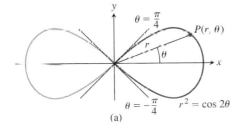

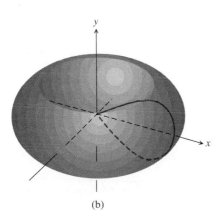

Figure 10.50 (a) The right-hand half of a lemniscate is (b) revolved about the y-axis to generate a surface whose area is calculated in Example 7.

Quick Review 10.6

Exercises 1–5 refer to the parametrized curve

$$x = 3 \cos t, \quad y = 5 \sin t, \quad 0 \leq t \leq 2\pi.$$

1. Find dy/dx.

2. Find the slope of the curve at $t = 2$.

3. Find the points on the curve where the slope is zero.

4. Find the points on the curve where the slope is not defined.

5. Find the length of the curve from $t = 0$ to $t = \pi$.

In Exercises 6–8, describe the portion of the graph of the polar curve $r = 1 + 2 \cos \theta$ traced in the given θ-interval.

6. $0 \leq \theta \leq \dfrac{2\pi}{3}$

7. $\dfrac{2\pi}{3} \leq \theta \leq \dfrac{4\pi}{3}$

8. $\dfrac{4\pi}{3} \leq \theta \leq 2\pi$

9. Find the area in the first quadrant under the curve $y = 6x - x^2$.

10. Find the area of the region enclosed by the curves $y = 2 \sin x$ and $y = x^2 - 2x + 1$.

Section 10.6 Exercises

In Exercises 1–4, find the slope of the curve at each indicated point.

1. $r = -1 + \sin \theta, \quad \theta = 0, \pi$

2. $r = \cos 2\theta, \quad \theta = 0, \pm\pi/2, \pi$

3. $r = 2 - 3 \sin \theta$

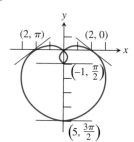

4. $r = 3(1 - \cos \theta)$

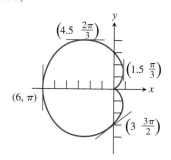

In Exercises 5–8, find the tangent lines at the pole.

5. $r = 3 \cos \theta, \quad 0 \leq \theta \leq 2\pi$

6. $r = 2 \cos 3\theta, \quad 0 \leq \theta \leq \pi$

7. $r = \sin 5\theta, \quad 0 \leq \theta \leq \pi$

8. $r = 2 \sin 2\theta, \quad 0 \leq \theta \leq 2\pi$

In Exercises 9–12, *work in groups of two or three*. Find equations for the horizontal and vertical tangent lines to the curve.

9. $r = -1 + \sin \theta, \quad 0 \leq \theta \leq 2\pi$

10. $r = 1 + \cos \theta, \quad 0 \leq \theta \leq 2\pi$

11. $r = 2 \sin \theta, \quad 0 \leq \theta \leq \pi$

12. $r = 3 - 4 \cos \theta, \quad 0 \leq \theta \leq 2\pi$

In Exercises 13–28, find the area of the region.

13. inside the oval limaçon $r = 4 + 2 \cos \theta$

14. inside the cardioid $r = a(1 + \cos \theta), \ a > 0$

15. inside the lemniscate $r^2 = 2a^2\cos 2\theta, \ a > 0$

16. inside one leaf of the four-leaved rose $r = \cos 2\theta$

17. inside one loop of the lemniscate $r^2 = 4 \sin 2\theta$

18. inside the six-leaved rose $r^2 = 2 \sin 3\theta$

19. shared by the circles $r = 2 \cos \theta$ and $r = 2 \sin \theta$

20. shared by the circles $r = 1$ and $r = 2 \sin \theta$

21. shared by the circle $r = 2$ and the cardioid $r = 2(1 - \cos \theta)$

22. shared by the cardioids $r = 2(1 + \cos \theta)$ and $r = 2(1 - \cos \theta)$

23. inside the circle $r = 3a \cos \theta$ and outside the cardioid $r = a(1 + \cos \theta), \ a > 0$

24. inside the lemniscate $r^2 = 6 \cos 2\theta$ and outside the circle $r = \sqrt{3}$

25. inside the circle $r = 2$ and outside the cardioid $r = 2(1 - \sin \theta)$

26. (a) inside the outer loop of the limaçon $r = 2 \cos \theta + 1$ (see Figure 10.46)

(b) inside the outer loop and outside the inner loop of the limaçon $r = 2 \cos \theta + 1$

27. inside the circle $r = 6$ above the line $r = 3 \csc \theta$

28. inside the lemniscate $r^2 = 6 \cos 2\theta$ to the right of the line $r = (3/2) \sec \theta$

29. (a) Find the area of the shaded region.

(b) Writing to Learn It looks as if the graph of $r = \tan \theta$, $-\pi/2 < \theta < \pi/2$, could be asymptotic to the lines $x = 1$ and $x = -1$. Is it? Give reasons for your answer.

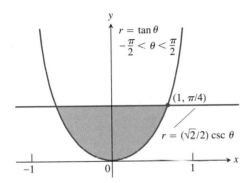

30. Writing to Learn The area of the region that lies inside the cardioid $r = \cos \theta + 1$ and outside the circle $r = \cos \theta$ is not

$$\frac{1}{2}\int_0^{2\pi}[(\cos \theta + 1)^2 - \cos^2 \theta]\,d\theta = \pi.$$

Why not? What is the area? Explain.

In Exercises 31–38, find the length of the curve.

31. the spiral $r = \theta^2$, $0 \le \theta \le \sqrt{5}$

32. the spiral $r = e^\theta/\sqrt{2}$, $0 \le \theta \le \pi$

33. the cardioid $r = 1 + \cos \theta$

34. the curve $r = a \sin^2 (\theta/2)$, $0 \le \theta \le \pi$, $a > 0$

35. the parabolic segment $r = 6/(1 + \cos \theta)$, $0 \le \theta \le \pi/2$

36. the parabolic segment $r = 2/(1 - \cos \theta)$, $\pi/2 \le \theta \le \pi$

37. the curve $r = \cos^3 (\theta/3)$, $0 \le \theta \le \pi/4$

38. the curve $r = \sqrt{1 + \sin 2\theta}$, $0 \le \theta \le \pi\sqrt{2}$

In Exercises 39–42, find the area of the surface generated by revolving the curve about the indicated axis.

39. $r = \sqrt{\cos 2\theta}$, $0 \le \theta \le \pi/4$, y-axis

40. $r = \sqrt{2}e^{\theta/2}$, $0 \le \theta \le \pi/2$, x-axis

41. $r^2 = \cos 2\theta$, x-axis

42. $r = 2a \cos \theta$, $a > 0$, y-axis

43. *Length of a Polar Curve* Assuming that the necessary derivatives are continuous, show how the substitutions

$$x = f(\theta) \cos \theta, \quad y = f(\theta) \sin \theta$$

(Equations 3 in the text) transform

$$L = \int_\alpha^\beta \sqrt{\left(\frac{dx}{d\theta}\right)^2 + \left(\frac{dy}{d\theta}\right)^2}\,d\theta$$

into

$$L = \int_\alpha^\beta \sqrt{r^2 + \left(\frac{dr}{d\theta}\right)^2}\,d\theta.$$

44. *Average Value* If f is continuous, the average value of the polar coordinate r over the curve $r = f(\theta)$, $\alpha \le \theta \le \beta$, with respect to θ is

$$r_{av} = \frac{1}{\beta - \alpha}\int_\alpha^\beta f(\theta)\,d\theta.$$

Use this formula to find the average value of r with respect to θ over the following curves ($a > 0$).

(a) the cardioid $r = a(1 - \cos \theta)$

(b) the circle $r = a$

(c) the circle $r = a \cos \theta$, $-\pi/2 \le \theta \le \pi/2$

45. Writing to Learn Can anything be said about the relative lengths of the curves

$$r = f(\theta), \quad \alpha \le \theta \le \beta,$$

and

$$r = 2f(\theta), \quad \alpha \le \theta \le \beta?$$

Give reasons for your answer.

46. Writing to Learn The curves

$$r = f(\theta), \quad \alpha \le \theta \le \beta,$$

and

$$r = 2f(\theta), \quad \alpha \le \theta \le \beta,$$

are revolved about the x-axis to generate surfaces. Can anything be said about the relative areas of these surfaces? Give reasons for your answer.

Explorations

47. *Videocassette Tape Length* The length of a tape wound onto a take-up reel as shown in the figure is

$$L = \int_0^\alpha \sqrt{r^2 + \left(\frac{b}{2\pi}\right)^2}\,d\theta,$$

where b is the tape thickness and

$$r = r_0 + \left(\frac{\alpha}{2\pi}\right)b$$

is the radius of the tape on the take-up reel. The initial radius of the tape on the take-up reel is r_0, and α is the angle in radians through which the wheel has turned.

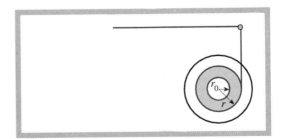

(a) Simulate the tape accumulating on the take-up reel using polar graphing with $r_0 = 1.75$ cm and $b = 0.06$ cm.

(b) Confirm the formula for L analytically.

(c) Determine the length of tape on the take-up reel if the reel has turned through an angle of 80π with $r_0 = 1.75$ cm and $b = 0.06$ cm.

(d) Assume that b is very small in comparison to r at any time. Show analytically that

$$L_a = \int_0^\alpha r\,d\theta$$

is an excellent approximation to the exact value of L.

(e) For the values given in part (c), compare L_a with L.

48. *(Continuation of Exercise 47)* Let n be the number of complete turns the take-up reel has made.

(a) Find a formula for n in terms of L, the tape length.

(b) When a VCR operates, the tape moves past the heads at a constant speed. Describe the speed of the take-up reel as time progresses.

(c) Suppose that the VCR tape counter is the number n of complete turns of the take-up reel. Describe the counter values as a function of time t. ■

Extending the Ideas

Centroids of Polar Regions (You might find it helpful to review Exercises 33–35 in Section 7.1 before you attempt Exercises 49 and 50.)

When thin flat plates are manufactured from material of constant density, the location of their centers of mass depends only on how the plates are shaped. Congruent plates of aluminum and steel, for instance, have their centers of mass in the same location. The location is thus a feature of the geometry of the plate and not of the material of which the plate is made. In such cases, engineers call the center of mass the **centroid** of the shape, as in "the centroid of a parabolic plate, triangle, or fan-shaped region."

Since the centroid of a triangle is located on each median, two-thirds of the way from the vertex to the opposite base, the lever arm for the moment about the x-axis of the thin triangular region in Figure 10.51 is about $(2/3)r\sin\theta$. Similarly, the lever arm for the moment of the triangular region about the y-axis is about $(2/3)r\cos\theta$.

These approximations improve as $\Delta\theta \to 0$ and lead to the following formulas for the coordinates $(\bar{x}, \bar{y})$ of the centroid of region *AOB*:

$$\bar{x} = \frac{\displaystyle\int_\alpha^\beta \frac{2}{3}r\cos\theta \cdot \frac{1}{2}r^2\,d\theta}{\displaystyle\int_\alpha^\beta \frac{1}{2}r^2\,d\theta} = \frac{\displaystyle\frac{2}{3}\int_\alpha^\beta r^3\cos\theta\,d\theta}{\displaystyle\int_\alpha^\beta r^2\,d\theta},$$

$$\bar{y} = \frac{\displaystyle\int_\alpha^\beta \frac{2}{3}r\sin\theta \cdot \frac{1}{2}r^2\,d\theta}{\displaystyle\int_\alpha^\beta \frac{1}{2}r^2\,d\theta} = \frac{\displaystyle\frac{2}{3}\int_\alpha^\beta r^3\sin\theta\,d\theta}{\displaystyle\int_\alpha^\beta r^2\,d\theta}.$$

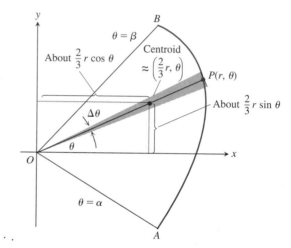

Figure 10.51 The moment of the thin triangular sector about the x-axis is approximately

$$\frac{2}{3}r\sin\theta\,dA = \frac{2}{3}r\sin\theta \cdot \frac{1}{2}r^2\,d\theta = \frac{1}{3}r^3\sin\theta\,d\theta.$$

49. Find the centroid of the region enclosed by the cardioid $r = a(1 + \cos\theta)$.

50. Find the centroid of the semicircular region $0 \le r \le a$, $0 \le \theta \le \pi$.

Chapter 10   Key Terms

acceleration vector (p. 533)
angle between vectors (p. 523)
antiderivative of a vector function (p. 535)
Archimedes spiral (p. 530)
area between polar curves (p. 563)
area differential (p. 562)
area in polar coordinates (p. 562)
cardioid (p. 554)
centroid (p. 568)
component form of a vector (p. 521)
component functions (p. 530)
component test for continuity (p. 531)
components of a vector (p. 530)
continuity of vector function at a point (p. 531)
continuous vector function (p. 531)
curve traced by vector function (p. 530)
cycloid (p. 515)
definite integral of vector function (p. 535)
derivative of parametric function at a point (p. 513)
derivative of vector function at a point (p. 532)
difference of vectors (p. 522)
differentiable parametrized curve (p. 513)
differentiable vector function (p. 532)
differentiation rules (p. 534)
directed line segment (p. 520)
direction of motion (p. 533)
distributive property of vectors (p. 528)
dot product of vectors (p. 524)
drag coefficient (p. 546)
drag force (p. 546)

equal vectors (p. 520)
equivalent directed line segments (p. 520)
graph of a vector function (p. 530)
horizontal component of a vector (p. 529)
Huygens's clock (p. 516)
ideal projectile motion (p. 539)
indefinite integral of vector function (p. 535)
initial point of directed line segment (p. 520)
inner product of vectors (p. 524)
launch angle (p. 540)
length of directed line segment (p. 520)
length of polar curve (p. 564)
length of smooth parametrized curve (p. 514)
length of vector (p. 520)
limit of vector function (p. 531)
linear combination of vectors (p. 529)
linear drag (p. 546)
magnitude of a vector (p. 521)
negative of a vector (p. 522)
normal vector (p. 525)
opposite of a vector (p. 522)
orthogonal vectors (p. 528)
parallelogram law (p. 522)
parametric equations for ideal projectile motion (p. 540)
particle's path (p. 530)
particle's position (p. 530)
particle's position vector (p. 530)
piecewise smooth curve (p. 532)
polar coordinates (p. 552)

pole (p. 552)
position vector (p. 521)
projectile's range (p. 541)
resultant vector (p. 522)
scalar (p. 520)
scalar function (p. 530)
scalar multiple of a vector (p. 522)
slope of a polar curve (p. 560)
slope of a vector (p. 522)
smooth parametrized curve (p. 514)
smooth vector curve (p. 532)
speed of a particle (p. 533)
standard position of a vector (p. 521)
standard unit vectors (p. 529)
sum of vectors (p. 522)
surface area of revolution for a parametrized curve (p. 517)
surface area of revolution for a polar curve (p. 565)
symmetry tests for polar graphs (p. 554)
tangent line to a vector curve (p. 532)
tangent vector (p. 525)
terminal point of directed line segment (p. 520)
unit vector (p. 522)
vector (p. 520)
vector equation for ideal projectile motion (p. 540)
vector (vector-valued) function (p. 530)
velocity vector (p. 533)
vertical component of a vector (p. 529)
zero vector (p. 522)

Chapter 10   Review Exercises

In Exercises 1–4, let $\mathbf{u} = \langle -3, 4 \rangle$ and $\mathbf{v} = \langle 2, -5 \rangle$. Find (a) the component form of the vector and (b) its magnitude.

1. $3\mathbf{u} - 4\mathbf{v}$

2. $\mathbf{u} + \mathbf{v}$

3. $-2\mathbf{u}$

4. $5\mathbf{v}$

In Exercises 5–8, find the component form of the vector.

5. the vector obtained by rotating $\langle 0, 1 \rangle$ through an angle of $2\pi/3$ radians

6. the unit vector that makes an angle of $\pi/6$ radian with the positive x-axis

7. the vector 2 units long in the direction $4\mathbf{i} - \mathbf{j}$

8. the vector 5 units long in the direction opposite to the direction of $(3/5)\mathbf{i} + (4/5)\mathbf{j}$

In Exercises 9 and 10, (a) find an equation for the tangent to the curve at the point corresponding to the given value of t, and (b) find the value of d^2y/dx^2 at this point.

9. $x = (1/2) \tan t$, $y = (1/2) \sec t$; $t = \pi/3$

10. $x = 1 + 1/t^2$, $y = 1 - 3/t$; $t = 2$

In Exercises 11–14, find the points at which the tangent to the curve is (a) horizontal; (b) vertical.

11. $x = (1/2) \tan t$, $y = (1/2) \sec t$

12. $x = -2 \cos t$, $y = 2 \sin t$

13. $x = -\cos t$, $y = \cos^2 t$

14. $x = 4 \cos t$, $y = 9 \sin t$

In Exercises 15 and 16, graph the set of points whose polar coordinates satisfy the inequality.

15. $0 \le r \le 6 \cos \theta$

16. $-4 \sin \theta \le r \le 0$

In Exercises 17–20, **(a)** graph the polar curve. **(b)** What is the smallest length θ-interval that will produce the graph?

17. $r = \cos 2\theta$

18. $r \cos \theta = 1$

19. $r^2 = \sin 2\theta$

20. $r = -\sin \theta$

In Exercises 21 and 22, *work in groups of two or three*. Find the tangent lines at the pole.

21. $r = \cos 2\theta, \quad 0 \le \theta \le 2\pi$

22. $r = 1 + \cos 2\theta, \quad 0 \le \theta \le 2\pi$

In Exercises 23 and 24, *work in groups of two or three*. Find equations for the horizontal and vertical tangent lines to the curve.

23. $r = 1 - \cos(\theta/2), \quad 0 \le \theta \le 4\pi$

24. $r = 2(1 - \sin \theta), \quad 0 \le \theta \le 2\pi$

25. Find equations for the lines that are tangent to the tips of the petals of the four-leaved rose $r = \sin 2\theta$.

26. Find equations for the lines that are tangent to the cardioid $r = 1 + \sin \theta$ at the points where it crosses the x-axis.

In Exercises 27–30, replace the polar equation by an equivalent Cartesian equation. Then identify or describe the graph.

27. $r \cos \theta = r \sin \theta$

28. $r = 3 \cos \theta$

29. $r = 4 \tan \theta \sec \theta$

30. $r \cos (\theta + \pi/3) = 2\sqrt{3}$

In Exercises 31–34, replace the Cartesian equation by an equivalent polar equation.

31. $x^2 + y^2 + 5y = 0$

32. $x^2 + y^2 - 2y = 0$

33. $x^2 + 4y^2 = 16$

34. $(x + 2)^2 + (y - 5)^2 = 16$

In Exercises 35–42, find the length of the curve.

35. $x = e^{2t} - t/8, \quad y = e^t, \quad 0 \le t \le \ln 2$

36. the loop of the curve $x = t^2, \quad y = (t^3/3) - t$, that starts at $t = -\sqrt{3}$ and ends at $t = \sqrt{3}$ as shown in the figure

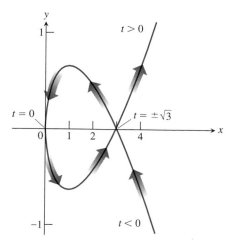

37. $r = -1 + \cos \theta, \quad 0 \le \theta \le 2\pi$

38. $r = 2 \sin \theta + 2 \cos \theta, \quad 0 \le \theta \le \pi/2$

39. $r = 8 \sin^3 (\theta/3), \quad 0 \le \theta \le \pi/4$

40. $r = \sqrt{1 + \cos 2\theta}, \quad -\pi/2 \le \theta \le \pi/2$

41. $\mathbf{r} = (2 \cos t)\mathbf{i} + (t^2)\mathbf{j}, \quad 0 \le t \le \pi/2$

42. $\mathbf{r} = (3 \sin t)\mathbf{i} + (2t^{3/2})\mathbf{j}, \quad 0 \le t \le 3$

In Exercises 43–46, find the area of the region.

43. enclosed by the limaçon $r = 2 - \cos \theta$

44. enclosed by one leaf of the three-leaved rose $r = \sin 3\theta$

45. inside the "figure eight" $r = 1 + \cos 2\theta$ and outside the circle $r = 1$

46. inside the cardioid $r = 2(1 + \sin \theta)$ and outside the circle $r = 2 \sin \theta$

In Exercises 47–50, find the area of the surface generated by revolving the curve about the indicated axis.

47. $x = t^2/2, \quad y = 2t, \quad 0 \le t \le \sqrt{5}; \quad x$-axis

48. $x = t^2 + 1/(2t), \quad y = 4t, \quad 1/\sqrt{2} \le t \le 1; \quad y$-axis

49. $r = \sqrt{\cos 2\theta}, \quad 0 \le \theta \le \pi/4; \quad x$-axis

50. $r^2 = \sin 2\theta; \quad y$-axis

In Exercises 51 and 52, $\mathbf{r}(t)$ is the position vector of a particle in the plane at time t.

(a) Find the velocity and acceleration vectors.

(b) Find the speed at the given value of t.

(c) Find the angle between the velocity and acceleration vectors at the given value of t.

51. $\mathbf{r}(t) = (4 \cos t)\mathbf{i} + (\sqrt{2} \sin t)\mathbf{j}, \quad t = \pi/4$

52. $\mathbf{r}(t) = (\sqrt{3} \sec t)\mathbf{i} + (\sqrt{3} \tan t)\mathbf{j}, \quad t = 0$

53. The position of a particle in the plane at time t is

$$\mathbf{r} = \frac{1}{\sqrt{1 + t^2}} \mathbf{i} + \frac{t}{\sqrt{1 + t^2}} \mathbf{j}.$$

Find the particle's greatest speed.

54. Writing to Learn Suppose that

$$\mathbf{r}(t) = (e^t \cos t)\mathbf{i} + (e^t \sin t)\mathbf{j}.$$

Show that the angle between $\mathbf{r}$ and the acceleration vector $\mathbf{a}$ never changes. What is the angle?

In Exercises 55 and 56, evaluate the integral.

55. $\displaystyle \int_0^1 [(3 + 6t)\mathbf{i} + (6\pi \cos \pi t)\mathbf{j}] \, dt$

56. $\displaystyle \int_e^{e^2} \left[\left(\frac{2 \ln t}{t} \right)\mathbf{i} + \left(\frac{1}{t \ln t} \right)\mathbf{j} \right] dt$

In Exercises 57–60, solve the initial value problem.

57. $\dfrac{d\mathbf{r}}{dt} = -(\sin t)\mathbf{i} + (\cos t)\mathbf{j}, \quad \mathbf{r}(0) = \mathbf{j}$

58. $\dfrac{d\mathbf{r}}{dt} = \dfrac{1}{t^2 + 1}\mathbf{i} + \dfrac{t}{\sqrt{t^2 + 1}}\mathbf{j}, \quad \mathbf{r}(0) = \mathbf{i} + \mathbf{j}$

59. $\dfrac{d^2\mathbf{r}}{dt^2} = 2\mathbf{j}, \quad \left.\dfrac{d\mathbf{r}}{dt}\right|_{t=0} = \mathbf{0}, \quad \mathbf{r}(0) = \mathbf{i}$

60. $\dfrac{d^2\mathbf{r}}{dt^2} = -2\mathbf{i} - 2\mathbf{j}, \quad \left.\dfrac{d\mathbf{r}}{dt}\right|_{t=1} = 4\mathbf{i}, \quad \mathbf{r}(1) = 3\mathbf{i} + 3\mathbf{j}$

61. *Particle Motion* A particle moves in the plane in such a manner that its coordinates at time t are

$$x = 3\cos\frac{\pi}{4}t, \quad y = 5\sin\frac{\pi}{4}t.$$

(a) Find the length of the velocity vector at $t = 3$.

(b) Find the x- and y-components of the acceleration of the particle at $t = 3$.

(c) Find a single equation in x and y for the path of the particle.

62. *Solid of Revolution* Let C be the curve

$$x = \frac{t - 2}{2}, \quad y = t\left(\frac{10 - t}{2}\right), \quad 0 \le t \le 10.$$

Let R be the region bounded by C and the x-axis.

(a) Find the length of C.

(b) Find the volume of the solid generated by revolving R about the x-axis.

(c) Find the surface area of the solid generated by revolving R about the x-axis.

63. *Particle Motion* At time t, $0 \le t \le 1$, the position of a particle moving along a path in the plane is given by the parametric equations

$$x = e^t \cos t, \quad y = e^t \sin t.$$

(a) Find the slope of the path of the particle at time $t = \pi$.

(b) Find the speed of the particle when $t = 3$.

(c) Find the distance traveled by the particle along the path from $t = 0$ to $t = 3$.

64. *Particle Motion* The position of a particle at any time $t \ge 0$ is given by

$$x(t) = t^2 - 2, \quad y(t) = \frac{2}{5}t^3.$$

(a) Find the magnitude of the velocity vector at $t = 4$.

(b) Find the total distance traveled by the particle from $t = 0$ to $t = 4$.

(c) Find dy/dx as a function of x.

65. *Navigation* An airplane, flying in the direction $80°$ east of north at 540 mph in still air, encounters a 55-mph tail wind acting in the direction $100°$ east of north. The airplane holds its compass heading but, because of the wind, acquires a different ground speed and direction. What are they?

66. *Combining Forces* A force of 120 lb pulls up on an object at an angle of $20°$ with the horizontal. A second force of 300 lb pulls down on the object at an angle of $-5°$. Find the direction and length of the resultant force vector.

67. *Shot Put* A shot leaves the thrower's hand 6.5 ft above the ground at a $45°$ angle at 44 ft/sec. Where is it 3 sec later?

68. *Javelin* A javelin leaves the thrower's hand 7 ft above the ground at a $45°$ angle at 80 ft/sec. How high does it go?

69. *Rolling Wheel* *Work in groups of two or three.* A circular wheel with radius 1 ft and center C rolls to the right along the x-axis at a half-turn per second (see figure). At time t seconds, the position vector of the point P on the wheel's circumference is

$$\mathbf{r}(t) = (\pi t - \sin \pi t)\mathbf{i} + (1 - \cos \pi t)\mathbf{j}.$$

(a) Graph the curve traced by P during the interval $0 \le t \le 3$.

(b) Find velocity and acceleration vectors $\mathbf{v}$ and $\mathbf{a}$ at $t = 0$, 1, 2, and 3.

(c) *Writing to Learn* At any given time, what is the forward speed of the topmost point of the wheel? of C? Give reasons for your answers.

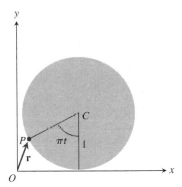

70. *The Dictator* The Civil War mortar Dictator weighed so much (17,120 lb) that it had to be mounted on a railroad car. It had a 13-in. bore and used a 20-lb powder charge to fire a 200-lb shell. The mortar was made by Mr. Charles Knapp in his ironworks in Pittsburgh, Pennsylvania, and was used by the Union army in 1864 in the siege of Petersburg, Virginia. How far did it shoot? Here we have a difference of opinion. The ordnance manual claimed 4325 yd, while field officers claimed 4752 yd. Assuming a 45° firing angle, what muzzle speeds are involved here?

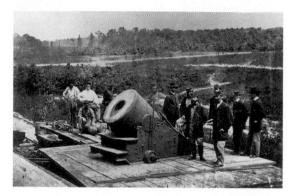

71. *World's Record for Popping a Champagne Cork*

(a) Until 1988, the world's record for popping a champagne cork, 109 ft 6 in., was held by Captain Michael Hill of the British Royal Artillery (of course). Assuming Captain Hill held the bottle at ground level at a 45° angle and the cork behaved like an ideal projectile, how fast was the cork going as it left the bottle?

(b) A new world record, 177 ft 9 in., was set on June 5, 1988, by Prof. Emeritus Heinrich of Rensselaer Polytechnic Institute, firing from 4 ft above ground level at a 45° angle at the Woodbury Vineyards Winery, New York. Assuming an ideal trajectory, what was the cork's initial speed?

72. *Javelin* In Potsdam in 1988, Petra Felke of (then) East Germany set a women's world record by throwing a javelin 262 ft 5 in.

(a) Assuming that Felke launched the javelin at a 40° angle to the horizontal from 6.5 ft above the ground, what was the javelin's initial speed?

(b) How high did the javelin go?

73. *Synchronous Curves* By eliminating α from the ideal projectile equations

$$x = (v_0 \cos \alpha)t, \quad y = (v_0 \sin \alpha)t - \frac{1}{2}gt^2,$$

show that $x^2 + (y + gt^2/2)^2 = v_0^2 t^2$. This shows that projectiles launched simultaneously from the origin at the same initial speed will, at any given instant, all lie on the circle of radius $v_0 t$ centered at $(0, -gt^2/2)$, regardless of their launch angle. These circles are the *synchronous curves* of the launching.

74. *Hitting a Baseball* A baseball is hit when it is 4 ft above the ground. It leaves the bat with an initial velocity of 155 ft/sec, making an angle of 18° with the horizontal. At the instant the ball is hit, an instantaneous 11.7 ft/sec gust of wind blows in the horizontal direction against the ball, adding a component of $-11.7\mathbf{i}$ to the ball's initial velocity. A 10-foot-high fence is 380 ft from home plate in the direction of the flight.

(a) Find vector and parametric forms for the path of the baseball.

(b) How high does the baseball go, and when does it reach maximum height?

(c) Find the range and flight time of the baseball.

(d) When is the baseball 25 ft high? How far (ground distance) is the baseball from home plate at that height?

(e) **Writing to Learn** Has the batter hit a home run? Explain.

75. *(Continuation of Exercise 74)* Consider the baseball problem of Exercise 74 again. This time, assume a linear drag model with a drag coefficient of 0.09.

(a) Find vector and parametric forms for the path of the baseball.

(b) How high does the baseball go, and when does it reach maximum height?

(c) Find the range and flight time of the baseball.

(d) When is the baseball 30 ft high? How far (ground distance) is the baseball from home plate at that height?

(e) Has the batter hit a home run? If "yes," find a drag coefficient that would have prevented a home run. If "no," find a drag coefficient that would have allowed the hit to be a home run.

76. *Parallelogram* The accompanying figure shows parallelogram *ABCD* and the midpoint *P* of diagonal *BD*.

(a) Express $\overrightarrow{BD}$ in terms of $\overrightarrow{AB}$ and $\overrightarrow{AD}$.

(b) Express $\overrightarrow{AP}$ in terms of $\overrightarrow{AB}$ and $\overrightarrow{AD}$.

(c) Prove that *P* is also the midpoint of diagonal *AC*.

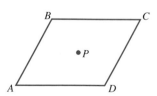

77. *Archimedes Spirals* The graph of an equation of the form $r = a\theta$, where a is a nonzero constant, is called an Archimedes spiral. Is there anything special about the widths between the successive turns of such a spiral?

Cumulative Review Exercises

In Exercises 1–8, determine the limit.

1. $\lim\limits_{x \to 1} \dfrac{2x^2 - x - 1}{x^2 + x - 12}$

2. $\lim\limits_{x \to 0} \dfrac{\sin 3x}{4x}$

3. $\lim\limits_{x \to 0} \dfrac{\dfrac{1}{x + 1} - 1}{x}$

4. $\lim\limits_{x \to \infty} \dfrac{x + e^x}{x - e^x}$

5. $\lim\limits_{t \to 0} \dfrac{t(1 - \cos t)}{t - \sin t}$

6. $\lim\limits_{x \to 0^+} \dfrac{\ln(e^x - 1)}{\ln x}$

7. $\lim\limits_{x \to 0} (e^x + x)^{1/x}$

8. $\lim\limits_{x \to 0} \left(\dfrac{3x + 1}{x} - \dfrac{1}{\sin x} \right)$

9. Let $f(x) = \begin{cases} 2x - x^2, & x \le 1 \\ 2 - x, & x > 1. \end{cases}$

   (a) Find $\lim_{x \to 1^-} f(x)$.

   (b) Find $\lim_{x \to 1^+} f(x)$.

   (c) Find $\lim_{x \to 1} f(x)$.

   (d) Is f continuous at $x = 1$?

   (e) Is f differentiable at $x = 1$?

10. Find all the points of discontinuity of

$$f(x) = \sqrt{\dfrac{1}{4 - x^2}}.$$

11. Identify all horizontal and vertical asymptotes of

$$y = \dfrac{\cos x}{2x^2 - x}.$$

12. Sketch a possible graph for a function $y = f(x)$ that satisfies:

$$\lim\limits_{x \to 2^-} f(x) = \infty, \quad \lim\limits_{x \to 2^+} f(x) = -1,$$

$$\lim\limits_{x \to -\infty} f(x) = -3, \quad \lim\limits_{x \to \infty} f(x) = 3.$$

13. Find the average rate of change of the function $f(x) = \sqrt{x + 4}$ over the interval $[0, 5]$.

In Exercises 14–28, find dy/dx.

14. $y = \dfrac{x + 1}{x - 2}$

15. $y = \cos(\sqrt{1 - 3x})$

16. $y = \sin x \tan x$

17. $y = \ln(x^2 + 1)$

18. $y = e^{x^2 - x}$

19. $y = x^2 \tan^{-1} x$

20. $y = x^{-3} e^x$

21. $y = \left(\dfrac{\csc x}{1 + \cos x} \right)^3$

22. $y = \cos^{-1} x - \cot^{-1} x$

23. $\cos(xy) + y^2 - \ln x = 1$

24. $y = \sqrt{|x|}$

25. $x = 1 + \cos t, \quad y = 1 - \sin t$

26. $y = (\cos x)^x, \quad -\dfrac{\pi}{2} < x < \dfrac{\pi}{2}$

27. $y = \displaystyle\int_0^x \sqrt{1 + t^3}\, dt$

28. $\displaystyle\int_{2x}^{x^2} \sin t\, dt$

29. Find $d^2 y/dx^2$ if $y^2 + 2y = \sec x$.

30. Suppose u and v are differentiable functions of x and that $u(0) = 2$, $u'(0) = -1$, $v(0) = -3$, and $v'(0) = 3$. Find

$$\dfrac{d}{dx} \left(\dfrac{u}{1 + v} \right) \Big|_{x=0}.$$

31. A particle moves along the x-axis with its position at time t in seconds given by $x = t^3 - 6t^2 + 9t, \ 0 \le t \le 5$, in meters.

   (a) Determine the velocity and acceleration of the particle at time t.

   (b) When is the particle at rest?

   (c) When is the particle moving to the right? left?

   (d) What is the velocity when the acceleration is zero?

In Exercises 32–36, find an equation for **(a)** the tangent line and **(b)** the normal line to the curve at the indicated point.

32. $y = 2x^3 - 6x^2 + 4x - 1$ at $x = 1$

33. $y = x \cos x$ at $x = \pi/3$

34. $\dfrac{x^2}{4} + \dfrac{y^2}{9} - 1$ at $\left(1, \dfrac{3\sqrt{3}}{2} \right)$

35. $x = 2 \cos t, \ y = 3 \sin t$, at $t = \pi/3$

36. $\mathbf{r}(t) = (\sec t)\mathbf{i} + (\tan t)\mathbf{j}$, at $t = \pi/4$

37. Sketch the graph of a continuous function f with

$$f(3) = 1 \quad \text{and} \quad f'(x) = \begin{cases} -1, & x < 3 \\ 2, & x > 3. \end{cases}$$

38. The graph of the function f over the interval $[-2, 3]$ is given. At what domain points does f appear to be

   (a) differentiable?

   (b) continuous but not differentiable?

   (c) neither continuous nor differentiable?

   (d) Identify any extreme values and where they occur.

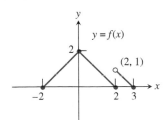

39. A driver handed in a ticket at a toll booth showing that in 1.5 h he had covered 111 mi on a toll road with speed limit of 65 mph. The driver was cited for speeding. Why?

40. Assume that f is continuous and differentiable on the interval $[-2, 2]$. The table gives some values of f'.

x	$f'(x)$	x	$f'(x)$
-2	-11	0.25	3.63
-1.75	-8.38	0.5	4
-1.5	-6	0.75	4.13
-1.25	-3.88	1	4
-1	-2	1.25	3.63
-0.75	-0.38	1.5	3
-0.5	1	1.75	2.13
-0.25	2.13	2	1
0	3		

(a) Estimate where f is increasing, decreasing, and has local extrema.

(b) Find a quadratic regression equation for the data in the table and superimpose its graph on a scatter plot of the data.

(c) Use the equation in (b) to find a formula for f that satisfies $f(0) = 1$.

41. Find the function f with $f'(x) = 2x - 3 + \sin x$ whose graph passes through the point $P(0, -2)$.

42. Suppose $f(x) = x^2\sqrt{4 - x^2}$. Find the intervals on which the graph of f is **(a)** increasing, **(b)** decreasing, **(c)** concave up, **(d)** concave down. Then find any **(e)** local extreme values and where they occur, and **(f)** any inflection points.

43. A function f is continuous on its domain $[-1, 3]$, $f(-1) = 1$, $f(3) = -2$, and f' and f'' have the following properties.

x	$-1 < x < 1$	$x = 1$	$1 < x < 2$	$x = 2$	$2 < x < 3$
f'	$+$	0	$-$	does not exist	$-$
f''	$-$	-1	$-$	does not exist	$+$

(a) Find where all absolute extrema of f occur.

(b) Find where the points of inflection of f occur.

(c) Sketch a possible graph of f.

44. A rectangle with base on the x-axis is to be inscribed under the upper half of the ellipse

$$\frac{x^2}{16} + \frac{y^2}{4} = 1.$$

What are the dimensions of the rectangle with largest area, and what is the largest area?

45. Find the linearization of $f(x) = \sec x$ at $x = \pi/4$.

46. The edge of a cube is measured as 8 cm with an error of 1%. The cube's volume is to be calculated from this measurement. Estimate the percentage error in the volume calculation.

47. A dinghy is pulled toward a dock by a rope from the bow through a ring on the dock 5 ft above the bow as shown in the figure. The rope is hauled in at the rate of 1.5 ft/sec.

(a) How fast is the boat approaching the dock when 8 ft of rope are out?

(b) At what rate is the angle θ changing at that moment?

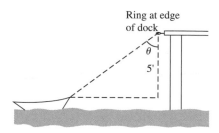

48. Coffee is draining from a conical filter into a cylindrical coffeepot at the rate of 9 in.^{3}/min as suggested in the figure.

(a) How fast is the level in the pot rising when the coffee in the cone is 5 in. deep?

(b) How fast is the level in the cone falling at that moment?

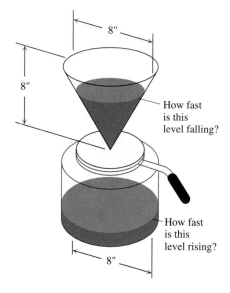

49. The table below shows the velocity of a model train engine moving along a track for 10 sec. Estimate the distance traveled by the engine, using 10 subintervals of length 1 with **(a)** LRAM and **(b)** RRAM.

Time (sec)	Velocity (in./sec)	Time (sec)	Velocity (in./sec)
0	0	6	28.8
1	1.8	7	29.4
2	6.4	8	25.6
3	12.6	9	16.2
4	19.2	10	0
5	25.0		

In Exercises 50–61, evaluate the integral analytically.

50. $\displaystyle\int_{-2}^{1} |x|\, dx$

51. $\displaystyle\int_{-2}^{2} \sqrt{4 - x^2}\, dx$

52. $\displaystyle\int_{1}^{3} \left(x^2 + \frac{1}{x}\right) dx$

53. $\displaystyle\int_{0}^{\pi/4} \sec^2 x\, dx$

54. $\displaystyle\int_{1}^{4} \frac{2 + \sqrt{x}}{\sqrt{x}}\, dx$

55. $\displaystyle\int_{e}^{2e} \frac{dx}{x\,(\ln x)^2}$

56. $\displaystyle\int_{1}^{3} [(3 - 2t)\mathbf{i} + (1/t)\mathbf{j}]\, dt$

57. $\displaystyle\int e^x \cot^2 (e^x + 1)\, dx$

58. $\displaystyle\int \frac{ds}{s^2 + 4}$

59. $\displaystyle\int \frac{\sin (x - 3)}{\cos^3 (x - 3)}\, dx$

60. $\displaystyle\int e^{-x} \cos 2x\, dx$

61. $\displaystyle\int \frac{x + 2}{x^2 - 5x - 6}\, dx$

62. A rectangular swimming pool is 25 ft wide and 40 ft long. The table below shows the depth $h(x)$ of the water at 5-ft intervals from one end of the pool to the other. Estimate the volume of the water in the pool using the Trapezoidal Rule with $n = 8$.

Position (ft) x	Depth (ft) $h(x)$	Position (ft) x	Depth (ft) $h(x)$
0	3	25	10.7
5	8.3	30	9.9
10	9.9	35	8.3
15	10.7	40	3
20	11		

In Exercises 63 and 64, solve the initial value problem.

63. $\dfrac{dy}{dt} = (t + 1)^{-2} + e^{-2t}, \quad y(0) = 2$

64. $\dfrac{d^2y}{d\theta^2} = \sin 2\theta - \cos \theta, \quad y(\pi/2) = y'(\pi/2) = 0$

65. Evaluate $\int x^2 \sin x\, dx$. Support your answer by superimposing the graph of one of the antiderivatives on a slope field of the integrand.

66. Evaluate $\int xe^x\, dx$. Confirm your answer by differentiation.

67. A colony of bacteria is grown under ideal conditions in a laboratory so that the population increases exponentially with time. At the end of 2 h there are 6,000 bacteria. At the end of 5 h there are 10,000 bacteria.

(a) Find a formula for the number of bacteria present at any time t.

(b) How many bacteria were present initially?

68. The temperature of an ingot of silver is 50°C above room temperature right now. Fifteen minutes ago, it was 65°C above room temperature.

(a) How far above room temperature will the silver be 2 hours from now?

(b) When will the silver be 5°C above room temperature?

In Exercises 69 and 70, solve the differential equation.

69. $\dfrac{dy}{dx} = 0.08y\left(1 - \dfrac{y}{500}\right)$

70. $\dfrac{dy}{dx} = (y - 4)(x + 3)$

71. Use Euler's method to solve the initial value problem
$$y' = y + \cos x, \quad y(0) = 0,$$
on the interval $0 \le x \le 1$ with $dx = 0.1$.

In Exercises 72–75, find the area of the region enclosed by the curves.

72. $y = \sin 2x, \quad y = 0, \quad x = -\pi, \quad x = \pi$

73. $y = 5 - x^2, \quad y = x^2 - 3$

74. $x = y^2 - 3, \quad y = x - 2$

75. $r = 3(1 + \cos \theta)$

In Exercises 76 and 77, find the volume of the solid generated by revolving the region bounded by the curves about the indicated axis.

76. $y = x^3/2, \quad y = 0, \quad x = -1, \quad x = 1; \quad x$-axis

77. $y = 4x - x^2, \quad y = 0; \quad y$-axis

78. Find the average value of $\sqrt{\sin x}$ on the interval $[0, \pi]$.

In Exercises 79–81, find the length of the curve.

79. $y = \tan x, \quad -\pi/4 \le x \le \pi/4$

80. $x = \sin t, \quad y = t + \cos t, \quad -\pi/2 \le t \le \pi/2$

81. $r = \theta, \quad 0 \le \theta \le \pi$

In Exercises 82–84, find the area of the surface generated by revolving the curve about the indicated axis.

82. $y = e^{-x/2}, \quad 0 \le x \le 2; \quad x$-axis

83. $x = \sin t, \quad y = t + \cos t, \quad 0 \le t \le \pi/2; \quad y$-axis

84. $r = \theta, \quad \pi/2 \le \theta \le \pi; \quad x$-axis

85. A solid lies between planes perpendicular to the x-axis at $x = 0$ and $x = 1$. The cross sections of the solid perpendicular to the x-axis between these planes are circular disks with diameters running from the parabola $y = x^2$ to the parabola $y = \sqrt{x}$. Find the volume of the solid.

86. Find the volume of the solid generated by revolving about the x-axis the region bounded by $y = 2 \tan x, \quad y = 0, \quad x = -\pi/4,$ and $x = \pi/4$. (The region lies in the first and third quadrants and resembles a bow tie.)

87. A force of 200 N will stretch a garage door spring 0.8 m beyond its unstressed length.

  (a) How far will a 300-N force stretch the spring from its unstressed length?

  (b) In (a), how much work was done in stretching the spring that far?

88. A right circular conical tank, point down, with top radius 5 ft and height 10 ft is filled with a liquid whose weight-density is 60 lb/ft^3.

  (a) To the nearest foot-pound, how much work will it take to pump the liquid to a point 2 ft above the tank?

  (b) If the pump is driven by a motor rated at 275 ft · lb/sec (1/2-hp), about how long will it take it to empty the tank?

89. You plan to store mercury (weight density 849 lb/ft^3) in a vertical right circular cylindrical tank of inside radius 1 ft whose interior side wall can withstand a total fluid force of 40,000 lb. About how many cubic feet of mercury can you store at any one time?

90. Does $f(x) = \ln x$ grow faster than, at the same rate as, or slower than $g(x) = \sqrt{x}$ as $x \to \infty$?

In Exercises 91–96, determine whether the integral converges or diverges.

91. $\displaystyle \int_3^\infty \frac{dt}{t^2 - 4}$

92. $\displaystyle \int_2^\infty \frac{dx}{\ln x}$

93. $\displaystyle \int_{-\infty}^\infty e^{-|x|}\, dx$

94. $\displaystyle \int_0^1 \frac{4r\, dr}{\sqrt{1 - r^2}}$

95. $\displaystyle \int_0^{10} \frac{dx}{1 - x}$

96. $\displaystyle \int_0^2 \frac{dx}{\sqrt[3]{x - 1}}$

97. Find a power series to represent

$$\frac{1}{1 + 2x}$$

and identify its interval of convergence.

98. **(a)** Find a power series for

$$F(x) = \int_0^x \cos(t^2)\, dt.$$

  (b) What is the interval of convergence of the series? Explain.

99. Find the Maclaurin series generated by $\ln(2 + 2x)$. What is its interval of convergence?

100. Find the Taylor series generated by $\sin x$ at $x = 2\pi$.

101. Find a polynomial that you know will approximate e^{-x} throughout the interval $[0, 1]$ with an error of magnitude less than 10^{-3}. Explain.

102. Find the Taylor series generated by $f(x) = \sqrt[3]{1 + x}$ at $x = 0$ and identify its radius of convergence.

In Exercises 103–106, determine whether the series converges or diverges.

103. $\displaystyle \sum_{n=0}^\infty \frac{2}{3^n}$

104. $\displaystyle \sum_{n=1}^\infty \frac{2}{\sqrt{n}}$

105. $\displaystyle \sum_{n=0}^\infty \frac{(-1)^n}{n + 1}$

106. $\displaystyle \sum_{n=0}^\infty \frac{3^n}{n!}$

In Exercises 107 and 108, **(a)** find the radius and interval of convergence. For what values of x is the convergence **(b)** absolute? **(c)** conditonal?

107. $\displaystyle \sum_{n=1}^\infty \frac{(-1)^n(x + 2)^n}{n}$

108. $\displaystyle \sum_{n=2}^\infty \frac{x^n}{n(\ln n)^2}$

109. Find the unit vector in the direction of $\langle 2, -3 \rangle$.

110. Find the component form of the unit vector that makes an angle of $\pi/3$ with the positive x-axis.

111. Find the unit vectors (four vectors in all) that are tangent and normal to the curve $x = 4 \sin t$, $y = 3 \cos t$, at $t = 3\pi/4$.

112. The position of a particle in the plane is given by

$$\mathbf{r}(t) = (1 - \sin t)\mathbf{i} + (t - \cos t)\mathbf{j}.$$

  (a) Find the velocity and acceleration of the particle.

  (b) Find the distance the particle travels along the path from $t = \pi/2$ to $t = 3\pi/2$.

113. A golf ball leaves the ground at a 45° angle at a speed of 100 ft/sec. Will it clear the top of a 35-ft tree 130 ft away? Explain.

114. Replace the polar equation $r \cos \theta - r \sin \theta = 2$ by an equivalent Cartesian equation. Then identify the graph.

115. Graph the polar curve $r = 1 + 2 \sin \theta$. What is the shortest length a θ-interval can have and still produce the graph?

116. Find equations for the horizontal and vertical tangents to the curve $r = 1 - \cos \theta$, $0 \le \theta \le 2\pi$.

APPENDICES

A1 Formulas from Precalculus Mathematics

Algebra • Geometry • Trigonometry

Algebra

1. Laws of Exponents

$$a^m a^n = a^{m+n}, \quad (ab)^m = a^m b^m, \quad (a^m)^n = a^{mn}, \quad a^{m/n} = \sqrt[n]{a^m}$$

If $a \neq 0$, $\qquad \dfrac{a^m}{a^n} = a^{m-n}, \quad a^0 = 1, \quad a^{-m} = \dfrac{1}{a^m}$

2. Zero Division by zero is not defined.

If $a \neq 0$: $\qquad \dfrac{0}{a} = 0, \quad a^0 = 1, \quad 0^a = 0$

For any number a: $\qquad a \cdot 0 = 0 \cdot a = 0$

3. Fractions

$$\frac{a}{b} + \frac{c}{d} = \frac{ad + bc}{bd}, \quad \frac{a}{b} \cdot \frac{c}{d} = \frac{ac}{bd}, \quad \frac{a/b}{c/d} = \frac{a}{b} \cdot \frac{d}{c}, \quad \frac{-a}{b} = -\frac{a}{b} = \frac{a}{-b},$$

$$\frac{(a/b) + (c/d)}{(e/f) + (g/h)} = \frac{(a/b) + (c/d)}{(e/f) + (g/h)} \cdot \frac{bdfh}{bdfh} = \frac{(ad + bc)fh}{(eh + fg)bd}$$

4. The Binomial Theorem
For any positive integer n,

$$(a + b)^n = a^n + na^{n-1}b + \frac{n(n-1)}{1 \cdot 2} a^{n-2}b^2$$

$$+ \frac{n(n-1)(n-2)}{1 \cdot 2 \cdot 3} a^{n-3}b^3 + \cdots + nab^{n-1} + b^n.$$

For instance, $\qquad (a + b)^1 = a + b,$

$\qquad\qquad\qquad (a + b)^2 = a^2 + 2ab + b^2,$

$\qquad\qquad\qquad (a + b)^3 = a^3 + 3a^2b + 3ab^2 + b^3,$

$\qquad\qquad\qquad (a + b)^4 = a^4 + 4a^3b + 6a^2b^2 + 4ab^3 + b^4.$

5. Differences of Like Integer Powers, $n > 1$

$$a^n - b^n = (a - b)(a^{n-1} + a^{n-2}b + a^{n-3}b^2 + \cdots + ab^{n-2} + b^{n-1})$$

For instance, $\qquad a^2 - b^2 = (a - b)(a + b),$

$\qquad\qquad\qquad a^3 - b^3 = (a - b)(a^2 + ab + b^2),$

$\qquad\qquad\qquad a^4 - b^4 = (a - b)(a^3 + a^2b + ab^2 + b^3).$

6. Completing the Square

If $a \neq 0$, we can rewrite the quadratic $ax^2 + bx + c$ in the form $au^2 + C$ by a process called completing the square:

$$ax^2 + bx + c = a\left(x^2 + \frac{b}{a}x\right) + c \qquad \text{Factor } a \text{ from the first two terms.}$$

$$= a\left(x^2 + \frac{b}{a}x + \frac{b^2}{4a^2} - \frac{b^2}{4a^2}\right) + c \qquad \text{Add and subtract the square of half the coefficient of } x.$$

$$= a\left(x^2 + \frac{b}{a}x + \frac{b^2}{4a^2}\right) + a\left(-\frac{b^2}{4a^2}\right) + c \qquad \text{Bring out the } -b^2/(4a^2).$$

$$= a\underbrace{\left(x^2 + \frac{b}{a}x + \frac{b^2}{4a^2}\right)}_{\text{This is } \left(x + \frac{b}{2a}\right)^2.} + \underbrace{c - \frac{b^2}{4a}}_{\substack{\text{Call this} \\ \text{part C.}}}$$

$$= au^2 + C \qquad\qquad u = x + \frac{b}{2a}$$

7. The Quadratic Formula

By completing the square on the first two terms of the equation

$$ax^2 + bx + c = 0$$

and solving the resulting equation for x (details omitted), we obtain

$$x = \frac{-b \pm \sqrt{b^2 - 4ac}}{2a}.$$

This equation is the **quadratic formula.**

The solutions of the equation $2x^2 + 3x - 1 = 0$ are

$$x = \frac{-3 \pm \sqrt{(3)^2 - 4(2)(-1)}}{2(2)} = \frac{-3 \pm \sqrt{9 + 8}}{4}$$

or

$$x = \frac{-3 + \sqrt{17}}{4} \quad \text{and} \quad x = \frac{-3 - \sqrt{17}}{4}.$$

Geometry

(A = area, B = area of base, C = circumference, h = height, S = lateral area or surface area, V = volume)

1. Triangle

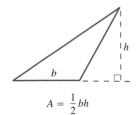

$$A = \frac{1}{2}bh$$

2. Similar Triangles

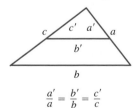

$$\frac{a'}{a} = \frac{b'}{b} = \frac{c'}{c}$$

3. Pythagorean Theorem

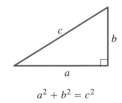

$$a^2 + b^2 = c^2$$

4. Parallelogram

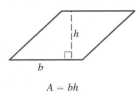

$$A = bh$$

5. Trapezoid

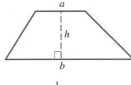

$$A = \frac{1}{2}(a + b)h$$

6. Circle

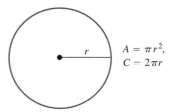

$$A = \pi r^2,$$
$$C = 2\pi r$$

7. Any Cylinder or Prism with Parallel Bases

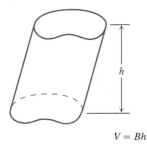 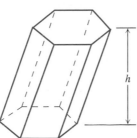

$$V = Bh$$

8. Right Circular Cylinder

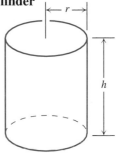

$$V = \pi r^2 h, \quad S = 2\pi rh$$

9. Any Cone or Pyramid

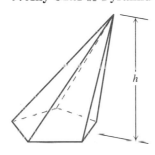 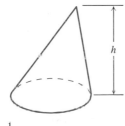

$$V = \frac{1}{3}Bh$$

10. Right Circular Cone

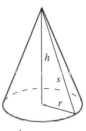

$$V = \frac{1}{3}\pi r^2 h, \quad S = \pi rs$$

11. Sphere

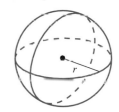

$$V = \frac{4}{3}\pi r^3, \quad S = 4\pi r^2$$

Trigonometry

1. Definitions of Fundamental Identities

$$\text{Sine:} \quad \sin \theta = \frac{y}{r} = \frac{1}{\csc \theta}$$

$$\text{Cosine:} \quad \cos \theta = \frac{x}{r} = \frac{1}{\sec \theta}$$

$$\text{Tangent:} \quad \tan \theta = \frac{y}{x} = \frac{1}{\cot \theta}$$

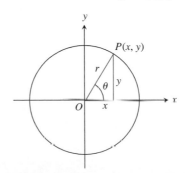

2. Identities

$$\sin(-\theta) = -\sin\theta, \quad \cos(-\theta) = \cos\theta$$

$$\sin^2\theta + \cos^2\theta = 1, \quad \sec^2\theta = 1 + \tan^2\theta, \quad \csc^2\theta = 1 + \cot^2\theta$$

$$\sin 2\theta = 2\sin\theta\cos\theta, \quad \cos 2\theta = \cos^2\theta - \sin^2\theta$$

$$\cos^2\theta = \frac{1 + \cos 2\theta}{2}, \quad \sin^2\theta = \frac{1 - \cos 2\theta}{2}$$

$$\sin(A + B) = \sin A \cos B + \cos A \sin B$$

$$\sin(A - B) = \sin A \cos B - \cos A \sin B$$

$$\cos(A + B) = \cos A \cos B - \sin A \sin B$$

$$\cos(A - B) = \cos A \cos B + \sin A \sin B$$

$$\tan(A + B) = \frac{\tan A + \tan B}{1 - \tan A \tan B}$$

$$\tan(A - B) = \frac{\tan A - \tan B}{1 + \tan A \tan B}$$

$$\sin\left(A - \frac{\pi}{2}\right) = -\cos A, \quad \cos\left(A - \frac{\pi}{2}\right) = \sin A$$

$$\sin\left(A + \frac{\pi}{2}\right) = \cos A, \quad \cos\left(A + \frac{\pi}{2}\right) = -\sin A$$

$$\sin A \sin B = \frac{1}{2}\cos(A - B) - \frac{1}{2}\cos(A + B)$$

$$\cos A \cos B = \frac{1}{2}\cos(A - B) + \frac{1}{2}\cos(A + B)$$

$$\sin A \cos B = \frac{1}{2}\sin(A - B) + \frac{1}{2}\sin(A + B)$$

$$\sin A + \sin B = 2\sin\frac{1}{2}(A + B)\cos\frac{1}{2}(A - B)$$

$$\sin A - \sin B = 2\cos\frac{1}{2}(A + B)\sin\frac{1}{2}(A - B)$$

$$\cos A + \cos B = 2\cos\frac{1}{2}(A + B)\cos\frac{1}{2}(A - B)$$

$$\cos A - \cos B = -2\sin\frac{1}{2}(A + B)\sin\frac{1}{2}(A - B)$$

3. Common Reference Triangles

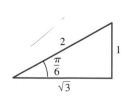
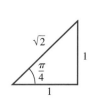
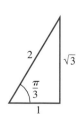

4. Angles and Sides of a Triangle

$$\text{Law of cosines:} \quad c^2 = a^2 + b^2 - 2ab \cos C$$

$$\text{Law of sines:} \quad \frac{\sin A}{a} = \frac{\sin B}{b} = \frac{\sin C}{c}$$

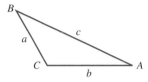

$$\text{Area} = \frac{1}{2}bc \sin A = \frac{1}{2}ac \sin B = \frac{1}{2}ab \sin C$$

A2　Mathematical Induction

Mathematical Induction Principle ● Other Starting Integers

Mathematical Induction Principle

Many formulas, like

$$1 + 2 + \cdots + n = \frac{n(n + 1)}{2},$$

can be shown to hold for every positive integer n by applying an axiom called the *mathematical induction principle*. A proof that uses this axiom is a *proof by mathematical induction* or a *proof by induction*.

The steps in proving a formula by induction are the following.

Step 1: Check that the formula holds for $n = 1$.

Step 2: Prove that if the formula holds for any positive integer $n = k$, then it also holds for the next integer, $n = (k + 1)$.

Once these steps are completed (the axiom says), we know that the formula holds for all positive integers n. By step 1 it holds for $n = 1$. By step 2 it holds for $n = 2$, and therefore by step 2 also for $n = 3$, and by step 2 again for $n = 4$, and so on. If the first domino falls, and the kth domino always knocks over the $(k + 1)$st when it falls, all the dominoes fall.

From another point of view, suppose we have a sequence of statements S_1, S_2, ..., S_n, ..., one for each positive integer. Suppose we can show that assuming any one of the statements to be true implies that the next statement in line is true. Suppose that we can also show that S_1 is true. Then we may conclude that the statements are true from S_1 on.

Example 1   SUM OF THE FIRST n POSITIVE INTEGERS

Show that for every positive integer n,

$$1 + 2 + \cdots + n = \frac{n(n+1)}{2}.$$

Solution We accomplish the proof by carrying out the two steps.

Step 1: The formula holds for $n = 1$ because

$$1 = \frac{1(1+1)}{2}.$$

Step 2: If the formula holds for $n = k,$ does it hold for $n = (k+1)$? The answer is yes, and here's why: If

$$1 + 2 + \cdots + k = \frac{k(k+1)}{2},$$

then

$$1 + 2 + \cdots + k + (k+1) = \frac{k(k+1)}{2} + (k+1)$$

$$= \frac{k^2 + k + 2k + 2}{2}$$

$$= \frac{(k+1)(k+2)}{2} = \frac{(k+1)((k+1)+1)}{2}.$$

The last expression in this string of equalities is the expression $n(n+1)/2$ for $n = (k+1)$.

The mathematical induction priniciple now guarantees the original formula for all positive integers n. All *we* have to do is carry out steps 1 and 2. The mathematical induction principle does the rest.

Example 2   SUMS OF POWERS OF 1/2

Show that for all positive integers n,

$$\frac{1}{2^1} + \frac{1}{2^2} + \cdots + \frac{1}{2^n} = 1 - \frac{1}{2^n}.$$

Solution We accomplish the proof by carrying out the two steps of mathematical induction.

Step 1: The formula holds for $n = 1$ because

$$\frac{1}{2^1} = 1 - \frac{1}{2^1}.$$

Step 2: If

$$\frac{1}{2^1} + \frac{1}{2^2} + \cdots + \frac{1}{2^k} = 1 - \frac{1}{2^k},$$

then

$$\frac{1}{2^1} + \frac{1}{2^2} + \cdots + \frac{1}{2^k} + \frac{1}{2^{k+1}} = 1 - \frac{1}{2^k} + \frac{1}{2^{k+1}}$$

$$= 1 - \frac{1 \cdot 2}{2^k \cdot 2} + \frac{1}{2^{k+1}}$$

$$= 1 - \frac{2}{2^{k+1}} + \frac{1}{2^{k+1}}$$

$$= 1 - \frac{1}{2^{k+1}}.$$

Thus, the original formula holds for $n = (k + 1)$ whenever it holds for $n = k$.

With these steps verified, the mathematical induction principle now guarantees the formula for every positive integer n.

Other Starting Integers

Instead of starting at $n = 1$, some induction arguments start at another integer. The steps for such an argument are as follows.

Step 1: Check that the formula holds for $n = n_1$ (the first appropriate integer).

Step 2: Prove that if the formula holds for any integer $n = k \geq n_1$, then it also holds for $n = (k + 1)$.

Once these steps are completed, the mathematical induction principle guarantees the formula for all $n \geq n_1$.

Example 3 FACTORIAL EXCEEDING EXPONENTIAL

Show that $n! > 3^n$ if n is large enough.

Solution How large is large enough? We experiment:

n	1	2	3	4	5	6	7
$n!$	1	2	6	24	120	720	5040
3^n	3	9	27	81	243	729	2187

It looks as if $n! > 3^n$ for $n \geq 7$. To be sure, we apply mathematical induction. We take $n_1 = 7$ in step 1 and try for step 2.

Suppose $k! > 3^k$ for some $k \geq 7$. Then

$$(k + 1)! = (k + 1)(k!) > (k + 1)3^k > 7 \cdot 3^k > 3^{k+1}.$$

Thus, for $k \geq 7$,

$$k! > 3^k \quad \Rightarrow \quad (k + 1)! > 3^{k+1}.$$

The mathematical induction principle now guarantees $n! > 3^n$ for all $n \geq 7$.

Section A2 Exercises

1. *General Triangle Inequality* Assuming that the triangle inequality $|a + b| \le |a| + |b|$ holds for any two numbers a and b, show that

$$|x_1 + x_2 + \cdots + x_n| \le |x_1| + |x_2| + \cdots + |x_n|$$

   for any n numbers.

2. *Partial Sums of Geometric Series* Show that if $r \ne 1$, then

$$1 + r + r^2 + \cdots + r^n = \frac{1 - r^{n+1}}{1 - r}$$

   for every positive integer n.

3. *Positive Integer Power Rule* Use the Product Rule,

$$\frac{d}{dx}(uv) = u\frac{dv}{dx} + v\frac{du}{dx},$$

   and the fact that

$$\frac{d}{dx}(x) = 1$$

   to show that

$$\frac{d}{dx}(x^n) = nx^{n-1}$$

   for every positive integer n.

4. *Products into Sums* Suppose that a function $f(x)$ has the property that $f(x_1 x_2) = f(x_1) + f(x_2)$ for any two positive numbers x_1 and x_2. Show that

$$f(x_1 x_2 \ldots x_n) = f(x_1) + f(x_2) + \cdots + f(x_n)$$

   for the product of any n positive numbers $x_1, x_2, \ldots, x_n$.

5. Show that

$$\frac{2}{3^1} + \frac{2}{3^2} + \cdots + \frac{2}{3^n} = 1 - \frac{1}{3^n}$$

   for all positive integers n.

6. Show that $n! > n^3$ if n is large enough.

7. Show that $2^n > n^2$ if n is large enough.

8. Show that $2^n \ge 1/8$ for $n \ge -3$.

9. *Sums of Squares* Show that the sum of the squares of the first n positive integers is

$$\frac{n\left(n + \dfrac{1}{2}\right)(n + 1)}{3}.$$

10. *Sums of Cubes* Show that the sum of the cubes of the first n positive integers is $(n(n + 1)/2)^2$.

11. *Rules for Finite Sums* Show that the following finite sum rules hold for every positive integer n.

   (a) $\displaystyle\sum_{k=1}^{n}(a_k + b_k) = \sum_{k=1}^{n} a_k + \sum_{k=1}^{n} b_k$

   (b) $\displaystyle\sum_{k=1}^{n}(a_k - b_k) = \sum_{k=1}^{n} a_k - \sum_{k=1}^{n} b_k$

   (c) $\displaystyle\sum_{k=1}^{n} ca_k = c \cdot \sum_{k=1}^{n} a_k$ (Any number c)

   (d) $\displaystyle\sum_{k=1}^{n} a_k = n \cdot c$

12. *Absolute Values* Show that $|x^n| = |x|^n$ for every positive integer n and every real number x.

A3   **Using the Limit Definition**

Limit Definition • Finding Deltas for Given Epsilons • Proving Limit Theorems

Limit Definition

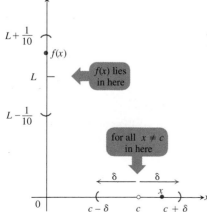

Figure A3.1 A preliminary stage in the development of the definition of limit.

We begin by setting the stage for the definition of limit. Recall that the limit of f of x as x approaches c equals L ($\lim_{x \to c} f(x) = L$) means that the values $f(x)$ of the function f approach or equal L as the values of x approach (but do not equal) c. Suppose we are watching the values of a function $f(x)$ as x approaches c (without taking on the value of c itself). Certainly we want to be able to say that $f(x)$ stays within one-tenth of a unit of L as soon as x stays within some distance δ of c (Figure A3.1). But that in itself is not enough, because as x continues on its course toward c, what is to prevent $f(x)$ from jittering about within the interval from $L - 1/10$ to $L + 1/10$ without tending toward L?

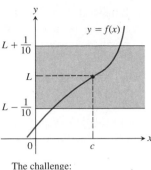

The challenge:
$$\text{Make } |f(x) - L| < \varepsilon = \frac{1}{10}$$

Figure A3.2 The first of a possibly endless sequence of challenges.

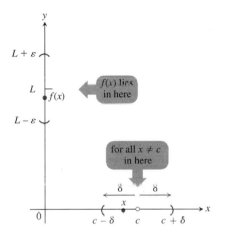

Figure A3.3 The relation of the δ and ε in the definition of limit.

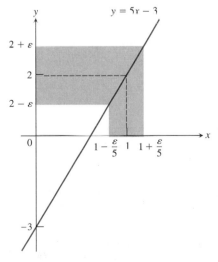

Figure A3.4 If $f(x) = 5x - 3$, then $0 < |x - 1| < \varepsilon/5$ guarantees that $|f(x) - 2| < \varepsilon$. (Example 1)

We can insist that $f(x)$ stay within 1/100 or 1/1000 or 1/100,000 of L. Each time, we find a new δ-interval about c so that keeping x within that interval keeps $f(x)$ within $\varepsilon = 1/100$ or 1/1000 or 1/100,000 of L. And each time the possibility exists that c jitters away from L at the last minute.

Figure A3.2 illustrates the problem. You can think of this as a quarrel between a skeptic and a scholar. The skeptic presents ε-challenges to prove that the limit does not exist or, more precisely, that there is room for doubt, and the scholar answers every challenge with a δ-interval around c.

How do we stop this seemingly endless sequence of challenges and responses? By proving that for every ε-distance that the challenger can produce, we can find, calculate, or conjure a matching δ-distance that keeps x "close enough" to c to keep $f(x)$ within that distance of L (Figure A3.3).

The following definition that we made in Section 2.1 provides a mathematical way to say that the closer x gets to c, the closer $f(x)$ must get to L.

Definition Limit

Let c and L be real numbers. The function f **has limit L as x approaches c** if, given any positive number ε, there is a positive number δ such that for all x
$$0 < |x - c| < \delta \quad \Rightarrow \quad |f(x) - L| < \varepsilon.$$
We write
$$\lim_{x \to c} f(x) = L.$$

Finding Deltas for Given Epsilons

From our work in Chapter 2 we know that $\lim_{x \to 1}(5x - 3) = 2$. In Example 1, we confirm this result using the definition of limit.

Example 1 USING THE DEFINITION OF LIMIT

Show that $\lim_{x \to 1}(5x - 3) = 2$.

Solution Set $\varepsilon = 1$, $f(x) = 5x - 3$, and $L = 2$ in the definition of limit. For any given $\varepsilon > 0$ we have to find a suitable $\delta > 0$ so that if $x \neq 1$ and x is within distance δ of $c = 1$, that is, if
$$0 < |x - 1| < \delta,$$
then $f(x)$ is within distance ε of $L = 2$, that is,
$$|f(x) - 2| < \varepsilon.$$

We find δ by working backwards from the ε-inequality:
$$|(5x - 3) - 2| = |5x - 5| < \varepsilon$$
$$5|x - 1| < \varepsilon$$
$$|x - 1| < \varepsilon/5$$

Thus we can take $\delta = \varepsilon/5$ (Figure A3.4). If $0 < |x - 1| < \delta = \varepsilon/5$, then
$$|(5x - 3) - 2| = |5x - 5|$$
$$= 5|x - 1| < 5(\varepsilon/5) = \varepsilon.$$

This proves that $\lim_{x \to 1}(5x - 3) = 2$.

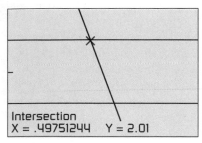

[0.48, 0.52] by [1.98, 2.02]

(a)

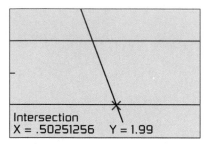

[0.48, 0.52] by [1.98, 2.02]

(b)

Figure A3.5 We can see from the two graphs that if $0.498 < x < 0.502$, then $1.99 < f(x) < 2.01$. (Example 2)

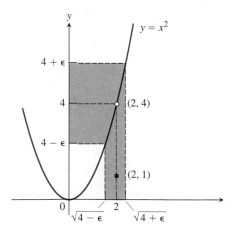

Figure A3.6 The function in Example 3.

The value of $\delta = \varepsilon/5$ is not the only value that will make $0 < |x - 1| < \delta$ imply $|f(x) - 2| = |5x - 5| < \varepsilon$ in Example 1. Any smaller positive δ will do as well. The definition does not ask for a "best" positive δ, just one that will work.

We can use graphs to find a δ for a specific ε as in Example 2.

Example 2 FINDING A δ GRAPHICALLY

For the limit $\lim_{x \to 0.5}(1/x) = 2$, find a δ that works for $\varepsilon = 0.01$. That is, find a $\delta > 0$ such that for all x

$$0 < |x - 0.5| < \delta \quad \Rightarrow \quad |f(x) - 2| < 0.01.$$

Solution Here $f(x) = 1/x$, $c = 0.5$, and $L = 2$. Figure A3.5 shows the graphs of f and the two horizontal lines

$$y = L - \varepsilon = 2 - 0.01 = 1.99 \quad \text{and} \quad y = L + \varepsilon = 2 + 0.01 = 2.01.$$

Figure A3.5a shows that the graph of f intersects the horizontal line $y = 2.01$ at about $(0.49751244, 2.01)$, and Figure A3.5b shows that the graph of f intersects the horizontal line $y = 1.99$ at about $(0.50251256, 1.99)$. It follows that

$$0 < |x - 0.5| < 0.002 \quad \Rightarrow \quad |f(x) - 2| < 0.01.$$

Thus, $\delta = 0.002$ works.

Example 3 FINDING δ ALGEBRAICALLY

Prove that $\lim_{x \to 2} f(x) = 4$ if

$$f(x) = \begin{cases} x^2, & x \neq 2 \\ 1, & x = 2. \end{cases}$$

Solution Our task is to show that given $\varepsilon > 0$ there exists a $\delta > 0$ such that for all x

$$0 < |x - 2| < \delta \quad \Rightarrow \quad |f(x) - 4| < \varepsilon.$$

Step 1: Solve the inequality $|f(x) - 4| < \varepsilon$ to find an open interval about $c = 2$ on which the inequality holds for all $x \neq c$.

For $x \neq c = 2$, we have $f(x) = x^2$, and the inequality to solve is $|x^2 - 4| < \varepsilon$:

$$|x^2 - 4| < \varepsilon$$
$$-\varepsilon < x^2 - 4 < \varepsilon$$
$$4 - \varepsilon < x^2 < 4 + \varepsilon$$
$$\sqrt{4 - \varepsilon} < |x| < \sqrt{4 + \varepsilon} \qquad \text{Assume } \varepsilon < 4.$$
$$\sqrt{4 - \varepsilon} < x < \sqrt{4 + \varepsilon} \qquad \begin{array}{l}\text{An open interval about 2} \\ \text{that solves the inequality}\end{array}$$

The inequality $|f(x) - 4| < \varepsilon$ holds for all $x \neq 2$ in the open interval $(\sqrt{4 - \varepsilon}, \sqrt{4 + \varepsilon})$ (Figure A3.6).

Step 2: Find a value of $\delta > 0$ that places the *centered* interval $(2 - \delta, 2 + \delta)$ inside the open interval $(\sqrt{4 - \varepsilon}, \sqrt{4 + \varepsilon})$.

Take δ to be the distance from $c = 2$ to the nearer endpoint of $(\sqrt{4 - \varepsilon}, \sqrt{4 + \varepsilon})$. In other words, take

$$\delta = \min \{2 - \sqrt{4 - \varepsilon}, \sqrt{4 + \varepsilon} - 2\},$$

the *minimum* (the smaller) of the two numbers $2 - \sqrt{4 - \varepsilon}$ and $\sqrt{4 + \varepsilon} - 2$. If δ has this or any smaller positive value, the inequality

$$0 < |x - 2| < \delta$$

will automatically place x between $\sqrt{4 - \varepsilon}$ and $\sqrt{4 + \varepsilon}$ to make

$$|f(x) - 4| < \varepsilon.$$

For all x,

$$0 < |x - 2| < \delta \quad \Rightarrow \quad |f(x) - 4| < \varepsilon.$$

This completes the proof.

Why was it all right to assume $\varepsilon < 4$ in Example 3? Because, in finding a δ such that, for all x, $0 < |x - 2| < \delta$ implied $|f(x) - 4| < \varepsilon < 4$, we found a δ that would work for any larger ε as well.

Finally, notice the freedom we gained in letting

$$\delta = \min \{2 - \sqrt{4 - \varepsilon}, \sqrt{4 + \varepsilon} - 2\}.$$

We did not have to spend time deciding which, if either, number was the smaller of the two. We just let δ represent the smaller and went on to finish the argument.

Proving Limit Theorems

We use the limit definition to prove parts 1, 3, and 5 of Theorem 1 (Properties of Limits) from Section 2.1.

Theorem 1 Properties of Limits

If L, M, c, and k are real numbers and

$$\lim_{x \to c} f(x) = L \quad \text{and} \quad \lim_{x \to c} g(x) = M, \quad \text{then}$$

1. *Sum Rule:* $\qquad\qquad\qquad\; \lim_{x \to c} (f(x) + g(x)) = L + M$

2. *Difference Rule:* $\qquad\quad\; \lim_{x \to c} (f(x) - g(x)) = L - M$

3. *Product Rule:* $\qquad\qquad\; \lim_{x \to c} (f(x) \cdot g(x)) = L \cdot M$

4. *Constant Multiple Rule:* $\quad \lim_{x \to c} k \cdot f(x) = k \cdot L$

5. *Quotient Rule:* $\qquad\qquad\; \lim_{x \to c} \dfrac{f(x)}{g(x)} = \dfrac{L}{M}, \quad M \neq 0$

6. *Power Rule:* $\qquad\qquad\;$ If r and s are integers, $s \neq 0$, then

$$\lim_{x \to c} (f(x))^{r/s} = L^{r/s}$$

provided $L^{r/s}$ is a real number.

Proof of the Limit Sum Rule We need to show that for any $\varepsilon > 0$, there is a $\delta > 0$ such that for all x in the common domain D of f and g,

$$0 < |x - c| < \delta \quad \Rightarrow \quad |f(x) + g(x) - (L + M)| < \varepsilon.$$

Regrouping terms, we get

$$
\begin{aligned}
|f(x) + g(x) - (L + M)| &= |(f(x) - L) + (g(x) - M)| \\
&\leq |f(x) - L| + |g(x) - M|. \qquad |a + b| \leq |a| + |b|
\end{aligned}
$$

Here we have applied the triangle inequality, which states that for all real numbers a and b, $|a + b| \leq |a| + |b|$. Since $\lim_{x \to c} f(x) = L$, there exists a number $\delta_1 > 0$ such that for all x in D

$$0 < |x - c| < \delta_1 \quad \Rightarrow \quad |f(x) - L| < \varepsilon/2.$$

Similarly, since $\lim_{x \to c} g(x) = M$, there exists a number $\delta_2 > 0$ such that for all x in D

$$0 < |x - c| < \delta_2 \quad \Rightarrow \quad |g(x) - M| < \varepsilon/2.$$

Let $\delta = \min \{\delta_1, \delta_2\}$, the smaller of δ_1 and δ_2. If $0 < |x - c| < \delta$ then

$$0 < |x - c| < \delta_1, \quad \text{so} \quad |f(x) - L| < \varepsilon/2,$$

and

$$0 < |x - c| < \delta_2, \quad \text{so} \quad |g(x) - M| < \varepsilon/2.$$

Therefore, $|f(x) + g(x) - (L + M)| < \dfrac{\varepsilon}{2} + \dfrac{\varepsilon}{2} = \varepsilon.$

This shows that $\lim_{x \to c} (f(x) + g(x)) = L + M$. ∎

Proof of the Limit Product Rule We show that for any $\varepsilon > 0$, there is a $\delta > 0$ such that for all x in the common domain D of f and g,

$$0 < |x - c| < \delta \quad \Rightarrow \quad |f(x)g(x) - LM| < \varepsilon.$$

Write $f(x)$ and $g(x)$ as $f(x) = L + (f(x) - L)$, $g(x) = M + (g(x) - M)$.

Multiply these expressions together and subtract LM:

$$
\begin{aligned}
f(x) \cdot g(x) - LM &= (L + (f(x) - L))(M + (g(x) - M)) - LM \\
&= LM + L(g(x) - M) + M(f(x) - L) + (f(x) - L)(g(x) - M) - LM \quad (1) \\
&= L(g(x) - M) + M(f(x) - L) + (f(x) - L)(g(x) - M)
\end{aligned}
$$

Since f and g have limits L and M as $x \to c$, there exist positive numbers $\delta_1, \delta_2, \delta_3$, and δ_4 such that for all x in D

$$
\begin{aligned}
0 < |x - c| < \delta_1 &\quad \Rightarrow \quad |f(x) - L| < \sqrt{\varepsilon/3} \\
0 < |x - c| < \delta_2 &\quad \Rightarrow \quad |g(x) - M| < \sqrt{\varepsilon/3} \\
0 < |x - c| < \delta_3 &\quad \Rightarrow \quad |f(x) - L| < \frac{\varepsilon}{3(1 + |M|)} \qquad (2) \\
0 < |x - c| < \delta_4 &\quad \Rightarrow \quad |g(x) - M| < \frac{\varepsilon}{3(1 + |L|)}.
\end{aligned}
$$

If we take δ to be the smallest of the numbers δ_1 through δ_4, the inequalities on the right-hand side of (2) will hold simultaneously for $0 < |x - c| < \delta$. Then, applying the triangle inequality to Equation 1, we have for all x in D, $0 < |x - c| < \delta$ implies

$$|f(x) \cdot g(x) - LM|$$

$$\leq |L||g(x) - M| + |M||f(x) - L| + |f(x) - L||g(x) - M|$$

$$\leq (1 + |L|)|g(x) - M| + (1 + |M|)|f(x) - L| + |f(x) - L||g(x) - M|$$

$$\leq \frac{\varepsilon}{3} + \frac{\varepsilon}{3} + \sqrt{\frac{\varepsilon}{3}}\sqrt{\frac{\varepsilon}{3}} = \varepsilon. \quad \text{Values from (2)}$$

This completes the proof of the Limit Product Rule.   ∎

Proof of the Limit Quotient Rule   We show that $\lim_{x \to c} (1/g(x)) = 1/M$. We can then conclude that

$$\lim_{x \to c} \frac{f(x)}{g(x)} = \lim_{x \to c} \left(f(x) \cdot \frac{1}{g(x)} \right) = \lim_{x \to c} f(x) \cdot \lim_{x \to c} \frac{1}{g(x)} = L \cdot \frac{1}{M} = \frac{L}{M}$$

by the Limit Product Rule.

Let $\varepsilon > 0$ be given. To show that $\lim_{x \to c} (1/g(x)) = 1/M$, we need to show that there exists a $\delta > 0$ such that for all x

$$0 < |x - c| < \delta \quad \Rightarrow \quad \left| \frac{1}{g(x)} - \frac{1}{M} \right| < \varepsilon.$$

Since $|M| > 0$, there exists a positive number δ_1 such that for all x

$$0 < |x - c| < \delta_1 \quad \Rightarrow \quad |g(x) - M| < \frac{|M|}{2}. \tag{3}$$

For any numbers A and B it can be shown that

$$|A| - |B| \leq |A - B| \quad \text{and} \quad |B| - |A| \leq |A - B|,$$

from which it follows that $||A| - |B|| \leq |A - B|$. With $A = g(x)$ and $B = M$, this becomes

$$||g(x)| - |M|| \leq |g(x) - M|,$$

which can be combined with the inequality on the right in (3) to get, in turn,

$$||g(x)| - |M|| < \frac{|M|}{2}$$

$$-\frac{|M|}{2} < |g(x)| - |M| < \frac{|M|}{2}$$

$$\frac{|M|}{2} < |g(x)| < \frac{3|M|}{2}$$

$$\frac{1}{|g(x)|} < \frac{2}{|M|} < \frac{3}{|g(x)|}. \quad \text{Multiply by } 2/(|M||g(x)|). \tag{4}$$

Therefore, $0 < |x - c| < \delta$ implies that

$$\left| \frac{1}{g(x)} - \frac{1}{M} \right| = \left| \frac{M - g(x)}{Mg(x)} \right| \le \frac{1}{|M|} \cdot \frac{1}{|g(x)|} \cdot |M - g(x)|$$

$$< \frac{1}{|M|} \cdot \frac{2}{|M|} \cdot |M - g(x)|. \quad \text{Inequality (4)} \quad (5)$$

Since $(1/2)|M|^2 \varepsilon > 0$, there exists a number $\delta_2 > 0$ such that for all x in D

$$0 < |x - c| < \delta_2 \quad \Rightarrow \quad |M - g(x)| < \frac{\varepsilon}{2} |M|^2. \tag{6}$$

If we take δ to be the smaller of δ_1 and δ_2, the conclusions in (5) and (6) both hold for all x such that $0 < |x - c| < \delta$. Combining these conclusions gives

$$0 < |x - c| < \delta_2 \quad \Rightarrow \quad \left| \frac{1}{g(x)} - \frac{1}{M} \right| < \varepsilon.$$

This completes the proof of the Limit Quotient Rule. ■

The last proof we give is of the Sandwich Theorem (Theorem 4) of Section 2.1.

Theorem 4 The Sandwich Theorem

If $g(x) \le f(x) \le h(x)$ for all $x \ne c$ in some interval about c, and

$$\lim_{x \to c} g(x) = \lim_{x \to c} h(x) = L,$$

then

$$\lim_{x \to c} f(x) = L.$$

Proof for Right-hand Limits Suppose that $\lim_{x \to c^+} g(x) = \lim_{x \to c^+} h(x) = L$. Then for any $\varepsilon > 0$ there exists a $\delta > 0$ such that for all x the inequality $c < x < c + \delta$ implies

$$L - \varepsilon < g(x) < L + \varepsilon \quad \text{and} \quad L - \varepsilon < h(x) < L + \varepsilon.$$

These inequalities combine with the inequality $g(x) \le f(x) \le h(x)$ to give

$$L - \varepsilon < g(x) \le f(x) \le h(x) < L + \varepsilon,$$

$$L - \varepsilon < f(x) < L + \varepsilon,$$

$$-\varepsilon < f(x) - L < \varepsilon.$$

Thus, for all x, the inequality $c < x < c + \delta$ implies $|f(x) - L| < \varepsilon$. Therefore, $\lim_{x \to c^+} f(x) = L$.

Proof for Left-hand Limits Suppose that $\lim_{x \to c^-} g(x) = \lim_{x \to c^-} h(x) = L$. Then for any $\varepsilon > 0$ there exists a $\delta > 0$ such that for all x the inequality $c - \delta < x < c$ implies

$$L - \varepsilon < g(x) < L + \varepsilon \quad \text{and} \quad L - \varepsilon < h(x) < L + \varepsilon.$$

We conclude as before that for all x, $c - \delta < x < c$ implies $|f(x) - L| < \varepsilon$. Therefore, $\lim_{x \to c^-} f(x) = L$.

Proof for Two-sided Limits If $\lim_{x \to c} g(x) = \lim_{x \to c} h(x) = L$, then $g(x)$ and $h(x)$ both approach L as $x \to c^+$ and $x \to c^-$; so $\lim_{x \to c^+} f(x) = L$ and $\lim_{x \to c^-} f(x) = L$. Hence $\lim_{x \to c} f(x)$ exists and equals L. ■

Section A3 Exercises

In Exercises 1 and 2, sketch the interval (a, b) on the x-axis with the point c inside. Then find a value of $\delta > 0$ such that for all x,

$$0 < |x - c| < \delta \;\Rightarrow\; a < x < b.$$

1. $a = 4/9, \quad b = 4/7, \quad c = 1/2$

2. $a = 2.7591, \quad b = 3.2391, \quad c = 3$

In Exercises 3 and 4, use the graph to find a $\delta > 0$ such that for all $x \;\; 0 < |x - c| < \delta \;\Rightarrow\; |f(x) - L| < \varepsilon$.

3.

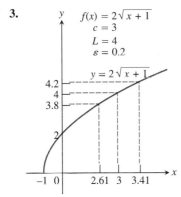

$f(x) = 2\sqrt{x + 1}$
$c = 3$
$L = 4$
$\varepsilon = 0.2$

$y = 2\sqrt{x + 1}$

NOT TO SCALE

4.

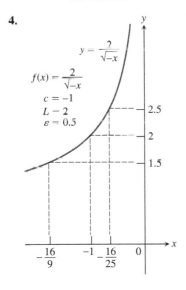

$y = \dfrac{2}{\sqrt{-x}}$

$f(x) = \dfrac{2}{\sqrt{-x}}$
$c = -1$
$L = 2$
$\varepsilon = 0.5$

Exercises 5–8 give a function $f(x)$ and numbers L, c, and ε. Find an open interval about c on which the inequality $|f(x) - L| < \varepsilon$ holds. Then give a value for $\delta > 0$ such that for all x satisfying $0 < |x - c| < \delta$ the inequality $|f(x) - L| < \varepsilon$ holds. Use algebra to find your answers.

5. $f(x) = 2x - 2, \quad L = -6, \quad c = -2, \quad \varepsilon = 0.02$

6. $f(x) = \sqrt{x + 1}, \quad L = 1, \quad c = 0, \quad \varepsilon = 0.1$

7. $f(x) = \sqrt{19 - x}, \quad L = 3, \quad c = 10, \quad \varepsilon = 1$

8. $f(x) = x^2, \quad L = 4, \quad c = -2, \quad \varepsilon = 0.5$

Exercises 9–12 give a function $f(x)$, a point c, and a positive number ε. **(a)** Find $L = \lim_{x \to c} f(x)$. Then **(b)** find a number $\delta > 0$ such that for all x

$$0 < |x - c| < \delta \;\Rightarrow\; |f(x) - L| < \varepsilon.$$

9. $f(x) = \dfrac{x^2 + 6x + 5}{x + 5}, \quad c = -5, \quad \varepsilon = 0.05$

10. $f(x) = \begin{cases} 4 - 2x, & x < 1, \\ 6x - 4, & x \ge 1, \end{cases} \quad c = 1, \quad \varepsilon = 0.5$

11. $f(x) = \sin x, \quad c = 1, \quad \varepsilon = 0.01$

12. $f(x) = \dfrac{x}{x^2 - 4}, \quad c = -1, \quad \varepsilon = 0.1$

In Exercises 13 and 14, use the definition of limit to prove the limit statement.

13. $\lim_{x \to 1} f(x) = 1 \quad$ if $\quad f(x) = \begin{cases} x^2, & x \ne 1 \\ 2, & x = 1 \end{cases}$

14. $\lim_{x \to \sqrt{3}} \dfrac{1}{x^2} = \dfrac{1}{3}$

15. *Relating to Limits* Given $\varepsilon > 0$, **(a)** find an interval $I = (5, 5 + \delta), \; \delta > 0$, such that if x lies in I, then $\sqrt{x - 5} < \varepsilon$. **(b)** What limit is being verified?

16. *Relating to Limits* Given $\varepsilon > 0$, **(a)** find an interval $I = (4 - \delta, 4), \; \delta > 0$, such that if x lies in I, then $\sqrt{4 - x} < \varepsilon$. **(b)** What limit is being verified?

17. Prove the Constant Multiple Rule for limits.

18. Prove the Difference Rule for limits.

19. *Generalized Limit Sum Rule* Suppose that functions $f_1(x)$, $f_2(x)$, and $f_3(x)$ have limits L_1, L_2, and L_3, respectively, as $x \to c$. Show that their sum has limit $L_1 + L_2 + L_3$. Use mathematical induction (Appendix 2) to generalize this result to the sum of any finite number of functions.

20. *Generalized Limit Product Rule* Use mathematical induction and the Limit Product Rule in Theorem 1 to show that if functions $f_1(x)$, $f_2(x)$, ..., $f_n(x)$ have limits L_1, L_2, ..., L_n, respectively, as $x \to c$, then

$$\lim_{x \to c} (f_1(x) \cdot f_2(x) \cdot \cdots \cdot f_n(x)) = L_1 \cdot L_2 \cdot \cdots \cdot L_n.$$

21. *Positive Integer Power Rule* Use the fact that $\lim_{x \to c} x = c$ and the result of Exercise 20 to show that $\lim_{x \to c} x^n = c^n$ for any integer $n > 1$.

22. *Limits of Polynomials* Use the fact that $\lim_{x \to c} k = k$ for any number k together with the results of Exercises 19 and 21 to show that $\lim_{x \to c} f(x) = f(c)$ for any polynomial function

$$f(x) = a_n x^n + a_{n-1} x^{n-1} + \cdots + a_1 x + a_0.$$

23. *Limits of Rational Functions* Use Theorem 1 and the result of Exercise 22 to show that if $f(x)$ and $g(x)$ are polynomial functions and $g(c) \neq 0$, then

$$\lim_{x \to c} \frac{f(x)}{g(x)} = \frac{f(c)}{g(c)}.$$

24. *Composites of Continuous Functions* Figure A3.7 gives the diagram for a proof that the composite of two continuous functions is continuous. Reconstruct the proof from the diagram. The statement to be proved is this: If f is continuous at $x = c$ and g is continuous at $f(c)$, then $g \circ f$ is continuous at c.

Assume that c is an interior point of the domain of f and that $f(c)$ is an interior point of the domain of g. This will make the limits involved two-sided. (The argument for the cases that involve one-sided limits are similar.)

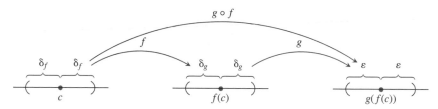

Figure A3.7 The continuity of composites holds for any finite number of functions. The only requirement is that each function be continuous where it is applied. Here, f is to be continuous at c and g at $f(c)$.

A4 Proof of the Chain Rule

Error in the Approximation $\Delta f \approx df$ • The Proof

Error in the Approximation $\Delta f \approx df$

Let $f(x)$ be differentiable at $x = a$ and suppose that Δx is an increment of x. We know that the differential $df = f'(a)\Delta x$ is an approximation for the change $\Delta f = (f(a + \Delta x) - f(a))$ in f as x changes from a to $(a + \Delta x)$. How well does df approximate Δf?

We measure the approximation error by subtracting df from Δf:

$$\text{Approximation error} = \Delta f - df = \Delta f - f'(a)\Delta x$$

$$= \underbrace{f(a + \Delta x) - f(a)}_{\Delta f} - f'(a)\Delta x$$

$$= \underbrace{\left(\frac{f(a + \Delta x) - f(a)}{\Delta x} - f'(a) \right)}_{\text{Call this part } \varepsilon.} \cdot \Delta x$$

$$= \varepsilon \cdot \Delta x$$

As $\Delta x \to 0$, the difference quotient $(f(a + \Delta x) - f(a))/\Delta x$ approaches $f'(a)$ (remember the definition of $f'(a)$), so the quantity in parentheses becomes a very small number (which is why we called it ε). In fact, $\varepsilon \to 0$ as $\Delta x \to 0$. When Δx is small, the **approximation error** $\varepsilon \Delta x$ is smaller still.

$$\underbrace{\Delta f}_{\substack{\text{true} \\ \text{change}}} = \underbrace{f'(a)\,\Delta x}_{\substack{\text{estimated} \\ \text{change}}} + \underbrace{\varepsilon\,\Delta x}_{\text{error}} \tag{1}$$

The Proof

Our goal is to show that if $f(u)$ is a differentiable function of u and $u = g(x)$ is a differentiable function of x, then the composite $y = f(g(x))$ is a differentiable function of x. More precisely, if g is differentiable at a and f is differentiable at $g(a)$, then the composite is differentiable at a and

$$\left.\frac{dy}{dx}\right|_{x=a} = f'(g(a)) \cdot g'(a).$$

Let Δx be an increment in x and let Δu and Δy be the corresponding increments in u and y. As you can see in Figure A4.1,

$$\left.\frac{dy}{dx}\right|_{x=a} = \lim_{\Delta x \to 0} \frac{\Delta y}{\Delta x},$$

so our goal is to show that the limit is $f'(g(a)) \cdot g'(a)$.

By Equation 1,

$$\Delta u = g'(a)\Delta x + \varepsilon_1 \Delta x = (g'(a) + \varepsilon_1)\Delta x,$$

where $\varepsilon_1 \to 0$ as $\Delta x \to 0$. Similarly, since f is differentiable at $g(a)$,

$$\Delta y = f'(g(a))\Delta u + \varepsilon_2 \Delta u = (f'(g(a)) + \varepsilon_2)\Delta u,$$

where $\varepsilon_2 \to 0$ as $\Delta u \to 0$. Notice also that $\Delta u \to 0$ as $\Delta x \to 0$. Combining the equations for Δu and Δy gives $\Delta y = (f'(g(a)) + \varepsilon_2)(g'(a) + \varepsilon_1)\Delta x$, so

$$\frac{\Delta y}{\Delta x} = f'(g(a))g'(a) + \varepsilon_2 g'(a) + f'(g(a))\varepsilon_1 + \varepsilon_2 \varepsilon_1.$$

Since ε_1 and ε_2 go to zero as Δx goes to zero, three of the four terms on the right vanish in the limit, leaving

$$\lim_{\Delta x \to 0} \frac{\Delta y}{\Delta x} = f'(g(a))g'(a).$$

This concludes the proof. ∎

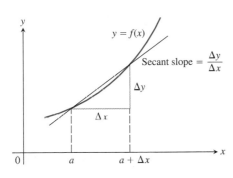

Figure A4.1 The graph of y as a function of x. The derivative of y with respect to x at $x = a$ is $\lim_{\Delta x \to 0} (\Delta y / \Delta x)$.

(Figure labels: y; $y = f(x)$; Secant slope $= \dfrac{\Delta y}{\Delta x}$; Δy; Δx; 0; a; $a + \Delta x$; x)

A5 Conic Sections

Overview

Conic sections are the paths traveled by planets, satellites, and other bodies (even electrons) whose motions are driven by inverse-square forces. Once we know that the path of a moving body is a conic section, we immediately have information about the body's velocity and the force that drives it. In this appendix, we study the connections between conic sections and quadratic equations and classify conic sections by eccentricity (Pluto's orbit is highly eccentric while Earth's is nearly circular).

A5.1 Conic Sections and Quadratic Equations

Circles • Interiors and Exteriors of Circles • Parabolas • Ellipses • Axes of an Ellipse • Hyperbolas • Asymptotes and Drawing • Reflective Properties • Other Applications

Circles

The Greeks of Plato's time defined conic sections as the curves formed by cutting through a double cone with a plane (Figure A5.1). Today, we define conic sections with the distance function in the coordinate plane.

Definition Circle

A **circle** is the set of points in a plane whose distance from a given fixed point in the plane is constant. The fixed point is the **center** of the circle; the constant distance is the **radius.**

If $a > 0$, the equation $x^2 + y^2 = a^2$ represents all the points (x, y) in the plane whose distance from the origin is

$$\sqrt{(x - 0)^2 + (y - 0)^2} = \sqrt{x^2 + y^2} = \sqrt{a^2} = a.$$

These are the points of the circle of radius a centered at the origin. If we shift the circle to place its center at the point (h, k), its equation becomes $(x - h)^2 + (y - k)^2 = a^2$.

Circle of Radius a Centered at (h, k)

$$(x - h)^2 + (y - k)^2 = a^2$$

Example 1   FINDING CENTER AND RADIUS

Find the center and radius of the circle

$$x^2 + y^2 + 4x - 6y - 3 = 0.$$

Solution We convert the equation to standard form by completing the squares in x and y:

$$x^2 + y^2 + 4x - 6y - 3 = 0$$

Start with the given equation.

$$(x^2 + 4x) + (y^2 - 6y) = 3$$

Gather terms. Move the constant to the right-hand side.

$$\left(x^2 + 4x + \left(\frac{4}{2}\right)^2\right) + \left(y^2 - 6y + \left(\frac{-6}{2}\right)^2\right) =$$
$$3 + \left(\frac{4}{2}\right)^2 + \left(\frac{-6}{2}\right)^2$$

Add the square of half the coefficient of x to each side of the equation. Do the same for y. The parenthetical expressions on the left-hand side are now perfect squares.

$$(x^2 + 4x + 4) + (y^2 - 6y + 9) = 3 + 4 + 9$$

$$(x + 2)^2 + (y - 3)^2 = 16$$

Write each quadratic as a squared linear expression.

With the equation now in standard form, we read off the center's coordinates and the radius: $(h, k) = (-2, 3)$ and $a = 4$.

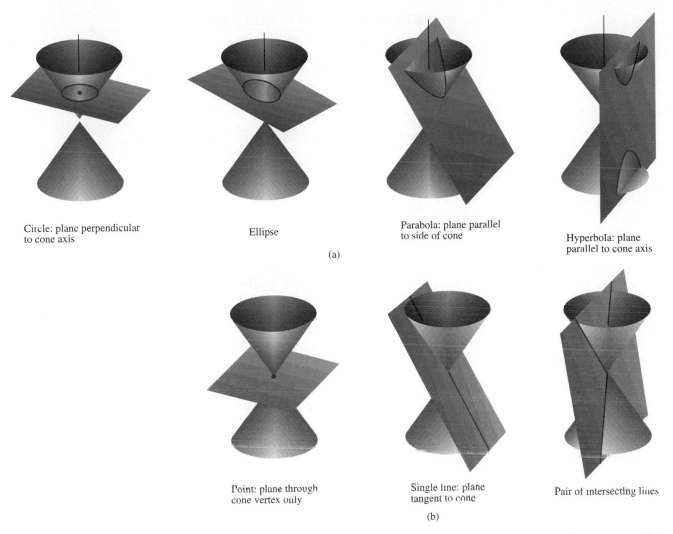

Circle: plane perpendicular to cone axis

Ellipse

Parabola: plane parallel to side of cone

Hyperbola: plane parallel to cone axis

(a)

Point: plane through cone vertex only

Single line: plane tangent to cone

Pair of intersecting lines

(b)

Figure A5.1 The standard conic sections (a) are the curves in which a plane cuts a double cone. Hyperbolas come in two parts, called *branches*. The point and lines obtained by passing the plane through the cone's vertex (b) are *degenerate* conic sections.

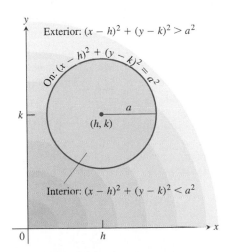

Figure A5.2 The interior and exterior of the circle $(x - h)^2 + (y - k)^2 = a^2$.

Interiors and Exteriors of Circles

The points that lie inside the circle $(x - h)^2 + (y - k)^2 = a^2$ are the points less than a units from (h, k). They satisfy the inequality

$$(x - h)^2 + (y - k)^2 < a^2.$$

They make up the region we call the **interior** of the circle (Figure A5.2).

The circle's **exterior** consists of the points that lie more than a units from (h, k). These points satisfy the inequality

$$(x - h)^2 + (y - k)^2 > a^2.$$

Example 2 INTERPRETING INEQUALITIES

Inequality	Region
$x^2 + y^2 < 1$	Interior of the unit circle
$x^2 + y^2 \leq 1$	Unit circle plus its interior
$x^2 + y^2 > 1$	Exterior of the unit circle
$x^2 + y^2 \geq 1$	Unit circle plus its exterior

Parabolas

> **Definition** **Parabola**
>
> A set that consists of all the points in a plane equidistant from a given fixed point and a given fixed line in the plane is a **parabola.** The fixed point is the **focus** of the parabola. The fixed line is the **directrix.**

If the focus F lies on the directrix L, the parabola is the line through F perpendicular to L. We consider this to be a degenerate case and assume henceforth that F does not lie on L.

A parabola has its simplest equation when its focus and directrix straddle one of the coordinate axes. For example, suppose that the focus lies at the point $F(0, p)$ on the positive y-axis and that the directrix is the line $y = -p$ (Figure A5.3). In the notation of the figure, a point $P(x, y)$ lies on the parabola if and only if $PF = PQ$. From the distance formula,

$$PF = \sqrt{(x - 0)^2 + (y - p)^2} = \sqrt{x^2 + (y - p)^2}$$
$$PQ = \sqrt{(x - x)^2 + (y - (-p))^2} = \sqrt{(y + p)^2}.$$

When we equate these expressions, square, and simplify, we get

$$y = \frac{x^2}{4p} \quad \text{or} \quad x^2 = 4py. \quad \text{Standard form} \tag{1}$$

These equations reveal the parabola's symmetry about the y-axis. We call the y-axis the **axis** of the parabola (short for "axis of symmetry").

The point where a parabola crosses its axis is the **vertex.** The vertex of the parabola $x^2 = 4py$ lies at the origin (Figure A5.3). The positive number p is the parabola's **focal length.**

If the parabola opens downward, with its focus at $(0, -p)$ and its directrix the line $y = p$, Equations 1 become

$$y = -\frac{x^2}{4p} \quad \text{or} \quad x^2 = -4py$$

(Figure A5.4). We obtain similar equations for parabolas opening to the right or to the left (Figure A5.5 and Table A5.1).

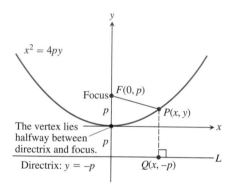

Figure A5.3 The parabola $x^2 = 4py$.

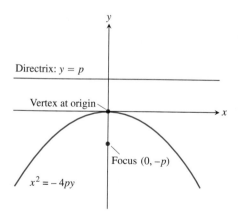

Figure A5.4 The parabola $x^2 = -4py$.

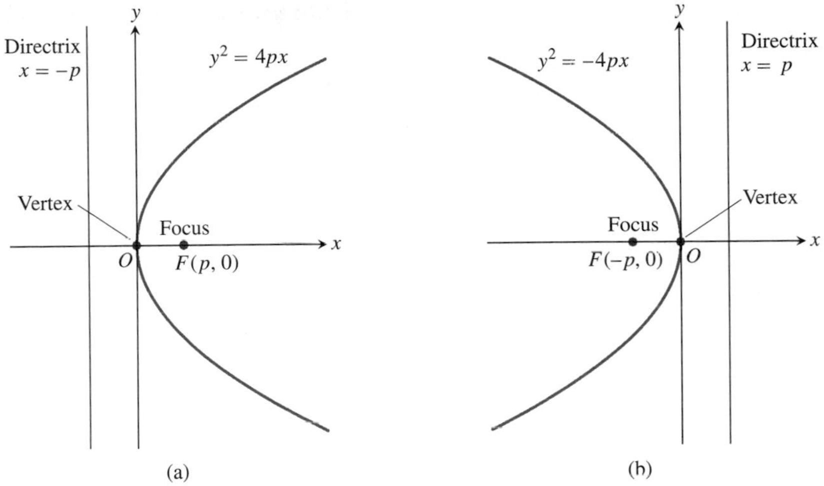

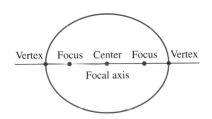

Figure A5.5 (a) The parabola $y^2 = 4px$. (b) The parabola $y^2 = -4px$.

Table A5.1 Standard-form equations for parabolas with vertices at the origin ($p > 0$)

Equation	Focus	Directrix	Axis	Opens
$x^2 = 4py$	$(0, p)$	$y = -p$	y-axis	Up
$x^2 = -4py$	$(0, -p)$	$y = p$	y-axis	Down
$y^2 = 4px$	$(p, 0)$	$x = -p$	x-axis	To the right
$y^2 = -4px$	$(-p, 0)$	$x = p$	x-axis	To the left

Example 3   FINDING FOCUS AND DIRECTRIX

Find the focus and directrix of the parabola $y^2 = 10x$.

Solution   We find the value of p in the standard equation $y^2 = 4px$:

$$4p = 10, \quad \text{so} \quad p = \frac{10}{4} = \frac{5}{2}.$$

Then we find the focus and directrix for this value of p:

$$\text{Focus:} \qquad (p, 0) = \left(\frac{5}{2}, 0\right)$$

$$\text{Directrix:} \quad x = -p \quad \text{or} \quad x = -\frac{5}{2}.$$

Ellipses

> ### Definition Ellipse
>
> An **ellipse** is the set of points in a plane whose distances from two fixed points in the plane have a constant sum. The fixed points are the **foci** of the ellipse. The line through the foci is the **focal axis.** The point on the axis halfway between the foci is the **center.** The points where the focal axis and ellipse cross are the **vertices** (Figure A5.6).

Figure A5.6 Points on the focal axis of an ellipse.

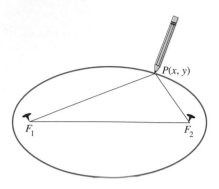

Figure A5.7 How to draw an ellipse.

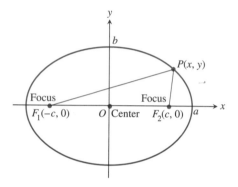

Figure A5.8 The ellipse defined by the equation $PF_1 + PF_2 = 2a$ is the graph of the equation $(x^2/a^2) + (y^2/b^2) = 1$.

The quickest way to construct an ellipse uses the definition. Put a loop of string around two tacks F_1 and F_2, pull the string taut with a pencil point P, and move the pencil around to trace a closed curve (Figure A5.7). The curve is an ellipse because the sum $PF_1 + PF_2$, being the length of the loop minus the distance between the tacks, remains constant. The ellipse's foci lie at F_1 and F_2.

If the foci are $F_1(-c, 0)$ and $F_2(c, 0)$ (Figure A5.8) and $PF_1 + PF_2$ is denoted by $2a$, then the coordinates of a point P on the ellipse satisfy the equation

$$\sqrt{(x + c)^2 + y^2} + \sqrt{(x - c)^2 + y^2} = 2a.$$

To simplify this equation, we move the second radical to the right-hand side, square, isolate the remaining radical, and square again, obtaining

$$\frac{x^2}{a^2} + \frac{y^2}{a^2 - c^2} = 1. \tag{2}$$

Since $PF_1 + PF_2$ is greater than the length F_1F_2 (triangle inequality for triangle PF_1F_2), the number $2a$ is greater than $2c$. Accordingly, $a > c$ and the number $a^2 - c^2$ in Equation 2 is positive.

The algebraic steps leading to Equation 2 can be reversed to show that every point P whose coordinates satisfy an equation of this form with $0 < c < a$ also satisfies the equation $PF_1 + PF_2 = 2a$. A point therefore lies on the ellipse if and only if its coordinates satisfy Equation 2.

If

$$b = \sqrt{a^2 - c^2}, \tag{3}$$

then $a^2 - c^2 = b^2$ and Equation 2 takes the form

$$\frac{x^2}{a^2} + \frac{y^2}{b^2} = 1. \tag{4}$$

Equation 4 reveals that this ellipse is symmetric with respect to the origin and both coordinate axes. It lies inside the rectangle bounded by the lines $x = \pm a$ and $y = \pm b$. It crosses the axes at the points $(\pm a, 0)$ and $(0, \pm b)$. The tangents at these points are perpendicular to the axes because

$$\frac{dy}{dx} = -\frac{b^2 x}{a^2 y} \qquad \text{Obtained from Eq. 4 by implicit differentiation}$$

is zero if $x = 0$ and infinite if $y = 0$.

Axes of an Ellipse

The **major axis** of the ellipse in Equation 4 is the line segment of length $2a$ joining the points $(\pm a, 0)$. The **minor axis** is the line segment of length $2b$ joining the points $(0, \pm b)$. The number a itself is the **semimajor axis,** the number b the **semiminor axis.** The number c found from Equation 3 as

$$c = \sqrt{a^2 - b^2},$$

is the **center-to-focus** distance of the ellipse.

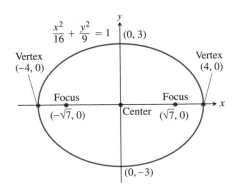

Figure A5.9 Major axis horizontal. (Example 4)

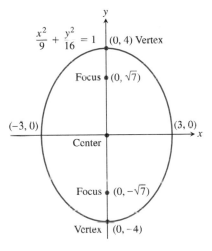

Figure A5.10 Major axis vertical. (Example 5)

Example 4   MAJOR AXIS HORIZONTAL

The ellipse

$$\frac{x^2}{16} + \frac{y^2}{9} = 1 \qquad (5)$$

(Figure A5.9) has

Semimajor axis:	$a = \sqrt{16} = 4$
Semiminor axis:	$b = \sqrt{9} = 3$
Center-to-focus distance:	$c = \sqrt{16 - 9} = \sqrt{7}$
Foci:	$(\pm c, 0) = (\pm\sqrt{7}, 0)$
Vertices:	$(\pm a, 0) = (\pm 4, 0)$
Center:	$(0, 0)$.

Example 5   MAJOR AXIS VERTICAL

The ellipse

$$\frac{x^2}{9} + \frac{y^2}{16} = 1, \qquad (6)$$

obtained by interchanging x and y in Equation 5, has its major axis vertical instead of horizontal (Figure A5.10). With a^2 still equal to 16 and b^2 equal to 9, we have

Semimajor axis:	$a = \sqrt{16} = 4$
Semiminor axis:	$b = \sqrt{9} = 3$
Center-to-focus distance:	$c = \sqrt{16 - 9} = \sqrt{7}$
Foci:	$(0, \pm c) = (0, \pm\sqrt{7})$
Vertices:	$(0, \pm a) = (0, \pm 4)$
Center:	$(0, 0)$.

There is never any cause for confusion in analyzing equations like (5) and (6). We simply find the intercepts on the coordinate axes; then we know which way the major axis runs because it is the longer of the two axes. The center always lies at the origin and the foci lie on the major axis.

Standard-Form Equations for Ellipses Centered at the Origin

Foci on the x-axis: $\dfrac{x^2}{a^2} + \dfrac{y^2}{b^2} = 1$ $(a > b)$

    Center-to-focus distance: $c = \sqrt{a^2 - b^2}$
    Foci: $(\pm c, 0)$
    Vertices: $(\pm a, 0)$

Foci on the y-axis: $\dfrac{x^2}{b^2} + \dfrac{y^2}{a^2} = 1$ $(a > b)$

    Center-to-focus distance: $c = \sqrt{a^2 - b^2}$
    Foci: $(0, \pm c)$
    Vertices: $(0, \pm a)$

In each case, a is the semimajor axis and b is the semiminor axis.

Hyperbolas

Definition Hyperbola

A **hyperbola** is the set of points in a plane whose distances from two fixed points in the plane have a constant difference. The two fixed points are the **foci** of the hyperbola.

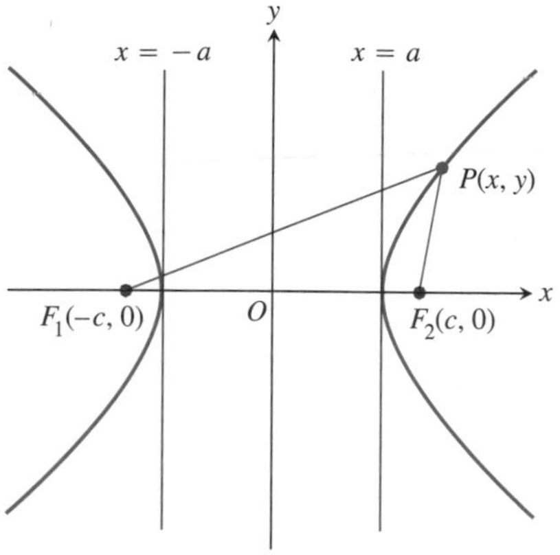

If the foci are $F_1(-c, 0)$ and $F_2(c, 0)$ (Figure A5.11) and the constant difference is $2a$, then a point (x, y) lies on the hyperbola if and only if

$$\sqrt{(x + c)^2 + y^2} - \sqrt{(x - c)^2 + y^2} = \pm 2a. \tag{7}$$

To simplify this equation, we move the second radical to the right-hand side, square, isolate the remaining radical, and square again, obtaining

$$\frac{x^2}{a^2} + \frac{y^2}{a^2 - c^2} = 1. \tag{8}$$

So far, this looks just like the equation for an ellipse. But now $a^2 - c^2$ is negative because $2a$, being the difference of two sides of triangle PF_1F_2 is less than $2c$, the third side.

The algebraic steps leading to Equation 8 can be reversed to show that every point P whose coordinates satisfy an equation of this form with $0 < a < c$ also satisfies Equation 7. A point therefore lies on the hyperbola if and only if its coordinates satisfy Equation 8.

If we let b denote the positive square root of $c^2 - a^2$,

$$b = \sqrt{c^2 - a^2}, \tag{9}$$

then $a^2 - c^2 = -b^2$ and Equation 8 takes the more compact form

$$\frac{x^2}{a^2} - \frac{y^2}{b^2} = 1. \tag{10}$$

The differences between Equation 10 and the equation for an ellipse (Equation 4) are the minus sign and the new relation

$$c^2 = a^2 + b^2. \quad \text{From Eq. 9}$$

Like the ellipse, the hyperbola is symmetric with respect to the origin and coordinate axes. It crosses the x-axis at the points $(\pm a, 0)$. The tangents at these points are vertical because

$$\frac{dy}{dx} = \frac{b^2 x}{a^2 y} \quad \begin{array}{l}\text{Obtained from Eq. 10 by}\\ \text{implicit differentiation}\end{array}$$

is infinite when $y = 0$. The hyperbola has no y-intercepts; in fact, no part of the curve lies between the lines $x = -a$ and $x = a$.

Definition Parts of a Hyperbola

The line through the foci of a hyperbola is the **focal axis.** The point on the axis halfway between the foci is the hyperbola's **center.** The points where the focal axis and hyperbola cross are the **vertices** (Figure A5.12).

Figure A5.11 Hyperbolas have two branches. For points on the right-hand branch of the hyperbola shown here, $PF_1 - PF_2 = 2a$. For points on the left-hand branch, $PF_2 - PF_1 = 2a$.

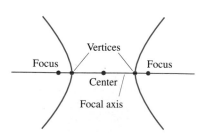

Figure A5.12 Points on the focal axis of a hyperbola.

Asymptotes and Drawing

The hyperbola

$$\frac{x^2}{a^2} - \frac{y^2}{b^2} = 1 \qquad (11)$$

has two asymptotes, the lines

$$y = \pm\frac{b}{a}x.$$

The asymptotes give the guidance we need to draw hyperbolas quickly. (See the drawing lesson.) The fastest way to find the equations of the asymptotes is to replace the 1 in Equation 11 by 0 and solve the new equation for *y:*

$$\underbrace{\frac{x^2}{a^2} - \frac{y^2}{b^2} = 1}_{\text{hyperbola}} \quad \Rightarrow \quad \underbrace{\frac{x^2}{a^2} - \frac{y^2}{b^2} = 0}_{\text{0 for 1}} \quad \Rightarrow \quad \underbrace{y = \pm\frac{b}{a}x.}_{\text{asymptotes}}$$

Standard-Form Equations for Hyperbolas Centered at the Origin

Foci on the x-axis: $\quad \dfrac{x^2}{a^2} - \dfrac{y^2}{b^2} = 1$

Center-to-focus distance: $\quad c = \sqrt{a^2 + b^2}$

Foci: $\quad (\pm c, 0)$

Vertices: $\quad (\pm a, 0)$

Asymptotes: $\quad \dfrac{x^2}{a^2} - \dfrac{y^2}{b^2} = 0 \quad$ or $\quad y = \pm\dfrac{b}{a}x$

Foci on the y-axis: $\quad \dfrac{y^2}{a^2} - \dfrac{x^2}{b^2} = 1$

Center-to-focus distance: $\quad c = \sqrt{a^2 + b^2}$

Foci: $\quad (0, \pm c)$

Vertices: $\quad (0, \pm a)$

Asymptotes: $\quad \dfrac{y^2}{a^2} - \dfrac{x^2}{b^2} = 0 \quad$ or $\quad y = \pm\dfrac{a}{b}x$

Notice the difference in the asymptote equations (b/a in the first, a/b in the second).

Drawing Lesson

How to Graph the Hyperbola $\dfrac{x^2}{a^2} - \dfrac{y^2}{b^2} = 1$

1. Mark the points $(\pm a, 0)$ and $(0, \pm b)$ with line segments and complete the rectangle they determine.

2. Sketch the asymptotes by extending the rectangle's diagonals.

3. Use the rectangle and asymptotes to guide your drawing.

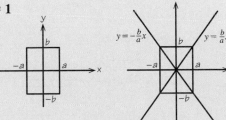 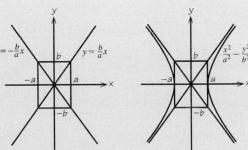

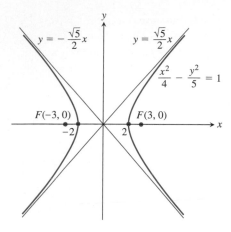

Figure A5.13 The hyperbola in Example 6.

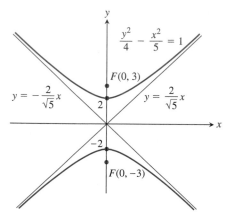

Figure A5.14 The hyperbola in Example 7.

Example 6 FOCI ON THE *x*-AXIS

The equation

$$\frac{x^2}{4} - \frac{y^2}{5} = 1 \tag{12}$$

is Equation 10 with $a^2 = 4$ and $b^2 = 5$ (Figure A5.13). We have

Center-to-focus distance: $c = \sqrt{a^2 + b^2} = \sqrt{4 + 5} = 3$

Foci: $(\pm c, 0) = (\pm 3, 0)$

Vertices: $(\pm a, 0) = (\pm 2, 0)$

Center: $(0, 0)$

Asymptotes: $\dfrac{x^2}{4} - \dfrac{y^2}{5} = 0$ or $y = \pm\dfrac{\sqrt{5}}{2}x.$

Example 7 FOCI ON THE *y*-AXIS

The hyperbola

$$\frac{y^2}{4} - \frac{x^2}{5} = 1,$$

obtained by interchanging x and y in Equation 12, has its vertices on the y-axis instead of the x-axis (Figure A5.14). With a^2 still equal to 4 and b^2 equal to 5, we have

Center-to-focus distance: $c = \sqrt{a^2 + b^2} = \sqrt{4 + 5} = 3$

Foci: $(0, \pm c) = (0, \pm 3)$

Vertices: $(0, \pm a) = (0, \pm 2)$

Center: $(0, 0)$

Asymptotes: $\dfrac{y^2}{4} - \dfrac{x^2}{5} = 0$ or $y = \pm\dfrac{2}{\sqrt{5}}x.$

Reflective Properties

The chief applications of parabolas involve their use as reflectors of light and radio waves. Rays originating at a parabola's focus are reflected out of the parabola parallel to the parabola's axis (Figure A5.15). This property is used by flashlight, headlight, and spotlight reflectors and by microwave broadcast

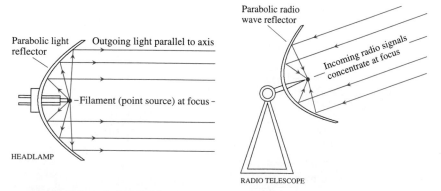

Figure A5.15 Two of the many uses of parabolic reflectors.

antennas to direct radiation from point sources into narrow beams. Conversely, electromagnetic waves arriving parallel to a parabolic reflector's axis are directed toward the reflector's focus. This property is used to intensify signals picked up by radio telescopes and television satellite dishes, to focus arriving light in telescopes, and to concentrate sunlight in solar heaters.

If an ellipse is revolved about its major axis to generate a surface (a surface called an *ellipsoid*) and the interior is silvered to produce a mirror, light from one focus will be reflected to the other focus (Figure A5.16). Ellipsoids reflect sound the same way, and this property is used to construct *whispering galleries,* rooms in which a person standing at one focus can hear a whisper from the other focus. Statuary Hall in the U.S. Capitol building is a whispering gallery. Ellipsoids also appear in instruments used to study aircraft noise in wind tunnels (sound at one focus can be received at the other focus with relatively little interference from other sources).

Light directed toward one focus of a hyperbolic mirror is reflected toward the other focus. This property of hyperbolas is combined with the reflective properties of parabolas and ellipses in designing modern telescopes. In Figure A5.17 starlight reflects off a primary parabolic mirror toward the mirror's focus F_P. It is then reflected by a small hyperbolic mirror, whose focus is $F_H = F_P$, toward the second focus of the hyperbola, $F_E = F_H$. Since this focus is shared by an ellipse, the light is reflected by the elliptical mirror to the ellipse's second focus to be seen by an observer.

As past experience with NASA's Hubble space telescope shows, the mirrors have to be nearly perfect to focus properly. The aberration that caused the malfunction in Hubble's primary mirror (now corrected with additional mirrors) amounted to about half a wavelength of visible light, no more than 1/50 the width of a human hair.

Other Applications

Water pipes are sometimes designed with elliptical cross sections to allow for expansion when the water freezes. The triggering mechanisms in some lasers are elliptical, and stones on a beach become more and more elliptical as they are ground down by waves. There are also applications of ellipses to fossil formation. The ellipsolith, once thought to be a separate species, is now known to be an elliptically deformed nautilus.

Hyperbolic paths arise in Einstein's theory of relativity and form the basis for the (unrelated) LORAN radio navigation system. (LORAN is short for "long range navigation.") Hyperbolas also form the basis for a new system the Burlington Northern Railroad developed for using synchronized electronic signals from satellites to track freight trains. Computers aboard Burlington Northern locomotives in Minnesota can track trains to within one mile per hour of their speed and to within feet of their actual location.

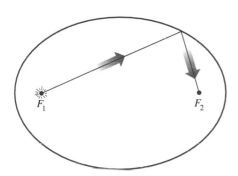

Figure A5.16 An elliptical mirror (shown here in profile) reflects light from one focus to the other.

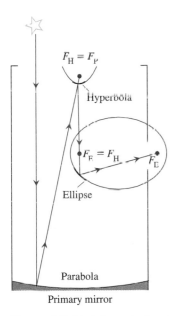

Figure A5.17 Schematic drawing of a reflecting telescope.

Section A5.1 Exercises

In Exercises 1 and 2, find an equation for the circle with center $C(h, k)$ and radius a. Sketch the circle in the xy-plane. Label the circle's center and x- and y-intercepts (if any) with their coordinate pairs.

1. $C(0, 2), \quad a = 2$ **2.** $C(-1, 5), \quad a = \sqrt{10}$

In Exercises 3 and 4, find the center and radius of the circle. Then sketch the circle.

3. $x^2 + y^2 + 4x - 4y + 4 = 0$ **4.** $x^2 + y^2 - 4x + 4y = 0$

In Exercises 5 and 6, describe the regions defined by the inequalities and pairs of inequalities.

5. $(x - 1)^2 + y^2 \le 4$

6. $x^2 + y^2 > 1, \quad x^2 + y^2 < 4$

Match the parabolas in Exercises 7–10 with the following equations:

$$x^2 = 2y, \quad x^2 = -6y, \quad y^2 = 8x, \quad y^2 = -4x.$$

Then find the parabola's focus and directrix.

7.

8.

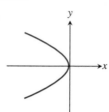

9.

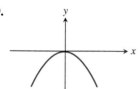

10.

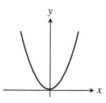

Match each conic section in Exercises 11–14 with one of these equations:

$$\frac{x^2}{4} + \frac{y^2}{9} = 1, \quad \frac{x^2}{2} + y^2 = 1,$$

$$\frac{y^2}{4} - x^2 = 1, \quad \frac{x^2}{4} - \frac{y^2}{9} = 1.$$

Then find the conic section's foci and vertices. If the conic section is a hyperbola, find its asymptotes as well.

11.

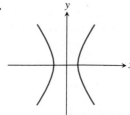

12.

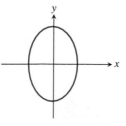

13.

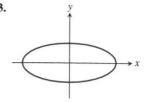

14.

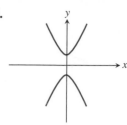

Exercises 15 and 16 give equations of parabolas. Find each parabola's focus and directrix. Then sketch the parabola. Include the focus and directrix in your sketch.

15. $y^2 = 12x$ **16.** $y = 4x^2$

Exercises 17 and 18 give equations for ellipses. Put each equation in standard form. Then sketch the ellipse. Include the foci in your sketch.

17. $16x^2 + 25y^2 = 400$ **18.** $3x^2 + 2y^2 = 6$

Exercises 19 and 20 give information about the foci and vertices of ellipses centered at the origin of the xy-plane. In each case, find the ellipse's standard-form equation from the given information.

19. Foci: $(\pm\sqrt{2}, 0)$ **20.** Foci: $(0, \pm 4)$

    Vertices: $(\pm 2, 0)$ Vertices: $(0, \pm 5)$

Exercises 21 and 22 give equations for hyperbolas. Put each equation in standard form and find the hyperbola's asymptotes. Then sketch the hyperbola. Include the asymptotes and foci in your sketch.

21. $x^2 - y^2 = 1$ **22.** $8y^2 - 2x^2 = 16$

Exercises 23 and 24 give information about the foci, vertices, and asymptotes of hyperbolas centered at the origin of the xy-plane. In each case, find the hyperbola's standard-form equation from the information given.

23. Foci: $(0, \pm\sqrt{2})$ **24.** Vertices: $(\pm 3, 0)$

    Asymptotes: $y = \pm x$ Asymptotes: $y = \pm\dfrac{4}{3}x$

25. The parabola $y^2 = 8x$ is shifted down 2 units and right 1 unit to generate the parabola $(y + 2)^2 = 8(x - 1)$.
  (a) Find the new parabola's vertex, focus, and directrix.
  (b) Plot the new vertex, focus, and directrix, and sketch in the parabola.

26. The ellipse $x^2/16 + y^2/9 = 1$ is shifted 4 units to the right and 3 units up to generate the ellipse

$$\frac{(x - 4)^2}{16} + \frac{(y - 3)^2}{9} = 1.$$

  (a) Find the foci, vertices, and center of the new ellipse.
  (b) Plot the new foci, vertices, and center, and sketch in the new ellipse.

27. The hyperbola $x^2/16 - y^2/9 = 1$ is shifted 2 units to the right to generate the hyperbola

$$\frac{(x-2)^2}{16} - \frac{y^2}{9} = 1.$$

(a) Find the center, foci, vertices, and asymptotes of the new hyperbola. **(b)** Plot the new center, foci, vertices, and asymptotes, and sketch in the hyperbola.

Exercises 28–31 give equations for conic sections and tell how many units up or down and to the right or left each is to be shifted. Find an equation for the new conic section and find the new vertices, foci, directrices, center, and asymptotes, as appropriate.

28. $y^2 = 4x$, left 2, down 3

29. $\dfrac{x^2}{6} + \dfrac{y^2}{9} = 1$, left 2, down 1

30. $\dfrac{x^2}{4} - \dfrac{y^2}{5} = 1$, right 2, up 2

31. $y^2 - x^2 = 1$, left 1, up 1

Find the center, foci, vertices, asymptotes, and radius, as appropriate, of each conic section in Exercises 32–36.

32. $x^2 + 4x + y^2 = 12$

33. $2x^2 + 2y^2 - 28x + 12y + 114 = 0$

34. $x^2 + 2x + 4y - 3 = 0$

35. $x^2 + 5y^2 + 4x = 1$

36. $x^2 - y^2 - 2x + 4y = 4$

37. *Archimedes' Formula for the Volume of a Parabolic Solid* The region enclosed by the parabola $y = (4h/b^2)x^2$ and the line $y = h$ is revolved about the y-axis to generate the solid shown here. Show that the volume of the solid is $3/2$ the volume of the corresponding cone.

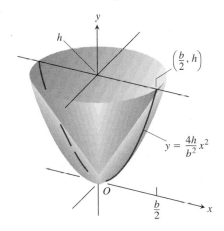

38. *Comparing Volumes* If lines are drawn parallel to the coordinate axes through a point P on the parabola $y^2 = kx$, $k > 0$, the parabola partitions the rectangular region bounded by these lines and the coordinate axes into two smaller regions, A and B.

(a) If the two smaller regions are revolved about the y-axis, show that they generate solids whose volumes have the ratio 4:1.

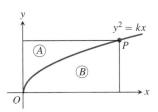

(b) What is the ratio of the volumes of the solids generated by revolving the regions about the x-axis?

39. *Perpendicular Tangents* Show that the tangents to the curve $y^2 = 4px$ from any point on the line $x = -p$ are perpendicular.

40. *Maximizing Area* Find the dimensions of the rectangle of largest area that can be inscribed in the ellipse $x^2 + 4y^2 = 4$ with its sides parallel to the coordinate axes. What is the area of the rectangle?

41. *Volume* Find the volume of the solid generated by revolving the region enclosed by the ellipse $9x^2 + 4y^2 = 36$ about the **(a)** x-axis, **(b)** y-axis.

42. *Volume* The "triangular" region in the first quadrant bounded by the x-axis, the line $x = 4$, and the hyperbola $9x^2 - 4y^2 = 36$ is revolved about the x-axis to generate a solid. Find the volume of the solid.

43. *Volume* The region bounded on the left by the y-axis, on the right by the hyperbola $x^2 - y^2 = 1$, and above and below by the lines $y = \pm 3$ is revolved about the y-axis to generate a solid. Find the volume of the solid.

Extending the Ideas

44. *Suspension Bridge Cables* The suspension bridge cable shown here supports a uniform load of w pounds per horizontal foot. It can be shown that if H is the horizontal tension of the cable at the origin, then the curve of the cable satisfies the equation

$$\frac{dy}{dx} = \frac{w}{H}x.$$

Show that the cable hangs in a parabola by solving this differential equation subject to the initial condition that $y = 0$ when $x = 0$.

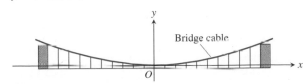

45. *Ripple Tank* Circular waves were made by touching the surface of a ripple tank, first at a point A and shortly thereafter at a nearby point B. As the waves expanded, their points of intersection appeared to trace a hyperbola. Did they really do that? To find out, we can model the waves with circles in the plane centered at nearby points labeled A and B as in the accompanying figure.

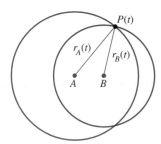

At time t, the point P is $r_A(t)$ units from A and $r_B(t)$ units from B. Since the radii of the circles increase at a constant rate, the rate at which the waves are traveling is

$$\frac{dr_A}{dt} = \frac{dr_B}{dt}.$$

Conclude from this equation that $r_A - r_B$ has a constant value, so that P must lie on a hyperbola with foci at A and B.

46. *How the Astronomer Kepler Used String to Draw Parabolas* Kepler's method for drawing a parabola (with more modern tools) requires a string the length of a T square and a table whose edge can serve as the parabola's directrix. Pin one end of the string to the point where you want the focus to be and the other end to the upper end of the T square. Then, holding the string taut against the T square with a pencil, slide the T square along the table's edge. As the T square moves, the pencil will trace a parabola. Why?

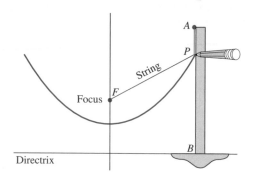

A5.2 Classifying Conic Sections by Eccentricity

Ellipses and Orbits • Hyperbolas • Focus-Directrix Equation

Ellipses and Orbits

We now associate with each conic section a number called the eccentricity. The eccentricity tells whether the conic is a circle, ellipse, parabola, or hyperbola, and, in the case of ellipses and hyperbolas, describes the conic's proportions. We begin with the ellipse.

Although the center-to-focus distance c does not appear in the equation

$$\frac{x^2}{a^2} + \frac{y^2}{b^2} = 1, \quad (a > b)$$

for an ellipse, we can still determine c from the equation $c = \sqrt{a^2 - b^2}$. If we fix a and vary c over the interval $0 \le c \le a$, the resulting ellipses will vary in shape (Figure A5.18). They are circles if $c = 0$ (so that $a = b$) and flatten as c increases. If $c = a$, the foci and vertices overlap and the ellipse degenerates into a line segment.

We use the ratio of c to a to describe the various shapes the ellipse can take. We call this ratio the ellipse's eccentricity.

Definition Eccentricity of Ellipse

The **eccentricity** of the ellipse $x^2/a^2 + y^2/b^2 = 1$ $(a > b)$ is

$$e = \frac{c}{a} = \frac{\sqrt{a^2 - b^2}}{a}.$$

Figure A5.18 The ellipse changes from a circle to a line segment as c increases from 0 to a.

The planets in the solar system revolve around the sun in elliptical orbits with the sun at one focus. Most of the orbits are nearly circular, as can be seen from the eccentricities in Table A5.2. Pluto has a fairly eccentric orbit, with $e = 0.25$, as does Mercury, with $e = 0.21$. Other members of the solar system have orbits that are even more eccentric. Icarus, an asteroid about 1 mile wide that revolves around the sun every 409 Earth days, has an orbital eccentricity of 0.83 (Figure A5.19).

Table A5.2 Eccentricities of planetary orbits

Mercury	0.21
Venus	0.01
Earth	0.02
Mars	0.09
Jupiter	0.05
Saturn	0.06
Uranus	0.05
Neptune	0.01
Pluto	0.25

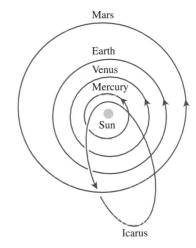

Figure A5.19 The orbit of the asteroid Icarus is highly eccentric. Earth's orbit is so nearly circular that its foci lie inside the sun.

Halley's comet

Edmund Halley (1656–1742; pronounced "*haw*-ley"), British biologist, geologist, sea captain, pirate, spy, Antarctic voyager, astronomer, adviser on fortifications, company founder and director, and the author of the first actuarial mortality tables, was also the mathematician who pushed and harried Newton into writing his *Principia*. Despite these accomplishments, Halley is known today chiefly as the man who calculated the orbit of the great comet of 1682: "wherefore if according to what we have already said [the comet] should return again about the year 1758, candid posterity will not refuse to acknowledge that this was first discovered by an Englishman." Indeed, candid posterity did not refuse—ever since the comet's return in 1758, it has been known as Halley's comet.

Last seen rounding the sun during the winter and spring of 1985–1986, the comet is due to return in the year 2062. The comet has made about 2000 cycles so far with about the same number to go before the sun erodes it away.

Example 1 FINDING THE ECCENTRICITY OF AN ORBIT

The orbit of Halley's comet is an ellipse 36.18 astronomical units long by 9.12 astronomical units wide. (One *astronomical unit* [AU] is 149,597,870 km, the semimajor axis of Earth's orbit.) Its eccentricity is

$$e = \frac{\sqrt{a^2 - b^2}}{a} = \frac{\sqrt{(36.18/2)^2 - (9.12/2)^2}}{(1/2)(36.18)}$$

$$= \frac{\sqrt{(18.09)^2 - (4.56)^2}}{18.09} \approx 0.97.$$

Example 2 LOCATING VERTICES

Locate the vertices of an ellipse of eccentricity 0.8 whose foci lie at the points $(0, \pm 7)$.

Solution Since $e = c/a$, the vertices are the points $(0, \pm a)$ where

$$a = \frac{c}{e} = \frac{7}{0.8} = 8.75,$$

or $(0, \pm 8.75)$.

Whereas a parabola has one focus and one directrix, each ellipse has two foci and two **directrices.** These are the lines perpendicular to the major axis at distances $\pm a/e$ from the center. The parabola has the property that

$$PF = 1 \cdot PD \tag{1}$$

for any point P on it, where F is the focus and D is the point nearest P on the directrix. For an ellipse, it can be shown that the equations that replace (1) are

$$PF_1 = e \cdot PD_1, \qquad PF_2 = e \cdot PD_2. \tag{2}$$

Here, e is the eccentricity, P is any point on the ellipse, F_1 and F_2 are the foci, and D_1 and D_2 are the points on the directrices nearest P (Figure A5.20).

In each equation in (2) the directrix and focus must correspond; that is, if we use the distance from P to F_1, we must also use the distance from P to the directrix at the same end of the ellipse. The directrix $x = -a/e$ corresponds to $F_1(-c, 0)$, and the directrix $x = a/e$ corresponds to $F_2(c, 0)$.

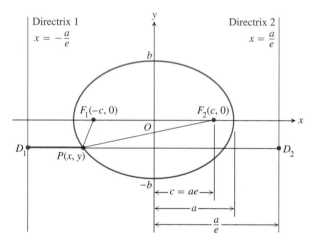

Figure A5.20 The foci and directrices of the ellipse $x^2/a^2 + y^2/b^2 = 1$. Directrix 1 corresponds to focus F_1, and directrix 2 to focus F_2.

Hyperbolas

The eccentricity of a hyperbola is also $e = c/a$, only in this case c equals $\sqrt{a^2 + b^2}$ instead of $\sqrt{a^2 - b^2}$. In contrast to the eccentricity of an ellipse, the eccentricity of a hyperbola is always greater than 1.

Definition Eccentricity of Hyperbola

The **eccentricity** of the hyperbola $x^2/a^2 - y^2/b^2 = 1$ is

$$e = \frac{c}{a} = \frac{\sqrt{a^2 + b^2}}{a}.$$

In both ellipse and hyperbola, the eccentricity is the ratio of the distance between the foci to the distance between the vertices (because $c/a = 2c/2a$).

$$\text{Eccentricity} = \frac{\text{distance between foci}}{\text{distance between vertices}}$$

In an ellipse, the foci are closer together than the vertices and the ratio is less than 1. In a hyperbola, the foci are farther apart than the vertices and the ratio is greater than 1.

Example 3 FINDING ECCENTRICITY

Find the eccentricity of the hyperbola $9x^2 - 16y^2 = 144$.

Solution We divide both sides of the hyperbola's equation by 144 to put it in standard form, obtaining

$$\frac{9x^2}{144} - \frac{16y^2}{144} = 1 \quad \text{or} \quad \frac{x^2}{16} - \frac{y^2}{9} = 1.$$

With $a^2 = 16$ and $b^2 = 9$, we find that $c = \sqrt{a^2 + b^2} = \sqrt{16 + 9} = 5$, so

$$e = \frac{c}{a} = \frac{5}{4}.$$

As with the ellipse, it can be shown that the lines $x = \pm a/e$ act as **directrices** for the hyperbola and that

$$PF_1 = e \cdot PD_1 \quad \text{and} \quad PF_2 = e \cdot PD_2. \tag{3}$$

Here P is any point on the hyperbola, F_1 and F_2 are the foci, and D_1 and D_2 are the points nearest P on the directrices (Figure A5.21).

Focus-Directrix Equation

To complete the picture, we define the eccentricity of a parabola to be $e = 1$. Equations 1–3 then have the common form $PF = e \cdot PD$.

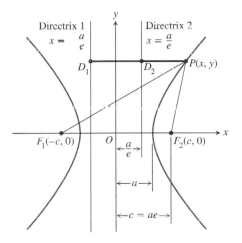

Figure A5.21 The foci and directrices of the hyperbola $x^2/a^2 - y^2/b^2 = 1$. No matter where P lies on the hyperbola, $PF_1 = e \cdot PD_1$, and $PF_2 = e \cdot PD_2$.

Definition Eccentricity of Parabola

The **eccentricity** of a parabola is $e = 1$.

The **focus-directrix** equation $PF = e \cdot PD$ unites the parabola, ellipse, and hyperbola in the following way. Suppose that the distance PF of a point P from a fixed point F (the focus) is a constant multiple of its distance from a fixed line (the directrix). That is, suppose

$$PF = e \cdot PD, \tag{4}$$

where e is the constant of proportionality. Then the path traced by P is

(a) a *parabola* if $e = 1$,

(b) an *ellipse* of eccentricity e if $e < 1$, and

(c) a *hyperbola* of eccentricity e if $e > 1$.

Equation 4 may not look like much to get excited about. There are no coordinates in it and when we try to translate it into coordinate form it translates in different ways, depending on the size of e. At least, that is what happens in Cartesian coordinates. However, in polar coordinates, the equation

$PF = e \cdot PD$ translates into a single equation regardless of the value of e, an equation so simple that it has been the equation of choice of astronomers and space scientists for nearly 300 years.

Given the focus and corresponding directrix of a hyperbola centered at the origin and with foci on the x-axis, we can use the dimensions shown in Figure A5.21 to find e. Knowing e, we can derive a Cartesian equation for the hyperbola from the equation $PF = e \cdot PD$, as in the next example. We can find equations for ellipses centered at the origin and with foci on the x-axis in a similar way, using the dimensions shown in Figure A5.20.

Example 4   USING FOCUS AND DIRECTRIX

Find a Cartesian equation for the hyperbola centered at the origin that has a focus at $(3, 0)$ and the line $x = 1$ as the corresponding directrix.

Solution We first use the dimensions shown in Figure A5.21 to find the hyperbola's eccentricity. The focus is

$$(c, 0) = (3, 0), \quad \text{so} \quad c = 3.$$

The directrix is the line

$$x = \frac{a}{e} = 1, \quad \text{so} \quad a = e.$$

When combined with the equation $e = c/a$ that defines eccentricity, these results give

$$e = \frac{c}{a} = \frac{3}{e}, \quad \text{so} \quad e^2 = 3 \quad \text{and} \quad e = \sqrt{3}.$$

Knowing e, we can now derive the equation we want from the equation $PF = e \cdot PD$. In the notation of Figure A5.22, we have

$$PF = e \cdot PD \qquad \text{Eq. 4}$$
$$\sqrt{(x - 3)^2 + (y - 0)^2} = \sqrt{3}\,|x - 1| \qquad e = \sqrt{3}$$
$$x^2 - 6x + 9 + y^2 = 3(x^2 - 2x + 1)$$
$$2x^2 - y^2 = 6$$
$$\frac{x^2}{3} - \frac{y^2}{6} = 1.$$

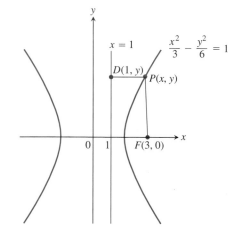

Figure A5.22 The hyperbola in Example 4.

Section A5.2 Exercises

In Exercises 1–4, find the eccentricity, foci, and directrices of the ellipse.

1. $16x^2 + 25y^2 = 400$ **2.** $2x^2 + y^2 = 2$

3. $3x^2 + 2y^2 = 6$ **4.** $6x^2 + 9y^2 = 54$

Exercises 5–8 give the foci or vertices and the eccentricities of ellipses centered at the origin of the xy-plane. In each case, find the ellipse's standard-form equation.

5. Foci: $(0, \pm 3)$
Eccentricity: 0.5

6. Foci: $(\pm 8, 0)$
Eccentricity: 0.2

7. Vertices: $(\pm 10, 0)$
Eccentricity: 0.24

8. Vertices: $(0, \pm 70)$
Eccentricity: 0.1

Exercises 9 and 10 give foci and corresponding directrices of ellipses centered at the origin of the xy-plane. In each case, use the dimensions in Figure A5.20 to find the eccentricity of the ellipse. Then find the ellipse's standard-form equation.

9. Focus: $(\sqrt{5}, 0)$
Directrix: $x = \dfrac{9}{\sqrt{5}}$

10. Focus: $(-4, 0)$
Directrix: $x = -16$

11. Draw an ellipse of eccentricity 4/5. Explain your procedure.

12. Draw the orbit of Pluto (eccentricity 0.25) to scale. Explain your procedure.

13. The endpoints of the major and minor axes of an ellipse are (1, 1), (3, 4), (1, 7), and (−1, 4). Sketch the ellipse, give its equation in standard form, and find its foci, eccentricity, and directrices.

14. Find an equation for the ellipse of eccentricity 2/3 that has the line $x = 9$ as a directrix and the point $(4, 0)$ as the corresponding focus.

In Exercises 15–18, find the eccentricity, foci, and directrices of the hyperbola.

15. $9x^2 - 16y^2 = 144$

16. $y^2 - x^2 = 8$

17. $8x^2 - 2y^2 = 16$

18. $8y^2 - 2x^2 = 16$

Exercises 19 and 20 give the eccentricities and the vertices or foci of hyperbolas centered at the origin of the xy-plane. In each case, find the hyperbola's standard-form equation.

19. Eccentricity: 3
    Vertices: $(0, \pm 1)$

30. Eccentricity: 3
    Foci: $(\pm 3, 0)$

Exercises 21 and 22 give foci and corresponding directrices of hyperbolas centered at the origin of the xy-plane. In each case, find the hyperbola's eccentricity. Then find the hyperbola's standard-form equation.

21. Focus: $(4, 0)$
    Directrix: $x = 2$

22. Focus: $(-2, 0)$
    Directrix: $x = -\dfrac{1}{2}$

23. A hyperbola of eccentricity 3/2 has one focus at $(1, -3)$. The corresponding directrix is the line $y = 2$. Find an equation for the hyperbola.

Explorations

24. *The Effect of Eccentricity on a Hyperbola's Shape* What happens to the graph of a hyperbola as its eccentricity increases? To find out, rewrite the equation $x^2/a^2 - y^2/b^2 = 1$ in terms of a and e instead of a and b. Graph the hyperbola for various values of e and describe what you find.

25. *Determining Constants* What values of the constants a, b, and c make the ellipse

$$4x^2 + y^2 + ax + by + c = 0$$

lie tangent to the x-axis at the origin and pass through the point $(-1, 2)$? What is the eccentricity of the ellipse? ■

Extending the Ideas

26. *The Reflective Property of Ellipses* An ellipse is revolved about its major axis to generate and ellipsoid. The inner surface of the ellipsoid is silvered to make a mirror. Show that a ray of light emanating from one focus will be reflected to the other focus. Sound waves also follow such paths, and this property is used in constructing "whispering galleries." (*Hint:* Place the ellipse in standard position in the xy-plane and show that the lines from a point P on the ellipse to the two foci make congruent angles with the tangent to the ellipse at P.)

27. *The Reflective Property of Hyperbolas* Show that a ray of light directed toward one focus of a hyperbolic mirror, as in the accompanying figure, is reflected toward the other focus. (*Hint:* Show that the tangent to the hyperbola at P bisects the angle made by segments PF_1 and PF_2.)

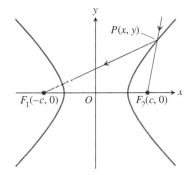

28. *A Confocal Ellipse and Hyperbola* Show that an ellipse and a hyperbola that have the same foci A and B, as in the accompanying figure, cross at right angles at their points of intersection. [*Hint:* A ray of light from focus A that met the hyperbola at P would be reflected from the hyperbola as if it came directly from B (Exercise 27). The same ray would be reflected off the ellipse to pass through B (Exercise 26).]

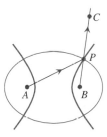

A5.3 Quadratic Equations and Rotations

Quadratic Curves • Cross Product Term • Rotating Axes to
Eliminate *Bxy* • Possible Graphs of Quadratic Equations •
Discriminant Test • Technology Application

Quadratic Curves

In this section, we examine one of the most amazing results in analytic geometry, which is that the Cartesian graph of any equation

$$Ax^2 + Bxy + Cy^2 + Dx + Ey + F = 0, \tag{1}$$

in which *A, B,* and *C* are not all zero, is nearly always a conic section. The exceptions are the cases in which there is no graph at all or the graph consists of two parallel lines. It is conventional to call all graphs of Equation 1, curved or not, **quadratic curves.**

Cross Product Term

You may have noticed that the term *Bxy* did not appear in the equations for the conic sections in Section A5.1. This happened because the axes of the conic sections ran parallel to (in fact, coincided with) the coordinate axes.

To see what happens when the parallelism is absent, let us write an equation for a hyperbola with $a = 3$ and foci at $F_1(-3, -3)$ and $F_2(3, 3)$ (Figure A5.23). The equation $|PF_1 - PF_2| = 2a$ becomes $|PF_1 - PF_2| = 2(3) = 6$ and

$$\sqrt{(x + 3)^2 + (y + 3)^2} - \sqrt{(x - 3)^2 + (y - 3)^2} = \pm 6.$$

When we transpose one radical, square, solve for the remaining radical and square again, the equation reduces to

$$2xy = 9, \tag{2}$$

a case of Equation 1 in which the cross product term is present. The asymptotes of the hyperbola in Equation 2 are the *x*- and *y*-axes, and the focal axis makes an angle of $\pi/4$ radians with the positive *x*-axis. As in this example, the cross product term is present in Equation 1 only when the axes of the conic are tilted.

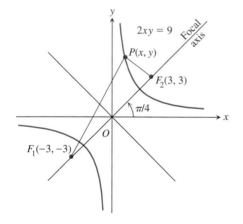

Figure A5.23 The focal axis of the hyperbola $2xy = 9$ makes an angle of $\pi/4$ radians with the positive *x*-axis.

Rotating Axes to Eliminate *Bxy*

To eliminate the *xy*-term from the equation of a conic, we rotate the coordinate axes to eliminate the "tilt" in the axes of the conic. The equations for the rotations we use are derived in the following way. In the notation of Figure A5.24, which shows a counterclockwise rotation about the origin through an angle α,

$$x = OM = OP \cos(\theta + \alpha) = OP \cos\theta \cos\alpha - OP \sin\theta \sin\alpha$$
$$y = MP = OP \sin(\theta + \alpha) = OP \cos\theta \sin\alpha + OP \sin\theta \cos\alpha. \tag{3}$$

Since

$$OP \cos\theta = OM' = x'$$

and

$$OP \sin\theta = M'P = y',$$

the equations in (3) reduce to the following.

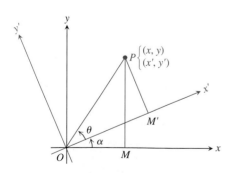

Figure A5.24 A counterclockwise rotation through angle α about the origin.

> ### Equations for Rotating Coordinate Axes
>
> $$x = x' \cos \alpha - y' \sin \alpha$$
>
> $$y = x' \sin \alpha + y' \cos \alpha \qquad (4)$$

Example 1 CHANGING AN EQUATION

The x- and y-axes are rotated through an angle of $\pi/4$ radians about the origin. Find an equation for the hyperbola $2xy = 9$ in the new coordinates.

Solution Since $\cos \pi/4 = \sin \pi/4 = 1/\sqrt{2}$, we substitute

$$x = \frac{x' - y'}{\sqrt{2}}, \qquad y = \frac{x' + y'}{\sqrt{2}}$$

from Equations 4 into the equation $2xy = 9$, obtaining

$$2\left(\frac{x' - y'}{\sqrt{2}}\right)\left(\frac{x' + y'}{\sqrt{2}}\right) = 9$$

$$x'^2 - y'^2 = 9$$

$$\frac{x'^2}{9} - \frac{y'^2}{9} = 1.$$

See Figure A5.25.

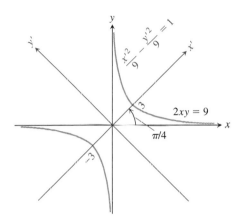

Figure A5.25 The hyperbola in Example 1 (x' and y' are the new coordinates).

If we apply Equations 4 to the quadratic Equation 1, we obtain a new quadratic equation

$$A'x'^2 + B'x'y' + C'y'^2 + D'x' + E'y' + F' = 0. \qquad (5)$$

The new and old coefficients are related by the equations

$$A' = A \cos^2 \alpha + B \cos \alpha \sin \alpha + C \sin^2 \alpha$$

$$B' = B \cos 2\alpha + (C - A) \sin 2\alpha$$

$$C' = A \sin^2 \alpha - B \sin \alpha \cos \alpha + C \cos^2 \alpha \qquad (6)$$

$$D' = D \cos \alpha + E \sin \alpha$$

$$E' = -D \sin \alpha + E \cos \alpha$$

$$F' = F.$$

Equations 6 show, among other things, that if we start with an equation for a curve in which the cross product term is present ($B \neq 0$), we can find a rotation angle α that produces an equation in which no cross product term appears ($B' = 0$). To find α, we set $B' = 0$ in the second equation in (6) and solve the resulting equation,

$$B \cos 2\alpha + (C - A) \sin 2\alpha = 0,$$

for α. In practice, this means determining α from one of the two equations.

> $$\cot 2\alpha = \frac{A - C}{B} \quad \text{or} \quad \tan 2\alpha = \frac{B}{A - C}. \qquad (7)$$

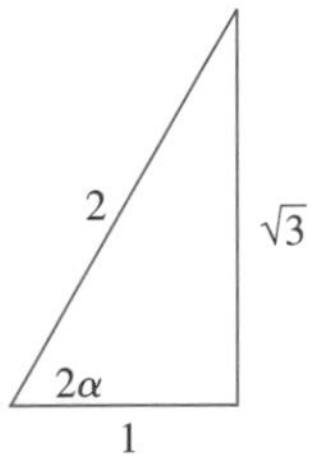

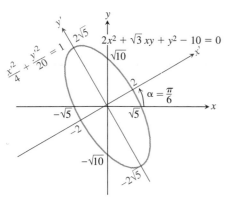

Figure A5.26 This triangle identifies $2\alpha = \cot^{-1}(1/\sqrt{3})$ as $\pi/3$. (Example 2)

Example 2 ELIMINATING A CROSS PRODUCT TERM

The coordinate axes are to be rotated through an angle α to produce an equation for the curve

$$2x^2 + \sqrt{3}xy + y^2 - 10 = 0$$

that has no cross product term. Find a suitable α and the corresponding new equation. Identify the curve.

Solution The equation $2x^2 + \sqrt{3}xy + y^2 - 10 = 0$ has $A = 2$, $B = \sqrt{3}$, and $C = 1$. We substitute these values into Equation 7 to find α:

$$\cot 2\alpha = \frac{A - C}{B} = \frac{2 - 1}{\sqrt{3}} = \frac{1}{\sqrt{3}}.$$

From the right triangle in Figure A5.26, we see that one appropriate choice of angle is $2\alpha = \pi/3$, so we take $\alpha = \pi/6$. Substituting $\alpha = \pi/6$, $A = 2$, $B = \sqrt{3}$, $C = 1$, $D = E = 0$, and $F = -10$ into Equations 6 gives

$$A' = \frac{5}{2}, \qquad B' = 0, \qquad C' = \frac{1}{2}, \qquad D' = E' = 0, \qquad F' = -10.$$

Equation 5 then gives

$$\frac{5}{2}x'^2 + \frac{1}{2}y'^2 - 10 = 0, \quad \text{or} \quad \frac{x'^2}{4} + \frac{y'^2}{20} = 1.$$

The curve is an ellipse with foci on the new y'-axis (Figure A5.27).

Possible Graphs of Quadratic Equations

We now return to the graph of the general quadratic equation.

Since axes can always be rotated to eliminate the cross product term, there is no loss of generality in assuming that this has been done and that the equation has the form

$$Ax^2 + Cy^2 + Dx + Ey + F = 0. \tag{8}$$

Equation 8 represents

(a) a *circle* if $A = C \neq 0$ (special cases: the graph is a point or there is no graph at all);

(b) a *parabola* if Equation 8 is quadratic in one variable and linear in the other;

(c) an *ellipse* if A and C are both positive or both negative (special cases: circles, a single point, or no graph at all);

(d) a *hyperbola* if A and C have opposite signs (special case: a pair of intersecting lines);

(e) a *straight line* if A and C are zero and at least one of D and E is different from zero;

(f) *one or two straight lines* if the left-hand side of Equation 8 can be factored into the product of two linear factors.

See Table A5.3 (on page 616) for examples.

Figure A5.27 The conic section in Example 2.

Discriminant Test

We do not need to eliminate the xy-term from the equation

$$Ax^2 + Bxy + Cy^2 + Dx + Ey + F = 0 \qquad (9)$$

to tell what kind of conic section the equation represents. If this is the only information we want, we can apply the following test instead.

As we have seen, if $B \neq 0$, then rotating the coordinate axes through an angle α that satisfies the equation

$$\cot 2\alpha = \frac{A - C}{B} \qquad (10)$$

will change Equation 9 into an equivalent form

$$A'x'^2 + C'y'^2 + D'x' + E'y' + F' = 0 \qquad (11)$$

without a cross product term.

Now, the graph of Equation 11 is a (real or degenerate)

(a) *parabola* if A' or $C' = 0$; that is, if $A'C' = 0$;

(b) *ellipse* if A' and C' have the same sign; that is, if $A'C' > 0$;

(c) *hyperbola* if A' and C' have opposite signs; that is, if $A'C' < 0$.

It can also be verified from Equations 6 that for any rotation of axes,

$$B^2 - 4AC = B'^2 - 4A'C'. \qquad (12)$$

This means that the quantity $B^2 - 4AC$ is not changed by a rotation. But when we rotate through the angle α given by Equation 10, B' becomes zero, so

$$B^2 - 4AC = -4A'C'.$$

Since the curve is a parabola if $A'C' = 0$, an ellipse if $A'C' > 0$, and a hyperbola if $A'C' < 0$, the curve must be a parabola if $B^2 - 4AC = 0$, an ellipse if $B^2 - 4AC < 0$, and a hyperbola if $B^2 - 4AC > 0$. The number $B^2 - 4AC$ is called the **discriminant** of Equation 9.

Discriminant Test

With the understanding that occasional degenerate cases may arise, the quadratic curve $Ax^2 + Bxy + Cy^2 + Dx + Ey + F = 0$ is

(a) a **parabola** if $B^2 - 4AC = 0$,

(b) an **ellipse** if $B^2 - 4AC < 0$,

(c) a **hyperbola** if $B^2 - 4AC > 0$.

Example 3 APPLYING THE DISCRIMINANT TEST

(a) $3x^2 - 6xy + 3y^2 + 2x - 7 = 0$ represents a parabola because

$$B^2 - 4AC = (-6)^2 - 4 \cdot 3 \cdot 3 = 36 - 36 = 0.$$

(b) $x^2 + xy + y^2 - 1 = 0$ represents an ellipse because

$$B^2 - 4AC = (1)^2 - 4 \cdot 1 \cdot 1 = -3 < 0.$$

(c) $xy - y^2 - 5y + 1 = 0$ represents a hyperbola because

$$B^2 - 4AC = (1)^2 - 4(0)(-1) = 1 > 0.$$

Technology Application

How Some Calculators Use Rotations to Evaluate Sines and Cosines
Some calculators use rotations to calculate sines and cosines of arbitrary angles. The procedure goes something like this: The calculator has, stored,

1. ten angles or so, say

$$\alpha_1 = \sin^{-1}(10^{-1}), \quad \alpha_2 = \sin^{-1}(10^{-2}), \quad \ldots, \quad \alpha_{10} = \sin^{-1}(10^{-10}),$$

and

2. twenty numbers, the sines and cosines of the angles $\alpha_1, \alpha_2, \ldots, \alpha_{10}$.

To calculate the sine and cosine of an arbitrary angle θ, we enter θ (in radians) into the calculator. The calculator substracts or adds multiples of 2π to θ to replace θ by the angle between 0 and 2π that has the same sine and cosine as θ (we continue to call the angle θ). The calculator then "writes" θ as a sum of multiples of α_1 (as many as possible without overshooting) plus multiples of α_2 (again, as many as possible), and so on, working its way to α_{10}. This gives

$$\theta \approx m_1\alpha_1 + m_2\alpha_2 + \cdots + m_{10}\alpha_{10}.$$

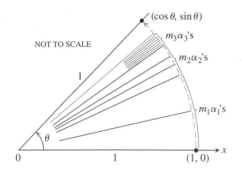

The calculator then rotates the point (1, 0) through m_1 copies of α_1 (through α_1, m_1 times in succession), plus m_2 copies of α_2, and so on, finishing off with m_{10} copies of α_{10} (Figure A5.28). The coordinates of the final position of (1, 0) on the unit circle are the values the calculator gives for $(\cos\theta, \sin\theta)$.

Figure A5.28 To calculate the sine and cosine of an angle θ between 0 and 2π, the calculator rotates the point (1, 0) to an appropriate location on the unit circle and displays the resulting coordinates.

Table A5.3 Examples of quadratic curves

$Ax^2 + Bxy + Cy^2 + Dx + Ey + F = 0$

	A	B	C	D	E	F	Equation	Remarks
Circle	1		1			−4	$x^2 + y^2 = 4$	$A = C$; $F < 0$
Parabola			1	−9			$y^2 = 9x$	Quadratic in y, linear in x
Ellipse	4		9			−36	$4x^2 + 9y^2 = 36$	A, C have same sign, $A \neq C$; $F < 0$
Hyperbola	1		−1			−1	$x^2 - y^2 = 1$	A, C have opposite signs
One line (still a conic section)	1						$x^2 = 0$	y-axis
Intersecting lines (still a conic section)		1		1	−1	−1	$xy + x - y - 1 = 0$	Factors to $(x - 1)(y + 1) = 0$, so $x = 1$, $y = -1$
Parallel lines (not a conic section)	1			−3		2	$x^2 - 3x + 2 = 0$	Factors to $(x - 1)(x - 2) = 0$, so $x = 1$, $x = 2$
Point	1		1				$x^2 + y^2 = 0$	The origin
No graph	1					1	$x^2 = -1$	No graph

Section A5.3 Exercises

Use the discriminant $B^2 - 4AC$ to decide whether the equations in Exercises 1–16 represent parabolas, ellipses, or hyperbolas.

1. $x^2 - 3xy + y^2 - x = 0$

2. $3x^2 - 18xy + 27y^2 - 5x + 7y = -4$

3. $3x^2 - 7xy + \sqrt{17}y^2 = 1$

4. $2x^2 - \sqrt{15}xy + 2y^2 + x + y = 0$

5. $x^2 + 2xy + y^2 + 2x - y + 2 = 0$

6. $2x^2 - y^2 + 4xy - 2x + 3y = 6$

7. $x^2 + 4xy + 4y^2 - 3x = 6$

8. $x^2 + y^2 + 3x - 2y = 10$

9. $xy + y^2 - 3x = 5$

10. $3x^2 + 6xy + 3y^2 - 4x + 5y = 12$

11. $3x^2 - 5xy + 2y^2 - 7x - 14y = -1$

12. $2x^2 - 4.9xy + 3y^2 - 4x = 7$

13. $x^2 - 3xy + 3y^2 + 6y = 7$

14. $25x^2 + 21xy + 4y^2 - 350x = 0$

15. $6x^2 + 3xy + 2y^2 + 17y + 2 = 0$

16. $3x^2 + 12xy + 12y^2 + 435x - 9y + 72 = 0$

In Exercises 17–26, rotate the coordinate axes to change the given equation into an equation that has no cross product (xy) term. Then identify the graph of the equation. (The new equations will vary with the size and direction of the rotation you use.)

17. $xy = 2$ **18.** $x^2 + xy + y^2 = 1$

19. $3x^2 + 2\sqrt{3}xy + y^2 - 8x + 8\sqrt{3}y = 0$

20. $x^2 - \sqrt{3}xy + 2y^2 = 1$

21. $x^2 - 2xy + y^2 = 2$

22. $3x^2 - 2\sqrt{3}xy + y^2 = 1$

23. $\sqrt{2}x^2 + 2\sqrt{2}xy + \sqrt{2}y^2 - 8x + 8y = 0$

24. $xy - y - x + 1 = 0$

25. $3x^2 + 2xy + 3y^2 = 19$

26. $3x^2 + 4\sqrt{3}xy - y^2 = 7$

27. Find the sine and cosine of an angle through which the coordinate axes can be rotated to eliminate the cross product term from the equation

$$14x^2 + 16xy + 2y^2 - 10x + 26{,}370y - 17 = 0.$$

Do not carry out the rotation.

28. Find the sine and cosine of an angle through which the coordinate axes can be rotated to eliminate the cross product term from the equation

$$4x^2 - 4xy + y^2 - 8\sqrt{5}x - 16\sqrt{5}y = 0.$$

Do not carry out the rotation.

The conic sections in Exercises 17–26 were chosen to have rotation angles that were "nice" in the sense that once we knew $\cot 2\alpha$ or $\tan 2\alpha$ we could identify 2α and find $\sin \alpha$ and $\cos \alpha$ from familiar triangles. The conic sections encountered in practice may not have such nice rotation angles, and we may have to use a calculator to determine α from the value of $\cot 2\alpha$ or $\tan 2\alpha$.

In Exercises 29–34, use a calculator to find an angle α through which the coordinate axes can be rotated to change the given equation into a quadratic equation that has no cross product term. Then find $\sin \alpha$ and $\cos \alpha$ to two decimal places and use Equations 6 to find the coefficients of the new equation to the nearest decimal place. In each case, say whether the conic section is an ellipse, hyperbola, or parabola.

29. $x^2 - xy + 3y^2 + x - y - 3 = 0$

30. $2x^2 + xy - 3y^2 + 3x - 7 = 0$

31. $x^2 - 4xy + 4y^2 - 5 = 0$

32. $2x^2 - 12xy + 18y^2 - 49 = 0$

33. $3x^2 + 5xy + 2y^2 - 8y - 1 = 0$

34. $2x^2 + 7xy + 9y^2 + 20x - 86 = 0$

35. *The Hyperbola* $xy = a$ The hyperbola $xy = 1$ is one of many hyperbolas of the form $xy = a$ that appear in science and mathematics.

   (a) Rotate the coordinate axes through an angle of $45°$ to change the equation $xy = 1$ into an equation with no xy-term. What is the new equation?

   (b) Do the same for the equation $xy = a$.

36. Writing to Learn Can anything be said about the graph of the equation $Ax^2 + Bxy + Cy^2 + Dx + Ey + F = 0$ if $AC < 0$? Give reasons for your answer.

37. Writing to Learn Does any nondegenerate conic section $Ax^2 + Bxy + Cy^2 + Dx + Ey + F = 0$ have all of the following properties?

   (a) It is symmetric with respect to the origin.

   (b) It passes through the point $(1, 0)$.

   (c) It is tangent to the line $y = 1$ at the point $(-2, 1)$.

Give reasons for your answer.

38. *When A = C* Show that rotating the axes through an angle of $\pi/4$ radians will eliminate the xy-term from Equation 1 whenever $A = C$.

Explorations

39. *90° Rotations* What effect does a 90° rotation about the origin have on the equations of the following conic sections? Give the new equation in each case.

(a) The ellipse $x^2/a^2 + y^2/b^2 = 1$ $(a > b)$

(b) The hyperbola $x^2/a^2 - y^2/b^2 = 1$

(c) The circle $x^2 + y^2 = a^2$

(d) The line $y = mx$

(e) The line $y = mx + b$

40. *180° Rotations* What effect does a 180° rotation about the origin have on the equations of the following conic sections? Give the new equation in each case.

(a) The ellipse $x^2/a^2 + y^2/b^2 = 1$ $(a > b)$

(b) The hyperbola $x^2/a^2 - y^2/b^2 = 1$

(c) The circle $x^2 + y^2 = a^2$

(d) The line $y = mx$

(e) The line $y = mx + b$ ∎

41. *Identifying a Conic Section*

(a) What kind of conic section is the curve $xy + 2x - y = 0$?

(b) Solve the equation $xy + 2x - y = 0$ for y and sketch the curve as the graph of a rational function of x.

(c) Find equations for the lines parallel to the line $y = -2x$ that are normal to the curve. Add the lines to your sketch.

42. *Sign of AC* Prove or find counterexamples to the following statements about the graph of $Ax^2 + Bxy + Cy^2 + Dx + Ey + F = 0$.

(a) If $AC > 0$, the graph is an ellipse.

(b) If $AC > 0$, the graph is a hyperbola.

(c) If $AC < 0$, the graph is a hyperbola.

Extending the Ideas

43. *Degenerate Conic Section*

(a) Decide whether the equation
$$x^2 + 4xy + 4y^2 + 6x + 12y + 9 = 0$$
represents an ellipse, a parabola, or a hyperbola.

(b) Show that the graph of the equation in (a) is the line $2y = -x - 3$.

44. *Degenerate Conic Section*

(a) Decide whether the conic section with equation
$$9x^2 + 6xy + y^2 - 12x - 4y + 4 = 0$$
represents a parabola, an ellipse, or a hyperbola.

(b) Show that the graph of the equation in (a) is the line $y = -3x + 2$.

45. *A Nice Area Formula for Ellipses* When $B^2 - 4AC$ is negative, the equation
$$Ax^2 + Bxy + Cy^2 = 1$$
represents an ellipse. If the ellipse's semi-axes are a and b, its area is πab (a standard formula). Show that the area is also $2\pi/\sqrt{4AC - B^2}$. (*Hint:* Rotate the coordinate axes to eliminate the xy-term and apply Equation 12 to the new equation.)

46. *Other Rotation Invariants* We describe the fact that $B'^2 - 4A'C'$ equals $B^2 - 4AC$ after a rotation about the origin by saying that the discriminant of a quadratic equation is an **invariant** of the equation. Use Equations 6 to show that the numbers (a) $A + C$ and (b) $D^2 + E^2$ are also invariants, in the sense that
$$A' + C' = A + C \quad \text{and} \quad D'^2 + E'^2 = D^2 + E^2.$$

We can use these equalities to check against numerical errors when we rotate axes. They can also be helpful in shortening the work required to find values for the new coefficients.

A6   # Hyperbolic Functions

Background • Definitions • Identities • Derivatives and Integrals • Inverse Hyperbolic Functions • Identities for $\text{sech}^{-1} x$, $\text{csch}^{-1} x$, $\coth^{-1} x$ • Derivatives of Inverse Hyperbolic Functions; Associated Integrals

Background

Suspension cables like those of the Golden Gate Bridge, which support a constant load per horizontal foot, hang in parabolas (Section A5.1, Exercise 44). Cables like power line cables, which hang freely, hang in curves called hyperbolic cosine curves.

Besides describing the shapes of hanging cables, hyperbolic functions describe the motions of waves in elastic solids, the temperature distributions in metal cooling fins, and the motions of falling bodies that encounter air resistance proportional to the square of the velocity. If a hanging cable were turned upside down (without changing shape) to form an arch, the internal forces, then reversed, would once again be in equilibrium, making the inverted hyperbolic cosine curve the ideal shape for a self-standing arch. The center line of the Gateway Arch to the West in St. Louis follows a hyperbolic cosine curve.

Definitions

The hyperbolic cosine and sine functions are defined by the first two equations in Table A6.1. The table also defines the hyperbolic tangent, cotangent, secant, and cosecant. As we will see, the hyperbolic functions bear a number of similarities to trigonometric functions after which they are named.

Pronouncing "cosh" and "sinh"

"Cosh" is often pronounced "kosh," rhyming with "gosh" or "gauche." "Sinh" is pronounced as if spelled "cinch" or "shine."

Table A6.1 The six basic hyperbolic functions

Hyperbolic cosine of x:	$\cosh x = \dfrac{e^x + e^{-x}}{2}$
Hyperbolic sine of x:	$\sinh x = \dfrac{e^x - e^{-x}}{2}$
Hyperbolic tangent:	$\tanh x = \dfrac{\sinh x}{\cosh x} = \dfrac{e^x - e^{-x}}{e^x + e^{-x}}$
Hyperbolic cotangent:	$\coth x = \dfrac{\cosh x}{\sinh x} = \dfrac{e^x + e^{-x}}{e^x - e^{-x}}$
Hyperbolic secant:	$\operatorname{sech} x = \dfrac{1}{\cosh x} = \dfrac{2}{e^x + e^{-x}}$
Hyperbolic cosecant:	$\operatorname{csch} x = \dfrac{1}{\sinh x} = \dfrac{2}{e^x - e^{-x}}$

See Figure A6.1 for graphs.

Identities

Hyperbolic functions satisfy the identities in Table A6.2. Except for differences in sign, these are identities we already know for trigonometric functions.

Derivatives and Integrals

The six hyperbolic functions, being rational combinations of the differentiable functions e^x and e^{-x}, have derivatives at every point at which they are defined (Table A6.3 on the following page). Again, there are similarities with trigonometric functions. The derivative formulas in Table A6.3 lead to the integral formulas seen there.

Table A6.2 Identities for hyperbolic functions

$$\sinh 2x = 2 \sinh x \cosh x$$

$$\cosh 2x = \cosh^2 x + \sinh^2 x$$

$$\cosh^2 x = \frac{\cosh 2x + 1}{2}$$

$$\sinh^2 x = \frac{\cosh 2x - 1}{2}$$

$$\cosh^2 x - \sinh^2 x = 1$$

$$\tanh^2 x = 1 - \operatorname{sech}^2 x$$

$$\coth^2 x = 1 + \operatorname{csch}^2 x$$

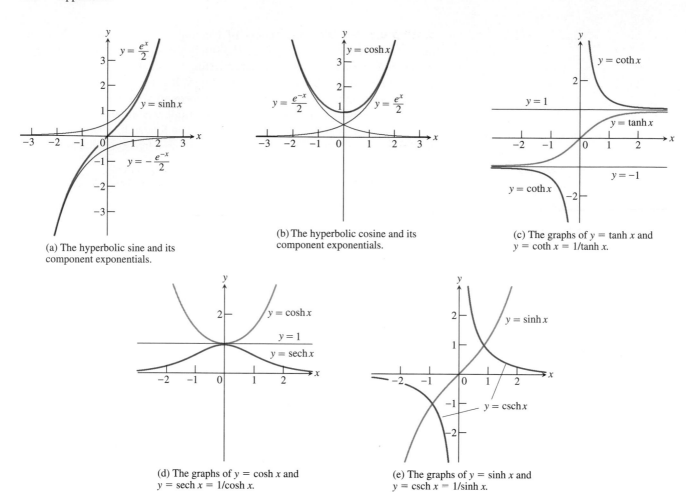

(a) The hyperbolic sine and its component exponentials.

(b) The hyperbolic cosine and its component exponentials.

(c) The graphs of $y = \tanh x$ and $y = \coth x = 1/\tanh x$.

(d) The graphs of $y = \cosh x$ and $y = \operatorname{sech} x = 1/\cosh x$.

(e) The graphs of $y = \sinh x$ and $y = \operatorname{csch} x = 1/\sinh x$.

Figure A6.1 The graphs of the six hyperbolic functions.

Table A6.3 Derivatives and companion integrals

$$\frac{d}{dx}(\sinh u) = \cosh u \frac{du}{dx} \qquad \int \sinh u \, du = \cosh u + C$$

$$\frac{d}{dx}(\cosh u) = \sinh u \frac{du}{dx} \qquad \int \cosh u \, du = \sinh u + C$$

$$\frac{d}{dx}(\tanh u) = \operatorname{sech}^2 u \frac{du}{dx} \qquad \int \operatorname{sech}^2 u \, du = \tanh u + C$$

$$\frac{d}{dx}(\coth u) = -\operatorname{csch}^2 u \frac{du}{dx} \qquad \int \operatorname{csch}^2 u \, du = -\coth u + C$$

$$\frac{d}{dx}(\operatorname{sech} u) = -\operatorname{sech} u \tanh u \frac{du}{dx} \qquad \int \operatorname{sech} u \tanh u \, du = -\operatorname{sech} u + C$$

$$\frac{d}{dx}(\operatorname{csch} u) = -\operatorname{csch} u \coth u \frac{du}{dx} \qquad \int \operatorname{csch} u \coth u \, du = -\operatorname{csch} u + C$$

Example 1 FINDING A DERIVATIVE

$$\frac{d}{dt}(\tanh \sqrt{1+t^2}) = \text{sech}^2 \sqrt{1+t^2} \cdot \frac{d}{dt}(\sqrt{1+t^2})$$

$$= \frac{t}{\sqrt{1+t^2}} \text{sech}^2 \sqrt{1+t^2}$$

Example 2 INTEGRATING A HYPERBOLIC COTANGENT

$$\int \coth 5x \, dx = \int \frac{\cosh 5x}{\sinh 5x} \, dx = \frac{1}{5} \int \frac{du}{u} \qquad \begin{array}{l} u = \sinh 5x, \\ du = 5 \cosh 5x \, dx \end{array}$$

$$= \frac{1}{5} \ln |u| + C = \frac{1}{5} \ln |\sinh 5x| + C$$

Example 3 USING AN IDENTITY TO INTEGRATE

Evaluate $\displaystyle\int_0^1 \sinh^2 x \, dx$.

Solution

Solve Numerically

To five decimal places,

$$\text{NINT}((\sinh x)^2, x, 0, 1) = 0.40672.$$

Confirm Analytically

$$\int_0^1 \sinh^2 x \, dx = \int_0^1 \frac{\cosh 2x - 1}{2} \, dx \qquad \text{Table A6.2}$$

$$= \frac{1}{2} \int_0^1 (\cosh 2x - 1) \, dx = \frac{1}{2} \left[\frac{\sinh 2x}{2} - x \right]_0^1$$

$$= \frac{\sinh 2}{4} - \frac{1}{2} \approx 0.40672$$

Inverse Hyperbolic Functions

We use the inverses of the six basic hyperbolic functions in integration. Since $d(\sinh x)/dx = \cosh x > 0$, the hyperbolic sine is an increasing function of x. We denote its inverse by

$$y = \sinh^{-1} x.$$

For every value of x in the interval $-\infty < x < \infty$, the value of $y = \sinh^{-1} x$ is the number whose hyperbolic sine is x (Figure A6.2a).

The function $y = \cosh x$ is not one-to-one, as we can see from the graph in Figure A6.1. But the restricted function $y = \cosh x$, $x \geq 0$, is one-to-one and therefore has an inverse, denoted by

$$y = \cosh^{-1} x.$$

For every value of $x \geq 1$, $y = \cosh^{-1} x$ is the number in the interval $0 \leq y < \infty$ whose hyperbolic cosine is x (Figure A6.2b).

Like $y = \cosh x$, the function $y = \text{sech } x = 1/\cosh x$ fails to be one-to-one, but its restriction to nonnegative values of x does have an inverse, denoted by

$$y = \text{sech}^{-1} x.$$

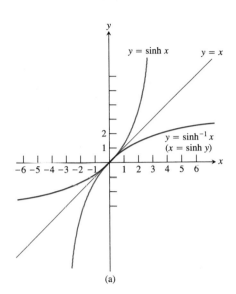

(a)

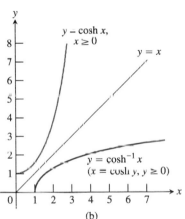

(b)

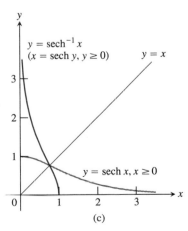

(c)

Figure A6.2 The graphs of the inverse hyperbolic sine, cosine, and secant of x. Notice the symmetries about the line $y = x$.

For every value of x in the interval $(0, 1]$, $y = \text{sech}^{-1} x$ is the nonnegative number whose hyperbolic secant is x (Figure A6.2c).

The hyperbolic tangent, cotangent, and cosecant are one-to-one on their domains and therefore have inverses, denoted by

$$y = \tanh^{-1} x, \quad y = \coth^{-1} x, \quad y = \text{csch}^{-1} x$$

(Figure A6.3).

Exploration 1 **Viewing Inverses**

Let $x_1(t) = t,$ $\quad y_1(t) = t,$

$\quad x_2(t) = t,$ $\quad y_2(t) = 1/\cosh t,$

$\quad x_3(t) = y_2(t), \quad y_3(t) = x_2(t).$

1. Graph the parametric equations simultaneously in a square viewing window that contains $0 \le x \le 6$, $0 \le y \le 4$. Set $t\text{Min} = 0$, $t\text{Max} = 6$, and t-step $= 0.05$. Explain what you see. Explain the domain of each function.
2. Let $x_4(t) = t$, $y_4(t) = \cosh^{-1}(1/t)$. Graph and compare (x_3, y_3) and (y_4, x_4). Predict what you should see, and explain what you do see.

Table A6.4 Identities for inverse hyperbolic functions

$$\text{sech}^{-1} x = \cosh^{-1} \frac{1}{x}$$

$$\text{csch}^{-1} x = \sinh^{-1} \frac{1}{x}$$

$$\coth^{-1} x = \tanh^{-1} \frac{1}{x}$$

Identities for $\text{sech}^{-1} x$, $\text{csch}^{-1} x$, $\coth^{-1} x$

We use the identities in Table A6.4 to calculate the values of $\text{sech}^{-1} x$, $\text{csch}^{-1} x$, and $\coth^{-1} x$ on calculators that give only $\cosh^{-1} x$, $\sinh^{-1} x$, and $\tanh^{-1} x$.

Derivatives of Inverse Hyperbolic Functions; Associated Integrals

The chief use of inverse hyperbolic functions lies in integrations that reverse the derivative formulas in Table A6.5.

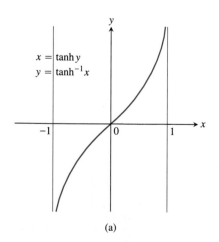

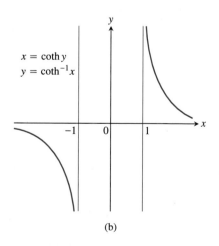

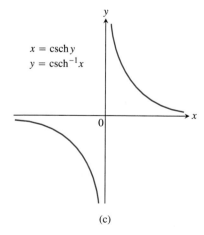

(a)   (b)   (c)

Figure A6.3 The graphs of the inverse hyperbolic tangent, cotangent, and cosecant of x.

Table A6.5 Derivatives of inverse hyperbolic functions

$$\frac{d(\sinh^{-1} u)}{dx} = \frac{1}{\sqrt{1 + u^2}} \frac{du}{dx}$$

$$\frac{d(\cosh^{-1} u)}{dx} = \frac{1}{\sqrt{u^2 - 1}} \frac{du}{dx}, \quad u > 1$$

$$\frac{d(\tanh^{-1} u)}{dx} = \frac{1}{1 - u^2} \frac{du}{dx}, \quad |u| < 1$$

$$\frac{d(\coth^{-1} u)}{dx} = \frac{1}{1 - u^2} \frac{du}{dx}, \quad |u| > 1$$

$$\frac{d(\operatorname{sech}^{-1} u)}{dx} = \frac{-du/dx}{u\sqrt{1 - u^2}}, \quad 0 < u < 1$$

$$\frac{d(\operatorname{csch}^{-1} u)}{dx} = \frac{-du/dx}{|u|\sqrt{1 + u^2}}, \quad u \neq 0$$

The restrictions $|u| < 1$ and $|u| > 1$ on the derivative formulas for $\tanh^{-1} u$ and $\coth^{-1} u$ come from the natural restrictions on the values of these functions. (See Figures A6.3a and b.) The distinction between $|u| < 1$ and $|u| > 1$ becomes important when we convert the derivative formulas into integral formulas. If $|u| < 1$, the integral of $1/(1 - u^2)$ is $\tanh^{-1} u + C$. If $|u| > 1$, the integral is $\coth^{-1} u + C$.

With appropriate substitutions, the derivative formulas in Table A6.5 lead to the integral formulas in Table A6.6.

Table A6.6 Integrals leading to inverse hyperbolic functions

1. $\displaystyle\int \frac{du}{\sqrt{a^2 + u^2}} = \sinh^{-1}\left(\frac{u}{a}\right) + C, \quad a > 0$

2. $\displaystyle\int \frac{du}{\sqrt{u^2 - a^2}} = \cosh^{-1}\left(\frac{u}{a}\right) + C, \quad u > a > 0$

3. $\displaystyle\int \frac{du}{a^2 - u^2} = \begin{cases} \dfrac{1}{a}\tanh^{-1}\left(\dfrac{u}{a}\right) + C & \text{if } u^2 < a^2 \\[2ex] \dfrac{1}{a}\coth^{-1}\left(\dfrac{u}{a}\right) + C & \text{if } u^2 > a^2 \end{cases}$

4. $\displaystyle\int \frac{du}{u\sqrt{a^2 - u^2}} = -\frac{1}{a}\operatorname{sech}^{-1}\left(\frac{u}{a}\right) + C, \quad 0 < u < a$

5. $\displaystyle\int \frac{du}{u\sqrt{a^2 + u^2}} = -\frac{1}{a}\operatorname{csch}^{-1}\left|\frac{u}{a}\right| + C, \quad u \neq 0$

Example 4 USING TABLE A6.6

Evaluate $\displaystyle\int_0^1 \frac{2\,dx}{\sqrt{3 + 4x^2}}$.

Solution

Solve Analytically

The indefinite integral is

$$\int \frac{2\,dx}{\sqrt{3 + 4x^2}} = \int \frac{du}{\sqrt{a^2 + u^2}} \qquad u = 2x, \quad du = 2\,dx, \quad a = \sqrt{3}$$

$$= \sinh^{-1}\left(\frac{u}{a}\right) + C \qquad \text{Formula from Table A6.6}$$

$$= \sinh^{-1}\left(\frac{2x}{\sqrt{3}}\right) + C.$$

Therefore,

$$\int_0^1 \frac{2\,dx}{\sqrt{3 + 4x^2}} = \sinh^{-1}\left(\frac{2x}{\sqrt{3}}\right)\Bigg]_0^1 = \sinh^{-1}\left(\frac{2}{\sqrt{3}}\right) - \sinh^{-1}(0)$$

$$= \sinh^{-1}\left(\frac{2}{\sqrt{3}}\right) - 0 \approx 0.98665.$$

Support Numerically

To five decimal places,

$$\text{NINT}\,(2/\sqrt{3 + 4x^2}, x, 0, 1) = 0.98665.$$

Section A6 Exercises

In Exercises 1–4, find the values of the remaining five hyperbolic functions.

1. $\sinh x = -\dfrac{3}{4}$

2. $\sinh x = \dfrac{4}{3}$

3. $\cosh x = \dfrac{17}{15}, \quad x > 0$

4. $\cosh x = \dfrac{13}{5}, \quad x > 0$

In Exercises 5–10, rewrite the expression in terms of exponentials and simplify the results as much as you can. Support your answers graphically.

5. $2\cosh(\ln x)$

6. $\sinh(2\ln x)$

7. $\cosh 5x + \sinh 5x$

8. $\cosh 3x - \sinh 3x$

9. $(\sinh x + \cosh x)^4$

10. $\ln(\cosh x + \sinh x) + \ln(\cosh x - \sinh x)$

11. Use the identities

$$\sinh(x + y) = \sinh x \cosh y + \cosh x \sinh y$$
$$\cosh(x + y) = \cosh x \cosh y + \sinh x \sinh y$$

to show that

(a) $\sinh 2x = 2\sinh x \cosh x$;

(b) $\cosh 2x = \cosh^2 x + \sinh^2 x$.

12. Use the definitions of $\cosh x$ and $\sinh x$ to show that

$$\cosh^2 x - \sinh^2 x = 1.$$

In Exercises 13–24, find the derivative of y with respect to the appropriate variable.

13. $y = 6\sinh\dfrac{x}{3}$

14. $y = \dfrac{1}{2}\sinh(2x + 1)$

15. $y = 2\sqrt{t}\,\tanh\sqrt{t}$

16. $y = t^2\tanh\dfrac{1}{t}$

17. $y = \ln(\sinh z)$

18. $y = \ln(\cosh z)$

19. $y = \text{sech } \theta(1 - \ln \text{sech } \theta)$

20. $y = \text{csch } \theta(1 - \ln \text{csch } \theta)$

21. $y = \ln \cosh x - \dfrac{1}{2} \tanh^2 x$

22. $y = \ln \sinh x - \dfrac{1}{2} \coth^2 x$

23. $y = (x^2 + 1) \text{ sech}(\ln x)$ (*Hint:* Before differentiating, express in terms of exponentials and simplify.)

24. $y = (4x^2 - 1) \text{ csch}(\ln 2x)$

In Exercises 25–36, find the derivative of y with respect to the appropriate variable.

25. $y = \sinh^{-1} \sqrt{x}$

26. $y = \cosh^{-1}(2\sqrt{x + 1})$

27. $y = (1 - \theta) \tanh^{-1} \theta$

28. $y = (\theta^2 + 2\theta) \tanh^{-1}(\theta + 1)$

29. $y = (1 - t) \coth^{-1} \sqrt{t}$

30. $y = (1 - t^2) \coth^{-1} t$

31. $y = \cos^{-1} x - x \text{ sech}^{-1} x$

32. $y = \ln x + \sqrt{1 - x^2} \text{ sech}^{-1} x$

33. $y = \text{csch}^{-1}\left(\dfrac{1}{2}\right)^{\theta}$

34. $y = \text{csch}^{-1} 2^{\theta}$

35. $y = \sinh^{-1}(\tan x)$

36. $y = \cosh^{-1}(\sec x), \quad 0 < x < \pi/2$

Verify the integration formulas in Exercises 37–40.

37. (a) $\displaystyle \int \text{sech } x \, dx = \tan^{-1}(\sinh x) + C$

(b) $\displaystyle \int \text{sech } x \, dx = \sin^{-1}(\tanh x) + C$

38. $\displaystyle \int x \text{ sech}^{-1} x \, dx = \dfrac{x^2}{2} \text{ sech}^{-1} x - \dfrac{1}{2}\sqrt{1 - x^2} + C$

39. $\displaystyle \int x \coth^{-1} x \, dx = \dfrac{x^2 - 1}{2} \coth^{-1} x + \dfrac{x}{2} + C$

40. $\displaystyle \int \tanh^{-1} x \, dx = x \tanh^{-1} x + \dfrac{1}{2}\ln(1 - x^2) + C$

Evaluate the integrals in Exercises 41–50.

41. $\displaystyle \int \sinh 2x \, dx$

42. $\displaystyle \int \sinh \dfrac{x}{5} \, dx$

43. $\displaystyle \int 6 \cosh\left(\dfrac{x}{2} - \ln 3\right) dx$

44. $\displaystyle \int 4 \cosh(3x - \ln 2) \, dx$

45. $\displaystyle \int \tanh \dfrac{x}{7} \, dx$

46. $\displaystyle \int \coth \dfrac{\theta}{\sqrt{3}} \, d\theta$

47. $\displaystyle \int \text{sech}^2\left(x - \dfrac{1}{2}\right) dx$

48. $\displaystyle \int \text{csch}^2(5 - x) \, dx$

49. $\displaystyle \int \dfrac{\text{sech } \sqrt{t} \tanh \sqrt{t} \, dt}{\sqrt{t}}$

50. $\displaystyle \int \dfrac{\text{csch}(\ln t) \coth(\ln t) \, dt}{t}$

Evaluate the integrals in Exercises 51–60 analytically and support with NINT.

51. $\displaystyle \int_{\ln 2}^{\ln 4} \coth x \, dx$

52. $\displaystyle \int_{0}^{\ln 2} \tanh 2x \, dx$

53. $\displaystyle \int_{-\ln 4}^{-\ln 2} 2e^{\theta} \cosh \theta \, d\theta$

54. $\displaystyle \int_{0}^{\ln 2} 4e^{-\theta} \sinh \theta \, d\theta$

55. $\displaystyle \int_{-\pi/4}^{\pi/4} \cosh(\tan \theta) \sec^2 \theta \, d\theta$

56. $\displaystyle \int_{0}^{\pi/2} 2 \sinh(\sin \theta) \cos \theta \, d\theta$

57. $\displaystyle \int_{1}^{2} \dfrac{\cosh(\ln t)}{t} \, dt$

58. $\displaystyle \int_{1}^{4} \dfrac{8 \cosh \sqrt{x}}{\sqrt{x}} \, dx$

59. $\displaystyle \int_{-\ln 2}^{0} \cosh^2\left(\dfrac{x}{2}\right) dx$

60. $\displaystyle \int_{0}^{\ln 10} 4 \sinh^2\left(\dfrac{x}{2}\right) dx$

In Exercises 61 and 62, find the volume of the solid generated by revolving the shaded region about the x-axis.

61.

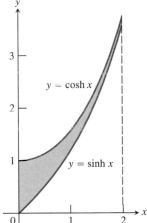

62.

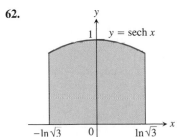

63. Find the volume of the solid generated by revolving the shaded region about the line $y = 1$.

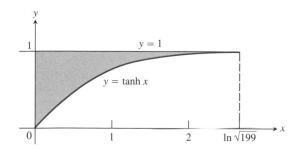

64. (a) Find the length of the curve $y = (1/2) \cosh 2x$, $0 \le x \le \ln \sqrt{5}$.

(b) Find the length of the curve $y = (1/a) \cosh ax$, $0 \le x \le b$.

Extending the Ideas

65. *Even-Odd Decompositions*

(a) Show that if a function f is defined on an interval symmetric about the origin (so that f is defined at $-x$ whenever it is defined at x), then

$$f(x) = \frac{f(x) + f(-x)}{2} + \frac{f(x) - f(-x)}{2}. \qquad (1)$$

Then show that

$$\frac{f(x) + f(-x)}{2} \quad \text{is even}$$

and

$$\frac{f(x) - f(-x)}{2} \quad \text{is odd.}$$

(b) In Equation 1, set $f(x) = e^x$. Identify the even and odd parts of f.

66. Writing to Learn *(Continuation of Exercise 65)* Equation 1 in Exercise 65 simplifies considerably if f itself is **(a)** even or **(b)** odd. What are the new equations? Explain.

67. *Skydiving* If a body of mass m falling from rest under the action of gravity encounters an air resistance proportional to the square of the velocity, then the body's velocity t seconds into the fall satisfies the differential equation

$$m\frac{dv}{dt} = mg - kv^2,$$

where k is a constant that depends on the body's aerodynamic properties and the density of the air. (We assume that the fall is short enough so that variation in the air's density will not affect the outcome.)

Show that

$$v = \sqrt{\frac{mg}{k}} \tanh\left(\sqrt{\frac{gk}{m}} \, t\right)$$

satisfies the differential equation and the initial condition that $v = 0$ when $t = 0$.

68. *Accelerations Whose Magnitudes Are Proportional to Displacement* Suppose that the position of a body moving along a coordinate line at time t is

(a) $s = a \cos kt + b \sin kt$,

(b) $s = a \cosh kt + b \sinh kt$.

Show in both cases that the acceleration d^2s/dt^2 is proportional to s but that in the first case it is directed toward the origin while in the second case it is directed away from the origin.

69. *Tractor Trailers and the Tractrix* When a tractor trailer turns into a cross street or driveway, its rear wheels follow a curve like the one shown here. (This is why the rear wheels sometimes ride up over the curb.)

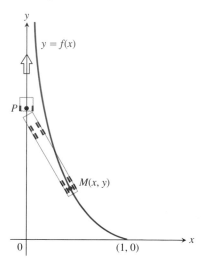

We can find an equation for the curve if we picture the rear wheels as a mass M at the point $(1, 0)$ on the x-axis attached by a rod of unit length to a point P representing the cab at the origin. As P moves up the y-axis, it drags M along behind it. The curve traced by M, called a *tractrix* from the Latin word *tractum* for "drag," can be shown to be the graph of the function $y = f(x)$ that solves the initial value problem

Differential equation: $\quad \dfrac{dy}{dx} = -\dfrac{1}{x\sqrt{1 - x^2}} + \dfrac{x}{\sqrt{1 - x^2}},$

Initial condition: $\quad y = 0 \quad \text{when} \quad x = 1.$

Solve the initial value problem to find an equation for the curve. (You need an inverse hyperbolic function.)

70. *A Minimal Surface* Find the area of the surface swept out by revolving the curve $y = 4 \cosh(x/4)$, $-\ln 16 \le x \le \ln 81$, about the x-axis.

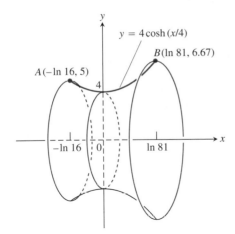

It can be shown that, of all continuously differentiable curves joining points A and B in the figure, the curve $y = 4 \cosh(x/4)$ generates the surface of least area. If you made a rigid wire frame of the end-circles through A and B and dipped them in a soap-film solution, the surface spanning the circles would be the one generated by the curve.

71. *Hanging Cables* Show that the function $y = a \cosh(x/a)$ solves the initial value problem

$$y'' = (1/a)\sqrt{1 + (y')^2}, \quad y'(0) = 0, \quad y(0) = a.$$

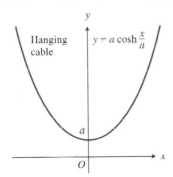

By analyzing the forces on hanging cables, we can show that the curves they hang in always satisfy the differential equation and initial conditions given here. That is how we know that hanging cables hang in hyperbolic cosines.

72. *The Hyperbolic in Hyperbolic Functions* In case you are wondering where the name *hyperbolic* comes from, here is the answer: Just as $x = \cos u$ and $y = \sin u$ are identified with points (x, y) on the unit circle, the functions $x = \cosh u$ and $y = \sinh u$ are identified with points (x, y) on the right-hand branch of the unit hyperbola $x^2 - y^2 = 1$ (Figure A6.4).

Another analogy between hyperbolic and circular functions is that the variable u in the coordinates $(\cosh u, \sinh u)$ for the points of the right-hand branch of the hyperbola $x^2 - y^2 = 1$ is twice the area of the sector AOP pictured in Figure A6.5. To see why, carry out the following steps.

(a) Let $A(u)$ be the area of sector AOP. Show that

$$A(u) = \frac{1}{2} \cosh u \sinh u - \int_1^{\cosh u} \sqrt{x^2 - 1} \, dx.$$

(b) Differentiate both sides of the equation in (a) with respect to u to show that

$$A'(u) = \frac{1}{2}.$$

(c) Solve the equation in (b) for $A(u)$. What is the value of $A(0)$? What is the value of the constant of integration C in your solution? With C determined, what does your solution say about the relationship of u to $A(u)$?

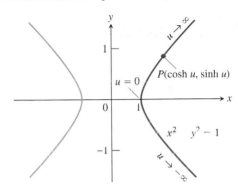

Figure A6.4 Since $\cosh^2 u - \sinh^2 u = 1$, the point $(\cosh u, \sinh u)$ lies on the right-hand branch of the hyperbola $x^2 - y^2 = 1$ for every value of u (Exercise 72).

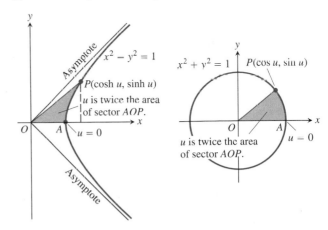

Figure A6.5 One of the analogies between hyperbolic and circular functions is revealed by these two diagrams (Exercise 72).

A7 A Brief Table of Integrals

The formulas below are stated in terms of constants a, b, c, m, n, and so on. These constants can usually assume any real value and need not be integers. Occasional limitations on their values are stated with the formulas. Formula 5 requires $n \neq -1$, for example, and Formula 11 requires $n \neq -2$. The formulas also assume that the constants do not take on values that require dividing by zero or taking even roots of negative numbers. For example, Formula 8 assumes $a \neq 0$, and Formula 13(a) cannot be used unless b is negative.

1. $\displaystyle\int u\, dv = uv - \int v\, du$

2. $\displaystyle\int a^u\, du = \frac{a^u}{\ln a} + C, \quad a \neq 1, \quad a > 0$

3. $\displaystyle\int \cos u\, du = \sin u + C$

4. $\displaystyle\int \sin u\, du = -\cos u + C$

5. $\displaystyle\int (ax + b)^n\, dx = \frac{(ax + b)^{n+1}}{a(n + 1)} + C, \quad n \neq -1$

6. $\displaystyle\int (ax + b)^{-1}\, dx = \frac{1}{a} \ln |ax + b| + C$

7. $\displaystyle\int x(ax + b)^n\, dx = \frac{(ax + b)^{n+1}}{a^2}\left[\frac{ax + b}{n + 2} - \frac{b}{n + 1}\right] + C, \quad n \neq -1, -2$

8. $\displaystyle\int x(ax + b)^{-1}\, dx = \frac{x}{a} - \frac{b}{a^2} \ln |ax + b| + C$

9. $\displaystyle\int x(ax + b)^{-2}\, dx = \frac{1}{a^2}\left[\ln |ax + b| + \frac{b}{ax + b}\right] + C$

10. $\displaystyle\int \frac{dx}{x(ax + b)} = \frac{1}{b} \ln \left|\frac{x}{ax + b}\right| + C$

11. $\displaystyle\int (\sqrt{ax + b})^n\, dx = \frac{2}{a} \frac{(\sqrt{ax + b})^{n+2}}{n + 2} + C, \quad n \neq -2$

12. $\displaystyle\int \frac{\sqrt{ax + b}}{x}\, dx = 2\sqrt{ax + b} + b\int \frac{dx}{x\sqrt{ax + b}}$

13. (a) $\displaystyle\int \frac{dx}{x\sqrt{ax + b}} = \frac{2}{\sqrt{-b}} \tan^{-1} \sqrt{\frac{ax + b}{-b}} + C, \quad \text{if } b < 0$

(b) $\displaystyle\int \frac{dx}{x\sqrt{ax + b}} = \frac{1}{\sqrt{b}} \ln \left|\frac{\sqrt{ax + b} - \sqrt{b}}{\sqrt{ax + b} + \sqrt{b}}\right| + C, \quad \text{if } b > 0$

14. $\displaystyle\int \frac{\sqrt{ax + b}}{x^2}\, dx = -\frac{\sqrt{ax + b}}{x} + \frac{a}{2}\int \frac{dx}{x\sqrt{ax + b}} + C$

15. $\displaystyle\int \frac{dx}{x^2\sqrt{ax + b}} = -\frac{\sqrt{ax + b}}{bx} - \frac{a}{2b}\int \frac{dx}{x\sqrt{ax + b}} + C$

16. $\displaystyle\int \frac{dx}{a^2 + x^2} = \frac{1}{a} \tan^{-1}\frac{x}{a} + C$

17. $\displaystyle\int \frac{dx}{(a^2 + x^2)^2} = \frac{x}{2a^2(a^2 + x^2)} + \frac{1}{2a^3} \tan^{-1}\frac{x}{a} + C$

18. $\displaystyle\int \frac{dx}{a^2 - x^2} = \frac{1}{2a} \ln \left|\frac{x + a}{x - a}\right| + C$

19. $\displaystyle\int \frac{dx}{(a^2 - x^2)^2} = \frac{x}{2a^2(a^2 - x^2)} + \frac{1}{2a^2}\int \frac{dx}{a^2 - x^2}$

20. $\displaystyle\int \frac{dx}{\sqrt{a^2 + x^2}} = \sinh^{-1}\frac{x}{a} + C = \ln\left(x + \sqrt{a^2 + x^2}\right) + C$

21. $\int \sqrt{a^2 + x^2} \, dx = \dfrac{x}{2}\sqrt{a^2 + x^2} + \dfrac{a^2}{2} \ln\left(x + \sqrt{a^2 + x^2}\right) + C$

22. $\int x^2\sqrt{a^2 + x^2} \, dx = \dfrac{x}{8}(a^2 + 2x^2)\sqrt{a^2 + x^2} - \dfrac{a^4}{8} \ln\left(x + \sqrt{a^2 + x^2}\right) + C$

23. $\int \dfrac{\sqrt{a^2 + x^2}}{x} \, dx = \sqrt{a^2 + x^2} - a \ln\left|\dfrac{a + \sqrt{a^2 + x^2}}{x}\right| + C$

24. $\int \dfrac{\sqrt{a^2 + x^2}}{x^2} \, dx = \ln\left(x + \sqrt{a^2 + x^2}\right) - \dfrac{\sqrt{a^2 + x^2}}{x} + C$

25. $\int \dfrac{x^2}{\sqrt{a^2 + x^2}} \, dx = -\dfrac{a^2}{2} \ln\left(x + \sqrt{a^2 + x^2}\right) + \dfrac{x\sqrt{a^2 + x^2}}{2} + C$

26. $\int \dfrac{dx}{x\sqrt{a^2 + x^2}} = -\dfrac{1}{a} \ln\left|\dfrac{a + \sqrt{a^2 + x^2}}{x}\right| + C$ **27.** $\int \dfrac{dx}{x^2\sqrt{a^2 + x^2}} = -\dfrac{\sqrt{a^2 + x^2}}{a^2 x} + C$

28. $\int \dfrac{dx}{\sqrt{a^2 - x^2}} = \sin^{-1}\dfrac{x}{a} + C$ **29.** $\int \sqrt{a^2 - x^2} \, dx = \dfrac{x}{2}\sqrt{a^2 - x^2} + \dfrac{a^2}{2} \sin^{-1}\dfrac{x}{a} + C$

30. $\int x^2\sqrt{a^2 - x^2} \, dx = \dfrac{a^4}{8} \sin^{-1}\dfrac{x}{a} - \dfrac{1}{8}x\sqrt{a^2 - x^2}\,(a^2 - 2x^2) + C$

31. $\int \dfrac{\sqrt{a^2 - x^2}}{x} \, dx = \sqrt{a^2 - x^2} - a \ln\left|\dfrac{a + \sqrt{a^2 - x^2}}{x}\right| + C$

32. $\int \dfrac{\sqrt{a^2 - x^2}}{x^2} \, dx = -\sin^{-1}\dfrac{x}{a} - \dfrac{\sqrt{a^2 - x^2}}{x} + C$ **33.** $\int \dfrac{x^2}{\sqrt{a^2 - x^2}} \, dx = \dfrac{a^2}{2} \sin^{-1}\dfrac{x}{a} - \dfrac{1}{2}x\sqrt{a^2 - x^2} + C$

34. $\int \dfrac{dx}{x\sqrt{a^2 - x^2}} = -\dfrac{1}{a} \ln\left|\dfrac{a + \sqrt{a^2 - x^2}}{x}\right| + C$ **35.** $\int \dfrac{dx}{x^2\sqrt{a^2 - x^2}} = -\dfrac{\sqrt{a^2 - x^2}}{a^2 x} + C$

36. $\int \dfrac{dx}{\sqrt{x^2 - a^2}} = \cosh^{-1}\dfrac{x}{a} + C = \ln\left|x + \sqrt{x^2 - a^2}\right| + C$

37. $\int \sqrt{x^2 - a^2} \, dx = \dfrac{x}{2}\sqrt{x^2 - a^2} - \dfrac{a^2}{2} \ln\left|x + \sqrt{x^2 - a^2}\right| + C$

38. $\int \left(\sqrt{x^2 - a^2}\right)^n dx = \dfrac{x\left(\sqrt{x^2 - a^2}\right)^n}{n + 1} - \dfrac{na^2}{n + 1}\int \left(\sqrt{x^2 - a^2}\right)^{n-2} dx, \quad n \neq -1$

39. $\int \dfrac{dx}{\left(\sqrt{x^2 - a^2}\right)^n} = \dfrac{x\left(\sqrt{x^2 - a^2}\right)^{2-n}}{(2 - n)a^2} - \dfrac{n - 3}{(n - 2)a^2}\int \dfrac{dx}{\left(\sqrt{x^2 - a^2}\right)^{n-2}}, \quad n \neq 2$

40. $\int x\left(\sqrt{x^2 - a^2}\right)^n dx = \dfrac{\left(\sqrt{x^2 - a^2}\right)^{n+2}}{n + 2} + C, \quad n \neq -2$

41. $\int x^2\sqrt{x^2 - a^2} \, dx = \dfrac{x}{8}(2x^2 - a^2)\sqrt{x^2 - a^2} - \dfrac{a^4}{8} \ln\left|x + \sqrt{x^2 - a^2}\right| + C$

42. $\int \dfrac{\sqrt{x^2 - a^2}}{x} \, dx = \sqrt{x^2 - a^2} - a \sec^{-1}\left|\dfrac{x}{a}\right| + C$

43. $\displaystyle\int \frac{\sqrt{x^2 - a^2}}{x^2}\, dx = \ln\left|x + \sqrt{x^2 - a^2}\right| - \frac{\sqrt{x^2 - a^2}}{x} + C$

44. $\displaystyle\int \frac{x^2}{\sqrt{x^2 - a^2}}\, dx = \frac{a^2}{2}\ln\left|x + \sqrt{x^2 - a^2}\right| + \frac{x}{2}\sqrt{x^2 - a^2} + C$

45. $\displaystyle\int \frac{dx}{x\sqrt{x^2 - a^2}} = \frac{1}{a}\sec^{-1}\left|\frac{x}{a}\right| + C = \frac{1}{a}\cos^{-1}\left|\frac{a}{x}\right| + C$

46. $\displaystyle\int \frac{dx}{x^2\sqrt{x^2 - a^2}} = \frac{\sqrt{x^2 - a^2}}{a^2 x} + C$

47. $\displaystyle\int \frac{dx}{\sqrt{2ax - x^2}} = \sin^{-1}\left(\frac{x - a}{a}\right) + C$

48. $\displaystyle\int \sqrt{2ax - x^2}\, dx = \frac{x - a}{2}\sqrt{2ax - x^2} + \frac{a^2}{2}\sin^{-1}\left(\frac{x - a}{a}\right) + C$

49. $\displaystyle\int \left(\sqrt{2ax - x^2}\right)^n dx = \frac{(x - a)\left(\sqrt{2ax - x^2}\right)^n}{n + 1} + \frac{na^2}{n + 1}\int \left(\sqrt{2ax - x^2}\right)^{n-2} dx$

50. $\displaystyle\int \frac{dx}{\left(\sqrt{2ax - x^2}\right)^n} = \frac{(x - a)\left(\sqrt{2ax - x^2}\right)^{2-n}}{(n - 2)a^2} + \frac{(n - 3)}{(n - 2)a^2}\int \frac{dx}{\left(\sqrt{2ax - x^2}\right)^{n-2}}$

51. $\displaystyle\int x\sqrt{2ax - x^2}\, dx = \frac{(x + a)(2x - 3a)\sqrt{2ax - x^2}}{6} + \frac{a^3}{2}\sin^{-1}\left(\frac{x - a}{a}\right) + C$

52. $\displaystyle\int \frac{\sqrt{2ax - x^2}}{x}\, dx = \sqrt{2ax - x^2} + a\sin^{-1}\left(\frac{x - a}{a}\right) + C$

53. $\displaystyle\int \frac{\sqrt{2ax - x^2}}{x^2}\, dx = -2\sqrt{\frac{2a - x}{x}} - \sin^{-1}\left(\frac{x - a}{a}\right) + C$

54. $\displaystyle\int \frac{x\, dx}{\sqrt{2ax - x^2}} = a\sin^{-1}\left(\frac{x - a}{a}\right) - \sqrt{2ax - x^2} + C$

55. $\displaystyle\int \frac{dx}{x\sqrt{2ax - x^2}} = -\frac{1}{a}\sqrt{\frac{2a - x}{x}} + C$

56. $\displaystyle\int \sin ax\, dx = -\frac{1}{a}\cos ax + C$

57. $\displaystyle\int \cos ax\, dx = \frac{1}{a}\sin ax + C$

58. $\displaystyle\int \sin^2 ax\, dx = \frac{x}{2} - \frac{\sin 2ax}{4a} + C$

59. $\displaystyle\int \cos^2 ax\, dx = \frac{x}{2} + \frac{\sin 2ax}{4a} + C$

60. $\displaystyle\int \sin^n ax\, dx = -\frac{\sin^{n-1} ax \cos ax}{na} + \frac{n - 1}{n}\int \sin^{n-2} ax\, dx$

61. $\displaystyle\int \cos^n ax\, dx = \frac{\cos^{n-1} ax \sin ax}{na} + \frac{n - 1}{n}\int \cos^{n-2} ax\, dx$

62. (a) $\displaystyle\int \sin ax \cos bx\, dx = -\frac{\cos(a + b)x}{2(a + b)} - \frac{\cos(a - b)x}{2(a - b)} + C, \quad a^2 \neq b^2$

(b) $\displaystyle\int \sin ax \sin bx\, dx = \frac{\sin(a - b)x}{2(a - b)} - \frac{\sin(a + b)x}{2(a + b)} + C, \quad a^2 \neq b^2$

(c) $\displaystyle\int \cos ax \cos bx\, dx = \frac{\sin(a - b)x}{2(a - b)} + \frac{\sin(a + b)x}{2(a + b)} + C, \quad a^2 \neq b^2$

63. $\int \sin ax \cos ax\, dx = -\dfrac{\cos 2ax}{4a} + C$

64. $\int \sin^n ax \cos ax\, dx = \dfrac{\sin^{n+1} ax}{(n+1)a} + C, \quad n \neq -1$

65. $\int \dfrac{\cos ax}{\sin ax}\, dx = \dfrac{1}{a} \ln |\sin ax| + C$

66. $\int \cos^n ax \sin ax\, dx = -\dfrac{\cos^{n+1} ax}{(n+1)a} + C, \quad n \neq -1$

67. $\int \dfrac{\sin ax}{\cos ax}\, dx = -\dfrac{1}{a} \ln |\cos ax| + C$

68. $\int \sin^n ax \cos^m ax\, dx = -\dfrac{\sin^{n-1} ax \cos^{m+1} ax}{a(m+n)} + \dfrac{n-1}{m+n} \int \sin^{n-2} ax \cos^m ax\, dx, \quad n \neq -m \quad$ (If $n = -m$, use No. 86.)

69. $\int \sin^n ax \cos^m ax\, dx = \dfrac{\sin^{n+1} ax \cos^{m-1} ax}{a(m+n)} + \dfrac{m-1}{m+n} \int \sin^n ax \cos^{m-2} ax\, dx, \quad m \neq -n \quad$ (If $m = -n$, use No. 87.)

70. $\int \dfrac{dx}{b + c \sin ax} = \dfrac{-2}{a\sqrt{b^2 - c^2}} \tan^{-1}\left[\sqrt{\dfrac{b-c}{b+c}} \tan\left(\dfrac{\pi}{4} - \dfrac{ax}{2}\right)\right] + C, \quad b^2 > c^2$

71. $\int \dfrac{dx}{b + c \sin ax} = \dfrac{-1}{a\sqrt{c^2 - b^2}} \ln \left|\dfrac{c + b \sin ax + \sqrt{c^2 - b^2} \cos ax}{b + c \sin ax}\right| + C, \quad b^2 < c^2$

72. $\int \dfrac{dx}{1 + \sin ax} = -\dfrac{1}{a} \tan\left(\dfrac{\pi}{4} - \dfrac{ax}{2}\right) + C$

73. $\int \dfrac{dx}{1 - \sin ax} = \dfrac{1}{a} \tan\left(\dfrac{\pi}{4} + \dfrac{ax}{2}\right) + C$

74. $\int \dfrac{dx}{b + c \cos ax} = \dfrac{2}{a\sqrt{b^2 - c^2}} \tan^{-1}\left[\sqrt{\dfrac{b-c}{b+c}} \tan\dfrac{ax}{2}\right] + C, \quad b^2 > c^2$

75. $\int \dfrac{dx}{b + c \cos ax} = \dfrac{1}{a\sqrt{c^2 - b^2}} \ln \left|\dfrac{c + b \cos ax + \sqrt{c^2 - b^2} \sin ax}{b + c \cos ax}\right| + C, \quad b^2 < c^2$

76. $\int \dfrac{dx}{1 + \cos ax} = \dfrac{1}{a} \tan \dfrac{ax}{2} + C$

77. $\int \dfrac{dx}{1 - \cos ax} = -\dfrac{1}{a} \cot \dfrac{ax}{2} + C$

78. $\int x \sin ax\, dx = \dfrac{1}{a^2} \sin ax - \dfrac{x}{a} \cos ax + C$

79. $\int x \cos ax\, dx = \dfrac{1}{a^2} \cos ax + \dfrac{x}{a} \sin ax + C$

80. $\int x^n \sin ax\, dx = -\dfrac{x^n}{a} \cos ax + \dfrac{n}{a} \int x^{n-1} \cos ax\, dx$

81. $\int x^n \cos ax\, dx = \dfrac{x^n}{a} \sin ax - \dfrac{n}{a} \int x^{n-1} \sin ax\, dx$

82. $\int \tan ax\, dx = \dfrac{1}{a} \ln |\sec ax| + C$

83. $\int \cot ax\, dx = \dfrac{1}{a} \ln |\sin ax| + C$

84. $\int \tan^2 ax\, dx = \dfrac{1}{a} \tan ax - x + C$

85. $\int \cot^2 ax\, dx = -\dfrac{1}{a} \cot ax - x + C$

86. $\int \tan^n ax\, dx = \dfrac{\tan^{n-1} ax}{a(n-1)} - \int \tan^{n-2} ax\, dx, \quad n \neq 1$

87. $\int \cot^n ax\, dx = \dfrac{\cot^{n-1} ax}{a(n-1)} - \int \cot^{n-2} ax\, dx, \quad n \neq 1$

88. $\int \sec ax\, dx = \dfrac{1}{a} \ln |\sec ax + \tan ax| + C$

89. $\int \csc ax\, dx = -\dfrac{1}{a} \ln |\csc ax + \cot ax| + C$

90. $\int \sec^2 ax\, dx = \dfrac{1}{a} \tan ax + C$

91. $\int \csc^2 ax\, dx = -\dfrac{1}{a} \cot ax + C$

92. $\displaystyle\int \sec^n ax\, dx = \frac{\sec^{n-2} ax \tan ax}{a(n-1)} + \frac{n-2}{n-1}\int \sec^{n-2} ax\, dx, \quad n \neq 1$

93. $\displaystyle\int \csc^n ax\, dx = -\frac{\csc^{n-2} ax \cot ax}{a(n-1)} + \frac{n-2}{n-1}\int \csc^{n-2} ax\, dx, \quad n \neq 1$

94. $\displaystyle\int \sec^n ax \tan ax\, dx = \frac{\sec^n ax}{na} + C, \quad n \neq 0$
 95. $\displaystyle\int \csc^n ax \cot ax\, dx = -\frac{\csc^n ax}{na} + C, \quad n \neq 0$

96. $\displaystyle\int \sin^{-1} ax\, dx = x \sin^{-1} ax + \frac{1}{a}\sqrt{1 - a^2x^2} + C$
 97. $\displaystyle\int \cos^{-1} ax\, dx = x \cos^{-1} ax - \frac{1}{a}\sqrt{1 - a^2x^2} + C$

98. $\displaystyle\int \tan^{-1} ax\, dx = x \tan^{-1} ax - \frac{1}{2a}\ln(1 + a^2x^2) + C$

99. $\displaystyle\int x^n \sin^{-1} ax\, dx = \frac{x^{n+1}}{n+1}\sin^{-1} ax - \frac{a}{n+1}\int \frac{x^{n+1}\, dx}{\sqrt{1 - a^2x^2}}, \quad n \neq -1$

100. $\displaystyle\int x^n \cos^{-1} ax\, dx = \frac{x^{n+1}}{n+1}\cos^{-1} ax + \frac{a}{n+1}\int \frac{x^{n+1}\, dx}{\sqrt{1 - a^2x^2}}, \quad n \neq -1$

101. $\displaystyle\int x^n \tan^{-1} ax\, dx = \frac{x^{n+1}}{n+1}\tan^{-1} ax - \frac{a}{n+1}\int \frac{x^{n+1}\, dx}{1 + a^2x^2}, \quad n \neq -1$

102. $\displaystyle\int e^{ax}\, dx = \frac{1}{a}e^{ax} + C$
 103. $\displaystyle\int b^{ax}\, dx = \frac{1}{a}\frac{b^{ax}}{\ln b} + C, \quad b > 0, \quad b \neq 1$

104. $\displaystyle\int xe^{ax}\, dx = \frac{e^{ax}}{a^2}(ax - 1) + C$
 105. $\displaystyle\int x^n e^{ax}\, dx = \frac{1}{a}x^n e^{ax} - \frac{n}{a}\int x^{n-1}e^{ax}\, dx$

106. $\displaystyle\int x^n b^{ax}\, dx = \frac{x^n b^{ax}}{a \ln b} - \frac{n}{a \ln b}\int x^{n-1}b^{ax}\, dx, \quad b > 0, \quad b \neq 1$

107. $\displaystyle\int e^{ax} \sin bx\, dx = \frac{e^{ax}}{a^2 + b^2}(a \sin bx - b \cos bx) + C$

108. $\displaystyle\int e^{ax} \cos bx\, dx = \frac{e^{ax}}{a^2 + b^2}(a \cos bx + b \sin bx) + C$
 109. $\displaystyle\int \ln ax\, dx = x \ln ax - x + C$

110. $\displaystyle\int x^n (\ln ax)^m\, dx = \frac{x^{n+1}(\ln ax)^m}{n+1} - \frac{m}{n+1}\int x^n (\ln ax)^{m-1}\, dx, \quad n \neq -1$

111. $\displaystyle\int x^{-1}(\ln ax)^m\, dx = \frac{(\ln ax)^{m+1}}{m+1} + C, \quad m \neq -1$
 112. $\displaystyle\int \frac{dx}{x \ln ax} = \ln|\ln ax| + C$

113. $\displaystyle\int \sinh ax\, dx = \frac{1}{a}\cosh ax + C$
 114. $\displaystyle\int \cosh ax\, dx = \frac{1}{a}\sinh ax + C$

115. $\displaystyle\int \sinh^2 ax\, dx = \frac{\sinh 2ax}{4a} - \frac{x}{2} + C$
 116. $\displaystyle\int \cosh^2 ax\, dx = \frac{\sinh 2ax}{4a} + \frac{x}{2} + C$

117. $\displaystyle\int \sinh^n ax\, dx = \frac{\sinh^{n-1} ax \cosh ax}{na} - \frac{n-1}{n}\int \sinh^{n-2} ax\, dx, \quad n \neq 0$

118. $\displaystyle\int \cosh^n ax\, dx = \frac{\cosh^{n-1} ax \sinh ax}{na} + \frac{n-1}{n}\int \cosh^{n-2} ax\, dx, \quad n \neq 0$

119. $\displaystyle\int x \sinh ax\, dx = \frac{x}{a}\cosh ax - \frac{1}{a^2}\sinh ax + C$

120. $\displaystyle\int x \cosh ax\, dx = \frac{x}{a}\sinh ax - \frac{1}{a^2}\cosh ax + C$

121. $\displaystyle\int x^n \sinh ax\, dx = \frac{x^n}{a}\cosh ax - \frac{n}{a}\int x^{n-1}\cosh ax\, dx$

122. $\displaystyle\int x^n \cosh ax\, dx = \frac{x^n}{a}\sinh ax - \frac{n}{a}\int x^{n-1}\sinh ax\, dx$

123. $\displaystyle\int \tanh ax\, dx = \frac{1}{a}\ln(\cosh ax) + C$

124. $\displaystyle\int \coth ax\, dx = \frac{1}{a}\ln|\sinh ax| + C$

125. $\displaystyle\int \tanh^2 ax\, dx = x - \frac{1}{a}\tanh ax + C$

126. $\displaystyle\int \coth^2 ax\, dx = x - \frac{1}{a}\coth ax + C$

127. $\displaystyle\int \tanh^n ax\, dx = -\frac{\tanh^{n-1} ax}{(n-1)a} + \int \tanh^{n-2} ax\, dx, \quad n \neq 1$

128. $\displaystyle\int \coth^n ax\, dx = -\frac{\coth^{n-1} ax}{(n-1)a} + \int \coth^{n-2} ax\, dx, \quad n \neq 1$

129. $\displaystyle\int \operatorname{sech} ax\, dx = \frac{1}{a}\sin^{-1}(\tanh ax) + C$

130. $\displaystyle\int \operatorname{csch} ax\, dx = \frac{1}{a}\ln\left|\tanh\frac{ax}{2}\right| + C$

131. $\displaystyle\int \operatorname{sech}^2 ax\, dx = \frac{1}{a}\tanh ax + C$

132. $\displaystyle\int \operatorname{csch}^2 ax\, dx = -\frac{1}{a}\coth ax + C$

133. $\displaystyle\int \operatorname{sech}^n ax\, dx = \frac{\operatorname{sech}^{n-2} ax \tanh ax}{(n-1)a} + \frac{n-2}{n-1}\int \operatorname{sech}^{n-2} ax\, dx, \quad n \neq 1$

134. $\displaystyle\int \operatorname{csch}^n ax\, dx = -\frac{\operatorname{csch}^{n-2} ax \coth ax}{(n-1)a} - \frac{n-2}{n-1}\int \operatorname{csch}^{n-2} ax\, dx, \quad n \neq 1$

135. $\displaystyle\int \operatorname{sech}^n ax \tanh ax\, dx = -\frac{\operatorname{sech}^n ax}{na} + C, \quad n \neq 0$

136. $\displaystyle\int \operatorname{csch}^n ax \coth ax\, dx = -\frac{\operatorname{csch}^n ax}{na} + C, \quad n \neq 0$

137. $\displaystyle\int e^{ax} \sinh bx\, dx = \frac{e^{ax}}{2}\left[\frac{e^{bx}}{a+b} - \frac{e^{-bx}}{a-b}\right] + C, \quad a^2 \neq b^2$

138. $\displaystyle\int e^{ax} \cosh bx\, dx = \frac{e^{ax}}{2}\left[\frac{e^{bx}}{a+b} + \frac{e^{-bx}}{a-b}\right] + C, \quad a^2 \neq b^2$

139. $\displaystyle\int_0^\infty x^{n-1} e^{-x}\, dx = (n-1)!, \quad n > 0$

140. $\displaystyle\int_0^\infty e^{-ax^2}\, dx = \frac{1}{2}\sqrt{\frac{\pi}{a}}, \quad a > 0$

141. $\displaystyle\int_0^{\pi/2} \sin^n x\, dx = \int_0^{\pi/2} \cos^n x\, dx = \begin{cases} \dfrac{1 \cdot 3 \cdot 5 \cdots (n-1)}{2 \cdot 4 \cdot 6 \cdots n} \cdot \dfrac{\pi}{2}, & \text{if } n \text{ is an even integer} \geq 2 \\[2ex] \dfrac{2 \cdot 4 \cdot 6 \cdots (n-1)}{3 \cdot 5 \cdot 7 \cdots n}, & \text{if } n \text{ is an odd integer} \geq 3 \end{cases}$

GLOSSARY

Absolute convergence: If the series $\sum |a_n|$ of absolute values converges, then $\sum a_n$ is said to converge absolutely. p. 502

Absolute error: $|(\text{true value}) - (\text{approximate value})|$ p. 221

Absolute maximum: The function f has an absolute maximum value $f(c)$ at a point c in its domain D if and only if $f(x) \leq f(c)$ for all x in D. p. 177

Absolute minimum: The function f has an absolute minimum value $f(c)$ at a point c in its domain D if and only if $f(x) \geq f(c)$ for all x in D. p. 177

Absolute, relative, and percentage change: As we move from $x = a$ to a nearby point $a + dx$, we can describe the corresponding change in the value of a function $f(x)$ three ways:

$\quad$ *Absolute change:* $\quad \Delta f = f(a + dx) - f(a)$

$\quad$ *Relative change:* $\quad \Delta f / f(a)$

$\quad$ *Percentage change:* $\quad (\Delta f / f(a)) \times 100$.

p. 226

Absolute value function: The function $f(x) = |x|$. See also *Absolute value of a number.* p. 15

Absolute value of a number: The absolute value of x is

$$|x| = \begin{cases} -x, & x < 0 \\ x, & x \geq 0. \end{cases}$$

Acceleration: The derivative of a velocity function with respect to time. p. 125

Acceleration vector: If $\mathbf{r}$ is the position vector of a particle moving along a smooth curve in the plane, then at any time t,

$$\mathbf{a} = \frac{d^2\mathbf{r}}{dt^2}$$

is the particle's acceleration vector. p. 533

Algebraic function: A function $y = f(x)$ that satisfies an equation of the form $P_n y^n + \cdots + P_1 y + P_0 = 0$ in which the P's are polynomials in x with rational coefficients. The function $y = 1/\sqrt{x + 1}$ is algebraic, for example, because it satisfies the equation $(x + 1)y^2 - 1 = 0$. Here, $P_2 = x + 1$, $P_1 = 0$, and $P_0 = -1$. All polynomials and rational functions are algebraic.

Alternating harmonic series: The series

$$\sum_{n=1}^{\infty} \frac{(-1)^{n-1}}{n}.$$

p. 500

Alternating series: A series in which the terms are alternately positive and negative. p. 500

Amplitude: Of a periodic function $f(x)$ continuous for all real x, the number

$$\frac{(\text{absolute max of } f) - (\text{absolute min of } f)}{2};$$

of the sine function

$$f(x) = A \sin\left[\frac{2\pi}{B}(x - C)\right] + D,$$

is the amplitude $|A|$.

Angle between curves: At a point of intersection of two differentiable curves, the angle between their tangent lines at the point of intersection.

Angle between vectors u and v: The angle

$$\theta = \cos^{-1}\left(\frac{\mathbf{u} \cdot \mathbf{v}}{|\mathbf{u}||\mathbf{v}|}\right).$$

p. 523

Antiderivative: A function $F(x)$ is an antiderivative of a function $f(x)$ if $F'(x) = f(x)$ for all x in the domain of f. p. 190

Antiderivative of a vector function: Vector function $\mathbf{R}(t)$ is an antiderivative of $\mathbf{r}(t)$ on an interval I if $d\mathbf{R}/dt = \mathbf{r}(t)$ at each point t of I. p. 535

Antidifferentiation: The process of finding an antiderivative. p. 190

Arbitrary constant: See *Constant of integration.*

Arc length: The length

$$\int_a^b \sqrt{1 + \left(\frac{dy}{dx}\right)^2}\, dx$$

of a smooth curve (y a function of x and dy/dx continuous) beginning at $x = a$ and ending at $x = b$. p. 397

Arccosine or inverse cosine function: The inverse of the cosine function with restricted domain $[0, \pi]$. p. 46

Arcsine or inverse sine function: The inverse of the sine function with restricted domain $[-\pi/2, \pi/2]$. p. 46

Arctangent or inverse tangent function: The inverse of the tangent function with restricted domain $(-\pi/2, \pi/2)$. p. 46

Asymptote: The line $y = b$ is a horizontal asymptote of the graph of a function $y = f(x)$ if either

$$\lim_{x\to\infty} f(x) = b \quad \text{or} \quad \lim_{x\to-\infty} f(x) = b.$$

The line $x = a$ is a vertical asymptote of the graph of a function $y = f(x)$ if either

$$\lim_{x\to a^+} f(x) = \pm\infty \quad \text{or} \quad \lim_{x\to a^-} f(x) = \pm\infty.$$

p. 65

Average rate of change of a quantity over a period of time: The amount of change divided by the time it takes. p. 86

Average value of a continuous function f on $[a, b]$:

$$\frac{1}{b-a}\int_a^b f(x)\,dx.$$

p. 271

Average velocity: Displacement (change in position) divided by time traveled. p. 123

Axis of revolution: The line about which a solid of revolution is generated. See *Solid of revolution*.

Base (exponential and logarithm): See *Exponential function with base a; Logarithm function with base a*.

Basic trigonometric functions: When an angle of measure θ is placed in standard position in the coordinate plane and (x, y) is the point in which its terminal ray intersects a circle with center $(0, 0)$ and radius r, the values of the six basic trigonometric functions of θ are

$$\text{Sine: } \sin\theta = \frac{y}{r} \qquad \text{Cosecant: } \csc\theta = \frac{r}{y}$$

$$\text{Cosine: } \cos\theta = \frac{x}{r} \qquad \text{Secant: } \sec\theta = \frac{r}{x}$$

$$\text{Tangent: } \tan\theta = \frac{y}{x} \qquad \text{Cotangent: } \cot\theta = \frac{x}{y}.$$

p. 41

Big-oh notation: If functions f and g are positive for x sufficiently large, then f is big-oh of g (written $f = O(g)$) if there exists a positive integer M such that

$$\frac{f(x)}{g(x)} \le M$$

for x sufficiently large. p. 429

Binomial series: The Maclaurin series for $f(x) = (1 + x)^m$. p. 479

Bounded: A function f is bounded on a given domain if there are numbers m and M such that $m \le f(x) \le M$ for any x in the domain of f. The number m is a lower bound and the number M is an upper bound for f. p. 266

Bounded (from) above: A function f is bounded (from) above on a given domain if there is a number M such that $f(x) \le M$ for all x in the domain. p. 270

Bounded (from) below: A function f is bounded (from) below on a given domain if there is a number m such that $m \le f(x)$ for all x in the domain. p. 270

Cardioid: Any of the heart-shaped polar curves $r = a(1 - \cos\theta)$ or $r = a(1 - \sin\theta)$, $a \ne 0$ a constant. p. 554

Center of a power series: See *Power series*.

Chain Rule: If $y = f(u)$ is differentiable at the point $u = g(x)$, and g is differentiable at x, then the composite function $(f \circ g)(x) = f(g(x))$ is differentiable at x, and

$$(f \circ g)'(x) = f'(g(x)) \cdot g'(x), \quad \text{or} \quad \frac{dy}{dx} = \frac{dy}{du} \cdot \frac{du}{dx},$$

where dy/du is evaluated at $u = g(x)$. p. 142

Circular functions: The functions $\cos x$ and $\sin x$ with reference to their function values corresponding to points $(\cos x, \sin x)$, x in radians, on the unit circle. More generally, the six basic trigonometric functions. p. 41

Closed interval $[a, b]$: The set of all real numbers x with $a \le x \le b$. p. 11

Comparison Test (Direct, for improper integrals): For f and g continuous on $[a, \infty)$ with $0 \le f(x) \le g(x)$ for all $x \ge a$,

(1) $\int_a^\infty f(x)\,dx$ converges if $\int_a^\infty g(x)\,dx$ converges;

(2) $\int_a^\infty g(x)\,dx$ diverges if $\int_a^\infty f(x)\,dx$ diverges. p. 438

Comparison Test (Direct, for infinite series): Let $\sum a_n$ be a series with no negative terms. Then

(1) $\sum a_n$ converges if there is a convergent series $\sum c_n$ with $a_n \le c_n$ for all $n > N$ for some integer N;

(2) $\sum a_n$ diverges if there is a divergent series $\sum d_n$ of nonnegative terms with $a_n \ge d_n$ for all $n > N$ for some integer N. p. 489

Comparison Test (Limit, for improper integrals): If the positive functions f and g are continuous on $[a, \infty)$ and if

$$\lim_{x \to \infty} \frac{f(x)}{g(x)} = L, \quad 0 < L < \infty,$$

then $\int_a^\infty f(x)\,dx$ and $\int_a^\infty g(x)\,dx$ both converge or both diverge. p. 439

Comparison Test (Limit, for infinite series): Suppose that $a_n > 0$ and $b_n > 0$ for all $n \ge N$, N an integer.

(1) If

$$\lim_{n \to \infty} \frac{a_n}{b_n} = c, \quad 0 < c < \infty,$$

then $\sum a_n$ and $\sum b_n$ both converge or both diverge;

(2) if

$$\lim_{n \to \infty} \frac{a_n}{b_n} = 0$$

and $\sum b_n$ converges, then $\sum a_n$ converges;

(3) if

$$\lim_{n \to \infty} \frac{a_n}{b_n} = \infty$$

and $\sum b_n$ diverges, then $\sum a_n$ diverges. p. 498

Complex number: An expression of the form $a + bi$ where a and b are real numbers and i is a symbol for $\sqrt{-1}$.

Component form of a vector: If **v** is a vector in the plane equal to the vector with initial point $(0, 0)$ and terminal point (v_1, v_2), the component form of **v** is $\mathbf{v} = \langle v_1, v_2 \rangle$. The number v_1 is the *horizontal component* of **v**; v_2 is the *vertical component.*   p. 521

Component functions: Of a parametrized curve $x = f(t)$, $y = g(t)$, $t \in I$, the functions f and g. Of a vector function $\mathbf{r}(t) = f(t)\mathbf{i} + g(t)\mathbf{j}$, the functions f and g.

Components of a vector: See *Component form of a vector.*

Composite function $f \circ g$: The function $(f \circ g)(x) = f(g(x))$.   p. 16

Concave down: The graph of a differentiable function $y = f(x)$ is concave down on an open interval I if y' is decreasing on I.   p. 197

Concave up: The graph of a differentiable function $y = f(x)$ is concave up on an open interval I if y' is increasing on I.   p. 197

Conditional convergence: An infinite series is conditionally convergent if it is convergent but not absolutely convergent.   p. 502

Constant function: A function that assigns the same value to every element in its domain.

Constant of integration: The arbitrary constant C in $\int f(x)\,dx = F(x) + C$, where F is any antiderivative of f.   p. 306

Continuity at an endpoint: A function $f(x)$ is continuous at a left endpoint a of its domain if $\lim_{x \to a^+} f(x) = f(a)$. The function is continuous at a right endpoint b of its domain if $\lim_{x \to b^-} f(x) = f(b)$.   p. 74

Continuity at an interior point: A function $f(x)$ is continuous at an interior point c of its domain if $\lim_{x \to c} f(x) = f(c)$.   p. 73

Continuity of vector functions: A vector function $\mathbf{r}(t)$ is continuous at $t = c$ if $\lim_{t \to c} \mathbf{r}(t) = \mathbf{r}(c)$ (use appropriate one-sided limits at endpoints). In terms of components, $\mathbf{r}(t) = f(t)\mathbf{i} + g(t)\mathbf{j}$ is continuous at $t = c$ if and only if f and g are continuous at $t = c$.   p. 531

Continuity on an interval: A function is continuous on an interval if and only if it is continuous at each point of the interval.   p. 81

Continuous extension of a function f: A function identical to f except that it is continuous at one or more points where f is not.   p. 77

Continuous function: A function that is continuous at each point of its domain.   p. 77

Convergent improper integral: An improper integral whose related limit(s) is (are) finite.   p. 434

Convergent sequence: A sequence converges if it has a limit. See *Limit at infinity.*

Convergent series: An infinite series converges if its sequence of partial sums has a finite limit.   p. 458

Cosecant function: See *Basic trigonometric functions.*

Cosine function: See *Basic trigonometric functions.*

Cotangent function: See *Basic trigonometric functions.*

Critical point (value): A point (value) in the interior of the domain of a function f at which $f' = 0$ or f' does not exist.   p. 180

Critical value: See *Critical point.*

Cross section area: The area of a cross section of a solid.   p. 383

Curve traced by a vector function: The curve traced by $\mathbf{r}(t) = f(t)\mathbf{i} + g(t)\mathbf{j}$, $t \in I$, is the parametrized curve $x = f(t)$, $y = g(t)$, $t \in I$.   p. 530

Decay models: See *Exponential growth and decay.*

Decreasing function: Let f be a function defined on an interval I. Then f decreases on I if, for any two points x_1 and x_2 in I,
$$x_1 < x_2 \;\Rightarrow\; f(x_1) > f(x_2).$$
p. 188

Decreasing on an interval: See *Decreasing function.*

Definite integral of a vector function $\mathbf{r}(t) = f(t)\mathbf{i} + g(t)\mathbf{j}$ from a to b: The vector
$$\int_a^b \mathbf{r}(t)\,dt = \left(\int_a^b f(t)\,dt\right)\mathbf{i} + \left(\int_a^b g(t)\,dt\right)\mathbf{j}.$$
p. 535

Definite integral of f over an interval $[a, b]$: For any partition P of $[a, b]$, let a number c_k be chosen arbitrarily in each subinterval $[x_{k-1}, x_k]$, and let $\Delta x_k = x_k - x_{k-1}$. If there exists a number I such that
$$\lim_{\|P\| \to 0} \sum_{k=1}^{n} f(c_k)\,\Delta x_k = I$$
no matter how P and the c_k's are chosen, then I is the definite integral of f over $[a, b]$.   p. 260

Delta notation (Δ): See *Increment.*

Derivative of a function f at a point a:
$$\lim_{h \to 0} \frac{f(a + h) - f(a)}{h},$$
provided the limit exists.   p. 95

Derivative of a function f with respect to x: The function f' whose value at x is
$$\lim_{h \to 0} \frac{f(x + h) - f(x)}{h},$$
provided the limit exists. Alternatively,
$$\lim_{x \to a} \frac{f(x) - f(a)}{x - a},$$
provided the limit exists.   p. 96

Difference of vectors:
$$\mathbf{v} - \mathbf{u} = \langle v_1, v_2 \rangle - \langle u_1, u_2 \rangle = \langle v_1 - u_1, v_2 - u_2 \rangle$$
p. 522

Difference quotient of the function *f* **at** *a:*

$$\frac{f(a + h) - f(a)}{h}.$$

Alternatively,

$$\frac{f(x) - f(a)}{x - a}.$$

p. 86

Differentiability: If $f'(x)$ exists, the function *f* is differentiable at *x*. A function that is differentiable at every point of its domain is a differentiable function.   p. 95

Differentiable curve: The graph of a differentiable function; also, a parametrized curve whose component functions are differentiable at every parameter value. pp. 95, 513

Differentiable vector function: $\mathbf{r}(t) = f(t)\mathbf{i} + g(t)\mathbf{j}$ is differentiable at *t* if *f* and *g* are differentiable at *t*, and is differentiable if *f* and *g* are differentiable. Its derivative is the vector

$$\frac{d\mathbf{r}}{dt} = \frac{df}{dt}\mathbf{i} + \frac{dg}{dt}\mathbf{j}.$$

Differential: If $y = f(x)$ is a differentiable function, the differential *dx* is an independent variable and the differential *dy* is $dy = f'(x)\, dx$.   p. 224

Differential calculus: The branch of mathematics that deals with derivatives.   p. 95

Differential equation: An equation containing a derivative. p. 303

Differentiation: The process of taking a derivative.   p. 95

Direction field: See *Slope field*.

Direction of motion: $\mathbf{v}/|\mathbf{v}|$, where **v** is the (nonzero) velocity vector of the motion.   p. 533

Discontinuity: If a function *f* is not continuous at a point *c*, then *c* is a point of discontinuity of *f*.   p. 75

Disk method: A method for finding the volume of a solid of revolution by evaluating $\int_a^b A(x)\, dx$, where $A(x)$ is the area of the disk cut by a cross section of the solid perpendicular to the axis of revolution at *x*.   p. 384

Distance traveled (from velocity): The integral of the absolute value of a velocity function with respect to time. p. 298

Divergent improper integral: An improper integral for which at least one of the defining limits does not exist. p. 434

Divergent sequence: An infinite sequence that has no limit as $n \to \infty$.   p. 468

Divergent series: An infinite series whose sequence of partial sums diverges.   p. 458

Domain of a function: See *Function*.

Domination: A function *f* dominates a function *g* on a domain *D* if $f(x) \geq g(x)$ for all *x* in *D*. A sequence $\{a_n\}$ dominates a sequence $\{b_n\}$ if $a_n \geq b_n$ for all *n*. An infinite series $\sum a_n$ dominates an infinite series $\sum b_n$ if $a_n \geq b_n$ for all *n*.

Dot (inner) product: Of vectors $\mathbf{u} = \langle u_1, u_2 \rangle$ and $\mathbf{v} = \langle v_1, v_2 \rangle$, the number $\mathbf{u} \cdot \mathbf{v} = u_1 v_1 + u_2 v_2$.   p. 524

Dummy variable of integration: In $\int_a^b f(x)\, dx$, the variable *x*. It could be any other letter without changing the value of the integral.   p. 262

$\dfrac{dy}{dx}$**:** The derivative of *y* with respect to *x*.   p. 97

e **(the number):** To 9 decimal places, $e = 2.718281828$. More formally,

$$e = \lim_{x \to \infty}\left(1 + \frac{1}{x}\right)^x.$$

p. 23

End behavior model: The function *g* is

(1) a right end behavior model for *f* if and only if

$$\lim_{x \to \infty}\frac{f(x)}{g(x)} = 1;$$

(2) a left end behavior model for *f* if and only if

$$\lim_{x \to -\infty}\frac{f(x)}{g(x)} = 1;$$

(3) an end behavior model for *f* if it is both a left and a right end behavior model for *f*.   p. 70

Euler's method: A method using linearizations to approximate the solution of an initial value problem. p. 350

Even function: A function *f* for which $f(-x) = f(x)$ for every *x* in the domain of *f*.   p. 13

Exponential change: See *Law of exponential change*.

Exponential function with base *a:* The function $f(x) = a^x$, $a > 0$ and $a \neq 1$.   p. 20

Exponential growth and decay: Growth and decay modeled by the functions $y = k \cdot a^x$, $k > 0$, with $a > 1$ for growth and $0 < a < 1$ for decay.   p. 20

Extreme value: See *Extremum*.

Extremum: A maximum or minimum value (extreme value) of a function on a set. See also *Absolute maximum; Absolute minimum; Local maximum; Local minimum*.

First derivative test (for local extrema): For a continuous function *f*,

(1) if f' changes sign from positive to negative at a critical point *c*, then *f* has a local maximum value at *c*;

(2) if f' changes sign from negative to positive at a critical point *c*, then *f* has a local minimum value at *c*;

(3) if f' does not change sign at a critical point *c*, then *f* has no local extreme value at *c*;

(4) if $f' < 0$ $(f' > 0)$ for $x > a$ where a is a left endpoint in the domain of f, then f has a local maximum (minimum) value at a;

(5) if $f' < 0$ $(f' > 0)$ for $x < b$ where b is a right endpoint in the domain of f, then f has a local minimum (maximum) value at b.   p. 195

Free fall equation: When air resistance is absent or insignificant and the only force acting on a falling body is the force of gravity, we call the way the body falls *free fall*. In a free fall short enough for the acceleration of gravity to be assumed constant, call it g, the position of a body released to fall from position s_0 at time $t = 0$ with velocity v_0 is modeled by the equation $s(t) = (1/2)gt^2 + v_0 t + s_0$.   p. 125

Frequency of a periodic function: The reciprocal of the period of the function, or the number of cycles or periods per unit time. The function $\sin x$, with x in seconds, has period 2π seconds and completes $1/2\pi$ cycles per second (has frequency $1/2\pi$).   p. 43

Function: A rule that assigns a unique element in a set R to each element in a set D. The set D is the *domain* of the function. The set of elements assigned from R is the *range* of the function.   p. 10

Fundamental Theorem of Calculus, Part 1: If f is continuous on $[a, b]$, then the function $F(x) = \int_a^x f(t)\, dt$ has a derivative with respect to x at every point in $[a, b]$ and

$$\frac{dF}{dx} = \frac{d}{dx} \int_a^x f(t)\, dt = f(x).$$

p. 277

Fundamental Theorem of Calculus, Part 2: If f is continuous on $[a, b]$, and F is any antiderivative of f on $[a, b]$, then

$$\int_a^b f(x)\, dx = F(b) - F(a).$$

p. 282

Gaussian curve: See *Normal curve*.

General linear equation: $Ax + By = C$ (A and B not both 0).

Geometric sequence: A sequence of the form a, ar, ar^2, …, ar^n, …, in which each term after the first is obtained from its preceding term by multiplying by the same number r. The number r is the *common ratio* of the sequence.

Geometric series: A series of the form

$$a + ar + ar^2 + \cdots + ar^n + \cdots, \quad \text{or} \quad \sum_{n=1}^{\infty} ar^{n-1},$$

in which each term after the first is obtained from its preceding term by multiplying by the same number r. The number r is the *common ratio* of the series.   p. 459

Global maximum: See *Absolute maximum*.

Global minimum: See *Absolute minimum*.

Graph of a function: The set of points (x, y) in the coordinate plane whose coordinates are the input-output pairs of the function.   p. 11

Growth models: See *Exponential growth and decay; Logistic growth*.

Growth rate: See *Relative growth rate*.

Half-life of a radioactive element: The time required for half of the radioactive nuclei present in a sample to decay.   p. 21

Harmonic series: The series

$$1 + \frac{1}{2} + \frac{1}{3} + \frac{1}{4} + \cdots + \frac{1}{n} + \cdots = \sum_{n=1}^{\infty} \frac{1}{n}.$$

p. 498

Hooke's Law: When a force is applied to stretch or compress a spring, the magnitude F of the force in the direction of motion is proportional to the distance x that the spring is stretched or compressed. In symbols, $F \sim x$ or $F = kx$, where k is the constant of proportionality. This relationship is *Hooke's Law*. If an elastic material is stretched too far, it becomes distorted and will not return to its original state. The distance beyond which distortion occurs is the material's *elastic limit*. Hooke's Law holds only as long as the material is not stretched past its elastic limit.   p. 369

Horizontal line: In the Cartesian coordinate plane, a line parallel to the x-axis.

Imaginary number: A complex number of the form $0 + bi$. See *Complex number*.

Implicit differentiation: A process for finding dy/dx when y is implicitly defined as a function of x by an equation of the form $f(x, y) = 0$.   p. 149

Improper integral: An integral on an infinite interval or on a finite interval containing one or more points of infinite discontinuity of the integrand. Its value is found as a limit or sum of limits.   p. 433

Increasing function: Let f be a function defined on an interval I. Then f increases on I if, for any two points x_1 and x_2 in I,

$$x_1 < x_2 \quad \Rightarrow \quad f(x_1) < f(x_2).$$

p. 188

Increasing on an interval: See *Increasing function*.

Increment: If coordinates change from (x_1, y_1) to (x_2, y_2), the increments in the coordinates are $\Delta x = x_2 - x_1$ and $\Delta y = y_2 - y_1$. The symbols Δx and Δy are read "delta x" and "delta y."   p. 1

Indefinite integral of a function f: The set of all antiderivatives of f, denoted by $\int f(x)\, dx$.   p. 306

Indefinite integral of a vector function $\mathbf{r}(t)$: The set of all antiderivatives of $\mathbf{r}$, denoted by $\int \mathbf{r}(t)\, dt$.   p. 535

Indeterminate form: A nonnumeric expression of the form $0/0$, ∞/∞, $0 \cdot \infty$, $\infty - \infty$, 1^∞, ∞^0, or 0^0 obtained when trying substitution to evaluate a limit. The expression reveals nothing about the limit, but does suggest that l'Hôpital's Rule may be applied to help find the limit.   p. 417

Infinite discontinuity: A point of discontinuity where one or both of the one-sided limits are infinite.   p. 76

Infinite limit: If the values of a function $f(x)$ outgrow all positive bounds as x approaches a finite number a, we say

$$\lim_{x \to a} f(x) = \infty.$$

If the values of f become large and negative, exceeding all negative bounds as $x \to a$, we say

$$\lim_{x \to a} f(x) = -\infty.$$

p. 68

Infinite series: An expression of the form

$$a_1 + a_2 + a_3 + \cdots + a_n + \cdots, \quad \text{or} \quad \sum_{k=1}^{\infty} a_k.$$

The numbers $a_1, a_2, \ldots$ are the *terms* or the series; a_n is the *n*th term. The *partial sums* of the series form a sequence

$$s_1 = a_1$$
$$s_2 = a_1 + a_2$$
$$s_3 = a_1 + a_2 + a_3$$
$$\vdots$$
$$s_n = a_1 + a_2 + a_3 + \cdots + a_n$$

of numbers, each defined as a finite sum. If the sequence of partial sums has a limit S as $n \to \infty$, the series *converges* to the *sum S*. A series that fails to converge *diverges*.   p. 458

Inflection point: A point where the graph of a function has a tangent line and the concavity changes.   p. 198

Initial condition: See *Initial value problem.*

Initial value problem: For a first order differential equation, the problem of finding the solution that has a particular value at a given point. The condition that the solution have this value at the point is the *initial condition* of the problem. For a second order differential equation, the problem of finding the solution given its value and the value of its first derivative at a point (*two* initial conditions).   p. 303

Inner product: See *Dot product.*

Instantaneous rate of change of f with respect to x at a: The derivative

$$f'(a) = \lim_{h \to 0} \frac{f(a + h) - f(a)}{h},$$

provided the limit exists.   p. 87

Instantaneous velocity: The derivative of a position function with respect to time.   p. 124

Integrable function on [a, b]: A function for which the definite integral over $[a, b]$ exists.   p. 260

Integral calculus: The branch of mathematics that deals with integrals.   p. 247

Integrand: $f(x)$ in $\int f(x)\, dx$ or in $\int_a^b f(x)\, dx$.   p. 261

Integration: The evaluation of a definite integral, an indefinite integral, or an improper integral.   p. 261

Integration by partial fractions: A method for integrating a rational function by writing it as a sum of proper fractions, called *partial fractions,* with linear or quadratic denominators.   p. 444

Integration by parts: A method of integration in which $\int u\, dv$ is rewritten as $uv - \int v\, du$.   p. 323

Integration by substitution: A method of integration in which $\int f(g(x)) \cdot g'(x)\, dx$ is rewritten as $\int f(u)\, du$ by substituting $u = g(x)$ and $du = g'(x)\, dx$.   p. 318

Intermediate Value Theorem for Continuous Functions: A function $y = f(x)$ that is continuous on a closed interval $[a, b]$ takes on every y-value between $f(a)$ and $f(b)$.   p. 79

Intermediate Value Theorem for Derivatives: If a and b are any two points in an interval on which f is differentiable, then f' takes on every value between $f'(a)$ and $f'(b)$.   p. 110

Interval: A subset of the number line formed by any of the following: (1) two points and the points in between; (2) only the points in between two points; (3) the points in between two points and one of the two points; (4) one point and the points to one side of it; (5) only the points to one side of a given point. The real line is also considered to be an interval. p. 11

Interval of convergence: The interval of x-values for which a power series converges. See *Power series.*

Inverse function f^{-1}: The function obtained by reversing the ordered pairs of a one-to-one function f.   p. 33

Jerk: The derivative of an acceleration function with respect to time.   p. 137

Jump discontinuity: A point of discontinuity where the one-sided limits exist but have different values. At such a point, the function jumps from one value to another.   p. 76

Lagrange error bound: A bound for truncation error obtained from the Lagrange form of the remainder for Taylor series.   p. 482

Lagrange form of the remainder: The formula

$$\frac{f^{n+1}(c)}{(n + 1)!}(x - a)^{n+1}$$

for the remainder in Taylor's Theorem.   p. 482

Law of exponential change: If a quantity y changes at a rate proportional to the amount present ($dy/dt = ky$) and $y = y_0$ when $t = 0$, then

$$y = y_0 e^{kt},$$

where $k > 0$ represents growth and $k < 0$ represents decay. The number k is the *rate constant*.   p. 330

Left end behavior model: See *End behavior model.*

Left-hand derivative: The derivative defined by a left-hand limit.   p. 100

Left-hand limit: The limit of f as x approaches c from the left, or $\lim_{x \to c^-} f(x)$.   p. 60

Length (magnitude) of a vector: The length (magnitude) of $\mathbf{v} = \langle v_1, v_2 \rangle$ is $|\mathbf{v}| = \sqrt{v_1^2 + v_2^2}$.   p. 521

L'Hôpital's Rule: Suppose that $f(a) = g(a) = 0$, that f and g are differentiable on an open interval I containing a, and that $g'(x) \neq 0$ on I if $x \neq a$. Then

$$\lim_{x \to a} \frac{f(x)}{g(x)} = \lim_{x \to a} \frac{f'(x)}{g'(x)},$$

provided the limit exists (or is $\pm\infty$). L'Hôpital's Rule also applies to quotients that lead to ∞/∞. If $f(x)$ and $g(x)$ both approach ∞ as $x \to a$, then

$$\lim_{x \to a} \frac{f(x)}{g(x)} = \lim_{x \to a} \frac{f'(x)}{g'(x)},$$

provided the latter limit exists (or is $\pm\infty$). In this case, a may itself be either finite or infinite.   p. 417

Limit: The function f has limit L as x approaches c if, given any positive number ε, there exists a positive number δ such that for all x,

$$0 < |x - c| < \delta \quad \Rightarrow \quad |f(x) - L| < \varepsilon.$$

This is represented as $\lim_{x \to c} f(x) = L$.   p. 57

Limit at infinity: The function f has limit L as x approaches ∞ if, given any positive number ε, there exists a positive number N such that for all $x > N$, $|f(x) - L| < \varepsilon$. This is represented as $\lim_{x \to \infty} f(x) = L$.

The function f has limit L as x approaches $-\infty$ if, given any positive number ε, there is a negative number N such that for all x with $x < N$, $|f(x) - L| < \varepsilon$. This is represented as $\lim_{x \to -\infty} f(x) = L$.

The sequence $f(n) = x_n$ has limit L, if $\lim_{n \to \infty} f(n) = L$. p. 468

Limit of a sequence: See *Limit at infinity*.

Limit of a vector function: If $\lim_{t \to c} f(t) = L_1$ and $\lim_{t \to c} g(t) = L_2$, and $\mathbf{r}(t) = f(t)\mathbf{i} + g(t)\mathbf{j}$, then $\lim_{t \to c} \mathbf{r}(t) = L_1\mathbf{i} + L_2\mathbf{j}$.   p. 531

Limits of integration: a and b in $\int_a^b f(x)\, dx$.   p. 261

Linear approximation (standard) of f at a: The approximation $f(x) \approx L(x)$ where $L(x)$ is the linearization of f at a.   p. 220

Linear combination of vectors u and v: Any vector $a\mathbf{u} + b\mathbf{v}$, where a and b are numbers, is a linear combination of $\mathbf{u}$ and $\mathbf{v}$.   p. 529

Linear equation: See *General linear equation*.

Linear function: A function that can be expressed in the form $f(x) = mx + b$.

Linearization of f at a: The approximating function $L(x) = f(a) + f'(a)(x - a)$ when f is differentiable at $x = a$.   p. 221

Little-oh notation: If functions f and g are positive for x sufficiently large, then f is little-oh of g (written $f = o(g)$) if $\lim_{x \to \infty} (f(x)/g(x)) = 0$.   p. 429

Local extrema: See *Local maximum; Local minimum*.

Local linearity: If a function $f(x)$ is differentiable at $x = a$, then, close to a, its graph resembles the tangent line at a. p. 107

Local linearization: See *Linearization of f at a*.

Local maximum: The function f has a local maximum value $f(c)$ at a point c in the interior of its domain if and only if $f(x) \leq f(c)$ for all x in some open interval containing c. The function has a local maximum value at an endpoint c if the inequality holds for all x in some half-open domain interval containing c.   p. 179

Local minimum: The function f has a local minimum value $f(c)$ at a point c in the interior of its domain if and only if $f(x) \geq f(c)$ for all x in some open interval containing c. The function has a local minimum value at an endpoint c if the inequality holds for all x in some half-open domain interval containing c.   p. 179

Logarithm function with base a: The function $y = \log_a x$ that is the inverse of the exponential function $y = a^x$, $a > 0$, $a \neq 1$.   p. 36

Logarithmic differentiation: The process of taking the natural logarithm of both sides of an equation, differentiating, and then solving for the desired derivative.   p. 169

Logistic curve: A solution curve of the logistic differential equation. It describes population growth that begins slowly when the population is small, speeds up as the number of reproducing individuals increases and nutrients are still plentiful, and slows down again as the population reaches the carrying capacity of its environment.   p. 199

Logistic differential equation:

$$\frac{dP}{dt} = \frac{k}{M} P(M - P)$$

where P is current population, t is time, M is the carrying capacity of the environment, and k is a positive proportionality constant.   p. 343

Logistic growth model: The solution

$$P = \frac{M}{1 + Ae^{-kt}},$$

to the logistic differential equation, A an arbitrary constant. This model assumes that the relative growth rate of a population is positive, but decreases as the population increases due to environmental and economic factors. See *Logistic differential equation*.   p. 343

Lower bound: See *Bounded*.

LRAM: Left-hand endpoint rectangular approximation method. The method of approximating a definite integral over an interval using the function values at the left-hand endpoints of the subintervals determined by a partition.   p. 249

Maclaurin series: See *Taylor series*.

Magnitude: Of a number, its absolute value; of a vector, its length.

Maximum: See *Absolute maximum; Local maximum.*

Mean value: See *Average value of a continuous function on [a, b].*

Mean Value Theorem for Definite Integrals: If f is continuous on $[a, b]$, then at some point c in $[a, b]$,

$$f(c) = \frac{1}{b - a} \int_a^b f(x)\, dx.$$

p. 272

Mean Value Theorem for Derivatives: If $y = f(x)$ is continuous at every point of the closed interval $[a, b]$ and differentiable at every point of its interior (a, b), then there is at least one point c in (a, b) at which

$$f'(c) = \frac{f(b) - f(a)}{b - a}.$$

p. 186

Minimum: See *Absolute minimum; Local minimum.*

Monotonic (monotone) function: A function f is *monotonic increasing* on a domain D if

$$x_1 < x_2 \implies f(x_1) < f(x_2)$$

for all x_1 and x_2 in D. The function is *monotonic nondecreasing* on D if

$$x_1 < x_2 \implies f(x_1) \le f(x_2)$$

for all x_1 and x_2 in D. Similarly, f is *monotonic decreasing* on a domain D if

$$x_1 < x_2 \implies f(x_1) > f(x_2)$$

for all x_1 and x_2 in D and *monotonic nonincreasing* on D if

$$x_1 < x_2 \implies f(x_1) \ge f(x_2)$$

for all x_1 and x_2 in D.   p. 188

Monotonic sequence: See *Monotonic function.*

MRAM: Midpoint rectangular approximation method. The method of approximating a definite integral over an interval using the function values at the midpoints of the subintervals determined by a partition.   p. 249

Natural logarithm: a is the natural logarithm of b if and only if $b = e^a$. The natural logarithm function $y = \ln x$ is the inverse of the exponential function $y = e^x$.   p. 36

NDER$(f(x), a)$: The numerical derivative of f at $x = a$. p. 108

Newton's second law (of motion): Most of the motion we observe undergoes changes that are the result of one or more applied forces. The overall net force, whether it be from a single force or a combination of forces, produces acceleration. The relationship of acceleration to force and mass is described by Newton's second law, which says that the acceleration of an object is directly proportional to the net force and inversely proportional to the object's mass:

$$\text{Acceleration} \sim \frac{\text{net force}}{\text{mass}} \quad \text{or} \quad a \sim \frac{F}{m}.$$

With appropriate units, the proportionality may be expressed as an exact equation: $F = ma$. To take direction into account as well as magnitude, we use the vector notation $\mathbf{F} = m\mathbf{a}$. p. 540

NINT$(f(x), x, a, b)$: The numerical integral of f with respect to x, from $x = a$ to $x = b$.   p. 281

Nonremovable discontinuity: A discontinuity that is not removable. See *Removable discontinuity.*

Norm of a partition: The longest subinterval length, denoted $\|P\|$, for a partition P.   p. 260

Normal curve: The graph of a normal probability density function.   p. 405

Normal line to a curve at a point of the curve: The line perpendicular to the tangent at that point.   p. 86

Normal probability density function: The normal probability density function for a population with *mean* μ and *standard deviation* σ is

$$f(x) = \frac{1}{\sigma\sqrt{2\pi}} e^{-(x - \mu)^2/(2\sigma^2)}.$$

The mean μ represents the average value of the variable x. The standard deviation σ measures the "scatter" around the mean.   p. 405

Normal vector: To a differentiable curve at a point P, a vector perpendicular to the tangent to the curve at P. To a curve traced by a vector function at a point P where the velocity is not zero, a vector perpendicular to the velocity vector at P.   p. 525

Numerical derivative: An approximation of the derivative of a function using a numerical algorithm.   p. 108

Numerical integration: Approximating the integral of a function using a numerical algorithm.   p. 280

Numerical method: A method for generating a numerical solution of a problem. For example, a method for estimating the value of a definite integral, for estimating the zeros of a function or solutions of an equation, or for estimating values of the function that solves an initial value problem.   p. 351

Numerical solution: Of an equation $f(x) = 0$, an estimate of one or more of its roots; of an initial value problem, a table of estimated values of the solution function.   p. 351

Odd function: A function f for which $f(-x) = -f(x)$ for every x in the domain of f.   p. 13

One-sided limit: See *Left-hand limit; Right-hand limit.*

One-to-one function: A function f for which $f(a) \ne f(b)$ whenever $a \ne b$.   p. 32

Open interval (a, b): All numbers x with $a < x < b$. p. 11

Optimization: In an application, maximizing or minimizing some aspect of the system being modeled.   p. 206

Order of a derivative: If y is a function of x, $y' = dy/dx$ is the first order, or first, derivative of y with respect to x; $y'' = d^2y/dx^2$ is the second order, or second, derivative of y; $y^{(n)} = dy^{(n-1)}/dx$ is the nth order, or nth, derivative of y. p. 119

Order of a differential equation: The order of the highest order derivative in the equation.   p. 303

Origin: The point $(0, 0)$ in the Cartesian coordinate plane; the point $(0, \theta)$ in the polar coordinate plane.   p. 552

Orthogonal curves: See *Perpendicular curves.*

Orthogonal vectors: Vectors making a 90° angle.   p. 528

Oscillating discontinuity: A point near which the function values oscillate too much for the function to have a limit. p. 76

Parallel curves: In the plane, curves that differ from one another by a vertical translation (shift).   p. 103

Parameter: See *Parametric equations.*

Parameter interval: See *Parametric equations.*

Parametric equations: If x and y are given as functions $x = f(t)$, $y = g(t)$, over an interval of t-values, then the set of points $(x, y) = (f(t), g(t))$ defined by these equations is a *parametric curve*, the equations are *parametric equations* for the curve, the variable t is the *parameter* for the curve, and the interval of allowable t-values is the *parameter interval*.   p. 26

Parametrization of a curve: The parametric equations and parameter interval describing a curve.   p. 27

Partial fractions: See *Integration by partial fractions.*

Partial sum: See *Infinite series.*

Particular solution: The unique solution of a differential equation satisfying given initial condition or conditions. p. 303

Partition of an interval $[a, b]$: A set

$$\{x_0 = a, x_1, x_2, \ldots, x_n = b\}$$

of points in $[a, b]$ numbered in order from left to right. p. 259

Percentage change: See *Absolute, relative, and percentage change.*

Percentage error:

$$\frac{|\text{approximate value} - \text{exact value}|}{|\text{exact value}|} \times 100.$$

Period of a periodic function f: The smallest positive number p for which $f(x + p) = f(x)$ for every value of x. See *Periodic function.*

Periodic function: A function f for which there is a positive number p such that $f(x + p) = f(x)$ for every value of x. p. 43

Perpendicular (orthogonal) curves: Two curves are said to be perpendicular (orthogonal) at a point of intersection if their tangents at that point are perpendicular.   p. 148

Piecewise-defined function: A function that is defined by applying different formulas to different parts of its domain. p. 15

Polar coordinates: Each point P in the polar coordinate plane has polar coordinates (r, θ) where r gives the directed distance from the origin O to P and θ gives a directed angle from the initial ray to ray OP.   p. 552

Pole: The polar coordinate origin.   p. 552

Polynomial: An expression of the form

$$a_n x^n + a_{n-1} x^{n-1} + \cdots + a_1 x + a_0.$$

Position function: A function f that gives the position $f(t)$ of a body on a coordinate axis at time t.   p. 86

Position vector: The vector $\langle a, b \rangle$ is the position vector of the point (a, b).   p. 530

Power series: An expression of the form

$$c_0 + c_1 x + c_2 x^2 + \cdots + c_n x^n + \cdots = \sum_{n=0}^{\infty} c_n x^n$$

is a power series centered at $x = 0$. An expression of the form

$$c_0 + c_1(x - a) + c_2(x - a)^2 + \cdots$$
$$+ c_n(x - a)^n + \cdots = \sum_{n=0}^{\infty} c_n(x - a)^n$$

is a power series centered at $x = a$. The number a is the *center* of the series.   p. 461

Prime notation $(f'(x))$: If $y = f(x)$, then both y' and $f'(x)$ denote the derivative of the function with respect to x.   p. 97

Probability: See *Probability density function.*

Probability density function (pdf): A function $f(x)$ such that $f(x) \geq 0$ for all x and $\int_{-\infty}^{\infty} f(x)\,dx = 1$. The *probability* associated with the interval $[a, b]$ is $\int_a^b f(x)\,dx$. p. 405

Product Rule: The product of two differentiable functions u and v is differentiable, and

$$\frac{d}{dx}(uv) = u\frac{dv}{dx} + v\frac{du}{dx}.$$

p. 115

p-series: A series of the form

$$\sum_{n=1}^{\infty} \frac{1}{n^p}, \qquad p \text{ a nonzero constant.}$$

p. 497

Quotient Rule: At a point where $v \neq 0$, the quotient $y = u/v$ of two differentiable functions is differentiable, and

$$\frac{d}{dx}\left(\frac{u}{v}\right) = \frac{v\dfrac{du}{dx} - u\dfrac{dv}{dx}}{v^2}.$$

p. 117

Radian measure: If a central angle of a circle of radius r intercepts an arc of length s on the circle, the radian measure of the angle is s/r.   p. 41

Radius of convergence: In general, the positive number R for which the power series

$$\sum_{n=0}^{\infty} c_n(x - a)^n$$

converges when $|x - a| < R$ and diverges when $|x - a| > R$. If the series converges only for $x = a$, then $R = 0$. If the series converges for every x, then $R = \infty$. p. 488

Range of a function: See *Function*.

Rate constant: See *Law of exponential change*.

Rate of change: See *Average rate of change of a quantity over a period of time; Instantaneous rate of change of f with respect to x at a.*

Ratio Test: Let $\sum a_n$ be a series with positive terms and with

$$\lim_{n \to \infty} \frac{a_{n+1}}{a_n} = L.$$

Then (1) the series converges if $L < 1$; (2) the series diverges if $L > 1$; (3) the test is inconclusive if $L = 1$. p. 491

Rational function: A function that can be expressed as the quotient of two polynomial functions.

Regular partition: A partition in which consecutive points are equally spaced. p. 260

Related rates: An equation involving two or more variables that are differentiable functions of time can be used to find an equation that relates the corresponding rates. p. 232

Relative change: See *Absolute, relative, and percentage change*.

Relative extrema: Same as *Local extrema*.

Relative growth rate: For a population modeled by a differentiable function $P(t)$ giving the number of individuals at time t, the quotient $(dP/dt)/P$. p. 342

Relative maximum: Same as *Local maximum*.

Relative minimum: Same as *Local minimum*.

Remainder of order n: The remainder

$$R_n(x) = \frac{f^{(n+1)}(c)}{(n + 1)!}(x - a)^{n+1}$$

in Taylor's Theorem. p. 482

Removable discontinuity: A discontinuity c of the function f for which $f(c)$ can be (re)defined so that $\lim_{x \to c} f(x) = f(c)$. p. 76

Resultant vector: The vector that results from adding or subtracting two vectors. p. 522

Riemann sum: A sum of the form

$$\sum_{k=1}^{n} f(c_k) \cdot \Delta x_k$$

where f is a continuous function on a closed interval $[a, b]$; c_k is some point in, and Δx_k the length of, the kth subinterval in some partition of $[a, b]$. p. 258

Right end behavior model: See *End behavior model*.

Right-hand derivative: The derivative defined by a right-hand limit. p. 100

Right-hand limit: The limit of f as x approaches c from the right, or $\lim_{x \to c^+} f(x)$. p. 60

Root of an equation: See *Zero of a function*.

Roundoff error: Error due to rounding. p. 115

RRAM: Right-hand endpoint rectangular approximation method. The method of approximating a definite integral over an interval using the function values at the right-hand endpoints of the subintervals determined by a partition. p. 249

Scalar: In the context of vectors, scalars are real numbers that behave like scaling factors. p. 520

Scalar function: Name used for real-valued functions to distinguish them from vector-valued functions. p. 530

Scalar multiple of a vector: $k\mathbf{u} = \langle ku_1, ku_2 \rangle$, k a scalar (real number) and $\mathbf{u} = \langle u_1, u_2 \rangle$. p. 522

Secant function: See *Basic trigonometric functions*.

Secant line to a curve: A line through two points on the curve. p. 82

Second derivative: If y is a function of x and $y' = dy/dx$ is the first derivative of y with respect to x, then $dy'/dx = y'' = d^2y/dx^2$ is the second derivative of y with respect to x. p. 119

Second derivative test (for local extrema): If $f'(c) = 0$ and $f''(c) < 0$, then f has a local maximum at $x = c$. If $f'(c) = 0$ and $f''(c) > 0$, then f has a local minimum at $x = c$. p. 200

Separable differential equation: A differential equation $y' = f(x, y)$ in which f can be expressed as a product of a function of x and a function of y. p. 320

Separating the variables: For a separable differential equation $y' = f(x, y)$, the process of combining all the y-terms with y' on one side of the equation and putting all the x-terms on the other side. p. 320

Sequence: A function whose domain is the set of positive integers.

Sigma notation: Notation using the Greek letter capital sigma, $\sum$, for writing lengthy sums in compact form. p. 258

Simple harmonic motion: Periodic motion, like the vertical motion of a weight bobbing up and down at the end of a spring, that can be modeled with a sinusoidal position function. p. 136

Sine function: See *Basic trigonometric functions*.

Slope field: A slope field for the first order differential equation $dy/dx = f(x, y)$ is a plot of short line segments with slopes $f(x, y)$ at a lattice of points (x, y) in the plane. p. 305

Slope of a curve: The slope of a curve $y = f(x)$ at the point $(a, f(a))$ is $f'(a)$, provided f is differentiable at a.   p. 85

Slope of a nonvertical vector: The common slope of the lines parallel to the vector.   p. 522

Smooth curve: The graph of a smooth function.   p. 396

Smooth function: A real-valued function $y = f(x)$ with a continuous first derivative. A vector function whose first derivative is continuous and never zero.   p. 396

Smooth parametrized curve: A curve $x = f(t)$, $y = g(t)$ for which the derivatives f' and g' are continuous and not simultaneously zero.   p. 513

Smooth vector curve: Curve traced by a smooth vector function. See *Smooth function.*

Solid of revolution: A solid generated by revolving a plane region about a line in the plane.   p. 385

Solution of a differential equation: Any function that together with its derivatives satisfies the equation. When we find all such functions, we have *solved* the differential equation.   p. 303

Speed: The absolute value or magnitude of velocity.   p. 124

Standard position of a vector: The representative vector with initial point at the origin.   p. 521

Standard unit vectors: See *Unit vector.*

Sum of a series: See *Infinite series.*

Sum of vectors:

$$\mathbf{v} + \mathbf{u} = \langle v_1, v_2 \rangle + \langle u_1, u_2 \rangle = \langle v_1 + u_1, v_2 + u_2 \rangle$$

p. 522

Symmetric difference quotient: The quotient

$$\frac{f(a + h) - f(a - h)}{2h}$$

that a graphing calculator uses to calculate $\text{NDER}(f(x), a)$, the numerical derivative of f at $x = a$.   p. 108

Symmetry: For a curve to have

(1) symmetry about the x-axis, the point (x, y) must lie on the curve if and only if $(x, -y)$ lies on the curve;

(2) symmetry about the y-axis, (x, y) must lie on the curve if and only if $(-x, y)$ lies on the curve;

(3) symmetry about the origin, (x, y) must lie on the curve if and only if $(-x, -y)$ lies on the curve.   p. 14

Tabular integration: A time-saving way to organize the work of repeated integrations by parts.   p. 327

Tangent function: See *Basic trigonometric functions.*

Tangent line: To the graph of a function $y = f(x)$ at a point $x = a$ where f' exists, the line through $(a, f(a))$ with slope $f'(a)$. To a parametrized curve at a point P where $dx/dt \neq 0$, the line through P with slope equal to the value of $(dy/dt)/(dx/dt)$ at P. To a curve traced by a vector function $\mathbf{r}(t)$ at a point P where $\mathbf{v} \neq \mathbf{0}$, the line through P parallel to $\mathbf{v}$.   p. 83

Tangent vector to a differentiable curve at a point P: A vector that is parallel to the line tangent to the curve at P.   p. 532

Taylor polynomial: Let f be a function with derivatives through order n throughout some open interval containing 0. Then

$$P_n(x) = \sum_{k=0}^{n} \frac{f^{(k)}(0)}{k!} x^k$$

is the Taylor polynomial of order n for f at $x = 0$. If f is a function with derivatives through order n throughout some open interval containing the point $x = a$, then

$$P_n(x) = \sum_{k=0}^{n} \frac{f^{(k)}(a)}{k!} (x - a)^k$$

is the Taylor polynomial of order n for f at $x = a$.   p. 470

Taylor series: Let f be a function with derivatives of all orders throughout some open interval containing 0. Then

$$f(0) + f'(0)x + \frac{f''(0)}{2!}x^2 + \cdots + \frac{f^{(n)}(0)}{n!}x^n + \cdots,$$

$$\text{or} \quad \sum_{k=0}^{\infty} \frac{f^{(k)}(0)}{k!} x^k,$$

is the Taylor series generated by f at $x = 0$. This series is also called the *Maclaurin series* generated by f.

If f is a function with derivatives of all orders throughout some open interval containing the point $x = a$, then

$$f(a) + f'(a)(x - a) + \frac{f''(a)}{2!}(x - a)^2 + \cdots$$
$$+ \frac{f^{(n)}(a)}{n!}(x - a)^n + \cdots,$$

$$\text{or} \quad \sum_{k=0}^{\infty} \frac{f^{(k)}(a)}{k!}(x - a)^k,$$

is the Taylor series generated by f at $x = a$.   p. 470

Term of a sequence or series: For the sequence

$$a_1, a_2, a_3, \ldots, a_n, \ldots,$$

or for the series

$$a_1 + a_2 + a_3 + \cdots + a_n + \cdots,$$

a_n is the nth term.   p. 458

Transcendental function: A function that is not algebraic. (See *Algebraic function.*) The six basic trigonometric functions are transcendental as are the inverse trigonometric functions and the exponential and logarithmic functions studied in this book.

Trapezoidal Rule: To approximate $\int_a^b f(x)\,dx$, use

$$T = \frac{h}{2}(y_0 + 2y_1 + 2y_2 + \cdots + 2y_{n-1} + y_n),$$

where $[a, b]$ is partitioned into n subintervals of equal length $h = (b - a)/n$ and y_i is the value of f at each partition point x_i.   p. 290

Trigonometric function: See *Basic trigonometric functions.*

Truncation error: The error incurred in using a finite partial sum to estimate the sum of an infinite series.   p. 481

Two-sided limit: Limit at an interior point of a function's domain. See *Limit*.

Unit circle: The circle of radius 1 centered at the origin.

Unit vector: A vector with magnitude 1. The vectors $\mathbf{i} = \langle 1, 0 \rangle$, $\mathbf{j} = \langle 0, 1 \rangle$ are the standard unit vectors.   p. 522

Upper bound: See *Bounded*.

u-substitution: See *Integration by substitution*.

Value of an improper integral: See *Improper integral*.

Variable of integration: In $\int f(x)\, dx$ or $\int_a^b f(x)\, dx$, the variable x.   p. 261

Vector in the plane: A directed line segment in the plane, with the understanding that two such vectors are equal if they have the same length and direction.   p. 520

Vector (vector-valued) function: The equation $\mathbf{r}(t) = f(t)\mathbf{i} + g(t)\mathbf{j}$, $t \in I$, defines $\mathbf{r}$ as a vector function of t on the interval I.   p. 530

Velocity: The rate of change of position with respect to time. See also *Average velocity; Instantaneous velocity; Velocity vector.*   p. 124

Velocity vector: If $\mathbf{r}$ is the position vector of a particle moving along a smooth curve in the plane, then at any time t, $\mathbf{v}(t) = d\mathbf{r}/dt$ is the particle's velocity vector.   p. 533

Vertical line: In the Cartesian coordinate plane, a line parallel to the y-axis.

Viewing window: On a graphing calculator, the portion of the coordinate plane displayed on the screen.

Volume by slicing: A method for finding the volume of a solid by evaluating $\int_a^b A(x)\, dx$ where $A(x)$ (assumed integrable) is the solid cross section area at x.   p. 384

x-intercept: The x-coordinate of the point where a curve intersects the x-axis.   p. 8

y-intercept: The y-coordinate of the point where a curve intersects the y-axis.   p. 4

Zero of a function: A solution of the equation $f(x) = 0$ is a zero of the function f or a *root* of the equation.

Zero vector: The vector $\mathbf{0} = 0\mathbf{i} + 0\mathbf{j}$.   p. 522

Chapter 1

Section 1.1 (pp. 1–9)

Quick Review

1. -2

2. -1

3. -1

4. $\dfrac{5}{4}$

5. (a) Yes   **(b)** No

6. (a) Yes   **(b)** No

7. $\sqrt{2}$

8. $\dfrac{5}{3}$

9. $y = \dfrac{4}{3}x - \dfrac{7}{3}$

10. $y = \dfrac{2}{5}x - \dfrac{3}{5}$

Exercises

1. $\Delta x = -2, \Delta y = -3$ **3.** $\Delta x = -5, \Delta y = 0$

5. (a) and **(c)**   **(b)** 3

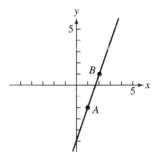

7. (a) and **(c)**   **(b)** 0

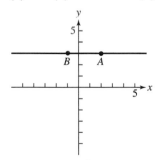

9. (a) $x = 2$ **(b)** $y = 3$

11. (a) $x = 0$ **(b)** $y = -\sqrt{2}$

13. $y = 1(x - 1) + 1$ **15.** $y = 2(x - 0) + 3$

17. $3x - 2y = 0$ **19.** $x = -2$

21. $y = 3x - 2$ **23.** $y = -\dfrac{1}{2}x - 3$

25. $y = \dfrac{5}{2}x$

27. (a) $-\dfrac{3}{4}$ **(b)** 3

(c)

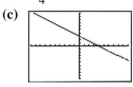

$[-10, 10]$ by $[-10, 10]$

29. (a) $-\dfrac{4}{3}$ **(b)** 4

(c)

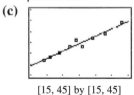

$[-10, 10]$ by $[-10, 10]$

31. (a) $y = -x$ **(b)** $y = x$

33. (a) $x = -2$ **(b)** $y = 4$

35. $m = \dfrac{7}{2}, b = -\dfrac{3}{2}$ **37.** $y = -1$

39. (a) $y = 0.680x + 9.013$

(b) The slope is 0.68. It represents the approximate average weight gain in pounds per month.

(c)

$[15, 45]$ by $[15, 45]$

(d) 29 pounds

41. $y = 1(x - 3) + 4$

$y = x - 3 + 4$

$y = x + 1$, which is the same equation.

43. (a) $k = 2$ **(b)** $k = -2$

45. 5.97 atmospheres ($k = 0.0994$)

47. (a) $y = 5632x - 11,080,280$

(b) The rate at which the median price is increasing in dollars per year

(c) $y = 2732x - 5,362,360$

(d) In the Northeast

49. The coordinates of the three missing vertices are (5, 2), (−1, 4) and (−1, −2).

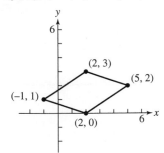

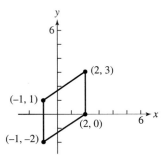

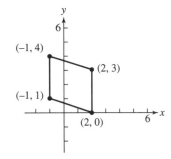

51. $y = -\frac{3}{4}(x - 3) + 4$ or $y = -\frac{3}{4}x + \frac{25}{4}$

Section 1.2 (pp. 9–19)

Quick Review

1. $[-2, \infty)$ **2.** $(-\infty, 0) \cup (2, \infty)$
3. $[-1, 7]$ **4.** $(-\infty, -3] \cup [7, \infty)$
5. $(-4, 4)$ **6.** $[-3, 3]$
7. Translate the graph of f 2 units left and 3 units downward.
8. Translate the graph of f 5 units right and 2 units upward.
9. (a) $x = -3, 3$ **(b)** No real solution
10. (a) $x = -\frac{1}{5}$ **(b)** No solution
11. (a) $x = 9$ **(b)** $x = -6$
12. (a) $x = -7$ **(b)** $x = 28$

Exercises

1. $A = \dfrac{\pi d^2}{4}$ **3.** $S = 6e^2$

5. (a) All reals **(b)** $(-\infty, 4]$
(c)

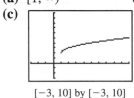

$[-5, 5]$ by $[-10, 10]$

(d) Symmetric about y-axis

7. (a) $[1, \infty)$ **(b)** $[2, \infty)$
(c)

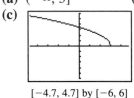

$[-3, 10]$ by $[-3, 10]$

(d) None

9. (a) $(-\infty, 3]$ **(b)** $[0, \infty)$
(c)

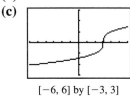

$[-4.7, 4.7]$ by $[-6, 6]$

(d) None

11. (a) All reals **(b)** All reals
(c)

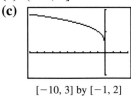

$[-6, 6]$ by $[-3, 3]$

(d) None

13. (a) $(-\infty, 0]$ **(b)** $[0, \infty)$
(c)

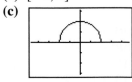

$[-10, 3]$ by $[-1, 2]$

(d) None

15. (a) $[-2, 2]$ **(b)** $[0, 2]$
(c)

$[-4.7, 4.7]$ by $[-3.1, 3.1]$

(d) Symmetric about y-axis

17. (a) $(-\infty, 0) \cup (0, \infty)$ **(b)** $(1, \infty)$

(c)

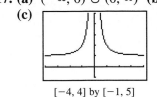

[−4, 4] by [−1, 5]

(d) Symmetric about y-axis

19. Even **21.** Neither

23. Even **25.** Odd

27. Neither

29. (a)

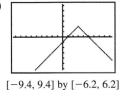

[−9.4, 9.4] by [−6.2, 6.2]

(b) All reals **(c)** $(-\infty, 2]$

31. (a)

[−4.7, 4.7] by [−1, 6]

(b) All reals **(c)** $[2, \infty)$

33. (a)

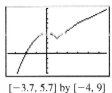

[−3.7, 5.7] by [−4, 9]

(b) All reals **(c)** All reals

35. Because if the vertical line test holds, then for each x-coordinate, there is at most one y-coordinate giving a point on the curve. This y-coordinate would correspond to the value assigned to the x-coordinate. Since there's only one y-coordinate, the assignment would be unique.

37. No **39.** Yes

41. $f(x) = \begin{cases} x, & 0 \le x \le 1 \\ 2 - x, & 1 < x \le 2 \end{cases}$

43. $f(x) = \begin{cases} 2 - x, & 0 < x \le 2 \\ \dfrac{5}{3} - \dfrac{x}{3}, & 2 < x \le 5 \end{cases}$

45. $f(x) = \begin{cases} -x, & -1 \le x < 0 \\ 1, & 0 < x \le 1 \\ \dfrac{3}{2} - \dfrac{x}{2}, & 1 < x < 3 \end{cases}$

47. $f(x) = \begin{cases} 0, & 0 \le x \le \dfrac{T}{2} \\ \dfrac{2}{T}x - 1, & \dfrac{T}{2} < x \le T \end{cases}$

49. (a) $x^2 + 2$ **(b)** $x^2 + 10x + 22$

(c) 2 **(d)** 22

(e) -2 **(f)** $x + 10$

51. (a) For $f \circ g$:

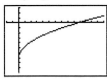

[−10, 70] by [−10, 3]

Domain: $[0, \infty)$; Range: $[-7, \infty)$

For $g \circ f$:

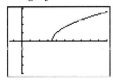

[−3, 20] by [−4, 4]

Domain: $[7, \infty)$; Range: $[0, \infty)$

(b) $(f \circ g)(x) = \sqrt{x} - 7$; $(g \circ f)(x) = \sqrt{x - 7}$

53. (a) For $f \circ g$:

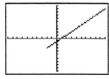

[−10, 10] by [−10, 10]

Domain: $[-2, \infty)$; Range: $[-3, \infty)$

For $g \circ f$:

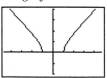

[−4.7, 4.7] by [−2, 4]

Domain: $(-\infty, -1] \cup [1, \infty)$; Range: $[0, \infty)$

53. continued

(b) $(f \circ g)(x) = (\sqrt{x+2})^2 - 3$

$\qquad = x - 1, x \geq -2$

$\qquad (g \circ f)(x) = \sqrt{x^2 - 1}$

55. Domain: $(-\infty, -2) \cup (2, \infty)$; Range: $(0, \infty)$

57. Domain: $(-\infty, -3) \cup (-3, 3) \cup (3, \infty)$

Range: $(-\infty, 0) \cup \left[\dfrac{2}{\sqrt[3]{9}}, \infty \right)$

59. (a)

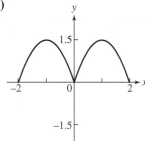

(b)

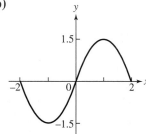

61. (a)

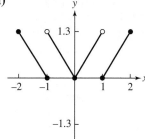

(b)

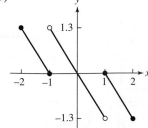

63. (a) $g(x) = x^2$ **(b)** $g(x) = \dfrac{1}{x - 1}$

(c) $f(x) = \dfrac{1}{x}$

(d) $f(x) = x^2$

(Note that the domain of the composite is $[0, \infty)$).

65. (a) Because the circumference of the original circle was 8π and a piece of length x was removed.

(b) $r = \dfrac{8\pi - x}{2\pi} = 4 - \dfrac{x}{2\pi}$

(c) $h = \sqrt{16 - r^2} = \dfrac{\sqrt{16\pi x - x^2}}{2\pi}$

(d) $V = \dfrac{1}{3}\pi r^2 h = \dfrac{(8\pi - x)^2 \sqrt{16\pi x - x^2}}{24\pi^2}$

67. (a)

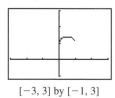

$[-3, 3]$ by $[-1, 3]$

(b) Domain of y_1: $[0, \infty)$

Domain of y_2: $(-\infty, 1]$

Domain of y_3: $[0, 1]$

(c) The results for $y_1 - y_2$, $y_2 - y_1$, and $y_1 \cdot y_2$ are the same as for $y_1 + y_2$ above.

Domain of $\dfrac{y_1}{y_2}$: $[0, 1)$

Domain of $\dfrac{y_2}{y_1}$: $(0, 1]$

(d) The domain of a sum, difference, or product of two functions is the intersection of their domains.

The domain of a quotient of two functions is the intersection of their domains with any zeros of the denominator removed.

Section 1.3 (pp. 20–26)

Quick Review

1. 2.924 **2.** 4.729

3. 0.192 **4.** 2.5713

5. 1.8882 **6.** ± 1.0383

7. \$630.58 **8.** \$1201.16

9. $x^{-18}y^{-5} = \dfrac{1}{x^{18}y^5}$ **10.** $a^2b^{-1}c^{-6} = \dfrac{a^2}{bc^6}$

Exercises

1. (a) **3.** (e)

5. (b)

7.

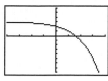

[−4, 4] by [−8, 6]

Domain: All reals
Range: $(-\infty, 3)$
x-intercept: ≈ 1.585
y-intercept: 2

9.

[−4, 4] by [−4, 8]

Domain: All reals
Range: $(-2, \infty)$
x-intercept: ≈ 0.405
y-intercept: 1

11. 3^{4x}

13. 2^{-6x}

15. $x \approx 2.3219$

17. $x \approx -0.6309$

19.

x	y	Δy
1	−1	
2	1	2
3	3	2
4	5	2

21.

x	y	Δy
1	1	
2	4	3
3	9	5
4	16	7

23. After 19 years

25. (a) $A(t) = 6.6\left(\dfrac{1}{2}\right)^{t/14}$

(b) About 38.1145 days later

27. ≈ 11.433 years

29. ≈ 11.090 years

31. ≈ 19.108 years

33. $2^{48} \approx 2.815 \times 10^{14}$

35. Since $\Delta x = 1$, the corresponding value of Δy is equal to the slope of the line. If the changes in x are constant for a linear function, then the corresponding changes in y are constant as well.

37. (a) Regression equation:
$P(x) = 6.033(1.030)^x$, where $x - 0$ represents 1900

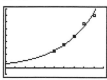

[0, 100] by [−10, 90]

(b) Approximately 6.03 million, which is not very close to the actual population

(c) The annual rate of growth is approximately 3%.

39. 7609.7 million

41. $a = 3, k = 1.5$

Section 1.4 (pp. 26–31)

Quick Review

1. $y = -\dfrac{5}{3}x + \dfrac{29}{3}$

2. $y = -4$

3. $x = 2$

4. x-intercepts: $x = -3$ and $x = 3$
y-intercepts: $y = -4$ and $y = 4$

5. x-intercepts: $x = -4$ and $x = 4$
y-intercepts: None

6. x-intercept: $x = -1$

y-intercepts: $y = -\dfrac{1}{\sqrt{2}}$ and $y = \dfrac{1}{\sqrt{2}}$

7. (a) Yes **(b)** No **(c)** Yes

8. (a) Yes **(b)** Yes **(c)** No

9. (a) $t = \dfrac{-2x - 5}{3}$ **(b)** $t = \dfrac{3y + 1}{2}$

10. (a) $a \geq 0$ **(b)** All reals **(c)** All reals

Exercises

1. Graph (c).
Window: $[-4, 4]$ by $[-3, 3]$, $0 \leq t \leq 2\pi$

3. Graph (d).
Window: $[-10, 10]$ by $[-10, 10]$, $0 \leq t \leq 2\pi$

5. (a) The resulting graph appears to be the right half of a hyperbola in the first and fourth quadrants. The parameter a determines the x-intercept. The parameter b determines the shape of the hyperbola. If b is smaller, the graph has less steep slopes and appears "sharper." If b is larger, the slopes are steeper and the graph appears more "blunt."

(b) This appears to be the left half of the same hyperbola.

(c) Because both sec t and tan t are discontinuous at these points. This might cause the grapher to include extraneous lines (the asymptotes to the hyperbola) in its graph.

(d) $\left(\dfrac{x}{a}\right)^2 - \left(\dfrac{y}{b}\right)^2 = (\sec t)^2 - (\tan t)^2 = 1$
by a standard trigonometric identity.

(e) This changes the orientation of the hyperbola. In this case, b determines the y-intercept of the hyperbola, and a determines the shape. The parameter interval $\left(-\dfrac{\pi}{2}, \dfrac{\pi}{2}\right)$ gives the upper half of the hyperbola. The parameter interval $\left(\dfrac{\pi}{2}, \dfrac{3\pi}{2}\right)$ gives the lower half. The same values of t cause discontinuities and may add extraneous lines to the graph.

7. (a)

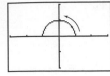

[−3, 3] by [−2, 2]

Initial point: (1, 0)
Terminal point: (−1, 0)

(b) $x^2 + y^2 = 1$; upper half (or $y = \sqrt{1 - x^2}$; all)

9. (a)

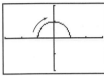

[−3, 3] by [−2, 2]

Initial point: (1, 0)
Terminal point: (0, 1)

(b) $x^2 + y^2 = 1$; upper half (or $y = \sqrt{1 - x^2}$; all)

11. (a)

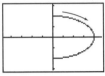

[−4.7, 4.7] by [−3.1, 3.1]

Initial point: (0, 2)
Terminal point: (0, −2)

(b) $\left(\dfrac{x}{4}\right)^2 + \left(\dfrac{y}{2}\right)^2 = 1$;
right half (or $x = 2\sqrt{4 - y^2}$; all)

13. (a)

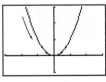

[−3, 3] by [−1, 3]

No initial or terminal point

(b) $y = x^2$; all

15. (a)

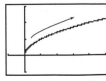

[−1, 5] by [−1, 3]

Initial point: (0, 0)
Terminal point: None

(b) $y = \sqrt{x}$; all (or $x = y^2$; upper half)

17. (a)

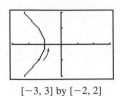

[−3, 3] by [−2, 2]

No initial or terminal point

(b) $x^2 - y^2 = 1$; left branch (or $x = -\sqrt{y^2 + 1}$; all)

19. (a)

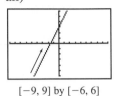

[−9, 9] by [−6, 6]

No initial or terminal point

(b) $y = 2x + 3$; all

21. (a)

[−3, 3] by [−2, 2]

Initial point: (0, 1)
Terminal point: (1, 0)

(b) $y = -x + 1$; (0, 1) to (1, 0)

23. (a)

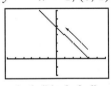

[−6, 6] by [−2, 6]

Initial point: (4, 0)
Terminal point: None

(b) $y = -x + 4$; $x \le 4$

25. (a)

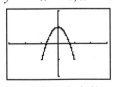

[−3, 3] by [−2, 2]

The curve is traced and retraced in both directions, and there is no initial or terminal point.

(b) $y = -2x^2 + 1$; $-1 \le x \le 1$

27. Possible answer:
$x = -1 + 5t, \; y = -3 + 4t, \; 0 \le t \le 1$

29. Possible answer:
$x = t^2 + 1, \; y = t, \; t \le 0$

31. Possible answer:
$x = 2 - 3t, \; y = 3 - 4t, \; t \ge 0$

33. $1 < t < 3$
35. $-5 \le t < -3$
37. Possible answer: $x = t, y = t^2 + 2t + 2, t > 0$
39. Possible answers:
   (a) $x = a \cos t, y = -a \sin t, 0 \le t \le 2\pi$
   (b) $x = a \cos t, y = a \sin t, 0 \le t \le 2\pi$
   (c) $x = a \cos t, y = -a \sin t, 0 \le t \le 4\pi$
   (d) $x = a \cos t, y = a \sin t, 0 \le t \le 4\pi$
41. $x = 2 \cot t, y = 2 \sin^2 t, 0 < t < \pi$

Section 1.5 (pp. 32–40)

Quick Review

1. 1   **2.** 5
3. $x^{2/3}$ **4.** $(x - 1)^{2/3} + 1$
5. Possible answer: $x = t, y = \dfrac{1}{t - 1}, t \ge 2$
6. Possible answer: $x = t, y = t, t < -3$
7. $(4, 5)$ **8.** $\left(\dfrac{8}{3}, -3\right) \approx (2.67, -3)$
9. (a) $(1.58, 3)$ **(b)** No intersection
10. (a) $(-1.39, 4)$ **(b)** No intersection

Exercises

1. No   **3.** Yes   **5.** Yes
7. Yes   **9.** No   **11.** No
13. $f^{-1}(x) = \dfrac{x - 3}{2}$
15. $f^{-1}(x) = (x + 1)^{1/3}$ or $\sqrt[3]{x + 1}$
17. $f^{-1}(x) = -x^{1/2}$ or $-\sqrt{x}$
19. $f^{-1}(x) = 2 - (-x)^{1/2}$ or $2 - \sqrt{-x}$
21. $f^{-1}(x) = \dfrac{1}{x^{1/2}}$ or $\dfrac{1}{\sqrt{x}}$
23. $f^{-1}(x) = \dfrac{1 - 3x}{x - 2}$

25. 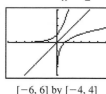   **27.**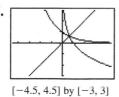
    $[-6, 6]$ by $[-4, 4]$ $[-4.5, 4.5]$ by $[-3, 3]$
29. 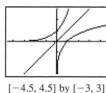   **31.**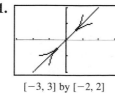
    $[-4.5, 4.5]$ by $[-3, 3]$ $[-3, 3]$ by $[-2, 2]$

33.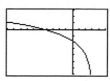
    $[-10, 5]$ by $[-7, 3]$ $[-3, 6]$ by $[-2, 4]$
    Domain: $(-\infty, 3)$;   Domain: $(-1, \infty)$;
    Range: all reals   Range: all reals
37. $t = \dfrac{\ln 2}{\ln 1.045} \approx 15.75$
39. $x = \ln\left(\dfrac{3 \pm \sqrt{5}}{2}\right) \approx -0.96$ or 0.96
41. $y = e^{2t+4}$
43. $f^{-1}(x) = \log_2\left(\dfrac{x}{100 - x}\right)$
45. (a) $f(f(x)) = \sqrt{1 - (f(x))^2}$
                $= \sqrt{1 - (1 - x^2)}$
                $= \sqrt{x^2}$
                $= x$, since $x \ge 0$
  (b) $f(f(x)) = f\left(\dfrac{1}{x}\right) = \dfrac{1}{1/x} = x$ for all $x \ne 0$
47. About 14.936 years. (If the interest is only paid
    annually, it will take 15 years.)
49. (a) $y = -2539.852 + 636.896 \ln x$
  (b) 209.94 million metric tons
  (c) When $x \approx 101.08$ or 2001
51. (a) Suppose that $f(x_1) = f(x_2)$. Then
        $mx_1 + b = mx_2 + b$, which gives $x_1 = x_2$
        since $m \ne 0$.
  (b) $f^{-1}(x) = \dfrac{x - b}{m}$; the slopes are reciprocals.
  (c) They are also parallel lines with nonzero
        slope.
  (d) They are also perpendicular lines with
        nonzero slopes.
53. If the graph of $f(x)$ passes the horizontal line test,
    so will the graph of $g(x) = -f(x)$ since it's the
    same graph reflected about the x-axis.
55. (a) Domain: All reals
        Range: If $a > 0$, then (d, ∞)
                 If $a < 0$, then $(-\infty, d)$
  (b) Domain: (c, ∞); Range: All reals

Section 1.6 (pp. 41–51)

Quick Review

1. $60°$ **2.** $-\left(\dfrac{450}{\pi}\right)° \approx -143.24°$
3. $-\dfrac{2\pi}{9}$ **4.** $\dfrac{\pi}{4}$

5. $x \approx 0.6435$, $x \approx 2.4981$

6. $x \approx 1.9823$, $x \approx 4.3009$

7. $x \approx 0.7854 \left(\text{or } \dfrac{\pi}{4} \right)$, $x \approx 3.9270 \left(\text{or } \dfrac{5\pi}{4} \right)$

8. $f(-x) = 2(-x)^2 - 3 = 2x^2 - 3 = f(x)$

The graph is symmetric about the y-axis because if a point (a, b) is on the graph, then so is the point $(-a, b)$.

9. $f(-x) = (-x)^3 - 3(-x) = -x^3 + 3x$
$= -(x^3 - 3x) = -f(x)$

The graph is symmetric about the origin because if a point (a, b) is on the graph, then so is the point $(-a, -b)$.

10. $x \geq 0$

Exercises

1. $\dfrac{5\pi}{4}$

3. $\dfrac{1}{2}$ radian or $\approx 28.65°$

5. Possible answers are:
  (a) $[0, 4\pi]$ by $[-3, 3]$
  (b) $[0, 4\pi]$ by $[-3, 3]$
  (c) $[0, 2\pi]$ by $[-3, 3]$

7. $\dfrac{\pi}{6}$ radian or $30°$

9. ≈ -1.3734 radians or $-78.6901°$

11. (a) π **(b)** 1.5
  (c) $[-2\pi, 2\pi]$ by $[-2, 2]$

13. (a) π **(b)** 3
  (c) $[-2\pi, 2\pi]$ by $[-4, 4]$

15. (a) 6 **(b)** 4
  (c) $[-3, 3]$ by $[-5, 5]$

17. (a) $\dfrac{2\pi}{3}$ **(b)** $x \neq \dfrac{k\pi}{3}$, for integers k
  (c) $(-\infty, -5] \cup [1, \infty)$
  (d)

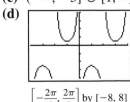

$\left[-\dfrac{2\pi}{3}, \dfrac{2\pi}{3} \right]$ by $[-8, 8]$

19. (a) $\dfrac{\pi}{3}$

  (b) $x \neq \dfrac{k\pi}{6}$, for odd integers k

  (c) All reals

(d)

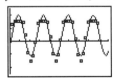

$\left[-\dfrac{\pi}{2}, \dfrac{\pi}{2} \right]$ by $[-8, 8]$

21. $\cos \theta = \dfrac{15}{17}$ $\sin \theta = \dfrac{8}{17}$ $\tan \theta = \dfrac{8}{15}$

$\sec \theta = \dfrac{17}{15}$ $\csc \theta = \dfrac{17}{8}$ $\cot \theta = \dfrac{15}{8}$

23. $\cos \theta = -\dfrac{3}{5}$ $\sin \theta = \dfrac{4}{5}$ $\tan \theta = -\dfrac{4}{3}$

$\sec \theta = -\dfrac{5}{3}$ $\csc \theta = \dfrac{5}{4}$ $\cot \theta = -\dfrac{3}{4}$

25. $x \approx 1.190$ and $x \approx 4.332$

27. $x = \dfrac{\pi}{6}$ and $x = \dfrac{5\pi}{6}$

29. $x = \dfrac{7\pi}{6} + 2k\pi$ and $x = \dfrac{11\pi}{6} + 2k\pi$, k any integer

31. $\dfrac{\sqrt{72}}{11} \approx 0.771$

33. (a) $y = 1.543 \sin (2468.635x - 0.494) + 0.438$

$[0, 0.01]$ by $[-2.5, 2.5]$

  (b) Frequency = 392.9, so it must be a "G."

35. (a) 37   **(b)** 365
  (c) 101   **(d)** 25

37. (a) $\cot (-x) = \dfrac{\cos (-x)}{\sin (-x)} = \dfrac{\cos (x)}{-\sin (x)} = -\cot (x)$

  (b) Assume that f is even and g is odd.
  Then $\dfrac{f(-x)}{g(-x)} = \dfrac{f(x)}{-g(x)} = -\dfrac{f(x)}{g(x)}$ so $\dfrac{f}{g}$ is odd.
  The situation is similar for $\dfrac{g}{f}$.

39. Assume that f is even and g is odd.
Then $f(-x) \, g(-x) = f(x)[-g(x)] = -f(x)g(x)$
so fg is odd.

41. (a) $y = 3.0014 \sin (0.9996x + 2.0012) + 2.9999$
  (b) $y = 3 \sin (x + 2) + 3$

43. (a) $\sqrt{2} \sin \left(ax + \dfrac{\pi}{4} \right)$
  (b) See part **(a)**.
  (c) It works.

(d) $\sin\left(ax + \dfrac{\pi}{4}\right)$

$= \sin(ax) \cdot \dfrac{1}{\sqrt{2}} + \cos(ax) \cdot \dfrac{1}{\sqrt{2}}$

$= \dfrac{1}{\sqrt{2}}(\sin ax + \cos ax)$

So, $\sin(ax) + \cos(ax) = \sqrt{2}\sin\left(ax + \dfrac{\pi}{4}\right)$.

45. Since $\sin(x)$ has period 2π,

$(\sin(x + 2\pi))^3 = (\sin(x))^3$. This function has period 2π. A graph shows that no smaller number works for the period.

47. One possible graph:

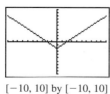

$\left[\dfrac{\pi}{60}, \dfrac{\pi}{60}\right]$ by $[-2, 2]$

Chapter Review (pp. 52–53)

1. $y = 3x - 9$ **2.** $y = -\dfrac{1}{2}x + \dfrac{3}{2}$

3. $x = 0$ **4.** $y = -2x$

5. $y = 2$ **6.** $y = -\dfrac{2}{5}x + \dfrac{21}{5}$

7. $y = -3x + 3$ **8.** $y = 2x - 5$

9. $y = -\dfrac{4}{3}x - \dfrac{20}{3}$ **10.** $y = -\dfrac{5}{3}x - \dfrac{19}{3}$

11. $y = \dfrac{2}{3}x + \dfrac{8}{3}$ **12.** $y = \dfrac{5}{3}x - 5$

13. $y = -\dfrac{1}{2}x + 3$ **14.** $y = -\dfrac{2}{7}x - \dfrac{6}{7}$

15. Origin **16.** y-axis

17. Neither **18.** y-axis

19. Even **20.** Odd

21. Even **22.** Odd

23. Odd **24.** Neither

25. Neither **26.** Even

27. (a) Domain: all reals **(b)** Range: $[-2, \infty)$

(c)

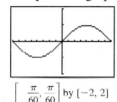

$[-10, 10]$ by $[-10, 10]$

28. (a) Domain: $(-\infty, 1]$ **(b)** Range: $[-2, \infty)$

(c)

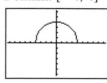

$[-9.4, 9.4]$ by $[-3, 3]$

29. (a) Domain: $[-4, 4]$ **(b)** Range: $[0, 4]$

(c)

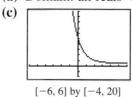

$[-9.4, 9.4]$ by $[-6.2, 6.2]$

30. (a) Domain: all reals **(b)** Range: $(1, \infty)$

(c)

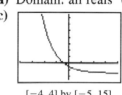

$[-6, 6]$ by $[-4, 20]$

31. (a) Domain: all reals **(b)** Range: $(-3, \infty)$

(c)

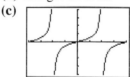

$[-4, 4]$ by $[-5, 15]$

32. (a) Domain: $x \neq \dfrac{k\pi}{4}$, for odd integers k

(b) Range: all reals

(c)

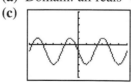

$\left[-\dfrac{\pi}{2}, \dfrac{\pi}{2}\right]$ by $[-8, 8]$

33. (a) Domain: all reals **(b)** Range: $[-3, 1]$

(c)

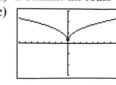

$[-\pi, \pi]$ by $[-5, 5]$

34. (a) Domain: all reals **(b)** Range: $[0, \infty)$

(c)

$[-8, 8]$ by $[-3, 3]$

35. (a) Domain: $(3, \infty)$ **(b)** Range: all reals
(c)

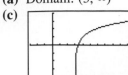

[−3, 10] by [−4, 4]

36. (a) Domain: all reals   **(b)** Range: all reals
(c)

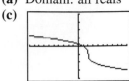

[−10, 10] by [−4, 4]

37. (a) Domain: $[-4, 4]$ **(b)** Range: $[0, 2]$
(c)

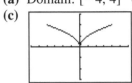

[−6, 6] by [−3, 3]

38. (a) Domain: $[-2, 2]$ **(b)** Range: $[-1, 1]$
(c)

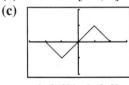

[−3, 3] by [−2, 2]

39. $f(x) = \begin{cases} 1 - x, & 0 \le x < 1 \\ 2 - x, & 1 \le x \le 2 \end{cases}$

40. $f(x) = \begin{cases} \dfrac{5x}{2}, & 0 \le x < 2 \\ -\dfrac{5}{2}x + 10, & 2 \le x \le 4 \end{cases}$

41. (a) 1

(b) $\dfrac{1}{\sqrt{2.5}} \left(= \sqrt{\dfrac{2}{5}} \right)$

(c) $x, x \ne 0$

(d) $\dfrac{1}{\sqrt{1/\sqrt{x+2} + 2}}$

42. (a) 2   **(b)** 1   **(c)** x

(d) $\sqrt[3]{\sqrt[3]{x + 1} + 1}$

43. (a) $(f \circ g)(x) = -x, x \ge -2$
$(g \circ f)(x) = \sqrt{4 - x^2}$
(b) Domain $(f \circ g)$: $[-2, \infty)$
Domain $(g \circ f)$: $[-2, 2]$
(c) Range $(f \circ g)$: $(-\infty, 2]$
Range $(g \circ f)$: $[0, 2]$

44. (a) $(f \circ g)(x) = \sqrt[4]{1 - x}$
$(g \circ f)(x) = \sqrt{1 - \sqrt{x}}$

(b) Domain $(f \circ g)$: $(-\infty, 1]$
Domain $(g \circ f)$: $[0, 1]$
(c) Range $(f \circ g)$: $[0, \infty)$
Range $(g \circ f)$: $[0, 1]$

45. (a)

[−6, 6] by [−4, 4]

Initial point: $(5, 0)$
Terminal point: $(5, 0)$

(b) $\left(\dfrac{x}{5}\right)^2 + \left(\dfrac{y}{2}\right)^2 = 1$; all

46. (a)

[−9, 9] by [−6, 6]

Initial point: $(0, 4)$
Terminal point: $(0, -4)$

(b) $x^2 + y^2 = 16$; left half

47. (a)

[−8, 8] by [−10, 20]

Initial point: $(4, 15)$
Terminal point: $(-2, 3)$

(b) $y = 2x + 7$; from $(4, 15)$ to $(-2, 3)$

48. (a)

[−8, 8] by [−4, 6]

Initial point: None
Terminal point: $(3, 0)$
(b) $y = \sqrt{6 - 2x}$; all

49. Possible answer:
$x = -2 + 6t, y = 5 - 2t, 0 \le t \le 1$

50. Possible answer:
$x = -3 + 7t, y = -2 + t, -\infty < t < \infty$

51. Possible answer:
$x = 2 - 3t, y = 5 - 5t, 0 \le t$

52. Possible answer:
$x = t,\ y = t(t - 4),\ t \le 2$

53. (a) $f^{-1}(x) = \dfrac{2 - x}{3}$

(b)

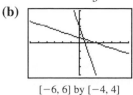

[−6, 6] by [−4, 4]

54. (a) $f^{-1}(x) = \sqrt{x} - 2$

(b)

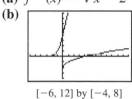

[−6, 12] by [−4, 8]

55. ≈ 0.6435 radians or $36.8699°$

56. ≈ -1.1607 radians or $-66.5014°$

57. $\cos \theta = \dfrac{3}{7}$ $\sin \theta = \dfrac{\sqrt{40}}{7}$ $\tan \theta = \dfrac{\sqrt{40}}{3}$

    $\sec \theta = \dfrac{7}{3}$ $\csc \theta = \dfrac{7}{\sqrt{40}}$ $\cot \theta = \dfrac{3}{\sqrt{40}}$

58. (a) $x \approx 3.3430$ and $x \approx 6.0818$

(b) $x \approx 3.3430 + 2k\pi$ and $x \approx 6.0818 + 2k\pi$, k any integer

59. $x = -5 \ln 4$

60. (a)

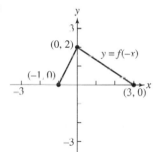

(b)

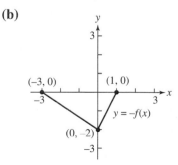

(c)

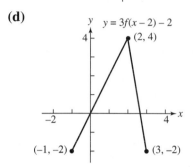

(d)

61. (a)

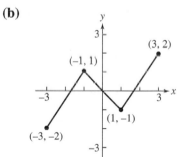

(b)

62. (a) $V = 100{,}000 - 10{,}000x,\ 0 \le x \le 10$

(b) After 4.5 years

63. (a) 90 units

(b) $90 - 52 \ln 3 \approx 32.8722$ units

(c)

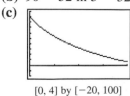

[0, 4] by [−20, 100]

64. After $\dfrac{\ln(10/3)}{\ln 1.08} \approx 15.6439$ years

(If the bank only pays interest at the end of the year, it will take 16 years.)

65. (a) $N = 4 \cdot 2^t$

(b) 4 days: 64; one week: 512

(c) After $\dfrac{\ln 500}{\ln 2} \approx 8.9658$ days, or after nearly 9 days.

(d) Because it suggests the number of guppies will continue to double indefinitely and become arbitrarily large, which is impossible due to the finite size of the tank and the oxygen supply in the water.

66. (a) $y = 20.627x + 338.622$

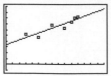

[0, 30] by [−100, 1000]

(b) Approximately 957

(c) Slope is 20.627. It represents the approximate arrival increase in number of doctorates earned by Hispanic Americans per year.

67. (a) $y = 14.60175 \cdot 1.00232^x$

(b) Sometime during the year 2132 (when $x \approx 232$)

(c) 0.232%

Chapter 2

Section 2.1 (pp. 55–65)

Quick Review

1. 0

2. $\dfrac{11}{12}$

3. 0

4. $\dfrac{1}{3}$

5. $-4 < x < 4$

6. $-c^2 < x < c^2$

7. $-1 < x < 5$

8. $c - d^2 < x < c + d^2$

9. $x - 6$

10. $\dfrac{x}{x+1}$

Exercises

1. (a) 3 **(b)** −2
  (c) No limit **(d)** 1

3. (a) −4 **(b)** −4
  (c) −4 **(d)** −4

5. (a) 4 **(b)** −3
  (c) No limit **(d)** 4

7. $-\dfrac{3}{2}$ **9.** −15   **11.** 0

13. 4   **15.** 1

17. Expression not defined at $x = -2$. There is no limit.

19. Expression not defined at $x = 0$. There is no limit.

21. $\dfrac{1}{2}$ **23.** $-\dfrac{1}{2}$

25. 12 **27.** −1 **29.** 0

31. (a) True **(b)** True **(c)** False
  (d) True **(e)** True **(f)** True
  (g) False **(h)** False **(i)** False
  (j) False

33. (c) **35.** (d) **37.** 0

39. 0 **41.** 1

43. (a) 6 **(b)** 0
  (c) 9 **(d)** −3

45. (a)

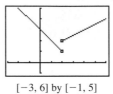

[−3, 6] by [−1, 5]

(b) Right-hand: 2
Left-hand: 1

(c) No, because the two one-sided limits are different.

47. (a)

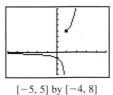

[−5, 5] by [−4, 8]

(b) Right-hand: 4
Left-hand: no limit

(c) No, because the left-hand limit doesn't exist.

49. (a)

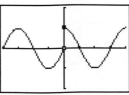

[−2π, 2π] by [−2, 2]

(b) $(-2\pi, 0) \cup (0, 2\pi)$

(c) $c = 2\pi$ **(d)** $c = -2\pi$

51. (a)

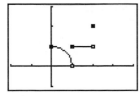

[−2, 4] by [−1, 3]

(b) $(0, 1) \cup (1, 2)$ **(c)** $c = 2$
(d) $c = 0$

53. 0   **55.** 0
57. (a) 14.7 m/sec   **(b)** 29.4 m/sec
59. (a)

x	−0.1	−0.01	−0.001	−0.0001
$f(x)$	−0.054402	−0.005064	−0.000827	−0.000031

(b)

x	0.1	0.01	0.001	0.0001
$f(x)$	−0.054402	−0.005064	−0.000827	−0.000031

The limit appears to be 0.

61. (a)

x	0.1	−0.01	−0.001	−0.0001
$f(x)$	2.0567	2.2763	2.2999	2.3023

(b)

x	0.1	0.01	0.001	0.0001
$f(x)$	2.5893	2.3293	2.3052	2.3029

The limit appears to be approximately 2.3.

63. (a) Because the right-hand limit at zero depends only on the values of the function for positive x-values near zero.

(b) Use: area of triangle $= \left(\dfrac{1}{2}\right)$(base)(height)

area of circular sector $= \dfrac{\text{(angle)(radius)}^2}{2}$

(c) This is how the areas of the three regions compare.

(d) Multiply by 2 and divide by sin θ.

(e) Take reciprocals, remembering that all of the values involved are positive.

(f) The limits for cos θ and 1 are both equal to 1. Since $\dfrac{\sin \theta}{\theta}$ is between them, it must also have a limit of 1.

(g) $\dfrac{\sin (-\theta)}{-\theta} = \dfrac{-\sin (\theta)}{-\theta} = \dfrac{\sin (\theta)}{\theta}$

(h) If the function is symmetric about the y-axis, and the right-hand limit at zero is 1, then the left-hand limit at zero must also be 1.

(i) The two one-sided limits both exist and are equal to 1.

65. (a) $f\left(\dfrac{\pi}{6}\right) = \dfrac{1}{2}$
(b) One possible answer: $a = 0.305$, $b = 0.775$
(c) One possible answer: $a = 0.513$, $b = 0.535$

Section 2.2 (pp. 65–73)

Quick Review

1. $f^{-1}(x) = \dfrac{x + 3}{2}$

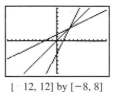

[−12, 12] by [−8, 8]

2. $f^{-1}(x) = \ln (x)$

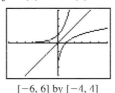

[−6, 6] by [−4, 4]

3. $f^{-1}(x) = \tan (x)$, $-\dfrac{\pi}{2} < x < \dfrac{\pi}{2}$

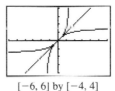

[−6, 6] by [−4, 4]

4. $f^{-1}(x) = \cot (x)$, $0 < x < \pi$

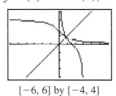

[−6, 6] by [−4, 4]

5. $q(x) = \dfrac{2}{3}$
$r(x) = -3x^2 - \left(\dfrac{5}{3}\right)x + \dfrac{7}{3}$

6. $q(x) = 2x^2 + 2x + 1$
$r(x) = -x^2 - x - 2$

7. (a) $f(-x) = \cos x$ **(b)** $f\left(\dfrac{1}{x}\right) = \cos \left(\dfrac{1}{x}\right)$

8. (a) $f(-x) = e^x$ **(b)** $f\left(\dfrac{1}{x}\right) = e^{-1/x}$

9. (a) $f(-x) = -\dfrac{\ln (-x)}{x}$

(b) $f\left(\dfrac{1}{x}\right) = -x \ln x$

10. (a) $f(-x) = \left(x + \dfrac{1}{x}\right) \sin x$

(b) $f\left(\dfrac{1}{x}\right) = \left(\dfrac{1}{x} + x\right) \sin\left(\dfrac{1}{x}\right)$

Exercises

1. (a) 1   **(b)** 1   **(c)** $y = 1$
3. (a) 0   **(b)** $-\infty$ **(c)** $y = 0$
5. (a) 3   **(b)** -3
   (c) $y = 3, y = -3$
7. (a) 1   **(b)** -1
   (c) $y = 1, y = -1$
9. ∞ **11.** $-\infty$
13. 0   **15.** ∞
17. (a) $x = -2, x = 2$
   (b) Left-hand limit at -2 is ∞.
      Right-hand limit at -2 is $-\infty$.
      Left-hand limit at 2 is $-\infty$.
      Right-hand limit at 2 is ∞.
19. (a) $x = -1$
   (b) Left-hand limit at -1 is $-\infty$.
      Right-hand limit at -1 is ∞.
21. (a) $x = k\pi$, k any integer
   (b) At each vertical asymptote:
      Left-hand limit is $-\infty$.
      Right-hand limit is ∞.
23. Both are 1   **25.** Both are 1
27. Both are 0   **29. (a)**
31. (d)
33. (a) $3x^2$ **(b)** None
35. (a) $\dfrac{1}{2x}$ **(b)** $y = 0$
37. (a) $4x^2$ **(b)** None
39. (a) e^x **(b)** $-2x$
41. (a) x **(b)** x
43. At ∞: ∞ **45.** At ∞: 0
   At $-\infty$: 0   At $-\infty$: 0
47. (a) 0   **(b)** -1
   (c) $-\infty$ **(d)** -1
49. One possible answer:

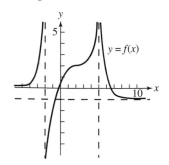

51. $\dfrac{f_1(x)/f_2(x)}{g_1(x)/g_2(x)} = \dfrac{f_1(x)/g_1(x)}{f_2(x)/g_2(x)}$

As x goes to infinity, $\dfrac{f_1}{g_1}$ and $\dfrac{f_2}{g_2}$ both approach 1.

Therefore, using the above equation, $\dfrac{f_1/f_2}{g_1/g_2}$ must

also approach 1.

53. (a) Using 1980 as $x = 0$:
      $y = -2.2316x^3 + 54.7134x^2 - 351.0933x$
         $+ 733.2224$

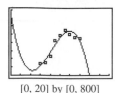

      [0, 20] by [0, 800]

(b) Again using 1980 as $x = 0$:
      $y = 1.458561x^4 - 60.5740x^3 + 905.8877x^2 -$
         $5706.0943x + 12967.6288$

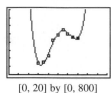

      [0, 20] by [0, 800]

(c) Cubic: approximately -2256 dollars
     Quartic: approximately 9979 dollars
(d) Cubic: End behavior model is $-2.2316x^3$.
           This model predicts that the grants
           will become negative by 1996.
     Quartic: End behavior model is $1.458561x^4$.
           This model predicts that the size of
           the grants will grow very rapidly
           after 1995.
   Neither of these seem reasonable. There is no
   reason to expect the grants to disappear
   (become negative) based on the data.
   Similarly, the data give no indication that a
   period of rapid growth is about to occur.
55. (a) This follows from $x - 1 < \text{int } x \le x$ which is
     true for all x. Dividing by x gives the result.
   (b) 1   **(c)** 1
57. This is because as x approaches infinity, $\sin x$
   continues to oscillate between 1 and -1 and
   doesn't approach any given real number.
59. Limit $= \ln(10)$,
   since $\dfrac{\ln x}{\log x} = \dfrac{\ln x}{\ln x/\ln 10} = \ln 10$.

Section 2.3 (pp. 73–81)

Quick Review

1. 2

2. (a) -2 **(b)** -1
   (c) No limit   **(d)** -1

3. (a) 1   **(b)** 2
   (c) No limit   **(d)** 2

4. $(f \circ g)(x) = \dfrac{x + 2}{6x + 1}, \ x \neq 0$

   $(g \circ f)(x) = \dfrac{3x + 4}{2x - 1}, \ x \neq -5$

5. $g(x) = \sin x, \ x \geq 0$
   $(f \circ g)(x) = \sin^2 x, \ x \geq 0$

6. $f(x) = \dfrac{1}{x^2} + 1, \ x > 0$

   $(f \circ g)(x) = \dfrac{x}{x - 1}, \ x > 1$

7. $x = \dfrac{1}{2}, -5$ **8.** $x \approx 0.453$

9. $x = 1$ **10.** Any c in $[1, 2)$

Exercises

1. $x = -2$, infinite discontinuity

3. None

5. All points not in the domain, i.e., all $x < -\dfrac{3}{2}$

7. $x = 0$, jump discontinuity

9. $x = 0$, infinite discontinuity

11. (a) Yes   **(b)** Yes
    (c) Yes   **(d)** Yes

13. (a) No   **(b)** No   **15.** 0

17. No, because the right-hand and left-hand limits are not the same at zero.

19. (a) $x = 2$
    (b) Not removable, the one-sided limits are different.

21. (a) $x = 1$
    (b) Not removable, it's an infinite discontinuity.

23. (a) All points not in the domain along with $x = 0, 1$
    (b) $x = 0$ is a removable discontinuity, assign $f(0) = 0$.
       $x = 1$ is not removable, the two-sided limits are different.

25. $y = x - 3$

27. $y = \begin{cases} \dfrac{\sin x}{x}, & x \neq 0 \\ 1, & x = 0 \end{cases}$ **29.** $y = \sqrt{x} + 2$

31. Assume $y = x$, constant functions, and the square root function are continuous.
Use the sum, composite, and quotient theorems.
Domain: $(-2, \infty)$

33. Assume $y = x$ and the absolute value function are continuous.
Use the product, constant multiple, difference, and composite theorems.
Domain: $(-\infty, \infty)$

35. Possible answer

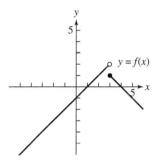

37. Possible answer

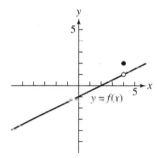

39. $x \approx -0.724$ and $x \approx 1.221$

41. $a = \dfrac{4}{3}$

43. (h)

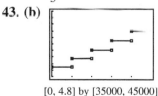

    $[0, 4.8]$ by $[35000, 45000]$

    Continuous at all points in the domain $[0, 5)$ except at $t = 1, 2, 3, 4$.

45. (a) Domain of f: $(-\infty, -1) \cup (0, \infty)$
    (b)

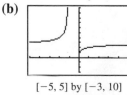

    $[-5, 5]$ by $[-3, 10]$

    (c) Because f is undefined there due to division by 0.

45. continued

(d) $x = 0$: removable, right-hand limit is 0
$x = -1$: not removable, infinite discontinuity

(e) 2.718 or e

47. Suppose not. Then f would be negative somewhere in the interval and positive somewhere else in the interval. So, by the Intermediate Value Theorem, it would have to be zero somewhere in the interval, which contradicts the hypothesis.

49. For any real number a, the limit of this function as x approaches a cannot exist. This is because as x approaches a, the values of the function will continually oscillate between 0 and 1.

Section 2.4 (pp. 82–90)

Quick Review

1. $\Delta x = 8, \Delta y = 3$

2. $\Delta x = a - 1, \Delta y = b - 3$

3. Slope $= -\dfrac{4}{7}$ **4.** Slope $= \dfrac{2}{3}$

5. $y = \dfrac{3}{2}x + 6$ **6.** $y = -\dfrac{7}{3}x + \dfrac{25}{3}$

7. $y = -\dfrac{3}{4}x + \dfrac{19}{4}$ **8.** $y = \dfrac{4}{3}x + \dfrac{8}{3}$

9. $y = -\dfrac{2}{3}x + \dfrac{7}{3}$ **10.** $b = \dfrac{19}{3}$

Exercises

1. (a) 19   (b) 1

3. (a) $\dfrac{1 - e^{-2}}{2} \approx 0.432$ (b) $\dfrac{e^3 - e}{2} \approx 8.684$

5. (a) $-\dfrac{4}{\pi} \approx -1.273$ (b) $-\dfrac{3\sqrt{3}}{\pi} \approx -1.654$

7. Using $Q_1 = (10, 225), Q_2 = (14, 375),$
$Q_3 = (16.5, 475), Q_4 = (18, 550),$ and
$P = (20, 650)$

(a)

Secant	Slope
PQ_1	43
PQ_2	46
PQ_3	50
PQ_4	50

Units are meters/second

(b) Approximately 50 m/sec

9. (a) -4 (b) $y = -4x - 4$

(c) $y = \dfrac{1}{4}x + \dfrac{9}{2}$

(d)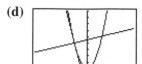

$[-8, 7]$ by $[-1, 9]$

11. (a) -1 (b) $y = -x + 3$

(c) $y = x - 1$

(d)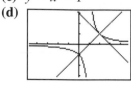

$[-4.7, 4.7]$ by $[-3.1, 3.1]$

13. (a) 1   (b) -1

15. No. Slope from the left is -2; slope from the right is 2. The two-sided limit of the difference quotient doesn't exist.

17. Yes. The slope is $-\dfrac{1}{4}$.

19. (a) $2a$

(b) The slope of the tangent steadily increases as a increases.

21. (a) $-\dfrac{1}{(a - 1)^2}$

(b) The slope of the tangent is always negative. The tangents are very steep near $x = 1$ and nearly horizontal as a moves away from the origin.

23. 19.6 m/sec   **25.** 6π in^2/in.

27. 3.72 m/sec   **29.** $(-2, -5)$

31. (a) At $x = 0$: $y = -x - 1$
At $x = 2$: $y = -x + 3$

(b) At $x = 0$: $y = x - 1$
At $x = 2$: $y = x - 1$

33. (a) 0.3 billion dollars per year

(b) 0.5 billion dollars per year

(c) $y = 0.0571x^2 - 0.1514x + 1.3943$

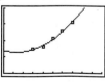

$[0, 10]$ by $[0, 4]$

(d) 1993 to 1995: 0.31 billion dollars per year
1995 to 1997: 0.53 billion dollars per year

(e) 0.65 billion dollars per year

35. (a) $\dfrac{e^{1+h} - e}{h}$

(b) Limit ≈ 2.718

(c) They're about the same.

(d) Yes, it has a tangent whose slope is about e.

37. No **39.** Yes

41. This function has a tangent with slope zero at the origin. It is sandwiched between two functions, $y = x^2$ and $y = -x^2$, both of which have slope zero at the origin.

Looking at the difference quotient,
$$-h \le \frac{f(0 + h) - f(0)}{h} \le h, \text{ so}$$
the Sandwich Theorem tells us that the limit is 0.

43. Slope ≈ 0.540

Chapter Review (pp. 91–93)

1. -15

2. $\frac{5}{21}$

3. No limit

4. No limit

5. $-\frac{1}{4}$

6. $\frac{2}{5}$

7. $+\infty, -\infty$

8. $\frac{1}{2}$

9. 2

10. 0

11. 6

12. 5

13. 0

14. 1

15. Limit exists

16. Limit exists

17. Limit exists

18. Doesn't exist

19. Limit exists

20. Limit exists

21. Yes

22. No

23. No

24. Yes

25. (a) 1 (b) 1.5 (c) No
(d) g is discontinuous at $x = 3$ (and points not in domain).
(e) Yes, can remove discontinuity at $x = 3$ by assigning the value 1 to $g(3)$.

26. (a) 1.5 (b) 0
(c) 0 (d) No
(e) k is discontinuous at $x = 1$ (and points not in domain).
(f) Discontinuity at $x = 1$ is not removable because the two one-sided limits are different.

27. (a) Vertical Asymp.: $x = -2$
(b) Left-hand limit $= -\infty$
Right-hand limit $= \infty$

28. (a) Vertical Asymp.: $x = 0$ and $x = -2$
(b) At $x = 0$:
Left-hand limit $= -\infty$
Right-hand limit $= -\infty$
At $x = -2$:
Left-hand limit $= \infty$
Right-hand limit $= -\infty$

29. (a) At $x = -1$:
Left-hand limit $= 1$
Right-hand limit $= 1$
At $x = 0$:
Left-hand limit $= 0$
Right-hand limit $= 0$
At $x = 1$:
Left-hand limit $= -1$
Right-hand limit $= 1$
(b) At $x = -1$:
Yes, the limit is 1.
At $x = 0$:
Yes, the limit is 0.
At $x = 1$:
No, the limit doesn't exist because the two one-sided limits are different.
(c) At $x = -1$:
Continuous because $f(-1) =$ the limit.
At $x = 0$:
Discontinuous because $f(0) \ne$ the limit.
At $x = 1$:
Discontinuous because limit doesn't exist.

30. (a) Left-hand limit $= 3$
Right-hand limit $= -3$
(b) No, because the two one-sided limits are different.
(c) Every place except for $x = 1$
(d) At $x = 1$

31. $x = -2$ and $x = 2$

32. There are no points of discontinuity.

33. (a) $\frac{2}{x}$ (b) $y = 0$ (x-axis)

34. (a) 2 (b) $y = 2$

35. (a) x^2 (b) None

36. (a) x (b) None

37. (a) e^x (b) x

38. (a) $\ln |x|$ (b) $\ln |x|$

39. $k = 8$ **40.** $k = \frac{1}{2}$

41. One possible answer:

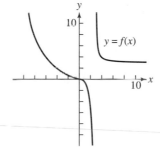

42. One possible answer:

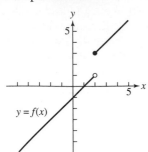

$y = f(x)$

43. $\dfrac{2}{\pi}$

44. $\dfrac{2}{3}\pi a H$

45. $12a$

46. $2a - 1$

47. (a) -1 (b) $y = -x - 1$
  (c) $y = x - 3$

48. $\left(\dfrac{3}{2}, -\dfrac{9}{4}\right)$

49. (a) 25. Perhaps this is the number of bears placed in the reserve when it was established.
  (b) 200
  (c) Perhaps this is the maximum number of bears which the reserve can support due to limitations of food, space, or other resources. Or, perhaps the number is capped at 200 and excess bears are moved to other locations.

50. (a)
$$f(x) = \begin{cases} 3.20 - 1.35 \cdot \mathrm{int}\,(-x + 1), & 0 < x \le 20 \\ 0, & x = 0 \end{cases}$$

  (b)

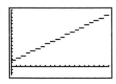

[0, 20] by [−5, 32]

  f is discontinuous at integer values of x: 0, 1, 2, ..., 19

51. (a) Cubic:
$$y = -1.644x^3 + 42.981x^2 - 254.369x + 300.232$$
  Quartic:
$$y = 2.009x^4 - 102.081x^3 + 1884.997x^2 - 14918.180x + 43004.464$$

  (b) Cubic: $-1.644x^3$, predicts spending will go to 0.
  Quartic: $2.009x^4$, predicts spending will go to ∞.

52. $\displaystyle\lim_{x \to c} f(x) = \dfrac{3}{2}$, $\displaystyle\lim_{x \to c} g(x) = \dfrac{1}{2}$

53. (a)

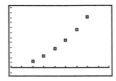

[3, 12] by [−2, 24]

  (b)

Year of Q	Slope of PQ
1995	3.48
1996	3.825
1997	4.1
1998	4.45
1999	4.9

  (c) Approximately 5 billion dollars per year
  (d) $y = 0.3214x^2 - 1.3471x + 1.3857$
  Predicted rate of change in 2000 is 5.081 billion dollars per year.

Chapter 3

Section 3.1 (pp. 95–104)

Quick Review

1. 4

2. $\dfrac{5}{2}$

3. -1

4. 8

5. 0

6. $(-\infty, 0]$ and $[2, \infty)$

7. $\displaystyle\lim_{x \to 1^+} f(x) = 0$; $\displaystyle\lim_{x \to 1^-} f(x) = 3$

8. 0

9. No, the two one-sided limits are different.

10. No. f is discontinuous at $x = 1$ because the limit doesn't exist there.

Exercises

1. (a) $y = 5x - 7$
  (b) $y = -\dfrac{1}{5}x + \dfrac{17}{5}$

3. $-\dfrac{1}{9}$

5. $\dfrac{dy}{dx} = 7$

7. (b)

9. (d)

11. $\dfrac{dy}{dx} = 4x - 13$, tangent line is $y = -x - 13$

13. (ii)

15. **(a)** Sometime around April 1. The rate then is approximately $\dfrac{1}{6}$ hour per day.

(b) Yes. Jan. 1 and July 1

(c) Positive: Jan. 1. through July 1
Negative: July 1 through Dec. 31

17. **(a)** 0 and 0
(b) 1700 and 1300

19. Graph of derivative:

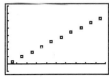

[0, 10] by [−10, 80]

(a) The speed of the skier
(b) Feet per second
(c) Approximately $D - 6.65t$

21.

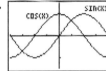

[−π, π] by [−1.5, 1.5]

Cosine could be the derivative of sine. The values of cosine are positive where sine is increasing, zero where sine has horizontal tangents, and negative where sine is decreasing.

23. $\displaystyle\lim_{h\to 0^+} \dfrac{f(0 + h) - f(0)}{h} = \lim_{h\to 0^+} \dfrac{\sqrt{h}}{h} = \lim_{h\to 0^+} \dfrac{1}{\sqrt{h}} = \infty$

Thus, the right-hand derivative at 0 does not exist.

25.

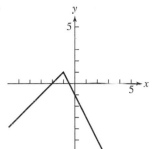

27. The y-intercept is $b - a$.

29. **(a)** 0.992   **(b)** 0.008

(c) If P is the answer to (b), then the probability of a shared birthday when there are four people is $1 - (1 - P)\dfrac{362}{365} \approx 0.016$.

(d) No. Clearly, February 29th is a much less likely birth date. Furthermore, census data do not support the assumption that the other 365 birth dates are equally likely. However, this simplifying assumption may still give us some insight into this problem even if the calculated probabilities aren't completely accurate.

Section 3.2 (pp. 105–112)

Quick Review

1. Yes   **2.** No   **3.** Yes
4. Yes   **5.** No   **6.** All reals
7. $[0, \infty)$ **8.** $[3, \infty)$ **9.** 3.2
10. 5

Exercises

1. Left-hand derivative = 0
Right-hand derivative = 1

3. Left-hand derivative = $\dfrac{1}{2}$
Right-hand derivative = 2

5. **(a)** All points in $[-3, 2]$
(b) None   **(c)** None

7. **(a)** All points in $[-3, 3]$ except $x = 0$
(b) None   **(c)** $x = 0$

9. **(a)** All points in $[-1, 2]$ except $x = 0$
(b) $x = 0$ **(c)** None

11. Discontinuity   **13.** Corner

15. Corner

17. All reals except $x = -1, 5$

19. All reals except $x = 0$

21. All reals except $x = 3$

23. **(a)** $x = 0$ is not in their domains, or, they are both discontinuous at $x = 0$.

(b) For $\dfrac{1}{x}$: 1,000,000

For $\dfrac{1}{x^2}$: 0

(c) It returns an incorrect response because even though these functions are not defined at $x = 0$, they are defined at $x = \pm 0.001$.

25.

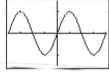

[−2π, 2π] by [−1.5, 1.5]

$\dfrac{dy}{dx} = \sin x$

27.

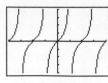

$[-2\pi, 2\pi]$ by $[-4, 4]$

$\dfrac{dy}{dx} = \tan x$

29. (a) $a + b = 2$

(b) $a = -3$ and $b = 5$

31. (a) Note that $-x \le x \sin \dfrac{1}{x} \le x$, for all x,

so $\displaystyle\lim_{x \to 0}\left(x \sin \dfrac{1}{x}\right) = 0$ by the Sandwich Theorem. Therefore, f is continuous at $x = 0$.

(b) $\dfrac{f(0 + h) - f(0)}{h} = \dfrac{h \sin \dfrac{1}{h} - 0}{h} = \sin \dfrac{1}{h}$

(c) The limit does not exist because $\sin \dfrac{1}{h}$ oscillates between -1 and 1 an infinite number of times arbitrarily close to $h = 0$.

(d) No

(e) $\dfrac{g(0 + h) - g(0)}{h} = \dfrac{h^2 \sin \dfrac{1}{h} - 0}{h}$

$= h \sin \dfrac{1}{h}$

As noted in part (a), the limit of this as x approaches zero is 0, so $g'(0) = 0$.

Section 3.3 (pp. 112–121)

Quick Review

1. $x + x^2 - 2x^{-1} - 2$ **2.** $x + x^{-1}$

3. $3x^2 - 2x^{-1} + 5x^{-2}$ **4.** $\dfrac{3}{2}x^2 - x + 2x^{-2}$

5. $x^{-3} + x^{-1} + 2x^{-2} + 2$ **6.** $x^2 + x$

7. Root: $x \approx 1.173$, $500x^6 \approx 1305$

Root: $x \approx 2.394$, $500x^6 \approx 94{,}212$

8. (a) 7 **(b)** 7

(c) 7 **(d)** 0

9. (a) 0 **(b)** 0

(c) 0

10. (a) $f'(x) = \dfrac{1}{\pi}$ **(b)** $f'(x) = -\pi x^{-2}$

Exercises

1. $\dfrac{dy}{dx} = -2x$, $\dfrac{d^2y}{dx^2} = -2$

3. $\dfrac{dy}{dx} = 2$, $\dfrac{d^2y}{dx^2} = 0$

5. $\dfrac{dy}{dx} = x^2 + x + 1$, $\dfrac{d^2y}{dx^2} = 2x + 1$

7. $\dfrac{dy}{dx} = 4x^3 - 21x^2 + 4x$, $\dfrac{d^2y}{dx^2} = 12x^2 - 42x + 4$

9. $\dfrac{dy}{dx} = -8x^{-3} - 8$, $\dfrac{d^2y}{dx^2} = 24x^{-4}$

11. (a) $3x^2 + 2x + 1$

(b) $3x^2 + 2x + 1$

13. $-\dfrac{19}{(3x - 2)^2}$ **15.** $\dfrac{3}{x^4}$

17. $\dfrac{x^4 + 2x}{(1 - x^3)^2}$ **19.** $\dfrac{12 - 6x^2}{(x^2 - 3x + 2)^2}$

21. $\dfrac{d}{dx}(c \cdot f(x)) = c \cdot \dfrac{d}{dx}f(x) + f(x) \cdot \dfrac{d}{dx}c$

$= c \cdot \dfrac{d}{dx}f(x) + 0 = c \cdot \dfrac{d}{dx}f(x)$

23. (a) 13 **(b)** -7

(c) $\dfrac{7}{25}$ **(d)** 20

25. (iii) **27.** $y = -\dfrac{1}{9}x + \dfrac{29}{9}$

29. $(-1, 27)$ and $(2, 0)$

31. At $(0, 0)$: $y = 4x$

At $(1, 2)$: $y = 2$

33. $-\dfrac{nRT}{(V - nb)^2} + \dfrac{2an^2}{V^3}$ **35.** $\dfrac{dR}{dM} = CM - M^2$

37. If the radius of a sphere is changed by a very small amount Δr, the change in the volume can be thought of as $(4\pi r^2)(\Delta r)$, which means that the change in the volume divided by the change in the radius is just $4\pi r^2$.

39. It is going down approximately 20 cents per year. (rate ≈ -0.201 dollars/year)

Section 3.4 (pp. 122–133)

Quick Review

1. Downward **2.** y-intercept $= -256$

3. x-intercepts $= 2, 8$ **4.** $(-\infty, 144]$

5. $(5, 144)$ **6.** $x = 3, 7$

7. $x = \dfrac{15}{8}$ **8.** $(-\infty, 5)$

9. 64 **10.** -32

Exercises

1. $3s^2$

3. (a) $\text{vel}(t) = 24 - 1.6t$, $\text{accel}(t) = -1.6$

(b) 15 seconds **(c)** 180 meters

(d) About 4.393 seconds

(e) 30 seconds

5. About 29.388 meters

7. For the moon:
$$x_1(t) = 3(t < 160) + 3.1(t \ge 160)$$
$$y_1(t) = 832t - 2.6t^2$$
t-values: 0 to 320
window: [0, 6] by [−10,000, 70,000]

For the earth:
$$x_1(t) = 3(t < 26) + 3.1(t \ge 26)$$
$$y_1(t) = 832t - 16t^2$$
t-values: 0 to 52
window: [0, 6] by [−1000, 11,000]

9. At the end of 10 minutes: 8000 gallons/minute
Average over first 10 minutes:
10,000 gallons/minute

11. (a)
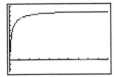
[0, 50] by [−500, 2200]
The values of x which make sense are the whole numbers, $x \ge 0$.

(b) $\dfrac{2000}{(x + 1)^2}$

(c) Approximately $55.56

(d) The limit is 0. This means that as x gets large, one reaches a point where very little extra revenue can be expected from selling more desks.

13. At $t = 1$: 0 m/sec; At $t = 2$: 1 m/sec

15. (a)
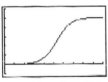
[0, 200] by [−2, 12]

(b) $x \ge 0$ (whole numbers)

(c)
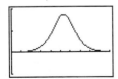
[0, 200] by [−0.1, 0.2]
P seems to be relatively sensitive to changes in x between approximately $x = 60$ and $x = 160$.

(d) The maximum occurs when $x \approx 106.44$. Since x must be an integer, $P(106) \approx 4.924$ thousand dollars or $4924.

(e) $13 per package sold, $165 per package sold, $118 per package sold, $31 per package sold, $6 per package sold, $P'(300) \approx 0$ (on the order of 10^{-6}, or $0.001 per package sold)

(f) The limit is 10. Maximum possible profit is $10,000 monthly.

(g) Yes. In order to sell more and more packages, the company might need to lower the price to a point where they won't make any additional profit.

17. Possible answer
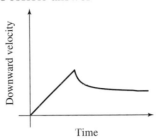

19. At $t \approx 2.83$

21. (a)
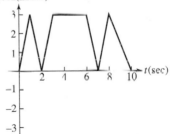

(b) $s'(1) = 18$, $s'(2.5) = 0$, $s'(3.5) = -12$

23. (a) At $t = 2$ and $t = 7$

(b) Between $t = 3$ and $t = 6$

(c)
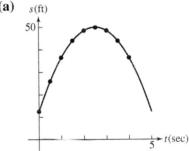

(d)
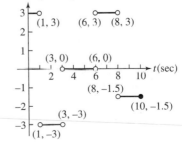

25. (a) Move forward: $0 \le t < 1$ and $5 < t < 7$
  move backward: $1 < t < 5$
  speed up: $1 < t < 2$ and $5 < t < 6$
  slow down: $0 \le t < 1$, $3 < t < 5$,
  and $6 < t < 7$
(b) Positive: $3 < t < 6$
  negative: $0 \le t < 2$ and $6 < t < 7$
  zero: $2 < t < 3$ and $7 < t \le 9$
(c) At $t = 0$ and $2 < t < 3$
(d) $7 < t \le 9$

27. (a) $\dfrac{4}{7}$ of a second.
  Average velocity = 280 cm/sec

(b) Velocity = 560 cm/sec;
  acceleration = 980 cm/sec^2
(c) About 28 flashes per second

29. Graph C is position, graph B is velocity, and graph A is acceleration.
B is the derivative of C because it is negative and zero where C is decreasing and has horizontal tangents, respectively.
A is the derivative of B because it is positive, negative, and zero where B is increasing, decreasing, and has horizontal tangents, respectively.

31. (a) 16π cubic feet of volume per foot of radius
(b) By about 11.092 cubic feet

33. Exit velocity ≈ 348.712 ft/sec ≈ 237.758 mi/h

35. Since profit = revenue − cost, using Rule 4 (the "difference rule"), and taking derivatives, we see that marginal profit
  = marginal revenue − marginal cost.

37. (a) Assume that f is even. Then,

$$f'(-x) = \lim_{h \to 0} \frac{f(-x + h) - f(-x)}{h}$$

$$= \lim_{h \to 0} \frac{f(x - h) - f(x)}{h},$$

and substituting $k = -h$,

$$= \lim_{k \to 0} \frac{f(x + k) - f(x)}{-k}$$

$$= -\lim_{k \to 0} \frac{f(x + k) - f(x)}{k} = -f'(x)$$

So, f' is an odd function.

(b) Assume that f is odd. Then,

$$f'(-x) = \lim_{h \to 0} \frac{f(-x + h) - f(-x)}{h}$$

$$= \lim_{h \to 0} \frac{-f(x - h) + f(x)}{h}, \text{ and}$$

substituting $k = -h$,

$$= \lim_{k \to 0} \frac{-f(x + k) + f(x)}{-k}$$

$$= \lim_{k \to 0} \frac{f(x + k) - f(x)}{k} = f'(x)$$

So, f' is an even function.

Section 3.5 (pp. 134–141)

Quick Review

1. $\dfrac{3\pi}{4} \approx 2.356$ **2.** $\left(\dfrac{306}{\pi}\right)^{\circ} \approx 97.403°$

3. $\dfrac{\sqrt{3}}{2}$

4. Domain: all reals; range: $[-1, 1]$

5. Domain: $x \ne k\dfrac{\pi}{2}$, where k is an odd integer; range: all reals

6. 0 **7.** $\pm\dfrac{1}{\sqrt{2}}$

8. Multiply by $\dfrac{1 + \cos h}{1 + \cos h}$ and use $1 - \cos^2 h = \sin^2 h$.

9. $y = 12x - 35$ **10.** 12

Exercises

1. $1 + \sin x$ **3.** $-\dfrac{1}{x^2} + 5 \cos x$

5. $-x^2 \cos x - 2x \sin x$ **7.** $4 \sec x \tan x$

9. $-\dfrac{\csc^2 x}{(1 + \cot x)^2} = -\dfrac{1}{(\sin x + \cos x)^2}$

11. $y = -x + \pi + 3$

13. Approximately $y = -8.063x + 25.460$

15. (a) $\dfrac{d}{dx} \tan x = \dfrac{d}{dx} \dfrac{\sin x}{\cos x}$

$$= \frac{(\cos x)(\cos x) - (\sin x)(-\sin x)}{(\cos x)^2}$$

$$= \frac{\cos^2 x + \sin^2 x}{\cos^2 x}$$

$$= \frac{1}{\cos^2 x} = \sec^2 x$$

(b) $\dfrac{d}{dx}\sec x = \dfrac{d}{dx}\dfrac{1}{\cos x}$

$\quad = \dfrac{(\cos x)(0) - (1)(-\sin x)}{(\cos x)^2}$

$\quad = \dfrac{\sin x}{(\cos x)^2} = \sec x \tan x$

17. $\dfrac{d}{dx}\sec x = \sec x \tan x$ which is 0 at $x = 0$, so the slope of the tangent line is 0.

$\dfrac{d}{dx}\cos x = -\sin x$ which is 0 at $x = 0$, so the slope of the tangent line is 0.

19. Tangent: $y = -x + \dfrac{\pi}{4} + 1$

Normal: $y = x + 1 - \dfrac{\pi}{4}$

21. (a) $y = -x + \dfrac{\pi}{2} + 2$

(b) $y = 4 - \sqrt{3}$

23. (a) Velocity: $-2\cos t$ m/sec
Speed: $|2\cos t|$ m/sec
Accel.: $2\sin t$ m/sec^2
Jerk: $2\cos t$ m/sec^3

(b) Velocity: $-\sqrt{2}$ m/sec
Speed: $\sqrt{2}$ m/sec
Accel.: $\sqrt{2}$ m/sec^2
Jerk: $\sqrt{2}$ m/sec^3

(c) The body starts at 2, goes to 0 and then oscillates between 0 and 4.
Speed: *Greatest* when $\cos t = \pm 1$ (or $t = k\pi$), at the center of the interval of motion.

Zero when $\cos t = 0$ $\left(\text{or } t = \dfrac{k\pi}{2}, k \text{ odd}\right)$, at the endpoints of the interval of motion.

Acceleration: *Greatest* (in magnitude) when $\sin t = \pm 1$ $\left(\text{or } t = \dfrac{k\pi}{2}, k \text{ odd}\right)$
Zero when $\sin t = 0$ (or $t = k\pi$)

Jerk: *Greatest* (in magnitude) when $\cos t = \pm 1$ (or $t = k\pi$)
Zero when $\cos t = 0$ $\left(\text{or } t = \dfrac{k\pi}{2}, k \text{ odd}\right)$

25. (a) The limit is $\dfrac{\pi}{180}$ because this is the conversion factor for changing from degrees to radians.

(b) This limit is still 0.

(c) $\dfrac{d}{dx}\sin x = \dfrac{\pi}{180}\cos x$

(d) $\dfrac{d}{dx}\cos x = -\dfrac{\pi}{180}\sin x$

(e) $\dfrac{d^2}{dx^2}\sin x = -\dfrac{\pi^2}{180^2}\sin x$

$\dfrac{d^3}{dx^3}\sin x = -\dfrac{\pi^3}{180^3}\cos x$

$\dfrac{d^2}{dx^2}\cos x = -\dfrac{\pi^2}{180^2}\cos x$

$\dfrac{d^3}{dx^3}\cos x = \dfrac{\pi^3}{180^3}\sin x$

27. $y'' = \dfrac{2 + 2\theta\tan\theta}{\cos^2\theta} = \dfrac{2\cos\theta + 2\theta\sin\theta}{\cos^3\theta}$

29. $\sin x$ 　　　　**31.** $y = x$

33. $\dfrac{d}{dx}\sin 2x = \dfrac{d}{dx}2\sin x\cos x$

$\quad = 2\,[(\sin x)(-\sin x) + (\cos x)(\cos x)]$

$\quad = 2\,[\cos^2 x - \sin^2 x] = 2\cos 2x$

35. $\lim\limits_{h\to 0}\dfrac{\cos h - 1}{h} = \lim\limits_{h\to 0}\dfrac{(\cos h - 1)(\cos h + 1)}{h(\cos h + 1)}$

$\quad = \lim\limits_{h\to 0}\dfrac{\cos^2 h - 1}{h(\cos h + 1)}$

$\quad = \lim\limits_{h\to 0}\dfrac{-\sin^2 h}{h(\cos h + 1)}$

$\quad = -\left(\lim\limits_{h\to 0}\dfrac{\sin h}{h}\right)\left(\lim\limits_{h\to 0}\dfrac{\sin h}{\cos h + 1}\right)$

$\quad = -(1)\left(\dfrac{0}{2}\right) = 0$

Section 3.6 (pp. 141–149)

Quick Review

1. $\sin(x^2 + 1)$ 　　　　**2.** $\sin(49x^2 + 1)$
3. $49x^2 + 1$ 　　　　**4.** $7x^2 + 7$
5. $\sin\dfrac{x^2 + 1}{7x}$ 　　　　**6.** $g(f(x))$
7. $g(h(f(x)))$ 　　　　**8.** $h(g(f(x)))$
9. $f(h(h(x)))$ 　　　　**10.** $f(g(h(x)))$

Exercises

1. $3\cos(3x + 1)$ 　　　　**3.** $-\sqrt{3}\sin(\sqrt{3}x)$
5. $\dfrac{10}{x^2}\csc^2\left(\dfrac{2}{x}\right)$ 　　　　**7.** $-\sin(\sin x)\cos x$
9. $-2(x + \sqrt{x})^{-3}\left(1 + \dfrac{1}{2\sqrt{x}}\right)$
11. $-5\sin^{-6}x\cos x + 3\cos^2 x\sin x$
13. $4\sin^3 x\sec^2 4x + 3\sin^2 x\cos x\tan 4x$
15. $-3(2x + 1)^{-3/2}$
17. $6\sin(3x - 2)\cos(3x - 2) = 3\sin(6x - 4)$

19. $-42(1 + \cos^2 7x)^2 \cos 7x \sin 7x$

21. $3 \sin\left(\dfrac{\pi}{2} - 3t\right)$ **23.** $\dfrac{4}{\pi} \cos 3t - \dfrac{4}{\pi} \sin 5t$

25. $-\sec^2 (2 - \theta)$ **27.** $\dfrac{\theta \cos \theta + \sin \theta}{2\sqrt{\theta \sin \theta}}$

29. $2 \sec^2 x \tan x$

31. $18 \csc^2 (3x - 1) \cot (3x - 1)$

33. $\dfrac{5}{2}$ **35.** $-\dfrac{\pi}{4}$ **37.** 0

39. (a) $-6 \sin (6x + 2)$ (b) $-6 \sin (6x + 2)$

41. $y = -x + 2\sqrt{2}$ **43.** $y = -\dfrac{1}{2}x - \dfrac{1}{2}$

45. $y = x + \dfrac{1}{4}$

47. $y = \sqrt{3}x + 2 - \dfrac{\pi}{\sqrt{3}}$

49. (a) $\dfrac{\cos t}{2t + 1}$

     (b) $\dfrac{d}{dt}\left(\dfrac{dy}{dx}\right) = -\dfrac{(2t + 1)(\sin t) + 2 \cos t}{(2t + 1)^2}$

     (c) $\dfrac{d}{dx}\left(\dfrac{dy}{dx}\right) = -\dfrac{(2t + 1)(\sin t) + 2 \cos t}{(2t + 1)^3}$

     (d) part (c)

51. 5   **53.** $\dfrac{1}{2}$

55. Tangent: $y = \pi x - \pi + 2$;

     Normal: $y = -\dfrac{1}{\pi}x + \dfrac{1}{\pi} + 2$

57. (a) 1   (b) 6   (c) 1

     (d) $-\dfrac{1}{9}$ (e) $-\dfrac{40}{3}$ (f) -6

     (g) $-\dfrac{4}{9}$

59. Because the symbols $\dfrac{dy}{dx}$, $\dfrac{dy}{du}$, and $\dfrac{du}{dx}$ are not fractions. The individual symbols dy, du, and dx do not have numerical values.

61. (a) On the 101st day (April 11th)
     (b) About 0.637 degrees per day

63. Acceleration $= \dfrac{dv}{dt} = \dfrac{dv}{ds}\dfrac{ds}{dt} = \dfrac{dv}{ds}v$

     $= \dfrac{k}{2\sqrt{s}}(k\sqrt{s}) = \dfrac{k^2}{2}$

65. Acceleration $= \dfrac{dv}{dt} = \dfrac{d f(x)}{dt}$

     $= \left[\dfrac{d f(x)}{dx}\right]\left[\dfrac{dx}{dt}\right]$

     $= f'(x)f(x)$

67. No, this does not contradict the Chain Rule. The Chain Rule states that if two functions are differentiable at the appropriate points, then their composite must also be differentiable.

69. As $h \to 0$, the second curve (the difference quotient) approaches the first $y = 2 \cos 2x$. This is because $2 \cos 2x$ is the derivative of $\sin 2x$, and the second curve is the difference quotient used to define the derivative of $\sin 2x$. As $h \to 0$, the difference quotient expression should be approaching the derivative.

71. (a) Let $f(x) = |x|$.

     Then $\dfrac{d}{dx} |u| = \dfrac{d}{dx} f(u) = f'(u)\dfrac{du}{dx}$

             $= f'(u)u' = \dfrac{u}{|u|} u'.$

     The derivative of the absolute value function is $+1$ for positive values, -1 for negative values, and undefined at 0. So $f'(u)$ should be $+1$ when $u > 0$ and -1 when $u < 0$. But this is exactly how the expression $\dfrac{u}{|u|}$ evaluates.

     (b) $f'(x) = \dfrac{(2x)(x^2 - 9)}{|x^2 - 9|}$

     $g'(x) = |x| \cos x + \dfrac{x \sin x}{|x|}$

Section 3.7 (pp. 149–157)

Quick Review

1. $y_1 = \sqrt{x}$, $y_2 = -\sqrt{x}$

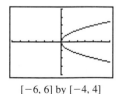

$[-6, 6]$ by $[-4, 4]$

2. $y_1 = \dfrac{2}{3}\sqrt{9 - x^2}$, $y_2 = -\dfrac{2}{3}\sqrt{9 - x^2}$

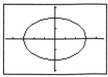

$[-4.7, 4.7]$ by $[-3.1, 3.1]$

3. $y_1 = \dfrac{x}{2}$, $y_2 = -\dfrac{x}{2}$

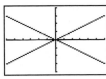

$[-6, 6]$ by $[-4, 4]$

4. $y_1 = \sqrt{9 - x^2}$, $y_2 = -\sqrt{9 - x^2}$

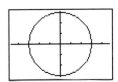

$[-4.7, 4.7]$ by $[-3.1, 3.1]$

5. $y_1 = \sqrt{2x + 3 - x^2}$, $y_2 = -\sqrt{2x + 3 - x^2}$

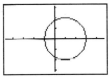

$[-4.7, 4.7]$ by $[-3.1, 3.1]$

6. $y' = \dfrac{4x - y + 2xy}{x^2}$

7. $y' = \dfrac{y + x \cos x}{\sin x - x}$

8. $y' = \dfrac{xy^2}{x^2 - y + x}$

9. $x^{3/2} - x^{5/6}$

10. $x^{-1/2} + x^{-5/6}$

Exercises

1. $\dfrac{9}{4}x^{5/4}$

3. $\dfrac{1}{3}x^{-2/3}$

5. $-(2x + 5)^{-3/2}$

7. $x^2(x^2 + 1)^{-1/2} + (x^2 + 1)^{1/2}$

9. $-\dfrac{2xy + y^2}{2xy + x^2}$

11. $\dfrac{1}{y(x + 1)^2}$

13. $-\dfrac{1}{4}(1 - x^{1/2})^{-1/2}x^{-1/2}$

15. $-\dfrac{9}{2}(\csc x)^{3/2} \cot x$

17. $\cos^2 y$

19. $-\dfrac{1}{x}\cos^2(xy) - \dfrac{y}{x}$

21. (b), (c), and (d)

23. $\dfrac{dy}{dx} = -\dfrac{x}{y}$

$\dfrac{d^2y}{dx^2} = -\dfrac{(x^2 + y^2)}{y^3} = -\dfrac{1}{y^3}$

25. $\dfrac{dy}{dx} = \dfrac{x + 1}{y}$

$\dfrac{d^2y}{dx^2} = \dfrac{y^2 - (x + 1)^2}{y^3} = -\dfrac{1}{y^3}$

27. (a) $y = \dfrac{7}{4}x - \dfrac{1}{2}$

(b) $y = -\dfrac{4}{7}x + \dfrac{29}{7}$

29. (a) $y = 3x + 6$

(b) $y = -\dfrac{1}{3}x + \dfrac{8}{3}$

31. (a) $y = \dfrac{6}{7}x + \dfrac{6}{7}$

(b) $y = -\dfrac{7}{6}x - \dfrac{7}{6}$

33. (a) $y = -\dfrac{\pi}{2}x + \pi$

(b) $y = \dfrac{2}{\pi}x - \dfrac{2}{\pi} + \dfrac{\pi}{2}$

35. (a) $y = 2\pi x - 2\pi$

(b) $y = -\dfrac{x}{2\pi} + \dfrac{1}{2\pi}$

37. (a) At $\left(\dfrac{\sqrt{3}}{4}, \dfrac{\sqrt{3}}{2}\right)$: Slope $= -1$;

at $\left(\dfrac{\sqrt{3}}{4}, \dfrac{1}{2}\right)$: Slope $= \sqrt{3}$

(b)

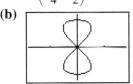

$[-1.8, 1.8]$ by $[-1.2, 1.2]$

Parameter interval:

$-1 \le t \le 1$

39. (a) $(-1)^3(1)^2 = \cos(\pi)$ is true since both sides equal: -1.

(b) The slope is $\dfrac{3}{2}$.

41. The points are $(\pm\sqrt{7}, 0)$.

$\dfrac{dy}{dx} = -\dfrac{2x + y}{2y + x}$

At both points, $\dfrac{dy}{dx} = -2$

43. First curve: $\dfrac{dy}{dx} = -\dfrac{2x}{3y}$

second curve: $\dfrac{dy}{dx} = \dfrac{3x^2}{2y}$

At $(1, 1)$, the slopes are $-\dfrac{2}{3}$ and $\dfrac{3}{2}$ respectively.

At $(1, -1)$, the slopes are $\dfrac{2}{3}$ and $-\dfrac{3}{2}$ respectively.

In both cases, the tangents are perpendicular.

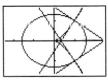

$[-2.4, 2.4]$ by $[-1.6, 1.6]$

45. Acceleration $= \dfrac{dv}{dt} = 4(s - t)^{-1/2}(v - 1)$

$= 32$ ft/sec^2

47. (a) At $(4, 2)$: $\dfrac{5}{4}$;

at $(2, 4)$: $\dfrac{4}{5}$

(b) At $(3\sqrt[3]{2}, 3\sqrt[3]{4}) \approx (3.780, 4.762)$

(c) At $(3\sqrt[3]{4}, 3\sqrt[3]{2}) \approx (4.762, 3.780)$

49. At $(-1, -1)$: $y = -2x - 3$;

at $(3, -3)$: $y = -2x + 3$

51. (a) $\dfrac{dy}{dx} = -\dfrac{b^2 x}{a^2 y}$

The tangent line is $y - y_1 = -\dfrac{b^2 x_1}{a^2 y_1}(x - x_1)$.

This gives: $a^2 y_1 y - a^2 y_1{}^2 = -b^2 x_1 x + b^2 x_1{}^2$,

$a^2 y_1 y + b^2 x_1 x = a^2 y_1{}^2 + b^2 x_1{}^2$.

But $a^2 y_1{}^2 + b^2 x_1{}^2 = a^2 b^2$ since (x_1, y_1) is on the ellipse.

Therefore, $a^2 y_1 y + b^2 x_1 x = a^2 b^2$, and dividing by $a^2 b^2$ gives $\dfrac{x_1 x}{a^2} + \dfrac{y_1 y}{b^2} = 1$.

(b) $\dfrac{x_1 x}{a^2} - \dfrac{y_1 y}{b^2} = 1$.

Section 3.8 (pp. 157–163)

Quick Review

1. Domain: $[-1, 1]$

Range: $\left[-\dfrac{\pi}{2}, \dfrac{\pi}{2}\right]$

At 1: $\dfrac{\pi}{2}$

2. Domain: $[-1, 1]$

Range: $[0, \pi]$

At 1: 0

3. Domain: all reals

Range: $\left(-\dfrac{\pi}{2}, \dfrac{\pi}{2}\right)$

At 1: $\dfrac{\pi}{4}$

4. Domain: $(-\infty, -1] \cup [1, \infty)$

Range: $\left[0, \dfrac{\pi}{2}\right) \cup \left(\dfrac{\pi}{2}, \pi\right]$

At 1: 0

5. Domain: all reals

Range: all reals

At 1: 1

6. $f^{-1}(x) = \dfrac{x + 8}{3}$

7. $f^{-1}(x) = x^3 - 5$

8. $f^{-1}(x) = \dfrac{8}{x}$

9. $f^{-1}(x) = \dfrac{2}{3 - x}$

10. $f^{-1}(x) = 3 \tan x$, $-\dfrac{\pi}{2} < x < \dfrac{\pi}{2}$

Exercises

1. $-\dfrac{2x}{\sqrt{1 - x^4}}$

3. $\dfrac{\sqrt{2}}{\sqrt{1 - 2t^2}}$

5. $\dfrac{1}{|2s + 1|\sqrt{s^2 + s}}$

7. $-\dfrac{2}{(x^2 + 1)\sqrt{x^2 + 2}}$

9. $-\dfrac{1}{\sqrt{1 - t^2}}$

11. $-\dfrac{1}{2\sqrt{t}(t + 1)}$

13. $-\dfrac{2s^2}{\sqrt{1 - s^2}}$

15. $0, x > 1$

17. $\sin^{-1} x$

19. (a) $y = 2x - \dfrac{\pi}{2} + 1$ **(b)** $y = \dfrac{1}{2}x - \dfrac{1}{2} + \dfrac{\pi}{4}$

21. (a) $f'(x) = 3 - \sin x$ and $f'(x) \neq 0$. So f has a differentiable inverse by Theorem 3.

(b) $f(0) = 1, f'(0) = 3$

(c) $f^{-1}(1) = 0, (f^{-1})'(1) = \dfrac{1}{3}$

23. (a) $v(t) = \dfrac{dx}{dt} = \dfrac{1}{1 + t^2}$ which is always positive.

(b) $a(t) = \dfrac{dv}{dt} = -\dfrac{2t}{(1 + t^2)^2}$ which is always negative.

(c) $\dfrac{\pi}{2}$

25. $\dfrac{d}{dx} \cot^{-1} x = \dfrac{d}{dx}\left(\dfrac{\pi}{2} - \tan^{-1} x\right)$

$= 0 - \dfrac{d}{dx} \tan^{-1} x$

$= -\dfrac{1}{1 + x^2}$

27. (a) $y = \dfrac{\pi}{2}$ **(b)** $y = -\dfrac{\pi}{2}$

(c) None

29. (a) $y = \dfrac{\pi}{2}$ **(b)** $y = \dfrac{\pi}{2}$

(c) None

31. (a) None **(b)** None

(c) None

33. (a)

$\alpha = \cos^{-1} x, \beta = \sin^{-1} x$

So $\dfrac{\pi}{2} = \alpha + \beta = \cos^{-1} x + \sin^{-1} x$.

(b)

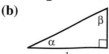

$\alpha = \tan^{-1} x, \beta = \cot^{-1} x$

So $\dfrac{\pi}{2} = \alpha + \beta = \tan^{-1} x + \cot^{-1} x$.

(c)

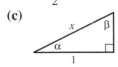

$\alpha = \sec^{-1} x, \beta = \csc^{-1} x$

So $\dfrac{\pi}{2} = \alpha + \beta = \sec^{-1} x + \csc^{-1} x$.

35. If s is the length of a side of the square, and let α, β, γ denote the angles labeled $\tan^{-1} 1$, $\tan^{-1} 2$, and $\tan^{-1} 3$, respectively.

$\tan \alpha = \dfrac{s}{s} = 1$, so $\alpha = \tan^{-1} 1$ and

$\tan \beta = \dfrac{s}{\dfrac{s}{2}} = 2$, so $\beta = \tan^{-1} 2$.

$\gamma = \pi - \alpha - \beta = \pi - \tan^{-1} 1 - \tan^{-1} 2$
$- \tan^{-1} 3$.

Section 3.9 (pp. 163–171)

Quick Review

1. $\dfrac{\ln 8}{\ln 5}$

2. $e^{x \ln 7}$

3. $\tan x$

4. $\ln (x - 2)$

5. $3x - 15$

6. $\dfrac{5}{4}$

7. $\ln (4x^4)$

8. $x = \dfrac{\ln 19}{\ln 3} \approx 2.68$

9. $x = \dfrac{\ln 18 - \ln (\ln 5)}{\ln 5} \approx 1.50$

10. $x = \dfrac{\ln 3}{\ln 2 - \ln 3} \approx -2.71$

Exercises

1. $2e^x$

3. $-e^{-x}$

5. $\dfrac{2}{3} e^{2x/3}$

7. $e^2 - e^x$

9. $\dfrac{e^{\sqrt{x}}}{2\sqrt{x}}$

11. $\pi x^{\pi - 1}$

13. $-\sqrt{2} x^{-\sqrt{2} - 1}$

15. $8^x \ln 8$

17. $-3^{\csc x}(\ln 3)(\csc x \cot x)$

19. $\dfrac{2 x^{\ln x} \ln x}{x}$

21. $\dfrac{2}{x}$

23. $-\dfrac{1}{x}, x > 0$

25. $\dfrac{1}{x + 2}, x > -2$

27. $\dfrac{\sin x}{2 - \cos x}$

29. $\dfrac{1}{x \ln x}$

31. $\dfrac{2}{x \ln 4} = \dfrac{1}{x \ln 2}$

33. $\dfrac{3}{(3x + 1) \ln 2}, x > -\dfrac{1}{3}$

35. $-\dfrac{1}{x \ln 2}, x > 0$

37. $\dfrac{1}{x}, x > 0$

39. $\dfrac{1}{\ln 10}$

41. $y = ex$

43. $(\sin x)^x [x \cot x + \ln (\sin x)]$

45.
$$\left(\dfrac{(x - 3)^4(x^2 + 1)}{(2x + 5)^3}\right)^{1/5}\left(\dfrac{4}{5(x - 3)} + \dfrac{2x}{5(x^2 + 1)} - \dfrac{6}{5(2x + 5)}\right)$$

47. rate ≈ 0.098 grams/day

49. (a) $\ln 2$ (b) $f'(0) = \lim\limits_{h \to 0} \dfrac{2^h - 1}{h}$

(c) $\ln 2$ (d) $\ln 7$

51. (a) The graph of y_4 is a horizontal line at $y = a$.

(b) The graph of y_3 is a horizontal line at $y = \ln a$.

(c) $\dfrac{d}{dx} a^x = a^x$ if and only if $y_3 = \dfrac{y_2}{y_1} = 1$.

So if $y_3 = \ln a$, then $\dfrac{d}{dx} a^x$ will equal a^x if and only if $\ln a = 1$, or $a = e$.

(d) $y_2 = \dfrac{d}{dx} a^x = a^x \ln a$. This will equal $y_1 = a^x$ if and only if $\ln a = 1$, or $a = e$.

53. (a) $y = \dfrac{1}{e} x$

(b) Because the graph of $\ln x$ lies below the graph of the line for all positive $x \neq e$.

(c) Multiplying by e, $e(\ln x) < x$, or $\ln x^e < x$.

(d) Exponentiate both sides of the inequality in (c).

(e) Let $x = \pi$ to see that $\pi^e < e^\pi$.

Chapter Review (pp. 172–175)

1. $5x^4 - \dfrac{x}{4} + \dfrac{1}{4}$

2. $-21x^2 + 21x^6$

3. $-2 \cos^2 x + 2 \sin^2 x = 2 \cos 2x$

4. $-\dfrac{4}{(2x - 1)^2}$

5. $2 \sin (1 - 2t)$

6. $\dfrac{2}{t^2} \csc^2 \dfrac{2}{t}$

7. $\dfrac{1}{2\sqrt{x}} - \dfrac{1}{2x^{3/2}}$

8. $\dfrac{3x + 1}{\sqrt{2x + 1}}$

9. $3 \sec (1 + 3\theta) \tan (1 + 3\theta)$

10. $-4\theta \tan (3 - \theta^2) \sec^2 (3 - \theta^2)$

11. $-5x^2 \csc 5x \cot 5x + 2x \csc 5x$

12. $\dfrac{1}{2x}, x > 0$

13. $\dfrac{e^x}{1 + e^x}$

14. $-xe^{-x} + e^{-x}$

15. e

16. cot x, where x is in an interval of the form $(k\pi, (k + 1)\pi)$, k even

17. $-\dfrac{1}{\cos^{-1} x \sqrt{1 - x^2}}$

18. $\dfrac{2}{\theta \ln 2}$

19. $\dfrac{1}{(t - 7) \ln 5}, t > 7$

20. $-8^{-t} \ln 8$

21. $\dfrac{2 (\ln x) x^{\ln x}}{x}$

22. $\dfrac{(2 \cdot 2^x)[x^3 \ln 2 + x \ln 2 + 1]}{(x^2 + 1)^{3/2}}$ or

$\dfrac{(2x)2^x}{\sqrt{x^2 + 1}}\left(\dfrac{1}{x} + \ln 2 + \dfrac{x}{x^2 + 1}\right)$

23. $\dfrac{e^{\tan^{-1}}}{1 + x^2}$

24. $-\dfrac{u}{\sqrt{u^2 - u^4}} = -\dfrac{u}{|u|\sqrt{1 - u^2}}$

25. $\dfrac{t}{|t|\sqrt{t^2 - 1}} + \sec^{-1}t - \dfrac{1}{2t}$

26. $-\dfrac{2 + 2t^2}{1 + 4t^2} + 2t \cot^{-1} 2t$

27. $\cos^{-1} z$

28. $-\dfrac{1}{x} + \dfrac{\csc^{-1}\sqrt{x}}{\sqrt{x - 1}}$

29. $-\dfrac{\sin x}{|\sin x|} = -\text{sign} (\sin x), x \neq \dfrac{\pi}{2}, \pi, \dfrac{3\pi}{2};$

or $\begin{cases} -1, & 0 \leq x < \pi, \quad x \neq \dfrac{\pi}{2} \\ 1, & \pi < x \leq 2\pi \quad x \neq \dfrac{3\pi}{2} \end{cases}$

30. $2\left(\dfrac{1 + \sin \theta}{1 - \cos \theta}\right)\left(\dfrac{\cos \theta - \sin \theta - 1}{(1 - \cos \theta)^2}\right)$

31. For all $x \neq 0$

32. For all real x

33. For all $x < 1$

34. For all $x \neq \dfrac{7}{2}$

35. $-\dfrac{y + 2}{x + 3}$

36. $-\dfrac{1}{3}(xy)^{-1/5}$

37. $-\dfrac{y}{x}$ or $-\dfrac{1}{x^2}$

38. $\dfrac{1}{2y(x + 1)^2}$

39. $-\dfrac{2x}{y^5}$

40. $-\dfrac{1 + 2xy^2}{x^4y^3}$

41. $-2\dfrac{(3y^2 + 1)^2 \cos x + 12y \sin^2 x}{(3y^2 + 1)^3}$

42. $\dfrac{2}{3}x^{-4/3}y^{1/3} + \dfrac{2}{3}x^{-5/3}y^{2/3} = \dfrac{8}{3}x^{-5/3}y^{1/3}$

43. $y' = 2x^3 - 3x - 1,$
$y'' = 6x^2 - 3,$
$y''' = 12x,$
$y^{(4)} = 12$, and the rest are all zero.

44. $y' = \dfrac{x^4}{24},$

$y'' = \dfrac{x^3}{6},$

$y''' = \dfrac{x^2}{2},$

$y^{(4)} = x,$

$y^{(5)} = 1$, and the rest are all zero.

45. (a) $y = \dfrac{2}{\sqrt{3}}x - \sqrt{3}$

   (b) $y = -\dfrac{\sqrt{3}}{2}x + \dfrac{5\sqrt{3}}{2}$

46. (a) $y = -x + \dfrac{\pi}{2} + 2$

   (b) $y = x - \dfrac{\pi}{2} + 2$

47. (a) $y = -\dfrac{1}{4}x + \dfrac{9}{4}$ **(b)** $y = 4x - 2$

48. (a) $y = -\dfrac{5}{4}x + 6$ **(b)** $y = \dfrac{4}{5}x - \dfrac{11}{5}$

49. $y = x - 2\sqrt{2}$ **50.** $y = \dfrac{4}{3}x + 4\sqrt{2}$

51. $y = \dfrac{10}{3}x - 5\sqrt{3}$

52. $y = (1 + \sqrt{2})x - \sqrt{2} - 1 - \dfrac{\pi}{4}$

   or $y \approx 2.414x - 3.200$

53. (a)

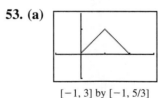

[−1, 3] by [−1, 5/3]

   (b) Yes, because both of the one-sided limits as $x \to 1$ are equal to $f(1) = 1$.
   (c) No, because the left-hand derivative at $x = 1$ is $+1$ and the right-hand derivative at $x = 1$ is -1.

54. (a) The function is continuous for all values of m, because the right-hand limit as $x \to 0$ is equal to $f(0) = 0$ for any value of m.
   (b) The left-hand derivative at $x = 0$ is 2, and the right-hand derivative at $x = 0$ is m, so in order for the function to be differentiable at $x = 0$, m must be 2.

55. (a) For all $x \neq 0$ **(b)** At $x = 0$
   (c) Nowhere

56. (a) For all x **(b)** Nowhere
   (c) Nowhere

57. (a) $[-1, 0) \cup (0, 4]$ **(b)** At $x = 0$
   (c) Nowhere in its domain

58. (a) $[-2, 0) \cup (0, 2]$ **(b)** Nowhere

(c) Nowhere in its domain

59.

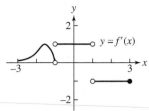

60.

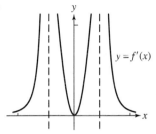

61. (a) iii **(b)** i

(c) ii

62.

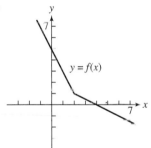

63.

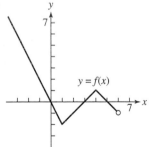

64. Answer is **D**: **i** and **iii** only could be true

65. (a)

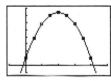

$[-1, 5]$ by $[-10, 80]$

(b)

t interval	avg. vel.
$[0, 0.5]$	56
$[0.5, 1]$	40
$[1, 1.5]$	24
$[1.5, 2]$	8
$[2, 2.5]$	-8
$[2.5, 3]$	-24
$[3, 3.5]$	-40
$[3.5, 4]$	-56

(c)

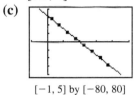

$[-1, 5]$ by $[-80, 80]$

(d) Average velocity is a good approximation to velocity.

66. (a) $-\dfrac{13}{10}$ **(b)** $-\dfrac{1}{3}$ **(c)** $\dfrac{1}{10}$

(d) -1 **(e)** $-\dfrac{2}{3}$ **(f)** -12

67. (a) 5 **(b)** 0 **(c)** 8

(d) 2 **(e)** 6 **(f)** -1

68. $\sqrt{3}$ **69.** $-\dfrac{1}{6}$

70. (a) One possible answer:

$$x(t) = 10 \cos\left(t + \frac{\pi}{4}\right)$$

$$y(t) = 1$$

(b) $5\sqrt{2}$ **(c)** $s = -10$ and $s = 10$

(d) At $t = \dfrac{\pi}{4}$:

velocity $= -10$

speed $= 10$

acceleration $= 0$

71. (a) $\dfrac{ds}{dt} = 64 - 32t$

$\dfrac{d^2s}{dt^2} = -32$

(b) 2 sec **(c)** 64 ft/sec

(d) $\dfrac{64}{5.2} \approx 12.3$ sec;

$s\left(\dfrac{64}{5.2}\right) \approx 393.8$ ft

72. (a) $\dfrac{4}{7}$ sec; 280 cm/sec

(b) 560 cm/sec; 980 cm/sec^2

73. $\pi(20x - x^2)$

74. (a) $r(x) = \left(3 - \dfrac{x}{40}\right)^2 x = 9x - \dfrac{3}{20}x^2 + \dfrac{1}{1600}x^3$

(b) 40 people; $4.00

(c) One possible answer:
Probably not, since the company charges less overall for 60 passengers than it does for 40 passengers.

75. (a) -0.6 km/sec

(b) $\dfrac{18}{\pi} \approx 5.73$ revolutions/min

76. Yes

77. $y'(r) = -\dfrac{1}{2r^2l}\sqrt{\dfrac{T}{\pi d}}$, so increasing r decreases the frequency.

$y'(l) = -\dfrac{1}{2rl^2}\sqrt{\dfrac{T}{\pi d}}$, so increasing l decreases the frequency.

$y'(d) = -\dfrac{1}{4rl}\sqrt{\dfrac{T}{\pi d^3}}$, so increasing d decreases the frequency.

$y'(T) = \dfrac{1}{4rl\sqrt{\pi Td}}$, so increasing T increases the frequency.

78. (a) $P(0) \approx 1.339$, so initially, one student was infected

(b) 200

(c) After 5 days, when the rate is 50 students/day

79. (a) $x \neq k\dfrac{\pi}{4}$, where k is an odd integer

(b) $\left(-\dfrac{\pi}{2}, \dfrac{\pi}{2}\right)$

(c) Where it's not defined, at $x = k\dfrac{\pi}{4}$, k an odd integer

(d) It has period $\dfrac{\pi}{2}$ and continues to repeat the pattern seen in this window.

80. $-\dfrac{1}{3\sqrt{3}}$

Chapter 4

Section 4.1 (pp. 177–185)

Quick Review

1. $\dfrac{-1}{2\sqrt{4-x}}$

2. $\dfrac{3}{4}x^{-1/4}$

3. $\dfrac{2x}{(9-x^2)^{3/2}}$

4. $\dfrac{-2x}{3(x^2-1)^{4/3}}$

5. $\dfrac{2x}{x^2+1}$

6. $-\dfrac{\sin(\ln x)}{x}$

7. $2e^{2x}$

8. 1

9. ∞

10. ∞

11. (a) 1   **(b)** 1   **(c)** Undefined

12. (a) $x \neq 2$

(b) $f'(x) = \begin{cases} 3x^2 - 2, & x < 2 \\ 1, & x > 2 \end{cases}$

Exercises

1. Maximum at $x = b$, minimum at $x = c_2$;
Extreme Value Theorem applies, and both the max and min exist.

3. Maximum at $x = c$, no minimum;
Extreme Value Theorem doesn't apply, since the function isn't defined on a closed interval.

5. Maximum at $x = c$, minimum at $x = a$;
Extreme Value Theorem doesn't apply, since the function isn't continuous.

7. Local minimum at $(-1, 0)$, local maximum at $(1, 0)$

9. Maximum at $(0, 5)$

11. Maximum value is $\dfrac{1}{4} + \ln 4$ at $x = 4$;
minimum value is 1 at $x = 1$;
local maximum at $\left(\dfrac{1}{2}, 2 - \ln 2\right)$

13. Maximum value is $\ln 4$ at $x = 3$;
minimum value is 0 at $x = 0$.

15. Maximum value is 1 at $x = \dfrac{\pi}{4}$;
minimum value is -1 at $x = \dfrac{5\pi}{4}$;
local minimum at $\left(0, \dfrac{1}{\sqrt{2}}\right)$;
local maximum at $\left(\dfrac{7\pi}{4}, 0\right)$

17. Maximum value is $3^{2/5}$ at $x = -3$;
minimum value is 0 at $x = 0$.

19. Minimum value is 1 at $x = 2$.

21. Local maximum at $(-2, 17)$;
local minimum at $\left(\dfrac{4}{3}, -\dfrac{41}{27}\right)$

23. Minimum value is 0 at $x = -1$ and $x = 1$.

25. Minimum value is 1 at $x = 0$.

27. Maximum value is 2 at $x = 1$;
minimum value is 0 at $x = -1$ and $x = 3$.

29. Maximum value is $\dfrac{1}{2}$ at $x = 1$;

minimum value is $-\dfrac{1}{2}$ at $x = -1$.

31. Maximum value is 11 at $x = 5$;
minimum value is 5 on the interval $[-3, 2]$;
local maximum at $(-5, 9)$

33. Maximum value is 5 on the interval $[3, \infty)$;
minimum value is -5 on the interval $(-\infty, -2]$.

35. (a) No

(b) The derivative is defined and nonzero for
$x \neq 2$. Also, $f(2) = 0$, and $f(x) > 0$ for all
$x \neq 2$.

(c) No, because $(-\infty, \infty)$ is not a closed interval.

(d) The answers are the same as (a) and (b) with
2 replaced by a.

37.

crit. pt.	derivative	extremum	value
$x = -\dfrac{4}{5}$	0	local max	$\dfrac{12}{25}10^{1/3} \approx 1.034$
$x = 0$	undefined	local min	0

39.

crit. pt.	derivative	extremum	value
$x = -2$	undefined	local max	0
$x = -\sqrt{2}$	0	minimum	-2
$x = \sqrt{2}$	0	maximum	2
$x = 2$	undefined	local min	0

41.

crit. pt.	derivative	extremum	value
$x = 1$	undefined	minimum	2

43.

crit. pt.	derivative	extremum	value
$x = -1$	0	maximum	5
$x = 1$	undefined	local min	1
$x = 3$	0	maximum	5

45. (c)   **47.** (d)

49. (a) Maximum value is 144 at $x = 2$.

(b) The largest volume of the box is 144 cubic
units and it occurs when $x = 2$.

51. (a) $f'(x) = 3ax^2 + 2bx + c$ is a quadratic, so it
can have 0, 1, or 2 zeros, which would be the
critical points of f. Examples:

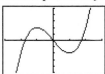

[−3, 3] by [−5, 5]

The function $f(x) = x^3 - 3x$ has two critical
points at $x = -1$ and $x = 1$.

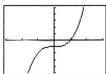

[−3, 3] by [−5, 5]

The function $f(x) = x^3 - 1$ has one critical
point at $x = 0$.

[−3, 3] by [−5, 5]

The function $f(x) = x^3 + x$ has no critical
points.

(b) Two or none

53. (a)

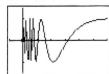

[−0.1, 0.6] by [−1.5, 1.5]

$f(0) = 0$ is not a local extreme value because
in any open interval containing $x = 0$, there
are infinitely many points where $f(x) = 1$ and
where $f(x) = -1$.

(b) One possible answer, on the interval $[0, 1]$:

$$f(x) = \begin{cases} (1 - x)\cos \dfrac{1}{1 - x}, & 0 \le x < 1 \\ 0, & x = 1 \end{cases}$$

This function has no local extreme value at
$x = 1$. Note that it is continuous on $[0, 1]$.

Section 4.2 (pp. 186–194)

Quick Review

1. $(-\sqrt{3}, \sqrt{3})$

2. $(-\infty, -\sqrt{2}) \cup (\sqrt{2}, \infty)$

3. $[-2, 2]$

4. For all x in its domain, or, $[-2, 2]$

5. On $(-2, 2)$ **6.** $x \neq \pm 1$

7. For all x in its domain, or, for all $x \neq \pm 1$

8. For all x in its domain, or, for all $x \neq \pm 1$

9. $C = 3$ **10.** $C = -4$

Exercises

1. **(a)** Local maximum at $\left(\dfrac{5}{2}, \dfrac{25}{4}\right)$

    (b) On $\left(-\infty, \dfrac{5}{2}\right]$ **(c)** On $\left[\dfrac{5}{2}, \infty\right)$

3. **(a)** None   **(b)** None

    (c) On $(-\infty, 0)$ and $(0, \infty)$

5. **(a)** None   **(b)** On $(-\infty, \infty)$

    (c) None

7. **(a)** Local maximum at $(-2, 4)$

    (b) None   **(c)** On $[-2, \infty)$

9. **(a)** Local maximum at
    $\approx (2.67, 3.08)$;
    local minimum at $(4, 0)$

    (b) On $\left(-\infty, \dfrac{8}{3}\right]$ **(c)** On $\left[\dfrac{8}{3}, 4\right]$

11. **(a)** Local maximum at $\left(-2, \dfrac{1}{4}\right)$;

    local minimum at $\left(2, -\dfrac{1}{4}\right)$

    (b) On $(-\infty, -2]$ and $[2, \infty)$

    (c) On $[-2, 2]$

13. **(a)** Local maximum at $\approx (-1.126, -0.036)$;
    local minimum at $\approx (0.559, -2.639)$

    (b) On $(-\infty, -1.126]$ and $[0.559, \infty)$

    (c) On $[-1.126, 0.559]$

15. **(a)** f is continuous on $[0, 1]$ and differentiable on $(0, 1)$.

    (b) $c = \dfrac{1}{2}$

17. **(a)** f is continuous on $[-1, 1]$ and differentiable on $[-1, 1]$.

    (b) $c \approx \pm 0.771$

19. **(a)** $y = \dfrac{5}{2}$ **(b)** $y = 2$

21. **(a)** Not differentiable at $x = 0$

    (b)
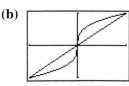
    $[-1, 1]$ by $[-1, 1]$

    (c) $c = \pm 3^{-3/2} \approx \pm 0.192$

23. **(a)** Not differentiable at $x = 0$

    (b)
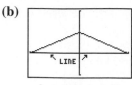
    $[-1, 1]$ by $[-1, 2]$

    (c) There are none

25. $\dfrac{x^2}{2} + C$ **27.** $x^3 - x^2 + x + C$

29. $e^x + C$ **31.** $\dfrac{1}{x} + \dfrac{1}{2}, x > 0$

33. $\ln (x + 2) + 3$

35. Possible answers:

    (a)
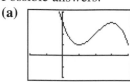
    $[-2, 4]$ by $[-2, 4]$

    (b)
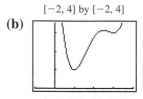
    $[-1, 4]$ by $[0, 3.5]$

    (c)
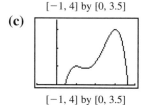
    $[-1, 4]$ by $[0, 3.5]$

37. One possible answer:
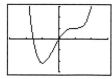
    $[-3, 3]$ by $[-15, 15]$

39. Because the trucker's average speed was 79.5 mph, and by the Mean Value Theorem, the trucker must have been going that speed at least once during the trip.

41. Because its average speed was approximately 7.667 knots, and by the Mean Value Theorem, it must have been going that speed at least once during the trip.

43. **(a)** 48 m/sec   **(b)** 720 meters

    (c) After about 27.604 seconds, and it will be going about 48.166 m/sec

45. Because the function is not continuous on $[0, 1]$.

47. $f(x)$ must be zero at least once between a and b by the Intermediate Value Theorem.
Now suppose that $f(x)$ is zero twice between a and b. Then by the Mean Value Theorem, $f'(x)$ would have to be zero at least once between the two zeros of $f(x)$, but this can't be true since we are given that $f'(x) \neq 0$ on this interval. Therefore, $f(x)$ is zero once and only once between a and b.

49. Let $f(x) = x + \ln(x + 1)$. Then $f(x)$ is continuous and differentiable everywhere on $[0, 3]$.

$f'(x) = 1 + \dfrac{1}{x + 1}$, which is never zero on $[0, 3]$.

Now $f(0) = 0$, so $x = 0$ is one solution of the equation. If there were a second solution, $f(x)$ would be zero twice in $[0, 3]$, and by the Mean Value Theorem, $f'(x)$ would have to be zero somewhere between the two zeros of $f(x)$. But this can't happen, since $f'(x)$ is never zero on $[0, 3]$. Therefore, $f(x) = 0$ has exactly one solution in the interval $[0, 3]$.

51. (a) Increasing: $[-2, -1.3]$ and $[1.3, 2]$;
decreasing: $[-1.3, 1.3]$;
local max: $x \approx -1.3$
local min: $x \approx 1.3$

(b) Regression equation: $y = 3x^2 - 5$

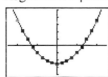

$[-2.5, 2.5]$ by $[-8, 10]$

(c) $f(x) = x^3 - 5x$

53. $\dfrac{f(b) - f(a)}{b - a} = \dfrac{\dfrac{1}{b} - \dfrac{1}{a}}{b - a} = -\dfrac{1}{ab}$

$f'(c) = -\dfrac{1}{c^2}$, so $-\dfrac{1}{c^2} = -\dfrac{1}{ab}$ and $c^2 = ab$.

Thus, $c = \sqrt{ab}$.

55. By the Mean Value Theorem,
$\sin b - \sin a = (\cos c)(b - a)$ for some c between a and b. Taking the absolute value of both sides and using $|\cos c| \leq 1$ gives the result.

57. Let $f(x)$ be a monotonic function defined on an interval D. For any two values in D, we may let x_1 be the smaller value and let x_2 be the larger value, so $x_1 < x_2$. Then either $f(x_1) < f(x_2)$ (if f is increasing), or $f(x_1) > f(x_2)$ (if f is decreasing), which means $f(x_1) \neq f(x_2)$. Therefore, f is one-to-one.

Section 4.3 (pp. 194–206)

Quick Review

1. $(-3, 3)$
2. $(-2, 0) \cup (2, \infty)$
3. f: all reals
   f': all reals
4. f: all reals
   f': $x \neq 0$
5. f: $x \neq 2$
   f': $x \neq 2$
6. f: all reals
   f': $x \neq 0$
7. $y = 0$
8. $y = 0$
9. $y = 0$ and $y = 200$
10. $y = 0$ and $y = 375$

Exercises

1. (a) Zero: $x = \pm 1$;
   positive: $(-\infty, -1)$ and $(1, \infty)$;
   negative: $(-1, 1)$
(b) Zero: $x = 0$;
   positive: $(0, \infty)$;
   negative: $(-\infty, 0)$

3. (a) $(-\infty, -2]$ and $[0, 2]$
(b) $[-2, 0]$ and $[2, \infty)$
(c) Local maxima: $x = -2$ and $x = 2$;
   local minimum: $x = 0$

5. (a) $[0, 1]$, $[3, 4]$, and $[5.5, 6]$
(b) $[1, 3]$ and $[4, 5.5]$
(c) Local maxima: $x = 1$, $x = 4$
   (if f is continuous at $x = 4$), and $x = 6$;
   local minima: $x = 0$, $x = 3$, and $x = 5.5$

7. (a) $\left[\dfrac{1}{2}, \infty\right)$
(b) $\left(-\infty, \dfrac{1}{2}\right]$
(c) $(-\infty, \infty)$
(d) Nowhere
(e) Local minimum at $\left(\dfrac{1}{2}, -\dfrac{5}{4}\right)$
(f) None

9. (a) $[-1, 0]$ and $[1, \infty)$
(b) $(-\infty, -1]$ and $[0, 1]$
(c) $\left(-\infty, -\dfrac{1}{\sqrt{3}}\right)$ and $\left(\dfrac{1}{\sqrt{3}}, \infty\right)$
(d) $\left(-\dfrac{1}{\sqrt{3}}, \dfrac{1}{\sqrt{3}}\right)$
(e) Local maximum: $(0, 1)$;
   local minima: $(-1, -1)$ and $(1, -1)$
(f) $\left(\pm\dfrac{1}{\sqrt{3}}, \dfrac{1}{9}\right)$

11. (a) $[-2, 2]$
(b) $[-\sqrt{8}, -2]$ and $[2, \sqrt{8}]$
(c) $(-\sqrt{8}, 0)$
(d) $(0, \sqrt{8})$
(e) Local maxima: $(-\sqrt{8}, 0)$ and $(2, 4)$;
   local minima: $(-2, -4)$ and $(\sqrt{8}, 0)$
(f) $(0, 0)$

13. (a) $(-\infty, -2]$ and $\left[-\dfrac{3}{2}, \infty\right)$

  (b) $\left[-2, -\dfrac{3}{2}\right]$ **(c)** $\left(-\dfrac{7}{4}, \infty\right)$

  (d) $\left(-\infty, -\dfrac{7}{4}\right)$

  (e) Local maximum: $(-2, -40)$;

  local minimum: $\left(-\dfrac{3}{2}, -\dfrac{161}{4}\right)$

  (f) $\left(-\dfrac{7}{4}, -\dfrac{321}{8}\right)$

15. (a) $(-\infty, \infty)$ **(b)** None

  (c) $(-\infty, 0)$ **(d)** $(0, \infty)$

  (e) None   **(f)** $(0, 3)$

17. (a) $(-\infty, \infty)$ **(b)** None

  (c) $(-\infty, 5 \ln 3) \approx (-\infty, 5.49)$

  (d) $(5 \ln 3, \infty) \approx (5.49, \infty)$

  (e) None

  (f) $\left(5 \ln 3, \dfrac{5}{2}\right) \approx (5.49, 2.50)$

19. (a) $(-\infty, 1)$ **(b)** $[1, \infty)$

  (c) None   **(d)** $(1, \infty)$

  (e) None   **(f)** None

21. (a) $(-\infty, -\sqrt{2}]$ and $[\sqrt{2}, \infty)$

  (b) $[-\sqrt{2}, 0)$ and $(0, \sqrt{2}]$

  (c) $(0, \infty)$ **(d)** $(-\infty, 0)$

  (e) Local maximum:
  $(-\sqrt{2}, -\sqrt{2e}) \approx (-1.41, -2.33)$;
  local minimum: $(\sqrt{2}, \sqrt{2e}) \approx (1.41, 2.33)$

  (f) None

23. (a) $(-\infty, \infty)$ **(b)** None

  (c) $(-\infty, 0)$ **(d)** $(0, \infty)$

  (e) None   **(f)** $(0, 0)$

25. (a) $[1, \infty)$ **(b)** $(-\infty, 1]$

  (c) $(-\infty, -2)$ and $(0, \infty)$

  (d) $(-2, 0)$

  (e) Local minimum: $(1, -3)$

  (f) $\approx (-2, 7.56)$ and $(0, 0)$

27. (a) Approximately $[0.15, 1.40]$ and $[2.45, \infty)$

  (b) Approximately $(-\infty, 0.15]$, $[1.40, 2)$,
  and $(2, 2.45]$

  (c) $(-\infty, 1)$ and $(2, \infty)$

  (d) $(1, 2)$

  (e) Local maximum: $\approx (1.40, 1.29)$;
  local minima: $\approx (0.15, 0.48)$ and $(2.45, 9.22)$

  (f) $(1, 1)$

29. (a) None   **(b)** At $x = 2$

  (c) At $x = 1$ and $x = \dfrac{5}{3}$

31.

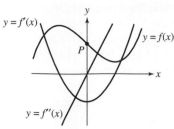

33. (a) Absolute maximum at $(1, 2)$;
  absolute minimum at $(3, -2)$

  (b) None

  (c) One possible answer:

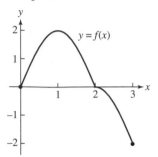

35.

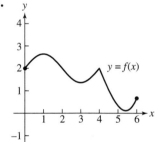

37. (a) $v(t) = 2t - 4$ **(b)** $a(t) = 2$

  (c) It begins at position 3 moving in a negative
  direction. It moves to position -1 when
  $t = 2$, and then changes direction, moving in
  a positive direction thereafter.

39. (a) $v(t) = 3t^2 - 3$

  (b) $a(t) = 6t$

  (c) It begins at position 3 moving in a negative
  direction. It moves to position 1 when $t = 1$,
  and then changes direction, moving in a
  positive direction thereafter.

41. (a) $t = 2.2, 6, 9.8$ **(b)** $t = 4, 8, 11$

43. No. f must have a horizontal tangent line at that
  point, but it could be increasing (or decreasing)
  on both sides of the point, and there would be no
  local extremum.

45. One possible answer:

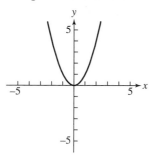

47. One possible answer:

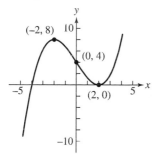

49. (a) Regression equation:

$$y = \frac{2161.4541}{1 + 28.1336e^{-0.8627x}}$$

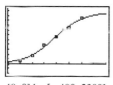

[0, 8] by [−400, 2300]

(b) At approximately $x = 3.868$ (late in 1996), when the sales are about 1081 million dollars/year

(c) 2161.45 million dollars/year

51. (a) $f'(x) = \dfrac{abce^{bx}}{(e^{bx} + a)^2}$, so the sign of $f'(x)$ is the same as the sign of the product abc.

(b) $f''(x) = -\dfrac{ab^2ce^{bx}(e^{bx} - a)}{(e^{bx} + a)^3}$. Since $a > 0$, this

changes sign when $x = \dfrac{\ln a}{b}$ due to the

$e^{bx} - a$ factor in the numerator, and there is a point of inflection at that location.

Section 4.4 (pp. 206–220)

Quick Review

1. None
2. Local maximum: $(-2, 17)$;
local minimum: $(1, -10)$

3. $\dfrac{200\pi}{3}$ cm^3

4. $r \approx 4.01$ cm and $h \approx 19.82$ cm, or,
$r \approx 7.13$ cm and $h \approx 6.26$ cm

5. $-\sin \alpha$ **6.** $\cos \alpha$
7. $\sin \alpha$ **8.** $-\cos \alpha$
9. $x = 1$ and $y = \sqrt{3}$, or, $x = -1$ and $y = -\sqrt{3}$
10. $x = 0$ and $y = 3$, or, $x = -\dfrac{24}{13}$ and $y = \dfrac{15}{13}$

Exercises

1. (a) As large as possible: 0 and 20;
as small as possible: 10 and 10

(b) As large as possible: $\dfrac{79}{4}$ and $\dfrac{1}{4}$;

as small as possible: 0 and 20

3. Smallest perimeter = 16 in., dimensions are 4 in. by 4 in.

5. (a) $y = 1 - x$ **(b)** $A(x) = 2x(1 - x)$

(c) Largest area $= \dfrac{1}{2}$, dimensions are 1 by $\dfrac{1}{2}$

7. Largest volume is $\dfrac{2450}{27} \approx 90.74$ in^3;

dimensions: $\dfrac{5}{3}$ in. by $\dfrac{14}{3}$ in. by $\dfrac{35}{3}$ in.

9. Largest area $= 80,000$ m^2;
dimensions: 200 m (perpendicular to river) by 400 m (parallel to river)

11. (a) 10 ft by 10 ft by 5 ft

(b) Assume that the weight is minimized when the total area of the bottom and the 4 sides is minimized.

13. 18 in. high by 9 in. wide

15. $\theta = \dfrac{\pi}{2}$ **17.** $\dfrac{8}{\pi}$ to 1

19. (a) $V(x) = 2x(24 - 2x)(18 - 2x)$

(b) Domain: $(0, 9)$

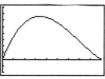

[0, 9] by [−400, 1600]

(c) Maximum volume ≈ 1309.95 in^3 when $x \approx 3.39$ in.

(d) $V'(x) = 24x^2 - 336x + 864$, so the critical point is at $x = 7 - \sqrt{13}$, which confirms the result in (c).

(e) $x = 2$ in. or $x = 5$ in.

19. continued

(f) The dimensions of the resulting box are $2x$ in., $(24 - 2x)$ in., and $(18 - 2x)$ in. Each of these measurements must be positive, so that gives the domain of $(0, 9)$.

21. Dimensions: width ≈ 3.44, height ≈ 2.61; maximum area ≈ 8.98

23. (a) At $x = 1$

(b)

a	b	A
0.1	3.71	0.33
0.2	2.86	0.44
0.3	2.36	0.46
0.4	2.02	0.43
0.5	1.76	0.38
0.6	1.55	0.31
0.7	1.38	0.23
0.8	1.23	0.15
0.9	1.11	0.08
1.0	1.00	0.00

(c)

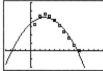

[0, 1.1] by [−0.2, 0.6]

(d) Quadratic:

$$A \approx -0.91a^2 + 0.54a + 0.34$$

[−0.5, 1.5] by [−0.2, 0.6]

Cubic:

$$A \approx 1.74a^3 - 3.78a^2 + 1.86a + 0.19$$

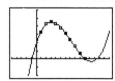

[−0.5, 1.5] by [−0.2, 0.6]

Quartic:

$$A \approx -1.92a^4 + 5.96a^3 - 6.87a^2 + 2.71a + 0.12$$

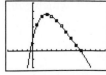

[−0.5, 1.5] by [−0.2, 0.6]

(e) Quadratic: $A \approx 0.42$; cubic: $A \approx 0.45$; quartic: $A \approx 0.46$

25. 18 in. by 18 in. by 36 in.

27. Radius $= \sqrt{2}$ m, height $= 1$ m, volume $\dfrac{2\pi}{3}$ m^3

29. $f'(x) = \dfrac{2x^3 - a}{x^2}$, so the only sign change in $f'(x)$ occurs at $x = \left(\dfrac{a}{2}\right)^{1/3}$, where the sign changes from negative to positive. This means there is a local minimum at that point, and there are no local maxima.

31. $\dfrac{32\pi}{3}$ cubic units

33. (a) 6 in. wide by $6\sqrt{3}$ in. deep

(b)

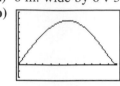

[0, 12] by [−2000, 8000]
$y = x(144 - x^2)^{3/2}$

(c)

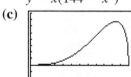

[0, 12] by [−2000, 8000]
$y = x^3(144 - x^2)^{1/2}$

Changing the value of k changes the maximum stiffness, but not the dimensions of the stiffest beam. The graphs for different values of k look the same except that the vertical scale is different.

35. $2\sqrt{2}$ amps

37. The minimum distance is 2.

39. (a) Because $f(x)$ is periodic with period 2π.

(b) No. It has an absolute minimum at the point $(\pi, 0)$.

41. (a) At $t = \dfrac{\pi}{3}$ sec and at $t = \dfrac{4\pi}{3}$ sec

(b) The maximum distance between particles is 1 m.

(c) Near $t = \dfrac{\pi}{3}$ sec and near $t = \dfrac{4\pi}{3}$ sec

43. (a) Answers will vary.

(b) $x = \dfrac{51}{8} = 6.375$ in.

(c) Minimum length ≈ 11.04 in.

45. $x = \dfrac{c + 100}{2} = 50 + \dfrac{c}{2}$

47. The rate v is maximum when $x = \dfrac{a}{2}$.

The rate then is $\dfrac{ka^2}{4}$.

49. 67 people

51. $p(x) = 6x - (x^3 - 6x^2 + 15x)$, $x \geq 0$. This function has its maximum value at the points $(0, 0)$ and $(3, 0)$.

53. (a) $y'(0) = 0$ **(b)** $y'(-L) = 0$

(c) $y(0) = 0$, so $d = 0$. $y'(0) = 0$, so $c = 0$.

Then $y(-L) = -aL^3 + bL^2 = H$ and

$y'(-L) = 3aL^2 - 2bL = 0$.

Solving, $a = 2\dfrac{H}{L^3}$ and $b = 3\dfrac{H}{L^2}$, which gives the equation shown.

55. (a) The x- and y-intercepts of the line through R and T are $x - \dfrac{a}{f'(x)}$ and $a - xf'(x)$ respectively.

The area of the triangle is the product of these two values.

(b) Domain: $(0, 10)$

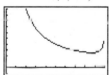

[0, 10] by [−100, 1000]

The vertical asymptotes at $x = 0$ and $x = 10$ correspond to horizontal or vertical tangent lines, which do not form triangles.

(c) Height $= 15$, which is 3 times the y-coordinate of the center of the ellipse.

(d) Part (a) remains unchanged.

The domain is $(0, C)$ and the graph is similar.

The minimum area occurs when $x^2 = \dfrac{3C^2}{4}$.

From this, it follows that the triangle has minimum area when its height is $3B$.

Section 4.5 (pp. 220–232)

Quick Review

1. $2x \cos(x^2 + 1)$

2. $\dfrac{1 - \cos x - (x + 1)\sin x}{(x + 1)^2}$

3. $x \approx -0.567$ **4.** $x \approx -0.322$

5. $y = x + 1$ **6.** $y = 2ex + e + 1$

7. (a) $x = -1$

(b) $x = -\dfrac{e + 1}{2e} \approx -0.684$

8.

x	$f(x)$	$g(x)$
0.7	−1.457	−1.7
0.8	−1.688	−1.8
0.9	−1.871	−1.9
1.0	−2	−2
1.1	−2.069	−2.1
1.2	−2.072	−2.2
1.3	−2.003	−2.3

9.

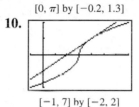

[0, π] by [−0.2, 1.3]

10.

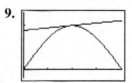

[−1, 7] by [−2, 2]

Exercises

1. (a) $L(x) = 10x - 13$

(b) Differs from the true value in absolute value by less than 10^{-1}

3. (a) $L(x) = 2$

(b) Differs from the true value in absolute value by less than 10^{-2}

5. (a) $L(x) = x - \pi$

(b) Differs from the true value in absolute value by less than 10^{-3}

7. $f(0) = 1$. Also, $f'(x) = k(1 + x)^{k-1}$, so $f'(0) = k$. This means the linearization at $x = 0$ is $L(x) = 1 + kx$.

9. The linearization is $1 + \dfrac{3x}{2}$. It is the sum of the two individual linearizations.

11. Center $= 1$, $L(x) = -5$

13. Center $= 1$, $L(x) = \dfrac{x}{4} + \dfrac{1}{4}$, or

Center $= 1.5$, $L(x) = \dfrac{4x}{25} + \dfrac{9}{25}$

15. $x \approx 0.682328$

17. $x \approx 0.386237, 1.961569$

19. (a) $dy = (3x^2 - 3)\, dx$
(b) $dy = 0.45$ at the given values

21. (a) $dy = (2x \ln x + x)\, dx$

(b) $dy = 0.01$ at the given values

23. (a) $dy = (\cos x)\, e^{\sin x}\, dx$
(b) $dy = 0.1$ at the given values

25. (a) $dy = \dfrac{dx}{(x + 1)^2}$

(b) $dy = 0.01$ at the given values

27. (a) 0.21   **(b)** 0.2   **(c)** 0.01
29. (a) $-\dfrac{2}{11}$ **(b)** $-\dfrac{1}{5}$ **(c)** $\dfrac{1}{55}$
31. $4\pi a^2\, dr$ **33.** $3a^2\, dx$ **35.** $2\pi ah\, dr$
37. (a) $x + 1$ **(b)** $f(0.1) \approx 1.1$
(c) The actual value is less than 1.1, since the derivative is decreasing over the interval $[0, 0.1]$.

39. The diameter grew $\dfrac{2}{\pi} \approx 0.6366$ in.
The cross section area grew about 10 in^2.

41. The side should be measured to within 1%.
43. The angle should be measured to within 0.76%. This is also ± 0.01 radian or $\approx \pm 0.57$ degree.
45. (a) Within 0.5%   **(b)** Within 5%
47. About 37.87 to 1
49. If $f'(x_1) \neq 0$, then x_2 and all later approximations are equal to x_1.
51. $x_2 = -2$, $x_3 = 4$, $x_4 = -8$, and $x_5 = 16$;
$|x_n| = 2^{n-1}$.

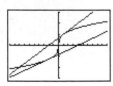

[−10, 10] by [−3, 3]
53. Just multiply the corresponding derivative formulas by dx.
55. $g(a) = c$, so if $E(a) = 0$, then $g(a) = f(a)$ and $c = f(a)$. Then
$E(x) = f(x) - g(x) = f(x) - f(a) - m(x - a)$.
Thus, $\dfrac{E(x)}{x - a} = \dfrac{f(x) - f(a)}{x - a} - m$.
$\displaystyle\lim_{x \to a} \dfrac{f(x) - f(a)}{x - a} = f'(a)$, so if the limit of $\dfrac{E(x)}{x - a}$ is zero, then $m = f'(a)$ and $g(x) = L(x)$.

Section 4.6 (pp. 232–241)

Quick Review

1. $\sqrt{74}$ **2.** $\sqrt{a^2 + b^2}$
3. $\dfrac{1 - 2y}{2x + 2y - 1}$ **4.** $-\dfrac{y + \sin y}{x + x \cos y}$
5. $2x \cos^2 y$ **6.** $2x + 2y - 1$
7. One possible answer:
$x = -2 + 6t$, $y = 1 - 4t$, $0 \le t \le 1$.
8. One possible answer:
$x = 5t$, $y = -4 + 4t$, $0 \le t \le 1$.

9. One possible answer:
$\dfrac{\pi}{2} \le t \le \dfrac{3\pi}{2}$

10. One possible answer:
$\dfrac{3\pi}{2} \le t \le 2\pi$

Exercises

1. $\dfrac{dA}{dt} = 2\pi r \dfrac{dr}{dt}$

3. (a) $\dfrac{dV}{dt} = \pi r^2 \dfrac{dh}{dt}$

(b) $\dfrac{dV}{dt} = 2\pi rh \dfrac{dr}{dt}$

(c) $\dfrac{dV}{dt} = \pi r^2 \dfrac{dh}{dt} + 2\pi rh \dfrac{dr}{dt}$

5. $\dfrac{ds}{dt} = \dfrac{x\dfrac{dx}{dt} + y\dfrac{dy}{dt} + z\dfrac{dz}{dt}}{\sqrt{x^2 + y^2 + z^2}}$

7. (a) 1 volt/sec

(b) $-\dfrac{1}{3}$ amp/sec

(c) $\dfrac{dV}{dt} = I\dfrac{dR}{dt} + R\dfrac{dI}{dt}$

(d) $\dfrac{dR}{dt} = \dfrac{3}{2}$ ohms/sec. R is increasing since $\dfrac{dR}{dt}$ is positive.

9. (a) $\dfrac{dA}{dt} = 14$ cm^2/sec   **(b)** $\dfrac{dP}{dt} = 0$ cm/sec

(c) $\dfrac{dD}{dt} = -\dfrac{14}{13}$ cm/sec

(d) The area is increasing, because its derivative is positive.
The perimeter is not changing, because its derivative is zero.
The diagonal length is decreasing, because its derivative is negative.

11. $\dfrac{dx}{dt} = \dfrac{3000}{\sqrt{51}}$ mph ≈ 420.08 mph

13. (a) 12 ft/sec   **(b)** $-\dfrac{119}{2}$ ft²/sec

    (c) -1 radian/sec

15. $\dfrac{19\pi}{2500} \approx 0.0239$ in³/min

17. (a) $\dfrac{32}{9\pi} \approx 1.13$ cm/min

    (b) $-\dfrac{80}{3\pi} \approx -8.49$ cm/min

19. $V = \dfrac{4}{3}\pi r^3$, so $\dfrac{dV}{dt} = 4\pi r^2 \dfrac{dr}{dt}$. But $S = 4\pi r^2$, so we

    are given that $\dfrac{dV}{dt} = kS = 4k\pi r^2$. Substituting,

    $4k\pi r^2 = 4\pi r^2 \dfrac{dr}{dt}$ which gives $\dfrac{dr}{dt} = k$.

21. (a) $\dfrac{5}{2}$ ft/sec   **(b)** $-\dfrac{3}{20}$ radian/sec

23. (a) $\dfrac{dc}{dt} = 0.3$, $\dfrac{dr}{dt} = 0.9$, $\dfrac{dp}{dt} = 0.6$

    (b) $\dfrac{dc}{dt} = -1.5625$, $\dfrac{dr}{dt} = 3.5$, $\dfrac{dp}{dt} = 5.0625$

25. $\dfrac{dy}{dt} = \dfrac{466}{1681} \approx 0.277$ L/min²

27. $\dfrac{2}{5}$ radian/sec   **29.** -3 ft/sec

31. In front: 2 radians/sec;

    Half second later: 1 radian/sec

33. 80 mph   **35.** -6 deg/sec

37. (a) $\dfrac{24}{5}$ cm/sec   **(b)** 0 cm/sec

    (c) $-\dfrac{1200}{160,801} \approx -0.00746$ cm/sec

39. (a) The point being plotted would correspond to a point on the edge of the wheel as the wheel turns.

    (b) One possible answer:
     $\theta = 16\pi t$, where t is in seconds.

    (c) Assuming counterclockwise motion, the rates are as follows.

     $\theta = \dfrac{\pi}{4}$: $\dfrac{dx}{dt} \approx -71.086$ ft/sec

           $\dfrac{dy}{dt} \approx 71.086$ ft/sec

     $\theta = \dfrac{\pi}{2}$: $\dfrac{dx}{dt} \approx -100.531$ ft/sec

           $\dfrac{dy}{dt} = 0$ ft/sec

     $\theta = \pi$: $\dfrac{dx}{dt} = 0$ ft/sec

           $\dfrac{dy}{dt} \approx -100.531$ ft/sec

41. (a) 9% per year

    (b) Increasing at 1% per year

Chapter Review (pp. 242–245)

1. Maximum: $\dfrac{4\sqrt{6}}{9}$ at $x = \dfrac{4}{3}$;
    minimum: -4 at $x = -2$

2. No global extrema

3. (a) $[-1, 0)$ and $[1, \infty)$
    (b) $(-\infty, -1]$ and $(0, 1]$
    (c) $(-\infty, 0)$ and $(0, \infty)$
    (d) None
    (e) Local minima at $(1, e)$ and $(-1, e)$
    (f) None

4. (a) $[-\sqrt{2}, \sqrt{2}]$
    (b) $[-2, -\sqrt{2}]$ and $[\sqrt{2}, 2]$
    (c) $(-2, 0)$ **(d)** $(0, 2)$
    (e) Local max: $(-2, 0)$ and $(\sqrt{2}, 2)$;
     local min: $(2, 0)$ and $(-\sqrt{2}, -2)$
    (f) $(0, 0)$

5. (a) Approximately $(-\infty, 0.385]$
    (b) Approximately $[0.385, \infty)$
    (c) None   **(d)** $(-\infty, \infty)$
    (e) Local maximum at $\approx (0.385, 1.215)$
    (f) None

6. (a) $[1, \infty)$ **(b)** $(-\infty, 1]$
    (c) $(-\infty, \infty)$ **(d)** None
    (e) Local minimum at $(1, 0)$
    (f) None

7. (a) $[0, 1)$ **(b)** $(-1, 0]$
    (c) $(-1, 1)$ **(d)** None
    (e) Local minimum at $(0, 1)$
    (f) None

8. (a) $(-\infty, -2^{-1/3}] \approx (-\infty, -0.794]$
    (b) $[-2^{-1/3}, 1) \approx [-0.794, 1)$ and $(1, \infty)$
    (c) $(-\infty, -2^{1/3}) \approx (-\infty, -1.260)$ and $(1, \infty)$
    (d) $(-1.260, 1)$
    (e) Local maximum at
     $\left(-2^{-1/3}, \dfrac{2}{3} \cdot 2^{-1/3}\right) \approx (-0.794, 0.529)$
    (f) $\left(-2^{1/3}, \dfrac{1}{3} \cdot 2^{1/3}\right) \approx (-1.260, 0.420)$

9. (a) None   **(b)** $[-1, 1]$
    (c) $(-1, 0)$ **(d)** $(0, 1)$
    (e) Local maximum at $(-1, \pi)$;
     local minimum at $(1, 0)$
    (f) $\left(0, \dfrac{\pi}{2}\right)$

10. (a) $[-\sqrt{3}, \sqrt{3}]$

(b) $(-\infty, -\sqrt{3}]$ and $[\sqrt{3}, \infty)$

(c) Approximately $(-2.584, -0.706)$ and $(3.290, \infty)$

(d) Approximately $(-\infty, -2.584)$ and $(-0.706, 3.290)$

(e) Local maximum at
$$\left(\sqrt{3}, \frac{\sqrt{3}-1}{4}\right) \approx (1.732, 0.183);$$
local minimum at
$$\left(-\sqrt{3}, \frac{-\sqrt{3}-1}{4}\right) \approx (-1.732, -0.683)$$

(f) $\approx (-2.584, -0.573)$, $(-0.706, -0.338)$, and $(3.290, 0.161)$

11. (a) $(0, 2]$ **(b)** $[-2, 0)$

(c) None

(d) $(-2, 0)$ and $(0, 2)$

(e) Local maxima at $(-2, \ln 2)$ and $(2, \ln 2)$

(f) None

12. (a) Approximately $[0, 0.176]$, $\left[0.994, \frac{\pi}{2}\right]$, $[2.148, 2.965]$, $\left[3.834, \frac{3\pi}{2}\right]$, and $\left[5.591, 2\pi\right]$

(b) Approximately $[0.176, 0.994]$, $\left[\frac{\pi}{2}, 2.148\right]$, $[2.965, 3.834]$, and $\left[\frac{3\pi}{2}, 5.591\right]$

(c) Approximately $(0.542, 1.266)$, $(1.876, 2.600)$, $(3.425, 4.281)$, and $(5.144, 6.000)$

(d) Approximately $(0, 0.542)$, $(1.266, 1.876)$, $(2.600, 3.425)$, $(4.281, 5.144)$, and $(6.000, 2\pi)$

(e) Local maxima at $\approx (0.176, 1.266)$, $\left(\frac{\pi}{2}, 0\right)$ and $(2.965, 1.266)$, $\left(\frac{3\pi}{2}, 2\right)$, and $(2\pi, 1)$; local minima at $\approx (0, 1)$, $(0.994, -0.513)$, $(2.148, -0.513)$, $(3.834, -1.806)$, and $(5.591, -1.806)$

Note that the local extrema at $x \approx 3.834$, $x = \frac{3\pi}{2}$, and $x \approx 5.591$ are also absolute extrema.

(f) $\approx (0.542, 0.437)$, $(1.266, -0.267)$, $(1.876, -0.267)$, $(2.600, 0.437)$, $(3.425, -0.329)$, $(4.281, 0.120)$, $(5.144, 0.120)$, and $(6.000, -0.329)$

13. (a) $\left(0, \dfrac{2}{\sqrt{3}}\right]$

(b) $(-\infty, 0]$ and $\left[\dfrac{2}{\sqrt{3}}, \infty\right)$

(c) $(-\infty, 0)$ **(d)** $(0, \infty)$

(e) Local maximum at
$$\left(\frac{2}{\sqrt{3}}, \frac{16}{3\sqrt{3}}\right) \approx (1.155, 3.079)$$

(f) None

14. (a) Approximately $[-0.578, 1.692]$

(b) Approximately $(-\infty, -0.578]$ and $[1.692, \infty)$

(c) Approximately $(-\infty, 1.079)$

(d) Approximately $(1.079, \infty)$

(e) Local maximum at $\approx (1.692, 20.517)$; local minimum at $\approx (-0.578, 0.972)$

(f) $\approx (1.079, 13.601)$

15. (a) $\left[0, \dfrac{8}{9}\right]$

(b) $(-\infty, 0]$ and $\left[\dfrac{8}{9}, \infty\right)$

(c) $\left(-\infty, -\dfrac{2}{9}\right)$

(d) $\left(-\dfrac{2}{9}, 0\right)$ and $(0, \infty)$

(e) Local maximum at $\approx (0.889, 1.011)$; local minimum at $(0, 0)$

(f) $\approx \left(\dfrac{2}{9}, 0.667\right)$

16. (a) Approximately $(-\infty, 0.215]$

(b) Approximately $[0.215, 2)$ and $(2, \infty)$

(c) Approximately $(2, 3.710)$

(d) $(-\infty, 2)$ and approximately $(3.710, \infty)$

(e) Local maximum at $\approx (0.215, -2.417)$

(f) $\approx (3.710, -3.420)$

17. (a) None   **(b)** At $x = -1$

(c) At $x = 0$ and $x = 2$

18. (a) At $x = -1$ **(b)** At $x = 2$

(c) At $x = \dfrac{1}{2}$

19. $f(x) = -\dfrac{1}{4}x^{-4} - e^{-x} + C$

20. $f(x) = \sec x + C$

21. $f(x) = 2\ln x + \dfrac{1}{3}x^3 + x + C$

22. $f(x) = \dfrac{2}{3}x^{3/2} + 2x^{1/2} + C$

23. $f(x) = -\cos x + \sin x + 2$

24. $f(x) = \dfrac{3}{4}x^{4/3} + \dfrac{x^3}{3} + \dfrac{x^2}{2} + x - \dfrac{31}{12}$

25. $s(t) = 4.9t^2 + 5t + 10$

26. $s(t) = 16t^2 + 20t + 5$

27. $L(x) = 2x + \dfrac{\pi}{2} - 1$

28. $L(x) = \sqrt{2}x - \dfrac{\pi\sqrt{2}}{4} + \sqrt{2}$

29. $L(x) = -x + 1$ **30.** $L(x) = 2x + 1$

31. Global minimum value of $\dfrac{1}{2}$ at $x = 2$

32. (a) T (b) P

33. (a) $(0, 2]$ (b) $[-3, 0)$
     (c) Local maxima at $(-3, 1)$ and $(2, 3)$

34. The 24th day

35.

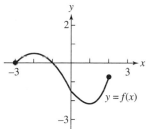

36. (a) Absolute minimum is -2 at $x = 1$;
        absolute maximum is 3 at $x = 3$
     (b) None
     (c)

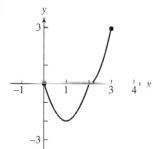

37. (a) $f(x)$ is continuous on $[0.5, 3]$ and differentiable on $(0.5, 3)$.
     (b) $c \approx 1.579$ (c) $y \approx 1.457x - 1.075$
     (d) $y \approx 1.457x - 1.579$

38. (a) $v(t) = -3t^2 - 6t + 4$
     (b) $a(t) = -6t - 6$
     (c) The particle starts at position 3 moving in the positive direction, but decelerating. At approximately $t = 0.528$, it reaches position 4.128 and changes direction, beginning to move in the negative direction. After that, it continues to accelerate while moving in the negative direction.

39. (a) $L(x) = -1$
     (b) Using the linearization, $f(0.1) \approx -1$
     (c) Greater than the approximation in (b), since $f'(x)$ is actually positive over the interval $(0, 0.1)$ and the estimate is based on the derivative being 0.

40. (a) $dy = (2x - x^2)e^{-x}\, dx$
     (b) $dy \approx 0.00368$

41. (a) $y = \dfrac{2701.73}{1 + 17.28e^{-0.36x}}$

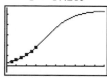

[0, 20] by [−300, 2800]

     (b) In 1998. There are approximately 1351 million transactions in that year.
     (c) Approximately 2702 million transactions per year

42. $x \approx 0.828361$ **43.** 1200 m/sec

44. 1162.5 m   **45.** $r = 25$ ft and $s = 50$ ft

46. 54 square units

47. Base is 6 ft by 6 ft, height $= 3$ ft

48. Base is 4 ft by 4 ft, height $= 2$ ft

49. Height $= 2$, radius $= \sqrt{2}$

50. $r = h = 4$ ft

51. (a) $V(x) = x(15 - 2x)(5 - x)$
     (b) $0 < x < 5$

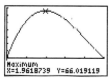

Maximum
X=1.9618739 Y=66.019119
[0, 5] by [−10, 70]

     (c) Maximum volume ≈ 66.019 in^3 when $x \approx 1.962$ in.
     (d) $V'(x) = 6x^2 - 50x + 75$ which is zero at $x = \dfrac{25 - 5\sqrt{7}}{6} \approx 1.962$.

52. 29.925 square units

53. $x = \dfrac{48}{\sqrt{7}} \approx 18.142$ mi and $y = \dfrac{36}{\sqrt{7}} \approx 13.607$ mi

54. $x = 100$ m and $r = \dfrac{100}{\pi}$ m

55. 276 grade A and 553 grade B tires

56. (a) 0.765 units
     (b) When $t = \dfrac{7\pi}{8} \approx 2.749$
        (plus multiples of π if they keep going)

57. Dimensions: base is 6 in. by 12 in., height $= 2$ in.; maximum volume $= 144$ in^3

58. -40m^2/sec   **59.** 5 m/sec

60. Increasing 1 cm/min   **61.** $\dfrac{dx}{dt} = 4$ units/second

62. (a) $h = \dfrac{5r}{2}$ (b) $\dfrac{125}{144\pi} \approx 0.276$ ft/min

63. 5 radians/sec

64. Not enough speed. Duck!

65. $dV \approx \dfrac{2\pi a h}{3}\, dr$

66. (a) Within 1% (b) Within 3%

67. (a) Within 4% (b) Within 8%

  (c) Within 12%

68. Height = 14 feet, estimated error = $\pm\dfrac{2}{45}$ feet

69. $\dfrac{dy}{dx} = 2 \sin x \cos x - 3$.

  Since $\sin x$ and $\cos x$ are both between 1 and -1,

  $2 \sin x \cos x$ is never greater than 2, and

  therefore $\dfrac{dy}{dx} \le 2 - 3 = -1$ for all values of x.

Chapter 5

Section 5.1 (pp. 247–257)

Quick Review

1. 400 miles **2.** 144 miles

3. 100 ft/sec $\approx$ 68.18 mph **4.** 9.46×10^{12} km

5. 28 miles **6.** 1200 gallons

7. $-3°$ **8.** 25,920,000 ft^3

9. 17,500 **10.** 176,400 times

Exercises

1. (a)

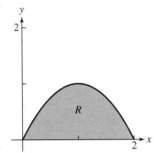

  (b)

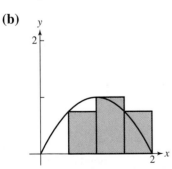

  LRAM = 1.25

3.

n	LRAM$_n$	MRAM$_n$	RRAM$_n$
10	1.32	1.34	1.32
50	1.3328	1.3336	1.3328
100	1.3332	1.3334	1.3332
500	1.333328	1.333336	1.333328

5. 13.5 **7.** 0.8821

9. $\approx$ 44.8, 6.7 L/min

11. (a) 5220 m (b) 4920 m

13. (a) 0.969 mi

  (b) 0.006 h = 21.6 sec

  116 mph

15.

n	error	% error
10	2.61799	0.5
20	0.65450	0.125
40	0.16362	3.12×10^{-2}
80	0.04091	7.8×10^{-3}
160	0.01023	2×10^{-3}

17. (a) $S_8 \approx 120.95132$ (b) 10%

  Underestimate

19. (a) 15,465 ft^3 (b) 16,515 ft^3

21. 39.26991

23. (a) 240 ft/sec

  (b) 1520 ft with RRAM and $n = 5$

25. (a) Upper: 60.9 tons;

  lower: 46.8 tons

  (b) By the end of October

27. (a) 2 (b) $2\sqrt{2} \approx 2.828$

  (c) $8 \sin\left(\dfrac{\pi}{8}\right) \approx 3.061$

  (d) Each area is less than the area of the circle, π. As n increases, the polygon area approaches π.

29. $\text{RRAM}_n f = \text{LRAM}_n f + f(x_n)\Delta x - f(x_0)\Delta x$

  Since $f(a) = f(b)$, or $f(x_0) = f(x_n)$, we have

  $\text{RRAM}_n f = \text{LRAM}_n f$.

Section 5.2 (pp. 258–268)

Quick Review

1. 55   **2.** 20   **3.** 5500

4. $\displaystyle\sum_{k=1}^{99} k$ **5.** $\displaystyle\sum_{k=0}^{25} 2k$ **6.** $\displaystyle\sum_{k=1}^{500} 3k^2$

7. $\displaystyle\sum_{x=1}^{50} (2x^2 + 3x)$ **8.** $\displaystyle\sum_{k=0}^{20} x^k$

9. $\displaystyle\sum_{k=0}^{n} (-1)^k = 0$ if n is odd.

10. $\displaystyle\sum_{k=0}^{n} (-1)^k = 1$ if n is even.

Exercises

1. $\displaystyle\int_0^2 x^2 \, dx$ **3.** $\displaystyle\int_1^4 \frac{1}{x} \, dx$

5. $\displaystyle\int_0^1 \sqrt{4 - x^2} \, dx$ **7.** 15

9. -480 **11.** 2.75   **13.** 21

15. $\dfrac{9\pi}{2}$ **17.** $\dfrac{5}{2}$ **19.** 3

21. $\dfrac{3\pi^2}{2}$ **23.** $\dfrac{1}{2}b^2$ **25.** $b^2 - a^2$

27. $\dfrac{3}{2}a^2$ **29.** 0   **31.** $\dfrac{1}{4}$

33. $\dfrac{3}{4}$ **35.** $\dfrac{1}{2}$ **37.** $-\dfrac{3}{4}$

39. 0.990501   **41.** $\dfrac{32}{3}$

43. (a) 0   (b) 1

45. (a) -1 (b) $-\dfrac{7}{2}$

47. (a) $f \to +\infty$

(b) Using right endpoints we have

$$\int_0^1 \frac{1}{x^2} \, dx = \lim_{n\to\infty} \sum_{k=1}^{n} \frac{1}{n}\frac{n^2}{k^2}$$

$$= \lim_{n\to\infty} \sum_{k=1}^{n} \frac{n}{k^2} = \lim_{n\to\infty} n\left[1 + \frac{1}{2^2} + \cdots + \frac{1}{n^2}\right].$$

$$n\left(1 + \frac{1}{2^2} + \cdots + \frac{1}{n^2}\right) > n \text{ and } n \to \infty$$

$$\text{so } n\left(1 + \frac{1}{2^2} + \cdots + \frac{1}{n^2}\right) \to \infty.$$

Section 5.3 (pp. 268–276)

Quick Review

1. $\sin x$ **2.** $\cos x$ **3.** $\tan x$

4. $\cot x$ **5.** $\sec x$ **6.** $\ln(x)$

7. x^n **8.** $-\dfrac{2^x \ln 2}{(2^x + 1)^2}$

9. $xe^x + e^x$ **10.** $\dfrac{1}{x^2 + 1}$

Exercises

1. (a) 0   (b) -8 (c) -12
     (d) 10   (e) -2 (f) 16

3. (a) 5   (b) $5\sqrt{3}$
     (c) -5 (d) -5

5. (a) 4   (b) 4

7. -14 **9.** 1

11. -1 **13.** $\dfrac{\pi}{2}$

15. $e^2 - 1 \approx 6.389$ **17.** $\dfrac{16}{3}$

19. $\dfrac{19}{3}$

21.

[0, 3] by [−1, 8]

     (a) 6   (b) $\dfrac{22}{3}$

23.

[0, 3] by [−3, 2]

     (a) 0   (b) $\dfrac{8}{3}$

25. 0, at $x = 1$ **27.** -2, at $x = \dfrac{1}{\sqrt{3}}$

29. $\dfrac{3}{2}$ **31.** 0

33. $\dfrac{1}{2} \le \displaystyle\int_0^1 \frac{1}{1 + x^4} \, dx \le 1$

35. $0 \le \displaystyle\int_0^1 \sin(x^2) \, dx \le \sin(1) < 1$

37. $0 \le \min f(b - a) \le \displaystyle\int_a^b f(x) \, dx$

39. $av(f) = \dfrac{1}{b - a} \displaystyle\int_a^b f(x) \, dx$ implies $\displaystyle\int_a^b f(x) \, dx$

$$= av(f)(b - a) = \int_a^b av(f) \, dx$$

41. Avg rate $= \dfrac{\text{total amount released}}{\text{total time}}$

$$= \frac{2000 \text{ m}^3}{100 \text{ min} + 50 \text{ min}} = 13\frac{1}{3} \text{ m}^3/\text{min}$$

43. $\dfrac{7}{6}$ **45.** $k \approx 2.39838$

Section 5.4 (pp. 277–288)

Quick Review

1. $2x \cos x^2$

2. $2 \sin x \cos x$

3. 0

4. 0

5. $2^x \ln 2$

6. $\dfrac{1}{2\sqrt{x}}$

7. $\dfrac{-x \sin x - \cos x}{x^2}$

8. $-\cot(t)$

9. $\dfrac{y+1}{2y-x}$

10. $\dfrac{1}{3x}$

Exercises

1. $5 - \ln 6 \approx 3.208$

3. 1

5. $\dfrac{5}{2}$

7. 2

9. $2\sqrt{3}$

11. 0

13. $\dfrac{8}{3}$

15. $\dfrac{5}{2}$

17. $\dfrac{1}{2}$

19. (a) No, $f(x) = \dfrac{x^2-1}{x+1}$ is discontinuous at $x = -1$.

(b) $-\dfrac{5}{2}$. $f(x)$ is bounded with only one discontinuity. Split it up at $x = -1$, or use area.

21. (a) No, $f(x) = \tan x$ is discontinuous at $x = \dfrac{\pi}{2}$ and $x = \dfrac{3\pi}{2}$.

(b) No, $\displaystyle\int_0^b \tan x\, dx \to \infty$ as $b \to \dfrac{\pi}{2}^-$.

23. (a) No, $f(x) = \dfrac{\sin x}{x}$ is discontinuous at $x = 0$.

(b) ≈ 2.55. Area is finite. $\dfrac{\sin x}{x}$ is bounded with only one discontinuity.

25. $\dfrac{5}{6}$

27. π

29. ≈ 3.802

31. ≈ 0.914

33. $x \approx 0.699$

35. $-\dfrac{3}{2}$

37. $\sqrt{1+x^2}$

39. $\dfrac{\sin x}{2\sqrt{x}}$

41. $3x^2 \cos(2x^3) - 2x \cos(2x^2)$

43. (d)

45. (b)

47. $x = a$

49. $L(x) = 2 + 10x$

51. $\displaystyle\int_0^{\pi/k} \sin kx\, dx = \dfrac{2}{k}$

53. (a) 0

(b) H is increasing on $[0,6]$ where $H'(x) = f(x) > 0$.

(c) H is concave up on $(9,12)$ where $H''(x) = f'(x) > 0$.

(d) $H(12) = \displaystyle\int_0^{12} f(t)\, dt > 0$ because there is more area above the x-axis than below for $y = f(x)$.

(e) $x = 6$ since $H'(6) = f(6) = 0$ and $H''(6) = f'(6) < 0$.

(f) $x = 0$ since $H(x) > 0$ on $(0, 12]$.

55. (a) $s'(3) = f(3) = 0$

(b) $s''(3) = f'(3) > 0$

(c) $s(3) = \displaystyle\int_0^3 f(x)\, dx = -\dfrac{1}{2}(3)(6) = -9$ units

(d) $s(t) = 0$ at $t = 6$ sec because $\displaystyle\int_0^6 f(x)\, dx = 0$

(e) $s''(t) = f'(t) = 0$ at $t = 7$ sec

(f) $0 < t < 3$: $s < 0$, $s' < 0 \Rightarrow$ away
$3 < t < 6$: $s < 0$, $s' > 0 \Rightarrow$ toward
$t > 6$: $s > 0$, $s' > 0 \Rightarrow$ away

(g) Positive side

57. (a) $\$9$

(b) $\$10$

59. (a) 300 drums

(b) $\$6.00$

61. Using area, $\displaystyle\int_0^x f(t)\, dt = -\int_{-x}^0 f(t)\, dt = \int_0^{-x} f(t)\, dt$

63. f odd $\to \displaystyle\int_0^x f(t)\, dt$ is even, but $\dfrac{d}{dx}\displaystyle\int_0^x f(t)\, dt = f(x)$ so f is the derivative of an even function.

Similarly for f even.

Section 5.5 (pp. 289–297)

Quick Review

1. Concave down

2. Concave up

3. Concave down

4. Concave down

5. Concave up

6. Concave down

7. Concave up

8. Concave up

9. Concave down

10. Concave down

Exercises

1. (a) 2 **(b)** Exact **(c)** 2

3. (a) 4.25 **(b)** Over **(c)** 4

5. (a) 5.146 **(b)** Under **(c)** $\dfrac{16}{3}$

7. $15{,}990$ ft^3

9. 0.9785 mi

11. The average of the 13 discrete temperatures gives equal weight to the low values at the end.

13. $S_{50} = 3.1379$, $S_{100} = 3.14029$

15. $S_{50} = 1.37066$, $S_{100} = 1.37066$ using $a = 0.0001$ as lower limit

$S_{50} = 1.37076$, $S_{100} = 1.37076$ using $a = 0.000000001$ as lower limit

17. (a) $T_{10} = 1.983523538$, $T_{100} = 1.999835504$
$T_{1000} = 1.999998355$

(b)

n	$|E_T|$
10	$0.016476462 = 1.6476462 \times 10^{-2}$
100	1.64496×10^{-4}
1000	1.645×10^{-6}

(c) $\left|E_{T_{10n}}\right| \approx 10^{-2} \times \left|E_{T_n}\right|$

(d) $\left|E_{T_n}\right| \le \dfrac{\pi^3 M}{12n^2}$. $\left|E_{T_{10n}}\right| \le \dfrac{\pi^3 M}{12(10n)^2}$
$= \dfrac{\pi^3 M}{12n^2} \times 10^{-2}$

19. (a) $f''(x) - 2\cos(x^2) - 4x^2\sin(x^2)$

(b)

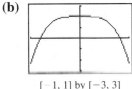

$[-1, 1]$ by $[-3, 3]$

(c) The graph shows that $3 \le f''(x) \le 2$ for $-1 \le x \le 1$.

(d) $|E_T| \le \dfrac{1 - (-1)}{12}(h^2)(3) = \dfrac{h^2}{2}$

(e) $|E_T| \le \dfrac{h^2}{2} \le \dfrac{0.1^2}{2} < 0.1$

(f) $n \ge 20$

21. 466.67 in^2

23. Each quantity is equal to
$\dfrac{h}{2}(y_0 + 2y_1 + 2y_2 + \cdots + 2y_{n-1} + y_n)$.

Chapter Review (pp. 298–301)

1.

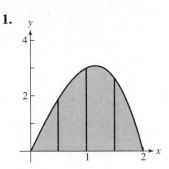

2. 3.75

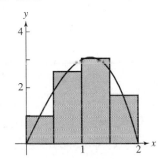

3. 4.125

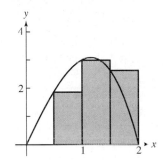

4. 3.75

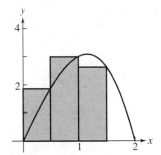

5. 3.75

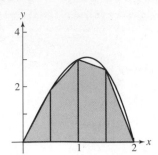

6. 4

7.

n	LRAM$_n$	MRAM$_n$	RRAM$_n$
10	1.78204	1.60321	1.46204
20	1.69262	1.60785	1.53262
30	1.66419	1.60873	1.55752
50	1.64195	1.60918	1.57795
100	1.62557	1.60937	1.59357
1000	1.61104	1.60944	1.60784

8. ln 5

9. (a) True **(b)** True **(c)** False

10. (a) $V = \lim\limits_{n\to\infty} \sum\limits_{i=1}^{n} \pi \sin^2(m_i)\, \Delta x$

(b) 4.9348

11. (a) 26.5 m

(b)

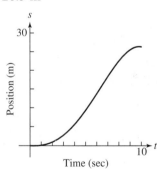

12. (a) $\int_0^{10} x^3\, dx$ **(b)** $\int_0^{10} x \sin x\, dx$

(c) $\int_0^{10} x(3x-2)^2\, dx$ **(d)** $\int_0^{10} (1+x^2)^{-1}\, dx$

(e) $\int_0^{10} \pi\left(9 - \sin^2\frac{\pi x}{10}\right) dx$

13. 10 **14.** 2 **15.** 20

16. 42 **17.** $\dfrac{\sqrt{2}}{2}$ **18.** 16

19. 3 **20.** 2 **21.** 2

22. 1 **23.** $\sqrt{3}$ **24.** 1

25. 8 **26.** 2 **27.** −1

28. 0 **29.** 2 ln 3 **30.** π

31. 40 **32.** 64π

33. (a) Upper = 4.392 L; **(b)** 4.2 L
lower = 4.008 L

34. (a) Lower = 87.15 ft; **(b)** 95.1 ft
upper = 103.05 ft

35. One possible answer:
The dx is important because it corresponds to actual physical quantity Δx in a Riemann sum. Without the Δx, our integral approximations would be way off.

36. $\dfrac{16}{3}$

37. $0 \le \sqrt{1+\sin^2 x} \le \sqrt{2}$

38. (a) $\dfrac{4}{3}$ **(b)** $\dfrac{2}{3}a^{3/2}$

39. $\sqrt{2+\cos^3 x}$ **40.** $14x\sqrt{2+\cos^3(7x^2)}$

41. $\dfrac{-6}{3+x^4}$ **42.** $\dfrac{2}{4x^2+1} - \dfrac{1}{x^2+1}$

43. \$230

44. $av(I) = 4800$ cases;
average holding cost = \$192.00 per day

45. $x = 1.63052$ or $x = -3.09131$

46. (a) True **(b)** True **(c)** True
(d) False **(e)** True **(f)** False
(g) True

47. $F(1) - F(0)$ **48.** $y = \int_5^x \dfrac{\sin t}{t}\, dt + 3$

49. Use the fact that $y' = 2x + \dfrac{1}{x}$.

50. (b); $\dfrac{dy}{dx} = 2x \to y = x^2 + c.\ y(1) = 4 \to c = 3$

51. (a) ≈ 2.42 gal **(b)** ≈ 24.83 mpg

52. (a) 6,144 ft **(b)** 4,296 ft **(c)** B

53. (a) $h(y_1 + y_3) + 2(2hy_2) = h(y_1 + 4y_2 + y_3)$

(b) Each expression is equal to
$$\frac{1}{3}[h(y_0 + 4y_1 + y_2) + h(y_2 + 4y_3 + y_4) + \cdots + h(y_{2n-2} + 4y_{2n-1} + y_{2n}).]$$

54. (a) 0 **(b)** −1
(c) $-\pi$ **(d)** $x = 1$
(e) $y = 2x + 2 - \pi$ **(f)** $x = -1, x = 2$
(g) $[-2\pi, 0]$

55. (a) NINT$(e^{-x^2/2}, x, -10, 10) \approx 2.506628275$
NINT$(e^{-x^2/2}, x, -20, 20) \approx 2.506628275$
(b) The area is $\sqrt{2\pi}$.

56. ≈ 1500 yd^3

57. (a) $(V^2)_{\text{av}} = \dfrac{(V_{\max})^2}{2}$ **(b)** 339 volts
$$V_{\text{rms}} = \frac{V_{\max}}{\sqrt{2}}$$

Chapter 6

Section 6.1 (pp. 303–315)

Quick Review

1. $106.00
2. $106.14
3. $106.17
4. $106.18
5. $3 \cos 3x$
6. $\dfrac{5}{2} \sec^2 \dfrac{5}{2}x$
7. $2Ce^{2x}$
8. $\dfrac{1}{x+2}$

9.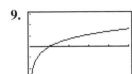

[0.01, 5] by [−3, 3]

The graphs appear to be the same.

10.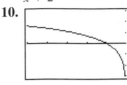

[−5, −0.01] by [−3, 3]

The graphs appear to be the same.

Exercises

1. $\dfrac{x^3}{3} - x^2 + x + C$

3. $\dfrac{x^3}{3} - \dfrac{8}{3}x^{3/2} + C$

5. $\dfrac{e^{4x}}{4} + C$

7. $\dfrac{x^6}{6} - 3x^2 + 3x + C$

9. $2e^{t/2} + \dfrac{5}{t} + C$

11. $\dfrac{x^4}{4} + \dfrac{1}{2x^2} + C$

13. $x^{1/3} + C$

15. $\sin\left(\dfrac{\pi}{2}x\right) + C$

17. $2 \ln|x-1| + \ln|x| + C$

19. $\tan 5r + C$

21. $\dfrac{x}{2} + \dfrac{\sin 2x}{4} + C$

23. $\tan\theta - \theta + C$

25. (a) Graph (b)
    (b) The slope is always positive, so (a) and (c) can be ruled out.

27. $y = x^2 - x - 2$

29. $y = \tan x - 2$

31. $y = 3x^3 - 2x^2 + 5x + 10$

33. $y = -2e^{-t} + 1$

35. $y = -\sin\theta + \theta - 3$

37. $y = \dfrac{1}{2}\ln|t| + \dfrac{5}{4}t^2 - \dfrac{1}{4}$

39. $s = 4.9t^2 + 5t + 10$

41. $s = 16t^2 + 20t$

43.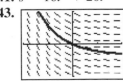

[−2, 2] by [−3, 3]

45–47. The derivative of the right hand side of the equation is equal to the integrand on the left hand side.

49. (a) $y = \dfrac{x^2}{2} + \dfrac{1}{x} + \dfrac{1}{2},\ x > 0$

(b) $y = \dfrac{x^2}{2} + \dfrac{1}{x} + \dfrac{3}{2},\ x < 0$

(c) $y' = \begin{cases} x - \dfrac{1}{x^2}, & x < 0 \\[2mm] x - \dfrac{1}{x^2}, & x > 0 \end{cases}$

(d) $C_1 = \dfrac{3}{2},\ C_2 = \dfrac{1}{2}$

(e) $C_1 = \dfrac{1}{2},\ C_2 = -\dfrac{7}{2}$

51. $c(x) = x^3 - 6x^2 + 15x + 400$

53. (a) $s = \dfrac{-kt^2}{2} + 88t$ (b) $t = \dfrac{88}{k}$
    (c) $k = 16$ ft/sec^2

55. ≈ 1.240 sec

57. $h = \left[-\dfrac{125}{48\pi}t + 10^{5/2}\right]^{2/5}$

$V = \dfrac{4\pi}{75}\left[-\dfrac{125}{48\pi}t + 10^{5/2}\right]^{6/5}$

59. (a) $y = 1200e^{0.0625t}$ (b) $t = \dfrac{\ln 3}{0.0625} \approx 17.6$ yr

61. (a) $\displaystyle\int_0^x te^t\,dt + C$ (b) $\displaystyle\int_0^x te^t\,dt + 1$

63.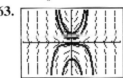

[−6, 6] by [−4, 4]

The concavity of each solution curve indicates the sign of y.

65.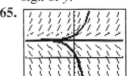

[−3, 3] by [−4, 10]

The concavity of each solution curve indicates the sign of y.

67. (a) $\dfrac{d}{dx}(\ln x + C) = \dfrac{1}{x}$ for $x > 0$

(b) $\dfrac{d}{dx}(\ln(-x) + C) = \dfrac{1}{x}$ for $x < 0$

(c) $\dfrac{d}{dx}\ln|x| = \dfrac{1}{x}$ for all x except 0.

(d) $\dfrac{dy}{dx} = \dfrac{1}{x}$ for all x except 0.

Section 6.2 (pp. 315–323)

Quick Review

1. $\dfrac{32}{5}$ 2. $\dfrac{16}{3}$

3. 3^x 4. 3^x

5. $4(x^3 - 2x^2 + 3)^3(3x^2 - 4x)$

6. $8 \sin (4x - 5) \cos (4x - 5)$

7. $-\tan x$ 8. $\cot x$

9. $\sec x$ 10. $-\csc x$

Exercises

1. $-\dfrac{1}{3} \cos 3x + C$ 3. $\dfrac{1}{2} \sec 2x + C$

5. $\dfrac{1}{3} \tan^{-1}\left(\dfrac{x}{3}\right) + C$ 7. $\dfrac{2}{3}\left(1 - \cos \dfrac{t}{2}\right)^3 + C$

9. $\dfrac{1}{1 - x} + C$ 11. $\dfrac{2}{3}(\tan x)^{3/2} + C$

13. $\ln (\ln 6)$ 15. $\dfrac{1}{3} \sin (3z + 4) + C$

17. $\dfrac{1}{7}(\ln x)^7 + C$ 19. $\dfrac{3}{4} \sin (s^{4/3} - 8) + C$

21. $\dfrac{1}{2} \sec (2t + 1) + C$ 23. 0

25. $\dfrac{1}{2} \ln 5 \approx 0.805$

27. $\dfrac{1}{3} \ln |\sec (3x)| + C = -\dfrac{1}{3} \ln |\cos (3x)| + C$

29. $\ln |\sec x + \tan x| + C$ 31. $\dfrac{14}{3}$

33. $-\dfrac{1}{2}$ 35. $\dfrac{10}{3}$

37. $2\sqrt{3}$ 39. $y = Ce^{(1/2)x^2 + 2x} - 5$

41. $y = -\ln (C - e^{\sin x})$ 43. $y = \dfrac{1}{x^2 + 3}$

45. (a) $\dfrac{d}{dx}\left(\dfrac{2}{3}(x + 1)^{3/2} + C\right) = \sqrt{x + 1}$

(b) Because $\dfrac{dy_1}{dx} = \sqrt{x + 1}$ and $\dfrac{dy_2}{dx} = \sqrt{x + 1}$

(c) $4\dfrac{2}{3}$

(d) $C = y_1 - y_2$

$= \displaystyle\int_0^x \sqrt{x + 1}\, dx - \int_3^x \sqrt{x + 1}\, dx$

$= \displaystyle\int_0^x \sqrt{x + 1}\, dx + \int_x^3 \sqrt{x + 1}\, dx$

$= \displaystyle\int_0^3 \sqrt{x + 1}\, dx$

47. (a) $\dfrac{1}{2}\sqrt{10} - \dfrac{3}{2} \approx 0.081$

(b) $\dfrac{1}{2}\sqrt{10} - \dfrac{3}{2} \approx 0.081$

49. Show $\dfrac{dy}{dx} = \tan x$ and $y(3) = 5$.

51. (a) $\sin^2 x + C$ (b) $-\cos^2 x + C$

(c) $-\dfrac{1}{2} \cos 2x + C$

(d) The derivative of each expression is $2 \sin x \cos x$.

Section 6.3 (pp. 323–329)

Quick Review

1. $2x^3 \cos 2x + 3x^2 \sin 2x$

2. $\dfrac{3e^{2x}}{3x + 1} + 2e^{2x} \ln (3x+1)$

3. $\dfrac{2}{1 + 4x^2}$ 4. $\dfrac{1}{\sqrt{1 - (x + 3)^2}}$

5. $x = \dfrac{1}{3}\tan y$ 6. $x = \cos y - 1$

7. $\dfrac{2}{\pi}$ 8. $y = \dfrac{1}{2}e^{2x} + C$

9. $y = \dfrac{1}{2}x^2 - \cos x + 3$

10. $\dfrac{d}{dx}\left[\dfrac{1}{2}e^x(\sin x - \cos x)\right] = e^x \sin x$

Exercises

1. $-x \cos x + \sin x + C$ 3. $\dfrac{1}{2}y^2 \ln y - \dfrac{1}{4}y^2 + C$

5. $x \tan x + \ln |\cos x| + C$

7. $(2 - t^2) \cos t + 2t \sin t + C$

9. $\dfrac{1}{4}x^4 \ln x - \dfrac{1}{16}x^4 + C$ 11. $(x^2 - 7x + 7)e^x + C$

13. $\dfrac{1}{2}e^y(\sin y - \cos y) + C$

15. $\dfrac{\pi^2}{8} - \dfrac{1}{2} \approx 0.734$

17. $\dfrac{1}{13}[e^6(2 \cos 9 + 3 \sin 9)$
$\qquad - e^{-4}(2 \cos 6 - 3 \sin 6)] \approx -18.186$

19. $y = \left(\dfrac{x^2}{4} - \dfrac{x}{8} + \dfrac{1}{32}\right)e^{4x} + C$

21. $y = \dfrac{\theta^2}{2}\sec^{-1} \theta - \dfrac{1}{2}\sqrt{\theta^2 - 1} + C$

23. (a) π (b) 3π (c) 4π

25. $\dfrac{1 - e^{-2\pi}}{2\pi} \approx 0.159$

27. $-2(\sqrt{x} \cos \sqrt{x} - \sin \sqrt{x}) + C$

29. $\dfrac{(x^6 - 3x^4 + 6x^2 - 6)e^{x^2}}{2} + C$

31. $u = x^n$, $dv = \cos x\, dx$ **33.** $u = x^n$, $dv = e^{ax}\, dx$

35. (a) Let $y = f^{-1}(x)$. Then $x = f(y)$,
so $dx = f'(y)\, dy$. Substitute directly.
  (b) $u = y$, $dv = f'(y)\, dy$

37. (a) $\displaystyle\int \sin^{-1} x\, dx = x \sin^{-1} x + \cos(\sin^{-1} x) + C$

  (b) $\displaystyle\int \sin^{-1} x\, dx = x \sin^{-1} x + \sqrt{1 - x^2} + C$

  (c) $\cos(\sin^{-1} x) = \sqrt{1 - x^2}$

39. (a) $\displaystyle\int \cos^{-1} x\, dx = x \cos^{-1} x - \sin(\cos^{-1} x) + C$

  (b) $\displaystyle\int \cos^{-1} x\, dx = x \cos^{-1} x - \sqrt{1 - x^2} + C$

  (c) $\sin(\cos^{-1} x) = \sqrt{1 - x^2}$

Section 6.4 (pp. 330–341)

Quick Review

1. $a = e^b$ **2.** $c = \ln(d)$

3. $x = e^2 - 3$ **4.** $x = \dfrac{1}{2}\ln 6$

5. $x = \dfrac{\ln 2.5}{\ln 0.85} \approx -5.638$ **6.** $k = \dfrac{\ln 2}{\ln 3 - \ln 2} \approx 1.710$

7. $t = \dfrac{\ln 10}{\ln 1.1} \approx 24.159$ **8.** $t = \dfrac{1}{2}\ln 4 = \ln 2$

9. $y = -1 + e^{2x+3}$ **10.** $y = -2 \pm e^{3t-1}$

Exercises

Most of the numerical answers in this section are approximations.

1. $y(t) = 100e^{1.5t}$ **3.** $y(t) = 50e^{(0.2 \ln 2)t}$

5. 8.06 yr doubling time; $13,197.10 in 30 yr

7. $600 initially; 13.2 yr doubling time

9. (a) 14.94 yr   **(b)** 14.62 yr
  (c) 14.68 yr   **(d)** 14.59 yr

11. (a) 2.8×10^{14} bacteria
  (b) The bacteria reproduce fast enough that even if many are destroyed there are enough left to make the person sick.

13. 0.585 days   **15.** $y \approx 2e^{0.4581t}$

17. $y = y_0 e^{-kt} = y_0 e^{-k(3/k)} = y_0 e^{-3} < 0.05 y_0$

19. (a) 17.53 minutes longer
  (b) 13.26 minutes

21. 6658 years

23. (a) 168.5 meters   **(b)** 41.13 seconds

25. 585.4 kg

27. (a) $p = 1013e^{-0.121h}$ **(b)** 2.383 millibars
  (c) 0.977 km

29. (a) $V = V_0 e^{-t/40}$ **(b)** 92.1 seconds

31. (b) $\displaystyle\lim_{t\to\infty} s(t) = \dfrac{v_0 m}{k}$

33. $s(t) = 1.32(1 - e^{-0.606t})$

35. (b) $T = T_s$ is a horizontal asymptote.

37. (a)

x	$\left(1 + \dfrac{1}{x}\right)^x$
10	2.5937
100	2.7048
1000	2.7169
10,000	2.7181
100,000	2.7183

$e \approx 2.7183$

  (c) As we compound more times the increment of time between compounding approaches 0. Continuous compounding is based on an instantaneous rate of change which is a limit of average rates as the increment in time approaches 0.

Section 6.5 (pp. 342–349)

Quick Review

1. All real numbers

2. $\displaystyle\lim_{x\to+\infty} f(x) = 50$; $\displaystyle\lim_{x\to-\infty} f(x) = 0$

3. $y = 0$ and $y = 50$ **4.** All real numbers

5.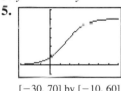

[−30, 70] hy [−10, 60]
no zeros

6.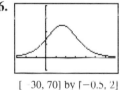

[−30, 70] by [−0.5, 2]
  (a) $(-\infty, \infty)$
  (b) None

7.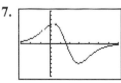

[−30, 70] by [−0.08, 0.08]
  (a) $(-\infty, 10 \ln 5) \approx (-\infty, 16.094)$
  (b) $(10 \ln 5, \infty) \approx (16.094, \infty)$

8. $(10 \ln 5, 25) \approx (16.094, 25)$

9. $A = 3$, $B = -2$ **10.** $A = -2$, $B = 4$

Exercises

1. (a) $\dfrac{dP}{dt} = 0.025P$ **(b)** $P(t) = 75,000e^{0.025t}$

1. continued

(c)

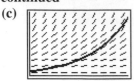

[0, 100] by [0, 1,000,000]

3. (a) $\dfrac{dP}{dt} = 0.00025P(200 - P)$

(b) $P(t) = \dfrac{200}{1 + 19e^{-0.05t}}$

(c)

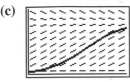

[0, 100] by [0, 250]

5. -30% **7.** $k = 0.04$; $M = 100$

9. (d)   **11.** (c)

13. (a) $k = 0.7$; $M = 1000$

(b) $P(0) \approx 8$; Initially there are 8 rabbits.

15. (a) 0.875%   **(b)** 275,980,017

17. (a) $P(t) = \dfrac{150}{1 + 24e^{-0.225t}}$

(b) About 17.21 weeks; 21.28 weeks

19. (a) $y = y_0 e^{kt}$ where $k = \ln 0.99^{(1/1000)} \approx -0.0001$

(b) $\approx 10{,}483$ yrs   **(c)** $\approx 81.8\%$

21. (a) $x = 11{,}000e^{0.1t} - 10{,}000$

(b) ≈ 23 yrs

23. (a) $y = \dfrac{18.70}{1 + 1.075e^{-0.0422x}}$

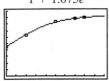

[0, 100] by [0, 20]

(b) 18.7 million

(c) $x \approx 1.7$ (year 1912);
population ≈ 9.35 million

25. Separate variables and rewrite $\dfrac{M}{P(M - P)}$ as
$\dfrac{1}{P} - \dfrac{1}{M - P}$ in order to integrate.

27. $y = e^{\sin x} - 1$ **29.** $y = \sqrt{x^2 + 4}$

31. (a) $\dfrac{dP}{dt}$ has the same sign as $(M - P)(P - m)$.

(b) $P(t) = \dfrac{1200Ae^{11kt/12} + 100}{1 + Ae^{11kt/12}}$

(c) $P(t) = \dfrac{300(8e^{11kt/12} + 3)}{9 + 2e^{11kt/12}}$

(d)

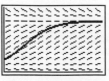

[0, 75] by [0, 1500]

(e) $P(t) = \dfrac{AMe^{(M-m)kt/M} + m}{1 + Ae^{(M-m)kt/M}}$ where

$A = \dfrac{P(0) - m}{M - P(0)}$

33. (a) $P(t) = \dfrac{P_0}{1 - kP_0 t}$

(b) Vertical asymptote at $t = \dfrac{1}{kP_0}$

Section 6.6 (pp. 350–356)

Quick Review

1. 9   **2.** $L(x) = 9x - 16$

3. 2   **4.** $L(x) = 2x - \dfrac{\pi}{2} + 1$

5. 0.4875   **6.** $y = 0.4875x + 0.9$

7. (a) 0.001762   **(b)** 0.061%

8. (a) 0.006976   **(b)** 0.236%

9. (a) 0.042361   **(b)** 1.351%

10. (a) 0.047321   **(b)** 1.783%

Exercises

1. $f'(x) = 1 - 2e^{-x} = x - f(x)$ and $f(0) = 1$.

3. $f'(x) = \dfrac{1}{5}(2e^{2x} - 2\cos x + \sin x) = 2f(x) + \sin x$
and $f(0) = 0$.

5. $y = 2e^x - 1$ **7.** $y = 2e^{x^2 + 2x}$

15. (a) $y = \dfrac{-1}{x^2 - 2x + 2}$, $y(3) = -0.2$

(b) -0.1851, error ≈ 0.0149

(c) -0.1929, error ≈ 0.0071

(d) -0.1965, error ≈ 0.0035

17. (a) -0.2024, error ≈ 0.0024

(b) -0.2005, error ≈ 0.0005

(c) -0.2001, error ≈ 0.0001

(d) As the step size decreases, the accuracy of
the method increases and the error decreases.

19.

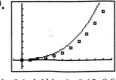

[−0.1, 1.1] by [−0.13, 0.88]

21.

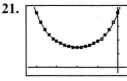

[−2.2, 0.2] by [−0.2, 2.2]

23.

x	y (Euler)	y (exact)	Error
0	1	1.0	0
-0.1	0.9000	0.9097	0.0097
-0.2	0.8200	0.8375	0.0175
-0.3	0.7580	0.7816	0.0236
-0.4	0.7122	0.7406	0.0284
-0.5	0.6810	0.7131	0.0321
-0.6	0.6629	0.6976	0.0347
-0.7	0.6566	0.6932	0.0366
-0.8	0.6609	0.6987	0.0377
-0.9	0.6748	0.7131	0.0383
-1.0	0.6974	0.7358	0.0384

25. (a)   **(b)**

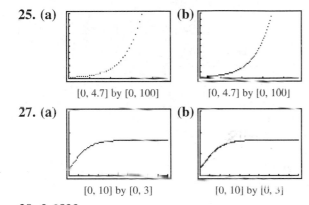

[0, 4.7] by [0, 100]   [0, 4.7] by [0, 100]

27. (a)   **(b)**

[0, 10] by [0, 3]   [0, 10] by [0, 3]

29. 2.6533, e

Chapter Review (pp. 358–361)

1. $\sqrt{3}$

2. 2

3. 8

4. 0

5. 2

6. $\dfrac{147}{8}$

7. $e - 1$

8. $\dfrac{2}{3}$

9. $-\ln|2 - \sin x| + C$

10. $\dfrac{1}{2}(3x + 4)^{2/3} + C$

11. $\dfrac{1}{2}\ln(t^2 + 5) + C$

12. $-\sec\dfrac{1}{\theta} + C$

13. $-\ln|\cos(\ln y)| + C$

14. $\ln|\sec(e^x) + \tan(e^x)| + C$

15. $\ln|\ln x| + C$

16. $\dfrac{-2}{\sqrt{t}} + C$

17. $x^3 \sin x + 3x^2 \cos x - 6x \sin x - 6 \cos x + C$

18. $\dfrac{x^5 \ln x}{5} - \dfrac{x^5}{25} + C$

19. $\left(\dfrac{3 \sin x}{10} - \dfrac{\cos x}{10}\right)e^{3x} + C$

20. $\left(-\dfrac{x^2}{3} - \dfrac{2x}{9} - \dfrac{2}{27}\right)e^{-3x} + C$

21. $y = \dfrac{x^3}{6} + \dfrac{x^2}{2} + x + 1$

22. $y = \dfrac{x^3}{3} + 2x - \dfrac{1}{x} - \dfrac{1}{3}$

23. $y = \ln(t + 4) + 2$

24. $y = -\dfrac{1}{2}\csc 2\theta + \dfrac{3}{2}$

25. $y = \dfrac{x^3}{3} + \ln x - x + \dfrac{2}{3}$

26. $r = \sin t - \dfrac{t^2}{2} - 2t - 1$

27. $y = 4e^x - 2$

28. $y = 2e^{x^2+x} - 1$

29. $-1 + \sqrt{x} + C$ or $\sqrt{x} + C$

30. $\dfrac{x^2}{2} + 1 - \sqrt{x} + C$ or $\dfrac{x^2}{2} - \sqrt{x} + C$

31. $-2\sqrt{x} - x + C$

32. $2 - 3x + C$ or $-3x + C$

33. (b)   **34.** (d)

35. iv, since the given graph looks like $y = x^2$, which satisfies $\dfrac{dy}{dx} = 2x$ and $y(1) = 1$.

36. Yes, $y = x$ is a solution.

37. (a) $v = 2t + 3t^2 + 4$ **(b)** 6 m

38.

[−10, 10] by [−10, 10]

39.

x	y
0	0
0.1	0.1000
0.2	0.2095
0.3	0.3285
0.4	0.4568
0.5	0.5946
0.6	0.7418
0.7	0.8986
0.8	1.0649
0.9	1.2411
1.0	1.4273
1.1	1.6241
1.2	1.8319
1.3	2.0513
1.4	2.2832
1.5	2.5285
1.6	2.7884
1.7	3.0643
1.8	3.3579
1.9	3.6709
2.0	4.0057

40.

x	y
-3	1
-2.9	0.6680
-2.8	0.2599
-2.7	-0.2294
-2.6	-0.8011
-2.5	-1.4509
-2.4	-2.1687
-2.3	-2.9374
-2.2	-3.7333
-2.1	-4.5268
-2.0	-5.2840
-1.9	-5.9686
-1.8	-6.5456
-1.7	-6.9831
-1.6	-7.2562
-1.5	-7.3488
-1.4	-7.2553
-1.3	-6.9813
-1.2	-6.5430
-1.1	-5.9655
-1.0	-5.2805

41. 0.9063

42. 4.4974

43. (a)

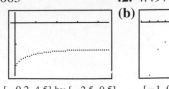

$[-0.2, 4.5]$ by $[-2.5, 0.5]$

(b)

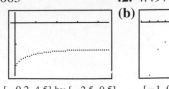

$[-1, 0.2]$ by $[-10, 2]$

44. (a)

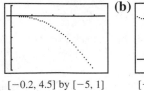

$[-0.2, 4.5]$ by $[-5, 1]$

(b)

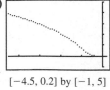

$[-4.5, 0.2]$ by $[-1, 5]$

45. (a) $k \approx 0.262059$ **(b)** About 3.81593 years

46. About 92 minutes   **47.** $-3°C$

48. About 41.2 years

49. About 18,935 years old

50. About 5.3%   **51.** About 59.8 ft

52. (a) $y = c + (y_0 - c)e^{-(kA/V)t}$

(b) c

53. (a) $k = 1$; carrying capacity $= 150$

(b) ≈ 2; Initially there were 2 infected students.

(c) About 6 days

54. Use the Fundamental Theorem of Calculus to obtain $y' = \sin(x^2) + 3x^2 + 1$. Then differentiate again and also verify the initial conditions.

55. $P = \dfrac{800}{1 + 15e^{-0.002t}}$

56. Method 1—Compare graph of $y_1 = x^2 \ln x$ with

$y_2 = \text{NDER}\left(\dfrac{x^3 \ln x}{3} - \dfrac{x^3}{9}\right)$.

Method 2—Compare graph of $y_1 = \text{NINT}(x^2 \ln x)$ with $y_2 = \dfrac{x^3 \ln x}{3} - \dfrac{x^3}{9}$.

57. (a) About 11.3 years **(b)** About 11 years

58. (a) $\dfrac{d}{dx}\displaystyle\int_0^x u(t)\, dt = u(x)$

$\dfrac{d}{dx}\displaystyle\int_3^x u(t)\, dt = u(x)$

(b) $C = \displaystyle\int_0^3 u(t)\, dt$

59. (a) $y = \dfrac{56.0716}{1 + 5.894e^{-0.0205x}}$

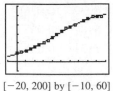

$[-20, 200]$ by $[-10, 60]$

(b) ≈ 56.0716 million **(c)** ≈ 1887, ≈ 28.0 million

60. (a) $T = 79.961(0.9273)^t$

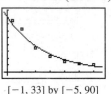

$[-1, 33]$ by $[-5, 90]$

(b) About 9.2 sec   **(c)** About 79.96°C

61. $s = 0.97(1 - e^{-0.8866t})$

Chapter 7

Section 7.1 (pp. 363–374)

Quick Review

1. Changes sign at $-\dfrac{\pi}{2}, 0, \dfrac{\pi}{2}$

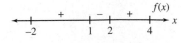

2. Changes sign at 1, 2

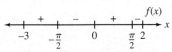

3. Always positive

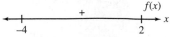

4. Changes sign at $-\dfrac{1}{2}$

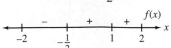

5. Changes sign at $\dfrac{\pi}{4}, \dfrac{3\pi}{4}, \dfrac{5\pi}{4}$

6. Always positive

7. Changes sign at 0

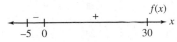

8. Changes sign at $-2, -\sqrt{2}, \sqrt{2}, 2$

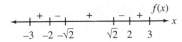

9. Changes sign at $0.9633 + k\pi$
$$2.1783 + k\pi$$
where k is an integer

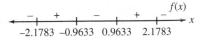

10. Changes sign at $\dfrac{1}{3\pi}, \dfrac{1}{2\pi}$

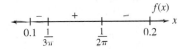

Exercises

1. (a) Right: $0 \le t < \dfrac{\pi}{2} \cup \dfrac{3\pi}{2} < t \le 2\pi$

Left: $\dfrac{\pi}{2} < t < \dfrac{3\pi}{2}$

Stopped: $t = \dfrac{\pi}{2}, \dfrac{3\pi}{2}$

(b) 0 **(c)** 20

3. (a) Right: $0 \le t < 5$

Left: $5 < t \le 10$

Stopped: $t = 5$

(b) 0 **(c)** 245

5. (a) Right: $0 < t < \dfrac{\pi}{2} \cup \dfrac{3\pi}{2} < t < 2\pi$

Left: $\dfrac{\pi}{2} < t < \pi, \pi < t < \dfrac{3\pi}{2}$

Stopped: $t = 0, \dfrac{\pi}{2}, \pi, \dfrac{3\pi}{2}, 2\pi$

(b) 0 **(c)** $\dfrac{20}{3}$

7. (a) Right: $0 \le t < \dfrac{\pi}{2} \cup \dfrac{3\pi}{2} < t \le 2\pi$

Left: $\dfrac{\pi}{2} < t < \dfrac{3\pi}{2}$

Stopped: $t = \dfrac{\pi}{2}, \dfrac{3\pi}{2}$

(b) 0 **(c)** $2e - \dfrac{2}{e} \approx 4.7$

9. (a) 63 mph **(b)** 344.52 feet

11. (a) -6 ft/sec **(b)** 5.625 sec

(c) 0 **(d)** 253.125 feet

13. 33 **15.** $t = a$

17. (a) 6 **(b)** 4 meters

19. (a) 5 **(b)** 7 meters

21. ≈ 332.965 billion barrels

23. (a) 2 miles **(b)** $2\pi r \Delta r$

(c) Population = Population density $\times$ Area

(d) Approximately 83,776

25. One possible answer:
Plot the speeds vs. time. Connect the points and find the area under the line graph. The definite integral also gives the area under the curve.

27. (a) 798.97 thousand

(b) The answer in (a) corresponds to the area of midpoint rectangles. Part of each rectangle is above the curve and part is below.

29. (a) 18 N **(b)** 81 N $\cdot$ cm

31. 0.04875

33. (a, b) Take $dm = \delta \, dA$ as m_k and letting $dA \to 0$, $k \to \infty$ in the center of mass equations.

35. $\bar{x} = \dfrac{4}{3}, \bar{y} = 0$

Section 7.2 (pp. 374–382)

Quick Review

1. 2

2. $\dfrac{1}{2}(e^2 - 1) \approx 3.195$

3. 2

4. 4

5. $\dfrac{9\pi}{2}$

6. $(6, 12); (-1, 5)$

7. $(0, 1)$

8. $(0, 0); (\pi, 0)$

9. $(-1, -1); (0, 0); (1, 1)$

10. $(-0.9286, -0.8008); (0, 0); (0.9286, 0.8008)$

Exercises

1. $\dfrac{\pi}{2}$ **3.** $\dfrac{1}{12}$ **5.** $\dfrac{128}{15}$

7. $\dfrac{5}{6}$ **9.** 16 **11.** $10\dfrac{2}{3}$

13. 4 **15.** $\dfrac{2}{3}a^3$ **17.** $21\dfrac{1}{3}$

19. $30\dfrac{3}{8}$ **21.** $\dfrac{8}{3}$ **23.** 8

25. $6\sqrt{3}$ **27.** $\dfrac{4 - \pi}{\pi} \approx 0.273$

29. $4 - \pi \approx 0.858$ **31.** $\dfrac{1}{2}$

33. $\sqrt{2} - 1 \approx 0.414$

35. (a) $(-\sqrt{c}, c)$; $(\sqrt{c}, c)$

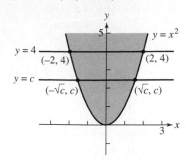

(b) $\int_0^c \sqrt{y}\, dy = \int_c^4 \sqrt{y}\, dy \Rightarrow c = 2^{4/3}$

(c) $\int_0^{\sqrt{c}} (c - x^2)\, dx$
$= (4 - c)\sqrt{c} + \int_{\sqrt{c}}^2 (4 - x^2)\, dx \Rightarrow c = 2^{4/3}$

37. $\dfrac{3}{4}$

39. Neither; both are zero

41. $\ln 4 - \dfrac{1}{2} \approx 0.886$

43. $k = 1.8269$

45. Since $f(x) - g(x)$ is the same for each region where $f(x)$ and $g(x)$ represent the upper and lower edges, area $= \int_a^b [f(x) - g(x)]\, dx$ will be the same for each.

Section 7.3 (pp. 383–394)

Quick Review

1. x^2

2. $\dfrac{x^2}{2}$

3. $\dfrac{\pi x^2}{2}$

4. $\dfrac{\pi x^2}{8}$

5. $\dfrac{\sqrt{3}}{4}x^2$

6. $\dfrac{x^2}{2}$

7. $\dfrac{x^2}{4}$

8. $\dfrac{\sqrt{15}}{4}x^2$

9. $6x^2$

10. $\dfrac{3\sqrt{3}}{2}x^2$

Exercises

1. (a) $\pi(1 - x^2)$ **(b)** $4(1 - x^2)$
  (c) $2(1 - x^2)$ **(d)** $\sqrt{3}(1 - x^2)$

3. 16

5. $\dfrac{16}{3}$

7. (a) $2\sqrt{3}$ **(b)** 8

9. 8π

11. (a) s^2h **(b)** s^2h

13. $\dfrac{2}{3}\pi$

15. $4 - \pi$

17. $\dfrac{32\pi}{5}$

19. 36π

21. $\dfrac{2}{3}\pi$

23. $\dfrac{117\pi}{5}$

25. $\pi^2 - 2\pi$

27. 2.301

29. 2π

31. $\dfrac{4}{3}\pi$

33. 8π

35. (a) 8π **(b)** $\dfrac{32\pi}{5}$
  (c) $\dfrac{8\pi}{3}$ **(d)** $\dfrac{224\pi}{15}$

37. (a) $\dfrac{16\pi}{15}$ **(b)** $\dfrac{56\pi}{15}$
  (c) $\dfrac{64\pi}{15}$

39. 8π **41.** $\dfrac{128\pi}{5}$

43. (a) $\dfrac{6\pi}{5}$ **(b)** $\dfrac{4\pi}{5}$
  (c) 2π **(d)** 2π

45. (a) $\dfrac{512\pi}{21}$ **(b)** $\dfrac{832\pi}{21}$

47. $\dfrac{11\pi}{48}$

49. (a) $\dfrac{36\pi}{5}$ cm^3 **(b)** 61.2π g

51. (a) 2.3, 1.6, 1.5, 2.1, 3.2, 4.8, 7.0, 9.3, 10.7, 9.3, 6.4, 3.2
  (b) $\dfrac{1}{4\pi} \displaystyle\int_0^6 C(y)^2\, dy$ **(c)** ≈ 34.7 in^3

53. (a) $\dfrac{32\pi}{3}$
  (b) The answer is independent of r.

55. 5   **57.** ≈ 13.614 **59.** ≈ 15.707

61. ≈ 53.226 **63.** ≈ 6.283

65. Hemisphere cross sectional area:
$$\pi(\sqrt{R^2 - h^2})^2 = A_1$$
Right circular cylinder with cone removed cross sectional area: $\pi R^2 - \pi h^2 = A_2$
Since $A_1 = A_2$, the two volumes are equal by Cavalieri's theorem. Thus,
volume of hemisphere
$$= \text{volume of cylinder} - \text{volume of cone}$$
$$= \pi R^3 - \frac{1}{3}\pi R^3 = \frac{2}{3}\pi R^3.$$

67. (a) $\dfrac{\pi h^2(3a - h)}{3}$ **(b)** $\dfrac{1}{120\pi}$ m/sec

Section 7.4 (pp. 395–401)

Quick Review

1. $x + 1$

2. $\dfrac{2 - x}{2}$

3. $\sec x$

4. $\dfrac{x^2 + 4}{4x}$

5. $\sqrt{2} \cos x$

6. 4

7. 0

8. -3

9. 2

10. $k\pi$, k any integer

Exercises

1. (a) $\displaystyle\int_{-1}^{2} \sqrt{1 + 4x^2}\, dx$

(b)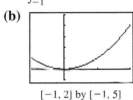

$[-1, 2]$ by $[-1, 5]$

(c) ≈ 6.126

3. (a) $\displaystyle\int_{0}^{\pi} \sqrt{1 + \cos^2 y}\, dy$

(b)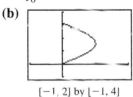

$[-1, 2]$ by $[-1, 4]$

(c) ≈ 3.820

5. (a) $\displaystyle\int_{-1}^{7} \sqrt{1 + \dfrac{1}{2x + 2}}\, dx$

(b)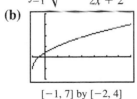

$[-1, 7]$ by $[-2, 4]$

(c) ≈ 9.294

7. (a) $\displaystyle\int_{0}^{\pi/6} \sqrt{1 + \tan^2 x}\, dx$

(b)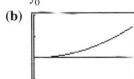

$\left[0, \dfrac{\pi}{6}\right]$ by $[-0.1, 0.2]$

(c) ≈ 0.549

9. (a) $\displaystyle\int_{-\pi/3}^{\pi/3} \sqrt{1 + \sec^2 x \tan^2 x}\, dx$

(b)

$\left[-\dfrac{\pi}{3}, \dfrac{\pi}{3}\right]$ by $[-1, 3]$

(c) ≈ 3.139

11. 12

13. $\dfrac{53}{6}$

15. $\dfrac{17}{12}$

17. 2

19. (a) $y = \sqrt{x}$

(b) Only one. We know the derivative of the function and the value of the function at one value of x.

21. 1

23. ≈ 21.07 inches

25. $(-19.909, 8.41)$

27. 2.1089

29. ≈ 1.623

31. Because the limit of the sum $\sum \Delta x_k$ as the norm of the partition goes to zero will always be the length $(b - a)$ of the interval (a, b).

33. (a) The fin is the hypotenuse of a right triangle with leg lengths Δx_k and
$$\left.\frac{df}{dx}\right|_{x = x_{k-1}} \Delta x_k = f'(x_{k-1})\, \Delta x_k.$$

(b) $\displaystyle\lim_{n \to \infty} \sum_{k=1}^{n} \sqrt{(\Delta x_k)^2 + (f'(x_{k-1})\, \Delta x_k)^2}$

$$= \lim_{n \to \infty} \sum_{k=1}^{n} \Delta x_k \sqrt{1 + (f'(x_{k-1}))^2}$$

$$\int_{a}^{b} \sqrt{1 + (f'(x))^2}\, dx$$

Section 7.5 (pp. 401–411)

Quick Review

1. $1 - \dfrac{1}{e}$

2. $e - 1$

3. $\dfrac{\sqrt{2}}{2}$

4. 15

5. $\dfrac{1}{3} \ln\left(\dfrac{9}{2}\right)$

6. $\displaystyle\int_{0}^{7} 2\pi(x + 2) \sin x\, dx$

7. $\displaystyle\int_{0}^{7} (1 - x^2)(2\pi x)\, dx$

8. $\displaystyle\int_{0}^{7} \pi \cos^2 x\, dx$

9. $\displaystyle\int_{0}^{7} \pi\left(\dfrac{y}{2}\right)^2 (10 - y)\, dy$

10. $\displaystyle\int_{0}^{7} \dfrac{\sqrt{3}}{4} \sin^2 x\, dx$

Exercises

1. ≈4.4670 J   **3.** 9 J
5. 4900 J   **7.** 1944 ft-lb
9. (a) 7238 lb/in.
   (b) ≈905 in.-lb, ≈2714 in.-lb
11. 780 J   **13.** 1123.2 lb
15. 3705 lb
17. (a) 1,497,600 ft-lb   **(b)** ≈100 min
   (d) 1,494,240 ft-lb, ≈100 min
     1,500,000 ft-lb, 100 min
19. Through valve: 84,687.3 ft-lb
   Over the rim: 98,801.8 ft-lb
   Through a hose attached to a valve in the bottom
   is faster, because it takes more time to do more
   work.
21. 53,482.5 ft-lb   **23.** ≈967,611 ft-lb, yes
25. (a) ≈209.73 lb
   (b) ≈838.93 lb; the fluid force doubles
27. (a) 0.5 (50%)   **(b)** ≈0.24 (24%)
   (c) ≈0.0036 (0.36%)
   (d) 0 if we assume a continuous distribution;
     ≈0.071 between 59.5 in. and 60.5 in.
29. Integration is a good approximation to the area.
31. 5.1446×10^{10} J

33. $F = m\dfrac{dv}{dt} = mv\dfrac{dv}{dx}$, so $w = \displaystyle\int_{x_1}^{x_2} F(x)\, dx$

$$= \int_{x_1}^{x_2} mv\frac{dv}{dx}\, dx = \int_{v_1}^{v_2} mv\, dv = \frac{1}{2}mv_2{}^2 - \frac{1}{2}mv_1{}^2$$

35. ≈85.1 ft-lb   **37.** 64.6 ft-lb
39. 110.6 ft-lb

Chapter Review (pp. 413–415)

1. 10.417 ft   **2.** 31.361 gal   **3.** 1464
4. 14 g   **5.** 14,400   **6.** 1
7. $\dfrac{9}{2}$ **8.** $\dfrac{1}{6}$ **9.** 18
10. 30.375   **11.** 0.0155   **12.** 4
13. 8.9023   **14.** 2.1043

15. $2\sqrt{3} - \dfrac{2}{3}\pi \approx 1.370$ **16.** $2\sqrt{3} + \dfrac{4}{3}\pi \approx 7.653$
17. 1.2956   **18.** 5.7312
19. 4π **20.** 2π
21. (a) $\dfrac{32\pi}{3}$ **(b)** $\dfrac{128\pi}{15}$
   (c) $\dfrac{64\pi}{5}$ **(d)** $\dfrac{32\pi}{3}$
22. (a) 4π **(b)** πk^2 **(c)** $\dfrac{1}{\pi}$
23. 276 in³   **24.** $\dfrac{\pi^2}{4}$
25. $\pi(2 - \ln 3)$ **26.** $\dfrac{28\pi}{3}$ ft³
27. 19.4942   **28.** 5.2454
29. 2.296 sec   **30. (a)** is true
31. 39
32. (a) 4000 J   **(b)** 640 J
   (c) 4640 J
33. 22,800,000 ft-lb   **34.** 12 J, 213.3 J
35. No, the work going uphill is positive, but the
   work going downhill is negative.
36. 113.097 in.-lb   **37.** 426.67 lbs
38. 6.6385 lb,   5.7726 lb,   9.4835 lb
     base   front   sides
                      back
39. 14.4   **40.** 0.2051 (20.5%)
41. Answer will vary.
42. (a) 0.6827 (68.27%)   **(b)** 0.9545 (95.45%)
   (c) 0.9973 (99.73%)
43. The probability that the variable has some value
   in the range of all possible values is 1.
44. π **45.** 3π
46. $2\pi^2$ **47.** $\dfrac{16\pi}{3}$
48. 9.7717
49. (a) $y = 5 - \dfrac{5}{4}x^2$ **(b)** 335.1032 in³
50. $f(x) = \dfrac{x^2 - 2\ln x + 3}{4}$ **51.** ≈3.84
52. ≈5.02

Chapter 8

Section 8.1 (pp. 417–425)

Quick Review

1. 1.105   **2.** 2.718   **3.** 1
4. ∞   **5.** 2   **6.** 2
7. 3   **8.** 1
9. $y = \dfrac{\sin h}{h}$ **10.** $y = (1 + h)^{1/h}$

Exercises

1. $\dfrac{1}{4}$ **3.** 1
5. $\dfrac{3}{11}$ **7.** e^2

9. (a)

x	10	10^2	10^3	10^4	10^5
$f(x)$	1.1513	0.2303	0.0354	0.0046	0.00058

Estimated limit $= 0$

(b) Note $\ln x^5 = 5 \ln x$.

$$\lim_{x \to \infty} \frac{5 \ln x}{x} = \lim_{x \to \infty} \frac{\frac{5}{x}}{1} = \frac{0}{1} = 0$$

11. $\dfrac{3}{4}$ **13.** 1   **15.** 0

17. -1 **19.** $\ln 2$ **21.** 1

23. 0   **25.** 1   **27.** 0

29. e^2 **31.** 0   **33.** 1

35. 1   **37.** e **39.** 1

41. e^{-1}

43. (a) L'Hôpital's rule does not help because applying l'Hôpital's rule to this quotient essentially "inverts" the problem by interchanging the numerator and denominator. It is still essentially the same problem and one is no closer to a solution. Applying l'Hôpital's rule a second time returns to the original problem.

(b, c) 3

45. Possible answers:

(a) $f(x) = 7(x - 3)$, $g(x) = x - 3$

(b) $f(x) = (x - 3)^2$, $g(x) = x - 3$

(c) $f(x) = x - 3$, $g(x) = (x - 3)^3$

47. $c = \dfrac{27}{10}$, because this is the limit of $f(x)$ as x approaches 0.

49. (a) $\ln\left(1 + \dfrac{r}{k}\right)^k = k \ln\left(1 + \dfrac{r}{k}\right)$. And, as $k \to \infty$,

$$\lim_{k \to \infty} k \ln\left(1 + \frac{r}{k}\right) = \lim_{k \to \infty} \frac{\ln\left(1 + \frac{r}{k}\right)}{\frac{1}{k}}$$

$$= \lim_{k \to \infty} \frac{\frac{-r}{k^2} / \left(1 + \frac{r}{k}\right)}{\frac{-1}{k^2}}.$$

$$= \lim_{k \to \infty} \frac{r}{1 + \frac{r}{k}} = r.$$

Therefore, $\lim_{k \to \infty}\left(1 + \dfrac{r}{k}\right)^k = e^r$.

Hence, $\lim_{k \to \infty} A_0\left(1 + \dfrac{r}{k}\right)^{kt} = A_0 e^{rt}$.

(b) Part (a) shows that as the number of compoundings per year increases toward infinity, the limit of interest compounded k times per year is interest compounded continuously.

51. (a) 1   **(b)** $\dfrac{\pi}{2}$ **(c)** π

53. (a) $(-\infty, -1) \cup (0, \infty)$

(b) ∞ **(c)** e

55. (a) $c = \dfrac{1}{3}$ **(b)** $c = \dfrac{\pi}{4}$

Section 8.2 (pp. 425–433)

Quick Review

1. 0   **2.** ∞ **3.** ∞

4. 0   **5.** $-3x^4$ **6.** $2x^2$

7. $\lim_{x \to \infty} \dfrac{f(x)}{g(x)} = \lim_{x \to \infty}\left(1 + \dfrac{\ln x}{x}\right) = 1 + 0 = 1$

8. $\lim_{x \to \infty} \dfrac{f(x)}{g(x)} = \lim_{x \to \infty} \sqrt{1 + \dfrac{5}{4x}} = 1$

9. (a) Local minimum at $(0, 1)$
Local maximum at $\approx(2, 1.541)$

(b) $[0, 2]$ **(c)** $(-\infty, 0]$ and $[2, \infty)$

10. f doesn't have an absolute maximum value. The values are always less than 2 and the values get arbitrarily close to 2 near $x = 0$, but the function is undefined at $x = 0$.

Exercises

1. Slower   **3.** Faster   **5.** Same rate

7. Slower   **9.** Slower   **11.** Same rate

13. Slower   **15.** Slower   **17.** Same rate

19. Slower   **21.** Faster

23. $e^{x/2}$, e^x, $(\ln x)^x$, x^x

25. $\lim_{x \to \infty} \dfrac{f_2(x)}{f_1(x)} = \sqrt{10}$ and $\lim_{x \to \infty} \dfrac{f_3(x)}{f_1(x)} = 1$, so f_2 and f_3 also grow at the same rate.

27. $\lim_{x \to \infty} \dfrac{f_2(x)}{f_1(x)} = 1$ and $\lim_{x \to \infty} \dfrac{f_3(x)}{f_1(x)} = 1$, so f_2 and f_3 also grow at the same rate.

29. (a) False   **(b)** False   **(c)** True
(d) True   **(e)** True   **(f)** True
(g) False   **(h)** True

31. $g = o(f)$

33. f and g grow at the same rate.

35. (a) The nth derivative of x^n is $n!$, which is a constant. Therefore n applications of l'Hôpital's rule give

$$\lim_{x \to \infty} \frac{e^x}{x^n} = \cdots = \lim_{x \to \infty} \frac{e^x}{n!} = \infty.$$

(b) In this case, n applications of l'Hôpital's rule give $\lim_{x \to \infty} \dfrac{a^x}{x^n} = \cdots = \lim_{x \to \infty} \dfrac{a^x (\ln a)^n}{n!} = \infty$

37. (a) $\lim_{x \to \infty} \dfrac{\ln x}{x^{1/n}} = \lim_{x \to \infty} \dfrac{\frac{1}{x}}{\frac{x^{(1/n)-1}}{n}} = \lim_{x \to \infty} \dfrac{n}{x^{1/n}} = 0$

(b) $\lim_{x \to \infty} \dfrac{\ln x}{x^a} = \lim_{x \to \infty} \dfrac{\frac{1}{x}}{ax^{a-1}} = \lim_{x \to \infty} \dfrac{1}{ax^a} = 0$

39. The one which is $O(n \log_2 n)$ is likely the most efficient, because of the three given functions, it grows the most slowly as $n \to \infty$.

41. This is the case because if $\lim_{x \to \infty} \dfrac{f(x)}{g(x)} = L$ where L is a nonzero finite real number, then for sufficiently large x, it must be the case that $\dfrac{f(x)}{g(x)} < L + 1 \le M$, for some integer M. Similarly for $g = O(f)$

43. (a) x^5 grows faster than x^2.
(b) They grow at the same rate.
(c) $m > n$ **(d)** $m = n$
(e) $m > n$ (or, degree of $g >$ degree of f)
(f) $m = n$ (or, degree of $g =$ degree of f)

45. (a) and (b) both follow from the fact that if f and g are negative, then

$$\lim_{x \to \infty} \frac{|f(x)|}{|g(x)|} = \lim_{x \to \infty} \frac{-f(x)}{-g(x)} = \lim_{x \to \infty} \frac{f(x)}{g(x)}$$

Section 8.3 (pp. 433–444)

Quick Review

1. $\ln 2$ **2.** 0
3. $\dfrac{1}{2} \tan^{-1} \dfrac{x}{2} + C$ **4.** $-\dfrac{1}{3} x^{-3} + C$
5. $(-3, 3)$ **6.** $(1, \infty)$
7. Because $-1 \le \cos x \le 1$ for all x
8. Because $\sqrt{x^2 - 1} < \sqrt{x^2} = x$ for $x > 1$
9. $\lim_{x \to \infty} \dfrac{4e^x - 5}{3e^x + 7} = \dfrac{4}{3}$ **10.** $\lim_{x \to \infty} \dfrac{\sqrt{2x - 1}}{\sqrt{x + 3}} = \sqrt{2}$

Exercises

1. (a) Because of an infinite limit of integration
(b) Converges   **(c)** $\dfrac{\pi}{2}$

3. (a) Because the integrand has an infinite discontinuity at $x = 0$
(b) Converges   **(c)** $-\dfrac{9}{2}$

5. (a) Because the integrand has an infinite discontinuity at $x = 0$
(b) Diverges   **(c)** No value

7. 1000 **9.** 4 **11.** $\dfrac{\pi}{2}$

13. $\ln 3$ **15.** $\sqrt{3}$ **17.** π

19. $\dfrac{\pi}{3}$ **21.** $2\pi^2$ **23.** -1

25. 2 **27.** Diverges   **29.** Converges
31. Converges   **33.** Converges   **35.** Diverges
37. Converges   **39.** Converges   **41.** Diverges
43. Converges   **45.** Converges   **47.** 1

49. (a) The integral in Example 1 gives the area of the region.
(b) ∞ **(c)** π
(d) Gabriel's horn has finite volume so it could only hold a finite amount of paint, but it has infinite surface area so it would presumably require an infinite amount of paint to cover itself.

51.(a) For $x \ge 6$, $x^2 \ge 6x$, and therefore, $e^{-x^2} \le e^{-6x}$. The inequality for the integrals follows. The value of the second integral is $\dfrac{e^{-36}}{6}$, which is less than 4×10^{-17}.
(b) The error in the estimate is the integral over the interval $[6, \infty)$, and we have shown that it is bounded by 4×10^{-17} in part (a).
(c) 0.13940279264 (This agrees with Figure 8.16.)
(d) $\displaystyle\int_0^\infty e^{-x^2}\, dx = \int_0^3 e^{-x^2}\, dx + \int_3^\infty e^{-x^2}\, dx$. The error in the approximation is
$$\int_3^\infty e^{-x^2}\, dx \le \int_3^\infty e^{-3x}\, dx \le 0.000042.$$

53. (a) It is divergent because as $x \to \infty$,
$$\lim_{x \to \infty} \ln(x^2 + 1) = \infty.$$
(b) *Both* the integral over $[0, \infty)$ and the integral over $(-\infty, 0]$ must converge in order for the integral over $(-\infty, \infty)$ to converge.

(c) Since this is an odd function, the integral over any interval of the form $[-b, b]$ equals 0. Therefore the limit as $b \to \infty$ is 0.

(d) Because the determination of convergence is not made using the method in part (c). In order for the integral to converge, there must be finite areas in both directions (toward ∞ and toward $-\infty$). In this case, there are infinite areas in both directions, but when one computes the integral over an interval $[-b, b]$, there is cancellation which gives 0 as the result.

55. From the properties of integrals, for any $b > a$,

$$\int_a^b f(x) \le \int_a^b g(x)\, dx.$$

If the infinite integral of g converges, then taking the limit in the above inequality as $b \to \infty$ shows that the infinite integral of f is bounded above by the infinite integral of g. Therefore, the infinite integral of f must be finite and it converges. If the infinite integral of f diverges, it must grow to infinity. So taking the limit in the above inequality as $b \to \infty$ shows that the infinite integral of g must also diverge to infinity.

57. (a) Although the values oscillate a bit, they appear to be approaching a limit of approximately 1.57.

(b) Yes, it converges.

Section 8.4 (pp. 444–453)

Quick Review

1. $A = -2, B = 1$
2. $A = -1, B = 2, C = 3$
3. $2x + 1 + \dfrac{x - 3}{x^2 - 3x - 4}$
4. $2 + \dfrac{3x - 4}{x^2 + 4x + 5}$ 5. $(x - 3)(x^2 + 1)$
6. $(y - 2)(y + 2)(y - 1)(y + 1)$
7. $-\dfrac{x + 13}{x^2 + x - 6}$ 8. $\dfrac{-x^2 + 12x - 15}{(x + 5)(x^2 - 4x + 5)}$
9. $-\dfrac{2t^3 + 5t^2 + 5t + 9}{(t^2 + 2)(t^2 + 1)}$ 10. $\dfrac{2x^2 - 7x + 6}{(x - 1)^5}$

Exercises

1. $\dfrac{2}{x - 1} + \dfrac{3}{x - 2}$ 3. $\dfrac{2}{t - 1} - \dfrac{2}{t} - \dfrac{1}{t^2}$
5. $1 - \dfrac{12}{x - 2} + \dfrac{17}{x - 3}$
7. $\dfrac{1}{2} \ln |x + 1| - \dfrac{1}{2} \ln |x - 1| + C$

9. $\dfrac{1}{4} \ln |y + 1| + \dfrac{3}{4} \ln |y - 3| + C$

11. $\dfrac{1}{6} \ln |t + 2| + \dfrac{1}{3} \ln |t - 1| - \dfrac{1}{2} \ln |t| + C$

13. $\dfrac{s^2}{2} - 2 \ln (s^2 + 4) + C$

15. $5x - \dfrac{5}{2} \ln (x^2 + x + 1) - \dfrac{5}{\sqrt{3}} \tan^{-1}\left(\dfrac{2x + 1}{\sqrt{3}}\right) + C$

17. $\dfrac{1}{4} \ln |x + 1| - \dfrac{\frac{1}{4}}{x + 1} - \dfrac{1}{4} \ln |x - 1| - \dfrac{\frac{1}{4}}{x - 1} + C$

19. $2 \tan^{-1} (r - 1) + C$
21. $\ln |x^2 + x + 1| - \ln |x - 1| + C$
23. $\dfrac{1}{x^2 + 4} + \dfrac{3}{2} \tan^{-1} \dfrac{x}{2} + C$
25. $1 - \ln 2$
27. $\ln |y - 1| - \ln |y| = e^x - 1 - \ln 2$
29. $y = \ln |x - 2| - \ln |x - 1| + \ln 2$

31. (a) Use $5 + 4x - x^2 = 9 - (x - 2)^2$, substitute $u = x - 2$, then use Formula 18 with $x = u$ and $a = 3$.

The integral is $\dfrac{1}{6} \ln \left| \dfrac{x + 1}{x - 5} \right| + C.$

(b) Rewrite as $\dfrac{1}{2a} \ln |x + a| - \ln |x - a|$. Differentiating gives $\dfrac{1}{2a}\left(\dfrac{1}{x + a} - \dfrac{1}{x - a}\right)$ which equals $\dfrac{1}{a^2 - x^2}.$

33. $6\pi \ln 5$
35. $\ln \left| \sqrt{9 + y^2} + y \right| + C$

37. $\dfrac{1}{2} \ln \left| 2x + \sqrt{4x^2 - 49} \right| + C$

39. $-\dfrac{x^2\sqrt{1 - x^2}}{3} - \dfrac{2}{3}\sqrt{1 - x^2} + C$

41. $\sqrt{16 - z^2} - 4 \ln \left| \dfrac{4}{z} + \dfrac{\sqrt{16 - z^2}}{z} \right| + C$

43. $y = \ln \left| \dfrac{x}{3} + \dfrac{\sqrt{x^2 - 9}}{3} \right|$ 45. ≈ 2.356

47. (a) $x(t) = \dfrac{1000e^{4t}}{e^{4t} + 499}$ or

$x(t) = \dfrac{1000}{1 + 499e^{-4t}}$

(b) After $t = \dfrac{\ln 499}{4} \approx 1.553$ days

(c) Since $\dfrac{dx}{dt} = kx(1000 - x)$, $\dfrac{dx}{dt}$ will have its maximum value where $x(1000 - x)$ is greatest, which is at $x = 500$.

49. (a) This can be seen geometrically in the figure.

(b) Using part (a), substitute $z = \dfrac{\sin x}{1 + \cos x}$ and then obtain a trigonometric identity.

Or, use the trigonometric identity
$$\frac{1 - \tan^2 \theta}{1 + \tan^2 \theta} = \cos 2\theta \text{ with } \theta = \frac{x}{2}.$$

(c) Using part (a), substitute $z = \dfrac{\sin x}{1 + \cos x}$ and then obtain a trigonometric identity.

Or, use the trigonometric identity
$$\frac{2 \tan \theta}{1 + \tan^2 \theta} = \sin 2\theta \text{ with } \theta = \frac{x}{2}.$$

(d) $dz = \left(\sec^2 \dfrac{x}{2} \right) \dfrac{1}{2} \, dx$

$\qquad = \left(1 + \tan^2 \dfrac{x}{2} \right) \dfrac{1}{2} \, dx$

$\qquad = \left(1 + z^2 \right) \dfrac{1}{2} \, dx$, then solve for dx.

51. $-\dfrac{1}{\tan \frac{x}{2}} + C$ **53.** $\ln \left| 1 + \tan \dfrac{t}{2} \right| + C$

Chapter Review (pp. 454–455)

Exercises

1. The limit doesn't exist.   **2.** $\dfrac{3}{5}$

3. 2   **4.** $\dfrac{1}{e}$ **5.** 1

6. e^3 **7.** 0   **8.** -1

9. $-\dfrac{1}{2}$ **10.** 1   **11.** 1

12. ∞ **13.** ∞ **14.** 0

15. Same rate, because $\lim\limits_{x \to \infty} \dfrac{f(x)}{g(x)} = \dfrac{1}{5}$

16. Same rate, because $\lim\limits_{x \to \infty} \dfrac{f(x)}{g(x)} = \dfrac{\ln 3}{\ln 2}$

17. Same rate, because $\lim\limits_{x \to \infty} \dfrac{f(x)}{g(x)} = 1$

18. Faster, because $\lim\limits_{x \to \infty} \dfrac{f(x)}{g(x)} = \infty$

19. Faster, because $\lim\limits_{x \to \infty} \dfrac{f(x)}{g(x)} = \infty$

20. Same rate, because $\lim\limits_{x \to \infty} \dfrac{f(x)}{g(x)} = 1$

21. Slower, because $\lim\limits_{x \to \infty} \dfrac{f(x)}{g(x)} = 0$

22. Slower, because $\lim\limits_{x \to \infty} \dfrac{f(x)}{g(x)} = 0$

23. Same rate, because $\lim\limits_{x \to \infty} \dfrac{f(x)}{g(x)} = \dfrac{1}{2}$

24. Slower, because $\lim\limits_{x \to \infty} \dfrac{f(x)}{g(x)} = 0$

25. Same rate, because $\lim\limits_{x \to \infty} \dfrac{f(x)}{g(x)} = 1$

26. Faster, because $\lim\limits_{x \to \infty} \dfrac{f(x)}{g(x)} = \infty$

27. (a) $\lim\limits_{x \to 0} f(x) = \ln 2$ **(b)** Define $f(0) = \ln 2$

28. (a) $\lim\limits_{x \to 0^+} f(x) = 0$ **(b)** Define $f(0) = 0$

29. True, $\lim\limits_{x \to \infty} \dfrac{\frac{1}{x^2} + \frac{1}{x^4}}{\frac{1}{x^2}} = 1$

30. False, $\lim\limits_{x \to \infty} \dfrac{\frac{1}{x^2} + \frac{1}{x^4}}{\frac{1}{x^4}} = \infty$

31. False, $\lim\limits_{x \to \infty} \dfrac{x}{x + \ln x} = 1$

32. True, $\lim\limits_{x \to \infty} \dfrac{\ln (\ln x)}{\ln x} = 0$

33. True, $\lim\limits_{x \to \infty} \dfrac{\tan^{-1} x}{1} = \dfrac{\pi}{2}$

34. True, $\lim\limits_{x \to \infty} \dfrac{\frac{1}{x^4}}{\frac{1}{x^2} + \frac{1}{x^4}} = 0$

35. True, $\lim\limits_{x \to \infty} \dfrac{\frac{1}{x^4}}{\frac{1}{x^2} + \frac{1}{x^4}} = 0$

36. True, $\lim\limits_{x \to \infty} \dfrac{\ln x}{x + 1} = 0$ **37.** True, $\lim\limits_{x \to \infty} \dfrac{\ln 2x}{\ln x} = 1$

38. True, $\lim\limits_{x \to \infty} \dfrac{\sec^{-1} x}{1} = \dfrac{\pi}{2}$ **39.** $\dfrac{\pi}{2}$

40. -1 **41.** 6   **42.** 0

43. $\ln 3$ **44.** $\ln \dfrac{3}{4} + 1$ **45.** 2

46. $-\dfrac{1}{9}$ **47.** Diverges   **48.** π

49. Diverges, by the limit comparison test, comparing with $\dfrac{1}{\theta}$ or directly from the antiderivative

50. Converges; directly from the antiderivative, value $\dfrac{1}{2}$

51. Diverges; by the direct comparison test with $\dfrac{1}{z}$ or directly from the antiderivative.

52. Converges; by the direct comparison test with e^{-t}

53. Converges; by the direct comparison test with e^{-x} on $[0, \infty)$ and with e^x on $(-\infty, 0]$ or directly from the antiderivative, value $\dfrac{\pi}{2}$.

54. Diverges; the problem is near $x = 0$. Compare with $\dfrac{1}{4x^2}$ there.

For $0 < x \leq 1$, $1 + e^x \leq 1 + e < 4$ and $\dfrac{1}{x^2(1 + e^x)} \geq \dfrac{1}{4x^2}$.

55. $9 \ln |x - 4| - 7 \ln |x - 3| + C$

56. $\ln |x| - \ln |x + 2| + \dfrac{2}{x} - \dfrac{2}{x^2} + C$

57. $4 \ln |t| - \dfrac{1}{2} \ln (t^2 + 1) + 4 \tan^{-1} t + C$

58. $\dfrac{1}{2} \tan^{-1} t - \dfrac{1}{2\sqrt{3}} \tan^{-1} \dfrac{t}{\sqrt{3}} + C$

59. $x + \ln |x - 1| - \ln |x| + C$

60. $\dfrac{x^2}{2} + \dfrac{3}{2} \ln |x + 1| - \dfrac{9}{2} \ln |x + 3| + C$

61. $y = \dfrac{500}{1 + 24e^{-x}}$

62. $y = \tan \left(\ln |x + 1| + \tan^{-1} \dfrac{\pi}{4} \right)$

63. $\ln |\sqrt{1 + 9y^2} + 3y| + C$

64. $\dfrac{1}{6} \sin^{-1} 3t + \dfrac{1}{2} t \sqrt{1 - 9t^2} + C$

65. $\ln |5x + \sqrt{25x^2 - 9}| + C$

66. $-\dfrac{4x}{\sqrt{1 - x^2}} - 4 \sin^{-1} x + C$

67. 2π **68.** 1

69. (a) $x = a - \dfrac{1}{kt + \dfrac{1}{a}}$ **(b)** $x = \dfrac{ab(e^{akt} - e^{bkt})}{ae^{akt} - be^{bkt}}$

Chapter 9

Section 9.1 (pp. 457–468)

Quick Review

1. $\dfrac{4}{3}, 1, \dfrac{4}{5}, \dfrac{2}{3}, \dfrac{1}{8}$ **2.** $-1, \dfrac{1}{2}, -\dfrac{1}{3}, \dfrac{1}{4}, \dfrac{1}{30}$

3. (a) 3 **(b)** $39{,}366$

    (c) $a_n = 2(3^{n-1})$

4. (a) $-\dfrac{1}{2}$ **(b)** $-\dfrac{1}{64}$

    (c) $a_n = 8\left(-\dfrac{1}{2}\right)^{n-1} = 8(-0.5)^{n-1}$

5. (a)

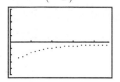

    $[0, 25]$ by $[-0.5, 0.5]$

    (b) 0

6. (a)

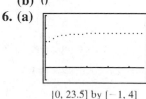

    $[0, 23.5]$ by $[-1, 4]$

    (b) e

7. (a)

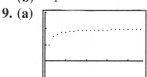

    $[0, 23.5]$ by $[-2, 2]$

    (b) The limit does not exist.

8. (a)

    $[0, 23.5]$ by $[-2, 2]$

    (b) -1

9. (a)

    $[0, 23,5]$ by $[-1, 3]$

    (b) 2

10. (a)

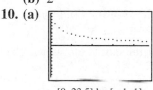

    $[0, 23.5]$ by $[-1, 1]$

    (b) 0

Exercises

1. (a) $* = n^2$ **(b)** $* = (n + 1)^2$
(c) $* = 3$

3. Different   **5.** Same

7. Converges; sum = 3   **9.** Converges; sum = $\dfrac{15}{4}$

11. Diverges
13. Converges; sum = $2 - \sqrt{2}$

15. Converges; sum = $\dfrac{e}{\pi - e}$

17. Interval: $-\dfrac{1}{2} < x < \dfrac{1}{2}$; function: $\dfrac{1}{1 - 2x}$

19. Interval: $1 < x < 5$; function: $\dfrac{2}{x - 1}$

21. Converges for all values of x except odd integer
multiples of $\dfrac{\pi}{2}$; function: $\dfrac{1}{1 - \sin x}$

23. (a) The partial sums tend toward infinity.
   (b) The partial sums are alternately 1 and 0.
   (c) The partial sums alternate between positive
   and negative while their magnitude increases
   toward infinity.

25. $x = \dfrac{19}{20}$

27. Assuming the series begins at $n = 1$:

(a) $\displaystyle\sum_{n=1}^{\infty} 2\left(\dfrac{3}{5}\right)^{n-1}$

(b) $\displaystyle\sum_{n=1}^{\infty} \dfrac{13}{2}\left(-\dfrac{3}{10}\right)^{n-1}$

29. Let $a = \dfrac{234}{1000}$ and $r = \dfrac{1}{1000}$, giving
$(0.234) + (0.234)(0.001) + (0.234)(0.001)^2$
   $+ (0.234)(0.001)^3 + \cdots$
So the sum is $\dfrac{26}{111}$.

31. $\dfrac{d}{9}$ **33.** $\dfrac{157}{111}$

35. $\dfrac{22}{7}$ **37.** ≈ 7.113 seconds

39. $\dfrac{\pi}{2}$

41. For $r \neq 1$, the result follows from:
If $|r| < 1$, $r^n \to 0$ as $n \to \infty$, and
if $|r| > 1$ or $r = -1$, r^n has no finite limit
as $n \to \infty$.
When $r = 1$, the nth partial sum is na, which goes
to $\pm\infty$.

43. Series: $x + 2x^2 + 4x^3 + \cdots + 2^{n-1}x^n + \cdots$
Interval: $-\dfrac{1}{2} < x < \dfrac{1}{2}$

45. Series: $1 - (x - 4) + (x - 4)^2 - (x - 4)^3 + \cdots$
   $+ (-1)^n(x - 4)^n + \cdots$
Interval: $3 < x < 5$

47. One possible series:
   $1 + (x - 1) + (x - 1)^2 + \cdots + (x - 1)^n + \cdots$
   Interval: $0 < x < 2$

49. (a) 2   **(b)** $t > -\dfrac{1}{2}$
   (c) $t > 9$

51. $(x - 1) - \dfrac{(x - 1)^2}{2} + \dfrac{(x - 1)^3}{3} - \cdots$
   $+ \dfrac{(-1)^{n-1}(x - 1)^n}{n} + \cdots$

53. (a) No, because if you differentiate it again, you
   would have the original series for f, but by
   Theorem 1, that would have to converge for
   $-2 < x < 2$, which contradicts the
   assumption that the original series converges
   only for $-1 < x < 1$.
   (b) No, because if you integrate it again, you
   would have the original series for f, but by
   Theorem 2, that would have to converge for
   $-2 < x < 2$, which contradicts the
   assumption that the original series converges
   only for $-1 < x < 1$.

55. Given an $\epsilon > 0$, by definition of convergence
   there corresponds an N such that for all $n < N$,
   $|L_1 - a_n| < \epsilon$ and $|L_2 - a_n| < \epsilon$. Now
   $|L_2 - L_1| = |L_2 - a_n + a_n - L_1|$
   $\leq |L_2 - a_n| + |a_n - L_1| < \epsilon + \epsilon = 2\epsilon$.
   $|L_2 - L_1| < 2\epsilon$ says that the difference between
   two fixed values is smaller than any positive
   number 2ϵ. The only nonnegative number smaller
   than every positive number is 0, so $|L_2 - L_1| = 0$
   or $L_1 = L_2$.

57. (a) $\displaystyle\lim_{n\to\infty} \dfrac{3n + 1}{n + 1} = 3$

   (b) The line $y = 3$ is a horizontal asymptote of
   the graph of the function $f(x) = \dfrac{3x + 1}{x + 1}$,
   which means $\displaystyle\lim_{x\to\infty} f(x) = 3$. Because $f(n) = a_n$
   for all positive integers n, it follows that
   $\displaystyle\lim_{n\to\infty} a_n$ must also be 3.

Section 9.2 (pp. 469–479)

Quick Review

1. $2^n e^{2x}$

2. $(-1)^n n!(x-1)^{-(n+1)}$

3. $3^x(\ln 3)^n$

4. $(-1)^{n-1}(n-1)!x^{-n}$

5. $n!$

6. $\dfrac{x^{n-1}}{(n-1)!}$

7. $\dfrac{2^n(x-a)^{n-1}}{(n-1)!}$

8. $\dfrac{(-1)^n x^{2n}}{(2n)!}$

9. $\dfrac{(x+a)^{2n-1}}{(2n-1)!}$

10. $-\dfrac{(1-x)^{n-1}}{(n-1)!}$

Exercises

1. $2x - \dfrac{4x^3}{3} + \dfrac{4x^5}{15} - \cdots + (-1)^n \dfrac{(2x)^{2n+1}}{(2n+1)!} + \cdots$
convergent for all real x

3. $x^2 - \dfrac{x^6}{3} + \dfrac{x^{10}}{5} - \cdots + (-1)^n \dfrac{x^{4n+2}}{(2n+1)} + \cdots$
convergent for $-1 \le x \le 1$

5. $(\cos 2) - (\sin 2)x - \dfrac{(\cos 2)x^2}{2} + \cdots$
$+ \dfrac{(-1)^A B x^n}{n!} + \cdots$, where $A = \text{int}\left(\dfrac{n+1}{2}\right)$, and
B is $\cos 2$ if n is even and $\sin 2$ if n is odd.

Alternately, the general term may be written as
$\left[\dfrac{1}{n!} \cos\left(2 + \dfrac{n\pi}{2}\right)\right]x^n$. The series converges for all
real x.

7. $x + x^4 + x^7 + \cdots + x^{3n+1} + \cdots$
convergent for $-1 < x < 1$

9. $P_0(x) = \dfrac{1}{2}$

$P_1(x) = \dfrac{1}{2} - \dfrac{x-2}{4}$

$P_2(x) = \dfrac{1}{2} - \dfrac{x-2}{4} + \dfrac{(x-2)^2}{8}$

$P_3(x) = \dfrac{1}{2} - \dfrac{x-2}{4} + \dfrac{(x-2)^2}{8} - \dfrac{(x-2)^3}{16}$

11. $P_0(x) = \dfrac{\sqrt{2}}{2}$

$P_1(x) = \dfrac{\sqrt{2}}{2} - \left(\dfrac{\sqrt{2}}{2}\right)\left(x - \dfrac{\pi}{4}\right)$

$P_2(x) = \dfrac{\sqrt{2}}{2} - \left(\dfrac{\sqrt{2}}{2}\right)\left(x - \dfrac{\pi}{4}\right) - \left(\dfrac{\sqrt{2}}{4}\right)\left(x - \dfrac{\pi}{4}\right)^2$

$P_3(x) = \dfrac{\sqrt{2}}{2} - \left(\dfrac{\sqrt{2}}{2}\right)\left(x - \dfrac{\pi}{4}\right) - \left(\dfrac{\sqrt{2}}{4}\right)\left(x - \dfrac{\pi}{4}\right)^2$
$\qquad + \left(\dfrac{\sqrt{2}}{12}\right)\left(x - \dfrac{\pi}{4}\right)^3$

13. (a) $4 - 2x + x^3$
(b) $3 + (x-1) + 3(x-1)^2 + (x-1)^3$

15. (a) 0
(b) $1 + 4(x-1) + 6(x-1)^2 + 4(x-1)^3$

17. (a) $P_3(x) = 4 - (x-1) + \dfrac{3}{2}(x-1)^2 + \dfrac{1}{3}(x-1)^3$
$\qquad f(1.2) \approx P_3(1.2) \approx 3.863$

(b) For f', $P_2(x) = -1 + 3(x-1) + (x-1)^2$
$\qquad f'(1.2) \approx P_2(1.2) = -0.36$

19. (a) $1 + \dfrac{x}{2} + \dfrac{x^2}{8} + \cdots + \dfrac{x^n}{2^n \cdot n!} + \cdots$

(b) $1 + \dfrac{x}{2!} + \dfrac{x^2}{3!} + \cdots + \dfrac{x^n}{(n+1)!} + \cdots$

(c) $g'(1) = 1$ and from the series,
$g'(1) = \dfrac{1}{2!} + \dfrac{2}{3!} + \dfrac{3}{4!} + \cdots + \dfrac{n}{(n+1)!} + \cdots$

21. (a) $1 + \dfrac{x}{2} - \dfrac{x^2}{8} + \dfrac{x^3}{16}$

(b) $1 + \dfrac{x^2}{2} - \dfrac{x^4}{8} + \dfrac{x^6}{16}$

(c) $5 + x + \dfrac{x^3}{6} - \dfrac{x^5}{40}$

23. 27 terms (or, up to and including the 52nd degree term)

25. (1) $\sin x$ is odd and $\cos x$ is even
(2) $\sin 0 = 0$ and $\cos 0 = 1$

27. $\dfrac{1}{24}$

29. (a)

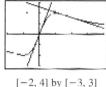

$[-2, 4]$ by $[-3, 3]$

(b) $f''(a)$ must be 0 because of the inflection point, so the second degree term in the Taylor series of f at $x = a$ is zero.

31. (a) $1 - \dfrac{x^2}{3!} + \dfrac{x^4}{5!} - \cdots + \dfrac{(-1)^n x^{2n}}{(2n+1)!} + \cdots$
(b) Because f is undefined at $x = 0$.
(c) $k = 1$

33. (a) Just differentiate 3 times.
(b) Differentiate k times and plug in $x = 0$.
(c) $\dfrac{m(m-1)(m-2)\cdots(m-k+1)}{k!}$
(d) $f(0) = 1$, $f'(0) = m$, and we're done by part (c).

Section 9.3 (pp. 480–487)

Quick Review

1. 2
2. 7
3. 1
4. $\dfrac{1}{2}$
5. 7
6. Yes
7. No
8. Yes
9. Yes
10. No

Exercises

1. $1 - 2x + 2x^2 - \dfrac{4}{3}x^3 + \dfrac{2}{3}x^4;\ f(0.2) \approx 0.6704$

3. $-5x + \dfrac{5}{6}x^3;\ f(0.2) \approx -0.9933$

5. $1 + 2x + 3x^2 + 4x^3 + 5x^4;\ f(0.2) \approx 1.56$

7. $\dfrac{x^5}{5!} - \dfrac{x^7}{7!} + \dfrac{x^9}{9!} - \cdots + (-1)^n\dfrac{x^{2n+5}}{(2n+5)!} + \cdots$

9. $x^2 - \dfrac{x^4}{3} + \dfrac{2x^6}{45} - \cdots + (-1)^n\dfrac{2^{2n+1}x^{2n+2}}{(2n+2)!} + \cdots$

11. Using the theorem, $-0.56 < x < 0.56$
    Graphically, $-0.57 < x < 0.57$

13. $|\text{Error}| < 1.67 \times 10^{-10}$

    $x < \sin x$ for negative values of x.

15. $|\text{Error}| < 1.842 \times 10^{-4}$

17. All of the derivatives of $\cosh x$ are either $\cosh x$ or $\sinh x$. For any real x, $\cosh x$ and $\sinh x$ are both bounded by $e^{|x|}$. So for any real x, let $M = e^{|x|}$ and $r = 1$ in the Remainder Estimation Theorem. It follows that the series converges to $\cosh x$ for all real values of x.

19. (a) 0
    (b) $-\dfrac{x^2}{2}$

    (c) The graphs of the linear and quadratic approximations fit the graph of the function near $x = 0$.

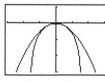

$[-3, 3]$ by $[-3, 1]$

21. (a) 1
    (b) $1 + \dfrac{x^2}{2}$

(c) The graphs of the linear and quadratic approximations fit the graph of the function near $x = 0$.

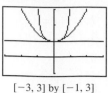

$[-3, 3]$ by $[-1, 3]$

23. (a) x (b) x

(c) The graphs of the linear and quadratic approximations fit the graph of the function near $x = 0$.

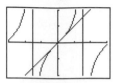

$[-3, 3]$ by $[-2, 2]$

25. $|\text{Error}| < 4.61 \times 10^{-6}$, by Remainder Estimation Theorem (actual maximum error is $\approx 4.251 \times 10^{-6}$)

27. (a) No

(b) Yes. $2 + x - \dfrac{x^3}{3} + \dfrac{x^5}{10} - \cdots$

$+ \dfrac{(-1)^{n+1}x^{2n-1}}{[(2n-1)(n-1)!]} + \cdots$

(c) For all real values of x. This is assured by Theorem 2 of Section 9.1, because the series for e^{-x^2} converges for all real values of x.

29. (a) $\tan x$ (b) $\sec x$

31. (a) It works.

(b) Let $P = \pi + x$ where x is the error in the original estimate. Then

$P + \sin P = (\pi + x) + \sin(\pi + x)$

$= \pi + x - \sin x$

But by the Remainder Theorem,
$|x - \sin x| < \dfrac{|x|^3}{6}$. Therefore, the difference between the new estimate $P + \sin P$ and π is less than $\dfrac{|x|^3}{6}$.

33. The derivative is
$(ae^{ax})(\cos bx + i \sin bx)$
$\quad + (e^{ax})(-b \sin bx + ib \cos bx)$
$= a[e^{ax}(\cos bx + i \sin bx)]$
$\quad + ib[e^{ax}(\cos bx + i \sin bx)]$
$= (a + ib)e^{(a+ib)x}.$

Section 9.4 (pp. 487–496)

Quick Review

1. $|x|$
2. $|x - 3|$
3. 0
4. $\dfrac{x^2}{16}$
5. $\dfrac{|2x + 1|}{2}$
6. $a_n = n^2, b_n = 5n, N = 6$
7. $a_n = 5^n, b_n = n^5, N = 6$
8. $a_n = \sqrt{n}, b_n = \ln n, N = 1$
9. $a_n = \dfrac{1}{10^n}, b^n = \dfrac{1}{n!}, N = 25$
10. $a_n = \dfrac{1}{n^2}, b_n = n^{-3}, N = 2$

Exercises

1. Diverges (*n*th-Term Test)
3. Converges (Ratio Test)
5. Converges (Ratio Test, Direct Comparison Test)
7. Converges (Ratio Test)
9. Converges (Ratio Test)
11. Converges (geometric series)
13. Diverges (*n*th-Term Test, Ratio Test)
15. Converges (Ratio Test)
17. One possible answer:

$\sum \dfrac{1}{n}$ diverges (see Exploration 1 in this section) even though $\lim\limits_{n\to\infty} \dfrac{1}{n} = 0$.

19. 1
21. $\dfrac{1}{4}$
23. 10
25. 3
27. 5
29. 3
31. $\dfrac{1}{2}$
33. 1

35. Interval: $-1 < x < 3$

Function: $-\dfrac{4}{x^2 - 2x - 3}$

37. Interval: $0 < x < 16$

Function: $\dfrac{2}{4 - \sqrt{x}}$

39. Interval: $-2 < x < 2$

Function: $\dfrac{3}{4 - x^2}$

41. Almost, but the ratio test won't determine whether there is convergence or divergence at the endpoints of the interval.

43. (a) For $k \le N$, it's obvious that

$$d_1 + \cdots + d_k$$
$$\le d_1 + \cdots + d_N + \sum_{n=N+1}^{\infty} a_n$$

For all $k > N$,

$$d_1 + \cdots + d_k$$
$$= d_1 + \cdots + d_N + d_{N+1} + \cdots + d_k$$
$$\le d_1 + \cdots + d_N + a_{N+1} + \cdots + a_k$$
$$\le d_1 + \cdots + d_N + \sum_{n=N+1}^{\infty} a_n.$$

(b) If $\sum a_n$ converged, that would imply that $\sum d_n$ was also convergent.

45. 1
47. 5
49. 1
51. $-\dfrac{\pi}{4}$

Section 9.5 (pp. 496–508)

Quick Review

1. Converges, $p > 1$
2. Diverges, limit comparison test with integral of $\dfrac{1}{x}$
3. Diverges, comparison test with integral of $\dfrac{1}{x}$
4. Converges, comparison test with integral of $\dfrac{2}{x^2}$
5. Diverges, limit comparison test with integral of $\dfrac{1}{\sqrt{x}}$
6. Yes
7. Yes
8. No
9. No
10. No

Exercises

1. Diverges
3. Diverges
5. Diverges
7. Diverges
9. Converges
11. Diverges
13. Diverges
15. Diverges
17. Converges absolutely
19. Converges absolutely
21. Diverges
23. Converges conditionally
25. Converges conditionally
27. (a) $(-1, 1)$ (b) $(-1, 1)$ (c) None
29. (a) $\left(-\dfrac{1}{2}, 0\right)$ (b) $\left(-\dfrac{1}{2}, 0\right)$ (c) None
31. (a) $(-8, 12)$ (b) $(-8, 12)$ (c) None
33. (a) $[-3, 3]$ (b) $[-3, 3]$ (c) None
35. (a) $(-8, 2)$ (b) $(-8, 2)$ (c) None
37. (a) $(-3, 3)$ (b) $(-3, 3)$ (c) None

39. (a) $\left(\dfrac{1}{2}, \dfrac{3}{2}\right)$ **(b)** $\left(\dfrac{1}{2}, \dfrac{3}{2}\right)$ **(c)** None

41. (a) $[-\pi - 1, -\pi + 1)$

    (b) $(-\pi - 1, -\pi + 1)$
    (c) At $x = -\pi - 1$

43. $40.554 <$ sum < 41.555

45.

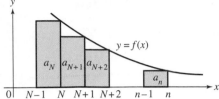

Comparing areas in the figures, we have for all

$$n \ge N, \int_N^{n+1} f(x)\, dx < a_N + \cdots + a_n < a_1$$
$$+ \int_N^n f(x)\, dx.$$

If the integral diverges, it must go to infinity, and the first inequality forces the partial sums of the series to go to infinity as well, so the series is divergent.
If the integral converges, then the second inequality puts an upper bound on the partial sums of the series, and since they are a nondecreasing sequence, they must converge to a finite sum for the series.

47. One possible answer: $\sum \dfrac{1}{n \ln n}$
This series diverges by the Integral Test, but its partial sums are roughly $\ln(\ln n)$, so they are much smaller than the partial sums for the harmonic series, which are about $\ln n$.

49. (a) Diverges

    (b) $S = \sum_{n=1}^{\infty} \dfrac{3n}{3n^3 + n} = \sum_{n=1}^{\infty} \dfrac{3}{3n^2 + 1}$ which

    converges.

51. Convergent for $-\dfrac{1}{2} \le x < \dfrac{1}{2}$.

    Use the Ratio Test, Direct Comparison Test, and

    Alternating Series Test.

53. Use the Alternating Series Test.

55. (a) It fails to satisfy $u_n \ge u_{n+1}$ for all $n \ge N$.

    (b) The sum is $-\dfrac{1}{2}$.

57. (a) Converges   **(b)** Converges
    (c) Converges

Chapter Review (pp. 509–511)

1. (a) ∞ **(b)** All real numbers
    (c) All real numbers   **(d)** None
2. (a) 3   **(b)** $[-7, -1)$
    (c) $(-7, -1)$ **(d)** At $x = -7$
3. (a) $\dfrac{3}{2}$ **(b)** $\left(-\dfrac{1}{2}, \dfrac{5}{2}\right)$

    (c) $\left(-\dfrac{1}{2}, \dfrac{5}{2}\right)$ **(d)** None
4. (a) ∞ **(b)** All real numbers
    (c) All real numbers   **(d)** None
5. (a) $\dfrac{1}{3}$ **(b)** $\left[0, \dfrac{2}{3}\right]$

    (c) $\left[0, \dfrac{2}{3}\right]$ **(d)** None
6. (a) 1   **(b)** $(-1, 1)$
    (c) $(-1, 1)$ **(d)** None
7. (a) 1   **(b)** $\left(-\dfrac{3}{2}, \dfrac{1}{2}\right)$

    (c) $\left(-\dfrac{3}{2}, \dfrac{1}{2}\right)$ **(d)** None
8. (a) ∞ **(b)** All real numbers
    (c) All real numbers   **(d)** None
9. (a) 1   **(b)** $[-1, 1)$
    (c) $(-1, 1)$ **(d)** At $x = -1$
10. (a) $\dfrac{1}{e}$ **(b)** $\left[-\dfrac{1}{e}, \dfrac{1}{e}\right]$

    (c) $\left[-\dfrac{1}{e}, \dfrac{1}{e}\right]$ **(d)** None
11. (a) $\sqrt{3}$ **(b)** $(-\sqrt{3}, \sqrt{3})$
    (c) $(-\sqrt{3}, \sqrt{3})$ **(d)** None
12. (a) 1   **(b)** $[0, 2]$
    (c) $(0, 2)$ **(d)** At $x = 0$ and $x = 2$
13. (a) 0   **(b)** $x = 0$ only
    (c) $x = 0$ **(d)** None
14. (a) $\dfrac{1}{10}$ **(b)** $\left[-\dfrac{1}{10}, \dfrac{1}{10}\right)$

    (c) $\left(-\dfrac{1}{10}, \dfrac{1}{10}\right)$ **(d)** At $x = -\dfrac{1}{10}$
15. (a) 0   **(b)** $x = 0$ only
    (c) $x = 0$ **(d)** None
16. (a) $\sqrt{3}$ **(b)** $(-\sqrt{3}, \sqrt{3})$
    (c) $(-\sqrt{3}, \sqrt{3})$ **(d)** None
17. $f(x) = \dfrac{1}{1 + x}$ evaluated at $x = \dfrac{1}{4}$. Sum $= \dfrac{4}{5}$.

18. $f(x) = \ln(1 + x)$ evaluated at $x = \dfrac{2}{3}$.
    Sum $= \ln\left(\dfrac{5}{3}\right)$.

19. $f(x) = \sin x$ evaluated at $x = \pi$. Sum $= 0$.

20. $f(x) = \cos x$ evaluated at $x = \dfrac{\pi}{3}$. Sum $= \dfrac{1}{2}$.

21. $f(x) = e^x$ evaluated at $x = \ln 2$. Sum $= 2$.

22. $f(x) = \tan^{-1} x$ evaluated at $x = \dfrac{1}{\sqrt{3}}$. Sum $= \dfrac{\pi}{6}$.

23. $1 + 6x + 36x^2 + \cdots + (6x)^n + \cdots$

24. $1 - x^3 + x^6 - \cdots + (-1)^n x^{3n} + \cdots$

25. $1 - 2x^2 + x^9$

26. $4x + 4x^2 + 4x^3 + \cdots + 4x^{n+1} + \cdots$

27. $\pi x - \dfrac{(\pi x)^3}{3!} + \dfrac{(\pi x)^5}{5!} - \cdots + (-1)^n \dfrac{(\pi x)^{2n+1}}{(2n+1)!} + \cdots$

28. $-\dfrac{2x}{3} + \dfrac{4x^3}{81} - \dfrac{4x^5}{3645} + \cdots$
$\qquad + \dfrac{(-1)^{n+1}}{(2n+1)!}\left(\dfrac{2x}{3}\right)^{2n+1} + \cdots$

29. $-\dfrac{x^3}{3!} + \dfrac{x^5}{5!} - \dfrac{x^7}{7!} + \cdots + (-1)^n \dfrac{x^{2n+1}}{(2n+1)!} + \cdots$

30. $1 + \dfrac{x^2}{2!} + \dfrac{x^4}{4!} + \cdots + \dfrac{x^{2n}}{(2n)!} + \cdots$

31. $1 - \dfrac{5x}{2!} + \dfrac{(5x)^2}{4!} - \cdots + (-1)^n \dfrac{(5x)^n}{(2n)!} + \cdots$

32. $1 + \dfrac{\pi x}{2} + \dfrac{\pi^2 x^2}{8} + \cdots + \dfrac{1}{n!}\left(\dfrac{\pi x}{2}\right)^n + \cdots$

33. $x - x^3 + \dfrac{x^5}{2!} - \dfrac{x^7}{3!} + \cdots + (-1)^n \dfrac{x^{2n+1}}{n!} + \cdots$

34. $3x - \dfrac{(3x)^3}{3} + \dfrac{(3x)^5}{5} - \cdots + (-1)^n \dfrac{(3x)^{2n+1}}{2n+1} + \cdots$

35. $-2x - 2x^2 - \dfrac{8x^3}{3} - \cdots - \dfrac{(2x)^n}{n} - \cdots$

36. $-x^2 - \dfrac{x^3}{2} - \dfrac{x^4}{3} - \cdots - \dfrac{x^{n+1}}{n} - \cdots$

37. $1 + (x - 2) + (x - 2)^2 + (x - 2)^3$
$\qquad + \cdots (x - 2)^n + \cdots$

38. $2 + 7(x + 1) - 5(x + 1)^2 + (x + 1)^3$

(Finite. general term $= 0$)

39. $\dfrac{1}{3} - \dfrac{x - 3}{9} + \dfrac{(x - 3)^2}{27} - \dfrac{(x - 3)^3}{81} + \cdots$
$\qquad + (-1)^n \dfrac{(x - 3)^n}{3^{n+1}} + \cdots$

40. $-(x - \pi) + \dfrac{(x - \pi)^3}{3!} - \dfrac{(x - \pi)^5}{5!} + \dfrac{(x - \pi)^7}{7!} - \cdots$
$\qquad + (-1)^{n+1} \dfrac{(x - \pi)^{2n+1}}{(2n+1)!} + \cdots$

41. Diverges. It's just -5 times the harmonic series.

42. Converges conditionally. Alternating Series Test and $p = \dfrac{1}{2}$.

43. Converges absolutely. Compare to $\dfrac{1}{n^2}$.

44. Converges absolutely. Ratio Test

45. Converges conditionally. Alternating Series Test and compare to $\dfrac{1}{n}$.

46. Converges absolutely. Integral Test

47. Converges absolutely. Ratio Test

48. Converges absolutely. nth-Root Test or Ratio Test

49. Diverges. nth-Term Test for Divergence

50. Converges absolutely. Compare to $\dfrac{1}{n^{3/2}}$.

51. Converges absolutely. Limit Comparison Test with $\dfrac{1}{n^2}$.

52. Diverges. nth-Term Test for Divergence

53. $\dfrac{1}{6}$ **54.** -1

55. (a) $P_3(x) = 1 + 4(x - 3) + 3(x - 3)^2 + 2(x - 3)^3$
$\qquad f(3.2) \approx P_3(3.2) = 1.936$
(b) For f: $P_2(x) = 4 + 6(x - 3) + 6(x - 3)^2$
$\qquad f'(2.7) \approx P_2(2.7) = 2.74$
(c) It underestimates the values, since the graph of f is concave up near $x = 3$.

56. (a) $f(4) = 7$ and $f'''(4) = -12$
(b) For f: $P_2(x) = -3 + 10(x - 4) - 6(x - 4)^2$
$\qquad f'(4.3) \approx P_2(4.3) = -0.54$
(c) $7(x - 4) - \dfrac{3}{2}(x - 4)^2 + \dfrac{5}{3}(x - 4)^3 - \dfrac{1}{2}(x - 4)^4$
(d) No. One would need the entire Taylor series for $f(x)$, and it would have to converge to $f(x)$ at $x = 3$.

57. (a) $\dfrac{5x}{2} - \dfrac{5x^3}{48} + \dfrac{x^5}{768} - \cdots$
$\qquad + (-1)^n \dfrac{5}{(2n+1)!}\left(\dfrac{x}{2}\right)^n + \cdots$

(b) All real numbers. Use the Ratio Test.

(c) Note that the absolute value of $f^{(n)}(x)$ is bounded by $\dfrac{5}{2^n}$ for all x and all $n = 1, 2, 3, \cdots$
So if $-2 < x < 2$, the truncation error using P_n is bounded by $\dfrac{5}{2^{n+1}} \cdot \dfrac{2^{n+1}}{(n+1)!} = \dfrac{5}{(n+1)!}$.
To make this less than 0.1 requires $n \geq 4$.

So, two nonzero terms (up through degree 4) are needed.

58. (a) $1 + 2x + 4x^2 + 8x^3 + \cdots + (2x)^n + \cdots$

(b) $\left(-\dfrac{1}{2}, \dfrac{1}{2}\right)$. The series for $\dfrac{1}{1-t}$ is known to converge for $-1 < t < 1$, so by substituting $t = 2x$, we find the resulting series converges for $-1 < 2x < 1$.

(c) Possible answer: $f\left(-\dfrac{1}{4}\right) = \dfrac{2}{3}$, so one percent is approximately 0.0067. It takes 7 terms (up through degree 6). This can be found by trial and error. Also, the series is an alternating series for $x = -\dfrac{1}{4}$. If you use the Alternating Series Estimation Theorem, it shows that 8 terms (up through degree 7) are sufficient. It is also a geometric series, and one could use the remainder formula for a geometric series to estimate the number of terms needed.

59. (a) $\dfrac{1}{e}$ **(b)** $-\dfrac{5}{18} \approx -0.278$

(c) By the Alternating Series Estimation Theorem, the error is bounded by the size of the next term, which is $\dfrac{32}{243}$, or about 0.132.

60. (a) $1 - (x - 3) + (x - 3)^2 - (x - 3)^3 + \cdots$
$+ (-1)^n(x - 3)^n + \cdots$

(b) $(x - 3) - \dfrac{(x - 3)^2}{2} + \dfrac{(x - 3)^3}{3} - \dfrac{(x - 3)^4}{4}$
$+ \cdots + (-1)^n\dfrac{(x - 3)^{n+1}}{n + 1} + \cdots$

(c) Evaluate at $x = 3.5$. This is an alternating series. By the Alternating Series Estimation Theorem, since the size of the third term is $\dfrac{1}{24} < 0.05$, the first two terms will suffice. The estimate for $\ln\left(\dfrac{3}{2}\right)$ is 0.375.

61. (a) $1 - 2x^2 + 2x^4 - \dfrac{4x^6}{3} + \cdots + (-1)^n\dfrac{2^nx^{2n}}{n!} + \cdots$

(b) All real numbers. Use the Ratio Test.

(c) This is an alternating series. The difference will be bounded by the magnitude of the fifth term, which is $\dfrac{(2x^2)^4}{4!} = \dfrac{2x^8}{3}$.
Since $-0.6 \le x \le 0.6$, this term is less than $\dfrac{2(0.6)^8}{3}$ which is less than 0.02.

62. (a) $x^2 - x^3 + x^4 - x^5 + \cdots + (-1)^nx^n + \cdots$

(b) No. The partial sums form the sequence 1, 0, 1, 0, 1, 0, ... which has no limit.

63. (a) $\dfrac{x^3}{3} - \dfrac{x^7}{7(3!)} + \dfrac{x^{11}}{11(5!)} + \cdots$
$+ \dfrac{(-1)^nx^{4n+3}}{(4n + 3)(2n + 1)!} + \cdots$

(b) The first two nonzero terms suffice (through degree 7).

(c) 0.31026830   **(d)** Within 1.5×10^{-7}

64. (a) 0.88566   **(b)** $\dfrac{41}{60} \approx 0.68333$

(c) Since f is concave up, the trapezoids used to estimate the area lie above the curve, and the estimate is too large.

(d) Since all the derivatives are positive (and $x > 0$), the remainder, $R_n(x)$, must be positive. This means that $P_n(x)$ is smaller than $f(x)$.

(e) $e - 2 \approx 0.71828$

65. (a) Because $[\$1000(1.08)^{-n}](1.08)^n = \$1000$ will be available after n years.

(b) Assume that the first payment goes to the charity at the end of the first year.
$1000(1.08)^{-1} + 1000(1.08)^{-2}$
$+ 1000(1.08)^{-3} + \cdots$

(c) This is a geometric series with sum equal to \$12,500. This represents the amount which must be invested today in order to completely fund the perpetuity forever.

66. \$16,666.67 [Again, assuming first payment at end of year.]

67. (a) $0\left(\dfrac{1}{2}\right) + 1\left(\dfrac{1}{2}\right)^2 + 2\left(\dfrac{1}{2}\right)^3 + 3\left(\dfrac{1}{2}\right)^4 + \cdots$

(b) $1 + 2x + 3x^2 + 4x^3 + \cdots$

(c) $\dfrac{x^2}{(1 - x)^2} = x^2 + 2x^3 + 3x^4 + 4x^5 + \cdots$

(d) The expected payoff of the game is \$1.

68. (a) $\dfrac{b^2\sqrt{3}}{4} + \dfrac{3b^2\sqrt{3}}{4^2} + \dfrac{3^2b^2\sqrt{3}}{4^3} + \cdots$

(b) $b^2\sqrt{3}$

(c) No, not every point is removed. But the remaining points are "isolated" enough that there are no regions and hence no area remaining.

69. $\dfrac{1}{(1 - x)^2} = 1 + 2x + 3x^2 + 4x^3 + 5x^4 + \cdots$
Substitute $x = \dfrac{1}{2}$ to get the desired result.

70. (b) Solve $x = \dfrac{2x^2}{(x - 1)^3}$. $x \approx 2.769$.

Chapter 10

Section 10.1 (pp. 513–520)

Quick Review

1. $(1, 0)$ **2.** $(0, -1)$
3. $x^2 + y^2 = 1$
4. The portion in the first three quadrants
5. $x = t, y = t^2 + 1, -1 \le t \le 3$
6. $x = 2 \cos t + 2, y = 2 \sin t + 3, 0 \le t \le 2\pi$
7. $\dfrac{3}{2}$ **8.** $y = \dfrac{3}{2}x + 3\sqrt{2}$
9. $y = -\dfrac{2}{3}x + \dfrac{5\sqrt{2}}{6}$ **10.** $\dfrac{31^{3/2} - 8}{27}$

Exercises

1. (a) $-\dfrac{1}{2} \tan t$ **(b)** $-\dfrac{1}{8} \sec^3 t$

3. (a) $-\sqrt{3 + \dfrac{3}{t}}$ **(b)** $-\dfrac{\sqrt{3}}{t^{3/2}}$

5. (a) $\dfrac{3t^2}{2t - 3}$ **(b)** $\dfrac{6t^2 - 18t}{(2t - 3)^3}$

7. (a) $(2, 0)$ and $(2, -2)$
   (b) $(1, -1)$ and $(3, -1)$

9. (a) At $t = \pm\dfrac{2}{\sqrt{3}}$, or $\sim (0.845, -3.079)$ and $(3.155, 3.079)$

   (b) Nowhere

11. 4   **13.** $\dfrac{2\sqrt{2} - 1}{3} \sim 0.609$

15. $\ln 2$ **17.** $8\pi^2$
19. ≈ 178.561
21. (a) $x(t) = 2t, y(t) = t + 1, 0 \le t \le 1$
   (b) $3\pi\sqrt{5}$ **(c)** $3\pi\sqrt{5}$
23. (a) π **(b)** π

25. Just substitute x for t and note that $\dfrac{dx}{dx} = 1$.

27. At $t = \sqrt{13} - 1$, or $\approx (3.394, 5.160)$
29. ≈ 144.513 **31.** $3\pi a^2$
33. (a) $x = \cos t + t \sin t, y = \sin t - t \cos t$
   (b) $2\pi^2$
35. (a) ~ 461.749 ft   **(b)** ≈ 41.125 ft

37. (a) ≈ 840.421 ft   **(b)** $\dfrac{16,875}{64} \approx 263.672$ ft

39. Just substitute x for t and note that $\dfrac{dx}{dx} = 1$.

41. ≈ 9.417

Section 10.2 (pp. 520–529)

Quick Review

1. $\sqrt{17}$ **2.** $\dfrac{1}{4}$
3. $b = 11$ **4.** $a = 4$
5. $b = 6$
6. (a) $120°$ **(b)** $\dfrac{2\pi}{3}$
7. (a) $-30°$ **(b)** $-\dfrac{\pi}{6}$
8. (a) $-45°$ **(b)** $-\dfrac{\pi}{4}$
9. $c \approx 2.832$
10. $\theta \approx 1.046$ radians or 59.935 degrees

Exercises

1. (a) $\langle 9, -6 \rangle$ **(b)** $3\sqrt{13}$
3. (a) $\langle 1, 3 \rangle$ **(b)** $\sqrt{10}$
5. (a) $\langle 12, -19 \rangle$ **(b)** $\sqrt{505}$
7. (a) $\left\langle \dfrac{1}{5}, \dfrac{14}{5} \right\rangle$ **(b)** $\dfrac{\sqrt{197}}{5}$
9. $\langle 1, -4 \rangle$ **11.** $\langle -2, -3 \rangle$
13. $\left(-\dfrac{1}{2}, \dfrac{\sqrt{3}}{2} \right)$ **15.** $\left(-\dfrac{\sqrt{3}}{2}, -\dfrac{1}{2} \right)$

17. The vector **v** is horizontal and 1 in. long. The vectors **u** and **w** are $\dfrac{11}{16}$ in. long. **w** is vertical and **u** makes a 45° angle with the horizontal. All vectors must be drawn to scale.

(a)

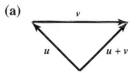

(b)

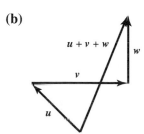

17. continued

(c)

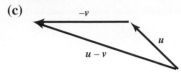

$u - v$

(d)

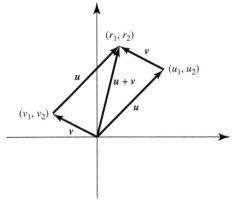

19. $\left\langle \dfrac{3}{5}, \dfrac{4}{5} \right\rangle$

21. $\left\langle -\dfrac{15}{17}, \dfrac{8}{17} \right\rangle$

23. Tangent: $\pm\left\langle \dfrac{1}{\sqrt{17}}, \dfrac{4}{\sqrt{17}} \right\rangle$

  Normal: $\pm\left\langle \dfrac{4}{\sqrt{17}}, -\dfrac{1}{\sqrt{17}} \right\rangle$

25. Tangent: $\pm\left\langle -\dfrac{12}{\sqrt{219}}, \dfrac{5}{\sqrt{73}} \right\rangle \approx \pm\langle -0.811, 0.585 \rangle$

  Normal: $\pm\left\langle \dfrac{5}{\sqrt{73}}, \dfrac{12}{\sqrt{219}} \right\rangle \approx \langle 0.585, 0.811 \rangle$

27. Angle at $A = \cos^{-1}\left(\dfrac{1}{\sqrt{5}}\right) \approx 63.435$ degrees

  Angle at $B = \cos^{-1}\left(\dfrac{3}{5}\right) \approx 53.130$ degrees

  Angle at $C = \cos^{-1}\left(\dfrac{1}{\sqrt{5}}\right) \approx 63.435$ degrees

29. (a) Both equal $u_1(v_1 + w_1) + u_2(v_2 + w_2)$.
  (b) Both equal $(u_1 + v_1)w_1 + (u_2 + v_2)w_2$.

31. $(\mathbf{u} + \mathbf{v}) \cdot (\mathbf{u} - \mathbf{v})$

$= (u_1 + v_1)(u_1 - v_1) + (u_2 + v_2)(u_2 - v_2)$

$= u_1^2 - v_1^2 + u_2^2 - v_2^2$

$= (u_1^2 + u_2^2) - (v_1^2 + v_2^2)$

$= |\mathbf{u}|^2 - |\mathbf{v}|^2$

33.

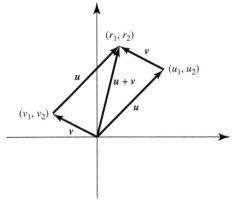

$r_1 - v_1 = u_1 \rightarrow r_1 = u_1 + v_1$
$r_2 - v_2 = u_2 \rightarrow r_2 = u_2 + v_2$

35. (a) Let $P = (a, b)$ and $Q = (c, d)$. Then

$$\left(\frac{1}{2}\right)\overrightarrow{OP} + \left(\frac{1}{2}\right)\overrightarrow{OQ} = \left(\frac{1}{2}\right)\langle a, b\rangle + \left(\frac{1}{2}\right)\langle c, d\rangle$$

$$= \left\langle \frac{(a + c)}{2}, \frac{(b + d)}{2} \right\rangle = \overrightarrow{OM}$$

(b) $\overrightarrow{OM} = \left(\dfrac{2}{3}\right)\overrightarrow{OP} + \left(\dfrac{1}{3}\right)\overrightarrow{OQ}$

(c) $\overrightarrow{OM} = \left(\dfrac{1}{3}\right)\overrightarrow{OP} + \left(\dfrac{2}{3}\right)\overrightarrow{OQ}$

(d) Possible answer:

  M is a fraction of the way from P to Q. Let d

  be this fraction. Then

$$\overrightarrow{OM} = d\overrightarrow{OQ} + (1 - d)\overrightarrow{OP}.$$

  Proof: $\overrightarrow{PM} = d\overrightarrow{PQ}$ and $\overrightarrow{MQ} = (1 - d)\overrightarrow{PQ}$, so

  $\overrightarrow{PQ} = \dfrac{1}{d}\overrightarrow{PM}$ and $\overrightarrow{PQ} = \dfrac{1}{1 - d}\overrightarrow{MQ}$.

  Therefore, $\dfrac{1}{d}\overrightarrow{PM} = \dfrac{1}{1 - d}\overrightarrow{MQ}$.

  But $\overrightarrow{PM} = \overrightarrow{OM} - \overrightarrow{OP}$ and $\overrightarrow{MQ} = \overrightarrow{OQ} - \overrightarrow{OM}$,

  so $\dfrac{1}{d}\overrightarrow{OM} - \dfrac{1}{d}\overrightarrow{OP} = \dfrac{1}{1 - d}\overrightarrow{OQ} - \dfrac{1}{1 - d}\overrightarrow{OM}$.

  Therefore,

$$\frac{1}{d}\overrightarrow{OM} + \frac{1}{1 - d}\overrightarrow{OM} = \frac{1}{d}\overrightarrow{OP} + \frac{1}{1 - d}\overrightarrow{OQ}.$$

$$\Rightarrow \overrightarrow{OM}\left(\frac{1}{d(1 - d)}\right) = \frac{1}{d}\overrightarrow{OP} + \frac{1}{1 - d}\overrightarrow{OQ}$$

$$\Rightarrow \overrightarrow{OM} = (1 - d)\overrightarrow{OP} + d\overrightarrow{OQ}.$$

37. Two adjacent sides of the rhombus can be given by two vectors of the same length, $\mathbf{u}$ and $\mathbf{v}$. Then the diagonals of the rhombus are $(\mathbf{u} + \mathbf{v})$ and $(\mathbf{u} - \mathbf{v})$. These two vectors are orthogonal since $(\mathbf{u} + \mathbf{v}) \cdot (\mathbf{u} - \mathbf{v}) = |\mathbf{u}|^2 - |\mathbf{v}|^2 = 0$.

39. Let two adjacent sides of the parallelogram be given by two vectors $\mathbf{u}$ and $\mathbf{v}$. The diagonals are then $(\mathbf{u} + \mathbf{v})$ and $(\mathbf{u} - \mathbf{v})$. So the lengths of the diagonals satisfy

$|\mathbf{u} + \mathbf{v}|^2 = (\mathbf{u} + \mathbf{v}) \cdot (\mathbf{u} + \mathbf{v})$

$\qquad = |\mathbf{u}|^2 + 2\mathbf{u} \cdot \mathbf{v} + |\mathbf{v}|^2$

and $|\mathbf{u} - \mathbf{v}|^2 = (\mathbf{u} - \mathbf{v}) \cdot (\mathbf{u} - \mathbf{v})$

$\qquad = |\mathbf{u}|^2 - 2\mathbf{u} \cdot \mathbf{v} + |\mathbf{v}|^2$.

The two lengths will be the same if and only if $\mathbf{u} \cdot \mathbf{v} = 0$, which means that $\mathbf{u}$ and $\mathbf{v}$ are perpendicular and the parallelogram is a rectangle.

41. The slopes are the same.

43. $\approx \langle -338.095, 725.046 \rangle$

45. Speed ≈ 346.735 mph

direction ≈ 14.266 degrees east of north

47. ≈ 39.337 lb

49. $\overrightarrow{AB} = \langle -3, 4 \rangle = \overrightarrow{CD}$

51. $\mathbf{u} = \langle u_1, u_2 \rangle$, $\mathbf{v} = \langle v_1, v_2 \rangle$, $\mathbf{w} = \langle w_1, w_2 \rangle$

(i) $\mathbf{u} + \mathbf{v} = \langle u_1 + v_1, u_2 + v_2 \rangle$
$= \langle v_1 + u_1, v_2 + u_2 \rangle = \mathbf{v} + \mathbf{u}$

(ii) $(\mathbf{u} + \mathbf{v}) + \mathbf{w}$
$= \langle u_1 + v_1, u_2 + v_2 \rangle + \langle w_1, w_2 \rangle$
$= \langle (u_1 + v_1) + w_1, (u_2 + v_2) + w_2 \rangle$
$= \langle u_1 + (v_1 + w_1), u_2 + (v_2 + w_2) \rangle$
$= \mathbf{u} + (\mathbf{v} + \mathbf{w})$

(iii) $\mathbf{u} + \mathbf{0} = \langle u_1, u_2 \rangle + \langle 0, 0 \rangle = \langle u_1 + 0, u_2 + 0 \rangle$
$= \langle u_1, u_2 \rangle = \mathbf{u}$

(iv) $\mathbf{u} + (-\mathbf{u}) = \langle u_1, u_2 \rangle + \langle -u_1, -u_2 \rangle$
$= \langle u_1 - u_1, u_2 - u_2 \rangle = \langle 0, 0 \rangle = \mathbf{0}$

(v) $0\mathbf{u} = 0\langle u_1, u_2 \rangle = \langle 0u_1, 0u_2 \rangle = \langle 0, 0 \rangle = \mathbf{0}$

(vi) $1\mathbf{u} = 1\langle u_1, u_2 \rangle = \langle 1u_1, 1u_2 \rangle = \langle u_1, u_2 \rangle = \mathbf{u}$

(vii) $a(b\mathbf{u}) = a(b \langle u_1, u_2 \rangle) = a\langle bu_1, bu_2 \rangle$
$= \langle abu_1, abu_2 \rangle = ab\langle u_1, u_2 \rangle = (ab)\mathbf{u}$

(viii) $a(\mathbf{u} + \mathbf{v}) = a\langle u_1 + v_1, u_2 + v_2 \rangle$
$= \langle au_1 + av_1, au_2 + av_2 \rangle$
$= \langle au_1, au_2 \rangle + \langle av_1, av_2 \rangle$
$= a\langle u_1, u_2 \rangle + a\langle v_1, v_2 \rangle$
$= a\mathbf{u} + a\mathbf{v}$

(ix) $(a + b)\mathbf{u} = (a + b)\langle u_1, u_2 \rangle$
$= \langle (a + b)u_1, (a + b)u_2 \rangle$
$= \langle au_1 + bu_1, au_2 + bu_2 \rangle$
$= \langle au_1, au_2 \rangle + \langle bu_1, bu_2 \rangle = a\mathbf{u} + b\mathbf{u}$

53. (a) $y = -x - 1$ (b) $y = x + 3$

Section 10.3 (pp. 529–539)

Quick Review

1. $y = \left(-\dfrac{1}{\sqrt{3}} \right) x + \dfrac{4}{\sqrt{3}}$, or approximately
$y = -0.577x + 2.309$

2. $y = \sqrt{3}x$ **3.** 0

4. Undefined; vertical tangent

5. $y = \left(-\dfrac{5\sqrt{3}}{4} \right) x + 10$ **6.** $y = \left(\dfrac{4\sqrt{3}}{15} \right) x + \dfrac{9}{10}$

7. $\dfrac{1}{4}$

8. ≈ 3.400

9. ≈ 2.958

10. $y = xe^x - e^x + 3$

Exercises

1. $6\mathbf{i} - 3\mathbf{j}$

3. (a) $-\mathbf{i} - \mathbf{j}$ (b) $7\mathbf{i} + 5\mathbf{j}$

5. (a)

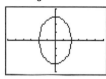

[−6, 6] by [−4, 4]

(b) $\mathbf{v}(t) = (-2 \sin t)\mathbf{i} + (3 \cos t)\mathbf{j}$
$\mathbf{a}(t) = (-2 \cos t)\mathbf{i} + (-3 \sin t)\mathbf{j}$

(c) Speed $= 2$
direction $= \langle -1, 0 \rangle$

(d) Velocity $= 2\langle -1, 0 \rangle$

7. (a)

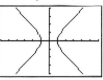

[−6, 6] by [−4, 4]

(b) $\mathbf{v}(t) = (\sec t \tan t)\mathbf{i} + (\sec^2 t)\mathbf{j}$
$\mathbf{a}(t) = (\sec t \tan^2 t + \sec^3 t)\mathbf{i} +$
$(2 \sec^2 t \tan t)\mathbf{j}$

(c) Speed $= \dfrac{2\sqrt{5}}{3}$

direction $= \left\langle \dfrac{1}{\sqrt{5}}, \dfrac{2}{\sqrt{5}} \right\rangle$

(d) Velocity $= \left(\dfrac{2\sqrt{5}}{3} \right) \left\langle \dfrac{1}{\sqrt{5}}, \dfrac{2}{\sqrt{5}} \right\rangle$

9. (a) $y = -1$ (b) $x = 0$

11. $-3\mathbf{i} + (4\sqrt{2} - 2)\mathbf{j}$

13. $(\sec t)\mathbf{i} + (\ln |\sec t|) \mathbf{j} + \mathbf{C}$

15. $\mathbf{r}(t) = ((t + 1)^{3/2} - 1)\mathbf{i} - (e^{-t} - 1)\mathbf{j}$

17. $\mathbf{r}(t) = (8t + 100)\mathbf{i} + (-16t^2 + 8t)\mathbf{j}$

19. $t = 0, \pi, 2\pi$ **21.** $t =$ all multiples of $\dfrac{\pi}{2}$

23. $\cos^{-1}\left(\dfrac{3}{5} \right) \approx 53.130$ degrees

25. (a) $3\mathbf{i}$ (b) $t \neq 0, -3$ (c) $t = 0, -3$

27. 2

29. (a) $\mathbf{v}(t) = (\cos t)\mathbf{i} - (2 \sin 2t)\mathbf{j}$

(b) $t = \dfrac{\pi}{2}, \dfrac{3\pi}{2}$

(c) $y = 1 - 2x^2, -1 \leq x \leq 1$. The particle starts
at $(0, 1)$, goes to $(1, -1)$, then goes to
$(-1, -1)$, and then goes to $(0, 1)$, tracing the
curve twice.

31. $\mathbf{r}(t) = \left(\dfrac{3}{2}t^2 + \dfrac{3\sqrt{10}}{5}t + 1 \right)\mathbf{i}$
$+ \left(-\dfrac{1}{2}t^2 - \dfrac{\sqrt{10}}{5}t + 2 \right)\mathbf{j}$

33. (a) 160 seconds   **(b)** 225 m

(c) $\dfrac{15}{4}$ meters per second

(d) At $t = 80$ seconds

35. (a) Referring to the figure, look at the circular arc from the point where $t = 0$ to the point "m".

On one hand, this arc has length given by $(r_0\theta)$, but it also has length given by (vt). Setting those two quantities equal gives the result.

(b) $\mathbf{a}(t) = -\dfrac{v^2}{r_0}\left[\left(\cos\dfrac{vt}{r_0}\right)\mathbf{i} + \left(\sin\dfrac{vt}{r_0}\right)\mathbf{j}\right]$

(c) From part (b) above, $\mathbf{a}(t) = -\left(\dfrac{v}{r_0}\right)^2 \mathbf{r}(t)$.

So, by Newton's second law, $\mathbf{F} = -m\left(\dfrac{v}{r_0}\right)^2 \mathbf{r}$.

Substituting for $\mathbf{F}$ in the law of gravitation gives the result.

(d) Set $\dfrac{vT}{r_0} = 2\pi$ and solve for vT.

(e) Substitute $\dfrac{2\pi r_0}{T}$ for v in $v^2 = \dfrac{GM}{r_0}$ and solve for T^2.

37. (a) Apply the Corollary to each component separately.

(b) Follows immediately from (a) since any two anti-derivatives of $\mathbf{r}(t)$ must have identical derivatives, namely $\mathbf{r}(t)$.

39. Let $\mathbf{C} = \langle C_1, C_2 \rangle$. $\dfrac{d\mathbf{C}}{dt} = \left\langle \dfrac{dC_1}{dt}, \dfrac{dC_2}{dt} \right\rangle = \langle 0, 0 \rangle$.

41. $\mathbf{u} = \langle u_1, u_2 \rangle$, $\mathbf{v} = \langle v_1, v_2 \rangle$

(a) $\dfrac{d}{dt}(\mathbf{u} + \mathbf{v}) = \dfrac{d}{dt}(\langle u_1 + v_1, u_2 + v_2 \rangle)$

$= \left\langle \dfrac{d}{dt}(u_1 + v_1), \dfrac{d}{dt}(u_2 + v_2) \right\rangle$

$= \langle u_1' + v_1', u_2' + v_2' \rangle$

$= \langle u_1', u_2' \rangle + \langle v_1', v_2' \rangle = \dfrac{d\mathbf{u}}{dt} + \dfrac{d\mathbf{v}}{dt}$

(b) $\dfrac{d}{dt}(\mathbf{u} - \mathbf{v}) = \dfrac{d}{dt}(\langle u_1 - v_1, u_2 - v_2 \rangle)$

$= \left\langle \dfrac{d}{dt}(u_1 - v_1), \dfrac{d}{dt}(u_2 - v_2) \right\rangle$

$= \langle u_1' - v_1', u_2' - v_2' \rangle$

$= \langle u_1', u_2' \rangle - \langle v_1', v_2' \rangle$

$= \dfrac{d\mathbf{u}}{dt} - \dfrac{d\mathbf{v}}{dt}$

43. $f(t)$ and $g(t)$ differentiable at $c \Rightarrow f(t)$ and $g(t)$ continuous at $c \Rightarrow \mathbf{r}(t) = f(t)\mathbf{i} + g(t)\mathbf{j}$ is continuous at c.

45. (a) Let $\mathbf{r}(t) = f(t)\mathbf{i} + g(t)\mathbf{j}$. Then

$\dfrac{d}{dt}\displaystyle\int_a^t \mathbf{r}(q)\, dq = \dfrac{d}{dt}\int_a^t [f(q)\mathbf{i} + g(q)\mathbf{j}]\, dq$

$= \dfrac{d}{dt}\left[\left(\int_a^t f(q)\, dq\right)\mathbf{i} + \left(\int_a^t g(q)\, dq\right)\mathbf{j}\right]$

$= \left(\dfrac{d}{dt}\int_a^t f(q)\, dq\right)\mathbf{i} + \left(\dfrac{d}{dt}\int_a^t g(q)\, dq\right)\mathbf{j}$

$= f(t)\mathbf{i} + g(t)\mathbf{j} = \mathbf{r}(t)$.

(b) Let $\mathbf{S}(t) = \displaystyle\int_a^t \mathbf{r}(q)\, dq$. Then part (a) shows that $\mathbf{S}(t)$ is an antiderivative of $\mathbf{r}(t)$. Let $\mathbf{R}(t)$ be any antiderivative of $\mathbf{r}(t)$. Then according to 37(b), $\mathbf{S}(t) = \mathbf{R}(t) + \mathbf{C}$.

Letting $t = a$, we have

$\mathbf{0} = \mathbf{S}(a) = \mathbf{R}(a) + \mathbf{C}$. Therefore,

$\mathbf{C} = -\mathbf{R}(a)$ and $\mathbf{S}(t) = \mathbf{R}(t) - \mathbf{R}(a)$.

The result follows by letting $t = b$.

Section 10.4 (pp. 539–552)

Quick Review

1. x-component $= 50 \cos 25° \approx 45.315$
y-component $= 50 \sin 25° \approx 21.131$

2. x-component $= 80 \cos 120° = -40$
y-component $= 80 \sin 120° = 40\sqrt{3} \approx 69.282$

3. x-intercepts: $\left(\dfrac{5}{2}, 0\right)$ and $(-8, 0)$
y-intercept: $(0, -40)$

4. $\left(-\dfrac{11}{4}, -\dfrac{441}{8}\right)$

5. x-intercepts: $(0, 0)$ and $(20, 0)$
y-intercept: $(0, 0)$

6. $(10, 100)$ **7.** $y = -\cos x + 2$

8. $y = \dfrac{t^3}{3} + 3t + \dfrac{25}{3}$ **9.** $y = 16 + 4e^{-t}$

10. $y = 2 - e^{-x^2}$

Exercises

1. 50 seconds

3. (a) After ≈ 72.154 seconds,
≈ 25.510 km downrange

(b) 4020 m   **(c)** ≈ 6377.55 m

5. After ≈ 2.135 seconds, ≈ 66.421 feet from the stopboard

7. (a) $7\sqrt{2} \approx 9.899$ m/sec
   (b) ≈ 18.435 or 71.565 degrees

9. ≈ 278.016 ft/sec, or ≈ 189.556 mph

11. No. When it has travelled 135 ft in the horizontal direction, it is only about 29.942 feet above the ground.

13. (a) ≈ 149.307 ft/sec   **(b)** ≈ 2.245 seconds

15. ≈ 39.261 and 50.739 degrees

17. ≈ 46.597 ft/sec

19. Integrating, $\dfrac{d}{dt}\mathbf{r}(t) = c_1\mathbf{i} + (-gt + c_2)\mathbf{j}$.

The initial condition on the velocity gives

$c_1 = v_0 \cos \alpha$ and $c_2 = v_0 \sin \alpha$. Integrating again,

$\mathbf{r}(t) = ((v_0 \cos \alpha)t + c_3)\mathbf{i}$
$\qquad + \left(-\dfrac{gt^2}{2} + (v_0 \sin \alpha)t + c_4\right)\mathbf{j}.$

The initial condition on the position gives

$c_3 = x_0$ and $c_4 = y_0$.

21. It takes about 1.924 seconds. The arrow passes about 3.698 feet above each rim. (It's 73.698 feet above the ground.)

23. Angle $\approx 62°$
Maximum height$- 4$ feet (independent of the measured angle)
Speed of engine≈ 8.507 ft/sec (changes with the angle)

25. (a)

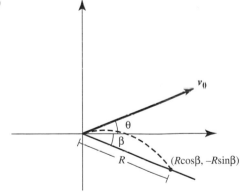

$x = (v_0 \cos \theta)t$
$y = (v_0 \sin \theta)t - \dfrac{1}{2}gt^2$

$x = R \cos \beta \Rightarrow R \cos \beta = (v_0 \cos \theta)t$

$\Rightarrow t = \dfrac{R \cos \beta}{v_0 \cos \theta}.$ Then $y = -R \sin \beta$

$\Rightarrow -R \sin \beta = \dfrac{(v_0 \sin \theta)\, R \cos \beta}{v_0 \cos \theta} - \dfrac{g}{2}\dfrac{R^2 \cos^2 \beta}{v_0^2 \cos^2 \theta}$

$\Rightarrow R = \dfrac{2v_0^2}{g \cos^2 \beta} \cos \theta \sin (\theta + \beta).$

Let $f(\theta) = \cos \theta \sin (\theta + \beta)$.
$f'(\theta) = \cos \theta \cos (\theta + \beta) - \sin \theta \sin (\theta + \beta)$
$f'(\theta) = 0 \Rightarrow \tan \theta \tan (\theta + \beta) = 1$
$\Rightarrow \tan \theta = \cot (\theta + \beta)$
$\Rightarrow \theta + \beta = 90° - \theta$. Note that $f''(\theta) < 0$, so R is maximum when $\theta + \beta = 90° - \theta$.

(b)

$R = \dfrac{2v_0^2}{g \cos^2 \beta} \cos \theta \sin (\theta - \beta)$ is maximum when $\tan \theta = \cot (\theta - \beta)$,

so $\theta - \beta = 90° - \theta$.

The initial velocity vector bisects the angle between the hill and the vertical for max range.

27. (a) (Assuming that "x" is zero at the point of impact.)
$\mathbf{r}(t) = (x(t))\mathbf{i} + (y(t))\mathbf{j}$, where
$x(t) = (35 \cos 27°)t$ and
$y(t) = 4 + (35 \sin 27°)t - 16t^2$.

(b) At $t \approx 0.497$ seconds, it reaches its maximum height of about 7.945 feet.

(c) Range ≈ 37.460 feet
flight time ≈ 1.201 seconds

(d) At $t \approx 0.254$ and $t \approx 0.740$ seconds, when it is ≈ 29.554 and ≈ 14.396 feet from where it will land.

(e) Yes. It changes things because the ball won't clear the net.

29. (a) $\mathbf{r}(t) = (x(t))\mathbf{i} + (y(t))\mathbf{j}$, where

$x(t) = \left(\dfrac{1}{0.08}\right)(1 - e^{-0.08t})(152 \cos 20° - 17.6)$ and

$y(t) = 3 + \left(\dfrac{152}{0.08}\right)(1 - e^{-0.08t})(\sin 20°)$
$\qquad + \left(\dfrac{32}{0.08^2}\right)(1 - 0.08t - e^{-0.08t})$

29. continued
  (b) At $t \approx 1.527$ seconds it reaches its maximum height of about 41.893 feet.
  (c) Range ≈ 351.734 feet
    Flight time ≈ 3.181 seconds
  (d) At $t \approx 0.877$ and $t \approx 2.190$ seconds, when it is about 106.028 and 251.530 feet from home plate.
  (e) No. The wind gust would need to be greater than 12.846 ft/sec in the direction of the hit in order for the ball to clear the fence for a home run.

31. The points in question are $(x, y) = \left(\dfrac{R}{2}, y_{\max}\right)$. So,

$$x = \frac{v_0^2 \sin \alpha \cos \alpha}{g}, \text{ and } y = \frac{(v_0 \sin \alpha)^2}{2g}.$$

Substituting these into the given equation for the ellipse yields an identity.

Section 10.5 (pp. 552–559)

Quick Review

1. $y = -x + 2$ **2.** $x^2 + y^2 = 9$
3. $(x + 2)^2 + (y - 4)^2 = 4$
4. (a) No   **(b)** No   **(c)** Yes
5. (a) No   **(b)** No   **(c)** No
6. (a) No   **(b)** Yes   **(c)** No
7. (a) Yes   **(b)** Yes   **(c)** Yes
8. Graph $y = (x - 2)^{1/2}$ and $y = -(x - 2)^{1/2}$
9. Graph $y = \left(\dfrac{4 - x^2}{3}\right)^{1/2}$ and $y = -\left(\dfrac{4 - x^2}{3}\right)^{1/2}$
10. $(x - 2)^2 + (y + 3)^2 = 4$, center $= (2, -3)$, radius $= 2$

Exercises

1. (a) and **(e)** are the same.
   (b) and **(g)** are the same.
   (c) and **(h)** are the same.
   (d) and **(f)** are the same.
3.

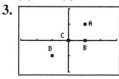

[−3, 3] by [−2, 2]
(a) $(1, 1)$ **(b)** $(1, 0)$
(c) $(0, 0)$ **(d)** $(-1, -1)$

5.

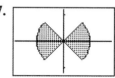

[−6, 6] by [−4, 4]
(a) $\left(\sqrt{2}, \dfrac{3\pi}{4}\right)$ or $\left(\sqrt{2}, -\dfrac{5\pi}{4}\right)$
(b) $\left(2, -\dfrac{\pi}{3}\right)$ or $\left(-2, \dfrac{2\pi}{3}\right)$
(c) $\left(3, \dfrac{\pi}{2}\right)$ or $\left(3, \dfrac{5\pi}{2}\right)$
(d) $(1, \pi)$ or $(-1, 0)$

7.

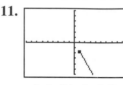

[−6, 6] by [−4, 4]

9.

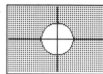

[−3, 3] by [−2, 2]

11.

[−9, 9] by [−6, 6]

13.

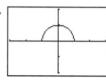

[−3, 3] by [−2, 2]

15.

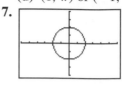

17.

[−1.8, 1.8] by [−1.2, 1.2]

19. $y = 0$, the x-axis
21. $y = 4$, a horizontal line
23. $x + y = 1$, a line (slope $= -1$, y-int. $= 1$)
25. $x^2 + y^2 = 4$, a circle (center $= (0, 2)$, radius $= 2$)
27. $xy = 1$ $\left(\text{or, } y = \dfrac{1}{x}\right)$, a hyperbola
29. $y = e^x$, the exponential curve
31. $y = \ln x$, the logarithmic curve
33. $x^2 + y^2 = -4x$, a circle (center $= (-2, 0)$, radius $= 2$)
35. $x^2 + y^2 = 2x + 2y$, a circle (center $= (1, 1)$, radius $= \sqrt{2}$)
37. $r \cos \theta = 7$ **39.** $\theta = \dfrac{\pi}{4}$
41. $r^2 = 4$ or $r = 2$
43. $r^2(4 \cos^2 \theta + 9 \sin^2 \theta) = 36$
45. $r \sin^2 \theta = 4 \cos \theta$ **47.** $r = 4 \sin \theta$

49. (a)

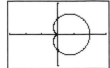

[−3, 3] by [−2, 2]

(b) Length of interval = 2π

51. (a)

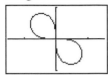

[−1.5, 1.5] by [−1, 1]

(b) Length of interval = $\dfrac{\pi}{2}$

53. (a)

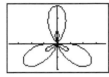

[−3.75, 3.75] by [−2, 3]

(b) Length of interval = 2π

55. (a)

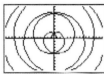

[−15, 15] by [−10, 10]

(b) Required interval = $(-\infty, \infty)$

57. (a)

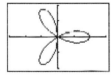

[−3, 3] by [−2, 2]

(b) Length of interval = π

59. x-axis, y-axis, origin   **61.** y-axis

63. (a) Because $r = a \sec \theta$ is equivalent to $r \cos \theta = a$ which is equivalent to the cartesian equation $x = a$.

(b) $r = a \csc \theta$ is equivalent to $y = a$.

65. (a) We have $x = r \cos \theta$ and $y = r \sin \theta$. By taking $t = \theta$, we have $r = f(t)$, so $x = f(t) \cos t$ and $y = f(t) \sin t$.

(b) $x = 3 \cos t, y = 3 \sin t$

(c) $x = (1 - \cos t) \cos t, y = (1 - \cos t) \sin t$

(d) $x = (3 \sin 2t) \cos t, y = (3 \sin 2t) \sin t$

67. $d = [(x_2 - x_1)^2 + (y_2 + y_1)^2]^{1/2}$

$= [(r_2 \cos \theta_2 - r_1 \cos \theta_1)^2$

$\qquad + (r_2 \sin \theta_2 - r_1 \sin \theta_1)^2]^{1/2}$

and then simplify using trigonometric identities.

69. (a)

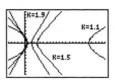

[−5, 25] by [−10, 10]

The graphs are hyperbolas.

(b) As $k \to 1^+$, the right branch of the hyperbola goes to infinity and "disappears". The left branch approaches the parabola $y^2 = 4 - 4x$.

Section 10.6 (pp. 559–568)

Quick Review

1. $-\dfrac{5}{3} \cot t$ \qquad **2.** $-\dfrac{5}{3} \cot 2 \approx 0.763$

3. $(0, 5)$ and $(0, -5)$ \qquad **4.** $(3, 0)$ and $(-3, 0)$

5. ≈ 12.763

6. The upper half of the outer loop

7. The inner loop

8. The lower half of the outer loop

9. 36 \qquad **10.** ≈ 2.403

Exercises

1. At $\theta = 0$: -1

At $\theta = \pi$: 1

3. At $(2, 0)$: $-\dfrac{2}{3}$

At $\left(1, \dfrac{\pi}{2}\right)$: 0

At $(2, \pi)$: $\dfrac{2}{3}$

At $\left(5, \dfrac{3\pi}{2}\right)$: 0

5. $\theta = \dfrac{\pi}{2}$ \quad $[x = 0]$

7. $\theta = 0$ \quad $[y = 0]$

$\theta = \dfrac{\pi}{5}$ \quad $\left[y = \left(\tan \dfrac{\pi}{5}\right)x\right]$

$\theta = \dfrac{2\pi}{5}$ \quad $\left[y = \left(\tan \dfrac{2\pi}{5}\right)x\right]$

$\theta = \dfrac{3\pi}{5}$ \quad $\left[y = \left(\tan \dfrac{3\pi}{5}\right)x\right]$

$\theta = \dfrac{4\pi}{5}$ \quad $\left[y = \left(\tan \dfrac{4\pi}{5}\right)x\right]$

9. Horizontal at: $\left(-\dfrac{1}{2}, \dfrac{\pi}{6}\right)$ $\left[y = -\dfrac{1}{4}\right]$,

$\left(-\dfrac{1}{2}, \dfrac{5\pi}{6}\right)$ $\left[y = -\dfrac{1}{4}\right]$,

$\left(-2, \dfrac{3\pi}{2}\right)$ $[y = 2]$

Vertical at: $\left(0, \dfrac{\pi}{2}\right)$ $[x = 0]$,

$\left(-\dfrac{3}{2}, \dfrac{7\pi}{6}\right)$ $\left[x = \dfrac{3\sqrt{3}}{4}\right]$,

$\left(-1.5, \dfrac{11\pi}{6}\right)$ $\left[x = -\dfrac{3\sqrt{3}}{4}\right]$

11. Horizontal at: $(0, 0)$ $[y = 0]$,

$\left(2, \dfrac{\pi}{2}\right)$ $[y = 2]$,

$(0, \pi)$ $[y = 0]$

Vertical at: $\left(\sqrt{2}, \dfrac{\pi}{4}\right)$ $[x = 1]$,

$\left(\sqrt{2}, \dfrac{3\pi}{4}\right)$ $[x = -1]$

13. 18π **15.** $2a^2$ **17.** 2

19. $\dfrac{\pi}{2} - 1$ **21.** $5\pi - 8$ **23.** $a^2\pi$

25. $8 - \pi$ **27.** $12\pi - 9\sqrt{3}$

29. (a) $\dfrac{3}{2} - \dfrac{\pi}{4}$

(b) Yes. $x = \tan\theta \cos\theta \Rightarrow x = \sin\theta$

$y = \tan\theta \sin\theta \Rightarrow y = \dfrac{\sin^2\theta}{\cos\theta}$

$\lim\limits_{\theta \to -\pi/2^+} x = -1, \quad \lim\limits_{\theta \to -\pi/2^+} = +\infty$

$\lim\limits_{\theta \to \pi/2^-} x = 1, \quad \lim\limits_{\theta \to \pi/2^-} y = +\infty$

31. $\dfrac{19}{3}$ **33.** 8

35. ≈ 6.887 **37.** $\dfrac{\pi + 3}{8}$

39. $\pi\sqrt{2} \approx 4.443$ **41.** $(4 - 2\sqrt{2})\pi \approx 3.681$

43. $\left(\dfrac{dx}{d\theta}\right)^2 + \left(\dfrac{dy}{d\theta}\right)^2$

$= (f'(\theta)\cos\theta - f(\theta)\sin\theta)^2$
$\quad + (f'(\theta)\sin\theta + f(\theta)\cos\theta)^2$
$= (f'(\theta)\cos\theta)^2 + (f(\theta)\sin\theta)^2 + (f'(\theta)\sin\theta)^2$
$\quad + (f(\theta)\cos\theta)^2$
$= (f(\theta))^2(\cos^2\theta + \sin^2\theta)$
$\quad + (f'(\theta))^2(\cos^2\theta + \sin^2\theta)$
$= (f(\theta))^2 + (f'(\theta))^2 = r^2 + \left(\dfrac{dr}{d\theta}\right)^2$

45. If $g(\theta) = 2f(\theta)$, then
$(g(\theta)^2 + g'(\theta)^2)^{1/2} = 2(f(\theta)^2 + f'(\theta)^2)^{1/2}$ so the
length of g is 2 times the length of f.

47. (a) Let $r = 1.75 + \dfrac{0.06\theta}{2\pi}, \ 0 \le \theta \le 12\pi$.

(b) Since $\dfrac{dr}{d\theta} = \dfrac{b}{2\pi}$, this is just equation (4) for
the length of the curve.

(c) ≈ 741.420 cm, or, ≈ 7.414 m

(d) $\left(\left(r^2 + \dfrac{b}{2\pi}\right)^2\right)^{1/2} = r\left(\left(1 + \dfrac{b}{2\pi r}\right)^2\right)^{1/2} \approx r$ since
$\left(\dfrac{b}{2\pi r}\right)^2$ is a very small quantity squared.

(e) $L \approx 741.420$ cm (from part (c)),
$L_a \approx 741.416$ cm

49. $\left(\dfrac{5a}{6}, 0\right)$

Chapter Review (pp. 569–572)

1. (a) $\langle -17, 32 \rangle$ **(b)** $\sqrt{1313}$
2. (a) $\langle -1, -1 \rangle$ **(b)** $\sqrt{2}$
3. (a) $\langle 6, -8 \rangle$ **(b)** 10
4. (a) $\langle 10, -25 \rangle$ **(b)** $\sqrt{725} = 5\sqrt{29}$
5. $\left(-\dfrac{\sqrt{3}}{2}, -\dfrac{1}{2}\right)$ [assuming counterclockwise]
6. $\left(\dfrac{\sqrt{3}}{2}, \dfrac{1}{2}\right)$ **7.** $\left(\dfrac{8}{\sqrt{17}}, -\dfrac{2}{\sqrt{17}}\right)$
8. $\langle -3, -4 \rangle$
9. (a) $y = \dfrac{\sqrt{3}}{2}x + \dfrac{1}{4}$ **(b)** $\dfrac{1}{4}$
10. (a) $y = -3x + \dfrac{13}{4}$ **(b)** 6
11. (a) $\left(0, \dfrac{1}{2}\right)$ and $\left(0, -\dfrac{1}{2}\right)$
(b) Nowhere
12. (a) $(0, 2)$ and $(0, -2)$ **(b)** $(-2, 0)$ and $(2, 0)$
13. (a) $(0, 0)$ **(b)** Nowhere
14. (a) $(0, 9)$ and $(0, -9)$ **(b)** $(-4, 0)$ and $(4, 0)$
15.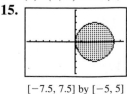
16.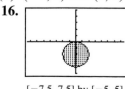

$[-7.5, 7.5]$ by $[-5, 5]$ $[-7.5, 7.5]$ by $[-5, 5]$

17. (a)

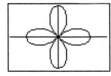

[−1.5, 1.5] by [−1, 1]

(b) 2π

18. (a)

[−3, 3] by [−2, 2]

(b) π

19. (a)

[−1.5, 1.5] by [−1, 1]

(b) $\dfrac{\pi}{2}$

20. (a)

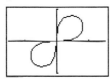

[−1.5, 1.5] by [−1, 1]

(b) π

21. Tangent lines at $\theta = \dfrac{\pi}{4}, \dfrac{3\pi}{4}, \dfrac{5\pi}{4}$, and $\dfrac{7\pi}{4}$
Cartesian equations are $y = \pm x$.

22. Tangent lines at $\theta = \dfrac{\pi}{2}$ and $\dfrac{3\pi}{2}$
Cartesian equation is $x = 0$.

23. Horizontal: $y = 0$, $y \approx \pm 0.443$, $y \approx \pm 1.739$
Vertical: $x = 2$, $x \approx 0.067$, $x \approx -1.104$

24. Horizontal: $y = \dfrac{1}{2}$, $y = -4$
Vertical: $x = 0$, $x \approx \pm 2.598$

25. $y = \pm x + \sqrt{2}$ and $y = \pm x - \sqrt{2}$
26. $y = x - 1$ and $y = -x - 1$
27. $x = y$, a line
28. $x^2 + y^2 = 3x$,
a circle $\left(\text{center} = \left(\dfrac{3}{2}, 0\right), \text{radius} = \dfrac{3}{2}\right)$
29. $x^2 = 4y$, a parabola
30. $x - \sqrt{3}y = 4\sqrt{3}$ or $y = \dfrac{x}{\sqrt{3}} - 4$, a line
31. $r = -5 \sin \theta$ **32.** $r = 2 \sin \theta$

33. $r^2 \cos^2 \theta + 4r^2 \sin^2 \theta = 16$, or,
$r^2 = \dfrac{16}{\cos^2 \theta + 4 \sin^2 \theta}$
34. $(r \cos \theta + 2)^2 + (r \sin \theta - 5)^2 = 16$
35. $\dfrac{\ln 2 + 24}{8} \approx 3.087$ **36.** $4\sqrt{3}$
37. 8   **38.** $\pi\sqrt{2}$
39. $\pi - 3$ **40.** $\pi\sqrt{2}$
41. ≈ 3.183 **42.** ≈ 12.363
43. $\dfrac{9\pi}{2}$ **44.** $\dfrac{\pi}{12}$
45. $\dfrac{\pi}{4} + 2$ **46.** 5π
47. $\dfrac{76\pi}{3}$ **48.** ≈ 10.110
49. ≈ 1.840 **50.** ≈ 12.566
51. (a) $\mathbf{v}(t) = (-4 \sin t)\mathbf{i} + (\sqrt{2} \cos t)\mathbf{j}$
$\mathbf{a}(t) = (-4 \cos t)\mathbf{i} + (-\sqrt{2} \sin t)\mathbf{j}$
(b) 3
(c) $\cos^{-1} \dfrac{7}{9} \approx 38.942$ degrees

52. (a) $\mathbf{v}(t) = (\sqrt{3} \sec t \tan t)\mathbf{i} + (\sqrt{3} \sec^2 t)\mathbf{j}$
$\mathbf{a}(t) = (\sqrt{3} \sec t \tan^2 t + \sec^2 t)\mathbf{i}$
$+ (2\sqrt{3} \sec^2 t \tan t)\mathbf{j}$
(b) $\sqrt{3}$ **(c)** 90 degrees
53. 1
54. $\mathbf{a}(t) = (-2 e^t \sin t)\mathbf{i} + (2 e^t \cos t)\mathbf{j}$, and
$\mathbf{r}(t) \cdot \mathbf{a}(t) = 0$ for all t. The angle between $\mathbf{r}$ and
$\mathbf{a}$ is always 90 degrees.
55. $6\mathbf{i}$ **56.** $3\mathbf{i} + (\ln 2)\mathbf{j}$
57. $\mathbf{r}(t) = (\cos t - 1)\mathbf{i} + (\sin t + 1)\mathbf{j}$
58. $\mathbf{r}(t) = (\tan^{-1} t + 1)\mathbf{i} + ((t^2 + 1)^{1/2})\mathbf{j}$
59. $\mathbf{r}(t) = \mathbf{i} + t^2\mathbf{j}$
60. $\mathbf{r}(t) = (-t^2 + 6t - 2)\mathbf{i} + (-t^2 + 2t + 2)\mathbf{j}$

61. (a) $\dfrac{\pi\sqrt{34}}{4\sqrt{2}} = \dfrac{\pi\sqrt{17}}{4} \approx 3.238$
(b) x-component: $\dfrac{3\pi^2}{16\sqrt{2}}$
y-component: $-\dfrac{5\pi^2}{16\sqrt{2}}$
(c) $\dfrac{x^2}{9} + \dfrac{y^2}{25} = 1$
62. (a) ≈ 25.874
(b) Volume $= \dfrac{1250\pi}{3}$
(c) Area ≈ 1040.728
63. (a) 1   **(b)** $e^3\sqrt{2}$
(c) $(e^3 - 1)\sqrt{2}$

64. (a) $\dfrac{104}{5}$ **(b)** $\dfrac{4144}{135}$

(c) $\dfrac{dy}{dx} = \dfrac{3}{5}\sqrt{x+2}$

65. Speed ≈ 591.982 mph
Direction ≈ 8.179 degrees north of east

66. Direction ≈ 2.073 degrees
Magnitude ≈ 411.891 lbs

67. It hits the ground ≈ 2.135 seconds later, approximately 66.421 feet from where it left the thrower's hand. Assuming it doesn't bounce or roll, it will still be there 3 seconds after it was thrown.

68. 57 feet

69. (a)

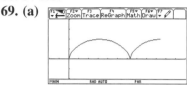

[−2, 12] by [−2, 4]

(b) $\mathbf{v}(0) = \langle 0, 0 \rangle$ $\mathbf{v}(1) = \langle 2\pi, 0 \rangle$
  $\mathbf{a}(0) = \langle 0, \pi^2 \rangle$ $\mathbf{a}(1) = \langle 0, -\pi^2 \rangle$
  $\mathbf{v}(2) = \langle 0, 0 \rangle$ $\mathbf{v}(3) = \langle 2\pi, 0 \rangle$
  $\mathbf{a}(2) = \langle 0, \pi^2 \rangle$ $\mathbf{a}(3) = \langle 0, -\pi^2 \rangle$

(c) Topmost point: 2π ft/sec
center of wheel: π ft/sec
Reasons: Since the wheel rolls half a circumference, or π feet every second, the center of the wheel will move π feet every second. Since the rim of the wheel is turning at a rate of π ft/sec about the center, the velocity of the topmost point relative to the center is π ft/sec, giving it a total velocity of 2π ft/sec.

70. For 4325 yds: $v \approx 644.360$ ft/sec
For 4752 yds: $v \approx 675.420$ ft/sec

71. (a) ≈ 59.195 ft/sec **(b)** ≈ 74.584 ft/sec

72. (a) ≈ 91.01 ft/sec **(b)** 59.97 ft

73. We have $x = (v_0 t)\cos\alpha$ and

$y + \dfrac{gt^2}{2} = (v_0 t)\sin\alpha$. Squaring and adding gives

$x^2 + \left(y + \dfrac{gt^2}{2}\right)^2 = (v_0 t)^2(\cos^2\alpha + \sin^2\alpha) = v_0^2 t^2$.

74. (a) $\mathbf{r}(t) = (155\cos 18° - 11.7)t\mathbf{i}$
   $+ (4 + 155\sin 18°t - 16t^2)\mathbf{j}$
  $x(t) = (155\cos 18° - 11.7)t$
  $y(t) = 4 + 155\sin 18°t - 16t^2$

(b) At ≈ 1.497 seconds, it reaches a maximum height of ≈ 39.847 ft.

(c) Range ≈ 417.307 ft
Flight time ≈ 3.075 secs

(d) At times $t \approx 0.534$ and $t \approx 2.460$ seconds, when it is ≈ 72.406 and ≈ 333.867 feet from home plate.

(e) Yes, the batter has hit a home run. When the ball is 380 feet from home plate (at $t \approx 2.800$ seconds), it is approximately 12.673 feet off the ground and therefore clears the fence by at least two feet.

75. (a)
$$\mathbf{r}(t) = \left[(155\cos 18° - 11.7)\left(\frac{1}{0.09}\right)(1 - e^{-0.09t})\right]\mathbf{i}$$
$$+\, [\,4 + \left(\frac{155\sin 18°}{0.09}\right)(1 - e^{-0.09t})$$
$$+\, \frac{32}{0.09^2}(1 - 0.09t - e^{-0.09t})]\mathbf{j}$$
$$x(t) = (155\cos 18° - 11.7)\left(\frac{1}{0.09}\right)(1 - e^{-0.09t})$$
$$y(t) = 4 + \left(\frac{155\sin 18°}{0.09}\right)(1 - e^{-0.09t})$$
$$+\, \frac{32}{0.09^2}(1 - 0.09t - e^{-0.09t})$$

(b) At ≈ 1.404 seconds, it reaches a maximum height of ≈ 36.921 feet.

(c) Range ≈ 352.520 ft
Flight time ≈ 2.959 secs

(d) At times $t \approx 0.753$ and $t \approx 2.068$ seconds, when it is ≈ 98.799 and ≈ 256.138 feet from home plate

(e) No, the batter has not hit a home run. If the drag coefficient k is less than ≈ 0.011, the hit will be a home run.

76. (a) $\overrightarrow{BD} = \overrightarrow{AD} - \overrightarrow{AB}$

(b) $\overrightarrow{AP} = \overrightarrow{AB} + \dfrac{1}{2}\overrightarrow{BD} = \dfrac{1}{2}\overrightarrow{AB} + \dfrac{1}{2}\overrightarrow{AD}$

(c) $\overrightarrow{AC} = \overrightarrow{AB} + \overrightarrow{AD}$, so by part (b), $\overrightarrow{AP} = \dfrac{1}{2}\overrightarrow{AC}$.

77. The widths between the successive turns are constant and are given by $2\pi a$.

Cumulative Review Exercises

Cumulative Review (pp. 573–576)

1. 0

2. $\dfrac{3}{4}$

3. -1

4. -1

5. 3

6. 1

7. e^2

8. 3

9. (a) 1   **(b)** 1

   (c) 1   **(d)** Yes

   (e) No

10. All $x \le -2$ and $x \ge 2$

11. Horizontal: $y = 0$

   Vertical: $x = 0$, $x = \dfrac{1}{2}$

12.

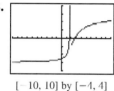

$[-10, 10]$ by $[-4, 4]$

13. $\dfrac{1}{5}$

14. $-\dfrac{3}{(x-2)^2}$

15. $\dfrac{3 \sin (\sqrt{1 - 3x})}{2\sqrt{1 - 3x}}$

16. $\sin x \sec^2 x + \cos x \tan x = \dfrac{(\sin x)(1 + \cos^2 x)}{\cos^2 x}$

17. $\dfrac{2x}{x^2 + 1}$

18. $(2x - 1)e^{x^2 - x}$

19. $2x \tan^{-1} x + \dfrac{x^2}{1 + x^2}$

20. $(x^{-3} - 3x^{-4})e^x$

21. $\dfrac{3 \csc^2 x}{(1 + \cos x)^4}(1 - \csc x \cot x - \cos x \csc x \cot x)$

   $= \dfrac{3(1 - 2 \cos x)}{(\sin^4 x)(1 + \cos x)^3}$

22. $-\dfrac{1}{\sqrt{1 - x^2}} + \dfrac{1}{1 + x^2}$

23. $\dfrac{1 + xy \sin (xy)}{2xy - x^2 \sin (xy)}$

24. $\dfrac{|x|}{2x\sqrt{|x|}}$

25. $\cot t = \dfrac{x - 1}{1 - y}$

26. $(\cos x)^{x-1}(\cos x \ln (\cos x) - x \sin x)$

27. $\sqrt{1 + x^3}$

28. $2x \sin (x^2) - 2 \sin (2x)$

29. $\dfrac{(2y + 2)^2(\sec^3 x + \sec x \tan^2 x) - 2 \sec^2 x \tan^2 x}{(2y + 2)^3}$

30. -1

31. (a) $v = 3t^2 - 12t + 9$, $a = 6t - 12$

   (b) $t = 1$ or $t = 3$

   (c) Right: $0 \le t < 1$, $3 < t \le 5$

     Left: $1 < t < 3$

   (d) -3 m/sec

32. (a) $y = -2x + 1$

   (b) $y = \dfrac{1}{2}x - \dfrac{3}{2}$

33. (a) $y = \left(\dfrac{3 - \pi\sqrt{3}}{6}\right)\left(x - \dfrac{\pi}{3}\right) + \dfrac{\pi}{6}$

   $\approx -0.407x + 0.950$

   (b) $y = \left(\dfrac{6}{\pi\sqrt{3} - 3}\right)\left(x - \dfrac{\pi}{3}\right) + \dfrac{\pi}{6}$

   $\approx 2.458x - 2.050$

34. (a) $y = -\dfrac{\sqrt{3}}{2}x + 2\sqrt{3} \approx -0.866x + 3.464$

   (b) $y = \dfrac{2}{\sqrt{3}}x + \dfrac{5}{2\sqrt{3}} \approx 1.155x + 1.443$

35. (a) $y = -\dfrac{\sqrt{3}}{2}x + 2\sqrt{3} \approx -0.866x + 3.464$

   (b) $y = \dfrac{2}{\sqrt{3}}x + \dfrac{5}{2\sqrt{3}} \approx 1.155x + 1.443$

36. (a) $y = \sqrt{2}x - 1 \approx 1.414x - 1$

   (b) $y = -\dfrac{1}{\sqrt{2}}x + 2 \approx -0.707x + 2$

37. $f(x) = \begin{cases} -x + 4, & x \le 3 \\ 2x - 5, & x > 3 \end{cases}$

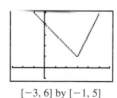

$[-3, 6]$ by $[-1, 5]$

38. (a) $x \ne 0, 2$ **(b)** $x = 0$

   (c) $x = 2$

   (d) Absolute maximum of 2 at $x = 0$

     Absolute minimum of 0 at $x = -2, 2, 3$

39. According to the Mean Value Theorem the driver speed at some time was $\dfrac{111}{1.5} = 74$ mph.

40. (a) Increasing in $[-0.7, 2]$, decreasing in $[-2, -0.7]$, and has a local minimum at $x = -0.7$.

(b) $y \approx -2x^2 + 3x + 3$

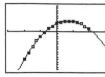

[−3, 3] by [−15, 10]

(c) $f(x) = -\frac{2}{3}x^3 + \frac{3}{2}x^2 + 3x + 1$

41. $f(x) = x^2 - 3x - \cos x - 1$

42. (a) $\left[-2, -\frac{2\sqrt{6}}{3}\right], \left[0, \frac{2\sqrt{6}}{3}\right]$

(b) $\left[-\frac{2\sqrt{6}}{3}, 0\right], \left[\frac{2\sqrt{6}}{3}, 2\right]$

(c) $\approx(-1.042, 1.042)$

(d) $\approx(-2, -1.042), \approx(1.042, 2)$

(e) Local max of approximately 3.079 at $x = -\frac{2\sqrt{6}}{3}$ and $x = \frac{2\sqrt{6}}{3}$ local min of 0 at $x = 0$

(f) Points of inflection at about $x = \pm 1.042$

43. (a) f has an absolute maximum at $x = 1$ and an absolute minimum at $x = 3$.

(b) f has a point of inflection at $x = 2$.

(c)

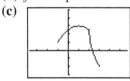

[−3.7, 5.7] by [−3,5]

44. Dimensions: $4\sqrt{2}$ by $\sqrt{2}$, area: 8

45. $y = \sqrt{2}\left(x - \frac{\pi}{4}\right) + \sqrt{2} \approx 1.414x + 0.303$

46. 3%

47. (a) About 1.9 ft/sec

(b) About 0.15 rad/sec

48. (a) $\frac{9}{16\pi} \approx 0.179$ in./min

(b) $\frac{36}{25\pi} \approx 0.458$ in./min

49. (a) 165 in. **(b)** 165 in.

50. 2.5 **51.** 2π

52. $\ln 3 + \frac{26}{3} \approx 9.765$

53. 1 **54.** 7

55. $\frac{\ln 2}{1 + \ln 2} \approx 0.409$

56. $-2\mathbf{i} + \ln(3)\mathbf{j}$

57. $-\cot(e^x + 1) - e^x + C$

58. $\frac{1}{2}\tan^{-1}\left(\frac{s}{2}\right) + C$

59. $\frac{1}{2\cos^2(x - 3)} + C$

60. $\frac{e^{-x}}{5}(2\sin 2x - \cos 2x) + C$

61. $\frac{8}{7}\ln|x - 6| - \frac{1}{7}\ln|x + 1| + C$

$= \frac{1}{7}\ln\frac{(x - 6)^8}{|x + 1|} + C$

62. 8975 ft³

63. $y = -\frac{1}{t + 1} - \frac{1}{2}e^{-2t} + \frac{7}{2}$

64. $y = -\frac{1}{4}\sin 2\theta + \cos \theta + \frac{1}{2}\theta - \frac{\pi}{4}$

65. $\int x^2 \sin x \, dx = (2 - x^2)\cos x + 2x \sin x + C$

The graph of the slope field of the differential equation $\frac{dy}{dx} = x^2 \sin x$ and the antiderivative $y = (2 - x^2)\cos x + 2x \sin x$ is shown below.

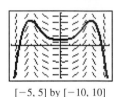

[−5, 5] by [−10, 10]

66. $\int xe^x \, dx = e^x(x - 1) + C$

$\frac{d}{dx}(e^x(x - 1) + C) = e^x$

67. (a) $y = 4268e^{kt}, k = \frac{\ln(5/3)}{3} \approx 0.170$

(b) About 4268

68. (a) About 6.13°C

(b) About 2 hours and 27 minutes after it was 65°C above room temperature, or about 2 hours and 12 minutes after it was 50°C above room temperature.

69. $y = \frac{500}{1 + Ce^{-0.08x}}$

70. $y = Ce^{x^2/2 + 3x} + 4$

71.

x	y
0	0
0.1	0.1
0.2	0.2095
0.3	0.3285
0.4	0.4568
0.5	0.5946
0.6	0.7418
0.7	0.8986
0.8	1.0649
0.9	1.2411
1.0	1.4273

72. 4

73. $\dfrac{64}{3}$

74. ≈ 16.039

75. ≈ 42.412

76. $\dfrac{\pi}{14} \approx 0.224$

77. $\dfrac{128\pi}{3} \approx 134.041$

78. ≈ 0.763

79. ≈ 2.556

80. 4

81. ≈ 6.110

82. ≈ 8.423

83. ≈ 3.470

84. ≈ 32.683

85. ≈ 0.101

86. ≈ 5.394

87. (a) 1.2 m   (b) 180 J

88. (a) 70,686 ft-lb   (b) 4 min, 17 sec

89. 12.166 ft³

90. $f(x) = \ln x$ grows slower than $g(x) = \sqrt{x}$.

91. Converges

92. Diverges

93. Converges

94. Converges

95. Diverges

96. Converges

97. $1 - 2x + 4x^2 - 8x^3 + \cdots + (-1)^n 2^n x^n + \cdots$,

$-\dfrac{1}{2} < x < \dfrac{1}{2}$

98. (a) $x - \dfrac{x^5}{5 \cdot 2!} + \dfrac{x^9}{9 \cdot 4!} - \dfrac{x^{13}}{13 \cdot 6!} + \cdots$

$+ (-1)^n \dfrac{x^{4n+1}}{(4n+1) \cdot (2n)!} + \cdots$

   (b) $-\infty < x < \infty$; Since the cosine series converges for all real numbers, so does the integrated series, by Theorem 2, the term-by-term integration theorem.

99. $\ln(2 + 2x) = \ln 2 + \ln(1 + x) = \ln 2 + x - \dfrac{x^2}{2} + \dfrac{x^3}{3} - \dfrac{x^4}{4} + \cdots + (-1)^{n-1}\dfrac{x^n}{n} + \cdots$, which converges for $-1 < x \le 1$.

100. $(x - 2\pi) - \dfrac{(x - 2\pi)^3}{3!} + \dfrac{(x - 2\pi)^5}{5!} - \cdots$

$+ (-1)^n \dfrac{(x - 2\pi)^{2n+1}}{(2n+1)!} + \cdots$

101. $P_6(x) = 1 - x + \dfrac{x^2}{2!} - \dfrac{x^3}{3!} + \dfrac{x^4}{4!} - \dfrac{x^5}{5!} + \dfrac{x^6}{6!}$

By the Alternating Series Estimation Theorem,

$|\text{error}| \le \left|\dfrac{x^7}{7!}\right| \le \dfrac{1}{7!} < 0.001.$

102. $1 + \dfrac{1}{3}x - \dfrac{2}{2! \cdot 3^2}x^2 + \dfrac{2 \cdot 5}{3! \cdot 3^3}x^3 - \dfrac{2 \cdot 5 \cdot 8}{4! \cdot 3^4}x^4 + \cdots$

$+ (-1)^{n-1}\dfrac{2 \cdot 5 \cdot \cdots \cdot (3n - 4)}{n! \cdot 3^n}x^n + \cdots$

$R = 1$

103. Converges   **104.** Diverges

105. Converges   **106.** Converges

107. (a) $R = 1$; $-3 < x \le -1$

   (b) $-3 < x < -1$

   (c) At $x = -1$

108. (a) $R = 1$; $-1 \le x \le 1$

   (b) $-1 \le x \le 1$

   (c) Nowhere

109. $\left\langle \dfrac{2}{\sqrt{13}}, -\dfrac{3}{\sqrt{13}} \right\rangle$

110. $\left\langle \dfrac{1}{2}, \dfrac{\sqrt{3}}{2} \right\rangle$

111. Tangent: $\left\langle -\dfrac{4}{5}, -\dfrac{3}{5} \right\rangle, \left\langle \dfrac{4}{5}, \dfrac{3}{5} \right\rangle$

normal: $\left\langle \dfrac{3}{5}, -\dfrac{4}{5} \right\rangle, \left\langle -\dfrac{3}{5}, \dfrac{4}{5} \right\rangle$

112. (a) $\mathbf{v}(t) = (-\cos t)\mathbf{i} + (1 + \sin t)\mathbf{j}$

   $\mathbf{a}(t) = (\sin t)\mathbf{i} + (\cos t)\mathbf{j}$

   (b) 4

113. Yes. When $x = 130$ ft,

$t = \dfrac{130}{100}\cos 45° \approx 1.838$ sec and $y \approx 75.9$ ft, high enough to easily clear the 35-ft tree.

114. $x - y = 2$; A line with slope 1 and y-intercept -2.

115.

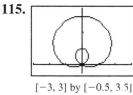

[−3, 3] by [−0.5, 3.5]

2π

116. Horizontal tangent line at:

$(0, 0) [y = 0], \left(\dfrac{3}{2}, \dfrac{2\pi}{3}\right)\left[y = \dfrac{3\sqrt{3}}{4}\right],$

$\left(\dfrac{3}{2}, \dfrac{4\pi}{3}\right)\left[y = -\dfrac{3\sqrt{3}}{4}\right], (0, 2\pi) [y = 0]$

Vertical tangent lines at:

$(2, \pi)[x = -2], \left(\dfrac{1}{2}, \dfrac{\pi}{3}\right)\left[x = \dfrac{1}{4}\right], \left(\dfrac{1}{2}, \dfrac{5\pi}{3}\right)\left[x = \dfrac{1}{4}\right]$

Appendix A3 (pp. 584–592)

1.

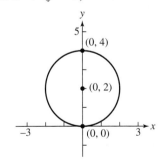

$\delta = \dfrac{1}{18}$

3. $\delta = 0.39$
5. $(-2.01, -1.99); \delta = 0.01$
7. $(3, 15); \delta = 5$
9. (a) -4 **(b)** $\delta = 0.05$
11. (a) $\sin 1 \approx 0.841$ **(b)** $\delta = 0.018$
13. $\delta = \min \{1 - \sqrt{1 - \epsilon}, \sqrt{1 + \epsilon} - 1\}$
15. (a) $I = (5, 5 + \epsilon^2)$
    (b) $\lim\limits_{x \to 5^+} \sqrt{x - 5} = 0$

Appendix A5.1 (pp. 593–606)

1. $x^2 + (y - 2)^2 = 4$

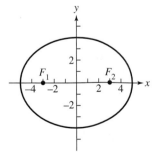

3. Center $= (-2, 2)$; radius $= 2$

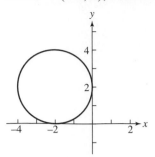

5. The circle with center at $(1, 0)$ and radius 2 plus its interior.
7. $y^2 = 8x$; focus is $(2, 0)$; directrix is $x = -2$
9. $x^2 = -6y$; focus is $\left(0, -\dfrac{3}{2}\right)$; directrix is $y = \dfrac{3}{2}$
11. $\dfrac{x^2}{4} - \dfrac{y^2}{9} = 1$; foci are $(\pm\sqrt{13}, 0)$;
vertices are $(\pm 2, 0)$; asymptotes are $y = \pm\dfrac{3}{2}x$

13. $\dfrac{x^2}{2} + y^2 = 1$; foci are $(\pm 1, 0)$;
vertices are $(\pm\sqrt{2}, 0)$

15. Focus is $(3, 0)$; directrix is $x = -3$

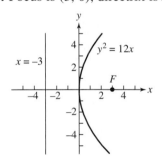

17. $\dfrac{x^2}{25} + \dfrac{y^2}{16} = 1$; foci are $(\pm 3, 0)$

19. $\dfrac{x^2}{4} + \dfrac{y^2}{2} = 1$

21. $x^2 - y^2 = 1$; asymptotes are $y = \pm x$; foci are $(\pm\sqrt{2}, 0)$

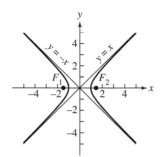

23. $y^2 - x^2 = 1$

25. (a) Vertex is $(1, -2)$; focus is $(3, -2)$; directrix is $x = -1$

(b)

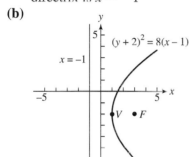

27. (a) Center is $(2, 0)$; foci are $(-3, 0)$ and $(7, 0)$; asymptotes are $y = \pm\dfrac{3(x-2)}{4}$; vertices are $(-2, 0)$ and $(6, 0)$

(b)

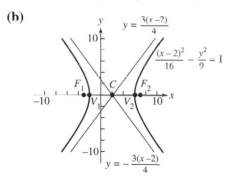

29. $\dfrac{(x+2)^2}{6} + \dfrac{(y+1)^2}{9} = 1$; vertices are $(-2, 2)$ and $(-2, -4)$; foci are $(-2, -1 \pm \sqrt{3})$; center is $(-2, -1)$

31. $(y-1)^2 - (x+1)^2 = 1$; vertices are $(-1, 2)$ and $(-1, 0)$; foci are $(-1, 1 \pm \sqrt{2})$; center is $(-1, 1)$; asymptotes are $y = \pm(x+1) + 1$

33. Circle; center is $(7, -3)$, radius is 1

35. Ellipse; center is $(-2, 0)$; foci are $(-4, 0)$ and $(0, 0)$; vertices are $(-2 \pm \sqrt{5}, 0)$

37. Volume of the parabolic solid is $V_1 = \dfrac{\pi h b^2}{8}$; volume of the cone is $V_2 = \dfrac{\pi h b^2}{12}$; $\dfrac{V_1}{V_2} = \dfrac{3}{2}$

39. The slopes of the two tangents to $y^2 = 4px$ from the point $(-p, a)$ are $m_1 = \dfrac{2p}{a + \sqrt{a^2 + 4p^2}}$ and $m_2 = \dfrac{2p}{a - \sqrt{a^2 + 4p^2}}$, and $m_1 m_2 = -1$.

41. (a) 24π **(b)** 16π

43. 24π

45. $\dfrac{dr_A}{dt} = \dfrac{dr_B}{dt} \Rightarrow \dfrac{d}{dt}(r_A - r_B) = 0$
$\Rightarrow r_A - r_B = $ a constant

Appendix A5.2 (pp. 606–611)

1. $e = \dfrac{3}{5}$; foci are $(\pm 3, 0)$; directrices are $x = \pm\dfrac{25}{3}$

3. $e = \dfrac{1}{\sqrt{3}}$; foci are $(0, \pm 1)$; directrices are $y = \pm 3$

5. $\dfrac{x^2}{27} + \dfrac{y^2}{36} = 1$ **7.** $\dfrac{x^2}{100} + \dfrac{y^2}{94.24} = 1$

9. $e = \dfrac{\sqrt{5}}{3}$; $\dfrac{x^2}{9} + \dfrac{y^2}{4} = 1$

11. Take $c = 4$ and $a = 5$, then $e = \dfrac{c}{a} = \dfrac{4}{5}$ and $b = 3$. The equation is $\dfrac{x^2}{25} + \dfrac{y^2}{9} = 1$.

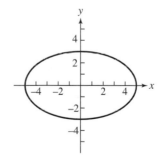

13. $\dfrac{(x-1)^2}{4} + \dfrac{(y-4)^2}{9} = 1$; foci are $(1, 4 \pm \sqrt{5})$;

$e = \dfrac{\sqrt{5}}{3}$; directrices are $y = 4 \pm \dfrac{9\sqrt{5}}{5}$.

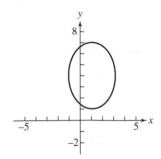

15. $e = \dfrac{5}{4}$; foci are $(\pm 5, 0)$; directrices are $x = \pm\dfrac{16}{5}$

17. $e = \sqrt{5}$; foci are $(\pm\sqrt{10}, 0)$; directrices are

$x = \pm\dfrac{\sqrt{10}}{5}$

19. $y^2 - \dfrac{x^2}{8} = 1$

21. $e = \sqrt{2}$; $\dfrac{x^2}{8} - \dfrac{y^2}{8} = 1$

23. $\dfrac{(y-6)^2}{36} - \dfrac{(x-1)^2}{45} = 1$

25. $a = 0$, $b = -4$, $c = 0$; $e = \dfrac{\sqrt{3}}{2}$

Appendix A5.3 (pp. 612–618)

1. Hyperbola   **3.** Ellipse

5. Parabola   **7.** Parabola

9. Hyperbola   **11.** Hyperbola

13. Ellipse   **15.** Ellipse

17. $(x')^2 - (y')^2 = 4$; hyperbola

19. $4(x')^2 + 16y' = 0$; parabola

21. $(y')^2 = 1$; parallel horiztonal lines

23. $(x')^2 + 4y' = 0$; parabola

25. $4(x')^2 + 2(y')^2 = 19$; ellipse

27. $\sin \alpha = \dfrac{1}{\sqrt{5}}$, $\cos \alpha = \dfrac{2}{\sqrt{5}}$; or $\sin \alpha = -\dfrac{2}{\sqrt{5}}$,

$\cos \alpha = \dfrac{1}{\sqrt{5}}$

29. $\sin \alpha \approx 0.23$, $\cos \alpha \approx 0.97$;

$A' \approx 0.88$, $B' \approx 0.00$, $C' \approx 3.12$, $D' \approx 0.74$,

$E' \approx -1.20$, $F' \approx -3$;

$0.88(x')^2 + 3.12(y')^2 + 0.74x' - 1.20y' - 3 = 0$;

ellipse

31. $\sin \alpha \approx 0.45$, $\cos \alpha \approx 0.89$;

$A' \approx 0.00$, $B' \approx 0.00$, $C' \approx 5.00$, $D' \approx 0$,

$E' \approx 0$, $F' \approx -5$;

$5.00(y')^2 - 5 = 0$ or $y' = \pm 1.00$; parallel lines

33. $\sin \alpha \approx 0.63$, $\cos \alpha \approx 0.77$;

$A' \approx 5.05$, $B' \approx 0.00$, $C' \approx -0.05$, $D' \approx -5.07$,

$E' \approx -6.19$, $F' \approx -1$;

$5.05(x')^2 - 0.05(y')^2 - 5.07x' - 6.18\,y' - 1 = 0$;

hyperbola

35. (a) $(x')^2 - (y')^2 = 2$

   (b) $(x')^2 - (y')^2 = 2a$

37. Yes, one example is $x^2 + 4xy + 5y^2 - 1 = 0$.

39. (a) $\dfrac{(x')^2}{b^2} + \dfrac{(y')^2}{a^2} = 1$ (b) $\dfrac{(y')^2}{a^2} - \dfrac{(x')^2}{b^2} = 1$

   (c) $(x')^2 + (y')^2 = a^2$ (d) $y' = -\dfrac{1}{m}x'$

   (e) $y' = -\dfrac{1}{m}x' + \dfrac{b}{m}$

41. (a) Hyperbola

   (b) $y = -\dfrac{2x}{x-1}$

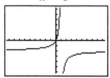

$[-9.4, 9.4]$ by $[-6.2, 6.2]$

   (c) At $(3, -3)$: $y = -2x + 3$

   At $(-1, -1)$: $y = -2x - 3$

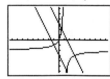

$[-9.4, 9.4]$ by $[-6.2, 6.2]$

43. (a) Parabola

   (b) The equation can be written in the form

    $(x + 2y + 3)^2 = 0$

Appendix A6 (pp. 618–627)

1. $\cosh x = \dfrac{5}{4}$; $\tanh x = -\dfrac{3}{5}$; $\coth x = -\dfrac{5}{3}$;

$\operatorname{sech} x = \dfrac{4}{5}$; $\operatorname{csch} x = -\dfrac{4}{3}$

3. $\sinh x = \dfrac{8}{15}$; $\tanh x = \dfrac{8}{17}$; $\coth x = \dfrac{17}{8}$;

$\operatorname{sech} x = \dfrac{15}{17}$; $\operatorname{csch} x = \dfrac{15}{8}$

5. $x + \dfrac{1}{x}$ **7.** e^{5x}

9. e^{4x}

13. $\dfrac{dy}{dx} = 2 \cosh \dfrac{x}{3}$

15. $\dfrac{dy}{dt} = \text{sech}^2 \sqrt{t} + \dfrac{\tanh \sqrt{t}}{\sqrt{t}}$

17. $\dfrac{dy}{dz} = \coth z$

19. $\dfrac{dy}{d\theta} = (\text{sech } \theta \tanh \theta)(\ln \text{sech } \theta)$

21. $\dfrac{dy}{dx} = \tanh^3 x$

23. $y = 2x;\ \dfrac{dy}{dx} = 2$

25. $\dfrac{dy}{dx} = \dfrac{1}{2\sqrt{x}(1 + x)}$

27. $\dfrac{dy}{d\theta} = \dfrac{1}{1 + \theta} - \tanh^{-1} \theta$

29. $\dfrac{dy}{dt} = \dfrac{1}{2\sqrt{t}} \coth^{-1} \sqrt{t}$ **31.** $\dfrac{dy}{dx} = -\text{sech}^{-1} x$

33. $\dfrac{dy}{d\theta} = \dfrac{\ln 2}{\sqrt{1 + \left(\frac{1}{2}\right)^{2\theta}}}$

35. $\dfrac{dy}{dx} = |\sec x|$

37. **(a)** $\dfrac{d}{dx}(\tan^{-1} (\sinh x) + C) = \text{sech } x$

(b) $\dfrac{d}{dx}(\sin^{-1} (\tanh x) + C) = \text{sech } x$

39. $\dfrac{d}{dx}\left(\dfrac{x^2 - 1}{2} \coth^{-1} x + \dfrac{x}{2} + C\right) = x \coth^{-1} x$

41. $\dfrac{\cosh 2x}{2} + C$

43. $12 \sinh \left(\dfrac{x}{2} - \ln 3\right) + C$

45. $7 \ln |e^{x/7} + e^{-x/7}| + C$

47. $\tan \left(x - \dfrac{1}{2}\right) + C$ **49.** $-2 \text{ sech } \sqrt{t} + C$

51. $\ln \left(\dfrac{5}{2}\right) \approx 0.916$ **53.** $\dfrac{3}{32} + \ln 2 \approx 0.787$

55. $e - e^{-1} \approx 2.350$ **57.** $\dfrac{3}{4}$

59. $\dfrac{3}{8} + \ln \sqrt{2} \approx 0.722$ **61.** 2π

63. $\left(2 \ln \dfrac{199}{100} - \dfrac{99}{100}\right)\pi \approx 1.214$

65. **(a)** If $g(x) = \dfrac{f(x) + f(-x)}{2}$, then

$g(-x) = \dfrac{f(-x) + f(x)}{2} = g(x)$. Thus,

$\dfrac{f(x) + f(-x)}{2}$ is even. If $h(x) = \dfrac{f(x) - f(-x)}{2}$,

then $h(-x) = \dfrac{f(-x) - f(x)}{2}$

$= -\dfrac{f(x) - f(-x)}{2} = -h(x).$

Thus $\dfrac{f(x) - f(-x)}{2}$ is odd.

(b) Even part: $\dfrac{e^x + e^{-x}}{2} = \cosh x$

Odd part: $\dfrac{e^x + e^{-x}}{2} = \sinh x$

69. $y = \text{sech}^{-1} (x) - \sqrt{1 - x^2}$

SELECTED SOLUTIONS

Chapter 1

Section 1.1 (pp. 1–9)

17. $m = \dfrac{3 - 0}{2 - 0} = \dfrac{3}{2}$

$y = \dfrac{3}{2}x$

$3x - 2y = 0$

27. $3x + 4y = 12$

$y = -\dfrac{3}{4}x + 3$

  (a) Slope: $-\dfrac{3}{4}$ **(b)** y-intercept: 3

  (c)

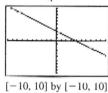

$[-10, 10]$ by $[-10, 10]$

31. **(a)** The desired line has slope -1 and passes through
$(0, 0)$: $y = -1(x - 0) + 0$ or $y = -x$.

  (b) The desired line has slope $\dfrac{-1}{-1} = 1$ and passes through
$(0, 0)$: $y = 1(x - 0) + 0$ or $y = x$.

35. $m = \dfrac{9 - 2}{3 - 1} = \dfrac{7}{2}$

$f(x) = \dfrac{7}{2}(x - 1) + 2 = \dfrac{7}{2}x - \dfrac{3}{2}$

Check: $f(5) = \dfrac{7}{2}(5) - \dfrac{3}{2} = 16$, as expected.

Since $f(x) = \dfrac{7}{2}x - \dfrac{3}{2}$, we have $m = \dfrac{7}{2}$ and $b = -\dfrac{3}{2}$.

39. **(d)** When $x = 30$, $y \approx 0.680(30) + 9.013 = 29.413$.
She weighs about 29 pounds.

Section 1.2 (pp. 9–19)

1. Since $A = \pi r^2 = \pi\left(\dfrac{d}{2}\right)^2$, the formula is $A = \dfrac{\pi d^2}{4}$, where A

represents area and d represents diameter.

7. **(a)** Since we require $x - 1 \geq 0$, the domain is $[1, \infty)$.

19. Even, since the function is an even power of x.

29. **(a)** Note that $f(x) = -|x - 3| + 2$, so its graph is the
graph of the absolute value function reflected across
the x-axis and then shifted 3 units right and 2 units
upward.

41. Line through $(0, 0)$ and $(1, 1)$: $y = x$
Line through $(1, 1)$ and $(2, 0)$: $y = -x + 2$

$$f(x) = \begin{cases} x, & 0 \leq x \leq 1 \\ -x + 2, & 1 < x \leq 2 \end{cases}$$

51. **(a)** Enter $y_1 = f(x) = x - 7$, $y_2 = g(x) = \sqrt{x}$,

$y_3 = (f \circ g)(x) = y_1(y_2(x))$,

and $y_4 = (g \circ f)(x) = y_2(y_1(x))$

$f \circ g$: $g \circ f$:

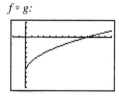 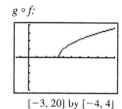

$[-10, 70]$ by $[-10, 3]$ $[-3, 20]$ by $[-4, 4]$

Domain: $[0, \infty)$ Domain: $[7, \infty)$

Range: $[-7, \infty)$ Range: $[0, \infty)$

55.

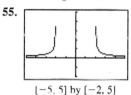

$[-5, 5]$ by $[-2, 5]$

We require $x^2 - 4 \geq 0$ (so that the square root is defined)
and $x^2 - 4 \neq 0$ (to avoid division by zero), so the domain
is $(-\infty, -2) \cup (2, \infty)$. For values of x in the domain,

$x^2 - 4 \left(\text{and hence } \sqrt{x^2 - 4} \text{ and } \dfrac{1}{\sqrt{x^2 - 4}}\right)$ can attain any

positive value, so the range is $(0, \infty)$. Note that grapher
failure may cause the range to appear as a finite interval on
a grapher.

Section 1.3 (pp. 20–26)

1. The graph of $y = 2^x$ is increasing from left to right and has
the negative x-axis as an asymptote. (a)

15.

$[-6, 6]$ by $[-2, 6]$

$x \approx 2.3219$

23. Let t be the number of years. Solving $500,000(1.0375)^t = 1,000,000$ graphically, we find that $t \approx 18.828$. The population will reach 1 million in about 19 years.

27. Let A be the amount of the initial investment, and let t be the number of years. We wish to solve $A(1.0625)^t = 2A$, which is equivalent to $1.0625^t = 2$. Solving graphically, we find that $t \approx 11.433$. It will take about 11.433 years. (If the interest is credited at the end of each year, it will take 12 years.)

Section 1.4 (pp. 26–31)

17. (a)

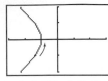

$[-3, 3]$ by $[-2, 2]$

No initial or terminal point. Note that it may be necessary to use a t-interval such as $[-1.57, 1.57]$ or use dot mode in order to avoid "asymptotes" showing on the calculator screen.

(b) $x^2 - y^2 = \sec^2 t - \tan^2 t = 1$

The parametrized curve traces the left branch of the hyperbola defined by $x^2 - y^2 = 1$ (or all of the curve defined by $x = -\sqrt{y^2 + 1}$).

19. (b) $y = 4t - 7 = 2(2t - 5) + 3 = 2x + 3$

The parametrized curve traces all of the line defined by $y = 2x + 3$.

23. (b) $y = \sqrt{t} = 4 - (4 - \sqrt{t}) = 4 - x = -x + 4$

The parametrized curve traces the portion of the line defined by $y = -x + 4$ to the left of $(4, 0)$, that is, for $x \leq 4$.

27. Using $(-1, -3)$ we create the parametric equations $x = -1 + at$ and $y = -3 + bt$, representing a line which goes through $(-1, -3)$ at $t = 0$. We determine a and b so that the line goes through $(4, 1)$ when $t = 1$.
Since $4 = -1 + a, a = 5$.
Since $1 = -3 + b, b = 4$.
Therefore, one possible parametrization is $x = -1 + 5t$, $y = -3 + 4t, 0 \leq t \leq 1$.

37. The graph of $y = x^2 + 2x + 2$ lies in Quadrant I for all $x > 0$. Substituting t for x, we obtain one possible parametrization: $x = t, y = t^2 + 2t + 2, t > 0$.

Section 1.5 (pp. 32–40)

7.

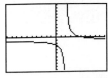

$[-10, 10]$ by $[-10, 10]$

Yes, the function is one-to-one since each horizontal line intersects the graph at most once, so it has an inverse function.

13. $y = 2x + 3$

$y - 3 = 2x$

$\dfrac{y - 3}{2} = x$

Interchange x and y.

$\dfrac{x - 3}{2} = y$

$f^{-1}(x) = \dfrac{x - 3}{2}$.

Verify.

$$(f \circ f^{-1})(x) = f\left(\frac{x - 3}{2}\right)$$
$$= 2\left(\frac{x - 3}{2}\right) + 3$$
$$= (x - 3) + 3$$
$$= x$$

$$(f^{-1} \circ f)(x) = f^{-1}(2x + 3)$$
$$= \frac{(2x + 3) - 3}{2}$$
$$= \frac{2x}{2}$$
$$= x$$

37. $(1.045)^t = 2$

$\ln(1.045)^t = \ln 2$

$t \ln 1.045 = \ln 2$

$t = \dfrac{\ln 2}{\ln 1.045} \approx 15.75$

Graphical support:

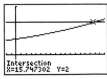

Intersection
X=15.747302 Y=2

$[-2, 18]$ by $[-1, 3]$

49. (a) $y = -2539.852 + 636.896 \ln x$

(b) When $x = 75, y \approx 209.94$. About 209.94 metric tons were produced.

(c) $-2539.852 + 636.896 \ln x = 400$

$636.896 \ln x = 2939.852$

$\ln x = \dfrac{2939.852}{636.896}$

$x = e^{\frac{2939.852}{636.896}} \approx 101.08$

According to the regression equation, Saudi Arabian oil production will reach 400 metric tons when $x \approx 101.08$, in about 2001. (Exactly when this is depends on whether $x = 0$ is January, June, or December in 1900.)

Section 1.6 (pp. 41–51)

1. Arc length $= \left(\frac{5\pi}{8}\right)(2) = \frac{5\pi}{4}$

5. (a) The period of $y = \sec x$ is 2π, so the window should have length 4π.
One possible answer: $[0, 4\pi]$ by $[-3, 3]$

   (b) The period of $y = \csc x$ is 2π, so the window should have length 4π.
One possible answer: $[0, 4\pi]$ by $[-3, 3]$

   (c) The period of $y = \cot x$ is π, so the window should have length 2π.
One possible answer: $[0, 2\pi]$ by $[-3, 3]$

7. Since $\frac{\pi}{6}$ is in the range $\left[-\frac{\pi}{2}, \frac{\pi}{2}\right]$ of $y = \sin^{-1} x$ and
$\sin \frac{\pi}{6} = 0.5$, $\sin^{-1}(0.5) = \frac{\pi}{6}$ radian or $\frac{\pi}{6} \cdot \frac{180°}{\pi} = 30°$.

11. (a) Period $= \frac{2\pi}{2} = \pi$
   (b) Amplitude $= 1.5$
   (c) $[-2\pi, 2\pi]$ by $[-2, 2]$

17. (b) Domain: Since $\csc(3x + \pi) = \frac{1}{\sin(3x + \pi)}$, we require $3x + \pi \neq k\pi$, or $x \neq \frac{(k-1)\pi}{3}$. This requirement is equivalent to $x \neq \frac{k\pi}{3}$ for integers k.

   (c) Since $\left|\csc(3x + \pi)\right| \geq 1$, the range excludes numbers between $-3 - 2 = -5$ and $3 - 2 = 1$. The range is $(-\infty, -5] \cup [1, \infty)$.

25. The angle $\tan^{-1}(2.5) \approx 1.190$ is the solution to this equation in the interval $-\frac{\pi}{2} < x < \frac{\pi}{2}$. Another solution in $0 \leq x < 2\pi$ is $\tan^{-1}(2.5) + \pi \approx 4.332$. The solutions are $x \approx 1.190$ and $x \approx 4.332$.

31. Let $\theta = \cos^{-1}\left(\frac{7}{11}\right)$. Then $0 \leq \theta \leq \pi$ and $\cos \theta = \frac{7}{11}$, so
$$\sin\left(\cos^{-1}\left(\frac{7}{11}\right)\right) = \sin \theta = \sqrt{1 - \cos^2 \theta} = \sqrt{1 - \left(\frac{7}{11}\right)^2}$$
$$= \frac{\sqrt{72}}{11} = \frac{6\sqrt{2}}{11} \approx 0.771.$$

Chapter 2

Section 2.1 (pp. 55–65)

7. $\lim\limits_{x \to -1/2} 3x^2(2x - 1) = 3\left(-\frac{1}{2}\right)^2\left[2\left(-\frac{1}{2}\right) - 1\right] = 3\left(\frac{1}{4}\right)(-2) = -\frac{3}{2}$

Graphical support:

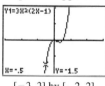

$[-3, 3]$ by $[-2, 2]$

21.

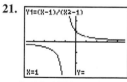

$[-4.7, 4.7]$ by $[-3.1, 3.1]$

$\lim\limits_{x \to 1} \frac{x - 1}{x^2 - 1} = \frac{1}{2}$

Algebraic confirmation:

$$\lim\limits_{x \to 1} \frac{x - 1}{x^2 - 1} = \lim\limits_{x \to 1} \frac{x - 1}{(x + 1)(x - 1)} = \lim\limits_{x \to 1} \frac{1}{x + 1} = \frac{1}{1 + 1} = \frac{1}{2}$$

33. $y_1 = \frac{x^2 + x - 2}{x - 1} = \frac{(x - 1)(x + 2)}{x - 1} = x + 2, x \neq 1$

   (c)

53.

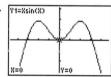

$[-4.7, 4.7]$ by $[-3.1, 3.1]$

$\lim\limits_{x \to 0} (x \sin x) = 0$

Confirm using the Sandwich Theorem, with $g(x) = -|x|$ and $h(x) = |x|$.

$|x \sin x| = |x| \cdot |\sin x| \leq |x| \cdot 1 - |x| - |x| \leq x \sin x \leq |x|$

Because $\lim\limits_{x \to 0} (-|x|) = \lim\limits_{x \to 0} |x| = 0$, the Sandwich Theorem gives $\lim\limits_{x \to 0} (x \sin x) = 0$.

Section 2.2 (pp. 65–73)

9.

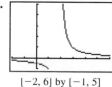

$[-2, 6]$ by $[-1, 5]$

$\lim\limits_{x \to 2^+} \frac{1}{x - 2} = \infty$

23. $y = \left(2 - \dfrac{x}{x+1}\right)\left(\dfrac{x^2}{5+x^2}\right) = \left(\dfrac{2(x+1) - x}{x+1}\right)\left(\dfrac{x^2}{5+x^2}\right)$

$= \left(\dfrac{x+2}{x+1}\right)\left(\dfrac{x^2}{5+x^2}\right) = \dfrac{x^3 + 2x^2}{x^3 + x^2 + 5x + 5}$

An end behavior model for y is $\dfrac{x^3}{x^3} = 1$.

$\lim\limits_{x\to\infty} y = \lim\limits_{x\to\infty} 1 = 1$

$\lim\limits_{x\to-\infty} y = \lim\limits_{x\to-\infty} 1 = 1$

39. (a) The function $y = e^x$ is a right end behavior model

because $\lim\limits_{x\to\infty} \dfrac{e^x - 2x}{e^x} = \lim\limits_{x\to\infty}\left(1 - \dfrac{2x}{e^x}\right) = 1 - 0 = 1$.

(b) The function $y = -2x$ is a left end behavior model

because $\lim\limits_{x\to-\infty} \dfrac{e^x - 2x}{-2x} = \lim\limits_{x\to-\infty}\left(-\dfrac{e^x}{2x} + 1\right) = 0 + 1 = 1$.

43.

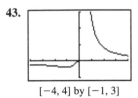

[−4, 4] by [−1, 3]

The graph of $y = f\left(\dfrac{1}{x}\right) = \dfrac{1}{x}e^{1/x}$ is shown.

$\lim\limits_{x\to\infty} f(x) = \lim\limits_{x\to0^+} f\left(\dfrac{1}{x}\right) = \infty$

$\lim\limits_{x\to\infty^-} f(x) = \lim\limits_{x\to0^-} f\left(\dfrac{1}{x}\right) = 0$

Section 2.3 (pp. 73–81)

1. The function $y = \dfrac{1}{(x+2)^2}$ is continuous because it is a

quotient of polynomials, which are continuous. Its only

point of discontinuity occurs where it is undefined. There

is an infinite discontinuity at $x = -2$.

19.

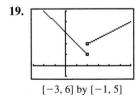

[−3, 6] by [−1, 5]

(a) $x = 2$

(b) Not removable, the one-sided limits are different.

25. For $x \neq -3$, $f(x) = \dfrac{x^2 - 9}{x+3} = \dfrac{(x+3)(x-3)}{x+3} = x - 3$.

The extended function is $y = x - 3$.

39.

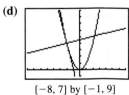

[−3, 3] by [−2, 2]

Solving $x = x^4 - 1$, we obtain the solutions
$x \approx -0.724$ and $x \approx 1.221$.

Section 2.4 (pp. 82–90)

1. (a) $\dfrac{\Delta f}{\Delta x} = \dfrac{f(3) - f(2)}{3 - 2} = \dfrac{28 - 9}{1} = 19$

(b) $\dfrac{\Delta f}{\Delta x} = \dfrac{f(1) - f(-1)}{1 - (-1)} = \dfrac{2 - 0}{2} = 1$

9. (a) $\lim\limits_{h\to0} \dfrac{y(-2+h) - y(-2)}{h} = \lim\limits_{h\to0} \dfrac{(-2+h)^2 - (-2)^2}{h}$

$= \lim\limits_{h\to0} \dfrac{4 - 4h + h^2 - 4}{h}$

$= \lim\limits_{h\to0} \dfrac{-4h + h^2}{h}$

$= \lim\limits_{h\to0}(-4 + h)$

$= -4$

(b) The tangent line has slope -4 and passes through
$(-2, y(-2)) = (-2, 4)$.
$y = -4[x - (-2)] + 4$
$y = -4x - 4$

(c) The normal line has slope $-\dfrac{1}{-4} = \dfrac{1}{4}$ and passes

through $(-2, y(-2)) = (-2, 4)$.

$y = \dfrac{1}{4}[x - (-2)] + 4$

$y = \dfrac{1}{4}x + \dfrac{9}{2}$

(d)

[−8, 7] by [−1, 9]

13. (a) Near $x = 2$, $f(x) = |x| = x$.

$\lim\limits_{h\to0} \dfrac{f(2+h) - f(2)}{h} = \lim\limits_{h\to0} \dfrac{(2+h) - 2}{h} = \lim\limits_{h\to0} 1 = 1$

(b) Near $x = -3$, $f(x) = |x| = -x$.

$\lim\limits_{h\to0} \dfrac{f(-3+h) - f(-3)}{h} = \lim\limits_{h\to0} \dfrac{(3-h) - 3}{h}$

$= \lim\limits_{h\to0} -1 = -1$

15. First, note that f is continuous at $x = 0$, since

$$\lim_{h \to 0^-} f(x) = \lim_{h \to 0^+} f(x) = f(0) = 2.$$

$$\lim_{h \to 0^-} \frac{f(0 + h) - f(0)}{h} = \lim_{h \to 0^-} \frac{(2 - 2h - h^2) - 2}{h}$$

$$= \lim_{h \to 0^-} \frac{-2h - h^2}{h}$$

$$= \lim_{h \to 0^-} (-2 - h)$$

$$= -2$$

$$\lim_{h \to 0^+} \frac{f(0 + h) - f(0)}{h} = \lim_{h \to 0^+} \frac{(2h + 2) - 2}{h}$$

$$= \lim_{h \to 0^+} 2$$

$$= 2$$

No, the slope from the left is -2 and the slope from the right is 2. The two-sided limit of the difference quotient does not exist.

19. (a) $\lim_{h \to 0} \dfrac{f(a + h) - f(a)}{h} = \lim_{h \to 0} \dfrac{[(a + h)^2 + 2] - (a^2 + 2)}{h}$

$$= \lim_{h \to 0} \frac{2ah + h^2}{h}$$

$$= \lim_{h \to 0} (2a + h) = 2a$$

(b) The slope of the tangent steadily increases as a increases.

23. Let $f(t) = 100 - 4.9t^2$.

$$\lim_{h \to 0} \frac{f(2 + h) - f(2)}{h}$$

$$= \lim_{h \to 0} \frac{[100 - 4.9(2 + h)^2] - [100 - 4.9(2)^2]}{h}$$

$$= \lim_{h \to 0} (-19.6 - 4.9h) = -19.6$$

The object is falling at a speed of 19.6 m/sec.

29. First, find the slope of the tangent at $x = a$.

$$\lim_{h \to 0} \frac{f(a + h) - f(a)}{h}$$

$$= \lim_{h \to 0} \frac{[(a + h)^2 + 4(a + h) - 1] - (a^2 + 4a - 1)}{h}$$

$$= \lim_{h \to 0} \frac{2ah + h^2 + 4h}{h} = \lim_{h \to 0} (2a + h + 4)$$

$$= 2a + 4$$

The tangent at $x = a$ is horizontal when $2a + 4 = 0$, or $a = -2$. The tangent line is horizontal at $(-2, f(-2)) = (-2, -5)$.

33. (e) $\lim_{h \to 0} \dfrac{y(7 + h) - y(7)}{h}$

$$= \lim_{h \to 0} \frac{1}{h} \left(\begin{array}{l} [0.0571(7 + h)^2 - 0.1514(7 + h) + 1.3943] \\ \quad - [0.0571(7)^2 - 0.1514(7) + 1.3943] \end{array} \right)$$

$$= \lim_{h \to 0} \frac{0.0571(14h + h^2) - 0.1514h}{h}$$

$$= \lim_{h \to 0} [0.0571(14) - 0.1514 + 0.0571h]$$

$$\approx 0.65$$

The funding was growing at a rate of about 0.65 billion dollars per year.

Chapter 3

Section 3.1 (pp. 95–104)

3. $f'(3) = \lim_{x \to 3} \dfrac{f(x) - f(3)}{x - 3}$

$$= \lim_{x \to 3} \frac{\dfrac{1}{x} - \dfrac{1}{3}}{x - 3}$$

$$= \lim_{x \to 3} \frac{3 - x}{(x - 3)(x)(3)}$$

$$= \lim_{x \to 3} \frac{1}{3x} = -\frac{1}{9}$$

7. The graph of $y = f_1(x)$ is decreasing for $x < 0$ and increasing for $x > 0$, so its derivative is negative for $x < 0$ and positive for $x > 0$. (b)

11. $\dfrac{dy}{dx} = \lim\limits_{h \to 0} \dfrac{y(x+h) - y(x)}{h}$

$= \lim\limits_{h \to 0} \dfrac{[2(x+h)^2 - 13(x+h) + 5] - (2x^2 - 13x + 5)}{h}$

$= \lim\limits_{h \to 0} \dfrac{2x^2 + 4xh + 2h^2 - 13x - 13h + 5 - 2x^2 + 13x - 5}{h}$

$= \lim\limits_{h \to 0} \dfrac{4xh + 2h^2 - 13h}{h}$

$= \lim\limits_{h \to 0} (4x + 2h - 13) = 4x - 13$

At $x = 3$, $\dfrac{dy}{dx} = 4(3) - 13 = -1$, so the tangent line has

slope -1 and passes through $(3, y(3)) = (3, -16)$.

$y = -1(x - 3) - 16$

$y = -x - 13$

13. Since the graph of $y = x \ln x - x$ is decreasing for $0 < x < 1$ and increasing for $x > 1$, its derivative is negative for $0 < x < 1$ and positive for $x > 1$. The only one of the given functions with this property is $y = \ln x$. Note also that $y = \ln x$ is undefined for $x < 0$, which further agrees with the given graph. (ii)

15. (a) The amount of daylight is increasing at the fastest rate when the slope of the graph is largest. This occurs about one-fourth of the way through the year, sometime around April 1. The rate at this time is approximately $\dfrac{4 \text{ hours}}{24 \text{ days}}$ or $\dfrac{1}{6}$ hour per day.

(b) Yes, the rate of change is zero when the tangent to the graph is horizontal. This occurs near the beginning of the year and halfway through the year, around January 1 and July 1.

(c) Positive: January 1 through July 1
Negative: July 1 through December 31

Section 3.2 (pp. 105–112)

1. Left-hand derivative:

$\lim\limits_{h \to 0^-} \dfrac{f(0+h) - f(0)}{h} = \lim\limits_{h \to 0^-} \dfrac{h^2 - 0}{h} = \lim\limits_{h \to 0^-} h = 0$

Right-hand derivative:

$\lim\limits_{h \to 0^+} \dfrac{f(0+h) - f(0)}{h} = \lim\limits_{h \to 0^+} \dfrac{h - 0}{h} = \lim\limits_{h \to 0^+} 1 = 1$

Since $0 \neq 1$, the function is not differentiable at the point P.

11. Since $\lim\limits_{x \to 0} \tan^{-1} x = \tan^{-1} 0 = 0 \neq y(0)$, the problem is a discontinuity.

17. Find the zeros of the denominator.

$x^2 - 4x - 5 = 0$

$(x + 1)(x - 5) = 0$

$x = -1 \text{ or } x = 5$

The function is a rational function, so it is differentiable for all x in its domain: all reals except $x = -1, 5$.

21. The function is piecewise-defined in terms of polynomials, so it is differentiable everywhere except possibly at $x = 0$ and at $x = 3$. Check $x = 0$:

$\lim\limits_{h \to 0^-} \dfrac{g(0+h) - g(0)}{h} = \lim\limits_{h \to 0^-} \dfrac{(h+1)^2 - 1}{h} = \lim\limits_{h \to 0^-} \dfrac{h^2 + 2h}{h}$

$= \lim\limits_{h \to 0^-} (h + 2) = 2$

$\lim\limits_{h \to 0^+} \dfrac{g(0+h) - g(0)}{h} = \lim\limits_{h \to 0^+} \dfrac{(2h+1) - 1}{h} = \lim\limits_{h \to 0^+} 2 = 2$

The function is differentiable at $x = 0$.

Check $x = 3$:

Since $g(3) = (4 - 3)^2 = 1$ and

$\lim\limits_{x \to 3^-} g(x) = \lim\limits_{x \to 3^-} (2x + 1) = 2(3) + 1 = 7$, the function is not continuous (and hence not differentiable) at $x = 3$.

The function is differentiable for all reals except $x = 3$.

Section 3.3 (pp. 112–121)

1. $\dfrac{dy}{dx} = \dfrac{d}{dx}(-x^2) + \dfrac{d}{dx}(3) = -2x + 0 = -2x$

$\dfrac{d^2y}{dx^2} = \dfrac{d}{dx}(-2x) = -2$

11. (a) $\dfrac{dy}{dx} = \dfrac{d}{dx}[(x+1)(x^2+1)]$

$= (x+1)\dfrac{d}{dx}(x^2+1) + (x^2+1)\dfrac{d}{dx}(x+1)$

$= (x+1)(2x) + (x^2+1)(1)$

$= 2x^2 + 2x + x^2 + 1$

$= 3x^2 + 2x + 1$

(b) $\dfrac{dy}{dx} = \dfrac{d}{dx}[(x+1)(x^2+1)]$

$= \dfrac{d}{dx}(x^3 + x^2 + x + 1)$

$= 3x^2 + 2x + 1$

13. $\dfrac{dy}{dx} = \dfrac{d}{dx}\dfrac{2x+5}{3x-2} = \dfrac{(3x-2)(2) - (2x+5)(3)}{(3x-2)^2} = -\dfrac{19}{(3x-2)^2}$

23. (a) At $x = 0$, $\dfrac{d}{dx}(uv) = u(0)v'(0) + v(0)u'(0)$

$= (5)(2) + (-1)(-3) = 13$

(b) At $x = 0$, $\dfrac{d}{dx}\left(\dfrac{u}{v}\right) = \dfrac{v(0)u'(0) - u(0)v'(0)}{[v(0)]^2}$

$= \dfrac{(-1)(-3) - (5)(2)}{(-1)^2} = -7$

(c) At $x = 0$, $\dfrac{d}{dx}\left(\dfrac{v}{u}\right) = \dfrac{u(0)v'(0) - v(0)u'(0)}{[u(0)]^2}$

$= \dfrac{(5)(2) - (-1)(-3)}{(5)^2} = \dfrac{7}{25}$

(d) At $x = 0$, $\dfrac{d}{dx}(7v - 2u) = 7v'(0) - 2u'(0)$

$= 7(2) - 2(-3) = 20$

25. $y'(x) = 2x + 5$
$y'(3) = 2(3) + 5 = 11$
The slope is 11. (iii)

27. $y'(x) = 3x^2 - 3$
$y'(2) = 3(2)^2 - 3 = 9$

The tangent line has slope 9, so the perpendicular line has

slope $-\dfrac{1}{9}$ and passes through $(2, 3)$.

$y = -\dfrac{1}{9}(x - 2) + 3$

$y = -\dfrac{1}{9}x + \dfrac{29}{9}$

Graphical support:

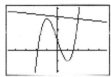

$[-4.7, 4.7]$ by $[-2.1, 4.1]$

31. $y'(x) = \dfrac{(x^2 + 1)(4) - 4x(2x)}{(x^2 + 1)^2} = \dfrac{-4x^2 + 4}{(x^2 + 1)^2}$

At the origin: $y'(0) = 4$

The tangent is $y = 4x$.

At $(1, 2)$: $y'(1) = 0$

The tangent is $y = 2$.

Graphical support:

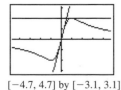

$[-4.7, 4.7]$ by $[-3.1, 3.1]$

Section 3.4 (pp. 122–133)

3. (a) Velocity: $v(t) = \dfrac{ds}{dt} = \dfrac{d}{dt}(24t - 0.8t^2) = 24 - 1.6t$

Acceleration: $a(t) = \dfrac{dv}{dt} = \dfrac{d}{dt}(24 - 1.6t) = -1.6$

(b) The rock reaches its highest point when
$v(t) = 24 - 1.6t = 0$, at $t = 15$. It took 15 seconds.

(c) The maximum height was $s(15) = 180$ meters.

(d) $s(t) = \dfrac{1}{2}(180)$

$24t - 0.8t^2 = 90$

$0 = 0.8t^2 - 24t + 90$

$t = \dfrac{24 \pm \sqrt{(-24^2) - 4(0.8)(90)}}{2(0.8)}$

$\approx 4.393, \ 25.607$

It took about 4.393 seconds to reach half its maximum

height.

(e) $s(t) = 0$
$24t - 0.8t^2 = 0$
$0.8t(30 - t) = 0$
$t = 0$ or $t = 30$
The rock was aloft from $t = 0$ to $t = 30$, so it was
aloft for 30 seconds.

11. (a)

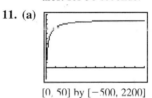

The values of x which make
sense are the whole
numbers, $x \geq 0$.

$[0, 50]$ by $[-500, 2200]$

(b) Marginal revenue $= r'(x) = \dfrac{d}{dx}\left[2000\left(1 - \dfrac{1}{x + 1}\right)\right]$

$= \dfrac{d}{dx}\left(2000 - \dfrac{2000}{x + 1}\right)$

$= 0 - \dfrac{(x + 1)(0) - (2000)(1)}{(x + 1)^2} = \dfrac{2000}{(x + 1)^2}$

(c) $r'(5) = \dfrac{2000}{(5 + 1)^2} = \dfrac{2000}{36} \approx 55.56$

The increase in revenue is approximately \$55.56.

(d) The limit is 0. This means that as x gets large, one
reaches a point where very little extra revenue can be
expected from selling more desks.

23. (a) The body reverses direction when v changes sign, at
$t = 2$ and at $t = 7$.

(b) The body is moving at a constant speed, $|v| = 3$ m/sec,
between $t = 3$ and $t = 6$.

23. continued

(c) The speed graph is obtained by reflecting the negative portion of the velocity graph, $2 < t < 7$, over the x-axis.

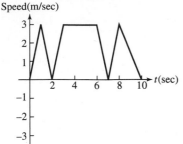

Speed(m/sec)

(d) For $0 \le t < 1$: $a = \dfrac{3-0}{1-0} = 3$ m/sec^2

For $1 < t < 3$: $a = \dfrac{-3-3}{3-1} = -3$ m/sec^2

For $3 < t < 6$: $a = \dfrac{-3-(-3)}{6-3} = 0$ m/sec^2

For $6 < t < 8$: $a = \dfrac{3-(-3)}{8-6} = 3$ m/sec^2

For $8 < t \le 10$: $a = \dfrac{0-3}{10-8} = -1.5$ m/sec^2

Acceleration (m/sec^2)

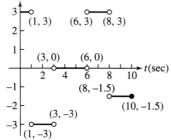

Section 3.5 (pp. 134–141)

1. $\dfrac{d}{dx}(1 + x - \cos x) = 0 + 1 - (-\sin x) = 1 + \sin x$

7. $\dfrac{d}{dx}\left(\dfrac{4}{\cos x}\right) = \dfrac{d}{dx}(4 \sec x) = 4 \sec x \tan x$

11. $y'(x) = \dfrac{d}{dx}(\sin x + 3) = \cos x$

$y'(\pi) = \cos \pi = -1$

The tangent line has slope -1 and passes through

$(\pi, \sin \pi + 3) = (\pi, 3).$

Its equation is $y = -1(x - \pi) + 3$, or $y = -x + \pi + 3$.

19. $y'(x) = \dfrac{d}{dx}(\sqrt{2} \cos x) = -\sqrt{2} \sin x$

$y'\left(\dfrac{\pi}{4}\right) = -\sqrt{2} \sin \dfrac{\pi}{4} = -\sqrt{2}\left(\dfrac{1}{\sqrt{2}}\right) = -1$

The tangent line has slope -1 and passes through

$\left(\dfrac{\pi}{4}, \sqrt{2} \cos \dfrac{\pi}{4}\right) = \left(\dfrac{\pi}{4}, 1\right)$, so its equation is

$y = -1\left(x - \dfrac{\pi}{4}\right) + 1$, or $y = -x + \dfrac{\pi}{4} + 1.$

The normal line has slope 1 and passes through $\left(\dfrac{\pi}{4}, 1\right)$, so

its equation is $y = 1\left(x - \dfrac{\pi}{4}\right) + 1$, or $y = x + 1 - \dfrac{\pi}{4}.$

21. $y'(x) = \dfrac{d}{dx}(4 + \cot x - 2 \csc x)$

$\qquad = 0 - \csc^2 x + 2 \csc x \cot x$

$\qquad = -\csc^2 x + 2 \csc x \cot x$

(a) $y'\left(\dfrac{\pi}{2}\right) = -\csc^2 \dfrac{\pi}{2} + 2 \csc \dfrac{\pi}{2} \cot \dfrac{\pi}{2}$

$\qquad = -1^2 + 2(1)(0) = -1$

The tangent line has slope -1 and passes through

$P\left(\dfrac{\pi}{2}, 2\right)$. Its equation is $y = -1\left(x - \dfrac{\pi}{2}\right) + 2$, or

$y = -x + \dfrac{\pi}{2} + 2.$

(b) $\qquad\qquad y'(x) = 0$

$-\csc^2 x + 2 \csc x \cot x = 0$

$-\dfrac{1}{\sin^2 x} + \dfrac{2 \cos x}{\sin^2 x} = 0$

$\dfrac{1}{\sin^2 x}(2 \cos x - 1) = 0$

$\cos x = \dfrac{1}{2}$

$x = \dfrac{\pi}{3}$ at point Q

$y\left(\dfrac{\pi}{3}\right) = 4 + \cot \dfrac{\pi}{3} - 2 \csc \dfrac{\pi}{3}$

$\qquad = 4 + \dfrac{1}{\sqrt{3}} - 2\left(\dfrac{2}{\sqrt{3}}\right)$

$\qquad = 4 - \dfrac{3}{\sqrt{3}} = 4 - \sqrt{3}$

The coordinates of Q are $\left(\dfrac{\pi}{3}, 4 - \sqrt{3}\right)$.

The equation of the horizontal line is $y = 4 - \sqrt{3}.$

29. Observe the pattern:

$\dfrac{d}{dx} \cos x = -\sin x$ $\qquad$ $\dfrac{d^5}{dx^5} \cos x = -\sin x$

$\dfrac{d^2}{dx^2} \cos x = -\cos x$ $\qquad$ $\dfrac{d^6}{dx^6} \cos x = -\cos x$

$\dfrac{d^3}{dx^3} \cos x = \sin x$ $\qquad$ $\dfrac{d^7}{dx^7} \cos x = \sin x$

$\dfrac{d^4}{dx^4} \cos x = \cos x$ $\qquad$ $\dfrac{d^8}{dx^8} \cos x = \cos x$

Continuing the pattern, we see that $\dfrac{d^n}{dx^n} \cos x = \sin x$ when $n = 4k + 3$ for any whole number k.

Since $999 = 4(249) + 3$, $\dfrac{d^{999}}{dx^{999}} \cos x = \sin x$.

Section 3.6 (pp. 141–149)

1. $\dfrac{dy}{dx} = \dfrac{d}{dx} \sin(3x + 1) = [\cos(3x + 1)]\dfrac{d}{dx}(3x + 1)$

$\quad = [\cos(3x + 1)](3) = 3 \cos(3x + 1)$

21. $\dfrac{ds}{dt} = \dfrac{d}{dt} \cos\left(\dfrac{\pi}{2} - 3t\right)$

$\quad = \left[-\sin\left(\dfrac{\pi}{2} - 3t\right)\right]\dfrac{d}{dt}\left(\dfrac{\pi}{2} - 3t\right)$

$\quad = \left[-\sin\left(\dfrac{\pi}{2} - 3t\right)\right](-3)$

$\quad = 3 \sin\left(\dfrac{\pi}{2} - 3t\right)$

29. $y' = \dfrac{d}{dx} \tan x = \sec^2 x$

$\quad y'' = \dfrac{d}{dx} \sec^2 x = (2 \sec x)\dfrac{d}{dx}(\sec x)$

$\qquad\qquad = (2 \sec x)(\sec x \tan x)$

$\qquad\qquad = 2 \sec^2 x \tan x$

33. $f'(u) = \dfrac{d}{du}(u^5 + 1) = 5u^4$

$\quad g'(x) = \dfrac{d}{dx}(\sqrt{x}) = \dfrac{1}{2\sqrt{x}}$

$\quad (f \circ g)'(1) = f'(g(1))g'(1) = f'(1)g'(1) = (5)\left(\dfrac{1}{2}\right) = \dfrac{5}{2}$

39. (a) $\dfrac{dy}{dx} = \dfrac{dy}{du}\dfrac{du}{dx}$

$\quad = \dfrac{d}{du}(\cos u)\dfrac{d}{dx}(6x + 2)$

$\quad = (-\sin u)(6)$

$\quad = -6 \sin u$

$\quad = -6 \sin(6x + 2)$

(b) $\dfrac{dy}{dx} = \dfrac{dy}{du}\dfrac{du}{dx}$

$\quad = \dfrac{d}{du}(\cos 2u)\dfrac{d}{dx}(3x + 1)$

$\quad = (-\sin 2u)(2) \cdot (3)$

$\quad = -6 \sin 2u$

$\quad = -6 \sin(6x + 2)$

41. $\dfrac{dx}{dt} = \dfrac{d}{dt}(2 \cos t) = -2 \sin t$

$\quad \dfrac{dy}{dt} = \dfrac{d}{dt}(2 \sin t) = 2 \cos t$

$\quad \dfrac{dy}{dx} = \dfrac{\dfrac{dy}{dt}}{\dfrac{dx}{dt}} = \dfrac{2 \cos t}{-2 \sin t} = -\cot t$

The line passes through $\left(2 \cos\dfrac{\pi}{4}, 2 \sin\dfrac{\pi}{4}\right) = (\sqrt{2}, \sqrt{2})$

and has slope $-\cot\dfrac{\pi}{4} = -1$. Its equation is

$y = -(x - \sqrt{2}) + \sqrt{2}$, or $y = -x + 2\sqrt{2}$.

51. $\dfrac{ds}{dt} = \dfrac{ds}{d\theta}\dfrac{d\theta}{dt} = \dfrac{d}{d\theta}(\cos \theta)\dfrac{d\theta}{dt}$

$\quad = (-\sin \theta)\left(\dfrac{d\theta}{dt}\right)$

When $\theta = \dfrac{3\pi}{2}$, $\dfrac{ds}{dt} = \left(-\sin\dfrac{3\pi}{2}\right)(5) = 5$.

Section 3.7 (pp. 149–157)

9. $\qquad\qquad x^2y + xy^2 = 6$

$\dfrac{d}{dx}(x^2y) + \dfrac{d}{dx}(xy^2) = \dfrac{d}{dx}(6)$

$x^2\dfrac{dy}{dx} + y(2x) + x(2y)\dfrac{dy}{dx} + y^2(1) = 0$

$x^2\dfrac{dy}{dx} + 2xy\dfrac{dy}{dx} = -(2xy + y^2)$

$(2xy + x^2)\dfrac{dy}{dx} = -(2xy + y^2)$

$\dfrac{dy}{dx} = -\dfrac{2xy + y^2}{2xy + x^2}$

11. $\qquad y^2 = \dfrac{x - 1}{x + 1}$

$\dfrac{d}{dx}y^2 = \dfrac{d}{dx}\dfrac{x - 1}{x + 1}$

$2y\dfrac{dy}{dx} = \dfrac{(x + 1)(1) - (x - 1)(1)}{(x + 1)^2}$

$2y\dfrac{dy}{dx} = \dfrac{2}{(x + 1)^2}$

$\dfrac{dy}{dx} = \dfrac{1}{y(x + 1)^2}$

21. (a) If $f(x) = \dfrac{3}{2}x^{2/3} - 3$, then

$f'(x) = x^{-1/3}$ and $f''(x) = -\dfrac{1}{3}x^{-4/3}$

which contradicts the given equation $f''(x) = x^{-1/3}$.

(b) If $f(x) = \dfrac{9}{10}x^{5/3} - 7$, then

$f'(x) = \dfrac{3}{2}x^{2/3}$ and $f''(x) = x^{-1/3}$,

which matches the given equation.

21. continued

(c) Differentiating both sides of the given equation

$f''(x) = x^{-1/3}$ gives $f'''(x) = -\frac{1}{3}x^{-4/3}$, so it *must* be true

that $f'''(x) = -\frac{1}{3}x^{-4/3}$.

(d) If $f'(x) = \frac{3}{2}x^{2/3} + 6$, then $f''(x) = x^{-1/3}$, which

matches the given equation.

Conclusion: (b), (c), and (d) could be true.

23. $\qquad x^2 + y^2 = 1$

$\frac{d}{dx}(x^2) + \frac{d}{dx}(y^2) = \frac{d}{dx}(1)$

$2x + 2yy' = 0$

$2yy' = -2x$

$y' = -\frac{x}{y}$

$y'' = \frac{d}{dx}\left(-\frac{x}{y}\right)$

$= -\frac{(y)(1) - (x)(y')}{y^2}$

$= -\frac{y - x\left(-\frac{x}{y}\right)}{y^2}$

$= -\frac{x^2 + y^2}{y^3}$

Since our original equation was $x^2 + y^2 = 1$, we may

substitute 1 for $x^2 + y^2$, giving $y'' = -\frac{1}{y^3}$.

27. $\qquad x^2 + xy - y^2 = 1$

$\frac{d}{dx}(x^2) + \frac{d}{dx}(xy) - \frac{d}{dx}(y^2) = \frac{d}{dx}(1)$

$2x + x\frac{dy}{dx} + (y)(1) - 2y\frac{dy}{dx} = 0$

$(x - 2y)\frac{dy}{dx} = -2x - y$

$\frac{dy}{dx} = \frac{-2x - y}{x - 2y} = \frac{2x + y}{2y - x}$

Slope at $(2, 3)$: $\frac{2(2) + 3}{2(3) - 2} = \frac{7}{4}$

(a) Tangent: $y = \frac{7}{4}(x - 2) + 3$ or $y = \frac{7}{4}x - \frac{1}{2}$

(b) Normal: $y = -\frac{4}{7}(x - 2) + 3$ or $y = -\frac{4}{7}x + \frac{29}{7}$

Section 3.8 (pp. 157–163)

1. $\frac{dy}{dx} = \frac{d}{dx}\cos^{-1}(x^2) = -\frac{1}{\sqrt{1-(x^2)^2}}\frac{d}{dx}(x^2)$

$\qquad = -\frac{1}{\sqrt{1-x^4}}(2x) = -\frac{2x}{\sqrt{1-x^4}}$

7. $\frac{dy}{dx} = \frac{d}{dx}\csc^{-1}(x^2 + 1)$

$\qquad = -\frac{1}{|x^2 + 1|\sqrt{(x^2 + 1)^2 - 1}}\frac{d}{dx}(x^2 + 1)$

$\qquad = -\frac{2x}{(x^2 + 1)\sqrt{x^4 + 2x^2}} = -\frac{2}{(x^2 + 1)\sqrt{x^2 + 2}}$

Note that the condition $x > 0$ is required in the last step.

11. $\frac{dy}{dt} = \frac{d}{dt}\cot^{-1}\sqrt{t} = -\frac{1}{1 + (\sqrt{t})^2}\frac{d}{dt}\sqrt{t}$

$\qquad = -\frac{1}{2\sqrt{t}(t + 1)}$

Section 3.9 (pp. 163–171)

7. $\frac{dy}{dx} = \frac{d}{dx}(xe^2) - \frac{d}{dx}(e^x) = e^2 - e^x$

17. $\frac{dy}{dx} = \frac{d}{dx}3^{\csc x} = 3^{\csc x}(\ln 3)\frac{d}{dx}(\csc x)$

$\qquad = 3^{\csc x}(\ln 3)(-\csc x \cot x)$

$\qquad = -3^{\csc x}(\ln 3)(\csc x \cot x)$

19. Use logarithmic differentiation.

$y = x^{\ln x}$

$\ln y = \ln x^{\ln x}$

$\ln y = \ln x \ln x$

$\frac{d}{dx}(\ln y) = \frac{d}{dx}(\ln x)^2$

$\frac{1}{y}\frac{dy}{dx} = (2 \ln x)\left(\frac{1}{x}\right)$

$\frac{dy}{dx} = \frac{2y \ln x}{x}$

$\frac{dy}{dx} = \frac{2x^{\ln x}\ln x}{x}$

29. $\frac{d}{dx}\ln(\ln x) = \frac{1}{\ln x}\frac{d}{dx}\ln x = \frac{1}{\ln x}\cdot\frac{1}{x} = \frac{1}{x \ln x}$

37. $\frac{dy}{dx} = \frac{d}{dx}(\ln 2 \cdot \log_2 x) = (\ln 2)\frac{d}{dx}(\log_2 x)$

$\qquad = (\ln 2)\left(\frac{1}{x \ln 2}\right) = \frac{1}{x}, x > 0$

39. $\dfrac{dy}{dx} = \dfrac{d}{dx}(\log_{10} e^x) = \dfrac{d}{dx}(x \log_{10} e) = \log_{10} e = \dfrac{\ln e}{\ln 10}$

$= \dfrac{1}{\ln 10}$

41. The line passes through (a, e^a) for some value of a and has slope $m = e^a$.

Since the line also passes through the origin, the slope is also given by $m = \dfrac{e^a - 0}{a - 0}$ and we have

$e^a = \dfrac{e^a}{a}$, so $a = 1$. Hence, the slope is e and the equation is $y = ex$.

Chapter 4

Section 4.1 (pp. 177–185)

11. The first derivative $f'(x) = -\dfrac{1}{x^2} + \dfrac{1}{x}$ has a zero at $x = 1$.

Critical point value: $f(1) = 1 + \ln 1 = 1$

Endpoint values: $f(0.5) = 2 + \ln 0.5 \approx 1.307$

$$f(4) = \dfrac{1}{4} + \ln 4 \approx 1.636$$

Maximum value is $\dfrac{1}{4} + \ln 4$ at $x = 4$;

minimum value is 1 at $x = 1$;

local maximum at $\left(\dfrac{1}{2}, 2 \quad \ln 2\right)$

21.

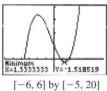

Minimum
X=1.3333333 Y=-1.518519

$[-6, 6]$ by $[-5, 20]$

To find the exact values, note that

$y' = 3x^2 + 2x - 8 = (3x - 4)(x + 2)$, which is zero when

$x = -2$ or $x = \dfrac{4}{3}$. Local maximum at $(-2, 17)$;

local minimum at $\left(\dfrac{4}{3}, -\dfrac{41}{27}\right)$

37.

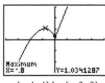

Maximum
X=-.8   Y=1.0341287

$[-4, 4]$ by $[-3, 3]$

$y' = x^{2/3}(1) + \dfrac{2}{3}x^{-1/3}(x + 2) = \dfrac{5x + 4}{3\sqrt[3]{x}}$

crit. pt.	derivative	extremum	value
$x = -\dfrac{4}{5}$	0	local max	$\dfrac{12}{25}10^{1/3} \approx 1.034$
$x = 0$	undefined	local min	0

45. Graph (c), since this is the only graph that has positive slope at c.

49. **(a)** $V(x) = 160x - 52x^2 + 4x^3$
$V'(x) = 160 - 104x + 12x^2 = 4(x - 2)(3x - 20)$
The only critical point in the interval $(0, 5)$ is at $x = 2$.
The maximum value of $V(x)$ is 144 at $x = 2$.

(b) The largest possible volume of the box is 144 cubic units, and it occurs when $x = 2$.

Section 4.2 (pp. 186–194)

1. **(a)** $f'(x) = 5 - 2x$
Since $f'(x) > 0$ on $\left(-\infty, \dfrac{5}{2}\right), f'(x) = 0$ at $x = \dfrac{5}{2}$, and $f'(x) < 0$ on $\left(\dfrac{5}{2}, \infty\right)$, we know that $f(x)$ has a local maximum at $x = \dfrac{5}{2}$. Since $f\left(\dfrac{5}{2}\right) = \dfrac{25}{4}$, the local maximum occurs at the point $\left(\dfrac{5}{2}, \dfrac{25}{4}\right)$. (This is also a global maximum.)

(b) Since $f'(x) > 0$ on $\left(-\infty, \dfrac{5}{2}\right), f(x)$ is increasing on $\left(-\infty, \dfrac{5}{2}\right]$.

(c) Since $f'(x) < 0$ on $\left(\dfrac{5}{2}, \infty\right), f(x)$ is decreasing on $\left[\dfrac{5}{2}, \infty\right)$.

9.

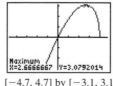

Maximum
X=2.6666667 Y=3.0792014

$[-4.7, 4.7]$ by $[-3.1, 3.1]$

9. continued

(a) $f'(x) = x \cdot \dfrac{1}{2\sqrt{4-x}}(-1) + \sqrt{4-x}$

$= \dfrac{-3x+8}{2\sqrt{4-x}}$

The local extrema occur at the critical point $x = \dfrac{8}{3}$ and

at the endpoint $x = 4$. There is a local (and absolute)

maximum at $\left(\dfrac{8}{3}, \dfrac{16}{3\sqrt{3}}\right)$ or approximately $(2.67, 3.08)$,

and a local minimum at $(4, 0)$.

(b) Since $f'(x) > 0$ on $\left(-\infty, \dfrac{8}{3}\right)$, $f(x)$ is increasing on

$\left(-\infty, \dfrac{8}{3}\right]$.

(c) Since $f'(x) < 0$ on $\left(\dfrac{8}{3}, 4\right)$, $f(x)$ is decreasing on $\left[\dfrac{8}{3}, 4\right]$.

15. (a) f is continuous on $[0, 1]$ and differentiable on $(0, 1)$.

(b) $f'(c) = \dfrac{f(1) - f(0)}{1 - 0}$

$2c + 2 = \dfrac{2 - (-1)}{1}$

$2c = 1$

$c = \dfrac{1}{2}$

19. (a) The secant line passes through $(0.5, f(0.5)) = (0.5, 2.5)$ and $(2, f(2)) = (2, 2.5)$, so its equation is $y = 2.5$.

(b) The slope of the secant line is 0, so we need to find c such that $f'(c) = 0$.

$1 - c^{-2} = 0$

$c^{-2} = 1$

$c = 1$

$f(c) = f(1) = 2$

The tangent line has slope 0 and passes through $(1, 2)$, so its equation is $y = 2$.

21. (a) Since $f'(x) = \dfrac{1}{3}x^{-2/3}$, f is not differentiable at $x = 0$.

(b)

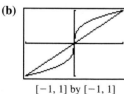

$[-1, 1]$ by $[-1, 1]$

(c) $f'(c) = \dfrac{f(1) - f(-1)}{1 - (-1)}$

$\dfrac{1}{3}c^{-2/3} = \dfrac{1 - (-1)}{2}$

$\dfrac{1}{3}c^{-2/3} = 1$

$c^{-2/3} = 3$

$c = \pm 3^{-3/2} \approx 0.192$

31. $f(x) = \dfrac{1}{x} + C, x > 0$

$f(2) = 1$

$\dfrac{1}{2} + C = 1$

$C = \dfrac{1}{2}$

$f(x) = \dfrac{1}{x} + \dfrac{1}{2}, x > 0$

43. (a) Since $v'(t) = 1.6$, $v(t) = 1.6t + C$. But $v(0) = 0$, so $C = 0$ and $v(t) = 1.6t$. Therefore, $v(30) = 1.6(30) = 48$. The rock will be going 48 m/sec.

(b) Let $s(t)$ represent position. Since $s'(t) = v(t) = 1.6t$, $s(t) = 0.8t^2 + D$. But $s(0) = 0$, so $D = 0$ and $s(t) = 0.8t^2$. Therefore, $s(30) = 0.8(30)^2 = 720$. The rock travels 720 meters in the 30 seconds it takes to hit bottom, so the bottom of the crevasse is 720 meters below the point of release.

(c) The velocity is now given by $v(t) = 1.6t + C$, where $v(0) = 4$. (Note that the sign of the initial velocity is the same as the sign used for the acceleration, since both act in a downward direction.) Therefore, $v(t) = 1.6t + 4$, and $s(t) = 0.8t^2 + 4t + D$, where $s(0) = 0$ and so $D = 0$. Using $s(t) = 0.8t^2 + 4t$ and the known crevasse depth of 720 meters, we solve $s(t) = 720$ to obtain the positive solution $t \approx 27.604$, and so $v(t) = v(27.604) = 1.6(27.604) + 4 \approx 48.166$. The rock will hit bottom after about 27.604 seconds, and it will be going about 48.166 m/sec.

Section 4.3 (pp. 194–206)

7. $y' = 2x - 1$

Intervals	$x < \dfrac{1}{2}$	$x > \dfrac{1}{2}$
Sign of y'	$-$	$+$
Behavior of y	Decreasing	Increasing

$y'' = 2$ (always positive: concave up)

Graphical support:

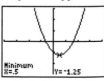

$[-4, 4]$ by $[-3, 3]$

(a) $\left[\dfrac{1}{2}, \infty\right)$ **(b)** $\left(-\infty, \dfrac{1}{2}\right]$

(c) $(-\infty, \infty)$ **(d)** Nowhere

(e) Local (and absolute) minimum at $\left(\dfrac{1}{2}, -\dfrac{5}{4}\right)$

(f) None

13. $y' = 12x^2 + 42x + 36 = 6(x + 2)(2x + 3)$

Intervals	$x < -2$	$-2 < x < -\dfrac{3}{2}$	$-\dfrac{3}{2} < x$
Sign of y'	$+$	$-$	$+$
Behavior of y	Increasing	Decreasing	Increasing

$y'' = 24x + 42 = 6(4x + 7)$

Intervals	$x < -\dfrac{7}{4}$	$-\dfrac{7}{4} < x$
Sign of y''	$-$	$+$
Behavior of y	Concave down	Concave up

Graphical support:

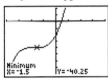

[−4, 4] by [−80, 20]

(a) $(-\infty, -2]$ and $\left[-\dfrac{3}{2}, \infty\right)$ **(b)** $\left[-2, -\dfrac{3}{2}\right]$

(c) $\left(-\dfrac{7}{4}, \infty\right)$ **(d)** $\left(-\infty, -\dfrac{7}{4}\right)$

(e) Local maximum: $(-2, -40)$;

local minimum: $\left(-\dfrac{3}{2}, -\dfrac{161}{4}\right)$

(f) $\left(-\dfrac{7}{4}, -\dfrac{321}{8}\right)$

19. $y' = \begin{cases} 2, & x < 1 \\ -2x, & x > 1 \end{cases}$

Intervals	$x < 1$	$1 < x$
Sign of y'	$+$	$-$
Behavior of y	Increasing	Decreasing

$y'' = \begin{cases} 0, & x < 1 \\ -2, & x > 1 \end{cases}$

Intervals	$x < 1$	$1 < x$
Sign of y''	0	$-$
Behavior of y	Linear	Concave down

Graphical support:

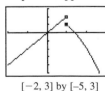

[−2, 3] by [−5, 3]

(a) $(-\infty, 1)$ **(b)** $[1, \infty)$

(c) None **(d)** $(1, \infty)$

(e) None **(f)** None

41. (a) The velocity is zero when the tangent line is horizontal, at approximately $t = 2.2$, $t = 6$, and $t = 9.8$.

(b) The acceleration is zero at the inflection points, approximately $t = 4$, $t = 8$, and $t = 11$.

Section 4.4 (pp. 206–220)

1. Represent the numbers by x and $20 - x$, where $0 \le x \le 20$.

(a) The sum of the squares is given by $f(x) = x^2 + (20 - x)^2 = 2x^2 - 40x + 400$. Then $f'(x) = 4x - 40$. The critical point and endpoints occur at $x = 0$, $x = 10$, and $x = 20$. Then $f(0) = 400$, $f(10) = 200$, and $f(20) = 400$. The sum of the squares is as large as possible for the numbers 0 and 20, and is as small as possible for the numbers 10 and 10.

Graphical support:

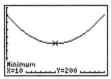

[0, 20] by [0, 450]

(b) The sum of one number plus the square root of the other is given by $g(x) = x + \sqrt{20 - x}$. Then $g'(x) = 1 - \dfrac{1}{2\sqrt{20 - x}}$. The critical point occurs when $2\sqrt{20 - x} = 1$, so $20 - x = \dfrac{1}{4}$ and $x = \dfrac{79}{4}$. Testing the endpoints and critical point, we find $g(0) = \sqrt{20} \approx 4.47$, $g\left(\dfrac{79}{4}\right) = \dfrac{81}{4} - 20.25$, and $g(20) = 20$.

The sum is as large as possible when the numbers are $\dfrac{79}{4}$ and $\dfrac{1}{4}$ $\left(\text{summing } \dfrac{79}{4} + \sqrt{\dfrac{1}{4}}\right)$, and is as small as possible when the numbers are 0 and 20 $(\text{summing } 0 + \sqrt{20})$.

Graphical support:

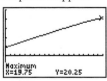

[0, 20] by [−10, 25]

5. (a) The equation of line AB is $y = -x + 1$, so the y-coordinate of P is $-x + 1$.

(b) $A(x) = 2x(1 - x)$

(c) Since $A'(x) = \dfrac{d}{dx}(2x - 2x^2) = 2 - 4x$, the critical point occurs at $x = \dfrac{1}{2}$. Since $A'(x) > 0$ for $0 < x < \dfrac{1}{2}$ and $A'(x) < 0$ for $\dfrac{1}{2} < x < 1$, this critical point corresponds to the maximum area. The largest possible area is $A\left(\dfrac{1}{2}\right) = \dfrac{1}{2}$ square unit, and the dimensions of the rectangle are $\dfrac{1}{2}$ unit by 1 unit.

Graphical support:

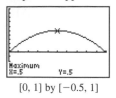

Maximum
X=.5   Y=.5

$[0, 1]$ by $[-0.5, 1]$

15. We assume that a and b are held constant. Then $A(\theta) = \dfrac{1}{2}ab \sin \theta$ and $A'(\theta) = \dfrac{1}{2}ab \cos \theta$. The critical point (for $0 < \theta < \pi$) occurs at $\theta = \dfrac{\pi}{2}$. Since $A'(\theta) > 0$ for $0 < \theta < \dfrac{\pi}{2}$ and $A'(\theta) < 0$ for $\dfrac{\pi}{2} < \theta < \pi$, the critical point corresponds to the maximum area. The angle that maximizes the triangle's area is $\theta = \dfrac{\pi}{2}$ (or $90°$).

17. Note that $\pi r^2 h = 1000$, so $h = \dfrac{1000}{\pi r^2}$. Then $A = 8r^2 + 2\pi rh = 8r^2 + \dfrac{2000}{r}$, so $\dfrac{dA}{dr} = 16r - 2000r^{-2} = \dfrac{16(r^3 - 125)}{r^2}$. The critical point occurs at $r = \sqrt[3]{125} = 5$ cm. Since $\dfrac{dA}{dr} < 0$ for $0 < r < 5$ and $\dfrac{dA}{dr} > 0$ for $r > 5$, the critical point corresponds to the least amount of aluminum used or wasted and hence the most economical can. The dimensions are $r = 5$ cm and $h = \dfrac{40}{\pi}$, so the ratio of h to r is $\dfrac{8}{\pi}$ to 1.

33. (a) Note that $w^2 + d^2 = 12^2$, so $d = \sqrt{144 - w^2}$. Then we may write $S = kwd^3 = kw(144 - w^2)^{3/2}$, so

$$\dfrac{dS}{dw} = kw \cdot \dfrac{3}{2}(144 - w^2)^{1/2}(-2w) + k(144 - w^2)^{3/2}(1)$$
$$= (k\sqrt{144 - w^2})(-3w^2 + 144 - w^2)$$
$$= (-4k\sqrt{144 - w^2})(w^2 - 36)$$

The critical point (for $0 < w < 12$) occurs at $w = 6$. Since $\dfrac{dS}{dw} > 0$ for $0 < w < 6$ and $\dfrac{dS}{dw} < 0$ for $6 < w < 12$, the critical point corresponds to the maximum stiffness. The dimensions are 6 in. wide by $6\sqrt{3}$ in. deep.

(b)

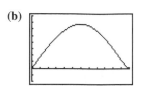

$[0, 12]$ by $[-2000, 8000]$
The graph of $S = w(144 - w^2)^{3/2}$ is shown. The maximum stiffness shown in the graph occurs at $w = 6$, which agrees with the answer to part (a).

(c)

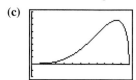

$[0, 12]$ by $[-2000, 8000]$
The graph of $S = d^3\sqrt{144 - d^2}$ is shown. The maximum stiffness shown in the graph occurs at $d = 6\sqrt{3} \approx 10.4$ agrees with the answer to part (a), and its value is the same as the maximum value found in part (b), as expected.

Changing the value of k changes the maximum stiffness, but not the dimensions of the stiffest beam. The graphs for different values of k look the same except that the vertical scale is different.

Section 4.5 (pp. 220–232)

1. (a) $f'(x) = 3x^2 - 2$
We have $f(2) = 7$ and $f'(2) = 10$.
$$L(x) = f(2) + f'(2)(x - 2)$$
$$= 7 + 10(x - 2)$$
$$= 10x - 13$$

(b) Since $f(2.1) = 8.061$ and $L(2.1) = 8$, the approximation differs from the true value in absolute value by less than 10^{-1}.

11. Center $= -1$

$f'(x) = 4x + 4$

We have $f(-1) = -5$ and $f'(-1) = 0$

$L(x) = f(-1) + f'(-1)(x - (-1))$

$\quad = -5 + 0(x + 1) = -5$

15. Let $f(x) = x^3 + x - 1$. Then $f'(x) = 3x^2 + 1$ and

$$x_{n+1} = x_n - \frac{f(x_n)}{f'(x_n)} = x_n - \frac{x_n^3 + x_n - 1}{3x_n^2 + 1}.$$

Note that f is cubic and f' is always positive, so there is exactly one solution. We choose $x_1 = 0$.

$\qquad x_1 = 0$

$\qquad x_2 = 1$

$\qquad x_3 = 0.75$

$\qquad x_4 \approx 0.6860465$

$\qquad x_5 \approx 0.6823396$

$\qquad x_6 \approx 0.6823278$

$\qquad x_7 \approx 0.6823278$

Solution: $x \approx 0.682328$

27. (a) $\Delta f = f(0.1) - f(0) = 0.21 - 0 = 0.21$

(b) Since $f'(x) = 2x + 2, f'(0) = 2$.
Therefore, $df = 2\ dx = 2(0.1) = 0.2$.

(c) $|\Delta f - df| = |0.21 - 0.2| = 0.01$

31. Note that $\dfrac{dV}{dr} = 4\pi r^2$, $dV = 4\pi r^2\ dr$. When r changes from

a to $a + dr$, the change in volume is approximately

$4\pi a^2\ dr$.

39. Let $A =$ cross section area, $C =$ circumference, and

$D =$ diameter. Then $D = \dfrac{C}{\pi}$, so $\dfrac{dD}{dC} = \dfrac{1}{\pi}$ and $dD = \dfrac{1}{\pi}\ dC$.

Also, $A = \pi\left(\dfrac{D}{2}\right)^2 = \pi\left(\dfrac{C}{2\pi}\right)^2 = \dfrac{C^2}{4\pi}$, so $\dfrac{dA}{dC} = \dfrac{C}{2\pi}$ and

$dA = \dfrac{C}{2\pi}\ dC$. When C increases from 10π in. to

$10\pi + 2$ in. the diameter increases by

$dD = \dfrac{1}{\pi}(2) = \dfrac{2}{\pi} \approx 0.6366$ in. and the area increases by

approximately $dA = \dfrac{10\pi}{2\pi}(2) = 10$ in^2.

41. Let $x =$ side length and $A =$ area. Then $A = x^2$ and

$\dfrac{dA}{dx} = 2x$, so $dA = 2x\ dx$. We want $|dA| \le 0.02A$, which

gives $|2x\ dx| \le 0.02x^2$, or $|dx| \le 0.01x$. The side length

should be measured with an error of no more than 1%.

Section 4.6 (pp. 232–241)

7. (a) Since V is increasing at the rate of 1 volt/sec,

$$\frac{dV}{dt} = 1 \text{ volt/sec}.$$

(b) Since I is decreasing at the rate of $\dfrac{1}{3}$ amp/sec,

$$\frac{dI}{dt} = -\frac{1}{3} \text{ amp/sec}.$$

(c) Differentiating both sides of $V = IR$, we have

$$\frac{dV}{dt} = I\frac{dR}{dt} + R\frac{dI}{dt}.$$

(d) Note that $V = IR$ gives $12 = 2 \cdot R$, so $R = 6$ ohms. Now

substitute the known values into the equation in (c).

$$1 = 2\frac{dR}{dt} + 6\left(-\frac{1}{3}\right)$$

$$3 = 2\frac{dR}{dt}$$

$$\frac{dR}{dt} = \frac{3}{2} \text{ ohms/sec}$$

R is changing at the rate of $\dfrac{3}{2}$ ohms/sec. Since this

value is positive, R is increasing.

13. Step 1:

$x =$ distance from wall to base of ladder

$y =$ height of top of ladder

$A =$ area of triangle formed by the ladder, wall, and ground

$\theta =$ angle between the ladder and the ground

Step 2:

At the instant in question, $x = 12$ ft and $\dfrac{dx}{dt} = 5$ ft/sec.

Step 3:

We want to find $-\dfrac{dy}{dt}, \dfrac{dA}{dt},$ and $\dfrac{d\theta}{dt}$.

Step 4, 5, and 6:

(a) $x^2 + y^2 = 169$

$$2x\frac{dx}{dt} + 2y\frac{dy}{dt} = 0$$

To evaluate, note that, at the instant in question,

$$y = \sqrt{169 - x^2} = \sqrt{169 - 12^2} = 5.$$

Then $2(12)(5) + 2(5)\dfrac{dy}{dt} = 0$

$\dfrac{dy}{dt} = -12$ ft/sec $\left(\text{or} -\dfrac{dy}{dt} = 12 \text{ ft/sec}\right)$

The top of the ladder is sliding down the wall at the

rate of 12 ft/sec. (Note that the *downward* rate of

motion is positive.)

13. continued

(b) $A = \frac{1}{2}xy$

$$\frac{dA}{dt} = \frac{1}{2}\left(x\frac{dy}{dt} + y\frac{dx}{dt}\right)$$

Using the results from step 2 and from part (a), we

have $\frac{dA}{dt} = \frac{1}{2}[(12)(-12) + (5)(5)] = -\frac{119}{2}$ ft/sec.

The area of the triangle is changing at the rate of

-59.5 ft²/sec.

(c) $\tan \theta = \frac{y}{x}$

$$\sec^2 \theta \frac{d\theta}{dt} = \frac{x\frac{dy}{dt} - y\frac{dx}{dt}}{x^2}$$

Since $\tan \theta = \frac{5}{12}$, we have $\left(\text{for } 0 \le \theta < \frac{\pi}{2}\right)$

$\cos \theta = \frac{12}{13}$ and so $\sec^2 \theta = \frac{1}{\left(\frac{12}{13}\right)^2} = \frac{169}{144}$.

Combining this result with the results from step 2 and

from part (a), we have $\frac{169}{144}\frac{d\theta}{dt} = \frac{(12)(-12) - (5)(5)}{12^2}$, so

$\frac{d\theta}{dt} = -1$ radian/sec. The angle is changing at the rate

of -1 radian/sec.

17. Step 1:

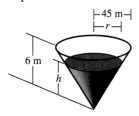

r = radius of top surface of water
h = depth of water in reservoir
V = volume of water in reservoir

Step 2:

At the instant in question, $\frac{dV}{dt} = -50$ m³/min and $h = 5$ m.

Step 3:

We want to find $-\frac{dh}{dt}$ and $\frac{dr}{dt}$.

Step 4:

Note that $\frac{h}{r} = \frac{6}{45}$ by similar cones, so $r = 7.5h$.

Then $V = \frac{1}{3}\pi r^2 h = \frac{1}{3}\pi (7.5h)^2 h = 18.75\pi h^3$

Step 5 and 6:

(a) Since $V = 18.75\pi h^3$, $\frac{dV}{dt} = 56.25\pi h^2 \frac{dh}{dt}$.

Thus $-50 = 56.25\pi(5^2)\frac{dh}{dt}$, and so

$$\frac{dh}{dt} = -\frac{8}{225\pi} \text{ m/min} = -\frac{32}{9\pi} \text{ cm/min}.$$

The water level is falling by $\frac{32}{9\pi} \approx 1.13$ cm/min.

(Since $\frac{dh}{dt} < 0$, the rate at which the water level is

falling is positive.)

(b) Since $r = 7.5h$, $\frac{dr}{dt} = 7.5\frac{dh}{dt} = -\frac{80}{3\pi}$ cm/min. The rate

of change of the radius of the water's surface is

$-\frac{80}{3\pi} \approx -8.49$ cm/min.

27. Step 1:

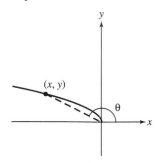

x = x-coordinate of particle's location
y = y-coordinate of particle's location
θ = angle of inclination of line joining the particle to the
 origin

Step 2:

At the instant in question, $\frac{dx}{dt} = -8$ m/sec and $x = -4$ m.

Step 3:

We want to find $\frac{d\theta}{dt}$.

Step 4:

Since $y = \sqrt{-x}$, we have $\tan \theta = \frac{y}{x} = \frac{\sqrt{-x}}{x} = -(-x)^{-1/2}$,

and so, for $x < 0$,

$\theta = \pi + \tan^{-1}[-(-x)^{-1/2}] = \pi - \tan^{-1}(-x)^{-1/2}$.

Step 5:

$$\frac{d\theta}{dt} = -\frac{1}{1 + [(-x)^{-1/2}]^2}\left(-\frac{1}{2}(-x)^{-3/2}(-1)\right)\frac{dx}{dt}$$

$$= -\frac{1}{1 - \left(\frac{1}{x}\right)}\frac{1}{2(-x)^{3/2}}\frac{dx}{dt}$$

$$= \frac{1}{2\sqrt{-x}(x-1)}\frac{dx}{dt}$$

Step 6:

$$\frac{d\theta}{dt} = \frac{1}{2\sqrt{4}(-4-1)}(-8) = \frac{2}{5} \text{ radian/sec}$$

The angle of inclination is increasing at the rate of

$\frac{2}{5}$ radian/sec.

29. Step 1:

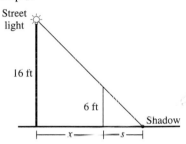

$x =$ distance from streetlight base to man
$s =$ length of shadow

Step 2:

At the instant in question, $\frac{dx}{dt} = -5$ ft/sec and $x = 10$ ft.

Step 3:

We want to find $\frac{ds}{dt}$.

Step 4:

By similar triangles, $\frac{s}{6} = \frac{s+x}{16}$. This is equivalent to

$16s = 6s + 6x$, or $s = \frac{3}{5}x$.

Step 5:

$$\frac{ds}{dt} = \frac{3}{5}\frac{dx}{dt}$$

Step 6:

$$\frac{ds}{dt} = \frac{3}{5}(-5) = -3\text{ft/sec}$$

The shadow length is changing at the rate of -3 ft/sec.

37. $\frac{dy}{dt} = \frac{dy}{dx}\frac{dx}{dt} = -10(1 + x^2)^{-2}(2x)\frac{dx}{dt} = -\frac{20x}{(1+x^2)^2}\frac{dx}{dt}$

Since $\frac{dx}{dt} = 3$ cm/sec, we have

$$\frac{dy}{dt} = -\frac{60x}{(1+x^2)^2} \text{ cm/sec.}$$

(a) $\frac{dy}{dt} = -\frac{60(-2)}{[1 + (-2)^2]^2} = \frac{120}{5^2} = \frac{24}{5}$ cm/sec

(b) $\frac{dy}{dt} = -\frac{60(0)}{(1+0^2)^2} = 0$ cm/sec

(c) $\frac{dy}{dt} = -\frac{60(20)}{(1+20^2)^2} = -\frac{1200}{160,801} \approx -0.00746$ cm/sec

Chapter 5

Section 5.1 (pp. 247–257)

5.

n	$LRAM_n$	$MRAM_n$	$RRAM_n$
10	12.645	13.4775	14.445
50	13.3218	13.4991	13.6818
100	13.41045	13.499775	13.59045
500	13.482018	13.499991	13.518018

Estimate the area to be 13.5.

9. LRAM:
Area $\approx f(2) \cdot 2 + f(4) \cdot 2 + f(6) \cdot 2 + \cdots + f(22) \cdot 2$
$= 2 \cdot (0 + 0.6 + 1.4 + \cdots + 0.5) = 44.8$ (mg/L) $\cdot$ sec
RRAM:
Area $\approx f(4) \cdot 2 + f(6) \cdot 2 + f(8) \cdot 2 + \cdots + f(24) \cdot 2$
$= 2(0.6 + 1.4 + 2.7 + \cdots 0) = 44.8$ (mg/L) $\cdot$ sec

MRAM.

Patient's cardiac output:

$$\frac{5 \text{ mg}}{44.8 \text{ (mg/L)} \cdot \text{sec}} \cdot \frac{60 \text{ sec}}{1 \text{ min}} \approx 6.7 \text{ L/min}$$

Note that estimates for the area may vary.

17. (a) Use RRAM with $\pi(16 - x^2)$
$S_8 \approx 120.9513$
S_8 is an underestimate because each rectangle is below the curve.

(b) $\frac{|V - S_8|}{V} \approx 0.10 = 10\%$

23. (a) 400 ft/sec $- (5 \text{ sec})(32 \text{ ft/sec}^2) = 240$ ft/sec

(b) Use RRAM with $400 - 32x$ on $[0, 5]$, $n = 5$.
$368 + 336 + 304 + 272 + 240 = 1520$ ft

Section 5.2 (pp. 258–268)

1. $\lim_{\|P\|\to 0} \sum_{k=1}^{n} c_k^2 \Delta x_k = \int_0^2 x^2 \, dx$ when P is any partition of $[0, 2]$

7. $\int_{-2}^{1} 5 \, dx = 5[1 - (-2)] = 15$

13. Graph the region under $y = \dfrac{x}{2} + 3$ for $-2 \le x \le 4$.

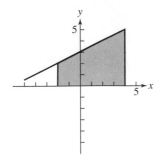

$\int_{-2}^{4} \left(\dfrac{x}{2} + 3 \right) dx = \dfrac{1}{2}(6)(2 + 5) = 21$

23. $\int_0^b x \, dx = \dfrac{1}{2}(b)(b) = \dfrac{1}{2}b^2$

29. Observe that the graph of $f(x) = x^3$ is symmetric with respect to the origin. Hence the area above and below the x-axis is equal for $-1 \le x \le 1$.

$\int_{-1}^{1} x^3 \, dx = -(\text{area below } x\text{-axis}) + (\text{area above } x\text{-axis}) = 0$

43. (a) The function has a discontinuity at $x = 0$.

(b)

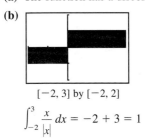

$[-2, 3]$ by $[-2, 2]$

$\int_{-2}^{3} \dfrac{x}{|x|} \, dx = -2 + 3 = 1$

Section 5.3 (pp. 268–276)

1. (c) $\int_1^2 3 f(x) \, dx = 3 \int_1^2 f(x) \, dx = 3(-4) = -12$

5. (a) $\int_3^4 f(z) \, dz = \int_3^0 f(z) \, dz + \int_0^4 f(z) \, dz$

$= -\int_0^3 f(z) \, dz + \int_0^4 f(z) \, dz$

$= -3 + 7 = 4$

7. An antiderivative of 7 is $F(x) = 7x$.

$\int_3^1 7 \, dx = F(1) - F(3) = 7 - 21 = -14$

17. Divide the shaded area as follows:

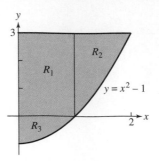

Note that an antiderivative of $x^2 - 1$ is $F(x) = \dfrac{1}{3}x^3 - x$.

Area of $R_1 = 3(1) = 3$

Area of $R_2 = (3)(1) - \int_1^2 (x^2 - 1) \, dx$

$= 3 - [F(2) - F(1)]$

$= 3 - \left[\left(\dfrac{2}{3} \right) - \left(-\dfrac{2}{3} \right) \right] = \dfrac{5}{3}$

Area of $R_3 = -\int_0^1 (x^2 - 1) \, dx$

$= -[F(1) - F(0)]$

$= -\left[\left(-\dfrac{2}{3} \right) - 0 \right] = \dfrac{2}{3}$

Total shaded area $= 3 + \dfrac{2}{3} + \dfrac{5}{3} = \dfrac{16}{3}$

21.

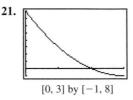

$[0, 3]$ by $[-1, 8]$

An antiderivative of $x^2 - 6x + 8$ is

$F(x) = \dfrac{1}{3}x^3 - 3x^2 + 8x.$

(a) $\int_0^3 (x^2 - 6x + 8) \, dx = F(3) - F(0) = 6 - 0 = 6$

(b) Area $= \int_0^2 (x^2 - 6x + 8) \, dx - \int_2^3 (x^2 - 6x + 8) \, dx$

$= [F(2) - F(0)] - [F(3) - F(2)]$

$= \left(\dfrac{20}{3} - 0 \right) - \left(6 - \dfrac{20}{3} \right) = \dfrac{22}{3}$

25. An antiderivative of $x^2 - 1$ is $F(x) = \frac{1}{3}x^3 - x$.

$$av = \frac{1}{\sqrt{3}}\int_0^{\sqrt{3}} (x^2 - 1)\, dx$$

$$= \frac{1}{\sqrt{3}}[F(\sqrt{3}) - F(0)]$$

$$= \frac{1}{\sqrt{3}}(0 - 0) = 0$$

Find $x = c$ in $[0, \sqrt{3}]$ such that $c^2 - 1 = 0$

$$c^2 = 1$$

$$c = \pm 1$$

Since 1 is in $[0, \sqrt{3}]$, $x = 1$.

29. The region between the graph and the x-axis is a triangle

of height 3 and base 6, so the area of region is $\frac{1}{2}(3)(6) = 9$

$$av(f) = \frac{1}{6}\int_{-4}^{2} f(x)\, dx = \frac{9}{6} = \frac{3}{2}$$

Section 5.4 (pp. 277–288)

1. $\displaystyle\int_{1/2}^{3}\left(2 - \frac{1}{x}\right) dx = \left[2x - \ln|x|\right]_{1/2}^{3}$

$$= (6 - \ln 3) - \left(1 - \ln\frac{1}{2}\right)$$

$$= 5 - \ln 3 - \ln 2 = 5 - \ln 6 \approx 3.208$$

15. Graph $y = 2 - x$.

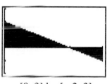

[0, 3] by [−2, 3]

Over $[0, 2]$: $\displaystyle\int_0^2 (2 - x)\, dx = \left[2x - \frac{1}{2}x^2\right]_0^2 = 2$

Over $[2, 3]$: $\displaystyle\int_2^3 (2 - x)\, dx = \left[2x - \frac{1}{2}x^2\right]_2^3 = \frac{3}{2} - 2 = -\frac{1}{2}$

Total area $= |2| + \left|-\frac{1}{2}\right| = \frac{5}{2}$

19. (b) $\dfrac{x^2 - 1}{x + 1} = x - 1$ for $x \neq -1$

The area between the graph of f and the x-axis over

$[-2, 1)$ where f is negative is $\frac{1}{2}(3)(3) = \frac{9}{2}$. The area

between the graph of f and the x-axis over $(1, 3]$ where

f is positive is $\frac{1}{2}(2)(2) = 2$.

$$\int_{-2}^{3} \frac{x^2 - 1}{x + 1}\, dx = -\frac{9}{2} + 2 = -\frac{5}{2}$$

25. First, find the area under the graph of $y = x^2$.

$$\int_0^1 x^2\, dx = \left[\frac{1}{3}x^3\right]_0^1 = \frac{1}{3}$$

Next find the area under the graph of $y = 2 - x$.

$$\int_1^2 (2 - x)\, dx = \left[2x - \frac{1}{2}x^2\right]_1^2 = 2 - \frac{3}{2} = \frac{1}{2}$$

Area of the shaded region $= \frac{1}{3} + \frac{1}{2} = \frac{5}{6}$

35. $\displaystyle\int_a^x f(t)\, dt + K = \int_b^x f(t)\, dt$

$$K = -\int_a^x f(t)\, dt + \int_b^x f(t)\, dt = \int_x^a f(t)\, dt + \int_b^x f(t)\, dt$$

$$= \int_b^a f(t)\, dt$$

$$K = \int_2^{-1} (t^2 - 3t + 1)\, dt$$

$$= \left[\frac{1}{3}t^3 - \frac{3}{2}t^2 + t\right]_2^{-1}$$

$$= \left[-\frac{1}{3} - \frac{3}{2} + (-1)\right] - \left[\frac{8}{3} - 6 + 2\right] = -\frac{3}{2}$$

Section 5.5 (pp. 289–297)

3. (a) $f(x) = x^3$, $h = \dfrac{2 - 0}{4} = \dfrac{1}{2}$

x	0	$\frac{1}{2}$	1	$\frac{3}{2}$	2
$f(x)$	0	$\frac{1}{8}$	1	$\frac{27}{8}$	8

$$T = \frac{1}{4}\left(0 + 2\left(\frac{1}{8}\right) + 2(1) + 2\left(\frac{27}{8}\right) + 8\right) = 4.25$$

(b) $f'(x) = 3x^2$, $f''(x) = 6x > 0$ on $[0, 2]$

The approximation is an overestimate.

(c) $\displaystyle\int_0^2 x^3\, dx = \left[\frac{1}{4}x^4\right]_0^2 = 4$

7. $\dfrac{5}{2}(6.0 + 2(8.2) + 2(9.1) + \cdots + 2(12.7) + 13.0)(30)$

$$= 15{,}990 \text{ ft}^3$$

21. $h = \dfrac{24''}{6} = 4''$

Estimate the area to be

$\frac{4}{3}[0 + 4(18.75) + 2(24) + 4(26) + 2(24) + 4(18.75) + 0]$

$$\approx 466.67 \text{ in}^2$$

Chapter 6

Section 6.1 (pp. 303–315)

11. $\int\left(x^3 - \dfrac{1}{x^3}\right) dx = \int(x^3 - x^{-3})\, dx$

$\qquad\qquad = \dfrac{x^4}{4} + \dfrac{x^{-2}}{2} + C$

$\qquad\qquad = \dfrac{x^4}{4} + \dfrac{1}{2x^2} + C$

21. $\int \cos^2 x\, dx = \int \dfrac{1 + \cos 2x}{2}\, dx$

$\qquad\qquad = \int\left(\dfrac{1}{2} + \dfrac{\cos 2x}{2}\right) dx$

$\qquad\qquad = \dfrac{x}{2} + \dfrac{\sin 2x}{4} + C$

31. $\qquad \dfrac{dy}{dx} = 9x^2 - 4x + 5$

$\qquad \int \dfrac{dy}{dx}\, dx = \int(9x^2 - 4x + 5)\, dx$

$\qquad\qquad y = 3x^3 - 2x^2 + 5x + C$

Initial condition: $y(-1) = 0$
$\qquad 0 = 3(-1)^3 - 2(-1)^2 + 5(-1) + C$
$\qquad 0 = -10 + C$
$\qquad 10 = C$
Solution: $y = 3x^3 - 2x^2 + 5x + 10$

Section 6.2 (pp. 315–323)

1. $u = 3x$

$du = 3\, dx$

$\dfrac{1}{3}du = dx$

$\int \sin 3x\, dx = \dfrac{1}{3}\int \sin u\, du$

$\qquad\qquad = -\dfrac{1}{3} \cos u + C$

$\qquad\qquad = -\dfrac{1}{3} \cos 3x + C$

Check: $\dfrac{d}{dx}\left(-\dfrac{1}{3} \cos 3x + C\right) = -\dfrac{1}{3}(-\sin 3x)(3) = \sin 3x$

21. Let $u = \cos(2t + 1)$

$du = -\sin(2t + 1)(2)\, dt$

$-\dfrac{1}{2}du = \sin(2t + 1)\, dt$

$\int \dfrac{\sin(2t + 1)}{\cos^2(2t + 1)}\, dt = -\dfrac{1}{2}\int u^{-2}\, du$

$\qquad\qquad = \dfrac{1}{2}u^{-1} + C$

$\qquad\qquad = \dfrac{1}{2 \cos(2t + 1)} + C$

$\qquad\qquad = \dfrac{1}{2}\sec(2t + 1) + C$

31. Let $u = y + 1$

$du = dy$

$\int_0^3 \sqrt{y + 1}\, dy = \int_1^4 u^{1/2}\, du$

$\qquad\qquad = \dfrac{2}{3}u^{3/2}\Big]_1^4$

$\qquad\qquad = \dfrac{2}{3}(4)^{3/2} - \dfrac{2}{3}(1)^{3/2}$

$\qquad\qquad = \dfrac{2}{3}(8) - \dfrac{2}{3} = \dfrac{14}{3}$

39. $\dfrac{dy}{dx} = (y + 5)(x + 2)$

$\dfrac{dy}{y + 5} = (x + 2)dx$

Integrate both sides.

$\int \dfrac{dy}{y + 5} = \int(x + 2)\, dx$

On the left, let $u = y + 5$

$du = dy$

$\int \dfrac{1}{u}\, du = \dfrac{1}{2}x^2 + 2x + C$

$\ln |u| = \dfrac{1}{2}x^2 + 2x + C$

$\ln |y + 5| = \dfrac{1}{2}x^2 + 2x + C$

$y + 5 = Ce^{(1/2)x^2+2x}$

$y = Ce^{(1/2)x^2+2x} - 5$

Section 6.3 (pp. 323–329)

1. Let $u = x \qquad\qquad dv = \sin x\, dx$

$\qquad du = dx \qquad\qquad v = -\cos x$

$\int x \sin x\, dx = -x \cos x + \int \cos x\, dx$

$\qquad\qquad = -x \cos x + \sin x + C$

5. Let $u = x \qquad\qquad dv = \sec^2 x\, dx$

$\qquad du = dx \qquad\qquad v = \tan x$

$\int x \sec^2 x\, dx = x \tan x - \int \tan x\, dx$

$\qquad\qquad = x \tan x - \int \dfrac{\sin x}{\cos x}\, dx$

$\qquad\qquad = x \tan x + \ln |\cos x| + C$

9. Let $u = \ln x$ $\qquad$ $dv = x^3\, dx$

$\qquad du = \dfrac{1}{x}\, dx$ $\qquad v = \dfrac{1}{4}x^4$

$\displaystyle\int x^3 \ln x\, dx = \dfrac{1}{4}x^4 \ln x - \dfrac{1}{4}\int x^4\!\left(\dfrac{1}{x}\right) dx$

$\qquad\qquad\qquad = \dfrac{1}{4}x^4 \ln x - \dfrac{1}{4}\int x^3\, dx$

$\qquad\qquad\qquad = \dfrac{1}{4}x^4 \ln x - \dfrac{1}{16}x^4 + C$

15. Use tabular integration with $f(x) = x^2$ and $g(x) = \sin 2x$.

$f(x)$ and its derivatives		$g(x)$ and its integrals
x^2	(+)	$\sin 2x$
$2x$	(−)	$-\dfrac{1}{2}\cos 2x$
2	(+)	$-\dfrac{1}{4}\sin 2x$
0		$\dfrac{1}{8}\cos 2x$

$\displaystyle\int x^2 \sin 2x\, dx = -\dfrac{1}{2}x^2 \cos 2x + \dfrac{1}{2}x \sin 2x + \dfrac{1}{4}\cos 2x + C$

$\qquad\qquad\qquad = \left(\dfrac{1 - 2x^2}{4}\right)\cos 2x + \dfrac{x}{2}\sin 2x + C$

$\displaystyle\int_0^{\pi/2} x^2 \sin 2x\, dx = \left[\left(\dfrac{1 - 2x^2}{4}\right)\cos 2x + \dfrac{x}{2}\sin 2x\right]_0^{\pi/2}$

$\qquad\qquad\qquad = \left(\dfrac{1 - 2\left(\dfrac{\pi}{2}\right)^2}{4}\right)(-1) + 0 - \dfrac{1}{4}(1) - 0$

$\qquad\qquad\qquad = \dfrac{\pi^2}{8} - \dfrac{1}{2} \approx 0.734$

Check: $\text{NINT}\!\left(x^2 \sin 2x, x, 0, \dfrac{\pi}{2}\right) \approx 0.734$

19. $y = \displaystyle\int x^2 e^{4x}\, dx$

Let $u = x^2$ $\qquad\qquad$ $dv = e^{4x}dx$

$\qquad du = 2x\, dx$ $\qquad\quad$ $v = \dfrac{1}{4}e^{4x}$

$y = (x^2)\!\left(\dfrac{1}{4}e^{4x}\right) - \displaystyle\int\!\left(\dfrac{1}{4}e^{4x}\right)(2x\, dx)$

$\quad = \dfrac{1}{4}x^2 e^{4x} - \dfrac{1}{2}\displaystyle\int xe^{4x}dx$

Let $u = x$ $\qquad\qquad$ $dv = e^{4x}\, dx$

$\qquad du = dx$ $\qquad\qquad$ $v = \dfrac{1}{4}e^{4x}$

$y = \dfrac{1}{4}x^2 e^{4x} - \dfrac{1}{2}\left[(x)\!\left(\dfrac{1}{4}e^{4x}\right) - \displaystyle\int\!\left(\dfrac{1}{4}e^{4x}\right)dx\right]$

$y = \dfrac{1}{4}x^2 e^{4x} - \dfrac{1}{8}xe^{4x} + \dfrac{1}{32}e^{4x} + C$

$y = \left(\dfrac{x^2}{4} - \dfrac{x}{8} + \dfrac{1}{32}\right)e^{4x} + C$

27. Let $w = \sqrt{x}$. Then $dw = \dfrac{dx}{2\sqrt{x}}$, so $dx = 2\sqrt{x}\, dw = 2w\, dw$.

$\displaystyle\int \sin \sqrt{x}\, dx = \int (\sin w)(2w\, dw) = 2\int w \sin w\, dw$

Let $u = w$ $\qquad\qquad$ $dv = \sin w\, dw$

$\qquad du = dw$ $\qquad\quad$ $v = -\cos w$

$\displaystyle\int w \sin w\, dw = -w \cos w + \int \cos w\, dw$

$\qquad\qquad\qquad = -w \cos w + \sin w + C$

$\displaystyle\int \sin \sqrt{x}\, dx = 2\int w \sin w\, dw$

$\qquad\qquad\qquad = -2w \cos w + 2 \sin w + C$

$\qquad\qquad\qquad = -2\sqrt{x} \cos \sqrt{x} + 2 \sin \sqrt{x} + C$

Section 6.4 (pp. 330–341)

5. Doubling time:

$\qquad A(t) = A_0 e^{rt}$

$\qquad 2000 = 1000e^{0.086t}$

$\qquad 2 = e^{0.086t}$

$\qquad \ln 2 = 0.086t$

$\qquad t = \dfrac{\ln 2}{0.086} \approx 8.06$ yr

Amount in 30 years:

$\qquad A = 1000e^{(0.086)(30)} \approx \$13{,}197.14$

13. $\qquad 0.9 = e^{-0.18t}$

$\qquad \ln 0.9 = -0.18t$

$\qquad t = -\dfrac{\ln 0.9}{0.18} \approx 0.585$ days

23. Note that the total mass is $66 + 7 = 73$ kg.

$\qquad v = v_0 e^{-(k/m)t}$

$\qquad v = 9e^{-3.9t/73}$

(a) $s(t) = \displaystyle\int 9e^{-3.9t/73}dt = -\dfrac{2190}{13}e^{-3.9t/73} + C$

$\qquad$ Since $s(0) = 0$ we have $C = \dfrac{2790}{13}$ and

$\qquad \displaystyle\lim_{t\to\infty} s(t) = \lim_{t\to\infty}\dfrac{2190}{13}(1 - e^{-3.9t/73}) = \dfrac{2190}{13} \approx 168.5$

$\qquad$ The cyclist will coast about 168.5 meters.

(b) $\qquad 1 = 9e^{-3.9t/73}$

$\qquad \dfrac{3.9t}{73} = \ln 9$

$\qquad t = \dfrac{73 \ln 9}{3.9} \approx 41.13$ sec

$\qquad$ It will take about 41.13 seconds.

29. (a) By the Law of Exponential Change, the solution is
$V = V_0 e^{-(1/40)t}$.

(b) $0.1 = e^{-(1/40)t}$

$\ln 0.1 = -\dfrac{t}{40}$

$t = -40 \ln 0.1 \approx 92.1$ sec

It will take about 92.1 seconds.

33. $\dfrac{v_0 m}{k} =$ coasting distance

$\dfrac{(0.80)(49.90)}{k} = 1.32$

$k = \dfrac{998}{33}$

We know that $\dfrac{v_0 m}{k} = 1.32$ and $\dfrac{k}{m} = \dfrac{998}{33(49.9)} = \dfrac{20}{33}$.

Using Equation 3, we have:

$s(t) = \dfrac{v_0 m}{k}(1 - e^{-(k/m)t})$

$= 1.32(1 - e^{-20t/33})$

$\approx 1.32(1 - e^{-0.606t})$

Section 6.5 (pp. 342–349)

1. (a) $\dfrac{dP}{dt} = 0.025P$

(b) Using the Law of Exponential Change from Section 6.4, the formula is $P = 75{,}000e^{0.025t}$.

(c)

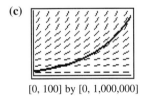

[0, 100] by [0, 1,000,000]

7. $\dfrac{dP}{dt} = 0.04P - 0.0004P^2$

$= 0.0004P(100 - P)$

$= \dfrac{0.04}{100}P(100 - P)$

$= \dfrac{k}{M}P(M - P)$

Thus, $k = 0.04$ and the carrying capacity is $M = 100$.

13. (a) $P(t) = \dfrac{1000}{1 + e^{4.8 - 0.7t}}$

$= \dfrac{1000}{1 + e^{4.8}e^{-0.7t}}$

$= \dfrac{M}{1 + Ae^{-kt}}$

This is a logistic growth model with $k = 0.7$ and $M = 1000$.

(b) $P(0) = \dfrac{1000}{1 + e^{4.8}} \approx 8$

Initially there are 8 rabbits.

23. (a) Note that the given years correspond to $x = 0$, $x = 20$, $x = 50$, $x = 70$, and $x = 80$.

$y = \dfrac{18.70}{1 + 1.075e^{-0.0422x}}$

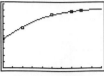

[0, 100] by [0, 20]

(b) Carrying capacity $= \lim\limits_{x\to\infty} y = 18.70$, representing 18.7 million people.

(c) Using NDER twice and solving graphically, we find that $y'' = 0$ when $x \approx 1.7$, corresponding to the year 1912. The population at this time was about $y(1.7) \approx 9.35$ million.

27. $\dfrac{dy}{dx} = (\cos x)e^{\sin x}$

$\displaystyle\int dy = \int (\cos x)e^{\sin x}\, dx$

$\displaystyle\int dy = \int e^u\, du$

$y = e^u + C$

$y = e^{\sin x} + C$

Initial value: $y(0) = 0$

$0 = e^{\sin 0} + C$

$-1 = C$

Solution: $y = e^{\sin x} - 1$

Section 6.6 (pp. 350–356)

1. Check the differential equation:

$y' = \dfrac{d}{dx}(x - 1 + 2e^{-x}) = 1 + 2e^{-x}(-1) = 1 - 2e^{-x}$

$x - y = x - (x - 1 + 2e^{-x}) = 1 - 2e^{-x}$

Therefore, $y' = x - y$.

Check the initial condition:

$y(0) = 0 - 1 + 2e^{-(0)} = -1 + 2 = 1$

5. Note that we are finding an exact solution to the initial value problem discussed in Examples 1–4.

$\dfrac{dy}{dx} = 1 + y$

$\displaystyle\int \dfrac{dy}{1 + y} = \int dx$

$\ln|1 + y| = x + C$

$|1 + y| = e^{x+C}$

$1 + y = \pm e^{x+C}$

$y = \pm e^C e^x - 1$

$y = Ae^x - 1$

Initial condition: $y(0) = 1$

$1 = Ae^0 - 1$

$2 = A$

Solution: $y = 2e^x - 1$

9. To find the approximate values, set $y_1 = 2y + \sin x$ and use EULERT with initial values $x = 0$ and $y = 0$ and step size 0.1 for 10 points. The exact values are given by

$$y = \frac{1}{5}(e^{2x} - 2\sin x - \cos x).$$

x	y (Euler)	y (exact)	Error
0	0	0	0
0.1	0	0.0053	0.0053
0.2	0.0100	0.0229	0.0129
0.3	0.0318	0.0551	0.0233
0.4	0.0678	0.1051	0.0374
0.5	0.1203	0.1764	0.0561
0.6	0.1923	0.2731	0.0808
0.7	0.2872	0.4004	0.1132
0.8	0.4090	0.5643	0.1553
0.9	0.5626	0.7723	0.2097
1.0	0.7534	1.0332	0.2797

15. (a)

$$\frac{dy}{dx} = 2y^2(x - 1)$$

$$\frac{dy}{y^2} = 2(x - 1)dx$$

$$\int y^{-2}\, dy = \int (2x - 2)\, dx$$

$$-y^{-1} = x^2 - 2x + C$$

Initial value: $y(2) = -\frac{1}{2}$

$$2 = 2^2 - 2(2) + C$$

$$2 = C$$

Solution: $-y^{-1} = x^2 - 2x + 2$ or $y = -\dfrac{1}{x^2 - 2x + 2}$

$$y(3) = -\frac{1}{3^2 - 2(3) + 2} = -\frac{1}{5} = 0.2$$

(b) To find the approximation, set $y_1 = 2y^2(x - 1)$ and use EULERT with initial values $x = 2$ and $y = -\frac{1}{2}$ and step size 0.2 for 5 points. This gives $y(3) \approx -0.1851$;

error ≈ 0.0149.

19. Set $y_1 = 2y + \sin x$ and use EULERG with initial values $x = 0$ and $y = 0$ and step size 0.1. The exact solution is $y = \frac{1}{5}(e^{2x} - 2\sin x - \cos x)$.

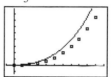

$[-0.1, 1.1]$ by $[-0.13, 0.88]$

25. Set $y_1 = y + e^x - 2$ and EULERG, with initial values $x = 0$ and $y = 2$ and step sizes 0.1 and 0.05.

(a)

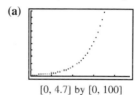

$[0, 4.7]$ by $[0, 100]$

(b)

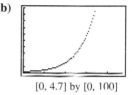

$[0, 4.7]$ by $[0, 100]$

Chapter 7

Section 7.1 (pp. 363–374)

1. (a) Right when $v(t) > 0$, which is when $\cos t > 0$, i.e., when $0 \le t < \frac{\pi}{2}$ or $\frac{3\pi}{2} < t \le 2\pi$. Left when $\cos t < 0$, i.e., when $\frac{\pi}{2} < t < \frac{3\pi}{2}$. Stopped when $\cos t = 0$, i.e., when $t = \frac{\pi}{2}$ or $\frac{3\pi}{2}$.

(b) Displacement

$$\int_0^{2\pi} 5\cos t\, dt = 5\Big[\sin t\Big]_0^{2\pi} = 5[\sin 2\pi - \sin 0] = 0$$

(c) Distance $= \int_0^{2\pi} |5\cos t|\, dt$

$$= \int_0^{\pi/2} 5\cos t\, dt + \int_{\pi/2}^{3\pi/2} -5\cos t\, dt + \int_{3\pi/2}^{2\pi} 5\cos t\, dt$$

$$= 5 + 10 + 5 = 20$$

9. (a) $v(t) = \int a(t)\, dt = t + 2t^{3/2} + C$, and since $v(0) = 0$,

$v(t) = t + 2t^{3/2}$. Then $v(9) = 9 + 2(27) = 63$ mph.

(b) First convert units:

$t + 2t^{3/2}$ mph $= \dfrac{t}{3600} + \dfrac{t^{3/2}}{1800}$ mi/sec. Then

Distance $= \displaystyle\int_0^9 \left(\dfrac{t}{3600} + \dfrac{t^{3/2}}{1800}\right) dt$

$= \left[\dfrac{t^2}{7200} + \dfrac{t^{5/2}}{4500}\right]_0^9 = \left[\left(\dfrac{9}{800} + \dfrac{27}{500}\right) - 0\right]$

$= 0.06525$ mi $= 344.52$ ft

15. At $t = a$, where $\dfrac{dv}{dt}$ is at a maximum (the graph is steepest

upward).

17. Distance $=$ Area under curve $= 4\left(\dfrac{1}{2} \cdot 1 \cdot 2\right) = 4$

(a) Final position $=$ Initial position $+$ Distance
$= 2 + 4 = 6$; ends at $x = 6$.

(b) 4 meters

21. $\displaystyle\int_0^{10} 27.08 \cdot e^{t/25}\, dt = 27.08\left[25e^{t/25}\right]_0^{10}$

$= 27.08[25e^{0.4} - 25] \approx 332.965$ billion barrels

23. (a) Solve $10{,}000(2 - r) = 0$: $r = 2$ miles.

(b) Width $= \Delta r$, Length $= 2\pi r$: Area $= 2\pi r \Delta r$

(c) Population $=$ Population density $\times$ Area

(d) $\displaystyle\int_0^2 10{,}000(2 - r)(2\pi r)\, dr = 20{,}000\pi \int_0^2 (2r - r^2)\, dr$

$= 20{,}000\pi\left[r^2 - \dfrac{1}{3}r^3\right]_0^2 = 20{,}000\pi\left[\left(4 - \dfrac{8}{3}\right) - 0\right]$

$= \dfrac{80{,}000}{3}\pi \approx 83{,}776$

29. $F(x) = kx$; $6 = F(3)$, so $k = 2$ and $F(x) = 2x$.

(a) $F(9) = 2(9) = 18$N

(b) $W = \displaystyle\int_0^9 F(x)\, dx = \int_0^9 2x\, dx = \left[x^2\right]_0^9 = 81$ N $\cdot$ cm

31. $\dfrac{12 - 0}{2(12)}[0.04 + 2(0.04) + 2(0.05) + 2(0.06) + 2(0.05)$

$+ 2(0.04) + 2(0.04) + 2(0.05) + 2(0.04)$

$+ 2(0.06) + 2(0.06) + 2(0.05) + 0.05] = 0.585$

The overall rate, then, is $\dfrac{0.585}{12} = 0.04875$.

Section 7.2 (pp. 374–382)

1. $\displaystyle\int_0^\pi (1 - \cos^2 x)\, dx = \left[\dfrac{1}{2}x - \dfrac{1}{4}\sin 2x\right]_0^\pi = \dfrac{\pi}{2}$

7. Integrate in two parts:

$\displaystyle\int_0^1 \left(x - \dfrac{x^2}{4}\right) dx + \int_1^2 \left(1 - \dfrac{x^2}{4}\right) dx$

$= \left[\dfrac{1}{2}x^2 - \dfrac{x^3}{12}\right]_0^1 + \left[x - \dfrac{x^3}{12}\right]_1^2$

$= \left[\left(\dfrac{1}{2} - \dfrac{1}{12}\right) - 0\right] + \left[\left(2 - \dfrac{2}{3}\right) - \left(1 - \dfrac{1}{12}\right)\right]$

$= \dfrac{5}{6}$

11. Solve $x^2 - 2 = 2$: $x^2 = 4$ and the curves intersect at

$x = \pm 2$.

$\displaystyle\int_{-2}^2 [2 - (x^2 - 2)]\, dx = \int_{-2}^2 (4 - x^2)\, dx$

$= \left[4x - \dfrac{1}{3}x^3\right]_{-2}^2 = \left(8 - \dfrac{8}{3}\right) - \left(-8 + \dfrac{8}{3}\right) = \dfrac{32}{3} = 10\dfrac{2}{3}$

21. Solve for x: $x = -y^2$ and $x = 2 - 3y^2$.

Now solve $-y^2 = 2 - 3y^2$: $y^2 = 1$ and the curves intersect

at $y = \pm 1$. $\displaystyle\int_{-1}^1 (2 - 3y^2 + y^2)\, dy = \int_{-1}^1 (2 - 2y^2)\, dy$

$= 2\displaystyle\int_{-1}^1 (1 - y^2)\, dy = 2\left[y - \dfrac{1}{3}y^3\right]_{-1}^1 = 2\left[\dfrac{2}{3} - \left(-\dfrac{2}{3}\right)\right] = \dfrac{8}{3}$

31. Solve for x: $x = y^3$ and $x = y$.

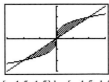

$[-1.5, 1.5]$ by $[-1.5, 1.5]$

The curves intersect at $x = 0$ and $x = \pm 1$. Use the area's

symmetry: $2\displaystyle\int_0^1 (y - y^3)\, dy = 2\left[\dfrac{1}{2}y^2 - \dfrac{1}{4}y^4\right]_0^1 = \dfrac{1}{2}$

41.

$[-1.5, 1.5]$ by $[-1.5, 1.5]$

The curves intersect at $x = 0$ and $x = \pm 1$. Use the area's

symmetry: $2\displaystyle\int_0^1 \left(\dfrac{2x}{x^2 + 1} - x^3\right) dx = 2\left[\ln\left(x^2 + 1\right) - \dfrac{1}{4}x^4\right]_0^1$

$= 2\ln 2 - \dfrac{1}{2} = \ln 4 - \dfrac{1}{2} \approx 0.886$

Section 7.3 (pp. 383–394)

1. In each case, the width of the cross section is
$w = 2\sqrt{1 - x^2}$.

   (a) $A = \pi r^2$, where $r = \dfrac{w}{2}$, so $A(x) = \pi\left(\dfrac{w}{2}\right)^2 = \pi(1 - x^2)$.

   (b) $A = s^2$, where $s = w$, so $A(x) = w^2 = 4(1 - x^2)$.

   (c) $A = s^2$, where $s = \dfrac{w}{\sqrt{2}}$, so $A(x) = \left(\dfrac{w}{\sqrt{2}}\right)^2 = 2(1 - x^2)$.

   (d) $A = \dfrac{\sqrt{3}}{4}w^2$ (see Quick Review Exercise 5), so
   $A(x) = \dfrac{\sqrt{3}}{4}(2\sqrt{1 - x^2})^2 = \sqrt{3}(1 - x^2).$

3. A cross section has width $w = 2\sqrt{x}$ and area
   $A(x) = s^2 = \left(\dfrac{w}{\sqrt{2}}\right)^2 = 2x$. The volume is
   $\displaystyle\int_0^4 2x\, dx = x^2\Big]_0^4 = 16.$

13. The solid is a right circular cone of radius 1 and height 2.
   $V = \dfrac{1}{3}Bh = \dfrac{1}{3}(\pi r^2)h = \dfrac{1}{3}(\pi 1^2)2 = \dfrac{2}{3}\pi.$

17.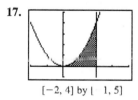

   $[-2, 4]$ by $[-1, 5]$

   A cross section has radius $r = x^2$ and area

   $A(x) = \pi r^2 = \pi x^4$. The volume is
   $\displaystyle\int_0^2 \pi x^4\, dx = \pi\left[\dfrac{1}{5}x^5\right]_0^2 = \dfrac{32\pi}{5}$

29.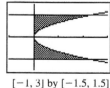

   $[-1, 3]$ by $[-1.5, 1.5]$

   A cross section has radius $r = \sqrt{5}y^2$ and area

   $A(y) = \pi r^2 = 5\pi y^4.$
   The volume is $\displaystyle\int_{-1}^1 5\pi y^4\, dy = \pi\left[y^5\right]_{-1}^1 = 2\pi.$

35.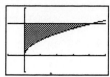

   $[-1, 5]$ by $[-1, 3]$

   The curved and horizontal line intersect at $(4, 2)$.

   (a) Use washer cross sections: a washer has inner radius
   $r = \sqrt{x}$, outer radius $R = 2$, and area
   $A(x) = \pi(R^2 - r^2) = \pi(4 - x)$. The volume is
   $\displaystyle\int_0^4 \pi(4 - x)\, dx = \pi\left[4x - \dfrac{1}{2}x^2\right]_0^4 = 8\pi$

   (b) A cross section has radius $r = y^2$ and area
   $A(y) = \pi r^2 = \pi y^4.$
   The volume is $\displaystyle\int_0^2 \pi y^4\, dy = \pi\left[\dfrac{1}{5}y^5\right]_0^2 = \dfrac{32\pi}{5}.$

   (c) A cross section has radius $r = 2 - \sqrt{x}$ and area
   $A(x) = \pi r^2 = \pi(2 - \sqrt{x})^2 = \pi(4 - 4\sqrt{x} + x).$

   The volume is
   $\displaystyle\int_0^4 \pi(4 - 4\sqrt{x} + x)\, dx = \pi\left[4x - \dfrac{8}{3}x^{3/2} + \dfrac{1}{2}x^2\right]_0^4$
   $\qquad = \dfrac{8\pi}{3}.$

   (d) Use washer cross sections: a washer has inner radius
   $r = 4 - y^2$, outer radius $R = 4$, and area
   $A(y) = \pi(R^2 - r^2) = \pi[16 - (4 - y^2)^2]$
   $\quad = \pi(8y^2 - y^4).$

   The volume is
   $\displaystyle\int_0^2 \pi(8y^2 - y^4)\, dy = \pi\left[\dfrac{8}{3}y^3 - \dfrac{1}{5}y^5\right]_0^2 = \dfrac{224\pi}{15}$

39.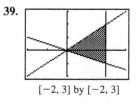

   $[-2, 3]$ by $[-2, 3]$

   A shell has radius x and height $x - \left(-\dfrac{x}{2}\right) = \dfrac{3}{2}x.$
   The volume is $\displaystyle\int_0^2 2\pi(x)\left(\dfrac{3}{2}x\right)\, dx = \pi\left[x^3\right]_0^2 = 8\pi.$

45.

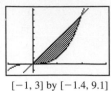

[−1, 3] by [−1.4, 9.1]

The functions intersect at (2, 8).

(a) Use washer cross sections: a washer has inner radius

$r = x^3$, outer radius $R = 4x$, and area

$A(x) = \pi(R^2 - r^2) = \pi(16x^2 - x^6)$. The volume is

$$\int_0^2 \pi(16x^2 - x^6)\, dx = \pi\left[\frac{16}{3}x^3 - \frac{1}{7}x^7\right]_0^2 = \frac{512\pi}{21}.$$

(b) Use cylindrical shells: a shell has a radius $8 - y$ and

height $y^{1/3} - \dfrac{y}{4}$. The volume is

$$\int_0^8 2\pi(8 - y)\left(y^{1/3} - \frac{y}{4}\right) dy$$

$$= 2\pi \int_0^8 \left(8y^{1/3} - 2y - y^{4/3} + \frac{y^2}{4}\right) dy$$

$$= 2\pi\left[6y^{4/3} - y^2 - \frac{3}{7}y^{7/3} + \frac{1}{12}y^3\right]_0^8 = \frac{832\pi}{21}.$$

57. $g'(y) = \dfrac{dx}{dy} = \dfrac{1}{2\sqrt{y}}$, and

$$\int_0^2 2\pi\sqrt{y}\sqrt{1 + \left(\frac{1}{2\sqrt{y}}\right)^2}\, dy = \int_0^2 \pi\sqrt{4y + 1}\, dy \text{ evaluates,}$$

using NINT, to ≈ 13.614.

Section 7.4 (pp. 395–401)

1. (a) $y' = 2x$, so

$$\text{Length} = \int_{-1}^2 \sqrt{1 + (2x)^2}\, dx = \int_{-1}^2 \sqrt{1 + 4x^2}\, dx$$

(b)

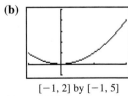

[−1, 2] by [−1, 5]

(c) ≈6.126

13. $x' = y^2 - \dfrac{1}{4y^2}$, so the length is $\displaystyle\int_1^3 \sqrt{1 + \left[y^2 - \frac{1}{4y^2}\right]^2}\, dy$

$$= \int_1^3 \sqrt{\left[y^2 + \frac{1}{4y^2}\right]^2}\, dy = \left[\frac{1}{3}y^3 - \frac{1}{4y}\right]_1^3 = \frac{53}{6}.$$

23. Find the length of the curve $y = \sin\dfrac{3\pi}{20}x$ for $0 \le x \le 20$.

$y' = \dfrac{3\pi}{20}\cos\dfrac{3\pi}{20}x$, so the length is

$$\int_0^{20} \sqrt{1 + \left(\frac{3\pi}{20}\cos\frac{3\pi}{20}x\right)^2}\, dx, \text{ which evaluate, using NINT,}$$

to ≈ 21.07 inches.

29. $y = (1 - \sqrt{x})^2,\ 0 \le x \le 1$

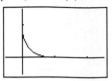

[−0.5, 2.5] by [−0.5, 1.5]

$y' = \dfrac{\sqrt{x} - 1}{\sqrt{x}}$, but NINT may fail using y' over the entire

interval because y' is undefined at $x = 0$. So, split the

curve into two equal segments by solving $\sqrt{x} + \sqrt{y} = 1$

with $y = x$: $x = \dfrac{1}{4}$. The total length is

$$2\int_{1/4}^1 \sqrt{1 + \left(\frac{\sqrt{x} - 1}{\sqrt{x}}\right)^2}\, dx, \text{ which evaluates, using NINT, to}$$

≈ 1.623.

Section 7.5 (pp. 401–411)

1. $\displaystyle\int_0^5 xe^{-x/3}\, dx = \left[9\left(-\frac{x}{3} - 1\right)e^{-x/3}\right]_0^5 = -\frac{24}{e^{5/3}} + 9 \approx 4.4670 \text{ J}$

5. When the bucket is x m off the ground, the water weighs

$$F(x) = 490\left(\frac{20 - x}{20}\right) = 490\left(1 - \frac{x}{20}\right) = (490 - 24.5x) \text{ N.}$$

Then

$$W = \int_0^{20} (490 - 24.5x)\, dx = \left[490x - 12.25x^2\right]_0^{20}$$

$$= 4900 \text{ J.}$$

9. (a) $F = kx$, so $21,714 = k(8 - 5)$ and $k = 7238$ lb/in.

(b) $F(x) = 7238x.$ $\displaystyle\int_0^{1/2} 7238x\, dx = \left[3619x^2\right]_0^{1/2}$

$$= 904.75 \approx 905 \text{ in.-lb, and } \int_{1/2}^1 7238x\, dx$$

$$= \left[3619x^2\right]_{1/2}^1 = 2714.25 \approx 2714 \text{ in.-lb.}$$

17. (a) Work to raise a thin slice $= 62.4(10 \times 12)(\Delta y)y$

$$\text{Total work} = \int_0^{20} 62.4(120)y\, dy = 62.4\left[60y^2\right]_0^{20}$$

$$= 1,497,600 \text{ ft-lb}$$

(b) (1,497,600 ft-lb) ÷ (250 ft-lb/sec) = 5990.4 sec
≈ 100 min

(c) Work to empty half the tank = $\int_0^{10} 62.4(120)y\, dy$

$= 62.4\left[60y^2\right]_0^{10} = 374{,}400$ ft-lb, and

374,400 ÷ 250 = 1497.6 sec ≈ 25 min

(d) The weight per ft^3 of water is a simple multiplicative

factor in the answers. So divide by 62.4 and multiply

by the appropriate weight-density

For 62.26:

$1{,}497{,}600\left(\dfrac{62.26}{62.4}\right) = 1{,}494{,}240$ ft-lb and $5990.4\left(\dfrac{62.26}{62.4}\right)$

$= 5976.96$ sec ≈ 100 min.

For 62.5:

$1{,}497{,}600\left(\dfrac{62.5}{62.4}\right) = 1{,}500{,}000$ ft-lb and

$5990.4\left(\dfrac{62.5}{62.4}\right) = 6000$ sec = 100 min.

23. The work to raise a thin disk is

$\pi r^2(56)h = \pi(\sqrt{10^2 - y^2})^2(56)(10 + 2 - y)\Delta y$

$= 56\pi(12 - y)(100 - y^2)\Delta y$. The total work is

$\int_0^{10} 56\pi(12 - y)(100 - y^2)\, dy$, which evaluates using

NINT to ≈967,611 ft-lb. This will come to

(967,611)($0.005) ≈ $4838, so yes, there's enough money

to hire the firm.

25. (a) The pressure at depth y is $62.4y$, and the area of a thin

horizontal strip is $2\Delta y$. The depth of water is $\dfrac{11}{6}$ ft,

so the total force on an end is

$\int_0^{11/6} (62.4y)(2\, dy) \approx 209.73$ lb.

(b) On the sides, which are twice as long as the ends, the

initial total force is doubled to ≈419.47 lb. When the

tank is upended, the depth is doubled to $\dfrac{11}{3}$ ft, and the

force on a side becomes

$\int_0^{11/3} (62.4y)(2\, dy) \approx 838.93$ lb, which means that the

fluid force doubles.

27. (a) 0.5 (50%), since half of a normal distribution lies
below the mean.

(b) Use NINT to find $\int_{63}^{65} f(x)\, dx$, where

$f(x) = \dfrac{1}{3.2\sqrt{2\pi}}e^{-(x - 63.4)^2/(2 \cdot 3.2^2)}$. The result is

≈0.24 (24%).

(c) 6 ft = 72 in. Pick 82 in. as a conveniently high upper

limit and with NINT, find $\int_{72}^{82} f(x)\, dx$. The result is

≈0.0036 (0.36%).

(d) 0 if we assume a continuous distribution. Between

59.5 in. and 60.5 in., the proportion is

$\int_{59.5}^{60.5} f(x)\, dx \approx 0.071$ (7.1%)

35. $\dfrac{0.3125 \text{ lb}}{32 \text{ ft/sec}^2} = 0.009765625$ slug, and

90 mph $= 90\left(\dfrac{5280 \text{ ft}}{1 \text{ mi}}\right)\left(\dfrac{1 \text{ hr}}{3600 \text{ sec}}\right) = 132$ ft/sec, so

Work = change in kinetic energy $= \dfrac{1}{2}(0.009765625)(132)^2$

≈ 85.1 ft-lb.

Chapter 8

Section 8.1 (pp. 417–425)

5. $\displaystyle\lim_{x\to 1} \frac{x^3 - 1}{4x^3 - x - 3} = \lim_{x\to 1} \frac{3x^2}{12x^2 - 1} = \frac{3}{11}$

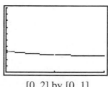

[0, 2] by [0, 1]

The graph supports the answer.

7. The limit leads to the indeterminate form 1^∞.

Let $\ln f(x) = \dfrac{\ln(e^x + x)}{x}$

$$\lim_{x \to 0^+} \frac{\ln(e^x + x)}{x} = \lim_{x \to 0^+} \frac{\frac{e^x + 1}{e^x + x}}{1} = \lim_{x \to 0^+} \frac{e^x + 1}{e^x + x} = \frac{2}{1} = 2$$

$$\lim_{x \to 0^+} (e^x + x)^{1/x} = \lim_{x \to 0^+} f(x) = \lim_{x \to 0^+} e^{\ln f(x)} = e^2$$

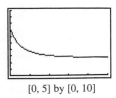

[0, 5] by [0, 10]

The graph supports the answer.

13. Let $f(x) = (1 + x)^{1/x}$.

x	10	10^2	10^3	10^4	10^5
$f(x)$	1.2710	1.0472	1.0069	1.0009	1.0001

Estimate the limit to be 1.

$$\ln f(x) = \frac{\ln(1 + x)}{x}$$

$$\lim_{x \to \infty} \frac{\ln(1 + x)}{x} = \lim_{x \to \infty} \frac{\frac{1}{1 + x}}{1} = \frac{0}{1} = 0$$

$$\lim_{x \to \infty} (1 + x)^{1/x} = \lim_{x \to \infty} f(x) = \lim_{x \to \infty} e^{\ln f(x)} = e^0 = 1$$

17. $\displaystyle \lim_{t \to 0} \frac{\cos t - 1}{e^t - t - 1} = \lim_{t \to 0} \frac{-\sin t}{e^t - 1} = \lim_{t \to 0} \frac{-\cos t}{e^t} = -1$

35. The limit leads to the indeterminate form 0^0.

Let $f(x) = (x^2 - 2x + 1)^{x-1}$

$$\ln(x^2 - 2x + 1)^{x-1} = (x - 1)\ln(x^2 - 2x + 1)$$

$$= \frac{\ln(x^2 - 2x + 1)}{\frac{1}{x - 1}}$$

$$\lim_{x \to 1} \frac{\ln(x^2 - 2x + 1)}{\frac{1}{x - 1}} = \lim_{x \to 1} \frac{\frac{2x - 2}{x^2 - 2x + 1}}{-\frac{1}{(x - 1)^2}}$$

$$= \lim_{x \to 1} \frac{\frac{2(x - 1)}{(x - 1)^2}}{-\frac{1}{(x - 1)^2}}$$

$$= \lim_{x \to 1} -2(x - 1) = 0$$

$$\lim_{x \to 1} (x^2 - 2x + 1)^{x-1} = \lim_{x \to 1} e^{\ln f(x)} = e^0 = 1$$

Section 8.2 (pp. 425–433)

7. $\displaystyle \lim_{x \to \infty} \frac{e^{\cos x}}{e^x} = 0$ since $e^{\cos x} \le e$ for all x.

$e^{\cos x}$ grows slower than e^x as $x \to \infty$.

11. $\displaystyle \lim_{x \to \infty} \frac{x^2 + 4x}{x^2} = \lim_{x \to \infty} \left(1 + \frac{4}{x}\right) = 1$

$x^2 + 4x$ grows at the same rate as x^2 as $x \to \infty$.

21. $\displaystyle \lim_{x \to \infty} \frac{x - 2\ln x}{\ln x} = \lim_{x \to \infty} \left(\frac{x}{\ln x} - 2\right) = \lim_{x \to \infty} \left(\frac{1}{1/x} - 2\right)$

$$= \lim_{x \to \infty} (x - 2) = \infty$$

$x - 2\ln x$ grows faster than $\ln x$ as $x \to \infty$.

23. Compare e^x to x^x.

$$\lim_{x \to \infty} \frac{e^x}{x^x} = \lim_{x \to \infty} \left(\frac{e}{x}\right)^x = 0$$

e^x grows slower than x^x.
Compare e^x to $(\ln x)^x$.

$$\lim_{x \to \infty} \frac{e^x}{(\ln x)^x} = \lim_{x \to \infty} \left(\frac{e}{\ln x}\right)^x = 0$$

e^x grows slower than $(\ln x)^x$.
Compare e^x to $e^{x/2}$.

$$\lim_{x \to \infty} \frac{e^x}{e^{x/2}} = \lim_{x \to \infty} e^{x/2} = \infty$$

e^x grows faster than $e^{x/2}$.
Compare x^x to $(\ln x)^x$.

$$\lim_{x \to \infty} \frac{x^x}{(\ln x)^x} = \lim_{x \to \infty} \left(\frac{x}{\ln x}\right)^x = \infty \text{ since } \lim_{x \to \infty} \frac{x}{\ln x} = \lim_{x \to \infty} \frac{1}{1/x} = \infty.$$

x^x grows faster than $(\ln x)^x$.
Thus, in order from slowest-growing to fastest-growing, we get $e^{x/2}$, e^x, $(\ln x)^x$, x^x.

27. Compare f_1 to f_2.

$$\lim_{x \to \infty} \frac{f_2(x)}{f_1(x)} = \lim_{x \to \infty} \frac{\sqrt{9^x - 2^x}}{3^x}$$

$$= \lim_{x \to \infty} \frac{\sqrt{9^x - 2^x}}{\sqrt{9^x}}$$

$$= \lim_{x \to \infty} \sqrt{1 - \left(\frac{2}{9}\right)^x} = 1$$

Thus f_1 and f_2 grow at the same rate.

Compare f_1 to f_3.

$$\lim_{x \to \infty} \frac{f_3(x)}{f_1(x)} = \lim_{x \to \infty} \frac{\sqrt{9^x - 4^x}}{3^x}$$

$$= \lim_{x \to \infty} \frac{\sqrt{9^x - 4^x}}{\sqrt{9^x}}$$

$$= \lim_{x \to \infty} \sqrt{1 - \left(\frac{4}{9}\right)^x} = 1$$

Thus f_1 and f_3 grow at the same rate.

By transitivity, f_2 and f_3 grow at the same rate, so all three functions grow at the same rate as $x \to \infty$.

29. (a) False, since $\lim\limits_{x \to \infty} \dfrac{x}{x} = 1 \neq 0$.

(b) False, since $\lim\limits_{x \to \infty} \dfrac{x}{x+5} = 1 \neq 0$

(c) True, since $\lim\limits_{x \to \infty} \dfrac{x}{x+5} = 1 \leq 1$.

(d) True, since $\lim\limits_{x \to \infty} \dfrac{x}{2x} = \dfrac{1}{2} \leq 1$.

(e) True, since $\lim\limits_{x \to \infty} \dfrac{e^x}{e^{2x}} = \lim\limits_{x \to \infty} \dfrac{1}{e^x} = 0$.

(f) True, since $\lim\limits_{x \to \infty} \dfrac{x + \ln x}{x} = \lim\limits_{x \to \infty} \dfrac{1 + \frac{1}{x}}{1} = 1 \leq 1$.

(g) False, since $\lim\limits_{x \to \infty} \dfrac{\ln x}{\ln 2x} = \lim\limits_{x \to \infty} \dfrac{1/x}{1/x} = 1 \neq 0$.

(h) True, since $\lim\limits_{x \to \infty} \dfrac{\sqrt{x^2 + 5}}{x} = \lim\limits_{x \to \infty} \sqrt{1 + \dfrac{5}{x^2}} = 1 \leq 1$.

31. From the graph, $\lim\limits_{x \to \infty} \dfrac{f(x)}{g(x)} = \infty$, so $\lim\limits_{x \to \infty} \dfrac{g(x)}{f(x)} = 0$.

Thus $g = o(f)$, so II is true.

Section 8.3 (pp. 433–444)

3 (a) The integral involves improper integrals because the integrand has an infinite discontinuity at $x = 0$.

(b) $\displaystyle\int_{-8}^{1} \dfrac{dx}{x^{1/3}} = \int_{-8}^{0} \dfrac{dx}{x^{1/3}} + \int_{0}^{1} \dfrac{dx}{x^{1/3}}$

$\displaystyle\int_{-8}^{0} \dfrac{dx}{x^{1/3}} = \lim_{b \to 0^-} \int_{-8}^{b} \dfrac{dx}{x^{1/3}}$

$\qquad = \lim\limits_{b \to 0^-} \left[\dfrac{3}{2}x^{2/3} \right]_{-8}^{b}$

$\qquad = \lim\limits_{b \to 0^-} \left(\dfrac{3}{2}b^{2/3} - 6 \right) = -6$

$\displaystyle\int_{0}^{1} \dfrac{dx}{x^{1/3}} = \lim_{b \to 0^+} \int_{b}^{1} \dfrac{dx}{x^{1/3}}$

$\qquad = \lim\limits_{b \to 0^+} \left[\dfrac{3}{2}x^{2/3} \right]_{b}^{1}$

$\qquad = \lim\limits_{b \to 0^+} \left(\dfrac{3}{2} - \dfrac{3}{2}b^{2/3} \right)$

$\qquad = \dfrac{3}{2}$

$\displaystyle\int_{-8}^{1} \dfrac{dx}{x^{1/3}} = -6 + \dfrac{3}{2} = -\dfrac{9}{2}$

The integral converges.

(c) $-\dfrac{9}{2}$

11. $\displaystyle\int_{0}^{1} \dfrac{dx}{\sqrt{1 - x^2}} = \lim_{b \to 1^-} \int_{0}^{b} \dfrac{dx}{\sqrt{1 - x^2}}$

$\qquad = \lim\limits_{b \to 1^-} \left[\sin^{-1} x \right]_{0}^{b}$

$\qquad = \lim\limits_{b \to 1^-} (\sin^{-1} b - 0) = \dfrac{\pi}{2}$

17. First integrate $\displaystyle\int \dfrac{dx}{(1 + x)\sqrt{x}}$ by letting

$u = \sqrt{x}$, so $du = \dfrac{1}{2\sqrt{x}} \, dx$.

$\displaystyle\int \dfrac{dx}{(1 + x)\sqrt{x}} = \int \dfrac{2 \, du}{1 + u^2}$

$\qquad = 2 \tan^{-1} u + C$

$\qquad = 2 \tan^{-1} \sqrt{x} + C$

Now evaluate the improper integral. Note that the integrand is infinite at $x = 0$.

$\displaystyle\int_{0}^{\infty} \dfrac{dx}{(1 + x)\sqrt{x}} = \int_{0}^{1} \dfrac{dx}{(1 + x)\sqrt{x}} + \int_{1}^{\infty} \dfrac{dx}{(1 + x)\sqrt{x}}$

$\quad = \lim\limits_{b \to 0^+} \int_{b}^{1} \dfrac{dx}{(1 + x)\sqrt{x}} + \lim\limits_{c \to \infty} \int_{1}^{c} \dfrac{dx}{(1 + x)\sqrt{x}}$

$\quad = \lim\limits_{b \to 0^+} \left[2 \tan^{-1} \sqrt{x} \right]_{b}^{1} + \lim\limits_{c \to \infty} \left[2 \tan^{-1} \sqrt{x} \right]_{1}^{c}$

$\quad = \lim\limits_{b \to 0^+} (2 \tan^{-1} 1 - 2 \tan^{-1} \sqrt{b}) +$

$\quad \lim\limits_{c \to \infty} (2 \tan^{-1} \sqrt{c} - 2 \tan^{-1} 1)$

$\quad = \left(\dfrac{\pi}{2} - 0 \right) + \left(\pi - \dfrac{\pi}{2} \right) = \pi$

23. Integrate $\displaystyle\int \theta e^{\theta} \, d\theta$ by parts.

$u = \theta \qquad dv = e^{\theta} \, d\theta$

$du = d\theta \qquad v = e^{\theta}$

$\displaystyle\int \theta e^{\theta} \, d\theta = \theta e^{\theta} - \int e^{\theta} \, d\theta = \theta e^{\theta} - e^{\theta} + C$

$\displaystyle\int_{-\infty}^{0} \theta e^{\theta} \, d\theta = \lim_{b \to -\infty} \int_{b}^{0} \theta e^{\theta} \, d\theta$

$\qquad = \lim\limits_{b \to -\infty} \left[\theta e^{\theta} - e^{\theta} \right]_{b}^{0}$

$\qquad = \lim\limits_{b \to -\infty} (-1 - be^{b} + e^{b}) = -1$

$\left(\text{Note that } \lim\limits_{b \to -\infty} be^{b} = \lim\limits_{c \to \infty} -ce^{-c} = \lim\limits_{c \to \infty} -\dfrac{c}{e^c} \right.$

$\left. = \lim\limits_{c \to \infty} -\dfrac{1}{e^c} = 0 \text{ and } \lim\limits_{b \to -\infty} e^{b} = \lim\limits_{c \to \infty} e^{-c} = 0. \right)$

35. $\int_0^2 \dfrac{dx}{1-x} = \int_0^1 \dfrac{dx}{1-x} + \int_1^2 \dfrac{dx}{1-x}$

$\int_0^1 \dfrac{dx}{1-x} = \lim\limits_{b \to 1^-} \int_0^b \dfrac{dx}{1-x}$

$\qquad = \lim\limits_{b \to 1^-} \left[-\ln|1-x| \right]_0^b$

$\qquad = \lim\limits_{b \to 1^-} (-\ln|1-b| + 0) = \infty$

Since this integral diverges, the given integral diverges.

39. Let $f(x) = \dfrac{\sqrt{x+1}}{x^2}$ and $g(x) = \dfrac{1}{x^{3/2}}$. Both are continuous on

$[1, \infty)$.

$\lim\limits_{x \to \infty} \dfrac{f(x)}{g(x)} = \lim\limits_{x \to \infty} \dfrac{\sqrt{x+1}}{\sqrt{x}} = \lim\limits_{x \to \infty} \sqrt{1 + \dfrac{1}{x}} = 1$

$\int_1^\infty \dfrac{1}{x^{3/2}}\, dx = \lim\limits_{b \to \infty} \int_1^b x^{-3/2}\, dx$

$\qquad = \lim\limits_{b \to \infty} \left[-2x^{-1/2} \right]_1^b$

$\qquad = \lim\limits_{b \to \infty} (-2b^{-1/2} + 2) = 2$

Since the integral converges, the given integral converges.

43. First rewrite $\dfrac{1}{e^x + e^{-x}}$.

$\dfrac{1}{e^x + e^{-x}} = \dfrac{1}{e^{-x}(e^{2x}+1)} = \dfrac{e^x}{1 + (e^x)^2}$

Integrate $\displaystyle\int \dfrac{e^x\, dx}{1 + (e^x)^2}$ by letting $u = e^x$ so $du = e^x\, dx$.

$\displaystyle\int \dfrac{dx}{e^x + e^{-x}} = \int \dfrac{e^x\, dx}{1 + (e^x)^2}$

$\qquad = \int \dfrac{du}{1 + u^2}$

$\qquad = \tan^{-1} u + C$

$\qquad = \tan^{-1} e^x + C$

$\displaystyle\int_{-\infty}^{\infty} \dfrac{dx}{e^x + e^{-x}} = \int_{-\infty}^{0} \dfrac{dx}{e^x + e^{-x}} + \int_0^{\infty} \dfrac{dx}{e^x + e^{-x}}$

$\displaystyle\int_{-\infty}^{0} \dfrac{dx}{e^x + e^{-x}} = \lim\limits_{b \to -\infty} \int_b^0 \dfrac{dx}{e^x + e^{-x}}$

$\qquad = \lim\limits_{b \to -\infty} \left[\tan^{-1} e^x \right]_b^0$

$\qquad = \lim\limits_{b \to -\infty} [\tan^{-1} 1 - \tan^{-1} e^b]$

$\qquad = \dfrac{\pi}{4} - 0 = \dfrac{\pi}{4}$

$\displaystyle\int_0^{\infty} \dfrac{dx}{e^x + e^{-x}} = \lim\limits_{b \to \infty} \int_0^b \dfrac{dx}{e^x + e^{-x}}$

$\qquad = \lim\limits_{b \to \infty} \left[\tan^{-1} e^x \right]_0^b$

$\qquad = \lim\limits_{b \to \infty} [\tan^{-1} e^b - \tan^{-1} 1]$

$\qquad = \dfrac{\pi}{2} - \dfrac{\pi}{4} = \dfrac{\pi}{4}$

Thus, the given integral converges.

Section 8.4 (pp. 444–453)

5. $x^2 - 5x + 6 \overline{\smash{\big)}\, x^2 \qquad\qquad + 8} \quad \overset{\textstyle 1}{}$

$\qquad \underline{x^2 - 5x + 6}$

$\qquad\qquad\qquad 5x + 2$

$\dfrac{x^2 + 8}{x^2 - 5x + 6} = 1 + \dfrac{5x + 2}{x^2 - 5x + 6}$

$x^2 - 5x + 6 = (x-3)(x-2)$

$\dfrac{5x+2}{x^2 - 5x + 6} = \dfrac{A}{x-3} + \dfrac{B}{x-2}$

$5x + 2 = A(x-2) + B(x-3)$

$\qquad\qquad = (A+B)x + (-2A - 3B)$

Equating coefficients of like terms gives

$A + B = 5$ and $-2A - 3B = 2$

Solving the system simultaneously yields

$A = 17,\ B = -12$

$\dfrac{x^2 + 8}{x^2 - 5x + 6} = 1 + \dfrac{17}{x-3} - \dfrac{12}{x-2}$

9. $y^2 - 2y - 3 = (y-3)(y+1)$

$\dfrac{y}{y^2 - 2y - 3} = \dfrac{A}{y-3} + \dfrac{B}{y+1}$

$y = A(y+1) + B(y-3)$

$\qquad = (A+B)y + (A - 3B)$

Equating coefficients of like terms gives

$A + B = 1$ and $A - 3B = 0$

Solving the system simultaneously yields $A = \dfrac{3}{4},\ B = \dfrac{1}{4}$.

$\displaystyle\int \dfrac{y\, dy}{y^2 - 2y - 3} = \int \dfrac{3/4}{y-3}\, dy + \int \dfrac{1/4}{y+1}\, dy$

$\qquad = \dfrac{3}{4} \ln|y-3| + \dfrac{1}{4} \ln|y+1| + C$

17. $(x^2 - 1)^2 = (x+1)^2(x-1)^2$

$\dfrac{1}{(x^2-1)^2} = \dfrac{A}{x+1} + \dfrac{B}{(x+1)^2} + \dfrac{C}{x-1} + \dfrac{D}{(x-1)^2}$

$1 = A(x+1)(x-1)^2 + B(x-1)^2 + C(x+1)^2(x-1)$

$\qquad + D(x+1)^2$

$\qquad = A(x^3 - x^2 - x + 1) + B(x^2 - 2x + 1)$

$\qquad\qquad + C(x^3 + x^2 - x - 1) + D(x^2 + 2x + 1)$

$\qquad = (A + C)x^3 + (-A + B + C + D)x^2$

$\qquad\qquad + (-A - 2B - C + 2D)x + (A + B - C + D)$

Equating coefficients of like terms gives
$A + C = 0, -A + B + C + D = 0,$
$-A - 2B - C + 2D = 0,$ and $A + B - C + D = 1$
Solving the system simultaneously yields

$$A = \frac{1}{4}, B = \frac{1}{4}, C = -\frac{1}{4}, D = \frac{1}{4}$$

$$\int \frac{dx}{(x^2 - 1)^2}$$

$$= \int \frac{1/4}{x + 1} \, dx + \int \frac{1/4}{(x + 1)^2} \, dx + \int \frac{-1/4}{x - 1} \, dx + \int \frac{1/4}{(x - 1)^2} \, dx$$

$$= \frac{1}{4} \ln |x + 1| - \frac{1}{4(x + 1)} - \frac{1}{4} \ln |x - 1| - \frac{1}{4(x - 1)} + C$$

21. $x^3 - 1 = (x - 1)(x^2 + x + 1)$

$$\frac{x^2 - 2x - 2}{x^3 - 1} = \frac{A}{x - 1} + \frac{Bx + C}{x^2 + x + 1}$$

$$x^2 - 2x - 2 = A(x^2 + x + 1) + (Bx + C)(x - 1)$$

$$= (A + B)x^2 + (A - B + C)x + (A - C)$$

Equating coefficients of like terms gives

$A + B = 1, A - B + C = -2,$ and $A - C = -2.$

Solving the system simultaneously yields

$A = -1, B = 2, C = 1$

$$\int \frac{x^2 - 2x - 2}{x^3 - 1} \, dx = \int \frac{-1}{x - 1} \, dx + \int \frac{2x + 1}{x^2 + x + 1} \, dx$$

$$= -\ln |x - 1| + \ln |x^2 + x + 1| + C$$

27.

$$\frac{1}{y^2 - y} \, dy = e^x \, dx$$

$$\int \frac{1}{y(y - 1)} \, dy = \int e^x \, dx = e^x + C$$

$$\frac{1}{y(y - 1)} = \frac{A}{y} + \frac{B}{y - 1}$$

$$1 = A(y - 1) + B(y)$$

$$= (A + B)y - A$$

Equating coefficients of like terms gives

$A + B = 0$ and $-A = 1$

Solving the system simultaneously yields $A = -1, B = 1.$

$$\int \frac{1}{y(y - 1)} \, dy = \int -\frac{1}{y} \, dy + \int \frac{1}{y - 1} \, dy$$

$$= -\ln |y| + \ln |y - 1| + C$$

$$-\ln |y| + \ln |y - 1| = e^x + C$$

Substitute $x = 0, y = 2.$

$-\ln 2 + 0 = 1 + C$ or $C = -1 - \ln 2$

The solution to the initial value problem is

$$-\ln |y| + \ln |y - 1| = e^x - 1 - \ln 2.$$

41. $z = 4 \sin \theta, dz = 4 \cos \theta \, d\theta, 0 < \theta < \frac{\pi}{2}$

$16 - z^2 = 16 - 16 \sin^2 \theta = 16 \cos^2 \theta$

$$\int \frac{\sqrt{16 - z^2}}{z} \, dz$$

$$= \int \frac{|4 \cos \theta| (4 \cos \theta) \, d\theta}{4 \sin \theta}$$

$$= \int \frac{4 \cos^2 \theta}{\sin \theta} \, d\theta$$

$$= \int \frac{4 - 4 \sin^2 \theta}{\sin \theta} \, d\theta$$

$$= \int (4 \csc \theta - 4 \sin \theta) \, d\theta$$

$$= -4 \ln |\csc \theta + \cot \theta| + 4 \cos \theta + C$$

$$= -4 \ln \left| \frac{4}{z} + \frac{\sqrt{16 - z^2}}{z} \right| + 4 \left| \frac{\sqrt{16 - z^2}}{4} \right| + C$$

$$= -4 \ln \left| \frac{4 + \sqrt{16 - z^2}}{z} \right| + \sqrt{16 - z^2} + C$$

Use Formula 89 with $a = 1$ and $x = \theta$. Use Figure 8.18(b)

from the text with $a = 4$ to get

$$\csc \theta = \frac{4}{|z|}, \cot \theta = \frac{\sqrt{16 - z^2}}{|z|} \text{ and } \cos \theta = \frac{\sqrt{16 - z^2}}{4}$$

43. $dy = \frac{dx}{\sqrt{x^2 - 9}}$

$x = 3 \sec \theta, dx = 3 \sec \theta \tan \theta \, d\theta, 0 < \theta < \frac{\pi}{2}$

$x^2 - 9 = 9 \sec^2 \theta - 9 = 9 \tan^2 \theta$

$$y = \int \frac{dx}{\sqrt{x^2 - 9}}$$

$$= \int \frac{3 \sec \theta \tan \theta \, d\theta}{|3 \tan \theta|}$$

$$= \int \sec \theta \, d\theta$$

$$= \ln |\sec \theta + \tan \theta| + C$$

$$= \ln \left| \frac{x}{3} + \frac{\sqrt{x^2 - 9}}{3} \right| + C$$

Substitute $x = 5, y = \ln 3.$

$\ln 3 = \ln \left(\frac{5}{3} + \frac{4}{3} \right) + C$ or $C = 0$

The solution to the initial value problem is

$$y = \ln \left| \frac{x}{3} + \frac{\sqrt{x^2 - 9}}{3} \right|$$

Chapter 9

Section 9.1 (pp. 457–468)

3. Different, since the terms of $\sum_{n=1}^{\infty}\left(-\dfrac{1}{2}\right)^{n-1}$ alternate between positive and negative, while the terms of $\sum_{n=1}^{\infty}-\left(\dfrac{1}{2}\right)^{n-1}$ are all negative.

9. Converges; $\sum_{n=0}^{\infty}\left(\dfrac{5}{4}\right)\left(\dfrac{2}{3}\right)^{n}=\dfrac{\frac{5}{4}}{1-\frac{2}{3}}=3\left(\dfrac{5}{4}\right)=\dfrac{15}{4}$

17. Since $\sum_{n=0}^{\infty}2^{n}x^{n}=\sum_{n=0}^{\infty}(2x)^{n}$, the series converges when $|2x|<1$ and the interval of convergence is $\left(-\dfrac{1}{2},\dfrac{1}{2}\right)$. Since the sum of the series is $\dfrac{1}{1-2x}$, the series represents the function $f(x)=\dfrac{1}{1-2x}$, $-\dfrac{1}{2}<x<\dfrac{1}{2}$.

33. $1.\overline{414}=1+0.414+0.414(0.001)+0.414(0.001)^{2}+\cdots$

$$=1+\sum_{n=0}^{\infty}0.414(0.001)^{n}$$

$$=1+\dfrac{0.414}{1-0.001}=1+\dfrac{46}{111}=\dfrac{157}{111}=1$$

43. Comparing $\dfrac{x}{1-2x}$ with $\dfrac{a}{1-r}$, the first term is $a=x$ and the common ratio is $r=2x$.

Series: $x+2x^{2}+4x^{3}+\cdots+2^{n-1}x^{n}+\cdots$

Interval: The series converges when $|2x|<1$, so the interval of convergence is $\left(-\dfrac{1}{2},\dfrac{1}{2}\right)$.

Section 9.2 (pp. 469–479)

1. Substitute $2x$ for x in the Maclaurin series for $\sin x$ shown at the end of Section 9.2.

$$\sin 2x=2x-\dfrac{(2x)^{3}}{3!}+\dfrac{(2x)^{5}}{5!}-\cdots+(-1)^{n}\dfrac{(2x)^{2n+1}}{(2n+1)!}+\cdots$$

$$=2x-\dfrac{4x^{3}}{3}+\dfrac{4x^{5}}{15}-\cdots+(-1)^{n}\dfrac{(2x)^{2n+1}}{(2n+1)!}+\cdots$$

This series converges for all real x.

9. $f(2)=\dfrac{1}{x}\Big|_{x=2}=\dfrac{1}{2}$

$f'(2)=-x^{-2}\Big|_{x=2}=-\dfrac{1}{4}$

$f''(2)=2x^{-3}\Big|_{x=2}=\dfrac{1}{4}$, so $\dfrac{f''(2)}{2!}=\dfrac{1}{8}$

$f'''(2)=-6x^{-4}\Big|_{x=2}=-\dfrac{3}{8}$, so $\dfrac{f'''(2)}{3!}=-\dfrac{1}{16}$

$P_{0}(x)=\dfrac{1}{2}$

$P_{1}(x)=\dfrac{1}{2}-\dfrac{x-2}{4}$

$P_{2}(x)=\dfrac{1}{2}-\dfrac{x-2}{4}+\dfrac{(x-2)^{2}}{8}$

$P_{3}(x)=\dfrac{1}{2}-\dfrac{x-2}{4}+\dfrac{(x-2)^{2}}{8}-\dfrac{(x-2)^{3}}{16}$

13. (a) Since f is a cubic polynomial, it is its own Taylor polynomial of order 3.

$$P_{3}(x)=x^{3}-2x+4 \text{ or } 4-2x+x^{3}$$

(b) $f(1)=x^{3}-2x+4\Big|_{x=1}=3$

$f'(1)=3x^{2}-2\Big|_{x=1}=1$

$f''(1)=6x\Big|_{x=1}=6$, so $\dfrac{f''(1)}{2!}=3$

$f'''(1)=6\Big|_{x=1}=6$, so $\dfrac{f'''(1)}{3!}=1$

$P_{3}(x)=3+(x-1)+3(x-1)^{2}+(x-1)^{3}$

17. (a) $P_{3}(x)=4+(-1)(x-1)+\dfrac{3}{2!}(x-1)^{2}+\dfrac{2}{3!}(x-1)^{3}$

$$=4-(x-1)+\dfrac{3}{2}(x-1)^{2}+\dfrac{1}{3}(x-1)^{3}$$

$f(1.2)\approx P_{3}(1.2)\approx 3.863$

(b) Since the Taylor series of $f'(x)$ can be obtained by differentiating the terms of the Taylor series of $f(x)$, the second order Taylor polynomial of $f'(x)$ is given by $P_{3}'(x)=-1+3(x-1)+(x-1)^{2}$.

$f'(1.2)\approx P_{3}'(1.2)=-0.36$

Section 9.3 (pp. 480–487)

1. $f(0) = e^{-2x}\big|_{x=0} = 1$

$f'(0) = -2e^{-2x}\big|_{x=0} = -2$

$f''(0) = 4e^{-2x}\big|_{x=0} = 4$, so $\dfrac{f''(0)}{2!} = 2$

$f'''(0) = -8e^{-2x}\big|_{x=0} = -8$, so $\dfrac{f'''(0)}{3!} = -\dfrac{4}{3}$

$f^{(4)}(0) = 16e^{-2x}\big|_{x=0} = 16$, so $\dfrac{f^{(4)}(0)}{4!} = \dfrac{2}{3}$

$P_4(x) = 1 - 2x + 2x^2 - \dfrac{4}{3}x^3 + \dfrac{2}{3}x^4$

$f(0.2) \approx P_4(0.2) = 0.6704$

9. $\sin^2 x - \dfrac{1}{2} - \dfrac{1}{2}\cos(2x)$

$\quad = \dfrac{1}{2} - \dfrac{1}{2}\left(1 - \dfrac{(2x)^2}{2!} + \dfrac{(2x)^4}{4!} - \dfrac{(2x)^6}{6!} + \cdots \right.$

$\qquad \left. + (-1)^n \dfrac{(2x)^{2n}}{(2n)!} + \cdots \right)$

$\quad = \dfrac{4x^2}{2 \cdot 2!} - \dfrac{16x^4}{2 \cdot 4!} + \dfrac{64x^6}{2 \cdot 6!} \quad \cdots$

$\qquad + (-1)^n + 1 \dfrac{2^{2n} x^{2n}}{2 \cdot (2n)!} + \cdots$

$\quad = x^2 - \dfrac{x^4}{3} + \dfrac{2x^6}{45} - \cdots + (-1)^{n-1} \dfrac{2^{2n-1} x^{2n}}{(2n)!} + \cdots$

Note: By replacing n with $n + 1$, the general term can be

written as $(-1)^n \dfrac{2^{2n+1} x^{2n+2}}{(2n+2)!}$.

11. Let $f(x) = \sin x$. Then $P_4(x) = P_3(x) = x - \dfrac{x^3}{6}$, so we use

the Remainder Estimation Theorem with $n = 4$. Since

$|f^{(5)}(x)| = |\cos x| \le 1$ for all x, we may use $M = r = 1$,

giving $|R_4(x)| \le \dfrac{|x|^5}{5!}$, so we may assure that

$|R_4(x)| \le 5 \times 10^{-4}$ by requiring $\dfrac{|x|^5}{5!} \le 5 \times 10^{-4}$, or

$|x| \le \sqrt[5]{0.06} \approx 0.5697$. Thus, the absolute error is no

greater than 5×10^{-4} when

$-0.56 < x < 0.56$ (approximately).

Alternate method: Using graphing techniques,

$\left|\sin x - \left(x - \dfrac{x^3}{6}\right)\right| \le 5 \times 10^{-4}$ when $-0.57 < x < 0.57$.

19. $f(0) = \ln(\cos x)\big|_{x=0} = \ln 1 = 0$

$f'(0) = \dfrac{1}{\cos x}(-\sin x)\big|_{x=0} = -\tan x\big|_{x=0} = 0$

$f''(0) = -\sec^2 x\big|_{x=0} = -1$ so $\dfrac{f''(0)}{2!} = -\dfrac{1}{2}$

(a) $L(x) = 0$

(b) $P_2(x) = -\dfrac{1}{2}x^2$

(c) The graphs of the linear and quadratic approximations fit the graph of the function near $x = 0$.

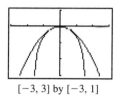

$[-3, 3]$ by $[-3, 1]$

Section 9.4 (pp. 487–496)

5. Converges by the Ratio Test, since

$\lim\limits_{n\to\infty} \dfrac{a_{n+1}}{a_n} = \lim\limits_{n\to\infty} \dfrac{2^{n+1}}{3^{n+1}+1} \cdot \dfrac{3^n+1}{2^n} = \dfrac{2}{3} < 1.$

Alternately, note that $\dfrac{2^n}{3^{n+1}+1} < \left(\dfrac{2}{3}\right)^n$ for all n.

Since $\sum\limits_{n=1}^{\infty} \left(\dfrac{2}{3}\right)^n$ converges, $\sum\limits_{n=1}^{\infty} \dfrac{2^n}{3^n+1}$ converges by the Direct

Comparison Test.

11. Converges, because it is a geometric series with $r = -\dfrac{2}{3}$, so $|r| < 1$.

21. This is a geometric series which converges only for

$|-(4x+1)| < 1$, or $\left|x + \dfrac{1}{4}\right| < \dfrac{1}{4}$, so the radius of

convergence is $\dfrac{1}{4}$.

31. $\lim\limits_{n\to\infty}\left|\dfrac{a_{n+1}}{a_n}\right| = \lim\limits_{n\to\infty} \dfrac{|(-2)^{n+1}|(n+2)|x-1|^{n+1}}{|-2^n|(n+1)|x-1|^n}$

$\qquad = \lim\limits_{n\to\infty} 2|x-1|\dfrac{n+2}{n+1}$

$\qquad = 2|x-1|$

The series converges for $|x - 1| < \dfrac{1}{2}$ and diverges for

$|x - 1| > \dfrac{1}{2}$, so the radius of convergence is $\dfrac{1}{2}$.

35. This is a geometric series with first term $a = 1$ and common ratio $r = \dfrac{(x-1)^2}{4}$. It converges only when $\left|\dfrac{(x-1)^2}{4}\right| < 1$, so the interval of convergence is $-1 < x < 3$.

$$
\begin{aligned}
\text{Sum} &= \frac{a}{1-r} = \frac{1}{1 - \dfrac{(x-1)^2}{4}} \\
&= \frac{4}{4 - (x-1)^2} \\
&= \frac{4}{-x^2 + 2x + 3} \\
&= -\frac{4}{x^2 - 2x - 3}
\end{aligned}
$$

Section 9.5 (pp. 496–508)

1. Diverges by the integral test, since $\displaystyle\int_1^\infty \frac{5}{x+1}\,dx$ diverges.

9. Converges by the Direct Comparison Test, since $\dfrac{\sqrt{n}}{n^2 + 1} < \dfrac{1}{n^{3/2}}$ for $n \geq 1$, and $\displaystyle\sum_{n=0}^\infty \frac{1}{n^{3/2}}$ converges as a p-series with $p = \dfrac{3}{2}$.

17. Converges absolutely, because it is a geometric series with $r = -0.1$.

33. $\displaystyle\lim_{n\to\infty}\left|\frac{a_{n+1}}{a_n}\right| = \lim_{n\to\infty}\frac{|x|^{n+1}}{(n+1)\sqrt{n+1}\cdot 3^{n+1}}\cdot\frac{n\sqrt{n}\,3^n}{|x|^n} = \frac{|x|}{3}$

The series converges absolutely for $|x| < 3$. Furthermore, when $|x| = 3$, $\displaystyle\sum_{n=1}^\infty\left|\frac{x^n}{n\sqrt{n}\,3^n}\right| = \sum_{n=1}^\infty\frac{1}{n^{3/2}}$, which also converges as a p-series with $p = \dfrac{3}{2}$.

(a) $[-3, 3]$ **(b)** $[-3, 3]$
(c) None

37. $\displaystyle\lim_{n\to\infty}\left|\frac{a_{n+1}}{a_n}\right| = \lim_{n\to\infty}\frac{\sqrt{n+1}\,|x|^{n+1}}{3^{n+1}}\cdot\frac{3^n}{\sqrt{n}\,|x|^n} = \frac{|x|}{3}$

The series converges absolutely for $|x| < 3$, or $-3 < x < 3$.

For $|x| = 3$, the series diverges by the nth-term test.

(a) $(-3, 3)$ **(b)** $(-3, 3)$
(c) None

53. $\ln(1+x) = \displaystyle\sum_{n=1}^\infty (-1)^{n+1}\frac{x^n}{n}$, so at $x = 1$, the series is $\displaystyle\sum_{n=1}^\infty\frac{(-1)^{n+1}}{n}$. This series converges by the Alternating Series Test.

Chapter 10

Section 10.1 (pp. 513–520)

1. (a) $\dfrac{dy}{dx} = y' = \dfrac{dy/dt}{dx/dt} = \dfrac{-2\sin t}{4\cos t} = -\dfrac{1}{2}\tan t$

(b) $\dfrac{d^2y}{dx^2} = \dfrac{dy'/dt}{dx/dt} = \dfrac{-\dfrac{1}{2}\sec^2 t}{4\cos t} = -\dfrac{1}{8}\sec^3 t$

7. $\dfrac{dy}{dx} = \dfrac{dy/dt}{dx/dt} = \dfrac{\cos t}{-\sin t} = -\cot t$

(a) $-\cot t = 0$ when $t = \dfrac{\pi}{2} + k\pi$ (k any integer). Then

$(x, y) = \left(2 + \cos\left(\dfrac{\pi}{2} + k\pi\right), -1 + \sin\left(\dfrac{\pi}{2} + k\pi\right)\right)$

$= (2, -1 \pm 1)$. The points are $(2, 0)$ and $(2, -2)$.

(b) $-\cot t$ is undefined when $t = k\pi$ (k any integer). Then

$(x, y) = (2 + \cos(k\pi), -1 + \sin(k\pi)) = (2 \pm 1, -1)$.

The points are $(1, -1)$ and $(3, -1)$.

11. $x' = -\sin t$, $y' = 1 + \cos t$, so

$$
\begin{aligned}
\text{length} &= \int_0^\pi \sqrt{(-\sin t)^2 + (1 + \cos t)^2}\,dt \\
&= \int_0^\pi \sqrt{2(1 + \cos t)}\,dt \\
&= \int_0^\pi \sqrt{4\cos^2\left(\frac{t}{2}\right)}\,dt \\
&= \int_0^\pi 2\cos\left(\frac{t}{2}\right)dt \\
&= 2\left[2\sin\left(\frac{t}{2}\right)\right]_0^\pi = 4
\end{aligned}
$$

17. $x' = -\sin t$, $y' = \cos t$, so

$$
\begin{aligned}
\text{Area} &= \int_0^{2\pi} 2\pi(2 + \sin t)\sqrt{(-\sin t)^2 + \cos^2 t}\,dt \\
&= 2\pi\int_0^{2\pi} (2 + \sin t)\,dt \\
&= 2\pi\left[2t - \cos t\right]_0^{2\pi} = 8\pi^2
\end{aligned}
$$

29.

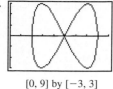

[0, 9] by [−3, 3]

Use the top half of the curve, and make use of the shape's symmetry.

$x' = 3 \cos t$, $y' = 6 \cos 2t$, so

$$\text{Area} = 2\int_0^{\pi/2} 2\pi(3 \sin 2t)\sqrt{(3 \cos t)^2 + (6 \cos 2t)^2}\, dt$$

which using NINT, evaluates to ≈ 144.513.

33. (a) $Q = (\cos t, \sin t)$, and since $\overrightarrow{QP}$ has length t, $P(x, y)$ is $t \sin t$ units right and $t \cos t$ units down from Q. So $x = \cos t + t \sin t$ and $y = \sin t - t \cos t$.

(b) $x' = t \cos t$, $y' = t \sin t$, so

$$\text{Length} = \int_0^{2\pi} \sqrt{(t \cos t)^2 + (t \sin t)^2}\, dt = \int_0^{2\pi} t\, dt$$

$$= 2\pi^2.$$

35. (a) The projectile hits the ground when $y = 0$.

$y = t(150 \sin 20° - 16t) = 0$

$t = 0$ or $t = \dfrac{75}{8} \sin 20° \approx 3.206$

$x' = 150 \cos 20°$, $y' = 150 \sin 20° - 32t$

Length =

$$\int_0^{(75 \sin 20°)/8} \sqrt{(150 \cos 20°)^2 + (150 \sin 20° - 32t)^2}\, dt$$

which, using NINT, evaluates to ≈ 461.749 ft

(b) The maximum height of the projectile occurs when

$y' = 0$, so $t = \dfrac{75}{16} \sin 20°$,

$y\left(\dfrac{75}{16} \sin 20°\right) \approx 41.125$ ft

41. $\dfrac{dy}{dx} = -\dfrac{1}{x^2}$, so Area $= \int_1^4 2\pi\left(\dfrac{1}{x}\right)\sqrt{1 + \left(-\dfrac{1}{x^2}\right)^2}\, dx$, which, using NINT, evaluates to ≈ 9.417.

Section 10.2 (pp. 520–529)

1. (a) $\langle 3(3), 3(-2)\rangle = \langle 9, -6\rangle$

  (b) $\sqrt{9^2 + (-6)^2} = \sqrt{117} = 3\sqrt{13}$

13. $\left\langle \cos \dfrac{2\pi}{3}, \sin \dfrac{2\pi}{3}\right\rangle = \left\langle -\dfrac{1}{2}, \dfrac{\sqrt{3}}{2}\right\rangle$

15. This is the unit vector which makes an angle of

$120 + 90 = 210°$ with the positive x-axis;

$\langle \cos 210°, \sin 210°\rangle = \left\langle -\dfrac{\sqrt{3}}{2}, -\dfrac{1}{2}\right\rangle$

19. $\sqrt{3^2 + 4^2} = 5$; $\dfrac{1}{5}\langle 3, 4\rangle = \left\langle \dfrac{3}{5}, \dfrac{4}{5}\right\rangle$

23. $x' = \dfrac{1}{2\sqrt{t}}$, $y' = 1 + \dfrac{1}{\sqrt{t}}$; for $t = 1$, $x' = \dfrac{1}{2}$, $y' = 2$, and

$\sqrt{(x')^2 + (y')^2} = \dfrac{\sqrt{17}}{2}$,

Tangent: $\pm\dfrac{2}{\sqrt{17}}\left\langle \dfrac{1}{2}, 2\right\rangle = \pm\left\langle \dfrac{1}{\sqrt{17}}, \dfrac{4}{\sqrt{17}}\right\rangle$,

Normal: $\pm\dfrac{2}{\sqrt{17}}\left\langle 2, -\dfrac{1}{2}\right\rangle = \pm\left\langle \dfrac{4}{\sqrt{17}}, -\dfrac{1}{\sqrt{17}}\right\rangle$.

27. $\overrightarrow{AB} = \langle 3, 1\rangle$, $\overrightarrow{BC} = \langle -1, -3\rangle$, and $\overrightarrow{AC} = \langle 2, -2\rangle$.

$\overrightarrow{BA} = \langle -3, -1\rangle$, $\overrightarrow{CB} = \langle 1, 3\rangle$, and $\overrightarrow{CA} = \langle -2, 2\rangle$.

$|\overrightarrow{AB}| = |\overrightarrow{BA}| = \sqrt{10}$, $|\overrightarrow{BC}| = |\overrightarrow{CB}| = \sqrt{10}$, and

$|\overrightarrow{AC}| = |\overrightarrow{CA}| = 2\sqrt{2}$.

Angle at $A = \cos^{-1}\left(\dfrac{\overrightarrow{AB} \cdot \overrightarrow{AC}}{|\overrightarrow{AB}||\overrightarrow{AC}|}\right)$

$= \cos^{-1}\left(\dfrac{3(2) + 1(-2)}{(\sqrt{10})(2\sqrt{2})}\right)$

$= \cos^{-1}\left(\dfrac{1}{\sqrt{5}}\right) \approx 63.435$ degrees,

Angle at $B = \cos^{-1}\left(\dfrac{\overrightarrow{BC} \cdot \overrightarrow{BA}}{|\overrightarrow{BC}||\overrightarrow{BA}|}\right)$

$= \cos^{-1}\left(\dfrac{(-1)(-3) + (-3)(-1)}{(\sqrt{10})(\sqrt{10})}\right)$

$= \cos^{-1}\left(\dfrac{3}{5}\right) \approx 53.130$ degrees, and

Angle at $C = \cos^{-1}\left(\dfrac{\overrightarrow{CB} \cdot \overrightarrow{CA}}{|\overrightarrow{CB}||\overrightarrow{CA}|}\right)$

$= \cos^{-1}\left(\dfrac{1(-2) + 3(2)}{(\sqrt{10})(2\sqrt{2})}\right)$

$= \cos^{-1}\left(\dfrac{1}{\sqrt{5}}\right) \approx 63.435$ degrees.

43. 25° west of north is $90 + 25 = 115°$ north of east. $800\langle \cos 115°, \sin 115°\rangle \approx \langle -338.095, 725.046\rangle$

Section 10.3 (pp. 529–539)

1. $[5 - (-1)]\mathbf{i} + (1 - 4)\mathbf{j} = 6\mathbf{i} - 3\mathbf{j}$

9. $\mathbf{v}(t) = (\cos t)\mathbf{i} + (2t + \sin t)\mathbf{j}$, $\mathbf{r}(0) = -\mathbf{j}$ and $\mathbf{v}(0) = \mathbf{i}$.
So the slope is zero (the velocity vector is horizontal).

(a) The horizontal line through $(0, -1)$: $y = -1$.

(b) The vertical line through $(0, -1)$: $x = 0$.

11. $\left(\int_1^2 (6 - 6t)\, dt\right)\mathbf{i} + \left(\int_1^2 3\sqrt{t}\, dt\right)\mathbf{j}$

$= \left[6t - 3t^2\right]_1^2 \mathbf{i} + \left[2t^{3/2}\right]_1^2 \mathbf{j}$

$= -3\mathbf{i} + (4\sqrt{2} - 2)\mathbf{j}$

15. $\mathbf{r}(t) = (t + 1)^{3/2}\mathbf{i} - e^{-t}\mathbf{j} + \mathbf{C}$, and
$\mathbf{r}(0) = \mathbf{i} - \mathbf{j} + \mathbf{C} = \mathbf{0}$, so $\mathbf{C} = -(\mathbf{i} - \mathbf{j}) = -\mathbf{i} + \mathbf{j}$
$\mathbf{r}(t) = ((t + 1)^{3/2} - 1)\mathbf{i} - (e^{-t} - 1)\mathbf{j}$

19. $\mathbf{v}(t) = (1 - \cos t)\mathbf{i} + (\sin t)\mathbf{j}$ and $\mathbf{a}(t) = (\sin t)\mathbf{i} + (\cos t)\mathbf{j}$.
Solve $\mathbf{v} \cdot \mathbf{a} = 0$: $(\sin t - \sin t \cos t) + (\sin t \cos t) = 0$
implies $\sin t = 0$, which is true for $t = 0$, π, or 2π.

23. $\mathbf{v}(t) = (-2 \sin t)\mathbf{i} + (\cos t)\mathbf{j}$, and

$\mathbf{a}(t) = (-2 \cos t)\mathbf{i} + (-\sin t)\mathbf{j}$. So

$\mathbf{v}\left(\dfrac{\pi}{4}\right) = (-\sqrt{2})\mathbf{i} + \left(\dfrac{1}{\sqrt{2}}\right)\mathbf{j}$, and

$\mathbf{a}\left(\dfrac{\pi}{4}\right) = (-\sqrt{2})\mathbf{i} + \left(-\dfrac{1}{\sqrt{2}}\right)\mathbf{j}$.

Then $|\mathbf{v}| = |\mathbf{a}| = \sqrt{\dfrac{5}{2}}$, $\mathbf{v} \cdot \mathbf{a} = \dfrac{3}{2}$, and

$\theta = \cos^{-1}\left(\dfrac{\mathbf{v} \cdot \mathbf{a}}{|\mathbf{v}||\mathbf{a}|}\right) = \cos^{-1}\left(\dfrac{3}{5}\right) \approx 53.130$ degrees.

Section 10.4 (pp. 539–552)

5. Use $y = (v_0 \sin \alpha)t - \dfrac{1}{2}gt^2 + 6.5$.

$16t^2 - 22\sqrt{2}t - 6.5 = 0$

$t = \dfrac{11\sqrt{2} + \sqrt{346}}{16} \approx 2.135$ seconds

(by the quadratic formula). Substitute that into

$x = (v_0 \cos \alpha)t = (44 \sin 45°)t$ to obtain $x \approx 66.4206$ feet

from the stopboard.

27. (a) (Assuming that "x" is zero at the point of impact.)
$\mathbf{r}(t) = (x(t))\mathbf{i} + (y(t))\mathbf{j}$, where
$x(t) = (35 \cos 27°)t$ and
$y(t) = 4 + (35 \sin 27°)t - 16t^2$.

(b) $y_{\max} = \dfrac{(v_0 \sin \alpha)^2}{2g} + 4 = \dfrac{(35 \sin 27°)^2}{64} + 4 \approx 7.945$ feet,

which is reached at

$t = \dfrac{v_0 \sin \alpha}{g} = \dfrac{35 \sin 27°}{32} \approx 0.497$ seconds.

(c) For the time, solve $y = 4 + (35 \sin 27°)t - 16t^2 = 0$

for t, using the quadratic formula:

$t = \dfrac{35 \sin 27° + \sqrt{(35 \sin 27°)^2 + 256}}{32} \approx 1.201$ seconds.

Then the range is about

$x(1.201) = (35 \cos 27°)(1.201) \approx 37.460$ feet.

(d) For the time, solve $y = 4 + (35 \sin 27°)t - 16t^2 = 7$

for t, using the quadratic formula:

$t = \dfrac{35 \sin 27° \pm \sqrt{(35 \sin 27°)^2 - 192}}{32} \approx 0.254$ and

0.740 seconds. At those times the ball is about

$x(0.254) = (35 \cos 27°)(0.254) \approx 7.906$ feet and

$x(0.740) = (35 \cos 27°)(0.740) \approx 23.064$ feet from the

impact point, or about $37.460 - 7.906 \approx 29.554$ feet

and $37.460 - 23.064 \approx 14.396$ feet from the landing

spot.

(e) Yes. It changes things because the ball won't clear the
net ($y_{\max} \approx 7.945$ ft).

29. (a) $\mathbf{r}(t) = (x(t))\mathbf{i} + (y(t))\mathbf{j}$, where

$x(t) = \left(\dfrac{1}{0.08}\right)(1 - e^{-0.08t})(152 \cos 20° - 17.6)$ and

$y(t) = 3 + \left(\dfrac{152}{0.08}\right)(1 - e^{-0.08t})(\sin 20°)$

$\quad + \left(\dfrac{32}{0.08^2}\right)(1 - 0.08t - e^{-0.08t})$

(b) Solve graphically: Enter $y(t)$ for Y_1 (where X stands in
for t), then use the maximum function to find that at
$t \approx 1.527$ seconds the ball reaches a maximum height
of about 41.893 feet.

(c) Use the zero function to find that $y = 0$ when the ball

has traveled for ≈ 3.181 seconds. The range is about

$x(3.181) = \left(\dfrac{1}{0.08}\right)(1 - e^{-0.08(3.181)})(152 \cos 20° - 17.6)$

$\quad \approx 351.734$ feet.

(d) Graph $Y_2 = 35$ and use the intersect function to find that $y = 35$ for $t \approx 0.877$ and 2.190 seconds, at which times the ball is about $x(0.877) \approx 106.028$ feet and $x(2.190) \approx 251.530$ feet from home plate.

(e) No; the range is less than 380 feet. To find the wind needed for a home run, first use the method of part (d) to find that $y = 20$ at $t \approx 0.376$ and 2.716 seconds.

Then define

$$x(w) = \left(\frac{1}{0.08}\right)(1 - e^{-0.08(2.716)})(152 \cos 20° + w), \text{ and}$$

solve $x(w) = 380$ to find $w \approx 12.846$ ft/sec. This is the speed of a wind gust needed in the direction of the hit for the ball to clear the fence for a home run.

Section 10.5 (pp. 552–559)

5.

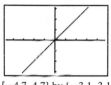

$[-6, 6]$ by $[-4, 4]$

(a) $r = \sqrt{(-1)^2 + 1^2} = \sqrt{2}$, $\tan \theta = \frac{1}{-1} = -1$ with θ in quadrant II. The coordinates are $\left(\sqrt{2}, \frac{3\pi}{4}\right)$ $\left(\sqrt{2}, -\frac{5\pi}{4}\right)$ also works, since r is the same and θ differs by 2π.

(b) $r = \sqrt{1^2 + (-\sqrt{3})^2} = 2$, $\tan \theta = -\frac{\sqrt{3}}{1} = -\sqrt{3}$ with θ in quadrant IV. The coordinates are $\left(2, -\frac{\pi}{3}\right)$. $\left(-2, \frac{2\pi}{3}\right)$ also works, since r has the opposite sign and θ differs by π.

(c) $r = \sqrt{0^2 + 3^2} = 3$, $\tan \theta = \frac{3}{0}$ is undefined with θ on the positive y-axis. The coordinates are $\left(3, \frac{\pi}{2}\right)$. $\left(3, \frac{5\pi}{2}\right)$ also works, since r is the same and θ differs by 2π.

(d) $r = \sqrt{(-1)^2 + 0^2} = 1$, $\tan \theta = \frac{0}{-1} = 0$ with θ on the negative x-axis. The coordinates are $(1, \pi)$. $(-1, 0)$ also works, since r has the opposite sign and θ differs by π.

25. $x^2 + y^2 = r^2$ and $y = r \sin \theta$, so the equation is $x^2 + y^2 = 4y \Rightarrow x^2 + (y - 2)^2 = 4$, a circle (center $= (0, 2)$, radius $= 2$).

31. $r \sin \theta = \ln r + \ln \cos \theta$
$\Rightarrow r \sin \theta = \ln (r \cos \theta)$
$\Rightarrow y = \ln x$, the logarithmic curve.

39. $x = y \Rightarrow r \cos \theta = r \sin \theta \Rightarrow \tan \theta = 1 \Rightarrow \theta = \frac{\pi}{4}$

More generally, $\theta = \frac{\pi}{4} + 2k\pi$ for any integer k.

The graph is a slanted line.

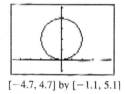

$[-4.7, 4.7]$ by $[-3.1, 3.1]$

47. $x^2 + (y - 2)^2 = 4 \Rightarrow$
$r^2 \cos^2 \theta + (r \sin \theta - 2)^2 = 4$
$r^2 \cos^2 \theta + r^2 \sin^2 \theta - 4r \sin \theta + 4 = 4$
$r^2 - 4r \sin \theta = 0$
$r = 4 \sin \theta$.
The graph is a circle centered at $(0, 2)$ with radius 2.

$[-4.7, 4.7]$ by $[-1.1, 5.1]$

Section 10.6 (pp. 559–568)

1. $\dfrac{dy}{dx} = \dfrac{f'(\theta) \sin \theta + f(\theta) \cos \theta}{f'(\theta) \cos \theta - f(\theta) \sin \theta}$

$= \dfrac{\cos \theta \sin \theta + (-1 + \sin \theta)\cos \theta}{\cos \theta \cos \theta - (-1 + \sin \theta)\sin \theta}$

$= \dfrac{2 \sin \theta \cos \theta - \cos \theta}{\cos^2 \theta - \sin^2 \theta + \sin \theta}$

$= \dfrac{\sin 2\theta - \cos \theta}{\cos 2\theta + \sin \theta}.$

$\left.\dfrac{dy}{dx}\right|_{\theta=0} = -\dfrac{1}{1} = -1$, $\left.\dfrac{dy}{dx}\right|_{\theta=\pi} = \dfrac{1}{1} = 1$

5.

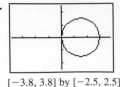

$[-3.8, 3.8]$ by $[-2.5, 2.5]$

The graph suggests a vertical tangent line with undefined

slope at the pole, with equation $\theta = \dfrac{\pi}{2}$ (i.e., $x = 0$).

Confirm analytically: $x = (3 \cos \theta) \cos \theta = 3 \cos^2 \theta$

$y = (3 \cos \theta) \sin \theta$

$\dfrac{dy}{d\theta} = (-3 \sin \theta)\sin \theta + (3 \cos \theta)\cos \theta = 3(\cos^2 \theta - \sin^2 \theta)$

and $\dfrac{dx}{d\theta} = 6 \cos \theta(-\sin \theta)$.

$\left(0, \dfrac{\pi}{2}\right)$ is a solution, $\dfrac{dx}{d\theta}\Big|_{\theta=\pi/2} = 0$, and

$\dfrac{dy}{d\theta}\Big|_{\theta=\pi/2} = 3(0^2 - 1^2) = -3$. So at $\left(0, \dfrac{\pi}{2}\right)$, $\dfrac{dx}{d\theta} = 0$

and $\dfrac{dy}{d\theta} \neq 0$.

13. The curve is complete for $0 \le \theta \le 2\pi$ (as can be verified

by graphing). The area is

$\displaystyle\int_0^{2\pi} \dfrac{1}{2}(4 + 2 \cos \theta)^2 \, d\theta$

$= 2 \displaystyle\int_0^{2\pi} (4 + 4 \cos \theta + \cos^2 \theta) \, d\theta$

$= 2 \left[4\theta + 4 \sin \theta + \dfrac{1}{2}\theta + \dfrac{1}{4} \sin 2\theta \right]_0^{2\pi} = 18\pi$

19.

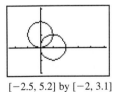

$[-2.5, 5.2]$ by $[-2, 3.1]$

The circles intersect at (x, y) coordinates $(0, 0)$ and

$(1, 1)$. The area shared is twice the area inside the

circle $r = 2 \sin \theta$ between $\theta = 0$ and $\theta = \dfrac{\pi}{4}$.

Shared area $= 2 \displaystyle\int_0^{\pi/4} \dfrac{1}{2}(2 \sin \theta)^2 \, d\theta$

$= \displaystyle\int_0^{\pi/4} 4 \sin^2 \theta \, d\theta$

$= 4 \left[\dfrac{1}{2}\theta - \dfrac{1}{4} \sin 2\theta \right]_0^{\pi/4}$

$= 4\left(\dfrac{\pi}{8} - \dfrac{1}{4}\right) = \dfrac{\pi}{2} - 1.$

31. $\dfrac{dr}{d\theta} = 2\theta$, so

Length $= \displaystyle\int_0^{\sqrt{5}} \sqrt{(\theta^2)^2 + (2\theta)^2} \, d\theta$

$= \displaystyle\int_0^{\sqrt{5}} \theta\sqrt{\theta^2 + 4} \, d\theta$

$= \left[\dfrac{1}{3}(\theta^2 + 4)^{3/2} \right]_0^{\sqrt{5}}$

$= \dfrac{1}{3}(27 - 8) = \dfrac{19}{3}.$

39. $\dfrac{dr}{d\theta} = \dfrac{1}{2\sqrt{\cos 2\theta}}(-\sin 2\theta)(2) = -\dfrac{\sin 2\theta}{\sqrt{\cos 2\theta}}$ so

Surface area

$= \displaystyle\int_0^{\pi/4} 2\pi \sqrt{\cos 2\theta} \cos \theta \sqrt{(\sqrt{\cos 2\theta})^2 + \left(-\dfrac{\sin 2\theta}{\sqrt{\cos 2\theta}}\right)^2} \, d\theta$

$= 2\pi \displaystyle\int_0^{\pi/4} \cos \theta \sqrt{\cos^2 2\theta + \sin^2 2\theta} \, d\theta$

$= 2\pi \displaystyle\int_0^{\pi/4} \cos \theta \, d\theta$

$= 2\pi \left[\sin \theta \right]_0^{\pi/4} = \pi\sqrt{2} \approx 4.443.$

APPLICATIONS INDEX

INDEX